Small Scale Processes *in* Geophysical Fluid Flows

This is Volume 67 in the
INTERNATIONAL GEOPHYSICS SERIES
A series of monographs and textbooks
Edited by RENATA DMOWSKA, JAMES R. HOLTON, and H. THOMAS ROSSBY
A complete list of books in this series appears at the end of this volume.

Small Scale Processes *in* Geophysical Fluid Flows

LAKSHMI H. KANTHA
University of Colorado
Boulder, Colorado

CAROL ANNE CLAYSON
Purdue University
West Lafayette, Indiana

San Diego San Francisco New York Boston London Sydney Tokyo

Front cover photograph: The cover image portrays backscatter from the NCAR scanning aerosol backscatter lidar taken during the 1996 Coastal Waves experiment. The image shows a hydraulic jump and internal waves in the marine boundary layer downstream of Point Sur, California (red color indicates cloud cover; flow is from right to left). Small scale processes such as these are the subject matter of this book. (courtesy of Dr. David Rogers of the Scripps Institute of Oceanography, La Jolla, CA)

This book is printed on acid-free paper. 

Academic Press
a division of Harcourt Brace & Company
525 B Street, Suite 1900, San Diego, California 92101-4495, USA
http://www.apnet.com

Academic Press
24-28 Oval Road, London NW1 7DX, UK
http://www.hbuk.co.uk/ap/

Library of Congress Catalog Card Number: 99-60586

International Standard Book Number: 0-12-434070-9

Printed and bound in the United Kingdom
Transferred to Digital Printing, 2011

To

The Office of Naval Research

on the occasion of its Fiftieth Anniversary

and

The U.S. Navy

Contents

Chapter 2
Oceanic Mixed Layer

Chapter 3
Atmospheric Boundary Layer

Chapter 6
Internal Waves

Chapter 7
Double-Diffusive Processes

Chapter 8
Lakes and Reservoirs

Appendix A
Units

Appendix B
Equations of State

Appendix C
Important Scales and Nondimensional Quantities

Appendix D
Wave Motions

Foreword

When I became a student of oceanography in the early 1940s, the physical mechanism of mixing had hardly been considered. The interior thermal time constant of a million years associated with a molecular conductivity of 10^{-7} m^2 s^{-1} was being challenged by observational data. Using profiles from Nansen bottle casts, eddy coefficients were being calculated and chosen to give agreement with observations, and more often than not, assumed to have a constant value. At that time, the principal focus was on the large scales at which the energy of the general circulation of the ocean and atmosphere is concentrated.

At the opposite end of the general circulation scale is the micro- (or dissipation) scale where energy is irreversibly converted into heat. We are talking about millimeters and centimeters, but just because the process scales are small does not mean that their importance is small. One may reasonably claim that successful model predictions on all scales, including the climate scale, require some understanding and realistic representation of these small-scale processes.

For this reason, I welcome the appearance of this ambitious survey of small-scale processes. The first chapter provides an extensive treatment of the fundamentals of statistical turbulence theory, with emphasis on the roles of stratification and rotation, the two ways geophysical turbulence differs from the ordinary laboratory situation. The next three chapters refer to the ocean and atmospheric boundary layers and surface exchange processes. This is followed by a discussion of surface and internal waves. The final chapters deal with double diffusive processes and the special situations in lakes and reservoirs.

The appendixes form an important component of this work. Appendix A has a compilation of frequently used conversion factors and of universal, physical, and geophysical constants. Appendix B summarizes equations of state for oceans, lakes, and atmosphere, and Appendix C gives a list of length, time, and velocity scales and nondimensional numbers. I look forward to taking advantage of the extensive effort that has gone into assembling this large array of information.

The juxtaposition of turbulent and wavelike processes is an essential feature. The distinction is subtle. Both are characterized as random processes with continuous spectra. In 1966, Owen Phillips spoke of the "promiscuous" interactions of the turbulent Fourier components, contrasting with the "weak, selective" interactions of gravity waves. In a 1981 survey of "Internal Waves and Small Scale Processes" I remarked: "The connection between internal waves and small scale processes—that is where the key is. I feel we are close to having these pieces fall into place, and I am uncomfortable with having attempted a survey at this time." The chapter on internal waves, particularly the section on abyssal mixing, was therefore of particular interest. Progress has been made in the past 15 years, but we are not there yet. I refer also to the section on breaking surface waves and their important role in the momentum transfer and exchange of gases across the air–sea interface.

The authors did not set out to write a monograph. Their stated, modest goal was to provide a modern overview of small-scale processes as an incentive to the reader to investigate these processes further. In doing so they have provided an important service to all those working in this active field.

Walter Munk
Scripps Institution of Oceanography
La Jolla, California

Preface

Human beings live on a thin crust of land, breathe the air from a thin layer of atmosphere, and drink water that ultimately comes from a thin layer of oceans. Because of their fluid nature, the atmospheric and oceanic envelopes contain a rich variety of processes; and while most human beings never experience more than their surface layers, a few tens of meters thick, they are profoundly affected by these processes and are in turn capable of influencing these envelopes, often in ways deleterious to the biosphere that sustains them. Fossil fuel burning and groundwater, atmospheric, and oceanic pollution are two such examples. An understanding of how these envelopes of fluid function and interact is therefore essential and necessarily involves small-scale processes, which play a very important role in shaping our environment and its characteristics.

We have always felt a need for a broad overview of small scale processes, which are ultimately responsible for mixing and dissipation, as well as for propagation of energy and momentum in geophysical flows. By this we mean small scale wave motions and turbulence. Both are ubiquitous in nature; there is no way one can avoid them when one deals with geophysical fluid media. However, we often dismiss them as too complex, too ill-understood, and therefore not worthy of the investment in time needed to gain a better understanding of their role in geophysical processes. But it is rather easy to make the point that a better understanding of geophysical processes requires a better understanding of small scale processes in geophysical flows.

This is not a monograph, but an elementary book on small scale processes. There is not much here that a diligent reader cannot look up among widely scattered scientific journals and monographs. Our principal objective has been to gather together in a single place such widely scattered material. We do not lay claim to expertise in all aspects of small scale processes in geophysical flows. The subject is too vast for any single person to attempt to gain expertise even in a single subtopic, let alone the entire field. Consequently, we owe a heavy debt of gratitude to the authors of various review articles and the latest literature on the topics, from whom we have borrowed liberally. These sources have been duly acknowledged. As far as the fundamentals are concerned, there is not much new here. What is different is some of the latest advances as seen by the authors. Here, heavy reliance is placed on scientific journals. The sole objective is to provide the reader with the basic principles of each subject and familiarity with selected topics of current interest in the field to give a modern flavor to the topic. We hope that these will provide sufficient incentives for the reader to investigate small scale processes further.

Also, with a subject this vast, it is futile to be all-inclusive. Therefore, the material inevitably exhibits personal biases. For this we apologize to the reader and the experts in the field whose vital contributions to the field may have been ignored. After all, this work is only one of many possible looks, no doubt biased, at the fascinating kaleidoscope that is Nature. Our intention has been to provide a broad overview of small scale processes and some insight into their role in geophysical flows. It is written so that a newcomer to the field, with the necessary mathematics and physics background, can learn the subject with ease and be introduced to some current research topics as well, without having to do an extensive literature survey on his or her own. Also, enough recent references are provided so that a single topic can be pursued further. Therefore, along with its companion volume, *Numerical Models of Oceans and Oceanic Processes*, (Academic Press, 2000), this book should prove useful to a novice as well as to a practicing expert. We hope that these books will inspire at least a few young people to take up careers in environmental science and engineering and thus contribute to a better understanding of our environment and possibly the betterment of a majority of the human population, which will be increasingly at the mercy of Nature for sustenance in the coming century.

We hope to make this book electronic. In addition to the regular hard-copy format, we hope to provide the student, eventually, with an electronic supplement that contains the source code, color graphics and animation packages, and sample runs that will allow more interactive use of the material. As far as is feasible, each chapter will be provided with project-like exercises to improve the student's skill and understanding. We hope this format will make it easier and more ''fun'' to learn topics that are normally quite dry and hard, and frankly turn off quite a few brilliant young people.

Finally, in a perhaps overambitious endeavor such as this, mistakes are inevitable, especially on topics on which we are not experts. The mere fact that we could put something like this together for so vast and intricate a field attests to our liberal borrowing (properly attributed of course) from experts in their individual areas of expertise. We thank them and apologize if we misquoted any of them. We would certainly appreciate hearing about any glaring errors that may have been inadvertently made.

It is our pleasure to acknowledge the contributions of many anonymous reviewers to this endeavor. Their comments have greatly improved this book. Particular thanks go to Dr. Eugene Terray of the Woods Hole Oceanographic Institution, whose thoughtful and thorough review of Chapter 5 is greatly appreciated. We would like to thank the many scientists who contributed by sending original figures from their work for inclusion in this text. We would also like to thank the following individuals for helping to prepare many of the final figures for this text: Tristan Johnson, Reed L. Clayson, Jason Hartz and Rebecca Priddy. Reed L. Clayson also provided valuable editorial assistance. It was our hope to complete these two books in time for the 50th birthday of the Office of Naval Research, but we severely underestimated the time involved in converting an initial draft to a final peer-reviewed set of chapters. Nevertheless, the principle "better late than never" governs our dedication of these books.

Last, but not least, we thank our very understanding, ever-patient, and tolerant spouses, Kalpana Kantha and Tristan Johnson, for their unflinching support and assistance. L.H.K. thanks Roshan, Vinod, and Kiran for putting up so patiently with an absentee father. CAC's infant son Johann did his part by consistently getting her up so she could devote productive pre-dawn hours to completing the text.

Lakshmi H. Kantha
Carol Anne Clayson

List of Acronyms

ABL	Atmospheric boundary layer
AIDJEX	Arctic Ice Dynamics Joint Experiment
AIWEX	Arctic Internal Wave Experiment
ALK	Alkalinity
ALPEX	Alpine Experiment
AMTEX	Air Mass Transformation Experiment
ARM	Atmospheric radiation measurements
ASTEX	Atlantic Stratocumulus Transition Experiment
ATEX	Atlantic Trade Wind Experiment
AVHRR	Advanced very high resolution radiometer
BBL	Bottom boundary layer
BL	Boundary layer
BOMEX	Barbados Oceanographic and Meteorological Experiment
CABL	Convective atmospheric boundary layer
CAPE	Convective available potential energy
CASP	Canadian Atlantic Storms Program
CEAREX	Coordinated Eastern Arctic Experiment
CEOF	Complex empirical orthogonal functions
CI	Cabelling instability
CNES	Centre National d'Etudes Spatiales
COADS	Consolidated Ocean Atmosphere Data Set
COARE	Coupled Ocean–Atmosphere Response Experiment
CODE	Coastal Ocean Dynamics Experiment
CoOP	Coastal Ocean Processes Program
COPE	Coastal Ocean Probe Experiment
COS	Carbonyl sulfide
CPSST	Cross-product sea surface temperature
CREAMS	Circulation Research in East Asian Marginal Seas

CrWT	Cross wavelet transform
C-SALT	Caribbean Sheets and Layers Experiment
CTBL	Cloud-topped atmospheric boundary layer
CWT	Continuous wavelet transform
CZCS	Coastal zone color scanner
DDC	Double-diffusive convection
DIA	Direct interaction approximation
DIC	Dissolved inorganic carbon
DMS	Dimethyl sulfide
DMSP	Defense Meteorological Satellite Program
DNS	Direct numerical simulation
DO	Dissolved oxygen
DOC	Dissolved organic carbon
DWT	Discrete wavelet transform
EAC	Ensemble average closure
EAPE	Evaporative available potential energy
ECMWF	European Center for Medium Range Weather Forecasting
EDQNM	Eddy damped quasi-normal Markovian
EEOF	Extended empirical orthogonal functions
EIC	Equatorial Intermediate Current
ENM	Empirical normal modes
ENSO	El Niño-Southern Oscillation
EOF	Empirical orthogonal functions
ERBE	Earth Radiation Budget Experiment
ERICA	Experiment on Rapidly Intensifying Cyclones over the Atlantic
ERS	Earth Resources Satellite
EUC	Equatorial Undercurrent
FASINEX	Frontal Air–Sea Interaction Experiment
FD	Fast-access data
FFT	Fast Fourier transform
FIRE	First International ISCCP Regional Experiment
FLIP	Floating instrument platform
FNMOC	Fleet Numerical Meteorology and Oceanography Center
FT	Fourier transform
GALE	Genesis of Atlantic Lows Experiment
GARP	Global Atmospheric Research Program
GATE	GARP Atlantic Tropical Experiment
GCM	General circulation model
GEOSAT	Geodetic satellite
GEWEX	Global Energy and Water Cycle Experiment
GFO	GEOSAT Follow-On

GM	Garrett-Munk
GOES	Geostationary orbiting environmental satellite
HEBBLE	High Energy Benthic Boundary Layer Experiment
HEXMAX	HEXOS Main Experiment
HEXOS	Humidity Exchange over the Sea Program
IBL	Internal boundary layer
IIOE	Indian Ocean Expedition
IOP	Intensive observation period
IR	Infrared
ISCCP	International Satellite Cloud Climatology Project
ITCZ	Intertropical Convergence Zone
IW	Internal wave
IWEX	Internal Wave Experiment
JASIN	Joint Air–Sea Interaction Experiment
JONSWAP	Joint North Sea Wave Project
JPDF	Joint probability density distribution
KC	Kantha-Clayson
LCL	Liquid condensation level
LEADEX	LEADs Experiment
LES	Large eddy simulation
LFC	Level of free convection
LHDIA	Lagrangian history direct interaction approximation
LHS	Left-hand side
LMD	Large-McWilliams-Doney
LOC	Limit of convection
LOTUS	Long-Term Upper Ocean Study
LSBS	Land–sea breeze system
LST	Lake surface temperature
LW	Longwave
MABL	Marine atmospheric boundary layer
MARSEN	Maritime Remote Sensing Experiment
MBLP	Marine Boundary Layer Project
MCSST	Multichannel sea surface temperature
METEOSAT	Meteorological satellite
MILE	Mixed Layer Experiment
MIZEX	Marginal Ice Zone Experiment
MKE	Mean kinetic energy
ML	Mixed layer
MLD	Mixed layer depth
MLML	Marine Light-Mixed Layer Experiment
M-O	Monin-Obukhoff
MY	Mellor-Yamada

NABL	Nocturnal atmospheric boundary layer
NASA	National Aeronautics and Space Administration
NCAR	National Center for Atmospheric Research
NCEP	National Centers for Environmental Prediction
NDBC	NOAA Data Buoy Center
NMC	National Meteorological Center
NOAA	National Ocean Atmosphere Administration
NOS	National Ocean Service
NPZ	Nutrient-phytoplankton-zooplankton
NPZD	Nutrient-phytoplankton-zooplankton-detritus
NWP	Numerical weather prediction
NWS	National Weather Service
OBL	Oceanic boundary layer
OML	Oceanic mixed layer
OPR	Operational products
OWS	Ocean weather station
PAR	Photosynthetically available radiation
PBL	Planetary boundary layer
PCA	Principal component analysis
PDF	Probability density distribution
PE	Potential energy
PIP	Principal interaction pattern
POD	Proper orthogonal decomposition
POP	Principal oscillation pattern
PWP	Price-Weller-Pinkel
RASEX	Riso Air–Sea Experiment
RHS	Right-hand side
RNG	Renormalization group analysis
RSC	Reynolds stress closure
SAR	Synthetic aperture radar
SEASAT	Sea satellite
SeaWIFS	Sea-viewing wide field-of-view sensor
SEMAPHORE	Structure des Echanges Mer-Atmosphere. Proprietes des Heterogeneites Oceaniques: Recherche Experimentale
SGS	Sub-grid scale
SHEBA	Surface Heat Budget of the Arctic Experiment
SMC	Second-moment closure
SMILEX	Shelf Mixed Layer Experiment
SMMR	Scanning Multichannel Microwave Radiometer
SOPHIA	Surface of the Ocean, Fluxes, and Interactions with the Atmosphere
SSA	Singular spectrum analysis

SSH	Sea surface height
SSM/I	Special sensor microwave imager
SST	Sea surface temperature
STP	Standard temperature and pressure
SW	Shortwave
SWADE	Surface Wave Dynamics Experiment
SWAMP	Surface Wave Modeling Project
SWAPP	Surface Wave Processes Program
TAO	Tropical atmosphere ocean array
TBI	Thermobaric instability
TC	TOGA/COARE
TFO	TOPEX Follow-On
TH	Tropic heat
TIWE	Tropical Instability Wave Experiment
TKE	Turbulence kinetic energy
TOGA	Tropical Ocean Global Atmosphere
TOPEX	Topography Experiment
TOVS	Tiros Operational Vertical Sounder
UV	Ultraviolet
WAM	Wave model
WAMDI	Wave Model Development and Implementation
WAVES	Water–Air Vertical Exchange Studies
WBL	Wave boundary layer
WD	Wavelet domain
WFT	Windowed Fourier transform
WKB	Wentzel–Kramers–Brillouin approximation in wave theory
WT	Wavelet transform
WWB	Westerly wind burst

Prologue

The subject of this treatise is small scale processes in the oceans and the atmosphere. While we will treat nongeophysical flows as needed, the emphasis will necessarily be on the oceans and the atmosphere, with particular focus on the immediate vicinity of the air–sea interface. By small scale processes, we mean principally the small spatial scales responsible for mass, momentum, and heat transfer in these fluids, scales much less than those responsible for large scale dynamical adjustment. In the atmosphere and the oceans, these scales normally range from a few millimeters to a few hundreds of meters, and in some cases, to a few kilometers. The corresponding time scales are also necessarily short, ranging from fractions of a second to fractions of an hour, and certainly less than a few hours in most cases. The fluid motions that fall into this part of the spatial and temporal spectrum are those associated with turbulent mixing and small scale waves. The former deals principally with the turbulent oceanic and atmospheric mixed layers adjacent to the air–sea interface, but includes mixing processes in the interior of the two geophysical fluid media as well. The latter include surface waves at the air–sea interface and internal waves in the interior. Processes primarily responsible for dynamical adjustments in the oceans (and to some extent the atmosphere) are dealt with in the companion volume, *Numerical Models of Oceans and Oceanic Processes* (Academic Press, 2000).

The subject is of great importance to understanding how the fluid flows behave in these media. Interpretation of observations in these geophysical flows, and modeling the oceans and the atmosphere with an eye to improving our understanding of their behavior and our capability to make accurate estimates of their state, requires attention to these small scale processes. Think about how the giant mid-ocean gyres and towering Hadley and Walker circulations in the atmosphere are generated and maintained and what their dissipation mechanisms are. These motions, one way or another, are deeply in debt to heat transfer processes in the oceans and the atmosphere. For example, the Walker circulation in the tropical atmosphere is maintained by heat transfer from the western Pacific warm pool. This transfer takes place through turbulent motions driven by convective transfer of heat from the ocean to the atmosphere. The giant oceanic gyres driven and maintained by marine surface winds owe their existence to poleward transfer of heat in the atmosphere. In fact it can be argued that the large scale motions that exist in both the atmosphere and the oceans (albeit altered by Earth's rotation quite dramatically) are there principally to reduce the large meridional gradients that would otherwise result from the net heating in the equatorial regions and the net cooling in the polar regions of the globe. If one remembers that heat and momentum are ultimately transferred from one fluid mass to another "molecule by molecule," albeit mediated by turbulent motions with a rich spectrum of spatial and temporal scales, it is not difficult to realize the important role small scale processes play in large scale motions in geophysical flows.

Turbulent motions are ubiquitous in nature. However, they are very hard to decipher and model. While one cannot but be thankful that fluid motions are almost always turbulent in the atmosphere and the oceans, the difficulty of the subject is quite daunting. Turbulence is an inherently nonlinear process and our ability to fathom its mysteries is quite limited, although the enormous gain in computing power in recent times has enabled us to finally tackle the problem in at least a brute-force approach, direct numerical solutions, if not in an intellectually pleasing and elegant fashion.

On the other hand, at first glance, one would expect wave motions to be less difficult, since the governing equations are often linear and simpler. This has not turned out to be the case, because of the essentially random nature of small scale wave motions in both the atmosphere and the oceans. Thus, for both turbulent motions and small scale wave motions, one has to appeal to statistical methods and spectral space. The energy transfer in spectral space from weak and strong nonlinear interactions is a crucial aspect of these processes. Herein lies another difficulty associated with these topics.

Of course, this is not to say that all is hopeless. Nature is indeed quite forgiving of our ignorance. Even a crude understanding of these physical processes enables us to model the large scale motions in the atmosphere and the oceans that so profoundly affect us all. No one argues the fact that parameterizations of subgrid scale processes in general circulation models (GCMs), both in the atmosphere and in the oceans, are at present rather crude. Yet, these models have been useful to many real-life applications such as short-range weather forecasting. The hope is, of course, that a better understanding of small scale processes would enable us to do better.

Dissipation of kinetic energy of fluid motions takes place at small viscous scales. There is thus a steady cascade of energy from large scales to small scales in both the atmosphere and the oceans. Now, look at the interaction between the atmosphere and the oceans. Differential heating of the atmosphere through small scale convective turbulent motions gives rise to winds. Winds blowing over the oceans transfer energy and momentum to currents and surface waves, which in turn transfer some of their energy to waves in the interior of the oceans. There is thus a cascade of energy not only from larger scales to smaller scales in a fluid mass but also across different types of fluid motions and across different media. It is therefore essential to understand the nature and mechanisms of these transfers and it is here that we need to appeal once again to small scale processes.

One more example of a potentially profound influence of small scale processes can be cited. The oceans are stably stratified for the most part, with water masses at a temperature of a few degrees Celsius in their abyssal regions. This stratification is the result of solar heating of the near-surface layers and wintertime formation of dense deep/intermediate water (DW/IW) in subpolar regions such as the Greenland Sea and around Antarctica. Averaged over the oceans, a slow upwelling on the order of about 2 m year^{-1} tends to bring cold water masses from the deep interior toward the surface. This is counteracted by turbulent mixing of heat downward from the near-surface layers into the interior so that a rough balance exists and the ocean stratification is maintained (Munk, 1966). The magnitude and sources of such mixing are poorly understood, although it is suspected that the deep-sea internal wave field fed by winds and internal tides may play a role (Munk and Wunsch, 1998). Mixing at the boundaries of the ocean basins and tidally driven mixing at the flanks of submerged seamounts are also thought to play an important role. In regions conducive to such instabilities, double diffusion may also contribute. In any case, whatever the source, turbulent mixing is essential to thermocline maintenance. A decrease in mixing (or an increase in DW/IW formation rate) would cause the oceans to fill up slowly with cold water over a time scale of several millennia, whereas an increase (or decrease in DW/IW formation rate) would tend to warm them up. Either could have profound climatic consequences, since the circulation and the CO_2 bearing capacity of the oceans depend very much on their thermal structure.

It is our goal to provide a broad survey of these small scale processes. We start off with a description of turbulence and turbulent mixing in Chapter 1. Unfortunately, this subject is highly mathematical and often obtuse. The reader is therefore encouraged to skip the difficult parts of this chapter for an initial reading. Chapter 2 deals with mixing processes in oceanic mixed layers. Chapter 3 addresses mixing processes in the atmospheric boundary layer. Chapter 4 concentrates on air–sea exchange processes crucial to understanding and modeling weather and climate. In Chapters 5 and 6, we describe surface and internal gravity waves and their role in the oceans. The reason for including and discussing these wave processes is their influence on momentum and energy transfer in the oceans and the atmosphere. Surface waves mediate the transfer of momentum between the atmosphere and the ocean. They radiate away some of the momentum flux from the region of wind forcing, and the surface gravity wave field determines the roughness felt by the atmosphere. When surface waves break, their energy is transferred to turbulence, and their momentum to the currents. Internal waves in the oceans are fed by winds and internal tides, and after cascading down the spectrum, the energy is lost to turbulence. Thus internal waves are a major source of mixing in the deep oceans. In the atmosphere, internal wave breaking is the main source of intermittent turbulence and mixing above the atmospheric boundary layer. Internal waves generated by mountains transfer their momentum to the mean flow when they break aloft. Internal gravity waves are ubiquitous in the upper troposphere and the stratosphere and are important to mixing there.

Chapter 7 is devoted to double-diffusive processes in the interior of the oceans that may also be a major source of mixing in the thermocline. In Chapter 8, we finish by describing mixing in lakes and reservoirs, simply because of the fascinating differences caused by the absence of salinity and the functional dependence of the density maximum of freshwater on temperature and depth.

In each chapter, we begin with a description of fundamental concepts before launching into detailed mathematical descriptions and developments. We have also tried hard to present advanced material on each topic. Given the rapid pace of scientific progress, it is likely that this material may be subject to change as new knowledge is acquired, but the elementary material, being invariant by definition, should endure. Recent references provided should enable the reader to pursue a particular topic further if need be. The level of treatment here is appropriate to graduate studies. Some elementary knowledge of fluid flows would, however, be desirable. The material laid out is useful for teaching a comprehensive, two-semester graduate-level course in small scale processes. However, selected topics could be taught in a single semester.

Chapter 1

Turbulence

We begin with turbulence and turbulent mixing, without a doubt, the most important small scale process in fluid flows. Geophysical boundary layers such as the oceanic mixed layer and the atmospheric boundary layer are characterized by their interaction with a boundary, such as the surface of the ocean, the land surface, or the ocean bottom. This interaction gives rise to flows that appear chaotic and dominated by eddies of various sizes. These seemingly random motions can be described as turbulence and the resulting variability can be seen in measurements of such properties as wind speed, temperature, and currents. Away from these regions, in the interior, the flow can be regarded as essentially inviscid, and for the most part, can therefore be described deterministically. However, even in the interior, there can be regions where intermittent turbulence caused by breaking internal waves and other mechanisms can exist. Clear-air turbulence in the atmosphere and double-diffusive convection in the oceanic thermocline are two such examples.

Turbulence is an outstanding, unsolved problem in physics, and is of interest to any field that involves the study of fluid flows. Consequently, there exists a vast literature, literally tens of thousands of papers, in almost all branches of science and engineering on the subject of turbulence. Several treatises exist that provide a more thorough description of turbulence in general (Batchelor, 1953; Leslie, 1973; Hinze, 1976; Townsend, 1976; Stanisic, 1985; Lesieur, 1990; McComb, 1990; Frisch, 1995), and in stratified fluids under the action of buoyancy forces in particular (Monin and Yaglom, 1971; Turner, 1973; Tennekes and Lumley, 1982), than we can provide in this chapter. In addition, there exist numerous monographs dealing with topics on turbulence in geophysical flows (Haugen, 1973; Phillips, 1977; Nieuwstadt and van Dop, 1982). Here we will provide a basic overview of turbulence and turbulent flows of interest to geophysics, with an emphasis on stratification effects in a gravitational field. For a more comprehensive treatment, the reader is referred to the treatises above. Of these, Monin and Yaglom (1973) provide a thorough and exhaustive survey of turbulence up to the 1960s and an extensive bibliography. Their discussion of isotropic, homogeneous turbulence is particularly detailed, and so is Stanisic's. Both Hinze (1976) and Stanisic (1985) provide detailed derivations that are useful for a beginner. Tennekes and Lumley (1982) is still an excellent starting point in learning turbulence. For more advanced treatments of current topics of interest, several of the specialized monographs cited above would be good starting points. The latest advances in turbulence research, including large eddy simulations (LES's), direct numerical simulations (DNS's), and renormalization group analysis (RNG) approaches, can be found in journals such as the *Journal of Fluid Mechanics* and *Physics of Fluids*. Recent articles on turbulence with geophysical applications, especially LES's, can be found in periodicals such as the *Journal of Geophysical Research,* the *Journal of Physical Oceanogaphy,* the *Journal of Atmospheric Sciences,* and *Boundary Layer Meterology,* the last of which is focused on the atmospheric boundary layer.

1.1 CHARACTERISTICS OF TURBULENT FLOWS

We begin our discussion with a qualitative description of turbulence, as do Tennekes and Lumley (1982), which will be followed by a more quantitative description. While it is easy enough to recognize a turbulent flow, it is hard to provide a succinct definition that is all-inclusive with respect to its properties. We can however list the essential aspects of a turbulent flow: it is inherently irregular (random), three-dimensional, strongly nonlinear, highly vortical, highly diffusive, and highly dissipative. It is these complex characteristics that make it a difficult task to understand and model turbulent processes.

1. Randomness (irregularity). Turbulent flows are highly irregular in both time and space. Observations at a point of any property in a turbulent flow show random fluctuations superposed on a secular or periodic trend. It is therefore difficult to describe a turbulent flow in detail at all its temporal and spatial scales. Turbulence is stochastic (random) in nature. Since randomness is involved, probabilistic/statistical methods have to be applied. While the flow is not deterministic, it is possible to discern distinct average properties such as the mean velocity, mean temperature, and mean shear stress using statistical approaches.

2. Three-dimensionality. All turbulent flows are three-dimensional. Turbulent fluctuations even in a two-dimensional flow have components in all three spatial dimensions, although the mechanism of vortex stretching that deforms the large eddies and transfers energy from large scales to successively smaller scales is absent in a strictly two-dimensional flow. This does not mean that turbulence at times cannot be close to being "two-dimensional." In strongly stratified flows or under strong rotation, there is a tendency for large scale eddies to be highly anisotropic, but they are never strictly two-dimensional. There is a considerable interest in a strictly two-dimensional "turbulence" from an academic point of view (e.g., Kraichnan, 1971, 1976; Leith, 1971), but the prevailing mechanisms are different. For example, the energy cascade is from smaller eddies to larger eddies in strictly two-dimensional turbulence, exactly the opposite of three-dimensional turbulence.

3. Vorticity. Turbulent flows are highly vortical, meaning that the deformation of a fluid particle involves rotation. The vorticity of the fluid is a measure of this rotation. While it is possible to have a flow that is random and hence looks like turbulence, if it is irrotational, then it is not turbulent. For example, surface and internal waves in the ocean are random processes requiring statistical methods for their description and modeling, but they are not turbulent in nature. Surface waves are irrotational and can be described by potential theory (Phillips, 1977, see Chapter 5). The flow in the vicinity of a turbulent jet or wake (Figure 1.1.1) is also random, but can be described quite well by potential flow theory (Phillips, 1955) and is not turbulent. Turbulence is characterized by strong, random vorticity fluctuations in all three spatial dimensions. Vorticity dynamics involving stretching of vortex lines in the flow by straining from eddies and mean flow is a central aspect of all turbulent flows.

4. Strong diffusivity. Turbulent flows are highly diffusive. Turbulent diffusivities of mass, momentum, heat, etc., are normally several orders of magnitude larger than molecular diffusivities. In fact, the turbulence Reynolds number, $R_t = q\ell/\nu$ (where q is the turbulence velocity scale, ℓ is the turbulence macroscale, and ν is the kinematic viscosity), can be thought of as the ratio of the turbulent diffusivity to the molecular diffusivity. For geophysical flows R_t is very large. For example, in the atmospheric boundary layer (ABL) R_t can be as high as 10^7. It is the strongly diffusive nature of turbulence that makes it highly useful in pollutant dispersal in the atmosphere and the oceans. However, unlike molecular diffusivity, which is a property of the fluid, turbulent diffusivity is a property of the flow. The flow however depends on the turbulent diffusivity and hence is not known a priori, and herein lies another difficulty associated with turbulence.

5. Strong dissipation. Turbulent flows are highly dissipative. Turbulence requires a steady supply of energy or it dies rapidly. The energy is extracted from the mean f low by turbulent shear stresses acting against the mean shear or

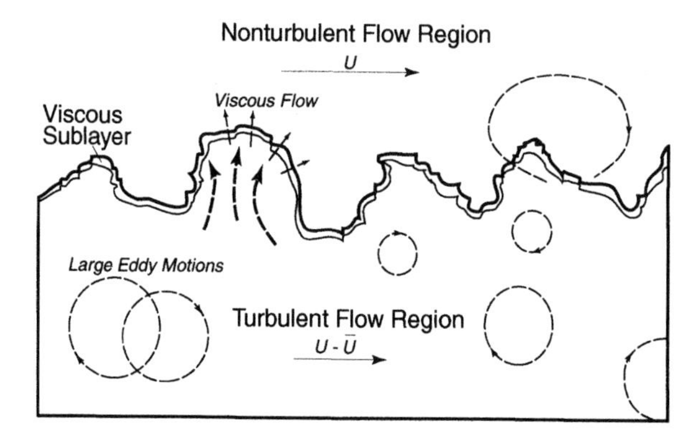

Figure 1.1.1. A sketch of the flow outside a turbulent boundary layer. The flow outside is irrotational and hence nonturbulent. Large eddies engulf this fluid and diffusion of vorticity at small scales completes its conversion to vortical rotational, turbulent fluid.

by buoyancy forces, into large eddies characterized by the turbulence macroscale ℓ. Energy then cascades continuously toward small eddies by nonlinear interactions. Dissipation of the turbulent kinetic energy occurs due to viscous forces at eddy scales comparable to the Kolmogoroff microscale, $\eta = (\nu^3/\varepsilon)^{1/4}$, where ε is the dissipation rate. The Kolmogoroff scale characterizes the smallest possible scales in a turbulent flow. However, these small scales are passive and the rate of dissipation in a high Reynolds number turbulent flow is independent of the viscosity. In fact, to a good approximation, the dissipation in a turbulent flow can be written as $\varepsilon \sim q^3/\ell$, an expression that does not involve the viscosity. The large eddies decay, not by the direct action of viscous forces acting on them, but by nonlinear transfer of their energy to smaller and smaller scales. The energy is then ultimately dissipated at Kolmogoroff scales. Turbulence decays rapidly once it is cut off from its energy source and the timescale involved is the eddy turnover timescale $t_e \sim \ell/q$. In contrast, random wave motions such as waves on the ocean surface are dissipated very slowly, and some are very nearly nondissipative.

6. Strong nonlinearity. Turbulent flows are highly nonlinear. In fact, it is the nonlinear terms in the Navier–Stokes equations that effect the cascade of energy from large eddies to small ones down the spectrum. Compared to wave motions, which are weakly nonlinearly interactive, there is a strong interaction between various scales, even though in both cases, the interaction and energy exchange appears to be local in spectral space. In other words, it is only the neighbors in the wavenumber spectrum that interact preferentially with one another. The exchange of energy between waves with different wavenumbers is a very slow process taking place over timescales on the order of many wave periods, but

takes place rapidly in turbulent flows, in a timescale on the order of the eddy turnover timescale.

7. Broad spectrum. The spectrum of turbulence is broad, but red, meaning the energy is concentrated in larger scales or lower wavenumbers. Turbulent fluctuations span a large spectrum in time and space, from the very large semipermanent eddies with a size characteristic of the mean flow to Kolmogoroff scales where energy dissipation is concentrated. The higher the Reynolds number of the flow, the smaller the smallest scales possible in the flow; that is why higher Reynolds number turbulent flows appear smaller grained. There are no distinct peaks in the spectrum except at the wavenumber corresponding to the energy-containing scale, the macroscale. Unlike tidal motions in the oceans which are characterized by peaks at distinct tidal frequencies, the turbulence spectrum is essentially smooth and broad.

8. Anisotropy of large scales. Turbulence is seldom isotropic, except at small scales at high Reynolds numbers. The large scales are invariably highly anisotropic and are continuously being oriented and elongated in the direction of the mean flow by the mean strain rate. Anisotropy is essential to the existence of shear stress; the shear stress is zero for a strictly isotropic flow. In isotropic turbulence, the turbulence characteristics are invariant to rotation and reflection of the coordinate system. Isotropy is distinct from homogeneity, in which the properties are invariant to translation in space. While the vast literature on isotropic, homogeneous turbulence is very helpful to understanding and characterizing certain universal properties of turbulence at small scales, progress in understanding and modeling turbulent flows has been hampered by the nonuniversal nature of the large anisotropic eddies that contain most of the turbulence energy and affect most of the turbulent transport, often against the prevailing gradient of the mean property. Characterization of these nonuniversal large eddies requires numerical models.

9. Loss of memory. An advecting fluid carries properties from one point to another in space. In this sense the large eddies in a turbulent flow have memory (in the Lagrangian sense). However, the memory in time from a Eulerian point of view is often short. In other words, because of the intense scrambling, initial conditions are quickly forgotten and the turbulence is often in "local equilibrium" temporally. This is often an important simplification, since memory effects are hard to deal with.

The smallest scales in turbulence, the Kolmogoroff dissipation scales, are still orders of magnitude larger than the molecular mean free path under most conditions. Turbulence can therefore be described by equations for a fluid continuum, the Navier–Stokes equations. However, its highly nonlinear and stochastic nature makes it one of the most fascinating, yet highly frustrating, processes to study in nature. The turbulence closure problem resulting directly from the nonlinearity of the governing equations makes any statistical or mechanical description an approximate "model" or postulate at best, not an exact theory. Turbulence research is akin to looking out into a beautiful valley through a frosted glass. One can only discern approximate shapes and what one might perceive to be beautiful deer grazing peacefully might turn out to be ugly boulders upon a closer examination or from a different point of view.

Fortunately, powerful supercomputers are enabling calculations to be made that render turbulence a little less mysterious, but there is still a long way to go. DNS and LES are adding far more to our knowledge of turbulence than pure observations were hitherto capable of. However, turbulence is a grand challenge problem that requires tera- or even petaflop computers to simulate realistic geophysical flow situations. Hopefully, the dawn of the new millennium will usher in new advances fueled by such computing capability. Nevertheless, many important findings in turbulence research have come from physical intuition and simple tools like scaling arguments, and not necessarily from sophisticated and complex theories. Discoveries like the universal law of the wall and the inertial subrange were made from incisive physical insight and simple dimensional analysis. Most descriptions of properties of turbulent flows assume that the governing Reynolds number is infinite in some sense and therefore we need not account for the Reynolds number in the parameterizations. Low Reynolds number turbulence is harder to characterize, since one can no longer invoke the asymptotic limit of zero molecular viscosity, and the Reynolds number enters explicitly as a parameter and the functional dependence on Reynolds number has to be prescribed.

1.2 ORIGIN AND TYPES OF TURBULENCE

Turbulence originates from the instability of laminar flows. At sufficiently large values of the characteristic dimensionless number governing the flow, the flow undergoes a transition from laminar flow to turbulent flow. This characteristic number is the Reynolds number, $R_N = UL/\nu$, where U is the characteristic flow velocity and L is its length scale, for sheared flows, and the Rayleigh number, $Ra = \Delta b\ L^3/k_T\nu$, where Δb is the buoyancy difference, for buoyant flows (Figure 1.2.1); ν and k_T are kinematic quantities—the molecular momentum and heat diffusivities. A flow can be intermittently turbulent at some transitional value of this parameter, but it has to be laminar or turbulent at any given time; it cannot be partly turbulent.

Since turbulent flows are highly dissipative, their method of generation and maintenance provides an important distinction of each flow. These methods fall into two categories: mixing caused by shear of the mean flow and mixing caused by convective gravitational instability (buoyancy forces). Both methods are

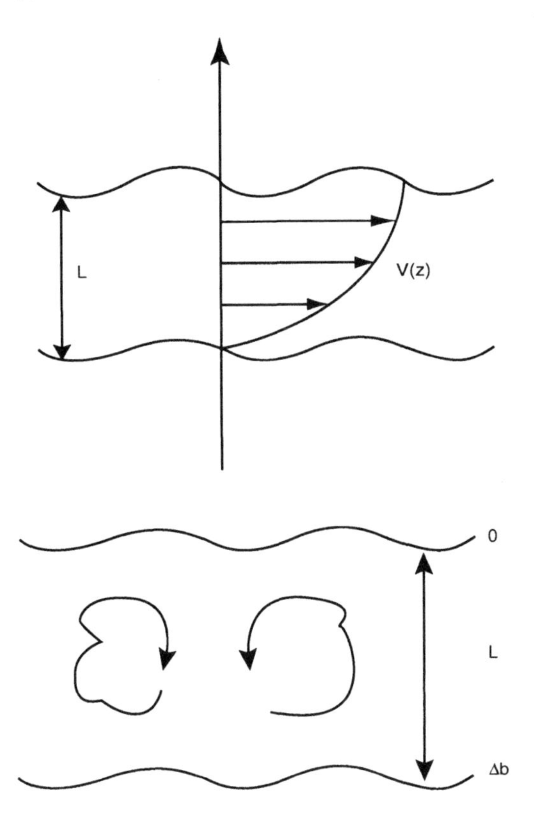

Figure 1.2.1. A sketch of a turbulent shear flow (top panel) and a turbulent convective flow (bottom panel).

important in the atmosphere and the oceans, and there are profound differences between the two mechanisms. The upper ocean is generally shear mixed from winds blowing at the surface (although nocturnal and wintertime cooling at the surface generates convective turbulence as well). The lower atmosphere experiences convective mixing due to solar heating of the ground and shear mixing due to the winds during the day, with convective mixing usually predominating, but only shear mixing at night. When convection drives the turbulence, the fluid is in a state of free convection, and when shear is the dominant process, the fluid is in a state of forced convection. Other sources of turbulent mixing in the bulk of the oceans and above the ABL are propagating internal waves that break and dissipate their energy into turbulence. Double diffusion is also an important mixing mechanism in the oceans. The turbulent flow at the bottom of the ocean in regions of strong bottom currents is driven by shear. Cloud-top radiative cooling and entrainment-induced instability produce mixing at the top of a cloud-topped ABL as well.

Turbulence can also be characterized as free turbulence or wall turbulence. Free turbulence occurs far away from any solid surfaces and is in general easier to characterize because of the absence of viscous sublayers and/or roughness elements, which are an important part of turbulent boundary layers. Computationally (for example, DNS and LES), free turbulence is easier to deal with than wall turbulence, because periodic boundary conditions can be employed and there is no need to deal with the damping of turbulence adjacent to a solid surface. Numerical simulations of wall turbulence, on the other hand, have to contend with the low Reynolds number effects adjacent to the boundary. There are however certain well-known aspects of wall turbulence, such as the asymptotic law of the wall, that can be appealed to in order to avoid facing such difficulties.

Traditionally, turbulence theory has also made a distinction between isotropic and nonisotropic turbulence. Isotropic, homogeneous turbulence is governed by a simpler subset of equations, which are nonetheless hard to solve because of their nonlinearity. Nevertheless, enormous efforts went into characterizing such turbulence in the early part of this century, both theoretically and experimentally (Taylor, 1935, 1938; Karman and Howarth, 1938; Batchelor, 1953; Corrsin and Kistler, 1954; Comte-Bellot and Corrsin, 1971; Champagne, 1978), and these studies have provided a considerable insight into turbulent processes. It is easy enough to generate and measure turbulence that is nearly isotropic and homogeneous in a wind tunnel using a grid. In fact, it has been argued recently (Sreenivasan, 1996) that even though the relevant Reynolds numbers are much smaller, a systematic study of grid turbulence might be useful in characterizing certain universal aspects of turbulence and in measuring, under controlled conditions, the relevant universal constants such as the Kolmogoroff constant that are hard to measure accurately in geophysical flows. Theoretical analyses are also simpler for such turbulence, although the closure problem and the basic necessity to postulate a model for the nonlinear transfer in spectral space make the analyses approximate at best.

In turbulence, there are no exact theories. There are only models based on one's concept of what the nonlinear transfer terms should be like. While isotropic, homogeneous turbulence theory has proved its use in understanding the nature of turbulence, it is difficult to find a practical turbulent flow that behaves that way. Most turbulent flows are basically anisotropic, especially at large scales characterized by the turbulence macroscale. It is only at the smaller scales that turbulence tends to be isotropic and somewhat universal in nature, and this proves useful in characterizing and modeling some aspects of turbulence. Other than this, there is not much use for isotropic, homogeneous turbulence theories in practical turbulent flows.

1.3 STATISTICAL DESCRIPTION OF TURBULENCE

Turbulence is stochastic; it can only be described statistically. Probability density or distribution functions can be used to describe turbulent quantities more completely than quantities such as statistical averages, which are essentially moments of the probability density functions. These moments contain useful information for quantifying and describing turbulence; however, they do not describe the turbulence completely. For this reason, we need to develop a more complete statistical representation, the probability density function (PDF).

The probability density function gives information about the probability of finding a random variable between two values. Figures 1.3.1 and 1.3.2 show PDFs for several different types of data sets. In the following, we will deal with only the fluctuating quantities, assuming that their mean is zero. Let u be a turbulent fluctuating quantity with a zero mean, and let P(u) be its probability density function. P(u) denotes the probability of finding the value of the random quantity between u and u + du. By definition, P(u) is positive definite and greater than zero, and the integral of P(u) over all u must be unity:

$$\int_{-\infty}^{\infty} P(u)\,du = 1 \tag{1.3.1}$$

Any ensemble average of a function f(u) of u can be obtained as follows, where the overbar indicates an average:

$$\overline{f(u)} = \int_{-\infty}^{\infty} f(u)P(u)\,du \tag{1.3.2}$$

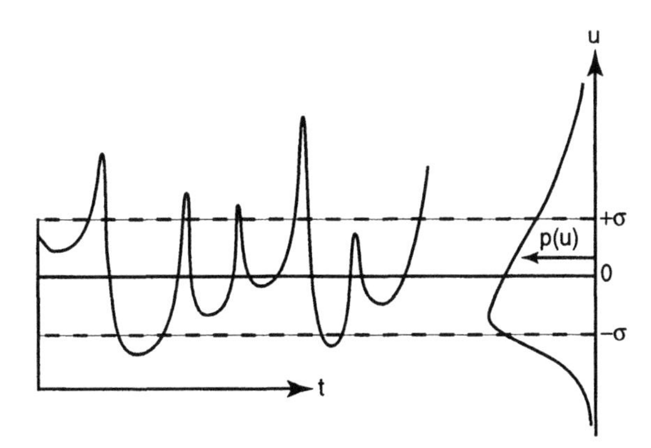

Figure 1.3.1. A function with positive skewness (from Tennekes and Lumley 1972).

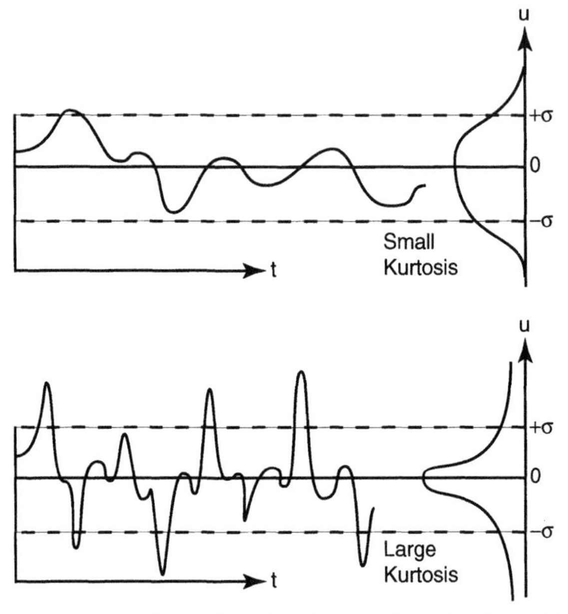

Figure 1.3.2. Functions with small and large kurtosis (from Tennekes and Lumley 1972).

This relationship applies to various powers of u as well. These are called its moments. There are an infinite number of them, but rarely does one go beyond the fourth moment. The first moment is the mean value of u and is, by virtue of the assumption above, zero:

$$\bar{u} = \int_{-\infty}^{\infty} u\, P(u)\, du = 0 \tag{1.3.3}$$

The second moment is the variance σ^2, the square root of which is the standard deviation, the root mean square (rms) value. It is a measure of the width of the probability density function P(u) and is essentially the intensity of the fluctuating quantity. It is defined as

$$\overline{u^2} = \sigma^2 = \int_{-\infty}^{\infty} u^2 P(u)\, du \tag{1.3.4}$$

For example, if u is the fluctuating velocity of the fluid in one spatial direction, then the second moment denotes twice the component of turbulence kinetic energy (TKE), or the intensity of turbulence, in that direction.

The second moment does not tell us whether the probability density function is symmetric about the origin or not. It just tells us what "energy" is contained. The third moment, on the other hand, tells us about the lack of symmetry and is called the skewness factor S when normalized by σ^3, and is defined as

$$S = \frac{\overline{u^3}}{\sigma^3} = \frac{1}{\sigma^3} \int_{-\infty}^{\infty} u^3 P(u) du \tag{1.3.5}$$

S is zero for a symmetric PDF. S can be positive or negative depending on whether large positive values of $\overline{u^3}$ are more frequent than large negative values or vice versa.

The fourth moment tells us whether the PDF is peaky or flat looking. When normalized by σ^4, it is called kurtosis or flatness factor K. K is smaller for a flatter PDF than a peaked one.

$$K = \frac{\overline{u^4}}{\sigma^4} = \frac{1}{\sigma^4} \int_{-\infty}^{\infty} u^4 P(u) du \tag{1.3.6}$$

It is also possible to describe a random variable completely by its probability distribution function (*PDF*), $P'(u)$, which is defined as the probability that the value of the random variable lies between $-\infty$ and u. By definition, $P'(u)$ is positive definite and greater than zero. It is the integral of the PDF between $-\infty$ and u:

$$P'(u) = \int_{-\infty}^{u} P(\tilde{u}) d\tilde{u} \tag{1.3.7}$$

Its value at ∞ must be unity by definition. While either the PDF or the *PDF* describes a random variable completely, it is a common practice to use the PDF. The central limit theorem states that the PDF of a random variable is a Gaussian in the limit, if the individual realizations of the random event are independent of one another. Gaussian distribution is one popular distribution in statistics, because many random events belong to this category. Skewness is zero and kurtosis is 3.0 for a Gaussian, and its PDF is defined as

$$P(u) = \frac{1}{(2\pi)^{1/2}} \cdot \frac{1}{\sigma} \exp\left(-u^2 / 2\sigma^2\right) \tag{1.3.8}$$

Unfortunately, many turbulent quantities are not normally distributed; they are not Gaussian. This is what makes them hard to deal with. The skewness and flatness factors are much different than Gaussian values and it appears that these departures from a normal distribution are the very essence of turbulence. For example, skewness of the velocity derivative for decaying homogeneous turbulence is –0.5 to –0.6.

Turbulence, of course, involves not one but several random variables dependent on one another. Therefore, it is necessary to define joint probability density functions (JPDFs) (an example is shown in Figure 1.3.3). For example, the JPDF $P_j(u,v)$ of variables u and v with zero means is the probability of finding the first random variable between u and u + du and the second one between v and v + dv. Once again a JPDF has a positive-definite value ≥ 0. The integral of P_j over the u,v two-dimensional space must be unity by definition:

$$\int_{-\infty}^{\infty}\int_{-\infty}^{\infty} P_j(u,v)\,du\,dv = 1 \tag{1.3.9}$$

We also get the PDF of u by integrating P_j over all values of v and PDF of v by integrating P_j over all u:

$$P(u) = \int_{-\infty}^{\infty} P_j(u,v)\,dv; \quad P(v) = \int_{-\infty}^{\infty} P_j(u,v)\,du \tag{1.3.10}$$

The moments of u and v can therefore be obtained from P_j as well. The joint first moment of u and v, $\overline{uv}$, can be written as

$$\overline{uv} = \int_{-\infty}^{\infty}\int_{-\infty}^{\infty} u\,v\,P_j(u,v)\,du\,dv \tag{1.3.11}$$

This is the covariance of u and v. If u and v are velocity components, the negative of this covariance denotes the kinematic turbulent shear stress. The covariance normalized by the rms values of u and v is called the correlation function, denoting the correlation between u and v. If the covariance or correlation function is zero, the two variables are uncorrelated, which, however, does not imply that the two variables are independent of each other. Statistical independence of the two variables requires that the JPDF be capable of being expressed as the product of the two PDFs.

$$P_j(u,v) = P(u)\,P(v) \tag{1.3.12}$$

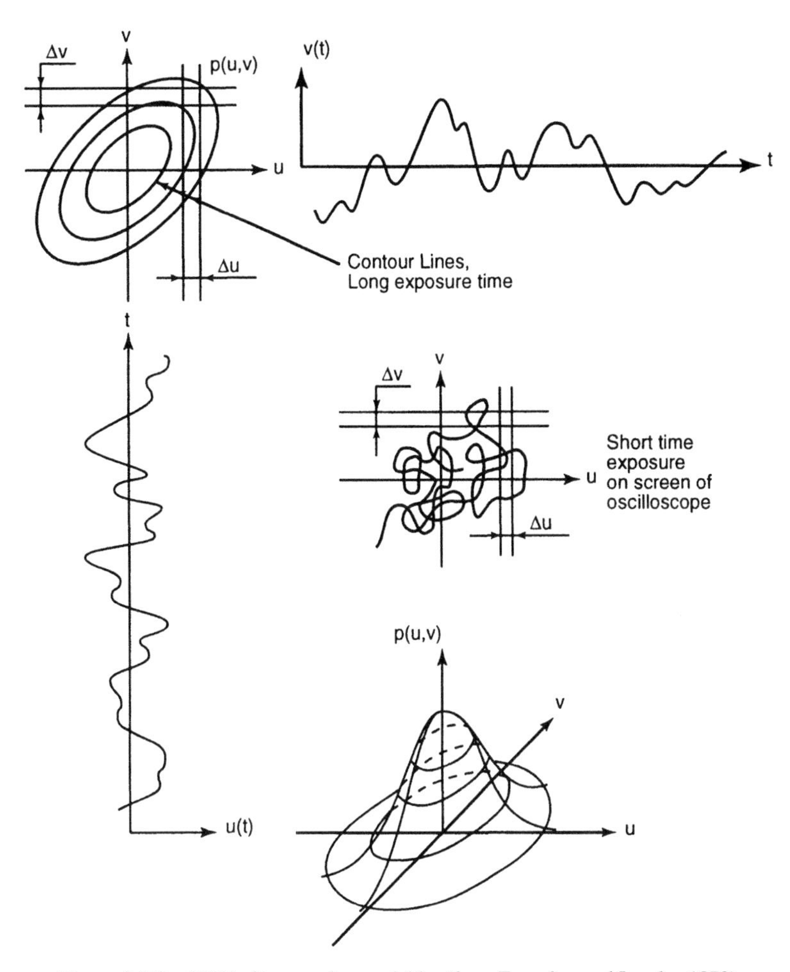

Figure 1.3.3. JPDF of two random variables (from Tennekes and Lumley 1972).

For perfectly correlated variables, the correlation function is ± 1. The covariance is a measure of the asymmetry of the JPDF. If $P_j(u,v)$ is equal to $P_j(-u,v)$, the covariance is zero.

It is, of course, possible to define joint probability distribution functions *(JPDFs)* as well. It is also possible to extend these concepts to three or more variables. Figure 1.3.4 (from Tennekes and Lumley, 1982) shows examples of JPDFs of two random variables that have negative, zero, and positive correlations. It also shows the JPDF of two uncorrelated variables that tend to inhibit each other.

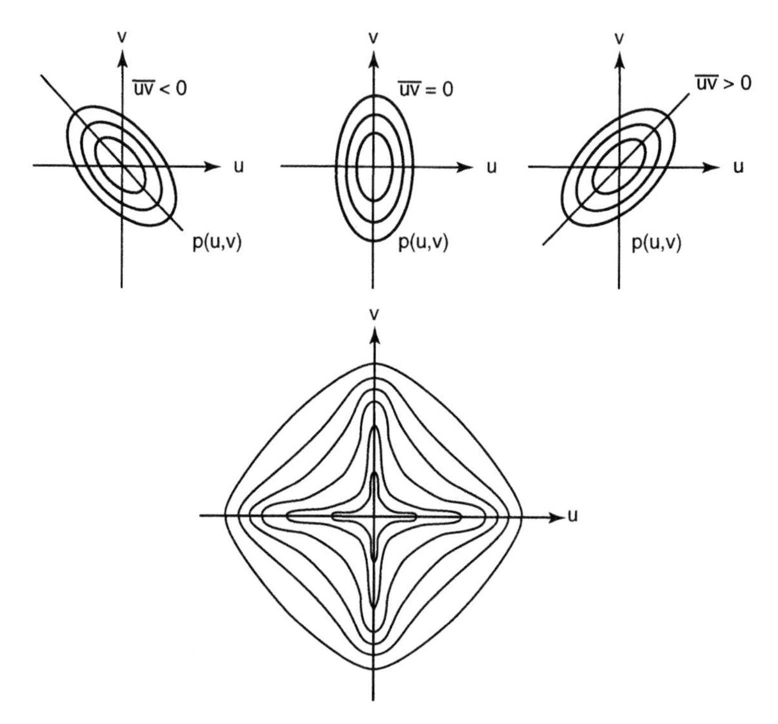

Figure 1.3.4. Examples of JPDFs with negative, zero and positive correlations, and JPDF of two uncorrelated variables (from Tennekes and Lumley 1972).

While PDFs and JPDFs are rather fundamental to turbulence, one seldom measures or uses these quantities. Most often, only the first and the second moments are measured and used to characterize a turbulent flow. These moments have definite physical meanings; for example, when u, v, and w denote the components of the fluctuating velocities, then the second moments of these quantities, $\overline{u^2}$, $\overline{v^2}$ and $\overline{w^2}$, are proportional to the components of TKE (they are also diagonal elements of the turbulent stress tensor), and the joint moments $-\overline{uv}$, $-\overline{uw}$ and $-\overline{vw}$ are components of the kinematic tangential Reynolds (turbulent) stresses. Thus these second moments and joint moments define the turbulent (or Reynolds) stress tensor that occurs in the governing Reynolds-averaged momentum equations for the mean flow, as we shall see shortly in Section 1.7.

All these descriptions assume that we are taking ensemble statistical averages, which requires taking averages over many, many ensembles or realizations, keeping the macro or overall flow conditions the same. These types of averages are the ones properly defined in probability theory. The number of samples or ensembles must be large enough for meaningful statistics. However, for statistically steady or stationary flows, the time average of a statistical quantity that is a function of time is independent of the time reference (i.e., translation along the time axis does not affect the results), and the time average is the same as that which would be obtained by ensemble averaging. Such flows are termed statistically stationary and all statistical quantities one cares to derive for such flows are independent of an arbitrary translation in time. The flow is called ergodic and the theorem equating temporal averages to ensemble averages is called the ergodicity theorem. It is a very powerful concept routinely invoked in the analysis of turbulent flows. A similar rule applies for spatial variables and averages as well.

Another statistical property of a time-dependent variable u is the autocovariance or autocorrelation function, the function that describes the self-variation of the variable in time. Autocovariance is equal to $\hat{\rho}(t,t') = \overline{u(t)u(t')}$, a measure of the relationship between values of u at two different times, t and t′. For a statistically steady or stationary variable, it is independent of time t, and depends only on the time difference $\tau = t - t'$. It is then also a symmetric function of τ: $\hat{\rho}(\tau) = \hat{\rho}(-\tau)$. The autocorrelation coefficient $\rho(\tau)$ is the normalized autocovariance:

$$\rho(\tau) = \overline{u(t)u(t+\tau)} / \overline{u^2}(t) \tag{1.3.13}$$

Since from Schwartz's inequality

$$\left|\overline{u(t)u(t+\tau)}\right| \le \left|\overline{u^2}(t)\overline{u^2}(t+\tau)\right|^{1/2} \tag{1.3.14}$$

for a stationary variable, $|\rho| \le 1$ and $\rho(0) = 1$. The autocorrelation coefficient can be used to define an integral timescale:

$$T = \int_0^\infty \rho(\tau)\, d\tau \tag{1.3.15}$$

This scale is an approximate measure of the time interval over which the random variable u(t) remains correlated with itself. If the variable is a function of distance x, then the integral scale defines a length scale, the macroscale of turbulence, which is a measure of the size of the large, energy-containing eddies in the flow (Figure 1.3.5).

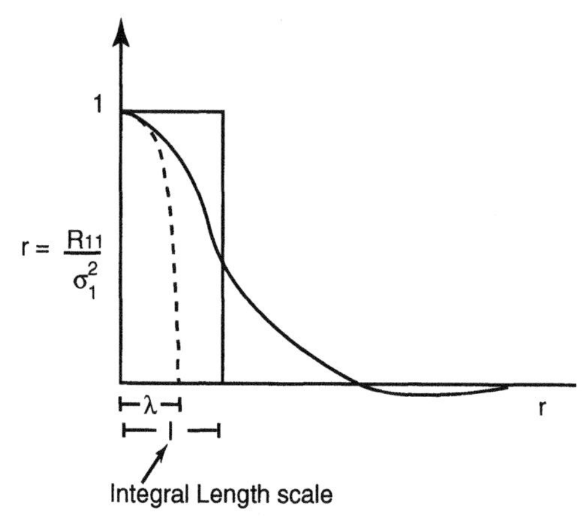

Figure 1.3.5. Autocorrelation of a random variable, showing the Taylor and integral scales.

1.4 THE IMPORTANT SCALES OF TURBULENCE

Most often, the crudest, bare minimum description of turbulence involves specifying its two scales: the length and velocity scales (and consequently the timescales and the various Reynolds numbers associated with these scales). There are three length scales of great importance in conventional turbulence in neutrally stratified flows. The first one is the macroscale ℓ, which is its integral length scale. The physical significance of this is that this denotes the scales over which turbulent quantities remain self-correlated. It also signifies the scale of the energy-containing or the most energetic eddies and corresponds to the peak of the three-dimensional turbulence spectrum E(k). This macroscale depends on the physical size of the turbulent region and is often proportional to it. It is also related to Prandtl's mixing length. For a boundary layer, ℓ is about half the boundary layer thickness. For a jet, it is proportional to the halfwidth of the jet.

The importance of the macroscale is that it essentially determines the dissipation rate of the flow. The large eddies continuously lose energy to smaller and smaller scales, with energy cascading eventually to dissipative viscous Kolmogoroff scales, where most of the dissipation is concentrated. It is this transfer rate down the spectrum that determines the dissipation rate, and not the molecular properties of the fluid. The macroscale therefore determines the dissipation rate. It is possible to define a macro Reynolds number or simply a Reynolds number of turbulence, $R_t = q\,\ell/\nu$, that is indicative of the intensity of turbulent mixing; it is simply proportional to the ratio of the turbulent viscosity to the molecular viscosity.

A timescale t_e can also be defined as $t_e \sim \ell/q$, which is often called the eddy turnover timescale. It is also proportional to K/ε (or q^2/ε), where K is the turbulence kinetic energy equal to $q^2/2$, and ε is its dissipation rate. It denotes the time over which the large energy-containing eddies lose a significant portion of their energy to dissipation through the cascade process. If the source of TKE is cut off, turbulence decays over a timescale on the order of t_e. It is also the timescale over which properties are mixed in a turbulent flow, because it involves the "mixing" length or the length scale of turbulent diffusion. In fact, the rate of dissipation of the large energy-containing eddies and hence the TKE can be looked upon as simply proportional to the ratio of TKE to the eddy turnover timescale, and hence proportional to $\left(q^3/\ell\right)$. It is important to realize that this energy loss occurs not by the direct action of viscous forces at these scales, which act very slowly over a long timescale on the order of ℓ^2/ν, but by the powerful nonlinear transfer mechanism by which their energy is handed down to smaller scales in a short timescale characterized by the eddy turnover timescale. The inverse of t_e is often called the frequency of turbulence f_t. This comes in handy in flows with their own natural frequency that entertain wave motions. For example, in stably stratified fluids, a fluid parcel when displaced vertically undergoes oscillations with a frequency equal to the buoyancy frequency or the Brunt–Vaisala frequency $N = (g/\rho_0)(d\,\bar{\rho}/dz)^{1/2}$ (see Section 1.7) and wave motions can exist only for frequencies less than the buoyancy frequency. Consequently, the ratio of f_t/N, the turbulence Froude number Fr_t (see below), is an important parameter in stably stratified turbulent flows.

The turbulence Reynolds number R_t can also be written as $(\ell/q)/\left(\nu/q^2\right)$. In other words, R_t can also be interpreted as the ratio of the turbulence timescale to the molecular timescale. Since the macroscale is often proportional to the size of the mean flow and the turbulence velocity macroscale to the mean velocity, the turbulence Reynolds number R_t is often proportional to the conventional mean flow Reynolds number, $R_N = UL/\nu$. However, R_N has a different meaning; it is the ratio of the inertial to viscous forces in the flow.

The second scale of importance is the Kolmogoroff microscale (Section 1.5 presents more discussion of this scale). This scale determines the size of the smallest possible eddies in the flow, the eddies which dissipate the TKE. The Kolmogoroff microscale is given by

$$\eta = \left(\nu^3/\varepsilon\right)^{1/4} \tag{1.4.1}$$

where ε is the dissipation rate, and the corresponding Kolmogoroff velocity scale is

$$v_\eta = (\nu\,\varepsilon)^{1/4} \tag{1.4.2}$$

A Kolmogoroff timescale can also be defined:

$$t_\eta \sim \eta / v_\eta = (\nu / \varepsilon)^{1/2} \tag{1.4.3}$$

The Reynolds number associated with η and v_η, R_η, is unity by definition. The viscous forces are of the same magnitude as the inertial forces at these scales of motion. This shows the importance of viscous forces at these scales. The significance of these scales of turbulence lies in the fact that, even though they are passive and adjust to the energy trickling down the spectrum, they are responsible for dissipating the TKE. They do so simply by becoming small enough so that the strain rates, and hence the viscous stresses and viscous dissipation, become just large enough to dissipate whatever amount of energy is being handed down the spectrum to them by the large energy-containing eddies! Since $\varepsilon \sim q^3/\ell$, it is easy to see that $\ell/\eta \sim R_t^{3/4}$. If one denotes k_e as the wavenumber of the spectral peak ($\sim 1/\ell$) and k_η as the wavenumber corresponding to the Kolmogoroff scale ($\sim 1/\eta$), then the ratio $k_\eta / k_e \sim R_t^{3/4}$. The separation between the two scales is therefore quite large for large turbulence Reynolds numbers R_t. A large R_t is essential for the existence of a well-defined inertial subrange in the turbulence spectrum (see Section 1.5). The inertial subrange consists of midsize eddies which are smaller than the largest eddies near ℓ that are gaining energy from the mean flow, and larger than the smallest eddies near the Kolmogoroff scale that are dissipating the energy. For a well-defined Kolmogoroff inertial subrange to exist, this Reynolds number must exceed 100, and for a well-defined Bachelor–Obukhoff–Corrsin subrange (the counterpart of the inertial subrange but for a passive scalar, or a property of the fluid which does not affect the mean dynamics, such as a pollutant) in a scalar spectrum, about 1000. Measurements by Grant, Stewart, and Molliet (1962) have shown that the dissipation spectrum is peaked at $k \sim 0.1/\eta$ at very high Reynolds numbers, close to the theoretical value (0.09) derived by Kraichnan (1964) using his Lagrangian History Direct Interaction Theory of turbulence. Measurements in the ocean by Gargett *et al.* (1984) and Moum (1990) indicate a value between 0.1 and 0.2.

The third scale in turbulence, made popular by G. I. Taylor (Taylor, 1935) and called the Taylor microscale λ, can also be related to dissipation. It is, however, more directly related to the curvature of the autocorrelation coefficient at the origin (Figure 1.3.5). The autocorrelation coefficient can be expanded in terms of powers of r/λ around the origin:

$$\rho(r) = 1 - \left(\frac{r}{\lambda}\right)^2 \tag{1.4.4}$$

For homogeneous turbulence, it can be shown that (e.g., Hinze, 1976)

$$\varepsilon = 5\nu\frac{q^2}{\lambda^2}, \quad \overline{\left(\frac{du}{dx}\right)^2} = \frac{2\overline{u^2}}{\lambda^2} \tag{1.4.5}$$

This expression is derived from the relationship between dissipation rate and the turbulence strain rate (defined below),

$$\varepsilon = 2\nu\overline{s_{ij}s_{ij}} = 15\frac{\nu u^2}{\lambda^2} = 5\frac{\nu q^2}{\lambda^2} \tag{1.4.6}$$

in isotropic turbulence, where s_{ij} is the turbulence strain rate (Hinze, 1976). It is then readily seen that the Taylor microscale is therefore related to dissipation, since the dissipation rate $\varepsilon = 2\nu\overline{s_{ij}\,s_{ij}}$ is equal to 15 ν $(du_1/dx_1)^2$ in isotropic turbulence (Hinze, 1977, p. 219), so that $\varepsilon = 15\ \nu\ u^2/\lambda^2 = 5\ \nu\ q^2/\lambda^2$ (since $u^2 = q^2/3$ for isotropic turbulence).

We can compare the Taylor microscale to the macroscale by examining the ratio $\ell/\lambda \sim R_t^{1/2}$, if we substitute the usual approximate expression $\varepsilon \sim q^3/\ell$. Thus one can see that the Taylor microscale is intermediate to the integral length or macroscale ℓ, and the viscous or Kolmogoroff microscale η. It marks the high wavenumber limit of the inertial subrange. It is not characteristic of the viscous dissipative scales, although it is related to the dissipation rate. The importance of the Taylor microscale is that the magnitude of the turbulent strain rate s_{ij} can be approximated as q/λ and therefore, λ is the scale of the vorticity fluctuations in the flow. Also, a Reynolds number $R_\lambda \sim u\lambda/\nu$ can be defined, which can be interpreted as the ratio of the large eddy turnover timescale t_e to the timescale of turbulent strain rate or vorticity fluctuations, λ/q. This Reynolds number is heavily used in interpretations of homogeneous, isotropic turbulence and laboratory grid-generated turbulence.

The following relationships between these three important turbulence scales in neutrally stratified flows are quite useful:

$$\frac{\eta}{\ell} \sim R_t^{-3/4}; \quad \frac{\lambda}{\ell} \sim R_t^{-1/2} \sim R_\lambda^{-1}; \quad \frac{\lambda}{\eta} \sim R_t^{1/4} \sim R_\lambda^{1/2} \tag{1.4.7}$$

These scales take the following typical values:

1. Small wind tunnel. Here $\nu \sim 10^{-5}$ m^2 s^{-1}. The flow section is about 1 m on average, so typically $\ell \sim 0.4$ m, and $q \sim 0.5$ m s^{-1}, so that

$$R_t \sim 2 \times 10^4, \eta \sim 2 \times 10^{-4} \text{ m}, \lambda \sim 10^{-2} \text{ m}, R_\lambda \sim 100$$

2. Oceans. Here $\nu \sim 10^{-6}$ m^2 s^{-1}. The oceanic mixed layer (OML) is ~50 m thick on average, so typically $\ell \sim 20$ m, and $q \sim 0.05$ m s^{-1}, so that

$$R_t \sim 10^6, \eta \sim 6 \times 10^{-4} \text{ m}, \lambda \sim 0.1 \text{ m}, R_\lambda \sim 10^3$$

3. Atmosphere. Here $\nu \sim 10^{-5}$ m^2 s^{-1}. The convective atmospheric boundary layer (CABL) is ~2 km thick, so typically $\ell \sim 800$ m, and $q \sim 0.5$ m s^{-1}, so that

$$R_t \sim 4 \times 10^7, \eta \sim 1.6 \times 10^{-3} \text{ m}, \lambda \sim 1 \text{ m}, R_\lambda \sim 10^4$$

For a nocturnal atmospheric boundary layer (NABL) about 200 m thick, $\ell \sim 80$ m, and $q \sim 0.1$ m s^{-1}, so that

$$R_t \sim 8 \times 10^5, \eta \sim 3.0 \times 10^{-3} \text{ m}, \lambda \sim 0.1 \text{ m}, R_\lambda \sim 10^3$$

The mean strain rate of the fluid, $S_{ij} = (1/2)\ (dU_i/dx_j + dU_j/dx_i)$, and the strain-rate fluctuations, $s_{ij} = (1/2)\ (du_i/dx_j + du_j/dx_i)$, are also of interest. Both have units of frequency and are important in understanding how turbulent eddies are strained. For unidirectional mean flow, dU/dz is the mean shear and is proportional to the mean strain rate, and its inverse is a timescale associated with shear, $t_S = (dU/dz)^{-1}$. This is the timescale over which the large eddies are strained and oriented by the mean flow. The ratio of this timescale to the turbulence timescale t_e is an important parameter in turbulent flows.

For fluctuations of a passive scalar in a turbulent fluid, a length scale analogous to the Kolmogoroff scale is the Batchelor scale,

$$\eta_b = \left(\nu k_c^2 / \varepsilon\right)^{1/4} \tag{1.4.8}$$

where k_c is the kinematic diffusivity of the scalar. This denotes the scale at which the steepening of the concentration gradient and smoothing by diffusion are in balance.

1.4.1 ADDITIONAL STRAIN RATES AND COMPLEX TURBULENCE

So far we have discussed only neutrally stratified, nonrotating flows with no streamline curvature. Geophysical flows are seldom neutrally stratified, and planetary rotation and streamline curvature are also inherent in geophysical flows. These effects introduce additional strain rates and the resulting turbulence is more complex. Density stratification introduces buoyancy forces that assist or act against motions in a gravitational field (magnetic body forces in electrically nonneutral fluids in the presence of a magnetic field is another example). Since work has to be done to displace a fluid parcel vertically against gravity, as in a stably stratified fluid, this comes at the expense of the KE of the eddies and hence the turbulence is weakened and in extreme cases extinguished. If buoyancy body forces assist the vertical motion, then turbulence is enhanced at the expense of the potential energy (PE) of the system. Therefore ambient density stratification introduces source and sink terms in the TKE equation, and this has an important effect on turbulence and its scales (e.g., Mellor, 1973; Kantha and Clayson, 1994).

Planetary rotation is another unique feature of geophysical flows. Rotational terms are fictitious body forces introduced into the momentum equations because of the noninertial nature of the reference coordinate system fixed to the rotating planet. As such they do no work, do not appear in energy equations, and have no role in the overall energy balance (no source or sink terms due to rotation appear in the TKE equation). However, they can redistribute TKE among its components, and since it is the vertical component that is the most important in many cases, rotation does affect turbulence and its scales (e.g., Kantha *et al.*, 1989).

Streamline curvature also introduces stabilizing and destabilizing terms into the momentum equations and hence source and sink terms appear in the TKE equation. It is important in flow over topography. If the flow is over a concave surface, turbulence is strengthened; if it is over a convex surface, it is weakened or even extinguished (see Section 3.8). Once again, turbulence characteristics are affected (e.g., Kaimal and Finnigan, 1994).

Often, two or more of these additional effects act in concert or against each other in geophysical (and other) flows. Of these additional interactions, the stratification effects are the most dominant and have been studied in great detail (e.g., Mellor, 1973). Rotational effects, important for large scale geophysical flows, are usually assumed to be negligible for small scales. This is true to a large extent. However, the horizontal component of rotation can make a difference of up to 10% in equatorial turbulence (Kantha *et al.*, 1989; Wang *et al.*, 1996). Curvature effects are only now being seriously studied in geophysical flows (Kaimal and Finnigan, 1994) and can be important when stratification effects are weak (Kantha and Rosati, 1990). Overall, gravitational stratification tends to overwhelm the rotation and curvature effects on small scale processes (see Section 2.10).

1.4.1.1 Stratification

Extensive microstructure measurements in the ocean over the past two decades by Gregg, Moum, and their collaborators have strengthened our understanding of oceanic turbulence in stratified conditions (e.g., Gregg, 1989; Peters *et al.,* 1995a,b; Moum *et al.,* 1995). These measurements are made by a dropped instrument (a few meters long) free-falling through the fluid, measuring the temperature, salinity, and velocity gradient fluctuations with depth. The resulting profile is a complex sample in time and space, and since it is impractical to make repeated drops in a small enough time to obtain meaningful ensemble averages, a series of assumptions must be made to extract useful turbulence quantities out of it. For example, the instrument is intended to sample the dissipative scales. However, the resolution attainable is such that it is impossible to resolve eddies of the order of the Kolmogoroff scales η and v_η (the dissipation rate spectrum peaks at scales $k \sim 0.1/\eta$), which are on the order of a millimeter and fractions of a millimeter per second in the main thermocline. Empirical spectral corrections are therefore necessary. The situation is worse for temperature and salinity fluctuations, since the molecular thermal and salinity diffusivities are respectively two and three orders of magnitude smaller than the momentum diffusivity. Also, since only 1 or 2 of the 12 components of the dissipation tensor is measured, assumption of local small scale isotropy is unavoidable. Determination of background mean values such as the ambient stratification from a single profile is also difficult. However, unlike the atmosphere, which is readily accessible and easily amenable to measurements by a wide variety of techniques, measurements in the ocean are highly constrained, expensive, and difficult, and the choice of platforms is few. Microstructure profilers are therefore invaluable in estimating the dissipation rates in the ocean.

The most important length scale in stably stratified flows representative of most geophysical flows is the buoyancy scale, $\ell_b = \left(\overline{w^2}\right)^{1/2} N^{-1}$, which is the maximum turbulence length scale associated with ambient stratification (a less rigorous definition is $\ell_b = q\, N^{-1}$). It is indicative of the vertical displacement at which all the vertical kinetic energy of the eddy is converted to potential energy by working against buoyancy forces and is therefore representative of the maximum vertical excursion a turbulent eddy can make.

Another estimate of the vertical excursions a parcel can theoretically make in a stratified fluid against buoyancy stratification in the absence of any overshoot is the length scale $\ell_d = \left(\overline{\rho^2}\right)^{1/2} (\partial\bar{\rho}/\partial z)^{-1}$. For cases in which temperature is the dominant stratifying agent, it can be defined as $\ell_d = \left(\overline{\theta^2}\right)^{1/2} (\partial\Theta/\partial z)^{-1}$. It can be estimated from the differences between the actual temperature (density) profile from the Thorpe reordered profile (see below).

An additional length scale is the Thorpe scale (Thorpe, 1977), which is easier to estimate from a vertical profile of potential density than the previous two scales. After binning the measured profile (for the oceans, a 1-m bin is a typical choice), the profile is reordered to restore monotonicity and hence static stability. The distance ζ_i over which each parcel has to be moved to obtain a monotonic profile is estimated. An individual overturn can then be defined as a section of the profile (ℓ_{ot}) over which the net displacement is zero: $\sum_i \zeta_i = 0$. The Thorpe scale has been found to be proportional to this overturn scale: $\ell_T \sim 0.6\,\ell_{ot}$. Averaging over 741 overturns, Peters *et al.* (1995b) found it also corresponds well to the buoyancy scale ℓ_b.

Most often in computing the Thorpe scale the potential temperature Θ is used instead of potential density for convenience. However, in places where salinity effects are also important, such as in thermohaline intrusions, potential density must be used. A judicious combination is more practical since the latter is affected by salinity spikes caused by insufficient vertical resolution of salinity fluctuations in microstructure measurements (Peters *et al.*, 1995b). Figure 1.4.1 (from Peters *et al.*, 1995b) shows an example of estimation of the overturn scales and the turbulence Froude number (see below).

The Ozmidov scale $\ell_O = \left(\varepsilon\, N^{-3}\right)^{1/2}$ is the scale at which buoyancy effects are felt strongly by turbulent eddies. It represents the vertical scale at which buoyancy and inertial forces are equal and therefore the largest vertical scale of motion that can exist in a stratified flow. The buoyancy scale ℓ_b, on the other hand, represents a buoyancy-limited length scale, which is the maximum vertical distance over which significant kinetic-to-potential energy conversions take place in turbulence under stratification. The ratio of either of these to the Kolmogoroff scale indicates the bandwidth of turbulence in a stably stratified fluid. The Ozmidov scale is related to both the Thorpe and the buoyancy length scales; it is proportional to them. This can be seen by putting $\varepsilon \sim q^3/\ell_O$ and rearranging to get $\ell \sim \ell_b$. However, Peters *et al.* (1995b) found that this relationship holds only when averaged over many overturns and not for a single overturn. This may have to do with the fact that a microstructure profiler samples an overturn at some random stage in its development phase and therefore a single measurement does not completely sample an overturn.

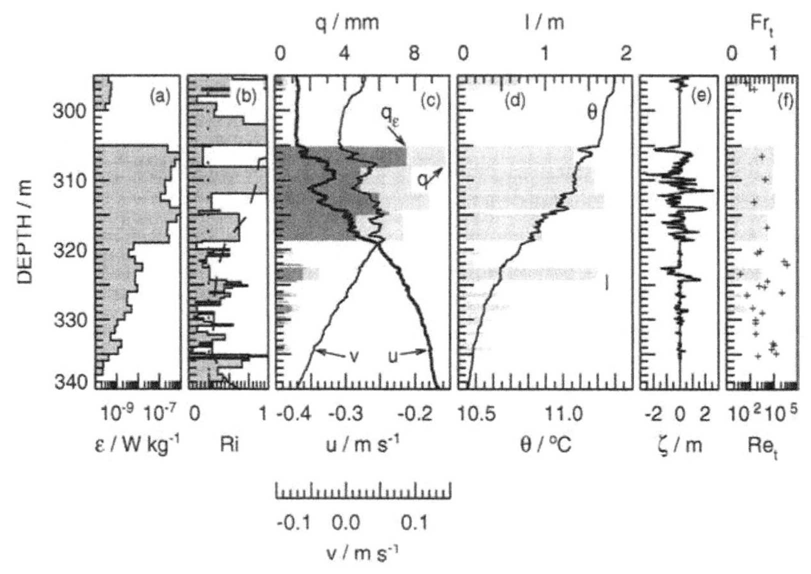

Figure 1.4.1. Plot showing various quantities measured by and estimated from microstructure profilers at the equator. a) dissipation rate, b) overturn Richardson number, c) velocity profiles and turbulent velocity, d) potential temperature and overturning scale estimated from Thorpe sorting, e) vertical displacements resulting from Thorpe sorting, f) resulting turbulence Reynolds number (shaded) and Froude number (pluses) (from Peters et al. 1995).

In all the definitions of length and timescales in stably stratified fluids, N refers to the background stratification that is often hard to determine from single realizations (as in microstructure profiles) because of the fluctuations present. Ensemble averages are the best, but are rather impractical. However, for microstructure profilers, one way to determine N is by the use of the Thorpe reordered profile. In all these measurements of turbulence-related quantities, care must be taken to exclude or limit the influence of internal wave contributions, which can produce temperature (and density) and velocity fluctuations even in the absence of turbulence. In principle, if dissipation rate can be unambiguously measured, the Ozmidov scale is the one that is most free from internal wave contamination.

All these length scales refer to the maximum turbulence length scale permissible in a stratified turbulent flow, and in the absence of internal wave contamination and in the limit of large Reynolds numbers, both seldom achieved in actual laboratory and field measurements. Observations show that within measurement uncertainties, the buoyancy scale ℓ_b, the Thorpe scale ℓ_T, and the Ozmidov scale ℓ_O are all proportional to each other. For example, Moum

(1996b) finds $\ell_T \sim \ell_b \sim 1.1\,\ell_O \sim 1.6\,\ell_d$. All these scales are used interchangeably in turbulence in stably stratified flows, although there is considerable uncertainty as to the magnitude of the proportionality constants. The latter two (and ℓ_d) can be estimated from microstructure profilers. Typical values for these scales are ~1 m in the main thermocline where N ~ 0.005 s^{-1} (Moum, 1996b). The buoyancy scale is particularly important in a strongly stratified but highly sheared flow such as in the vicinity of the Equatorial Undercurrent (EUC) core, where it is on the order of a meter or two (Peters *et al.,* 1995b). Its estimate by microstructure profilers involves a series of assumptions and approximations so that there is considerable quantitative uncertainty. Nevertheless, even a rough idea of its magnitude is helpful.

There is considerable confusion as to the relationship of these length scales to the length scale often used in second-moment closure models to define the dissipation rate $\varepsilon = c_\varepsilon\, q^3/\ell$, where the constant of proportionality $c_\varepsilon \sim 0.06$. If the dissipation rate is defined as $\varepsilon = c_\varepsilon\, q^3/\ell_b$, then $c_\varepsilon \sim 0.7$ (Moum, 1996a,b). However, the buoyancy scale is indicative of the upper bound on the length scale in a stably stratified flow, and the actual length scale that enters the dissipation rate definition might be a fraction of this scale. Also, the use of different velocity scales and length scales greatly complicates the comparison and reconciling of various measurements of the constant of proportionality c_ε; consequently the estimates for this constant from oceanic, atmospheric, and laboratory measurements vary widely, from 0.04 to 5, although most values cluster around 0.4–0.8 (Moum, 1996b).

The buoyancy timescale is the most important in a stably stratified fluid: $t_b \sim N^{-1}$. It is the natural period of oscillation of a parcel of fluid displaced vertically in a stably stratified fluid. A timescale associated with the dissipation of temperature fluctuations is $t_\theta = \overline{\theta^2}/\varepsilon_\theta$; timescales can be defined analogously for salinity and density. This is the timescale for diffusive (turbulent) smoothing of scalar fluctuations. If the source of turbulence is cut off, this is the timescale needed for the scalar fluctuations to die down. A velocity scale can also be defined as $q_s = \left(\varepsilon\, N^{-1}\right)^{1/2}$. This scale can often be used as a surrogate for vertical velocity fluctuations. It is also possible to define timescales based on $\left(\overline{w^2}\right)^{1/2}$ (or equivalently q if the vertical velocity fluctuations are not seriously suppressed by stratification effects) and the above length scales—ℓ_b/q, for example.

Nondimensional numbers based on the above scales are also of interest. Of these, the turbulence Froude number is of fundamental importance since it is proportional to the ratio of the turbulence frequency ($f_t \sim 1/t_e$) to the buoyancy frequency ($N \sim 1/t_b$):

$$Fr_t = \left(\frac{q}{N\ell}\right) = \frac{t_b}{t_e} \sim \frac{f_t}{N} \qquad (1.4.7)$$

This number can also be defined by using the Thorpe or buoyancy scale instead. For Froude numbers less than a critical value (about 1/3), turbulence appears to collapse (Hopfinger, 1987; Etling, 1993). Since stratified fluids permit internal waves at frequencies lower than N, it is easy to see that eddies with a Fr_t less than a certain value can lose their energy to radiating internal waves. This mechanism competes with the nonlinear transfer down the spectrum. In fact, it is not unreasonable to think that there should be a lower bound on Fr_t, beyond which eddies become increasingly inefficient in participating in the cascade process and hence cannot really be considered turbulent in aspects related to TKE dissipation, mixing, etc. This upper bound is in fact determined by the buoyancy or Ozmidov or Thorpe length scales, and can be thought to delineate the "turbulence" and "internal wave" parts of the spectrum. However, this transition from turbulence to internal waves in the turbulence spectrum is likely to be gradual. The turbulence Froude number is also proportional to q_s^2 / q^2, the ratio of the kinetic energy associated with the buoyancy velocity scale q_s to that associated with turbulence velocity scale q.

The ratio t_e/t_θ is larger than unity. Moum (1996b) indicates a value of about 3 from microstructure measurements, but this appears to be high compared to a value of 1.6 from classical laboratory measurements using passive scalars, for neutral stratification. It is possible that this ratio is a function of ambient stratification.

Peters *et al.* (1995b) have found that quantities related to mixing in an intermittently turbulent environment are more closely related to the turbulence Froude number than the gradient Richardson number (Figure 1.4.2). If confirmed, this has serious implications as to how the turbulence away from the actively turbulent regions such as the OML should be parameterized. However, Fr_t is an internal parameter and it is not clear how one can make use of it. They also found that the turbulence Reynolds number R_t has a more direct interpretation than the turbulence activity, $A_t = \left(\frac{\varepsilon}{\nu N^2}\right) = (t_b / t_\eta)^2$, which is often used to characterize fossil or decaying turbulence in stably stratified flows.

Moum (1998) presents an interesting plot that summarizes the range of turbulence measurements in the upper layers of the ocean, reproduced here as Figure 1.4.3. It plots on a loglog scale, the dissipation rate ε as a function of the buoyancy frequency. Lines of constant energy-containing scale (the Ozmidov scale) and the diffusive scales $(\nu k^2/\varepsilon)^{1/4}$ (where k is the diffusivity for velocity,

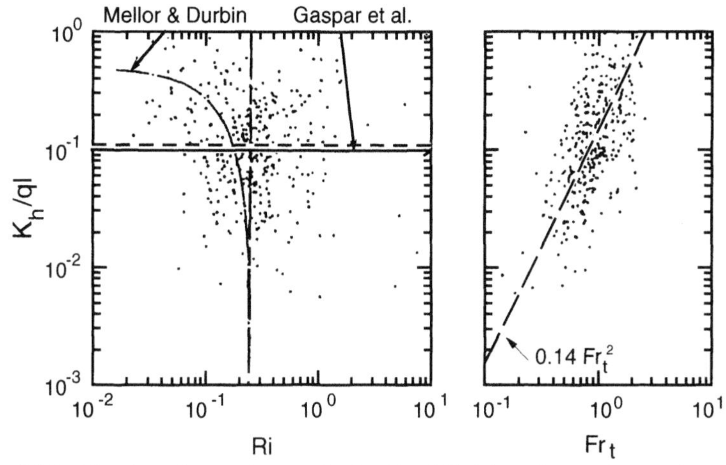

Figure 1.4.2. The dependence of normalized eddy diffusivity of a scalar on the gradient Richardson and turbulent Froude numbers. Correlation is better with the latter. Comparisons are with models of Mellor and Durbin (1975) and Gaspar et al. (1990) (from Peters et al. 1995).

heat, and salinity) are also shown to illustrate the range of scales that exist in the upper ocean. The ratio of the two scales is $(\varepsilon/\nu N^2)^{3/4} \sim 10^5$–$10^6$, and its cube determines the computing power needed to resolve oceanic turbulence. He makes the point that the large range makes it impossible to model oceanic turbulence using DNS in the foreseeable future (except for weak, highly stratified turbulence with $\ell_O \sim 10$ cm), while the LES approach is severely limited by the strong stable stratification. The reader is referred to Moum (1998) for a succinct description of the state of the art of oceanic turbulent flux measurements.

1.4.1.2 Rotational Effects

The parameter that characterizes rotational effects is the Coriolis parameter in either the vertical (f) or the horizontal (f_y) direction. In fact, it is possible to define scales similar to the above scales defined for stratification effects by replacing N, the buoyancy frequency, by one of these two parameters. However, the most important scales are the rotational length scale, $\ell_r = \left(\frac{q}{f, f_y}\right)$, and the rotational turbulent Froude number, $Fr_t^r = \left(\frac{q}{(f, f_y)\ell}\right) = \frac{t_r}{t_e} \sim \frac{f_t}{f, f_y}$. The latter

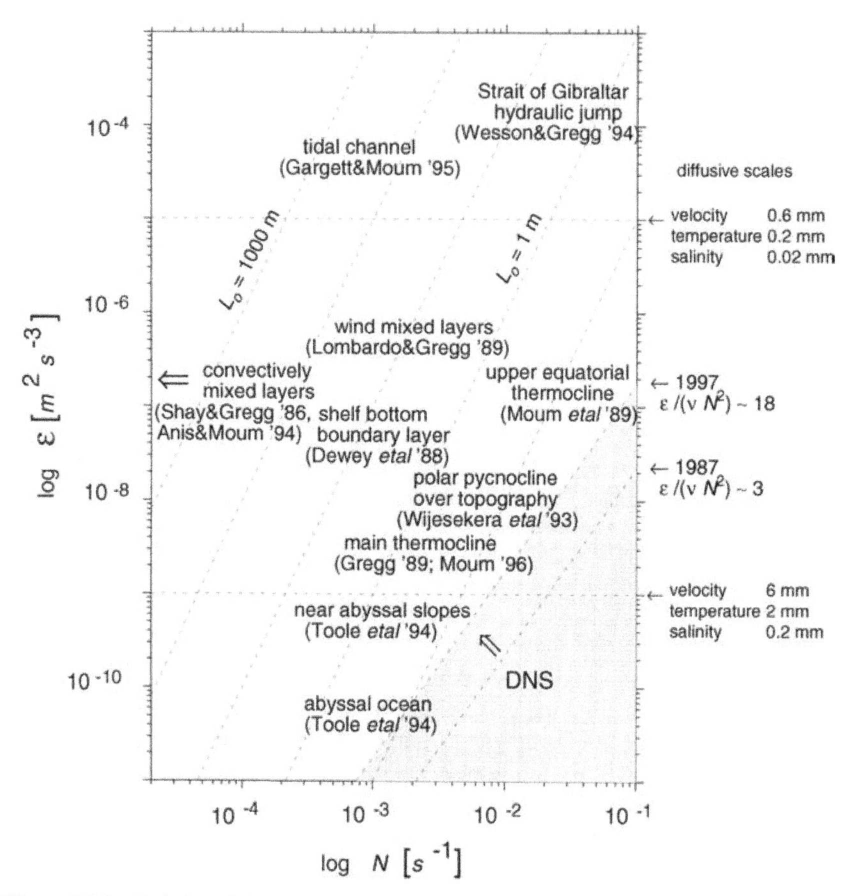

Figure 1.4.3. Variation of the dissipation rate with buoyancy frequency from Moum (1998). The references are listed in Moum (1998). The dashed lines indicate constant values of the energy containing scale L_0. The diffusive scales are indicated by dotted lines and those for velocity, temperature and salinity for dissipation rates of 10^{-5} and 10^{-9} $m^2 s^{-1}$ are shown at right. The range that can be currently simulated by DNS is shown shaded.

must be bounded, since the length scale is bounded on the upper side in neutral and stably stratified flows by the Ekman scale.

A rotational gradient Richardson number analogous to the gradient Richardson number in stratified flows (see Section 1.8) can also be defined, $Ri_R = \left(f, f_y\right)\left(\frac{\partial U_j}{\partial z}\frac{\partial U_j}{\partial z}\right)^{-1/2} = \left(\frac{f, f_y}{f_s}\right)$, and the magnitude of this appears to govern the influence of rotation on small scale turbulence (Kantha *et al.*, 1989).

1.4.1.3 Curvature Effects

The curvature effects depend on the magnitude of the streamline curvature R, or more appropriately the ratio of frequency associated with flow curvature to that associated with mean shear. One parameter that characterizes the magnitude of the curvature effects is the curvature gradient Richardson number: $Ri_C = \left(\frac{U}{R} \left(1 + \frac{z}{R} \right)^{-1} \right) \left(\frac{\partial U}{\partial z} \right)^{-1}$. U is the flow speed dependent on the distance z perpendicular to the flow streamlines. The centrifugal acceleration (deceleration) tends to stabilize (destabilize) the flow and this is characterized by Ri_C, or a similar related parameter (Kantha and Rosati, 1990).

1.5 UNIVERSAL EQUILIBRIUM RANGE AND THE INERTIAL SUBRANGE

A. N. Kolmogoroff made the most seminal contribution to turbulence theory when he postulated certain universal spectral properties of turbulence (Kolmogoroff, 1941a,b). He did this without appealing to any of the complex statistical theories of G. I. Taylor (1938) or simple mechanistic theories of Prandtl (1925). All he used was physical intuition and scaling arguments. Yet, the Kolmogoroff law stands as one of the cornerstones in the theory of turbulence. In fact, sophisticated analytical and numerical solutions of turbulence test their validity by appealing to this law, and observationalists always check for the presence of the Kolmogoroff inertial subrange in their measurements to assure themselves that the turbulence is at a sufficiently high Reynolds number to be characterized by asymptotic similarity and invariance principles. Kolmogoroff postulated:

1. At sufficiently high Reynolds numbers, there is a range of high wavenumbers where the turbulence is statistically in equilibrium and uniquely determined by the parameters ε and ν. This state of equilibrium is universal.
2. If the Reynolds number is infinitely large, the energy spectrum in the subrange of wavenumbers far from the wavenumbers characterizing both the energy-containing scales and the viscous scales is independent of viscosity ν and dependent only on one parameter: ε, the dissipation rate.

These postulates lead to some profound consequences as shown below. First we will discuss the manner in which these postulates can be deduced.

The straining rate of large energy-containing eddies is q/ℓ, roughly similar in magnitude to the strain rate $|S_{ij}|$ of the mean flow. This also means that the anisotropy of these eddies is maintained by straining by the mean flow, which orients them preferentially, for example, in the flow direction in a jet or a boundary layer. The straining rate of small eddies $|s_{ij}|$, or the straining rate of "turbulence," is q/λ. This is quite large compared to the strain rate of large eddies, because ℓ/λ is large, and therefore the anisotropy at these small scales tends to be rapidly destroyed. Another way to look at this is that for large ℓ/λ, the nonlinear interactions handing down the energy from necessarily anisotropic large scales to these scales are many, since the eddies interact principally with those of similar size (and are simply advected, not strained significantly, by eddies much larger than themselves, and in turn advect eddies that are much smaller than themselves). During the course of these interactions, all information on orientation is lost. In other words, the turbulence is "scrambled" at scales much smaller than the energy-containing scales. Thus local isotropy prevails and because the timescale associated with these scales is small enough for these scales to adjust "instantaneously" to changes in the large scales and mean flow, they are always close to equilibrium, even though the macro conditions might be changing. They then become "independent" of the large eddies; all they recognize is the rate at which energy is handed down to them by the large eddies, which under equilibrium conditions is equal to the average dissipation rate ε.

Thus, as stated by Kolmogoroff in his first postulate, the parameters ε and ν define all of the high wavenumber statistical characteristics of the flow at these high Reynolds numbers. These two parameters can then be used (see Section 1.4) to determine the internal scales of length (η), velocity (v_η), and time (t_η) [Eqs. (1.4.1)–(1.4.3)]. These are the typical dimensions, velocities, and lifetimes of eddies at the smallest turbulence scales. They are universal in that all turbulent flows that are fully developed (high enough Reynolds number) will have similar statistical characteristics. One convenient way to easily encapsulate the characteristics of the flow is by examining the spectrum of turbulent kinetic energy (see Section 1.7). This spectrum, in wavenumber space, can be written as $E = E(k, \nu, \varepsilon, \ell)$. E is also called the three-dimensional spectrum. Due to Kolmogoroff's first postulate, the parameter ℓ drops out and the spectrum E becomes a function of only k, ν, and ε. By dimensional analysis then

$$E(k) \sim \nu^{5/4}\, \varepsilon^{1/4}\, f(k\eta) \tag{1.5.1}$$

This is the famous Kolmogoroff law for the equilibrium range of the turbulence spectrum. The functional form for f is unknown, but is expected to be of the form $\exp(-ak\eta)$ for large $k\eta$, with constant $a \geq 1$. Also it is nonmonotonic, increasing up to roughly $k\eta \sim 1$ before decreasing at higher values of $k\eta$. A similar relationship holds for the one-dimensional spectrum, but with a reduced value for the constant. Observations (Grant *et al.*, 1962) show there exists a broad range in the one-dimensional turbulence spectrum $E_{11}(k)$ where this law holds (Figure 1.5.1). This includes the high wavenumber end of the spectrum, where the dissipative Kolmogoroff scales reside (Figure 1.5.2).

If the Reynolds number is sufficiently high, there exists an intermediate range in the wavenumber spectrum, in between the wavenumbers corresponding to the viscous and energy-containing scales (Figure 1.5.2), where the eddies are "independent" of both the large scales and the viscous scales. In other words, both ℓ and ν drop out in the above relationship and E can be written by dimensional analysis as

$$E(k) = \alpha\, \varepsilon^{2/3}\, k^{-5/3} \tag{1.5.2}$$

This is the famous Kolmogoroff law for the inertial subrange of the turbulence spectrum. α, called the Kolmogoroff constant (~1.6–1.7), is one of the very few universal constants in turbulence, the other being the constant associated with the universal logarithmic law of the wall in a turbulent boundary

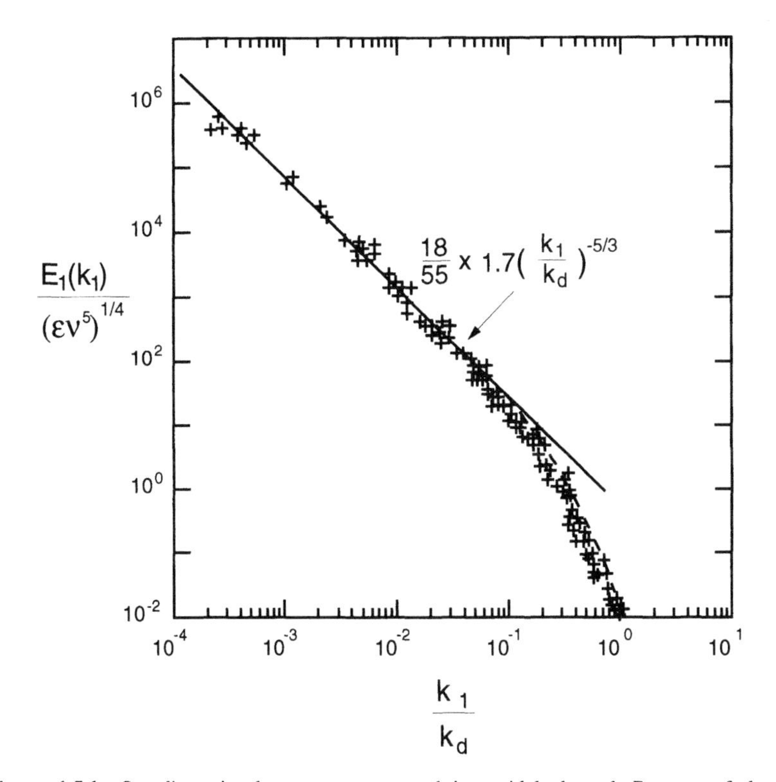

Figure 1.5.1. One-dimensional spectrum measured in a tidal channel. Because of the large Reynolds numbers typical of such flows, an extensive inertial subrange can be seen (adapted from Grant et al. 1962).

layer—the von Karman constant—which has a value of 0.40–0.41. The two constants are related to each other. Both these laws were derived from simple physical intuition and dimensional analysis.

It was only after Grant, Stewart, and Molliet's (1962) observations in a tidal channel that the existence of the inertial subrange was confirmed unequivocally. Earlier work involving laboratory flows did not achieve Reynolds numbers large enough to display unambiguously a broad range of the spectrum characterized by the inertial subrange. Only now are computers powerful enough that direct numerical solutions have been made at large enough Reynolds numbers to display an incipient inertial subrange. DNS's use spectral methods to solve the Navier–Stokes equations, directly resolving the important dissipative scales given by Kolmogoroff scales. These simulations will be discussed in Section 1.13. Kraichnan's direct interaction theory of turbulence (Kraichnan, 1964) had to be modified when in its original form it failed to produce the Kolmogoroff inertial subrange law.

Strictly speaking, Kolmogoroff postulates hold for a globally (not locally) isotropic and homogeneous flow. Sreenivasan and Antonia (1997) provide a timely and authoritative review of many aspects of small scale turbulence (for earlier ones see Nelkin, 1994; Frisch, 1995). It is impossible here to reproduce the depth of their inquiry. The universality of the Kolmogoroff constant has been both questioned (Praskovsky and Oncley, 1994) and vigorously defended (Sreenivasan, 1995) recently. Careful measurements by Praskovsky and Oncley (1994) show a weak dependence of the Kolmogoroff constant on Reynolds number, even at what one would normally regard as large Reynolds numbers. Therefore, the existence of asymptotic invariance and fully developed turbulence has been doubted (Barenblatt and Goldenfeld, 1995). But Sreenivasan (1995) has made a systematic survey of hundreds of available measurements of turbulence spectra in laboratory grid turbulence and shear flows, as well as geophysical flows, and concludes that despite the inevitable scatter, observations suggest that at sufficiently high Reynolds numbers the Kolmogoroff constant is indeed approximately constant independent of both the flow and the Reynolds number. However, Sreenivasan and Antonia (1997) state that if one looks at all the higher order moments of the probability distribution, for example, the skewness and flatness factors for the velocity derivatives, which appear not to be constants, but functions of Reynolds numbers, Kolmogoroff universality might not hold strictly. They also conclude that unequivocal measurements at $R_\lambda \sim 10^4$, or equivalently $R_t \sim 10^8$, under well-controlled conditions may be needed to resolve the question of asymptotic invariance.

Monin and Yaglom (1973) use the name quasi-equilibrium range to denote the Kolmogoroff universal range (Figure 1.5.2). The dissipation rate used in the Kolmogoroff law is the mean value of energy flux down the spectrum. However, there is reason to believe that there exist strong fluctuations at small scales and

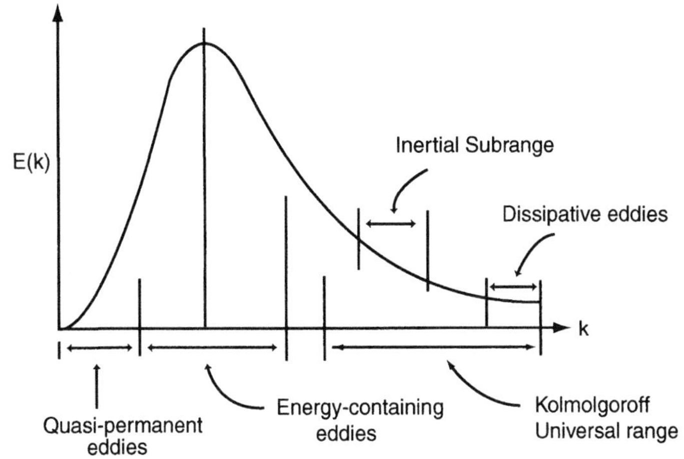

Figure 1.5.2. Three-dimensional spectrum showing the inertial subrange nested between the energy-containing eddies and the dissipative Kolmogoroff eddies.

hence in the dissipation rate itself. The nature of this fluctuation in the dissipation rate itself may depend on the large scale flow and thereby introduce a Reynolds number dependence. If so, it introduces corrections to the Kolmogoroff law that depend on the probability distribution of the dissipation rate fluctuations, which is likely to be highly non-Gaussian in nature. These are hard to measure and formulate theoretical ideas about, since it is then necessary to postulate the spatial characteristics of high vorticity fluctuations in the flow (vortex sheet-like or vortex tube-like structures of turbulence at small scales). Kolmogoroff himself modified his law somewhat by taking into account the intermittency (spatial and temporal) of turbulence at small scales (Kolmogoroff, 1962; see also Obukhoff, 1962). Nevertheless, both theoretical and observational evidence suggests that the correction, if any, is rather small. Scatter in the observational data makes it hard to discern any systematic deviations, and for all practical purposes, the Kolmogoroff -5/3 law can be considered to be a rather universal behavior of small scales in a turbulent flow.

The parameter most often measured in the laboratory and in the field is the one-dimensional (longitudinal) spectrum defined by

$$E_{11}(k_1) = \alpha_1\, \varepsilon^{2/3}\, k_1^{-5/3} \qquad (1.5.3)$$

where the subscript 1 indicates the flow direction. Sreenivasan (1995) finds $\alpha_1 \sim 0.53$ as the average value of all the data he has examined. Mostly based on a compilation by Yaglom (1981), Hogstrom (1996) concluded from a careful examination of 32 data sets from the laboratory, the atmosphere, and the ocean that this value is 0.52 ±0.01. Figure 1.5.3 shows α_1 plotted against the Reynolds number based on the Taylor microscale. Local isotropy implies $\alpha = 3.66\ \alpha_1$ and therefore $\alpha \sim 1.91$. Since $\overline{u_1^2} = \int_0^\infty E_{11}(k_1)dk_1$ (see Section 1.10), the transverse spectra are related to the longitudinal one by $E_{22}(k_1) = E_{33}(k_1) = (4/3)\ E_{11}(k_1)$, if local isotropy prevails.

Since the spectrum of the longitudinal velocity fluctuations in the surface layer of the atmospheric boundary layer (ABL) obeys similarity rules (see Section 3.3), and the von Karman constant (κ) is part of the expression for the spectrum, it is possible to derive the value of the Kolmogoroff constant from the well-known value of the von Karman constant. Kaimal and Finnigan (1994) find that $\alpha_1 \approx \kappa^{2/3}$, and for $\kappa = 0.4$, $\alpha_1 = 0.54$.

Kolmogoroff's theories assume the overall flow of energy in spectral space is one way—from large eddies down the spectrum to small eddies. While this is true overall, there is often a reverse flow from small eddies to larger ones locally. This is called scattering, as well as negative viscosity. Some LES models try to parameterize this effect. The overall impact of this scattering on spectral characteristics of turbulence is still rather uncertain.

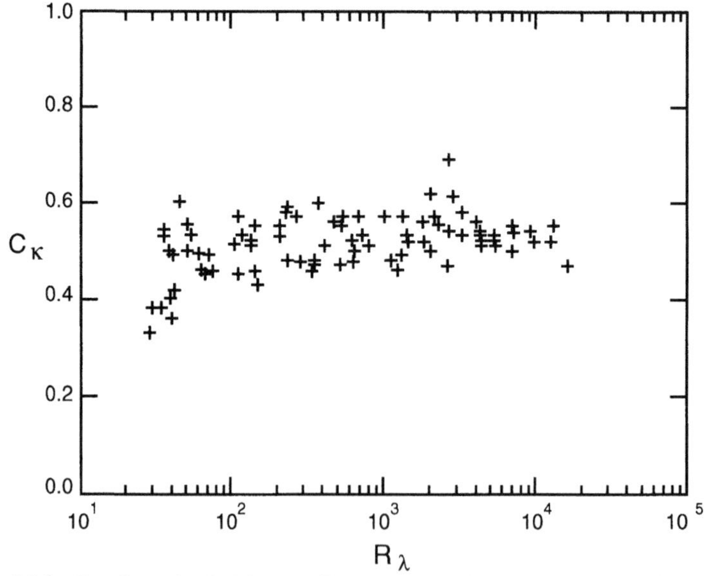

Figure 1.5.3. One-dimensional Kolmogoroff constant plotted against the Taylor microscale for a large number of spectrum measurements. Note the apparent universality of the constant in spite of the large scatter (from Sreenivasan 1995).

1.6 VON KARMAN LOGARITHMIC LAW OF THE WALL

Another very important "universal" turbulence law is the one related to the behavior of turbulence close to a boundary. This important law can also be derived from dimensional analysis. Adjacent to an unbounded wall, if the spatial scales of variability along the wall are much larger than those perpendicular to the wall, then in a region close to the wall the various fluxes (the momentum flux, the heat flux, etc.) remain approximately constant with distance from the wall. This region is called the constant flux layer. In the constant flux region of a neutrally stratified turbulent boundary layer (see Chapter 3 for a generalization to stratified flows), the momentum flux, or the shear stress τ, is constant. It is then possible to define a velocity scale $u_* = (\tau/\rho)^{1/2}$, called the frictional velocity, which is the scale of the turbulent velocity fluctuations in the layer. Sufficiently far away from the outer regions of the boundary layer, where the thickness of the boundary layer δ is the relevant length scale, and far away from the immediate vicinity of the wall, where the viscous scale ν/u_* is the relevant length scale (for a smooth wall—for a rough wall it is z_0, the roughness scale), the only relevant length scale is the distance from the wall, z. Thus, the external parameter dU/dz, the mean shear, has to be scaled by u_* and z,

$$\frac{dU}{dz} \sim \frac{u_*}{z} \tag{1.6.1}$$

which can be readily integrated to yield

$$\frac{U}{u_*} = \frac{1}{\kappa} \ln z + C \tag{1.6.2}$$

where κ is the von Karman constant (0.40–0.41) and C is the integration constant. This is the famous logarithmic law of the wall, derived independently by both von Karman and Prandtl and called the von Karman (often the Karman–Prandtl) universal logarithmic law of the wall (Karman, 1930; Prandtl, 1932). The value of this constant has been measured in a wide variety of flows in the laboratory that indicate a universal value of ~0.41. Some early measurements in the atmosphere (Businger *et al.*, 1971) suggested a much smaller value of 0.35, and this led to speculation for a while that the constant may not be universal, but instead a function of the salient nondimensional number in the flow (for example, the Rossby number—see Section 3.2). Careful reexamination of the errors involved (Purtell *et al.,* 1981; Hogstrom, 1996) and more recent observations (Zhang *et al.,* 1988) and laboratory measurements (Raupach, 1981; Raupach *et al.,* 1991) indicate that the constant is indeed a constant with a value of around 0.40±0.01.

Both the Kolmogoroff law for the inertial subrange and the von Karman law of the wall can also be obtained by appealing to asymptotic matching principles expounded originally by the Nobel laureate Clark Millikan to derive the logarithmic law of the wall (Millikan, 1939). The argument goes as follows. Far away from the wall, the normalizing length scale is the thickness of the boundary layer δ, the normalizing velocity is the friction velocity u_*, and therefore the mean velocity (or, more appropriately, the defect or the departure of the mean velocity from the free stream velocity) can be written as

$$(U-U_o)/u_* = F(Z), \text{ where } Z = z/\delta \tag{1.6.3}$$

This is the outer law. In fact, all variables in the flow, such as the dissipation rate, must scale with u_* and δ. In the viscous layer adjacent to the wall, on the other hand, the normalizing variables are u_* and the viscous scale ν/u_*, so that

$$U/u_* = f(z_+), \text{ where } z_+ = u_* z/\nu \tag{1.6.4}$$

This is the inner law. For a rough wall, the length scale is z_0 and the law of the wall can be derived using the same arguments. In the region far from the wall and far from the edge of the boundary layer, both laws must be valid, provided the limits $z_+ \rightarrow \infty$ and $Z \rightarrow 0$ are taken simultaneously. In other words, there exists an overlap region where both laws must hold simultaneously. The matching requires that the velocity gradients be the same in that limit:

$$dU/dz = (u_*/\delta)(dF/dZ) \tag{1.6.5}$$

$$dU/dz = (u_*^2/\nu)(df/dz_+) \tag{1.6.6}$$

Equating the two expressions and rearranging,

$$Z\, dF/dZ = z_+\, df/dz_+ \tag{1.6.7}$$

The left-hand side (LHS) is a function of only Z and the right-hand side (RHS) is a function of only z_+, and therefore equality demands that both be equal to a constant = κ^{-1}. Integrating the resulting expressions,

$$F(Z)=(1/\kappa)\ \ln(Z) + C_1$$

and
$$f(z_+) = (1/\kappa)\ \ln(z_+) + C \tag{1.6.8}$$

both valid for $z_+ >> 1$ and $Z << 1$. These expressions give

$$U/u_* = (1/\kappa)\ \ln(z_+) + C \tag{1.6.9}$$

the logarithmic law of the wall (Figure 1.6.1), and

$$(U-U_o)/u_* = (1/\kappa)\ \ln(z/\delta) + C_1 \tag{1.6.10}$$

the velocity-defect law (Coles, 1956) or Cole's law of the wake (Figure 1.6.2). Thus, by appealing to matched asymptotic expansions (singular perturbation theory), we have been able to derive both the outer wake law and the logarithmic

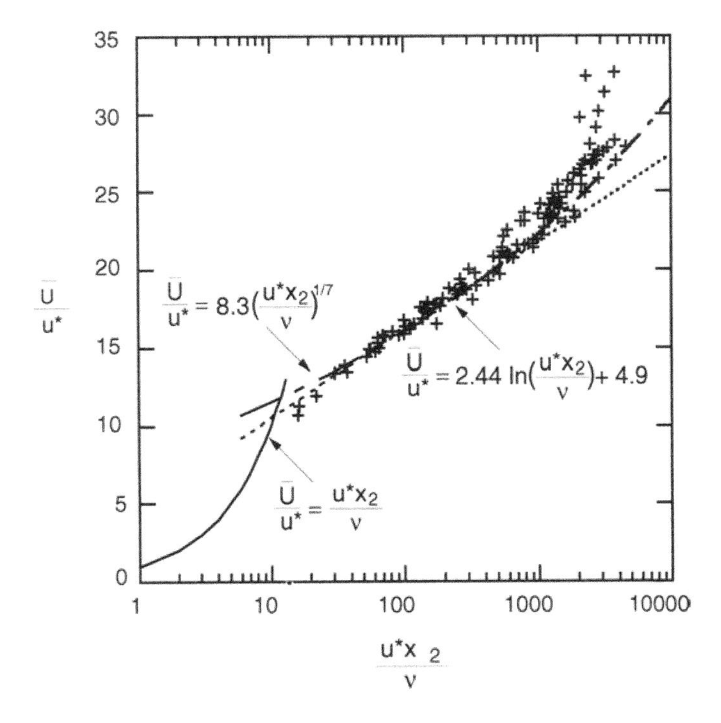

Figure 1.6.1. Mean-velocity distribution adjacent to a smooth wall, showing the logarithmic distribution away from the viscous region close to the wall and the wake region near the edge of the turbulent flow (note: x_2 is z; theoretical curves from Hinze 1976; data sources are listed in Hinze, 1976).

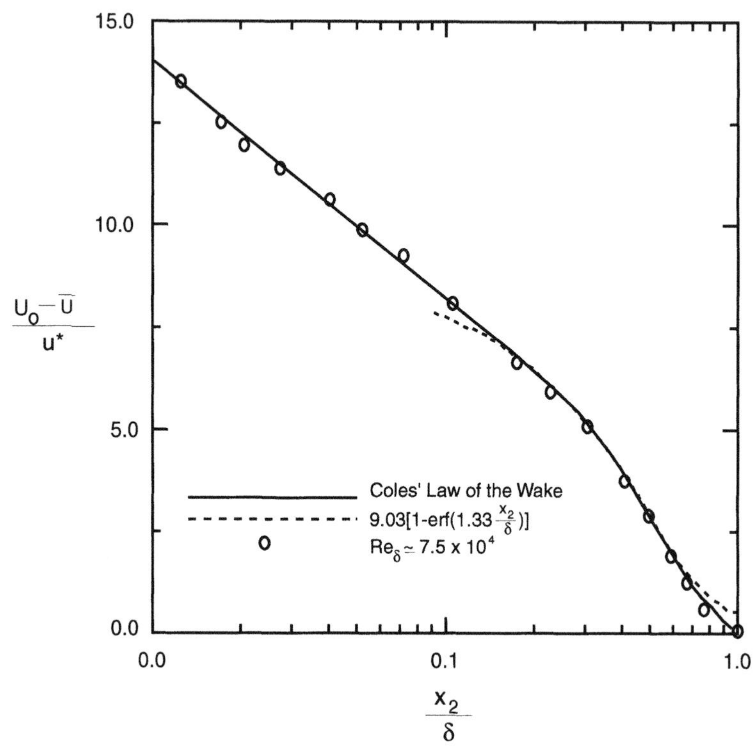

Figure 1.6.2. Velocity distribution in a turbulent shear flow showing the Coles law of the wake (from Hinze 1976). Note: x_2 is z.

law of the wall. Appealing to similarity theory or dimensional reasoning gave us only the law of the wall. Clark Millikan was the one who developed the matching procedure in 1939, but a rigorous mathematical technique of matched asymptotic expansions was not formalized until the fifties.

The corresponding "outer" law for turbulence in spectral space is

$$\frac{E(k)}{q^2 \ell} \sim F(k\ell) \tag{1.6.11}$$

This functional relationship is not universal and is different for different classes of flows, since it is characterized by the macroscale which depends on the mean flow. The important thing is that the functional relationship does not involve viscosity ν, since we postulate that the direct action of viscous forces on the large eddies is negligibly small. In the dissipative range of the spectrum (toward the high wavenumber end of the spectrum), E should be independent of the large scales and depend only on the viscosity and the dissipation rate. In other words,

Kolmogoroff microscales are the normalizing variables (ℓ does not occur in the relationship):

$$\eta E / \nu^2 = f(k\eta) \quad (1.6.12)$$

This is the "inner" law, the Kolmogoroff law for the equilibrium range. Now imagine an intermediate range of the wavenumber spectrum ($k_e << k << k_d$). Here the two laws must hold simultaneously in the asymptotic sense, that is, as $k\eta \to 0$ and $k\ell \to \infty$ simultaneously. There must exist an overlap or matched region where both the inner (dissipative) and the outer (macroscale) laws must have proper limits. Matching (e.g., Tennekes and Lumley, 1982) gives

$$F(kl) = \alpha\,(k\ell)^{-5/3}$$

$$f(k\eta) = \alpha\,(k\eta)^{-5/3} \quad (1.6.13)$$

The latter, along with Eqs. (1.6.12) and (1.4.1), gives $E(k) = \alpha\varepsilon^{2/3}k^{-5/3}$, the Kolmogoroff inertial subrange law.

1.7 GOVERNING EQUATIONS

In order to determine the present and future state of a turbulent fluid, it is necessary to apply the laws of fluid mechanics and thermodynamics to the flow. However, since turbulence is random, a statistical method for determining the mean properties of the flow is required. We will first discuss the governing equations of the flow and then describe Reynolds averaging, the statistical procedure used to derive the governing equations for turbulence.

Turbulent flows are governed by one or another variant of the Navier–Stokes equations, since continuum approximation is quite adequate to characterize turbulent flows, because even at the very high Reynolds numbers characteristic of geophysical flows, the smallest turbulence microscales are several orders of magnitude larger than the mean molecular free path. Only in the very rarefied regions of the upper reaches of the Earth's atmosphere does one have to relax the continuum approximation. At any instant in time, the mass, momentum, and scalar are conserved in the flow and these conservation equations can be written in tensorial notation as

$$\frac{\partial \tilde{u}_i}{\partial x_i} = 0 \quad (1.7.1)$$

$$\frac{\partial \tilde{u}_j}{\partial t} + \frac{\partial}{\partial x_k}\left(\tilde{u}_k \tilde{u}_j\right) + \varepsilon_{jkl} f_k \tilde{u}_l = -\frac{1}{\rho_0}\frac{\partial \tilde{p}}{\partial x_j} + \frac{\partial}{\partial x_k}\left(\tilde{\sigma}_{kj}\right) - g_j \beta \tilde{\theta} \qquad (1.7.2)$$

$$\frac{\partial \tilde{\theta}}{\partial t} + \frac{\partial}{\partial x_k}\left(\tilde{u}_k \tilde{\theta}\right) = \frac{\partial}{\partial x_k}\left(k_T \frac{\partial \tilde{\theta}}{\partial x_k}\right) \qquad (1.7.3)$$

Note that in these equations, Einstein's summation convention applies, meaning a term involving repeated indices implies summation over all possible combinations of index values. For example, $\tilde{u}_i \tilde{u}_i = \tilde{u}_1^2 + \tilde{u}_2^2 + \tilde{u}_3^2$ (e.g., Batchelor, 1967). The quantity ε_{jkl} is a third-order, skew-symmetric tensor, called the alternating tensor. It is equal to +1 if i, j, k are in cyclic order, −1 if i, j, k are in anticyclic order, and 0 if any two of the i, j, k indices are equal. For example,

$$\varepsilon_{123} = \varepsilon_{231} = 1, \quad \varepsilon_{321} = \varepsilon_{213} = -1, \quad \varepsilon_{112} = \varepsilon_{232} = 0$$

We have used the Boussinesq approximation in these equations, meaning that the density of the fluid is considered constant as far as its mass is concerned, but density variations are retained in buoyancy considerations. In other words, density is regarded as constant unless it is multiplied by gravity. This is a very good approximation for most geophysical flow phenomena.

These equations are for an incompressible flow with a single property $\tilde{\theta}$, temperature, affecting the buoyancy of the fluid. Extension to geophysical fluids, where usually two properties govern the density (temperature $\tilde{\theta}$ and salinity $\tilde{S}$ in the oceans, and temperature and specific humidity in the atmosphere), is straightforward but requires an additional conservation equation as well as the equation of state relating density to these properties (see Appendix B for general equations of state for oceans, freshwater lakes, and the atmosphere). For the oceans, a linearized equation of state can be written as

$$\rho = \rho_r + \left[\beta(\tilde{\theta} - \tilde{\theta}_r) + \beta_S (\tilde{S} - \tilde{S}_r)\right] \qquad (1.7.4)$$

Subscript r denotes reference values and β and β_S are coefficients of expansion and are functions of $\tilde{\theta}$ and $\tilde{S}$ in general, but a good approximation at normal $\tilde{\theta}$ and $\tilde{S}$ values is $\beta \sim -0.16$ kg m^{-3} °C^{-1} and $\beta_S \sim 0.75$ kg m^{-3} psu^{-1}. In general, a complicated equation of state is needed for the oceans (see Appendix B for the widely used UNESCO equation of state). For the atmosphere, specific humidity is important for density considerations. The water vapor-laden atmosphere can be treated as a mixture of ideal gases and the Clasius–Clapeyron equation can be used to obtain the necessary equation of state (Appendix B).

The instantaneous velocity and pressure are denoted by $\tilde{u}_i$ and $\tilde{p}$. We use a right-handed coordinate system with x_3 directed positive upward, and x_1 in the zonal and x_2 in the meridional directions. Note that

$$g_j = -\delta_{j3}\, g \tag{1.7.5}$$

where the quantity δ_{ij} is the Kronecker delta (equal to 1 if i = j, but 0 if i ≠ j). $\tilde{\sigma}_{ij}$ is the second-order, symmetric viscous stress sensor,

$$\tilde{\sigma}_{ij} = 2\nu \cdot \frac{1}{2}\left(\frac{\partial \tilde{u}_i}{\partial x_j} + \frac{\partial \tilde{u}_j}{\partial x_i}\right) = 2\nu \tilde{S}_{ij} \tag{1.7.6}$$

where $\tilde{S}_{ij}$ is the instantaneous rate of strain; ν is the kinematic viscosity. As can be seen from Eq. (1.7.6), viscous stress occurs because of shear in the fluid. This shear acts to deform the fluid, and can act in any of the three Cartesian directions on any of the three sides of the fluid, creating nine components, of which only six are independent.

Note that as far as possible, we will deal with kinematic quantities in this book, meaning quantities normalized by the density ρ of the fluid. This simplifies dimensional analysis and scaling arguments, since the units are then length and time only (meter and second). Therefore, $\tilde{\sigma}_{ij}$ is the kinematic stress tensor equal to the viscous stress tensor divided by density ρ. Note also that kinematic viscosity $\nu = \mu/\rho$, where μ is the molecular viscosity. Similarly, thermal diffusivity $k_T = K_T/\rho\, c_p$, where K_T is the thermal conductivity and c_p is the specific heat. Both ν and k_T have units of $m^2\ s^{-1}$. We have assumed a Newtonian fluid in Eq. (1.7.6), implying a linear constitutive relationship— the viscous stress in a fluid medium is proportional to the rate of strain in the fluid (the stress is however proportional to the strain in elastic media, the so-called Hooke's Law).

The stress tensor $\tilde{\sigma}_{ij}$ is often defined to include pressure terms:

$$\tilde{\sigma}_{ij} = -\tilde{p}\,\delta_{ij} + 2\nu\,\tilde{s}_{ij} \tag{1.7.7}$$

The deformation of a fluid parcel as it is being advected around consists of two components, the rotation and the strain; only the latter contributes to the stress.

The deformation rate $\partial \tilde{u}_i / \partial x_j$ can therefore be written as

$$\frac{\partial \tilde{u}_i}{\partial x_j} = \tilde{s}_{ij} + \tilde{r}_{ij} \tag{1.7.8}$$

where $\tilde{s}_{ij}$ is the symmetric strain rate tensor and $\tilde{r}_{ij}$ is the skew-symmetric rotation tensor:

$$\tilde{s}_{ij} = \frac{1}{2}\left(\frac{\partial \tilde{u}_i}{\partial x_j} + \frac{\partial \tilde{u}_j}{\partial x_i}\right); \quad \tilde{r}_{ij} = \frac{1}{2}\left(\frac{\partial \tilde{u}_i}{\partial x_j} - \frac{\partial \tilde{u}_j}{\partial x_i}\right) \tag{1.7.9}$$

Vorticity is an important aspect of fluid flows. It is the curl of the velocity and figures prominently in turbulence, since turbulent flows are highly vortical:

$$\tilde{\omega}_i = \varepsilon_{ijk} \frac{\partial \tilde{u}_k}{\partial x_j} \tag{1.7.10}$$

The vorticity vector and rotation tensor are related:

$$\tilde{\omega}_i = \varepsilon_{ijk}\, \tilde{r}_{kj}\,; \quad \tilde{r}_{ij} = -\varepsilon_{ijk}\, \tilde{\omega}_k \tag{1.7.11}$$

Vorticity dynamics plays a major role in fluid flows, because of the angular momentum conservation arguments related to a fluid parcel. In geophysics, vorticity plays an even more important role due to the existence of a background vorticity f_k, called planetary vorticity, that dominates all aspects of large scale flows on the rotating Earth and is important often in small scale processes as well.

Quantity f_k is the rotation vector, and the terms involving f_k in Eq. (1.7.2) are the Coriolis acceleration or Coriolis force terms. Coriolis forces are fictitious forces that appear when the equations are written in a rotating frame of reference appropriate to geophysical fluid flows. They do not do any work and therefore f_k does not appear in any energy equation (for example, the TKE equation); they drop out. They are there in momentum equations simply because an unaccelerated motion in an inertial frame of reference appears accelerated in a rotating coordinate system which is noninertial, and vice versa. This requires introducing fictitious force terms in the equations of motion in a rotating coordinate frame. On any point on the the surface of a sphere, such as the Earth, f_k has three components, zonal f_1, meridional f_2 and vertical f_3. Term f_1, the zonal component of rotation, is by definition zero. For most dynamical considerations,

f_2 can be ignored $(f_2 = 2\Omega \cos\theta)$. Therefore, only the vertical component need be retained,

$$f_k = \delta_{k3}\, 2\Omega \sin\theta \tag{1.7.12}$$

where Ω is the angular velocity of Earth's rotation ($0.707 \times 10^{-4}\ s^{-1}$) and θ is the latitude. However, near the equator, f_2 has to be retained.

1.7.1 Reynolds Averaging

Equations (1.7.1)–(1.7.3) are for instantaneous quantities and we are usually more interested in averaged quantities in a turbulent flow, such as the average shear stress and the average velocity. There are three types of averaging procedures possible:

1. Time averaging:

$$U_i(x_i, t) = \mathrm{Lim}_{T\to\infty} \frac{1}{2T} \int_{-T}^{T} \tilde{u}_i(x_i, t)\, dt \tag{1.7.13}$$

2. Spatial averaging (or volume averaging in some considerations):

$$U_i^s(x_i, t) = \mathrm{Lim}_{x_1\to\infty} \frac{1}{2x_1} \int_{-x_1}^{x_1} \tilde{u}_i(x_i, t)\, dx_1 \tag{1.7.14}$$

3. Ensemble averaging:

$$U_i^e(x_i, t) = \mathrm{Lim}_{N\to\infty} \frac{1}{N} \sum_{1}^{N} \tilde{u}_i^n(x_i, t) \tag{1.7.15}$$

For statistically stationary turbulence, $U_i = U_i(x_i \text{ only})$ in Eq. (1.7.13). In Eq. (1.7.14) we have defined an average only in the x_1 direction. For homogeneous turbulence, the x_1 dependence drops out from the left-hand side of Eq. (1.7.14), just as time drops out of the left-hand side of Eq. (1.7.13) for statistically steady or stationary turbulence. Equation (1.7.15) describes averages over ensembles of N realizations and involves the PDF of $\tilde{u}_i$. Since

$$\int_{-\infty}^{\infty} P(\tilde{u}_i)\, d\tilde{u}_i = 1 \tag{1.7.16}$$

Eq. (1.7.15) can be written as

$$U_i^e(x_i, t) = \int_{-\infty}^{\infty} \tilde{u}_i(x_i, t)\, P(\tilde{u}_i)\, d\tilde{u}_i \tag{1.7.17}$$

No summation is implied in either Eq. (1.7.16) or Eq. (1.7.17).

For a stationary, homogeneous turbulent flow, by virtue of the ergodic hypothesis,

$$U_i(x_i) = U_i^s(x_i) = U_i^e(x_i) \tag{1.7.18}$$

While stochastic methods imply ensemble averages, most of the time we appeal to the ergodic hypothesis and consider only time averages. Also, we cannot, strictly speaking, take the limit $T \to \infty$ in Eq. (1.7.13) for time-dependent flows, which most geophysical flows are. We then imply that the averaging interval is large in some sense compared to the timescale of the turbulent fluctuations, yet small compared to the timescale of change of the mean flow itself.

Equations (1.7.1)–(1.7.3) can be averaged subject to the following rules: If $A=\overline{A}+a$, $B=\overline{B}+b$, where bars denote averaged quantities (by definition, $\overline{a}=\overline{b}=0$), then

$$\begin{aligned} &\overline{\overline{A}} = \overline{A} \\ &\overline{\overline{A}\,\overline{B}} = \overline{A}\,\overline{B} \\ &\overline{\overline{A}\,a} = 0 \\ &\overline{A\,B} = \overline{A}\,\overline{B} + \overline{a\,b} \neq \overline{A}\,\overline{B} \end{aligned} \tag{1.7.19}$$

so that the average of a product of two averaged quantities does not change the product, the average of an averaged and fluctuating quantity is zero, and the average of the product of fluctuating quantities is nonzero. Let us now write each instantaneous quantity in Eqs. (1.7.1)–(1.7.3) as the sum of an average quantity and a fluctuating quantity:

$$\begin{aligned} &\tilde{u}_i = U_i + u_i \\ &\tilde{\theta} = \Theta + \theta \\ &\tilde{p} = P + p \\ &\tilde{\sigma}_{ij} = \Sigma_{ij} + \sigma_{ij} \end{aligned} \tag{1.7.20}$$

Throughout the remainder of this book we will use capital letters to denote averaged quantities and small letters for fluctuating or turbulent quantities.

We now use Reynolds averaging to separate the equations for the mean flow quantities from those for the turbulent flow quantities. We substitute Eq. (1.7.20) in Eqs. (1.7.1)–(1.7.3) and take an ensemble average, resulting in equations for mean flow quantities. The equations for the mean flow, that is, the equations for averaged quantities, are

$$\frac{\partial U_i}{\partial x_i} = 0 \tag{1.7.21}$$

$$\begin{aligned}\frac{\partial U_j}{\partial t} + \frac{\partial}{\partial x_k}\left(U_k\, U_j\right) + \varepsilon_{jkl}\, f_k\, U_l &= -\frac{1}{\rho_0}\frac{\partial P}{\partial x_j} + \frac{\partial}{\partial x_k}\Sigma_{kj} \\ &\quad - g_j\, \beta\Theta - \frac{\partial}{\partial x_k}\left(\overline{u_k\, u_j}\right)\end{aligned} \tag{1.7.22}$$

$$\frac{\partial \Theta}{\partial t} + \frac{\partial}{\partial x_k}\left(U_k\, \Theta\right) = \frac{\partial}{\partial x_k}\left(k_T \frac{\partial \Theta}{\partial x_k}\right) - \frac{\partial}{\partial x_k}\left(\overline{u_k \theta}\right) \tag{1.7.23}$$

These are called Reynolds-averaged equations. The process of taking ensemble averages to derive these is called Reynolds averaging. It is named after Osborne Reynolds, a fluid dynamicist who performed the very first experiments to demonstrate the onset of turbulence in a pipe flow by using a dye injected at the center line. Above a certain flow rate (or more appropriately, a Reynolds number), the dye line becomes chaotic and dispersive. Reynolds called this "sinuous motion" (Reynolds, 1895), not "turbulent motion." Who first coined the word turbulent is not known.

The quantity Σ_{ij} is the mean viscous stress tensor and involves the kinematic molecular viscosity ν. The quantity $\tau_{ij} = -\rho_0\, \overline{u_i u_j}$ is the turbulent stress tensor or the Reynolds stress tensor. The Reynolds stress tensor exists only when the fluid is in turbulent motion. The transfer of momentum by turbulent eddies acts to deform the fluid exactly as we would observe if we applied a force; hence the use of "stress" in the designation. τ_{ij} is a symmetric tensor ($\tau_{ij} = -\tau_{ij}$), as all stress tensors are. It is determined solely by the fluctuating velocity components of the flow. The diagonal terms are normal stresses (pressures) and are related to the turbulence kinetic energy (TKE):

$$\text{TKE} = \frac{1}{2}\tau_{ii} = \frac{\rho_0}{2}\overline{u_i u_i} = \frac{\rho_0}{2} q^2 \tag{1.7.24}$$

Quantity q^2, the trace of τ_{ij}, indicates the intensity (or violence) of turbulence. It also determines the turbulence velocity scale q. The stress tensor has six components: three normal Reynolds stresses, $-\rho_0\left(\overline{u_1^2}, \overline{u_2^2}, \overline{u_3^2}\right)$, and three tangential Reynolds stresses, $-\rho_0\left(\overline{u_1u_2}, \overline{u_2u_3}, \overline{u_3u_1}\right)$.

The quantity $Q_i = -\rho_0\, c_p\, \overline{u_i\theta}$ is the turbulent heat flux vector or the Reynolds heat flux. As with the momentum flux, this is a transfer of heat due to eddy motions in a turbulent fluid. For stratified fluids, it is more common to consider turbulent buoyancy fluxes, because these are the quantities that affect flow dynamics:

$$q_{bi} = \beta g\, \overline{u_i\theta} \tag{1.7.25}$$

For a fluid with an equation of state involving two variables (temperature $\tilde{\theta}$ and salinity $\tilde{S}$, for example), it is more appropriate to talk about density fluctuations and buoyancy fluctuations,

$$q_{bi} = \frac{g}{\rho_0}\overline{u_i\rho} = \overline{u_i b} \tag{1.7.26}$$

where $b = \rho/\rho_0$ is the buoyancy. The corresponding term in Eq. (1.7.22) that replaces $g_j \beta \Theta$ is $g_j\, \bar{\rho}/\rho_0$. Note that it is the density, or more appropriately the buoyancy fluctuations that affect the flow dynamically, that too only when a gravitational field is present, and not the temperature or salinity or humidity fluctuations individually.

Fluid flows involve conversion of energy from one form to the other, while the total energy of the flow is conserved. Molecular friction converts the kinetic energy of both the mean flow (MKE) and the fluctuations (TKE) into heat energy—it acts to dissipate the kinetic energy of the fluid. In a gravitational field, there exists potential energy as well. For stably stratified flows, which most geophysical flows are, work has to be done to displace a fluid particle in the vertical. This usually comes at the expense of TKE and results in an increase in the potential energy (PE) of the system. In fact, the fraction of TKE converted to PE can be regarded as the efficiency of turbulent mixing. It is therefore desirable to look at the equations governing the MKE (K_M) and TKE (K). First we need to derive the governing equations for turbulent (fluctuating) quantities. This can be done by simply subtracting Eqs. (1.7.21)–(1.7.23) (the equations for averaged quantities) from Eqs. (1.7.1)–(1.7.3) (the equations for instantaneous quantities),

$$\frac{\partial u_i}{\partial x_i} = 0 \tag{1.7.27}$$

$$\frac{\partial u_j}{\partial t} + \frac{\partial}{\partial x_k}\left(U_k u_j + U_j u_k + u_k u_j - \overline{u_k u_j}\right) + \varepsilon_{jkl} f_k u_l = -\frac{1}{\rho_0}\frac{\partial p}{\partial x_j} + \frac{\partial}{\partial x_k}\sigma_{kj} - g_j \beta \theta \tag{1.7.28}$$

$$\frac{\partial \theta}{\partial t} + \frac{\partial}{\partial x_k}\left(\Theta u_k + U_k \theta + u_k \theta - \overline{u_k \theta}\right) = \frac{\partial}{\partial x_k}\left(k_T \frac{\partial \theta}{\partial x_k}\right) \tag{1.7.29}$$

$$\sigma_{ij} = \rho_0 \cdot 2\nu s_{ij} \tag{1.7.30}$$

where

$$s_{ij} = \frac{1}{2}\left(\frac{\partial u_i}{\partial x_j} + \frac{\partial u_j}{\partial x_i}\right) \tag{1.7.31}$$

is the strain rate of fluctuations. The mean kinetic energy and turbulence kinetic energy are defined by

$$K_M = \frac{1}{2} U_j U_j, \quad K = \frac{1}{2}\overline{u_j u_j} \tag{1.7.32}$$

It is a simple matter to multiply Eq. (1.7.22) by U_i, replace index j by i in Eq. (1.7.22) and multiply by U_j, add the two equations to derive the equation for the rate of change of $U_i U_j$, Reynolds average, and sum up terms following the Einstein summation convention to derive the equation for the MKE:

$$\frac{\partial}{\partial t} K_M + \frac{\partial}{\partial x_k}(U_k K_M) = -\frac{1}{\rho_0}\overset{\mathrm{I}}{\frac{\partial}{\partial x_j}}(U_j P) + \frac{\partial}{\partial x_j}\left[2\nu \overset{\mathrm{II}}{U_i} S_{ij} - \overset{\mathrm{III}}{\overline{u_i u_j}}\; U_i\right] + \underset{\mathrm{IV}}{\overline{u_i u_j}\, S_{ij}} - \underset{\mathrm{V}}{2\nu S_{ij} S_{ij}} - \underset{\mathrm{VI}}{g_j \beta U_j \Theta} \tag{1.7.33}$$

Note that the Coriolis terms drop out. As mentioned earlier, this is due to the fact that these are fictitious forces that do no work and therefore do not contribute to the kinetic energy. Because of incompressibility, deformation work done by pressure forces PS_{ii} vanish and do not appear in Eq. (1.7.33).

The rate of change of MKE as the fluid particle is advected past a point consists of changes due to work done by pressure (I), transport of MKE by viscous stresses (II), transport of MKE by Reynolds stresses (III), extraction of energy by turbulence (IV), viscous dissipation (V), and work done by or against gravitational force (VI). Terms I, II, and III are flux divergence terms and cause transfer from one part of the fluid to another. Their volume integral inside a closed surface must be zero. Viscous dissipation results from the action of viscous stresses $\Sigma_{ij} S_{ij}$ and is related only to the strain-rate portion of the deformation rate of a fluid parcel, and not its rotation rate or vorticity. Viscous terms II and V are usually negligible.

The equation for TKE can be derived by a similar procedure but using Eq. (1.7.28) instead, resulting in

$$\frac{\partial K}{\partial t}+\frac{\partial}{\partial x_k}(U_k K)+\frac{\partial}{\partial x_k}\left(\overset{\text{I}}{\frac{1}{2}\overline{u_k u_j u_j}}\right)=-\frac{1}{\rho_0}\frac{\partial}{\partial x_k}\left(\overset{\text{II}}{\overline{u_k p}}\right)+\frac{\partial}{\partial x_k}\left(\overset{\text{III}}{2\nu\,\overline{u_k s_{kj}}}\right)$$
$$\overset{\text{IV}}{-\overline{u_i u_j}\,S_{ij}}\ \overset{\text{V}}{-2\nu\,\overline{s_{ij}s_{ij}}}\ \overset{\text{VI}}{-g_j\,\beta\,\overline{u_j\theta}} \qquad (1.7.34)$$

The most important equation governing the evolution of turbulence is Eq. (1.7.34) for its TKE. Here, term I represents the turbulent transport of TKE by eddies; term II represents the transport due to pressure fluctuations; and term III represents the transport of TKE by viscous stresses. These are all flux divergence terms and so do not constitute a net source or sink; they just redistribute TKE from one part of space to another. They are also called diffusion terms. Usually term III is much smaller than term I and can be neglected; turbulent diffusion dominates by far the viscous diffusion.

The most important terms in Eq. (1.7.34) for TKE are IV, V, and VI. The most important term in Eq. (1.7.33) for MKE is the work done by the Reynolds stresses (IV) that extracts energy from the flow and deposits it into turbulence. It represents a sink term for MKE, but a source term for TKE. If Eqs. (1.7.33) and (1.7.34) are added to get an equation for total KE of the flow, this term would vanish. Term IV is called the shear production term. Reynolds stresses, acting on the mean strain rate or mean shear, extract energy from the mean flow and put it into fluctuations. The shear production of turbulence or just shear production is

$$P_s = -\rho_0\,\overline{u_i u_j}\,S_{ij} = -\rho_0\,\overline{u_i u_j}\,\frac{\partial U_i}{\partial x_j} \qquad (1.7.35)$$

It is an important source term for TKE. Term V is the work done by viscous stresses acting on the fluctuation strain rate. This is the viscous dissipation term ε:

$$\varepsilon = \rho_0 \, 2\nu \, \overline{s_{ij} s_{ij}} \tag{1.7.36}$$

This is an important sink term for TKE. Note that despite the appearance of ν in the term, it is *not* small enough to be neglected. Turbulence dynamics is such that no matter how small ν is, s_{ij}, the strain rate of fluctuations, becomes large enough to dissipate whatever energy needs to be dissipated. As we saw earlier, the dissipation of TKE is independent of ν and can be written as

$$\varepsilon \sim q^3 / \ell \tag{1.7.37}$$

Term VI is the buoyancy production (destruction) term. In the presence of a gravitational field, fluid stratification implies conversions between potential and kinetic energies in the fluid. For unstably stratified fluids, the buoyancy flux term VI adds to the KE of turbulence; it constitutes a source term. For stably stratified fluids, flow has to work against gravitational forces. Energy is extracted from turbulence and converted to potential energy by the buoyancy flux term; it then becomes a sink term for TKE.

Most often, the principal balance in Eq. (1.7.34) is between terms IV, V, and VI. There exists an approximate balance between production and destruction of turbulence, that is, between the source and sink terms. This is called the local equilibrium or superequilibrium condition. If we recall that only g_3 is nonzero ($g_3 = -g$) and write Eq. (1.7.34) in terms of buoyancy fluxes, the equilibrium approximation yields (putting $u_3 = w$)

$$-\overline{u_i u_j} \, S_{ij} - 2\nu \, \overline{s_{ij} s_{ij}} + \beta g \, \overline{w\theta} = 0 \tag{1.7.38}$$

or

$$P_s - \varepsilon - q_b = 0 \tag{1.7.39}$$

where $q_b = \overline{bw} = -\beta g \overline{w\theta} = (g/\rho_0)\overline{\rho w}$ is the buoyancy flux. For stable stratification, $\overline{w\theta} < 0$ ($q_b > 0$, $\overline{\rho w} > 0$), and for unstable stratification, $\overline{w\theta} > 0$ ($q_b < 0$, $\overline{\rho w} < 0$). The ratio

$$Ri_f = \frac{q_b}{P_s} = \frac{(g/\rho_0)\overline{\rho w}}{-\overline{u_i u_j} \, S_{ij}} = \frac{-g\beta \overline{w\theta}}{-\overline{u_i u_j} \left(\partial U_i / \partial x_j \right)} \tag{1.7.40}$$

is called the flux Richardson number (If the horizontal axis is oriented along the mean flow direction, then the denominator can be approximated as $-\overline{uw}(\partial U/\partial z)$. For stably stratified fluids, Ri_f determines the efficiency of mixing q_b/P_s, since it is the turbulent eddies that sacrifice some of their TKE to entrain and lift heavier fluid (in the oceans) or lower lighter fluid (in the atmosphere) against gravitational forces into the mixed layer, often past a stable buoyancy interface. Since most of the TKE is normally just dissipated, this efficiency is usually small; Ri_f seldom exceeds 0.2 in geophysical flows.

However, it is traditional to define (erroneously) the mixing efficiency as $\gamma_m = -g\beta\overline{w\theta}/\varepsilon = (g/\rho_0)\overline{\rho w}/\varepsilon = \overline{bw}/\varepsilon$, the ratio of the buoyancy flux to the dissipation rate, the knowledge of which enables inference of heat flux from dissipation rate measurements. For turbulence in local equilibrium (production balancing destruction), it is related to Ri_f by

$$\gamma_m = Ri_f\,(1 + Ri_f)^{-1} \tag{1.7.41}$$

Moum (1996a) recommends that it be called the dissipation flux coefficient; however, traditionally entrenched notations are hard to change.

Stably stratified fluids are characterized by a stable density gradient:

$$\frac{\partial \bar{\rho}}{\partial z} < 0 \quad \text{or} \quad \frac{\partial \Theta}{\partial z} > 0 \tag{1.7.42}$$

It is possible to define a frequency N (equivalently, a period T_b) associated with stable stratification:

$$N^2 = -\frac{g}{\rho_0}\frac{\partial \bar{\rho}}{\partial z} \quad \left(\text{or} \quad \beta g \frac{\partial \Theta}{\partial z}\right) \tag{1.7.43}$$

$$T_b = 2\pi / N \tag{1.7.44}$$

N is the buoyancy frequency, also called the Brunt–Vaisala frequency, that characterizes free oscillations in stably stratified fluids. A fluid particle, if displaced vertically, undergoes vertical oscillations with a frequency N (or period T_b), before viscous damping returns it eventually to its equilibrium position. The significance is that internal wave motions of frequency ω can exist in stably stratified fluid only if $\omega < N$ or the wave period is larger than T_b. N is imaginary and hence undefined for unstably stratified fluids. Another important number in stratified fluids is the gradient Richardson number Ri_g:

$$Ri_g = \frac{-(g/\rho_0)\dfrac{\partial\bar{\rho}}{\partial z}}{\left(\dfrac{\partial U_j}{\partial z}\dfrac{\partial U_j}{\partial z}\right)} = \frac{g\beta\dfrac{\partial\Theta}{\partial z}}{\left(\dfrac{\partial U_j}{\partial z}\dfrac{\partial U_j}{\partial z}\right)} \qquad (j=1,2) \qquad (1.7.45)$$

It can also be written as N^2/f_s^2, the ratio of the square of the ratio of the buoyancy frequency to the frequency associated with mean shear. Its significance is that it determines the stability of sheared stably stratified flows. For $Ri_g > 0.25$, the inviscid theory of Miles (1961, 1963) and Howard (1961) says the flow should be stable. Ri_g being everywhere greater than 0.25 ensures the flow is stable and laminar. $Ri_g < 0.25$ is necessary for instability. The energy needed for fluid motions against gravity is provided by the mean shear. For turbulent flows in stably stratified flows, $Ri_g < 0.25$.

For completeness, we will also include an equation for the variance of temperature by multiplying Eq. (1.7.29) by θ and taking ensemble averages:

$$\frac{\partial\overline{\theta^2}}{\partial t} + \frac{\partial}{\partial x_k}\left(U_k\overline{\theta^2}\right) + \frac{\partial}{\partial x_k}\overbrace{\left(\overline{u_k\theta^2}\right)}^{\text{I}}$$

$$= k_T\overbrace{\frac{\partial\overline{\theta^2}}{\partial x_k\partial x_k}}^{\text{II}} - 2\overbrace{\overline{u_k\theta}}^{\text{III}}\frac{\partial\Theta}{\partial x_k} - \left(2k_T\overbrace{\overline{\frac{\partial\theta}{\partial x_k}\frac{\partial\theta}{\partial x_k}}}^{\text{IV}}\right) \qquad (1.7.46)$$

Term I is the transport of the variance by turbulence and term II is the very small and negligible diffusion by molecular processes. The most important terms are III, production of temperature variance P_θ, and IV, its dissipation, often denoted by N_θ or ε_θ. There exists a rough balance between these two terms: $P_\theta = N_\theta$. Similar equations can be written for specific humidity variance in the atmosphere and salinity variance in the oceans, and there also there exists a rough balance between production and dissipation of these variances: $P_q = N_q$ and $P_S = N_S$. Microstructure measurements in the oceans can provide the value of ε and N_θ (measurement of N_S is much harder), thus enabling the production terms and hence turbulent fluxes of momentum and heat to be inferred. In both the atmosphere and the oceans, P_θ can be well approximated by $P_\theta \sim -2\overline{w\theta}(\partial\Theta/\partial z)$. If one assumes local small scale isotropy, then

$\varepsilon_\theta \sim 6k_T\overline{(\partial\theta/\partial z)^2}$. Putting $-\overline{w\theta} = K_H(\partial\Theta/\partial z)$, where K_H is the turbulent heat diffusivity, and assuming turbulence to be in local equilibrium so that the production and destruction of temperature variance balance, $P_\theta = \varepsilon_\theta$, one gets

$$K_H = 3k_T\text{Co}; \quad \text{Co} = \overline{\left(\frac{\partial\theta}{\partial z}\right)^2}\left(\frac{\partial\Theta}{\partial z}\right)^{-2} \tag{1.7.47}$$

where Co is the one-dimensional Cox number, one-third the value of the three-dimensional Cox number. The significance of this is that Co can be estimated from microstructure measurements in the ocean (Osborn and Cox, 1972). On the other hand (Osborn, 1980),

$$K_\rho = \frac{-\overline{w\rho}}{(\partial\bar{\rho}/\partial z)} = \frac{\overline{wb}}{N^2} = \frac{\gamma_m\varepsilon}{N^2} = \frac{Ri_f\,\varepsilon}{(1-Ri_f)N^2} \tag{1.7.48}$$

using the definition of γ_m and Eq. (1.7.41). By assuming a value or upper bound for Ri_f, it is possible to deduce the vertical scalar diffusivity in the ocean from measurements of the dissipation rate of TKE by microstructure profiler measurements. These indicate a nearly universal value of 10^{-5} m^2 s^{-1} for K_ρ in the abyssal oceans.

Assuming turbulence in local equilibrium, the temperature variance equation (1.7.46) gives $-2\overline{w\theta}(\partial\Theta/\partial z) = \varepsilon_\theta$, so that the mixing efficiency is $\gamma_m = g\beta\varepsilon_\theta\left[2\varepsilon(\partial\Theta/\partial z)\right]^{-1}$. Since both ε and ε_θ can be measured by microstructure profilers in the ocean, γ_m can be measured. However, the values are widely scattered with the mean values around 0.1–0.4 (Moum 1990), although most recently Moum (1996a) indicates a value around 0.25–0.33, somewhat higher than the value most often used (0.2) from laboratory measurements.

The free convective limit of turbulent flows is frequently attained in geophysical flows. In this case, shear is absent and the turbulence in the fluid is driven by the heating from the bottom or cooling from the top. In this case, the term IV is zero in (1.7.34) and the principal balance in the TKE equation is between the buoyancy production term VI and viscous dissipation term V. The salient governing quantity is the Rayleigh number, $Ra = g\beta\Delta\Theta h^3/(\nu k_T)$, where $\Delta\Theta$ is the temperature difference between the top and the bottom of the fluid layer of height h (a heat flux-based Rayleigh number can also be defined—see Chapter 4). In the ABL and OML, Ra can reach values as high as 10^{19}–10^{22}. Enormous literature exists on free convection, a classic fluid mechanical problem, but most of the theoretical and observational studies pertain to rather low Ra values (10^4–10^7) clustered around the critical value for transition. Investigations of very high Ra turbulent convection are relatively few. An excellent review of high Ra convection can be found in Siggia (1994).

Another important nondimensional quantity in free turbulent convection is the Nusselt number (Nu), which is the ratio of the transport of heat by turbulence to the transport of heat by molecular diffusion. The traditional relationship between Nu and Ra ($Nu \sim Ra^{1/3}$), which is defined by the heat flux between two flat plates in the high Rayleigh number free convection limit, can be obtained through classical scaling arguments. The argument goes as follows. Let

$$Nu = \frac{Q}{k_T\left(\Delta\Theta/h\right)} \sim \left(\frac{g\beta\Delta\Theta h^3}{\nu k_T}\right)^n = Ra^n \qquad (1.7.49)$$

be the asymptotic power law for $Ra \to \infty$, where Q is the kinematic heat flux and $\Delta\Theta$ is the difference in temperature between the plates placed a distance h apart. The thermal layer adjacent to the plates is exceedingly thin at high Ra values, and therefore it is argued that the two boundary layers in thermal convection do not communicate with each other and hence the distance between the two plates is irrelevant to the expression for the heat flux in the asymptotic limit. For h to drop out of the power law relationship, n must be 1/3.

Recent laboratory observations have shown departures from this classical asymptotic power law relationship between Nu and Ra. The relationship appears to be $Nu \sim Ra^{2/7}$ instead. This is due to the persistent coherent large scale buoyancy-driven circulation at Ra values as high as 10^{14}. The shear due to this circulation modifies the thermal boundary layers, whose behavior is ultimately responsible for the heat flux. The existence of this circulation also invalidates the basic assumption underlying the argument that leads to the 1/3 law. The two thermal layers do communicate via the large eddies, and this introduces the distance between the plates as an additional length scale. While modification to the scaling arguments can provide the 2/7 power law (see Siggia, 1994), it is not yet clear if this power law is really obtained at the very high Ra numbers (10^{19}) typical of geophysical flows, which are also free of the lateral boundaries that constrain convective flows in a laboratory apparatus. Most work in geophysical flows related to high Rayleigh number turbulent convection relies on the 1/3 law.

1.8 TURBULENCE CLOSURE

Equations (1.7.21) to (1.7.23), the governing equations for the mean quantities, contain second moments $\overline{u_i u_j}$ and $\overline{u_j \theta}$. Therefore, these equations are not closed; there are more unknowns than equations, as there are six components of the Reynolds stress tensor and three components of the Reynolds heat flux vector that are unknown. In order to obtain the governing equations for these second moments using Eqs. (1.7.27)–(1.7.29), the following procedure is used. First, multiply Eq. (1.7.28) by u_i to derive an equation for $u_i \, \partial u_j / \partial t$, switch indices i and j, add the two equations to get an equation for $\partial \left(u_i u_j\right) / \partial t$, and then take an ensemble average to get an equation for the rate of change of the negative of the Reynolds stress $\partial\left(-\overline{u_i u_j}\right)/\partial t$. Similarly, from Eqs. (1.7.28) and (1.7.29), it is possible to derive an equation for the rate of change of the negative of the turbulent heat flux $\partial\left(-\overline{u_j \theta}\right)/\partial t$. This equation involves the term $\overline{\theta}^2$, for which an equation can once again be derived from Eq. (1.7.29) in a similar fashion. The resulting equations are

$$\frac{\partial}{\partial t}\left(\overline{u_i u_j}\right)+\frac{\partial}{\partial x_k}\left(U_k \, \overline{u_i u_j}\right)+\frac{\partial}{\partial x_k}\left[\underset{\mathrm{I}}{\overline{u_k u_i u_j}}-\underset{\mathrm{II}}{\nu \frac{\partial}{\partial x_k} \overline{u_i u_j}}\right]$$
$$+\underset{\mathrm{III}}{\left[\frac{\partial}{\partial x_j}\overline{p u_i}+\frac{\partial}{\partial x_i}\overline{p u_j}\right]}+f_k\underset{\mathrm{IV}}{\left(\varepsilon_{jkl}\overline{u_l u_i}+\varepsilon_{ikl}\overline{u_l u_j}\right)}$$
$$=\underset{\mathrm{V}}{\left[-\overline{u_k u_i}\frac{\partial U_j}{\partial x_k}-\overline{u_k u_j}\frac{\partial U_i}{\partial x_k}\right]}-\underset{\mathrm{VI}}{\beta\left(g_j \overline{u_i \theta}+g_i \overline{u_j \theta}\right)} \tag{1.8.1}$$
$$+\underset{\mathrm{VII}}{\overline{p\left(\frac{\partial u_i}{\partial x_j}+\frac{\partial u_j}{\partial x_i}\right)}}-\underset{\mathrm{VIII}}{2\nu\overline{\frac{\partial u_i}{\partial x_k}\frac{\partial u_j}{\partial x_k}}}$$

$$\frac{\partial}{\partial t}\left(\overline{u_j \theta}\right)+\frac{\partial}{\partial x_k}\left(U_k \, \overline{u_j \theta}\right)+\frac{\partial}{\partial x_k}\left[\underset{\mathrm{I}}{\overline{u_k u_j \theta}}-\underset{\mathrm{II}}{k_T \, \overline{u_j \frac{\partial \theta}{\partial x_k}}}-\nu\,\overline{\theta \frac{\partial u_j}{\partial x_k}}\right] \tag{1.8.2}$$

$$+\underset{\text{III}}{\overline{\theta\frac{\partial p}{\partial x_j}}}+\underset{\text{IV}}{\varepsilon_{jkl}\,f_k\,\overline{u_l\theta}}=\left[\underset{\text{V}}{-\overline{u_ju_k}\,\frac{\partial\Theta}{\partial x_k}-\overline{\theta u_k}\,\frac{\partial U_j}{\partial x_k}}\right]$$
$$-\underset{\text{VI}}{\beta g_j\overline{\theta^2}}-\underset{\text{VIII}}{(k_T+\nu)\overline{\frac{\partial u_j}{\partial x_k}\frac{\partial\theta}{\partial x_k}}} \tag{1.8.2}$$

$$\frac{\partial\overline{\theta^2}}{\partial t}+\frac{\partial}{\partial x_k}\left(U_k\,\overline{\theta^2}\right)+\frac{\partial}{\partial x_k}\left[\underset{\text{I}}{\overline{u_k\theta^2}}-\underset{\text{II}}{k_T\,\frac{\partial\overline{\theta^2}}{\partial x_k}}\right]=\underset{\text{V}}{-2\overline{u_k\theta}\,\frac{\partial\Theta}{\partial x_k}}$$
$$\underset{\text{VIII}}{-2k_T\,\overline{\frac{\partial\theta}{\partial x_k}\frac{\partial\theta}{\partial x_k}}} \tag{1.8.3}$$

In these equations, terms labeled VIII are dissipation terms; terms labeled V are production from the action of second moments against mean gradients; and terms labeled VI are buoyancy production (destruction) terms. There are also rotational terms that are normally negligible (IV), and flux divergence terms that represent spatial redistribution by turbulence transport (I), molecular transport (II), and pressure transport (III). The last three terms are also called turbulent, molecular, and pressure diffusion terms, respectively. Molecular diffusion terms can be neglected for high Reynolds number turbulence.

The term VII involving pressure–strain rate covariance $2\overline{p s_{ij}}$ deserves special attention. Note that this term vanishes on contraction of Eq. (1.8.1) and hence it does not appear in the TKE equation (1.7.34), which can also be written as

$$\frac{\partial}{\partial t}q^2+\frac{\partial}{\partial x_k}\left(U_k\,q^2\right)+\frac{\partial}{\partial x_k}\left(\overset{\text{I}}{\overline{u_k\,q^2}}-\overset{\text{II}}{\nu\,\frac{\partial}{\partial x_k}q^2}\right)+\frac{\partial}{\partial x_k}\overset{\text{III}}{\left(\overline{p\,u_k}\right)}$$
$$=\underset{\text{V}}{-2\,\overline{u_ku_j}\,\frac{\partial U_j}{\partial x_k}}-\underset{\text{VI}}{2\beta g_j\,\overline{u_j\theta}}-\underset{\text{VIII}}{2\nu\overline{\frac{\partial u_j}{\partial x_k}\frac{\partial u_j}{\partial x_k}}} \tag{1.8.4}$$

Thus the pressure–strain rate covariance does not contribute to the TKE but rather acts to redistribute energy among the three components of q^2 or TKE (TKE = $q^2/2$). Also, rotational terms are absent in Eq. (1.8.4) since they perform no net work due to their fictitious nature. The pressure–strain rate covariance term tends to make the turbulence isotropic, without, of course, ever succeeding.

Equations (1.8.1)–(1.8.3) contain unknown terms as well as third moments (e.g., $\overline{u_i u_j u_k}$) and are therefore not closed. These unknown terms and third moments can be "modeled." In other words, postulates can be made to close the equations at this level. This is called second-moment closure. Or equations can be derived for third moments $\overline{u_i u_j u_k}$ in a similar manner to the derivation of the second-moment equations. However, these equations contain fourth moments and thus are also not closed. Closure can be made at the third-moment level by means of postulates for the fourth moments. In any case, the closure problem of turbulence persists no matter to what level one is willing to proceed. This is why turbulence remains an unsolved problem in physics.

An equation for turbulent dissipation rate $\varepsilon = \nu \overline{\frac{\partial u_j}{\partial x_k}\frac{\partial u_j}{\partial x_k}}$ can also be derived by differentiating Eq. (1.7.28) w.r.t. x_k, multiplying it by

$$
\begin{aligned}
\frac{\partial \varepsilon}{\partial t} + \frac{\partial}{\partial x_k}(U_k \varepsilon) = {} & 2\frac{\partial}{\partial x_k}\left(\nu \frac{\partial \varepsilon}{\partial x_k}\right) - 2\nu \overline{\frac{\partial u_i}{\partial x_j}\frac{\partial u_k}{\partial x_j}}\frac{\partial U_i}{\partial x_k} \\
& - 2\nu \overline{\frac{\partial u_j}{\partial x_i}\frac{\partial u_j}{\partial x_k}}\frac{\partial U_i}{\partial x_k} - 2\nu \overline{u_k \frac{\partial u_i}{\partial x_j}}\frac{\partial^2 U_i}{\partial x_k \partial x_j} - 2\nu \overline{\frac{\partial u_i}{\partial x_k}\frac{\partial u_i}{\partial x_m}\frac{\partial u_k}{\partial x_m}} \\
& - 2\nu \frac{\partial}{\partial x_k}\left(\overline{u_k \frac{\partial u_i}{\partial x_m}\frac{\partial u_i}{\partial x_m}}\right) - 2\nu \frac{\partial}{\partial x_k}\overline{\frac{\partial p}{\partial x_m}\frac{\partial u_k}{\partial x_m}} - 2\nu \varepsilon_{jkl} f_k \overline{\frac{\partial u_j}{\partial x_k}\frac{\partial u_l}{\partial x_k}} \\
& - 2\nu^2 \overline{\frac{\partial^2 u_i}{\partial x_k \partial x_m}\frac{\partial^2 u_i}{\partial x_k \partial x_m}} - 2\nu g_j \beta \overline{\frac{\partial \theta}{\partial x_k}\frac{\partial u_j}{\partial x_k}}
\end{aligned}
\tag{1.8.5}
$$

$\nu \frac{\partial u_j}{\partial x_k}$, and taking an ensemble average (note that $\varepsilon = \frac{1}{2}\varepsilon_{ii}$):

This equation often serves as a surrogate for the turbulence macroscale, since asymptotically at very high Reynolds numbers, it is possible to write $\varepsilon \sim q^3/\ell$. Unfortunately, to make use of this equation, almost all the terms on the RHS have to be modeled. For future reference and completeness, we will also present the equation for dissipation of variance of scalar fluctuations (Eq. 1.7.46), $\varepsilon_\theta = 2k_T \overline{\frac{\partial \theta}{\partial x_k}\frac{\partial \theta}{\partial x_k}}$ (sometimes denoted by N_θ):

$$\frac{\partial \varepsilon_\theta}{\partial t} + \frac{\partial}{\partial x_k}(U_k \varepsilon_\theta) = \frac{\partial}{\partial x_k}\left(k_T \frac{\partial \varepsilon_\theta}{\partial x_k}\right) - 2k_T \overline{\frac{\partial \theta}{\partial x_j}\frac{\partial u_k}{\partial x_j}}\frac{\partial \Theta}{\partial x_k}$$

$$-2k_T \overline{\frac{\partial \theta}{\partial x_i}\frac{\partial \theta}{\partial x_k}}\frac{\partial U_i}{\partial x_k} - 2k_T \overline{u_k \frac{\partial \theta}{\partial x_j}}\frac{\partial^2 \Theta}{\partial x_k \partial x_j} - 2k_T \overline{\frac{\partial \theta}{\partial x_k}\frac{\partial \theta}{\partial x_m}\frac{\partial u_k}{\partial x_m}}$$

$$-2k_T \frac{\partial}{\partial x_k}\left(\overline{u_k \frac{\partial \theta}{\partial x_m}\frac{\partial \theta}{\partial x_m}}\right) - 2k_T^2 \overline{\frac{\partial^2 \theta}{\partial x_k \partial x_m}\frac{\partial^2 \theta}{\partial x_k \partial x_m}} \tag{1.8.6}$$

First attempts at turbulence closure involved closure at the first-moment level. Using eddy viscosity and mixing length approaches (see below), Eq. (1.7.22) was closed, assuming

$$\tau_{ij} = -\overline{u_i u_j} = -\frac{2}{3}K\,\delta_{ij} - 2\nu_t\,S_{ij} \tag{1.8.7}$$

where ν_t is the eddy viscosity. When the scales of fluctuations are smaller than the scales in the mean flow $(T_e << T_m, \ell << L)$, a localized (in time and space) expression such as Eq. (1.8.7) is possible. The basic problem in turbulence closure is that the governing length scale (integral length scale) is comparable to the scale of the flow (for instance the BL thickness). So there is no reason to expect local closures to work, unlike in laminar flow, where the mean free path is several orders of magnitude smaller and hence local constitutive laws hold very well. The inequalities are not satisfied since T_e can be $O(T_m)$. Still, local closures work remarkably well, at least from an engineering point of view.

The original Boussinesq eddy viscosity approach assumed ν_t to be a constant, but much larger than ν. But ν_t is a property of the flow and not the fluid, and can hardly be considered constant. However, ν_t can be approximated as $q\ell$. Since q and ℓ need to be specified, the zero equation models such as Prandtl's mixing length theories assumed $\ell \sim \ell_m$ and $q \sim \ell_m \left|\frac{\partial U}{\partial z}\right|$ by making use of the loose analogy between mixing length ℓ_m and the mean free path in the kinetic theory of gases. With ℓ_m chosen judiciously, Prandtl's mixing length theory was able to do a remarkable job of collapsing turbulence data on simple flows such as plane jets, wakes, and boundary layers [see Schlichting (1977) for examples]. This approach was extended to three-dimensional flows by

Smagorinsky (1963), using

$$\nu_t = \ell_m^{\ 2} \left(2 S_{ij} S_{ij}\right)^{1/2} \tag{1.8.8}$$

or equivalently,

$$\nu_t = \ell_m^{\ 2} \left(\bar{\omega}_i \bar{\omega}_i\right)^{1/2} \tag{1.8.9}$$

where $\bar{\omega}_i = \varepsilon_{ijk} \, \partial U_k / \partial x_j$ is the mean vorticity.

The basic conceptual problem behind the mixing length approach is that the analogy with the kinetic theory of gases fails miserably in turbulent flows, and ν_t should really be a function of the turbulence field, not the mean flow. Prandtl himself recognized this and suggested obtaining q in $\nu_t \sim q \, \ell$ using the TKE equation (1.7.34) (Prandtl, 1945). But further closure assumptions are needed to do this. Postulating a gradient transport hypothesis for the turbulent transport terms,

$$\left(\frac{1}{2}\overline{u_k u_j u_j} + \overline{p u_k}\right) \sim -\nu_t \frac{\partial K}{\partial x_k} \tag{1.8.10}$$

where

$$\nu_t = K^{1/2} \ell \tag{1.8.11}$$

and modeling dissipation ε as proportional to $K^{3/2}/\ell$, it is possible to derive an equation for K. Prescribing ℓ suitably enables ν_t to be determined then and the flow problem to be solved. This is the so-called one-equation model of turbulence. The equation for the TKE is closed and the length scale is prescribed empirically.

The most popular turbulence models are two-equation models, where an additional equation for a quantity involving ℓ is also written and closed suitably. This second equation can be for ε, the dissipation, since $\ell \sim K^{3/2}/\varepsilon$ (Spaziale, 1991),

$$\frac{\partial \varepsilon}{\partial t} + \frac{\partial}{\partial x_k}\left(U_k \varepsilon\right) = P_\varepsilon + D_\varepsilon - \varepsilon_\varepsilon + \nu \frac{\partial^2 \varepsilon}{\partial x_k \partial x_k} \tag{1.8.12}$$

where P_ε, D_ε, and ε_ε are the production, turbulent transport (diffusion), and dissipation of dissipation ε. Once again, assuming localness,

$$D_\varepsilon = \frac{\partial}{\partial x_k}\left(\frac{\nu_t}{c_2}\frac{\partial \varepsilon}{\partial x_k}\right) \tag{1.8.13}$$

$$P_\varepsilon = -2c_2\varepsilon\, b_{ik}\frac{\partial U_i}{\partial x_k} = -c_3\frac{\varepsilon}{K}\tau_{ik}\frac{\partial U_i}{\partial x_k} \tag{1.8.14}$$

$$\varepsilon_\varepsilon = c_4\frac{\varepsilon^2}{K} \tag{1.8.15}$$

where

$$b_{ij} = \left(\tau_{ij} - \frac{2}{3}K\,\delta_{ij}\right)/\,2K \tag{1.8.16}$$

is the anisotropy tensor. This is based on the assumption that while production must depend on the anisotropy of the Reynolds stress tensor, dissipation must depend only on turbulence length and velocity scales (Hanjalic and Launder, 1972). These closure postulates yield the popular two-equation K–ε model of turbulence (Spaziale, 1991; Rodi, 1987),

$$\tau_{ij} = \frac{2}{3}K\,\delta_{ij} - \nu_t\left(\frac{\partial U_i}{\partial x_j} + \frac{\partial U_j}{\partial x_i}\right) \tag{1.8.17}$$

$$\nu_t = c_1\frac{K^2}{\varepsilon} \tag{1.8.18}$$

$$\frac{\partial K}{\partial t} + \frac{\partial}{\partial x_K}(U_K K) = -\tau_{ij}\frac{\partial U_i}{\partial x_j} - \varepsilon + \frac{\partial}{\partial x_k}\left(\frac{\nu_t}{c_2}\frac{\partial K}{\partial x_K}\right) + P_b \tag{1.8.19}$$

$$\frac{\partial \varepsilon}{\partial t} + \frac{\partial}{\partial x_K}(U_K\varepsilon) = -c_3\frac{\varepsilon}{K}\tau_{ik}\frac{\partial U_i}{\partial x_K} - c_4\frac{\varepsilon^2}{K} + \frac{\partial}{\partial x_k}\left(\frac{\nu_t}{c_5}\frac{\partial \varepsilon}{\partial x_K}\right) + c_6\frac{\varepsilon}{K}P_b \tag{1.8.20}$$

where $P_b = -\beta g_j \overline{u_j\theta}$ is the buoyancy production/destruction term and ν_t is the eddy viscosity. Note in these expressions that molecular terms are omitted. The constants are $c_1 = 0.09, c_2 = 1.0, c_3 = 1.44,\ c_4 = 1.92, c_5 = 1.3$ (Spaziale, 1991), and c_6 = 1.5 (Sommer and So, 1995). Spaziale (1991) provides examples to show that this model does quite well when compared to turbulence flows such as that over a backward facing step. However, the K–ε model cannot be applied close a solid wall. Matching to the law of the wall or damping functions must be used, as for all turbulence models invoking asymptotic invariance at large Reynolds numbers.

The two-equation K–ε model has worked reasonably well for neutrally stratified flows, but a better model for application to buoyancy-dominated flows is the four-equation $K–\varepsilon–\overline{\theta^2}-\varepsilon_\theta$ model. The K–ε models used for buoyant flows assume that the ratio of the turbulence dynamic timescale $T_e \sim K/\varepsilon$ to the scalar timescale $T_\theta \sim \overline{\theta^2}/2\varepsilon_\theta$ is constant (as do many second-moment closure models). Use of the $\overline{\theta^2}-\varepsilon_\theta$ equations helps avoid this (Sommer and So, 1995). An equation for $\overline{\theta^2}$, and one for ε_θ, similar to Eq. (1.8.20) for ε, can be derived,

$$\frac{\partial\overline{\theta^2}}{\partial t}+\frac{\partial}{\partial x_k}\left(U_k\overline{\theta^2}\right)=\frac{\partial}{\partial x_i}\left(-c_7\frac{K}{\varepsilon}\tau_{ik}\frac{\partial\overline{\theta^2}}{\partial x_k}\right)-2\overline{u_k\theta}\frac{\partial\Theta}{\partial x_k}-2\varepsilon_\theta$$

$$\frac{\partial\varepsilon_\theta}{\partial t}+\frac{\partial}{\partial x_k}(U_k\varepsilon_\theta)=-c_8\frac{\varepsilon_\theta}{K}\tau_{ik}\frac{\partial U_i}{\partial x_k}-(c_9\frac{\varepsilon}{K}+c_{10}\frac{\varepsilon_\theta}{\overline{\theta^2}})\varepsilon_\theta \tag{1.8.21}$$

$$+\frac{\partial}{\partial x_i}\left(-c_{11}\frac{K}{\varepsilon}\tau_{ik}\frac{\partial\varepsilon_\theta}{\partial x_k}\right)-(c_{12}\frac{\varepsilon}{K}+c_{13}\frac{\varepsilon_\theta}{\overline{\theta^2}})\overline{u_k\theta}\frac{\partial\Theta}{\partial x_k}$$

with $c_7 = c_{11} = 0.11$, $c_8 = 0.72$, $c_9 = 0.8$, $c_{10} = 2.2$, $c_{12} = 0$, and $c_{13} = 1.8$ (Sommer and So, 1995). An equation for $\overline{u_j\theta}$ is also needed to close the problem and second-moment closure must be attempted (Sommer and So, 1995). The result is a considerable increase in complexity. Nevertheless, this model is capable of simulating the countergradient transport in flows such as decaying homogeneous turbulence in stably stratified flows (Sommer and So, 1995).

Two-equation models of the K–ε type (and others) are well suited for simple flows. They do not however do a good job in turbulent flows that are complex, meaning when additional strain rates such as those due to rotation, streamline curvature, buoyancy, and other body forces are present in the flow. For example, for turbulence in a strongly rotating frame of reference, there are profound differences in the characteristics of turbulence. Yet the K–ε model has no rotational terms in the equations and therefore provides results as if rotation were zero. Note that c_1 above is a constant and not a function of the ambient stratification. Unless c_1 is made a function of stability, K–ε models cannot be expected to perform well in geophysical flows where buoyancy effects on turbulence are important. It is therefore necessary to appeal to the full second-moment equations, which, for example, retain the rotational terms for the study of turbulence in complex flows. The K–ε model also ignores nonlocal and memory effects on Reynolds stresses. While it is possible to overcome some of these problems by a nonlinear eddy viscosity formulation, second-moment closure models with explicit consideration of transport equations for Reynolds stresses offer a better solution. They are much better suited to modeling complex turbulent flows. An additional advantage of second-moment closure models compared to first-moment closure is that since closure is performed at the second-moment level, possible inaccuracies in modeling third moments still tend to yield a better solution for mean quantities than closing equations at the first-moment level. One also determines a solution for the turbulence field as well, instead of just the mean flow field.

Rotta (1951) made pioneering contributions to second-moment closure by proposing to model pressure–strain rate covariance terms as proportional to the degree of anisotropy in the turbulence field. In other words, he modeled these terms as tendency-toward-isotropy terms. Research on rapidly strained homogenous, isotropic turbulence showed, however, the necessity for a term proportional to the mean strain rate also (Crow, 1968). Thus,

$$\overline{ps_{ij}} \sim \frac{q^3}{\ell} b_{ij} + 2c_1 q^2 S_{ij} \tag{1.8.22}$$

Using concepts of isotropy of small scales in turbulence at which dissipation occurs, the dissipative terms in Eqs. (1.8.1)–(1.8.3) can be written as

$$2\nu \overline{\frac{\partial U_i}{\partial x_k} \frac{\partial U_j}{\partial x_k}} \sim \frac{2}{3} \frac{q^3}{\ell} \delta_{ij} \tag{1.8.23}$$

$$\left(k_T + \nu\right) \overline{\frac{\partial U_i}{\partial x_k} \frac{\partial \theta}{\partial x_k}} = 0 \tag{1.8.24}$$

$$2k_T \overline{\frac{\partial \theta}{\partial x_k} \frac{\partial \theta}{\partial x_k}} \sim \frac{q}{\ell} \overline{\theta^2} \tag{1.8.25}$$

These equations, along with suitable models for turbulent and pressure transport terms, enable the equations to be closed. Recent large eddy simulations of turbulence have confirmed the form given by Eq. (1.8.22) for $\overline{ps_{ij}}$ and provided valuable guidance on modeling other unknown terms as well (Andren and Moeng, 1993; Moeng and Wyngaard, 1986). With resulting modifications to second-moment closure, such models appear to be reasonably good for application to geophysical boundary layers (Kantha and Clayson, 1994).

In its complete form, second-moment closure involves solving 10 additional transport equations for second moments in addition to 5 first-moment equations. This additional burden is most often unjustified in routine applications to computing geophysical flows. Some simplifications are possible by a perturbation expansion in terms of the degree of anisotropy indicated by tensor b_{ij} (Mellor and Yamada, 1974; Galperin *et al.,* 1988). Retaining only the lowest order terms results in a quasi-equilibrium model for turbulence, which consists of a TKE equation (an equation for length scale is added as well), along with a set of algebraic relations for second moments (second moment closure will be dealt with in detail in Chapter 2). These algebraic relations can be reduced for simple, stratified geophysical boundary layers to

$$-\overline{uw} = q\ell S_M \frac{\partial U}{\partial z}$$
$$-\overline{w\theta} = q\ell S_H \frac{\partial \Theta}{\partial z} \qquad (1.8.26)$$

where S_M and S_H are stability-dependent functions of the flux Richardson number in the limit of local equilibrium. The variation of S_M and S_H with Ri_f is shown in Kantha and Clayson (1994) (see also Section 2.10). For the general case (meaning the principal balance is not production equal to dissipation), S_M and S_H are functions of $\overline{Ri_t} = N^2\ell^2/q^2$. This parameter can be looked upon as a turbulence gradient Richardson number [similar to $Ri_g = \frac{N^2}{(dU/dz)^2}$] with shear characterized by q/ℓ. Another interpretation is that it is the square of the ratio of the buoyancy frequency to the turbulence frequency.

The weakest link in second-moment closure is the equation for the length scale of turbulence. This is where future research can help. At this point, there are four different ad hoc models for quantities involving the length scale: (1) $\varepsilon \sim q^3/\ell$, (2) $q^2\ell$, (3) $q\ell$, and (4) q/ℓ. But all these models require most if not all of the terms on the RHS to be modeled. Often the physical basis for such models is obscure and the modeling is more in analogy with the TKE equation rather than founded on any rigorous basis. Production, dissipation, and transport terms are modeled based mostly on dimensional analysis. All except the q/ℓ formulations require ad hoc terms introduced to yield the law of the wall, namely, the linear behavior of a length scale close to a boundary. This diminishes confidence in the rigor of such models.

Another major deficiency in current turbulence models (including second moment and LES ones) is the inability of these models to be integrated down to a solid boundary. Current closures are essentially asymptotic theories independent of the Reynolds number R_t and assume $R_t \to \infty$. This assumption is violated close to the wall. Rigorous extension to finite R_t is needed, but appears far more difficult to achieve than asymptotic models.

Since DNS is still limited to low Reynolds numbers (relatively speaking), even with the projected increase in computing power over the coming decades, simulation of geophysical turbulent flows by DNS is rather unlikely. LES offers the best solution. However, there are still unresolved problems in LES, especially as related to simulations close to a solid boundary or a buoyancy interface. These have to do with subgrid scale closure problems, which are not unlike problems with ensemble average closure (EAC) methods such as the second-moment closure (SMC) discussed above.

There are several flow situations that are simple enough to be simulated in the laboratory and analytically by theories of turbulence. Of these, two have been extensively used for verifying turbulence models. One is the decay of homogenous, isotropic turbulence and the other is the growth of turbulence in a homogeneous shear flow. The first one is governed by the following simple form of K–ε equations:

$$\frac{dK}{dt} = -\varepsilon$$
$$\frac{d\varepsilon}{dt} = -c_4 \frac{\varepsilon^2}{K} \tag{1.8.27}$$

These equations can be readily integrated to obtain an asymptotic $(t \to \infty)$ decay law for TKE,

$$K \sim t^{-a} \tag{1.8.28}$$

where $a = 1/(1-c_4) \sim 1.2$ from observations in wind tunnel experiments (although theoretically a value of unity is rather appealing). Equivalently, $c_4 \sim 11/6 = 1.833$.

For homogenous shear flows the equations are [from Eqs. (1.8.17)–(1.8.20)]

$$\frac{dK}{dt} = -\tau \frac{dU}{dz} + \frac{d}{dz}\left(\frac{\nu_t}{c_2}\frac{dK}{dz}\right) - \varepsilon \tag{1.8.29}$$

$$\frac{d\varepsilon}{dt} = -c_3 \frac{\varepsilon}{K} \tau \frac{dU}{dz} - c_4 \frac{\varepsilon^2}{K} + \frac{d}{dz}\left(\frac{\nu_t}{c_5} \frac{\partial \varepsilon}{dz}\right)$$

$$\nu_t = c_1 \frac{K^2}{\varepsilon} \tag{1.8.29}$$

$$\tau = -\nu_t \frac{dU}{dz}$$

An equation for $\eta = \frac{SK}{E}$ can be derived, where $S=2^{1/2}\frac{dU}{dz}$ is the strain rate of the shear flow. Substituting $\tau = S\,t$,

$$\frac{d\eta}{dt} = -c_1(c_3 - 1)\eta^2 + (c_4 - 1)$$

$$\frac{dK}{d\tau} = K\left(c_1\eta - \frac{1}{\eta}\right) \tag{1.8.30}$$

Asymptotic solutions for K and ε of the form $K \sim e^{\lambda\tau}$, $E \sim e^{\lambda\tau}$, yield $\frac{d\eta}{d\tau} = 0$, and for $\tau \gg 1$,

$$\lambda = \left[\frac{c_1(c_4 - c_3)^2}{(c_3 - 1)(c_4 - 1)}\right]^{1/2}, \eta \to \eta_\infty = \left[(c_4 - 1)/c_1(c_3 - 1)\right]^{1/2} \tag{1.8.31}$$

is the solution. $\lambda \sim 0.13 - 0.15$ as obtained from subjecting an initially isotropic turbulence to a constant shear in a wind tunnel. η has an asymptotically constant $\eta_\infty\ (\tau \to \infty)$ value of 6.08. For the $K-\varepsilon$ model discussed above, $\lambda = -0.22$, $\eta_\infty = 4.82$.

1.9 DESCRIPTION IN SPECTRAL SPACE

The minimum description of turbulence involves specifying its macro length scale ℓ (the integral scale, the energy-containing eddy scale, or the scale corresponding to the peak of the 3D spectrum) and its velocity scale q (which also is a measure of its intensity or "violence"). Specifying these also specifies the Reynolds number R_t of turbulence, its dissipation rate $\varepsilon = q^3/\ell$, and several other Reynolds numbers associated with it, as well as the Kolmogoroff and

Taylor microscales described earlier. This is equivalent to specifying the overall properties of the turbulence energy spectrum, the overall energy content and the location of the peak. While this is a reasonably good description of turbulence, it is not a complete description by any means. As discussed earlier, a complete description utilizes PDFs and JPDFs. However, an intermediate step is to describe the spectrum of turbulence, namely, the distribution of its energy in wavenumber (or equivalently frequency) space. This makes sense since turbulence contains an entire range of scales all the way from large quasi-permanent eddies, to energy-containing eddies, down to Kolmogoroff microscales, and it is of interest to know how the energy is distributed in the wavenumber space. In order to be able to do this, we need to discuss how to describe turbulence in spectral space.

The correlation tensor R_{ij} is the covariance of $u_i(\tilde{x},t)$ and $u_j(\tilde{x}+\tilde{r},t)$, a function of $\tilde{x}$ and $\tilde{r}$ in general (Figure 1.9.1):

$$R_{ij}(\tilde{x},\tilde{r}) = \overline{u_i(\tilde{x},t)u_j(\tilde{x}+\tilde{r},t)} \tag{1.9.1}$$

For homogeneous turbulence, R_{ij} is a function only of the separation vector $\tilde{r}$. The Fourier transform of R_{ij} is called the spectrum tensor F_{ij}. R_{ij} and F_{ij} are related thusly:

$$\begin{aligned} R_{ij}(\tilde{r}) &= \int_{-\infty}^{\infty}\int_{-\infty}^{\infty}\int_{-\infty}^{\infty} F_{ij}(\tilde{k})\, e^{i\tilde{k}\cdot\tilde{r}} d\tilde{k} \\ F_{ij}(\tilde{k}) &= \frac{1}{(2p)^3}\int_{-\infty}^{\infty}\int_{-\infty}^{\infty}\int_{-\infty}^{\infty} R_{ij}(\tilde{r})\, e^{-i\tilde{k}\cdot\tilde{r}} d\tilde{r} \end{aligned} \tag{1.9.2}$$

Clearly, $F_{ii}(\tilde{k}) = F_{11}(\tilde{k}) + F_{22}(\tilde{k}) + F_{33}(\tilde{k})$ is proportional to the TKE at wavenumber k, since

$$R_{ii}(0) = \overline{u_i u_i} = q^2 = \int_{-\infty}^{\infty}\int_{-\infty}^{\infty}\int_{-\infty}^{\infty} F_{ii}(\tilde{k})\, d\tilde{k} \tag{1.9.3}$$

where the integral is in the three-dimensional k space, in other words, over a spherical shell of radius $k = |\tilde{k}| = (k_i k_i)^{1/2}$ (Figure 1.9.2). If ds is the elemental

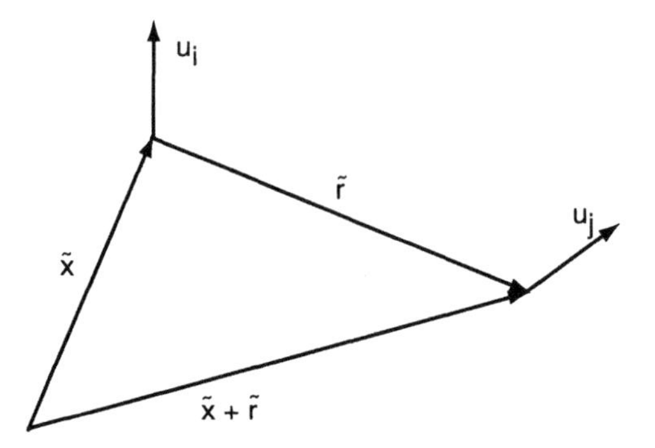

Figure 1.9.1. Definition of quantities involved in defining the correlation tensor.

surface area of a sphere of radius k, then

$$E(k)=\frac{1}{2}\iint F_{ii}(\tilde{k})\,ds \tag{1.9.4}$$

where E(k) is known as the three-dimensional spectrum of turbulence, or simply the spectrum, and is one of the most important quantities in a turbulent flow. Then TKE/unit mass (or simply TKE henceforth) can be written as

$$\text{TKE}=\frac{1}{2}\overline{u_i u_i}=\frac{1}{2}q^2=\int_0^\infty E(k)\,dk=\frac{1}{2}\int_{-\infty}^{\infty}\int_{-\infty}^{\infty}\int_{-\infty}^{\infty}F_{ii}(\tilde{k})\,d\tilde{k} \tag{1.9.5}$$

The shape of E(k) tells us how the TKE is distributed in the wavenumber space, or equivalently, the energy contained in eddies of various scales. Note that k is a scalar and so is E(k). While the spectrum tensor F_{ij} is a more complete description of the stochastic variable $u_i(x,t)$, the spectrum often provides enough information on the energetics of the variable. The one-dimensional spectra are related to E(k) by

$$\begin{aligned}E_{11}(k_1)&=\int_{k_1}^{\infty}\left(1-\frac{k_1^2}{k^2}\right)\frac{E(k)}{k}dk\\E_{22}(k_1)=E_{33}(k_1)&=\frac{1}{2}\left(E_{11}(k_1)-k_1\frac{d}{dk_1}E_{11}(k_1)\right)\end{aligned} \tag{1.9.6}$$

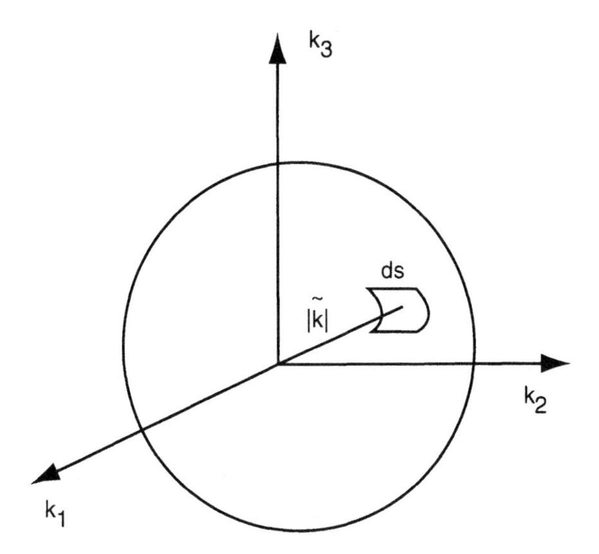

Figure 1.9.2. The spherical shell over which the spectrum tensor has to be integrated to obtain the three-dimensional spectrum E(k).

1.10 GOVERNING EQUATIONS IN SPECTRAL SPACE

Consider the TKE equation (1.7.34). For simplicity, consider the case of a horizontally homogeneous turbulence. This condition is very well approximated in the oceanic and atmospheric mixed layers, where the horizontal scales are much larger than the vertical scales in most situations. Thus, for a horizontally homogeneous turbulent flow,

$$\frac{\partial K}{\partial t}+\frac{\partial}{\partial z}\left[\overline{w\left(p+K\right)}\right]=-\overline{uw}\,\frac{\partial U}{\partial z}+\overline{wb}-\varepsilon \tag{1.10.1}$$

where $b = g\rho/\rho_0$ is the buoyancy. This equation tells us the overall energy balance in turbulent motions. It has, however, no information on how the TKE is distributed among the various scales in the turbulence spectrum. For this information, one has to look at the spectral energy balance. Such an equation can be derived as follows. Take the equation for the turbulence velocity $u_j\left(x_i\right)$, Eq. (1.7.28), and multiply it by $u_j\,(x_i')$ to form a scalar product. Interchange x_i and x_i', add the two equations, and take an ensemble average to derive an equation of the form

$$\frac{\partial}{\partial t}\left[\overline{u_j(x_i)\,u_j(x_i')}\right] = \cdots$$

Substitution of $x_i = y_i - \frac{1}{2} r_i$, $x_i' = y_i + \frac{1}{2} r_i$, gives an equation for the autocovariance $R_{ii}(y_i, r_i)$. Taking the Fourier transform with respect to the separation vector r_i gives an equation for the spectrum tensor $F_{ii}\left(y_i, \underset{\sim}{k_i}\right)$. Integrate this over a spherical shell of radius $k = (k_i k_i)^{1/2}$ to get an expression for E(k), the scalar three-dimensional spectrum. The result for the horizontally homogeneous case is

$$\frac{\partial}{\partial t} E(k) = T(k) - \frac{\partial}{\partial z} Q(k) + \tau(k)\frac{\partial U}{\partial z} + B(k) - 2\nu k^2 E(k) \quad (1.10.2)$$

Equation (1.10.2) also assumes that the vertical scales of turbulence are much smaller than the scale of the mean flow and its shear. T(k) is the net energy transfer due to nonlinear interactions to wavenumbers between k and k + dk from all other wavenumbers in spectral space. It denotes transport in spectral space:

$$\int_0^\infty T(k)\, dk = 0 \quad (1.10.3)$$

If F(k) is the net energy transfer from wavenumbers less than k to those larger than k, the energy acquired in the wavenumber band k, k + dk, is $F(k) - F(k+dk)$. Therefore,

$$T(k) = -\frac{d}{dk} F(k) \quad (1.10.4)$$

F(k) is the energy flux in spectral space. Q(k) denotes transport in physical space and hence redistribution in physical space. It involves Fourier transforms of triple covariances. τ(k) is the Fourier transform of the Reynolds stress. The term τ(k) dU/dz denotes turbulence production by shear at wavenumber k. The last term, $2\nu k^2 E(k) \equiv \varepsilon(k)$, is the viscous dissipation at wavenumber k. The approximate shapes of E(k) and $2\nu k^2 E(k)$ are shown in Figure 1.10.1.

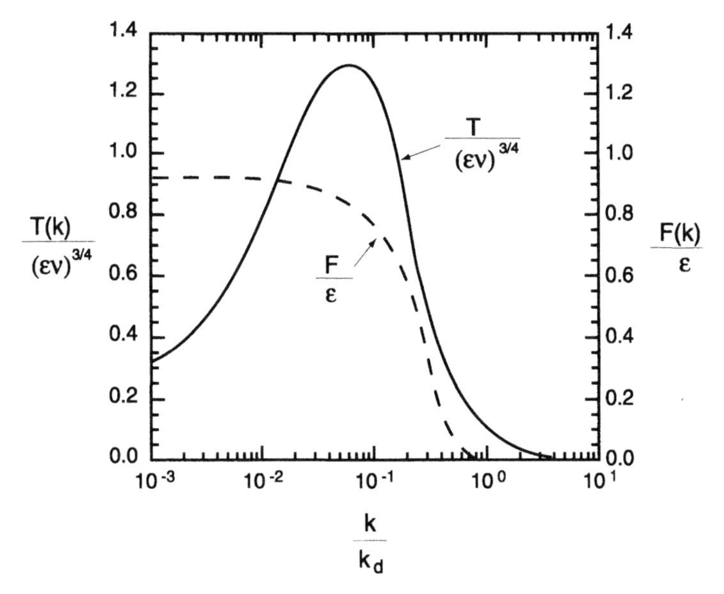

Figure 1.10.1. Functions F(k) and T(k) in spectral space.

The buoyancy flux spectrum B(k) denotes the distribution of buoyancy flux in wavenumber space, just as E(k) denotes the distribution of TKE in wavenumber space. B(k) dk is the contribution to the buoyancy flux in the wavenumber range k and k + dk:

$$\int_0^\infty B(k)\,dk = \overline{wb}$$
$$\int_0^\infty E(k)\,dk = K \qquad (1.10.5)$$

The term B(k) can be either a source term or a sink term, depending on the ambient stratification. For stable stratification, B(k) is negative and extracts energy from turbulence. B(k) is important at large scales or low wavenumbers. The contribution to τ(k), the Reynolds stress spectrum, is also at large scales. Thus E(k), τ(k), and B(k) are large at energy-containing scales, although τ(k) drops off much faster with k than with E(k), with a –7/3 power law dependence (see Figure 1.10.2). Dissipation ε(k) is, however, negligible at energy-containing scales and peaks at dissipative, Kolmogoroff microscales, where the energy density itself is small. F(k), denoting transfer in spectral space from energy-

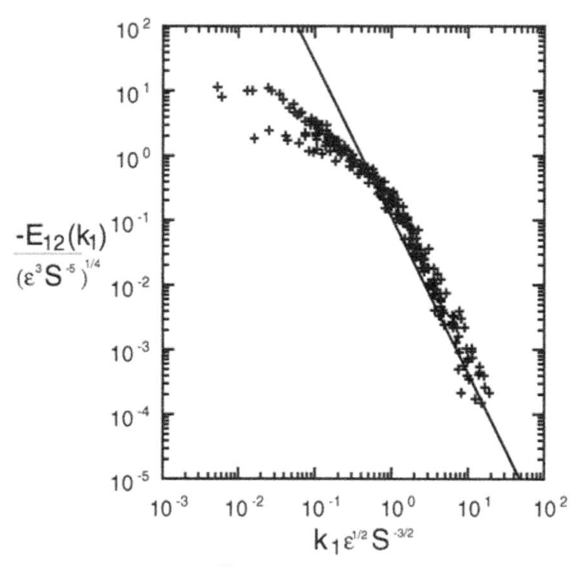

Figure 1.10.2. One-dimensional $k_1^{-7/3}$ Reynolds stress spectrum from observations (from Saddoughi and Veeravalli 1994).

containing eddies to microscales, is significant at all wavenumbers, i.e., over the entire spectrum. Unfortunately, it is also the most difficult term to model.

Consider a statistically steady, homogeneous, isotropic turbulence in a neutrally stratified fluid (see, e.g., Karman and Howarth, 1938). Equation (1.10.2) becomes

$$\frac{\partial}{\partial t} E(k) = -\frac{\partial F(k)}{\partial k} - 2\nu k^2 E(k) \qquad (1.10.6)$$

This deceptively simple looking equation has attracted talents like Heisenberg, Obukhoff, von Karman, Saffman, and Kovazsnay (see Monin and Yaglom, 1973, Section 17). The literature on homogeneous isotropic turbulence research initiated by G. I. Taylor (1935) is vast (Leslie, 1973; Hinze, 1975; Monin and Yaglom, 1973; Stanisic, 1985; Kraichnan, 1964, 1971). The basic difficulty is that Eq. (1.10.6) is one equation with two unknowns, the classic problem of turbulence closure, but in spectral space. F(k) is the principal difficulty arising from the very strong and nonlinear interactions between wavenumbers in spectral space. One has to formulate a model for the spectral energy transfer to close the equation, and herein lies the difficulty, since all such models/theories are postulates, albeit some are better than others. There is one limit where it can be neglected—in the final stages of decay of a homogenous isotropic turbulence, when spectral transfer of energy among eddies is very small, and Eq. (1.10.6) becomes

$$\frac{\partial}{\partial t} E(k) = -2\nu k^2 E(k) \tag{1.10.7}$$

which can be readily integrated to yield

$$E(k) = E_0(k) \exp\left[-2\nu k^2 (t-t_0)\right] \tag{1.10.8}$$

This shows that high wavenumbers or small scales decay at a much higher rate than low wavenumbers or large eddies. This makes sense since the large eddies are no longer feeding the small scales through the spectrum, and viscosity is rapidly dissipating whatever little energy they have left.

For the general case of Eq. (1.10.6), we will consider a postulate by Pao that yields a reasonable solution for the spectrum (Hinze, 1976). Pao assumes that the energy flux F(k) is proportional to the local energy density, $F(k) = \gamma E(k)$, where γ, the rate at which energy is transported by the cascade process, is a function only of ε and k. From dimensional analysis,

$$F(k) = -B\,\varepsilon^{1/3} k^{5/3} E(k) \tag{1.10.9}$$

Therefore, Eq. (1.10.6) can be readily integrated to yield

$$E(k) = \alpha\, \varepsilon^{2/3} k^{-5/3} \exp\left[-\frac{3}{2B}\frac{\nu}{\varepsilon^{1/3}} k^{4/3}\right]$$

Constant B is obtained by noting that $\varepsilon = 2\nu \int_k^\infty k^2 E(k)\, dk$, since dissipation in wavenumber range 0 to k is negligible. The result is called the Pao spectrum, which agrees well with observations,

$$E(k) = \alpha\, \varepsilon^{2/3} k^{-5/3} \exp\left[-\frac{3}{2}\alpha \frac{\nu}{\varepsilon^{1/3}} k^{4/3}\right] \tag{1.10.10}$$

or in Kolmogoroff variables,

$$\frac{E(k)}{\varepsilon^{1/4}\nu^{5/4}} = \alpha\, (k\eta)^{-5/3} \exp\left[-\frac{3}{2}\alpha (k\eta)^{4/3}\right] \tag{1.10.11}$$

This is the spectral shape in the universal equilibrium range, which includes the inertial subrange and the dissipative subrange. It yields the Kolmogoroff law in the inertial subrange, not a major achievement since agreement was forced by dimensional considerations alone. $\alpha \sim 1.7$ according to the measurements of Grant, Stewart, and Molliet (1962). Kraichnan's analytical theory predicts an upper limit of 1.77 for α. A more recently determined value is 1.62 (see Section 1.5)

If C is the concentration of a passive scalar in the flow, it is governed by an equation similar to that for temperature, Eq. (1.7.23):

$$\frac{\partial C}{\partial t}+\frac{\partial}{\partial x_K}(U_K C)=\frac{\partial}{\partial x_K}\left(k_c\frac{\partial C}{\partial x_K}\right)-\frac{\partial}{\partial x_K}\left(\overline{u_K c}\right) \tag{1.10.12}$$

The only difference is that the passive scalar does not affect the momentum of the fluid and $\overline{u_K c}$ does not occur in the momentum equations. The scalar is passively transported and mixed by the fluid. A heated jet is a typical example, when the degree of heating is too small to affect the buoyancy of the fluid. Then the temperature can be used as a "tracer" to study mixing (Corrsin and Kistler, 1954).

Equations for $\overline{u_j c}$ and $\overline{c^2}$ can also be written along the same lines starting from the equation for concentration fluctuations c,

$$\frac{\partial c}{\partial t}+\frac{\partial}{\partial x_k}\left(Cu_k+U_k c+u_k c-\overline{u_k c}\right)=\frac{\partial}{\partial x_k}\left(k_c\frac{\partial c}{\partial x_k}\right) \tag{1.10.13}$$

from which

$$\frac{\partial \overline{c^2}}{\partial t}+\frac{\partial}{\partial x_k}\left[U_k\overline{c^2}+\overline{u_k c^2}-k_c\frac{\partial \overline{c^2}}{\partial x_k}\right]=-2\overline{u_k c}\frac{\partial C}{\partial x_k}-2k_c\overline{\frac{\partial c}{\partial x_k}\frac{\partial c}{\partial x_k}} \tag{1.10.14}$$

$$\frac{\partial}{\partial t}\left(\overline{u_j c}\right)+\frac{\partial}{\partial x_k}\left[U_k\overline{u_j c}+\overline{u_k u_j c}-k_c\overline{u_j\frac{\partial c}{\partial x_k}}-\nu\overline{c\frac{\partial u_j}{\partial x_k}}\right]+f_k\varepsilon_{jkl}\overline{u_l c}$$
$$+\frac{\partial}{\partial x_j}\overline{pc}=-\overline{u_j u_k}\frac{\partial C}{\partial x_k}-\overline{cu_k}\frac{\partial U_j}{\partial x_k}+\overline{p\frac{\partial c}{\partial x_j}}-(k_c+\nu)\overline{\frac{\partial u_j}{\partial x_k}\frac{\partial c}{\partial x_k}} \tag{1.10.15}$$

These equations come in handy for dealing with the transport of passive scalars such as pollutants in the ABL. When their influence on buoyancy is not large, these equations apply to both temperature and humidity in the atmosphere.

Obukhoff (1949) in Russia and Corrsin (1951) in the United States independently postulated the existence of a subrange (the so-called inertial-convective subrange) in the spectrum of a passive scalar that is very much similar to the inertial subrange in the TKE spectrum. Once again, as long as the wavenumber k is such that $\frac{1}{\ell} \ll k \ll \frac{1}{\eta_c}$, where η_c is the microscale corresponding to the dissipation of the variance of c, $\overline{c}^2$, then in this inertial-convective range of the spectrum, the spectral density E_c should be independent of both ℓ and $1/\eta_c$, and a function only of the viscous dissipation rate ε of TKE, the molecular dissipation rate ε_c of $\overline{c}^2$, and the local wavenumber k,

$$E_c = E_c\left(k,\varepsilon,\varepsilon_c\right) \tag{1.10.16}$$

where $\varepsilon_c = 2k_c \overline{\frac{\partial c}{\partial x_k} \frac{\partial c}{\partial x_k}}$ (note that k_c is the kinematic diffusivity of the scalar, not a wavenumber). Simple dimensional analysis gives the Bachelor–Obukhoff–Corrsin scalar spectrum:

$$E_c\left(k\right) = \alpha_B \varepsilon_c \varepsilon^{-1/3} k^{-5/3} \tag{1.10.17}$$

The –5/3 power law is the same as for the TKE spectrum. Constant α_B is called the Batchelor constant (or the Obukhoff–Corrsin constant by some) and has a value ~ 1.52. In fact, it is related to the Kolgomoroff constant α by

$$\alpha_B = Pr_t \ \alpha \tag{1.10.18}$$

where Pr_t is the turbulent Prandtl number and has a value of about 0.8. Experimental values for Pr_t range from 0.7 to 0.9. This relationship is an asymptotic relationship and hence valid only at very high Reynolds numbers, since only then can one expect the values of eddy viscosity and eddy diffusivity to be similar. This is the so-called Reynolds similarity law.

Sreenivasan (1996) has carefully analyzed most available data from laboratory and geophysical observations on the scalar spectrum, and finds that for sufficiently large values of the Reynolds number, the values of the one-dimensional Obukhoff–Corrsin (Batchelor) constant is scattered around a value

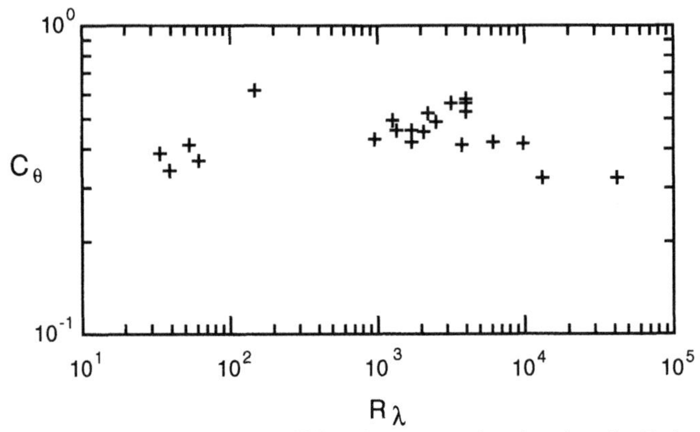

Figure 1.10.3. One-dimensional Obukhoff-Corrsin constant plotted against the Taylor microscale for a large number of passive scalar spectrum measurements. Note the apparent universality of the constant in spite of the large scatter (from Sreenivasan 1996).

of 0.4 (Figure 1.10.3), corresponding to a turbulent Prandtl number of nearly 0.8. This also corresponds to a three-dimensional Batchelor constant of 1.52. The Reynolds number based on the Taylor microscale R_λ must be above 1000 for a well-defined inertial-convective subrange to exist (Sreenivasan, 1996). Figure 1.10.4 shows two passive scalar spectra from recent measurements that show a well-defined inertial-convective subrange.

When buoyancy forces are involved, the temperature is no longer a passive scalar and the temperature spectrum shows three distinct slopes. For low wavenumbers where internal wave motions may predominate, the spectrum has a k^{-2} behavior, and for large wavenumbers in the viscous subrange $\eta < k < \eta_b$, k^{-1} is the result. In between these two lies the Obukhoff–Corrsin $k^{-5/3}$ inertial subrange (Tennekes and Lumley 1982).

1.11 WAVELET TRANSFORMS IN TURBULENCE

Interpreting turbulence measurements and gaining insight into the underlying physics is a complex task, often requiring a detailed analysis of the characteristics of the temporal variations of turbulence variables. As we have seen, traditionally, Fourier transforms have been the centerpiece of turbulence time series analysis. However, Fourier transforms provide only a limited amount of information on the characteristics of a temporally varying signal. They provide only its frequency content over the record length, making it possible to identify the magnitude and frequency (or equivalently the period) of various events embedded in the time series. However, it is impossible to localize each of the

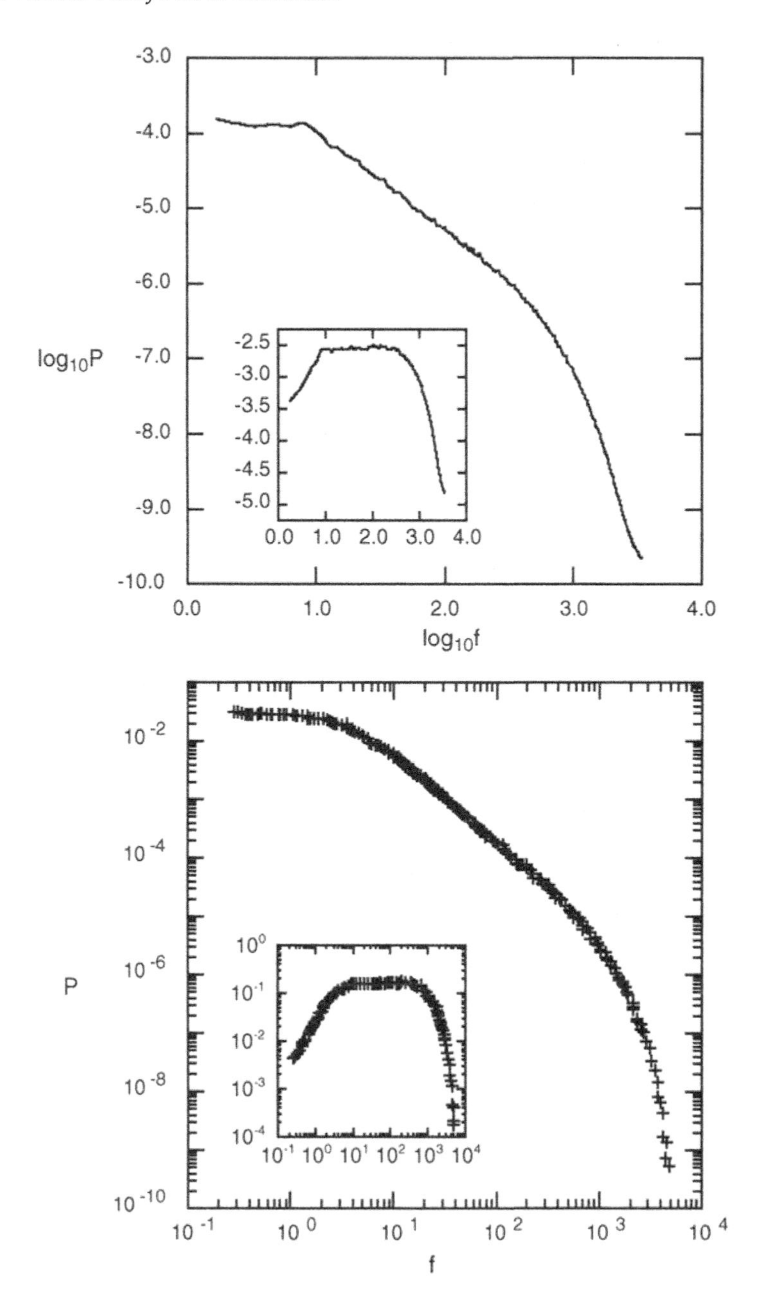

Figure 1.10.4. One-dimensional Obukhoff-Corrsin passive scalar spectra from recent measurements. The so-called compensated spectra are shown in the inset. Note the large inertial-convective subrange. P, power spectral density; f, frequency in Hz. (From Sreenivasan 1996).

events in the time domain, in other words, to identify when a particular event is occurring in the time series and what its characteristics are.

The wavelet transform (WT) is a powerful time series analysis tool that is capable of localizing the signal variability simultaneously in both time and scale (frequency) domains. It is ideally suited for analyzing records containing episodic events and multiple scales. It complements the traditional Fourier transforms and empirical orthogonal functions in time series analysis. The WT is a far more complete characterization of the time series than a traditional fast Fourier transform (FFT), and in fact provides a complete picture of scale (equivalently, frequency)—time evolution of various events composing the signal. Unlike an FFT, it is well suited for analysis of nonstationary time series as well.

We will not consider WTs in any great detail here. See Appendix B of Kantha and Clayson (2000) for a compact review of the method. Kumar and Foufoula-Georgiou (1994, 1997) provide an excellent and timely review of wavelet analysis for geophysical applications (see also Hubbard, 1996). Farge (1992a,b) and Farge *et al.* (1996) deal with its applications specifically to turbulence. Suffice it to say that the WT uses compactly supported basis functions called wavelets to decompose the time series in the time–frequency domain, so that events can be localized in time. FFT techniques can be employed to compute discrete wavelet transforms (DWTs) using either an orthogonal wavelet basis or a nonorthogonal one. The technique is readily extended to two dimensions. Consequently, the method is well suited to identification, isolation, and analysis of coherent spatial structures in turbulent flows. Farge (1992b) finds that large coherent structures correspond to strong wavelet coefficients, and filamental structures to weak ones. In addition, Fourier transforms could be replaced by wavelet transforms in solving turbulent flow equations in LES and DNS approaches. In spite of the resulting additional complexities, the method is promising, because of the ability of WTs to localize turbulence events such as bursting near solid boundaries in both time and frequency domains.

We will present just one example of a WT application to analysis of turbulence data. Howell and Mahrt (1994) applied orthogonal Haar wavelets to transform a time series record of the longitudinal component of velocity in the lower atmospheric boundary layer. They were thus able to separate the large eddies transporting finer scale features around (see Figure 1.11.1). See Hudgins *et al.* (1993) for another application of WTs to atmospheric turbulence.

1.12 LARGE EDDY SIMULATIONS

The large eddy simulation approach to modeling turbulence was pioneered by atmospheric scientists (Smagorinsky, 1963; Lilly, 1967; Deardorff, 1973). Doug

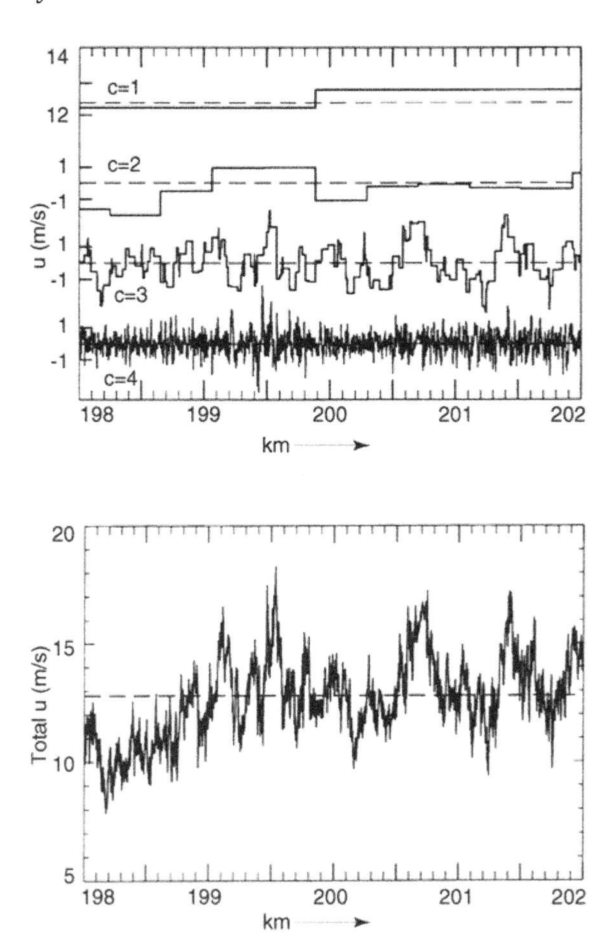

Figure 1.11.1 The four constituent modes: mesoscale, large eddy, transporting eddy, and fine scale (top panel) from Haar orthogonal wavelet transforms of the longitudinal component of velocity, whose time series is shown in the bottom panel (from Howell and Mahrt 1994).

Lilly at NCAR was the first to propose the idea and also formulated the two widely used subgrid scale parameterizations (Lilly, 1967). However, it was James Deardorff's implementation of the idea using the relatively primitive computers of the late sixties that launched this area of research. With the advent of multigigaflop computers in the nineties, LES has become a very powerful tool for numerical modeling of turbulent flows and obtaining turbulence "data" on quantities that are hard to measure in the laboratory and in the field. For example, recent work on convective ABLs has shed light on the pressure–strain rate and other covariances central to second-moment closure models of turbulence (Moeng and Wyngaard, 1986). The maturity of the field prompted a conference to be held recently, the proceedings of which (Galperin and Orszag, 1993) constitute an excellent summary of the current state of research in the field. Particularly noteworthy is the review article by Wyngaard and Moeng (1993; see also Herring, 1979; Wyngaard, 1982). More recent reviews can be found in Lesieur and Metais (1996) and Kosovic (1996). There appears to be a convergence in the field in the sense that different groups around the world have now constructed LES models that are quite similar in performance to one another, at least for convectively driven, clear ABLs (Nieuwstadt *et al.,* 1991), and shear-driven, neutral ABLs (Andren *et al.,* 1994). The major differences in performance appear to emanate from the different subgrid scale (SGS) parameterizations used in different models. In this section we will provide a brief overview of the topic and some pertinent results of LES in geophysical boundary layers. For a more detailed description, the reader is referred to the original paper of Deardorff (1973) and recent reviews cited above.

Although the energetic large eddies dominate the various aspects of turbulent mixing and transport, they are so dependent on the flow that there is no "universal" character to them and they are therefore difficult to categorize and study. The smaller scales, on the other hand, are universal in character and have been treated successfully by approaches such as the inertial subrange of the turbulence spectrum. They are, however, more passive and simply adjust to the energy cascading down the spectrum. Conventional ensemble averaging averages over all scales, from the largest quasi-permanent ones to the Kolmogoroff scales. In the process, one loses the ability to explicitly simulate the critically important large eddies. Another approach (DNS—see Section 1.13), in an attempt to resolve all scales of turbulence from the Kolmogoroff scales up, uses efficient spectral techniques and therefore simulates all scales explicitly. This task is monumental and it will be impossible to achieve large Reynolds numbers characteristic of practical geophysical turbulent flows in the foreseeable future.

An acceptable compromise is to explicitly simulate the large eddies in the flow and parameterize the small or subgrid scales. The theory is that if the resolution is such that most of the important energy-containing scales can be resolved, then the inaccuracies in modeling the small, less energy-containing subgrid scales cannot have a major impact on the derived properties of the turbulent flow. The explicit calculation of the "large" eddies also enables averages of various quantities to be obtained numerically that are very difficult, impossible, or expensive to obtain through observations. It is the promise of LES to provide the badly needed "data" on turbulence that can in turn be used to calibrate, verify, and improve the simpler, less computing-intensive techniques that is currently the most appealing aspect of the approach, and not its explicit and routine use in geophysical predictions such as numerical weather prediction

(NWP). The computing resources needed will remain too demanding for the latter to happen for the near future. Even the current multigigaflop computers enable simulations such as the convective ABL to be made only in a box a few tens of kilometers on its side at best.

In LES, the simulations are necessarily time dependent and on a 3D grid, while in ensemble average closure (EAC), neither of these is essential. The model grid resolution is chosen (in ideal circumstances) such that the subgrid scales are in the Kolmogoroff inertial subrange and therefore their dissipative properties can be parameterized (Deardorff, 1973; Herring, 1979; Moeng, 1984). The resolved variables are defined by the application of Leonard's spatial filter, whose effect is to remove the subgrid scales from the equations. The effect of these small scales is then parameterized and numerical simulations are carried out in time until a statistically steady state is reached and time averages of various resolved turbulent quantities computed. The technique has worked remarkably well for the convective ABL, which is ideally suited to this approach because of its domination by large buoyant eddies. It has been less successful when applied to shear-generated turbulence and other situations such as the ocean mixed layer where the largest eddies in the flow tend to be quite small and large computing resources are needed to resolve these eddies. A case in point is the turbulence close to a solid boundary, where the scale of the large eddies is the distance from the boundary (law of the wall region). It has been quite difficult to extend these models down to the boundary, and appeal is most often made to matching the calculations to the law of the wall to get over this difficulty. Also, because asymptotic theory of the inertial subrange is relied upon to model subgrid scales, the method is inherently flawed at the lower Reynolds numbers characteristic of flow adjacent to a solid boundary. Like most other approaches, including EAC and DNS, turbulence close to a boundary in the presence of mean shear is hard to model by LES.

The spatial averages are determined by a low-pass filter that eliminates fluctuations on scales smaller than the grid size Δx:

$$\bar{G}(\tilde{x}) = \iiint F(\tilde{x}-\tilde{x}')\, G(\tilde{x}')\, d\tilde{x}' \qquad (1.12.1)$$

$F(\tilde{x}-\tilde{x}')$ is the filter function; $F=\Delta^{-3}$ (where D is a length scale) inside the grid element, and 0 outside for a top hat distribution, and

$$F = \left(\frac{\sqrt{\gamma/\pi}}{\Delta} \right)^{n} \exp\left[-\left(\sqrt{\gamma}/\Delta \right)^{n} (\tilde{x}-\tilde{x}')^{2} \right] \qquad (1.12.2)$$

for a Gaussian. Moeng uses $\gamma = 6$ and $\Delta = 2\Delta x$, where Δx is the grid spacing in the horizontal; n is the number of dimensions of the filter. The application of the Gaussian or top hat filter to the initial conservation equations [(1.7.1)–(1.7.3)] leads to equations that look very similar to the equations obtained by Reynolds averaging; however, a Leonard average is not the same as a Reynolds average, as can be seen by these resulting continuity and momentum equations,

$$\frac{\partial U_i}{\partial x_i} = 0$$
$$\frac{\partial U_j}{\partial t} + \frac{\partial}{\partial x_k}(U_j U_k) = -\frac{1}{\rho_0}\frac{\partial P}{\partial x_j} + g_j \beta \Theta + \frac{\partial}{\partial x_k}(2\nu S_{kj} + T_{kj}) \tag{1.12.3}$$

where

$$T_{ij} = \left[(U_i U_j - \overline{U_i U_j}) - (\overline{U_i u_j} + \overline{U_j u_i} + \overline{u_i u_j})\right] = 2\nu_t S_{ij} + \frac{\delta_{ij}}{3} T_{ii}$$

The overbar denotes the filtering operation, the upper case letters denote the resolved (filtered) variables, and the lower case ones denote the subgrid scales. Note that unlike Reynolds averages, the average of the resolved and unresolved variables does not vanish. T_{ij} is called the subgrid scale stress tensor, the second term of which involves its trace and is usually absorbed into the pressure term. Upon taking the divergence of the momentum equation, one gets an elliptic equation for the modified pressure. The term in the first parentheses is known as the Leonard stress tensor. Although the Leonard stress tensor can be computed from the filtered field, the terms in the second parentheses are unknown and thus the entire SGS stress tensor T_{ij} is usually modeled.

The temperature equation is similarly

$$\frac{\partial \Theta}{\partial t} + \frac{\partial}{\partial x_k}(U_k \Theta) = \frac{\partial}{\partial x_k}\left[(\nu \,\mathrm{Pr} + \nu_t \,\mathrm{Pr}_t)\frac{\partial \Theta}{\partial x_k}\right] \tag{1.12.4}$$

where Pr_t and ν_t are the subgrid scale turbulent Prandtl number and viscosity.

Closing the equations thus requires modeling the subgrid scale quantities, Reynolds stresses, and heat fluxes. A good representation of these quantities is necessary for accurate simulations, and is a problem faced by atmospheric scientists in NWP. In this case it was necessary to remove the energy piling up at the subgrid scales due to cascade from the resolved scales so that computations of future states of the weather could be made without a numerical blowup. In this sense, it was purely a numerical artifact. To a large extent, even in LES, the function of the subgrid scale model is still principally to "dissipate" this cascading energy. It is the complexity of the subgrid model that distinguishes one LES model from the other [and to some extent the filter shape—finite difference approaches use a top hat function (Deardorff, 1973), while pseudo-spectral approaches based on fast Fourier transforms use a Gaussian (Moeng, 1984)].

One approach to the subgrid scale model was developed by Smagorinsky (1963), who used the extension of mixing length concepts to relate the subgrid scale eddy viscosity to the mean strain rate by setting

$$\nu_t = (c_s \Delta x)^2 |S| \qquad \text{and} \qquad |S| = (2S_{ij}S_{ij})^{1/2} \tag{1.12.5}$$

where $|S|$ is the strain rate and c_s is a constant. This formulation was originally for quasi-two-dimensional motions in the atmosphere, since only the horizontal components of the strain rate were used. However, Lilly (1967) showed that the formulation is valid for three-dimensional flows and the constant c_s can be obtained by assuming the cutoff wavenumber $k_c = (2\pi / 2\Delta x)$ lies in the inertial subrange and appealing to the Kolmogoroff inertial subrange law. Since

$$\frac{\partial}{\partial t} \int_0^{k_c} E(k)dk = \varepsilon = \int_0^{k_c} 2\nu_t k^2 E(k)dk \tag{1.12.6}$$

using Kolmogoroff inertial subrange expression $E(k) = \alpha \varepsilon^{2/3} k^{-5/3}$ (Eq. 1.5.2), one gets $\varepsilon = (1.5\alpha)^3 k_c^4 \nu_t^3$. But $\varepsilon = \nu_t |S|^2$, and equating the two expressions produces $c_s = \pi^{-1}(1.5\alpha)^{-3/4} \approx 0.17$, although the most commonly used value is around 0.1.

This parameterization does not account for the effects of stratification on subgrid scales. Nevertheless, it works reasonably well in NWP models and in neutrally stratified flows. It is inaccurate when applied to rotating, stratified fluids in geophysics as Deardorff (1973) discovered when he applied it to a stably stratified planetary boundary layer (PBL). An additional problem is that the subgrid scale stresses are a function of the unresolved subgrid field, not of the mean flow. This is very much similar to the situation in Reynolds or ensemble averaging approaches. Lilly also proposed using a subgrid scale turbulence-based parameterization,

$$\nu_t = \hat{c}_s \, q_{sgs} \Delta x \tag{1.12.7}$$

where q_{sgs} is obtained from an SGS TKE equation. Since

$$q_{sgs}^2 = 2\int_{k_c}^{\infty} E(k)dk = 3\alpha\varepsilon^{2/3}k_c^{-2/3} \tag{1.12.8}$$

substituting the expressions above for ε gives $\hat{c}_s = 2^{-1/2}(1.5\alpha)^{-3/2}\pi^{-1} \approx 0.12$. This approach is similar to the one-equation (for TKE) model in ensemble average closures and accounts for stratification effects through their influence on q_{sgs}. Deardorff (1973) proposed writing equations for the second moments of subgrid scale quantities and closing them in manner very much similar to the SMC approach. The only difference between SMC and LES is then the prescription of the turbulence length scale; in SMC it is modeled by an additional equation, while in LES it is usually tied to the grid size and a separate equation is not written for it. The full SMC approach, but for subgrid scales, is followed by some LES groups, although the use of full SMC is quite expensive for LES. An upper bound is put on the length scale for stably stratified cases (Moeng, 1984), similar to the bound used in ensemble average closures:

$$\ell = \min\left(\Delta x, 0.76 q_{sgs} N^{-1}\right) \tag{1.12.9}$$

Most groups (Nieuwstadt *et al.,* 1991; Moeng *et al.,* 1996) however use only a one-equation model for subgrid scales, the approach proposed by Lilly (1967). This is a good compromise since it permits buoyancy effects on subgrid scales to be included indirectly through the TKE equation, yet does not require the many additional equations of a full SMC to be solved.

Traditionally, LES has been quite successful when applied to turbulence away from boundaries, and in geophysical flows such as the dry convective ABL, where the solution is relatively insensitive to the exact form of the SGS parameterization used. With some tuning of the constant, simple linear Smagorinsky-type eddy viscosity formulations by Lilly (1967) and Deardorff (1970) are realistic (see, e.g., Moin and Kim, 1982). On the other hand, SGS parameterization is turning out to be the key to the success or failure of the LES approach when applied to more complex cases such as a sheared flow near a boundary or the stably stratified geophysical boundary layer. The principal difference between a convective ABL and a stably stratified shear-driven ABL is the much smaller scales at which TKE production takes place in the latter and the much higher anisotropy of turbulence under stable stratification and vertical shear. Suppression of turbulence by stable stratification leads to intermittency, gravity wave excitation, and generally low Reynolds numbers, and these effects are difficult to take into account in closure schemes based essentially on asymptotic theories of continuous, high Reynolds number turbulence. As mentioned earlier, the smaller and smaller scales of the energy-containing eddies as one approaches a boundary (whether it is a wall or an inversion) make resolving them explicitly in a LES more and more difficult and hence parameterizing these highly anisotropic subgrid scales accurately by SGS closure more and more critical. It is for these reasons that the details of the SGS parameterization become crucial to the results of the LES itself, a situation not much different from the classical EAC that LES was supposed to replace. As long as the energy resident in the resolved turbulence scales is at least an order of magnitude larger than that in the subgrid scales, LES retains its appeal and advantage. Once the two become comparable, and this is the case for sheared near-wall turbulence and stably stratified turbulence, the distinction between LES and EAC becomes rather blurred and the computational expense of the former a bit harder to justify.

In sheared wall turbulence and stably stratified turbulence, simple Smagorinsky hypothesis-based SGS formulations do not even yield correct first-order statistics such as the mean profiles (Kosovic, 1996). This has prompted searches for alternative formulations (e.g., Germano *et al.,* 1991; Germano, 1992; Lilly, 1992; Brown *et al.,* 1994; Sullivan *et al.,* 1994; Lesieur and Metais, 1996). The linear eddy viscosity formulations turn out to be far too dissipative close to a boundary, a salient characteristic of the underlying down-the-gradient momentum diffusion approximation. As a consequence, they have difficulty simulating phenomena such as the transition from laminar to turbulent flow near a wall (Lesieur and Metais, 1996). There is too much drainage from resolved to unresolved scales, which cannot be corrected by merely choosing a lower value for c_s, as Deardorff discovered. In reality, the flow of energy in the turbulence spectrum is not always from large scales to small scales. There is occasional transfer of energy from small scales to large ones, as demonstrated by DNS's of turbulence (Domaradzki *et al.,* 1993), although, overall, the transfer is still down the spectrum from the red end to the violet, consistent with the Kolmogoroff hypothesis. Some approaches have therefore attempted to allow for this backscatter of energy by permitting negative viscosities, but often at the risk of numerical instabilities. Other approaches have resorted to closure in spectral space instead of the physical space (see Lesieur, 1990; Lesieur and Metais, 1996). Using the Eddy-Damped-Quasi-Normal-Markovian (EDQNM) hypothesis in the equation for the kinetic energy spectrum for homogeneous isotropic turbulence [closure effected in spectral space–see Orszag (1970) and Lesieur (1990)], and assuming k_c lies in the Kolmogoroff inertial subrange, one gets an expression for eddy viscosity that allows for some backscatter in spectral space:

$$\nu(k,k_c) = 0.441\alpha^{-3/2}\left[k_c^{-1}\left(k_c E(k_c)\right)^{1/2}\right] f\left(\frac{k}{k_c}\right) \quad (1.12.10)$$

The term in the square brackets is the normalizing value of eddy viscosity, a product of the characteristic length scale (k_c^{-1}) and the velocity scale, and f is a nondimensional function, determinable from EDQNM, that increases from 1 at low wavenumbers up to $k = k_c$ (the so-called cusp behavior). When energy is injected at wavenumber k_I in spectral space, this form allows for transfer of energy to wavenumber space $k < k_I$ from nonlinear resonant interactions between neighboring energy-containing modes near k_I. Thus this form produces some backscatter of energy (Mason, 1994) to scales larger than the grid scale, even though the overall eddy viscosity is positive.

Transforming to physical space and requiring that SGS dissipation equal ε yields

$$\nu_t = \frac{2}{3}\alpha^{-3/2}\left[k_c^{-1}\left(k_c E(k_c)\right)^{1/2}\right] \quad (1.12.11)$$

The structure function approach (see Lesieur and Metais, 1996) uses this expression, with $E(k_c)$ determined from the local second-order velocity structure function of the filtered field. It appears to work well for isotropic turbulence, yielding a reasonable Kolmogoroff spectrum for TKE as well as Batchelor's $E_p(k) = 1.32\alpha^2\varepsilon^{4/3}k^{-7/3}$ pressure spectrum. However, just like the Smagorinsky approach, it is still too dissipative for some applications to near-wall turbulence, although some modifications have been suggested (see Lesieur and Metais, 1996).

Germano *et al.* (1991) pioneered the dynamic model, where the use of two filters of differing cutoff wavelengths allows the local Smagorinsky constant at each point in time and space to be calculated rather than prescribed a priori. While this works well for near-wall low Reynolds number turbulence, the system is overdetermined. Lilly's (1992) least-squares approach to overcome this indeterminacy leads to excessive backscatter, leading to large negative viscosities and hence numerical instabilities (Lesieur and Metais, 1996). Also, the dynamic models appear to have difficulties at high Reynolds numbers (Sullivan and Moeng, 1993).

Sullivan *et al.* (1994) followed Schumann's (1975) approach of decomposing SGS stresses into mean and fluctuating components but used an isotropy factor so that the Smagorinsky constant becomes a function of the flow shear,

$$T_{ij} - \frac{\delta_{ij}}{3} T_{ii} = 2\gamma v_t S_{ij} + 2\bar{v}_t \bar{S}_{ij} \tag{1.12.12}$$

where γ is the isotropy factor defined as that ratio of the magnitude of the fluctuating strain rate to the magnitude of the total strain rate. This anisotropic model was used by Andren (1995) to simulate a stably stratified ABL.

Based on the nonlinear constitutive relationship suggested by Spaziale (1991) for K–ε closure models and similar to the model used by Pope (1975) for the Reynolds stress, Kosovic (1996) used a combination of the linear SGS term and two additional nonlinear terms to model the SGS tensor,

$$T_{ij} = \hat{c}_s q_{sgs} \Delta x\ S_{ij} + (c_s \Delta x)^2 \left\{ c_1 \left[\left(S_{ik} S_{kj} - \frac{1}{3} \frac{|S|^2}{2} \delta_{ij} \right) \right] + c_2 \left(S_{ik} \Omega_{kj} - \Omega_{ik} S_{kj} \right) \right\} \tag{1.12.13}$$

where Ω_{ij} is the resolved rotation rate tensor. Using spectral properties of homogeneous isotropic turbulence, he related the constants c_s and c_1 to the backscatter parameter c_b first determined by Leslie and Quarini (1979) using the EDQNM turbulence model and the skewness parameter. Based on empirical data on homogeneous sheared flows, he also put $c_2 = c_1$ and determined c_b by sensitivity studies for the isotropic case. He applied his model to a neutrally stratified ABL and compared it to the results of the linear model of Sullivan *et al.* (1994) and the stochastic backscatter model of Andren *et al.* (1994), who have compared three different LES approaches to modeling the neutral boundary layer. Kosovic (1996) asserts that his nonlinear model is superior in performance to both the linear Smagorinsky model and the stochastic backscatter models, in terms of the ability to reproduce the similarity profiles in the constant flux layer.

For application to stably stratified flows, it is necessary to introduce stability dependence of the eddy viscosity, whether directly through Richardson-number-dependent Smagorinsky eddy viscosity (Mason and Derbyshire, 1990) or indirectly through the SGS approach. However, it is also necessary to include backscatter (Brown *et al.,* 1994; Kosovic, 1996). Kosovic (1996) applied his nonlinear model to the stratified ABL and strong inversion-capped clear-air ABL observed by aircraft during the 1994 Beaufort and Arctic Sea Experiment (BASE). His simulations appear to reproduce reasonably well the surface layer similarity relationships in both cases.

The major difficulty in LES arises from the difficulty of accurately modeling the unresolved SGS component. It is also here that LES models differ from one another, and this is reflected in their performance as applied to situations where SGS is important or even dominant. The various intercomparisons of different LES codes suggest that numerics and other aspects related to resolving the large eddies do not have the same degree of impact as the SGS schemes employed and parameterizations of physics involved, such as those associated with liquid water condensation and evaporation effects (Nieuwstadt *et al.,* 1991; Andren *et al.,* 1994; Moeng *et al.,* 1996). Moeng *et al.* (1996) describe intercomparisons between 10 LES models applied to the nocturnal stratocumulus-topped PBL, in which the cloud-top radiative cooling provides the principal source of TKE. They conclude that while overall the performance with respect to mean quantities such as the PBL structure is roughly similar, the turbulence second-moment quantities differ widely (Figure 1.12.1). This is natural since the SGS parameterization is where the models differ and this has a major effect on the near-cloud-top processes, especially the turbulent entrainment rates, which were found to vary significantly among the LES models compared.

The various intercomparisons mentioned above have highlighted both the major strength and the major deficiency of the LES approach. Since LES is designed to directly simulate large eddies in the flow that contain most of the TKE and do most of the transport, it does well where the turbulence physics is dominated by such large eddies. This explains its phenomenal success in simulating the clear-air convective atmospheric boundary layer (CABL). But its Achilles' heel is the SGS parameterization, since this is most often the principal source of error. Where small scales and hence SGS parameterization have little effect on the flow overall, such as a clear CABL, LES performs well; where they are dominant or even important, as in a cloud-topped ABL, or a stably stratified PBL (and OML), the success very much depends on the details of the SGS model incorporated. Shear production at small scales, the larger degree of anisotropy in a stably stratified ABL (and OML), and the small scale entrainment processes that affect cloud-top dynamics are all processes that must be parameterized accurately enough for accurate LES results. Since shear production at small scales is involved, LES performance close to solid boundaries in the surface layer is poor. Since the dynamics of a shear-dominated PBL depends very much on what happens in the surface layer, SGS has a major

---

Figure 1.12.1. Vertical profiles of mean and turbulence quantities in a nocturnal cloud-topped ABL from several LES models (from Moeng et al. 1996).

∅ – UKMO LES • – UW LES •– MPI LES • – UMIST LES

● – NCAR LES ⊗ – WVU LES ⊕ – UOK LES × – CSU LES

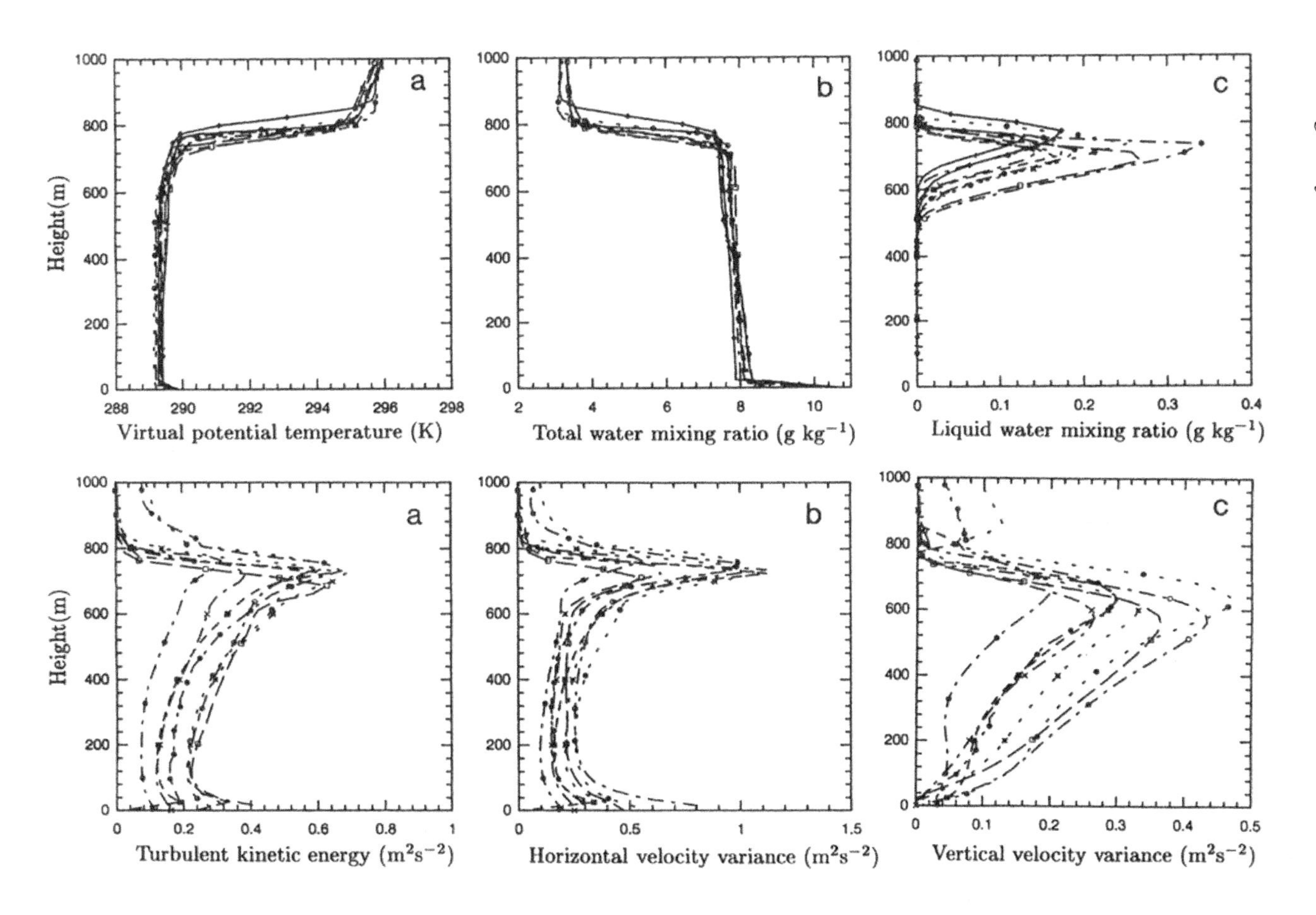
a
b
c
Height(m)
Virtual potential temperature (K)
Total water mixing ratio (g kg^{-1})
Liquid water mixing ratio (g kg^{-1})
a
b
c
Height(m)
Turbulent kinetic energy (m^{2}s^{-2})
Horizontal velocity variance (m^{2}s^{-2})
Vertical velocity variance (m^{2}s^{-2})

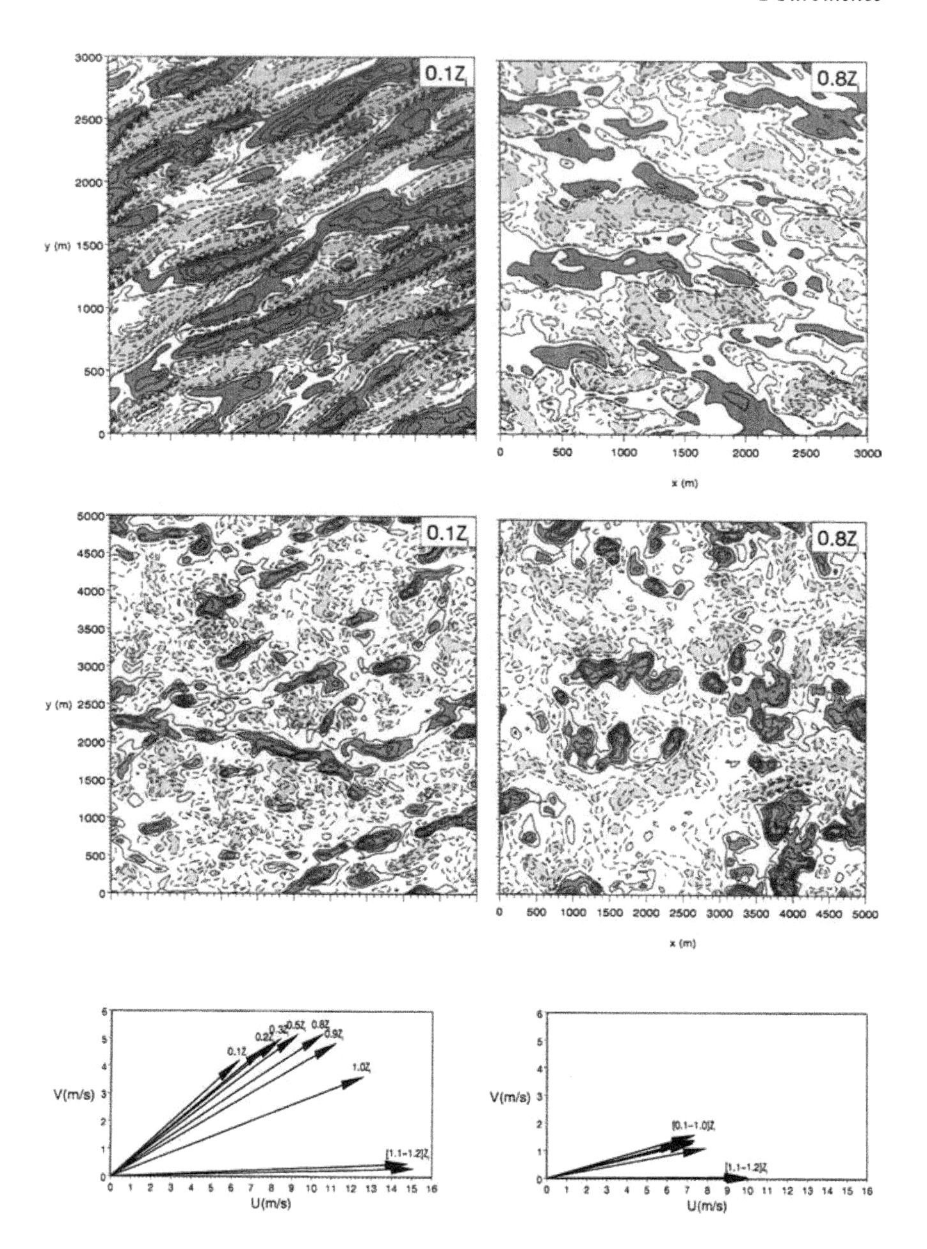

Figure 1.12.2. Structures in the CABL at two different heights in the PBL (0.1 and 0.8 z_i, where z_i is the inversion height, for the shear-dominated case (top) and buoyancy-dominated case (bottom). Hodographs for the two cases are also shown, with the shear case at left and the buoyancy case at right (from Moeng and Sullivan 1994).

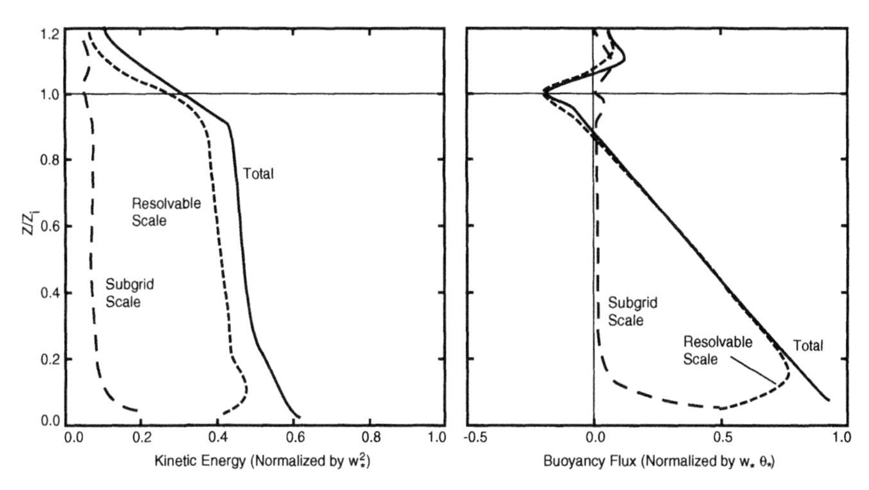

Figure 1.12.3. The TKE and the buoyancy flux in the CABL for the shear-dominated case (left) and buoyancy-dominated case (right) (from Moeng and Wyngaard 1989).

impact on the performance of the LES in the entire PBL. Since it is prohibitively expensive to try and resolve the small scales close to a solid boundary, appeal must be made to empirical laws such as the Karman law of the wall to provide the needed boundary condition (Deardorff, 1973). Intermittent turbulence, characteristic of a NABL, is also a major problem common to all turbulence models.

Normally the principal role of SGS in LES is to provide an energy sink to

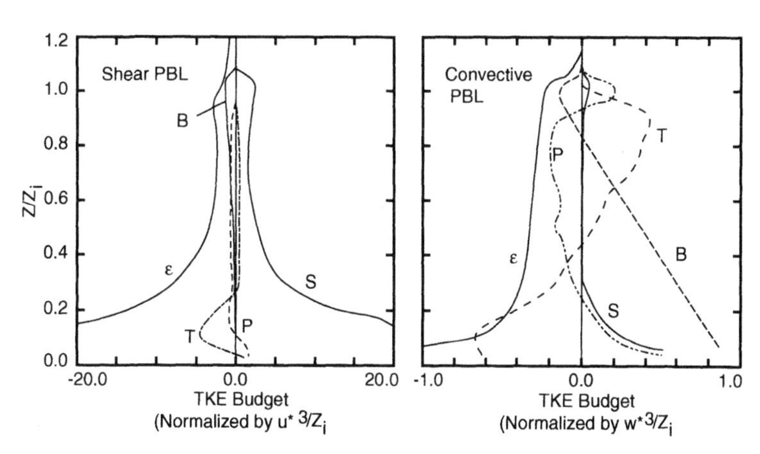

Figure 1.12.4. The TKE budget in the CABL for the shear-dominated case (left) and buoyancy-dominated case (right) (from Moeng and Sullivan 1994).

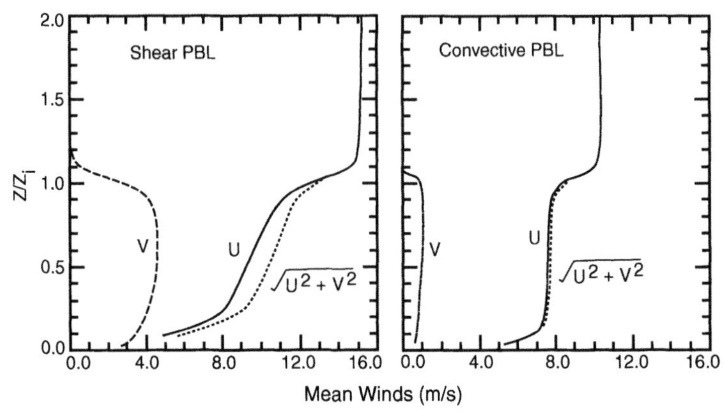

Figure 1.12.5. The mean velocity profiles in the CABL for the shear-dominated case (left) and buoyancy-dominated case (right). Note the near-homogeneity of velocity for buoyancy-dominated mixing (from Moeng and Sullivan 1994).

absorb the transfer of TKE from resolved scales and thus simulate the effect of small scales on resolved eddies. When SGS is important to the accurate simulation of large scales, such as in the surface layer, the deficiencies begin to matter. Therefore, most of the effort in recent years in LES has been to improve SGS parameterization, and this has led to some successful simulations of PBL in recent years. Moeng and Sullivan (1994) describe one such application of LES that provides a good comparison between shear- and buoyancy-driven PBL. Figure 1.12.2 shows the difference in the turbulence structure between shear-dominated PBL and convection-dominated PBL at two different heights in the PBL. The streaky structures readily visible in the former at low levels are noticeably absent in the latter. At larger heights, the two appear more similar. The hodographs show that the turning of the wind vector is more gradual for the shear-dominated case than for the buoyancy-dominated case, in which most of the turning occurs across the inversion and the flow within the ABL shows very little turning. Figure 1.12.3 shows the TKE and buoyancy flux distributions in the two cases. Close to the surface, the TKE and buoyancy flux associated with the SGS become comparable to those of the resolved scales, a problem discussed earlier. Figure 1.12.4 shows the TKE budget. In the shear case, the major balance is between shear production and dissipation, the turbulent transport playing a very minor role in the bulk of the PBL. However, in the buoyancy-dominated case, turbulent transport is an important component of the budget. It is for this reason down-gradient parameterizations of turbulence fail in the CABL, and why LES, which explicitly computes the large eddies responsible for this transport, does so well. Figure 1.12.5 shows the mean velocity profiles; note

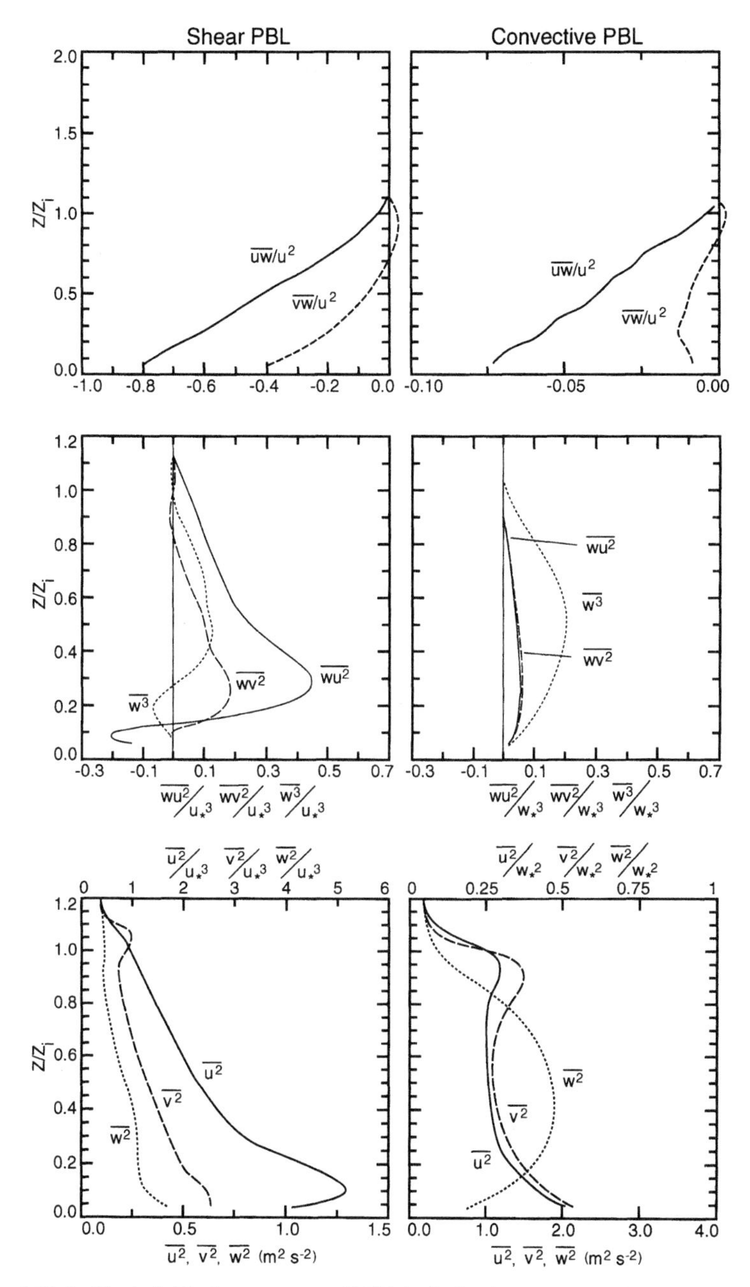

Figure 1.12.6. The individual components of TKE (top), the shear stresses (middle) and the third moments (bottom) in the CABL for the shear-dominated case (left) and buoyancy-dominated case (right) (from Moeng and Sullivan 1994).

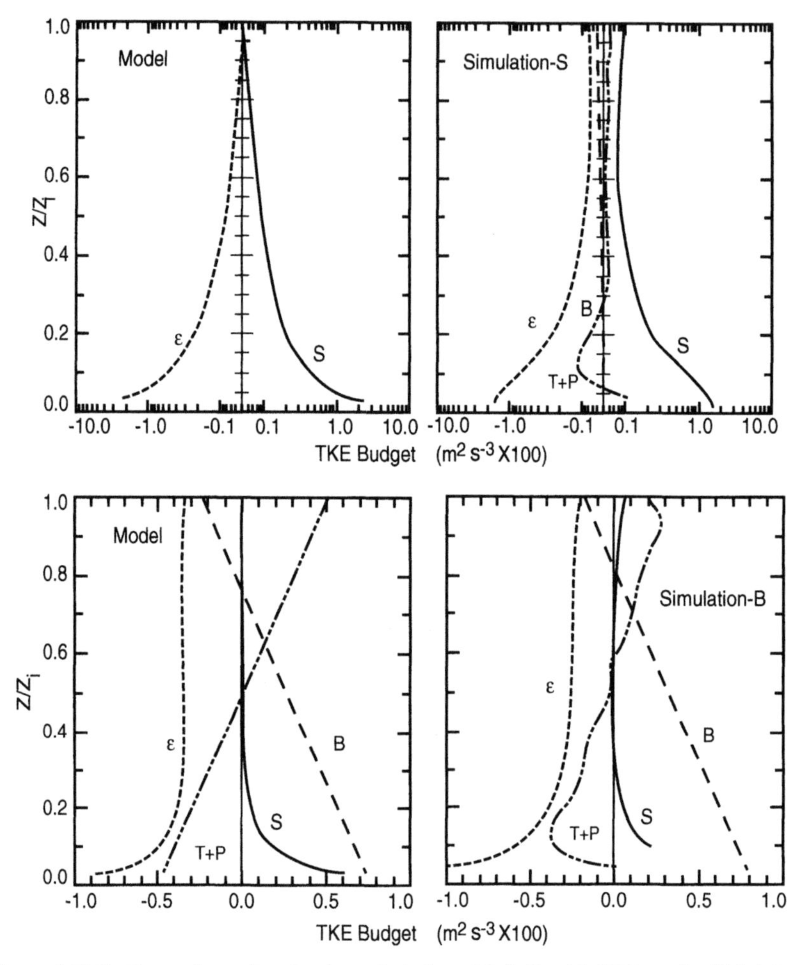

Figure 1.12.7. Comparison of a simple analytical model (left) with LES results (right) in the CABL for the shear-dominated case (top) and buoyancy-dominated case (bottom) (from Moeng and Sullivan 1994).

the near-homogeneity indicating efficient mixing for the buoyancy-dominated case. Figure 1.12.6 shows the distribution of various second and third moments in the PBL. It is this kind of information, that is hard to obtain otherwise, that makes LES so valuable to turbulence parameterizations in atmospheric and oceanic models. LES results also enable one to postulate simple analytical models for the PBL. Figure 1.12.7 shows the TKE distribution from one such model (Moeng and Sullivan, 1994) for both the shear- and the buoyancy-dominated PBLs, compared to the LES results. Here the shear production, buoyancy production, and dissipation are modeled as

$$P_s = u_*^3 \left(1 - \frac{z}{z_i}\right) \frac{\phi_M}{\kappa z}$$

$$P_b = \frac{w_*^3}{z_i}\left(1 - 1.2\frac{z}{z_i}\right) - \frac{u_*^3}{z_i}\frac{z}{z_i} \tag{1.12.14}$$

$$\varepsilon = -0.4\frac{w_*^3}{z_i} - u_*^3\left(1 - \frac{z}{z_i}\right)\frac{\phi_M}{\kappa z}$$

with the turbulent and pressure transport terms constituting the remainder of the TKE balance. Here z_i is the inversion height, ϕ_M is the nondimensional shear, and w_* is the Deardorff velocity scale equal to $(q_b\ z_i)^{1/3}$, with q_b being the buoyancy flux See Chapters 2 and 3). This analytical model compares quite well with the full LES model, except close to the surface, where the results from the latter are susceptible to large errors.

LES results have been helpful in understanding and modeling turbulence in geophysical flows. For example, LES results on second moments of turbulence quantities and pressure–scalar covariances in the CABL (Moeng and Wyngaard, 1986, 1989) have been useful in second-moment closure modeling (e.g., Kantha and Clayson, 1994). LES has clarified the nature of countergradient diffusion and bottom–up and top–down scalar diffusion in the CABL (Holtslag and Moeng, 1991), and has proved to be a valuable tool in understanding and modeling turbulence in geophysical flows. Consequently, there has been a recent upsurge in LES applications to geophysical flows. Examples of application to the ocean include equatorial currents (Skyllingstad and Denbo, 1994), Langmuir circulations (Skyllingstad and Denbo, 1995) (see also Section 2.4), deep convection in the Greenland Sea (Denbo and Skyllingstad, 1996), and diurnal variability in the OML (Wang *et al.,* 1996). A recent review of LES and DNS related to stably stratified shear flows can be found in Schumann (1996; see also Kosovic, 1996).

1.13 RENORMALIZATION GROUP ANALYSIS (RNG)

There are essentially three approaches to dealing with turbulence. One is primarily a semianalytical method based on turbulence properties in physical space. Examples are EAC models such as the K–ε model and SMC and LES models. The second approach involves numerical solutions without any approximations whatsoever made to the governing equations (DNS). The third involves statistical methods based on the governing equations in spectral space. Classical theories of isotropic, homogeneous turbulence of Heisenberg, von Karman, etc., are based on this latter approach. Other examples are Kraichnan's Lagrangian History Direct Interaction Approximation (LHDIA) theory (Kraichnan, 1964), Orszag's EDQNM model (Orszag, 1970; Chollet and Lesieur, 1981; Chasnov, 1991), and RNG.

The RNG technique originated in statistical physics and was successful in describing phase transitions in quantum field theory (Wilson, 1971). It was first applied to the fluid flow problem by Foster, Nelson, and Stephens (1977), who studied fluctuations in a randomly stirred fluid at rest (with a Gaussian random force included in the Navier–Stokes equations) for application to cases where the influence of small eddies on large eddies can be treated by eddy viscosity concepts. They showed that the classical power law spectral relationship in the inertial subrange that had been derived by Kolmogoroff in the forties can be reproduced by this approach. The renormalized eddy viscosity was found to scale with the cutoff wavenumber at the ultraviolet end of the spectrum and the amplitude of the random forcing introduced into the Navier–Stokes equations. Yakhot and Orszag (1986) were able to obtain the constants in this relationship, so that classical turbulence constants such as the Kolmogoroff constant could be computed. They went further to compute other classical constants such as the Batchelor–Obukhoff–Corrsin constant in the passive scalar variance spectrum, the turbulent Prandtl number, the skewness factor, the von Karman constant, and the power law in the decay law of isotropic turbulence. By stopping at a particular point in the process of removing small scales, they could also derive equations applicable to LES. Equations equivalent to the K–ε theory of turbulence could be obtained by eliminating all small scales, thus yielding equations applicable to EAC models of turbulence (Spaziale, 1991).

This amazing ability to consistently predict universal turbulence constants in the scaling laws which were in reasonable agreement with observations—simply by a systematic removal of small scales and renormalizing the equations to obtain an effective eddy viscosity that accounts for the influence of removed scales on the remaining scales, and without appealing to any turbulence observations—established RNG as a highly promising approach to the classical problem of turbulence closure. This feat had never been accomplished before. Seminal contributions to turbulence theory, such as the Kolmogoroff law for the inertial subrange in the turbulence spectrum and von Karman's law of the wall, had established classical relationships between turbulence quantities, but the constants had to be derived from observations. Laboratory measurements of the Kolmogoroff constant showed large uncertainties in its value (1.3–2.3), and it was not until observations were made in geophysical flows characterized by large Reynolds numbers (which are required for the existence of a broad unambiguous inertial subrange in the spectrum) was the value finally determined to have a possible asymptotic value of about 1.62. In contrast, the RNG theory of Yakhot and Orszag (YO) gave a value of 1.617 for this constant and no appeal was made to any observations! YO went on to show that it was possible to apply the technique to other aspects of turbulence and derive values for other turbulent constants, such as the Prandtl number, that were close to observed values again. Kraichnan (1971) had earlier derived both the Kolmogoroff inertial range power law $E \sim \alpha k^{-5/3}$ and a value for the constant (1.77) using his Lagrangian History Direct Interaction Approximation theory of turbulence, but had failed to reproduce a correct value for the turbulent Prandtl number. When applied to the problem of diffusion of a passive scalar in a turbulent flow, he obtained a value of 0.14 for Pr_t, instead of the observed value in the range 0.7–0.9. In contrast, RNG gives a value of 0.718 for Pr_t.

The success of RNG was a source of both amazement and skepticism. Various groups around the country have begun to look closely at the basic RNG procedures (e.g., Lam, 1992; Smith and Reynolds, 1992; Eynk, 1994), which are so mathematically complex as to deter most researchers. YO's procedure has been duplicated and many algebraic errors in the original paper corrected (Smith and Reynolds, 1992). For example, the skewness factor was shown to be –0.59, not YO's –0.49. The most important finding, however, was that because of several algebraic errors, RNG [originally thought to yield good agreement for the constant n (~1.331) in the power law for the decay of isotropic turbulence ($K \sim t^{-n}$, $n \sim 1.111$)] failed to reproduce by a wide margin ($n \sim 0.215$) this classic turbulent flow (Smith and Reynolds, 1992; Lam, 1992), which has traditionally been used to benchmark turbulence models (see Section 1.10). In this sense the failure was akin to Kraichnan's LHDIA theory. Spaziale (1991) remarks that one of the constants in the K–ε theory of turbulence derived from RNG is also close to a singular value, thus yielding unrealistically high growth rates for TKE in homogeneous shear flow.

This has prompted the authors of the RNG theory to reexamine some of the procedures and attempt to correct some of the shortcomings. Yakhot and Smith (1992) have redone the RNG calculations to obtain constants that yield a somewhat better agreement with observations. Nevertheless, the expansion procedure failed for the very important shear term when applied to the dissipation rate equation (Yakhot and Orszag, 1992). Instead, Yakhot and Orszag (1992) make use of a double expansion technique for proper application of RNG to shear flows, the salient parameter being the ratio of turbulence timescale to the mean shear timescale ($r \sim SK/\varepsilon \sim t_e/t_s$, where S is the mean strain rate dU/dz). However, the theory in its present form requires empirical determination of one of its constants by forcing it to yield the accepted value of 0.41 for the von Karman constant. Thus while some of the flaws in RNG have been eliminated, its original appeal in not relying on empiricism for derived constants has diminished somewhat.

Lam (1992) has also reexamined RNG theory, especially its controversial ε-expansion procedure, and has offered a different interpretation for the random stirring force in the Navier–Stokes equation postulated by RNG theory, concluding that ε is essentially an adjustable parameter.

It is impossible to give a thorough treatment of the highly complex mathematical theory behind RNG in this rather elementary book. Instead, we refer the reader to the original paper by Yakhot and Orszag (1986) and its recent modifications (Yakhot and Smith ,1992; Yakhot *et al.,* 1992). Smith and Reynolds (1992) is an excellent tutorial on RNG; it follows the same methodology step-by-step and provides far more mathematical details than the original paper. However, the algebra is extremely messy. Lam (1992) provides an alternative look at the original RNG theory. A more recent and timely review of RNG can be found in Smith and Woodruff (1998). The interested reader is encouraged to follow up via the references cited above, which in turn contain extensive bibliographies of recent work on the subject. Here we will confine ourselves to a general description of the original procedure, based on the papers cited above.

RNG is based on the so-called correspondence principle: A turbulent flow characterized by scaling laws in the inertial range (and hence at high Reynolds numbers) is assumed to be describable in this inertial range by the addition of a random "stirring force" in the Navier–Stokes (N-S) equation that generates the velocity fluctuations that obey the inertial range scaling in the original unmodified, unforced system (Yakhot and Orszag, 1986). This equation is then the basis for a systematic iterative elimination of small scales and the representation of their effects on the retained scales by renormalized transport coefficients, for example, the eddy viscosity. It is thus strictly an asymptotic theory. Heavy reliance is placed on the existence of the inertial subrange and, according to Lam (1992), indirectly the Kolmogoroff scaling law for "closure." The forced Navier–Stokes equation is

$$\partial v_i / \partial x_i = 0$$

$$\frac{\partial v_j}{\partial t} + \frac{\partial}{\partial x_k}\left(v_k v_j\right) = -\frac{1}{\rho}\frac{\partial p}{\partial x_j} + \nu \frac{\partial^2 v_j}{\partial x_k \partial x_k} + f_j \tag{1.13.1}$$

where f_i is a Gaussian random stirring force that leads to a statistically steady state. This equation is then postulated to yield solutions statistically equivalent to those of the original unforced equations in the inertial subrange. The correlation function for f in frequency–wavenumber or Fourier space is given by

$$\overline{\tilde{f}_i(k_i,\omega)\tilde{f}_j(k_i',\omega')} = 2D_0^*(\varepsilon,k)k^{-d}(2\pi)^{d+1}\left(\delta_{ij}-\frac{k_ik_j}{k^2}\right)\delta(k_i+k_i')\delta(\omega+\omega') \quad (1.13.2)$$

$$D_0^*(\varepsilon,k) = D_0(\varepsilon)k^{4-\varepsilon}, k = |k_ik_i|^{1/2} \quad (1.13.3)$$

Quantity $\tilde{f}_i$ is the space–time Fourier transform of f_i, d is the spatial dimension (3 for three-dimensional turbulence), and ε is a dimensionless parameter. $D_0(\varepsilon)$ is assumed to be independent of any length scale and this scale invariance leads to $\varepsilon = 4$ (Lam, 1992) and $D_o(\varepsilon, k) = D_o(\varepsilon)$. k is the wavenumber vector, and ω is the frequency. Operator $(D_{ij} - k_ik_j/k^2)$ renders the random force isotropic and divergence-free. The Dirac delta function in wavenumber and frequency ensure statistical homogeneity in space and time (white noise in time). f is put in to simulate turbulence generated by hydrodynamic processes in the inertial subrange. A recent modification to the forcing function assumes an infrared cutoff as well: $\langle f_if_j \rangle = 0$ for $k < A_L$. This is not important for renormalizing the momentum equation, but it is important for the dissipation equation.

In addition, equations for TKE and dissipation rate ε, as well as the equation for scalar conservation (see Section 1.7), are written down and RNG procedure applied to all these equations.

RNG procedure is carried out in Fourier space and consists of eliminating modes $v^>(k)$ $(A_o + \delta A < k < A_o)$ from the equation for retained modes $v^<(k)$ $(0 < k < A_oe^{-r})$. A_o is the ultraviolet cutoff, equivalent to the wavenumber corresponding to the Kolmogoroff dissipation scale. These are the smallest scales possible in a turbulent flow. Averaging or filtering is done over an infinitesimal band of small scales and the removal process is iterated to obtain finite changes (Yakhot *et al.,* 1992). Now, this removal alters the "filtered" solution significantly, which no longer satisfies the original N-S equation, since additional terms are generated by this removal/filtering process. This is taken into account by modifying the viscosity ν_0 by a correction $\delta\nu$. By recursive filtering, the ultraviolet cutoff wavenumber A can be moved lower and lower into the inertial subrange, while viscosity is continuously corrected to yield an eddy viscosity ν_t. The residuals not absorbed by the increased viscosity are ignored as higher order in ε in the ε-expansion procedure [which has been challenged by Lam (1992)]. Skipping the extremely messy details (refer to

Yakhot and Orszag, 1986; Smith and Reynolds, 1992; Lam, 1992), one gets an equation for the effective viscosity,

$$\frac{\partial \nu_t}{\partial \Lambda} = -\frac{\nu_t}{\Lambda} A_d \, \bar{\lambda}^2 \, F(\bar{\lambda}^2) \tag{1.13.4}$$

$$A_d(\varepsilon_0) = (d^2 - d - \varepsilon_0) / [2d(d+2)] \tag{1.13.5}$$

$$\bar{\lambda}^2 = \frac{D_0(\varepsilon_0) S_d}{(2\pi)^d} \frac{1}{\nu_0^3 \Lambda^\varepsilon} \tag{1.13.6}$$

$$F(\bar{\lambda}^2) = 1 + C_1(\varepsilon_0, d)(\bar{\lambda}^2) + \cdots C_n(\varepsilon_0, d)(\bar{\lambda}^2)^n + \cdots \tag{1.13.7}$$

where $S_d = 2\pi^{d/2} / \Gamma(d/2)$ is the area of unit sphere of dimension d, and $c_n(\varepsilon_o, d)$ are dimensionless numbers. The initial condition is

$$\nu_t = \nu_0 \quad \text{at} \quad \Lambda = \Lambda_0 \tag{1.13.8}$$

Assuming $D_0(\varepsilon_0)$ to be independent of Λ, the equation was integrated, using the leading approximation $F(\bar{\lambda}^2) \sim 1$. YO showed that $\bar{\lambda}^2$ quickly approached a fixed point value $\bar{\lambda}^2{}_*$ independent of ν_0 as Λ is decreased into the inertial range. Lam (1992) shows the same conclusion holds for arbitrary $F(\bar{\lambda}^2)$. The scaling law for ν_t is obtained by equating $\bar{\lambda}^2{}_* = \bar{\lambda}^2$ in Eq. (1.13.6).

By eliminating all modes > k, one obtains for three-dimensional turbulence (d = 3)

$$\nu_t(k) = \left[\frac{3}{8} A_d(\varepsilon_0) \frac{2 D_0 S_d}{(2\pi)^d} \right]^{1/3} k^{-4/3} \tag{1.13.9}$$

To compute the value of A_d, YO use an ε-expansion procedure and put $\varepsilon = 0$ to lowest order in Eq. (1.13.5) to get $A_d = 0.2$ for $d = 3$:

$$\nu_t(k) = 0.422 \left[\frac{2 D_0 S_d}{(2\pi)^d} \right]^{1/3} k^{-4/3} \tag{1.13.10}$$

A similar expansion procedure gives the energy spectrum to lowest order:

$$E(k) = 1.186 \left[\frac{2D_0 S_d}{(2\pi)^d} \right]^{2/3} k^{-5/3} \tag{1.13.11}$$

The coefficients are still undetermined. If one now appeals to Kolmogoroff scaling in the inertial subrange, dimensional analysis gives

$$\begin{aligned} \nu_t(k) &= \alpha_1 \varepsilon^{1/3} k^{-4/3} \\ E(k) &= \alpha\, \varepsilon^{2/3} k^{-5/3} \end{aligned} \tag{1.13.12}$$

If one further recognizes that ε, the dissipation rate, can be written as

$$\varepsilon = \lim_{\Lambda \to \infty} \left[2\nu_t(\Lambda) \int_0^{\Lambda} k^2\, E(k)\, dk \right] \tag{1.13.13}$$

then

$$\alpha_1 \alpha = 2/3 \tag{1.13.14}$$

This immediately provides the value of the undetermined coefficient,

$$\frac{2D_0 S_d}{(2\pi)^d} = 1.549\, \varepsilon \tag{1.13.15}$$

leading to a value of 1.617 for α ($\alpha_1 = 0.512$). In effect, Kolmogoroff scaling, Eq. (1.13.12), is the effective closure in RNG.

A similar RNG procedure applied to a passive scalar gives a value of 0.718 for the turbulent Prandtl number Pr_t. The Batchelor number can also be obtained since

$$\varepsilon_c = \lim_{\Lambda \to \infty} \left[2\nu_{ct}(\Lambda) \int_0^{\Lambda} k^2 E_c(k)\, dk \right] \tag{1.13.16}$$

where

$$E_c(k) = \alpha_B\, \varepsilon_c\, \varepsilon^{-1/3} k^{-5/3} \tag{1.13.17}$$

The Batchelor constant can be shown to be

$$\alpha_B = \alpha \, Pr_t \tag{1.13.18}$$

by simply taking the ratio of Eq. (1.13.17) to Eq. (1.13.12), where

$$Pr_t = \frac{\nu_t(k)}{\nu_{ct}(k)}$$

is the ratio of the two turbulent diffusivities. Therefore, $\alpha_B = 1.161$ (Yakhot and Orszag, 1987). In addition, YO derive values for the skewness factor, –0.59, after the algebraic error is corrected. Since

$$K(\Lambda) = \int_{\Lambda}^{\infty} E(k)\, dk \tag{1.13.19}$$

it can be shown that $K(L) = \frac{3\alpha}{2} \varepsilon^{2/3} L^{-2/3}$, but from Eq.(1.13.13),

$$\varepsilon^{1/3} = \frac{3}{2} \alpha \, \nu_t(\Lambda) \Lambda^{4/3} \tag{1.13.20}$$

Combining the two,

$$\nu_t(\Lambda) = \left(\frac{2}{3\alpha} \right)^3 \frac{K^2}{\varepsilon} \tag{1.13.21}$$

The RNG procedure was also applied to the equation for the dissipation rate to obtain for homogeneous isotropic turbulence

$$\frac{D\varepsilon}{Dt} = -c_4 \frac{\varepsilon^2}{K} \tag{1.13.22}$$

where $c_4 = 1.722$. Combined with $\frac{DK}{Dt} = -\varepsilon$, this gives a decay law

$$K \sim t^{\frac{-1}{c_4 - 1}} \sim t^{-1.331} \tag{1.13.23}$$

However, Smith and Reynolds (1992) corrected YO's algebraic errors to yield $c_4 = 5.65$, thus leading to $K \sim t^{-0.215}$, in very poor agreement with experimental values. Recent recalculations yield $c_4 \sim 1.68$ and $K \sim t^{-1.47}$ (Yakhot and Smith, 1992).

Removal of only the smallest scales by RNG gives rise to subgrid scale models useful for LES. YO showed that in this case, the Smagorinsky model results,

$$\gamma_t = (c\Delta)^2 \, S_{ij}^{<} \tag{1.13.24}$$

where $S_{ij}^{<}$ is the strain rate of the resolved scales; $c \sim 0.11$. Once again, these values are of the same order as used in many LES calculations.

Removal of all significant scales from the TKE and dissipation equation results in the $K-\varepsilon$ model (see Section 1.10). In other words, ensemble averaged equations result. YO's values for the constants in the $K-\varepsilon$ model are [c_4 corrected by Smith and Reynolds (1992)]

$$c_1 = 0.084;\ c_2 = 0.718;\ c_3 = 1.063;\ c_4 = 5.65;\ c_5 = 0.718 \tag{1.13.25}$$

Contrast these with values determined empirically to yield good agreement with simple flows. As discussed earlier, the value of c_4 is too large to be consistent with observations on homogeneous, isotropic turbulence. For homogeneous shear flow, TKE increases as (Spaziale, 1991)

$$K \sim e^{\lambda t} \tag{1.13.26}$$

where

$$\lambda \sim \left[\frac{c_1 \left(c_4 - c_3\right)^2}{\left(c_3 - 1\right)\left(c_4 - 1\right)} \right]^{1/2} \tag{1.13.27}$$

Here the value of c_3 is too close to the singular value and the RNG model yields unrealistically large growth rates. However, the recomputed values of $c_3 = 1.42$ and $c_4 = 1.68$ yield much better agreement: $\lambda = 0.142$.

Lam (1992) provides an alternative explanation of the RNG method. He shows that the ε-expansion scheme used by YO is inconsistent since $\varepsilon = 4$ is needed to derive the Kolmogoroff constant from the expression for ε using the Kolmogoroff scaling law, whereas elsewhere small expansions in ε are used. He shows that the stirring force introduced is a surrogate for a new body force introduced by the filtering process of eliminating small scales and extrapolating; ε is an adjustable parameter in the problem and not an expansion parameter. He mentions several alternative RNG approaches developed recently.

Yakhot *et al.* (1992) extend the $K-\varepsilon$ model of YO to shear flows using expansion in η, the ratio of the timescale of turbulence t_e to the timescale of the mean strain rate,

$$\eta = \left(2 S_{ij} S_{ij}\right)^{1/2} K/\varepsilon \tag{1.13.28}$$

They also present recalculated values for various constants. c_3 has been recalculated to be 1.420, and c_4 to 1.68. An additional term R has been added to the dissipation rate equation to account for strong shear,

$$\frac{\partial \varepsilon}{\partial t} + \frac{\partial}{\partial x_k}\left(U_K \varepsilon\right) = -c_3 \frac{\varepsilon}{K} \tau_{ij} S_{ij} - c_4 \frac{\varepsilon^2}{K} + \frac{\partial}{\partial x_k}\left(\frac{\nu_t}{c_5} \frac{\partial \varepsilon}{\partial x_k}\right) - R \tag{1.13.29}$$

where

$$R = 2\nu S_{ij} \overline{\frac{\partial U_k}{\partial x_i} \frac{\partial U_k}{\partial x_j}} \tag{1.13.30}$$

For strong shear, η is large and R cannot be neglected ($\eta \sim 3$ in the logarithmic region). Yakhot *et al.* (1992) postulate a model for R motivated by a Padé approximation and scaling arguments that yield the universal value for $\eta = \eta_\infty$ obtained in homogeneous shear flows:

$$R = \frac{\left(-\tau_{ij} S_{ij}\right) \eta \left(1 - \frac{\eta}{\eta_\omega}\right)}{\left(1 + c_6 \eta^3\right)} \frac{\varepsilon}{K} \tag{1.13.31}$$

For homogeneous shear flows, $-\tau_{ij} S_{ij} = \gamma_t S^2$, and thus

$$R = \frac{c_1 \eta^3 \left(1 - \frac{\eta}{\eta_\infty}\right)}{1 + c_6 \eta^3} \frac{\varepsilon^2}{K} \tag{1.13.32}$$

For weak shear ($\eta \rightarrow 0$), $R \rightarrow 0$, and for strong shear ($\eta \rightarrow \infty$), $R \rightarrow -\frac{c_1}{c_6}\frac{\eta}{\eta_\infty}\frac{\varepsilon^2}{K}$. The constant c_6 is undetermined and RNG has to appeal to the law of the wall and the empirical value of the von Karman constant κ. $\eta = c_1^{-1/2}$, $\varepsilon = -\tau_{ij} S_{ij}$, in the wall region, and since only $\partial / \partial z$ terms remain in the ε equation, it can be shown that

$$\kappa^2 = \left[c_4 - c_3 + \frac{c_1 \eta^3 \left(1 - \frac{\eta}{\eta_\infty}\right)}{1 + c_6 \eta^3} \right] c_1^{1/2} c_5 \qquad (1.13.33)$$

Using $\kappa \sim 0.4$, one gets $c_6 \sim 0.012$.

This completes the part of the RNG-based generalized $K-\varepsilon$ model applicable to strong shear flows. It is also important to account for the time lag in the response of turbulence to sudden application of strain by including relaxation terms in the Reynolds stress transport terms to account for the initial evolution of turbulence (Yakhot *et al.*, 1992). For asymptotic growth rate considerations, the relaxation effects are not important.

A more recent development is an improved RNG model by Canuto and Dubovikov (1996a,b,c) in which nonlinear interactions introduce both a turbulent stirring force that is ultimately responsible for the spectral energy flux and a wavenumber-dependent viscosity that can be renormalized and summed up, as described above, by RNG analysis. Locality of transfer in the wavenumber spectrum is postulated because of the nonrenormalizable infrared divergence of the stirring force. There is considerable evidence that nonlocal interactions between disparate wavenumbers mostly cancel one another (Kraichnan, 1971), and the interactions that contribute to local spectral transfer are principally local in nature (Eynk, 1994). This is also confirmed by DNS and LES calculations (Domaradzki, 1988; Domaradzki and Rogallo, 1990). Canuto and Dubovikov (1996a,b,c) derive Kolmogoroff and Obukhov–Corrsin spectra, the associated constants (5/3 and 1.2), and skewness for freely decaying turbulence (0.5), and compare their model to DNS, LES, and laboratory data.

1.14 DIRECT NUMERICAL SIMULATIONS (DNS)

Direct numerical solutions exploit the computing power of modern high performance computers to explicitly solve the Navier–Stokes equations, resolving all scales from the Kolmogoroff dissipative scales to the largest eddies in the e.g., Moin and Kim, 1982; Domaradzki, 1988; Domaradzki *et al.*, 1993, 1994; Coleman *et al.*, 1992; Schumann, 1996). In this sense, they are the other extreme to EAC models. DNS has enabled various properties of turbulence to be calculated numerically, without any approximations or "models" whatsoever. Herein lies the attraction of DNS. It is not a turbulence model; it is an exact calculation. These calculations tell us the true nature of many aspects of turbulence, including its vortex tube-like structure at small scales, that are hard or impossible to measure or model accurately. The technique is especially well suited to the simulation of unbounded turbulent flows, but simulations near a wall are more difficult. For a timely review of DNS, see Moin and Mahesh (1998).

The main drawback of DNS [see the review by Reynolds (1990)] is that to resolve the microscales in the dissipative range of the spectrum, since $\ell/\eta \sim R_t^{3/4}$, the number of grid points needed is $O(R_t^{9/4})$ (Moin and Mahesh (1998) point out that it is only necessary to represent dissipative scales accurately and it is seldom necessary to resolve the Kolmogoroff microscale). In addition, to resolve the associated timescales and do calculations to statistically steady state, $O(R_t^{3/4})$ resources are needed. So the magnitude of the problem scales like R_t^3. Just doubling the Reynolds number requires a computer an order of magnitude more powerful. DNS turbulence research therefore rightfully belongs to the category of grand challenge problems that require computing capabilities beyond a teraflop machine. So far, even with the most efficient techniques of solution and optimization, the method is impractical beyond a turbulent Reynolds number (Re_t) value of about 2000. Since geophysical boundary layers have R_t values of over 10^7, there is therefore little likelihood of DNS being used for accurate simulation of geophysical turbulent flow problems such as the ABL and the OML in the foreseeable future. Nevertheless, these calculations are being done at the low Reynolds numbers currently attainable in the hope that they can shed some light on the otherwise intractable problem of modeling turbulence (Coleman *et al.*, 1988, 1992; Domaradzki *et al.*, 1993, 1994; Piomelli *et al.*, 1996; Kerr *et al.*, 1996; Schumann, 1996), and some are beginning to display an incipient inertial subrange, a broad range of which exists in the spectrum at only large Reynolds numbers (Moin and Kim, 1982; Rogallo and Moin, 1984). DNS is also beginning to provide detailed "data" on laboratory turbulence that is valuable for understanding turbulence. More recently DNS has been applied to geophysical-type flow situations, but at Reynolds numbers much lower than those typical of such flows (e.g., Coleman *et al.*, 1988, 1992).

However, as Moin and Mahesh (1998) point out, the real utility of DNS is not in simulating geophysical and other turbulent layers, but more for obtaining a better understanding of turbulence dynamics, which would in turn be useful in modeling and understanding turbulent layers.

DNS obtained its start in the simulation of the classical turbulence problem of homogeneous isotropic turbulence, long studied by turbulence theorists. There has however been a significant increase in applications of DNS to practical flow problems in geophysics in the 1990s (Coleman *et al.*, 1992; Schumann, 1996). The Reynolds numbers achievable in DNS are severely limited by computing memory and time requirements. In calculations of homogeneous stratified turbulent flows, early supercomputers of the 1980s had a maximum feasible grid size of about 64^3 (Gerz *et al.*, 1989), but a multigigaflop computers such as the Cray C90 allow computations to be done over a 128^3 grid (Schumann, 1996), leading to a Re_t of about 120 (and a Re_λ of about 50). The Kolmogoroff scale η in these calculations is comparable to the grid size, and the integral scale ℓ is usually a fraction of the domain size. The ratio ℓ/η, proportional to $R_t^{3/4}$, is therefore a fraction of the number of grid points in a single direction. Thus the number of grid points in each direction determines the Reynolds number of the simulation. As of 1999, on current high performance computers, efficient pseudo-spectral techniques enable homogeneous isotropic turbulence to be simulated (e.g., Kerr *et al.*, 1996) with a numerical grid of 1024^3, and a turbulence Reynolds number Re_t (based on the integral length scale) of about 2400 (Re_λ based on Taylor microscale of about 800). A 1024^3 calculation enables a reasonably broad Kolmogoroff inertial subrange to be realized. Even higher Reynolds numbers would become feasible on massively parallel teraflop computers of the next century.

The biggest advantage of DNS is that it is an exact calculation and not a model. Therefore it is possible to determine the various turbulence quantities of interest to both LES and EAC, although care must be exercised in interpreting and extrapolating the low Reynolds numbers results of DNS to the high Reynolds number flows simulated by LES and EAC. This applies especially to interpreting transfers in spectral space and resolving questions related to local vs distant scale interactions, since in DNS, the range of scales is small and the interactions are therefore mostly local.

In EAC, the question is how best to model unknown terms such as the pressure–strain covariances, and once again it is here that DNS (and LES) is providing some guidance. In LES, the question is how best to parameterize the subgrid scales, which can be explored by filtering the DNS results and examining the resolved and subgrid scales. It is indeed such studies that revealed the importance of including backscatter (Piomelli *et al.*, 1991, 1996) in parameterizing the subgrid scales in LES. Traditional Smagorinsky-type eddy viscosity formulations for subgrid scales have proven inadequate for LES of all but the simplest of flows. Simulations of homogeneous isotropic turbulence (Domaradzki *et al.*, 1993) and wall-bounded turbulence (Domaradzki *et al.*, 1994) have shown that the net eddy viscosity has a cusp-like behavior near the cutoff wavenumber resulting from the balance between forward transfer of energy from resolved scales to subgrid scales and the backscatter from subgrid scales to resolved scales. Piomelli *et al.*'s (1996) DNS results have shown that in the buffer region of a turbulent boundary layer, the forward scatter is associated predominantly with ejections, and the backscatter with sweeps. Furthermore, subgrid scale transfer is local, with the scales immediately above the cutoff interacting with those immediately below. It may therefore be necessary to model the local and nonlocal transfers separately. Linear formulations such as the Smagorinsky eddy viscosity consider only the forward transfer of energy from resolved scales to subgrid scales and hence are absolutely dissipative. This often leads to excessive damping of turbulence. Nevertheless, their success is related to the fact that they do predict overall dissipation reasonably well even though they fail to predict the details of the subgrid scale stresses (Piomelli *et al.,* 1996). Dynamic models that utilize a dual-filtering approach can account for the backscatter. A variety of other approaches to accounting for the backscatter from subgrid scales to resolved scales have also been proposed, leading to better LES's in certain flow situations such as boundary layer flows (e.g., Kosovic, 1996).

DNS employs numerical methods that can be classified into two categories: the pseudo-spectral method and the finite-difference-based method. The former is the preferred method when feasible, since the equations can be transformed into spectral space for a more efficient solution. Tests with isotropic turbulence show that the same accuracy can be achieved by spectral methods at half the resolution of the finite-difference-based schemes (Gerz *et al.,* 1989). The assumption of horizontal homogeneity in the flow enables the Navier–Stokes equations to be transformed from physical to wavenumber space and pseudo-spectral techniques to be used for solution. These techniques use simplified periodic boundary conditions. For homogeneous isotropic turbulence far away from any boundaries, periodic boundary conditions can be applied in all three directions. However, when shear and/or stable stratification is involved, as is the case in the ABL, periodic conditions fail for one of the three directions in the Eulerian frame of reference (the same is true of LES). Either time-dependent coordinate transformation in which the flow can be assumed periodic (Holt *et al.,* 1992) or finite differences must be used (Gerz *et al.,* 1989; Schumann, 1996); both give similar results according to Holt *et al.* (1992). The use of spectral decomposition in the horizontal enables conversion of PDEs to ODEs and efficient techniques such as the Runge–Kutta schemes to be employed for solution. Such numerical details can be found in the original papers and are too complicated to describe here.

There is considerable similarity between numerical techniques employed for DNS and those for LES. The principal difference is that the former solves the full N-S equations, whereas the latter solves the filtered equations. The former needs no subgrid scale model; the latter needs one. Both use homogeneity and periodic boundary conditions in the horizontal directions, and often pseudo-spectral techniques in the horizontal. In fact, Kaltenbach and co-workers extended their DNS method (Kaltenbach *et al.*, 1991) to LES (Kaltenbach *et al.*, 1994). Using the Smagorinsky model for subgrid scales they found that they could reproduce the LES results for weakly stratified shear flows in DNS by replacing its constant molecular viscosity by the initial mean turbulent viscosity of LES. LES can therefore be looked upon as a DNS with temporally and spatially varying viscosity (Schumann, 1996). At present, because of the inability of DNS to achieve high Reynolds numbers, LES gives more accurate results for practical flow situations, provided the subgrid scales are parameterized correctly. In any numerical simulation of turbulence, energy must be removed from the smallest resolved scales (Coleman *et al.*, 1992) by either molecular viscosity (in DNS), eddy viscosity (in LES), or a hyperviscosity [in LES, where the viscous term is multiharmonic (∇^n, n > 6) instead of the biharmonic (∇^2) type, for purely numerical reasons], and the underlying assumption is always that the larger energy-containing scales are not very sensitive to exactly how the removal is done!

Schumann (1996) summarizes recent developments in DNS and LES of stratified homogeneous shear flows. DNS of the neutrally stratified Ekman layer has been performed by Coleman *et al.* (1990), as well as that of a stably stratified Ekman layer (Coleman *et al.*, 1992). The latter collaborators (using a 96 × 96 × 45 computational grid) found some agreement between their DNS results and the LES results of Mason and Derbyshire (1990). They also found support for the z-less scaling of Nieuwstadt (1984) (see Chapter 3). A salient feature of the stable stratification is the limitation on the eddy scale related to dissipation due to buoyancy effects (Brost and Wyngaard, 1978). This scale is bounded by the Ozmidov scale in stably stratified flows. Therefore the length scale distribution in the boundary layer adjacent to the wall is the lesser of the distance from the wall and this length scale. When DNS results were corrected for low Reynolds number effects, the results on the length scale in the stable boundary layer agreed with this. They also agreed well with gradient closure approximations for the temperature variance and the heat flux. Figure 1.14.1 shows the variation of coefficients a and b as a function of Ri_g, where

$$\sigma_\theta = a\left(Ri_g\right)\frac{w}{N}\frac{\partial\Theta}{\partial z}; \ \overline{w\theta} = b\left(Ri_g\right)\frac{w^2}{N}\frac{\partial\Theta}{\partial z} \tag{1.14.1}$$

We stop here in our rather limited exploration of turbulence and turbulence modeling. The subject is too vast and complex to describe fully in a single chapter. The reader is however encouraged to explore specific approaches via the references cited. Hopefully, this chapter provides the necessary background.

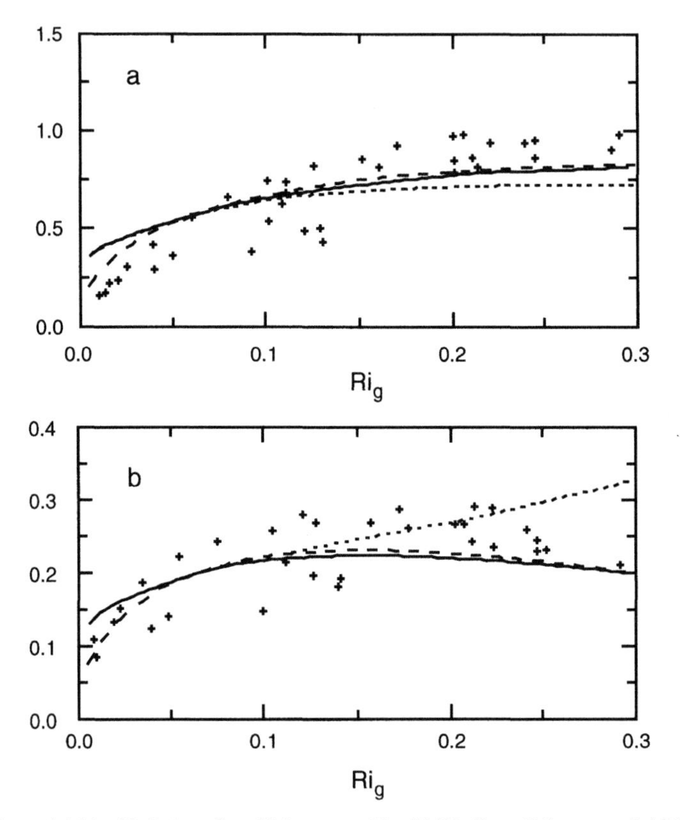

Figure 1.14.1. Variation of coefficients a and b with Ri_g (from Coleman et al. 1992).

LIST OF SYMBOLS

α, α_1	Kolmogoroff constants in the universal inertial subrangeof the three-dimensional and one-dimensional turbulence kinetic energy spectrum
α_B	Batchelor constant in the universal inertial-convective subrange of the Obukhoff–Corrsin passive scalar spectrum
β, β_S	Coefficient of expansion (thermal and due to salinity)
δ	Boundary layer thickness
ε, ε_θ, ε_c	Dissipation rate of TKE, temperature, and scalar variance
γ_m	Mixing efficiency
φ_M	Monin–Obukhoff similarity function
η	Kolmogoroff microscale (also the ratio of the turbulence timescale to the timescale of mean strain rate)

η_b	Batchelor scale
η_c	Microscale corresponding to the dissipation of the variance of c
κ	Von Karman constant
λ	Taylor microscale
ν, ν_t	Kinematic viscosity (molecular and turbulent)
θ	Fluctuating temperature
ρ	Fluctuating fluid density (also correlation coefficient)
ρ_0, ρ_r	Mean and reference density
σ	Standard deviation
σ_{ij}	Turbulent stress tensor
ω_i	Fluctuating component of vorticity
τ	Shear stress value (also time delay in correlations and covariances)
$\tau(k)$	Shear stress spectrum
τ_{ij}	Shear stress tensor
Δ	Grid size
Δb	Buoyancy change
Θ	Mean temperature
Ω_i	Mean vorticity (also rotation rate of the Earth)
Σ_{ij}	Mean stress tensor
b	Buoyancy
c	Fluctuating concentration of a passive scalar
f, f_y, f_k	Coriolis parameter (also frequency), twice the horizontal component of rotation, and twice the rotation vector
f_t, f_S	Turbulence frequency and frequency associated with vertical shear
g	Acceleration due to gravity
k, k_1	Wavenumber and longitudinal component of the wavenumber
k_e, k_η	Wavenumbers corresponding to the spectral peak and Kolmogoroff scales
k_T, k_S, k_c	Diffusivity of heat and salt, and a passive scalar (kinematic)
k_H, k_ρ	Turbulent heat diffusivity, turbulent buoyancy diffusivity
ℓ	Integral microscale of turbulence
ℓ_b, ℓ_O, ℓ_T	Buoyancy length scale, Ozmidov length scale, and Thorpe length scale
ℓ_d	Alternative buoyancy length scale
ℓ_m	Mixing length
ℓ_r	Rotational length scale
p	Pressure
q	Turbulence velocity scale

q_b	Buoyancy flux
r	Radius of curvature
r_{ij}	Turbulence rotation tensor
s_{ij}	Turbulence strain-rate tensor
t	Time
t_s	Timescale over which large eddies are strained and oriented
t_θ	Timescale associated with dissipation of temperature fluctuations
u, v, w	Turbulence velocity components
u^*	Friction velocity
v_η	Kolmogoroff velocity scale
x	Horizontal coordinate
x_i	Coordinates
z	Vertical coordinate
z_i	Inversion height
$z_)$	Roughness scale
A_t	Turbulence activity
B(k)	Buoyancy flux spectrum
C	Mean concentration of a passive scalar
D	Diffusive transport
E(k), $E_c(k)$	Three-dimensional spectrum of TKE and a passive scalar
E_{11}, E_{22}, E_{33}	One-dimensional spectra in three directions (E_{11} is longitudinal)
F_{ij}	Spectrum tensor
F(k)	Energy flux in spectral space
Fr, Fr_t	Froude number and turbulence Froude number
Fr'_t	Rotational turbulence Froude number
K, K_M	Turbulence kinetic energy (also kurtosis) and mean kinetic energy
L	Length scale
N	Buoyancy (Brunt–Vaisala frequency)
Nu	Nusselt number
N_θ	Dissipation rate of temperature variance
P	Mean pressure (also probability density function)
P_S, P_b	Shear and buoyancy production of TKE
Pr_t	Turbulent Prandtl number
R_{ij}	Correlation tensor
R_N	Flow Reynolds number
R_t, R_λ	Turbulence Reynolds number and Taylor microscale Reynolds number
Ra	Rayleigh number
Ri_g, Ri_f	Gradient Richardson number and Flux Richardson number
Ri_C	Curvature gradient Richardson number
Ri_R	Rotational gradient Richardson number

Ri_t	Turbulence gradient Richardson number
S	Salinity (also skewness)
S_{ij}	Mean strain-rate tensor
S_M, S_H	Stability functions in turbulent mixing coefficients
T_e, T_η, T_b	Eddy turnover timescale, Kolmogoroff timescale, and buoyancy timescale
U	Flow velocity

Chapter 2

Oceanic Mixed Layer

In this chapter, we discuss the salient characteristics of oceanic mixed layers (OMLs), and how we go about modeling them and obtaining a better understanding of the OML for many practical applications. We concentrate on the mixed layer adjacent to the air–sea interface, and describe the various processes affecting it. The bottom/benthic layer (BBL) is then described, and finally, methods for modeling OMLs are outlined.

There are very few existing reviews of the OML, one of them being the now classic but somewhat dated monograph by Phillips (1977). Most of the observational and other research work done on the OML is published in oceanography journals such as the *Journal of Geophysical Research (Oceans)* of the American Geophysical Union, and the *Journal of Physical Oceanography* of the American Meteorological Society, although related articles can often be found in journals on the atmosphere. Journals such as the *Journal of Marine Research, Deep-Sea Research,* and the *Journal of Continental Shelf Research* are also relevant to the OML. The *Journal of Hydraulics* of the American Society of Civil Engineers publishes research, mostly engineering-related, on the coastal oceans. The reader is referred to these for the latest advances related to the OML.

2.1 IMPORTANCE

In matters related to geophysical flows, turbulent boundary layers and the associated small scale processes play a vital role. The atmospheric boundary layer (ABL) and the OML adjacent to the air–sea interface mediate the exchange of mass, momentum, energy, and heat between the atmosphere and the ocean. This is the principal reason why the OML is so central to air–sea exchange, a problem of importance to long-term weather and climate.

Since the atmosphere is to a large extent transparent to solar radiation, the surface of the Earth is mainly responsible for most of the solar heating. In addition, because of the high heat capacity of water (2.5 m of the upper ocean has the same heat capacity as the entire troposphere), and because the oceans compose over two-thirds of the surface of the globe, most of the solar heating on Earth passes through the oceans, or more specifically the OML, first (roughly 60% is absorbed in the upper 20 m). Oceans act as heat reservoirs, gaining heat during spring–summer and losing it slowly during fall–winter. Most of this heat loss is through the atmospheric column and therefore oceans act like a flywheel in matters related to weather on timescales of weeks and longer. In contrast, the seasonal heat storage in the ground is for most practical purposes quite negligible. The heat gained by the oceans, primarily in the tropics, is transported poleward by western boundary currents to subpolar regions, where there is a net heat loss during the course of a year. It is the poleward heat transport by the oceans and the atmosphere (in roughly equal amounts) that keeps high midlatitudes habitable.

The OML also plays an important role in the oceanic food chain, which supplies a large fraction of the protein needs of the human population. Primary production by phytoplankton is the first link in this chain. The need for an energy source in producing biomass restricts primary production to the upper few tens of meters (the euphotic or photic zone), in which the solar insolation is strong enough to assist carbon fixation. Because of the absorption characteristics of seawater, the blue-green part of the spectrum of electromagnetic radiation important to biology can penetrate only a few tens of meters in clear waters and even less in murkier waters (the extinction is faster in murky waters due to sediments or high phytoplankton concentrations—see Section 2.3). Therefore it is primarily in the OML that most primary productivity occurs. The characteristics of the OML affect primary productivity, and to some extent, through its influence on solar extinction, there is a feedback on the OML as well.

The mixing at the base of the OML is also crucial to biological productivity. The OML is normally nutrient-poor and it is the injection of nutrients from the nutrient-rich waters below the seasonal thermocline that permits higher levels of primary productivity. Wind forcing is crucial to this since the upwelling and downwelling produced as a consequence of wind forcing affect the nutrient balance in the OML. In fact, it is the upwelling regions (which compose just a few percent of the world's oceans and are located mostly along coasts such as Peru), where nutrient-rich waters are forced into the OML and brought into the photic zone, that provide most of the fish catch around the world.

In addition to nutritional needs, biological productivity is important from a climatic point of view over timescales of decades or more. Carbon fixing that occurs in the OML essentially sequesters the CO_2 in the atmosphere, since the fecal pellets of zooplankton raining down onto the deep ocean floor constitute a sink mechanism for CO_2. Thus there exists a biological pathway for removing some of the anthropogenic CO_2 introduced into the atmosphere. It is likely that the ocean acts as an important CO_2 sink on the globe and accounts for a significant fraction of the "missing" anthropogenic CO_2 input to the atmosphere. The inorganic pathway is also likely to be important since it is in the cold subpolar oceans that there is a significant uptake of CO_2, and some of these regions are also regions of deep and intermediate water formation.

Air–sea exchange involves not only momentum, heat, and water vapor, but some photochemically produced and other "greenhouse" gases as well. For example, N_2O and carbonyl sulfide (CoS) are two examples of such gases for which the ocean is a nonnegligible source. CoS is produced through photochemistry by the action of the UV part of the spectrum on dissolved organic matter in the upper few meters of the ocean (the extinction scale of the UV radiation) and then outgassed to the atmosphere. N_2O is produced by anaerobic bacteria primarily in oxygen-poor waters away from the ventilated OML and then brought into the OML during winter by the deepening of the seasonal thermocline. It is then outgassed to the atmosphere, eventually ending up in the stratosphere, where it has an impact on ozone concentrations. CoS is also a greenhouse gas of some importance to climate. The concentrations and therefore the rates of exchange of these gases to the atmosphere depend on the mixing in the OML.

Acoustic propagation in the oceans is affected by the vertical gradients of temperature (and to some extent salinity). The OML determines acoustic transmission characteristics important to sound transmission from an acoustic source in the water column.

Finally, the OML constitutes the first link in the chain of oceanic pollution. Most of the pollution in the global oceans takes place along the coast in the coastal oceans through the OML, and therefore the fate of any pollutants accidentally or intentionally deposited in the OML depends on the mixing and dispersion in the OML.

2.2 SALIENT CHARACTERISTICS

The oceanic mixed layer is the region adjacent to the air-sea interface, which responds directly to surface forcing. The OML is typically tens of meters deep, and due to the fact that it is well mixed, the temperature and salinity (and therefore the density) are fairly uniform (see Figure 2.6.1 for some typical temperature profiles). The rapidly changing regions below these uniform regimes of temperature, salinity, and density are called the thermocline, halocline, and pycnocline, respectively. The mixing is primarily shear driven, since wind stress at the surface is the primary mixing agent, although at night significant convective mixing takes place. Convective mixing is described in Section 2.5. Various mixing processes which occur at different latitudes are discussed in Sections 2.6 to 2.8.

The OML is heated both at the top and deeper in the water column from solar radiation in the visible part of the spectrum penetrating into the OML. This solar heating produces a diurnal cycle that varies in importance and magnitude at different latitudes. The cooling, however, is driven from heat and evaporative losses at the surface. At high latitudes, the diurnal cycle of heating is overshadowed by dynamical effects due to the presence of sea-ice cover, as discussed in Section 2.8. The diurnal modulation is, however, quite important at lower latitudes. Seasonal variation of the OML due to radiative heating is also important, although the importance depends on the latitude.

Heating and cooling of the ocean surface occurs across the skin of the ocean (see Figures 4.6.1 and 4.7.1). Within this skin layer, which is on the order of a millimeter in thickness, there can be a sharp drop in temperature of a few tenths of a degree Celsius. Exchanges of heat, momentum, and mass through this region are by molecular processes. This cool skin plays an important role in air–sea transfer processes. The details of the cool skin and the exchanges are found in Section 4.7. Transport of dissolved gases also occurs across a molecular sublayer of similar thickness to the thermal sublayer. There is an equivalent layer on the atmospheric side of the ocean–atmosphere interface called the interfacial sublayer that is governed by similar physics (Section 4.6).

Below this thin layer, turbulent processes dominate, driven by momentum and energy exchanges from the atmosphere to the ocean, which involve wave motions at the ocean surface. Wind stress acting at the ocean surface creates currents as well as waves, the fraction of the momentum input to each depending critically on the degree of development of the wave field and hence the fetch. At early stages of wind-wave development (equivalently, short fetches), a large fraction of the momentum input goes into waves. At late stages, when the wave field is mature or nearly fully developed, waves over most of the wavenumber/frequency spectrum lose any excess momentum and energy gained from the wind and remain "saturated." This excess energy is dissipated by wave breaking and the excess momentum is transferred to ocean currents. Thus a larger fraction of the momentum input goes into ocean currents. It is the dependence of the momentum transfer from winds on the degree of wave field development that is partly responsible for the large scatter found in the measurements of the drag coefficient in classical bulk formulas.

The presence of the wave sublayer and a mobile interface, whose roughness as felt by both the ocean and the atmosphere is dynamically determined, provides the most important distinction between an ABL over land and that over the ocean. Consequently, surface layer similarity laws derived from the ABL over land are at best loose analogies when applied to the ABL over water and the OML. This is more so for the OML where the fraction of the OML affected directly by wave orbital velocities and hence wave dynamics can be a large. In contrast, the fraction of the ABL affected directly by wave motions is small. At anemometric height, the fraction of the momentum flux from the atmosphere to the ocean supported by waves is small. Most of that flux is carried by turbulent eddies. However, as the air–sea interface is approached, an increasing part of this flux is supported by waves. The fraction of the flux carried by waves depends critically on the wave age. The atmosphere transfers more momentum and energy to young waves compared to more mature ones; the result is that the roughness and the drag felt by the atmosphere is higher for the former.

A further effect of wind-induced currents is the creation of Ekman spirals. The mechanics of the Ekman spiral are discussed in Section 2.8; the importance of the Ekman spiral to the OML is in the transport of mixed layer water due to these currents. These currents cause convergences and divergences in this upper layer (called Ekman pumping) which link the OML to circulations in the deeper ocean.

The importance of organized motions in the OML such as those due to large eddies and Langmuir circulations is being realized increasingly. Langmuir cells (Figures 2.2.1 and 2.4.1) are unique to the oceans since they result from a subtle interaction of the wind-driven turbulence and the Stokes drift current produced by surface gravity waves (see Section 5.2 for a description of the Stokes drift current). These motions are not only capable of injecting additional energy into the OML for mixing, but also are capable of transporting floating particles such as phytoplankton deep into the OML. Observational programs and advanced computer models such as large eddy simulations (LES's) are helping us understand such large scale features of the OML. These cellular motions have their counterparts in the ABL, the horizontal rolls that extend over the entire ABL, but the mechanism is totally different.

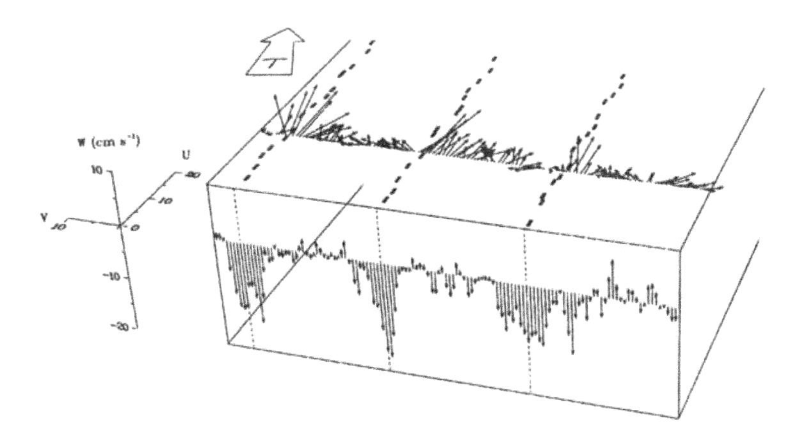

Figure 2.2.1. Velocity fields associated with Langmuir circulations. Notice the strong vertical and horizontal velocities close to the surface in the region of convergence of these cells (from Weller *et al.*, 1985).

The ocean is also driven by the exchange of longwave radiation, and the turbulent sensible and latent heat fluxes. These heat fluxes, in addition to the freshwater flux (a combination of evaporation and precipitation), together make up the buoyancy flux. Rain events alter the buoyancy of the upper layers by decreasing their salinity, changing the surface temperature, and promoting enhanced heating and cooling at the air–sea interface.

A major factor in OML dynamics in the equatorial regions is the presence of strong background currents in the vicinity of the OML. The Equatorial Undercurrent in the Equatorial Pacific is a typical example. It exists at depths ranging from 50 to 200 m and is an eastward-flowing current that produces a strong vertical shear, which has a major influence on mixing in the upper water column. In contrast, in midlatitude oceans, the principal balance is between the Coriolis terms and the stress divergence, and the currents are not continuously accelerated by a steady wind; instead a steady state is reached and an Ekman-like spiral is produced.

In ice-covered oceans, the ice mediates the exchange of momentum betweenthe atmosphere and the OML. The principal balance in ice is between the Coriolis force, the wind stress at the top, and the shear stress on the ocean at the bottom. In addition, ice growth and melting causes buoyancy fluxes that affect the OML below the ice. A significant difference from ice-free oceans is the absence of a surface wave field (apart from the propagating swell near the ice edge). Stirring by deep ice keels is an important factor.

The active turbulent mixed layer in the upper ocean is usually capped below by a strong buoyancy interface, in the form of a layer with either a sharp decrease in temperature (seasonal thermocline) or a sharp increase in salinity (halocline). In either case, this layer (called a pycnocline) is stably stratified and hence turbulence is damped by buoyancy forces. The transition region from active turbulent mixing to mostly quiescent layers below can be called a turbucline. Normally, the turbucline coincides with the seasonal thermocline or halocline, but not necessarily both. During a rain event, a shallow brackish layer can form and the halocline and turbucline are at similar depth but the thermocline is much deeper. In the tropical western Pacific, a similar situation can exist, leading to the so-called barrier layer.

An OML is mixed from both the top and the bottom. At the top, it is the winds, waves, and buoyancy fluxes that stir the fluid. At the bottom, it is the entrainment driven by large turbulent eddies in the OML that mixes the denser fluid from below into the OML (Kraus and Turner, 1967; Kantha *et al.,* 1977). Wind-driven current in the OML also causes strong shear at the base of the mixed layer; shear instability ensues, inducing Kelvin–Helmholtz (K-H) billows, which thicken the buoyancy interface and hence decrease its resistance to erosion by turbulent eddies. In deep OMLs, it is these mechanisms at the bottom that are responsible for a majority of the deepening of the OML. In shallow OMLs, the surface-stirring processes are also important. Note that turbulent erosion tends to sharpen the pycnocline, while K-H billows tend to make it more diffuse (Woods, 1968; see also Woods and Wiley, 1972).

Thus, an OML can be divided into four parts, the very thin but important molecular sublayer, a few millimeters thick; the wave sublayer, normally 2 to 6 m thick; the main bulk of the OML, 10–40 m thick; and the interfacial layer or the entrainment sublayer of about 5–10 m thickness. In deep convective OMLs, where the mixed layer depth is a few hundred meters, the fractions of the wave and entrainment sublayers are small. In a shallow diurnal OML, a few meters thick, the wave sublayer can be a large fraction. An active gravity wave field can damp out the diurnal modulation of sea surface temperature (SST) by wave-driven mixing, through Langmuir cells or wave breaking processes.

2.2.1 Differences with the ABL

There are striking similarities between the ABL and the OML. Figure 2.2.2 shows the variation of the ratio of the dissipation rate to the buoyancy flux with depth/height in the OML and ABL under convective conditions. Similar scaling laws should hold in both turbulent layers and this is indeed the case. For example, under neutral stratification, it is possible to find a region where the universal law of the wall scaling would apply: $q \sim u_*$, $l \sim z$ and $\varepsilon \sim z^{-1}$, $\nu_t \sim z$. Indeed this scaling is found in the upper part of the OML, except close to the surface. Close to the surface, under strong wind conditions, modern measurements (e.g., Agarwal *et al.,* 1992) have found that the dissipation rate is one to two orders of magnitude larger than that given by the law of the wall. In the upper few meters, recent measurements (Terray *et al.,* 1996; Drennan *et al.,* 1996) indicate a region where the dissipation rate scales as z^{-2}. This near-surface elevated dissipation rate is due to the influence of surface waves and wave breaking. Wave breaking generates intermittent, shear-free turbulence somewhat akin to the turbulence generated by a stirring grid in a fluid (whose scaling follows $q \sim z^{-1}$, $l \sim z$, $\varepsilon \sim z^{-4}$, $\nu_t \sim$ constant). The turbulence intensity drops off sharply away from the source. Figure 2.2.3 shows that under stormy conditions, where one might expect to see extensive wave breaking, the above scaling applies, whereas during calm conditions, the dissipation scales more like the law of the wall. Figure 2.2.4 (from Melville, 1994) shows elevated dissipation levels in the near-surface layers, as much as 50 times that expected from law of the wall scaling. However, while the turbulence intensities are elevated above the usual levels during extensive wind-wave breaking, this turbulence is important only to a depth on the order of the amplitude of the breaking waves, since the resulting turbulence drops off rapidly with depth. Below these depths, the law of the wall can often be found once again. Wave breaking and associated turbulence are likely to be important for the dynamics of OMLs, especially shallow ones; because of the elevated near-surface dissipation rates, they may bring about a higher exchange of gas and heat across the air–sea interface. If it were not for the surface waves, the turbulence near the surface of an OML would behave roughly similar to that adjacent to a solid boundary, such as the ABL.

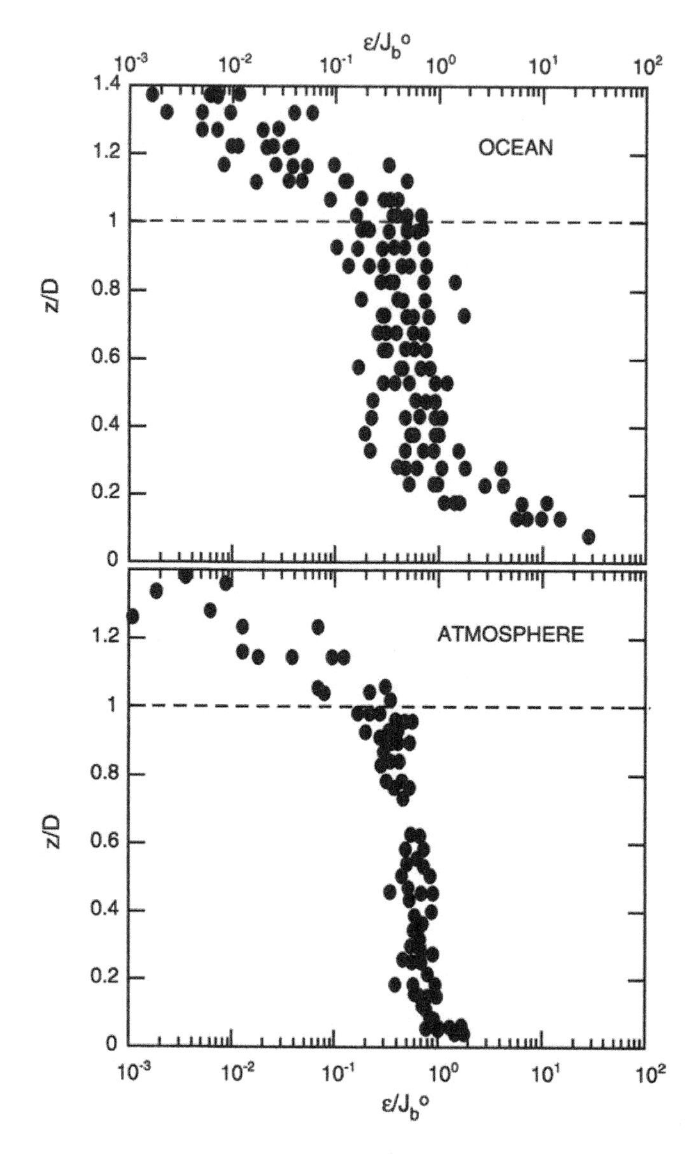

Figure 2.2.2. Comparison of the ratio of the dissipation rate to the buoyancy flux as a function of depth between the OML and the ABL under convective conditions (from Shay and Gregg, 1986).

The most important difference between the OML and the ABL is the depth. The OML is typically tens of meters deep, while the ABL is hundreds of meters deep. The ABL is principally convectively mixed during the day and shear mixed at night. The OML is principally shear mixed, since wind stress at the surface is the primary mixing agent, although at night significant convective

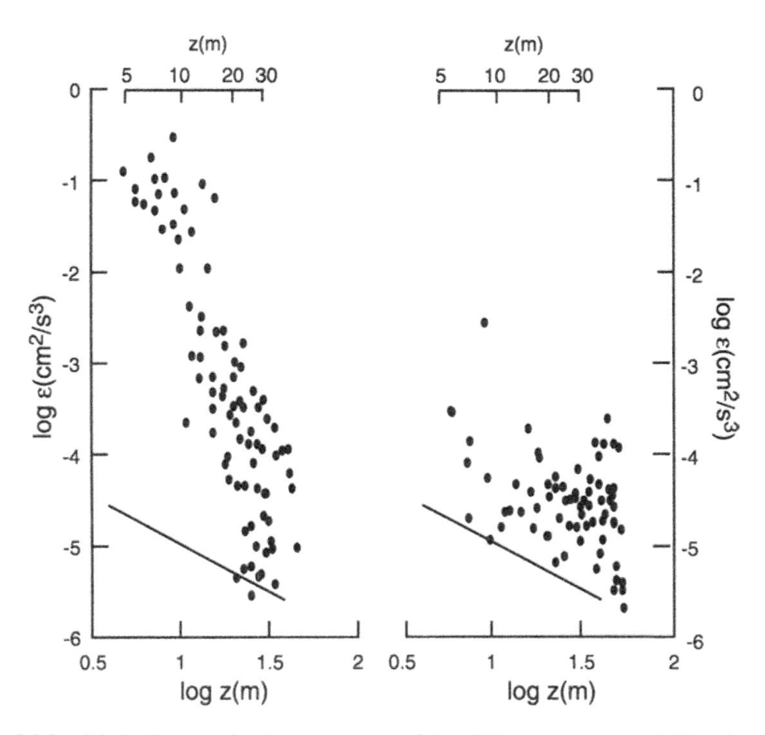

Figure 2.2.3. Dissipation rate in the upper parts of the OML under stormy (left) and calm conditions. Classical law of the wall scaling prevails in the latter, whereas surface wave breaking contributes significantly to the departure from this law during the former (from Gargett, 1989, with permission from the *Annual Review of Fluid Mechanics,* Volume 21, Copyright 1989, by Annual Reviews).

mixing takes place. The ABL is heated primarily from below. The OML is, on the other hand, heated both from the top and from solar radiation in the visible part of the spectrum penetrating into the OML. The cooling is however driven from heat and evaporative losses at the surface, while the entire atmospheric column cools during the night. The seasonal modulation of the OML is quite important, whereas seasonal modulations in the ABL are of interest mostly in the polar atmosphere during the long polar days and nights. But perhaps the most important distinguishing feature of the OML is the presence of surface waves at the air–sea interface that play an active role in its dynamics. The dynamical influence of the ground surface on the ABL is determined by its roughness and topography, which are invariant, whereas it is the effective roughness of the mobile sea surface that is constantly changing with the winds that is important.

The duration of daylight at a given latitude as a function of the season is quite important to the diurnal modulation of the ABL. Long polar days and nights play a crucial role in mixing in the ABL. In the OML, this is however

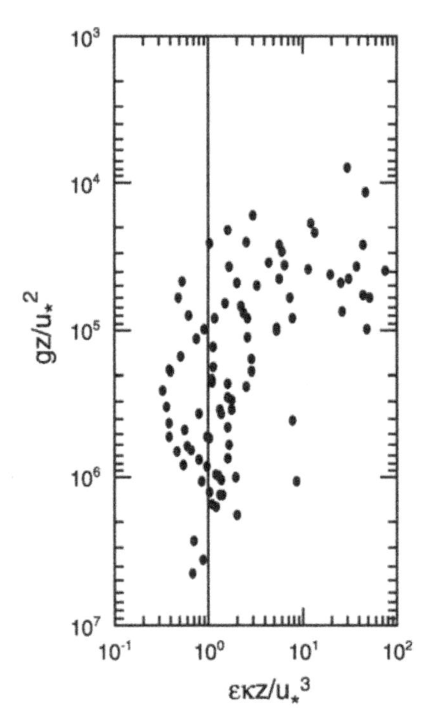

Figure 2.2.4. Measured dissipation rate in the wall layer of the OML. The dissipation rate is more than an order of magnitude higher than that expected from the classical law of the wall scaling (from Melville, 1994).

overshadowed by other dynamical effects due to the presence of sea-ice cover in the polar regions. The diurnal modulation is however quite important at lower latitudes. There are significant latitudinal variations in OML dynamics.

Rain causes a freshening of the upper ocean, which in turn acts to stabilize the upper ocean and reduce mixing. This effect, in combination with the opposite effect due to evaporation, provides the second component of the buoyancy flux to the ocean (the first being the surface heat flux). Rain also affects the ABL, as raindrops falling through this layer evaporate, moistening the ABL. However, the effects of rain on the structure of the ABL, similarity theory, etc., is not well-known in part due to the difficulty of obtaining accurate turbulence measurements during precipitation events.

The condensation and evaporation of liquid water has a major effect on ABL dynamics and thermodynamics. Such phase conversions are absent in the OML, but dissolved substances (salt, for example) play an important role in the dynamics of the OML. Unlike the ABL, statistics of turbulence quantities such as the turbulence kinetic energy (TKE) budget are hard to measure and hence largely unknown in the OML.

2.3 PENETRATIVE SOLAR HEATING

The solar heating of the ABL is mostly from the bottom; heating in the bulk is negligible for the most part, at least in the absence of clouds. The oceans, on the other hand, are heated at the surface (by both shortwave and longwave radiation) and below to depths of up to approximately 100 m by shortwave components that are not absorbed close to the surface. The degree of heating in the body of the upper water column, combined with the net heat balance at the surface, affects the mixing in the OML. Since heating is stabilizing, it tends to suppress turbulence in the water column and to confine mixing to a shallower layer near the surface. Consequently, the momentum transferred by the wind is confined to shallower layers, resulting in stronger currents.

The solar radiation incident at the ocean surface can be divided into three components: short wavelengths in the ultraviolet part of the spectrum (<350 nm), the wavelengths available for photosynthesis [photosynthetically available radiation (PAR), 350–700 nm], and the infrared and near-infrared wavelengths (>700 nm). The UV portion is roughly 2%, PAR 53%, and IR 45% of the total solar insolation. PAR coincides roughly with the visible portion of the spectrum and is the most important of the three portions for biological aspects of the upper ocean. Primary productivity and fixing of carbon by phytoplankton take place only in the euphotic zone, defined usually by the depth at which PAR decreases to 1% of its surface value. The ultraviolet part is important to the production of certain photochemicals such as CoS.

The depth of the euphotic zone depends greatly on water clarity, which is in turn determined by dissolved matter, suspended sediments, and chlorophyll concentration in the water (mostly the latter two). The infrared and near-infrared parts of the solar insolation are absorbed within the top 10 cm of the surface and the e-folding scale for UV radiation is less than 5 m. It is the visible part that penetrates much deeper, the extent of penetration depending very critically on the density of phytoplankton or pigment concentration, and in some coastal waters (near river outflows), the sediment concentration. Away from the coast, it is principally the chlorophyll content that affects water clarity. Ocean waters are usually divided into broad classes and types depending on optical clarity (Jerlov, 1976). Waters in midocean gyres tend to be the most optically clear (Figure 2.3.1).

However, there is variability in water clarity at various timescales all the way from biological timescales to interannual ones. For the oceanic Case One waters (which include Types I to III), the chlorophyll concentration can change over several orders of magnitude (0.01 to 20 mg m^{-3}) and the corresponding euphotic depth can vary from ~100 m to 10 m (Morel and Antoine, 1994). This is of importance not only to biology but also to dynamics, since the heat deposition in the water column suppresses mixing. During the TOGA/COARE experiment in the western Pacific, a large increase in chlorophyll concentration was detected immediately after the onset of a westerly wind burst. This increase in concen-

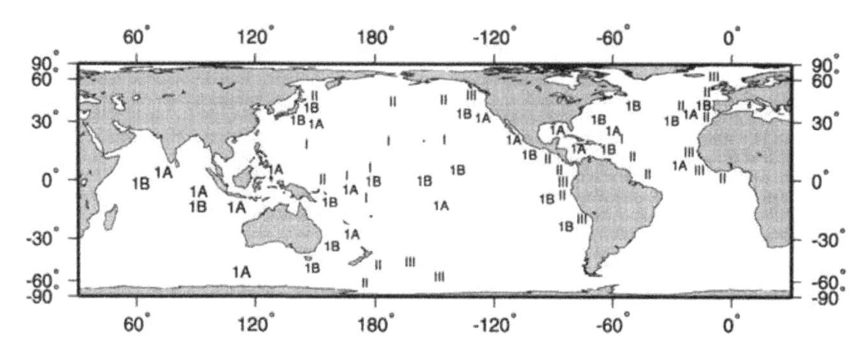

Figure 2.3.1. Water types (from Jerlov, 1976) in the global oceans.

tration was due to increased mixing in the water column and a consequent increase of nutrient concentration in the euphotic zone.

It is clear that we need to understand and realistically model the solar extinction in the water column. An excellent and succinct summary of the solar radiation in the ocean can be found in Morel and Antoine (1994), which we follow closely here. Ivanoff (1977) and Bird (1973) are also good but dated references.

Solar irradiance in the upper ocean can be modeled as

$$I(z)=I_0 \sum_{n=1}^{N} \alpha_n \exp\left(z / L_n\right) \tag{2.3.1}$$

where I_0 is the insolation immediately below the ocean surface, N is the number of bands, and α_n and L_n are the fraction of insolation resident in the band and its extinction length scale. Traditionally, a two-band model of solar extinction has been used in mixed layer applications. Paulson and Simpson (1977; see also Simpson and Dickey, 1981a,b) have tabulated the values of α_n and L_n for various Jerlov water types. The two bands correspond roughly to the red and near-infrared and the visible parts of the spectrum. Obviously, for biological applications, a much finer division is essential, especially in the PAR band. But Morel and Antoine (1994) have shown that even for dynamical applications, the two-band model is quite deficient. Application to the cool-skin aspects of the ocean surface may require as many as nine. Simpson and Dickey (1981b) present a nine-band parameterization for clear water. Also, since the Jerlov classification is based on optical clarity and not the pigment concentration, many practical applications such as consideration of the biological feedback on the dynamics of mixing cannot use the Jerlov approach, and explicit functional dependence of α_n and L_n on chlorophyll concentration (C) is needed. Morel (1988) analyzed the extinction properties in the visible part of the spectrum

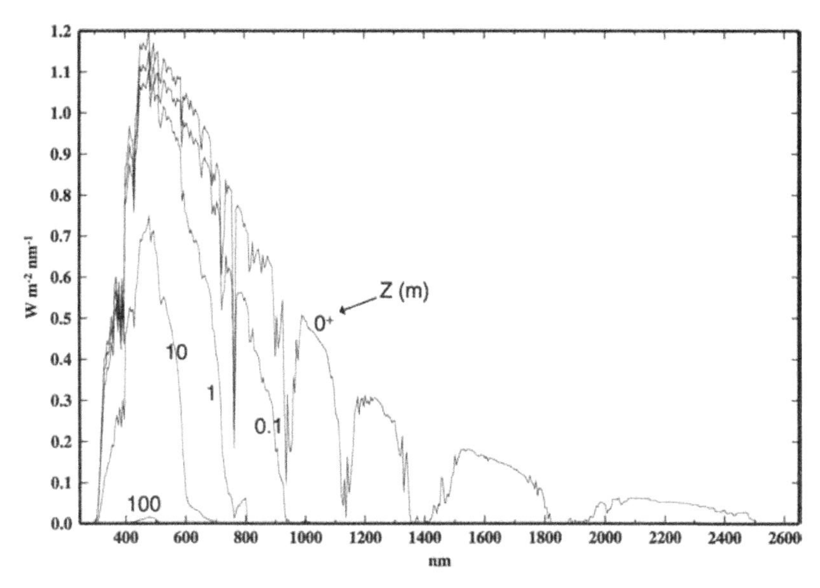

Figure 2.3.2. Spectral characteristics of the solar radiation impinging on the sea surface. Note the absorption of certain wavebands due to the intervening atmosphere. Solar radiation at 0.1-, 1-, 10-, and 100-m depths for a chlorophyll concentration of 0.2 mg m^{-3} is also shown. Note that only the visible part is left at a 10-m depth (from Morel and Antoine, 1994).

(400–700 nm) in 61 wavelength bands for uniform pigment distribution. While this fine division may be indispensable for the analysis of photosynthesis, for dynamical calculations, Morel and Antoine (1994) recommend a simpler three-band parameterization, with polynomial expressions for α_n and L_n.

The solar insolation spectrum at the surface is shown in Figure 2.3.2. Morel and Antoine (1994) point out that less than 0.5% extraterrestrial solar radiation lies outside the 250- to 4250-nm spectral band. Due to efficient absorption by principally H_2O and CO_2, less than 1% of the radiation remains beyond 2500 nm at the ocean surface. Also N, O, O_2, and O_3 absorb the radiation below 300 nm, so that for oceanic applications only the 300- to 2500-nm band is relevant. The insolation at the ocean upper surface is a complex function of atmospheric properties: the vertically integrated water, ozone and aerosol contents (visibility), and the amount of cloud cover. For clear skies, the sun elevation angle and atmospheric properties control the amount and distribution across the spectrum of solar radiation incident on the ocean surface, and for cloudy skies, the extent of cloud cover and the liquid water content are of great importance. Fortunately, most of this complexity can be handled by radiation models with appropriate properties for the atmospheric column as input. Figure 2.3.3 shows the fraction of the radiation transmitted to the ocean surface as a function of visibility, sun angle, and water vapor content.

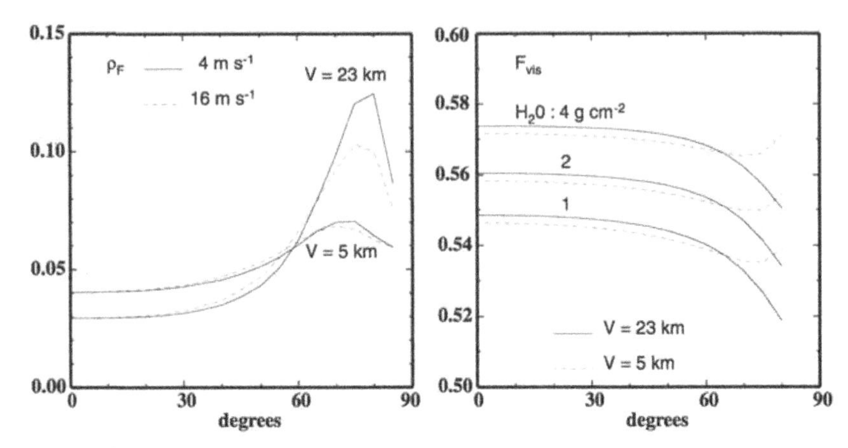

Figure 2.3.3. Fraction of the solar radiation at the sea surface (right) as a function of the sun angle and water vapor content for clear and hazy conditions and albedo (left) as a function of Sun angle, visibility, and wind speed (from Morel and Antoine, 1994).

Figure 2.3.2 shows the absorption characteristics for C = 0.2 mg m^{-3} from Morel and Antoine (1994). The rapid absorption of wavelengths beyond the visible in the upper meter of the water column can be seen. The solar insolation at the bottom of the ocean surface is a fraction of that incident at the upper surface, the fraction depending on the albedo, which is in turn a function of the Sun's zenith angle and the sea state (and, to some extent, the wavelength). The Fresnel reflection at the air–sea interface is apparently more important than the diffuse reflection from upper layers. The former leaves the spectrum essentially unchanged so that for all practical purposes, the spectral shape immediately below the interface is the same as the one immediately above it (Morel and Antoine, 1994). Normally (for clear skies) the total albedo when the Sun is directly above is about 0.03, but can be as high as 0.3 for low Sun angles and calm seas (diffuse reflection is less than 0.7%). For overcast skies, Morel and Antoine (1994) recommend a value of 0.066. Payne (1972) tabulates the albedo as a function of Sun elevation angle and atmospheric transmittance. Rough seas reduce the albedo at low Sun angles. Figure 2.3.3, from Morel and Antoine (1994), shows the albedo from Fresnel reflectance for different conditions.

The absorption characteristics of seawater are such that the spectrum can be demarcated by a 750-nm wavelength. Infrared wavelengths that lie above this limit are essentially independent of water turbidity and so pure water values can be used. For dynamical purposes, the infrared region can be characterized by a single band throughout the OML, except for the skin region. In this shallow region (see Section 4.7) the IR absorption close to the surface is so important that at least two and frequently more bands are needed. Fortunately, dependence on chlorophyll and sediment concentrations can be ignored and pure water values can be used (Simpson and Dickey, 1981b) (see Table 2 also). The two-band and one-band models for this part of the spectrum can be written as

$$
\begin{aligned}
& I_{IR}(z) = I_{IR}(0) \sum_n \alpha_n \exp(z/L_n) \\
& n=1, \quad \alpha_1 = 1, \quad L_1 = 0.267 \cos\alpha \\
& n=2, \quad \alpha_1 = 0.805, \quad \alpha_2 = 0.195 \\
& L_1 = 0.043 \cos\alpha, \quad L_2 = 0.42 (\cos\alpha)^{1/2}
\end{aligned}
\tag{2.3.2}
$$

Absorption of the solar radiation in the spectrum below 750 nm (especially the blue-green PAR range) depends crucially on the chlorophyll content. Figure 2.3.4, from Morel and Antoine (1994), shows the absorption as a function of chlorophyll concentration C (and also the IR absorption). Table 2.3.1, from the same reference, tabulates the parameters needed to compute the extinction scale L in the entire 300- to 2600-nm spectral band of importance to the oceans (at a 5-nm resolution) from the relationship

$$
1/L_n = \lambda_n = \lambda_{on} + x_n C^{e_n} \tag{2.3.3}
$$

where λ_n is the inverse of the extinction scale and subscript 0 denotes the value for pure water. This relationship is valid for C between 0.02 and 20 mg m^{-3}. These numbers can be used to compute the solar insolation at any depth in the entire spectral range of interest and for most pigment concentrations of practical importance. Figure 2.3.5 shows results for various values of C.

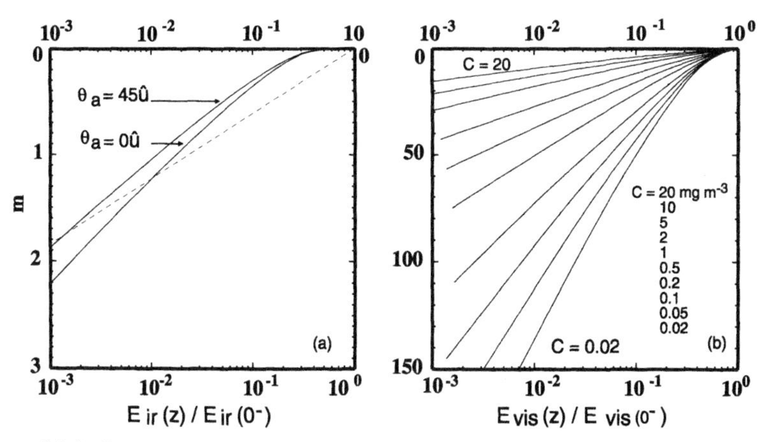

Figure 2.3.4. Downwelling solar radiation in the near-IR band (left) at two solar zenith angles and in the visible (and UV) band (right) for various chlorophyll concentrations (from Morel and Antoine, 1994).

TABLE 2.3.1
Scalar Extinction Parameters for Computing the Extinction Scale L for Various Wavelengths

	λ (nm)	$K_w(\lambda)$ (m^{-1})	$\chi(\lambda)$	$e(\lambda)$	λ (nm)	$K_w(\lambda)$ (m^{-1})	$\chi(\lambda)$	$e(\lambda)$	λ (nm)	$a(\lambda)$ (m^{-1})	
	300	.154	.196	.889	530	0.052	.047	.670	775	2.400	
	305	.135	.191	.878	535	0.054	.045	.665	800	1.963	
	310	.116	.187	.867	540	0.057	.044	.660	825	2.772	
	315	.105	.183	.856	545	0.062	.043	.655	850	4.331	
UV	320	.094	.179	.845	550	0.064	.041	.650	875	5.615	
	325	.085	.174	.834	555	0.068	.040	.645	900	6.786	
	330	.076	.170	.823	560	0.072	.039	.640	925	14.401	
	335	.070	.166	.812	565	0.076	.038	.630	950	38.764	
	340	.064	.161	.801	570	0.081	.036	.623	975	44.857	IR
	345	.058	.157	.790	575	0.094	.034	.615	1000	36.317	
	350	.053	.153	.778	580	0.107	.033	.610	1200	103.568	
	355	.048	.149	.767	585	0.128	.033	.614	1400	1238.685	
	360	.044	.144	.756	590	0.157	.032	.618	1600	671.515	
	365	.040	.140	.737	595	0.200	.033	.622	1800	802.851	
	370	.035	.136	.720	600	0.253	.034	.626	2000	6911.504	
	375	.031	.131	.700	605	0.279	.035	.630	2200	1650.764	
	380	.027	.127	.685	610	0.296	.036	.634	2400	5005.604	
	385	.025	.123	.673	615	0.303	.038	.638	2600	15321.306	
	390	.023	.119	.670	620	0.310	.038	.642			
	395	.022	.114	.668	625	0.315	.040	.647			
	400	.021	.110	.668	630	0.320	.042	.653			

420	.018	.113	.693	650	0.350	.045	.672
425	.018	.110	.701	655	0.370	.046	.677
430	.017	.108	.707	660	0.405	.047	.682
435	.017	.106	.708	665	0.418	.049	.687
440	.017	.104	.707	670	0.430	.052	.695
445	.017	.100	.704	675	0.440	.052	.697
450	.017	.097	.701	680	0.450	.051	.693
455	.017	.094	.699	685	0.470	.044	.665
460	.017	.090	.700	690	0.500	.039	.640
465	.017	.086	.703	695	0.550	.034	.620
470	.018	.082	.703	700	0.650	.030	.600
475	.018	.079	.703	705	0.742	.025	.600
480	.019	.075	.703	710	0.834	.020	.600
485	.020	.073	.704	715	1.002	.015	.600
490	.022	.069	.702	720	1.170	.010	.600
495	.024	.066	.700	725	1.485	.007	.600
500	.027	.064	.700	730	1.800	.005	.600
505	.032	.060	.695	735	2.090	.002	.600
510	.038	.058	.690	740	2.380	.000	.600
515	.045	.054	.685	745	2.420	.000	.600
520	.049	.050	.680	750	2.470	.000	.600
525	.051	.047	.675				

Data from Morel and Antoine, 1994.

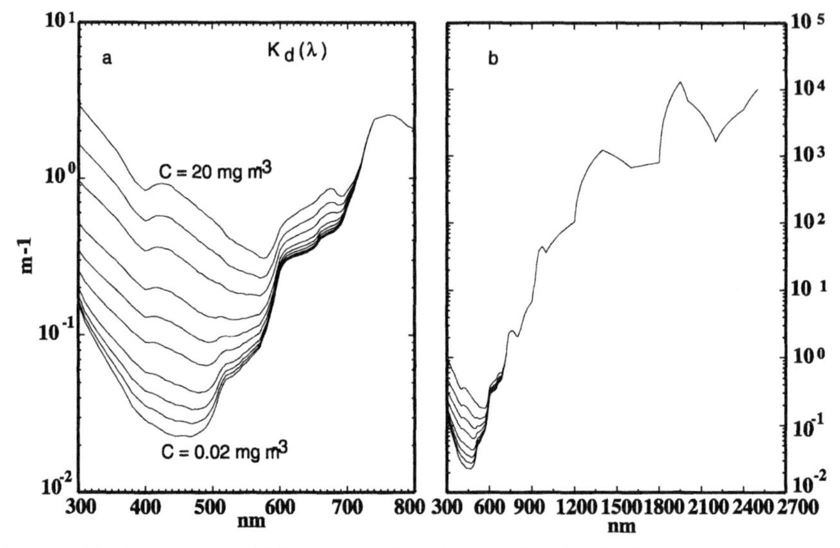

Figure 2.3.5. Attenuation coefficient as a function of the wavelength and chlorophyll concentration. Values for the 300- to 750-nm domain are shown in the right panel, and the full solar spectral domain up to 2.5 mm (including the curves from the right panel) is shown in the left panel (from Morel and Antoine, 1994)

For dynamical applications, it is adequate to consider only two bands in the visible–near-UV band:

$$I_V(z) = I_V(0) \sum_{n=1}^{2} \alpha_n \exp(z/L_n)$$
$$\alpha_n, L_n \sim \alpha_n, L_n(C) \tag{2.3.4}$$

Table 2.3.2, from Morel and Antoine (1994), provides the coefficients in the polynomial expansions for the extinction scale L_n and the fraction of the visible radiation α_n resident in the band in terms of $x = \log_{10}(C)$

$$Y = a_0 + a_1 x + a_2 x^2 + a_3 x^3 + a_4 x^4 + a_5 x^5 \tag{2.3.5}$$

The dynamics of a mixed layer are quite sensitive to the parameterization of solar extinction in the water column. OML modelers have therefore tried to incorporate better parameterizations of absorption of solar radiation in the water column (Ivanoff, 1977; Simpson and Dickey, 1981a,b; Siegel and Dickey, 1987; Kantha and Clayson, 1994). Most OML models now have at least a two-band model, representing the longwave IR and shortwave visible parts of the spectrum that have different extinction length scales, following the recomen-

TABLE 2.3.2
Coefficients for Polynomial Expansions for Eq. (2.3.5)

	a_0	a_1	a_2	a_3	a_4	a_5
V_1	0.353	−0.047	0.083	0.047	−0.011	−0.009
V_2	0.647	0.047	−0.083	−0.047	0.011	0.009
Z_1	1.662	−0.605	0.128	0.033	−0.051	−0.004
Z_2	8.541	−8.924	4.020	0.077	−0.536	0.055
V_1	0.321	0.008	0.132	0.038	−0.017	−0.007
V_2	0.679	−0.008	−0.132	−0.038	0.017	0.007
Z_1	1.540	−0.197	0.166	−0.252	−0.055	0.042
Z_2	7.925	−6.644	3.662	−1.815	−0.218	0.502

dations of Paulson and Simpson (1977) for various water types. However, as discussed above, it has become increasingly clear that an even more accurate extinction parameterization is essential for a better simulation of many OML properties, including those pertaining to the dynamics (Martin and Allard, 1993). Since photosynthesis is sensitive to the spectral properties of shortwave solar radiation in the water column, it is even more important for applications to biological productivity (Sathyendranath and Platt, 1988).

The feedback effects of chlorophyll-induced turbidity in the water column on the solar extinction profile itself are being recognized as an important aspect that affects the primary productivity in the upper ocean (Martin and Allard, 1993). Since carbon uptake by the oceans through organic pathways is central to anthropogenic carbon dioxide-induced global warming, understanding and modeling primary biological productivity in the upper ocean has become an important issue in global change (Smith, 1993). With the advent of satellite-borne ocean color sensors that provide a global estimate of chlorophyll concentrations (and hence primary productivity) and solar extinction in the water column [see Martin and Allard (1993) for a detailed discussion of satellite-sensed ocean color], the need for OML models that can better simulate solar extinction has grown. Ramp *et al.* (1991) observed a patch of turbid waters off California, where the 4-cm temperatures were as much as 4.7°C elevated above the 2-m values due to strong attenuation of solar fluxes in the turbid upper layer.

Martin and Allard (1993) have recently made a thorough investigation of the effects of chlorophyll concentration on turbidity and hence solar extinction properties. They found that in both long-term and short-term OML simulations, increasing the turbidity of the water affects the SST considerably. For example, increasing the chlorophyll concentrations from 0 to 2 mg m^{-3} (roughly equivalent to changing the Jerlov water type from Type I to Type III) causes the SST to warm up by as much as 4°C in summer at Ocean Weather Ship (OWS) November and 2°C at OWS Papa. Even on timescales of a few days, the SST change can reach 1°C. They also find that the popular two-band parameterization of extinction is inaccurate in that it underestimates solar irradiance in the upper meter or so, and overestimates it in the rest of the water column. Such inaccuracies might be detrimental to modeling photochemical and biological production in the upper ocean (especially the former), since most photochemical production is concen-trated in the upper few meters (the extinction scale for UV radiation is 2–5 m).

2.4 LANGMUIR CIRCULATION

Langmuir (1938) was the first to observe and study the phenomena of organized counterrotating vortices at the surface of a lake, with axes aligned roughly with the wind, and associated with a three-dimensional, cell-like circulation. These have come to be known as Langmuir cells or windrows. Their presence is manifest by the surface convergence at the boundary of counterrotating cells (Figure 2.4.1). Windrows are often visible to the naked eye because seaweed and flotsam accumulate at the surface in these convergence regions. On a blustery day with a vigorous surface wave field, the convergence region is made visible by whitecapping and bubble entrainment due to breaking of small scale waves in the convergence regions resulting in parallel white lines roughly aligned with the wind and roughly uniformly spaced. Langmuir carried out a series of observations confirming the existence of cell-like circulation associated with windrows.

Bubble clouds (Thorpe, 1984; Zedel and Farmer, 1991; Smith *et al.,* 1987) seen in side-scan sodar observations of the near-surface layers of the ocean are also associated with Langmuir cells (Thorpe, 1992a). Bubble plumes are manifest in the form of streaks with a variety of scales, merging at characteristic Y junctions to form large circulation cells, with the streaks aligned in the wind direction and drifting to the right of the wind. Smith *et al.* (1987) showed that these bubble plumes are concentrated at surface convergence zones. Bubble clouds are important to air–sea gas transfer, and because they are efficient volume backscatterers, important also to ocean acoustics at high frequencies. The depth of penetration of the bubble cloud depends very much on the strength of the surface convergence and hence the strength of Langmuir circulations.

It has long been suspected that Langmuir cells might play a role in the mixing in the upper ocean, and other near-surface processes such as air–sea exchange of gases, entrainment of bubbles, photochemical production in the upper layers, and vertical transfer of momentum and other properties into the interior from the air–sea interface. This is simply because these cells can be quite vigorous and the downward vertical velocity immediately below the convergence region can be as high as a few tens of cm s^{-1}, leading to bubble entrainment and transport to greater depths (Zedel and Farmer, 1991). The cells also provide a mechanism for

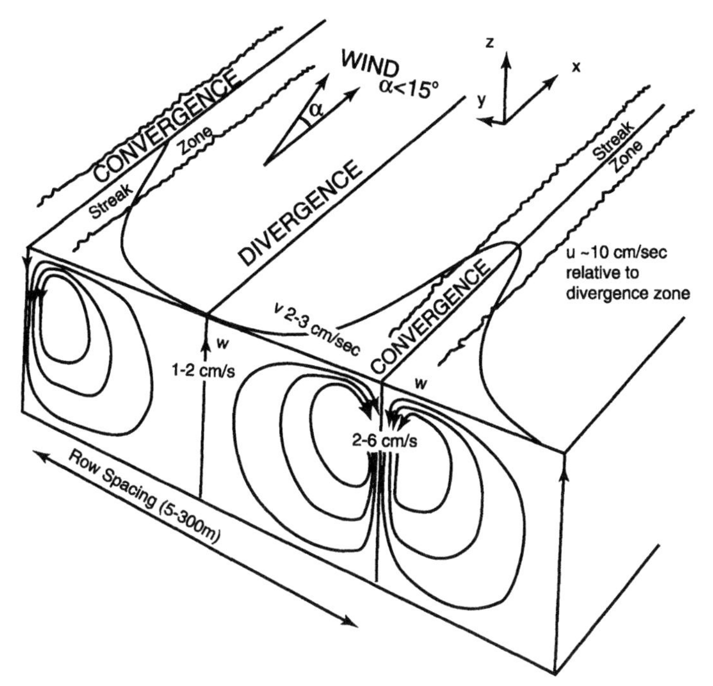

Figure 2.4.1. A schematic of the circulation induced by Langmuir cells (reprinted from Pollard, *A Voyage of Discovery,* Copyright 1977, with permission from Elsevier Science).

distribution of properties in the vertical in the mixed layer and transport of phytoplankton and zooplankton in the vertical. However, Langmuir cells are inherently transient events and not much is known about their generation and decay characteristics. Therefore it has been difficult to assess their importance and significance (Weller and Price, 1988). Even the exact mechanism for generation of these cells was controversial until Craik (1970, 1977), Craik and Leibovich (1976), and Leibovich (1977) showed that these cells result from the interaction between the Stokes drift due to surface waves and the vertical shear of the wind-induced currents. The prevailing theory is that due to Craik (1977) and Leibovich (1977), known as the Craik–Leibovich theory, which hypothesizes that the instability brought on by the vortex force term that appears in the momentum equations due to the interaction of the Stokes drift with the mean shear leads to the formation of cellular patterns, aligned in the direction parallel to the wind. The vertical vorticity associated with small lateral variations in vertical shear is twisted by the Stokes drift into horizontal vorticity and Langmuir cells. Leibovich (1983; see also Pollard, 1977) has reviewed earlier work on Langmuir cells.

Weller *et al.* (1985) and Smith *et al.* (1987) observed vertical velocities as high as 25 cm s^{-1} in Langmuir circulations off California. Smith (1992) reports on acoustic Doppler velocity measurements also off California indicating vertical velocities of ~10 cm s^{-1}. Weller and Price (1988) made very careful measurements of Langmuir cells from the ocean platform FLIP off the coast of California, including measurements of vertical velocity using an instrument capable of measuring all three components of velocity. Their observations have greatly enhanced our knowledge of the characteristics of these cells. They found much larger vertically downward velocities below the convergence zone (up to 27 cm s^{-1}) than had been suspected, with the maximum located subsurface in the upper part of the mixed layer. There was a jet-like horizontal flow in the convergence zone in the downwind direction (Figure 2.4.1), once again with a subsurface maximum. Overall, the conceptual picture presented by Pollard (1977) is essentially correct, although the downwelling downwind flow in the convergence zone appears to be confined in the vertical and in the cross-wind directions.

Weller and Price (1988) found the cells to be highly transient and were unable to determine the conditions for their onset, growth, and decay. They also found no definite dependence of downwelling velocity on wind speed (Figure 2.4.2). Their most important finding, however, was that the Langmuir cells were able to rapidly destroy the surface thermal stratification in shallow diurnal mixed layers. However, there was no evidence of their direct participation in mixing in the deeper parts of the OML. This is not surprising, since Langmuir

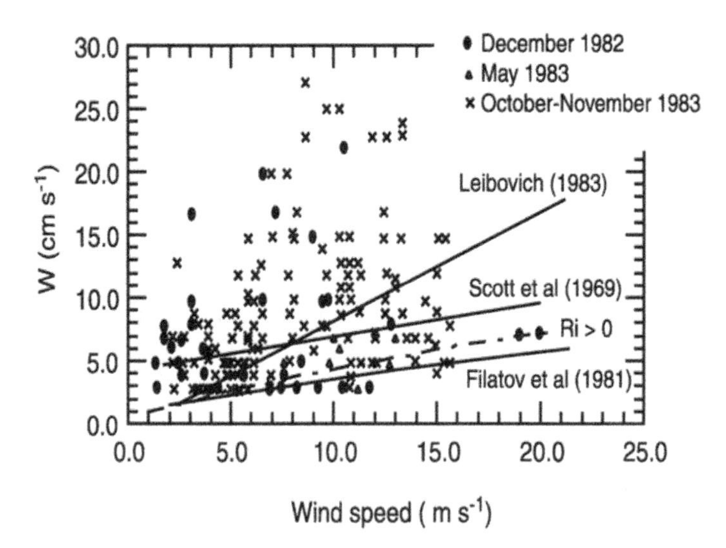

Figure 2.4.2. Vertical velocity in Langmuir cells as a function of wind speed (reprinted from *Deep Sea Research,* 35, Weller and Price, Langmuir circulation within the oceanic mixed layer, 711–747, Copyright 1988 with permission from Elsevier Science).

cells are intimately coupled to the Stokes drift from surface waves, which whilestrong near the surface, decays exponentially with depth. The vertical velocities below convergence zones, while quite strong immediately below the surface, also decay rapidly with depth. Thus one can expect the vertical scale of the cells to be essentially the vertical decay scale of the wave effects (1/2k) and this is often much smaller than the depth of the OML. Thus, while Langmuir cells are undoubtedly important to mixing in the near-surface layers, they might not be as significant to the overall mixing. Their intimate association with the wave field also makes them inherently transient.

Finally, numerical solutions are also adding to our understanding of Langmuir circulations. Skyllingstad and Denbo (1995) have performed LES's of Langmuir circulations under a variety of conditions, including wind- and convection-driven mixing with and without Stokes drift to highlight their importance in the structure of the upper layers. They demonstrate the rapid growth in the scale of Langmuir circulations brought on by increased wind forcing evident in the observations of Smith (1992). Unlike the two-dimensional Craik–Leibovich theories, they were also able to simulate successfully the multiple cell scales characteristic of Thorpe's measurements of Langmuir cells (Thorpe, 1992a). Their simulations indicate the limited effect of surface heating and cooling on the near-surface structure of Langmuir circulation. Figure 2.4.3 shows the near-surface vertical velocity structure for a variety of cases, illustrating the importance of the Stokes drift in producing the characteristic elongated cellular structures and Y junctions in Langmuir circulations. Only their cases WS, CWS, and HWS, the wind and wave forced simulations without any heat flux and with cooling and heating, respectively, demonstrate this organized structure; cases C, CW, and HW, the corresponding runs without surface wave forcing, do not. Also, these characteristic elongated structures were not evident at greater depths in the OML. Figure 2.4.4, from their simulation, confirms the tendency of surface floaters to congregate in convergence zones. Figure 2.4.5 shows the enhancement of vertical velocity variance and the increased entrainment heat flux in the CWS case relative to the CW case, leading to the conclusion that Langmuir circulations might be important to the dynamics of the OML.

Langmuir circulations (Craik and Leibovich, 1976) are formed due to the instability of the wind-driven flow by the interaction of the wind-driven surface shear with the Stokes drift of the surface waves. This instability is due to the vortex force term that appears in the momentum equations. Neglecting viscous terms for simplicity, the governing equations for momentum can be written as

$$\frac{\partial U_j}{\partial t} + \frac{\partial}{\partial x_k}\left(U_k U_j\right) + \varepsilon_{jkl} f_k U_l = -\frac{1}{\rho_0}\frac{\partial P}{\partial x_j} - g_j \beta \Theta - \frac{\partial}{\partial x_k}\left(\overline{u_k u_j}\right) + \varepsilon_{jpl} V_{Sp} \Omega_l \qquad (2.4.1)$$

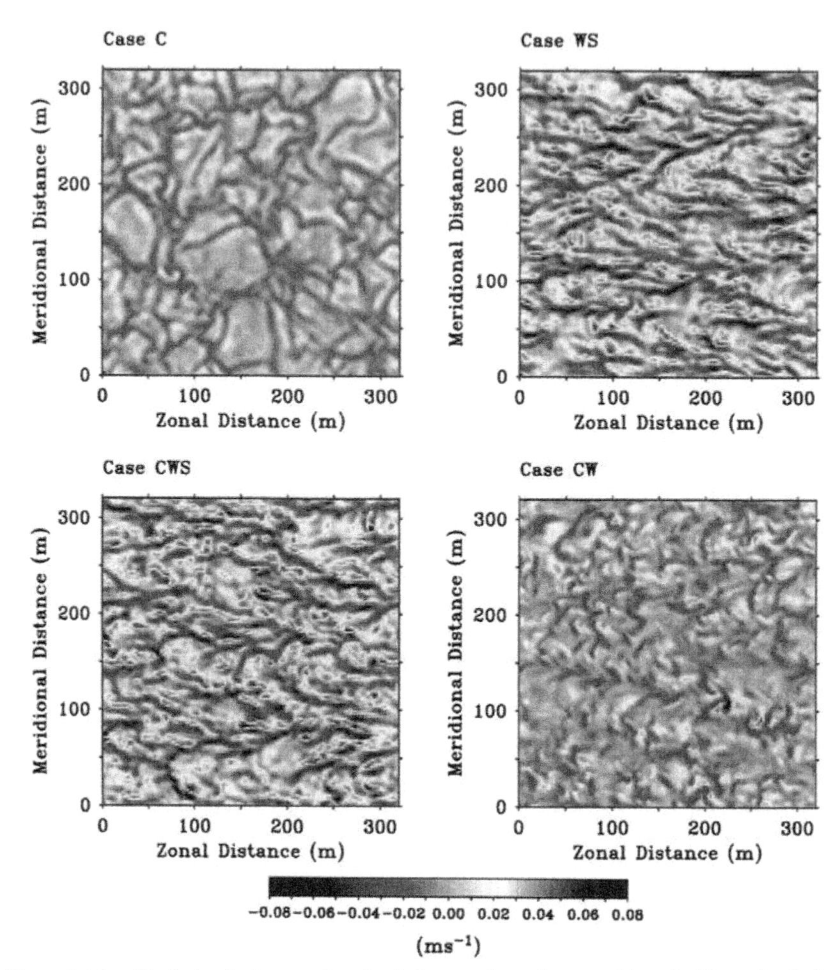

Figure 2.4.3. Vertical velocity at a 5-m depth for a variety of cases with and without the Stokes drift from an LES model of the OML. Note the characteristic cellular structure elongated in the downstream direction whenever Stokes drift is included (WS, CWS, and HWS cases) and its absence otherwise (from Skyllingstad and Denbo, 1995).

where

$$\Omega_l = \varepsilon_{lmn} \frac{\partial U_n}{\partial x_m} \tag{2.4.1}$$

is the vorticity and V_{Sp} (p = 1, 2) is the Stokes drift velocity due to surface waves, whose magnitude is given by

$$|V_S| = \left(V_{Sp} V_{Sp}\right)^{1/2} = V_{S0} \exp\left(2kz\right) = C\left(ka\right)^2 \exp\left(2kz\right) \tag{2.4.2}$$

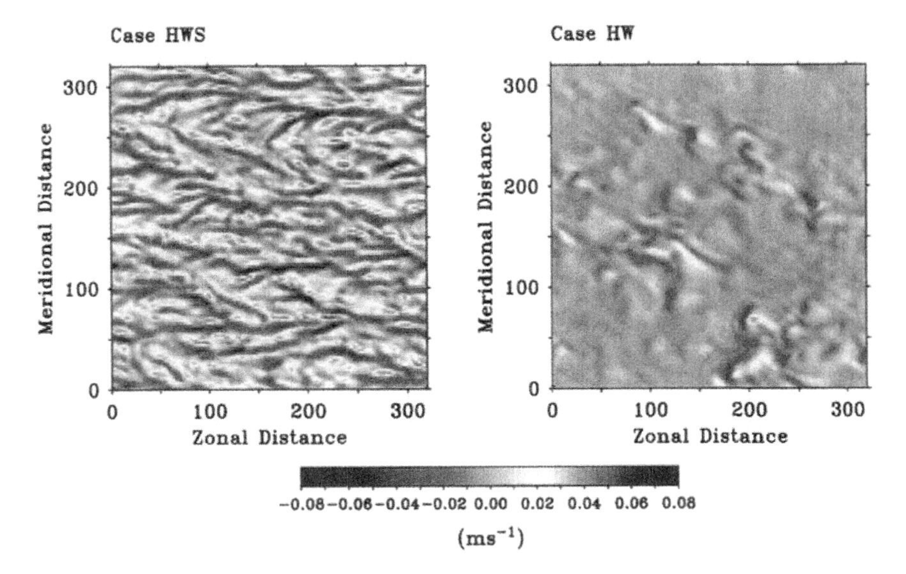

Figure 2.4.3. *(Continued)*

C is the wave phase speed, k is the wavenumber, and a is the amplitude. This vortex force acts like a buoyancy force term in the vertical momentum equation. The horizontally averaged KE equation can be written as

$$\frac{\partial}{\partial t} q^2 = -2\,\overline{u_k u_j}\,\frac{\partial U_j}{\partial x_k} + 2V_{Sp}\varepsilon_{ipl}\,\overline{u_i \omega_l} + \cdots \tag{2.4.3}$$

where only the Reynolds stress terms and vortex force terms are retained for simplicity. The most important thing to note is the appearance of the term due to the Stokes drift. With no loss of generality, we can align the x_1-axis in the direction of wave propagation and put $V_{S1} = V_S$ and $V_{S2} = 0$. It is instructive to look at the individual components of KE. For simplicity we will write only the shear production and Langmuir circulation terms:

$$\begin{aligned}
\frac{\partial}{\partial t}\overline{u_1^2} &= -2\overline{u_1 u_3}\frac{\partial U_1}{\partial x_3} + \cdots \\
\frac{\partial}{\partial t}\overline{u_2^2} &= -2\overline{u_2 u_3}\frac{\partial U_2}{\partial x_3} + 2V_S\overline{u_1\frac{\partial u_3}{\partial x_3}} + \cdots \\
\frac{\partial}{\partial t}\overline{u_3^2} &= 2V_S\overline{\frac{\partial (u_1 u_3)}{\partial x_3}} - 2V_S\overline{u_1\frac{\partial u_3}{\partial x_3}} + \cdots
\end{aligned} \tag{2.4.4}$$

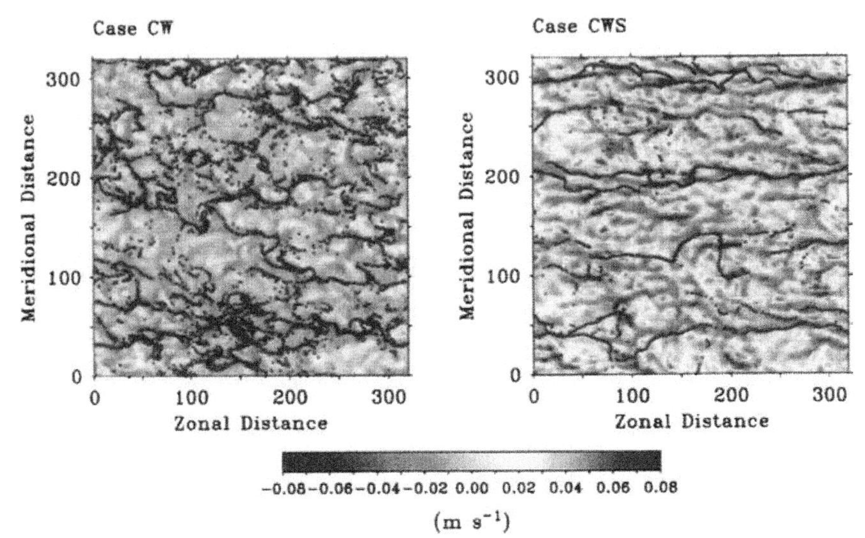

Figure 2.4.4. Vertical velocity and location of surface floats without and with Stokes drift, illustrating the elongated convergence zones in the latter (from Skyllingstad and Denbo, 1995).

The vortex force has introduced additional terms into the vertical component and that perpendicular to the wave direction (i.e., in the cross-cell direction). The component in the direction of the wind is unchanged. The second term in the second and third equations is just a redistribution term; it redistributes energy between the vertical and cross-cell directions. The first term in the vertical component is the extra term that survives when all the components are summed. This term is

$$P_L \sim -2V_S \frac{\partial}{\partial x_3}\left(\overline{u_1 u_3}\right) \qquad (2.4.5)$$

When the cells are fully developed, one can expect that this energy input is lost to small scale turbulence and hence acts as a source term in the TKE equation. The ratios of this term to the usual shear production and buoyancy production terms therefore denote the relative importance of Langmuir circulation to the mixed layer

$$R_S = \frac{P_L}{P_S} = \frac{-V_S \dfrac{\partial}{\partial x_3}\left(\overline{u_1 u_3}\right)}{-\overline{u_1 u_3}\dfrac{\partial U_1}{\partial x_3} - \overline{u_2 u_3}\dfrac{\partial U_2}{\partial x_3}} \qquad (2.4.6)$$

$$R_B = \frac{P_L}{P_B} = \frac{-V_S \dfrac{\partial}{\partial x_3}\left(\overline{u_1 u_3}\right)}{\beta g \overline{u_3 \theta}}$$

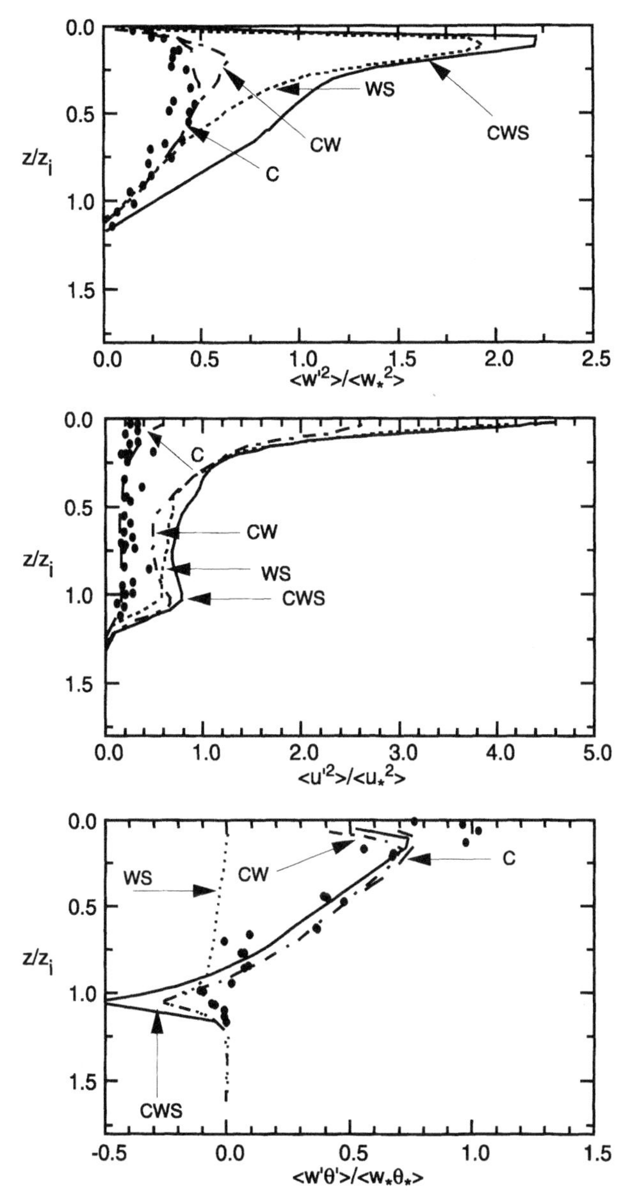

Figure 2.4.5. Profiles of vertical and horizontal velocity variance and heat flux from observations (points) and model results (lines) for various cases with and without the Stokes drift, showing the enhancement of vertical mixing (from Skyllingstad and Denbo, 1995).

There are several points worth noting. First of all, it is clear that both the Stokes drift and the vertical gradient of the shear stress are essential. For simplicity, if we assume a linear variation of the shear stress with depth d of the mixed layer, it is possible to integrate Eq. (2.4.5) in the vertical to show that the total input into Langmuir circulations is

$$V_{S0}u_*^2\left[1-e^{-2kd}\right](2kd)^{-1} \tag{2.4.7}$$

For shallow mixed layers, where kd < 1, this reduces to $V_{S0}u_*^2$, and for deep mixed layers where kd > 1, $V_{S0}u_*^2(kd)^{-1}$. Clearly, the term is important for shallow mixed layers. Also, the kinematic wind stress in the direction of wave motion and the Stokes drift appear together as a product and therefore the appropriate velocity scale for Langmuir circulations is

$$V_L \sim (u_*^2 V_{S0}\cos\theta)^{1/3} \sim \left[u_*^2(ka)^2 C\cos\theta\right]^{1/3} \tag{2.4.8}$$

Equation (2.4.7) suggests that the strength of the Langmuir circulations depends on the combined effects of the Stokes drift and wind stress, so that strong winds and small waves can have an influence similar to that of weak winds and large waves. But both are essential. This fact is underscored by Plueddemann *et al.* (1996), who could not scale their sonar observations of the near-surface rms convergent velocities during the Surface Wave Processes Program (SWAPP) (Weller *et al.*, 1991) by the friction velocity alone and had to use a velocity scale based on the combination of the friction and Stokes velocities. They, however, use $(u_*V_{S0})^{1/2}$ as the characteristic velocity scale for Langmuir circulations.

It is the action of the vortex force terms in the momentum equation that is crucial and hence it is not necessary to have wind-generated shear to have Langmuir circulations. Any ambient shear would do, such as the shear of the inertial currents. Plueddemann *et al.* (1996), in fact, observed Langmuir circulations for up to a day after abrupt reductions in wind stress. The surface wave field, which decayed more slowly than the wind, interacted with ambient shear to maintain the circulation. However, such shear is usually much weaker than the wind-produced ones and so the circulation may not be as strong.

Equation (2.4.8) can also be derived from dimensional analysis. If we take V_{S0} and the wave slope (ka) to characterize the wave influence, and u_* to characterize the wind stress, then a velocity scale that can be formed from this combination is given by Eq. (2.4.8), taking into account the fact that the velocity scale must also depend on the angle between the wind and wave directions (usually zero). Stokes drift decays rapidly with depth and so does the vortex force responsible for maintaining the cells. The relevant length scale is $(2k)^{-1}$. The drift is only 10% of its value at the surface at a depth of 0.18λ, where λ is the wavelength. Beyond this depth, one can expect the cell circulation to be quite weak. For a typical λ value of 50 m, this depth is 9 m. Thus, the importance of the Langmuir cell contribution to TKE decreases rapidly with depth, but since $C/u^* \sim 200$ and wave slope $ka \sim 0.05$ are quite typical values, the ratio is O(1) close to the surface and the enhancement of production there is quite significant. A related nondimensional number,

$$N_S = kV_{S0}d/u_* = (ka)^2\,(kd)\,(C/u_*) \tag{2.4.9}$$

that characterizes the importance of Langmuir circulation relative to wind-induced shear effects is noteworthy. Note that C/u_* is the wave age (see Chapter 5) and ka is the wave slope in a wind-wave field. The nondimensional number characterizing the relative importance of Langmuir forcing to buoyancy forcing at the surface is

$$N_B \sim (u_*^2\ V_{so}\ k)/(-\beta g\overline{w\theta}) \tag{2.4.10}$$

which is the inverse of the Hoenikker number (Skyllingstad and Denbo, 1995). This number can be larger than unity in magnitude, as described by Li and Garrett (1995). For convective cooling at the surface, this number can be written in terms of the ML depth d and the Deardorff convective velocity scale w_* as

$$N_B \sim u_*^2\ V_{so}\ (kd)/w_*^3 \tag{2.4.11}$$

Dramatic evidence of the capability of Langmuir cells to effect mixing in the near-surface layers comes from Figure 2.4.6, which shows the temperature difference in the upper OML as a function of time as well as an index of the strength of Langmuir circulation. Note the absence, during Langmuir circulation episodes, of diurnal peaks in the SST due to strong solar heating and of shallow diurnal mixed layers, that are normally present. It is apparent that these cells are capable of affecting shallow diurnal mixed layers.

For mixed layer modeling, it is important to look at the contribution of Langmuir circulation to the TKE balance. Since this is proportional to $u_*^2V_{S0}\cos\theta$, the ratio of this to the production by wind shear near the surface u_*^3 (and buoyancy production), plus the shear production at the base of the mixed layer, delineates the importance of Langmuir circulation to overall ML dynamics.

Gnanadesikan (1994) has done a thorough study of Langmuir cells and their influence on mixing, both experimentally and numerically. He shows that the near-surface shear in neutrally stratified cases in the presence of Langmuir cells is much less than that indicated by log-law scaling (Figure 2.4.7). The downwelling below the convergence zone, the cellular motions in the upper

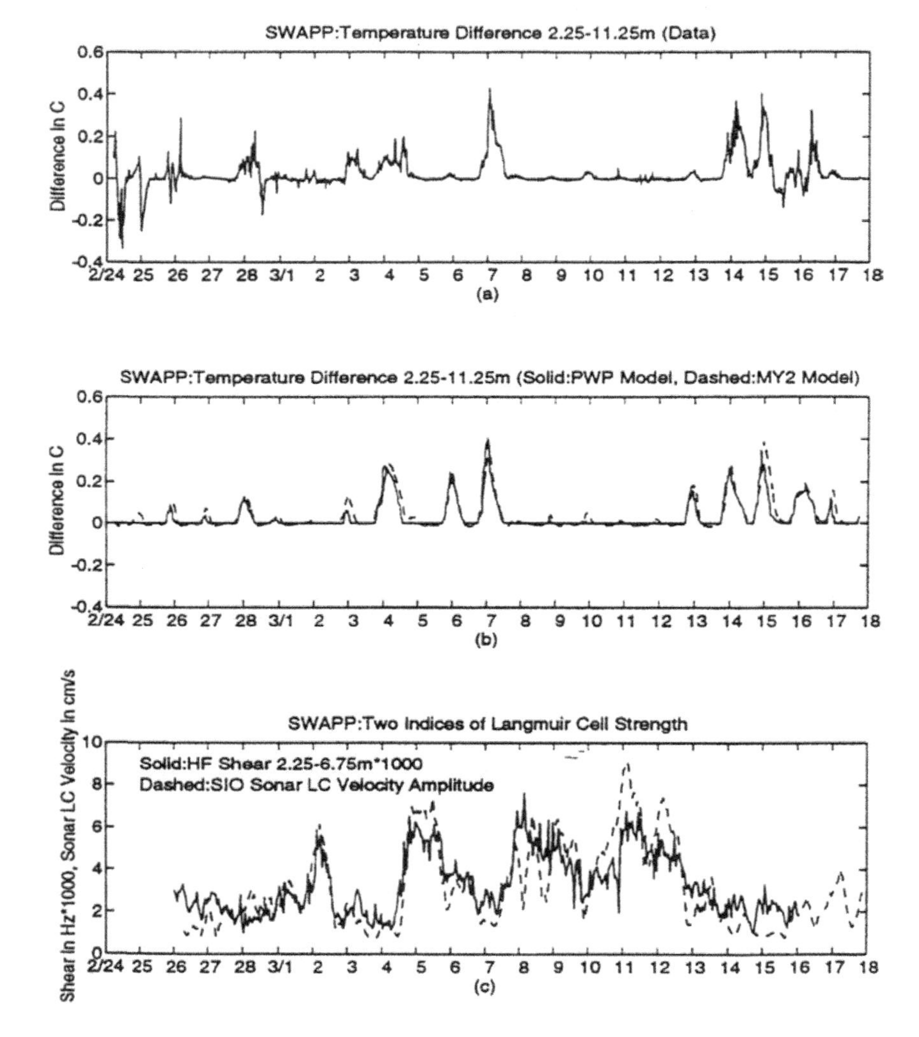

Figure 2.4.6. Temperature difference in the upper OML, along with indexes of Langmuir circulation during SWAPP. Note the absence of strong near-surface diurnal heating when Langmuir circulation is strong. (from Gnanadesikan, 1994).

layers, and the associated mixing would tend to decrease the shear according to

$$\frac{kz}{u_*}\frac{\partial U}{\partial z} = f\left(\frac{V_{s0}}{u_*}, kz\right) \tag{2.4.12}$$

C is the phase speed, k the wavenumber, and a the amplitude at the spectral peak, assuming that the wave field can be characterized by the peak of the

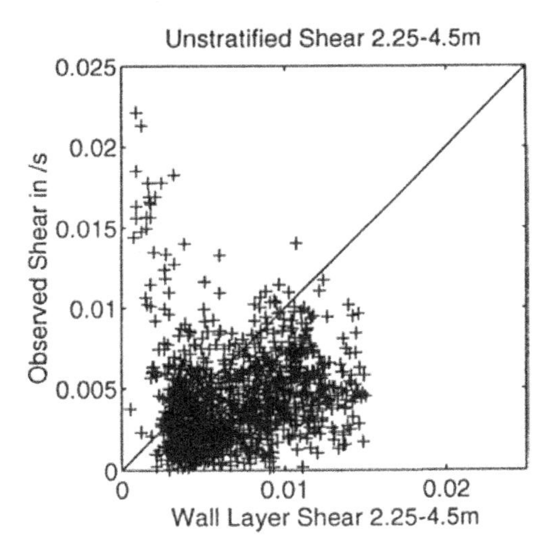

Figure 2.4.7. Observed near-surface vertical shear plotted against that expected from wall layer scaling for unstratified conditions during SWAPP. Note the reduced shears presumably due to Langmuir circulations (from Gnanadesikan, 1994).

spectrum. If the relationship is taken to be linear to a first approximation, then

$$\frac{kz}{u_*}\frac{\partial U}{\partial z}=1-\frac{V_{s0}}{u_*}\cdot kz \qquad (2.4.13)$$

Integrating between levels z_1 and z_2, the ratio of shear with and without Langmuir cells is

$$\frac{S_L}{S_5}=1-\frac{V_{s0}}{u_*}k\Delta z\ln\frac{z_1}{z_2} \qquad (2.4.14)$$

where $\Delta z = z_1 - z_2$.

2.5 CONVECTIVE MIXING

Penetrative convection, a process involving turbulent erosion of the bounding buoyancy interface and the consequent increase in the depth of the convectively mixed layer, is common in the oceans and the atmosphere and important to the daily and seasonal cycles of mixing in the ocean and lakes, and the diurnal evolution of the ABL. Convective mixing is in many ways more efficient than shear mixing. For example, in midlatitudes, there is an inherent limit to the depth a mixed layer can achieve under the action of a shear stress applied at the surface, even when there is neutral stratification, whether it is the OML forced by the wind stress or the ABL with winds aloft. This is the classical Ekman depth $D_e \sim 0.4\, u^*/|f|$. Mixed layers cannot be deeper than this under the action of a shear stress, and in stratified fluids, the depth is even less. No such limit exists for convective mixing. In fact, in subpolar latitudes, in regions of formation of deep/dense water in the oceans such as the Greenland and Labrador seas, OML depths can reach values as large as 1000–1500 m in local regions called chimneys. The comparative Ekman depth is a few hundred meters even under very strong wind forcing. The convective ABL can reach heights of a few kilometers (whereas shear-driven nocturnal ABLs are seldom more than a few hundred meters high), and in regions of cumulonimbus convection, convection can span the entire troposphere!

Convective mixing is therefore quite important to the deepening phases of mixed layers, nocturnal and wintertime deepening of the OML, and the daytime CABL. It is also quite important to the seasonal cycle in freshwater lakes during both heating and cooling phases because of the density maximum of fresh water at 4°C. During winter cooling, the lakes are cooled from the surface and convective mixing erodes the summer thermocline with typical depths of 10–20 m, until the water column reaches a value of 4°C. Any further cooling is confined to layers near the surface because the resulting water is lighter than water at 4°C. The surface layers cool to 0°C, at which point ice begins to form and cover the surface of the lake. Any further cooling results in ice formation, the heat loss being made up by the latent heat of fusion released during the phase change. Consequently, the waters below the thermocline during winter have higher temperatures than the freezing water column above it and even shallow freshwater lakes are seldom frozen solid and support life in the bottom layers during winter. When spring heating commences, the ice at the surface melts and the upper layers warm above freezing temperature. Due to the density maximum at 4°C, heating at the top destabilizes the water column and drives convection once again (convection can also start even with ice at the top from penetrative solar heating), until the entire water column reaches 4°C. Further evolution of the temperature structure involves stabilization by heating and the behavior is quite similar to spring–summer heating of the OML. Thus in freshwater lakes, convection occurs during both heating and cooling phases of the seasonal cycle (Figure 2.5.1), and this is quite important to the limnology of shallow dimictic freshwater lakes in redistribution of oxygen and nutrients in the water column.

In polar oceans and fjords, wintertime formation of ice at the surface drives convective mixing in the water column. Convection is not driven by heat loss per se, because the upper layers are already at the freezing point and the heat loss is made up by the phase change and ice formation. Instead, it is driven by the salt extruded by growing ice. The salinity of freshly formed sea ice is typically 10–15 psu and the salinity of the water column is in the thirties; the excess salt rejected by ice in the form of dense brine drives convective mixing in

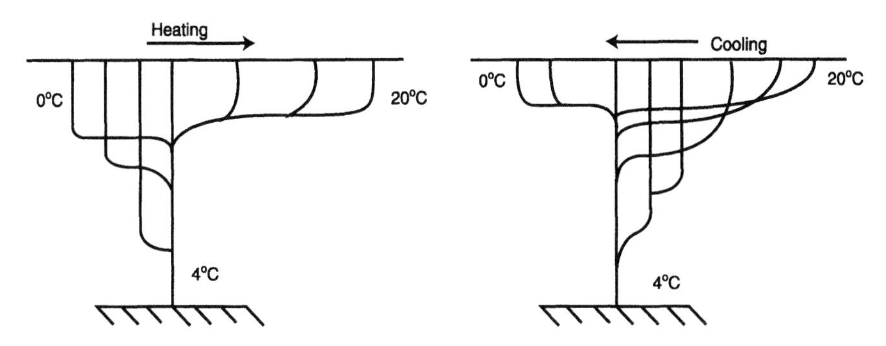

Figure 2.5.1. Evolution of temperature in shallow dimictic freshwater lakes during spring heating and winter cooling seasons.

the water column. On Siberian shelves, this wintertime ice formation and brine rejection leads to the formation of dense water that ends up forming a barrier layer between the cold (–1.7°C) near-surface layers and the warm (0.5–1.5°C) Atlantic waters at a depth of a few hundred meters in the Arctic Basin, thus preventing the release of heat from the Atlantic waters, and helping to maintain the perennial ice covering the Arctic.

In the atmosphere, daytime deepening is predominantly due to convective mixing and helps decrease the concentration of pollutants injected into the shallow nocturnal ABL during the night. The penetrative convection in the ABL is driven by surface fluxes. On days in which the surface warms rapidly in the morning, there is a rapid increase in the height of the ABL, as thermals driven by the transfer of heat from the surface to the atmosphere rise through the boundary layer.

In all these cases of convective mixing, deepening of the mixed layer is resisted by the buoyancy forces due to the presence of a strongly stratified region adjacent to the mixed layer, which is the capping inversion in the ABL and the thermocline or halocline (or more appropriately, pycnocline) in the oceans. Turbulent eddies generated in the mixed layer by convection expend some of their energy in entraining the heavier (lighter) fluid from below (above) the OML (ABL) and convection penetrates into the adjoining stratified fluid—hence the name penetrative convection. Deepening of the mixed layer can, but seldom does, occur without penetrative convection (Figure 2.5.2). The resulting entrainment fluxes at the top of the ABL have a major influence on dispersion of passive scalars such as contaminants on the ABL. Nonpenetrative convection is not a good approximation in the OML either.

Because of the importance of penetrative convection, a large number of studies have been done both in the laboratory and in the field. An excellent summary of the state of the subject can be found in a monograph by Zilitinkevich (1991). We will describe a few salient points here.

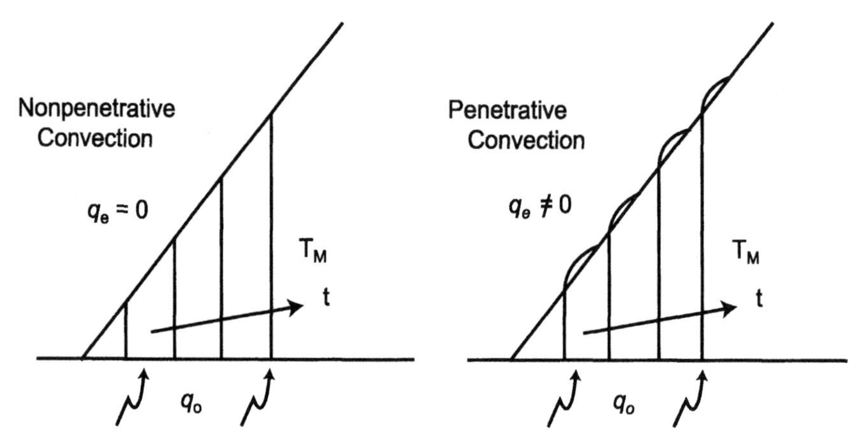

Figure 2.5.2. Mixed layer deepening due to convective mixing: nonpenetrative (left) and penetrative (right). Note the absence of entrainment in the former case.

The temperature (more generally buoyancy) profile in the fluid column can be idealized as shown in Figure 2.5.3. Assuming a homogeneous distribution of properties in the mixed layer is a very good approximation for convective mixing. The buoyancy interface is assumed to be sharp for simplicity, while in practice it is usually more diffuse. Extension to diffuse interfaces can be found in Deardorff (1980) and Zilitinkevich (1991). A stably stratified layer with buoyancy frequency N (the Brunt–Vaisala frequency) exists on the other side of the interface. If we consider the conditions to be horizontally homogeneous, then the mean equation reduces to

$$\frac{\partial T}{\partial t} = -\frac{\partial Q_T}{\partial z} \tag{2.5.1}$$

where Q_T is the total heat flux.

This has to be supplemented by the initial temperature profile. While it is more appropriate to write this equation in terms of buoyancy for general applications, there is no loss of generality in doing this and it is easy enough to replace temperatures by buoyancy of the fluid. Integration of this equation in the vertical yields

$$\frac{d}{dt}(D\Delta b) = D\frac{d}{dt}(\Delta b) + \Delta b\, u_e = D\frac{d}{dt}(\Delta b) + q_e = q_0 \tag{2.5.2}$$

where D = D(t) is the mixed layer depth. In particular u_e = dD/dt, the rate of deepening of the mixed layer, is unknown and hence the equation is not closed. q_0 is the applied heat flux (kinematic) that drives convection in the mixed layer.

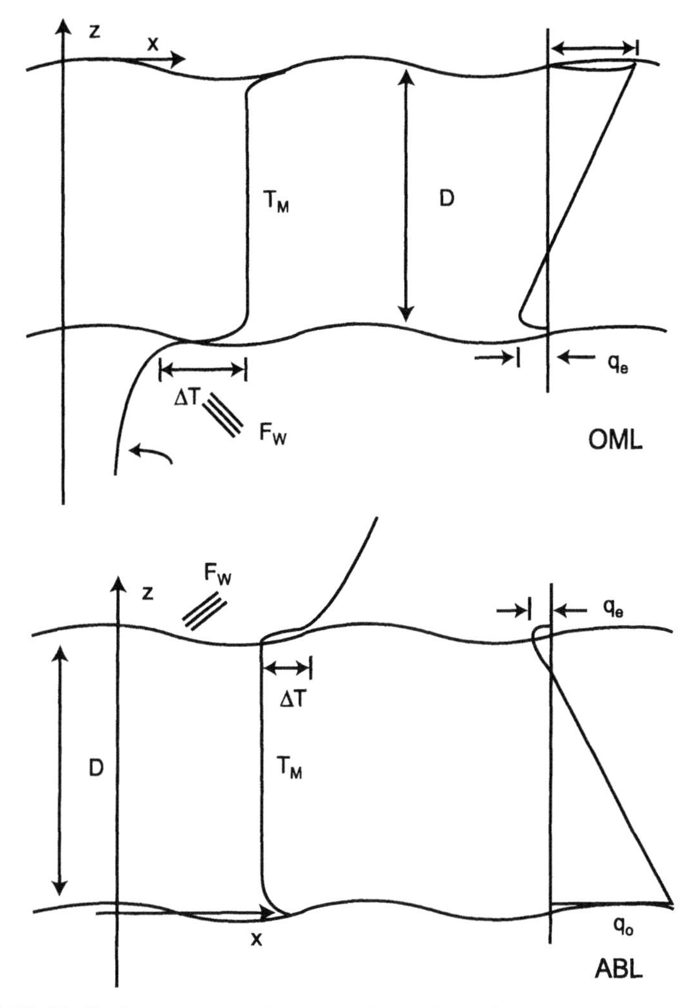

Figure 2.5.3. Idealized temperature and buoyancy flux profile in the upper ocean (top), showing a well-mixed OML, a strong buoyancy interface at its base, and a stably-stratified region below. The corresponding features are shown for the ABL in the bottom panel.

The problem in penetrative convection simply boils down to determining the value of u_e, the entrainment velocity (or equivalently, the entrainment heat flux q_e) in a wide range of the governing parameters. Since it is a process involving turbulence, there are no analytical solutions and appeal has to be made to observations and/or numerical calculations. The characteristic turbulence velocity scale in the problem is the convective or the Deardorff scale:

$$w_* = \left(\beta g q_0 D\right)^{1/3} \tag{2.5.3}$$

Using w_* we can define a nondimensional entrainment velocity:

$$E=u_e/w_* \tag{2.5.4}$$

A nondimensional flux ratio can also be defined:

$$R=q_e/q_0 \tag{2.5.5}$$

The important stability parameters in the problem that characterize the work to be done by turbulent eddies in effecting entrainment are the bulk Richardson numbers Ri and Ri_N,

$$Ri=\frac{D\Delta b}{w_*^2}, \quad Ri_N=\frac{N^2D^2}{w_*^2} \tag{2.5.6}$$

so that

$$E=E(Ri, Ri_N), \quad R=E\,Ri=R(Ri, Ri_N) \tag{2.5.7}$$

Figure 2.5.4 shows vertical profiles from the classical experiments of Deardorff *et al.* (1969) in an initially linearly stratified fluid which show the evolution of temperature structure with time. Note the nonpenetrative nature of convection ($R \sim 0.12$). In these experiments, the entrainment rate is initially large and decreases with time, because of the increase in the value of Ri with time. Figure 2.5.5 shows the variation of R with Ri in Kantha's two-layer experiments (Kantha, 1980b). Here the entrainment increases with time with eventual overturning and $Ri_N = 0$. Kantha (1980b) argued that the heat flux ratio must be a function of Ri. $R \sim Ri$ at very low values of Ri (equivalent to $E \sim$ const), when the buoyancy interface is weak enough that Ri has little influence on entrainment. In the moderate range of Ri, typical of most geophysical situations, $R \sim$ const (equivalent to $E \sim Ri^{-1}$), implying that the rate of change of potential energy of the system due to entrainment ($u_e\Delta b$) must be proportional to the rate of work done (w_*^3). Here the entrainment is due to the energetic impact of large turbulent eddies against the interface and the resulting sweeping up of the heavier fluid from below between interfacial convolutions. At large Ri values, the interface is flat and entrainment is due more to sporadic local shear enhancements by eddies. Here R should fall off with increases in Ri [Zilitinkevich (1991) argues against the decrease in the value of R below 0.2 at high Ri values]. However, for most applications, Kantha's results indicate that a constant value of $R \sim 0.2$ is quite a reasonable choice. As $Ri \rightarrow 0$, Kantha's results indicate that $E \rightarrow 0.15$, while Deardorff and Willis (1985) results in their two-layer experiments indicate $E \rightarrow 0.24$. An average value of 0.20 is a reasonable choice.

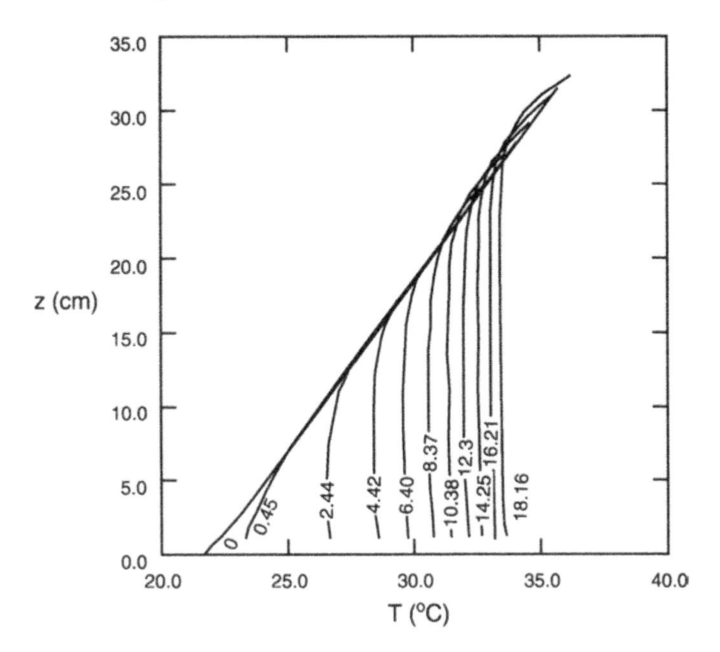

Figure 2.5.4. Deepening of the mixed layer in an initially linearly stratified fluid due to convective heating at the bottom (from Deardorff *et al.*, 1969). Numbers on the curves indicate time in minutes.

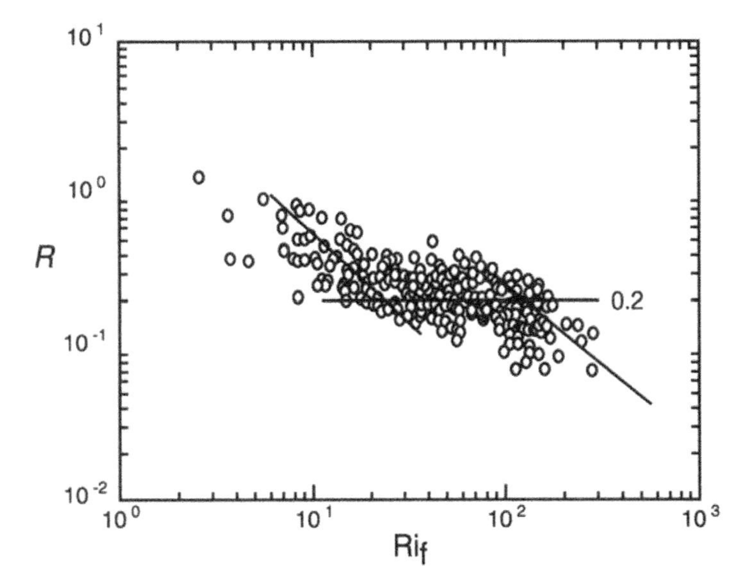

Figure 2.5.5. Entrainment ratio as a function of the bulk Richardson number (from Kantha, 1980).

It is difficult to make measurements of E or R in the field under controlled conditions. Consequently, the scatter in the deduced values of R is quite large, with the values ranging anywhere from 0.1 to 0.35. Table 2.5.1, taken from Zilitinkevich (1991), summarizes the values of R from laboratory and field observations. The theoretical limits for R are R = 0 (Zubov, 1945) and R = 1 (Ball, 1960).

The closure problem in penetrative convection then is to develop functional relationships for E or equivalently R. To do this, we need to appeal at least to the

TABLE 2.5.1
Values of R from Field and Laboratory Observations (from Zilitinkevich, 1991)

Source	Data type[a]	*A*
Koprov and Zwang (1965)	M	0.35
Deardorff (1967)	M	0.1–0.3
Lenschow and Johnson (1963)	M	0.9
Deardorff *et al.* (1969)	L	0.12
Lenschow (1970)	M	0.1
Deardorff (1972b)	M	0.1
Lenschow (1973)	M	−0.4–0.17
Stull (NCAR Report, 1973; see Stull, 1976)	M	< 0.1
Deardorff (1973)	M	0.2
Carson (1973b)	M	0.25
Betts (1973)	M	0.25
Cattle and Weston (1973)	M	0.25
Lenschow (1974)	M	< 0.12
Pennel and Le Mone (1974)	M	< 0.1
Rayment and Readings (1974)	M	0.25
Deardorff *et al.* (1974)	L	0.23
Betts (1974)	M	0.30
Willis and Deardorff (1974)	L	0.11–0.23
Cattle and Weston (1975)	M	0.29; 0.32
Farmer (1975)	H	0.2
Heidt (1977)	L	0.18
Yamamoto *et al.* (1977)	M	0.2
Kantha (1979a)	L	0.2 ± 0.1
Coultman (1980)	M	0.25
Dubosclard (1980)	M	0.9
Varfolomeyev and Sutyrin (1981)	L	0.21
Denton and Wood (1981)	L	0.2–0.5
Driedonks and Tennekes (1984)	M	0.2

[a] Laboratory (L), meteorological (M), and hydrological (H) data.

TKE equation [Eq. (1.7.34)] in a vertically integrated form:

$$\frac{d}{dt}\int_0^{-D} K\,dz = \frac{D}{2}\left[\beta g \Delta q_0 - \beta g \Delta T \frac{dD}{dt}\right] - [F_D - F_0] - \int_0^{-D} \varepsilon\,dz - F_w \qquad (2.5.8)$$

F_w denotes the energy flux due to internal waves radiated into the stratified interior. This is often quite substantial and cannot be ignored. Also this mechanism is important in its own right as far as internal waves in the atmosphere and the ocean are concerned. It can be parameterized as (Kantha, 1977)

$$F_w \sim \rho A^2 N^2 Cg_z \qquad (2.5.9)$$

where A is the amplitude of the waves and Cg_z is the vertical group velocity of the waves. The problem is how to scale A and the wave frequency n. Kantha (1977) suggested that the maximum value for F_w can be written as

$$F_w \sim \rho A^2 N^3 \lambda \qquad (2.5.10)$$

where λ is the horizontal scale of the waves (see also Stull, 1976; Kantha, 1977; Zilitinkevich, 1991). By energetics arguments (simply balancing the kinetic energy of eddies impinging at the base by the increase in potential energy due to the indentations produced),

$$\beta g \Delta T \cdot A + \frac{N^2 A^2}{2} \sim \frac{w_*^2}{2} \qquad (2.5.11)$$

where ΔT is the buoyancy "jump" at the base of the mixed layer and N the buoyancy frequency in the stratified fluid below. Kantha takes the horizontal scale of the waves to be proportional to mixed layer depth D so that

$$\frac{F_w}{\rho w_*^3} \sim Ri^{-2}\, Ri_N^3 \left(Ri_N \to 0\right), \sim Ri_N \left(Ri \to 0\right) \qquad (2.5.12)$$

If one considers a diffuse thermocline model, then the latter expression is valid if N is regarded as the stratification in the thermocline.

A different expression results if one parameterizes A and λ slightly differently. Zilitinkevich (1991) takes

$$A \sim \frac{R}{1+R} \cdot D, \quad \lambda \sim R\,D \qquad (2.5.13)$$

so that

$$F_w \sim \rho\, w_*^3\, Ri_N^{3/2} \left(\frac{E R_i}{1 + E R_i} \right)^3 \qquad (2.5.14)$$

after substituting R = E Ri. F_D and F_o in Eq. (2.5.8) denote turbulence energy fluxes at the buoyancy interface and the surface ($F_o = 0$). Both F_D and integrated dissipation terms scale with w_*^3, and the TKE (K) scales like w_*^2, so that

$$(C_2 + Ri)E + C_3\, Ri^{3/2} \left(\frac{E\, Ri}{1 + E\, Ri} \right)^3 = C_1 \qquad (2.5.15)$$

When N ~ 0, this relationship reduces to

$$E(C_2 + Ri) = C_1 \qquad (2.5.16)$$

which is valid for low to moderate Ri values. The best values for C_1, C_2, and C_3 are $C_1 \sim 0.2$, $C_2 \sim 1.0$, and $C_3 \sim 0.1$ (Zilitinkevich, 1991). Zilitinkevich (1991) argues that the form of the term due to internal wave radiation is consistent with laboratory observations. Equation (2.5.15) can be used to parameterize penetrative convection to a fairly good degree of approximation. It proves quite useful in semianalytical models of the seasonal cycle of density structure in the oceans and lakes (Zilitinkevich, 1991).

However, the expression (2.5.14) for F_w is far from certain. From Chapter 5, it can be seen that

$$F_w = \rho A^2 N^2 C_{g_z} = \rho A^2 N^2 \frac{n}{k_h} \left(1 - \frac{n^2}{N^2} \right) \qquad (2.5.17)$$

For internal waves to be generated and radiated, n < N. The horizontal scale of the waves should be scaled by the depth of the mixed layer because it is proportional to the size of the energy-containing eddies. The frequency must be the inverse of the eddy turnover timescale, $k_h \sim \frac{1}{D}$, $n \sim \frac{w_*}{D}$, so that

$$F_w \sim \rho\, w_*^3 \frac{A^2}{D^2} (Ri_N - C_4) \qquad (2.5.18)$$

If we take A ~ D, assuming a weak buoyancy jump ($\Delta T \to 0$), then we get

$$F_w \sim \rho\, w_*^2\, Ri_N \qquad (2.5.19)$$

The penetrative convection processes described above occur in the daytime CABL and nocturnal OML. However, often shallow convection can exist in the daytime OML, under calm no-wind conditions. This is because there is usually a heat loss immediately at the surface even during daytime solar heating that can drive convection. However, the subsurface bulk heating due to penetrative shortwave (SW) radiation damps the convection caused by surface cooling and limits it to shallow depths. An analogous situation exists in the cloud-topped ABL, where radiative longwave (LW) cooling at the cloud top drives convection, but bulk SW heating of the cloud layer limits the depth of convection. Woods (1980) considered the damping of convective instability by bulk heating and proposed that when the Rayleigh number Ra_q falls below a critical value (~1700), convection ceases and this defines the depth of the shallow convective layer,

$$Ra_q = \frac{\beta g[I(z) - I(D_c)]z^4}{\nu\, k_T^2} \tag{2.5.20}$$

where D_c is the compensation depth defined as the depth at which the heat loss at the surface equals the net heat absorption above that depth:

$$I_s - I(D_c) = q_s \tag{2.5.21}$$

Since the maximum Ra_q occurs at a depth z_{max} given by

$$z_{max} \left.\frac{dI}{dz}\right|_{z_{max}} = 4[I_s - I(z_{max}) - q_s] \tag{2.5.22}$$

the condition

$$Ra_q^{max} = \frac{\beta g\left(I_s - I(z_{max}) - q_s\right)z_{max}^4}{\nu\, k_T^2} < 1700 \tag{2.5.23}$$

provides the depth of shallow convection that results, usually 0.7–0.9 D_c (Soloviev and Schlussel, 1996). Dalu and Purini (1981) suggest that the convection depth is that depth at which the heating rate by insolation equals the heating rate of the layer above:

$$\left.\frac{dI}{dz}\right|_{z_{max}} = \frac{[I_s - I(z_{max}) - q_s]}{z_{max}} \tag{2.5.24}$$

This simple parameterization has been used in the Price–Weller–Pinkel (1986) bulk ML model and works well under conditions of strong solar insolation (see Section 2.10).

2.5.1 Deep Convection in the Ocean

Convective processes outlined thus far are applicable to a majority of the oceans and the atmosphere. The associated vertical length scale is usually a fraction of the fluid column (generally less than 200 m deep in the ocean and less than 1–3 km in the atmosphere), and therefore such convection can be more appropriately called shallow convection. The associated timescale is mostly diurnal, and therefore short enough for rotational effects to be ignored. But there exist a few regions in the high latitude oceans, where convection is both deep and long-lasting. Under the present climatic conditions, in the Greenland Sea (Schott *et al.,* 1993), the Labrador Sea (Lab Sea Group, 1998), and the western Mediterranean Sea (Schott *et al.,* 1996) in the northern hemisphere, and in the Weddell Sea in the southern hemisphere, strong, prolonged wintertime cooling occurs, leading to deep convective layers that extend over most of the water column. Deep convection in the open ocean is the means by which the deep ocean is ventilated and its thermal structure maintained (Munk, 1966). The resulting meridional thermohaline circulation and the poleward oceanic heat transport from the low latitudes to the mid–high latitudes associated with it are a major factor in the climate at these latitudes. In these few deep convection regions, the stable stratification in the water column that normally isolates the abyss from the atmosphere is broken down violently by strong convective cooling at the surface. The associated timescales are much larger than the inertial period, and hence deep convection occurs under the influence of Earth's rotation. An analogous process in the atmosphere is a similarly violent cumulonimbus convection that can span the entire troposphere; however, the associated timescale is short enough for rotational effects to be neglected. Here we have attempted to provide a brief overview of deep convection in the ocean. For a more detailed one, see the excellent reviews by Killworth (1983) and Marshall and Schott (1998) and the references therein.

Deep convection was first observed in the Gulf of Lions in the western Mediterranean in the 1960s. Measurements detected rapid deepening of the mixed layer down to 2 km, with vertical currents exceeding 10 cm s^{-1}. Since then, evidence of deep convection reaching down to 2.5–3 km has been gathered in both the Labrador and the Greenland seas. In all these cases, deep convection occurs in a region adjacent to continents, less than a hundred kilometers in size, driven by cold wintertime continental air masses blowing from nearby land masses over the relatively warmer ocean at high speeds, causing heat losses of as much as 1000 W m^{-2} and strong convection in the water column. However, there is tremendous variability. Deep convection does not occur every winter. In the Labrador Sea, strong deep convection was present in the 1970s. It was weak during the 1980s, but the 1996–1997 Labrador Sea Deep Convection Experiment sponsored by the Office of Naval Research (Lab Sea Group, 1998) has detected strong deep convective activity once again in the 1990s. Evidence of deep convection in the Greenland Sea in the 1980s exists. In the Mediterranean Sea, deep convection leading to the formation of the western Mediterranean deep water is more common, and instead of decadal variability that seems to be a feature of the Greenland and Labrador sea deep convection, interannual variability appears to be more prevalent (Mertens and Scott, 1998).

Marshall and Schott (1998) identify three phases of deep convection: (1) The preconditioning phase, where the prevailing large scale circulation brings the weakly stratified deep water masses closer to the surface for the stratification to be gradually eroded by strong sustained surface cooling during early winter. This phase is crucial to the whole process. Deep convection in the open ocean is found only in regions with cyclonic circulation that causes an upward doming of the isotherms. In the Labrador Sea, the cyclonic circulation is due to the West Greenland and Labrador currents hugging the continental slope. In the Mediterranean, the cyclonic Lions Gyre provides the preconditioning. Strong, sustained cooling is also essential to break down the stratification built up in the upper layers during the previous spring and summer. It is interesting that even stronger heat losses (~1000 W m^{-2}) occur in the oceans during wintertime cold air outbreaks off the east coasts of continents leading to strong cyclogenesis in the atmosphere, but not deep water formation because of the brevity of the event. In the polar oceans, the air–sea temperature differences during off-ice wind conditions can reach 30–40°C, and if sustained long enough can lead to intermediate and deep water formation, examples being the Weddell Sea in the Antarctic, the Sea of Okhotsk, and the Arctic shelves. (2) Eventually the stratification breaks down. A strong cooling event lasting several days with heat losses of 500–1000 W m^{-2}, brought on by air–sea temperature differences of 8–12°C, and strong wind bursts are common. Deep convection ensues with intense plumes a few kilometers in size reaching down to 2–3 km. (3) The cooling weakens, and the well-mixed water mass in the convective "chimney" spreads laterally, undergoing baroclinic instabilities in the process, and mixes with the ambient waters. Stratification is restored and the stage is set for the next cycle.

The most relevant parameters are the buoyancy flux (due to both sensible and evaporative heat losses) B_o that can reach values of $1\text{–}3 \times 10^{-7}$ m^2 s^{-3}, the inertial frequency f, the depth of the mixed layer D, 1000–3000 m, and the Rossby radius of deformation a, typically a few kilometers. Under conditions where the rotational effects dominate, the relevant length and velocity scales are $l_{dc} = (B_o/f^3)^{1/2}$ and $u_{dc} = (B_0/f)^{1/2}$ (Jones and Marshall, 1993; Maxworthy and Narimousa, 1994; Fernando and Ching, 1994; Raasch and Etling, 1998). The associated Rossby number is unity. The relevant Rayleigh number is $Ra = B_0D^4/(\nu k^2)$, on the order of 10^{26} (compared to laboratory experiments on convection under rotation of $<10^{13}$). Note that in the atmosphere, where rotational effects are not important, the relevant length scale is D, and the relevant velocity scale is the Deardorff scale $w_*=(B_0D)^{1/3}$. In the oceans, since the temperature and salinity changes at the bottom of the mixed layer tend to compensate each other, leading to negligible density change (Marshall and Schott, 1998), it is difficult to define a value of $c = (g'D)^{1/2}$ for use in the Rossby radius calculation. Instead the relevant g' might be $u_{dc}^2/l_{dc} = (B_0 f)^{1/2}$, so that a ~ $(B_0 D^2/f^3)^{1/4}$. The velocity scale u_{dc} ~ 2–6 cm s^{-1}, of the same order as the Deardorff velocity scale, 5–8 cm s^{-1} [observed vertical velocities in the plumes are 6–10 cm s^{-1} (Marshall and Schott, 1998)]. In any case, l_{dc}, typically 0.4–0.7 km, dictates the scale of the convective plumes (plumes observed in the three seas are 0.5–1.5 km in diameter) and the depth below the surface where rotational effects begin to dominate. Several laboratory studies (Fernando *et al.*, 1991; Maxworthy and Narimousa, 1994; Coates *et al.*, 1995) suggest that the critical depth beyond which rotation effects begin to dominate the flow is an order of magnitude larger than l_{dc}. If this is proven to be correct, the depth at which this occurs in the ocean would be on the order of 4 km or more. Since the depth of deep convection is typically 1–4 km, this would suggest that deep convection in the ocean is affected by rotation, but not necessarily dominated by it.

The ratio $l_{dc}/D = [B_0/(f^3 D^2)]^{1/2}$, the ratio of the rotational timescale to the convective timescale, is a Rossby number, typically 0.1–0.4 (0.1 in the Labrador and Greenland seas and 0.4 in the Mediterranean), and therefore in the upper part of the deeply convective mixed layer, convection appears similar to shallow convection, and in the lower part, it may take the form of sinewy rope-like convective plumes with a diameter on the order of l_{dc} reaching down to the bottom of the mixed layer. In contrast, in cumulonimbus convection in the atmosphere, D ~ 10 km and the ratio l_{dc}/D ~10, and hence rotational effects are not strongly felt. The ratio $a/D = (l_{dc}/D)^{1/2}$ and is proportional to the aspect ratio of the patches of mixed layer fluid that result during the third spreading phase of the deep convection. The size of these patches is typically 2–3 km. The buoyancy difference Δb (typically 3–4 × 10^{-6} m s^{-2}) between the top and bottom of the deeply convective layer is another parameter. A buoyancy flux based on Δb and D can also be defined, $B_p \sim (\Delta b^3 D)^{1/2}$; this is a measure of the buoyancy flux in the plumes. The ambient stratification N defines a radius of deformation $a_a = ND/f$, and if the size of the convecting chimney is larger than this, it is susceptible to baroclinic instability. The ambient stratification is weak in regions susceptible to deep convection with N/f ~ 5–10, so that a_a is typically 10–30 km.

Nonhydrostatic models are being increasingly used to study deep convection. Jones and Marshall (1993) and Marshall *et al.* (1997a) are noteworthy examples. The former studied the problem of deep convection, whereas the latter was a coarse resolution (1°) study of the global oceans using the nonhydrostatic model. While the global model captures the preferred sites of deep convection, it does not simulate their characteristics well (Marshall and Schott, 1998). Legg *et al.* (1998) have performed simulations of deep convection in a preconditioned, cold core, cyclonic eddy under uniform surface forcing, rather than the disk-shaped surface cooling of a horizontally homogeneous ocean (e.g., Jones and Marshall, 1993; Visbeck *et al.*, 1996). This allows for baroclinic instabilities to develop more naturally. Recent LES models of deep convection include Julien *et al.*

(1996) who studied rotating penetrative convection, and Garwood *et al.* (1994) and Denbo and Skyllingstad (1996) who included the effect of thermobaricity on convection. The rotational effects decrease entrainment at a stable density interface at the base of the convecting layer, while thermobaricity appears to do the opposite. Raasch and Etling (1998) set out to simulate the laboratory experimental results of Coates *et al.* (1995) who simulated unsteady convection in an initially linearly stratified fluid heated from below in a localized region in the presence of rotation. Model results show that due to the horizontal temperature gradient between the plume and the surroundings, a strong rim current with a size of about 2 a_a develops around the convective plume. This current becomes baroclinically unstable and breaks up after a few inertial periods into eddies that then spread out horizontally, leading to a strong slowdown in the growth of the mixed layer. Schott *et al.* (1996) report a rim current about 20 km in width in the Gulf of Lions, and Morawitz *et al.* (1996) report a 50-km convective chimney of 1-km depth that broke up in 3–6 days. LES's of Raasch and Etling (1998) appear to be consistent with these observations. Finally, Kampf and Backhaus (1998) have applied a 3D non-hydrostatic model to the equally important problem of brine-driven convection in shallow Arctic and Antarctic coastal polynyas.

2.6 MIXING PROCESSES IN MIDLATITUDE OCEANS

Perhaps the most salient aspect of the OML in midlatitudes is its diurnal and seasonal variability. Figure 2.6.1 shows the typical seasonal cycle in mid-latitudes. This seasonal variability in ML depth and temperature, and hence the heat content of the ML, is a prime factor in the air–sea exchange at these latitudes. The onset of spring warming restratifies the water column and once the shallow spring–summertime thermocline forms, its depth stays roughly the same. However, the formation period is heavily influenced by wind events at the time. Similarly, wind forcing controls the deepening of the ML at the onset of autumn cooling. During this time, the ML deepens episodically during intense storms that pass through the region with significant assistance from cooling at the surface. Cold air outbreaks during winter along the east side of continents lead to rapid ML deepening, a large heat loss from the ocean, and cyclogenesis in the atmosphere. At higher latitudes, the ML deepens more due to penetrative convection, and in subpolar regions, deep convection occurs. Both sensible and latent heat fluxes are important in cooling the ocean at midlatitudes, whereas it is principally the evaporative losses that dominate the air–sea exchange at warmer low latitudes and sensible heat loss at colder high latitudes. Precipitation events also play a role in mixing. The ML structure is affected by both its Salinity and temperature structures, whereas in subpolar and polar oceans,

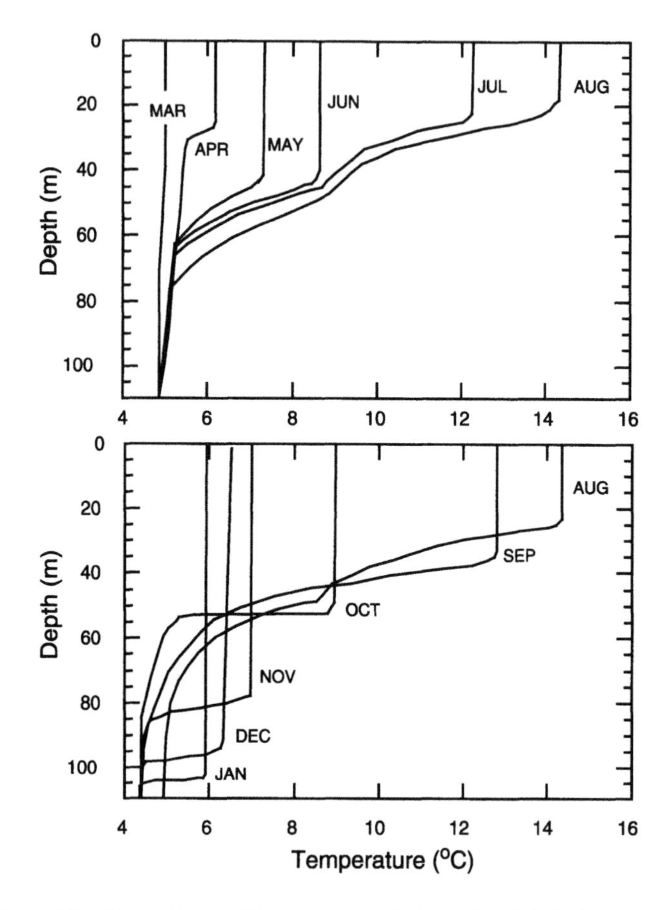

Figure 2.6.1 Seasonal cycle of temperature evolution in the midlatitude upper ocean.

salinity plays an overwhelmingly important role, and in the tropics, the thermal structure in general predominates.

Diurnal variability affects the heat exchange on shorter timescales and may play a role over long timescales as well. The intensity of diurnal modulation of the ML depth and temperature depends on the season. Generally, the modulation is stronger if the solar insolation is strong and winds weak. This situation is typical of summer. Insolations as high as 1000 W m^{-2} are possible during clear summer days. This combined with very low winds gives rise to a diurnal modulation of as much as 2–4°C. Part of the heat built up in the ML during the day is lost by nocturnal cooling, which drives a vigorous convection and mixing in the water column that normally mixes some of the heat gained into the seasonal mixed layer. Figure 2.6.2 shows observations made during LOTUS (Price *et al.*, 1986) that show this diurnal modulation quite well.

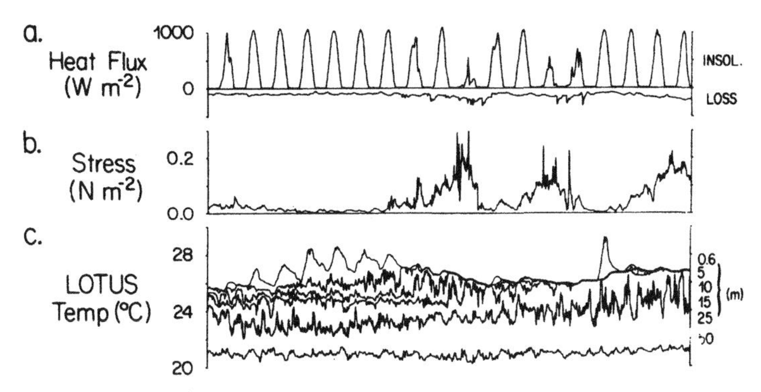

Figure 2.6.2. Diurnal modulation of the OML. (a) The observed heat flux, (b) wind stress, and (c) temperature at various depths during LOTUS (from Stramma *et al.*, 1986).

Stratification and mixing in the upper layers affects biological processes, as demonstrated by the Marine Light-Mixed Layer (MLML) program (Plueddemann *et al.,* 1995; Stramska *et al.,* 1995). During these observations, springtime restratification occurred and this led to a strong bloom. The bloom was, however, erased by the passage of a strong storm, which deepened the ML and transported the primary producers to depths where the light was low. The productivity recovered after the storm, but not to the same levels as before, even though storm-induced mixing imports nutrients into the ML. Diurnal modulation has a similar strong impact on biological processes. Strong daytime stratification confines the phytoplankton to upper layers with abundant sunlight, and if the nutrient supply is not limiting, high productivity ensues. However, this also depletes the nutrients, unless replenished from below. Nocturnal cooling does that to some extent, but it also transports phytoplankton deep into the ML, where the light levels are lower during the day. Figure 2.6.3, shows the impact of mixing events on a neutrally stratified float in the ML. Note the marked excursions in the vertical during the night, when the float reaches depths of 50–60 m. The vertical velocities resulting from turbulent eddies in the ML are generally too strong even for the "swimmers" and they are transported up and down by the resulting vertical motions. Processes such as Langmuir circulations also transport phytoplankton and zooplankton in the vertical and hence have an effect on biological processes.

One of the features of the ABL and OML in mid- and upper latitudes is the classic Ekman spiral, first described by Ekman to explain ice motion (see Section 2.8 for a full description). Simplistically, the Ekman spiral occurs as the surface water receives momentum from the winds in the direction of the wind. This surface current then by friction transfers a portion of this momentum to the water underneath; however, due to Coriolis effects, the current is turned to the right (left) in the northern (southern) hemisphere. This transfer of momentum continues through the water column, leading to the classic Ekman spiral shown

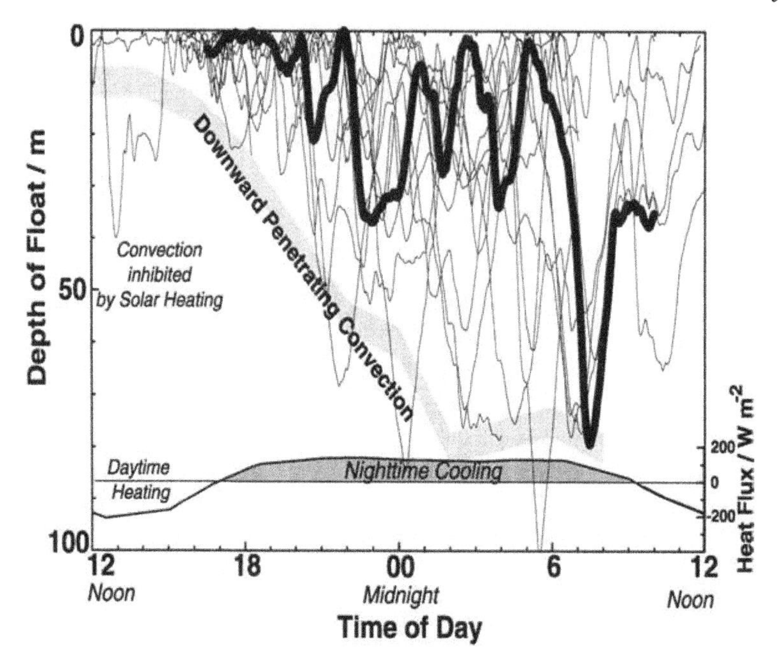

Figure 2.6.3. Track of a neutrally buoyant float in the OML, portraying dramatically the shallow diurnal and deep nocturnal mixed layer (from Denman and Gargett, 1995, with permission from the *Annual Review of Fluid Mechanics*, Volume 27, Copyright 1995, by Annual Reviews).

in Figure 2.6.4 (Brown, 1990). The situation is complicated by the coexistence of many physical processes that affect currents, such as surface wave motions and Langmuir circulations. The existence of the Ekman spiral in the ocean was not demonstrated conclusively until careful processing by Price *et al.* (1985) of observed currents from R/V Flip to remove extraneous effects. Wind vector rotation in the lower ABL is readily observed in the atmosphere.

2.7 EQUATORIAL MIXING PROCESSES

The mixed layer is important to the dynamics of the equatorial oceans. Nowhere else in the global oceans does the OML exert a primary influence on the dynamics of the underlying current system. This is principally because the stress divergence due to turbulent mixing is a dominant term in the force balance right at the equator (Dillon *et al.*, 1989), whereas elsewhere the primary balance is nearly geostrophic and the stress divergence is a smaller term. The mixed layer in the equatorial regions differs from the midlatitude ML in that although there is still strong diurnal variability, there is little seasonal variability. In addition, some areas of the world's tropical ocean regions receive an abundance

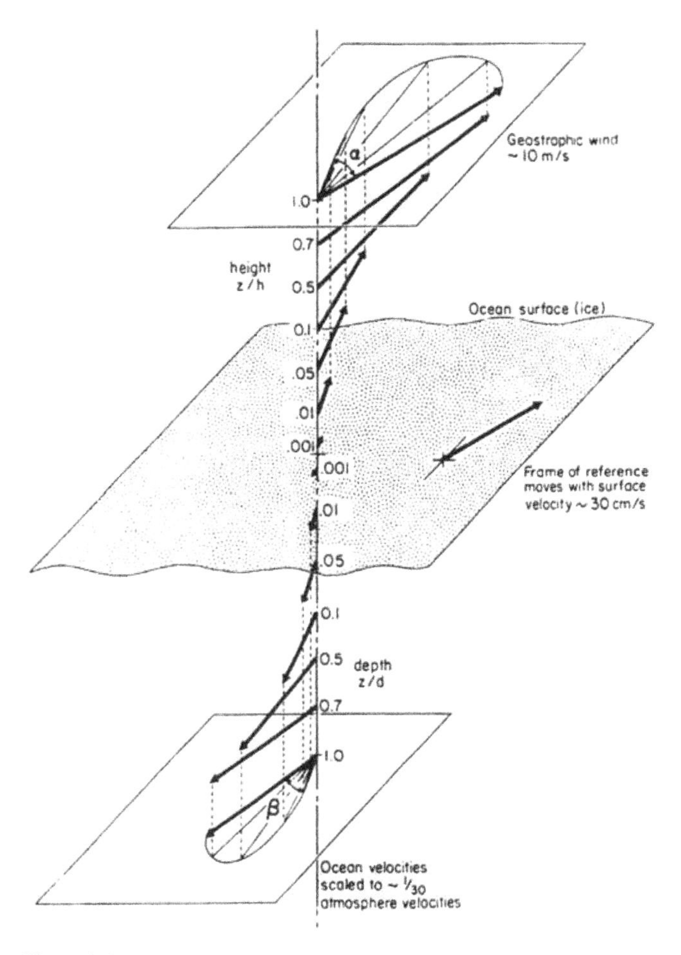

Figure 2.6.4. The Ekman spiral in the atmosphere and the ocean (from Brown, 1990).

of rain, enough in some cases to offset losses due to evaporation. Thus although the strong diurnal variability produced by the cycle of solar heating is the main influence on the upper ocean density structure (and consequently a control on the depth of mixing), strong precipitation events also play a role in mixing processes in the tropical mixed layer.

Since $f = 0$ at the equator, the Coriolis acceleration cannot balance the vertical divergence of the wind-induced shear stress and therefore a steady wind cannot lead to a steady Ekman spiral as in midlatitudes. Instead, a steady zonal wind causes steadily accelerating currents in the zonal direction in the upper layers on short timescales (a few days), until pressure gradients are set up to balance the wind stress (which takes large scale flow adjustments in the equatorial waveguide and hence involves timescales typically larger than those of synoptic variability of the winds). In addition, the presence of a strong current in the form of the eastward-flowing Equatorial Undercurrent (EUC) at depths ranging from 50 to 100 m in the central Pacific to 150–200 m in the western Pacific causes a strong vertical shear in the upper layers above the core of the EUC that has profound consequences on mixing above the EUC core, as we shall see below. As a consequence, for steady easterlies, the vertical current shear due to the EUC in the equatorial Pacific is considerably enhanced in the upper water column. This tends to enhance the mixing rates in the upper layers. By the same token, steady westerlies tend to reduce the EUC-induced vertical shear and suppress shear-induced mixing. Changing wind conditions in the equatorial waveguide constantly accelerate and decelerate the upper layer currents, and the diurnal heating and cooling cycle modulates the ML depth. There is considerable long timescale variability in both the strength and the location of the EUC itself. The net effect is a pronounced variability in shear generation of turbulence in the upper layers up to the depth of the EUC core.

Our understanding of the turbulent mixing processes in the global oceans has greatly increased due to microstructure measurements over the past two decades (Gregg, 1987; Peters *et al.,* 1988). Intensive observations of turbulent mixing at the equator during the Tropic Heat 1984 (Moum and Caldwell, 1985; Gregg *et al.,* 1985) and Tropic Heat 1987 (Moum *et al.,* 1992a,b; Hebert *et al.,* 1992; McPhaden and Peters, 1992) field experiments have greatly changed our concepts of mixing in the upper layers of the equatorial oceans. During Tropic Heat 1984 a strong diurnal variability was found in the turbulent dissipation rate in the upper layers of the equatorial Pacific at 140° W, *below* the surface mixed layer, with dissipation rates varying over two orders of magnitude from 10^{-8} to 10^{-6} W kg^{-1} (Peters *et al.,* 1988; Moum *et al.,* 1989). What was surprising was not the diurnal variability of the dissipation rate (a factor of 2 to 3) in the well-mixed surface layer 10–35 m deep (which is to be expected because of the modulation of mixing by strong solar heating during the day and convective mixing caused by cooling at night), but the variability deep in the stably stratified, but highly sheared, low Richardson number region above the core of the Equatorial Undercurrent, to depths of 80–90 m, and over two orders of magnitude. Enhanced dissipation occurred typically at night and persisted for several hours after sunrise. Mixing occurred in bursts lasting 2–3 hr, with overturning scales of typically 10 m in the vertical. The exact mechanism of periodic penetration of high turbulent dissipation deep into the thermocline during Tropic Heat was not clear, although it was speculated (Gregg *et al.,* 1985; Peters *et al.,* 1988; Moum *et al.,* 1989; Brainard *et al.,* 1994; Peters *et al.,* 1994) that the internal waves generated at the base of the convectively mixed nocturnal mixed layer were responsible for the sporadic bursts of turbulence and elevated nocturnal dissipation rates. Dillon *et al.* (1989) even suggested that such internal waves may be important to the large scale zonal momentum balance at the equator.

More observations in this area were made during Tropic Heat 1987 (Figure 2.7.1) (Peters *et al.*, 1991; McPhaden and Peters, 1992; Moum *et al.*, 1992a,b; Hebert *et al.*, 1992). While the conditions were much different during 1987 (measurements were made in April with weak easterly winds and a warm SST, whereas 1984 measurements were in November during strong easterlies and a colder SST), the same kind of diurnal variability was found (Figure 2.7.1). Moum *et al.* (1992a) found evidence for a diurnal cycle of internal wave activity from towed thermistor chain data, and Hebert *et al.* (1992) were able to intercept a mixing event (Figure 2.7.2) during the cruise that showed the strong dissipative character of the event. The dissipation rate during this event was three orders of magnitude above the background values. McPhaden and Peters (1992) observed a diurnal cycle in the variance of temperature and isotherm displacements in the moored time series at 140° W, with elevated variances during the night (Figure 2.7.3). They found pronounced variability at horizontal scales of 100–300 m, far larger than the mixed layer depth, and at periods of 10–30 cph, far larger than the local buoyancy frequency, which they attributed to Doppler shifting by the EUC. They suggested that their observations were consistent with internal waves generated at the base of the nocturnal mixed layer and propagating down into the thermocline, helping trigger Kelvin–Helmholtz shear instabilities and consequently turbulence and elevated dissipation rates.

Moum *et al.* (1992a) noted that the "internal wave" episodes occur only during periods of sustained westward winds and exclusively at night or during early morning hours (see, for example, Figure 2.7.4). This prompted them to postulate that these internal waves are generated either by convectively

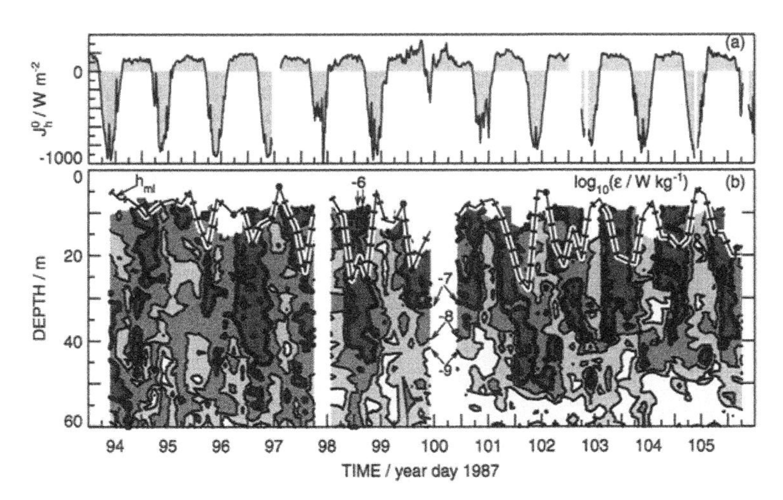

Figure 2.7.1. Diurnal fluctuations in the dissipation rate and the mixed layer depth at 140° W on the equator. Note the large diurnal modulations in the dissipation rate below the mixed layer that are correlated with nocturnal cooling (from Peters *et al.*, 1994).

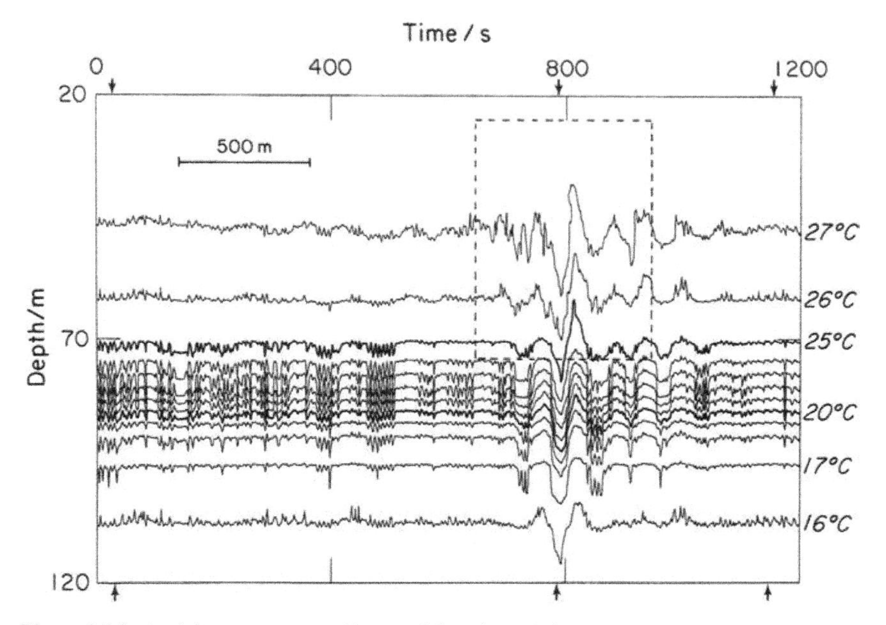

Figure 2.7.2. A mixing event captured in towed thermistor chain measurements during TH87 (from Hebert *et al.*, 1992).

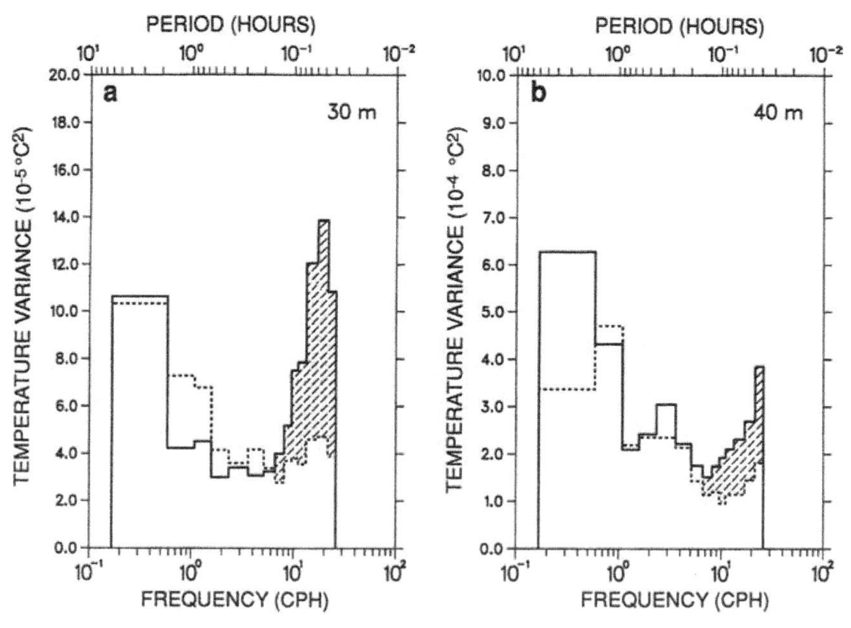

Figure 2.7.3. Elevated temperature (from McPhaden and Peters, 1992) and isotherm displacement variances (from Moum *et al.*, 1992) showing large diurnal variability.

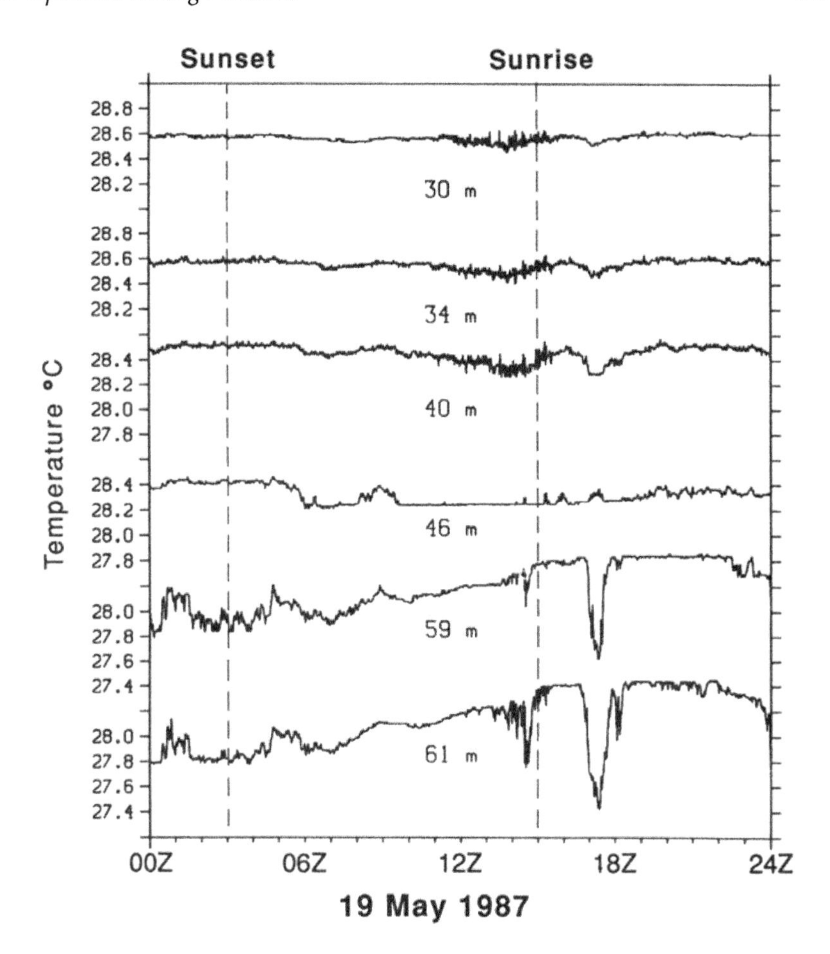

Figure 2.7.4. Moored temperature observations showing high-frequency internal wave activity in the early morning hours (from McPhaden and Peters, 1992).

generated eddies at the mixed layer base or by shear instabilities, propagate deep, and become unstable and break, generating the episodic high dissipation events. However, they did note that these waves could be forced modes resulting from shear instabilities. While there is definitely a link between the "internal wave" activity and the high dissipation rate events (Peters *et al.,* 1994), it is not clear that the latter result from the former. There is currently speculation about the cause of these enhanced dissipation rate events.

For instance, Clayson and Kantha (1999), using their one-dimensional mixed layer model (Kantha and Clayson, 1994), have reproduced deep penetration of dissipation by simply including the mixing from the vertical shear of the eastward-flowing EUC enhanced by, and interacting with, the shear stress due to westward (easterly) winds. It is present whether or not there is nocturnal cooling, thus suggesting that it has little to do with nocturnal cooling or internal waves below the mixed layer. The phenomenon is absent during eastward (westerlies) winds and then the behavior is more like high-latitude mixed layers. Their results for a simulation with an easterly wind stress of 0.2 Pa, a peak solar insulation of 1000 W m^{-2}, and a constant cooling rate of 200 W m^{-2} are shown in Figures 2.7.5 and 2.7.6. The mixed layer depth, determined as the depth at which the density increases from the surface value by 10^{-2} kg m^{-3}, consistent with the definition used by microstructure investigators (e.g., Peters *et al.*, 1994), is also shown. Periodic penetration of turbulent dissipation well below the OML is quite evident.

The mechanism that is responsible for generation of turbulence in this model at the base of the ML is also that responsible for the observed deep penetration. The turbulent shear stress acting on the mean velocity shear is the source of turbulent energy at the base of the ML and this causes elevated dissipation rates. The turbulence thus generated mixes the fluid at the base, entrains the quiescent fluid below, and advances slowly into deeper layers. This advance of turbulence into deeper parts of the water column is slow but steady, the rate depending on the ambient stratification and the strength of the turbulence source. The penetration is stopped by the strong stratification (and weak shear) present near the EUC core.

Shear generation of turbulence in the deep layers is a self-limiting process. Once the shear stress acting on local vertical shear generates mixing, this tends to mix the water column locally, reduce or eliminate the vertical shear, and therefore reduce the strength of the source of turbulence. When turbulence decays, diffusive processes tend to reestablish the shear in the water column and the whole process can repeat once again. It is important to remember that it is the shear stress acting against the vertical shear that is the source of turbulence and it is necessary to have both to produce mixing and elevate dissipation rates by shear generation. It is not surprising then to see some randomness superimposed in deeper layers even in the presence of steady forcing at the top.

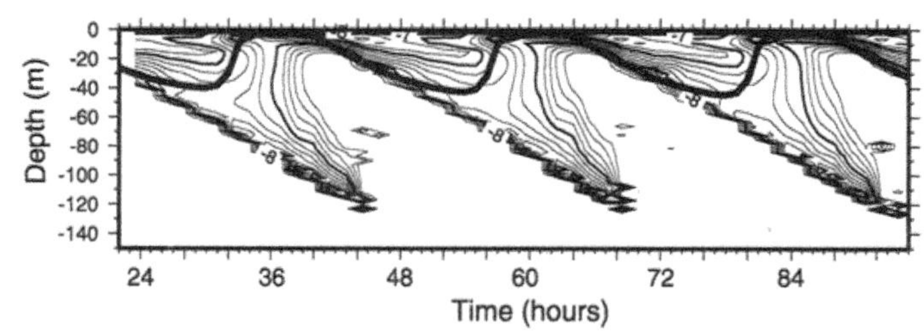

Figure 2.7.5. Diurnal penetration of elevated dissipation rates for easterly winds in one-dimensional simulations of Clayson and Kantha (1999). Note the wavelike progression to a 120-m depth, close to the EUC core. The mixed layer depth is also shown.

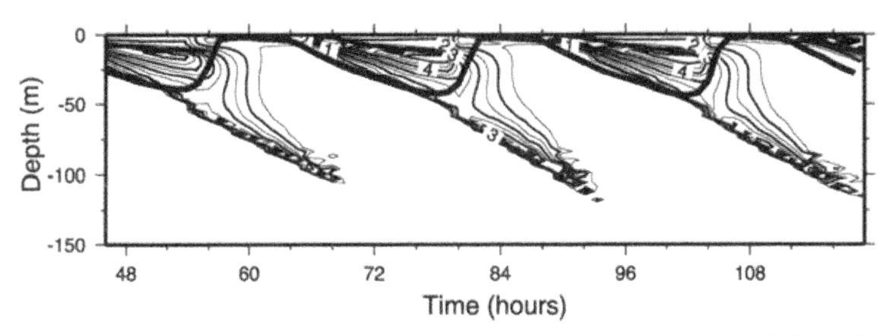

Figure 2.7.6. Diurnal modulation of TKE for easterly winds in one-dimensional simulations of Clayson and Kantha (1999). Note the wavelike progression to 120-m depth, close to the EUC core. The mixed layer depth is also shown.

These results can be compared to results obtained from a simulation with a westerly wind stress, in which the deep penetration of dissipation was not evident, suggesting that it is primarily the enhancement of EUC-induced shear by the easterly wind stress that causes this progressively deeper penetration. This is consistent with the findings of Skyllingstad and Denbo (1994), who in a numerical simulation of internal wave generation in the stratified, sheared region above the EUC core, using a nonhydrostatic model, discovered that when they switched the winds from westward to eastward, internal wave activity ceased within a day of the reversal. These are forced waves generated by shear instabilities in the water column and hence enhancement of EUC-induced shear by easterlies is crucial to their generation, unlike the free waves generated by turbulent eddies in the mixed layer, which are generated by both easterlies and westerlies.

Several other experiments in the tropical Pacific include the TOGA/COARE (TC) Intensive Observation Period (IOP) and the 1991 Tropical Instability Wave Experiment (TIWE) (Lien *et al.,* 1995). TOGA/COARE was a major international program (Webster and Lukas, 1992) conducted in the early 1990s to understand the dynamics and thermodynamics of the tropical western Pacific warm pool, central to the ENSO (El Niño–Southern Oscillation) process. Wijesekera and Gregg (1996) present microstructure measurements from R/V Moana Wave during the program and discuss various aspects such as the ML response to winds and rain squalls. Weller and Anderson (1997) discuss the variability of the ocean ML in this region as observed during TOGA/COARE.

The current and density structure of the ML in the tropics is also subject to changes in wind speed, leading to equatorially trapped waves, which play a role in modulating thermocline depth. One such burst in westerly winds in the western tropical Pacific is discussed in detail by McPhaden *et al.* (1988). During this period, winds as high as 10 m s^{-1} lasted for 10 days, and in their wake, the SST dropped by 0.3–0.4°C and the South Equatorial Current accelerated to speeds exceeding 1 m s^{-1}. These westerly wind bursts generate eastward-propagating equatorial Kelvin waves that are responsible for redistributing the water masses (e.g., see McPhaden *et al.*, 1992) and causing profound changes in the coupled ocean–atmosphere system in the tropical Pacific.

As mentioned above, rainfall also plays a role in the upper ML in the tropics. Moum and Caldwell (1994; see also Smyth *et al.*, 1996b) describe the formation and evolution of a rainwater lens during the TC IOP. An intense rain event can depress the salinity in the upper few meters by 0.2–0.4 psu for several hours, but this signature is quickly erased if winds are strong enough and the rain water rapidly mixed into the water column. Only if the winds are sustainedly weak do the rain waters tend to form a halocline that can give rise to a barrier layer below. Barrier layers separating the deep thermocline from a shallower halocline are a salient feature of the western Pacific warmpool (Lucas and Lindstrom, 1991) and have an important effect on the air-sea exchange in the tropical Pacific.

2.8 MIXING UNDER SEA-ICE COVER

Although sea ice covers a small fraction of the global oceans (7% on the average), it is an important influence on the global climate. There are two factors responsible for the inordinately large role (relative to its size) that sea ice plays: (1) The sea-ice cover insulates the oceans from the bitterly cold winter atmosphere and reduces the heat loss from the oceans by two orders of magnitude. (2) Because of its much higher albedo compared to open water, it provides a positive feedback to any climatic cooling or warming. The latter makes it a sensitive indicator of any small but steady changes in the environmental conditions. In fact, most GCM simulations of anthropogenic carbon dioxide-induced global warming indicate much larger changes in the polar regions than in lower latitudes. There are indications from recent expeditions that there have been some significant changes in the ice cover in the Arctic in the 1990's. Whether the thinning of the ice cover and decrease in its extent are due to natural variability in climate or some other causes is hard to say. In any case, the thermodynamics and dynamics of sea ice are closely coupled to those of the ocean, in particular, the OML underneath.

The OML under ice is in some ways both easier and harder to study than the OML without the sea-ice cover. It is easier to observe, since the large ice floes that form the interior ice pack provide a stable platform for the deployment of instrumentation to more accurately measure its various properties (McPhee and Smith, 1976; McPhee, 1980, 1986, 1992, 1994). The near absence of surface waves makes it much simpler to observe and study. In fact, well-defined Ekman spirals (Figure 2.8.1) are found more readily under ice (e.g., Hunkins, 1975) than in open water, because of the complications due to other influences such as waves and Langmuir circulations in the latter case (Price *et al.*, 1985). Similarities to the ABL can be exploited in scaling observational data (McPhee and Smith, 1976; McPhee, 1994) and forming theories (McPhee, 1991). However,

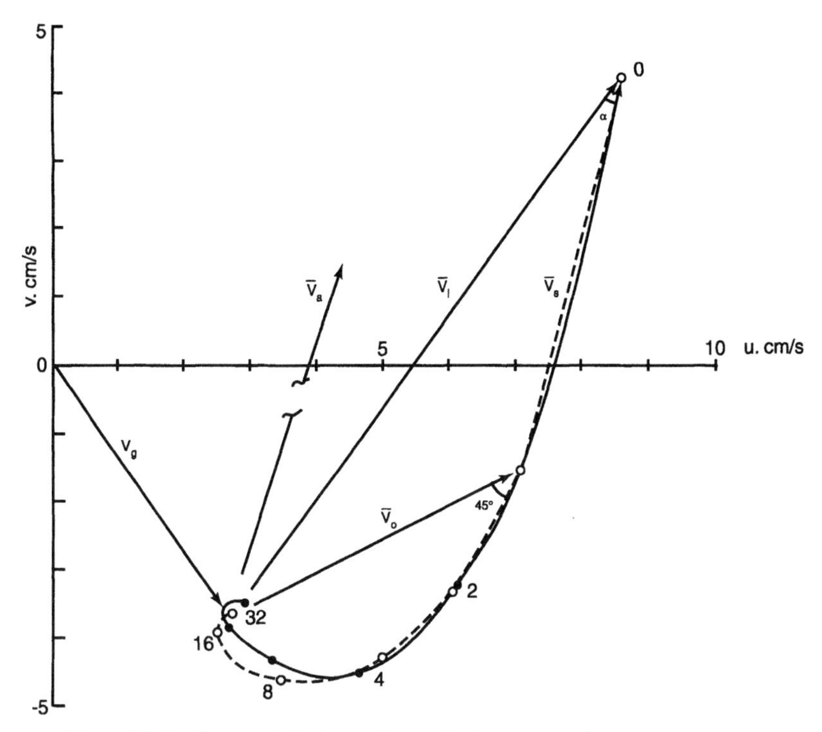

Figure 2.8.1. Ekman spiral observed in the OML under ice (from Hunkins, 1975).

the mediation of sea ice in the exchange of heat and momentum with the atmosphere complicates the picture. The presence of leads and deep ice pressure keels produce inhomogeneities that compound the difficulty. Long polar nights and persistent cloudiness hamper any remote sensing other than that based on microwaves. The harsh ambient environment and its remoteness render observations costly and even hazardous. Nevertheless, its importance has made the ice-covered ocean the subject of intensive study through programs such as the Arctic Ice Dynamics Joint Experiment (AIDJEX) in the 1970s, the Marginal Ice Zone Experiment (MIZEX, covered in special sections of the *Journal of Geophysical Research* **88,** C5, 1983; **92,** C7, 1987; and **96,** C3, 1991) and the Coordinated Eastern Arctic Experiment (CEAREX) in the 1980s, the Leads Experiment (LEADEX) in the early 1990s, and the Surface Heat Budget of the Arctic (SHEBA). Sea-ice-covered oceans have fascinated many for decades, and in fact, it is the observations by Nansen of the drift of his ice-bound ship Fram and the ice pack consistently 20–40° to the right of the wind that provided the inspiration to Ekman (1905) to formulate his theory of wind-driven currents that forms a basis for our understanding of wind-driven ocean circulation (Pedlosky, 1996).

The stress transmitted by sea ice from the winds to the water depends on the prevailing momentum balance on the ice floe. Figure 2.8.2 shows the typical steady-state balance of forces in the central Arctic (Hibler, 1979). The interior ice pack in the Arctic consists of large ice floes, a few kilometers to tens of kilometers in size. Internal ice stresses form an important part of this momentum balance. The presence of sea ice can either enhance or diminish the stress applied by the wind to the water. Most of the momentum flux is absorbed by ice and little is transmitted to the water if internal ice stresses are large. Close to the shore (for example, in landfast ice), internal stresses are capable of balancing the wind stress entirely so that no ice motions or momentum transfer to water results. Close to the edge of the ice pack, near the marginal ice zone (MIZ), the ice pack consists of small floes that are pretty much in free drift, where the principal balance is between the wind stress, the water stress, and the Coriolis force,

$$\vec{\tau}_{IO} = \vec{\tau}_{AI} - D_I f \vec{U}_I; \vec{\tau}_{AI} = \rho_A c_{dAI} \left| \vec{U}_A - \vec{U}_I \right| \left(\vec{U}_A - \vec{U}_I \right) \tag{2.8.1}$$

where the subscripts A, O, and I refer to the atmosphere, the ocean, and the ice, and D_I is the mass per unit area of ice. There have been extensive measurements of c_{dAI} (e.g., Overland, 1985; see Brown, 1990) that show that the drag coefficient c_{dAI} depends very much on the roughness of the ice surface. Ice is effectively much smoother in the interior of the ice pack ($c_{dAI} \sim 0.002$) than toward the ice edge, where rafting processes can make the ice surface appear much rougher to the atmosphere ($c_{dAI} \sim 0.004$). Nevertheless, because of the roughness of ice, in the interior of the ice pack due to pressure ridges, and near the ice edge due to rafting and the presence of large areas of open water, the wind stress is usually larger than it would be in the absence of ice cover. The stress applied to the water is also larger as a result.

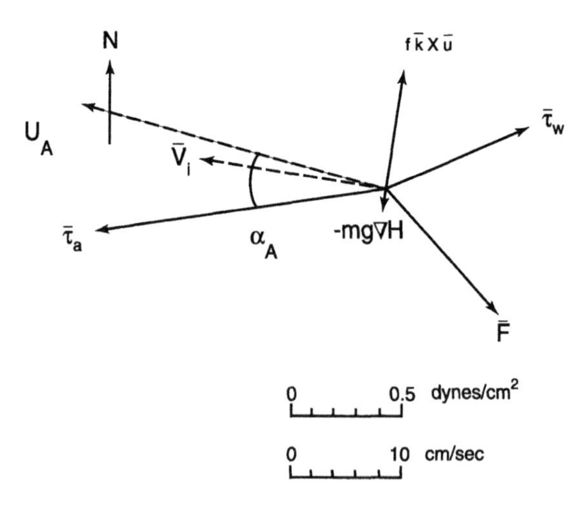

Figure 2.8.2. Stress balance in sea ice. Note the importance of internal stress $\vec{F}$.

Another complication in ML dynamics under ice is that due to the presence of either open water or thin ice in linear rift zones in ice called leads (Kantha, 1995b). While the ice insulates the OML from excessive heat loss during winter, its high albedo makes it a less efficient absorber of solar heating during summer. In fact, a major fraction of the summer heating (Maykut and McPhee, 1995) and winter cooling (Morison *et al.,* 1992) of the ice-covered OML occurs through open water in leads surrounding pack ice. During winter, leads (which compose only 1–2% of the pack ice in area) account for nearly half of the heat loss from the Arctic Ocean to the atmosphere above. During summer, solar heating of the open water and thin ice areas, which are now much higher in percentage, provides the energy to melt the ice laterally. This SW radiation entering the ice pack, which can reach peak values of 50 W m^{-2}, is thought to constitute the bulk of the ocean-to-ice heat flux (annual average of 5 W m^{-2}) in the Arctic. Without this heat flux, the ice cover in the Arctic would be much thicker than the prevailing average of 3 m, since the presence of a strong halocline at an average depth of 50 m prevents the warmer but deeper north Atlantic waters underneath (Figure 2.8.3) from providing a much larger heat flux. The situation is much different in the sea ice surrounding the Antarctic, where the ice is much thinner since mixing processes enable the OML underneath to provide large amounts of heat directly to the ice. A rough balance between the heat flux from the ocean and transfer through ice out to the atmosphere keeps the ice thickness around 0.5–0.8 m.

Ice growth and decay also play an important role in mixing. Formation of ice extrudes concentrated brine from the sea ice formed, and this provides the buoyancy flux to drive free convection below the growing ice. Since the water is close to the freezing point at ambient salinity underneath the ice (average of –1.7°C), the heat loss to the atmosphere does not cool the OML much; instead, this heat loss is provided by phase conversion. Fresh ice forms rapidly during winter in the leads and polynyas around the ice pack, and the resulting extrusion of salt provides the energy for mixing. Conversely, during spring–summer heating, the melting of ice laterally from heat absorbed in the surrounding open water, and the water running down from the melt ponds formed on the ice surface, provides the freshwater flux that tends to inhibit mixing in the OML underneath. Therefore, the surface heat and freshwater balances are also an integral part of the ML dynamics in ice-covered seas.

The morphology of sea ice also plays a role. The interior of the ice pack in the Arctic consists of pressure ridges and rubble created by collision between floes, and these increase the roughness of the otherwise smooth ice surface. The accompanying deep pressure keels provide an efficient stirring mechanism when the ice pack is in motion. They are also a source of internal waves. Toward the edge of the ice pack, in the MIZ, melting ice creates a strong but shallow halocline that confines any momentum transferred by ice to the water. Consequently, the "slippery-water" effect causes the pack ice to move rapidly in response to applied wind stress. Internal waves generated by ice motion also account for part of the momentum balance (McPhee and Kantha, 1989).

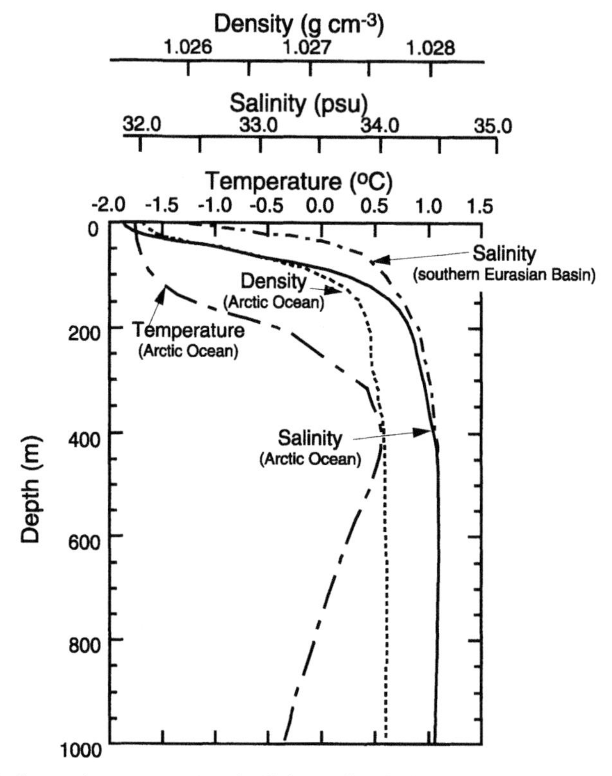

Figure 2.8.3. Composite temperature and salinity profiles in the Arctic. Note the warm temperatures at 300- to 400-m depths (from Barry *et al.*, 1993).

Wintertime convection in leads is not only central to the air–sea exchange but also to the ML physics (Figure 2.8.4). Leads are by their very nature transient (Smith *et al.,* 1990; Morison *et al.,* 1992; Kantha, 1995a), with timescales of interest on the order of hours to days. They are created by the divergence in ice, often storm-induced, but rapid ice growth or convergence of floes closes them quickly. Growth rates of 12–15 cm day^{-1} have been observed during LEADEX (see special section on leads and polynyas in the *Journal of Geophysical Research* **100,** C3, 1995). Brine expulsion from the lead ice causes vigorous convection in the ML underneath that has been modeled by Kantha (1995a), Smith and Morison (1993, 1998), Potts (1998) and others.

The large scale circulation also plays a role in ML dynamics. The large scale ice motion, and hence, to some extent, the upper layer circulation in the Arctic driven by prevailing winds, consists of the anticyclonic Beaufort Gyre in the Canadian Basin and the broad Transpolar Drift Stream (TDS) in the Eurasian Basin (Figure 2.8.5). Typical average velocities are 3–5 cm s^{-1}, although the velocities in the TDS can exceed 10 cm s^{-1} around the Fram Strait. The Arctic

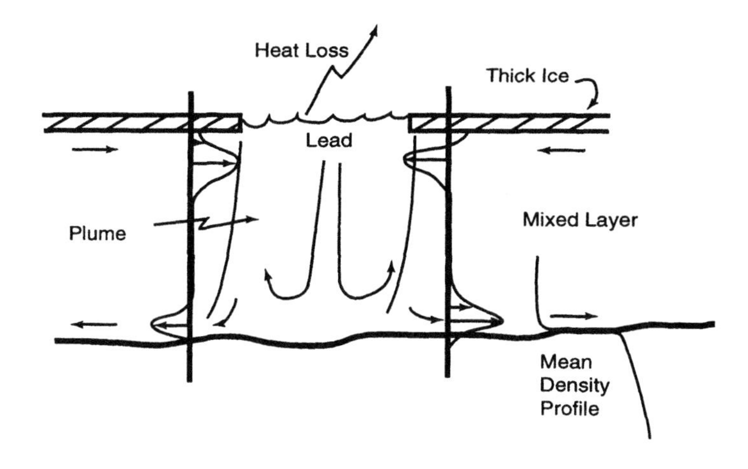

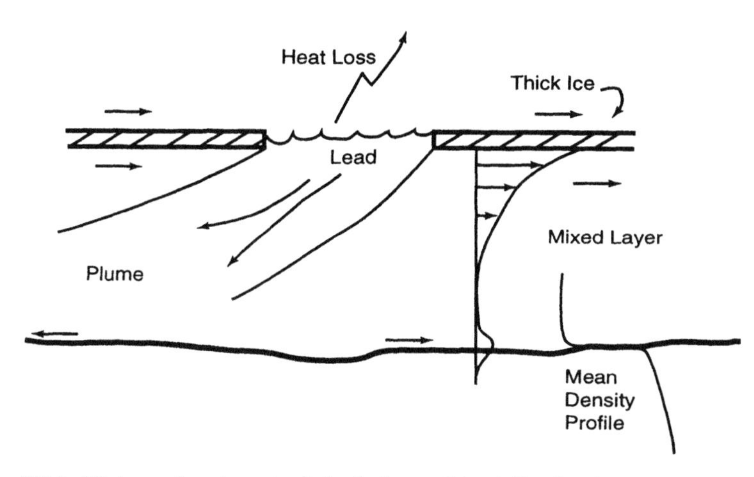

Figure 2.8.4. Mixing under winter Arctic leads for small ice drift velocities (top) and strong ice drifts (bottom). Free convection induced by salt extrusion dominates in the former.

exports 0.1–0.2 Sv of ice through the Fram Strait. Sea ice is created during previous winters, on the shallow shelves around the Arctic. Roughly 15% of the inflow of water into the Arctic, about 1 Sv, occurs through the Bering Strait (45 m deep, 85 km wide), and outflow occurs through the Canadian Archipelago and Fram Strait. The major influence on the basin is the large inflow (roughly 75% of the total) of relatively warm (~1°C) North Atlantic waters at depths of a few hundred meters (Figure 2.8.3) through the deep (2600-m deep) Fram Strait. Slow vertical diffusion of heat from this flow limits ice growth in the Arctic. However, the strong cold halocline immediately below the upper ML (25–50 m

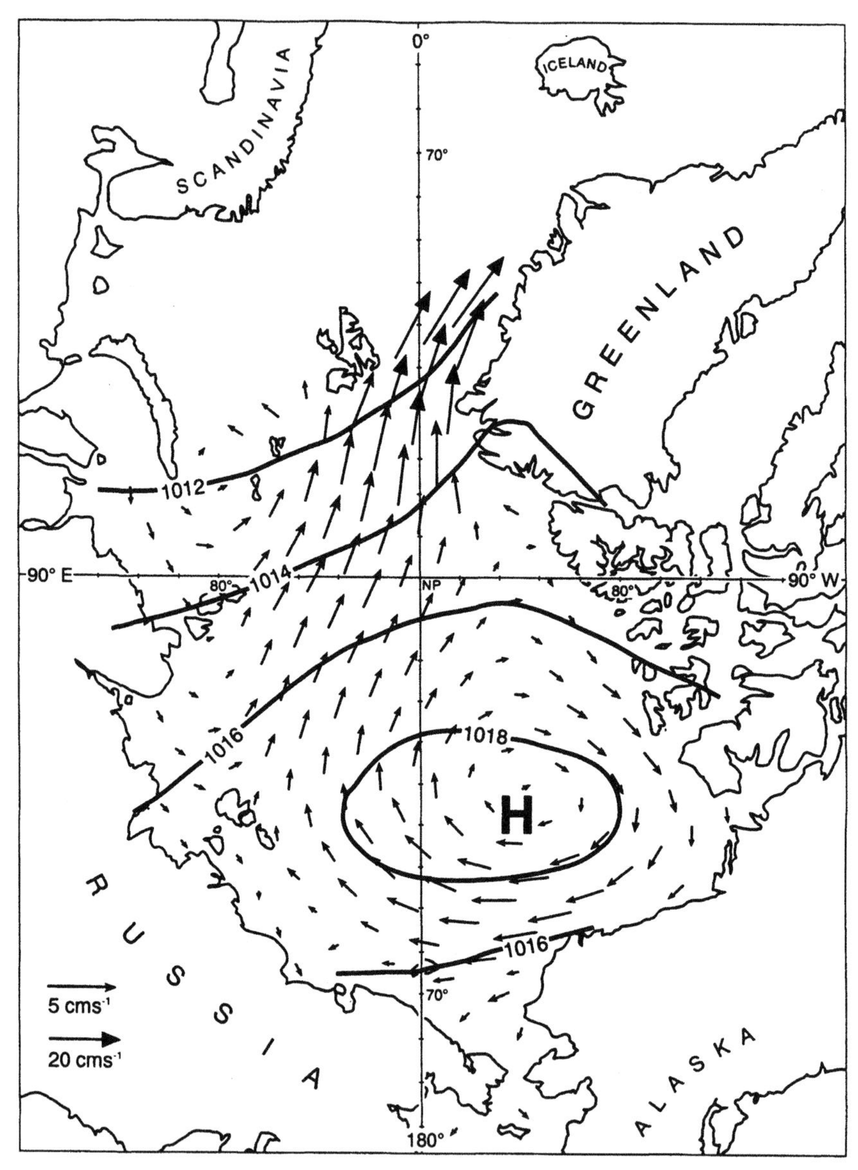

Figure 2.8.5. Ice circulation in the Arctic. Note the Beaufort Gyre in the Canadian Basin and the transpolar drift stream (from Barry *et al.*, 1993).

thick) isolates the warm Atlantic waters from contact with ice and keeps the central Arctic perennially ice-covered. The formation mechanism for this layer is uncertain, but the wintertime dense water formation on Siberian shelves and the Bering Strait inflow might be responsible. The freshwater flow from the large rivers in Siberia (~0.1 Sv) also plays a role in the circulation (Carmack, 1990). Holland *et al.* (1996) have performed realistic simulations of the coupled ice-ML in the Arctic, using a coarse resolution isopycnal ocean model (Oberhuber, 1993a,b) coupled to a realistic sea ice model (Holland *et al.,* 1993), and conclude that the near-surface salinity controlled by ice melt/growth, evaporation/precipitation, and river runoff is the key factor in controlling the circulation in the Arctic ML. In the Antarctic, the presence of the strong Circumpolar Current stream also plays an important role in sea-ice cover.

Sea ice in polar and subpolar oceans advances and retreats in harmony with the seasons. During the peak of winter, the entire Arctic Ocean is covered by ice, as well as parts of marginal seas, such as the Bering Sea, The Sea of Okhotsk, Hudson Bay, the Labrador Sea, the Greenland Sea, and even the Baltic Sea and the Sea of Japan (Barry *et al.,* 1993). During summer, all the marginal seas and most of the shelves around the Arctic are ice-free (Figure 2.8.6). The ice area nearly doubles from boreal summer to boreal winter (from 8 to 15 million km^2). Around Antarctica also, sea ice quintuples from astral summer to astral winter (from 4 to 21 million km^2). Roughly 7% of the global ocean (25 million km^2) is periodically covered and uncovered by sea ice, more than two-thirds of it in the Southern Ocean around the Antarctic (Niebauer, 1991). There are interannual variations, as well, of a few million km^2.

The dynamic edge of the ice pack is the MIZ, and it is one of the regions of pronounced variability in the oceans, as well as a highly biologically productive one. The ice pack in the MIZ responds vigorously to winds and a large fraction of ice created and destroyed during the year occurs in the MIZ. The biological productivity is quite low in the polar pack ice, but increases dramatically toward and in the MIZ, where spring blooms driven by melt-water stratification are common. The broad shelf in the Bering Sea, in particular, is characterized by nutrient-rich, well-mixed waters with low phytoplankton concentrations and ice cover during winter. As the ice begins to melt and retreat during spring, the restratification, combined with abundant nutrients and sunlight, and the low grazing pressure characteristic of cold waters, causes a dramatic increase in biological productivity that consumes most of the nutrients, thus limiting the bloom to a few weeks in spring. Most of this new production ends up on the benthos (Niebauer, 1991). In contrast, the nutrient concentrations remain high even after the bloom in regions like the Weddell Sea around Antarctica, due partly to the steady upwelling of deep nutrient-rich waters.

The Siberian shelves around the Arctic are not only responsible for most of the first-year ice in the Arctic but are also regions of dense intermediate water formation. During winter, winds often blow ice away from the coast, exposing water to cold winter air. The resulting temperature differences of as much as 30–

Figure 2.8.6. Extent of sea-ice cover in the Arctic during the peaks of summer and winter (from Barry *et al.*, 1993).

40°C drive a large heat loss, ice production, and brine extrusion. The intermediate water so formed is thought to sink to the bottom of the shelf and spread around the basin, forming a strong cold halocine at the bottom of the Arctic ML.

The Greenland and Labrador seas in the northern hemisphere and the Weddell Sea in the southern hemisphere are regions of formation of dense, deep waters central to the long timescale thermohaline circulation in the global oceans. Observational studies indicate approximately 1.0, 3.9, and 2.4 Sv as the current best estimates for the amount of dense water formed each year in the Greenland, Labrador, and Weddell seas, respectively. Strong wintertime storms cause deep convection in cyclonic gyres that create deep mixed layers that often exceed 1000 m in depth. Based on LES studies, Denbo and Skyllingstad (1996) suggest that thermobaric effects (see Chapter 8) and rotation play an important role in the deep penetrative convection processes occurring in these seas.

From the above discussion, it is clear that mixing under sea-ice cover covers a whole gamut of processes that require proper consideration of sea-ice dynamics and thermodynamics. It is the ice motion and the net heat loss (gain) at the sea–ice/air–sea interface that drive the circulation in the ML, the latter through brine (melt water) formation. The dynamical and thermodynamic coupling to ice distinguishes ML modeling in ice-covered seas from that in open water. Examples can be found in Parkinson and Washington (1979), Hibler and Bryan (1979), Kantha and Mellor (1989b), Mellor and Kantha (1989), Hakkinen *et al.* (1992), Hakkinen and Mellor (1990, 1994), Kantha (1995a), and Holland *et al.* (1996). Recent reviews of ice–ocean coupled models can be found in Hakkinen (1990) and Hakkinen and Mellor (1994).

The sea ice can be considered to be a continuum governed by the equations

$$\frac{\partial A_I}{\partial t} + \frac{\partial}{\partial x_k}(U_{Ik} A_k) = \frac{\rho_w A_I}{\rho_I D_I}(So_A - Si_A)$$
$$\frac{\partial D_I}{\partial t} + \frac{\partial}{\partial x_k}(U_{Ik} D_k) = \frac{\rho_w}{\rho_I}(So_D - Si_D)$$
$$\frac{\partial (D_I U_{Ij})}{\partial t} + \frac{\partial}{\partial x_k}(D_I U_{Ik} U_{Ij}) - \varepsilon_{ijk} D_I f_i U_{Ik} = D_I g \frac{\partial \zeta}{\partial x_j} + \frac{1}{\rho_I}\frac{\partial \sigma_{ij}}{\partial x_i} + \frac{A_I}{\rho_I}(\tau_{AIj} - \tau_{IOj}) \tag{2.8.2}$$

where A_I, D_I, and U_I are the area concentration (fractional area covered by ice), mass per unit area, and velocity of ice. Note the source and sink terms in the equations of conservation of area fraction and mass of ice. These terms are dependent on the thermodynamics of air–sea and ice–sea interactions; they parameterize the growth and decay of ice in terms of the net heat loss or gain (Hibler, 1979; Mellor and Kantha, 1989). Note also the second term on the RHS of the momentum conservation equation; this term represents the internal ice stresses. It is modeled traditionally by considering the ice to be a viscous-plastic continuum (Hibler, 1979; see also Flato and Hibler, 1996). The first term is the pressure gradient due to sea level slope. The solution of Eq. (2.8.2) coupled to the ML equations provides the necessary fluxes of momentum, heat, and salt needed to solve for the evolution of the OML. The momentum flux from ice to the water is the simplest of these to address. Tables 1 and 2 show the measurements of roughness length z_0 of the undersurface of ice (from McPhee, 1990) and c_{d1}, the drag coefficient referred to a level 1 m from the ice (Shirasawa and Ingram, 1991). The two are related to each other by

$$z_0 = z_1 \exp\left(-\kappa c_d^{-1/2}\right); z_1 = 1\,\mathrm{m} \tag{2.8.3}$$

The parameterization of scalar fluxes at the rough ice–ocean interface requires careful consideration to the disparity in roughness scales for momentum, heat, and salt (see also Chapter 4). While momentum can be transmitted by pressure forces, no similar mechanism exists for scalar transfers, and therefore the heat and salt fluxes at the ice–ocean interface must pass through a thin molecular sublayer on the ice underside. This leads to the peculiar situation where the momentum exchange is independent of molecular properties of the fluid; heat and mass exchanges are not, even in the asymptotic limit of infinite Reynolds numbers. The heat and salt fluxes at the ice–ocean interface can be written as (McPhee *et al.*, 1987; Mellor and Kantha, 1989)

$$-\overline{wq} = w_0 L_F + H_c ; H_c = -k_I \frac{dT}{dz}$$
$$-\overline{ws} = (w_0 + w_p)(S_0 - S_I) \tag{2.8.4}$$

where w_0 is the freezing (if negative, but melting if positive) rate at the ice underside, w_p is the rate of percolation of water from above the interface, k_I is the thermal conductivity of ice divided by the water density and specific heat at constant pressure of water, and L_F is the latent heat of fusion of ice divided by the specific heat of water. S_0 is the salinity at the interface and S_I is the ice salinity. Equation (2.8.4) expresses the balance between the turbulent heat flux from the ocean through the thin ice–ocean interface and the heat conducted out through the ice and the heat flux due to any phase change such as freezing or melting. It also expresses the salt flux due to melting or freezing of ice at the interface. The problem now is to determine w_0 in terms of known properties in the OML. Writing

$$\overline{w\theta}, \overline{ws} = -k_{H,S} \frac{\partial}{\partial z}(T, S) \tag{2.8.5}$$

and integrating Eq. (2.8.4),

$$\frac{u_* \left[T(z) - T_0\right]}{w_0 L_F + H_c} = \Phi_T = \int_z^0 \frac{u_*}{k_H} dz'$$
$$\frac{u_* \left[S(z) - S_0\right]}{(w_0 + w_p)(S_0 - S_I)} = \Phi_S = \int_z^0 \frac{u_*}{k_S} dz' \tag{2.8.6}$$

where $k_{H,S}$ are heat and salt diffusivities, principally turbulent values, and u_* is the friction velocity at the interface. Note that T_0 and S_0 are the temperature and salinity values at the interface related to each other by the equation for the freezing line ($T_0 = -m\, S_0$; $m \sim 0.054$). Using this and Eqs. (2.8.4) to (2.8.6), one can solve for S_0 (from McPhee, 1990):

$$\begin{aligned} & mS_0^2+\left\{\left[T(z)-(\Phi_T/u_*)H_c\right]+\left[1+(\Phi_S w_p/u_*)\right](\Phi_T H_c/\Phi_S)-mS_I\right\}S_0 \\ & -\left\{\left[T(z)-(\Phi_T/u_*)H_c\right]+(\Phi_T H_c/\Phi_S)\left[S(z)+(\Phi_S w_p/u_*)S_I\right]\right\}=0 \end{aligned} \tag{2.8.7}$$

The ablation velocity is given by

$$w_0=\left(\frac{S(z)-S_0}{\Phi_S(S_0-S_I)}\right)u_*-w_p \tag{2.8.8}$$

From laboratory measurements of Yaglom and Kader (1974), it is possible to write (e.g., Mellor and Kantha, 1989)

$$\Phi_{T,S}=\Phi_{turb}+b\left(\frac{u_* z_0}{\nu}\right)^{1/2}\left(\frac{\nu}{k_{T,S}}\right)^{2/3} \tag{2.8.9}$$

where b ~ 3.2. To within a few percent accuracy, it is possible to ignore the turbulent contribution, since the second term, the change in temperature and salinity across the molecular sublayer, is the most dominant. From measurements under ice, McPhee (1987) suggests b ~ 1.6. The above equations can be used to compute the ablation/freezing rate of ice.

A simple instructive analytical model for the OML underneath ice is possible if some simplifications can be accepted (see McPhee, 1986, 1990, 1991, 1992). Consider a steady-state horizontally homogeneous OML (the same as Ekman considered) with coordinates fixed to the moving ice. Removing the $\partial/\partial t$ terms in the momentum equations filters out inertial oscillations [which are quite vigorous, especially when the ice is in free drift and internal ice stresses are negligible—see Figure 2.8.7, from McPhee (1986)] and so we are considering motions of much longer timescales. Using complex variables, the momentum equation becomes

$$\begin{aligned} & i\hat{U}=\frac{\partial\hat{\tau}}{\partial\zeta} \\ & \hat{U}=\frac{fD}{u_*}\frac{\vec{U}}{u_*},\ \hat{\tau}=\frac{\vec{\tau}}{u_*^2},\ \zeta=\frac{z}{D};\ \vec{U}=U_1+iU_2;\ \vec{\tau}=\tau_1+i\tau_2 \end{aligned} \tag{2.8.10}$$

Using the eddy viscosity approach

$$\hat{\tau}=k_m^*\frac{\partial\hat{U}}{\partial\zeta};\ k_m^*=\frac{k_m}{fD^2} \tag{2.8.11}$$

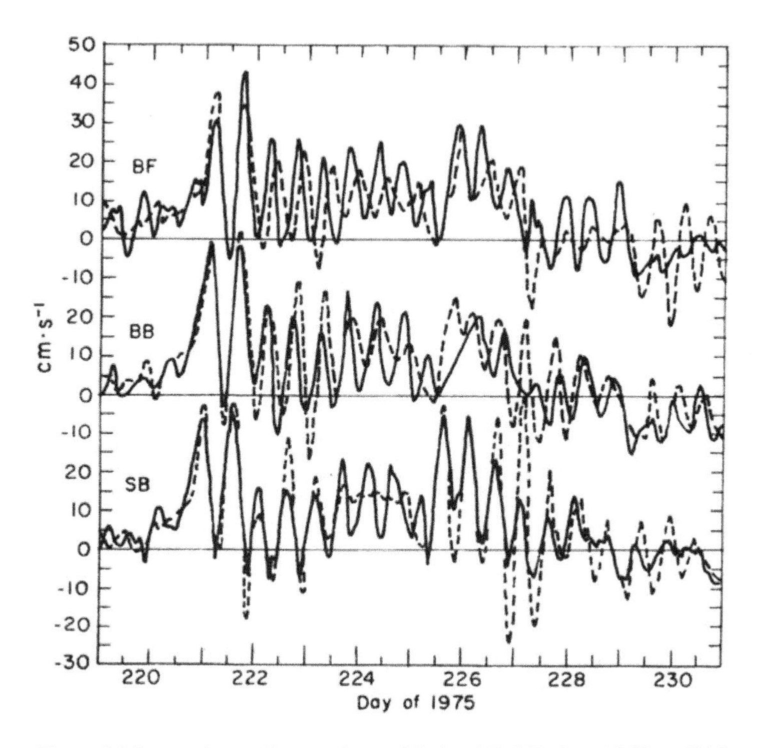

Figure 2.8.7. Inertial oscillations observed during AIDJEX (from McPhee, 1986).

and differentiating the momentum equation,

$$i\frac{\hat{\tau}}{k_m^*} = \frac{\partial^2 \hat{\tau}}{\partial \zeta^2} \quad (2.8.12)$$

The question is what is the functional form of $k_m^*(\zeta)$? Only for simple forms is the analytical solution possible. Now, there are four relevant length scales in the problem: z, $\underline{D}$, u_*/f (the neutral PBL scale) and L, the Monin–Obukhoff scale $[L = u_*^3/(\kappa\ \overline{wb})]$ that is indicative of the stratification and the scaling length in the constant flux layer (see Chapter 4). The turbulence scale in the inner constant flux layer is $l \sim z$, and in the bulk of the ML under neutral stratification, $l \sim u_*/f$. Now, for stable stratification, McPhee argues that L sets the scale, $l \sim Ri_c L$. A simple scaling that is valid for both is

$$l = a\,(u_*/f)\left[1 + \frac{a(u_*/f)}{Ri_c L}\right]^{-1} \;;\; a\sim 0.05,\ Ri_c \sim 0.2 \quad (2.8.13)$$

If one assumes k^*_m is a constant (= κa) irrespective of stratification and location in the OML, Eq. (2.8.6) can be readily integrated under the condition $\hat{\tau}(0)=1; \hat{\tau} \to 0, \hat{U} \to 0$ as $\zeta \to -\infty$:

$$\hat{\tau} = \exp\left(\frac{\zeta}{(k_m^*/i)^{1/2}}\right)$$
$$\hat{U} = (ik_m^*)^{-1/2} \exp\left(\frac{\zeta}{(k_m^*/i)^{1/2}}\right) \tag{2.8.14}$$

The solution is shown as hodographs in Figure 2.8.8, from McPhee (1986). Note that the implicit assumption here is that the depth scale D for the OML is proportional to the geometric mean of l and u_*/f: $D = [l\,(au_*/f)]^{1/2}/a$. Note that the e-folding scale of the stress in Eq. (2.8.8) is $(\kappa a)^{1/2}D$; for neutral stratification, this is equal to $0.14(u_*/f)$. Note also that since $i^{1/2} = \exp(i\pi/4)$, we get the classic 45° for the angle between the stress and velocity vectors at the ice surface. This is typical of all theories that postulate a constant eddy viscosity in the PBL. Strictly speaking, this is not correct and observations show that this angle varies between 20° and 40°. Nevertheless, this simple model extends the similarity approach and integrates the stratified PBL into the Ekman theory and appears to often provide an approximate fit to observations, for example, drifting station observations in the Weddell Sea [Figure 2.8.9, from McPhee and Martinson (1994)]. Note that the approach fails for free convection since then the length scale in the OML is the ML depth itself and u_*/f is irrelevant in this limit (see Chapter 3). Measurements presented in McPhee (1994) during LEADEX suggests that l increases to a maximum of κD_{ML} in the convectively driven OML.

Ice melt creates shallow MLs conducive to internal wave (IW) generation by moving ice floes. The IWs radiate energy away and therefore represent a sink of momentum flux and hence additional drag on sea ice. By characterizing the underside relief of ice by a peak wavenumber k_0 and amplitude h_0, and considering an idealized stratification consisting of a buoyancy jump Δb at the ML base followed by uniform stratification with buoyancy frequency N, McPhee and Kantha (1989) were able to calculate the IW drag on ice. The IW drag coefficient is given by

$$c_{dIW} = \frac{\tau_{IW}}{\rho_w U_I^2} = \frac{1}{2}(k_0 h_0)^2 \left[R_N^2 - 1\right]^{1/2}$$
$$\left\{\sinh^2(k_0 D)\left[\left(\coth(k_0 D) - R_b\right)^2 + R_N^2 - 1\right]\right\}^{-1} \tag{2.8.15}$$
$$R_N = \frac{N}{U_I k_0}; R_b = \frac{\Delta b}{k_0 U_I^2}$$

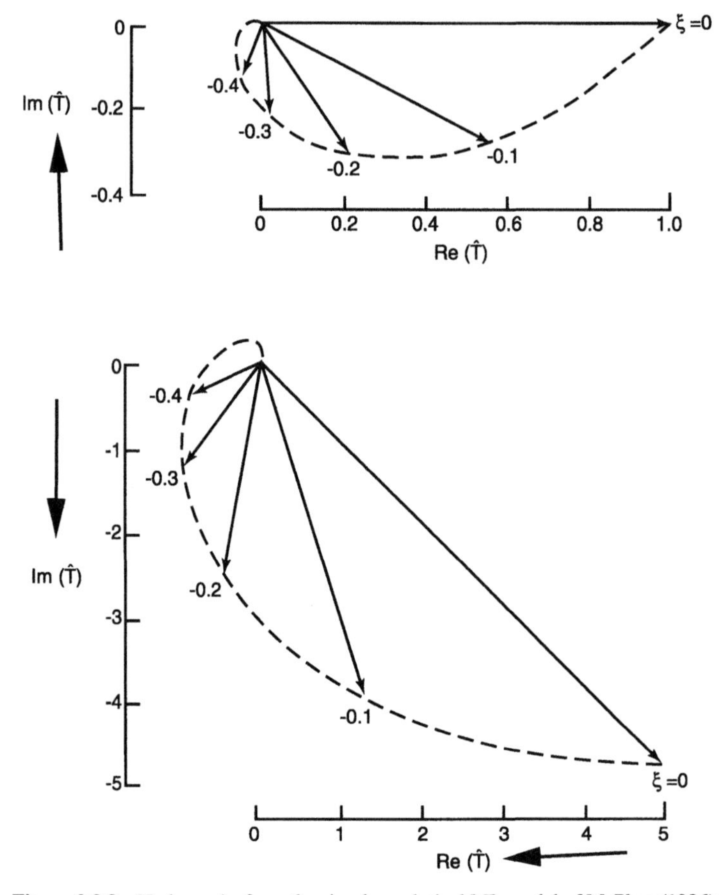

Figure 2.8.8. Hodographs from the simple analytical ML model of McPhee (1986).

where D is the ML depth, and R_N and R_b are parameters indicative of the stratification in the water column. The curly bracket term is the attenuation factor that accounts for the ML depth and the buoyancy change at its base. It decreases rapidly with the increasing ML depth, so that for D values greater than say 10 m, the IW drag is small. However, the ML depth is often much smaller during rapid melting at the ice edge and IW drag cannot be ignored. This attenuation factor multiplies the term that indicates the drag that would be felt if the stable stratification extended all the way to the ice undersurface, without a strong pycnocline intervening. Equation (2.8.4) can also be generalized to a spectrum of underice relief (McPhee and Kantha, 1989). This parameterization explained the otherwise anomalous ice-drift behavior in the Greenland Sea MIZ (McPhee and Kantha, 1989; Morison *et al.*, 1987).

Figure 2.8.10 shows an example of application of an ice–ocean coupled

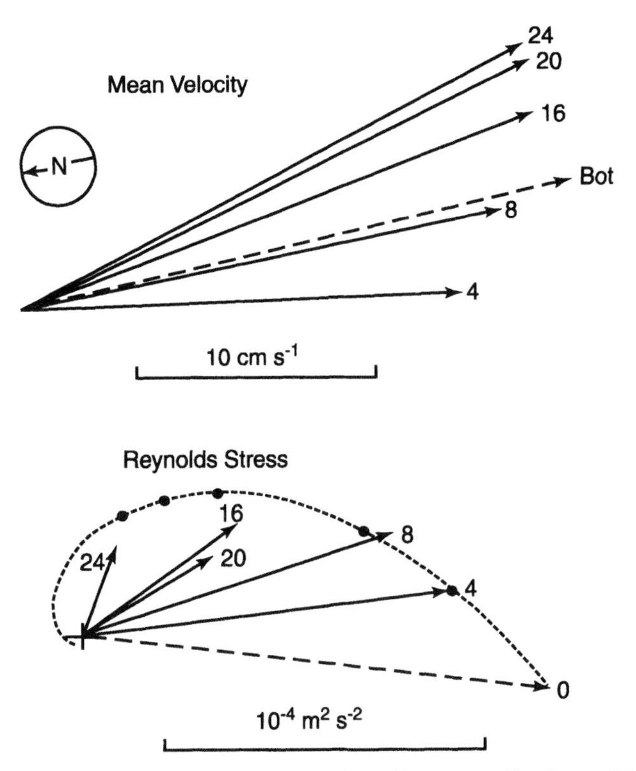

Figure 2.8.9. Observed mean velocity (top) and Reynolds stress profiles (bottom) in the Weddell Sea (from McPhee and Martinson, 1994).

model to determine the structure of the ML in the Bering Sea MIZ during winter (Kantha and Mellor, 1989). Sea ice is created in the northern sections of the Bering Sea and transported southward by winds, where it encounters warm (2–3°C) Bering Sea water and melts. This leads to a strong freshwater-induced haline front at the ice edge, and a layered structure underneath the ice. Strong tidal mixing near the bottom and mixing by ice motion near the top give rise to a characteristic two-layer structure of the water column with the OML separated by the benthic boundary layer (BBL) by a strong halocline. This structure is common on the mid-Bering shelf, while on the inner shelf a single-layer structure, and on the outer shelf a three-layer structure, is observed. This characteristic one/two/three-layer stratification pattern has strong implications for biological production on the Bering shelf (Overland *et al.*, 1999). The layered structure can also be seen in the distribution of TKE in the water column in Figure 2.8.10. The ocean model includes a second-moment closure submodel for calculating the turbulence parameters under ice, and the ice model includes the internal ice stresses.

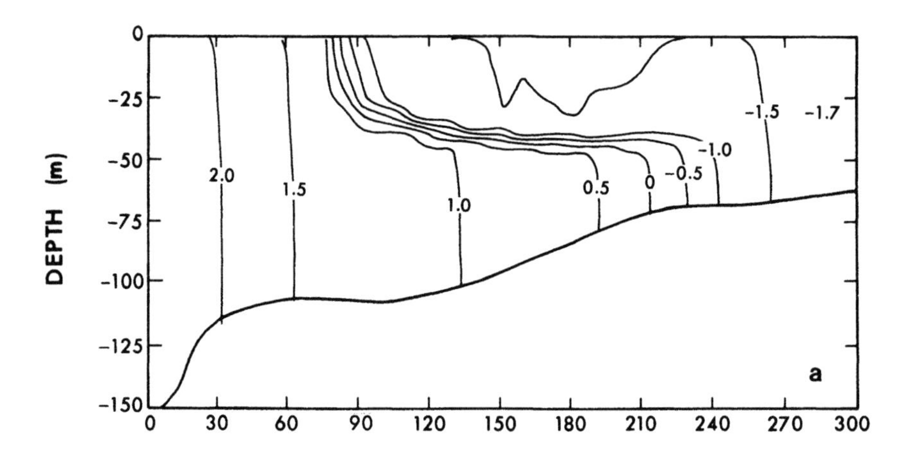

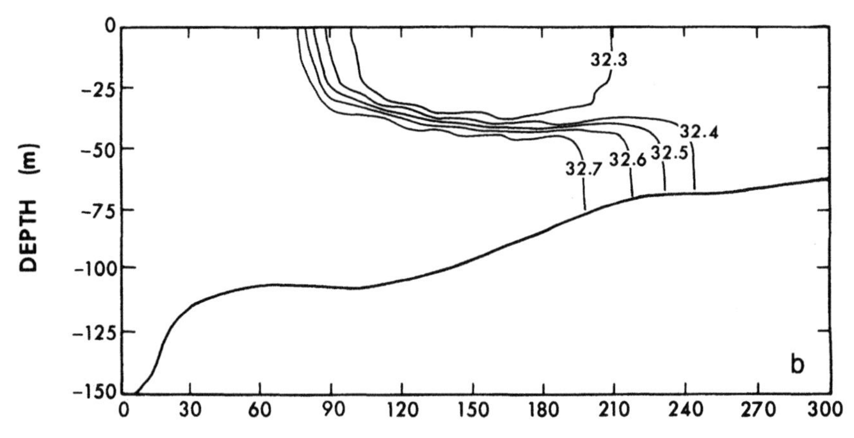

Figure 2.8.10. (a) Temperature, (b) salinity, (c) along-ice edge velocity and (d) TKE in simulations of the Bering Sea MIZ by Kantha and Mellor (1989).

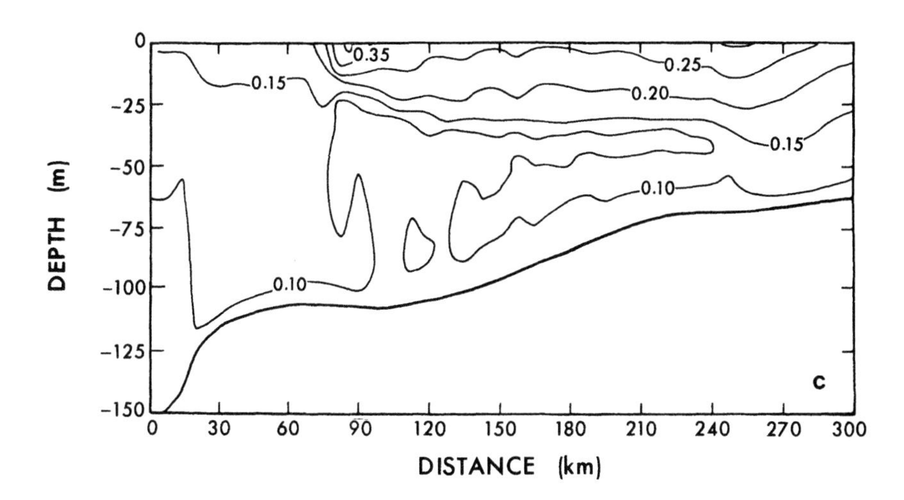

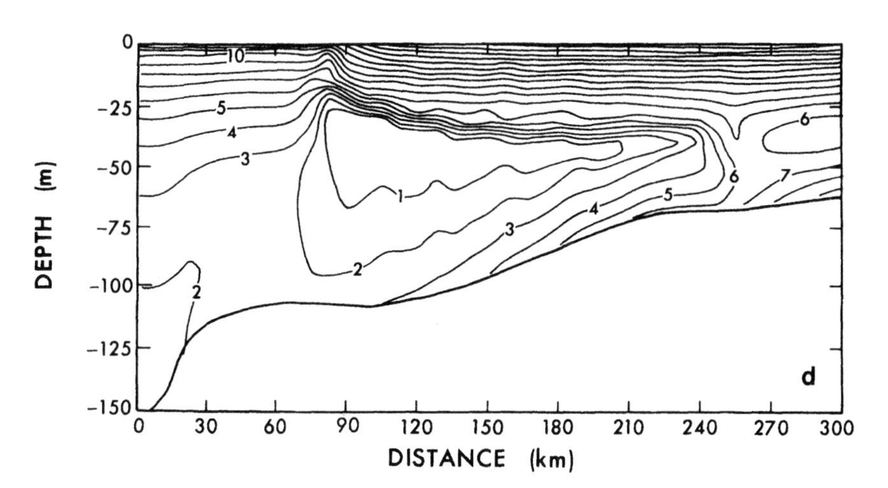

Figure 2.8.10. *(Continued).*

2.9 BOTTOM/BENTHIC BOUNDARY LAYER

Oceanic currents flowing over the bottom give rise to well-mixed layers. These layers are called benthic boundary layers (Armi and Millard, 1976) in the deep ocean, and bottom boundary layers in the shallow water columns on the continental shelf. These bottom/benthic boundary layers (BBLs) provide an important dissipation mechanism for the currents and are therefore important to the dynamics of bottom currents. They are also important from the point of view of sediment suspension (resuspension) and transport, and hence to the bottom-dwelling (benthic) organisms. In this section, we will briefly review such boundary layers.

Consider first the benthic boundary layer (Figure 2.9.1). In regions of strong currents in the deep ocean such as the Mediterranean outflow plume (Johnson *et al.,* 1994), and the western boundary undercurrents that are part of the thermohaline circulation of the global oceans, one often observes a well-mixed region a few tens of meters thick near the bottom, in which, if the bottom consists of easily suspended sediments, there is considerable sediment suspen-sion and transport. Like all energetic currents (typical value for current intensity ~ 20–40 cm s^{-1}), there is considerable temporal and spatial variability in these currents and this causes episodic sediment suspension and settling. These processes are important to bottom-dwelling organisms and the resulting bioturbation of the sediment layer on the deep ocean bottom. Normally, currents at the ocean bottom are quite weak (a few cm s^{-1} at the most) and the associated mixed layers, if any, are quite thin. The characteristic scaling under neutral stratification is the turbulent Ekman layer thickness (u_{*b}/f), where u_{*b} is the bottom friction velocity. The density stratification near the ocean bottom is normally quite weak and close to neutral stratification, and neutral Ekman scaling may therefore be fairly accurate. In regions of strong bottom currents, the BBL can be a few tens of meters thick (20–40 m). The benthic heat flux to the water column is also small (typically 0.1 W m^{-2}) and therefore quite unimportant to the dynamics of the benthic boundary layer and benthic currents, except in a few regions of strong heat flow (regions near midocean ridges are typical examples). In these regions, a convective boundary layer ensues and the ambient stratification, no matter how weak, becomes important to the dynamics. Scaling laws from the well-known ABL, which also evolves under the action of geostrophic flow aloft and heating at the bottom, can be readily applied to the oceanic benthic boundary layer. In regions of strong stratification or weak rotation (for example, near the equator), the BBL thickness scales with U_b/N_b, where N_b is the background buoyancy frequency and U_b is the current magnitude.

Temporal variability is an important aspect of most BBLs. Observations in HEBBLE (High Energy Benthic Boundary Layer Experiment) showed a bottom

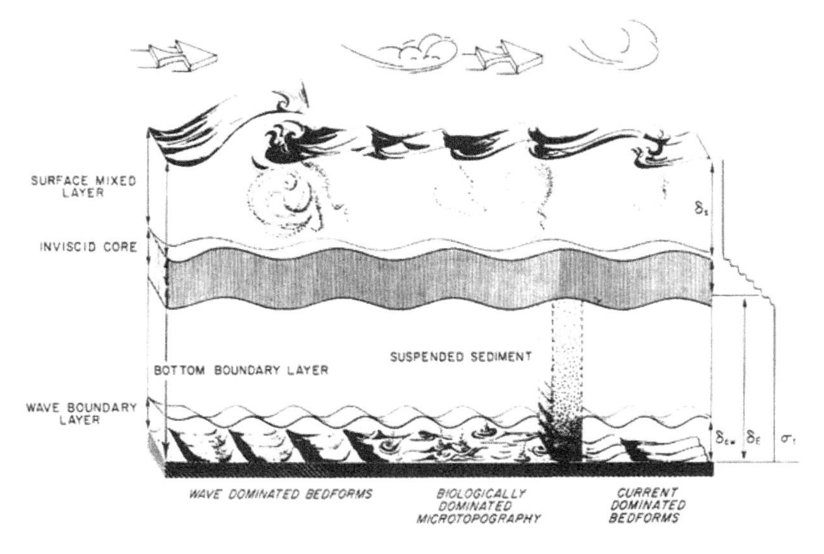

Figure 2.9.1. A sketch of the BBL on the continental shelf showing the various processes influencing its state (from Grant and Madsen 1986 with permission from the *Annual Review of Fluid Mechanics,* Volume 18, Copyright 1986, by Annual Reviews).

mixed layer that had been apparently evolving with time, eroding the ambient stratification and mixing the water column in the process. This aspect of dynamics is readily dealt with using knowledge from observations in the field and the laboratory on similar mixing processes in other stratified fluids such as the ABL. The salient difference with the ABL is the sediment suspension during episodes of strong currents. Above a particular threshold of bottom stress, whose value depends critically on the nature and density of sediments, the sediments are suspended into the water column. Sediment concentration, if large enough, can alter the dynamics significantly, since the turbulence has to act against the gravitational forces tending to resettle the sediments. Therefore they are akin to stable stratification in the water column. These aspects are important to bioturbation and the fate of $CaCO_3$ rained onto the ocean floor from the upper ocean.

Bottom boundary layers on continental shelves, on the other hand, are almost always important to the dynamics of the shelf. Vigorous currents are set up on the continental shelf by along-shore winds, along-shore pressure gradients set up by the large scale basin processes, tidal action, surface gravity waves, density differences due to freshwater input and midshelf density fronts, and even atmospheric pressure gradients. There is a plethora of forcing mechanisms operating on the continental shelf, the relative importance of each depending most crucially on the depth of the water column. The continental shelf can usually be divided into three regions: the inner shelf, the middle shelf, and the outer shelf, and the dynamics of each of these regions are dependent on a slightly different balance among the driving forces. The inner shelf is delineated by water depths of a few tens of meters (typically 25–30 m); the outer shelf is that beyond water depths of about 75–100 m; and in between is the middle shelf. This demarcation is not quite strict. It depends on the dynamical balances prevailing and could be temporally variable. Nevertheless, this artificial division is useful for discussing the BBL aspects on the continental shelf.

Shallow waters on the inner shelf (where one is likely to observe a single well-mixed water column most often) are driven vigorously by winds and they respond in kind on timescales similar to those of wind forcing [see Allen (1980) and Winant (1980) for reviews of wind-driven currents on the shelf]. The strong currents resulting from wind driving "feel" the bottom and one finds more often than not a well-mixed water column over the inner shelf. This is often also a region affected by vigorous tidal currents, wind-generated surface waves, and fresh river water runoff from continents. It is likely to be affected by the entire spectrum of wind-generated waves, because waves in the <6-second band typical of wind waves can "feel" the bottom and therefore affect the dynamics of the BBL.

On the midshelf, one is likely to find distinct upper and bottom boundary layers and here the influence of wind waves is confined to the long period swell components of the surface wave field generated in the deep water and propagating onshore (a 12-second wave can feel the bottom in 100-m waters). This region is subject to both wind driving and the influence of off-shelf pressure gradients set up by the large scale flow in the basin. The BBL is due to the bottom currents resulting from the various forcing mechanisms. There is usually a column of stratified fluid in the interior between the upper OML and the lower BBL. There is often a density front that delineates the well-mixed inner-shelf waters from the midshelf water masses. This front is often found in the very outward regions of the continental shelf during winter, when strong cooling and vigorous storms tend to mix the entire water column to depths of up to 100 m.

On the outer shelf, the surface wave influence is negligible and the principal driving forces are the pressure gradients set up by the off-shelf flow. The wind action is confined to the upper OML (as it is in the deep ocean) and the bottom currents are less vigorous than on the inner and middle shelves. The flow on the outer shelf is more vigorous and more variable, and these spatial and temporal scales affect the BBL as well.

The BBL is important on the entire shelf. This is where strong bottom dissipation of flow energy takes place and the bottom stress is an important part of the dynamical balance. It is here that vigorous exchanges between the seabed and the water column take place. Sediment suspension and resettling is a very important aspect of BBLs on the continental shelf.

For our purposes, the most salient aspect of the BBL on the continental shelf is the coexistence of a thin surface, wave-induced, oscillatory boundary layer at the bottom with a thicker planetary Ekman boundary layer due to the mean flow.

This makes for a fascinating interplay that has interesting consequences. For example, the bottom frictional effects are enhanced, and because of the variability of the surface wave field across the shelf, there is corresponding variability in the shelf circulation also. Sediment suspension is facilitated by the presence of vigorous oscillatory motion above the seabed. Most of the work in this area has been done by Grant and his colleagues, and an excellent review of the BBL (Grant and Madsen, 1986) is the source of the following summary. Bowden (1978), Wimbush and Munk (1970), Soulsby (1983), Nowell (1983), and Smith (1977) provide reviews of various aspects of benthic boundary layers. Figure 2.9.1, from Grant and Madsen (1986), summarizes the processes related to the BBL.

The planetary boundary layer that exists on the bottom due to currents is very similar to other Ekman layers such as those in the atmosphere. The thickness of this PBL, under neutral stratification, is

$$\delta_p = \kappa u_{*b} / f \tag{2.9.1}$$

The typical value is 20–40 m. There exists a region ($z_0 \ll z \ll \delta_p$) where the velocity profile is logarithmic. This result can be obtained and the drag law derived by asymptotically matching the inner law close to the wall and the outer law of the wake near the edge of the PBL (Tennekes, 1973; Grant and Madsen, 1986).

The surface, wave-induced, oscillatory boundary layer (WBL) is 5–30 cm thick. The WBL thickness is a function of the wave frequency n and a representative average friction velocity u_{*b} related to the orbital velocity of the fluid particles near the seabed. Once again for $z_0 \ll z \ll \delta_w$, logarithmic law holds. Also, because of the much higher frequencies of waves, the wave-induced shear stress is much larger than that due to geostrophic flow,

$$\delta_w = \kappa u_{*w} / n$$

$$\frac{u_*}{U_g} = \left(\frac{f_g}{2}\right)^{1/2}, \quad \frac{u_{*w}}{U_b} = \left(\frac{f_w}{2}\right)^{1/2} \tag{2.9.2}$$

$$\frac{1}{4\left(f_{g,w}\right)^{1/2}} + \log\frac{1}{4\left(f_{g,w}\right)^{1/2}} = \log\left(\frac{U_g}{f z_0}, \frac{U_b}{n z_0}\right) - 1.65$$

where U_g is the geostrophic flow velocity and U_b is the orbital velocity of the wave near the bottom. The velocity profile in the WBL is (Grant and Madsen, 1986)

$$U(z) = \frac{\ker 2\sqrt{z/l} + i \operatorname{kei} 2\sqrt{z/l}}{\ker 2\sqrt{z/l} + i \operatorname{kei} 2\sqrt{z/l}} U_b \exp(int) \tag{2.9.3}$$

where ker and kei are Kelvin functions of zeroth order and l is the turbulence length scale: $l = \kappa u_{*w}/n$. This reduces to the log law for $z < 0.1\ \delta_w$. These results are in good agreement with observations (Figure 2.9.2).

When both waves and currents are present, the WBL is embedded in the PBL and the nonlinear interaction between the two affects both. However, there is a wide disparity in the two vertical scales; the WBL thickness is two orders of magnitude smaller than the PBL thickness. Grant and Madsen (1986) show that for $z < \delta_{cw}$ the scaling velocity is u_{*cw}, and for $z > \delta_{cw}$ it is simply u_{*c}. Therefore the WBL solution ($z < \delta_{cw}$) holds if u_{*w} is replaced by u_{*cw},

$$u_{*cw} = u_{*wm}\left[1+2\left(u_{*c}/u_{*wm}\right)^2\right]\cos\alpha + \left[\left(u_{*c}/u_{*wm}\right)^4\right]^{1/4} = C_R^{1/2}\, u_{*wm} \quad (2.9.4)$$

$$u_{*wm} = C_R^{1/2}\sqrt{\frac{f_w}{2}}\, U_b \quad (2.9.5)$$

where α is the angle between the waves and the current. C_R is slightly above 1 for $u_{*c} < u_{*wm}$, the usual condition on the shelf. The drag law becomes

$$0.25\left(f_w\right)^{-1/2} + \log\left[0.25\left(f_w\right)^{-1/2}\right] = \log\left(\frac{C_R U_b}{n z_0}\right) - 1.65 + 0.48 f_w^{1/2} \quad (2.9.6)$$

The current profiles are given by

$$U_c = \begin{cases} \dfrac{u_c}{\kappa}\left(\dfrac{u_{*c}}{u_{*cw}}\right)\ln\dfrac{z}{z_0} & z \le \delta_{cw} \\[2ex] \dfrac{u_{*c}}{\kappa}\ln\dfrac{z}{z_{0c}} & z \ge \delta_{cw} \end{cases} \quad (2.9.7)$$

z_{0c} is the apparent roughness scale felt by the PBL because of the wave motions and is the appropriate quantity to be used to scale PBL drag law and velocity profiles outside the WBL. By matching the current from the WBL and PBL at $z = \delta_{cw}$,

$$U_c = \frac{u_{*c}}{\kappa}\left(\frac{u_{*c}}{u_{*cw}}\ln\frac{\delta_{cw}}{z_0} + \ln\frac{z}{\delta_{cw}}\right) \quad (2.9.8)$$

for the current velocity outside the WBL. In the limit $u_{*c} << u_{*cw}$; this reduces

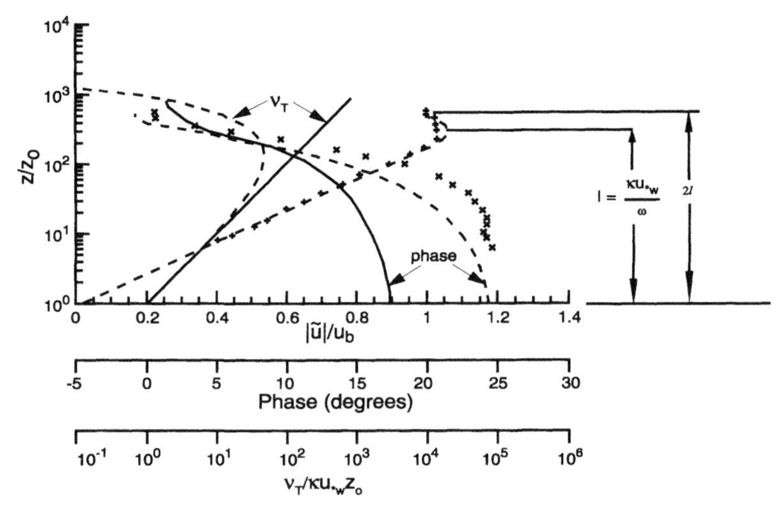

Figure 2.9.2. Comparison of velocity profiles in the wave boundary layer between model (lines) and observations (scatter points) (from Grant and Madsen 1986 with permission from the *Annual Review of Fluid Mechanics,* Volume 18, Copyright 1986, by Annual Reviews).

to a log law with an apparent roughness scale of δ_{cw} and herein is the large effect of the WBL on the PBL properties, including the drag. The current speeds decrease with increase in wave motion and hence δ_{cw}. For a geostrophic current of a given magnitude, the average bottom stress felt by the PBL in the presence of the WBL is therefore much higher. Grant and Madsen (1986) suggest that within the various uncertainties in the problem, it is reasonable to use $\delta_{cw} \sim (1-2)l$.

In practice, the above scenario is considerably complicated by aspects related to bottom sediments. The oscillatory shear stress due to waves is quite efficient in suspending sediments in the water column. The sediment load acts to stabilize the BBL. The sediment-induced stability effects can be treated using the classical Monin–Obukhov approach adapted to sediment suspension. This approach shows that it is not the total sediment concentration that determines the stable stratification effect but the integral of the product of the concentration and settling velocity. In addition, there are complications brought on by sediment motion and the resulting variability in the bed forms that affect the roughness felt by the flow. Also, there are the usual additional effects due to stratification in the water column, internal wave motions, etc. It is easy to see why the BBL on a continental shelf is richer and far more complex in structure compared to the oceanic BBL. For more details, the reader is referred to Grant and Madsen (1986).

Accurate numerical modeling of the BBL is especially important on the continental shelf and in the shallow water of the coastal oceans. Most bulk ML models apply to the open ocean and are not designed to handle the shallow water MLs where not only the BBL might exist, but its structure might be complex due to its interaction with the upper ML. Diffusion-based models, on the other hand, such as the Mellor–Yamada (1982) and Kantha–Clayson (1994) models (see Section 2.10), model the BBL also, provided sufficient resolution is allocated to adequately resolve the BBL. However, so far, none of these models have incorporated the surface wave effects discussed above.

Analytical models of the benthic boundary layer take the stream tube approach, where the flow is confined to a tube with properties such as temperature, salinity, and velocity constant across the tube (Smith, 1975). The flow is retarded by bottom friction and entrainment occurs at the top surface of the tube, and by proper choice of parameters, such models can be made to perform well (See Killworth, 1977; Price and Baringer, 1994). However they suffer from the disadvantage that they allow only for a growing BBL, since detrainment is not allowed. Mellor–Yamada- and Kantha–Clayson-type diffusion models avoid this, and can calculate the BBL explicitly provided enough resolution is allocated, which may not always be possible in an ocean circulation model. For this reason, a slab model akin to a slab upper mixed layer model (see Section 2.10.1 below) is desirable. Killworth and Edwards (1997) present such a slab model. The approach is very much similar to a slab ABL model discussed in Section 3.11, although there are differences in details, especially with regard to parameterizing the BBL thickness, which is taken as $\sim(u_{*b}/f)\ [1+(u_{*b}/f)^2(N_b/U_b)^2]^{-1/2}$. This BBL model allows for both entrainment and detrainment and is suitable for inclusion in oceanic GCMs (e.g., Webb, 1996).

2.10 MODELING THE OCEANIC MIXED LAYER

The air–sea interface acts as a significant barrier to the exchange of heat and gases between the ocean and the atmosphere, because the transfer of these scalar quantities is mediated by a molecular sublayer at the air–sea interface and the associated molecular diffusion. However, the sea surface is invariably covered with surface waves, and under certain conditions, these waves break. When they do, additional transfer mechanisms come into play that are more efficient in effecting the transfer across the interface. For example, breaking waves entrain air in the form of small bubbles that are propelled into the water to depths on the order of the depth of breaking and therefore more efficiently transfer properties to the water. Similarly, spray and droplets ejected into the air during breaking are an efficient mechanism for transferring water vapor, heat, and dissolved gases from the ocean to the atmosphere. Breaking waves also create additional turbulence and mixing. This has a considerable effect on the near-surface distributions of mixed layer properties, such as the velocity, temperature, salinity, and concentrations of suspended matter, phytoplankton, and dissolved gases. This aspect is of particular importance to air–sea exchange of greenhouse-active photochemicals in the water column such as carbonyl sulfide. These chemicals are produced principally in the upper few meters by the ultraviolet component of the solar insolation penetrating into the water column, and are then mixed downward by turbulent eddies in the mixed layer. For all these reasons, it may be important to include the effect of wave breaking in modeling the upper layers of the OML (Craig and Banner, 1994; Stacey and Pond, 1997; Clayson and Kantha, 1999).

As we saw earlier, surface waves are also responsible for generating large scale coherent motions in the upper layers. These take the form of counter rotating horizontal cells with horizontal dimensions scaling roughly with the depth of the mixed layer and with their axes roughly aligned with the wind. These Langmuir cells have the capability to transport suspended matter such as phytoplankton and zooplankton vertically downward deep into the mixed layer. The near-surface vertical velocities in the convergence and divergence zones of these cells can reach values on the order of 10 cm s^{-1}, and therefore Langmuir circulation, when present, can have a dramatic effect on near-surface properties in the mixed layer. These cells can also transfer some of their kinetic energy directly to turbulence and this must be taken into account in mixed layer models. Gemmrich and Farmer (1999) discuss wave-breaking-enhanced turbulence near the surface and the effects of subsurface advection of tracers by Langmuir circulation.

In addition, surface waves can extract energy from mean motions (Phillips, 1977). They can also transfer energy directly to turbulence, without any breaking (Kitaigorodskii *et al.,* 1983). These wave-mean flow-turbulence interactions are not very well understood at present, but could nevertheless be important to the overall dynamics of the upper layers of the ocean.

Traditionally, most OML models have been patterned after the atmospheric boundary layer (e.g., Mellor and Yamada, 1974, 1982; Kantha and Clayson, 1994) and have therefore ignored the fact that the air–sea interface is a nonrigid, mobile surface capable of sustaining gravity wave motions with all their attendant complex dynamics. The net result is that traditional concepts such as the Karman–Prandtl law of the wall in shear flows adjacent to a rigid boundary have been assumed to be valid near the more complex air–sea interface as well. In the presence of a vigorous sea state the principal balance in the turbulence kinetic energy equation in the immediate vicinity of the air–sea interface is not between local shear production and dissipation, but is instead between downward diffusion of turbulence produced by breaking waves and dissipation. The influence of coherent motions such as Langmuir cells is also excluded in this approach. The principal justification has been that these wave effects are confined to the upper few meters of the water column and in deep mixed layers, at least, ignoring these effects is not very detrimental to the overall mixed layer dynamics. Both wave breaking and Langmuir circulations are "surface-trapped" phenomena and their influence decays rapidly with distance from the air–sea interface. As we shall see, whatever limited comparisons with observational data that have been possible (see, e.g., Mellor and Yamada, 1982; Kantha and Clayson, 1994; Clayson *et al.,* 1997) have generally supported this approach. Nevertheless, one particular situation where this approach is particularly uncomfortable is that of shallow mixed layers such as the diurnal mixed layer generated by weak winds and strong solar insolation, and a rain-mixed layer generated by heavy precipitation. These mixed layers are typically less than 10 m deep, and therefore in the presence of a vigorous sea state, gravity wave influence cannot be justifiably ignored.

There have been a few attempts to incorporate the effect of wave breaking in mixed layer models. Some bulk ML models (see, e.g., Niiler and Kraus, 1977) have recognized the fact that a mixed layer is mixed from both the top and the bottom, and in addition to incorporating the shear-produced TKE at the ML base, have parameterized the TKE input at the surface by a term proportional to u_*^3, where u_* is the water-side friction velocity. On the other hand, except for a few exceptions, second-moment closure-based models (Mellor and Yamada, 1982; Kantha and Clayson, 1994) have traditionally ignored this input. Those that have (Kundu, 1980; Kantha, 1985; Craig and Banner, 1994; Stacey and Pond, 1997, Clayson and Kantha 1999) have confirmed the notion that the influence of this input is confined to the near-surface layers on the order of a few meters. All but one (Kantha, 1985) of these assume that the wave field is fully developed and corresponds to that generated by the local wind, and parameterize the TKE input at the surface as proportional to u_*^3. While the capillary-gravity range of the wind-wave spectrum adjusts rather quickly to changes in magnitude and direction of local winds, this is not generally true for the low wavenumber range, especially near the spectral peak. Fetch effects may therefore be important and could account for the large scatter in the near-surface dissipation rate (see Figures 7 and 8 of Craig and Banner, 1994). Clayson and Kantha (1999) explore the effect of fetch on the TKE input and ML dynamics.

Except for a large eddy simulation of the ML (Skyllingstad and Denbo, 1995), there has been no attempt to include the effect of Langmuir circulations in ML models. Clayson and Kantha (1999) parameterize Langmuir cells in conventional ML models of the bulk and second-moment closure type, concentrating on the direct input of energy from Langmuir circulations into ML turbulence.

Be that as it may, as a minimum, a mixed layer model applied to the upper ocean mixed layer must simulate or parameterize the turbulent mixing processes that occur at the base of the mixed layer, leading to its deepening, and the extinction of turbulence there by stable stratification that leads to the shallowing of the mixed layer. Almost all mixed layer models do so.

Mixed layer models can be grouped into roughly two categories: bulk (or slab) models and diffusion models. We summarize here the characteristics of each kind. For a detailed discussion of the ML models in the upper ocean, including the governing equations and model constraints and constants, the reader is referred to Martin (1985, 1986), Kantha and Clayson (1994), and Large *et al.* (1994).

2.10.1 SLAB MODELS

Bulk or slab models attempt to model the ML in an integral sense (Niiler, 1975; Niiler and Kraus, 1977; Garwood, 1977). The governing equations are integrated over the mixed layer so that the momentum and heat balance of the entire mixed layer, under the action of momentum and buoyancy fluxes at the ocean surface, can be considered. The major problem in bulk mixed layer modeling arises from the necessity to parameterize the advance and retreat of the ML under the action of surface momentum and buoyancy fluxes. The entrainment rate at the base of the ML, determined by turbulence processes, governs the deepening of the ML. This has been a subject of both laboratory (e.g., Kantha *et al.*, 1977) and field experiments (Price *et al.*, 1978; Price, 1979). It is also necessary to know the depth to which turbulence generated at the surface can penetrate under the action of a stabilizing buoyancy flux at the surface, as, for example, during a rainstorm or strong solar heating (Kantha, 1980a). Bulk models parameterize the entrainment (ML deepening) and "detrainment" (ML retreat) in terms of surface fluxes of momentum and buoyancy, using well-known properties of turbulence in geophysical mixed layers and/or observational evidence (Niiler, 1975; Garwood, 1977) (see also Niiler and Kraus, 1977). However, these parameterizations are far from universal, and it is often necessary to tune the entrainment coefficients for different situations. This means that there is often the danger of being able to compensate for model deficiencies and inaccurate surface fluxes by proper tuning of these constants.

The governing equations for a slab ML model can be derived by assuming a universal form for profiles of velocity, temperature, and salinity, and vertically integrating the momentum and conservation equations, Eqs. (1.7.22) and (1.7.23). For the purposes of illustrating the principles of mixed layer modeling, it is adequate to consider horizontally homogeneous conditions. The governing equations for such a one-dimensional slab ML model, assuming vertically homogeneous distribution of properties, are

$$
\begin{aligned}
&D_M \frac{\partial U_{Mj}}{\partial t} + \varepsilon_{j3k} f U_{Mk} + \Delta U_{Mj} \frac{\partial}{\partial t} D_M = \tau_j^w - \nu \left. \frac{\partial U_j}{\partial z} \right|_b \\
&D_M \frac{\partial T_M}{\partial t} + \Delta T_M \frac{\partial}{\partial t} D_M = Q_{Ts} + I_s - I_b - k_T \left. \frac{\partial T}{\partial z} \right|_b \\
&D_M \frac{\partial S_M}{\partial t} + \Delta S_M \frac{\partial}{\partial t} D_M = Q_{Ss} - k_S \left. \frac{\partial S}{\partial z} \right|_b
\end{aligned}
\tag{2.10.1}
$$

where T_M, S_M, and U_{Mj} (j = 1, 2) are ML averaged temperature, salinity, and velocity, and D_M is the ML depth. The quantity τ_j^w; denotes the kinematic wind stress (wind stress divided by density) at the surface, Q_{Hs} and Q_{Ss} are kinematic heat flux (heat flux divided by ρc_p) and salinity flux at the surface, and Δ denotes the "jump" in the properties at the base of the mixed layer between the values in the ML and those just below its base. It is the buoyancy jump at the base that tends to inhibit mixing in the vertical. The last terms denote the molecular diffusion at the base of the ML; k_T and k_S are molecular diffusion coefficients for heat and salt. Solar insolation I (kinematic value, meaning it has been divided by ρc_p) constitutes a volume heating source, with subscripts s and b denoting values at the surface and just below the ML base.

Equation (2.10.1) comprises three equations for four unknowns: T_M, S_M, and U_{Mj}, and thus an auxiliary condition is needed. This is obtained by vertically integrating the TKE equation (1.7.34),

$$\frac{1}{2}\frac{\Delta\rho g D_M}{\rho_w}\frac{\partial D_M}{\partial t} = P_{Ss} + P_{Sb} + P_{Bs} + P_{Bv} - D_v \tag{2.10.2}$$

where P_{Ss} indicates the vertically integrated value of TKE production at the surface by the applied wind stress, P_{Sb} the shear production at the base of the mixed layer, P_{Bs} the production (destruction) due to buoyancy flux at the surface, P_{Bv} the contribution (negative) from volumetric heating by penetrative solar insolation, and D_v the dissipation in the ML. P_{Ss} can include the production due to wave breaking. The LHS indicates the increase in the potential energy of the system, and hence this equation just allocates part of the net TKE production by various sources to increasing the potential energy of the system. Equation (2.10.2) constitutes an additional equation and provides the closure needed. Some of the terms on the RHS of Eq. (2.10.2) need to be parameterized. This introduces empirical constants into the picture,

$$\frac{1}{2}\frac{\Delta\rho g D_M}{\rho_w}\frac{\partial D_M}{\partial t} = m_0 A + m_1 B + C + m_2 D \tag{2.10.3}$$

where

$$A = u_*^3; B = \frac{1}{2}\left[\left(\Delta U_{Mj}\Delta U_{Mj}\right)\frac{\partial D_M}{\partial t} + \nu\, \Delta U_{Mj}\left.\frac{\partial U_j}{\partial z}\right|_b\right]$$

$$C = \alpha g\int_{-D_M}^{0} I\, dz' - \frac{1}{2}D_M\left[B_s + \alpha g I_b + g\left(\alpha k_T\left.\frac{\partial T}{\partial z}\right|_b - \beta k_S\left.\frac{\partial S}{\partial z}\right|_b\right)\right] \tag{2.10.4}$$

$$D = \frac{1}{4}D_M\left(B_s - \left|B_s\right|\right)$$

where α and β are volumetric coefficients and B_s is the buoyancy flux at the surface given by

$$B_s = \alpha g\left(H_s + I_s\right) - \beta g S_M (\dot{E} - \dot{P}) \tag{2.10.5}$$

where the last term denotes the difference between evaporation and precipitation rates. Note that the dissipation rates of TKE produced by different mechanisms have been assumed to be proportional to the respective production rates, and the constants m_0, m_1, and m_2 are therefore fractions that account for net production near the surface, at the base of the mixed layer due to shear, and due to the buoyancy flux at the surface, when it is negative (convective mixing). See Niiler (1975) and Niiler and Kraus (1977) for a detailed derivation of the above equations. Martin (1985) indicates that the best empirical values for these constants are $m_0 \sim 0.39$, $m_1 \sim 0.48$, and $m_2 \sim 0.83$. Note, however, that these values might need to be tuned to a particular application, since in general they could be functions of a stratification parameter such as the bulk Richardson number, $Ri_B = \Delta\rho g D_M / \rho_w$.

The application of Eq. (2.10.2) requires that the ML be deepening. In those situations where the turbulence in the ML is extinguished by a positive buoyancy flux, and the ML shallows, the LHS is assumed to be zero, and the resulting algebraic equation is solved for the ML depth:

$$D_M = \frac{m_0 u_*^3 + \alpha g \int_{-D_M}^{0} I\, dz'}{\frac{1}{2}\left[B_s + \alpha g I_b\right]} \tag{2.10.6}$$

Note the absence of the shear production term at the ML base.

The equations above are solved from specified initial conditions on the temperature, salinity, and velocity in the water column, with wind stress, heat flux, solar insolation, and precipitation (and evaporation most often) prescribed as a function of time. An alternative method is to compute from the model the wind stress and the heat and evaporative fluxes, given atmospheric parameters such as air temperature, humidity, and wind velocity as functions of time. The water column below the ML can be considered to be quiescent or the following equations that involve only molecular diffusivities can be solved:

$$\begin{aligned}
&\frac{\partial U_j}{\partial t} + \varepsilon_{j3k} f U_j = \frac{\partial}{\partial z}\left(\nu \frac{\partial U_j}{\partial z}\right) \\
&\frac{\partial T}{\partial t} = \frac{\partial}{\partial z}\left(k_T \frac{\partial T}{\partial z}\right) + \frac{\partial I}{\partial z} \\
&\frac{\partial S}{\partial t} = \frac{\partial}{\partial z}\left(k_S \frac{\partial S}{\partial z}\right)
\end{aligned} \tag{2.10.7}$$

A slight variation of the above bulk model is Garwood's bulk ML model (Garwood, 1977). Instead of a single integrated TKE equation, he uses separate budgets for TKE and the vertical component of the TKE, and parameterizes the ML deepening in terms of the TKE and its vertical component. This approach is appealing, since it is principally the vertical fluctuations that effect entrainment at the base of the ML.

$$0 = m_0 A - m_3\left(q^2 - 3\overline{w^2}\right)q - \frac{2}{3}\left(m_4 q^3 + m_5 f D_M q^2\right)$$

$$0 = \frac{1}{2}\frac{\Delta\rho g D_M}{\rho_w}\frac{\partial D_M}{\partial t} + m_3\left(q^2 - 3\overline{w^2}\right)q - \frac{1}{3}\left(m_4 q^3 + m_5 f D_M q^2\right) + E$$

(2.10.8)

$$E = \alpha g \int_{-D_M}^{0} I\, dz' - \frac{1}{2} D_M \left[B_s + \alpha g I_b\right]$$

$$\frac{\Delta\rho g D_M}{\rho_w}\frac{\partial D_M}{\partial t} = m_6 q^2 \left(\overline{w^2}\right)^{1/2}$$

where $m_0 = 4.5$, $m_3 = m_4 = m_6 = 1$, and $m_5 = 4.6$. Note that terms containing m_3 denote redistribution of TKE between horizontal and vertical components, and m_4 and m_5, the dissipation terms. Since the m_0 term does not account for dissipation, the value of m_0 is higher than that in the Niiler and Kraus model above. For a retreating ML case, the tendency terms are put to zero in Eq. (2.10.8) and the resulting algebraic equations are solved for D_M. One advantage of Garwood's approach is that it is possible to extend the model to include the effect of the horizontal component of rotation on redistribution of energy among the different components of TKE. Since the rotation terms vanish in the TKE equation, without separating the vertical component, this cannot be done. The horizontal component of rotation can play a role in mixing in the equatorial region (Garwood *et al.*, 1985a,b; Kantha *et al.*, 1989; Wang *et al.*, 1996). Figure 2.10.1 illustrates its most recent application to Station Papa in the Pacific (McClain *et al.*, 1996).

Another model of the bulk variety is the shear (dynamical) instability model, originally proposed by Pollard *et al.* (1973), and enhanced by Price *et al.* (1986). It is assumed that the sheared buoyancy interface at the base of the ML is close to neutral stability, as dictated by a bulk Richardson number criterion. The hypothesis is that when strongly sheared, the interface becomes unstable, turbulence is generated, and the ML deepens just enough to restore neutral stability to the interface. The modeling problem here is to select the appropriate value of the bulk Richardson number, and this is usually set at anywhere from 0.6 to 1.0. Price *et al.* (1986) use

$$Ri_b = \frac{\Delta\rho g D_M}{\rho_w\left(\Delta U_j \Delta U_j\right)} \geq 0.65 \qquad (2.10.9)$$

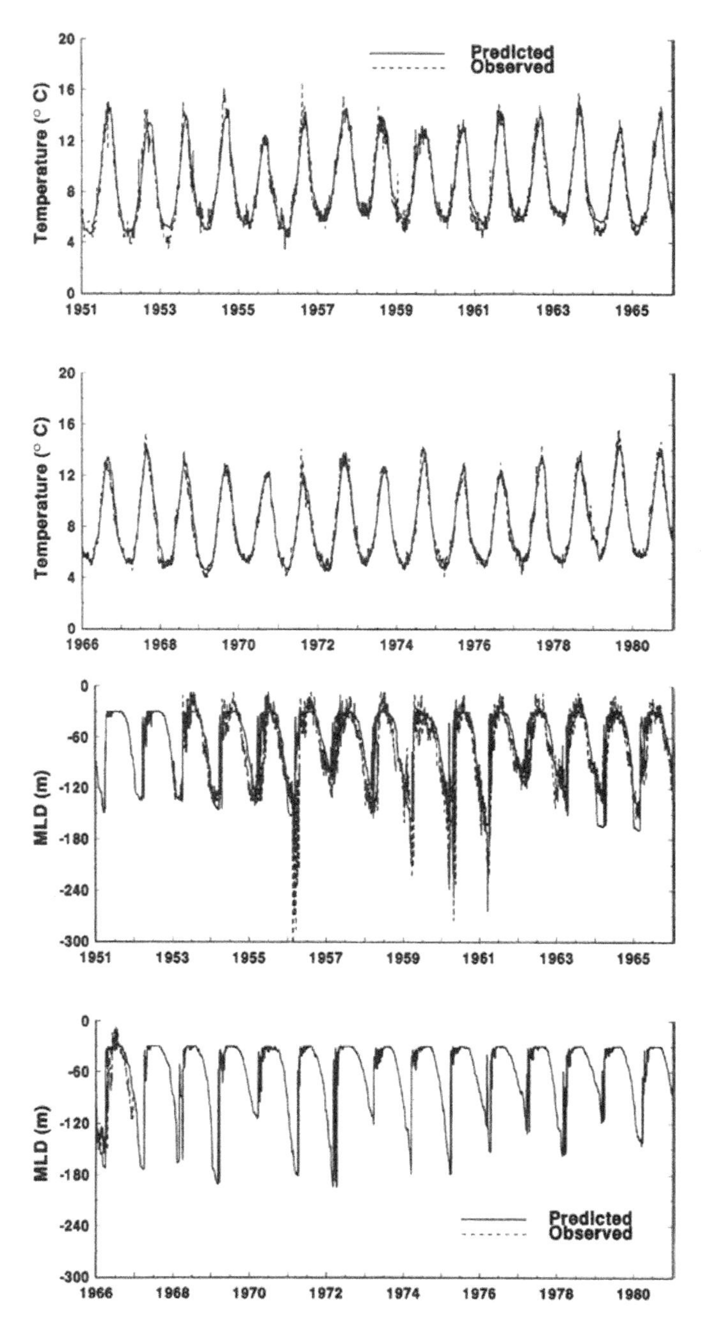

Figure 2.10.1. Application of Garwood slab model to Station Papa. Time series of observed and predicted temperatures (upper panels) and ML depths (lower panels) are shown (from McClain *et al.*, 1996).

In addition, the ML and the sheared layers below the ML are forced to neutral stability as governed by a gradient Richardson number criterion. This also affects mixing in the strongly stratified but also strongly sheared fluid below the mixed layer:

$$Ri_b = \frac{\frac{g}{\rho_w}\frac{\partial \rho}{\partial z}}{\left(\frac{\partial U_j}{\partial z}\frac{\partial U_j}{\partial z}\right)} \geq 0.25 \qquad (2.10.10)$$

This is the classic Miles–Howard (Miles, 1961, 1963; Howard, 1961) criterion for stability of a sheared stably stratified flow and is particularly important during storm-induced ML deepening, when inertial currents can produce large shears in the fluid below the active ML (see below). It also smooths out the sharp discontinuity at the base of the ML that the application of Eq. (2.10.9) alone would produce and makes the resulting vertical profiles more realistic. Such a criterion has also been used by other bulk ML modelers.

Price *et al.* (1986) enhanced the utility of the dynamic instability bulk mixed layer models by also including the influence of stabilizing and destabilizing buoyancy fluxes at the surface. The instability model is unable to account for convective deepening which takes the form of penetrative convection (Stull, 1973; Kantha, 1980b) (see Section 2.5). This problem is usually overcome by convective adjustment (Price *et al.*, 1986), an iterative procedure that restores neutral stability to the fluid in the ML under the action of destabilizing surface buoyancy fluxes.

$$\frac{\partial \rho}{\partial z} \geq 0 \qquad (2.10.11)$$

However, this procedure is equivalent to nonpenetrative convection, and it has been evident from both laboratory (Kantha, 1980b) and field observations in the ABL (Stull, 1988; Zilitinkevich, 1991) that penetrative heat flux at a convectively driven interface is at least 20% of the driving convective heat flux (Deardorff, 1980a) and hence cannot be ignored (see also Large *et al.*, 1994). In addition, not all properties (such as momentum) that can be mixed are adjusted by this procedure.

The opposite case of a strong stabilizing buoyancy flux due to penetrative solar heating is parameterized by an empirical model (Price *et al.*, 1986; Schudlich and Price, 1992) that sets the ML depth to the convection depth D_c, which is the depth at which the heating rate by insolation equals the heat absorbed above that depth given by

$$D_M = D_c = \frac{I_s - I_c + H_s}{\left.\frac{\partial I}{\partial z}\right|_c} \qquad (2.10.12)$$

Since this relates the resulting ML depth to the gradient of the solar insolation, the model displays some sensitivity to the parameterization of penetrative solar radiation in the upper layers. Nevertheless, the inclusion of these additional processes provided excellent model results for diurnal MLs observed from R/P Flip (Price *et al.*, 1986). The usual procedure is to apply Eq. (2.10.11) after computing the density profile resulting from absorption of insolation and surface fluxes. Then Ri_b is computed and ML deepened if needed until Eq. (2.10.9) is satisfied. Then condition (2.10.10) is applied to the water column. For strong solar heating and weak wind mixing cases, Eq. (2.10.12) is applied.

This model (PWP henceforth) has since been applied to other data sets, including the Long-Term Upper Ocean Study (LOTUS), Ocean Storms, and Tropic Heat, with differing degrees of success (see Gaspar *et al.*, 1990; Large *et al.*, 1994; Crawford, 1993; Schudlich and Price, 1992; see also Large and Crawford, 1995). Still, given the simplifications in ML physics implied by Eqs. (2.10.9) to (2.10.12), the PWP model has done remarkably well. Its recent applications are to simulating the MLML observations (Plueddemann *et al.*, 1995) and TOGA/COARE observations (Fairall *et al.*, 1996a; Anderson *et al.*, 1996; Weller and Anderson, 1996), where it has done remarkably well. Figure 2.10.2 shows a comparison between observed and modeled ML evolution during spring restratification measured during the 1991 MLML experiment in the Atlantic. Nevertheless, since no turbulence properties are ever explicitly calculated in this model (unlike, for example, Garwood's model above and the Kantha–Clayson model described below), related quantities such as the dissipation rate have to be inferred indirectly. Also, there exists a conceptual difficulty, namely, the well mixedness of all properties in a slab ML implies essentially infinite diffusivities. Observationally, while scalars do appear to have very small gradients in the ML proper, velocities seldom do.

With proper selection of the empirical constants, most bulk ML models can be made to agree reasonably well with observations.

2.10.2 Diffusion Models

The second kind of models are diffusion models that attempt to directly parameterize the turbulent mixing and diffusion in the ML. Either this parameterization is drawn from the abundant theoretical and observational knowledge of the surface layers and hence contains empirical or semiempirical formulations for turbulent diffusion in the ABL (Troen and Mahrt, 1986) or the OML (Large *et al.*, 1994), or it can be based on actual modeling of turbulence quantities by appealing to turbulence closure at the second (Mellor and Yamada, 1982; Gaspar *et al.*, 1990) or third moment (Andre and Lacarrere, 1985) level. The former builds upon our knowledge of Monin–Obukhov similarity relations in the constant flux layer that is adjacent to a boundary and part of the planetary

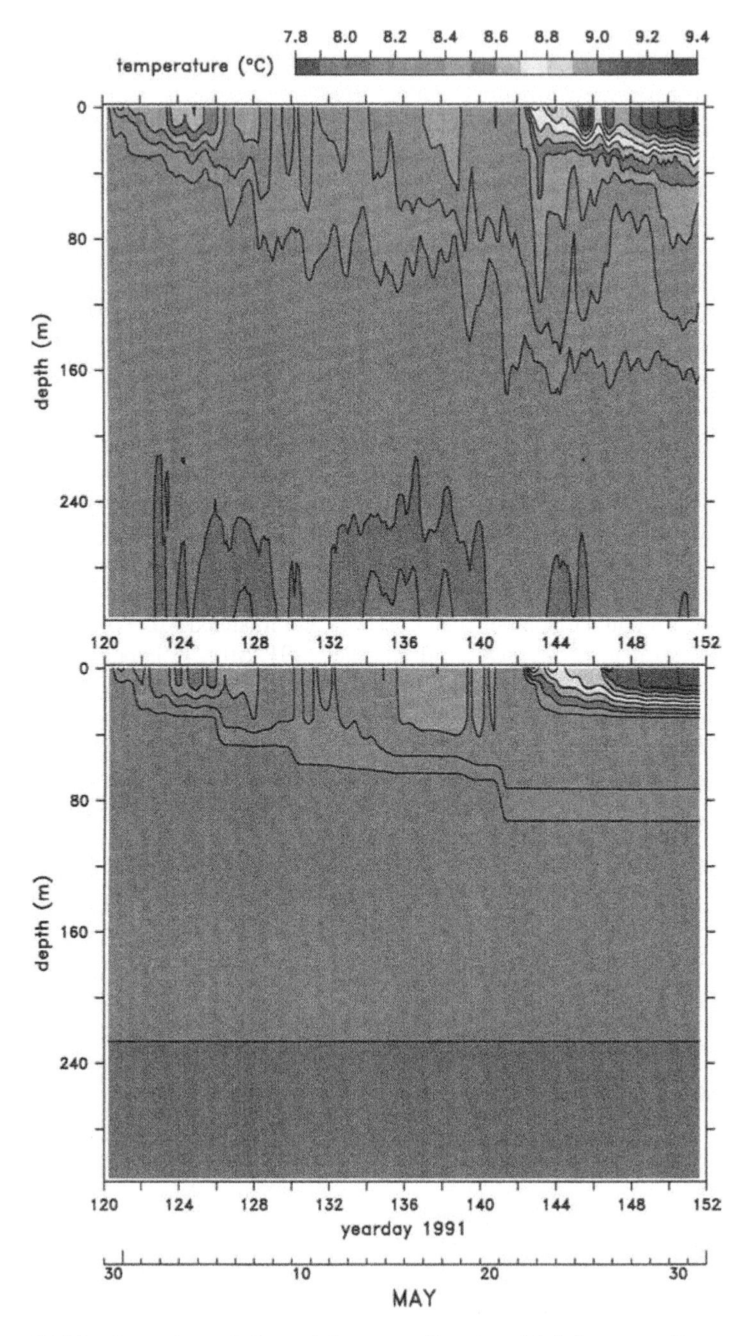

Figure 2.10.2. Comparison of observed (upper panel) and modeled (lower panel) temperatures using the PWP model during the MLML experiment (from Plueddemann *et al.*, 1995).

boundary layer (PBL). The well-known flux profile relationships (Businger *et al.,* 1971; Hogstrom, 1996) (see Section 3.3) in the ABL are judiciously extended into the entire PBL and are used to parameterize diffusion in the entire ABL/OML. However, this K profile approach also needs an equation for the ML depth that accounts for the deepening and shallowing of the ML due to convective and stabilizing surface buoyancy fluxes. In this sense, they are a hybrid of bulk and diffusion models. The reader is referred to Large *et al.* (1994) who describe this kind of model (LMD henceforth) and its application to several oceanic data sets, including OWS Papa, LOTUS, and Ocean Storms. The Large *et al.* model also includes several other attractive features such as a gradient Richardson-number-dependent first-order-closure-based mixing below the active ML, inclusion of nonlocal effects on mixing during free convection (Deardorff, 1972a), and the turbulence contribution to the vertical shear used to calculate the bulk Richardson number that determines the ML depth.

As in bulk ML models, the ML depth in the LMD model is determined by a critical bulk Richardson number criterion for deepening situations,

$$\mathrm{Ri_b} = \frac{\frac{(\rho_b - \rho_r)}{\rho_w} g D_M}{\left((U_{rj} - U_{bj})(U_{rj} - U_{bj})\right) + u_{tb}^2} = \mathrm{Ri_c} \sim 0.3$$

$$u_{tb}^2 = \frac{0.21 N w_t D_M}{\kappa \mathrm{Ri_c}} \qquad (2.10.13)$$

where subscript r refers to reference values (instead of surface values) that are averages over the surface layer, that is, the upper 10% of the ML; subscript b refers to the values below the ML. Large *et al.* (1994) argue that use of these reference values makes the model results less sensitive to strong variations in the surface layer. The second term denotes the contribution of large turbulent eddies, and is especially important for convective situations with little or no shear. Its form has been chosen to yield the empirical ratio (–0.2) of entrainment buoyancy flux to the surface buoyancy flux in pure convection (see Section 2.5). N is the local buoyancy frequency; w_t is the turbulence velocity scale.

For stable buoyancy flux, the ML depth is required to be less than both the Monin–Obukhoff length scale L and the neutral Ekman scale, taken by Large *et al.* (1994) to be 0.7 u_*/f, the constant being somewhat higher than the traditional value of 0.4.

The normalized diffusivities in the ML are assumed to be a cubic function of the normalized depth,

$$K_{M,H} = D_M w_{M,H}(\hat{z}) G(\hat{z})$$

$$G(\hat{z}) = a_0 + a_1\hat{z} + a_2\hat{z}^2 + a_3\hat{z}^3; \quad \hat{z} = z / D_M \qquad (2.10.14)$$

where the constants are determined by matching the values and the gradients of diffusivities at the top and bottom of the ML: $a_0 = 0$, $a_1 = 1$, and a_2 and a_3 are determined by matching the interior diffusivity and its gradient to the ML values at the base of the ML. The parameter $w_{M,H}$ denotes the turbulence velocity scale for momentum and scalar quantities, which depend on the cube root of a linear combination of u_*^3 and w_*^3, the velocity scales for shear mixing and convective mixing (see Section 2.5). In the limit of free convection, these become proportional to w_*, the Deardorff convective velocity scale.

By far the most salient aspect of the LMD model is the countergradient mixing (Deardorff, 1966), which is quite important under convective conditions. This is accomplished by adding a nonlocal transport term to turbulent fluxes under convective conditions:

$$\overline{w\theta}(z'),\overline{ws}(z') = -K_H\left\{\frac{\partial}{\partial z}(T,S) - 6.33\left[\frac{\left(\overline{w\theta_0} - (I_s - I_b)\right),\left(\overline{ws_0}\right)}{w_H(z')D_M}\right]\right\} \qquad (2.10.15)$$

No such countergradient transport is implemented for momentum.

Mixing below the ML is taken to be the sum of shear instability mixing, internal wave mixing, and double-diffusive mixing. Mixing due to shear instabilities is parameterized by a gradient Richardson number-dependent parameterization similar to that of Pacanowski and Philander (1981),

$$\begin{aligned} K_{M,H}/K_0 &= 1 & & Ri_g < 0 \\ K_{M,H}/K_0 &= \left[1-(Ri_g/Ri_{gc})^2\right]^3 & & 0<Ri_g<Ri_{gc} \\ K_{M,H}/K_0 &= 0 & & Ri_g > Ri_{gc} \end{aligned} \qquad (2.10.16)$$

where Ri_{gc} = 0.7 and $K_0 = 50 \times 10^{-4}\ m^2\ s^{-1}$. Internal waves modulate the background shear and this often leads to episodic lowering of the Richardson number below the critical value leading to turbulence and mixing. This is parameterized by constant values:

$$K_{M,H} = 1.0, 0.1 \text{ x } 10^{-4}\ m^2 s^{-1} \qquad (2.10.17)$$

Double-diffusive mixing occurs in statically stable fluids due to differential diffusivity of heat and salt under appropriate conditions represented by R_ρ that is indicative of the relative stratifications due to heat and salt (see Chapter 7). It is assumed to be, for the salt finger type of double diffusion,

$$\begin{aligned} K_S/K_0 &= 0 & & R_\rho \geq 1.9 \\ K_S/K_0 &= \left[1-\left(\frac{R_\rho - 1}{1.9-1}\right)^2\right]^3 & & 1.0<R_\rho<1.9 \\ K_H &= 0.7K_S & & \end{aligned} \qquad (2.10.18)$$

where $R_\rho = (\alpha\partial T/\partial z)/(\beta\partial S/\partial z)$ and K_0 = 10 × 10^{-4} m^2 s^{-1}. For double-diffusive convection,

$$K_H = 0.909\nu \exp\left\{4.6\exp\left[-0.54\left(R_\rho^{-1}-1\right)\right]\right\}$$

$$\begin{aligned} K_S &= K_H(1.85-0.85R_\rho^{-1})R_\rho \qquad && 0.5 \le R_\rho < 1.0 \\ &= K_H(0.15R_\rho) && R_\rho < 0.5 \end{aligned} \qquad (2.10.19)$$

This completes a basic description of the LMD model. It is, however, impossible to go into all the details, and the reader is referred to the original paper for more details, and a thorough review of the state of the ML modeling. The model appears to be remarkably successful in modeling the OML on a variety of timescales. In applications to the LOTUS diurnal mixed layer, the model accurately depicted the diurnal modulation of the SST. Wind and inertial current-induced ML deepening and warming of the thermocline below by rapidly moving storm systems in high latitudes, as depicted by the Ocean Storms data set, are also reproduced well (Crawford, 1993). The model also demonstrates a capability for remarkably accurate simulation of interannual variability at Station Papa, when proper attention is paid to the heat and salt balances in the mixed layer as well as cold water advection during winter. The principal deficiency is the assumption that the Monin-Obukhoff similarity scaling from the lower ABL over land is applicable to the upper OML, where the presence of a surface wave field on the air-sea interface and associated processes question its validity.

2.10.3 SECOND-MOMENT CLOSURE MIXED LAYER MODELS

ML models based on higher moments of governing equations close the governing equations for turbulence quantities at some level by judicious modeling of the unknown higher moments and other terms (Section 1.11) (Mellor and Yamada, 1982; Markatos, 1986; Rodi, 1987). Once turbulence is thus quantified, it is a straightforward matter to deduce the mixing intensity. In second-moment closure models, the turbulence equations are closed at the second-moment level; conservation equations for turbulence Reynolds stresses, heat fluxes, and scalar variances are solved by modeling the unknown third-moment turbulence quantities and pressure–strain rate and pressure–scalar gradient covariance terms by appealing to physical intuition and/or observational evidence (and lately to LES and DNS numerical models of turbulence). While the model constants are empirical and most often chosen by an optimization procedure by application to different turbulent flows in the laboratory, they are also regarded as nontunable from one turbulent flow to the other. Herein lies the attractiveness of this approach.

However, the complexity of this approach is at least an order of magnitude more than that of the simpler models cited above, since there is now a need to solve 11 partial differential equations governing second moments in addition to the usual 5 for mass, momentum (for U and V), and scalar (for T and S) conservation. Attempts have therefore been made to simplify the set (Mellor, 1973; Mellor and Yamada, 1974, 1982; Galperin *et al.,* 1988) by once again utilizing certain aspects of turbulence such as its departure from the state of local isotropy (somewhat equivalently from local equilibrium) (see Galperin *et al.,* 1988). The result is a hierarchy of models, of which the most useful for geophysical applications is the model that consists of one conservation equation for the turbulence kinetic energy (TKE, half the square of the turbulence velocity macroscale q) and a set of algebraic equations for turbulence second-moment quantities. The resulting simplicity and the potential "universality" of application are of particular interest in modeling geophysical MLs. Since the most basic description of turbulence is incomplete without quantifying its macro length scale l, this set is supplemented either by auxiliary information on the turbulence length scale such as the Blackadar formula (Blackadar, 1962), or by utilizing an equation for a quantity that includes the turbulence length scale such as dissipation (q^3/l) or, as in the Mellor–Yamada model, the product of the turbulence length scale and twice the TKE (q^2l). The reader is referred to Mellor and Yamada (1982) for a detailed description of the Mellor–Yamada model (MY henceforth) and its application to a wide variety of atmospheric and oceanic flows.

The MY model is actually a hierarchy of models ranging from those governed by the full set of second-moment equations to the superequilibrium model governed by an algebraic set. The so-called 2 1/2-level model consisting of a conservation equation for TKE has been used widely in geophysical applications. The hierarchy itself was originally derived by Mellor and Yamada (1974) using a systematic expansion procedure, which was amended recently by Galperin *et al.* (1988), who demonstrated that a more robust quasi-equilibrium model results from a slightly different expansion. We will deal with the latter quasi-equilibrium model and its modifications, as well as comparison with observations, in the rest of this chapter.

The constants in the MY-type second-moment closure model have been chosen from various laboratory turbulent flows, mostly in neutrally stratified and nonrotating fluids. There is therefore some uncertainty as to its validity in geophysics, where stratification, acceleration, and rotation effects on turbulence become quite important. These effects give rise to additional strain rates, leading to what is often called complex turbulence. The universality of constants in going from laboratory flows to geophysical flows has been the basic tenet of this kind of model, and the model loses some appeal if fine tuning of constants ever becomes necessary. In its application to geophysical MLs (Martin, 1985, 1986; Gaspar *et al.,* 1988; Large *et al.,* 1994; Crawford, 1993), it has become clear that while the model has been found quite skillful, it has a major deficiency, namely, an underestimation of mixing leading to a systematic warm sea surface temperature (SST) bias that has so far been difficult to correct without altering the constants. This underestimation has been cited as one major reason (simplicity being the other) for the development of the new similarity-theory-based K-profile LMD ML model described above. With additional improvements to model parameterizations and the addition of parameterization of mixing below the ML, Kantha and Clayson (1994) showed that the second-moment closure models perform reasonably well for OMLs and, by implication, for ABLs as well. This model will be discussed below as an example of this type of model.

Closure is also possible at the third-moment level, which is the basis for models such as that of Andre and Lacarrere (1985). These models are much more complex, since they involve a large number of conservation equations for third moments, that require rather tenuous closure models for fourth moments. The reader is referred to the literature for a full treatment.

The Kantha–Clayson model (Kantha and Clayson, 1994) (KC model henceforth) is an improved second-moment closure model that performs well in comparison with numerous ML data. We will describe this model in great detail here. The modifications made to the Galperin *et al.* (1988) version of the MY mixed layer model to obtain the KC model are twofold. First, recent findings of the LES turbulence research group at NCAR (Wyngaard and Moeng, 1993; Moeng and Sullivan, 1994) are incorporated in modeling the unknown terms in the second-moment equations. Second, a mixing model of the often highly active stably stratified but strongly sheared thermocline/halocline region immediately below the ML is added to remedy the major shortcoming of the MY model, namely, inadequate mixing. The inclusion of an empirical Richardson-number-based model for the episodic mixing below the ML is the most significant improvement to the model.

The turbulent diffusion in the PBL has been known for some time to be more complex than that approximated by a down-the-gradient parameterization in second-moment closure models. In fact, gradient transport models in turbulence in general have been under criticism by turbulence researchers (Corrsin, 1974; Lumley, 1978, 1983) for a long time. These models use only local properties and hence do not require information on the nonlocal integral properties of the mixed layer as a whole. However, in some situations such as convective mixing, large turbulent eddies bring about mixing against the prevailing mean gradient. However, at least in the case of simple turbulence in local equilibrium where dissipation and production very nearly balance, turbulent transport terms are of secondary importance, and errors in modeling these terms may be inconsequential.

Countergradient transport terms are especially important in a convective PBL, as indeed has been pointed out by Deardorff (1966), and is known from his own laboratory experiments on turbulent convection and observations in the ABL (e.g., Caughey and Wyngaard, 1979). However, while the turbulent transport and the pressure transport of TKE are undoubtedly large, they are of equal and opposite sign, and hence nearly cancel each other out. For neutral and stable situations, the turbulent transport terms do not appear to be as important, and the basic balance is between turbulence production by shear and its destruction by dissipation and buoyancy forces.

There is also more evidence in recent times from numerical LES models of turbulence in the ABL (Moeng and Wyngaard, 1986, 1989; Wyngaard and Moeng, 1993), where the large scale eddies are calculated directly, while small scale eddies are parameterized [see Galperin and Orzag (1993) for a recent review of LES's]. Though computationally intensive, LES calculations in the neutral (Andren and Moeng, 1993) and convective (Moeng and Wyngaard, 1986, 1989) ABL have enabled quantities such as turbulent transport and pressure–scalar (and pressure–strain rate) covariances to be explicitly calculated from LES's (it is difficult to measure the latter). This provides valuable guidance in modeling these terms in turbulence closure models.

Equations (1.7.21) to (1.7.23) and (1.8.1) to (1.8.3) are the basic equations for mean and turbulence quantities in a rotating, incompressible, stratified turbulent flow under the Boussinesq approximation under tensorial notation (Mellor, 1973; Mellor and Yamada, 1982; Kantha and Rosati, 1990). The pioneering works of Kolmogoroff (1942) and Rotta (1951) have made it possible to model some of the unknown terms on the right-hand side of the equations for second moments and hence made second-moment closure feasible. They enable us to postulate relations for the third moments and other unknown quantities on the right-hand side, based on principles of tensorial invariance and on physical reasoning derived from measurements of classic turbulent flows. Unfortunately, the models for the various terms, even with these constraints, are by no means necessarily unique. The hope is that, whatever the model used, it is "universal" and applicable to all turbulent flows without any changes whatsoever. Unlike closure at the first-moment level (for example, Prandtl's mixing length approach), higher-moment closures provide a means of estimating the turbulence field also. This is especially attractive in view of the microstructure measurements that are now being routinely made in the global oceans and from which the dissipation rate in the water column can be deduced.

The pressure–strain rate covariance terms in Eq. (1.8.1) are modeled as proportional to departure from isotropy, following Rotta (1951). This return to isotropy approximation has been essentially confirmed by LES, and there is consensus among modelers as to the form of this term. However, it is also necessary to add a second term accounting for the response of turbulence to rapid distortion. The particular form used is that due to Crow (1968). The pressure–strain rate covariance is therefore modeled as

$$\overline{p\left(\frac{\partial u_i}{\partial x_j}+\frac{\partial u_j}{\partial x_i}\right)} = -\frac{q}{3A_1 l}\left[\overline{u_i u_j} - \frac{\delta_{ij}}{3}q^2\right] + C_1 q^2\left(\frac{\partial U_i}{\partial x_j}+\frac{\partial U_j}{\partial x_i}\right). \tag{2.10.20}$$

The values used for A_1 and C_1 are 0.92 and 0.08. Strict adherence to the results of the response of isotropic turbulence to rapid distortion (Crow, 1968) suggests a higher value (0.2) for C_1. However, C_1 is not an independent constant and is related to A_1 and B_1, and strict adherence to these findings deemphasizes the importance of return-to-isotropy terms and degrades model performance. This is confirmed also by the experience of the Launder group of second-moment modelers, who use an alternative form, and who originally chose their constants to conform to the rapid distortion findings but lately have reduced them to give more prominence to return to isotropy (Gibson and Younis, 1986). This appears to have brought their value of constant A_1 closer to the MY value.

Rotational terms (see Section 2.10.5) do not appear in Eq. (2.10.20). This is consistent with the LES calculations of a neutral PBL by Andren and Moeng (1993). They recommend a value of 0.12 for C_1 instead of the value of 0.08 used by Mellor and Yamada (1982). They also found B_2/B_1 to be 0.42, whereas the MY value is 0.6. Observations on grid-generated turbulence yield a value of 0.8, while those on shear-generated turbulence yield 0.5, bounding the MY value of 0.6. In consequence, the Andren and Moeng value appears to be too low. Similarly, Andren and Moeng indicate a value of 3.0 for q/u_*, whereas the MY value is 2.55, which implies a value of 27.0 for B_1, instead of the MY value of 16.6. In view of the large departures of Andren and Moeng's values from generally accepted ones and in light of the experience accumulated by MY models, earlier MY values are to be preferred. However, it may be appropriate to reevaluate them in the near future.

LES findings (Andren and Moeng, 1993; Moeng and Wyngaard, 1986) have pointed out the inadequacies in the original MY modeling of the pressure–scalar gradient covariance term in Eq. (1.8.2). Mellor (1973) modeled it as

$$\overline{p\frac{\partial\theta}{\partial x_j}} = -\frac{q}{3A_2 l}\overline{u_j\theta} \qquad \overline{p\theta} = 0. \tag{2.10.21}$$

It appears that it is essential to include a shear term in the formulation (Andren and Moeng, 1993) and another term proportional to the scalar variance (Moeng and Wyngaard, 1986). KC therefore model this term as

$$\overline{\theta\frac{\partial p}{\partial x_i}} = \frac{\partial}{\partial x_i}\left(\overline{p\theta}\right) - \overline{p\frac{\partial\theta}{\partial x_i}}$$

$$= \frac{q}{3A_2 l}\overline{u_j\theta} + C_3\beta g_j\overline{\theta^2} + C_2\overline{\theta u_k}\frac{\partial U_j}{\partial x_k}. \tag{2.10.22}$$

Moeng and Wyngaard recommend a value of 0.75 for C_2, while Launder (1975) uses 0.5; we choose 0.7. We put C_3 equal to 0.2, instead of the value of 0.5 recommended by Launder. The reason for this becomes clear once we look at the Monin–Obukhov similarity relations derived from the resulting model.

The diffusion terms in Eqs. (1.8.1)–(1.8.3) are the most controversial. Despite their conceptual inelegance, they have traditionally been modeled in MY models using the down-the-gradient form,

$$\overline{u_i u_j u_k} = -\frac{3}{5} q l S_q \left[\frac{\partial}{\partial x_k} \overline{u_i u_j} + \frac{\partial}{\partial x_j} \overline{u_i u_k} + \frac{\partial}{\partial x_i} \overline{u_j u_k} \right] \tag{2.10.23}$$

$$\overline{u_k u_j \theta} = -q l S_{u\theta} \left[\frac{\partial}{\partial x_j} \overline{u_k \theta} + \frac{\partial}{\partial x_k} \overline{u_j \theta} \right] \tag{2.10.24}$$

and for the diffusion of temperature variance,

$$\overline{u_k \theta^2} = -q l S_\theta \frac{\partial}{\partial x_k} \overline{\theta^2}. \tag{2.10.25}$$

It might, however, be useful to add the countergradient terms for convective PBL situations (see discussion in connection with the LMD model above). When the Mellor and Yamada (1974) expansion procedure is used, the diffusion terms drop out of all the equations except the TKE equation, where a term of the form suggested by the third-moment closure models of the ABL and OML by Andre *et al.* (1978) and Andre and Lacarrere (1985) can be used. Finally, from tensorial considerations, it is possible to model the pressure diffusion terms (Mellor and Yamada, 1982) as

$$\overline{p u_j} = q l S'_q \frac{\partial q^2}{\partial x_j}. \tag{2.10.26}$$

It is possible to absorb pressure diffusion terms into velocity diffusion terms and put $S'_q = 0$. Laboratory and field observations (Caughey and Wyngaard, 1979) show that the pressure diffusion counteracts velocity diffusion and nearly cancels it in the convective ABL.

The model is now self-complete and closure has been effected. There remains the task of selecting the various constants. It is worth pointing out that the above closure assumes that turbulence can be completely characterized by a single velocity scale q and a single length scale l. The latter is certainly an approximation, because in the presence of external influences such as stratification, rotation, streamline curvature, and a wall, the correlation scale is not isotropic. In fact, there is a preferred direction in all these cases, and the length scale should be a vector quantity. However, we do not have the empirical base to make the length scale a vector, and in application to geophysical BLs, which are nearly horizontally homogeneous on the scales of the PBL thickness, the length scale l can be considered indicative of the direction perpendicular to the surface. Some modelers have used more than one length scale to characterize turbulence (Gaspar *et al.,* 1990).

It is also a common practice to put $S_q = S_{u\theta} = S_\theta$ and regard the constants A_1, B_1, C_1, A_2, B_2, C_2, C_3, and S_q as universal constants. However, it is not contrary to the principles of closure to have these be universal functions of relevant nondimensional numbers such as Reynolds, Rossby, and Richardson numbers. Fortunately, it appears that these are indeed constants (or imperceptibly weak functions) and therefore can be determined from laboratory measurements of simple, classical, neutral, nonrotating turbulent flows, and then used to model complex turbulent flows such as those affected by rotation, stratification, and streamline curvature (Kantha and Rosati, 1990; Kantha *et al.,* 1989), without readjustment.

With the above modeling assumptions, with the use of the expansion procedure of Galperin *et al.* (1988), the turbulence field is described by a differential equation for q^2 and algebraic relations for the second moments:

$$\overline{u_i u_j} = \frac{\delta_{ij}}{3} q^2 - \frac{3A_1 l}{q} \left[\begin{array}{c} \overline{u_i u_k} \dfrac{\partial U_j}{\partial x_k} + \overline{u_j u_k} \dfrac{\partial U_i}{\partial x_k} + \dfrac{2}{3} \delta_{ij} \left(\dfrac{q^3}{B_1 l} \right) \\ - C_1 q^2 \left(\dfrac{\partial U_i}{\partial x_j} + \dfrac{\partial U_j}{\partial x_i} \right) + \beta \left[g_j \overline{u_i \theta} + g_i \overline{u_j \theta} \right] \\ + f_k \left[\varepsilon_{jkl} \overline{u_l u_i} + \varepsilon_{ikl} \overline{u_l u_j} \right] \end{array} \right] \tag{2.10.27}$$

$$\overline{u_j \theta} = - \frac{3A_2 l}{q} \left[\begin{array}{c} \overline{u_j u_k} \dfrac{\partial \Theta}{\partial x_k} + (1 - C_2) \overline{\theta u_k} \dfrac{\partial U_j}{\partial x_k} \\ + (1 - C_3) \beta g_j \overline{\theta^2} + f_k \varepsilon_{jkl} \overline{u_l \theta} \end{array} \right] \tag{2.10.28}$$

$$\overline{\theta^2} = - \frac{B_2 l}{q} \overline{u_k \theta} \frac{\partial \Theta}{\partial x_k} \tag{2.10.29}$$

$$\frac{D}{Dt}(q^2) - \frac{\partial}{\partial x_k} \left[q l S_q \frac{\partial}{\partial x_k}(q^2) \right] = 2 \left[- \overline{u_i u_j} \frac{\partial U_i}{\partial x_j} - \beta g_j \overline{u_j \theta} - \frac{q^3}{B_1 l} \right] . \tag{2.10.30}$$

For level 2 1/2, the third term in Eq. (2.10.27) is replaced by

$$\frac{2}{3} \delta_{ij} \left[- \overline{u_i u_j} \frac{\partial U_i}{\partial x_j} - \beta g_j \overline{u_j \theta} \right] .$$

For application to boundary layers (BLs), all gradients except those in the direction perpendicular to the boundary can be neglected. Also, rotational effects appear to be small in most geophysical BLs (Kantha *et al.*, 1989). We will therefore neglect all rotational terms, although it is straightforward to retain them. Then the mixing coefficients are scalars, and it is possible to align the coordinates in the direction of the local mean flow and put V = 0 without any loss of generality. The set of equations for turbulence quantities then become in their component form

$$\overline{u^2} = \frac{q^2}{3}\left(1 - \frac{6A_1}{B_1}\right) - \frac{6A_1 l}{q}\,\overline{uw}\,\frac{\partial U}{\partial z}$$

$$\overline{v^2} = \frac{q^2}{3}\left(1 - \frac{6A_1}{B_1}\right)$$

$$\overline{w^2} = \frac{q^2}{3}\left(1 - \frac{6A_1}{B_1}\right) + \frac{6A_1 l}{q}\beta g\,\overline{w\theta}$$

$$\overline{uv} = 0 \quad \overline{vw} = 0 \quad \overline{v\theta} = 0$$

$$\overline{uw} = -\frac{3A_1 l}{q}\left[\left(\overline{w^2} - C_1 q^2\right)\frac{\partial U}{\partial z} - \beta g\overline{u\theta}\right] \qquad (2.10.31)$$

$$\overline{u\theta} = -\frac{3A_2 l}{q}\left[\overline{uw}\frac{\partial \Theta}{\partial z} + (1 - C_2)\,\overline{w\theta}\frac{\partial U}{\partial z}\right]$$

$$\overline{w\theta} = -\frac{3A_2 l}{q}\left[\overline{w^2}\frac{\partial \Theta}{\partial z} - (1 - C_3)\,\beta g\,\overline{\theta^2}\right]$$

$$\overline{\theta^2} = -\frac{B_2 l}{q}\,\overline{w\theta}\frac{\partial \Theta}{\partial z}.$$

The values of the constants chosen are

$$(A_1, A_2, B_1, B_2, C_1, C_2, C_3, S_q) = (0.92, 0.74, 16.6, 10.1, 0.08, 0.7, 0.2, 0.2). \qquad (2.10.32)$$

As noted earlier, C_1 is not independent and is related to A_1 and B_1:

$$A_1\left(1 - 6\frac{A_1}{B_1} - 3C_1\right) = B_1^{-1/3}. \qquad (2.10.33)$$

It is possible to derive from this set of equations expressions for the shear stress and vertical heat flux:

$$-\overline{uw} = -\frac{3A_1 l}{q}\left\{\begin{aligned}&\left[\frac{q^3}{3}\left(1-\frac{6A_1}{B_1}\right)+\frac{6A_1 l}{q}\beta g\overline{w\theta}-C_1 q^2\right]\frac{\partial U}{\partial z}\\&+\frac{3A_2 l}{q}\beta g\left[-\overline{uw}\frac{\partial \Theta}{\partial z}-(1-C_2)\overline{w\theta}\frac{\partial U}{\partial z}\right]\end{aligned}\right\} \tag{2.10.34}$$

$$-\overline{w\theta} = -\frac{3A_2 l}{q}\left\{\begin{aligned}&\left[\frac{q^3}{3}\left(1-\frac{6A_1}{B_1}\right)+\frac{6A_1 l}{q}\beta g\overline{w\theta}\right]\frac{\partial \Theta}{\partial z}\\&-(1-C_3)\beta g\frac{B_2 l}{q}\overline{w\theta}\frac{\partial \Theta}{\partial z}\end{aligned}\right\} \tag{2.10.35}$$

If we now define

$$-\overline{uw} = qlS_M\frac{\partial U}{\partial z} \tag{2.10.36}$$

$$-\overline{w\theta} = qlS_H\frac{\partial \Theta}{\partial z} = ql\tilde{S}_H\frac{\partial \Theta}{\partial z}\left(1-\frac{\gamma}{\partial\Theta/\partial z}\right), \tag{2.10.37}$$

we obtain the following expressions for S_M and S_H, stability factors in the mixing coefficients K_M and K_H:

$$S_H = \frac{A_2\left(1-\frac{6A_1}{B_1}\right)}{1-3A_2G_H\left[6A_1+B_2(1-C_3)\right]} \tag{2.10.38}$$

$$S_M = A_1\left\{\frac{\left(1-\frac{6A_1}{B_1}-3C_1\right)+9\left[2A_1+A_2(1-C_2)\right]S_H G_H}{(1-9A_1A_2G_H)}\right\}. \tag{2.10.39}$$

The countergradient term, originally suggested by Deardorff (1966, 1972a), is included in the alternative definition of the stability parameter in Eq. (2.10.37). The quantity g is a constant. The expressions for S_M and S_H involve only the buoyancy gradient, with consequent simplifications in the prescription of realizability constraints (Galperin *et al.*, 1988). Also,

$$G_H = -\frac{l^2}{q^2}\beta g\frac{\partial \Theta}{\partial z}. \tag{2.10.40}$$

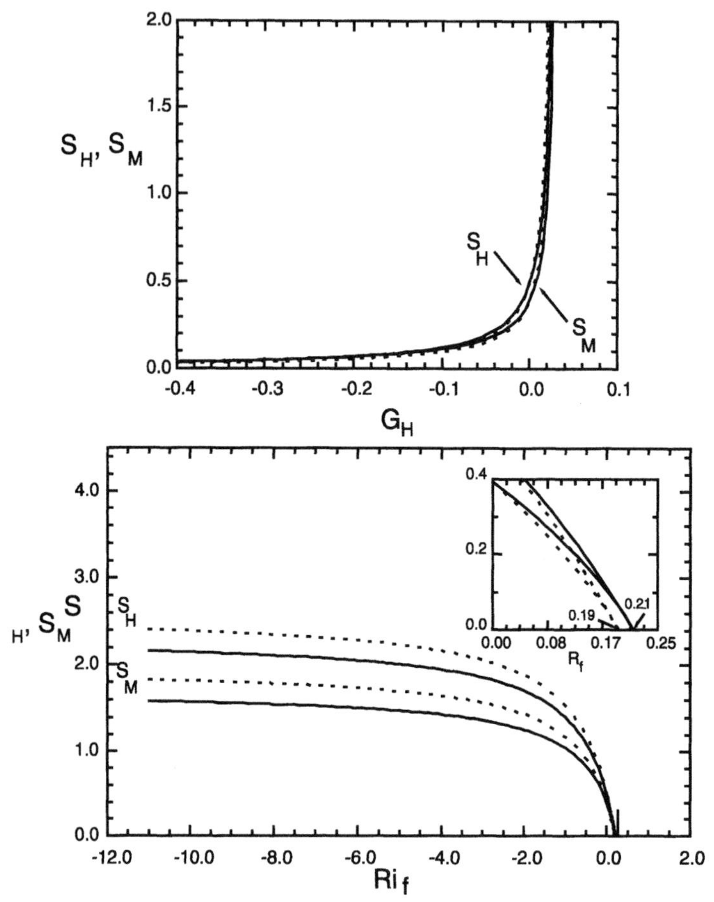

Figure 2.10.3. Stability factors S_M and S_H from the KC second-moment closure model (from Kantha and Clayson, 1994).

Figure 2.10.3 shows the variation of S_M and S_H with G_H, for both nonzero and zero values of constants C_2 and C_3. If we further make the superequilibrium (level 2) approximation

$$S_M G_M + S_H G_H = \frac{1}{B_1}, \tag{2.10.41}$$

where

$$G_M = \frac{l^2}{q^2}\left(\frac{\partial U}{\partial z}\right)^2, \tag{2.10.42}$$

we get

$$S_H = 3A_2\left[\frac{\gamma_1 - (\gamma_1 + \gamma_2)Ri_f}{1 - Ri_f}\right] \tag{2.10.43}$$

$$S_M = \frac{A_1}{A_2}\left\{\frac{B_1(\gamma_1+C_1) - [B_1(\gamma_1+C_1) + 6A_1 + 3A_2(1 - C_2)]Ri_f}{B_1\gamma_1 - [B_1(\gamma_1+\gamma_2) - 3A_1]Ri_f}\right\}S_H \tag{2.10.44}$$

where

$$\gamma_1 = \frac{1}{3} - \frac{2A_1}{B_1} \qquad \gamma_2 = \frac{B_2}{B_1}(1 - C_3) + \frac{6A_1}{B_1} \tag{2.10.45}$$

where Ri_f is the flux Richardson number, the ratio of the destruction rate of TKE by buoyancy forces to the production rate by shear:

$$Ri_f = -\beta g\overline{w\theta} \Big/ \left(-\overline{uw}\frac{\partial U}{\partial z}\right). \tag{2.10.46}$$

Negative values of Ri_f correspond to the unstably stratified situation, where buoyancy forces enhance TKE production, while positive values pertain to stable stratification, which suppresses turbulence. Ri_f is related to the gradient Richardson number Ri by

$$Ri_f = \frac{S_H}{S_M} Ri , \tag{2.10.47}$$

where

$$Ri = \frac{\beta g \dfrac{\partial \Theta}{\partial z}}{\left(\dfrac{\partial U}{\partial z}\right)^2} = -\frac{G_H}{G_M}. \tag{2.10.48}$$

Figure 2.10.3 also shows the variation of S_M and S_H with Ri_f for the super-equilibrium approximation, and compares it to the case of zero values for C_2 and C_3. The critical value of Ri_f for extinction of turbulence by stable stratification is 0.213 as opposed to 0.191 for the old MY value. This is in the right direction of promoting increased mixing, even though the increase is rather modest. Martin (1985) points out that a much higher value of critical flux Richardson number for extinction of turbulence improves the performance of MY level 2 models for Stations Papa and November.

The constant flux region of the ABL, where the shear stress and the buoyancy flux are approximately constant (Monin and Yaglom, 1971) and the wind turning angle small, lends itself to a simple analysis, because there the turbulence is in local equilibrium. Therefore the classical Monin–Obukhov similarity relations in the surface layer can be derived by appealing to the superequilibrium limit of second-moment closure (Mellor, 1973):

$$\frac{1}{3A_1 q_*} = \phi_M \left(\gamma_1 - \frac{6A_1}{q_*^3}\zeta - \frac{3A_2(1 - C_2)}{q_*^3}\zeta - C_1 \right) - \frac{3A_2\zeta}{q_*^3}\phi_H \tag{2.10.49}$$

$$\frac{1}{3A_2 q_*} = \phi_H \left(\gamma_1 - \frac{6A_1}{q_*^3}\zeta - \frac{B_2(1 - C_3)}{q_*^3}\zeta \right) \tag{2.10.50}$$

$$q_*^3 = B_1 \left(\phi_M - \zeta \right), \tag{2.10.51}$$

where

$$\phi_M = \frac{l}{u_*}\frac{\partial U}{\partial z} \tag{2.10.52}$$

$$\phi_H = \frac{l}{\theta_*}\frac{\partial \Theta}{\partial z} \tag{2.10.53}$$

are the Monin–Obukhoff similarity functions for velocity and temperature, which are functions of the Monin–Obukhov similarity variable ζ (Monin and Yaglom, 1971). They provide information on the velocity and temperature profiles in the constant flux layer of a PBL and have proved to be immensely useful to ABL modelers. These flux profiles are in fact the foundation for the ABL model of Troen and Mahrt (1986) and the LMD (Large *et al.*, 1994) model of the OML.

Figure 2.10.4 shows the Monin–Obukhov similarity functions ϕ_M and ϕ_H as functions of ζ, for zero and nonzero values of C_2 and C_3. The observational data of Businger *et al.* (1971) are also shown. Optimum values of C_2 and C_3 were chosen from the best agreement possible with Businger *et al.* (1971) data. The figure also shows results for $C_2 = 0.5$ and $C_3 = 0.5$. In general, higher values of C_2 tend to raise the curves in the negative region, while the slope on the stable side diminishes. We found the best compromise to be $C_2 = 0.7$ and $C_3 = 0.2$. Values of $C_2 = 0.5$ and $C_3 = 0.5$ lead to unrealistic lowering of the slope of ϕ_H on the stable side, while raising the values of ϕ_H above the observed values on the unstable side, but does not improve significantly the values of ϕ_M on the unstable side.

In applications of the MY (and other) closure models, one needs to impose realizability conditions on quantities to prevent nonphysical results. For example, G_H should be bounded so that G_M does not become negative on the unstable side. This condition can be derived from Eq. (2.10.45):

$$G_H \le \left\{ A_2 \left[B_1 + 12A_1 + 3B_2(1 - C_3) \right] \right\}^{-1}. \tag{2.10.54}$$

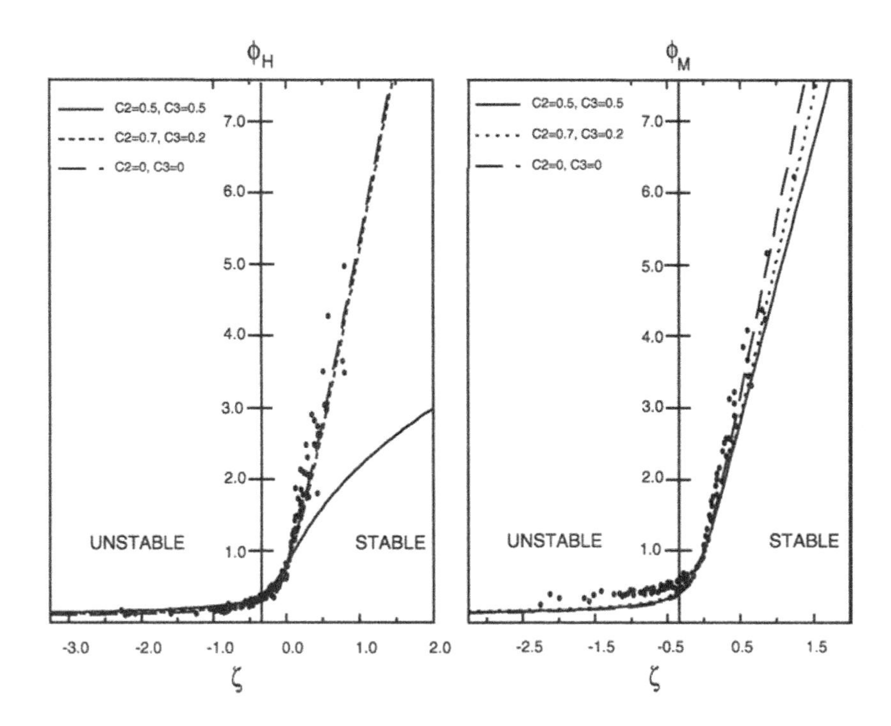

Figure 2.10.4 Comparison of Monin–Obukhoff similarity functions from observations and the KC model (from Kantha and Clayson, 1994).

This also ensures positive definiteness of the vertical velocity variance $\overline{w^2}$. Mellor and Yamada (1982) imposed an upper bound of 0.0326 in their level 2 1/2 model to achieve this, while a somewhat lower bound of 0.0288 ensures positive definiteness for all velocity variances. It is also necessary to ensure that none of the components of TKE fall below a certain fraction of the total. Mellor and Yamada (1982) ensure that $\overline{w^2}/q^2 > 0.12$ by requiring $G_M \leq (0.48 - 15.0\ G_H)$. Hassid and Galperin (1983) impose a slightly different set of realizability conditions. Their upper bound on G_M is different from that above, but nevertheless ensures that an increase in the velocity gradient does not cause a decrease in the Reynolds stress. In the quasi-equilibrium model presented here, S_M and S_H depend only on G_H, and therefore it is necessary to impose realizability conditions only on G_H. In this sense, the KC model, akin to that of Galperin *et al.* (1988), is more robust than the original MY model. We impose an upper bound on G_H of 0.029, slightly lower than that given by Eq. (2.10.54), to ensure positive definiteness of all velocity variances.

There is also a need for a lower bound on G_H, which has not been widely recognized (see Galperin *et al.,* 1988). Under strong stable stratification conditions, there is a limit to the size beyond which eddies are incapable of overturning and hence lose their ability to contribute to the cascade of TKE down the spectrum to dissipative scales. Velocity fluctuations beyond this limit are more indicative of internal wave activity. This bound is dictated by the Ozmidov length scale, $L_o = (\varepsilon/N^3)^{1/2}$, where ε is the dissipation rate. This is an important parameter in stably stratified flows. Imposing a bound on the turbulence length scale that is proportional to the Ozmidov scale leads to the following result:

$$\frac{Nl}{q} \leq C_4 \, . \tag{2.10.55}$$

C_4 is 0.53 (Galperin *et al.*, 1988), which leads to a lower bound on G_H of –0.28. The above relationship can also be written as

$$\frac{N}{(q/l)} \leq C_4. \tag{2.10.56}$$

The denominator is the turbulence frequency, the inverse of the characteristic macro timescale of turbulent motions, which is the eddy turnover timescale. Only below the local buoyancy frequency N are internal wave motions permitted by the stable stratification. The upper bound on turbulence length scale therefore ensures that the turbulent motions do not "intrude" into the internal wave portion of the frequency spectrum. The closure model deals only with turbulent motions that participate in the cascade and in the mixing processes, and not with internal wave motions, which do not.

The practical utility of this limit arises in regions like the equatorial mixed layer, where immediately below the upper mixed layer lies the strong equatorial undercurrent. Without the bound, the undercurrent tends to get excessively smeared. Smith and Hess (1993) have reported on the use of this bound and demonstrated its usefulness in models of the equatorial oceans.

Finally, the model requires a prescription for the turbulence length scale. It has been traditional to use either a Blackadar (1962) type formulation (Martin, 1985) or a conservation equation for a quantity that includes the length scale (Mellor and Yamada, 1982). The former postulates a length scale proportional to the distance from the surface near the bounding surface, but asymptoting to a constant value as the edge of the boundary layer is approached. It is valid only for MLs adjacent to a surface such as the ABL and OML, but fails for multiple turbulent regions separated from each other (for example, the upper layers of the equatorial region). Solving a prognostic relationship for the length scale overcomes this problem as long as adequate care is given to the proper bounds on the length scale as discussed above.

Different formulations are possible for the length scale equation and various methods have been tried (Mellor and Yamada, 1982; Rodi, 1987). Kantha and Clayson (1994) and Mellor and Yamada (1982) use an equation for the term $q^2 l$ which includes a wall proximity term to force the model toward a log-law behavior near a surface,

$$\frac{D}{D_t}(q^2) - \frac{\partial}{\partial z}\left[lqS_q\frac{\partial}{\partial z}(q^2)\right] = -2\overline{uw}\frac{\partial U}{\partial z} + 2\beta g\overline{w\theta} - \frac{2q^3}{B_1 l} \tag{2.10.57}$$

$$\frac{D}{Dt}(q^2 l) - \frac{\partial}{\partial z}\left[lqS_l\frac{\partial}{\partial z}(q^2 l)\right]$$

$$= E_1 l\left[-\overline{uw}\frac{\partial U}{\partial z} + E_3\beta g\overline{w\theta}\right] - \frac{q^3}{B_1}\left[1 + E_2\left(\frac{l}{\kappa l_w}\right)^2\right] \tag{2.10.58}$$

where $S_l = S_q$. The last term in Eq. (2.10.58) is the wall proximity correction term. The length scale l_w denotes the distance from a surface; this term ensures log-law behavior near a surface. The values for constants in these equations are

$$(E_1, E_2, E_3, E_4) = (1.8, 1.33, 1.0, 1.0)\,. \tag{2.10.59}$$

It is also possible to add a term to the length scale equation to account for the influence of rotation (Kantha *et al.,* 1989) and countergradient terms to the equations above, patterned after Andre and Lacarrere (1985). However, the values of additional empirical constants that would result are not known with certainty. As the inclusion of the countergradient term made little difference to their results overall, the countergradient terms in both the q^2 and $q^2 l$ equations can be ignored.

An alternative formulation popular in aerodynamic flow calculations is to use an equation for the dissipation rate to provide information on the length scale. Whatever the formulation, many ad-hoc assumptions are involved and the proper prescription of the length scale remains one of the major unresolved issues in turbulence modeling in general. Resolution of this issue would have some bearing on the simulation of the bulk skin sea surface temperature difference in the upper ocean (Schluessel *et al.,* 1987; Wick *et al.,* 1992).

Although the S_M, S_H, and other expressions above were derived by orienting the coordinates in the local flow direction (V = 0), the expressions are of general applicability. For example, by replacing the shear dU/dz by $[(dU/dz)^2 + (dV/dz)^2]^{1/2}$, the coordinates can be chosen more generally.

The KC model also includes mixing in the strongly sheared, strongly stable thermocline/halocline region immediately below the ML. As discussed above, second-moment closure applies only to the asymptotic limit of high Reynolds number turbulence, in other words, to a fully turbulent region like the ML. In the stably stratified region below, the turbulence is intermittent and episodic. There exists at present no reliable closure model for intermittent turbulence, although attempts have been made to treat this problem by analogy to two-phase fluids, the two phases being a laminar and a turbulent fluid (Markatos, 1986). The best that can be done at present are empirical relationships from laboratory and field observations. It is traditional to present these observations in terms of gradient Richardson-number-dependent mixing coefficients, and these can be used for parameterizations of mixing below the OML.

As originally formulated, MY models did not account for mixing below the OML. Attempts have since been made to remedy this shortcoming (Mellor, 1989). However, there has been a failure to recognize the unusually high levels of mixing that exist in the region, and the parameterizations used have been such that the added mixing below made little difference to the outcome. KC follow the work of Large *et al.* (1994), who demonstrate the need for strong mixing in the region to remedy this deficiency in a ML model.

Mixing in the strongly stable but highly sheared region below the OML is caused by episodic Kelvin–Helmholtz shear instabilities. In addition, this region happens to be a region of considerable variability on short timescales due to significant internal wave activity, especially in the equatorial regions. Internal waves generated by turbulence in the ML (Kantha, 1977, 1979a; McPhee and Kantha, 1989) as well as those propagating into the region from elsewhere modulate the shear significantly and enhance the instabilities episodically. The usual scenario is for the shear to be intensified by some process, causing a lowering of the gradient Richardson number, which leads to shear instability and intense mixing that increases the Richardson number locally and stabilizes the fluid column. Such episodic mixing has been difficult to comprehend and model. While mixing in a fully turbulent region such as the ML is better understood and, arguably, reasonably well parameterized, a good understanding and modeling of intermittent turbulence have proved elusive. The best we can do at present are ad hoc parameterizations of mixing obtained from knowledge culled from observations in the laboratory and in the field (Peters *et al.*, 1988).

There are a variety of Richardson-number-dependent parameterizations possible (Pacanowski and Philander, 1981; Peters *et al.*, 1988; Gregg, 1987), but KC chose to follow Large *et al.* (1994) in parameterizing the shear-induced diffusivities as a strongly decreasing function of the gradient Richardson number Ri and constant values for internal-wave-induced diffusion,

$$K_{MR} = 10^{-4} \; K_{HR} = 5 \times 10^{-5} \qquad \mathrm{Ri} > 0.7$$

$$\left.\begin{aligned} K_{MB} &= K_{MR} + 5\times10^{-3}\left[1-\left(\frac{Ri}{0.7}\right)^2\right]^3 \\ K_{HB} &= K_{HR} + 5\times10^{-3}\left[1-\left(\frac{Ri}{0.7}\right)^2\right]^3 \end{aligned}\right\} 0 < \mathrm{Ri} < 0.7 \qquad (2.10.60)$$

$$\left.\begin{aligned} K_{MB} &= K_{MR} + 5\times10^{-3} \\ K_{HB} &= K_{HR} + 5\times10^{-3} \end{aligned}\right\} \qquad \mathrm{Ri} < 0.$$

The unit for all diffusivities in Eq. (2.10.60) is $m^2\ s^{-1}$. K_{MR} and K_{HR} denote diffusivities due to internal wave (IW) activity in the pycnocline, and K_{MB} and K_{HB} denote the total diffusivities for momentum and scalars. There is a high level of residual diffusion inherent in this formulation that is quite appropriate for the highly sheared and active region immediately below the ML. The value KC chose for K_{HR} is, however, different from the value used by Large *et al.* (1994) (5×10^{-5} as compared to their value of $10^{-5}\ m^2\ s^{-1}$). Experience indicates the necessity for the high levels of mixing, especially on short and event timescales represented by the diurnal response and storm-induced mixing episodes. In the former, the mixed layer is quite shallow, and the proximity to the ocean surface might be responsible for enhanced variability in the strongly sheared and stable region below. For the latter, vigorous eddy activity as well as induced, strong, inertial currents may be responsible for enhanced mixing events below the ML proper. In the new model, the diffusivities are put to the above values immediately below the fully turbulent ML. The base of the active ML is diagnosed by q^2 values dropping to a low value ($10^{-6}\ m^2\ s^{-2}$ or less), indicating extinction of turbulence. There is little sensitivity to the choice of this value, since extinction near the ML base is rather abrupt.

It appears that ML models cannot be made to work properly without a strong Ri-dependent mixing in the thermocline. The strong heating in the thermocline below the active ML observed during storm events (Crawford, 1993) cannot be simulated without this term. Lack of this capability in earlier second-moment closure models is partly responsible for the poor performance of those models in the past.

The KC model has in addition a weak but constant background diffusion with values appropriate to the deep ocean ($K_{MD} = K_{HD} = 10^{-5}\ m^2\ s^{-1}$). Provision has also been made for buoyancy-frequency-dependent deep mixing (Gargett and Holloway, 1984, 1992).

The surface boundary condition used in the q^2 equation is that on q^2, a Dirichlet condition. It is possible to use instead Neumann conditions that prescribe its vertical derivative. This is equivalent to prescribing the flux of TKE (instead of the TKE itself), and is well suited to impose the influence of surface wave breaking. Kantha (1985), Craig and Banner (1994), Stacey and Pond (1997) and Clayson and Kantha (2000) have shown that this alternative prescription is important only in the upper few meters, where the properties in the mixed layer such as the mean velocity change significantly because of the increased mixing brought on by injection of turbulence by wave breaking. It is therefore important in modeling shallow mixed layers. However, since a mixed layer is stirred by turbulence from both the top and the bottom, and the shear-induced turbulence at the bottom is more effective in deepening, the alternative prescription does not change the deepening rate of the mixed layer significantly, unless it is rather shallow.

2.10.4 Comparison with Oceanic Data Sets

The mixed layer models described above have been applied to various data sets available from field observations in the global oceans. A few of these comparisons are given below, in order to demonstrate the utility of the various ML models and other issues associated with ML model simulations. Further comparisons, in addition to those discussed in this chapter, have been performed by Stramma *et al.* (1986), Gaspar (1988), Gaspar *et al.* (1990), Therry and Lacarrere (1983), and Schudlich and Price (1992).

One important issue in dealing with ML model simulations is related to surface fluxes imposed. Since all relevant surface fluxes are seldom measured directly in the field, it is necessary to parameterize at least some. It becomes difficult, therefore, to separate the test of model skill from that of the accuracy of forcing parameterization. Verification data for assessing model skill in detail are also scarce. Most often only the SST is available for comparison, and properties such as the TKE and its dissipation rate are not available for testing the turbulence field predicted by the model. There is only one data set [Tropical Ocean and Global Atmosphere/Coupled Ocean–Atmosphere Response Experiment (TOGA/COARE) data], where all the surface fluxes were measured, and measured accurately and concurrently with the properties in the OML (Andersen *et al.,* 1996; Weller and Andersen, 1996). This data set therefore serves well the task of assessing the skill of a ML model.

Additional complications in using and validating ML models are advection and upwelling/downwelling processes. In general, validation of ML models is performed by running the models in a one-dimensional version, precluding the inclusion of changes in the water column due to three-dimensional effects. Some of the ways that researchers have dealt with this difficulty are demonstrated below.

The dynamics of a mixed layer are sensitive to the parameterization of solar extinction in the water column. ML modelers have therefore tried to incorporate better parameterizations of absorption of solar radiation in the water column (Ivanoff, 1977; Simpson and Dickey, 1981a,b; Siegel and Dickey, 1987). Most ML models now have at least a two-band model, representing the longwave IR and shortwave visible parts of the spectrum that have different extinction length scales, following recommendations of Paulson and Simpson (1977) for various Jerlov (1976) water types. However, it has become increasingly clear (see Morel and Antoine, 1994) that an even more accurate extinction parameterization is essential for a better simulation of many ML properties, including those pertaining to the dynamics (Martin and Allard, 1993). Since photosynthesis is sensitive to the spectral properties of shortwave solar radiation in the water column, it is even more important for applications to biological productivity (Sathyendranath and Platt, 1988). Solar extinction characteristics were described in Section 2.3.

Modelers must also consider how best to validate their models. Traditionally modelers have used only surface temperatures and temperature profiles for validation. There are several reasons for this; one of these is that for many cases temperature data are the only data available. Salinity data are still rather scarce, outside of a few isolated cases. In addition, turbulence data are even scarcer than salinity data. As opposed to the ABL, where the various terms of the TKE equation have been measured during field experiments for many years, the only turbulence data available for the OML are dissipation rates deduced from microstructure profiles (see Section 2.7 for examples of calculated and modeled dissipation rates). Thus, models cannot be compared with rates of TKE production, turbulent transport, etc., but only with the end product of the mixing processes, temperature and salinity profiles, both of which can be affected by other processes like advection. Not all ML models simulate the turbulent quantities either (bulk models, for example). Finally, in terms of air–sea coupling and climate modeling, it is the sea surface temperature to which the ABL responds (due to its effect on surface fluxes). It is therefore natural that a premium is placed on a model's ability to reproduce this parameter.

A mixed layer model suitable for geophysical applications must be able to simulate processes on a variety of timescales from diurnal and event scales to multiyear timescales. It must perform reliably under strong daytime heating and nighttime cooling. However, this ability is necessary but not sufficient. A mixed layer model must also perform well under strong event scale forcing such as that due to rapidly moving storms and transient wind events lasting from a few hours to days throughout the global oceans. Wintertime ML deepening at midlatitudes occurs principally during strong winter storm events. Often the divergence of ML model predictions (such as SST) from reality can be traced directly back to inadequate performance during an earlier mixing event [see, for example, the MY model simulations presented by Martin (1985) for MILE]. In this section, we present results from several studies emphasizing a range of timescales and events.

One of the few places in the global oceans where long-term observations are available is the ocean weather Station (OWS) Papa located in the North Pacific at 145° W, 50° N, where every 3 hr meteorological and upper ocean temperature profiles were collected from the 1940s to the early 1980s. Due to minimal advection effects [except during the winter months—see Large *et al.* (1994)] many comparisons of one-dimensional mixed layer models have utilized this data set, beginning with Denman (1973). Martin (1985, 1986) performed extensive comparisons of the Mellor and Yamada (1982), Niiler (1975), and

Therry and Lacarrere (1983) mixed layer models at Station Papa. More recently, Large *et al.* (1994) have a made a thorough investigation of the heat balances and fluxes there, in addition to making a very long (20-year) simulation of Papa data with their model, with excellent results on interannual variability. Both Martin (1985, 1986) and Large *et al.* (1994) present results for year 1961, and it is this year that we will also concentrate on.

For these simulations of seasonal variability, all the ML models were forced by heat fluxes which are computed using the same radiation and bulk formulas as Martin. However, evaporation was included, and precipitation was prescribed, using the mean annual cycle derived by Tabata (1965) as in the simulation by Large *et al.* (1994). The salinity flux was not included in Martin's simulations. As Large *et al.* have shown, for long-term simulations it is essential to consider the salinity fluxes at the surface, as well as the net heat balances over the year and Ekman-divergence-driven upwelling from below the thermocline, to produce proper interannual variability. At Station Papa there appears to exist advection of cold water in the seasonal thermocline during winter that essentially balances the 29 W m^{-2} excess heat input to the upper mixed layer. For the year 1961, this imbalance appears to be 34 W m^{-2} (Large *et al.,* 1994). However, advection effects were not included in either the KC, the PWP, or the Garwood model simulations, contributing to the overestimation of SST by a degree or so at the end of the year. As in Martin's simulations, the model calculations are started at the beginning of the year (Large *et al.* begin their simulations in the middle of March). The model is initialized by the observed temperature profile. The strong halocline that exists below 125 m is reproduced by initializing the model with climatological salinity data. Solar extinction corresponding to the Type II waters of Jerlov (1976) was used for this study. Martin (1985, 1986) also used Type II water for his model simulations. Large *et al.* cite evidence that the water quality varies with seasons, from Type IA during winter to Type II during summer.

Figure 2.10.5 shows the observed SST, and the modeled SST from the KC, PWP, and Garwood models, with the same initialization and forcing data. The observed and modeled mixed layer depths are shown in Figure 2.10.6. There is a small cold bias that may be attributable more to surface flux parameterization than to model performance, as the simulations have a deficit of 25 W m^{-2} for the year. In addition, because of the 34 W m^{-2} net heat input during the year and lack of cold water advection to balance this, the model SST ends up about 1°C higher than the observed value at the end of the model year. As Large *et al.* (1994) showed, this bias can be eliminated by allowing for wintertime advection effects in multiyear simulations. Figure 2.10.7 shows modeled SST for Papa from the LMD model. It appears from these simulations that with careful attention to the heat budget considerations, one can simulate the seasonal cycle of SST to within about 1 K.

While simulations at PAPA focused on long-term temperature changes, other experiments have focused on diurnal variability. Price *et al.* (1986) reported on observations and model simulations, using the data collected at R/P Flip in 1980 at a location 4° west of San Diego at 30° N latitude. CTD observations were made every hour during the deployment, in addition to the usual suite of meteorological measurements. The data between May 7 and May 11 have been found well suited for testing 1D ML models because of the absence of advection and the presence of very strong insolation that taxes the ability of ML models to

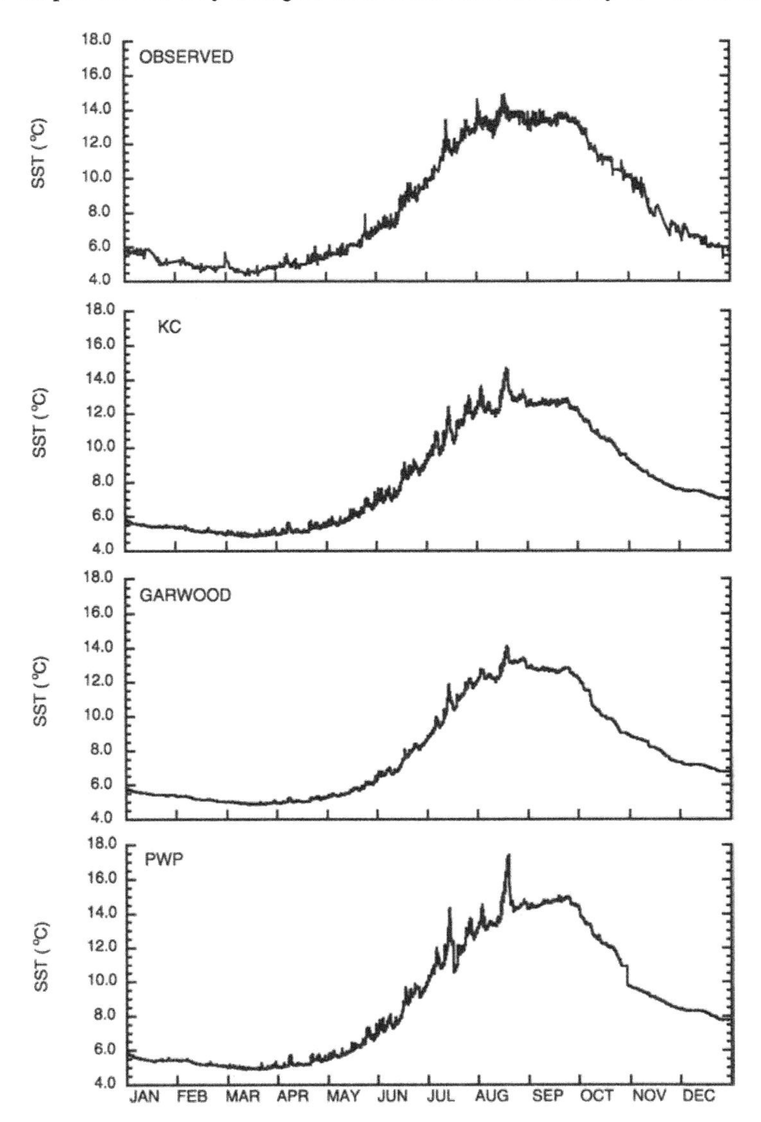

Figure 2.10.5. Comparison of SST observations with simulations from KC, Garwood, and PWP models for Station Papa. Note the seasonal modulation of SST (from Kantha and Clayson, 1994).

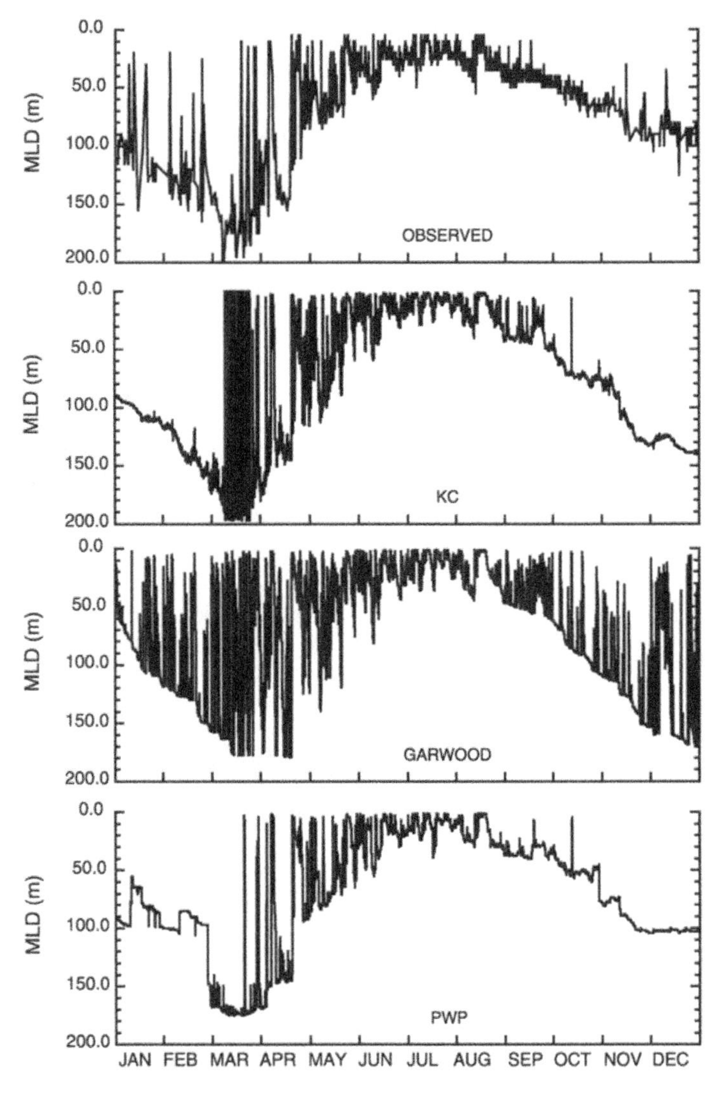

Figure 2.10.6. Comparison of observations of ML depth with simulations from KC, Garwood, and PWP models for Station Papa (from Kantha and Clayson, 1994).

simulate the diurnal ML. The magnitudes of the diurnal SST peaks over several days are a critical test of a model's ability to simulate conditions pertaining to strong positive buoyancy flux at the surface, as well as the ability to simulate convective cooling the following night. Martin (1986) also tested several ML models against Flip data from May 9 to May 10, and reported that the PWP and Therry and Lacarrere (1983) models performed best, while the MY model

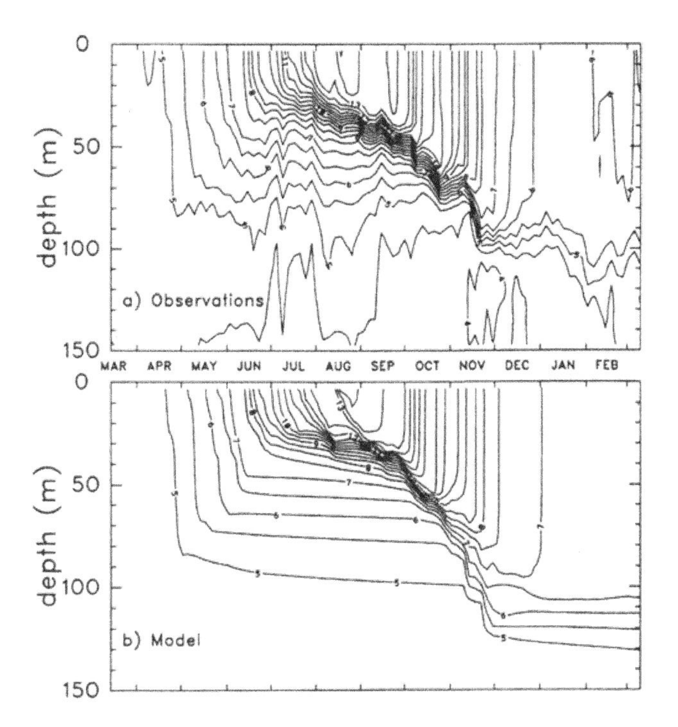

Figure 2.10.7. Comparison of observations of ML temperature with simulations from the LMD model for Station Papa (from Large *et al.*, 1994).

overpredicted SST buildup due to insufficient mixing. Here we present results for Flip of the KC, PWP, and Garwood models.

The forcing data used in these simulations are the same as those Martin (1986) calculated from half-hourly meteorological observations and used in his simulations (solar insolation was measured). The models were initialized with BT data taken during the morning of May 7 (Day 128) and integrated forward to the end of May 12 (Day 133). Salinity was taken to be uniform in the upper layer. Figure 2.10.8 compares the modeled temperatures at the surface and at 5-, 10-, 15-, 20-, 25-, and 30-m levels to the observed values. The KC model appears to estimate the peak SSTs quite well. The warm bias in earlier MY second-moment closure models (Martin, 1986) is not present. The performance vis-â-vis the SST peaks is comparable to the PWP model, while the Garwood model tends to underpredict the diurnal variation. No results are available for the LMD model. However, Kantha and Clayson (1994) and Large *et al.* (1994) simulated the diurnal mixed layer evolution during the LOTUS experiment, and from those results one can conclude that given sufficiently accurate surface fluxes, modern ML models appear to be able to simulate the amplitude of the diurnal cycle of SST to within about 0.5 K.

Data taken during the Ocean Storms experiment (D'Asaro, 1985) provide an excellent observation of the mixed layer behavior during several storm events (see Crawford, 1993) [see Large and Crawford (1995) for a detailed discussion of the observations and processing procedures]. Crawford (1993) performed detailed ML simulations using the MY, PWP, and LMD models. He reports in

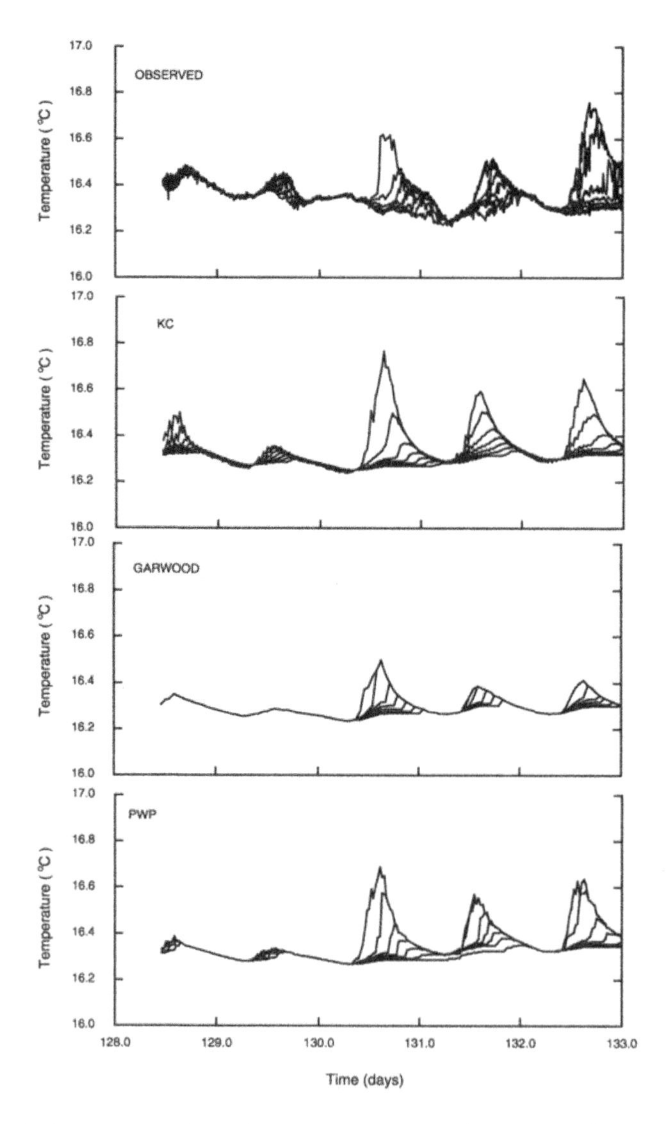

Figure 2.10.8. Comparison of observations of temperature at various depths with simulations from KC, Garwood, and PWP models for FLIP. Note the strong diurnal modulation of near-surface temperatures (from Kantha and Clayson, 1994).

detail his simulations for the storm event that he designated Event 3 on Day 277 The SST dropped by about 1.1°C between Days 277.1 and 278.0, even though the ML depth did not increase dramatically. Instead, the most remarkable feature was the considerable lowering of isotherms below the seasonal thermocline (by as much as 20 m) as a result of the storm, indicating substantial heating there (Large *et al.,* 1986). Crawford (1993) finds that the resonant ocean response to inertially rotating winds is responsible for this (Large and Crawford, 1995; Large *et al.,* 1994). This suggests that considerable mixing occurs below the conventional mixed layer, which conditions the water column for subsequent mixing events. Thus it is important to accurately model the mixing not only in the ML but also in the strongly sheared region of the seasonal thermocline immediately below the ML.

Simulations for Storm Event 3 of Crawford (1993) will be presented. Crawford discovered that while the LMD model tended to slightly overestimate mixing and predict cooler SSTs, the MY and PWP models tended to underpredict mixing during storm events and therefore to underpredict the degree of cooling. For example, for Storm Event 3 the observed SST drop was 1.15°C, while the LMD model indicates a change of 1.3°C, and the MY and PWP models indicate changes of around 0.65 and 0.85°C, respectively. Also, the warming of the deep layers was more realistic in the LMD model. The results are shown in Figure 2.10.9, which compares the KC model temperature profile with the observed one. LMD, PWP, and MY results are also shown for comparison. The KC model indicates a drop in SST of about 1.0°C, and the depression of isotherms by the storm also appears realistic, although the extent of penetration is slightly lower than that observed. Note that the storm commences on Day 277.1 and is essentially over by Day 278.0, although the SST continues to decrease.

Perhaps the most critical and stringent test of a ML model is the tropical ML in the equatorial waveguide. Since it appears that the pycnocline here is a region of highly elevated turbulent dissipation and intense internal wave activity at night (Brainard *et al.,* 1994; Peters *et al.,* 1994), it is likely that the shear-instability-induced mixing below the active ML is even more important for equatorial mixing (Moum and Caldwell, 1989; Peters *et al.,* 1988; Garwood *et al.,* 1989). In addition, the presence of strong currents such as the Equatorial Undercurrent (EUC) and hence shears in the immediate vicinity of the ML play an important role in ML deepening. Some results of modeling in this region were discussed in Section 2.7.

During the TOGA/COARE Intensive Operations Phase (IOP), the 3-m WHOI Improved Meteorological (IMET) buoy was equipped with a redundant set of meteorological sensors and deployed at 1° 45′ S, 156° E, in the warm pool region of the western tropical Pacific. The data collected from this buoy were described by Weller and Anderson (1996). This data were used by Anderson *et al.* (1996) to force and validate the PWP model. In order to alleviate the effects of advection on the resultant simulation, the model was run over the entire IOP

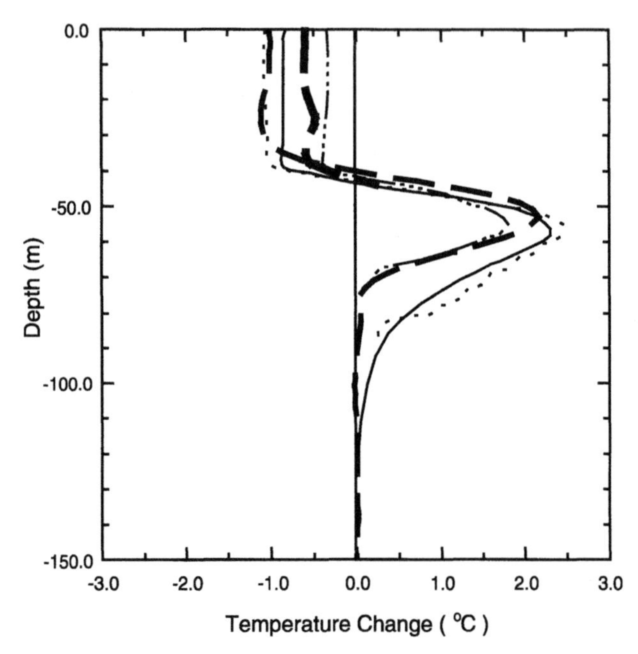

Figure 2.10.9. Comparison of the observed (solid line) and modeled temperature changes due to a storm (from Kantha and Clayson, 1994).

with an initialization of the mean temperature, salinity, and velocity profiles over this time period. Anderson *et al.* (1996) show a consistent positive bias in their PWP model simulations of about 0.8–1.0°C throughout the IOP period, while the amplitude of the diurnal SST modulation is well represented. The KC model was also run over this time period, using the same forcing and initialization profile, and the results of the observed versus model sea surface temperature are shown in Figure 2.10.10. Model SSTs agree well with measured SSTs, although there is a divergence between the model and the observed SST of about 0.5°C after Day 382 (January 17th). However, this excellent agreement is misleading since the prescribed initial conditions are averaged over the entire IOP, thus minimizing any influence of advection events that the 1D model cannot account for. When the model is initialized by the profiles observed at the beginning of the IOP, its performance deteriorates quite a bit, as can be seen from the fact that the model begins to diverge around Day 318 (November 14th) and differs by nearly 0.6°C until Day 375 (January 10th), but toward the end of the IOP (Day 430, March 3rd) the difference is nearly 1.5°C (Figure 2.10.11). This is presumably due to nonlocal (advective) influences. Observed temperature and salinity profiles show strong advection-like events around Days 318, 340, 370, 400, and 410. This highlights the risky nature of using a 1D model to simulate the upper ocean structure over a long period in a region like

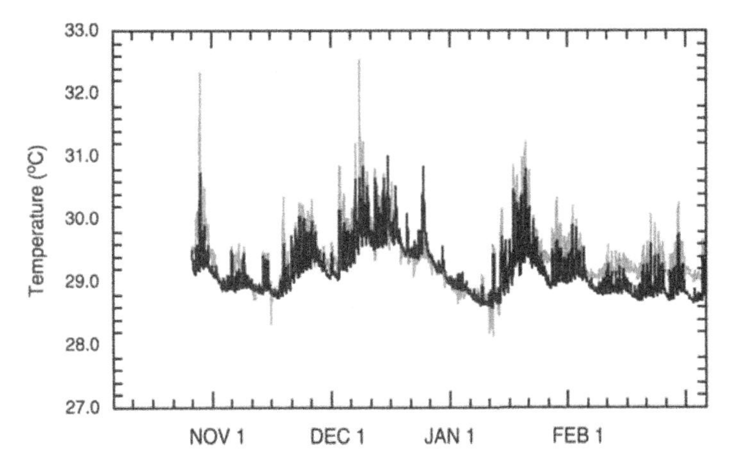

Figure 2.10.10. Observed SST (gray line) compared with SST simulated by the KC model (black line) initialized using mean profiles of temperature, salinity, and velocity.

the tropical waveguide, where nonlocal events are bound to affect the ML over time scales of weeks to months. However, simulations restricted to simulation periods, when advective effects are small, suggests that modern ML models are capable of accurately simulating the ML properties (for example, SST to accuracies better than 0.5°C), as long as accurate surface fluxes (as in TOGA/COARE observations above) are used to drive these models (see Clayson *et al.* 1997).

Temperature and salinity are not the only features of the mixed layer that ML models can be used to reproduce. The upper ocean is also an important source of certain photochemicals important to stratospheric and tropospheric atmospheric chemistry and hence to the global radiative and ozone budget. The reader is referred to Najjar *et al.* (1994) for both an in-depth review and a thorough analysis of the problem. Carbonyl sulfide (CoS) and carbon monoxide (CO) are examples of such photochemicals that are produced in the upper ocean by solar insolation and then outgassed to the atmosphere. These photochemicals are also destroyed by bacterial activity or chemical reactions; hence their concentrations near the ocean surface depend on the balance between the production and destruction rates. However, mixing plays a very important role in determining the concentration of these photochemicals, since they are produced only near the surface and can be transported to depths by turbulence.

CoS and CO are both produced by photolysis of dissolved organic matter in seawater by UV radiation and hence exhibit strong diurnal variability. While UV radiation is only 2% of the solar insolation (Martin and Allard, 1993), it is important to the photochemical question. CoS and CO may often be undersaturated during the night and become supersaturated during the day. CoS is lost by hydrolysis; hence its concentration is a function of the balance between

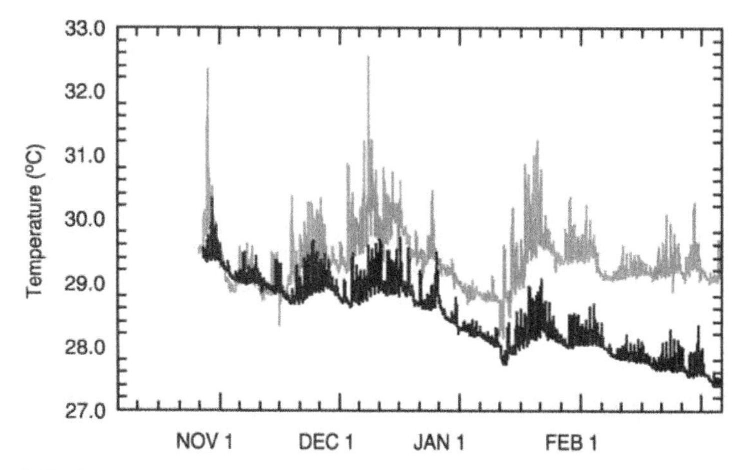

Figure 2.10.11. Observed SST (gray line) compared with SST simulated by the KC model (black line) initialized using the observed initial profiles of temperature, salinity, and velocity

photolytic and hydrolytic reactions, and of mixing processes. To model the concentration of such a photochemical compound well, it is essential to model the mixing in the upper ocean well. In fact, the ability to model photochemical compounds well may be another stringent test of a mixed layer model (and its solar extinction profile parameterization).

From the few measurements on photochemicals that have been made (see the review by Najjar *et al.,* 1994), a few qualitative aspects emerge: (1) The photochemicals can become highly supersaturated during the day with supersaturation ratios reaching 10–15. (2) The concentration decreases rapidly once the sun sets, often to near-saturation or undersaturation values during the night. (3) There may be a significant lag (typically a few hours) between the peak of solar insolation and the peak concentration of the photochemical. These features can be tested at least in a qualitative sense in a mixing model.

The KC ML model was applied to the problem of modeling CoS concentrations (Kantha and Clayson 1994). An equation for the CoS concentration that is very similar to the scalar conservation equation (such as for salinity) is used but with appropriate source and sink terms on the right-hand side in addition to the diffusion terms; the photochemical production term S_o and the hydrolytic destruction term S_i have been described by Najjar *et al.* (1994). Air–sea transfer of CoS is parameterized by a standard bulk formula with proper accounting for the solubility of CoS in seawater. The sink term S_i is K_hC, where K_h is modeled as an exponential function of temperature with appropriate constants for a pCO_2 of 350 ma, a salinity of 34.6 psu, and an alkalinity of 2320 meq kg^{-1}, from information compiled by R. G. Najjar (personal communication, 1993):

$$K_h = 2.2 \times 10^{-2} \exp (T/6.3) \qquad (2.10.65)$$

The solubility and diffusivity of CoS are also modeled as exponential functions of temperature. The extinction scale for UV radiation is taken as 5 m, and the CoS production rate is made proportional to the shortwave solar insolation, since the UV component at the top of the atmosphere is a constant fraction of the shortwave insolation. The proportionality constant is, however, quite arbitrary. The dependence of UV component intensity at the surface on the zenith angle can also be taken into account.

Sensitivity studies of the diurnal modulation of CoS concentration with the model show that at high temperature values (~25°C) typical of summer at midlatitudes and low latitudes, peak CoS concentration at the sea surface lags the peak of solar insolation by 1–2 hr. However, since the peak of the SST lags by 2–3 hr and since the CoS saturation value is a strong decreasing function of temperature, a double peak is often observed in the CoS saturation ratio, one near the peak of solar insolation and another near the SST peak, separated by 1/2 to 1 hr. The surface concentrations are a strong function of the solar insolation, as can be expected. The peak CoS values at deeper levels do not lag the surface value significantly.

Salient aspects such as those revealed by these sensitivity studies can be examined by application to different field data sets. These studies show that while this leads to an overestimation of photochemical concentrations by up to 10%, the behavior around the peak solar insolation is not much different. The simulations for LOTUS and Papa locations in which accurate solar insolation measurements present a good contrast in behavior of CoS under different solar insolation values, SSTs, and mixing. At the LOTUS site, the models can be tested on synoptic and diurnal timescales at midlatitude conditions, while at Papa, seasonal simulations are possible.

Figure 2.10.12 shows measured solar insolation, SST, and CoS concentration as a function of time for the LOTUS data set. The strong diurnal variability shown by the model is consistent with the few observations that exist. The peak CoS saturation ratios reach a value of 4–6 on most days, in good agreement with the values observed typically during the day, and dip below 1.0 during the night. A double peak in saturation ratio is not evident due to the more shallow depth of the diurnal mixed layer and hence the level to which CoS is mixed out. The peak CoS saturation ratio lags an hour behind the peak insolation, whereas the SST peak lags by nearly 4 hr.

In Chapter 1, we discussed the two-equation K-ε models that are also based on the turbulence equations: the kinetic energy and dissipation equations. These models have been popular in the fluid mechanics community. Rodi (1987) has provided a detailed discussion of their capabilities. Burchard *et al.* (1998) have compared such models with MY second-moment closure models described above in the context of both neutrally stratified and stably stratified flows. They find that in neutrally stratified turbulent Couette and open channel flows, K-ε and MY models do equally well, provided the prescription of the length scale l_W in the q^2l equation (2.10.58) in MY-type closure is modified slightly. The usual

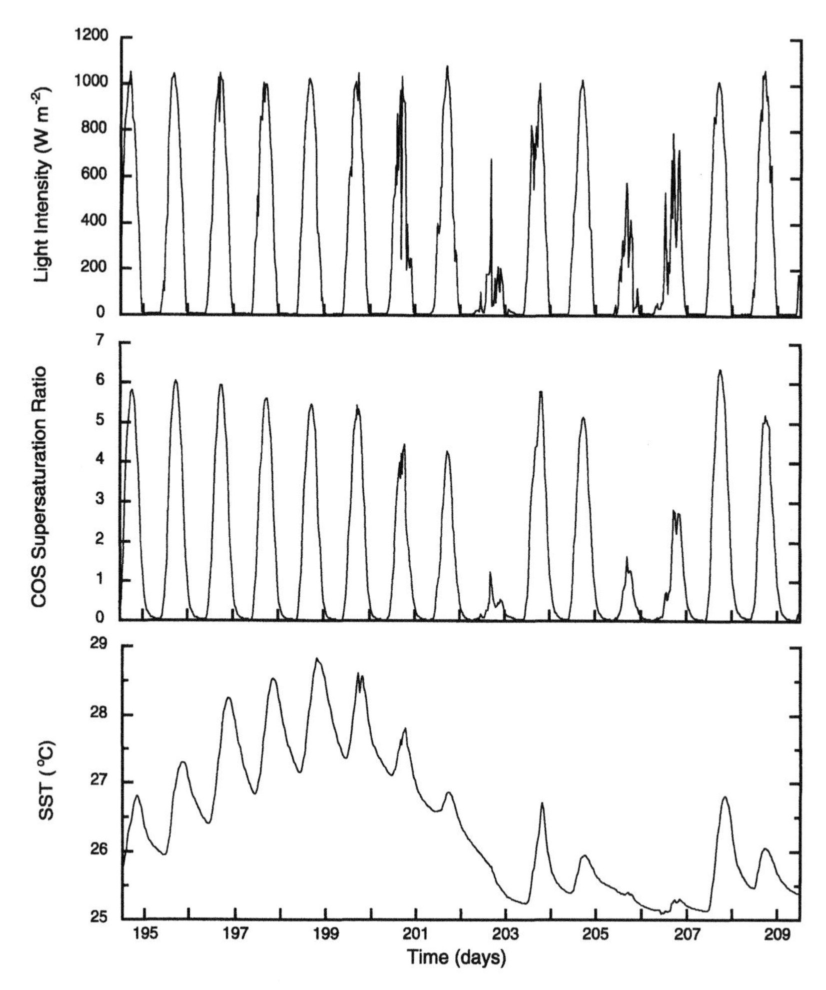

Figure 2.10.12. Simulation of the diurnal cycle in CoS at LOTUS (from Kantha and Clayson, 1994).

prescription in the presence of two boundaries enclosing the flow is $l_w = (d_s^{-1} + d_b^{-1})^{-1}$, where d_s and d_b are distances from the two boundaries. Burchard *et al.* (1998) find that the prescription $l_w = \min(d_s, d_b)$ gives better comparisons with experimental results. Figure 2.10.13 shows the comparisons. The constant c_1 in the K-ε model in these comparisons is kept constant at approximately 0.56. For stably stratified flows, the K-ε model performs poorly, unless c_1 (which is proportional to the stability function S_M in second-moment closure) is made a function of the ambient stability. Based on Galperin *et al.* (1988) derivations of the quasi-equilibrium forms for the stability functions, Burchard *et al.* (1998)

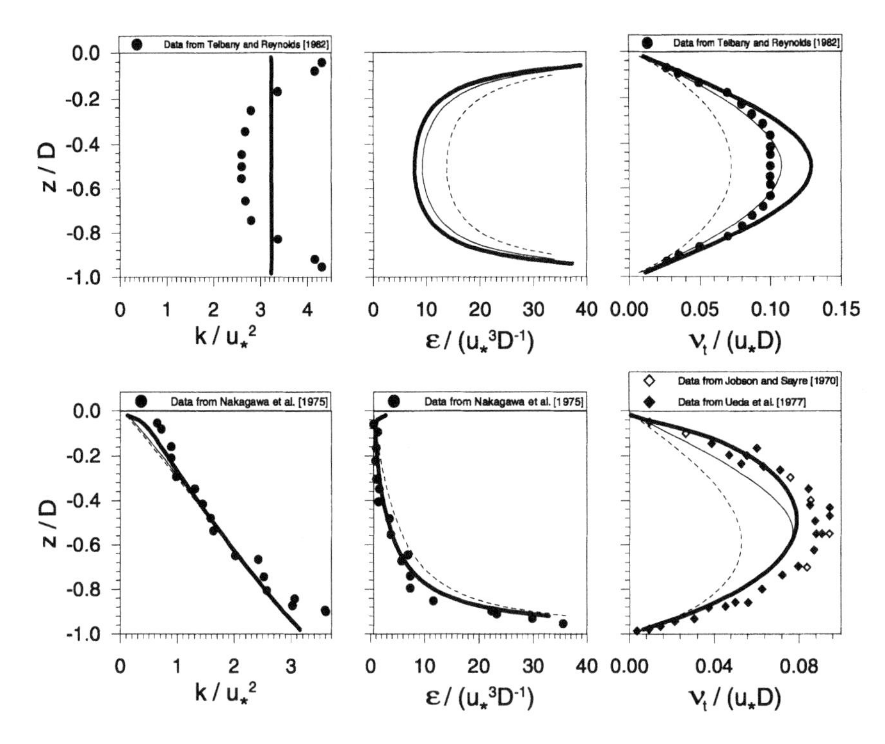

Figure 2.10.13. Comparison of turbulence closure model simulations with data in neutrally stratified (top) stress-driven Couette flow and (bottom) pressure-gradient-driven open channel flow from Burchard (1998). Couette flow TKE and eddy viscosity data are from Telbany and Reynolds (1982). Open channel TKE and dissipation rate data are from Nakagawa *et al.* (1975), and eddy viscosity data from Jobson and Sayre (1970) and Ueda *et al.* (1977). The thick line denotes the K-• model, the dashed line the MY second-moment closure model with the usual length scale prescription, and the thin line the MY model with a modified turbulence length scale.

prescribed a stability-dependent form for c_1 and found that this improved the performance of K-ε models to a level roughly similar to that of second-moment closure, when applied to stratified flows. Figure 2.10.14 shows comparison of the two types of models with turbulence dissipation measurements in the Irish Sea by Simpson *et al.* (1996). Note that there is very little difference between the two alternative prescriptions for length scale l_w in MY models. This is not surprising since what is more important is the stable-stratification-imposed upper bound on the turbulence length scale, $l \leq 0.53q/N$, as discussed earlier. Following Galperin *et al.* (1988) and Kantha and Clayson (1994), Burchard *et al.* (1998) also impose such a limit in K-ε models, which translates to a lower bound on the dissipation rate: $\varepsilon \geq 0.15qN$. See Clayson and Kantha (1998) for a comparison of the computed dissipation rates in the equatorial oceanic mixed layers with TKE dissipation rates measured with microstructure profilers.

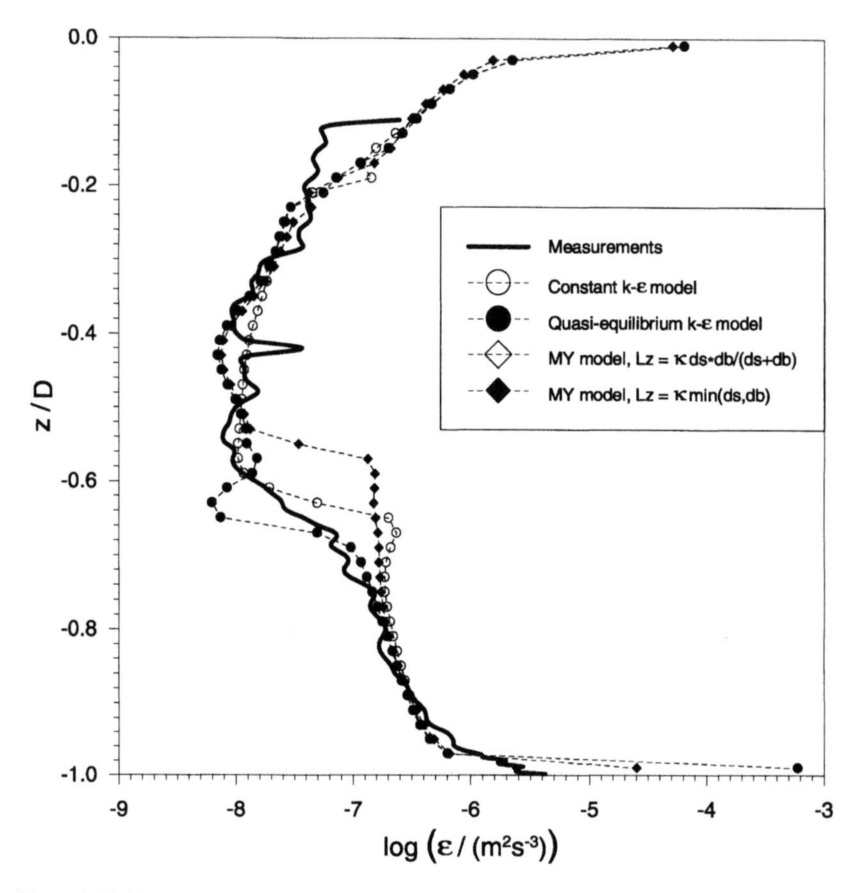

Figure 2.10.14. Comparison of dissipation rates measured in the Irish Sea by Simpson *et al.* (1996) with various turbulence closure model simulations (from Burchard, 1998).

It is important to realize that the models discussed above do not account directly for surface wave breaking effects and Langmuir circulations. Efforts have been made to include the former, Craig and Banner (1994), and Stacey and Pond (1996) being examples. Clayson and Kantha (2000) have also included the additional input of energy into turbulence from the latter. Their results reinforce the notion that both these effects are important for shallow mixed layers, and their overall influence decreases with increase in ML depth.

It should be noted that in aerodynamics literature, a turbulence model similar to the two-equation models discussed above is used in modeling turbulence in internal and external flows, but with an equation for the dissipation rate instead of the equation for q^2l. It is called the algebraic closure model there.

2.10.5 Effect of Rotation and Curvature

A primary characteristic of geophysical flows is, of course, the dominant influence of planetary rotation and gravitational stratification. For small scale processes, the latter is overwhelming and it has been traditional to ignore rotation effects. However, more accurate parameterization of mixing in geophysical flows has focused attention on rotational effects on turbulence often acting in concert with western tropical Pacific might be due to the influence of the horizontal component of rotation. Rotational terms do tend to redistribute energy among the different components of turbulence, although because of their fictitious nature, they do no work and hence drop out of the TKE equation. Since the vertical component of turbulence is the most important in changing the KE of turbulence to the PE of the system, rotation can therefore be expected to affect the mixing in the OML and ABL. The second effect of rotation is on the turbulence length scales. This is often more profound, although we do not know much about this because of the dearth of suitable observations (but see Bardina *et al.*, 1985; Launder *et al.*, 1987). It turns out that the deep MLs in the western tropical Pacific are thermal in nature and the dynamical ML is not necessarily anomalously deep (Lukas and Lindstrom, 1991), because of the presence of a shallow halocline caused by excessive precipitation in the region. Kantha *et al.* (1989; see also Galperin and Kantha, 1989; Galperin *et al.*, 1989), using second-moment closure, showed that the effect of rotation on small scale (not geostrophic, planetary scale) turbulence is at the most a few percent, except for neutral stratification conditions. This result has also been confirmed recently by the LES model runs of Wang *et al.* (1996) of the OML. Nevertheless, it is instructive to consider how rotation affects small scale processes.

The effect of modeling rotational terms in the various covariance terms in the RHS of the equations for second moments, Eqs. (1.8.1) to (1.8.3), is simply to multiply the rotational terms in Eq. (2.10.27) for the Reynolds stresses by a constant A_3 and those in Eq. (2.10.28) for Reynolds heat fluxes by a constant A_4. The precise values of these two constants is not known. However, Kantha *et al.* (1989) put these constants equal to unity, since this then means that no rotational terms in the equations need be modeled and no additional constants are introduced. Confirmation of their conclusions by the LES model of Wang *et al.* (1996) suggests that their assumption is basically correct. Now, note that the TKE equation (2.10.57) is unchanged, but it is possible to add a rotation term to the RHS of the length scale equation (2.10.58) of the form $E_5 \left(f_k f_k\right)^{1/2} q^2 l$, but the value of the constant is unknown. In component form, the equations for second-moment quantities (Kantha *et al.*, 1989), replacing Eq. (2.10.31),

$$\overline{u^2} = \frac{q^2}{3}\left(1 - \frac{6A_1}{B_1}\right) - \frac{6A_1 l}{q}\overline{uw}\frac{\partial U}{\partial z} + \frac{6A_1 l}{q}\left(f\,\overline{uv} - f_y\,\overline{uw}\right)$$

$$\overline{v^2} = \frac{q^2}{3}\left(1 - \frac{6A_1}{B_1}\right) - \frac{6A_1 l}{q}\overline{vw}\frac{\partial V}{\partial z} - \frac{6A_1 l}{q}\left(f\,\overline{uv}\right)$$

$$\overline{w^2} = \frac{q^2}{3}\left(1 - \frac{6A_1}{B_1}\right) + \frac{6A_1 l}{q}\beta g\overline{w\theta} + \frac{6A_1 l}{q}\left(f_y\,\overline{uw}\right)$$

$$\overline{uv} = \frac{3A_1 l}{q}\left[-\overline{uw}\frac{\partial V}{\partial z} - \overline{vw}\frac{\partial U}{\partial z} - f\left(\overline{u^2} - \overline{v^2}\right) - f_y\,\overline{vw}\right]$$

$$\overline{uw} = \frac{3A_1 l}{q}\left[-\left(\overline{w^2} - C_1 q^2\right)\frac{\partial U}{\partial z} + \beta g\overline{u\theta} - f_y\left(\overline{w^2} - \overline{u^2}\right) + f\,\overline{vw}\right]$$

$$\overline{vw} = \frac{3A_1 l}{q}\left[-\left(\overline{w^2} - C_1 q^2\right)\frac{\partial V}{\partial z} + \beta g\overline{v\theta} + f_y\,\overline{uv} - f\,\overline{uw}\right]$$

$$\overline{u\theta} = \frac{3A_2 l}{q}\left[-\overline{uw}\frac{\partial \Theta}{\partial z} - (1 - C_2)\overline{w\theta}\frac{\partial U}{\partial z} + f\,\overline{v\theta} - f_y\,\overline{w\theta}\right]$$

$$\overline{v\theta} = \frac{3A_2 l}{q}\left[-\overline{vw}\frac{\partial \Theta}{\partial z} - (1 - C_2)\overline{w\theta}\frac{\partial V}{\partial z} - f\,\overline{u\theta}\right] \qquad (2.10.66)$$

$$\overline{w\theta} = \frac{3A_2 l}{q}\left[-\overline{w^2}\frac{\partial \Theta}{\partial z} + (1 - C_3)\beta g\overline{\theta^2} + f_y\,\overline{u\theta}\right]$$

$$\overline{\theta^2} = \frac{B_2 l}{q}\left[-\overline{w\theta}\frac{\partial \Theta}{\partial z}\right]$$

The presence of f and f_y in Eq. (2.10.66) imparts tensorial properties to the mixing coefficients for momentum and it is necessary now to distinguish between S_{MU} and S_{MV} (Kantha *et al.,* 1989). The solutions are straightforward but algebraically tedious. We will therefore refer the reader to Kantha *et al.* (1989) and present only some salient results.

The introduction of rotational terms introduces two additional parameters, the turbulence Rossby numbers Ro and Ro_y that govern the behavior of turbulence under rotation:

$$Ro = \frac{fl}{q}; Ro_y = \frac{f_y l}{q} \qquad (2.10.67)$$

Two cases can be considered. For $Ro_y = 0$, the x-axis can be aligned in the direction of the local stress (therefore $S_{MV} = 0$). In this case the governing nondimensional number is the Rotation Richardson number:

$$Ri_R = f\left(\frac{\partial U}{\partial z}\right)^{-1} \tag{2.10.68}$$

Figure 2.10.15 shows the effect of Ri_R on the momentum mixing coefficient as derived by Kantha *et al.* (1989), assuming C_2 and C_3 equal 0, for various stratifications as indicated by the flux Richardson number Ri_f. Note that the vertical component of rotation has no effect on the scalar stability parameter S_H.

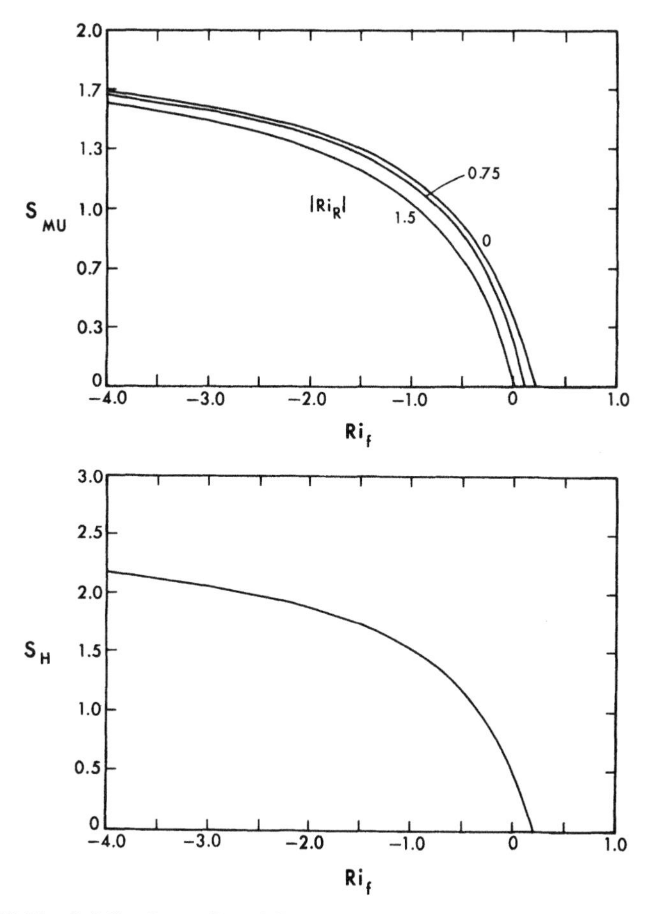

Figure 2.10.15. Stability factors S_M and S_H as functions of flux and rotation Richardson numbers for nonzero f (from Kantha *et al.*, 1989)

The correction for rotation is on the order of Ri_R^2, and since Ri_R rarely exceeds 0.1, the rotational effects can be safely ignored.

The case of nonzero Ro_y is more interesting from the point of view of equatorial mixing processes. Here the direction of the mean flow is important. The relevant nondimensional number is now

$$Ri_{Ry} = f_y \left[\left(\frac{\partial U}{\partial z} \right)^2 + \left(\frac{\partial V}{\partial z} \right)^2 \right]^{-1/2} \quad (2.10.69)$$

Figure 2.10.16 shows the effect of Ri_{Ry} on the stability factors S_{MU} and S_H for various Ri_f and for purely zonal flows. Increasing rotation tends to counteract

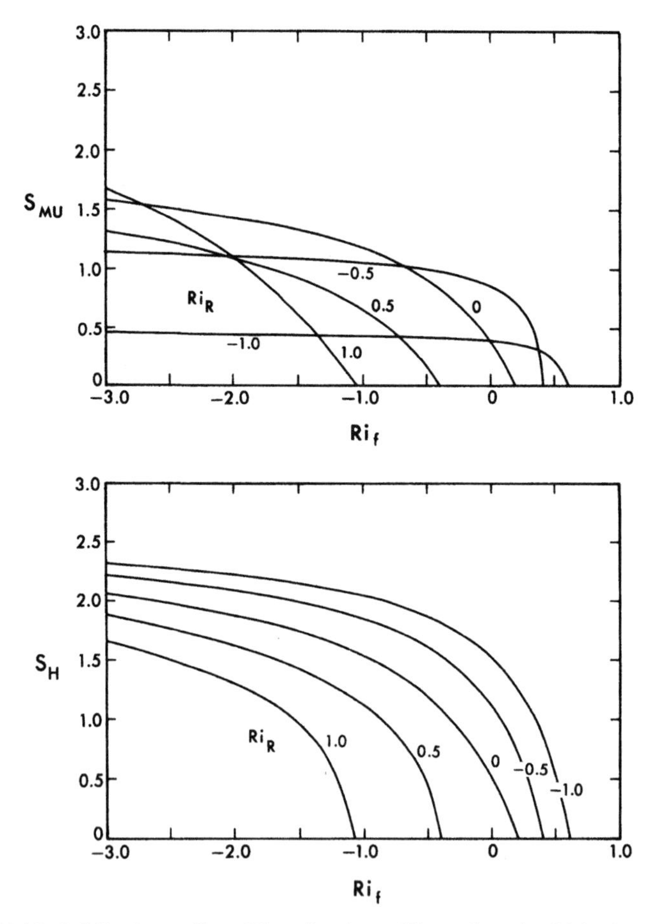

Figure 2.10.16. Stability factors S_M and S_H as functions of flux and rotation Richardson numbers for nonzero f_y (from Kantha *et al.*, 1989).

the effect of unstable stratification and suppress turbulence for an eastward flow, while the effect is exactly opposite for a westward zonal flow. However, turbulence can be kept alive by destabilizing rotation for much higher values of stable stratification. Figure 2.10.17 shows Ri_f^c, the critical value for extinction of turbulence by stable gravitational stratification for various values of Ri_{Ry}. Outside the domain delineated by the curve, turbulence is extinguished. Negative values of Ri_{Ry} permit turbulence to exist at a much higher value of Ri_f than the value of 0.191 for zero rotation (note that C_2 and C_3 are zero here unlike in the KC model; the value would be 0.213 for nonzero values in the KC model). For large values of Ri_{Ry}, the effect can be quite dramatic. Yet the value of Ri_{Ry} rarely exceeds 0.1 and hence the effects of horizontal rotation are still confined to be less than 10% or so.

Wang *et al.* (1996) simulated the equatorial ML with and without the rotational terms, using a LES model, for a variety of conditions, including the diurnal cycle of heating and cooling. The largest difference they observed was for the case of zero surface heat flux. The ML was deeper when the rotational terms were included, but the difference was less than 10%, confirming the above conclusions. But if the thermocline is absent, the effect of horizontal rotation on an equatorial ML could be larger under neutral and unstable stratification conditions, consistent with the study of Hassid and Galperin (1994). This condition, however, is seldom attained in the equatorial oceans.

As mentioned earlier, the effect of rotation on the length scale and eddy structure is quite substantial. In numerical simulations of mixing in rotating, unstratified fluids, it is essential to impose a limit on the length scale based on

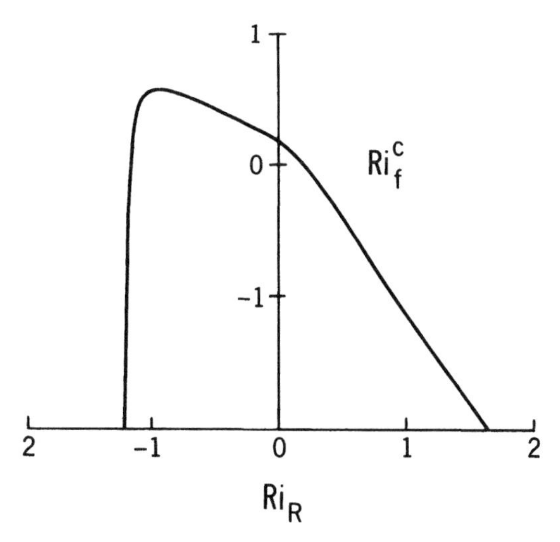

Figure 2.10.17. Critical flux Richardson number for extinction of turbulence as a function of the rotation Richardson number for nonzero f_y (from Kantha *et al.*, 1989).

rotation, analogous to the limit imposed by stratification, Eq. (2.10.55):

$$\frac{fl}{q} \le C_5; \frac{f_y l}{q} \le C_6 \qquad (2.10.70)$$

Since q ~ u*, and u*/f is the Ekman scale, the first inequality in Eq. (2.10.70) says that the turbulence length scale must be bounded by a fraction of the Ekman scale in a neutrally stratified, rotating PBL ($C_5 \sim 0.1$) ; this is physically consistent with the fact that in a rotating fluid, the influence of shear mixing cannot penetrate below the Ekman layer. The physical interpretation of the second inequality is not that straightforward.

The effect of curvature on small scale mixing was also studied by Kantha and Rosati (1990), using similar procedures. The governing equations were cast in orthogonal curvilinear coordinates and simplifications for small boundary layer thickness vis-à-vis the radius of curvature were invoked to simplify the problem. The algebra is quite tedious, but it is possible to illustrate the effects by looking at the component form for the stresses and heat fluxes as before. Aligning the x-coordinate in the direction of the mean flow and ignoring the advection terms in the second-moment equations (Kantha and Rosati, 1990),

$$\begin{aligned}
\overline{u^2} &= \frac{q^2}{3}\left(1-\frac{6A_1}{B_1}\right)-\frac{6A_1 l}{q}\overline{uw}\left(\frac{\partial U}{\partial z}+C\right) \\
\overline{v^2} &= \frac{q^2}{3}\left(1-\frac{6A_1}{B_1}\right) \\
\overline{w^2} &= \frac{q^2}{3}\left(1-\frac{6A_1}{B_1}\right)+\frac{6A_1 l}{q}\left(2C\overline{uw}+\beta g\overline{w\theta}\right) \\
\overline{uv} &= 0 \\
\overline{uw} &= \frac{3A_1 l}{q}\left[2C\overline{u^2}-\left(\overline{w^2}-C_1 q^2\right)\frac{\partial U}{\partial z}-\left(\overline{w^2}+C_1 q^2\right)C+\beta g\overline{u\theta}\right] \\
\overline{vw} &= 0 \\
\overline{u\theta} &= \frac{3A_2 l}{q}\left[-\overline{uw}\frac{\partial \Theta}{\partial z}-(1-C_2)\overline{w\theta}\left(C+\frac{\partial U}{\partial z}\right)\right] \\
\overline{v\theta} &= 0 \\
\overline{w\theta} &= \frac{3A_2 l}{q}\left[-\overline{w^2}\frac{\partial \Theta}{\partial z}+(1-C_3)\beta g\overline{\theta^2}+2C\overline{u\theta}\right] \\
\overline{\theta^2} &= \frac{B_2 l}{q}\left[-\overline{w\theta}\frac{\partial \Theta}{\partial z}\right]
\end{aligned} \qquad (2.10.71)$$

The relevant nondimensional number is the curvature Richardson number Ri_C given by

$$Ri_C = C\left(\frac{\partial U}{\partial z}\right)^{-1} = \left[\frac{\left(\frac{U}{R}\right)}{\left(1+\frac{z}{R}\right)}\right]\left(\frac{\partial U}{\partial z}\right)^{-1} \tag{2.10.72}$$

Figure 2.10.18 shows the dependence of S_M and S_H on Ri_C for various values of the stratification parameter Ri_f under local equilibrium approximation. It is

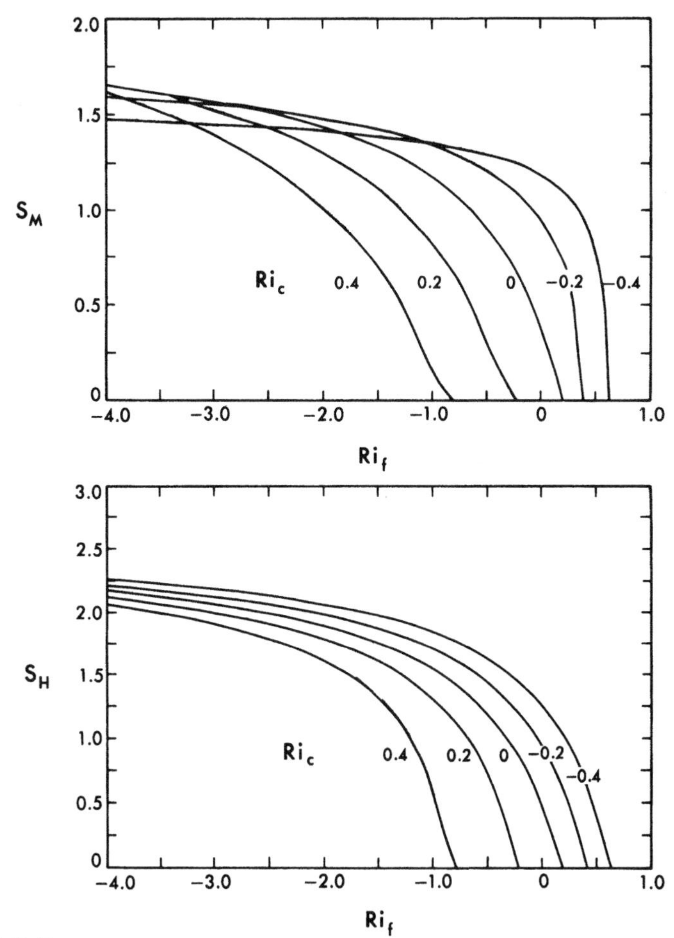

Figure 2.10.18. Stability factors S_M and S_H as functions of flux and the curvature Richardson number (from Kantha and Rosati, 1990).

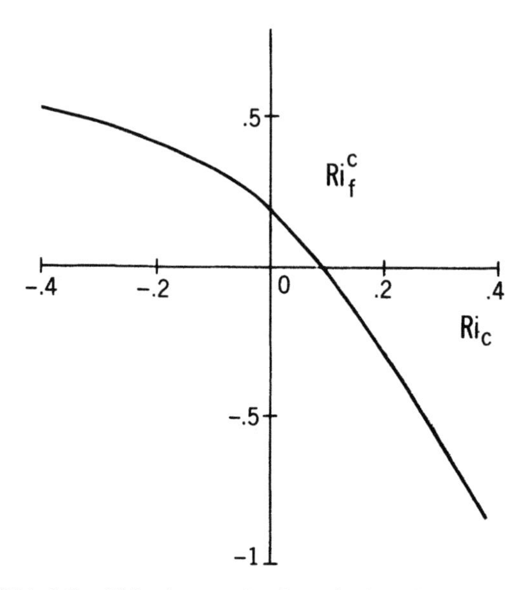

Figure 2.10.19. Critical flux Richardson number for extinction of turbulence as a function of the curvature Richardson number (from Kantha and Rosati, 1990).

clear that for positive Ri_C, stabilizing curvature (convex upward for positive z being upward), turbulence can be suppressed even when the stratification is unstable. Conversely, under strong stable stratification, destabilizing curvature (concave upward) can keep the turbulence alive much longer. Figure 2.10.19 shows the extinction flux Richardson number as a function of Ri_C. The Ri_C values needed, for example, for 50% enhancement or curtailment of mixing is on the order of 0.1, a value readily attained in geophysical BL flows, such as in the ABL over mountains. It may therefore be important to account for curvature effects on mixing in flow over topography (see Chapter 3).

2.11 CHEMICAL AND BIOLOGICAL MIXING MODELS

In the late 1970s and the early 1980s, a remarkable event in the history of biological oceanography occured. NASA orbited on a NOAA satellite a sensor called the Coastal Zone Color Scanner (CZCS) to sense the "color" of the oceans (Esias *et al.,* 1986). This sensor provided for the first time a global perspective of the chlorophyll concentration and primary productivity in the upper layers of the ocean on a variety of timescales. Figure 2.11.1 presents an example pertaining to the change in primary productivity with seasons in the North Atlantic. By the end of summer, biological activity depletes the nutrients in the euphotic zone (the zone in which solar insolation is strong enough to

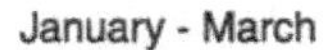

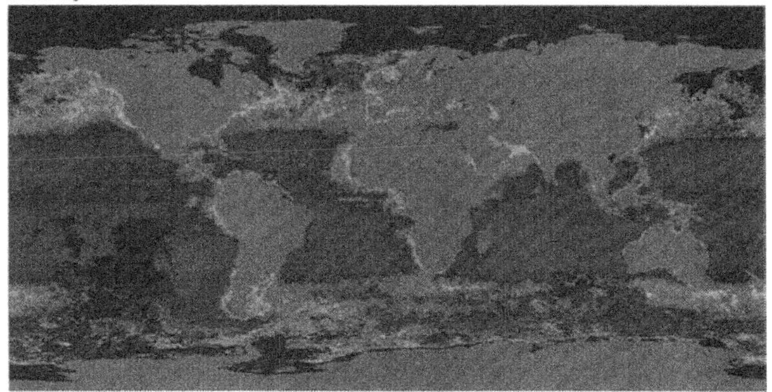

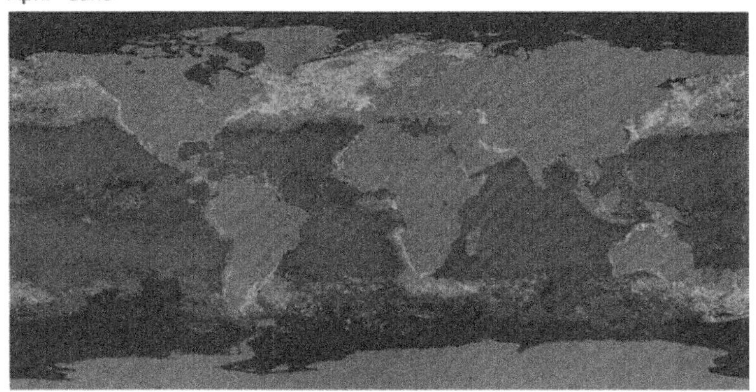

Figure 2.11.1. CZCS false color image showing the spring bloom in the North Atlantic Ocean (courtesy of NASA).

assist carbon fixation) and the onset of autumn/winter brings increasingly less sunlight into the upper ocean. This combination slows down the primary production considerably. During autumn/winter, storms manage to stir the upper layers vigorously, bring the nutrients from below, and mix them into the euphotic zone. However, the light levels and ambient temperature are not conducive to biological activity. With the onset of spring and increased sunlight, increasing ambient temperatures and the abundance of nutrients in the euphotic zone cause an explosion in biological productivity. This spring bloom is clearly seen in Figure 2.11.1. CZCS also highlighted the few areas in the global oceans, which are upwelling-favorable and hence regions of continuous input of nutrients into the euphotic zone, where most of the primary productivity is concentrated.

Overall, this color sensor has revolutionized the way we see our oceans in terms of its productivity. For the first time, biological oceanographers were able to get a rough estimate of the primary production in the global oceans, the first link in the oceanic food chain. Until then, they were restricted to very sparse, isolated samples collected here and there, and lacked the global perspective of this problem, important to the protein needs of a burgeoning global population and long-term climate change. Primary production also constitutes a primary pathway, biological, for sequestering of the excess anthropogenic carbon dioxide in the atmosphere, since the phytoplankters are consumed by zooplankters, whose shells rain down onto the deep seafloor, where they reside for centuries before being recycled back into the upper ocean and again coming into contact with the atmosphere. It is estimated that 18–40% of the total anthropogenic CO_2 emissions (Najjar, 1992; Sarmiento, 1992) have been absorbed by the oceans (about rmain in the atmosphere), and this has led to a slow-down in greenhouse warming.

The color scanner accomplishes the task of diagnosing the primary productivity by detecting the small change in the spectral characteristics of the upwelling shortwave radiation from the upper ocean due to increased chlorophyll concentration brought on by the abundance of phytoplankton in the upper layers. This it does by sensing the spectrum with multiple bands, and removing by far the most predominant part of the signal, the one due to the intervening atmospheric column. It is then able to sense the change from blue to bluish green in the upwelling radiation of the upper layers and provide an estimate of the biological activity. The follow-on to CZCS, the NASA SeaWIFS sensor, which has been in operation since the middle of 1997, uses more bands and is designed to more effectively remove signal corruption by dissolved organic matter and sediments in coastal waters. These remotely sensed data, in combination with reliable global ocean biological models of primary productivity, have the potential to revolutionize our understanding of the upper ocean as a biological factory. The only other color sensor, the Ocean Color and Temperature Sensor (OCTS) orbited on the Japanese satellite ADEOS, met an untimely demise when a massive power failure terminated the ADEOS mission.

However, there are plans to orbit even more sensitive color sensors. The spectral characteristics of remotely sensed reflectance (ocean color) are significantly different in coastal waters compared to the open ocean. Higher concentrations of phytoplankton along with suspended sediments and dissolved organic matter in the usually more turbid coastal waters contribute to a much richer coastal spectra, which in turn requires muvh higher spatial and spectral (especially in the 600-750 nm red region) resolutions than those required to characterize the open ocean waters. Consequently, While multispectral sensors such as SeaWiFS launched in 1997 (1.1 km spatial resolution and 8 channels in 402-885 nm band) are adequate for mapping the variability in the open ocean, hyperspectral sensors such as the Coastal Ocean Imaging Spectrometer (COIS, 30 m spatial resolutiontribution of various constituents to the spectral signature of the coastal waters. MODIS on, 210 channels in 400-2500 nm band)) to be orbited in 2000 under the Navy Earth Map Observer (NEMO) mission would provide the ability to evaluate the cto be orbited on the NASA EOS-AM mission also has a higher spectral resolution (36 channels in 405-14385 nm band) than SeaWiFS. For more details on these developments, see Arnone and Gould (1998).

Primary productivity is critically dependent on the availability of nutrients and sunlight, and it is in the former that mixing in the upper ocean plays a dominant role. Nutrients have to be brought up into the normally depleted euphotic zone from deeper nutrient-rich waters by mixing across the stable pycnocline, which also coincides most often with the nutricline. This requires energy input from the surface either by winds or by convective cooling. Because of this heavy dependence on mixing processes, biological models have to be coupled to physical models to infer primary productivity in the upper ocean. Figure 2.11.2 shows the carbon cycle in the ocean, the organic part of which depends on the primary productivity in the upper layers.

Air–sea exchange of dissolved gases in the upper ocean, such as CO_2, methane, dimethyl sulfide, carbonyl sulfide, and nitrous oxide, plays an important role in the global climate. Carbon dioxide, methane, and nitrous oxide are among the primary greenhouse gases. The oceans constitute by far the largest reservoir of CO_2 (65 times that of the atmosphere), which is resident not only as a dissolved gas but also in the form of dissolved calcium carbonate and bicarbonate. The uptake of CO_2 in the cold subpolar oceans and its outgassing in the warmer tropics and subtropics depend very much on the mixing processes in the upper layers of the ocean. The long-term (timescale ~ 1000 yr) thermohaline circulation in the global oceans is also dependent on wintertime deep convection and mixing in subpolar oceans. Overall, it is estimated that the CO_2 equivalent of the entire atmosphere is cycled through the oceans once every 6 to 7 years (Sarmiento, 1992). The oceans act as a net sink of CO_2, but are a source of methane (2% of total from all sources, 15% contribution to greenhouse effect) and N_2O (~30% of total from all sources, ~6% contribution to greenhouse effect). They are also a source of dimethyl sulfide, an important source of marine aerosols that tend to cool the Earth (Najjar, 1992).

The upper ocean also constitutes an important source of certain photochemical trace gases important to stratospheric and tropospheric chemistry, and hence to the global radiative and ozone budget (Najjar *et al.*, 1994; Doney *et al.*, 1995). Carbonyl sulfide (CoS), carbon monoxide (CO), and hydrogen peroxide (H_2O_2) are examples of such gases photochemically produced in the upper few meters of the ocean from dissolved organic matter by the UV part (280–350 nm) of the solar insolation and outgassed to the atmosphere (sinks are species-specific—microbial uptake for CO and inorganic hydrolysis for CoS). However, the rate of exchange depends on their concentration, which in turn depends on the mixing processes in the OML (Kantha and Clayson, 1994). Nitrous oxide (N_2O) is another example of a trace gas also produced in the oceans which

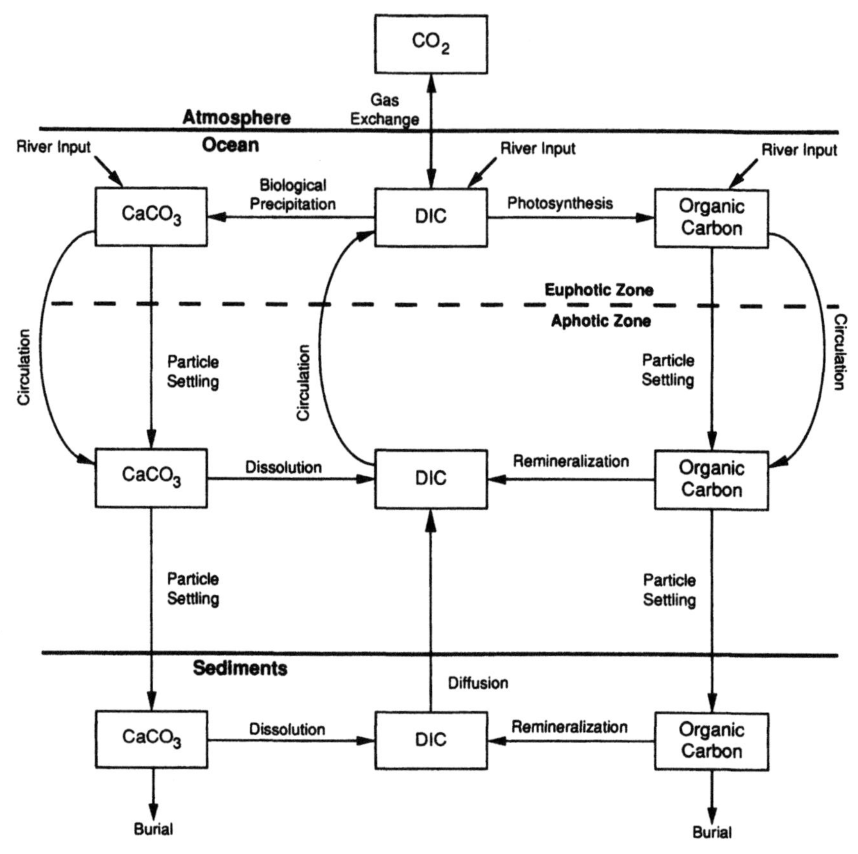

Figure 2.11.2. Carbon cycle in the ocean (from Najjar, 1992; reprinted with the permission of Cambridge University Press and the University Corporation for Atmospheric Research).

destroys ozone and has long residence times in the stratosphere, and whose concentrations and seasonal evolution in the upper ocean depend on mixing in the OML (Nevison *et al.*, 1995). It is also a greenhouse gas.

Ever since their formation, approximately 3.5 billion years ago, the ocean has played an important role in the geochemical evolution of the Earth. Almost all of the oxygen found in the Earth's atmosphere was created by oxygenic photosynthesis in the ocean by unicellular phytoplankton, with oxygen levels reaching the present day levels roughly 2.2 billion years ago. A massive amount of organic carbon (15 Pt) was simultaneously produced and sequestered in sedimentary rocks. Satellites have made it possible to estimate the global annual net primary production (NPP) of carbon, the amount of photosynthetically fixed carbon available to other trophic levels. From ocean color data gathered by the CZCS sensor, Field *et al.* (1998) estimate that in the present day oceans, NPP is

~49 Gt, roughly one third of which is exported to the deep ocean and hence sequestered from the atmosphere for centuries to millenia (Falkowski *et al.* 1998). The corresponding terrestrial value (deduced from the land vegetation index from AVHRR) is ~56 Gt, comparable to the oceanic value, even though phytoplnkton biomass (~1 Gt) accounts for only 0.2% of the global photosynthetically active primary producer biomass (~500 Gt). This means that the average turnover time scale of biomass is roughly a week in the ocean, compared to 19 years on land (Field *et al.* 1998). This rapid turnover implies that increased NPP will not result in substantial changes in carbon stored in the phytoplankton biomass, but rather in carbon sequestered through transport of carbon into the oceanic interior. Phytoplankton absorb only 7% of the PAR incident at the ocean surface, while terrestrial plants absorb 31% of the PAR incident on land without permanent ice cover, and the production per unit surface area over land is three times that over the ocean. Maximum NPP is however, similar (1-1.5 kg C m^{-2} yr^{-1}), with upwelling regions being high NPP regions in the ocean and humid tropics on land.

Primary biological production in the ocean is conditional upon the simultaneous availability of solar insolation (the dominant energy source available and essential for photosynthesis), and inorganic nutrients dissolved in the water (to form plant tissue). The latter include inorganic elements carbon, nitrogen, phosphorus and sulfur available in the form of carbon dioxide, nitrates (or molecular nitrogen), phosphates and sulfates dissolved in the water. Primary poductivity is therefore limited to the euphotic zone, where the solar insolation is adequate for photosynthesis. Solar insolation is the primary limiting factor in high latitudes and responsible for seasonal modulations in productivity. In near-surface waters at low latitudes, the sunlight is usually not a limiting factor, and the growth rate of phytoplankton and hence primary productivity is dictated instead by the scarcest of the inorganic nutrients. Which nutrient is limiting depends on the region under consideration, but in most of the oceans it is nitrogen and secondarily phosphorus. However, certain trace minerals such as dissolved iron are also essential for photosynthesis and the availability of these trace mineralconstituents in often limits productivity even when the bulk nutrients are plentiful. Vast areas of the global ocean, including Southern Ocean, and the eastern equatorial Pacific are high-nutrient, low-chlorophyll (HNLC) regions, where despite the relative abundance of bulk nutrients, biological productivity is small because of the scarcity of iron (Falkowski *et al.* 1998, Behrenfeld and Kilber 1999). This opens up the possibility that the productivity in these waters can be increased to the level limited only by the availability of bulk nutrients such as nitrogen, by an artificial augmentation of trace nutrients. This was confirmed by experiments in the equatorial Pacific, where artificial iron enrichment gave rise to spectacular, but temporary phytoplankton blooms (Mullineaux 1999).

Finally, primary productivity can have a direct impact on dynamical processes in the OML, since chlorophyll-induced turbidity affects the solar extinction in the upper layers, and hence the stratification and mixing aspects of the OML. This biological feedback has been mostly ignored thus far in ML models, since it requires coupling to biological models of primary productivity. However, its importance to the dynamics of the OML itself is being increasingly realized. TOGA/COARE researchers discovered a several-fold increase in turbidity of the normally clear waters in the western tropical Pacific due to a westerly wind burst, which enhanced the mixing and brought nutrients into the euphotic zone and increased the turbidity significantly.

2.11.1 Modeling Chemistry—The Inorganic Carbon Cycle

Modeling the fate of dissolved chemicals and gases and biological constituents in the ocean is straightforward in principle, but quite complicated in practice. The governing equation is the conservation equation for the concentration C of the constituent,

$$\frac{\partial C}{\partial t}+\frac{\partial}{\partial x_k}(U_k C)=k_C\nabla^2 C-\frac{\partial}{\partial x_k}(\overline{u_k c})+So_C-Si_C \tag{2.11.1}$$

where k_C is the molecular diffusivity of the constituent and $-\overline{u_k c}$ is the turbulent flux of the quantity. Equation (2.11.1) needs to be solved in conjunction with the equations for physical variables such as temperature, velocity, and turbulence (see Section 2.10), because of the coupling to physical processes brought on by the presence of dynamical quantities in the equation. The principal effect of physical processes is due to advection and mixing by the fluid, although the influence of physical variables such as temperature and salinity on the source and sink terms has to be accounted for also. By far, the principal problem is the nonconservative nature of the constituent implied by the presence of the source and sink terms So_C and Si_C. In many cases, these are very poorly known and not easy to parameterize. Nevertheless, it is possible to solve equations like this starting from known initial conditions (equivalent in a one-dimensional model to prescribing the initial vertical profile of C),

$$C=C(x_k) \quad \text{at} \quad t=0 \tag{2.11.2}$$

and integrate forward in time (in conjunction with the equations for physical variables) under prescribed surface and bottom boundary conditions:

$$\begin{aligned} Q_C &= -(\overline{wC})_s \quad \text{at} \quad z=0 \\ Q_C &= -(\overline{wC})_b \quad \text{at} \quad z=-H \end{aligned} \tag{2.11.2}$$

or

$$C = C_s \quad \text{at} \quad z = 0$$
$$C = C_b \quad \text{at} \quad z = -H \tag{2.11.3}$$

For chemical constituents, it is essential to know the rates of their production and destruction (or conversion to other constituents) due to chemical reactions [see recent reviews by Najjar (1992) and Sarmiento (1992), and the references cited therein]. For dissolved gases, the air–sea exchange rate, which is a function of not only the surface concentrations but also the turbulent processes on both sides of the air–sea interface, must be known or parameterized. This is another principal difficulty, since air–sea exchange of these quantities is often poorly known (see a recent review by Wanninkhof, 1992).

We will now describe a simple, one-dimensional model of the fate of inorganic CO_2 to illustrate the principles of modeling chemical constituents in the ocean. The following comes from Najjar (1992). More details on marine biogeochemistry and its modeling can be found in excellent tutorials by Najjar (1992) and Sarmiento (1992).

CO_2 dissolved in water forms carbonic acid H_2CO_3, which dissociates to form H^+ and a bicarbonate ion HCO_3^-, which in turn dissociates into H^+ and a carbonate ion CO_3^{2-}:

$$CO_2 + H_2O \Leftrightarrow H^+ + HCO_3^- \tag{2.11.4}$$

$$HCO_3^- \Leftrightarrow H^+ + CO_3^{2-} \tag{2.11.5}$$

The superscripts denote the electrical charge carried; the carbonate ion carries two negative charges. Since these and other chemical reactions associated with CO_2 occur rapidly (in a few minutes time), chemical equilibrium is normally present and the reactions must obey the chemical equilibrium relations

$$k_1 = \frac{[H^+][HCO_3^-]}{[CO_2]}$$
$$k_2 = \frac{[H^+][CO_3^{2-}]}{[HCO_3^-]} \tag{2.11.6}$$

where k_1 and k_2 are the first and second dissociation constants of carbonic acid, and the brackets denote concentrations expressed usually as micromoles per kilogram of water (μmol kg^{-1}), where 1 mol is the molecular weight in grams (for example, 1 mol of CO_2 is 44 g). Water also ionizes to form

$$H_2O \Leftrightarrow \left[H^+\right] + \left[OH^-\right] \tag{2.11.7}$$

$$k_w = \left[H^+\right]\left[OH^-\right] \tag{2.11.8}$$

where k_w is the ionization constant of water. All these equilibrium constants are functions of temperature, salinity, and pressure. Dissolution of CO_2 in water also involves conservation of electrical neutrality so that in pure water

$$\left[H^+\right] = \left[OH^-\right] + \left[HCO_3^-\right] + 2\left[CO_3^{2-}\right] \tag{2.11.9}$$

This just states that to maintain electrical neutrality, the concentration of positive H^+ ions must equal the sum of concentrations of negative hydroxyl, bicarbonate, and carbonate ions. Since the carbonate ion carries two charges, its concentration must be multiplied by two in Eq. (2.11.9), which along with Eqs. (2.11.6) and (2.11.8), with the equilibrium constants known, provides four equations for the five unknowns CO_2, H^+, OH^-, HCO_3^-, and CO_3^{2-}. It is more useful to deal with the total (or dissolved) inorganic carbon ΣCO_2 (DIC) instead of CO_2:

$$[DIC] = [CO_2] + \left[HCO_3^-\right] + \left[CO_3^{2-}\right] \tag{2.11.10}$$

If there is no addition to or removal of CO_2 from the system, DIC is conserved, although there might be interchanges among the different carbon species. DIC is usually expressed in μmol kg^{-1}.

In seawater, there are a lot more ions due to dissolved salts (and associated acids and bases) and the charge balance is much more complex than Eq. (2.11.9). There is a slight excess of positive charge as a result and this is called alkalinity (ALK),

$$[ALK] = \left[OH^-\right] + \left[HCO_3^-\right] + 2\left[CO_3^{2-}\right] - \left[H^+\right] + \left[B(OH)_4^-\right] \tag{2.11.11}$$

where the borate ion enters the relationship. Alkalinity is usually expressed in microequivalents per kilogram of seawater (μeq kg^{-1}). The borate ion concentration can be calculated from the reaction

$$[H_3BO_3]+[H_2O] \Leftrightarrow \left[H^+\right]+\left[B(OH)_4^-\right] \quad (2.11.12)$$

Borate species is conserved for a given salinity so that

$$[H_3BO_3]+\left[B(OH)_4^-\right]=nS \quad (2.11.13)$$

where n is a constant and S is salinity. The reaction rate is governed by

$$k_b = \frac{\left[H^+\right]\left[B(OH)_4^-\right]}{[H_3BO_3]} \quad (2.11.14)$$

where k_b is the dissociation constant of boric acid. Equations (2.11.6), (2.11.7), (2.11.11), (2.11.13), and (2.11.14) provide six equations for the eight unknowns DIC, ALK, HCO_3^-, CO_3^{2-}, H^+, OH^-, $B(OH)_4^-$, and H_3BO_3. Specifying any two of these, usually DIC and ALK, along with ambient conditions, temperature, salinity, and pressure, determines all of them. It is therefore enough to carry two conservation equations, one for DIC and the other for ALK, to infer inorganic carbon dioxide chemistry in the ocean. However, since change in salinity due to precipitation and evaporation affects their values, it is traditional to deal with equations for values normalized to 35 psu instead,

$$\frac{\partial C_T}{\partial t}+\frac{\partial}{\partial x_k}(U_k C_T)=k_C \nabla^2 C_T - \frac{\partial}{\partial x_k}(\overline{u_k c_T})+So_C - Si_C \quad (2.11.15)$$

$$\frac{\partial A}{\partial t}+\frac{\partial}{\partial x_k}(U_k A)=k_a \nabla^2 A - \frac{\partial}{\partial x_k}(\overline{u_k a})+So_a - Si_a \quad (2.11.16)$$

where C_T = (DIC) (S/35) and A = (ALK) (S/35). Normally, the source and sink terms in Eq. (2.11.16) for alkalinity are zero. Those in Eq. (2.11.15), the equation for total carbon, correspond to exchanges at the air–sea interface (which can be a source or a sink) and biological uptake (sink) and recycling (source). These equations are solved from known initial conditions on C_T and A with prescribed fluxes of these quantities at the surface and the bottom. For air–sea exchange purposes, it is important to know the partial pressure of dissolved carbon dioxide (CO_2), pCO_2, which is given by

$$pCO_2 = [CO_2]/\alpha \quad (2.11.17)$$

where α is the solubility of CO_2. The flux of alkalinity is zero at the sea surface and the flux of C_T depends on the difference in the partial pressures of CO_2 in the atmosphere p_a and the ocean p_c. The resistance on the atmospheric side to CO_2 transfer can be neglected (see Jahne and Haußecker, 1998, for a recent review of processes affecting air–sea gas exchange; high solubility means atmospheric side dominates the resistance, while low solubility implies that the water side resistance is more dominant) so that the flux can be written as

$$Q_C = -(\overline{wc_T})_s = u_p([CO_2]_w - \alpha p_a) = u_p \alpha (p_w - p_a) \qquad (2.11.18)$$

where u_p has units of velocity and is called the piston velocity. Its value is a function of parameters affecting air–sea exchange, principally the wind velocity and the surface wave field. It is also a function of the Schmidt number $Sc = \nu/k_c$, where k_c is the diffusion coefficient of CO_2 in water. Enormous efforts have been expended in determining the piston velocity and its functional dependences (Wanninkhof, 1992), because of the climatic implications of CO_2 uptake by the oceans. Q_c is positive for outgassing of CO_2 as in equatorial oceans, and negative for uptake of CO_2 as in subpolar seas.

Figure 2.11.3 shows a typical distribution of C_T and A in the global oceans. There is a marked deficit in the upper 500 to 1000 m. Figure 2.11.4, from Najjar (1992), shows the dependence of pCO_2 on T and S, for typical near-surface values of DIC (2000 μmol kg^{-1}) and ALK (2300 μeq kg^{-1}), and on DIC and ALK for T = 20°C and S = 35 psu (pressure is 1 atm). pCO_2 can also be written as

$$pCO_2 = \frac{k_2}{\alpha k_1} \frac{\left[HCO_3^-\right]^2}{\left[CO_3^{2-}\right]} \qquad (2.11.19)$$

pCO_2 in the atmosphere was about 300 μatm in preindustrial times and is 360 μatm at present, presently increasing at a rate of 1.3 $\mu atm yr^{-1}$.

The bottom boundary conditions are once again zero fluxes of C_T and A. However, there may be sediments of $CaCO_3$ on the seafloor and it may be necessary to account for its dissolution. Now, the principal effect of adding CO_2 to the oceans is to consume a CO_3^{2-} ion,

$$[CO_2] + \left[CO_3^{2-}\right] + H_2O \Leftrightarrow 2\left[HCO_3^-\right] \qquad (2.11.20)$$

whereas the effect of dissolving $CaCO_3$ is to increase it:

$$CaCO_3 \rightarrow \left[Ca^{2+}\right] + \left[CO_3^{2-}\right] \qquad (2.11.21)$$

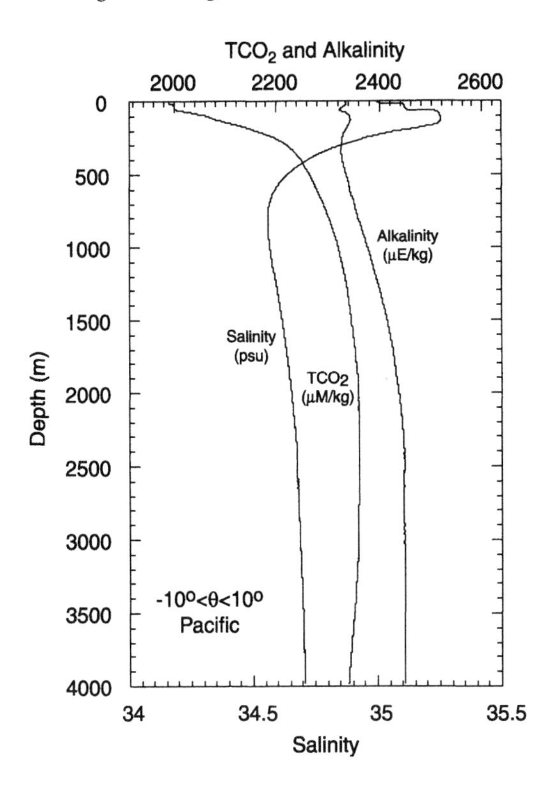

Figure 2.11.3. Typical distribution of total carbon and alkalinity in the global oceans (data from Takahashi *et al.*, 1979).

Thus every micromole of $CaCO_3$ dissolved increases C_T by 1 μ mol and A by 2 μ eq. If (CO_3^{2-}) falls below its saturation value, dissolution of $CaCO_3$ occurs. If above, $CaCO_3$ can precipitate. The water has to be supersaturated for those living organisms that form shells to function. The saturation value is proportional roughly to the ratio of the solubility product k_{sp} of $CaCO_3$, which is a function of temperature, salinity, and pressure, to the salinity. $CaCO_3$ is found in two forms in the ocean, aragonite and calcite. Figure 2.11.5 shows the typical variation of the saturation values for these two, as well as the typical concentration of carbonate ion with depth. Aragonite is more soluble than calcite in the oceans. The depth at which the CO_3^{2-} concentration equals the saturation value determines the depth below which $CaCO_3$ dissolution takes place. This depth is around 500–1500 m for aragonite, but 3000–4000 m for calcite. The saturation value for aragonite is given by (Broecker and Takahashi, 1978)

$$\left[CO_3^{2-}\right]_s = 120 \exp\left[0.15(0.001H - 4)\right] \ \mu\text{mol kg}^{-1} \qquad (2.11.22)$$

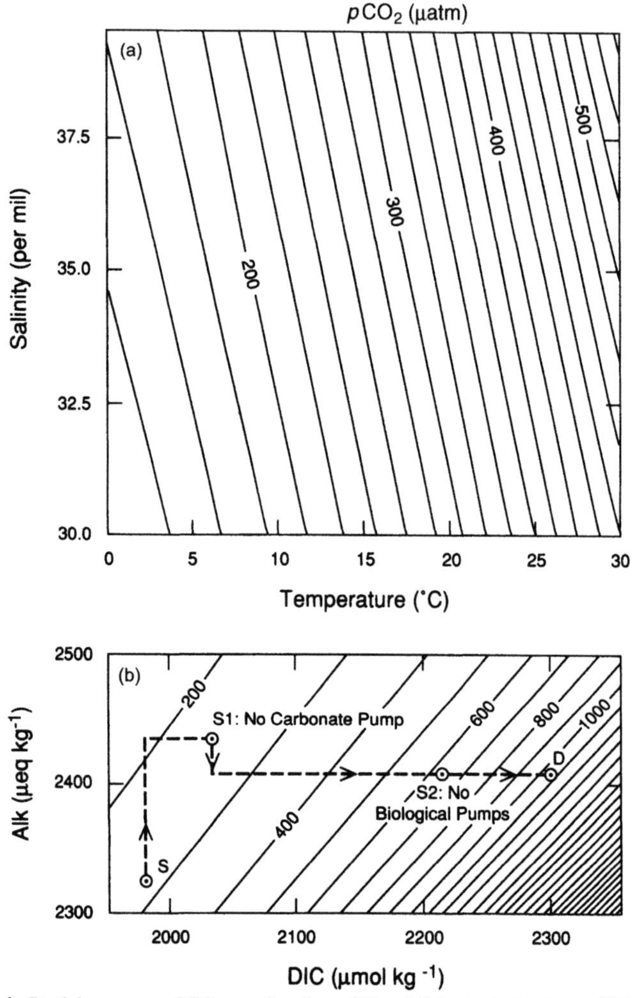

Figure 2.11.4. Partial pressure of CO_2 as a function of T and S for typical values of DIC and ALK, and as a function of DIC and ALK for typical T and S values (from Najjar, 1992; reprinted with the permission of Cambridge University Press and the University Corporation for Atmospheric Research).

The dissolution rate is proportional to the power of the degree of under-saturation. The bottom boundary condition must then account for the flux of C_T and A due to the flux of the carbonate ion. However, complications involving bioturbation and other effects make the problem complex. One simple way to account for the dissolution is to assume that as long as there are sediments, the value of carbonate ion is at the saturation value. This depletes the sediments and once the sediments are gone, the fluxes of both drop to zero.

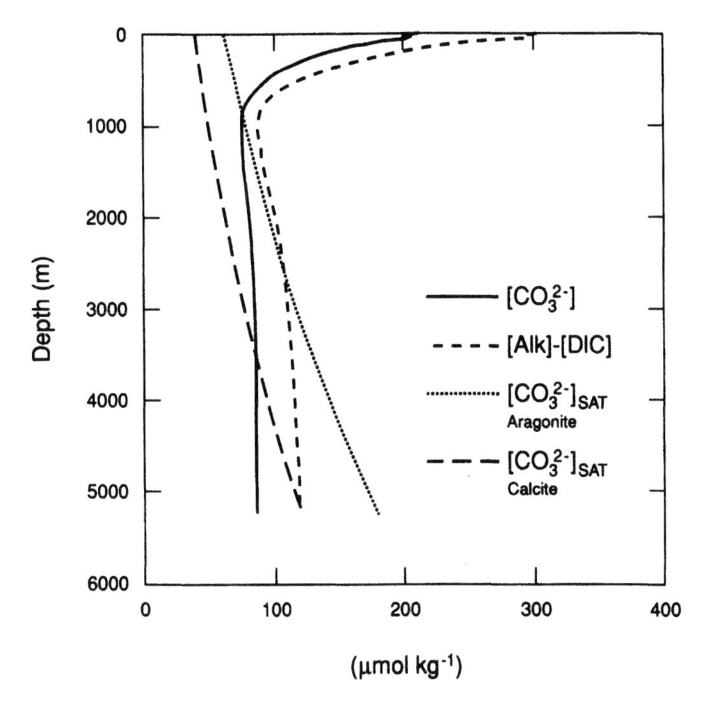

Figure 2.11.5. Typical carbonate ion concentration profiles in the ocean, showing the different depths of aragonite and calcite saturations (from Najjar, 1992; reprinted with the permission of Cambridge University Press and the University Corporation for Atmospheric Research).

The capacity of the oceans to absorb anthropogenic CO_2 in the atmosphere is large (Brewer, 1983; Broecker and Peng, 1982; Najjar, 1992). There are two factors involved. First, DIC in the water can increase. However, this is tied to the thermohaline cycle, since absorption occurs in colder latitudes and dense water forms there. The turnover timescale for the deep oceans is about 1000 years. The second factor involves $CaCO_3$. Increase in DIC decreases (CO_3^{2-}) since

$$\left[CO_3^{2-}\right] \sim [ALK]-[DIC] \tag{2.11.23}$$

which causes shallowing of the saturation horizon and dissolution of $CaCO_3$ in the sediments, an increase in alkalinity as a result, and therefore lower CO_2.

When the various constants in the above chemical reactions are properly prescribed, Eqs. (2.11.15) and (2.11.16) enable the fate of inorganic carbon in the oceans to be modeled. However, modeling the biological pathway of carbon (DOC, dissolved organic carbon) requires an ecosystem model that accounts for primary biological productivity of the upper ocean is needed. A recent example of the state-of-the-art in carbon cycle modeling in the upper ocean is Walsh *et al.* (1999), who model the fate of both inorganic and organic carbon in the water column to examine the carbon-nitrogen cycling in the Cariaco Basin off Venezuela during spring uwelling events. They incorporate biological-chemical model into a two-layer ocean model and apply it to the Cariaco Basin. In addition to the biological variables involved in primary productivity, they solve additional conservation equations for dissolved inorganic carbon and two forms of dissolved organic carbon. The inorganic nitrogen and carbon are tied together using the Redfield ratio for C/N uptake. They ignore however issues related to alkalinity and solve only for the DIC.

2.11.2 MODELING BIOLOGY—THE NITROGEN CYCLE

Equation (2.11.1) can also be applied to biological quantities such as phytoplankton and zooplankton concentrations, provided the source and sink terms due to biological production, and destruction by mortality and grazing by herbivores and carnivores, are accounted for. Production in turn depends on the availability of nutrients and the reproduction rate, which is dependent on physiology and ambient parameters such as temperature. Considerable effort has been expended in determining these sources and sinks, but there are still large uncertainties (Denman and Gargett, 1995). Biological processes are extremely complex and affected by a myriad of factors of which we know very little, especially on scales of importance to primary productivity in the upper ocean. The reader is referred to Denman and Gargett (1995) for a recent review of the problems. Nevertheless, considerable progress is being made as can be seen from the application of a coupled mixed layer-biological model to simulate phytoplankton dynamics in the North Atlantic observed in 1991 (Stramska *et al.*, 1995), biological processes at Station Papa in the North Pacific (McClain *et al.*, 1996), spring bloom in the North Atlantic (McGillicuddy *et al.*, 1995a,b), the annual cycle of primary productivity in the central Black Sea (Oguz *et al.*, 1996), and the influence of trace element iron on primary productivity in the central equatorial Pacific (Leonard *et al.* 1999).

Considerable insight into biological processes is also being gained by organized observations such as the BioWatt program in the early 1980s and the Marine Light-Mixed Layers (MLML) experiment in the North Atlantic in the early 1990s (Marra, 1995; Plueddemann *et al.*, 1995; Marra *et al.*, 1995). Of these, MLML is noteworthy, since for the very first time sophisticated instruments capable of measuring biological parameters were deployed in conjunction with physical oceanographic instruments in order to measure simultaneously all the physical oceanographic, meteorological, and biological parameters needed to obtain a complete picture of biological productivity. The reader is referred to this collection of papers in a special section of the *Journal of Geophysical Research,* **100,** which is a good description of our current capability to observe and model coupled physical-biological processes in the upper ocean. It also has a good compilation of a modern references on the topic.

Stramska and Dickey (1994), Plueddemann *et al.* (1995), and Stramska *et al.* (1995) describe the observations collected at a heavily instrumented mooring in the North Atlantic (59°36′ N, 20°58′ W), near OWS India, from April 30 to July 19, 1991. The stratification was weak and phytoplankton concentration small at the start, but spring stratification and increase of the SST, accompanied by a phytoplankton bloom, occurred immediately after. During one particular period (year days 128–140), there was a marked diurnal variability in the MLD, with large phytoplankton concentrations that were mixed downward deep into the nocturnal ML, leading to concentrations much larger than expected from productivity alone. This observation once again illustrates the importance of mixing processes in primary productivity. Just like photochemicals (which are produced in the upper few meters of the oceans and mixed down deep, affecting their concentrations in the ML and hence their rate of outgassing at the air–sea interface), phytoplankton produced by the abundance of light in, and confinement to, the shallow upper layers during the day is mixed down deep during the night into regions with much smaller irradiance and productivity. This diurnal modulation (Stramska *et al.,* 1995) is shown in Figure 2.11.6, which shows the winds, the surface heat flux, the MLD, chlorophyll a, and temperature at 10- and 50-m depths. Note the large diurnal modulations, and the high chlorophyll a concentrations at 50 m. In contrast, a period with little diurnal variability in MLD is also shown in Figure 2.11.7, which has much lower chlorophyll a levels at 50 m.

In addition to capturing the early spring bloom and the diurnal variability, MLML also managed to capture the response of the system to a storm passage from year days 140 to 143. Phytoplankton concentrations decreased drastically during the storm, but water restratification and recovery of phytoplankton to stronger concentrations occurred shortly thereafter (Figure 2.11.8).

Denman and Gargett (1995) emphasize the important role of the vertical velocities associated with gyre scale flows and mesoscale features such as jets, fronts, and eddies on biological processes. They also discuss the role of large vertical excursions of phytoplankton and zooplankton produced by the diurnal modulation of the MLD, and as a result of the large eddies and any Langmuir circulations present. These excursions take the primary producers and their predators in and out of the euphotic zone and thus modulate primary productivity. The phenomenon is complex and not well understood; nevertheless, the application of coupled physical-biological LES models might provide a better understanding. For the present, we will describe and discuss only simple, one-dimensional models for primary productivity, simply to illustrate the principles of coupled physical-biological modeling.

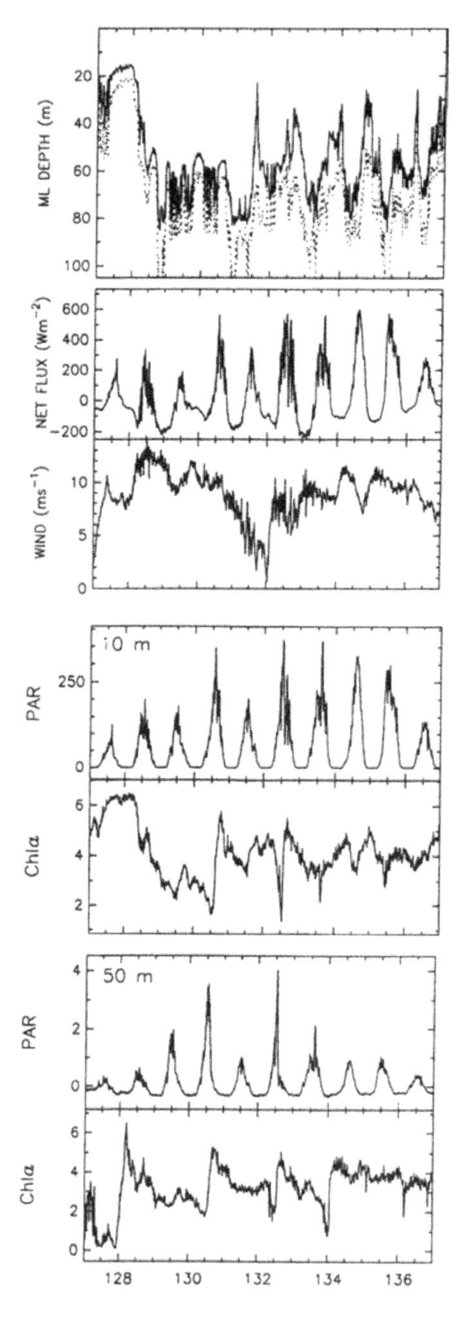

Figure 2.11.6. Observations of meteorological parameters and chlorophyll concentration and PAR at 10- and 50-m depths from MLML experiments showing large diurnal modulations (from Stramska *et al.*, 1995).

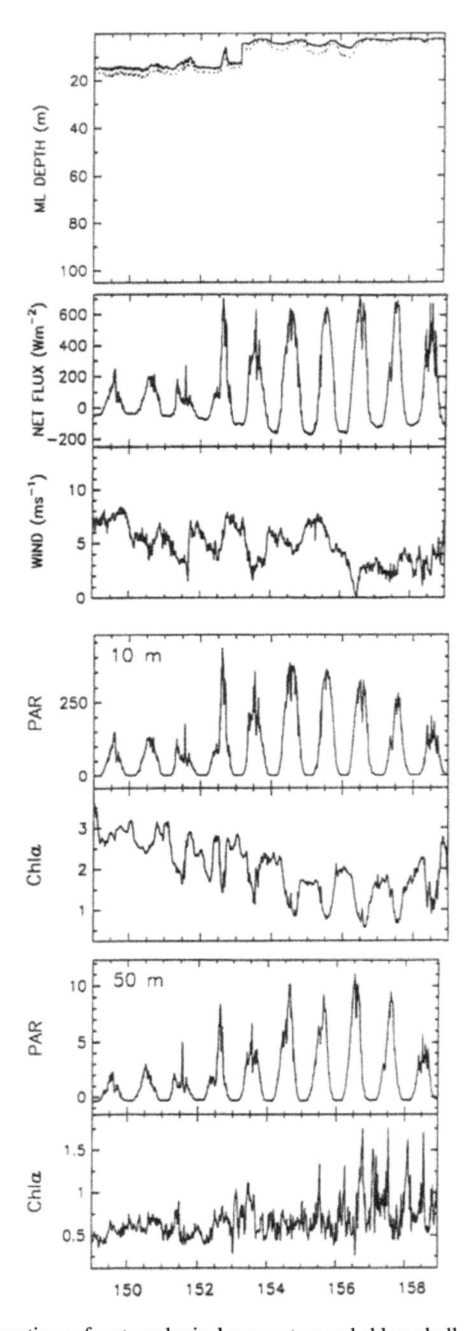

Figure 2.11.7. Observations of meteorological parameters and chlorophyll concentration and PAR at 10- and 50-m depths from MLML experiments showing very little diurnal modulation (from Stramska *et al.*, 1995).

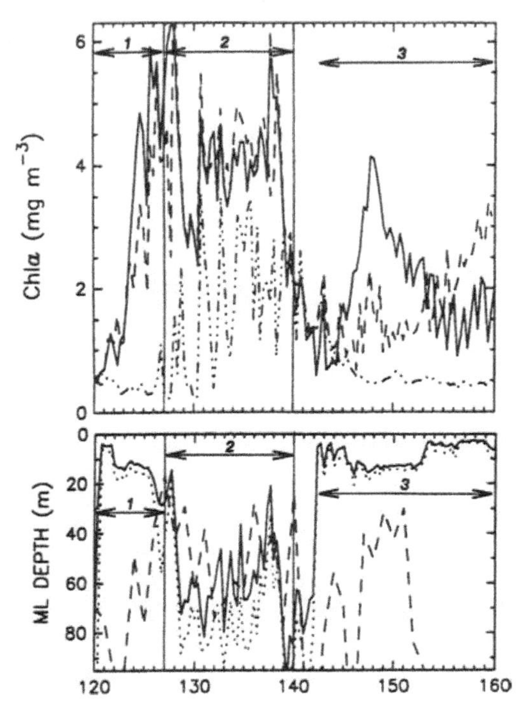

Figure 2.11.8. Observations of column-integrated chlorophyll concentration showing the effects of mixing and restratification (from Stramska *et al.*, 1995).

Biological, or more appropriately, ecosystem models are made up of several "boxes" (or components), each box corresponding to a particular constituent. For example, a four-box NPZD model consists of nutrient, phytoplankton, zooplankton, and detritus components interacting with one another. The nutrient considered is nitrate and the phytoplankton, zooplankton, and detritus biomasses are all expressed in nitrogen concentrations (for example, μmol N m^{-3}). It is assumed that no other nutrient is a limiting factor in the biological production (in some areas of the ocean such as the North Pacific, iron deficiency limits the primary productivity). The most general form of the conservation equations for any of these constituents takes the form

$$\frac{\partial \Phi}{\partial t}+\frac{\partial}{\partial x_j}(U_j\Phi)+\frac{\partial}{\partial z}(W\Phi)+\frac{\partial}{\partial z}(W_S\Phi)=$$

$$\nu_\Phi \frac{\partial^2 \Phi}{\partial x_k \partial x_k}-\frac{\partial}{\partial x_k}(\overline{u_k\phi})+So_\phi - Si_\phi \qquad (j=1,2;\ \ k=1,3) \tag{2.11.24}$$

where Φ is the mean concentration of the constituent (μmol N m^{-3}) and ϕ is the fluctuating value. The LHS contains the tendency term, the horizontal advection term, the vertical advection term, and the sinking term, respectively, where W is the vertical advection velocity (positive upward), W_S is the sinking velocity (negative, since it is always downward) of the constituent, and U_j is the horizontal component of velocity (z is the vertical coordinate, positive upward). The first term on the RHS is the molecular diffusion term, ν_ϕ being the kinematic molecular diffusion coefficient (units of m^2 s^{-1}). The second term accounts for turbulent diffusion, with u_k denoting the fluctuating value of the velocity component. The last two terms are the source and sink terms, equivalently the production and destruction terms for the constituent in question. It is the parameterization of these source and sink terms that determines the skill of the biological model. These terms also couple the equations for various constituents to one another. Provided the turbulent diffusion terms can be modeled appropriately, it is quite straightforward to solve the coupled set of equations to obtain the evolution of the concentrations of the various constituents from known initial conditions.

In a one-dimensional model, all the horizontal advection and horizontal diffusion terms vanish by definition (due to the condition of horizontal homogeneity), so that only tendency terms and vertical advection and diffusion terms are left. The vertical advection term is of importance in upwelling and downwelling scenarios, but the profile of the upwelling/downwelling vertical velocity has to be prescribed in a 1-D model. Similarly it is possible to include sinking of components such as phytoplankton in its equation (McClain *et al.*, 1996). The surface and boundary conditions are also quite simple. Normally, zero flux conditions for these quantities both at the surface and at the bottom suffice.

Denman and Gargett (1995) considered a simple one-dimensional NPZD model where the vertical advection and sinking terms were also neglected so that the equations for the nutrient concentration (N), phytoplankton concentration (P), zooplankton concentration (Z), and detritus concentration (D) become

$$
\begin{aligned}
\frac{\partial N}{\partial t} &= \frac{\partial}{\partial x_3}\left(\nu_N \frac{\partial N}{\partial x_3}\right) - \frac{\partial}{\partial x_3}\left(\overline{wn}\right) + So_N - Si_N \\
\frac{\partial P}{\partial t} &= \frac{\partial}{\partial x_3}\left(\nu_P \frac{\partial P}{\partial x_3}\right) - \frac{\partial}{\partial x_3}\left(\overline{wp}\right) + So_P - Si_P \\
\frac{\partial Z}{\partial t} &= \frac{\partial}{\partial x_3}\left(\nu_Z \frac{\partial Z}{\partial x_3}\right) - \frac{\partial}{\partial x_3}\left(\overline{wz}\right) + So_Z - Si_Z \\
\frac{\partial D}{\partial t} &= \frac{\partial}{\partial x_3}\left(\nu_P \frac{\partial D}{\partial x_3}\right) - \frac{\partial}{\partial x_3}\left(\overline{wD}\right) + So_D - Si_D
\end{aligned}
\qquad (2.11.25)
$$

So and Si are source–sink terms in these equations that couple each of the equations to others. The turbulent diffusion term can be written as

$$-(\overline{u_3\phi}) = K_\phi \frac{\partial \Phi}{\partial x_3} \tag{2.11.26}$$

where Φ and ϕ stand for any of the four quantities above. K_ϕ, the turbulent diffusivity, comes from the physical model and is usually taken to be the same for all four quantities.

Now, the source and sink terms can be written as

$$\begin{aligned}
&So_N = \text{Recycle}; \; Si_N = \text{Uptake} \\
&So_P = \text{Uptake}; \; Si_P = \text{Grazing} + \text{Mortality}_P \\
&So_Z = f_G \cdot \text{Grazing}; \; Si_Z = \text{Mortality}_Z \\
&So_D = \text{Mortality}_{P+Z} + (1 - f_G)\text{Grazing}; \; Si_D = \text{Recycle}
\end{aligned} \tag{2.11.27}$$

Note that there is global conservation of nitrogen:

$$\frac{\partial}{\partial t}(N + P + Z + D) = 0 \tag{2.11.28}$$

However, there is no consensus on how each of these terms is to be parameterized. Therefore, unlike the case for modeling physical processes, even the equations to be used for modeling biological processes are uncertain to some extent. One possibility is (Denman and Gargett, 1995)

$$\begin{aligned}
&\text{Uptake} = \left[\frac{N}{(N^h + N)}\right]\left[\frac{I}{(I^h + I)}\right] p_g P \\
&\text{Grazing} = r_m Z[1 - \exp(-\gamma P)] \\
&\text{Mortality}_P = p_m P \\
&\text{Mortality}_Z = z_m Z^2 \\
&\text{Recycle} = d_m (p_m P + z_m Z)
\end{aligned} \tag{2.11.29}$$

Here, the uptake indicates the growth rate of phytoplankton. It depends only on the nutrient concentration and solar insolation, and is a function of both. Another factor may be the ambient temperature that determines the metabolic rate; yet another may be the trace catalytic nutrients such as iron. Both are ignored here. The maximum growth rate of phytoplankton is parameterized as proportional to its concentration ($p_g P$), and the terms in the square brackets denote the dependence of productivity on nutrient and light availability. Superscript h denotes half-saturation values; I is the solar insolation. Following Ivlev (1955), grazing is usually parameterized as proportional to the zooplankton concentration ($r_m Z$), but with an asymptotic value for large phytoplankton concentrations; γ is the Ivlev constant. Constant p_g is the maximum phytoplankton growth rate and r_m is the maximum zooplankton growth rate, both in units of day^{-1}. Constants p_m and z_m determine the mortality rates (unit, day^{-1}). Note that the mortality rate for zooplankton has a quadratic dependence on its concentration and this is appropriate for predation by carnivores. The factor f_G is the grazing efficiency of zooplankters, and d_m determines the transfer rate between the detritus and nutrient pools. Observational data to determine the various constants and even to verify the assumed form of various source and sink terms are rather scant.

Increased sophistication is usually achieved by more comprehensive parameterizations of the sources and sinks and finer differentiation of the various nutrient pools via addition of more zooplankton, nutrients, and other components. However, this adds more parameters that are even more uncertain. Therefore, while it is clear that more accurate biological simulations must include more trophic levels, this comes at a cost. Simple models are useful at least for obtaining a better understanding of the fundamental nature of biological-physical interactions and that is our focus here.

An even more simplified version is the NPZ model (D = 0) in which the phytoplankton and zooplankton losses are immediately converted to nutrients so that

$$So_N = \text{Mortality}_{P+Z} + (1 - f)\,\text{Grazing} \tag{2.11.30}$$

Denman and Gargett (1995) have used this model coupled to a highly simplified mixing model to simulate an annual cycle of phytoplankton concentration (Figure 2.11.9). They show that a decreased uptake efficiency (increased value of I^h) leads to a smaller overwintering phytoplankton population, that in turn supports a smaller overwintering zooplankton population, which cannot consume phytoplankton fast enough to prevent a spring bloom when increased insolation and nutrients promote rapid phytoplankton growth. On the other hand, a larger overwintering phytoplankton population supporting a correspondingly larger overwintering zooplankton population tends not to produce a spring bloom. Eventually, increased grazing from increasing zooplankton eliminates any spring bloom.

A slightly different version of the biological model is one in which the nutrient pool is divided into ammonia (N_1) and nitrate (N_2) pools and treated separately. This permits new production to be monitored and has been quite popular (McGillicuddy *et al.*, 1995a,b; McClain *et al.*, 1996; Oguz *et al.*, 1996). Such a one-dimensional N^2PZ model coupled to a one-dimensional bulk mixed layer model (Garwood, 1977) was applied by McGillicuddy *et al.* (1995a,b) to the spring bloom in the North Atlantic and McClain *et al.* (1996) to model the biological processes at Station PAPA in the North Pacific (50°° N, 145° W). A

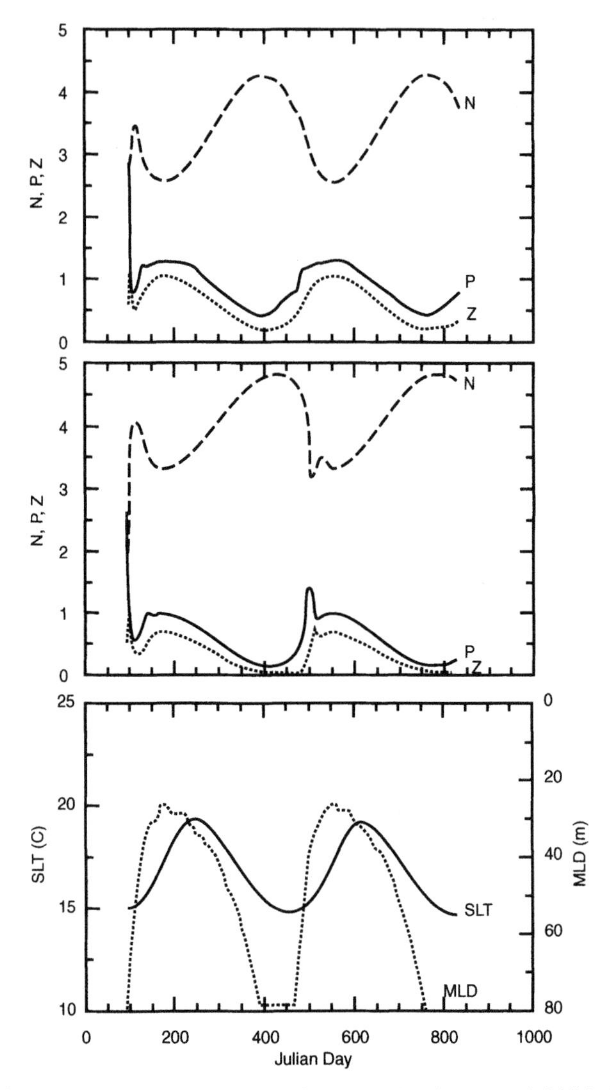

Figure 2.11.9. Model simulations showing the absence (top) and presence (middle) of spring bloom in phytoplankton. Bottom panel shows the MLD and SST. (from Denman and Gargett 1995 with permission from the *Annual Review of Fluid Mechanics,* Volume 27, Copyright 1995, by Annual Reviews).

N^2PZD model coupled to a second-moment closure-based physical model was applied by Oguz *et al.* (1996) to the simulation of the annual cycle in the central Black Sea. To our knowledge, these are comprehensive, realistic, physical-biological model–observation comparisons and therefore we will describe them briefly. The reader is urged to consult the original references for more details and justifications for the parameterizations, and for the latest references on the subject of physical-biological models and observations. Consider first the McClain *et al.* model. The source and sink terms used by McClain *et al.* (1996) were

$$
\begin{aligned}
&So_{N_1} = f_M \cdot \text{Mortality}_{P+Z} + \text{Respiration}_{P+Z};\ Si_{N_1} = \text{Uptake}_1 \\
&So_{N_2} = 0;\ Si_{N_2} = \text{Uptake}_2 \\
&So_P = \text{Uptake}_{1+2};\ Si_P = \text{Grazing} + \text{Mortality}_P + \text{Respiration}_P \\
&So_Z = f_G \cdot \text{Grazing};\ Si_Z = \text{Mortality}_Z + \text{Respiration}_Z
\end{aligned}
\tag{2.11.31}
$$

where N_1 refers to ammonia (NH_4) and N_2 to nitrate (NO_3) (P, Z, N_1, and N_2 are all in μmol N m^{-3} units). The mortality rate for the zooplankton is taken to be proportional to its concentration ~ $z_m Z$, with z_m ~ 0.05–0.1 day^{-1}, and the mortality rate for phytoplankton is $p_m P$, with p_m ~ 0–0.1 day^{-1}. Respiration is also parameterized by linear terms $z_r Z$ and $p_r P$, respectively, with the coefficients a strong function of ambient temperature:

$$
\begin{aligned}
&p_r = p_{r0}\exp(k_{rp}T);\ p_{r0} = 0.020\ d^{-1},\ k_{rp} = 0.0633\ C^{-1} \\
&z_r = z_{r0}\exp(k_{rz}T);\ z_{r0} = 0.019\ d^{-1},\ k_{rz} = 0.1500\ C^{-1}
\end{aligned}
\tag{2.11.32}
$$

Grazing is slightly different from Denman and Gargett (1995) so as to define an asymptotic upper limit to grazing due to food acclimatization,

$$
\text{Grazing} = r_m Z[1 - \exp(-\gamma P)] \cdot \gamma P \tag{2.11.33}
$$

where γ is the Ivlev constant ~ 0.8 (μmol N $m^{-3})^{-1}$ and r_m ~ 1.6 day^{-1}. The difference from the usual form of grazing due to Ivlev (1955) is that instead of asymptoting to a constant value at large phytoplankton concentrations, grazing becomes proportional to the phytoplankton concentration. This food acclimatization is regarded as important by McClain *et al.* (1996).

The efficiency of assimilation by zooplankton is f_G ~ 0.7. The fraction of phytoplankton and zooplankton converted to ammonia, N_1, is f_M ~ 0.33–0.5. By far the most complex is the uptake, and its partition between ammonia and nitrate, parameterized quite differently from that in Denman and Gargett (1995):

$$\text{Uptake} = p_g P$$

$$\text{Uptake}_1 = \pi_1 p_g P; \; \text{Uptake}_2 = \pi_2 p_g P$$

$$\pi_{1,2} = \frac{N_{1,2}^l}{N_1^l + N_2^l}; \; \pi_1 + \pi_2 = 1 \tag{2.11.34}$$

$$N_1^l = \frac{N_1}{N_1^h + N_1}; \; N_2^l = \frac{N_2}{N_2^h + N_2} \exp(-\alpha N_1)$$

$$f \;\; \text{ratio} = N_1^l / (N_1^l + N_2^l) = \pi_2$$

Defining separate uptakes this way permits total primary production ($p_g P$) to be partitioned into new ($\pi_2 p_g P$) and regenerated ($\pi_1 p_g P$) values. Note that the f ratio is the nitrate portion of the nitrogen-based growth and hence denotes new production. Subscript 1 denotes the limiting concentrations and superscript h denotes concentrations where growth reaches half-maximal (half-saturation) values; $N_1^h \sim 0.1$–0.5, and $N_2^h \sim 0.2$–1.0 μmol N m^{-3}. Inhibition of N_2 intake due to the presence of N_1 is represented by the exponential term with $\alpha \sim 1.5-3.0$ (μmol N $m^{-3})^{-1}$. The growth rate p_g is taken to be a strong function of temperature,

$$p_g = \beta p_{gm} = \beta p_{g0} \exp(k_g T); \; p_{g0} \sim 0.851 \; \text{day}^{-1}; \; k_g \sim 0.0633 \; C^{-1} \tag{2.11.35}$$

where p_{gm} is the maximum growth rate, which is temperature dependent. The growth-limiting factor β is determined by the least of the two limits, the nutrient and insolation limits, the former taken as the least limiting of the two nutrient limits,

$$\beta = \min\left[\max(N_1^l, N_2^l), I^l\right]$$

$$I^l = 1 - \exp\left[-\frac{I}{I^k}\left(1 + 10 \exp(-\delta I^d\right)\right] \tag{2.11.36}$$

$$\delta = \exp\left[1.089 - 2.12 \log I^k\right]$$

where I^k is the maximum photoacclimation irradiance (25 microeinsteins per square meter per second, essentially W m^{-2}, in upper layers < 60 m in depth and 250 m elsewhere), and I^d is the 1-day moving average irradiance. This definition accounts for the complex dependence of growth of phytoplankton on irradiance.

Note that an equation for detritus can also be written so that there is a conservation of nitrogen overall, but no recycling back to nutrients occurs; in other words, only the source terms are nonzero, but sink terms are zero (and there is no corresponding source term in the equations for nutrients due to recycling):

$$So_N = (1 - f_M)\ Mortality_{P+Z} + (1 - f_G)\ Grazing \qquad (2.11.37)$$

Also, with the chlorophyll a to nitrogen ratio taken as constant (~1), quantities of interest to optical clarity and carbon concentration can be derived from the quantities above.

McClain *et al.* (1996) also considered the phytoplankton sinking rate W_S (m day^{-1}) and parameterized it as being dependent on nutrient concentration so as to account for the accelerated sinking when the nutrients are depleted:

$$W_s = W_s^m\left[1 - \tanh(2.16 N_2 N_2^1)\right];\ W_s^m \sim 1\ \mathrm{m\ day^{-1}} \qquad (2.11.38)$$

Another aspect of biology is the effect of chlorophyll concentration on the extinction coefficient of PAR, essentially SW solar insolation in the visible band (the near-infrared component absorbed very near the surface is essentially unaffected). It is desirable to account for this effect, since it affects both biology as well as mixing. McClain accounts for this by assuming the extinction coefficient of PAR to be modified by self-shading of phytoplankton according to

$$k = k_w + k_c (chl)^{0.428};\ k_w \sim 0.027\mathrm{m^{-1}};k_c \sim 0.0518 \qquad (2.11.39)$$

where k_w is the extinction coefficient of clear water.

Using extensive observations of physical parameters such as the SST, the MLD, temperature profiles, insolation, and wind forcing, and biological parameters such as chlorophyll a, McClain *et al.* (1996) performed a 30-year simulation at OWS PAPA. Upwelling velocity W was computed using the known wind stress curl values at PAPA. Figure 2.11.10 shows the time series of the average chlorophyll a and zooplankton concentrations in the upper 50 m, which shows an autumn peak of 0.34 mg-atom N m^{-3} to a late spring minimum of 0.20 mg-atom N m^{-3} for chlorophyll a and corresponding values of 0.6 and 0.2 mg-atom N m^{-3} for zooplankton, but a lag of about 1–2 months. A secondary early spring peak also appears in chlorophyll a. Figure 2.11.11 shows comparisons of predicted and observed annual cycles of chlorophyll a, primary production, and nitrate concentration in the upper 50 m. The agreement is remarkably good, given the various uncertainties in coupled, physical-biological models.

McGillicuddy *et al.* (1995a) also applied a similar coupled NPZ physical-biological model (with ammonia and nitrate pools) to observations made in the North Atlantic during the 1989 JGOFS North Atlantic Bloom Experiment

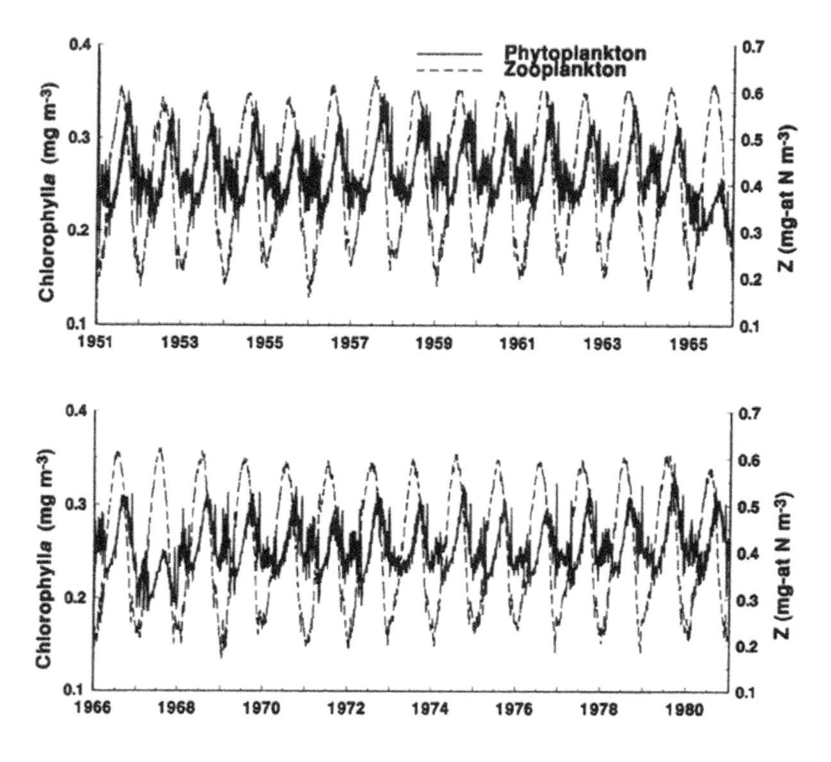

Figure 2.11.10. Model simulations of phytoplankton and zooplankton biomass in the ML at Station Papa (from McClain *et al.*, 1996).

(NABE), but with the parameterizations

$$
\begin{aligned}
&So_{N1} = (1-f_G)*\text{Grazing} + f_{M1}\text{Mortality}_{Z1} + f_{M2}\text{Mortality}_{Z2};\ Si_{N1} = \text{Uptake}_1 \\
&So_{N2} = 0\ ;\ Si_{N2} = \text{Uptake}_2 \\
&So_P = \text{Uptake};\ Si_P = \text{Grazing} \\
&So_Z = f_G * \text{Grazing};\ Si_Z = \text{Mortality}_{Z1} - \text{Mortality}_{Z2} \\
&\text{Uptake}{=}p_g P\ (N_1^1{+}N_2^1);\ \text{Uptake}_{1,2} = \pi_{1,2}\cdot\text{Uptake} \\
&p_g = p_{gm}\left[1-\exp(-\alpha_1 I)\right]\exp(-\alpha_2 I) \\
&p_{gm} = 0.66\ \text{d}^{-1},\ \alpha_{1,2} = 0.0019, 0.0\ (\text{Wm}^{-2})^{-1} \\
&f_{M1} \sim 0.25,\ f_{M2} \sim 0.50;\ f_G \sim 0.75 \\
&\text{Grazing} = r_m Z[1-\exp(-\gamma P)];\ r_m \sim 0.69\ \text{d}^{-1},\ \gamma{\sim}1.0\ \left(\mu\text{mol N m}^{-3}\right)^{-1} \\
&\text{Mortality}_{Z1} = z_{m1}Z;\ \text{Mortality}_{Z1} = z_{m2}Z^2;\ z_{m1} \sim 0.11\ \text{d}^{-1},\ z_{m2} \sim 0.52\ \text{d}^{-1}
\end{aligned}
\tag{2.11.40}
$$

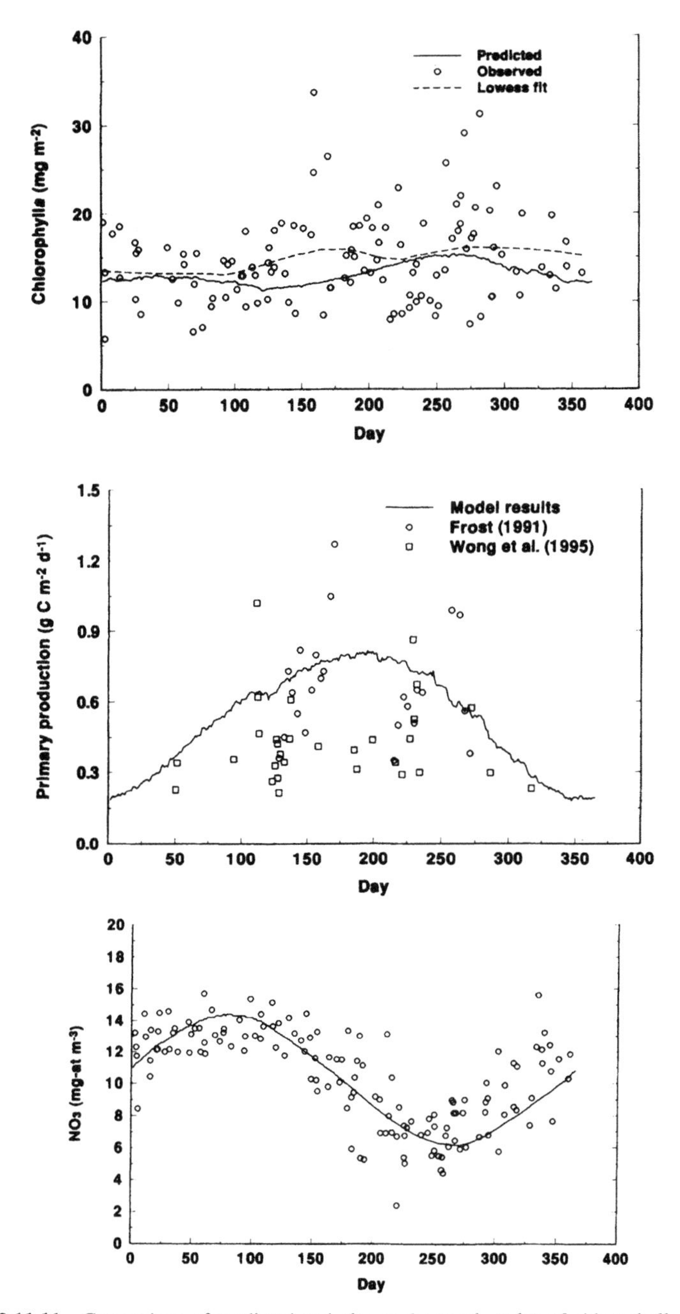

Figure 2.11.11. Comparison of predicted and observed annual cycles of chlorophyll a, primary production, and nitrate concentration in the ML at Station Papa (from McClain *et al.*, 1996).

where α_1 is the initial slope of photosynthetic response, α_2 is the photoinhibition parameter, and p_{gm} is the maximum growth rate. The phytoplankton growth is limited by both light and nutrients. The ratio of uptakes π_1 and π_2 are the same as in McClain *et al.* (1996) but with values $N_1^h \sim 0.05\ \mu\text{mol N m}^{-3}$, $N_2^h \sim 0.2\mu\text{mol N m}^{-3}, \alpha \sim 27.2\left(\mu\text{mol N m}^{-3}\right)^{-1}$. The noninteracting detritus pool has

$$So_D = (1 - f_{M1})\ \text{Mortality}_{Z1} + (1 - f_{M2})\ \text{Mortality}_{Z2} \quad (2.11.41)$$

Note that the zooplankton mortality rate has both linear and quadratic terms, unlike the model of McClain *et al.* (1996). The quadratic term is an effective way to incorporate predation, and apparently is essential to maintain stability of the phytoplankton and zooplankton biomasses when the phytoplankton growth rate nears grazing. For PAR attenuation, McGillicuddy *et al.* (1995a) use

$$k = k_w + k_P P;\ k_w \sim 0.05\text{m}^{-1}; k_P \sim 0.04 \quad (2.11.42)$$

McGillicuddy *et al.* (1995a) determined the values of the various parameters in the model by experimentation and applied their model to spring bloom in the North Atlantic using both a one-dimensional (McGillicuddy *et al.*, 1995a) and a three-dimensional model (McGillicuddy *et al.*, 1995b) incorporating Garwood's bulk mixed layer model. They accounted for the effect of phytoplankton on light extinction in the mixed layer also. Their 1D model results show good qualitative agreement with observations of nitrate distribution and evolution. The spring bloom is also qualitatively well simulated by the model.

Oguz *et al.* (1996) applied a N^2PZD model to the Black Sea,

$$\begin{aligned}
&So_{N1} = \text{Remineralization} + \text{Respiration}_Z;\ Si_{N1} = \text{Uptake}_1 + \text{Oxidation}\\
&So_{N2} = \text{Oxidation};\ Si_{N2} = \text{Uptake}_2\\
&So_P = \text{Uptake}_{1+2};\ Si_P = \text{Grazing} + \text{Mortality}_P\\
&So_Z = f_G \cdot \text{Grazing};\ Si_Z = \text{Mortality}_Z + \text{Respiration}_Z\\
&So_D = (1 - f_G) \cdot \text{Grazing+Mortality}_{P+Z};\ Si_D = \text{Remineralization}
\end{aligned} \quad (2.11.43)$$

where

$$\begin{aligned}
&\text{Grazing} = r_m Z\left(\frac{P}{P + P^h}\right);\ r_m \sim 0.8\ \text{d}^{-1},\ P^h \sim 0.5\ \mu\text{mol N m}^{-3}\\
&\text{Mortality}_P = p_m P;\ \text{Mortality}_Z = z_m Z;\ p_m, z_m \sim 0.04\ \text{d}^{-1}\\
&\text{Respiration}_z = z_r Z;\ z_r \sim 0.07\ \text{d}^{-1}; \quad f_G \sim 0.75\\
&\text{Remineralization} = d_m D;\ d_m \sim 0.1\ \text{d}^{-1}\\
&\text{Oxidation} = d_o N_1;\ d_o \sim 0.05\ \text{d}^{-1}
\end{aligned} \quad (2.11.44)$$

$$\mathrm{Uptake} = p_{gm} P \cdot \min\left[I^l, (N_1^l + N_2^l) \right]; \ \mathrm{Uptake}_{1,2} = \pi_{1,2} \cdot \mathrm{Uptake}$$

$$p_{gm} \sim 1.5\ \mathrm{d}^{-1}, N_1^h, N_2^h \sim 0.2, 0.5\ \mu\mathrm{mol\ N\ m}^{-3} \qquad (2.11.45)$$

$$I^l = \tanh\left(\frac{I}{I_k}\right); \ I_k \sim 100\ \mathrm{Wm}^{-2}$$

Note that the fractional uptake of nutrients N_1 and N_2 is modeled similarly by all three models above, but with different half-saturation values. The uptake itself is modeled slightly differently, although all three formulations consider both the light and the nutrient limitations. Also both McGillicuddy *et al.* (1995a) and Oguz *et al.* (1996) ignore temperature dependence of phytoplankton growth rate. Oguz *et al.* also consider self-shading of the same form as McGillicuddy *et al.*, but with $k_w \sim 0.08\ \mathrm{m}^{-1}$ and $k_P \sim 0.07$.

Oguz *et al.* (1996) show that the spring bloom depends critically on the vertical stability of the water column, occurring just before the seasonal stratification is set up. Once spring–summer stratification is established, the productivity in the upper layers decreases, while the layers below the thermocline receive enough light and nutrients to produce a subsurface phytoplankton maximum. Figure 2.11.12 shows the measured and modeled evolution of average chlorophyll a and primary productivity in the euphotic layer during the annual cycle.

Stramska and Dickey (1994) and Stramska *et al.* (1995) applied the Mellor–Yamada (1982) ML model coupled to a simple light pigment productivity model (Kiefer and Mitchell, 1983) to simulate their observations during MLML. Only a single equation for chlorophyll a concentration is solved:

$$\frac{\partial C_a}{\partial t} = k_c \frac{\partial^2}{\partial x_3 \partial x_3} C_a - \frac{\partial}{\partial x_3}(\overline{wc_a}) + So_c - Si_c \qquad (2.11.46)$$

$$So_c = \left(\frac{I}{I^h + I}\right) c_g C_a, \ Si_c = c_m C_a; \quad c_g, c_m \sim 0.112, 0.12\ \mathrm{d}^{-1}, I^h \sim 116\ \mu\mathrm{E\ m}^{-2}\mathrm{s}^{-1}$$

No equation for nutrients or zooplankton is solved. Yet, even this highly simplified model appears to show fairly good agreement with observations.

Moisan and Hofmann (1996) describe a 1-D nine-box coupled physical-biological model applied to simulate the primary productivity in the California coastal transition zone (CTZ). It has three nutrient pools, nitrate, ammonia, and silicate; two phytoplankters (small and large); three zooplankters (copepods, doliolids, and euphausiids); and a detritus pool. The large phytoplankton is supposed to represent silicate and nitrate-dependent diatom that grows best under high nutrient and light conditions, and the smaller one only nitrate-dependent flagellate that grows best under low nutrient and light conditions.

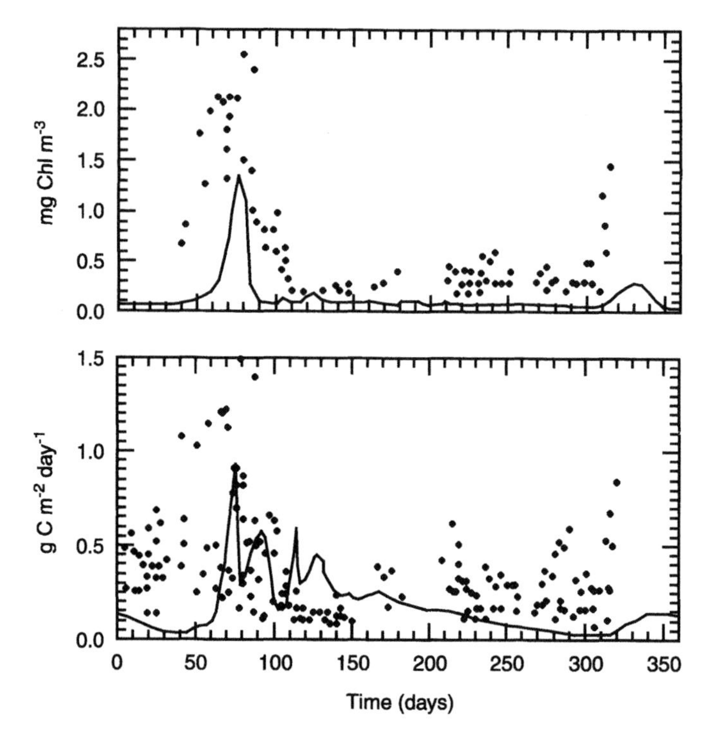

Figure 2.11.12. Comparison of eutrophic layer (top) average chlorophyll concentration and (bottom) integrated primary production from the model compared to observations (from Oguz *et al.*, 1996).

Complex source and sink functions were used for phytoplankters, including combined silicate and nitrate limitation, and light limitation on growth rate. The three zooplankters include herbivores and omnivores. Nitrate and silicate concentrations are governed by the phytoplankton uptake. The vertical diffusion coefficient is set at 10^{-4} m^2 s^{-1} and vertical velocity is prescribed to be linear over the top 100 m. The irradiance model of Sathyendranath and Platt (1988) was used and the effect of chlorophyll concentration on light extinction was taken into account. Vertical sinking rates were set for detritus and the two phytoplankters. Extensive sensitivity runs were made to determine the optimum values for various parameters in the complex coupled model. When applied to the CTZ, the model was able to simulate the development and maintenance of the subsurface chlorophyll maximum. The latter was primarily due to *in situ* primary production. The PAR extinction scale was affected by zooplankton grazing activity. Figure 2.11.13 shows the temporal evolution of the components of the ecosystem during the model simulation.

Another example of an ecosystem model is that by Leonard *et al.* (1999) of the central equatorial Pacific, which is a HNLC region. They apply a 1-D nine-

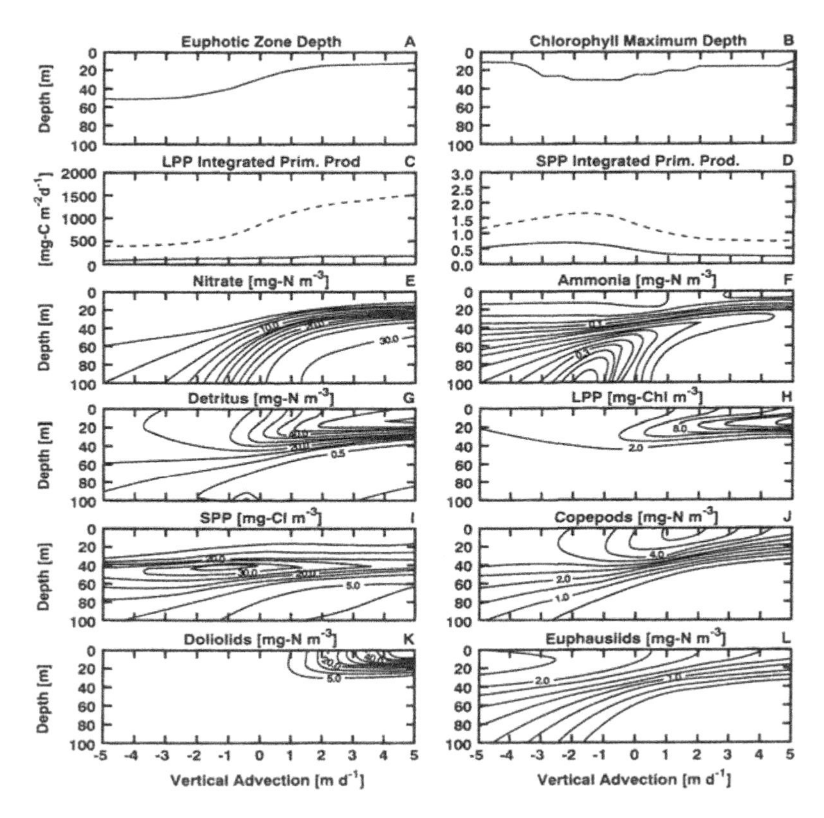

Figure 2.11.13. Temporal evolution of the various components of the nine-box coupled physical-biological model applied to the California coastal transition zone (from Moisan and Hofmann, 1996). Note the subsurface phytoplankton maximum maintained primarily by in situ primary productivity.

box ecosystem model, comprising two phytoplankton, two zooplankton and two detrital categories in addition to a nitrate, ammonium and dissolved iron boxes, to show that the chlorophyll concentrations are both iron-limited and grazing-controlled. The conservation equation for dissolved iron makes use of specified C:Fe ratios to parameterize the source and sink terms on the RHS.

Clearly, the above models show the level of sophistication attainable in biological ecosystem models. However, it is difficult to determine with certainty the large number of parameters that describe such multibox models. Thus, although in some cases the level of complexity introduced by multicomponent models of phytoplankters and zooplankters is unavoidable, it is not clear if these complex models are preferable in general to simple ecosystem models with fewer components but with better-determined parameters. This is precisely the reason cited by Stoens *et al.* (1999) in using a single nitrate box model to simulate the nitrate variability in the equatorial Pacific during the 1992-95 El Niño. There is still much to learn in coupled physical-biological models of primary productivity in the upper ocean.

2.11.3 SCALING ARGUMENTS

Since chemical and biological processes in the OML are intimately coupled to physical processes, a natural question to ask concerns what governs the relative importance of physics vis-à-vis chemistry and biology. To answer this, we need to look at the timescales and length scales involved. Let us consider a simple one-dimensional conservation equation for a photochemical constituent whose production rate depends on the intensity of insolation,

$$\frac{\partial C}{\partial t} = \frac{\partial}{\partial z}\left(K_c \frac{\partial C}{\partial z}\right) + \frac{\gamma I_s(t)}{T_g}\exp(z/D_e) - \frac{C}{T_d} \quad (2.11.40)$$

where T_g is the timescale associated with production, and T_d is the timescale associated with decay rate or destruction of the constituent ($t_d \sim 5$ hr for CoS). It is also called the turnover time (Doney *et al.,* 1995). D_e is the extinction scale for the photochemically (UV part) active radiation, and γ is a constant associated with production. In addition to this, for dissolved gases, we need to consider the flux at the ocean surface,

$$F_{SC} = v_p DC; DC = C_s - C_a \quad (2.11.41)$$

where v_p is the piston velocity, and we will assume that the air-side concentration is constant so that ΔC depends just on C_s. We have lumped all the air–sea exchange physics into one unknown v_p.

Now, ignore tendency and mixing terms, assume I_s is constant, and look at the equilibrium solution, where the balance between production and destruction prevails at every depth. The equilibrium profile is given by

$$C(z) = \frac{T_d}{T_g}\gamma I_s \exp(z/D_e) = C_s \exp(z/D_e) \quad (2.11.42)$$

where C_s is the surface concentration, which is a function of the chemical parameters and surface insolation. However, this assumes that concentration of dissolved biological matter (for production of photochemicals such as CoS) is uniform and not limiting. Let D_m be the limit of mixing, usually the ML depth. Now the depth at which the concentration reaches the depth-averaged concentration defines a length scale D_t given by

$$D_t = D_e \ln\left[\frac{D_e}{D_m}\left(1 - \exp(D_m/D_e)\right)\right] \quad (2.11.43)$$

The significance of this depth is that for a given mixed layer depth and insolation and chemical characteristics, this depth will determine approximately the depth above which mixing tends to reduce the concentration (acts as a sink) and below which it will increase the concentration (acts as a source). In essence, C_s and D_t determine the character of the equilibrium solution. For a given chemical model and ML depth, D_m, C_s, D_t, and D_e characterize the system completely.

The physics of mixing is tied up with the mixing coefficient K_c, which is proportional to the characteristic mixing velocity scale q_* (equal to u_* for shear-driven turbulence, w_* for convection-driven turbulence, and a combination for both) and a characteristic length scale, which is proportional to D_m in the bulk of the ML. A characteristic length scale can be defined out of K_c and t_d: $D_c = (K_c t_d)^{1/2}$. This is the e-folding length scale if the mixing coefficient were constant and the balance were between the mixing and decay. This depth is called the eddy consumption depth (Doney *et al.,* 1995). A characteristic timescale for air–sea transfer of the constituent is simply $t_v = D_m/v_p$, the ventilation timescale for the mixed layer. Efficiency of mixing is indicated by the eddy turnover timescale $t_e = D_m/q_*$. There are therefore three timescales, t_d, t_e, and t_v, and four length scales, D_c, D_m, D_c, and D_e, in the problem. By dimensional analysis,

$$\frac{C(z)}{C_s} = f\left(\frac{z}{D_c}, \frac{t_v}{t_d}, \frac{t_e}{t_d}, \frac{D_c}{D_m}, \frac{D_e}{D_m}, \frac{D_t}{D_m}\right) \quad (2.11.44)$$

The first timescale ratio in Eq. (2.11.44) determines the relative importance of air–sea transfer. The smaller this ratio, the more important the air–sea transfer (see, e.g., Gnanadesikan, 1996). The second ratio determines the efficiency of mixing: the smaller this number, the more efficient the mixing is. These five ratios define a complex phase space in different portions of which prevail different mechanisms and balances.

A rapidly deepening ML introduces another parameter, the entrainment rate at the base of the ML, $\partial D_m/\partial t$ (or the nondimensional parameter $q_*^{-1}\partial D_m/\partial t$). For a biological parameter such as phytoplankton concentration, the situation is even more complicated because of coupling to other parameters such as nutrient concentration and predator population, and dependence on factors such as metabolic rate and grazing pressure. Nevertheless, an equilibrium solution of the coupled system can be obtained without mixing and tendency terms. Both D_t and C_e thus obtained will be complex functions of biological parameters such as N_m/N_H and I_s/I_H (see above). Note that D_e refers now to photosynthetically active radiation. Also for phytoplankton, $t_v = 0$, but it is replaced by the timescale associated with its sinking rate. With these caveats, the above equations should be applicable to biological variables, also.

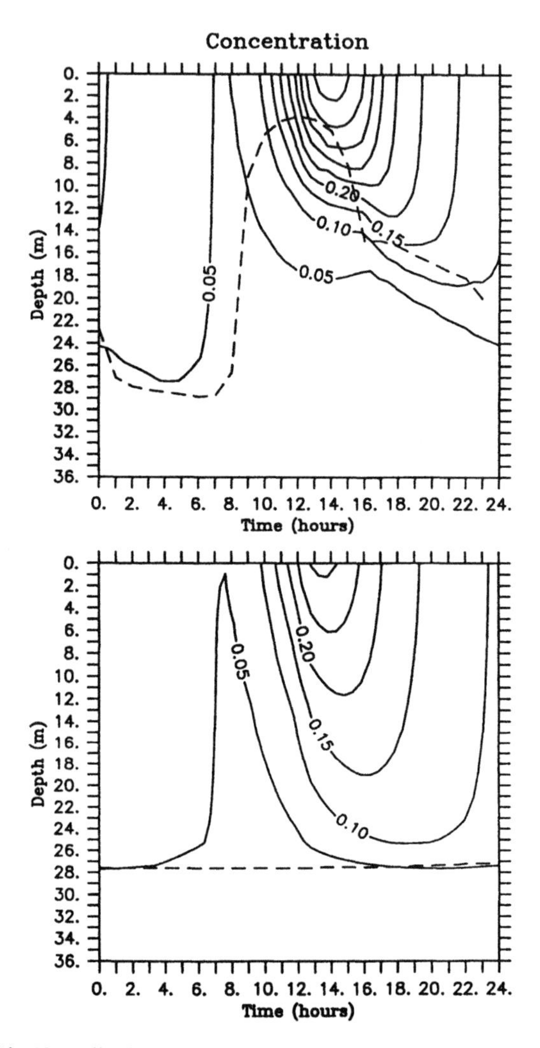

Figure 2.11.14. Normalized concentration of the photochemical species with (top) and without (bottom) the diurnal cycle. The dotted line shows the mixed layer depth (from Doney *et al.*, 1995).

Diurnal variability introduces further complications in photochemical and biological production. The relevant parameter is the ratio t_d/t_h, where t_h is the length of the day (~1/2 day). For small values of this parameter, the delay between the peak of solar insolation and the peak concentration is small. For values of this ratio approaching unity, large lags can be expected (Doney *et al.*, 1995). Secondly, because of the suppression of mixing and shallow ML depths during the day, large concentrations are built up near the surface during the day, which are erased during the night by deep nocturnal mixing. Both these effects can be seen in the numerical model results of Doney *et al.* (1995), which show the temporal evolution of a typical photochemical with and without diurnal variability (Figure 2.11.14).

Overall, the inclusion of realistic mixing physics and complex biology/chemistry makes the problem complex enough that the numerical solutions discussed above become more and more attractive (see Kantha and Clayson, 1994; Doney *et al.*, 1995; Gnanadesikan, 1996).

LIST OF SYMBOLS

α_n	The fraction of solar radiation in wavelength band n
α	The Sun angle
β, β_S	Coefficients of expansion (thermal and due to salinity)
δ_p, δ_w	Planetary boundary layer and wave boundary layer thicknesses
ε	Dissipation rate of TKE
φ_M	Monin–Obukhoff similarity function
κ	Von Karman constant
λ_n	Inverse extinction length scale
ν, ν_t	Kinematic viscosity (molecular and turbulent)
θ	Fluctuating temperature (also angle of the waves to wind direction)
ρ	Fluctuating fluid density (also correlation coefficient)
ρ_w, ρ_A, ρ_I	Density of water, air, and ice
σ_{ij}	Ice stress tensor (also turbulent stress tensor)
ζ	Sea surface height
τ_{AI}, τ_{IO}, τ_{AO}	Shear stress at air–ice, ice–ocean, and air–ocean interfaces
τ_{ij}	Shear stress tensor
ΔT, ΔS, ΔU	Changes in temperature, salinity, and velocity
Θ	Mean temperature
a	Amplitude of the wave
b	Buoyancy
c	Fluctuating scalar concentration
c_d	Drag coefficient
f	Coriolis parameter
g	Acceleration due to gravity
k	Wavenumber
k_1, k_2, ...	Chemical reaction constants
k_T, k_S, k_c	Diffusivity of heat, salt, and a passive scalar (kinematic)

l	Integral microscale of turbulence
n	Wave frequency
q, q_*	Turbulence velocity scales
q_b	Buoyancy flux
q_0, q_e	Applied convective heat flux and entrainment buoyancy flux
t	Time
t_e, t_η, t_b	Eddy turnover timescale, Kolmogoroff timescale, and buoyancy timescale
u, v, w	Turbulence velocity components
u^*, u_*	Friction velocity
u_e	Entrainment velocity
w_*	Deardorff convective velocity scale
x	Horizontal coordinate
x_i	Coordinates
z	Vertical coordinate
z_i	Inversion height
z_0	Roughness scale

A	Alkalinity
A_I	Ice concentration
C	Chlorophyll or another scalar mean concentration (also the wave phase speed)
C_T	Total dissolved carbon
C_g	Group velocity
D_M, D_e, D_I	Mixed layer depth, Ekman layer depth, and ice thickness
E	Entrainment ratio
F_w	Energy flux due to internal waves
H_c	Conductive heat flux
$I(z)$, I_0, I_{IR}, I_v	Solar insolation at depth z and the surface, and the infrared and visible fractions
L_n	Extinction length scale
L_F	Latent heat of fusion
N	Buoyancy frequency (also nitrogen concentration)
P	Mean pressure
P_S, P_b, P_L	Shear, buoyancy, and Langmuir circulation production of TKE
Pr_t	Turbulent Prandtl number
R	Flux ratio in a convective ML
Ri_g, Ri_f	Gradient Richardson number and flux Richardson number
Ri, Ri_N	Bulk Richardson number and stratification Richardson number
S, S_I	Salinity of water and ice
S_M, S_H	Stability functions in turbulent mixing coefficients
So, Si	Sources and sinks in the conservation equations
S_M	Mixed layer salinity

T_M	Mixed layer temperature
U_A, U_I	Wind velocity and ice velocity.
U_g, U_b	Geostrophic and bottom velocities
U_M	Mixed layer velocity
V_S	Stokes drift velocity due to surface waves

Chapter 3

Atmospheric Boundary Layer

In this chapter, we discuss the salient characteristics of the atmospheric boundary layer (ABL) and how we go about parameterizing and modeling the associated processes. We will discuss important mixing processes in the ABL and the causative agents. The literature on the ABL is vast and all we can do here is summarize a few salient features of importance in the context of small scale processes in geophysics. There are several excellent monographs on the topic, including some that are dated but still useful (Sutton, 1953; Lumley and Panofsky, 1964; Plate, 1971; Oke, 1978; Panofsky and Dutton, 1984). More recent ones by Arya (1988), Stull (1988), Sorbjan (1989), Garratt (1992), and Kaimal and Finnigan (1994) are excellent reference sources for the ABL. Kaimal and Finnigan (1994) also provide a survey of observational techniques for the ABL and data processing methods. There is also a collection of articles such as Haugen (1973), Wyngaard (1980), Nieuwstadt and van Dop (1982), and Lenschow (1986). Recent advances in the field can be found in journals such as the *Journal of Geophysical Research of the American Geophysical Union,* the *Journal of Atmospheric Sciences,* the *Journal of Applied Climatology,* the *Journal of Atmospheric and Oceanic Technology of the American Meteorological Society,* and *Boundary Layer Meteorology,* among others. Kraus and Businger (1994) and Kagan (1995) deal with air–sea interactions, a topic that necessarily involves the ABL. Some of the introductory material in this chapter follows Stull (1988), Sorbjan (1989), Garratt (1992), and Kaimal and Finnigan (1994).

3.1 SALIENT CHARACTERISTICS

The atmospheric boundary layer is quite important to many aspects of human life. It is in a very thin layer of the lower ABL that most people live and as such, many aspects of the ABL affect daily life. Dispersion of pollutants is one such example. The characteristics of turbulent mixing in the ABL determine the concentration levels of pollutants. Most importantly, the momentum, heat, and moisture exchanges between the atmosphere, Earth, and oceans takes place through the ABL, and it is for this reason that ABL has received intense attention.

The ABL over land can be grouped into two types—the daytime convective ABL (CABL) and the nighttime or nocturnal ABL (NABL). The mixing in the CABL and NABL is basically different. In the former, it is primarily due to solar heating at the lower boundary, and in the latter, it is primarily due to winds. The CABL is extensively studied, while the NABL has not received that much attention. The presence or absence of condensation of water vapor and clouds is another important distinction.

The CABL can be divided into three parts—the lower part close to the surface (composing perhaps 10% of its depth), the middle layer (constituting the bulk of the ABL), and the upper part near the capping inversion layer, which makes up another 10–20% of the ABL. Turbulence in the entire CABL is driven by the heat flux from the ground. However, the salient processes, characteristics of mixing that goes on, and scaling properties are all different in these regions. In the layers close to the surface, fluxes of various properties such as momentum, heat, water vapor, and trace gas concentrations are nearly constant. In this constant-flux surface layer, the important nondimensional variable is the distance z from the surface normalized by the Monin–Obukhoff length that characterizes the stability conditions. The surface layer is the part that is the most well-measured and well-understood part of the ABL.

In the bulk of the ABL, the scaling is determined by the depth of the inversion z_i. Nevertheless, it is also driven by the convective, destabilizing heat flux to the atmosphere near the ground. Mixing processes are so efficient that the gradients of properties in the bulk of the CABL are small, and the transport is mainly due to large eddies and thermals. Near the top, close to the capping inversion, the dynamics are different. The ambient stratification is such that stable, lighter air masses from above the ABL are entrained into the ABL by the large eddies. The intensity of this entrainment determines the properties in the upper parts of the CABL. It is, in fact, possible to divide the turbulent diffusion of a passive scalar such as a pollutant in the CABL into two parts, top-down and bottom-up. The top–down diffusion can be scaled using the entrainment characteristics at the top of the CABL. The bottom–up diffusion depends on the characteristics of the turbulence driven by heat flux near the ground. The timescale for mixing a scalar released at the ground throughout the CABL is typically on the order of an hour. Therefore, the vertical profile of the scalar through the CABL depends on the ratio of its lifetime (determined by chemical reaction rates) to this mixing time. If this ratio is not large, the profile will be different than that of a conserved scalar. Often fair weather cumulus clouds form at the top of CABL from condensation of moisture transferred upward from near the ground surface by convective eddies. If these are extensive, they affect the further evolution of the CABL by modifying the fluxes through it.

The NABL is of a fundamentally different character than the CABL. Here the source of mixing is the shear from winds aloft, and not the destabilizing heat flux to the atmosphere from the bottom. In fact, the heat flux is normally from the atmosphere to the ground, and turbulence has to work against gravity, and therefore the NABL is also often called a stable boundary layer (SBL). Since winds are the sole source of mixing, their variability affects the NABL. Consequently, mixing in the NABL can be weak and very intermittent, and accompanied by internal wave motions. Continuous turbulence is often confined to the lower parts of the NABL, with intermittent turbulence characterizing the upper part. Strong stable stratification and gradient Richardson numbers close to critical are salient features. Because of the relative inefficiency of shear-induced mixing vis-à-vis convective mixing, properties often show strong gradients in the NABL, whereas the CABL tends to be more uniformly mixed, except perhaps close to the inversion. NABL depths are also smaller, several tens of meters to a few hundred meters at the most, whereas the CABL is typically one to a few kilometers deep. Very stable NABLs occur under light winds and strong surface cooling under clear skies. Unlike CABL, where simple similarity laws suffice and numerical models can be easily constructed, here the prevailing processes are complex, and difficult to characterize by any simple theory and to model: layered structures, intermittent turbulence, internal waves, and low level jets above the surface inversion.

The normal sequence of events in the ABL in the midlatitudes is as follows. Over the previous night, a low level inversion has formed close to the ground, typically 100–200 m high, and the atmospheric column has also cooled by radiation into space during the night. The heat flux is to the ground and on the order of a few tens of W m^{-2}. When solar heating commences in the morning, the layers close to the ground get heated, convective instability ensues, and the resulting turbulence begins to erode the inversion above. Gradually, over the course of the day, the CABL deepens and by mid-afternoon begins to approach an asymptotic value of 1–2 km. Maximum temperatures are reached about mid-afternoon. All this time, vigorous entrainment takes place at the top of the ABL. Once the solar heating ceases, turbulence in the bulk of the CABL collapses in a matter of a few tens of minutes and it is then confined to the vicinity of the ground. The depth of this turbulent mixed layer depends on the strength of the winds aloft, since it is the wind-induced shear that is the principal source of mixing. Throughout the night, the NABL is fairly constant in depth, although in many cases, there are some changes; the NABL may grow slowly through the night. Nocturnal inversion strength normally increases during the night and the stage is set for the diurnal cycle to start all over again in the morning. There are many other factors that influence the course of these events, such as the presence of clouds, which tend to decrease the cooling during the night, and the presence or absence of phase conversion. Wyngaard (1992) presents an excellent review of processes that occur in the CABL and NABL.

A similar diurnal cycle occurs in the oceanic mixed layer (OML) as well, although its character is somewhat different. Strong solar heating during the day gives rise to a shallow, diurnal mixed layer a few meters deep, especially during summer and when the winds are low. The turbulence in the bulk of the seasonal OML is extinguished. A strong temperature gradient builds up at the bottom from the strong heating of this shallow, mixed layer. But cessation of solar heating and nocturnal cooling begins to erode the "inversion" built up during the day and if the winds are not light, mixing penetrates to the depth of the seasonal thermocline and mixes the heat gained during the day by the shallow upper layers into the rest of the OML.

Seasonal modulation due to changing solar insolation is normally not as important as the predominant diurnal variations, as far as mixing in the ABL in midlatitudes is concerned. In polar regions, however, the seasonal changes are quite strong. In the Arctic, during the long polar night, the principal source of turbulence in the ABL is the wind. Arctic winds can be strong and sustained. The heat flux from leads and polynyas is usually not a major factor, except locally. Consequently, generally speaking, the ABL is quite shallow, 50–300 m deep (Andreas, 1996). In contrast, the ABL during an Arctic summer can be deeper due to the steady and incessant insolation. The much longer timescales of principal variability are the most distinguishing feature of the polar ABL, relative to the diurnally modulated ABL at lower latitudes.

A very important aspect of the ABL that distinguishes itself from its counterpart in the oceans is the presence of water in both vapor and condensed liquid form, and the associated phase conversions. Condensation releases the latent heat of vaporization and this internal source of heating is quite important to the fate of the ABL. Conversely, evaporation acts as a heat sink. Low level convective clouds are typically found in the upper part of the ABL and their flat bottoms often characterize the lower boundary of the capping inversion. Cloud-top cooling is an important source of energy for mixing in the ABL. Clouds also affect the heat balance at the air–sea and air–ground interfaces and are therefore important to the dynamics of both the ABL and the OML. Cloud-topped ABL (CTBL) is therefore quite complex and difficult to understand and model. Our lack of familiarity with the subject prevents us from dealing with the subject of phase conversions of water in the ABL and cloud dynamics, and its impact on the ABL, except cursorily. An excellent reference for this important topic is Houze (1993).

The ABL can also be grouped into that over land and that over sea. The ABL over the sea is fundamentally different. Because of the large heat capacity of the oceans, which constitute a strong source of heat and moisture, the temperature at the bottom, the sea surface temperature (SST), changes by a much smaller amount, and the diurnal cycle in the marine ABL (MABL) is therefore much different. The abundance of moisture in the MABL leads most often to cloud formation there and this has a large effect on the MABL structure and dynamics. Also under clear conditions the MABL is directly heated by the sun, leading to less stable conditions than in a NABL over land that suffers from strong IR cooling (Dabberdt *et al.,* 1993). In mid to high latitudes, the strong seasonal cycle of the OML affects the MABL. Warm air advection over cooler waters leads to strongly stable MABL, whereas cold air masses transiting over warm waters give rise to strong convection and an unstable convectively mixed MABL.

The ABL over sea ice is affected by the insulating properties of sea-ice cover. The heat flux is usually small or negligible so that most often the ABL over ice is close to neutral stratification with winds being the major source of turbulent mixing. Consequently, the ABL is shallow, a few hundred meters in most cases. Advection of air masses from relatively warmer water on to ice causes suppression of turbulence in the ABL and strong inversions can develop over ice under these conditions.

Extensive measurements from aircraft, towers, and tethered balloons over the past two decades (Kaimal *et al.,* 1976; Caughey, 1982; Lenschow, 1979; Lenschow *et al.,* 1988; Caughey and Wyngaard, 1979) have increased our knowledge of the CABL considerably. For those quantities that are hard to observe and measure accurately, LES's (Moeng and Wyngaard, 1986, 1989; Moeng and Sullivan, 1994) have filled the gap (see Chapter 1). As a result, a very accurate picture of mixing in the CABL has emerged. As a result, it is also possible to postulate simple models of mixing in the ABL (Moeng and Sullivan, 1994). In contrast, the picture for the NABL and CTBL is not that complete. Also, most of our knowledge comes from measurements in a relatively horizontally homogeneous ABL. Only recently has increasing attention been given to orographic effects and horizontal inhomogeneities such as those across the land–sea boundary.

In the seventies and eighties, the team of James Deardorff and Glen Willis performed laboratory simulations of convection in a laboratory convection tank (Deardorff *et al.,* 1969; Willis and Deardorff, 1974; Deardorff, 1980; see also Kantha, 1980b) that shed some light on the structure of the CABL, and the nature of entrainment processes near the capping inversion. The entrainment rate was measured and used to establish simple models (Deardorff, 1980). These experiments have been very useful, even though the Reynolds numbers that could be achieved were less than the atmospheric values (R_t is several orders of magnitude less, typically less than 10^4). The main problem has been the finite aspect ratio effects, brought on by practical limits on the size of the tank (Deardorff and Willis, 1985; Kantha, 1980b). Nevertheless, the knowledge gained from laboratory simulations has been invaluable. Turbulent entrainment across a buoyancy interface (Kantha *et al.,* 1977; Kantha, 1980a,b) has long been a subject of laboratory studies (see Fernando, 1991, for a recent summary; see also Nokes, 1988; Gregg, 1987), and has added to our understanding of mixing processes in the ABL. Needless to say, the various careful field observations made over the past two decades (see, for example, Sorbjan, 1989; Kaimal and Finnigan, 1994) has helped further our understanding of the ABL.

3.1.1 Ground-Based Remote Sensing

While *in situ* measurements from towers, balloons, kites, sondes, and aircraft have yielded a wealth of information on the ABL structure and dynamics over the past few decades, their obvious limitations w.r.t. spatial and temporal coverage have spurred the development of satellite-orbited and ground-based remote sensors. There has been rapid development in the latter over the past 25 years, and ground-based radars, sodars (SOund Detection And Ranging), and lidars (LIght Detection And Ranging) have matured enough to be routinely employed to probe the ABL structure (Lenschow, 1986; Atlas, 1990; Wilczak *et al.*, 1996) and add significantly to our knowledge of ABL processes, such as drainage flows, nocturnal jets, internal waves, land–sea breezes, flow convergences, and pollutant transport. We draw upon a compact review of the progress to date in Wilczak *et al.* (1996).

The principal advantage of most ground-based sensors is that not only can they monitor continuously the vertical profiles of properties in the ABL but they can also scan a large volume of the ABL continuously in time to provide horizontal distributions as well. A classic example is the Doppler radar, an operational network (NEXRAD) of which is now routinely used in the United States for assisting regional weather forecasts around the country.

Frequency-modulated (FM) continuous wave (CW) radars (frequencies of a few hundred MHz to a GHz) depend on backscattering of microwave energy from point scatterers in the ABL such as raindrops, snow particles, and insects, as well as from refractive index inhomogeneities. The Radio Acoustic Sounding System (RASS) that uses both acoustic and microwave energies enables monitoring of both wind and temperature profiles in the ABL. Sodars, on the other hand, depend principally on backscattering of acoustic energy by refractive index changes in the ABL, although they are sensitive to point scatterers as well. Time–height images of backscattered acoustic intensity have provided a wealth of information on the ABL structure, such as convective plumes, temperature inversions, and thermal fronts. Doppler techniques have enabled mean wind profiles to be measured in the lower ABL. Lidars employ backscattering by aerosol particles and hydrometeors of energy in the visible range of the electromagnetic spectrum to probe the ABL structure. Once again Doppler techniques enable the velocity to be profiled. Lidar ceilometers have been used to measure the height of the cloud base, and lidars have been useful for monitoring the vertical aerosol structure in the ABL and tracking pollutant plumes. Use of Raman backscatter enables water vapor profile measurements to be made.

All three techniques have advantages and limitations; for example, only radars can probe through clouds. Because of the presence of large turbulent fluctuations in temperature and humidity and particulates in the entrainment zone at the top, all three can be used to monitor the depth and evolution of the daytime CABL readily, but the absence of strong discontinuities makes it hard to use them for the NABL. Appropriate use of temporal averaging provides a better estimate of the ABL depth than a single sonde profile can, in spite of the resolution limitations (a few tens of meters typically). Mean wind profiles can be measured to within 1 m s^{-1} from both Doppler radars and sodars. Adding an acoustic source with a wavelength half that of the radar enables the radar to sense Bragg backscattering from the acoustic wave and measure its velocity, which is a function of the ambient temperature. A 915-MHz profiler system (RASS) appears to be able to profile the temperature to within 1 K to a height of 500 m to 1 km (Wilczak *et al.*, 1996). Most importantly, radars and sodars appear to hold the promise of measuring turbulence-related properties in the ABL. The techniques are well-suited to neutral and convective ABL (not NABL). Figure 3.1.1 shows a vertical profile of momentum flux measured from a dual-Doppler radar. If the radar or sodar wavelength lies within the Kolmogoroff inertial subrange, it is possible to deduce the dissipation rate of TKE. Thus, it is in principle possible to use ground-based remote sensors, to profile both the mean properties such as velocity, temperature and humidity, and turbulence properties (Gal-Chen *et al.*, 1992) such as the fluxes of momentum, heat, and water vapor, and the dissipation rate in the daytime ABL. The accuracies involved, the limitations in resolution and range, and the difficulties are discussed in detail by Wilczak *et al.* (1996). He also provides examples of applications to studies of processes in the ABL such as sea breezes, convergence boundaries, drainage flows, nonprecipitating cloud systems, and pollutant dispersion.

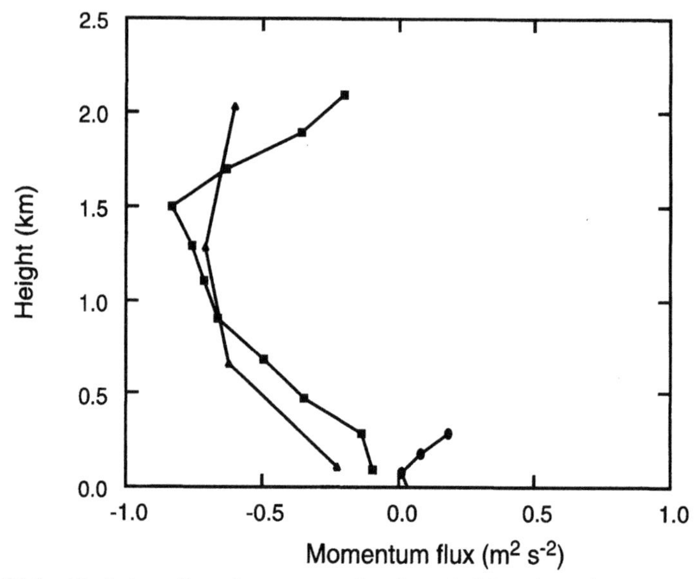

Figure 3.1.1. Vertical profiles of momentum flux from dual-Doppler radar (squares), aircraft (triangles), and towers (circles) (from Wilczak *et al.*, 1996 with kind permission from Kluwer Academic Publishers).

3.2 GEOSTROPHIC DRAG LAWS

It is possible to derive functional relationships for geostrophic drag laws in the planetary boundary layer (or ABL) by applying principles of asymptotic matching, without the necessity to solve the complex turbulent flow in the ABL. These laws relate the geostrophic velocity immediately outside the boundary layer to the friction velocity or the shear stress. The procedure is the same as the one due to Millikan for neutrally stratified, nonrotating turbulent boundary layers (Yaglom, 1979) that occur in the laboratory (see Section 1.6). Consider first a neutrally stratified, horizontally homogeneous, turbulent boundary layer in a rotating fluid. The governing equations are

$$
\begin{aligned}
-f\left(U_2 - G_2\right) &= \frac{\partial}{\partial z}\left(-\overline{u_1 u_3}\right) \\
f\left(U_1 - G_1\right) &= \frac{\partial}{\partial z}\left(-\overline{u_2 u_3}\right)
\end{aligned}
\tag{3.2.1}
$$

G_1 and G_2 are components of the geostrophic flow aloft:

$$-f\,G_2 = -\frac{\partial p}{\partial x_1}$$
$$f\,G_1 = -\frac{\partial p}{\partial x_2} \qquad (3.2.2)$$

Let G be the magnitude of the geostrophic velocity: $G = (G_1 + G_2)^{1/2}$. The velocities are brought to zero at the surface by friction and the resulting profile depends very much on the resulting turbulent shear stresses. Unfortunately, these stresses are part of the solution to the above equations and unless turbulence closure is effected, it is not possible to solve for the flow in the ABL. Solutions are readily obtained when the flow is laminar. The resulting profile is the laminar Ekman profile. If one assumed constant eddy viscosity in the ABL, one would get a similar profile (although this assumption is hard to justify),

$$U_1 = G_1\left[1-\exp\left(-\frac{z}{\delta_1}\right)\cos\left(\frac{z}{\delta_1}\right)\right] - G_2\exp\left(-\frac{z}{\delta_1}\right)\sin\left(\frac{z}{\delta_1}\right)$$

$$U_2 = G_1\exp\left(-\frac{z}{\delta_1}\right)\sin\left(\frac{z}{\delta_1}\right) + G_2\left[1-\exp\left(-\frac{z}{\delta_1}\right)\cos\left(\frac{z}{\delta_1}\right)\right] \qquad (3.2.3)$$

$$\delta_1 = (2A_e/f)^{1/2}$$

where A_e is the eddy viscosity. The Ekman layer thickness is taken as $\pi\delta_1$. If the x_1-axis is aligned with the geostrophic velocity vector, $G_2 = 0$, $G_1 = G$. Taking the derivatives w.r.t. z of U_1 and U_2, the surface stress components are

$$\frac{\tau_1}{\rho} = -\overline{u_1 u_3} = A_e\frac{G}{\delta_1};\quad \frac{\tau_2}{\rho} = -\overline{u_2 u_3} = -A_e\frac{G}{\delta_1} \qquad (3.2.4)$$

The angle between the stress at the surface and the geostrophic velocity vector is 45°, with the stress to the left (right) of the geostrophic wind in the northern (southern) hemisphere. The magnitude of the surface stress from Eq. (3.2.4) gives

$$u_*^2 = 2^{1/2}A_e\frac{G}{\delta_1} \text{ and } \delta_1 = 2^{1/2}\frac{u_*^2}{Gf} \qquad (3.2.5)$$

Alternatively, if we align the x_1-axis in the direction of the surface stress, then for $\tau_2 = 0$, $G_2 = -G_1 = -2^{-1/2}G$, and Eq. (3.2.3) gives the corresponding velocity profiles. The angle of the geostrophic wind is again 45° w.r.t. the surface stress. In either case, both components of velocity overshoot slightly above their geostrophic values around $z/\delta = \pi/2$ before asymptoting to them beyond $z = \pi\delta$; the velocities are therefore (slightly) supergeostrophic.

Eddy viscosity is hardly a constant in the turbulent Ekman layer and solutions are hard to obtain analytically. Generally, Ekman turning is observed even when the layer is turbulent, but the angle between the surface stress and the geostrophic wind is much less, roughly 20°–30° instead of 45°. Also, limited solutions are possible without having to resort to solving the governing equations with turbulence closure approximations.

The important parameters in the problem are u_*, the friction velocity (also called shear velocity), G, the geostrophic velocity, and f, the Coriolis parameter ($= 2\Omega \sin\theta$, where Ω is the angular rotation of the Earth and θ is the latitude). The relevant length scales in the problem are the boundary layer thickness δ and the roughness scale z_0, characterizing the roughness of the underlying surface(or for smooth surfaces, the viscous scale ν/u_*; real ABLs and PBLs are seldom smooth, so z_0 is the relevant scale). The boundary layer thickness δ scales as (u_*/f) and invariably $\delta \gg z_0$. Two Rossby numbers can be defined, one based on the geostrophic velocity G, $Ro = (G/fz_0)$, and the other based on friction velocity u_*, $Ro_* = (u_*/fz_0)$. Ro_* is the friction Rossby number, and $Ro > Ro_* \ggg 1$.

Close to the boundary, since $\delta \gg z_0$, the solutions should depend only on the inner variable z/z_0 and should be independent of δ. In this so-called inner layer, the flow should be more or less aligned with the shear stress at the boundary, and therefore, aligning the x_1-axis in the direction of the shear stress at the surface,

$$\frac{U_1}{u_*} = f_0(z_+), \quad \frac{U_2}{u_*} = 0; \quad z_+ = \frac{z}{z_0} \tag{3.2.6}$$

Far away from the surface, inner scale z_0 is irrelevant, and the flow scales with the outer scale δ. The departure of the flow velocity from geostrophic conditions (the velocity defect) should therefore be a function of only the outer variable z/δ and independent of z_0:

$$\frac{U_1 - G_1}{u_*} = f_1(\eta), \quad \frac{U_2 - G_2}{u_*} = f_2(\eta); \quad \eta = \frac{z}{\delta} \tag{3.2.7}$$

Now for $Ro_* \to \infty$, there exists an overlap region ($z_0 \ll z \ll \delta$), where both the inner and the outer laws are simultaneously valid asymptotically ($z/z_0 \rightarrow \infty$ and $z/\delta \rightarrow 0$ simultaneously). Matching the profiles in the overlap region (Yaglom, 1979) requires equating the gradients:

$$z_+ \frac{df_0}{dz_+} = \frac{z}{u_*} \frac{\partial U_1}{\partial z} = \eta \frac{df_1}{d\eta} \quad (3.2.8)$$

Since each of these terms should be a constant, this results in logarithmic relationships for the velocity defect in the outer layer and the velocity in the inner region,

$$\frac{U_1 - G_1}{u_*} = \frac{1}{\kappa} \ln \eta - A' = \frac{1}{\kappa} \ln \frac{z}{\delta} - A' \quad (z >> z_0)$$

$$(3.2.9)$$

$$\frac{U_1}{u_*} = \frac{1}{\kappa} \ln z_+ = \frac{1}{\kappa} \ln \frac{z}{z_0} \quad (z << \delta)$$

so that (Tennekes and Lumley, 1982)

$$\frac{G_1}{u_*} = \frac{1}{\kappa} \ln \frac{\delta}{z_0} - A' \quad (3.2.10)$$

Note that for the U_2 component, since

$$\eta \frac{df_2}{d\eta} = \frac{z}{u_*} \frac{dU_2}{dz} \quad (3.2.11)$$

the functional form f_2 cannot be determined. However,

$$\frac{G_2}{u_*} = -f_2(0) = -B' \quad (3.2.12)$$

Substituting $\delta \sim u_* / f$, the geostrophic drag law becomes

$$\frac{G_1}{u_*} = \frac{1}{\kappa} \left[\ln \frac{u_*}{f z_0} - A \right], \quad \frac{G_2}{u_*} = -\frac{B}{\kappa} \quad (3.2.13)$$

The angle between the surface velocity and the geostrophic velocity vectors is given by

$$\tan\alpha = -\frac{G_2}{G_1} = B\left[\ln\left(\frac{u_*}{f z_0}\right) - A\right]^{-1} \tag{3.2.14}$$

Figures 3.2.1 and 3.2.2 show the rotation of the velocity vector for the atmosphere and the ocean, respectively. Observations indicate A ~ 1.6 and B ~ 4.8 (Tennekes and Lumley, 1973). These relationships enable determination of the surface stress, provided the value of the roughness length scale is known. Table 3.2.1 derived from Stull (1988) and Garratt (1992), lists values for z_0 for typical surface conditions in the ABL, shown graphically in Figure 3.2.3. For the marine ABL, z_0 is a function of the surface wave field, but when the wave field is in equilibrium with the wind, $z_0 \sim u_*^2/g$ (see Chapter 4).

The thickness of the neutral ABL, $\delta \sim u_*/f$. There is considerable disagreement as to the value of the proportionality constant. Values range from 0.2 to 0.6, depending on how the outer edge of the boundary layer is defined [see Table 3.2.2 from Garratt (1992)]. It is, however, attractive to consider this constant to be the von Karman constant so that

$$\delta = \kappa \frac{u_*}{f} \tag{3.2.15}$$

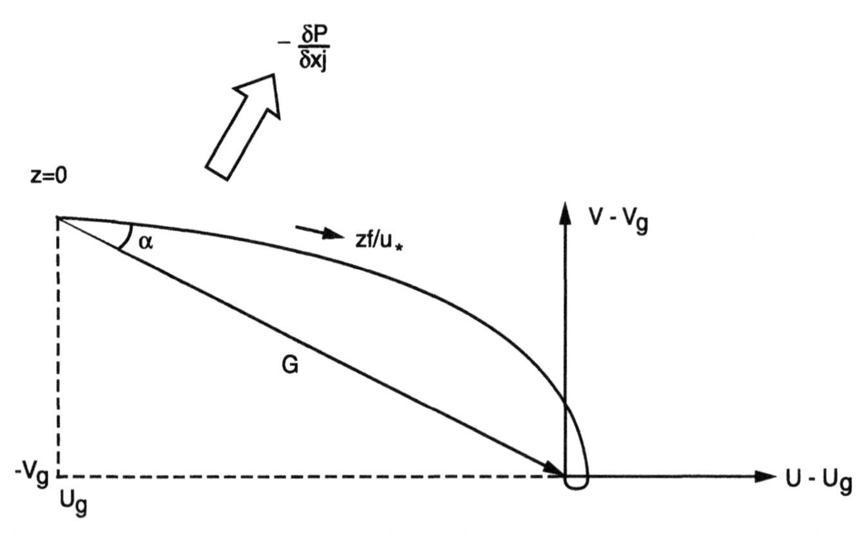

Figure 3.2.1. The Ekman spiral in the ABL. The shear stress at the ground is to the left of the geostrophic wind aloft in the northern hemisphere.

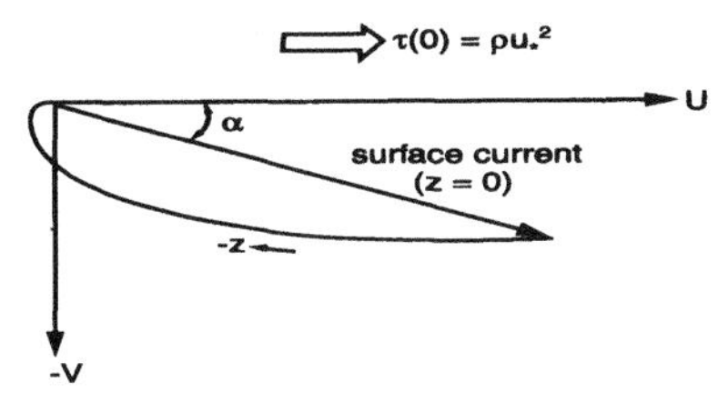

Figure 3.2.2. The Ekman spiral in the ocean. The surface current is to the right of the wind stress in the northern hemisphere.

TABLE 3.2.1
Values of Aerodynamic Roughness Length and Zero-Plane Displacement for a Range of Natural Surfaces (from Stull, 1988; Garratt, 1992)

Surface	Reference	h_c(m)	z_0(m)	d/h_c
Soils			0.001–0.01	
Grass				
Thick	Sutton (1953)	0.1	0.023	
Thin	Sutton (1953)	0.5	0.05	
Sparse	Clarke *et al.* (1971)	0.025	0.0012	
	Deacon (1953)	0.015	0.002	
		0.45	0.018	
		0.65	0.039	
Crops				
Wheat stubble	Izumi (1971)	0.18	0.025	
Wheat	Garratt (1977b)	0.25	0.005	
		0.4	0.015	
		1.0	0.05	
Corn	Kung (1961)	0.8	0.064	
Beans	Thom (1971)	1.18	0.077	
Vines	Hicks (1973)	0.9	0.023[a]	
		1.4	0.12[b]	
Vegetation	Fichtl and McVehil (1970)	1–2	0.2	
Woodland				
Trees	Fichtl and McVehil (1970)	10–15	0.4	
Savannah	Garratt (1980)	8	0.4	0.6
		9.5	0.9	0.75
Forests				
Pine	Hicks *et al.* (1975)	12.4	0.32	
Pine	Thom *et al.* (1975)	13.3	0.55	
		15.8	0.92	
Coniferous	Jarvis *et al.* (1976)	10.4[c]–27.5	0.28–3.9	0.61–0.92
Tropical	Thomson and Pinker (1975)	32	4.8	
Tropical	Shuttleworth (1989)	35	2.2	0.85

[a] Flow parallel to rows.
[b] Flow normal to rows.
[c] Range in h_c for 11 sites; the mean z_0/h_c is 0.076 and the mean d/h_c is 0.78.

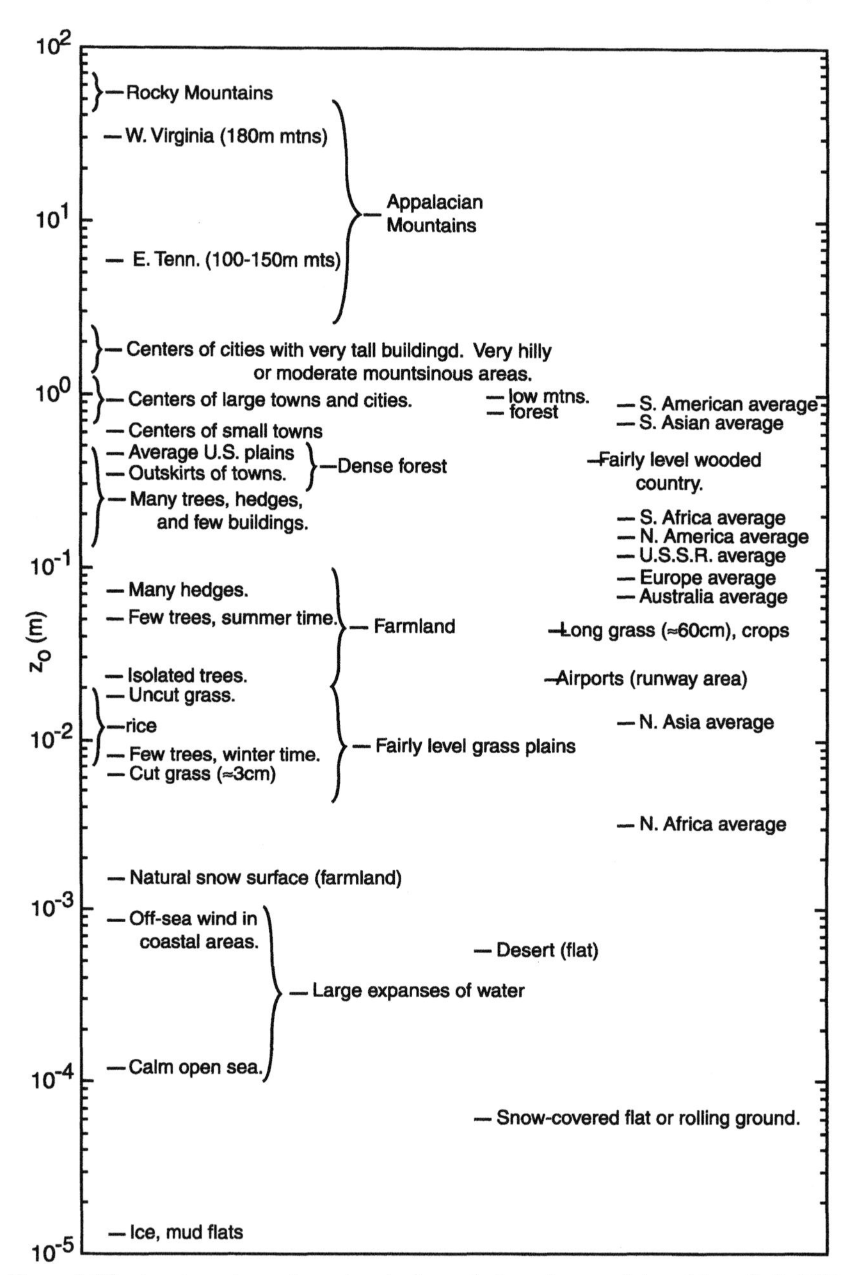

Figure 3.2.3. Aerodynamic roughness lengths for typical terrain types (adapted from Stull, 1988; from Garratt 1977, Hicks *et al.* 1975, Kondo and Yamazawa 1986, Nappo 1977, Smedman-Hogstrom and Hogstrom 1978, and Thompson 1978, with kind permission from Kluwer Academic Publishers).

TABLE 3.2.2
Values of the Proportionality Constant for Determining Thickness of the Neutral ABL (from Garratt, 1992)[a]

Reference	c	Comments
Csanady (1967)		Introduces a scaling depth ($\alpha\ u_*/f$) but gives no value for c
Blackadar and Tennekes (1968)	0.25	Quote, but do not derive
Hanna (1969)	0.2	Quotes three references that utilize numerical simulations
Clarke (1970)	0.2	Evaluates from observations, and interpolation to neutral ABL
Plate (1971)	0.185	Analytical derivation
Wyngaard (1973)	0.25	Quotes, but does not derive
Tennekes (1973)	0.3	Quotes, but does not derive
Clarke and Hess (1973)	0.3	Based on an Ekman spiral assumption and observations
Deardorff (1974)	0.33	Neutral limit, from an ABL growth formula
Zilitinkevich and Monin (1974)	0.4	Thickness of neutral Ekman layer, with $c = k$
Zilitinkevich and Deardorff (1974)	0.3	Quote, but do not derive
Yanmada (1976)	0.3	Quotes, but does not derive
Brutsaert (1982)	0.15 to 0.3	Quotes a typical range
Panofsky and Dutton (1984)	≈ 0.2	Quote, but do not derive

[a]References can be found in these volumes.

Note that neither the constant stress assumption nor the constant flux layer (as is traditional) was invoked to derive the logarithmic law of the wall close to the boundary. From the governing equations, it is easy to show that

$$-\overline{u_1 u_3} = u_*^2 \left[1 - \frac{B}{\kappa} \frac{f z}{u_*} \right]$$

$$-\overline{u_2 u_3} = u_*^2 \frac{f z}{u_*} \frac{1}{\kappa} \left[\ln \frac{f z}{u_*} + A\text{-}1 \right] \tag{3.2.16}$$

In nonrotating boundary layers, logarithmic law of the wall is valid for about 10% of the boundary layer thickness. Applied to the rotating PBL, this implies that the log layer thickness is 0.04 u_*/f, and therefore, the stress in the x_1 direction has decreased by ~40% and turned by about 30°. The constant stress layer is usually taken to be the thickness over which the stress has decreased by less than 10% and the turning angle is less than 10°. Thus, the validity of the logarithmic law of the wall extends well beyond the constant stress region.

A similar matching procedure is valid for any scalar such as temperature, humidity, or gas concentration and leads to

$$\begin{aligned} \frac{F-F_R}{F_*} &= \frac{Pr_t}{\kappa} \ln \frac{z}{z_0} \\ \frac{F-F_\infty}{F_*} &= \frac{Pr_t}{\kappa} \ln \frac{z}{\delta} + C' \\ \frac{F_\infty - F_R}{F_*} &= \frac{Pr_t}{\kappa} \left[\ln \frac{u_*}{f z_0} - C \right] \end{aligned} \tag{3.2.17}$$

where F_R is the reference value. This value is not the same as the surface value F_S, since there exists a molecular sublayer at the surface, across which scalar transfer takes place. Unlike momentum, which can be transferred across the boundary by pressure forces acting on roughness elements, scalar transfer across the surface can only take place through the action of molecular diffusion. This leads to a substantial change in the scalar value across the molecular sublayer and it is the value at the edge of the sublayer that is the reference value. In practice, it is possible to define a scalar roughness scale z_{0F} that is different from the momentum roughness scale z_0, so that in the wall layer

$$\frac{F-F_S}{F_*} = \frac{Pr_t}{\kappa} \ln \frac{z}{z_{0F}} \tag{3.2.18}$$

z_{0F} depends, of course, on the thickness and the change across the molecular sublayer (see Chapter 4).

When stratification effects are important, the problem is complicated by the appearance of another length scale in the problem, the Monin–Obukhoff length scale (Monin and Obukhoff, 1954). A similar matching procedure can be conducted, but the profiles are no longer simply logarithmic. The above constants A, B, and C in the geostrophic drag laws now become functions of the ratio of the boundary layer thickness to the Monin–Obukhoff length scale or

u_*/fL. The parameter

$$\mu = \frac{\kappa u_*}{fL} \tag{3.2.19}$$

is known as the Monin–Kazanski parameter and the functional forms for A (μ), B (μ), and C (μ) are known from observations (see, for example, Yardanov, 1976). These are reproduced in Figure 3.2.4. However, the scatter in the data is quite large and this raises some doubt as to whether the Monin–Kazanski parameter (Kazanski and Monin, 1960), which is simply the ratio of the Ekman scale to the Monin–Obukhoff scale, characterizes the stratified boundary layer, since the Ekman scale is most often irrelevant under stratification conditions. The unstably stratified ABL is invariably capped by an inversion that defines the mixed layer height h, which is the more relevant length scale in this situation, since the Ekman scale becomes irrelevant in the limit of free convection. Observations show that various parameters in the CABL indeed scale with its height. For stably stratified flows also, such as in the NABL, the Ekman scale is not as relevant as the height of the boundary layer itself. In both cases, the BL height h is the relevant scale and therefore the Monin–Kazanski parameter should be defined as h/L.

$$\mu = \frac{h}{L} \tag{3.2.20}$$

When this is done, Yamada (1976) has shown (see Figure 3.2.5 that the data on A, B, and C collapse quite well. These can be approximated by

$$\left.\begin{aligned} A(\mu) &= 10.0 - 8.145(1 - 0.008376\mu)^{-1/3} \\ B(\mu) &= 3.020(1 - 3.290\mu)^{-1/3} \\ C(\mu) &= 12.0 - 8.355(1 - 0.03106\mu)^{-1/3} \end{aligned}\right\} \quad (\mu \le 0)$$

$$\left.\begin{aligned} A(\mu) &= 1.855 - 0.380\mu \\ B(\mu) &= 3.020 + 0.300\mu \end{aligned}\right\} \quad (0 \le \mu \le 35)$$

$$C(\mu) = 3.665 - 0.829\mu \qquad (0 \le \mu \le 18) \tag{3.2.21}$$

$$\left.\begin{aligned} A(\mu) &= -2.940(\mu - 19.940)^{1/2} \\ B(\mu) &= 2.850(\mu - 12.470)^{1/2} \end{aligned}\right\} \quad (\mu \ge 35)$$

$$C(\mu) = -4.320(\mu - 11.210)^{1/2} \qquad (\mu \ge 18)$$

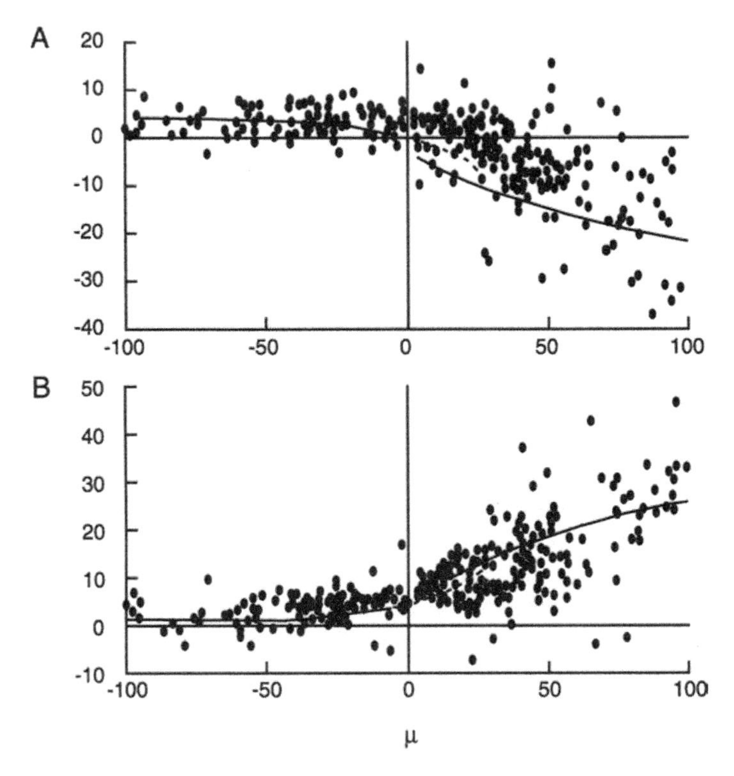

Figure 3.2.4. Stratification similarity functions A and B as functions of the Monin–Kazanski stratification parameter based on the Ekman scale (from Yardanov, 1976).

There are other formulations as well, but Yamada's are the most frequently used. These relations depend on being able to define the boundary layer height correctly. For unstably stratified ABL, this is not a problem, since the well-defined base of the inversion defines the turbulent boundary layer. For a stably stratified ABL, such as the NABL or polar ABL, the turbulent layer may or may not coincide with the inversion (see Chapter 3.5 on the NABL), and using the inversion height, even if one is able to define it properly, overestimates h. Andreas (1996) suggests that it is best to define h as the height of the core of the jet found near the top of the ABL. G/u_* and the turning angle α are given by (Andreas, 1996)

$$\frac{G}{u_*} = \frac{1}{\kappa}\left\{\left[\ln\left(h/z_0\right) - A(\mu)\right]^2 + B(\mu)^2\right\}^{1/2}$$
$$\tan\alpha = \frac{-B(\mu)}{\ln\left(h/z_0\right) - A(\mu)} \tag{3.2.22}$$

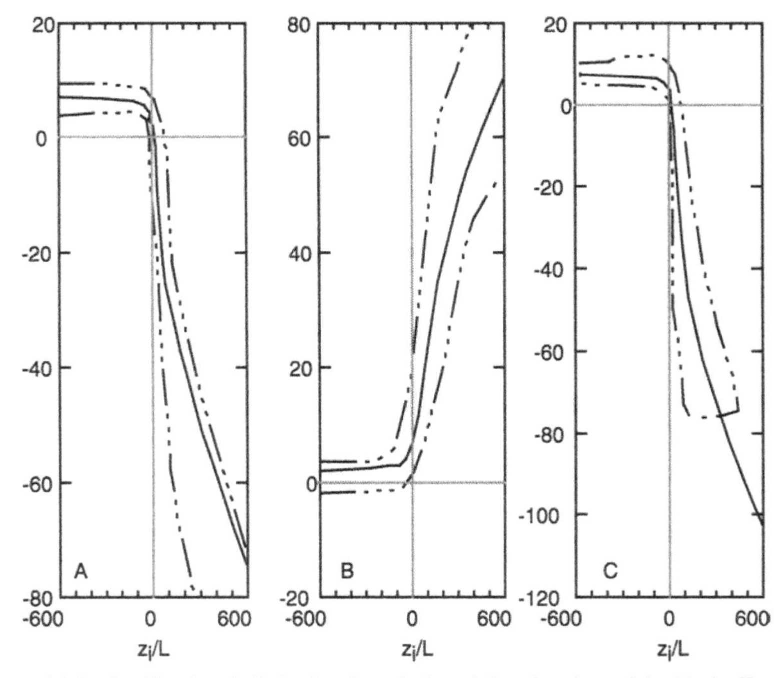

Figure 3.2.5. Stratification similarity functions A, B, and C as functions of the Monin–Kazanski stratification parameter based on the PBL thickness (from Yamada, 1976). Dashed-dotted lines denote extremes of observations.

Since the turning angle is proportional to B, which is small in the CABL, it is clear that there is not much turning in the CABL, but turning is large in the NABL. However, the ever-present horizontal gradients of density lead to vertical shear in the geostrophic winds called thermal wind, whose presence makes it difficult to see a well-defined Ekman spiral in the atmosphere. Sorbjan (1989) suggests ways to incorporate thermal wind into the Ekman solutions.

A note on the von Karman constant that is ubiquitous in the boundary layer relationships: This constant is named, of course, after von Karman, who discovered the logarithmic law of the wall in laboratory turbulent boundary layers. Engineers have always used a value of 0.41 for this constant (Schlichting, 1977; Hinze, 1975), even though the Reynolds numbers in laboratory flows are much smaller than those in geophysical flows. Yet, when Businger *et al.* (1971) made their famous Kansas measurements, they suggested 0.35 as the best value. This led to a raging controversy about whether this constant is really a constant. Some even suggested it should be a function of parameters such as the Reynolds number in the laboratory (Tennekes, 1968) and the roughness Rossby number in the atmosphere (Tennekes, 1973). However careful reanalysis of all available surface layer measurements (Hogstrom, 1996) and new observations have led to the conclusion that this constant is indeed a constant equal to 0.40±0.01 (Zhang *et al.*, 1988; see also Hogstrom, 1988). Thus the two so-called constants in turbulence theory, the von Karman constant and the Kolmogoroff constant, appear to be true universal constants (in the asymptotic limit of large Reynolds numbers), even though this still comes under attack occasionally.

3.3 SURFACE LAYER (MONIN–OBUKHOFF SIMILARITY THEORY)

As we saw in Section 3.2, there exists a region (usually a few tens of meters thick) near the ground but sufficiently far from the roughness elements on the ground, where the outer and inner scaling laws can be matched to obtain universal shapes for the profiles of various quantities. For neutral stratification, this yields the famous von Karman logarithmic law of the wall. When the atmosphere is stratified, as is invariably the case (even the marine ABL, often close to neutral stratification, is seldom neutrally stratified), the profiles depart significantly from logarithmic behavior, but still exhibit similarity. Similarity laws in the surface layer of the ABL were proposed by Monin and Obukhoff, and are known as Monin–Obukhoff similarity relationships, or often, simply Obukhoff similarity laws, since Obukhoff derived them first. These laws are one of the finest achievements in ABL turbulence and enormous effort has been directed toward determining the exact form of these functional relationships, as described below.

Monin–Obukhoff similarity laws can be derived quite simply by dimensional analysis. Assume the conditions in the ABL are nearly steady. If $\partial F/\partial z$ denotes the gradient of any property F at a point z in the surface layer that is far away from the roughness elements and from the edge of the boundary layer ($z_0 >> z >> \delta$), then this gradient should be independent of both z_0 and δ, and be a function of only z. We can postulate such relationships for only the gradients since the properties themselves can be referenced to any arbitrary reference value (for example, velocity can be referenced to any arbitrary frame of reference; the requirement of Galilean invariance to arbitrary translation is satisfied by its gradient, but not the velocity itself).

In the surface layer, the scale of variations in the vertical direction is much smaller than the scale of variations in the horizontal direction and it is therefore reasonable to assume conditions of horizontal homogeneity. Under these conditions, the momentum, scalar, and other fluxes through the surface layer are roughly constant with height (or independent of z) and are the internal parameters that govern the flow and its properties in the layer. This is the reason that the surface layer is often called the constant flux layer. These internal flux parameters can be expressed as kinematic quantities: for example, the friction velocity at the surface u_* characterizing the momentum flux ($\tau = \rho u_*^2$), the kinematic heat flux ($Q_H = H/\rho c_p$), and the kinematic flux of any other scalar property, such as water vapor or gas concentrations. The dynamics of the surface layer are influenced by stratification effects, which can be characterized by the buoyancy flux through the layer which is equal to the buoyancy flux at the surface Q_b ($\overline{wb}$), and therefore the gradient of any property in the surface layer should obey

$$dF/dz \sim f(z, u_*, Q_F, Q_b) \tag{3.3.1}$$

where Q_F is the kinematic flux of that property in the surface layer. Remember that the flux divergences are assumed to be zero (or nearly so) for all properties, and therefore flux gradients do not enter into this relationship. Equation (3.3.1) can be rewritten in nondimensional form as

$$\phi_F = \frac{\kappa z}{F_*}\frac{\partial F}{\partial z} = \phi_F\left(\frac{z}{L}\right) = \phi_F(\zeta) \tag{3.3.2}$$

where

$$F_* = Q_F / u_* = -\overline{wf} / u_* \tag{3.3.3}$$

is the friction value of the property (the ratio of its kinematic flux to the friction velocity, for example, friction temperature $\theta_* = Q_H / u_*$), and

$$L = \frac{u_*^3}{\kappa Q_b} \tag{3.3.4}$$

is the Monin–Obukhoff (or simply Obukhoff) length scale that characterizes the effects of stratification.

$$\zeta = \frac{z}{L} \tag{3.3.5}$$

is the celebrated Monin–Obukhoff similarity variable. It is traditional to retain the von Karman constant κ in these relationships, although it is not necessary to do so. Equation (3.3.2) is referred to as the Monin–Obukhoff (M-O) similarity law for the constant flux or the surface layer. It is one shining example of the success of scaling arguments in their application to complex turbulent flows in nature, and has turned out to be extremely useful for many practical applications (see Chapter 4). Enormous effort has been expended in determining the precise form of the M-O similarity functions empirically (for example, Businger *et al.*, 1971) and analytically (for example, Mellor, 1973; Kantha and Clayson, 1994) and as a result, the functional forms are reasonably well known and of considerable utility.

Note that for both the marine and continental (unless it is dry) ABL, water vapor flux in the surface layer influences the buoyancy flux and cannot be ignored. Q_b can be written as

$$Q_b = \overline{wb} = -\frac{g}{T_v}\overline{w\theta_v} \tag{3.3.6}$$

where the subscript v denotes virtual quantities (for example, T_v is the absolute virtual temperature),

$$\begin{aligned} T_v &= T\left(1{+}0.61q\right) \\ \overline{w\theta_v} &= \overline{w\theta}\left(1+0.61q\right)+0.61T\overline{wq} \end{aligned} \tag{3.3.7}$$

where q is the specific humidity and $-\overline{wq}$ is the humidity (moisture) flux. The Monin–Obukhoff scale can then be written as

$$L = \frac{u_*^3 T_v}{\kappa g \theta_{v*}} = \frac{u_*^3 T\left(1+0.61q\right)}{\kappa g\left[\theta_*\left(1+0.61q\right)+0.61Tq_*\right]} \tag{3.3.8}$$

The ratio of the sensible heat flux to the latent heat flux,

$$B_r = \frac{c_p\,\overline{w\theta}}{L\overline{wq}} \tag{3.3.9}$$

is called the Bowen ratio (c_p/L is called the psychrometric constant).

TKE balance is the most important aspect of any turbulent flow, and this is true for the ABL and the surface layer as well. Turbulence is most often close to local equilibrium so that the production (and destruction) closely balances dissipation. The ratio of buoyancy production (destruction) to shear production of TKE, the flux Richardson number Ri_f, is an important parameter characterizing buoyancy effects and is related to the gradient Richardson number Ri_g,

$$Ri_f = \frac{-\frac{g}{T_v}\overline{w\theta_v}}{-\overline{uw}\frac{\partial U}{\partial z} - \overline{vw}\frac{\partial V}{\partial z}} = \frac{K_H}{K_M} Ri_g$$

$$Ri_g = \frac{\frac{g}{T_V}\frac{\partial T_V}{\partial z}}{\left(\frac{\partial U}{\partial z}\right)^2 + \left(\frac{\partial V}{\partial z}\right)^2} \quad (3.3.10)$$

where K_M and K_H are turbulent mixing coefficients (eddy diffusivities) for momentum and heat. The shear production term in the denominator of the first equation in Eq. (3.3.10) is always positive-definite. For an unstable ABL and surface layer ($\partial T_v / \partial z < 0, \overline{w\theta} > 0, \overline{wb} > 0$), the buoyancy production term is also positive, so that both shear and buoyancy generate turbulence, and both Ri_f and Ri_g are negative. For a stable ABL and surface layer ($\partial T_v / \partial z > 0, \overline{w\theta} < 0, \overline{wb} < 0$), the buoyancy term is negative and hence tends to suppress turbulence by destroying TKE, and both Ri_f and Ri_g are positive. The ratio K_H/K_M is the turbulent Prandtl number Pr_t, believed to be a universal constant for neutral stratification (at least in an asymptotic sense) with a value of about 0.86–0.95 (Hogstrom, 1996). Note that in the surface layer, the convention is to align the coordinate in the direction of the mean flow and therefore $V = 0$.

In the surface or the constant flux layer, all flow variables suitably normalized must be universal functions of the Monin–Obukhoff similarity variable ζ. Thus the velocity, temperature, humidity, and gas concentration profiles can be written as

$$\begin{aligned}
\phi_M &= \frac{\kappa z}{u_*}\frac{\partial U}{\partial z} = \phi_M(\zeta) \\
\phi_H &= \frac{\kappa z}{\theta_*}\frac{\partial T}{\partial z} = \phi_H(\zeta) \\
\phi_E &= \frac{\kappa z}{q_*}\frac{\partial q}{\partial z} = \phi_E(\zeta) \\
\phi_G &= \frac{\kappa z}{c_*}\frac{\partial C}{\partial z} = \phi_G(\zeta)
\end{aligned} \quad (3.3.11)$$

Observations confirm this. Figure 3.3.1 shows observations on the variation of the mean momentum and scalar profiles in the surface layer. This applies to turbulence quantities as well. For second-moment quantities,

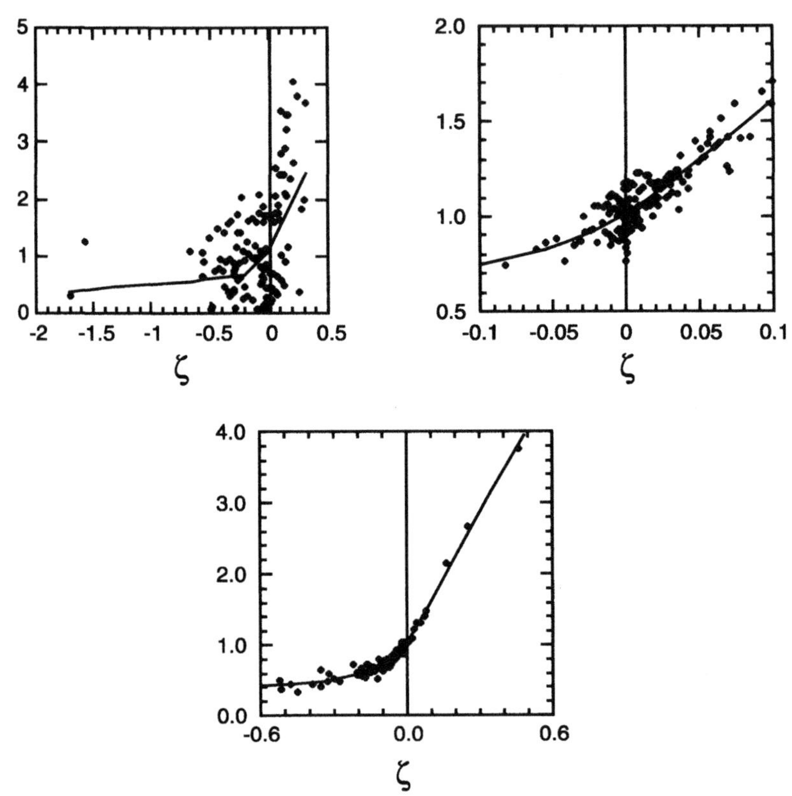

Figure 3.3.1. Monin–Obukhoff similarity functions as functions of Monin–Obukhoff similarity variable ζ for momentum (top left from Zeller 1992), wind shear (top right from Zhang *et al.* 1988), and TKE production (bottom, from Frenzen and Vogel, 1992).

$$
\begin{aligned}
\phi_u &= \overline{u^2}/u_*^2 = \phi_u(\zeta) \\
\phi_v &= \overline{v^2}/u_*^2 = \phi_v(\zeta) \\
\phi_w &= \overline{w^2}/u_*^2 = \phi_w(\zeta) \\
\phi_K &= \frac{\overline{u^2} + \overline{v^2} + \overline{w^2}}{2u_*^2} = \phi_K(\zeta) \\
\phi_\theta &= \overline{\theta^2}/\theta_*^2 = \phi_\theta(\zeta) \\
\phi_q &= \overline{q^2}/q_*^2 = \phi_q(\zeta) \\
\phi_c &= \overline{c^2}/c_*^2 = \phi_c(\zeta)
\end{aligned}
\tag{3.3.12}
$$

In particular, the TKE equation can be written as

$$\phi_M - \zeta - \phi_\varepsilon - \phi_t = 0 \tag{3.3.13}$$

The first term is the normalized shear production, the second term the normalized buoyancy production, and the third term the normalized dissipation rate:

$$\phi_\varepsilon = \kappa z \varepsilon / u_*^3 \tag{3.3.14}$$

The last term is the flux divergence term. Normally, this term is small in the surface layer, where the turbulence is close to local equilibrium, meaning that the production and dissipation terms balance approximately. However, ϕ_t has been observed to be nonzero in some recent observations (Vogel and Frenzen, 1993). ϕ_t and ϕ_ε are also universal functions of ζ. Thus

$$\phi_\varepsilon = \frac{\kappa z \varepsilon}{u_*^3} = \phi_\varepsilon(\zeta) \tag{3.3.15}$$

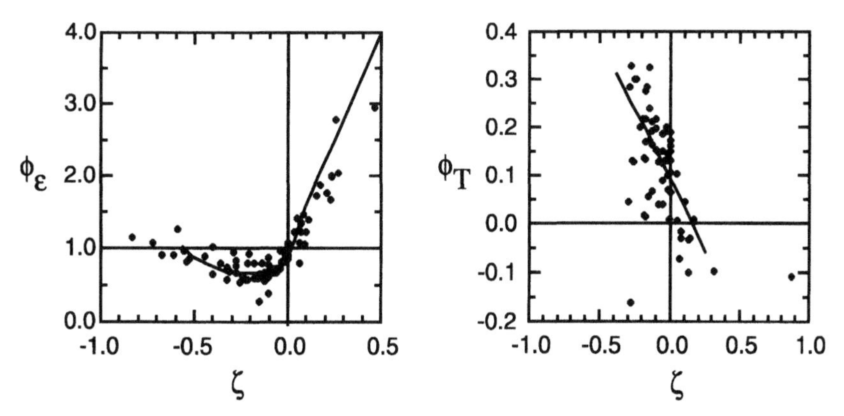

Figure 3.3.2. Monin–Obukhoff similarity functions ϕ_t, ϕ_ε as functions of the Monin–Obukhoff similarity variable ζ.

Figure 3.3.2 shows the observed variation of ϕ_t and ϕ_ε. Similarly the flux and gradient Richardson numbers and Pr_t should also be universal functions of ζ:

$$Ri_g = \zeta \frac{\phi_H}{\phi_M^2} = Ri_g(\zeta)$$

$$Ri_f = \frac{K_H}{K_M} Ri_g = \frac{\phi_M}{\phi_H} Ri_g = \frac{\zeta}{\phi_M} = Ri_f(\zeta) \tag{3.3.16}$$

$$Pr_t = \frac{K_M}{K_H} = \frac{\phi_M}{\phi_H} = Pr_t(\zeta)$$

Similarity scaling applies to curvatures of the mean profiles as well,

$$D_M = \frac{-z\left(\partial^2 U / \partial z^2\right)}{(\partial U / \partial z)} = 1 - \frac{\zeta}{\phi_M(\zeta)} \frac{\partial\left[\phi_M(\zeta)\right]}{\partial \zeta} = D_M(\zeta)$$

$$D_F = \frac{-z\left(\partial^2 F / \partial z^2\right)}{(\partial F / \partial z)} = 1 - \frac{\zeta}{\phi_F(\zeta)} \frac{\partial\left[\phi_F(\zeta)\right]}{\partial \zeta} = D_F(\zeta) \tag{3.3.17}$$

where F stands for any scalar. These are called Deacon numbers.

The M-O length scale L is positive for stably stratified flows and negative for unstable stratification so that $\zeta > 0$ for stable stratification and $\zeta < 0$ for unstable stratification. $\zeta = 0$ for neutrally stratified flows. L denotes the height at which the buoyancy production of TKE is of the same magnitude as the shear production of TKE. For $z > L$, the buoyancy production (destruction) dominates, and for $z < L$, the shear production dominates.

The neutral values of various similarity functions are of some interest:

$$\phi_M(0) = 1.0; \; \phi_H(0) = 0.86 - 0.95 \quad (0.90)$$

$$\phi_E(0) = 0.86 - 0.95 \quad (0.90); \; \phi_G(0) = 0.86 - 0.95 \quad (0.90)$$

$$\phi_u(0) = 3.7 - 6.6; \; \phi_v(0) = 2.4 - 5.0; \; \phi_w(0) = 1.3 - 2.0 \tag{3.3.18}$$

$$\phi_\theta(0) = 4.0; \; \phi_q(0) = 4.0; \; \phi_c(0) = 4.0$$

$$\phi_K(0) = 3.7 - 6.8; \; \phi_\varepsilon(0) = 1.0; \; \phi_t(0) = 0.08$$

The constant values for the similarity functions imply that the profiles for the mean quantities are all logarithmic,

$$\frac{U}{u_*} = \frac{1}{\kappa} \ln \frac{z}{z_0}$$

$$\frac{F - F_s}{f_*} = \frac{Pr_t}{\kappa} \ln \frac{z}{z_{0f}} \tag{3.3.19}$$

where F stands for any scalar and f_* for its friction quantity; z_{0f} is the scalar roughness scale. The utility of these relations lies in that by using measured profiles of mean quantities, the roughness scales can be determined empirically. For example, since by definition z_0 is the value of the height z above the surface at which U = 0, plotting U vs ln z and extrapolating linearly in the surface layer toward the surface to find the intercept along the ln z axis at which U = 0 gives z_0. Since z_0 is usually a fraction of the actual height of roughness elements, measurements must be made much above the roughness heights (for the log law to be valid). Theoretical determination of z_0 is virtually impossible because of the complicated dependence on the height, shape, and distribution of the roughness elements.

It is difficult to derive the form of M-O similarity functions over the entire range of ζ analytically. Instead, observational data are most often relied upon to characterize their variation with ζ for diabatic (nonneutral) conditions (see Figure 3.3.1). The most common form used for the unstable surface layer (Paulson, 1970) is

$$\left.\begin{aligned} \phi_M(\zeta) &= (1 - \gamma_1 \zeta)^{-1/4} \\ \phi_H(\zeta) &= Pr_t^0 (1 - \gamma_2 \zeta)^{-1/2} \end{aligned}\right\} \quad -5 < \zeta < 0 \tag{3.3.20}$$

and for the stable layer

$$\left.\begin{aligned} \phi_M(\zeta) &= 1 + \beta_1 \zeta \\ \phi_H(\zeta) &= Pr_t^0 + \beta_2 \zeta \end{aligned}\right\} \quad \zeta > 0 \tag{3.3.21}$$

The usual values for the constants are (see Table 3.3.1)

$$\gamma_1 \cong \gamma_2 \sim 16, \quad \beta_1 \cong \beta_2 \sim 5, \quad Pr_t^0 \sim 0.86 \tag{3.3.22}$$

TABLE 3.3.1
Values of Surface-Layer Constants, where κ Is the von Karman Constant, P_{tN} is the Neutral Turbulent Prandtl Number, and β and γ Are Experimental Constants Appearing in the Expressions for the Monin–Obukhov Stability Functions Φ_M and Φ_H

	k	P_{tN}	γ_1	γ_2	β_{1M}	β_{1H}
Observations						
W70	—	—	—	—	5.2	5.2
DH70	0.41	1	16	—	—	—
B71	0.35	0.74	15	9	4.7	4.7
G77	0.41	—	—	—	—	—
W80	0.41	1	22	13	6.9	9.2
DB82	0.40	1	28	14	—	—
W82	—	1	20.3	12.2	—	—
H85	0.40	1	—	—	4	—
H88	0.40	0.95	19	11.6	6.0	7.8
Z88	0.40	—	—	—	—	—
Review						
D74	0.41	1	16	16	5	5

Sources: W70, Webb (1970); DH70, Dyer and Hicks (1970); B71, Businger *et al.* (1971); G77, Garratt (1977); W80, Wieringa (1980); DB82, Dyer and Bradley (1982); W82, Webb (1982); H85, Hogstrom (1985); H88, Hogstrom (1988); Z88, Zhang *et al.* (1988); D74, Dyer (1974).

Recently, in a review of surface layer characteristics, after careful examination of various observations, Hogstrom (1996) recommends $Pr_t \sim 0.95$, $\gamma_1 \sim 19$, and $\gamma_2 \sim 11.6$ for $-\zeta < 2$, and $\beta_1 \sim 5.3$ and $\beta_2 \sim 8.0$ for $0 < \zeta < 0.5$. It appears that despite the enormous amount of effort put into characterizing the surface layer, there still exist uncertainties in flux profile relationships and constants.

There are several interesting facts to note in these relationships. The slope of the functions ϕ_M and ϕ_H are not the same at $\zeta = 0$ on the unstable and stable sides. This is not inconsistent with the differing nature of turbulence for convectively driven and buoyancy-stabilized cases. For the stable case, the linear dependence of the similarity functions leads to zero Deacon numbers and a log-linear profile for the mean quantities:

$$\frac{U}{u_*} = \frac{1}{\kappa}\left(\ln\frac{z}{z_0} + \beta_1\frac{z - z_0}{L}\right)$$

$$\frac{F - F_s}{f_*} = \frac{Pr_t}{\kappa}\left(\ln\frac{z}{z_{0f}} + \beta_2\frac{z - z_{0f}}{L}\right) \tag{3.3.23}$$

The profiles for unstable stratification are not consistent with the free convective scaling that should prevail in the asymptotic limit of $\zeta \rightarrow -\infty$. In the free convection limit, u_* is no longer the scaling parameter, and hence the velocity and temperature scales are the free convection scales,

$$u_{*f} = \left(-\frac{g}{T_v} \overline{w\theta_v}\, z \right)^{1/3} = \left(\overline{wb}\, z \right)^{1/3}$$
$$\theta_{*f} = -\overline{w\theta_v} / u_{*f} \tag{3.3.24}$$

so that

$$\frac{z}{\theta_{*f}} \frac{\partial \overline{\theta_v}}{\partial z} = -\alpha_1 ; \quad \alpha_1 \sim 0.7 \tag{3.3.25}$$

and therefore

$$\frac{\partial \theta_v}{\partial z} = -\alpha_1 \left(\overline{w\theta_v} \right)^{2/3} \left(\frac{g}{T_v} \right)^{-1/3} z^{-4/3} \tag{3.3.26}$$

The –4/3 law relationships result from the classical Nu ~ $Ra^{1/3}$ relationship for pure convection, which demands that the heat flux be independent of the layer depth asymptotically. In other words, the height z is the only relevant length scale. Observational evidence for this classical relationship has been fairly solid until recently, when experiments at very large Ra appear to indicate that there may be significant departures from the 1/3 law. This is thought to be due to the influence of local, persistent shear near the surface introduced by large convective eddies in the flow, even though the mean shear averaged over a long timescale is zero. Nevertheless, for free convective scaling to hold, M-O similarity functions must have the following asymptotic form:

$$\phi_M(\zeta) \rightarrow \kappa^{4/3} (\zeta)^{-1/3}$$
$$\phi_H(\zeta) \rightarrow \alpha_1 \kappa^{4/3} (\zeta)^{-1/3} \tag{3.3.27}$$

This scaling law for the free convection limit was also derived by Prandtl (1945). As far as is known, that was his only contribution to geophysical flow problems. Observational difficulties and uncertainties under free convective conditions and very low wind speeds do not enable us to verify the correct form for M-O functions in the free convective limit. Therefore, the forms suggested above (often called Businger–Dyer forms when the turbulence Prandtl number is put to unity) are to be regarded as essentially empirical approximations (Dyer, 1974;

Businger, 1988). These forms are also convenient in the sense that they allow simple analytical forms to be realized for integrals of M-O functions, which are needed for obtaining the velocity and scalar profiles in the surface layer. For example, the velocity profile can be written as

$$\frac{U}{u_*} = \frac{1}{\kappa}\left[\ln\frac{z}{z_0} - \psi_M(\zeta)\right]$$

$$\psi_M(\zeta) = 2\ln\left(\frac{1+x}{2}\right) + \ln\left(\frac{1+x^2}{2}\right) - 2\tan^{-1}x + \frac{\pi}{2} \qquad (\zeta < 0) \tag{3.3.28}$$

$$= -\beta_1\zeta \qquad (\zeta > 0)$$

where

$$x = 1/\phi_M(\zeta) = (1-\gamma_1\zeta)^{1/4} \tag{3.3.29}$$

and the temperature profile is given by

$$\frac{\theta_v - \theta_0}{\theta_{*v}} = \frac{Pr_t^0}{\kappa}\left[\ln\frac{z}{z_{0T}} - \psi_H(\zeta)\right]$$

$$\psi_H(\zeta) = 2\ln\frac{1+y}{2} \qquad (\zeta < 0) \tag{3.3.30}$$

$$= -\beta_2\zeta \qquad (\zeta > 0)$$

where

$$y = 1/\phi_H(\zeta) = (1-\gamma_2\zeta)^{1/2} \tag{3.3.31}$$

Table 3.3.1, from Garratt (1992), shows the values of the constants in the above equations from different sources. All other scalar profiles can be taken to be similar in form to the temperature profile.

Note that the roughness scales for heat and other scalars are not the same as that for momentum z_0. This is due to the fact that the momentum transfer across an aerodynamically rough surface occurs principally through form drag across roughness elements. Governing equations for mean velocities permit this because of the presence of pressure gradient terms in the momentum equations. However, because of the absence of similar terms in the heat and scalar conservation equations, heat and scalar transfer across rough surfaces has to be effected by molecular diffusion through embedded molecular sublayers. These molecular layers are thin enough for the gradients to be large enough to drive the desired fluxes. This leads to the fact that the roughness scales for scalars are functions of molecular properties and often smaller than the momentum roughness scale by an order of magnitude or more. There is also a "jump" in properties across these molecular layers so that the reference value for the scalar in the logarithmic profile relationship would be different than the value at the surface if the same roughness scale as momentum were used. In fact, the ratio of the scalar roughness scale to the momentum roughness scale is a function of this jump:

$$\frac{\Delta\theta}{\theta_*} = \frac{Pr_t^0}{k} \ln\frac{z_0}{z_{0T}} \tag{3.3.32}$$

M-O similarity functions appear in the expressions for the drag coefficient C_D and the bulk heat (moisture) transfer coefficient C_H (C_E) (see Chapter 4):

$$\begin{aligned} C_D &= \kappa^2\left[\ln\frac{z}{z_0} - \psi_M(\zeta)\right]^{-2} \\ C_{H,E} &= \kappa^2\left[\ln\frac{z}{z_0} - \psi_M(\zeta)\right]^{-1}\left[\ln\frac{z}{z_{0T,0E}} - \psi_{H,E}(\zeta)\right]^{-1} \end{aligned} \tag{3.3.33}$$

The ratio C_{DN}/C_{HN} of the bulk transfer coefficients for neutral stratification is a function of the roughness scales:

$$\frac{C_{DN}}{C_{HN}} = \left(\ln\frac{z}{z_{0T}}\right)\left(\ln\frac{z}{z_0}\right)^{-1} \tag{3.3.34}$$

Since $z_{0T} < z_0$ and $C_{DN} > C_{HN}$, the momentum transfer is carried out more efficiently. This is simply due to the fact that form drag is an efficient mechanism for momentum transfer, but molecular diffusion of heat through a molecular sublayer is not.

The forms for the similarity functions for stable stratification are controversial. The linear relationships also yield a constant gradient Richardson number in the limit of strong stability ($\zeta \rightarrow \infty$), a value of 0.20. Normally, turbulence is extinguished when Ri_g increases beyond about this value. The theoretical Miles–Howard stability condition (Miles, 1961, 1963; Howard, 1961) for an infinite inviscid stably stratified shear flow is $Ri_g = 0.25$, above which flow is stable and laminar. However, observations indicate turbulence existing intermittently even when Ri_g is near unity. It is possible that hysteresis might be responsible for this discrepancy, with laminar flow becoming unstable only when Ri_g falls below 0.25, but turbulence not being extinguished until Ri_g attains much higher values (Andreas, 1996). Nonstationarity of the flow might also permit turbulence to exist at Richardson numbers much higher 0.25. The ratio of buoyancy frequency to turbulence frequency Nl/q also determines if turbulence can exist or not. Above a certain value for this parameter (~3), turbulence collapses (Hopfinger, 1987; Etling, 1993). In light of this, other expressions have been offered for stable conditions that have a higher (or no) critical Ri_g for strongly stable conditions. Lettau (1979), in light of his measurements at the South Pole, suggests

$$\phi_M(\zeta) = (1+4.5\zeta)^{3/4}$$
$$\phi_H(\zeta) = (1+4.5\zeta)^{3/2} \tag{3.3.35}$$

which yield Deacon numbers of 1/4 and −1/2 for momentum and heat and a Ri_g that is linear in ζ and unbounded (Andreas, 1996).

The turbulence is necessarily intermittent under strongly stable conditions with internal waves and turbulence alternating or coexisting, and it is still not clear which formulation is the most realistic. A further complication is that the gravitational forces see to it that the vertical dimensions of eddies are small and hence z may no longer be the relevant scale, leading to the so-called z-less scaling in the bulk of the ABL (see Chapter 3.5).

Similarity theory also implies that any turbulence quantity normalized appropriately (by u_*, θ_*, q_*, ...) must also be a function of ζ (Figure 3.3.3). The vertical component of turbulence is observed to scale well according to M-O theory and is well represented by

$$\frac{\overline{w^2}}{u_*^2} = 1.74\,(1-3\zeta)^{2/3} \tag{3.3.36}$$

This behavior provides a proper asymptotic free convection limit, where the appropriate scales are given by Eqs. (3.3.23):

$$\frac{\overline{w^2}}{u_*^2} \to 3.24\,(-\zeta)^{2/3} \tag{3.3.37}$$

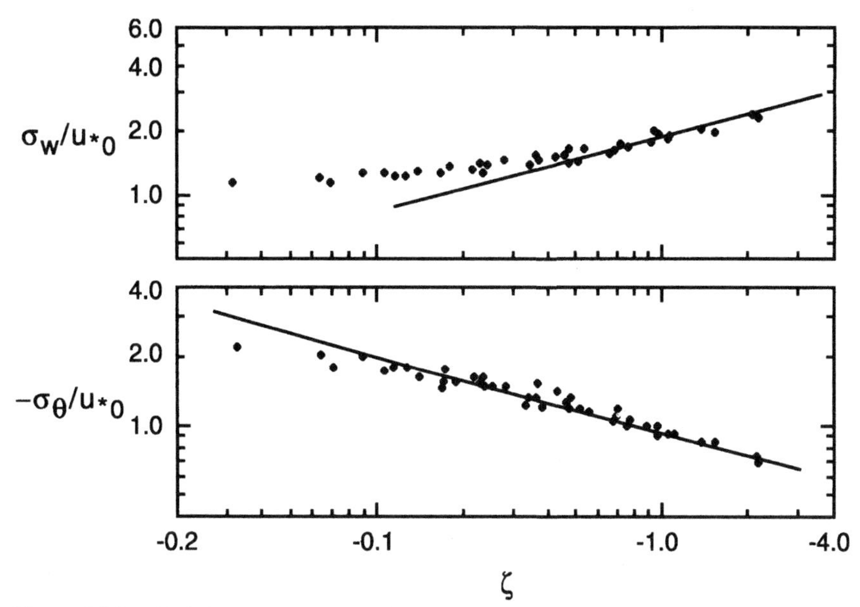

Figure 3.3.3. Turbulence intensities as functions of the similarity variable under unstable stratification, showing free convective scaling in the asymptotic limit (from Wyngaard *et al.* 1971).

Temperature variance behaves similarly. The horizontal components of turbulence, however, do not follow the M-O surface layer scaling. Instead, they are also dependent on the depth of the ABL. The existence of large eddies in the ABL appears to influence the horizontal components even in the surface layer and invalidate the surface layer scaling.

Kolmogoroff theory indicates that for locally isotropic turbulence, the spectra of various turbulence quantities in the inertial subrange should be functions of only the corresponding dissipation rate and the wavenumber. Thus from dimensional analysis, the one-dimensional longitudinal velocity spectrum should be

$$nS_u(n) = \alpha_u \varepsilon^{2/3} k^{-2/3} = \frac{\alpha_u}{(2\pi)^{2/3}} \varepsilon^{2/3} n^{-2/3} U^{2/3} \tag{3.3.38}$$

where ε is the dissipation rates of TKE and Taylor's frozen turbulence hypothesis has been used to convert the frequency to the wavenumber $k = 2\pi n/U$, where U is the mean velocity (Hogstrom, 1996; see also Kaimal and Finnigan, 1994; Serra *et al.,* 1997). A similar reasoning yields the spectra for scalars in the inertial subrange. For temperature and humidity, these are

$$\begin{aligned} nS_\theta(n) &= \frac{\alpha_\theta \varepsilon_\theta}{(2\pi)^{2/3}} \varepsilon^{-1/3} n^{-2/3} U^{2/3} \\ nS_q(n) &= \frac{\alpha_q \varepsilon_q}{(2\pi)^{2/3}} \varepsilon^{-1/3} n^{-2/3} U^{2/3} \end{aligned} \tag{3.3.39}$$

where ε_θ and ε_q are the corresponding dissipation rates of half the temperature and humidity variances, and α_θ and α_q are the corresponding Kolmogoroff constants. Based on data from Yaglom (1981) and others, Hogstrom (1996) concludes that $\alpha_u \sim 0.51 \pm 0.01$ and $\alpha_\theta = \alpha_q = 0.80 \pm 0.01$. α_u is related to the Karman constant (Kaimal and Finnigan, 1994) by $\alpha_u \kappa^{-2/3} \sim 1$, and if a value of 0.4 is chosen for κ, then $\alpha_u \sim 0.55$. Nondimensionalizing these expressions by the relevant frictional quantities and using the length scale z as the normalizing scale in the surface layer, the longitudinal velocity, temperature, and humidity frequency spectra in the surface layer become

$$\begin{aligned} \frac{nS_u(n)}{u_*^2} &= \frac{\alpha}{(2\pi)^{2/3}} \varphi^{2/3} \hat{n}^{-2/3} \\ \frac{nS_\theta(n)}{\theta_*^2} &= \frac{\alpha_\theta}{(2\pi)^{2/3}} \varphi^{-1/3} \varphi_\theta \hat{n}^{-2/3} \\ \frac{nS_q(n)}{q_*^2} &= \frac{\alpha_q}{(2\pi)^{2/3}} \varphi^{-1/3} \varphi_q \hat{n}^{-2/3} \end{aligned} \tag{3.3.40}$$

where

$$\hat{n} = \frac{zn}{U} = \frac{z}{\lambda}, \varphi = \frac{\varepsilon z}{u_*^3}, \varphi_\theta = \frac{\varepsilon_\theta z}{u_* \theta_*^2}, \varphi_q = \frac{\varepsilon_q z}{u_* q_*^2} \tag{3.3.41}$$

Note that for isotropic turbulence, the vertical and lateral velocity spectra are related to the longitudinal spectrum by $S_v = S_w = (4/3)\, S_u$, and all these nondimensional spectra must be functions of only $\zeta = z/L$, where L is the Monin–Obukhoff length scale. See Kaimal and Finnigan (1994) for various spectra in the surface layer obtained from measurements over land surfaces in Kansas and a lucid discussion of their properties in stable and unstable conditions. Briefly, the peak of the spectrum corresponds to the most energetic eddy and is therefore of great interest. Going from stable to neutral conditions, the various spectra show a systematic shift of the peak of the spectrum to lower and lower frequencies. The wavelength corresponding to the peak scales with ζ, except at large values of ζ (>2) where z-less scaling prevails. Under unstable conditions, the depth of the CABL (z_i) becomes an important parameter, and the wavelength at the peak scales with z_i. The Monin–Obukhoff scaling breaks down for u and v spectra in unstable conditions. For the w spectrum, the wavelength at the peak stops scaling with ζ, scaling with z instead with a constant of proportionality of $(0.17)^{-1}$. This is the so-called free convective limit, where the Monin–Obukhoff length scale loses its significance. Note that the expressions in Eq. (3.3.40) differ from that in Kaimal and Finnigan (1994) in that the von Karman constant does not multiply z in Eq. (3.3.41).

Figure 3.3.4, from Serra *et al.* (1997), compares the various spectra measured over the ocean during recent field campaigns in the tropical Pacific COARE and CEPEX studies with those obtained from the Kansas experiment. We present this only to show that systematic differences exist between spectra in the surface layer measured over land and over humid cloudy conditions over the tropical oceans. Serra *et al.* (1997) point out that even in stable and near-stable conditions, the presence of clouds and the associated circulations within tend to bring cool dry air down, forcing warm moist air up. While free convective conditions are dominated by small scale processes, large scale processes originating above the surface layer affect forced convective conditions. Serra *et al.* (1997) find the free convective limit corresponding to $\hat{n} \sim 0.17$ in their w spectra, in agreement with Kansas data.

In the inertial subrange, in the absence of significant buoyancy effects, the cospectra of turbulence quantities must also depend only on the wavenumber, the dissipation rate, and the local vertical gradient of the mean property. This means that when suitably normalized, the cospectra must be functions of only the normalized frequency and ζ. For example, the uw, wθ, and wq cospectra can be written as

$$
\begin{aligned}
-\frac{nC_{uw}(n)}{u_*^2} &= \gamma G(\zeta)\hat{n}^{-4/3} \\
-\frac{nC_{w\theta}(n)}{u_*\theta_*} &= \gamma_\theta G(\zeta)\hat{n}^{-4/3}; -\frac{nC_{wq}(n)}{u_*q_*} = \gamma_q G(\zeta)\hat{n}^{-4/3}
\end{aligned}
\tag{3.3.42}
$$

Kaimal and Finnigan (1994) present the uw and wθ cospectra from land surface measurements. The behavior is similar to that of the spectra. The cospectra also show a systematic dependence of the location of the spectral peak with ζ in stable conditions. Also, under unstable conditions the cospectra cluster around that for the neutral value. Figure 3.3.5, from Serra *et al.* (1997), compares the uw, wθ, and wq cospectra with those obtained in Kansas under neutral conditions. The higher frequency of the cospectral peaks is due to the dominance

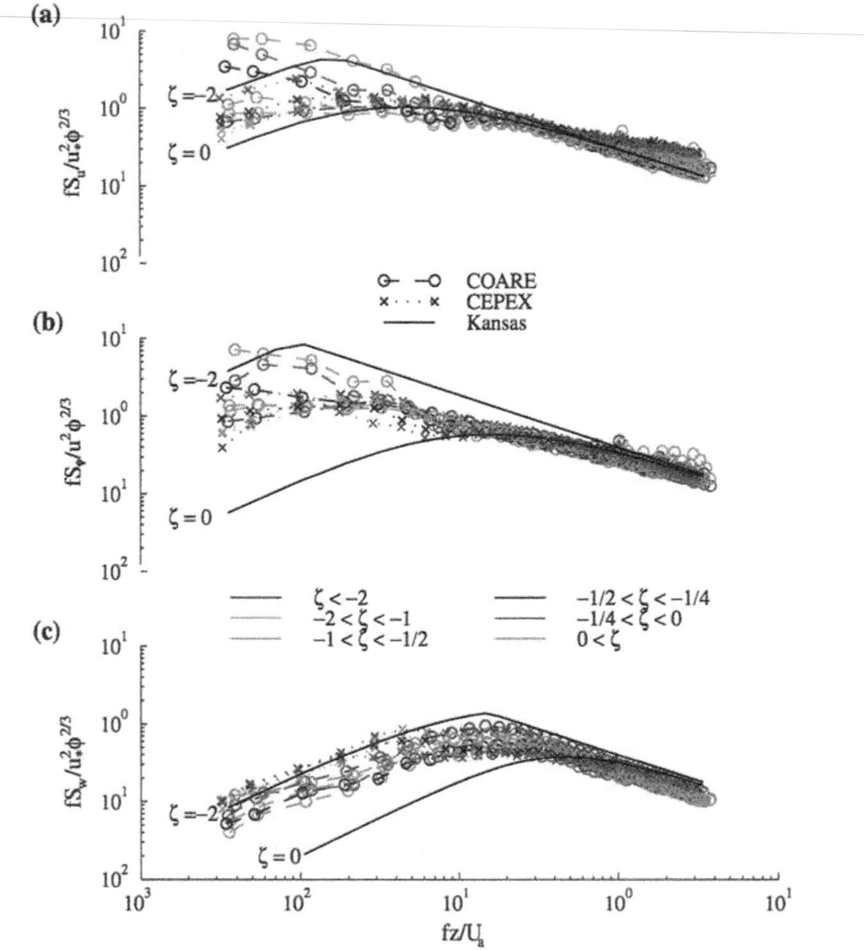

Figure 3.3.4. Normalized logarithmic spectra of (a) along-wind velocity, (b) cross-wind velocity, (c) vertical velocity, (d) potential temperature, and (e) specific humidity (from Serra *et al.*, 1997).

of the vertical velocity variance over the horizontal one in transporting momentum and scalars in the surface layer over the tropical oceans.

Structure function parameter C_θ^2 (Wyngaard *et al.*, 1971; Kaimal and Finnigan, 1994) of temperature θ defined as

$$C_\theta^2 r^{2/3} = \overline{[\theta(x+r)-\theta(x)]^2} \tag{3.3.43}$$

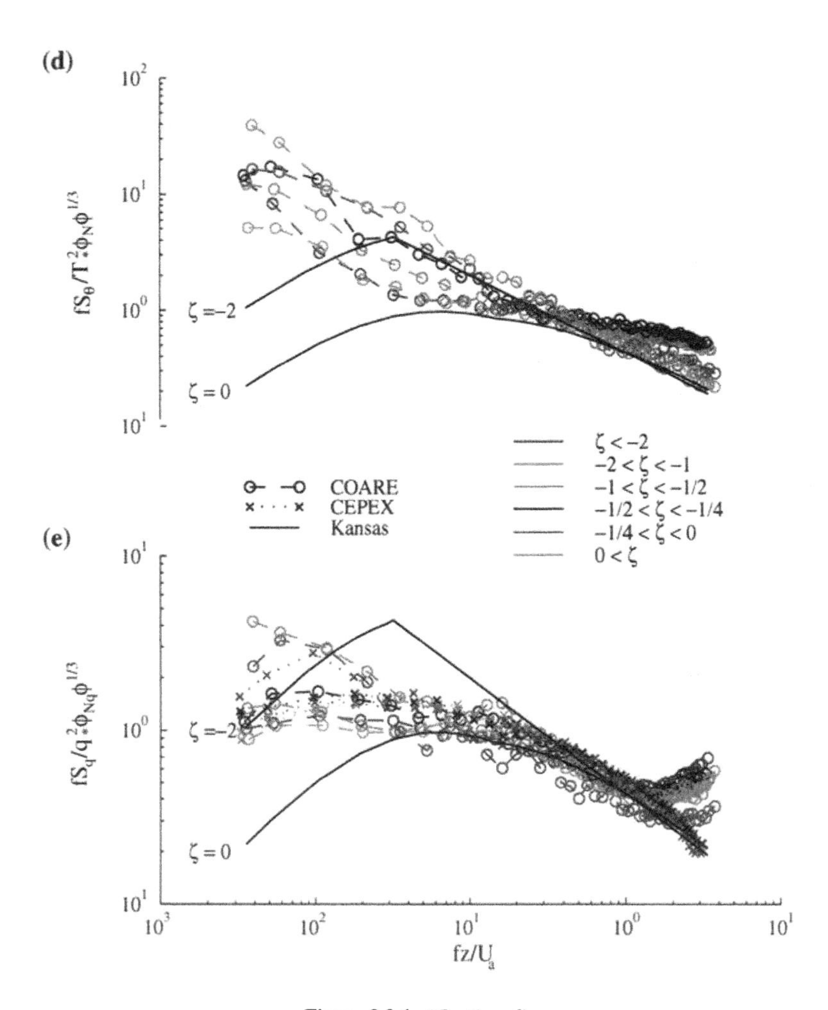

Figure 3.3.4. *(Continued)*

is of importance in wave propagation through a turbulent medium (the structure parameter for refractive index fluctuations important to remote sensing using electromagnetic and acoustic waves is closely related). It is related to the one-dimensional wavenumber spectrum of temperature and dissipation rate of half the temperature variance by

$$C_\theta^2 \sim 4k_1{}^{5/3}S_\theta(k_1) \sim 4\alpha_\theta\varepsilon_\theta\varepsilon^{-1/3} \tag{3.3.44}$$

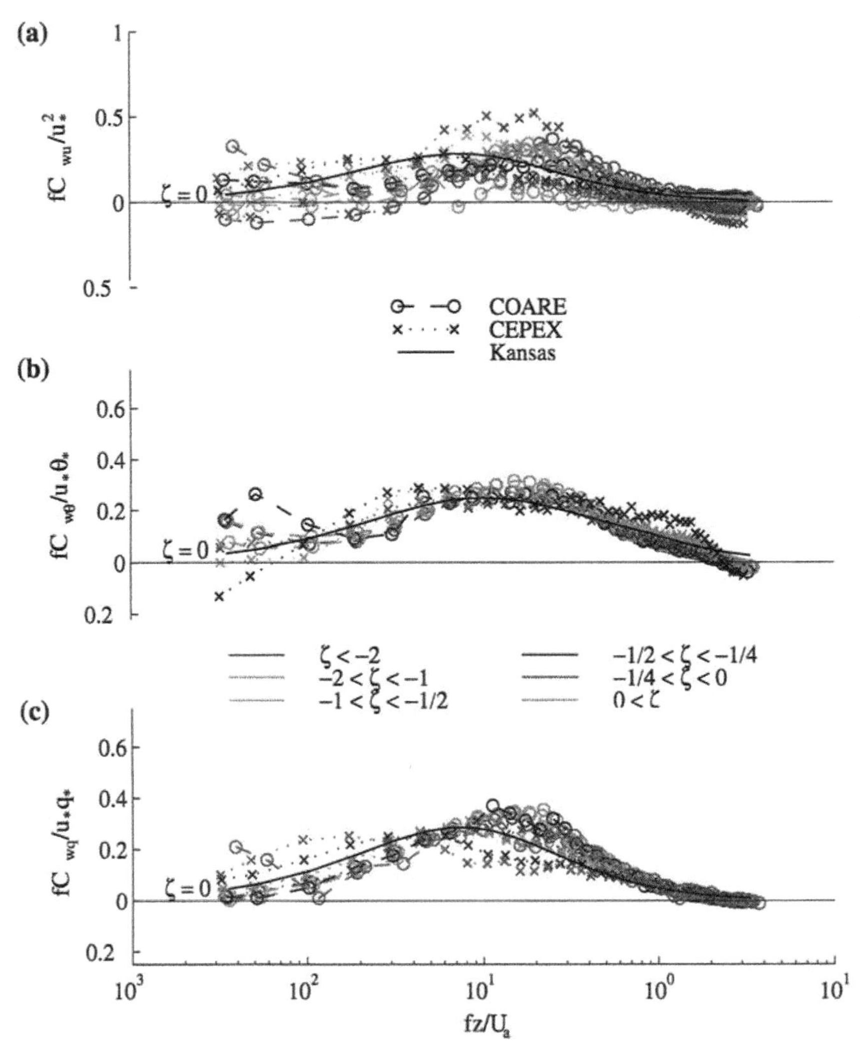

Figure 3.3.5. Normalized logarithmic cospectra of (a) vertical velocity and along-wind component of velocity (b) vertical velocity and potential temperature, and (c) vertical velocity and specific humidity (from Serra *et al.* 1997).

Similar structure function parameters can be defined for other variables such as humidity and velocity components and obey similar relationships. Structure function parameters of all variables including the velocity must obey M-O similarity laws in the surface layer. For example,

$$C_\theta^2 = \theta_*^2 z^{-2/3} f_\theta(\zeta) \tag{3.3.45}$$

The spectra of various properties obey M-O similarity quite well as indicated by the plots in Figure 3.3.6 although scatter is large.

Normalized equations for TKE, half the temperature, and humidity variances in the surface layer are (invoking steady state and horizontal homogeneity)

$$\begin{aligned}
&\phi_T(\zeta)+\phi_P(\zeta)=\phi_M(\zeta)-\zeta-\phi_\varepsilon(\zeta)\\
&\phi_{T\theta}(\zeta)=\phi_H(\zeta)-\phi_{\varepsilon_\theta}(\zeta)\\
&\phi_{Tq}(\zeta)=\phi_E(\zeta)-\phi_{\varepsilon_q}(\zeta)
\end{aligned}\tag{3.3.46}$$

where ϕ_T denotes the corresponding normalized turbulent transport, and ϕ_ε the corresponding normalized dissipation rates. ϕ_P is the pressure transport term that appears only in the TKE budget:

$$\begin{aligned}
&\phi_T=\frac{\kappa z}{2u_*^3}\frac{\partial}{\partial z}\left[\overline{w\left(u^2+v^2+w^2\right)}\right];\phi_P=\frac{\kappa z}{\rho u_*^3}\frac{\partial}{\partial z}\left(\overline{wp}\right);\phi_\varepsilon=\frac{\kappa z\varepsilon}{u_*^3}\\
&\phi_{T\theta}=\frac{\kappa z}{2u_*\theta_*^2}\frac{\partial}{\partial z}\left(\overline{w\theta^2}\right);\phi_{\varepsilon_\theta}=\frac{\kappa z\varepsilon_\theta}{u_*^2\theta_*}\\
&\phi_{Tq}=\frac{\kappa z}{2u_*q_*^2}\frac{\partial}{\partial z}\left(\overline{wq^2}\right);\phi_{\varepsilon_q}=\frac{\kappa z\varepsilon_q}{u_*^2q_*}
\end{aligned}\tag{3.3.47}$$

These equations form the basis of the inertial dissipation method for computing the fluxes of various properties in the surface layer (see Chapter 2). For neutral conditions, it is normal to assume local equilibrium and ignore all the transport terms, although there are indications (Hogstrom, 1996, among others) that the terms on the LHS of the first equation in Eq. (3.3.46), the TKE budget, are nonzero. For unstable conditions, the pressure transport and turbulent transport terms are roughly equal in magnitude and opposite in sign and are important not only in the surface layer but in the bulk of the CABL as well. For stable conditions, these terms may be approximately zero. Note that we have throughout here used budgets and spectra involving half the variances of scalar properties. This is not always the case in literature, and often budgets for variances are used. This has led to considerable confusion about various constants in the spectral and budget relationships.

A few comments with respect to the applicability of M-O similarity to the ABL and OML with mobile surface elements are in order. Sand grains are set in motion by wind above a certain threshold wind stress, and the motion takes the form of creep as well as saltation. Creeping motion consists of sand grains rolling along the surface, while during saltation, the impact of another sand

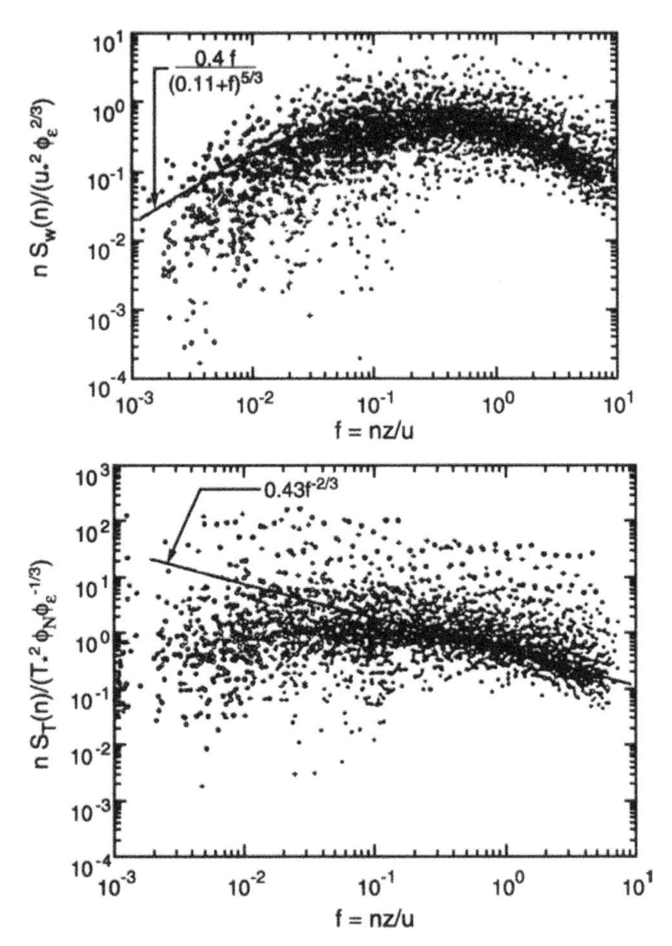

Figure 3.3.6. Scaled vertical velocity (top) and temperature (bottom) spectra for unstable conditions. Note the Kolmogoroff and Corrsin–Obukhoff inertial subranges. The lines are derived from similarity theory for the unstable surface layer (from Gaynor *et al.*, 1992).

particle launches a sand grain into the air and when it falls back, it sets off another in motion. Clearly, saltation involves momentum flux to the ground, but the mechanism is different. However, as long as the effective shear stress is used, M-O similarity should still hold.

The application of M-O similarity to the OML is not so unambiguous because of the influence of surface waves, the high wavenumber end of their spectrum contributing to the roughness of the sea surface and hence determining the shear stress. Nevertheless, away from the wave-influenced upper layers, M-O scaling is approximately valid. One potential modification is during strong rainstorms, when the added momentum and buoyancy fluxes due to rain must be considered (see Section 4.1) in defining the M-O length scale.

Another aspect worthy of note is the behavior of the drag coefficient over ice-covered surfaces in polar seas. Overland (1985) reviews observations of the drag coefficients over sea ice (see also Overland and Davidson, 1992). He showed that the drag coefficients are smaller over smooth interior pack ice compared to those over rafted marginal sea ice. Invariably the surfaces of pack ice are covered by snow that drifts and is blown around by winds above 6–8 m s^{-1}. The roughness felt by the ABL depends very much on whether the snow drifts (sastrugis) are piled up parallel or perpendicular to the mean wind (with the latter leading to an increased, and the former a decreased, value for C_D), the angle the wind makes to the sastrugis being an important factor (Andreas, 1996). Andreas (1987b) reviews observations on the values of the heat transfer coefficient over snow-covered ice surfaces, and Andreas (1996) comments on the large scatter in the observations and suggests empirical relationships, based on surface-renewal theory (see Chapter 4), for $z_{0T,0E}$ as functions of the roughness Reynolds number $u_* z_0/\nu$. According to his model, C_{HN10} increases with C_{DN10} for a given wind speed, but decreases with increasing wind speed. The observational data are, however, too scattered to verify this behavior.

3.4 CONVECTIVE ATMOSPHERIC BOUNDARY LAYER (CABL)

The convective ABL can be defined as the ABL dominated by heating near the ground. This does not mean that the mixing effects due to the mean shear of the geostrophic winds are negligible. It just means that convective mixing dominates over the bulk of the ABL and determines its characteristics. The CABL over land is the most extensively studied, both observationally and numerically. Its salient characteristics are now quite well known. We will consider only the cloud-free ABL for simplicity and defer discussion of cloud-topped ABL (CTBL) and marine ABL (MABL). In the absence of liquid water and clouds that add additional complexity to the thermodynamics and dynamics of the ABL, it is adequate to make use of the virtual potential temperature to take into account the water vapor effects on the dynamics and thermodynamics. With this modification, the dynamics are essentially similar to that of a dry ABL.

The most salient characteristic of the CABL in midlatitudes is its strong diurnal evolution. Figure 3.4.1 shows the evolution of the ABL over a typical day. At sunrise, the shallow shear-driven nocturnal ABL (NABL) that formed around the previous sunset and developed over the previous night (Figure 3.4.2) sets the stage for the evolution of the CABL. We will discuss the NABL in detail later, but suffice it to say that radiative cooling of the atmospheric column during

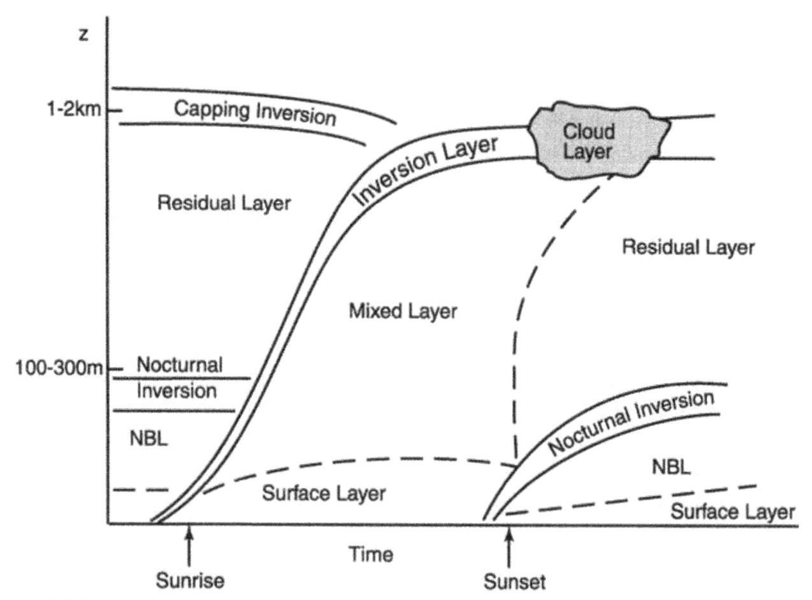

Figure 3.4.1. The diurnal evolution of the ABL. Conditions are shown for a typical ABL over land (from Wyngaard, 1992).

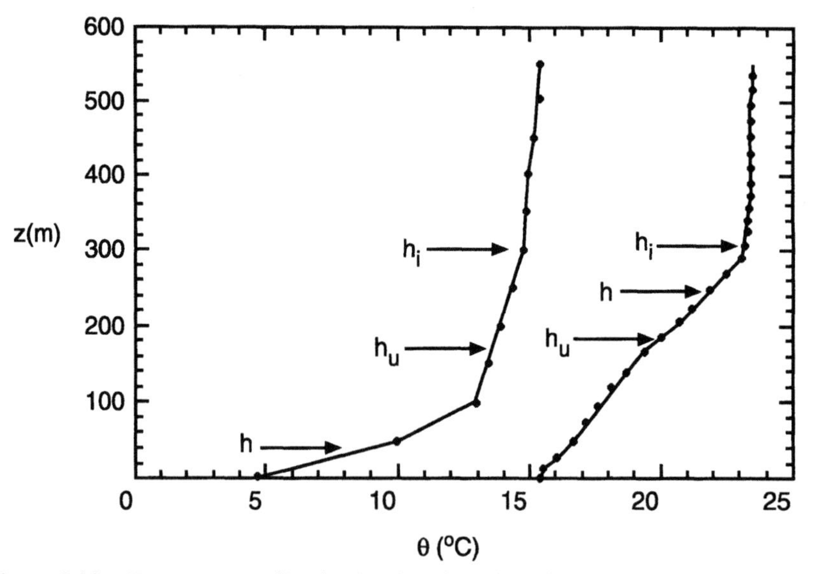

Figure 3.4.2. Temperature profiles in the clear-sky NABL from WANGARA and VOVES observations, showing the heights of inversion (h_i), low-level wind maximum (h_u), and NABL (h) (from Andre and Mahrt, 1982).

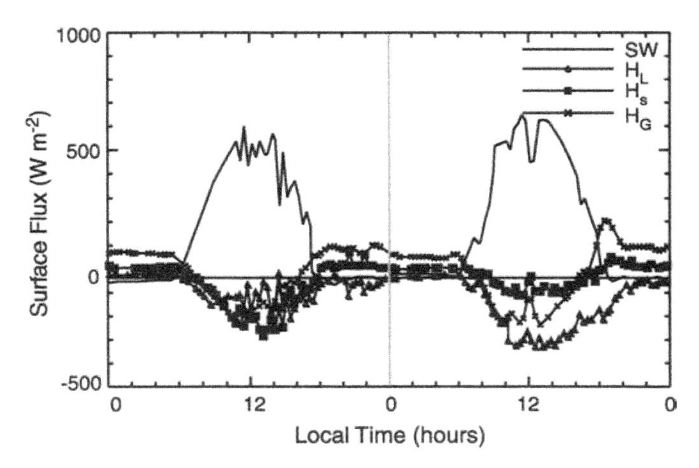

Figure 3.4.3. Typical flux balance indicating the half-sinusoidal form of heating of the CABL from below (from Kustas *et al.* 1992). Data was collected over semiarid rangeland; several rain events occurred between the day shown on the left and the day shown on the right.

the night (typically 0.1–0.2° C per hour on clear cloudless nights, although it can be as high as 2° C per hour near the ground) produces a typical nocturnal inversion with a nearly linear potential temperature and strong static stability in the 100 or 200 m near the ground (Figure 3.4.2). The exact conditions at sunrise are very much dependent on the evolution of the NABL during the previous night, including the intensity of cooling and the strength of the geostrophic winds aloft. The typical variation of the net heat flux at the ground during the day can be approximated by a half-sinusoid (Figure 3.4.3) for cloud-free conditions, with the heat flux increasing rapidly to a maximum around local noon and then diminishing to near-zero values around sunset (Figure 3.4.4). The peak value is typically 400 to 700 W m^{-2}, depending on the season and intensity of solar insolation. The heat flux during the night is typically 50 to 100 W m^{-2}, but negative, that is downward, from the atmosphere to the ground. The turbulent heat flux (sum of the sensible and latent fluxes) at the ground–air interface is the result of a complex thermodynamic balance involving shortwave (SW) and longwave (LW) radiation impinging on the ground, LW backradiation from the ground, and conduction into/from the ground surface (see Section 4.2). The situation is even more complex in the presence of vegetation due to evapotranspiration effects.

For many practical purposes, it is possible to consider a local equilibrium approximation in which the ground temperature adjusts instantaneously to effect a balance between the net incoming SW and LW solar radiation, the turbulent sensible and latent heat fluxes at the ground–air interface, and the heat flux into the ground. In many cases, the net heat flux to the ground averaged over a day is

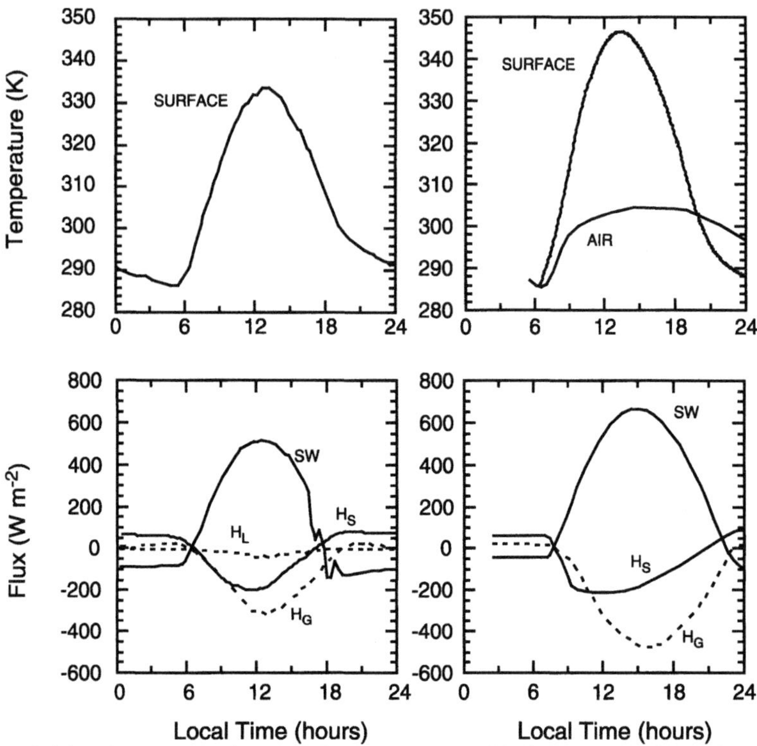

Figure 3.4.4. Diurnal cycle of net radiation, latent and sensible heat fluxes, and heat flux to the ground for clear skies for a semiarid rangeland (left, from Kustas *et al.*, 1991) and over crops (right) (from Garratt, 1992).

very nearly zero. The heating during the day is nearly balanced by cooling during the night. This is a reasonably good approximation when looking at diurnal evolution of the ABL over land. This does not mean that the ground does not store heat during summer and lose it over winter. There is definitely a modulation of below-ground temperatures, especially in the absence of vegetation, over the year, but compared to the OML, the heat storage capacity of land is small, and its influence on the ABL is consequently small, also. Over the global oceans, the OML manages to store considerable heat during spring-summer and transfer it to the atmosphere during winter, thus limiting the temperature extremes in the ABL over the year. The lack of a similar moderating influence of the ground is responsible for the larger temperature extremes in the ABL over land in the interior of the continents.

For purposes of discussing the evolution of the ABL, it is adequate to consider a half-sinusoid heat flux into the ABL during the day, and a nearly constant heat flux from the ABL during the night. During early morning, heating erodes the inversion developed during the night. Once the inversion is broken down, the depth of the CABL increases rapidly until mixing reaches the vicinity of the capping inversion around noon at typically a height of 1–2 km. The rate of increase of the CABL depth slows down dramatically at this point and in many cases, the depth does not increase much further during the afternoon. The CABL is in a quasi-steady state during most of the afternoon. The ABL temperature, on the other hand, continues to increase during the afternoon, reaching a maximum around mid-afternoon. The precise manner in which the CABL evolves during the day depends on the potential temperature profile in the atmospheric column at sunrise—especially the location and strength of the capping inversion. In some cases, such as over deserts during summer, strong heating can produce CABL depths of a few kilometers by mid-afternoon.

Even before the sun sets and the heat flux at the ground reverses, the rapid decrease of heat flux into the ABL reduces convective production of turbulence, and the turbulence in the bulk of the ABL begins to decay. Given the fact that the eddy turnover timescale (h/w^*) is typically 20–30 min near sunset, the decay of turbulence in the bulk of the ABL takes about an hour or so. But close to the ground, turbulence is often kept alive by shear generation due to prevailing geostrophic winds, and this continues into the night even in the presence of the stable stratification due to reversal of heat flux at the air–ground interface.

Convective generation of turbulence dominates the CABL and the shear generation is generally unimportant except close to the ground. Similarly the radiative effects are small in the CABL and can be ignored for the most part during the day, whereas the LW radiative balance is quite central to the nocturnal evolution of the atmospheric column. Extensive observations and LES's (the CABL is ideally suited for LES's—see Chapter 1.11) have given us a very clear picture of mixing processes in the CABL. In general, the CABL can be divided into three parts:

1. The layer near the ground composing perhaps 10% of the height of the CABL ($0 < z/h < 0.1$), where local free convection scaling (with the length scale determined by z, and the velocity and temperature scales are u_f and θ_f) applies (except in the surface layer). In this layer is embedded the surface layer, a few tens of meters deep, in which Monin–Obukhoff similarity scaling holds and all quantities except horizontal turbulence velocity components scale with z/L.

2. The bulk of the ABL ($0.1 < z/h < 1.0$), where strong turbulent mixing driven by convective eddies and plumes prevails. Here the turbulence quantities scale with the depth of the CABL, h, and the convective velocity scale, w^* (and temperature scale Q_0/w^*, where Q_0 is the kinematic heat flux at the ground), which is typically 1–2 m s^{-1}. The velocity and temperature gradients are small in this layer and the Ekman turning in the bulk of the CABL is also small, with most of the turning taking place in the upper portions of the CABL. The heat flux decreases nearly linearly with height. For nonpenetrative convection, the heat flux should be zero at the top of the CABL. However, this is rarely the case,

requiring very strong, nearly impenetrable capping inversions. Normally, the value of heat flux becomes negative at the top of the CABL due to energetic entrainment by turbulent eddies, the ratio of the kinematic heat flux at the top (Q_t) to the kinematic heat flux at the ground (Q_0) being typically 0.05–0.3 (0.2 is a good approximation in most cases). It is the presence of this entrainment heat flux at the top that gives rise to the bottom–up/top–down mixing characteristics of the CABL.

3. The inversion layer above the mean CABL top ($1 < z/h < 1.1$ to 1.5), where quantities are governed by the characteristics of the capping inversion and the stable region above. The size of this region depends very much on the prevailing static stability and the strength of convection in the CABL. The region is characterized by hummocks due to penetrating thermal plumes and internal waves generated by such quasi-periodic intrusions. In addition, there are Kelvin–Helmholtz billows driven by shear instabilities due to large local vertical shear induced by motion in convective plumes.

It is the properties in the bulk of the CABL that are of principal interest in many cases. Figures 3.4.5 and 3.4.6 show the turbulence velocity components and temperature variance from observations. The normalized spectra (normalized by mixed layer scales h and w_*) are shown in Figure 3.4.7; a nice inertial subrange can be seen.

Typical values for the CABL scales are $h \sim 1000$ m, $H_0 \sim 400$ W m^{-2}, $H_t = -100$ W m^{-2}, $w_* \sim 2$ m s^{-1}, $T_* \sim 0.2$°C, and $t_e \sim 500$ s. Turbulence in the CABL is highly coherent and dominated by buoyant updrafts and downdrafts. Convective updrafts or plumes in the bulk of the CABL result from the merging of small thermals that originate in the superadiabatic layer near the ground. These are embedded in a slightly stable, surrounding air mass that includes air entrained into the mixed layer by the action of turbulent eddies at the top of the CABL. Descending air masses take the form of downdrafts surrounding ascending plumes, and the warm air from above the CABL is drawn and entrained into the CABL around hummocks created by the ascending plumes. The wispy downdrafts and columnar ascending plumes are clearly seen in radar and sodar records (Figure 3.4.8) and in laboratory experiments (Deardorff and Willis, 1974; Kantha, 1980b).

The horizontal structure of the CABL depends very much on the relative strength of the wind. LES's (Moeng and Sullivan, 1994) show that in very light winds, the structures are typical of free convection with irregular, polygon-shaped, cell-like features and buoyant plumes at their edges (see Chapter 1.11). If these plumes reach the lifting condensation level, cumulus clouds form and display the cellular structure of the convective flow. However, linear features aligned roughly in the direction of the mean velocity characterize the flow at moderate to high wind speeds. These roll-vortices have also been observed in the

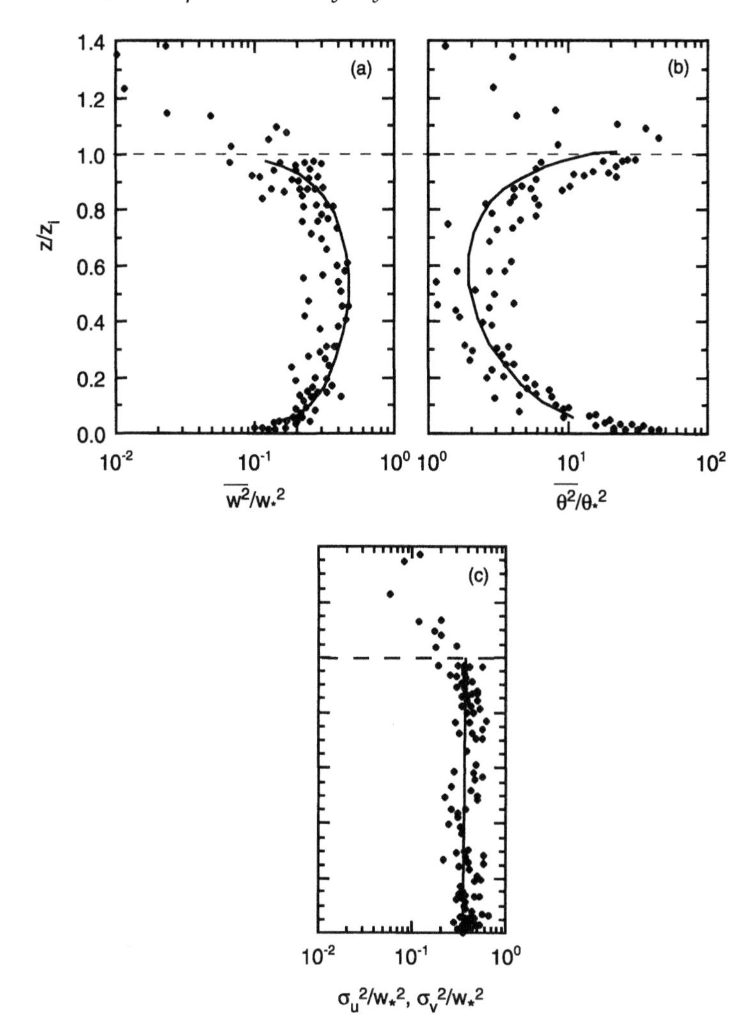

Figure 3.4.5. Normalized (a) vertical velocity variance, (b) temperature variance, and (c) horizontal velocity variances in CABL. (a) and (b) from Sorbjan 1991; (c) from Garratt, 1992, reprinted with the permission of Cambridge University Press.

field and give rise to cloud streets at convergence zones of the rolls, aligned in the along-wind direction.

The CABL grows by both encroachment (nonpenetrative convection) and by turbulent entrainment of quiescent air above the CABL. Turbulent entrainment requires expenditure of TKE. Energetic turbulent eddies created by convection convert some of their kinetic energy into potential energy in this manner and the rate of growth of the CABL (or any ML, for that matter) is intimately related to

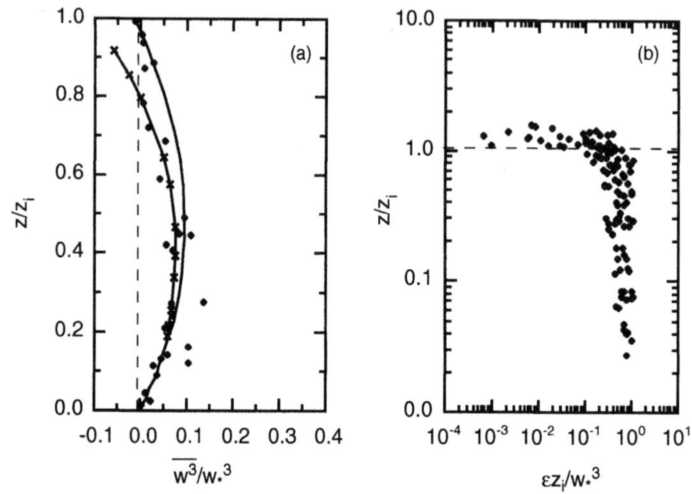

Figure 3.4.6. Normalized (a) third moment of vertical velocity and (b) dissipation rate profiles in the CABL (from Sorbjan 1991).

the energy balance at the buoyancy interface capping the CABL (see Chapter 3.4). There is often a strong shear across the top of the CABL, as well, due to prevailing winds, and this is an additional source of TKE which is often ignored. In any case, the turbulent region grows by incorporating stable quiescent adjacent air masses into it and advancing into the quiescent region. The resulting entrainment heat and buoyancy fluxes at the top of the CABL are germane to the scaling as we will see below. They are also important to cloud formation and dissipation at the top of the CABL as we will see later.

In the absence of entrainment fluxes at the top of CABL, that is, for nonpenetrative convection (encroachment), the only scaling possible is that due to the heat and buoyancy flux at the bottom of the CABL. The bottom–up mixing prevails, Q_0 is the governing parameter, and quantities in the CABL scale with w_*, T_*, and z/h. The presence of heat and buoyancy fluxes at the top of the CABL gives rise to top–down mixing with scaling determined by Q_t, and the relevant distance scale is $(1- z/h)$. Observations have shown that in the vicinity of the ground, bottom–up scaling dominates even for penetrative convection and the conditions are not far different from the nonpenetrative convection situation. However, close to the top, top–down scaling must prevail. It is possible then to regard all quantities in the entire CABL to scale according to a linear superposition of nonpenetrative components scaled by bottom–up mixing scales and the residual penetrative components scaled by top–down mixing scales. Both observations and LES's show this is indeed a very good approximation. Following Sorbjan (1991), define then two sets of local distance, velocity, and

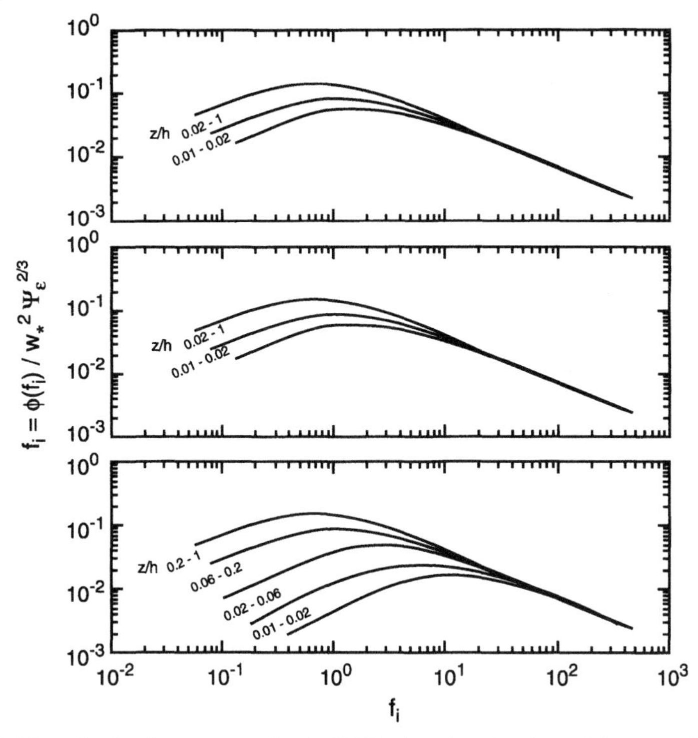

Figure 3.4.7. Normalized velocity spectra in the CABL plotted as functions of the normalized frequency f_i = nh/u. U-component spectrum (top), v-component spectrum (middle), and w-component spectrum (bottom) (from *Atmospheric Boundary Layers* by J. C. Kaimal and J. J. Finnigan. Copyright 1994 Oxford University Press, Inc. Used by permission of Oxford University Press, Inc.).

temperature scales in the CABL: z and

$$u_{bu} = \left[\beta z \overline{w\theta}_{bu}(z)\right]^{1/3}, \qquad T_{*bu} = \overline{w\theta}_{bu}(z) / u_{bu} \tag{3.4.1}$$

for the bottom–up component of mixing, and h–z and

$$u_{td} = \left[\beta (h\text{-}z) \overline{w\theta}_{td}(z)\right]^{1/3}, \qquad T_{*td} = \overline{w\theta}_{td}(z) / u_{td} \tag{3.4.2}$$

for the top–down component. Since the local heat flux is

$$\begin{aligned} -\overline{w\theta}(z) &= -\overline{w\theta}_{bu}(z) - \overline{w\theta}_{td}(z) \\ &= Q_0\left(1 - \frac{z}{h}\right) + Q_t \frac{z}{h} \end{aligned} \tag{3.4.3}$$

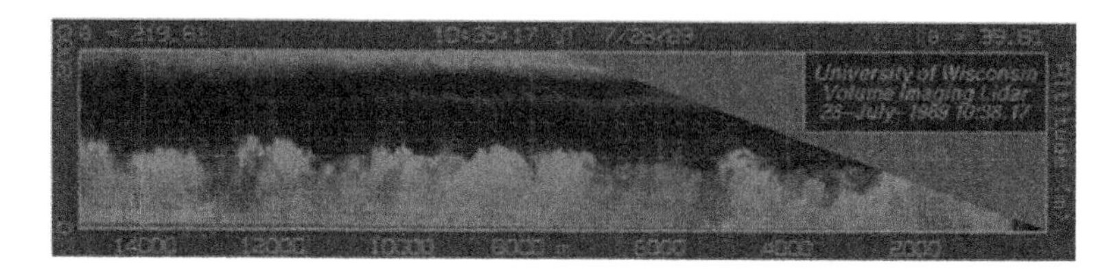

Figure 3.4.8. Spatial variability of the CABL from lidar data during the FIFE II field experiment, showing the entrainment layer and the downdrafts as measred by the University of Wisconsin Lidar Group (figure provided by E. Eloranta).

we get

$$
\begin{aligned}
u_{bu} &= w_* \left(\frac{z}{h}\right)^{1/3} \left(1-\frac{z}{h}\right)^{1/3} \\
T_{*bu} &= T_* \left(\frac{z}{h}\right)^{-1/3} \left(1-\frac{z}{h}\right)^{2/3} \\
u_{td} &= w_* R^{1/3} \left(\frac{z}{h}\right)^{1/3} \left(1-\frac{z}{h}\right)^{1/3} \\
T_{*td} &= T_* R^{2/3} \left(\frac{z}{h}\right)^{2/3} \left(1-\frac{z}{h}\right)^{-1/3}
\end{aligned}
\tag{3.4.4}
$$

where

$$
w_* = (\beta Q_0 h)^{1/3}, \qquad T_* = -Q_0 / w_* \tag{3.4.5}
$$

$R = H_t/H_0 = Q_t/Q_0,$ the ratio of heat fluxes, and w_* and T_* are the usual free convection scales that hold for nonpenetrative convection. Now each turbulent quantity can be written as the sum of the nonpenetrative and residual penetrative components. For example,

$$
\overline{\theta^2} = \overline{\vartheta_{bu}^2} + \overline{\theta_{td}^2} = c_1 \overline{T_{*bu}^2} + c_1' \overline{T_{*td}^2} \tag{3.4.6}
$$

so that

$$
\frac{\overline{\theta^2}}{T_*^2} = 2\left(\frac{z}{h}\right)^{-2/3} \left(1-\frac{z}{h}\right)^{4/3} + 8R\left(\frac{z}{h}\right)^{4/3} \left(1-\frac{z}{h}\right)^{-2/3} \tag{3.4.7}
$$

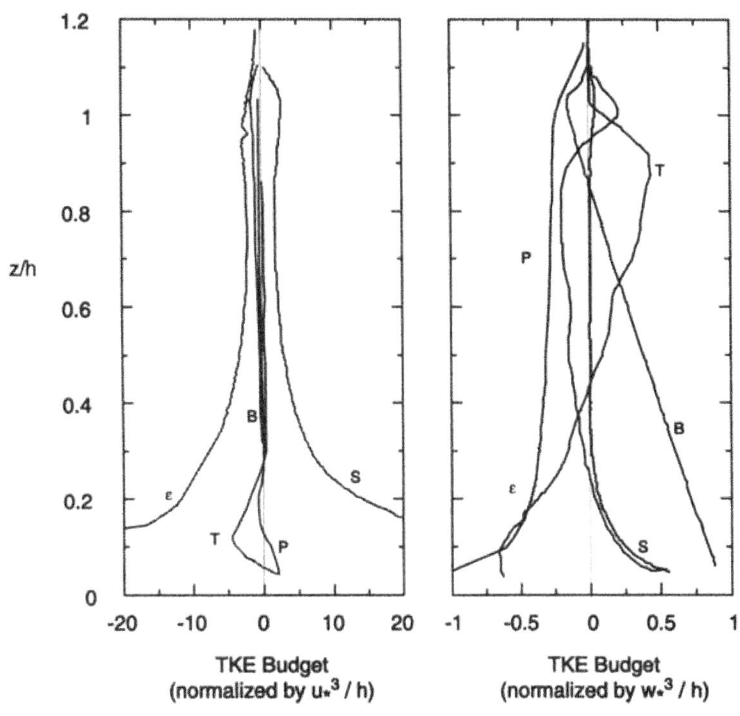

Figure 3.4.9. The TKE budget for the CABL from observations, showing the dissipation, buoyancy and shear production, and transport for both a shear-driven (left) and a convectively-driven CABL (from Moeng and Sullivan 1994).

All other quantities can be expressed similarly:

$$\frac{\overline{w^2}}{w_*^2} = \left(\frac{z}{h}\right)^{2/3}\left(1-\frac{z}{h}\right)^{2/3} + 0.5R^{2/3}\left(\frac{z}{h}\right)^{2/3}\left(1-\frac{z}{h}\right)^{2/3}$$

$$\frac{h}{w_*^3}\frac{\partial}{\partial z}\left(\overline{w^3}\right) = 0.4\left(1-\frac{z}{h}\right) + 2R\left(\frac{z}{h}\right) \tag{3.4.8}$$

$$\frac{h}{w_*^3}\varepsilon = 0.8\left(1-\frac{z}{h}\right) + 2.5R\left(\frac{z}{h}\right)$$

LES studies have given us insight into the turbulent mixing processes in the CABL. The CABL is ideally suited to LES and it has enabled aspects such as bottom–up/top–down mixing to be studied (Moeng and Wyngaard, 1989). LES has also enabled studies of aspects such as the relative importance of shear and

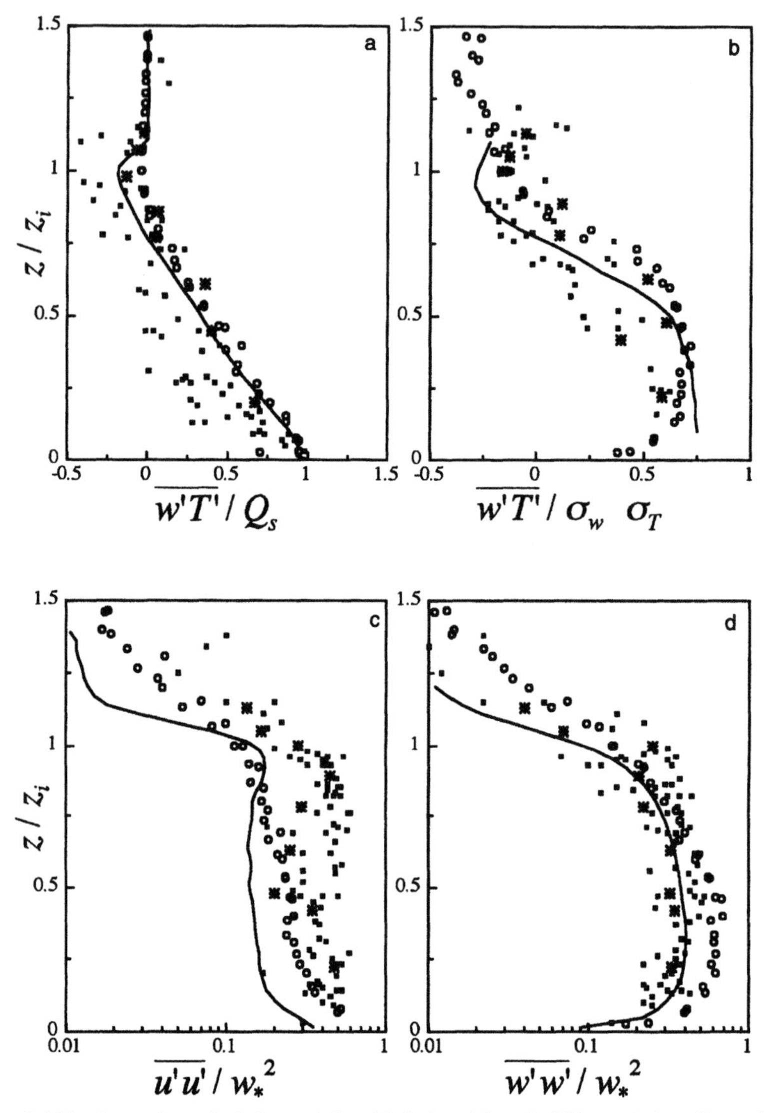

Figure 3.4.10. Comparison of wind tunnel data (circles) on (a) vertical kinematic heat flux and (b) the correlation coefficient between vertical velocity and temperature fluctuations with the results of atmosphere measurements (filled squares) from Caughey and Palmer (1979) (a) and from Sorbjan (1991) (b). The water tank simulations of Deardorff and Willis (1985), asterisks, and LES results of Schmidt and Schumann (1989) line are also shown. Profiles of (c) horizontal and (d) vertical velocity variances in the simulated boundary layer (circles). Other symbols as previously (from Federovich *et al.,* 1996). Vertical distribution of (e) the dimensionless temperature variance and (f) and (g) the third moment of the vertical velocity fluctuations from wind tunnel simulations (circles); other measurements as described before; and (h) the third moment of temperature fluctuations. Sumbols are as previously (from Federovich *et al.,* 1996).

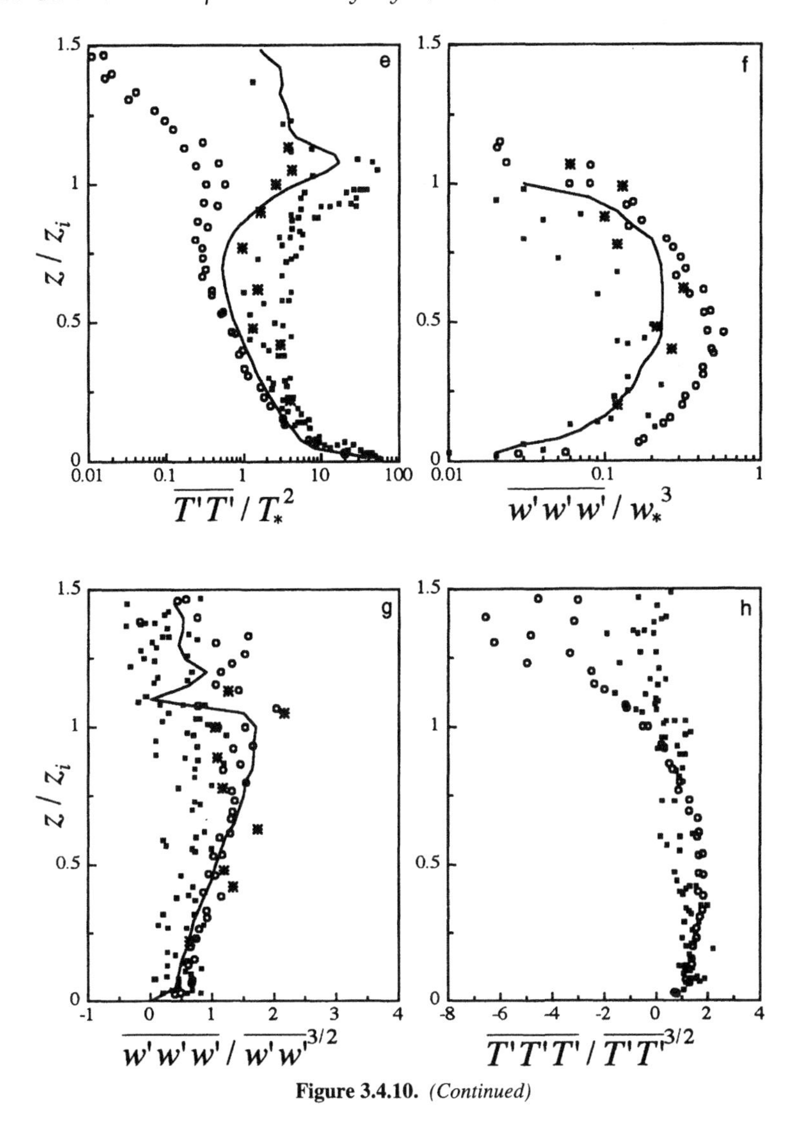

Figure 3.4.10. *(Continued)*

convective mixing. Moeng and Sullivan (1994) have investigated the properties of shear-dominated and convection-dominated CABL using LES (see Section 1.11 for a comparison). From these results it is easy to see that while the shear stress profile in the flow direction is linear in both cases, convection is more efficient at mixing the momentum than shear. Also the TKE budget shows that the shear production and dissipation nearly balance each other and the turbulent transport terms are small for the shear-dominated case. The transport terms are quite large for the convection-dominated CABL and shear production is important only close to the ground. Figure 3.4.9 shows the TKE budget inferred from observations, which confirm this behavior for the principally convection-dominated CABL.

Most of our knowledge on convective processes comes from laboratory measurements of unsteady convection in a stationary convection tank (for example, Deardorff *et al.,* 1969; Willis and Deardorff, 1974; Kantha, 1980a; Deardorff and Willis, 1985) and from measurements in the atmosphere (Clarke *et al.,* 1971; Kaimal *et al.,* 1976; Caughey and Palmer, 1979; Lenschow *et al.,* 1980). Measurements in the ocean are comparatively few (Anis and Moum, 1992). LES studies are also mostly for the atmosphere (for example, Schmidt and Schumann, 1989). Recently, Federovich *et al.* (1996) have performed experiments on steady convection with shear in a temperature-stratified wind tunnel at the University of Karlsruhe under conditions corresponding to weak capping inversion and weak stability above the inversion ($Ri \sim 10$, $Ri_N \sim 20$) and $u_*/w_* \sim 0.2$–0.5. Figure 3.4.10 shows the profiles of various turbulence quantities from this experiment compared to earlier measurements. Shown also are LES results of Schmidt and Schumann (1989). This plot shows the difficulty of making these measurements whether in the laboratory or in the atmosphere or on a computer. The scatter is large, but the general trends are encouraging.

More recent examples of ABL measurement campaigns are the Flatland Boundary Layer Experiments (FlaBLE) (Angevine *et al.,* 1998) conducted over a very flat agricultural cropland region of Illinois, and the Montreal-96 Experiment on Regional Mixing and Ozone (MERMOZ) (Mailhot *et al.,* 1998). Both observations stress the importance of entrainment processes at the top in modeling the ABL and interpreting the dynamics and chemistry of the ABL. FlaBLE employed 915-MHz wind profilers able to also measure virtual temperature profiles. The profiler reflectivity highlights the refractive index fluctuations and therefore displays the ABL structure due to rising thermals and sinking downdrafts. Figure 3.4.11 shows the CABL evolution during the day. The bottom panel shows the net solar radiation, the latent heat flux, and the sensible heat flux (respectively from top to bottom) over the crops. The incoming solar radiation was seen to be partitioned more into latent heat flux than sensible heat flux.

Comparisons of the MERMOZ observations of the ABL evolution with an ABL model (Benoit *et al.,* 1997) show that while the inclusion of nonlocal mixing parameterization (for example, Troen and Mahrt, 1986) leads to slight improvement in the vertical temperature and humidity profiles, the mixed layer deepening rate does not change significantly. Instead, the mixing length parameterization appears to be more crucial.

Stably stratified ABL also occurs when warm air from a land mass passes over a cold sea. This situation is quite typical of large semienclosed seas in high latitudes surrounded by land masses (except during winter) and is therefore of

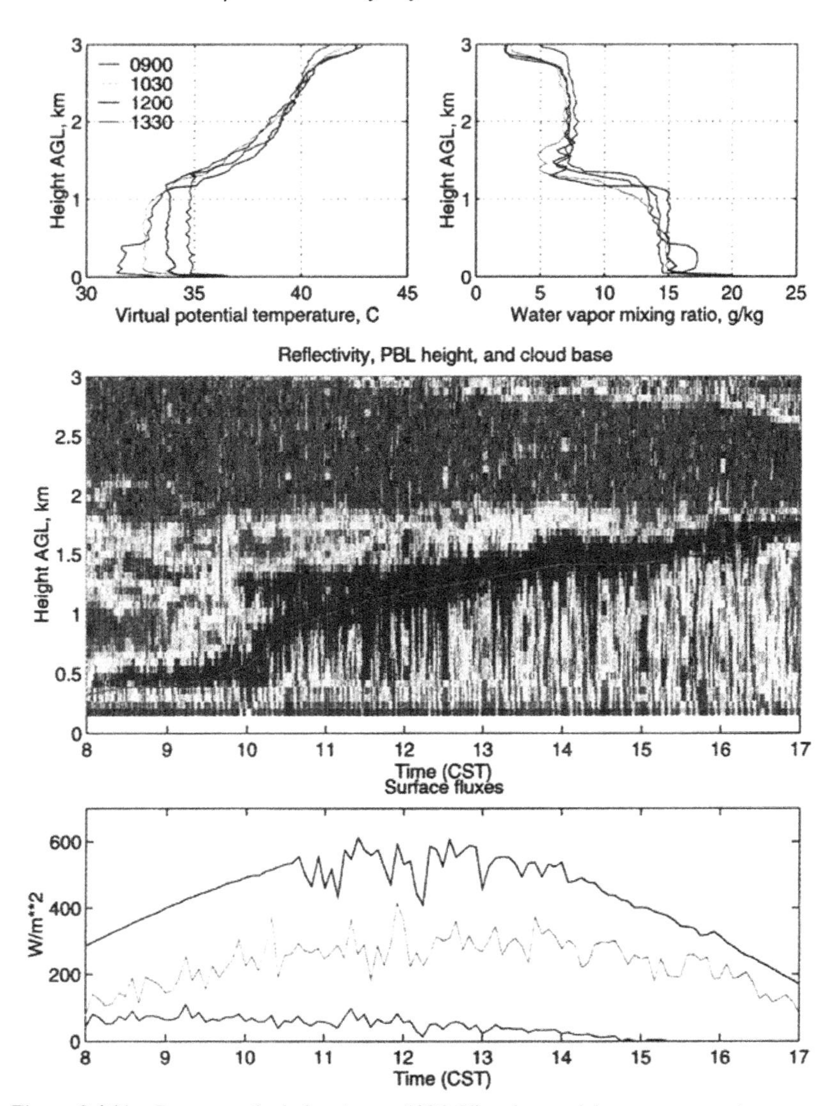

Figure 3.4.11. Case example during August 1996. Virtual potential temperature and water vapor mixing ratio from four soundings at the times shown are plotted in the upper panels. The middle panel shows the profiler reflectivity (arbitrary scale). The bottom panel shows the net solar radiation (upper line), the latent heat flux (middle line), and sensible heat flux (bottom line) (from Angevine *et al.*, 1998).

considerable importance to the atmospheric forcing of these water bodies. In the Baltic Sea, a semienclosed sea surrounded by the European land mass and with the only outlet to the North Sea through the Denmark Strait, this situation occurs 66% of the time on the average (Smedman *et al.*, 1997). Observations during the

Baltic Sea Experiment (BALTEX) indicate that as the air mass passes over the sea, the turbulence in the ABL is extinguished and an internal layer begins to grow with distance from the coastline. The temperature in this layer tends asymptotically to the sea temperature from its initial land temperature value, and eventually a neutrally stratified mixed layer capped by a strong density interface at the top with a temperature change roughly equal to the difference between the land and sea temperature values forms (Smedman *et al.*, 1997). Earlier work on this topic includes Csanady (1974), who studied a similar situation off Lake Ontario, Mulhearn (1981), Garratt and Ryan (1989), Melas (1989), and Bergstrom and Smedman (1994).

3.4 NOCTURNAL ATMOSPHERIC BOUNDARY LAYER (NABL)

The nocturnal ABL over land is characterized by strong stable stratification. Since the heat flux is reversed at night and is from the air to the ground, it tends to suppress turbulence and the main source of turbulence in the NABL is the shear due to geostrophic winds aloft. In addition, stable stratification gives rise to internal wave motions which often break and give rise to intermittent turbulence. Generally, episodic wave motions appear to coexist with turbulence. The evolution of the NABL itself depends very much, therefore, on a variety of factors, including the strength of the winds aloft and the degree of radiative cooling in the layers near the ground. The flow in a stable boundary layer is quite sensitive to even small ground slopes (Mahrt, 1982). This and their generally nonstationary nature makes NABLs hard to study and model. Fortunately, stable boundary layers over ice in polar regions are excellent analogues and observations about polar ABLs can be brought to bear upon the problem (Andreas, 1996). NABLs are particularly important to the study of pollutant dispersal since their shallow depths and relatively lower mixing levels tend to trap any pollutants (and pollen) closer to the ground, leading to much higher ground level concentrations during the night. They are also important to forecasting nocturnal temperatures and any associated hazard such as low level fog. Excellent reviews of stably stratified ABLs can be found in Hunt *et al.* (1996) and Andreas (1996), the latter pertaining to polar regions (see also Derbyshire, 1994). Schumann (1996) compares DNS and LES models for stratified shear flows.

Around sunset, turbulence in the bulk of the CABL begins to decay. Active turbulence is then confined to the vicinity of the ground, where shear generation prevails. The sudden cessation of turbulent mixing can often give rise to vigorous inertial oscillations in the residual layer above the NABL. As the night evolves, the air column cools, generally by 0.1–0.2°C hr^{-1}, except near the ground where the cooling rate can be as high as 1–2°C hr^{-1} in the bottom few meters. LW radiative heat transfer, quite inconsequential to the daytime CABL, is central to the evolution of the NABL. The rapid decrease in temperature near the ground is associated with a negative heat flux near the ground of typically 50–100 W m^{-2} and this begins to counteract shear mixing. In spite of the suppression of mixing by stable stratification, the NABL tends to evolve in height during the night. Generally, the timescale for evolution is a few hours and therefore the NABL is continuously evolving and seldom reaches a steady state (unlike the CABL, which reaches a quasi-steady state by mid-afternoon), except perhaps during long winter nights in polar regions. The NABL over the Arctic sea ice and over the Antarctic continent, although subject to variability due to changing wind conditions, is often in quasi-steady-state conditions. In addition, the intermittent nature of turbulence, the coexistence of wave motions with turbulence, the importance of LW radiation cooling, and the presence of surface inversion (and, often, a strong nocturnal jet) all combine to make the NABL much more complex than the CABL and harder to observe and model. The only "advantage" is that it is quite shallow, 100 or 200 m in depth, and is therefore more easily amenable to observations from meteorological towers and soundings.

Even the depth of the NABL is difficult to define. LW cooling leads often to a surface inversion (see Figure 3.4.2) under low winds and the resulting temperature profile defines a height that is different than the height defined by continuous turbulence. The position of the low level wind maximum defines another height. This definition is favored by Andreas (1996) in the strongly stable ABL over polar seas during late fall and winter. However, for most practical applications, it is the height of the shallow turbulent layer that matters. Above this layer, the shear stress is low, and while the heat flux is also small, the gradient Richardson number is large, indicating static stability, except intermittently during periods of wave breaking.

The NABL can be classified into principally two different turbulent regimes (Sorbjan, 1989). For $h/L < 2$, the turbulence in the NABL is stable-continuous, with continuous turbulence in the entire NABL, whereas for $h/L > 2$, it is stable-sporadic, with continuous turbulence in layers adjacent to the ground ($z < 2$ L) and patchy turbulence above.

Figure 3.5.1 shows the evolution of the NABL at Sprakensehl in Germany on a cloudless fair-weather night, with an easterly geostrophic wind of about 4 m s^{-1}. Figure 3.5.2 shows the various mean profiles. With $L \sim 11$ m and $h \sim 185$ m, this NABL is stable-sporadic, with continuous turbulence up to a 40-m height. A low level nocturnal jet typical of weak to moderate geostrophic winds on clear nights can also be seen (Wittich, 1991). Figure 3.5.2 also shows the various mean profiles from the SESAME 1979 experiment in the Great Plains (Lenschow *et al.,* 1988). A maximum in Ri_g appears to better define the NABL

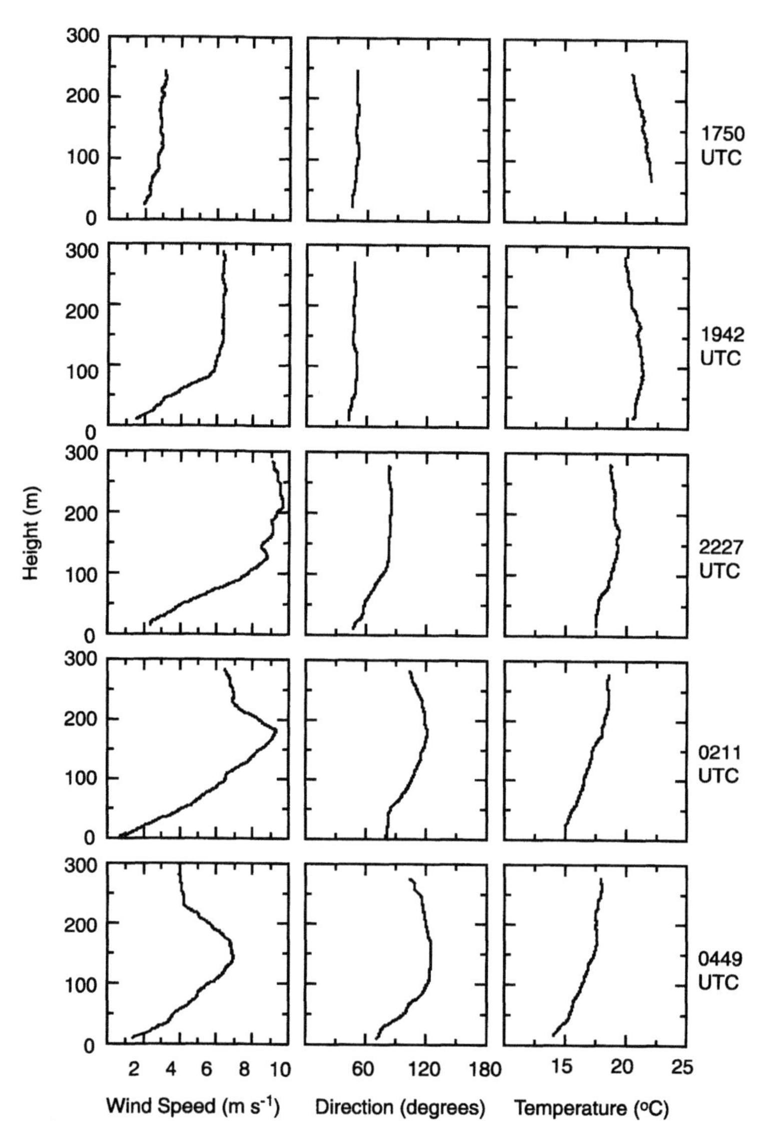

Figure 3.5.1. The evolution of the NABL at Sprakensehl. Wind speed and direction, and temperature show the formation of nocturnal inversion and its growth during the night (from Wittich 1991, with kind permission from Kluwer Academic Publishers).

top in both of these observations. Vigorous inertial oscillations were present at the top of the NABL and above in the residual layer (see Figure 3.5.8). Cooling due to turbulent heat flux divergence was maximum near the ground and zero at the top.

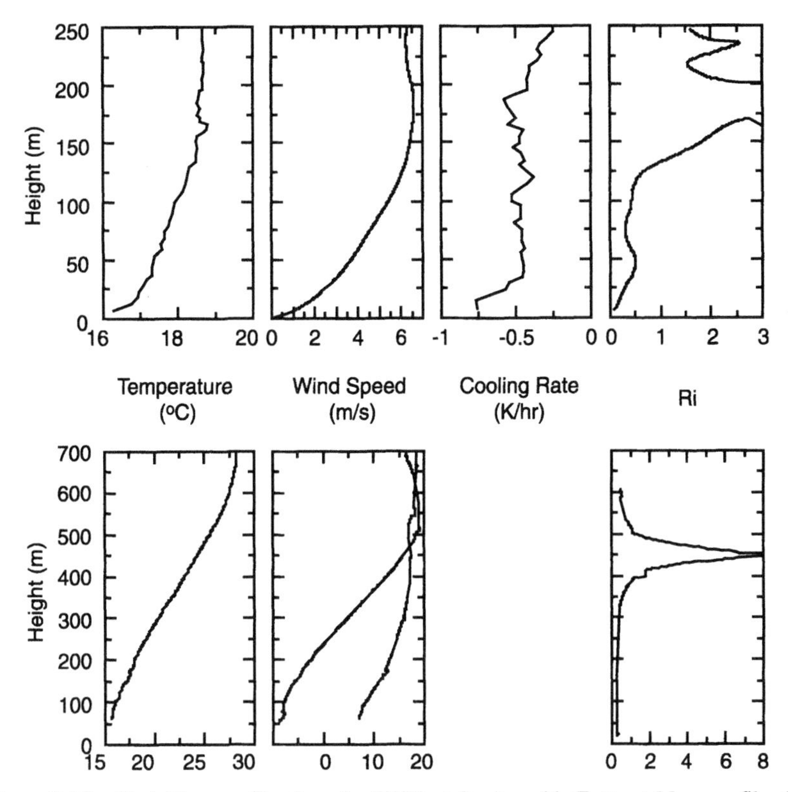

Figure 3.5.2. (Top) Mean profiles from the NABL at Sprakensehl. (Bottom) Mean profiles from SESAME 1979 observations (from Lenschow *et al.*, 1988 with kind permission from kluwer Academic Publishers).

Nieuwstadt (1984; see also Derbyshire, 1990, 1994) argued that in strongly stable boundary layers, the flux and gradient Richardson numbers are close to their critical values (around 0.20). This appears to be the case, especially when strong inversion develops and caps the NABL. Observations from Hunt *et al.* (1996) appear to support this.

Concept of local z-less scaling (Nieuwstadt, 1984; Sorbjan, 1993) applies to the NABL proper, while the M-O similarity applies to the surface layer itself. The reasoning is that under strong stable stratification conditions, the turbulence length scales are necessarily small (see Chapter 1) and independent of the distance from the surface. Therefore, the region well above the surface layer must be independent of the surface fluxes. Then the only relevant scales are the local values of u_{*l}, T_{*l}, and h_{*l}, where $u_{*l} = u_{*l}(z)$, $T_{*l} = \overline{w\theta}(z) / u_{*l}(z)$, and $h_{*l} = u_{*l}^2(z) / (\kappa\beta T_{*l}(z))$. Then all quantities scaled with these local values must become constants in the z-less turbulence state of the NABL. The values of these constants can be obtained by matching the NABL values with the observed surface layer values and therefore (Sorbjan, 1993)

$$\frac{\overline{w^2}}{u_{*l}^2} \sim 2.5, \quad \frac{\theta^2}{T_{*l}^2} \sim 6.0, \quad \frac{\overline{u^2}+\overline{v^2}}{u_{*l}^2} \sim 9.3;$$

$$\frac{\kappa h_{*l}}{u_{*l}} \frac{\partial}{\partial z}\left(U^2+V^2\right)^{1/2} \sim \frac{1}{\zeta}\left(1+\beta_1\zeta\right); \quad \beta_1 \sim 5 \tag{3.5.1}$$

$$\frac{\kappa h_{*l}}{T_{*l}} \frac{\partial \Theta}{\partial z} \sim \frac{1}{\zeta}\left(Pr_t^o+\beta_2\zeta\right); \quad Pr_t^o \sim 0.86, \beta_2 \sim 5$$

$$\frac{h_{*l}\varepsilon}{u_{*l}^3} \sim \frac{c_\varepsilon}{\zeta}\left(1+\alpha_1\zeta\right); \quad \alpha_1 \sim 3.7$$

$$\frac{h_{*l}\chi}{u_{*l}T_{*l}^2} \sim \frac{c_\chi}{\zeta}\left(Pr_t^o+\alpha_2\zeta\right); \quad \alpha_2 \sim 3.7$$

The values of c_ε and c_x depend on the stage of NABL development (Sorbjan, 1988a, 1993). For steady state, $c_\varepsilon = 1$. SESAME-79 data show $c_\varepsilon \sim 2.7$; Minnesota data show $c_\varepsilon \sim 1.25$. In general, the values for the various similarity constants for the NABL exhibit a much greater degree of variability than for the CABL, because of the sensitivity of its properties to a variety of factors. For example, even a gentle terrain slope can lead to significant nocturnal drainage flows and cause significant departures from flat terrain values. Also, the turbulence is often intermittent and coexists with wave motions. Under these conditions, asymptotic turbulence scaling laws, which assume high Reynolds number continuous turbulence, are of doubtful validity.

The above relationships cannot yield the variation of various quantities with distance above the ground, unless the variation of the scaling variables with height is known and this is usually quite difficult to model. Observationally,

$$\tau/\rho = u_*^2\left(1-\frac{z}{h}\right)^a$$

$$H = H_0\left(1-\frac{z}{h}\right)^b$$

so that

$$
\begin{aligned}
u_{*l} &= u_* \left(1-\frac{z}{h}\right)^{a/2} \\
T_{*l} &= \theta_* \left(1-\frac{z}{h}\right)^{b-\frac{a}{2}} \\
h_{*l} &= L\left(1-\frac{z}{h}\right)^{\frac{3a}{2}-b}
\end{aligned}
\tag{3.5.2}
$$

Sorbjan (1988a, 1993) points out that a must be less than b to avoid a singularity in the temperature profile at the top of the ABL. Generally, these coefficients differ widely among different data sets, because the state of turbulent mixing in a NABL depends on a variety of factors including surface inhomogeneities, terrain slope, baroclinicity, and the temporal evolution. Figure 3.5.3, from Wittich (1991), presents several data sets, including Sprakensehl data taken during the middle of the night, with a composite value of a ~ 1.5 and b ~ 2. According to Sorbjan (1993), Cabauw data (Nieuwstadt, 1984) in later stages of NABL development beginning 3 hr after sunset indicate a ~ 1 and b ~ 1, Yokoyama *et al.* (1979) data indicate b ~ 1–3, and Minnesota observations

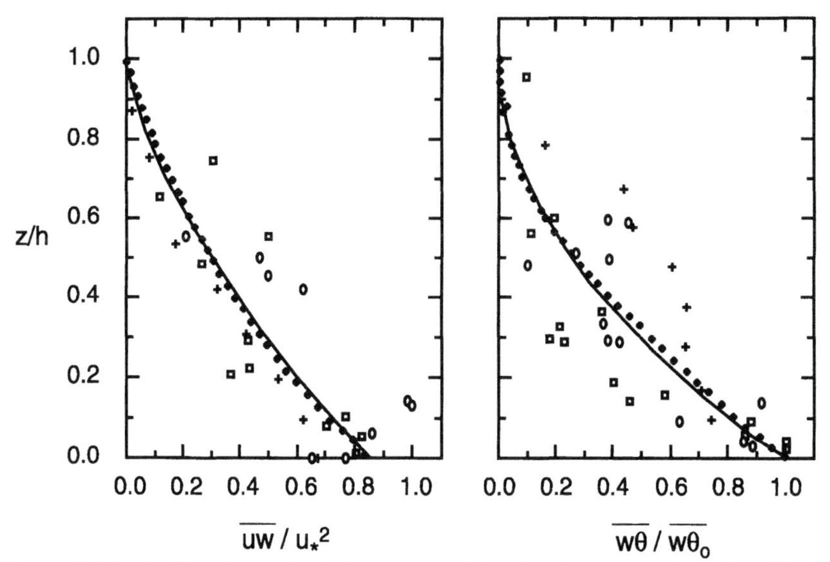

Figure 3.5.3. Profiles of normalized shear stress and heat flux from observations (from Wittich, K.-P., 1991 with kind permission from Kluwer Academic Publishers).

(Caughey *et al.,* 1979) indicate a ~ 2 and b ~ 3 near sunset. Using Minnesota values,

$$\frac{\overline{w^2}}{u_*^2} \sim 2.5\left(1-\frac{z}{h}\right)^2$$
$$\frac{\overline{\theta^2}}{\theta_*^2} \sim 6.0\left(1-\frac{z}{h}\right)^4 \tag{3.5.4}$$
$$\frac{h\varepsilon}{u_*^3} \sim 3.6\,(1+\alpha_1\zeta)\left(1-\frac{z}{h}\right)^3\left(\frac{z}{h}\right)^{-1}$$

These are compared with observations in Figure 3.5.4. SESAME-1979 data suggest (Sorbjan, 1988a; Lenschow *et al.,* 1988) a ~ 1.5 and b ~ 2, so that

$$u_{*l} = u_*\left(1-\frac{z}{h}\right)^{0.75}$$
$$T_{*l} = \theta_*\left(1-\frac{z}{h}\right)^{1.25} \tag{3.5.5}$$
$$h_{*l} = L\left(1-\frac{z}{h}\right)^{0.25}$$
$$\frac{\overline{u^2}}{u_{*l}^2} \sim 4, \quad \frac{\overline{v^2}}{u_{*l}^2} \sim 4.5, \quad \frac{\overline{w^2}}{u_{*l}^2} \sim 3.0, \quad \frac{\overline{\theta^2}}{T_{*l}^2} \sim 3.0$$

and therefore

$$\overline{u^2} \sim 4u_*^2\left(1-\frac{z}{h}\right)^{1.5}$$
$$\overline{v^2} \sim 4.5u_*^2\left(1-\frac{z}{h}\right)^{1.5}$$
$$\overline{w^2} \sim 3.0u_*^2\left(1-\frac{z}{h}\right)^{1.5} \tag{3.5.6}$$
$$\overline{\theta^2} \sim 3.0T_*^2\left(1-\frac{z}{h}\right)^{1.5}$$

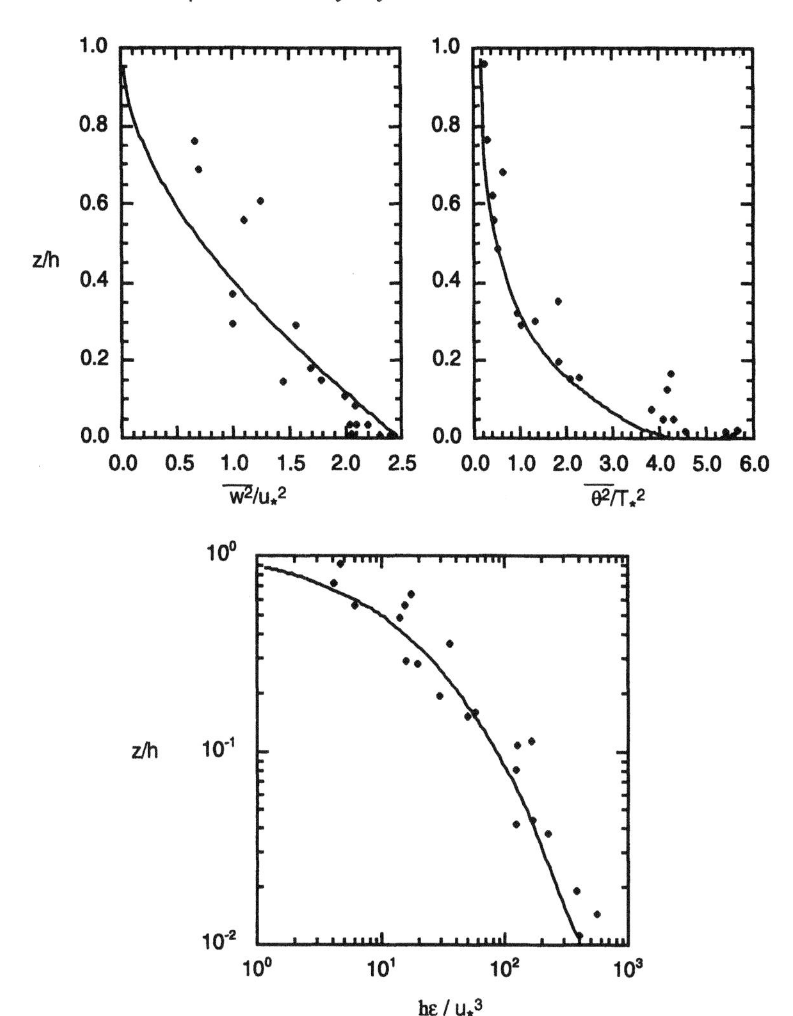

Figure 3.5.4. Profiles of vertical velocity variance (top left), temperature variance (top right) and dissipation rate (bottom) from Minnesota observations (from Caughey *et al.,* 1979).

These profiles are shown in Figure 3.5.5. Also

$$\frac{h_{*l}}{u_{*l}^3}\varepsilon \sim \frac{2.7}{\zeta}(1+3.7\zeta) \tag{3.5.7}$$

Local similarity applies to spectra and cospectra as well. For example, the

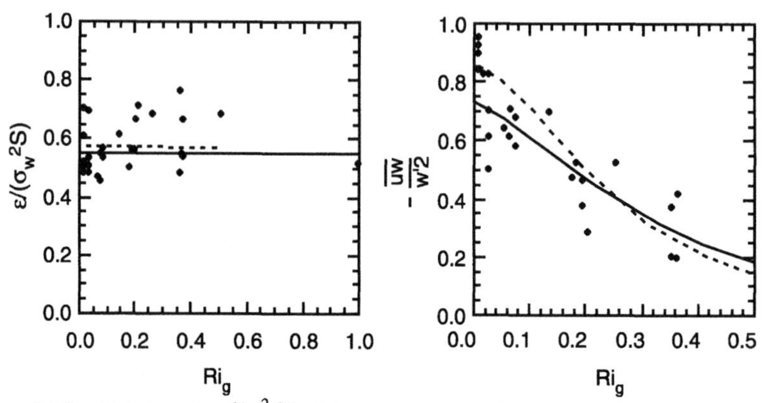

Figure 3.5.5. Variation of $\varepsilon/(\sigma_w^2 S)$ (left) and the Reynolds stress normalized by the vertical velocity variance (right) with gradient Richardson number Ri_g from Schumann (1996).

spectrum of the ith component of velocity is (Sorbjan, 1988a)

$$\begin{aligned} \frac{kS_i}{u_{*l}^2} &= \frac{\sigma_i^2}{u_{*l}^2}\frac{0.644\,\overline{z}/\overline{z}_m}{1+1.5\left(\overline{z}/\overline{z}_m\right)^{5/3}} \\ \overline{z}_m &= c_i\left(\frac{h_{*l}\varepsilon}{u_{*l}^3}\right)\left(\frac{\sigma_i^2}{u_{*l}^2}\right)^{-3/2}; \quad c_1 \sim 0.57, \quad c_{2,3} \sim 0.88 \end{aligned} \tag{3.5.8}$$

where $\overline{z} = z/\lambda$, and λ is the wavelength $(2\pi/k)$.

The height of the NABL can be defined based on several different criteria. The surface inversion layer depth is easily determined as the height at which dT/dz vanishes. However, the NABL depth can be defined as the height at which TKE falls to 5% of its surface value, the shear stress falls to 5% of the surface value, where the wind speed and Ri_g are maximum, or where the lapse rate vanishes or becomes adiabatic (Stull, 1988). Figure 3.4.2 shows the different possibilities. The height of the stationary NABL was first derived by Zilitinkevich (see Zilitinkevich, 1991) to be

$$h_e \sim c\left(\frac{u_* L}{f}\right)^{1/2}; \quad c \sim 0.4 = \kappa \tag{3.5.9}$$

This equilibrium depth is seldom reached. NABL is seldom in a steady state, but evolves slowly in time roughly according to (Nieuwstadt and Tennekes, 1981; Garratt, 1992)

$$\frac{\partial h}{\partial t} = \frac{h_e - h}{T_r} \tag{3.5.10}$$

where T_r is a relaxation timescale, which depends on the surface cooling rate

$$T_r \sim (\theta_h - \theta_s) / \left(\frac{\partial \theta_s}{\partial t} \right) \tag{3.5.11}$$

and ranges from a few hours early in the night to nearly half a day toward sunrise (Garratt, 1992). With a typical NABL depth of 100–200 m, this means that the rate of change of NABL height is typically a few m hr^{-1} to a few tens of m hr^{-1}. This implies that the NABL adjusts very slowly to a change in external conditions and the night period is generally too short for it to reach an equilibrium value (Nieuwstadt and Tennekes, 1981). This also means that the initial state of the NABL at the time of the sunset dominates its subsequent evolution during the night. This initial state is critically dependent on the characteristics of the decay of turbulence in the bulk of the CABL around sunset and is difficult to model.

Zilitinkevich's (1972) formula predicts that ($f\, h_e/u_*$) varies inversely with the square root of (u_*/fL). However, this behavior is by no means conclusively proven by observations. In fact, often the observations are contradictory. For example, sodar observations (Koracin and Berkowicz, 1988) of NABL heights indicate that ($f\, h/u_*$) is instead close to a constant (0.07). However, the difficulty of defining NABL height unambiguously and determining it by acoustic backscatter from small scale temperature fluctuations that provides the thickness of the layer with temperature fluctuations (not necessarily the same as the thickness determined by dynamical considerations) make this scaling of marginal utility.

The turbulence length scale under strong stable stratification typical of the NABL has been subject to considerable uncertainty. The basic question is whether the turbulence scale is determined by the strong shear that prevails or by the strong stable stratification. Hunt *et al.* (1988) suggest a bound on the length scale based on mean shear $S = dU/dz$,

$$\ell^{-1} = \frac{0.27}{z} + 0.45 \frac{S}{\sigma_w}; \; \varepsilon = \frac{\sigma_w^3}{\ell} \tag{3.5.12}$$

whereas Brost and Wyngaard (1978) suggested

$$\ell^{-1} = \frac{1}{z} + 0.6 \frac{N}{\sigma_w}; \; \varepsilon = \frac{q^3}{7.2\ell} \tag{3.5.13}$$

Both use the vertical velocity variance although the former is shear-based. The justification for shear-based scaling ignoring stratification is that this scale is smaller than the buoyancy scale and therefore it is the mean strain rate that sets the dissipation rate, not the stratification. DNS results of Coleman *et al.* (1992) found support for both. Based on their LES's of stably stratified flows, Schumann (1996) suggest that a form similar to Eq. (3.5.12) holds for $Ri_g < 1$, but without the z term and with the constant equal to 0.50. Figure 3.5.5 shows the variation of $\varepsilon/(\sigma_w^2 S)$ with Ri_g from their LES results and available measurements, and the variation of the vertical velocity variance with Ri_g.

The evolution of the NABL during the night is critically dependent on the relative magnitudes of turbulent and radiative cooling. Under calm conditions, radiative cooling predominates. It is also dominant near the ground and near the top under most conditions. With moderate to strong mixing, in the bulk of the NABL it is the cooling due to turbulent heat flux divergence that dominates. Averaged over the entire NABL, both are important. It is the influence of radiative cooling that leads to departures of the total heat flux profile from the linear one that would prevail if turbulent flux divergence were the only or predominant mechanism (as in CABL). This also has interesting consequences as far as the temperature profile is concerned (Garratt, 1992). With strong wind mixing, the profile tends to be more homogeneous (for example, the stable marine ABL) whereas the dominance of radiative cooling such as in Wangara data produces a negative curvature (see Figure 3.5.6).

The development of surface inversion is of importance to subsequent evolution of the daytime CABL. The height of this inversion depends very much,

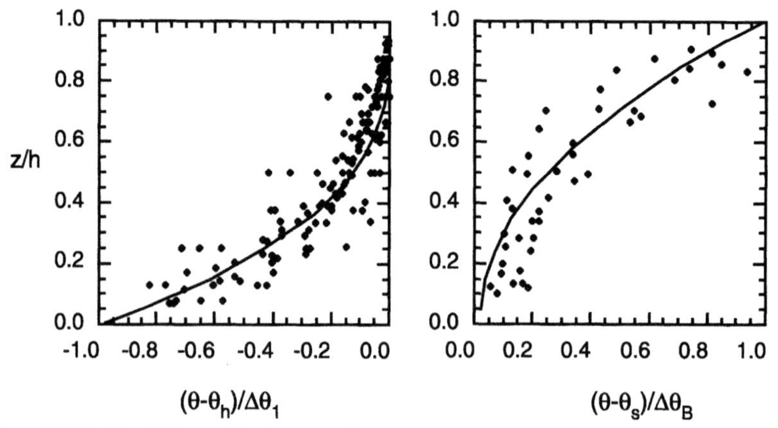

Figure 3.5.6. The different distributions of the temperature in the stable NABL (left) and a stable MABL (right) (from Garratt, 1992).

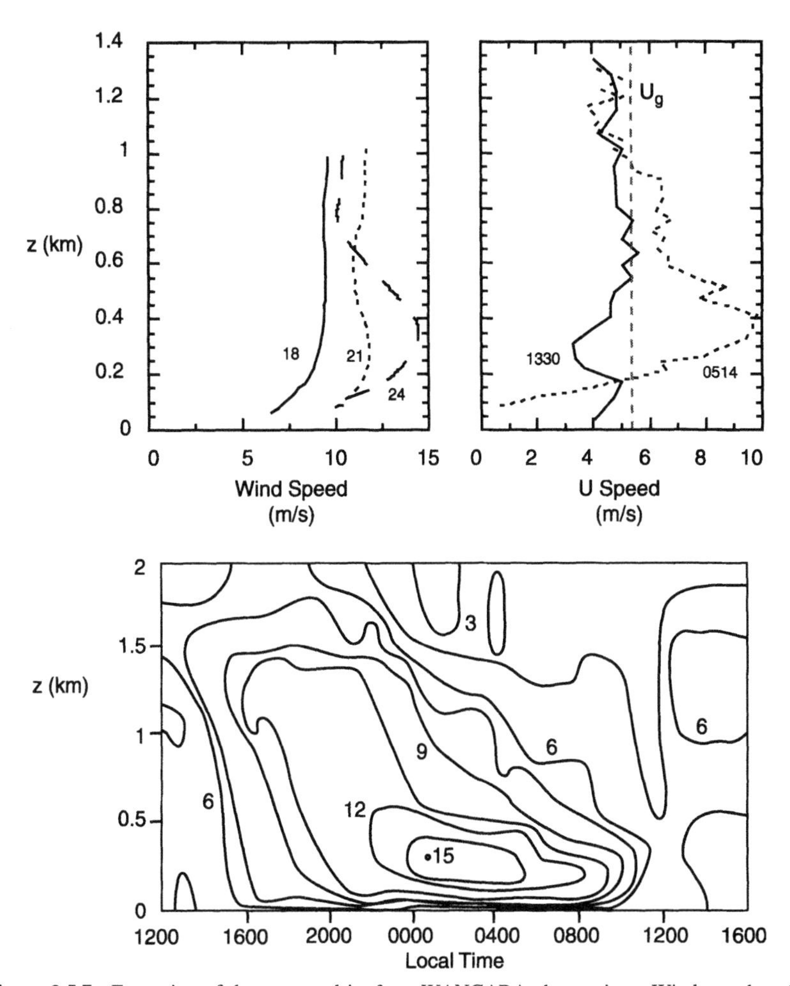

Figure 3.5.7. Formation of the nocturnal jet from WANGARA observations. Wind speed profiles on Day 13 (top left); profiles of the u-component of the wind velocity, with the x-axis along the geostrophic wind direction, for mid-afternoon and early morning (top right); and height-time cross-section of wind speed on days 13/14 (bottom) near Ascot, England (from Garratt 1992).

once again, on both radiative and turbulent cooling processes. During a cloud-free night, a surface inversion of a few hundred meters deep and 10–15°C strength can develop by sunrise (Garratt, 1992). Within this inversion exists the NABL, often with a low level nocturnal jet near the top of the NABL. The nocturnal jet is particularly pronounced during cloudless nights with moderate to strong geostrophic winds. Figure 3.5.7 shows the development of such a jet during Day 13 of WANGARA observations and near Ascot, England. The

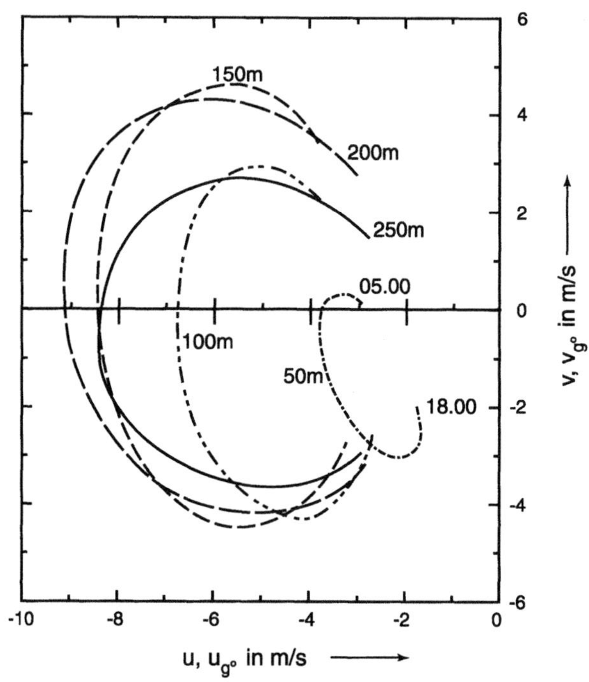

Figure 3.5.8. Strong inertial oscillations near and above the NABL shown by hodographs of the wind vector (from Wittich 1991, with kind permission from Kluwer Academic Publishers).

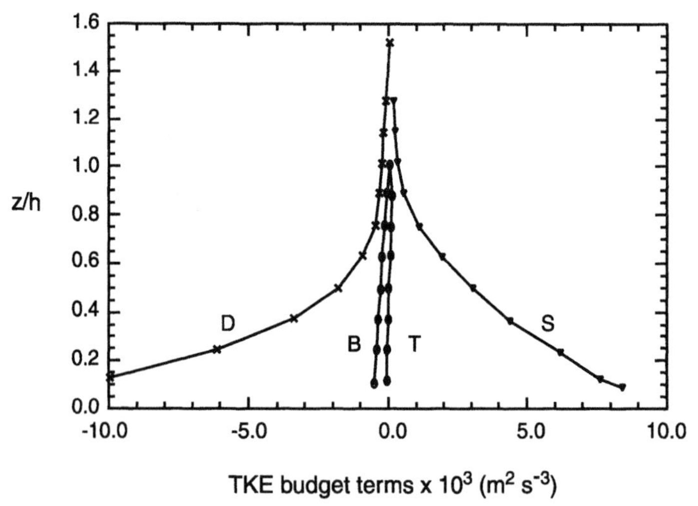

Figure 3.5.9. The TKE budget for the NABL. Note the near-balance between the shear production and dissipation, and the relative insignificance of turbulent transport (from Lenschow, *et al.*, 1988 with kind permission from Kluwer Academic Publishers).

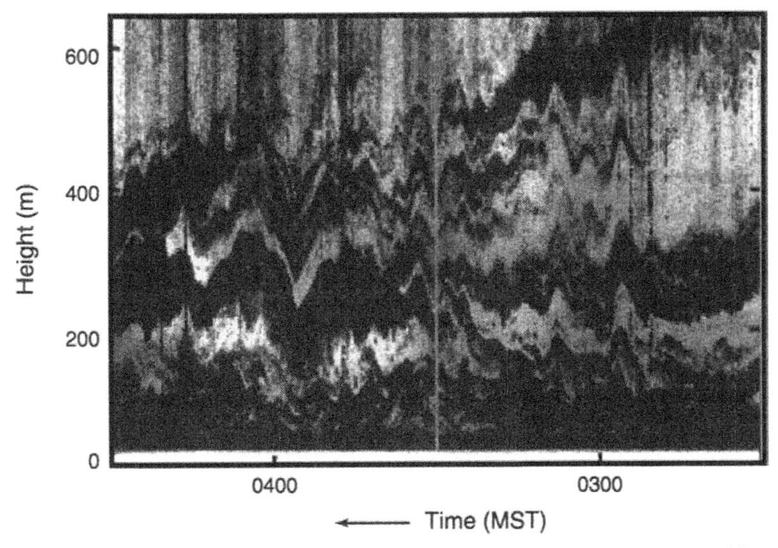

Figure 3.5.10. NABL structure including vigorous internal waves from sodar probing near Boulder, Colorado (from Hooke and Jones, 1986).

dynamics of this jet is closely tied to the inertial oscillations that develop with the cessation of turbulent mixing in the bulk of the CABL around sunset. During the day, the presence of turbulent friction makes the velocities in the CABL subgeostrophic, but the "sudden" removal of friction in the bulk of the CABL makes the pressure gradients tend to accelerate the winds toward the geostrophic value. The result is inertial oscillations and supergeostrophic winds near and above the top of the NABL (Figure 3.5.8).

The TKE budget in a typical NABL is of interest simply because its characteristics are a contrast to that in the convection-dominated CABL. Figure 3.5.9 shows that unlike the CABL, the shear production and dissipation very nearly balance and the turbulent transport is, for the most part, negligible. Contrast this to the CABL, where the turbulent transport is a dominant term, except in situations where shear generation dominates.

Also, due to the stable stratification that develops during the night, the top of the NABL is characterized by strong internal wave motions, and these are manifest in most remote sensing of the NABL. For example, acoustic probing of the NABL shows vigorous fluctuations in the NABL due to internal waves (see Figure 3.5.10. Finally, Figure 3.5.11 shows normalized spectra and cospectra for various quantities in the stable ABL (Caughey *et al.,* 1979). These spectral intensities are normalized using local scaling, and the frequency is normalized by the frequency of the spectral maximum. It is clear from these plots that the local z-less scaling applies very well indeed to the stable boundary layer.

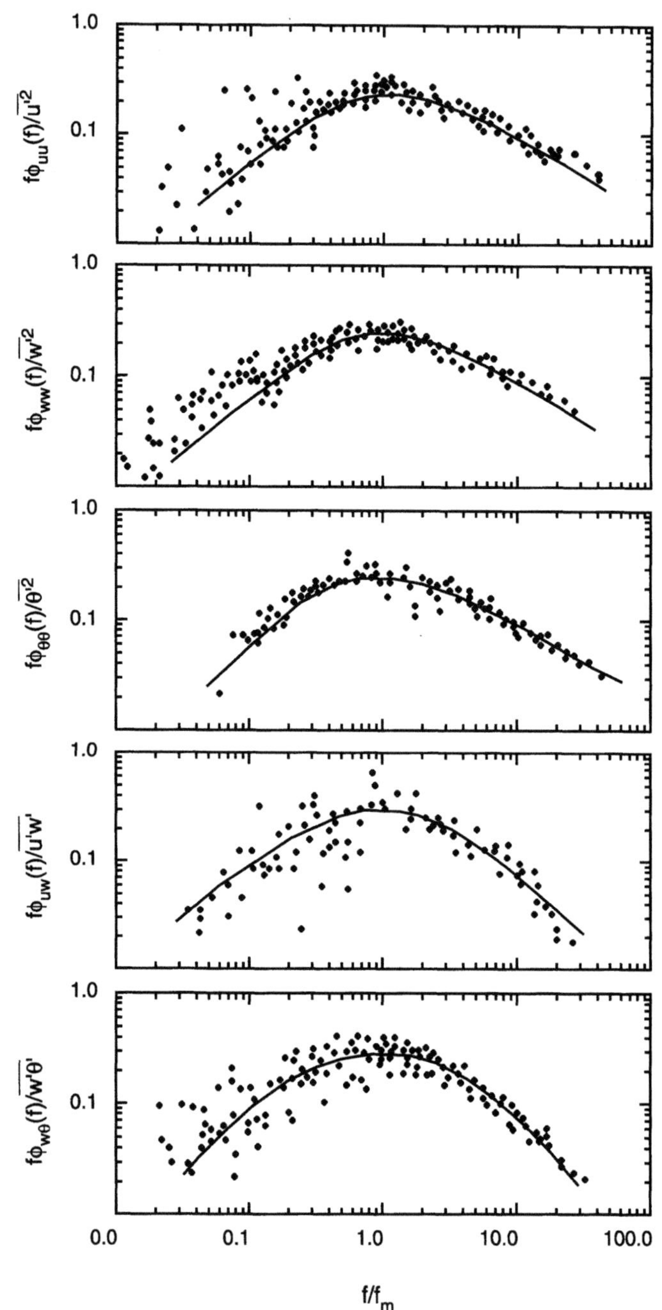

Figure 3.5.11. Normalized spectra (top three panels) and cospectra (bottom two panels) in the NABL (from Caughey *et al.*, 1979).

3.6 MARINE ATMOSPHERIC BOUNDARY LAYER (MABL)

The ABL over the oceans is significantly different from that over land. One important difference is the presence of the ocean below the MABL forming an essentially infinite source of water vapor. This means that the MABL is invariably humid and is often associated with clouds, usually either of cumulous or of stratocumulus variety. The roughness of the underlying ocean surface is itself determined by the MABL winds, which drive the surface wave motions that in turn determine the surface roughness. Also the MABL tends to be more horizontally homogeneous than the ABL over land, which necessarily has topographic variations on a variety of spatial scales, even in relatively flat terrain.

The diurnal variations in the SST and hence the properties of the MABL are much less pronounced because of the large heat capacity of the oceans. The heat capacity of the OML ($\rho c_p h$) is typically two orders of magnitude higher than that of the MABL (2.5 m of the water column is equivalent to the entire 10 km of the troposphere), and therefore the OML serves as a heat reservoir for most of the MABL over the global oceans. The SW radiation that manages to impinge upon the ocean surface at the bottom of the MABL, unlike in the ABL over land, does not take part in direct heating of the MABL. Instead, a significant fraction penetrates the ocean and heats up the upper few tens of meters and consequently contributes to the heat flux to the MABL only indirectly. The SST difference between day and night is typically less than 1°C, except in regions and periods of calm winds and high solar insolation. Most often, especially in the tropics, it is the latent heat flux from the ocean to the atmosphere that dominates the evolution and fate of the MABL. A similar situation exists during cold-air outbreaks on east sides of continents during winter, when cold, dry continental air masses flow over warm ocean waters and often cause explosive cyclogenesis.

Figure 3.6.1 shows aircraft observations of the coastal MABL over the Baltic Sea (Tjernstrom and Smedman, 1993). The summertime conditions under which the ocean temperatures were much cooler than the air temperatures over land produced a stratification that was slightly stable but close to neutral stratification, and the profiles of turbulence quantities scale well with the neutral stratification scaling of Townsend (1976), except near the MABL top, because of the presence of a low level jet there creating additional turbulence.

Oceans cover two-thirds of the surface area of the Earth; hence, the MABL is a major component of the Earth's climatic system and marine boundary layer clouds have an important effect on the Earth's radiation balance. Satellites show that mid- and low-latitude clouds cover a large part of the Earth's surface at any given time, and because of their large horizontal extent and high albedo, they

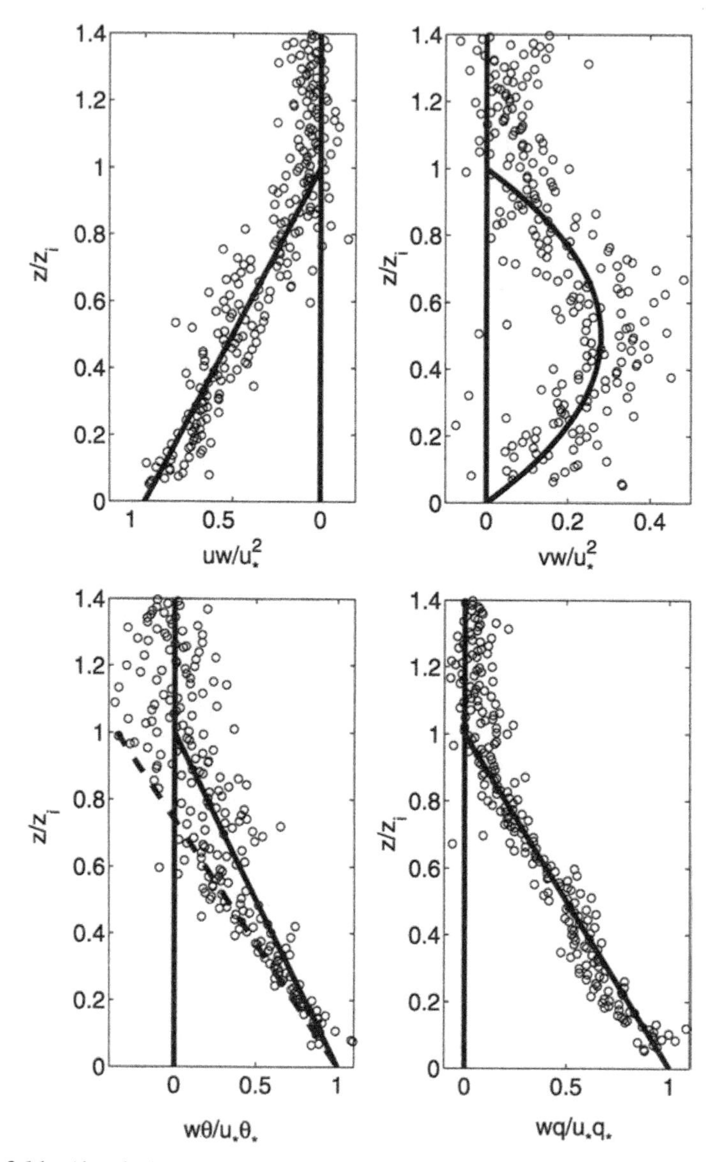

Figure 3.6.1. Aircraft observations of the turbulent quantities in the MABL over the Baltic Sea (from Tjernstrom and Smedman, 1993). Scaled turbulent fluxes of momentum, temperature, and water vapor plotted against normalized altitude are shown in the first four panels; scaled variances of the three wind component plotted against normalized altitude are shown in the last three panels.

play a very important role in the shortwave radiation budget (Randall *et al.*, 1984; Ramanathan *et al.*, 1989). Extensive stratocumulus cloud decks are

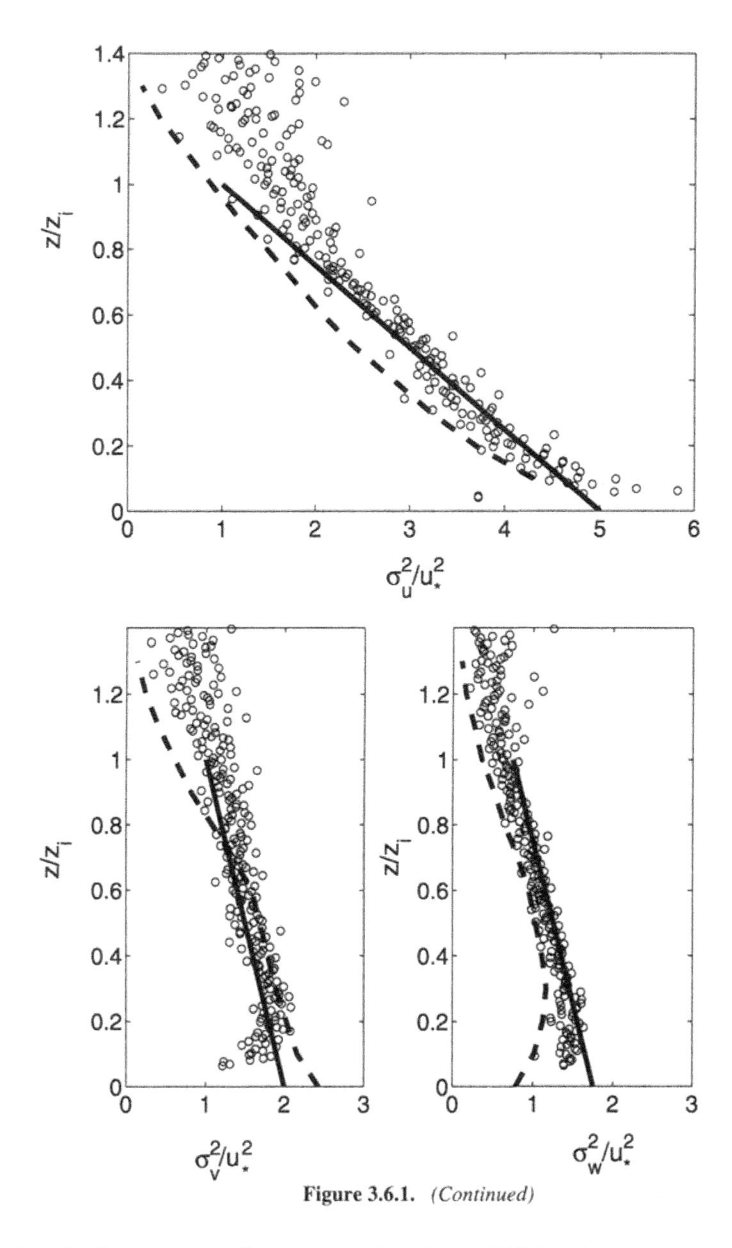

Figure 3.6.1. *(Continued)*

prevalent in the eastern and equatorward regions of high subsidence subtropical high pressure zones over cooler waters near western coasts of continents—for example, off the California coast. But as the atmospheric column encounters increasingly warmer sea surfaces and smaller subsidence, the planetary boundary layer (PBL) deepens, the cloud cover decreases drastically, and there is a transition from overcast stratus to scattered cumulus, with a corresponding change in the PBL structure and circulation from a well-mixed boundary layer to a decoupled two-layer structure with a cloud layer and a well-mixed subcloud layer. This meridional transition from midlatitude stratocumulus to tropical trade cumulus was the subject of extensive observations during the Atlantic Stratocumulus Transition Experiment (ASTEX) (Albrecht *et al.*, 1995) in the Azores area of the eastern Atlantic in 1992. Midlatitude stratocumulus itself was the subject of observations during the 1987 First International Satellite Cloud Climatology Regional Experiment (FIRE) (Albrecht *et al.*, 1988). During both field programs, extensive satellite, aircraft, and surface-based observations were made of PBL structure and a variety of parameters affecting clouds in the PBL.

It is estimated that marine stratocumulus clouds cover about 25% of the world ocean at any given time (Charleson *et al.*, 1987), and because they increase the local planetary albedo by 30–50%, they have a large effect on the shortwave radiation budget. On the other hand, the stratocumulus cloud top is only slightly cooler (a few degrees) than the ocean surface; they have a minimal impact on the longwave radiation budget (Kogan *et al.*, 1995; Miller and Albrecht, 1995). The net effect of MABL clouds is to cool the ocean surface, primarily by reflecting a larger portion of incoming SW solar radiation back out into space. It is estimated that the cloud radiative forcing, the difference between the planetary radiation balance with and without clouds, is typically 100 W m^{-2}, and an increase in cloud cover or cloud albedo of a few percent can offset the anthropogenic global warming. By the same token, decreases of similar magnitude can double the effect. Consequently, considerable effort has been devoted to understanding and accurately modeling the cloud–radiation–ocean feedback, since biases or errors of even a few percent in modeling the effect of clouds in climate models are intolerable.

The structure of stratocumulus clouds has been studied extensively (for example, Nicholls, 1984; Driedonks and Duynkerke, 1989; Paluch and Lenschow, 1991). Here the cloud-top longwave radiative cooling, confined to a few tens of meters near the cloud top, creates convective instability and drives convective mixing throughout the MABL, maintaining it in a well-mixed state. The cloud-top cooling rate depends on the cloud liquid water content and can be as high as 40 K hr^{-1}, although the usual rate is 10 K hr^{-1} (Rogers *et al.*, 1995). Convective flux at the sea surface also assists in mixing. This convection takes the form of cold, descending downdrafts compensated by warm, ascending updrafts. The turbulence in the boundary layer entrains warm, dry, stable air from above the cloud top into the boundary layer, which might lead to drying and hence thinning of the cloud layer. This mechanism could be responsible for the breakup of stratocumulus into scattered cumulus. The entrainment also weakens the convection driven by cloud-top cooling.

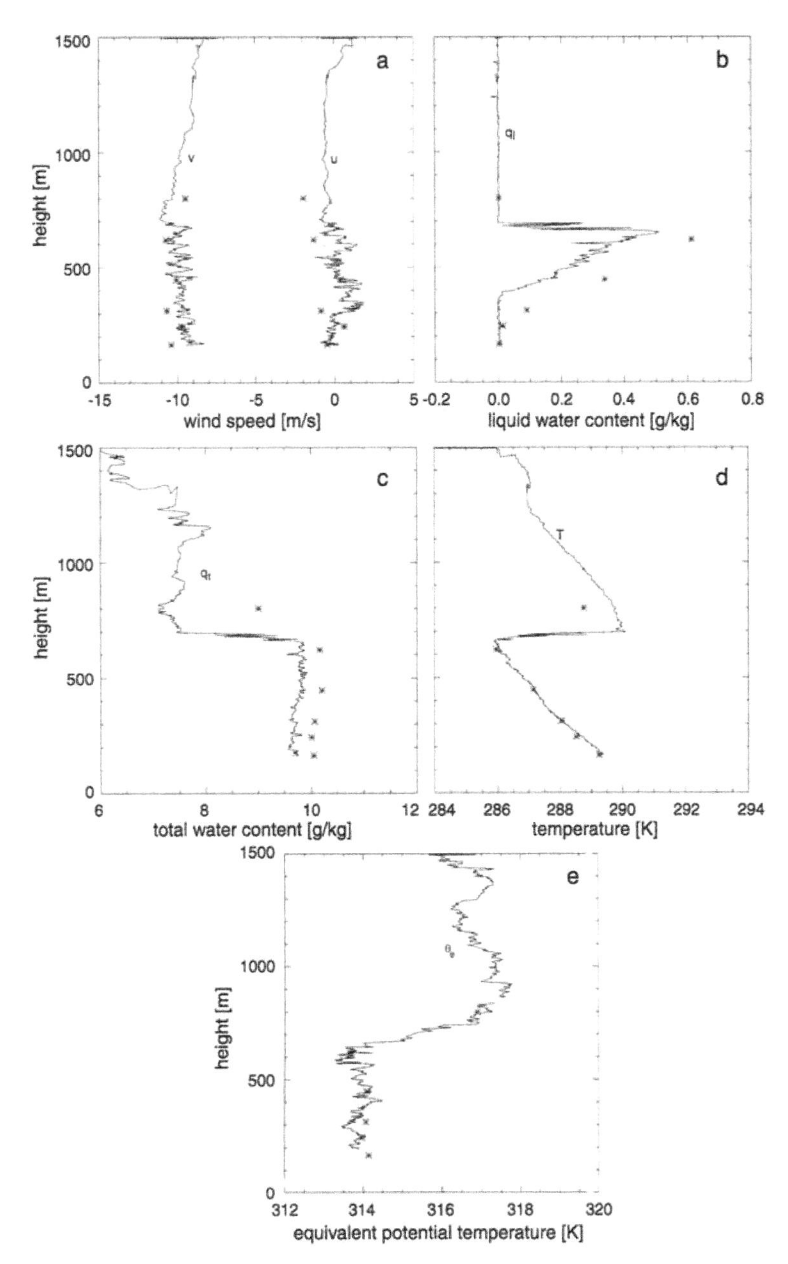

Figure 3.6.2. The structure of the nocturnal stratocumulus in the MABL observed during ASTEX (from Duynkerke *et al.,* 1995). (a) Horizontal wind speed, (b) liquid water content, (c) total water content, (d) temperature, and (e) equivalent potential temperature during run 1 (points) and profile P1 (line).

Figure 3.6.2 shows the structure of the nocturnal stratocumulus observed during ASTEX (Duynkerke *et al.*, 1995). Both profiles and horizontal averages measured from aircraft are shown. The cloud base was between 240 and 300 m, and the cloud top between 690 and 780 m. These profiles show very little variation in liquid water and equivalent potential temperature below the cloud top. The cloud top is clearly seen from a sharp decrease in liquid water content at the cloud top. There is also a strong jump in the equivalent potential temperature at the cloud top, but very little turbulence there. Figure 3.6.3 shows upwelling and downwelling longwave radiative fluxes. Two jumps in net longwave flux can be seen—one at the cloud top and one at the cloud base, the latter being much smaller than the former. The cloud-top jump of ~70 W m^{-2} leads to strong radiative cooling and vigorous convection, while the one at the cloud base of ~ −3 W m^{-2} causes heating. The net LW flux is zero in the cloud itself since the cloud acts as a black body. Duynlerke *et al.* (1995) found that turbulence quantities were best scaled by

$$w_{*s} = \left[2.5(g/T_0) \int_0^{D_M} \overline{w\theta_v} dz \right]^{1/3}$$

$$\theta_{v*} = w_{*s}^2 T_0 /(g D_M) \tag{3.6.1}$$

$$q_{l*} = (\overline{wq_v} + \overline{wq_l})\Big|_0 / w_{*s}$$

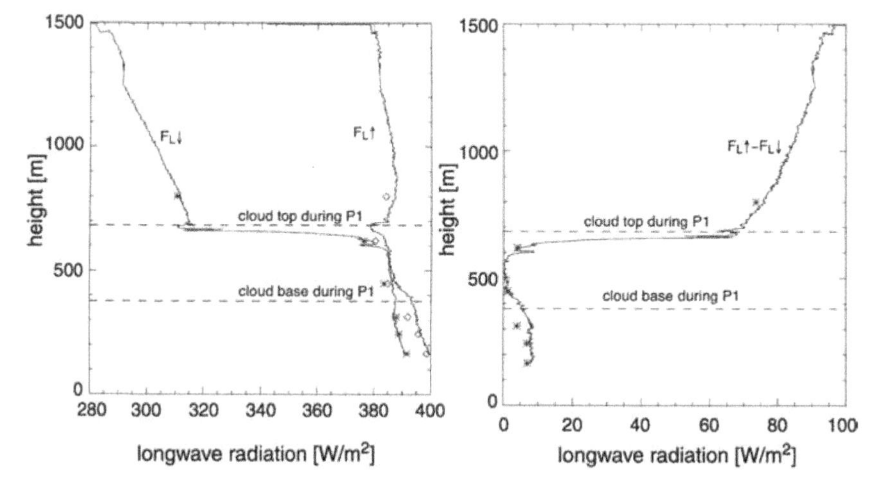

Figure 3.6.3. Downwelling and upwelling LW radiative fluxes observed in the stratocumulus-topped MABL during ASTEX (from Duynkerke *et al.*, 1995).

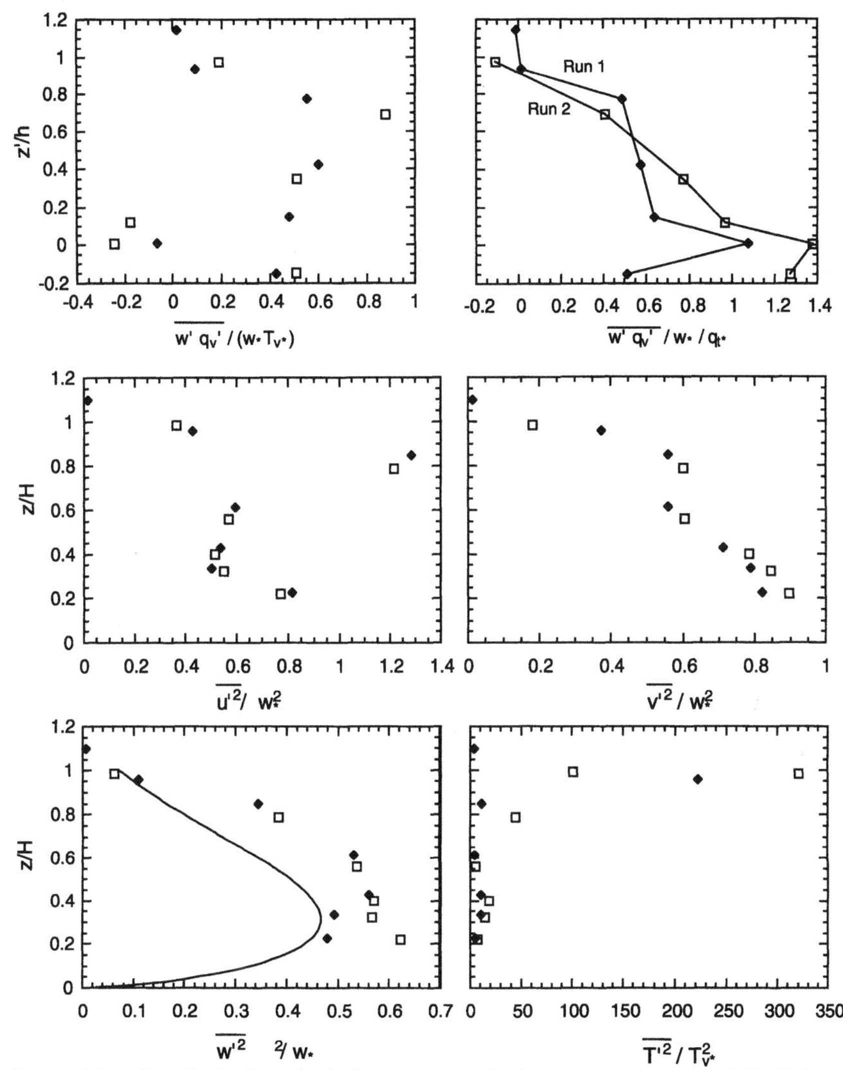

Figure 3.6.4. The distribution of turbulence quantities in the stratocumulus-topped MABL during ASTEX (from Duynkerke *et al.*, 1995). Different symbols indicate different runs.

where w_{*s} is the convective velocity scale, θ_{v*} is the virtual friction temperature, and q_{l*} is the friction liquid water mixing ratio; D_M is the ML depth (height of the cloud top). Figure 3.6.4 shows the distribution of some of the variances and fluxes normalized by the above scales (h is the cloud depth). Notice that the vertical velocity variance reaches a maximum near the cloud base and then decreases with height. It is zero above the cloud top. Instead of the maximum being in the upper part of the cloud, as would be the case if convection were driven mainly by cloud-top cooling, the maximum is at a location similar to observations over land. This is due to the fact that in this particular case, convection is being driven both by heat flux at the surface and by cooling at the top. Temperature variance reaches a maximum near the cloud top.

One difference between a dry CABL and a MABL is noteworthy. In the former, the buoyancy flux driving convection is at the bottom and positive, whereas entrainment is at the top, causing a negative stabilizing buoyancy flux. In a stratocumulus-topped MABL, both the negative entrainment buoyancy flux and the positive radiative cooling flux occur near the cloud top, the former balancing the latter somewhat. The net effect is a positive buoyancy flux (in this case about 20%), which drives convection in the MABL. The largest buoyancy fluxes are near the cloud top and at the surface; there is little buoyancy flux through the cloud base. The relative strength of these fluxes depends on the entrainment rate and the air–sea difference. For high entrainment rates, the resulting cloud-top radiative cooling-driven convection is weak; for low values, it is strong.

This simple situation is complicated by the effect of evaporation on entrained air in the cloud layer, the so-called cloud-top entrainment instability (Lilly, 1968; Randall, 1980; Deardorff, 1980b). It is presumed that the dry air from above entrained into the cloud layer by turbulent processes can be evaporatively cooled by mixing with the cloudy air, become virtually colder than the surrounding cloudy air, and hence reinforce convection driven by cloud-top cooling. It is this process that can lead to rapid dissipation of the cloud layer, since a positive feedback mechanism can exist under certain conditions. If the entrained air is sufficiently dry, then mixing and evaporative cooling strengthen turbulence, which in turn can cause a higher entrainment at the cloud top. However, observations do not support the Lilly–Randall–Deardorff hypothesis. Since this mechanism depends critically on the mixing of entrained air with the surrounding air, it is a strong function of the turbulence dynamics and cloud microphysics in the cloud layer. Strong preexisting turbulence is a necessary condition for this mechanism to operate. Cloud-top cooling is ultimately the driving mechanism for entrainment, and therefore the degree of such cooling might be a critical factor in the balance between the conflicting effects of entrainment per se and of the evaporative cooling of entrained air. These processes are hard to parameterize in numerical models, since they depend on small scale processes and cloud microphysics (Moeng *et al.,* 1996; Kogan *et al.,* 1995), and herein lies the major difficulty in accurate simulation of cloud-topped ABL, whether marine or continental.

During daytime, the situation is further complicated by shortwave radiative heating of the cloud, which stabilizes the cloud layer and tends to suppress

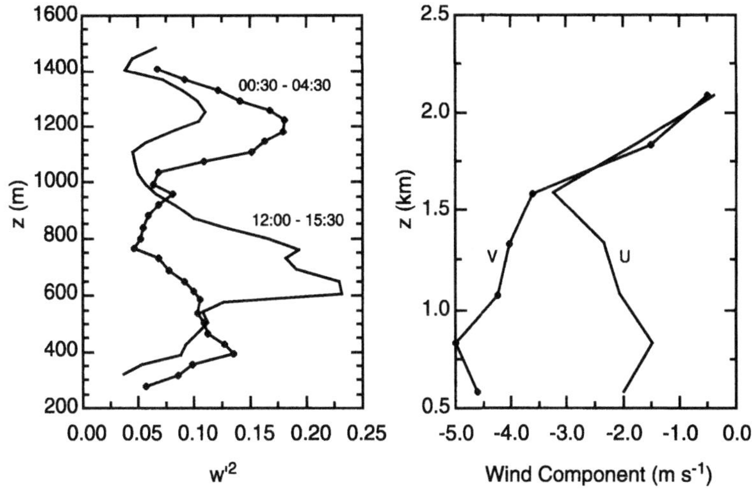

Figure 3.6.5. Doppler radar measurements of vertical velocity variances in the stratocumulus-topped MABL during ASTEX, showing the maximum in the upper part of the cloud layer for a daytime and nighttime case (left); wind profiler wind components for the daytime case from near cloud base through cloud top (right) (from Frisch *et al.*, 1995).

turbulence. The situation is somewhat akin to an OML, mixed from the top and stabilized by bulk heating due to penetrative SW radiation, the difference being the absence of entrainment at the top. Doppler radar measurements during ASTEX (Frisch *et al.*, 1995) showed that the vertical velocity variance had a distinct maximum in the upper part of the cloud during both day and night, but the daytime value was about half that of the nighttime one (Figure 3.6.5), illustrating the effect of daytime SW warming in the cloud layer. This SW heating often leads to a cloud layer that is decoupled from the subcloud layer. Decoupled cloud layers are commonly observed during the daytime. Nocturnal cloud-top cooling, however, manages to reinvigorate convection and hence mix the entire MABL.

During ASTEX, surface-based observations (Rogers *et al.*, 1995) showed a distinct surface-based well-mixed layer decoupled from the cloud base during the day by a layer which is stable to dry turbulent mixing. This layer is formed by SW heating of the cloud layer, which causes evaporation of the cloud from the base up. It isolates the cloud layer from the well-mixed layer below and restricts transfer of heat and moisture between the two. Cumulus clouds often form in this transition layer and were observed during ASTEX. The decoupling of the surface-based mixed layer from the stratocumulus layer by a dry transition layer causes moisture and heat buildup in the surface layer, lowering the lifting condensation level (LCL), allowing air parcels that overshoot the ML top to saturate, become unstable with respect to the surrounding dry air in the transition layer, and develop into cumulus clouds [see Figure 3.6.6, from Martin *et al.*

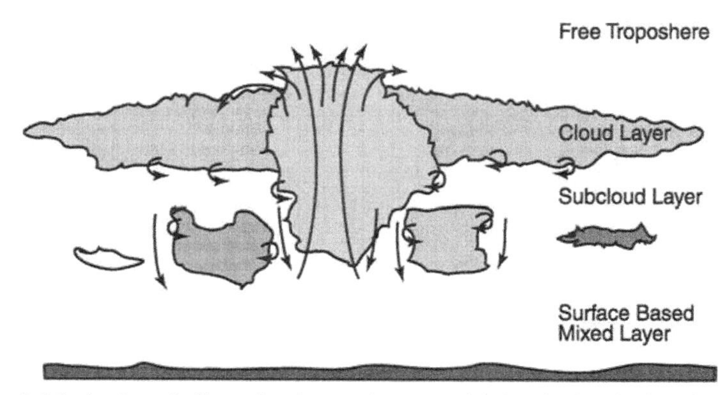

Figure 3.6.6. A schematic illustrating the cumulus convection that develops in the subcloud layer under the stratus.

(1995)]. Their tops may remain below the stratocumulus base, or as observed during ASTEX (Rogers *et al.,* 1995), can penetrate into the stratocumulus deck and occasionally through the capping inversion, thus providing a source of moisture that can offset the drying near the stratocumulus base due to SW heating and entrainment and thus help maintain the stratocumulus layer. Cumulus clouds beneath a stratocumulus layer are observed commonly in extratropical MABL, for example, during the Joint Air–Sea Interaction (JASIN) experiment in the northeast Atlantic in 1979. Rogers (1989) presents observations (Figure 3.6.7) during the Frontal Air–Sea Interaction Experiment (FASINEX) near Bermuda that show a two-cloud-layer structure. ASTEX observations clearly emphasized the importance of the cumulus underneath the stratocumulus in the thermodynamics and microphysics of the stratocumulus (Martin *et al.,* 1995; Rogers *et al.,* 1995).

Tropical oceans are dominated by cumulus convective clouds that show a large diurnal variability. These were the subject of intense observations during the 1992 TOGA/COARE observations in the western Pacific warm pool region (Webster and Lucas, 1992).

Finally, the two most important parameters that measure the effect of a cloud layer on the radiation balance and hence the climatic effect are the optical thickness and the droplet size, and fortunately both can be measured remotely. Nonabsorbing wavelengths (visible part of the spectrum) reflected by cloud tops contain information on optical thickness, whereas the absorbing wavelengths near infrared contain information on the droplet size. Observations during ASTEX confirmed the feasibility of satellite retrieval of these parameters (Platnick and Valero, 1995). This enlists satellite remote sensing techniques in routine monitoring and developing a better understanding of the role of clouds in the Earth's radiation balance.

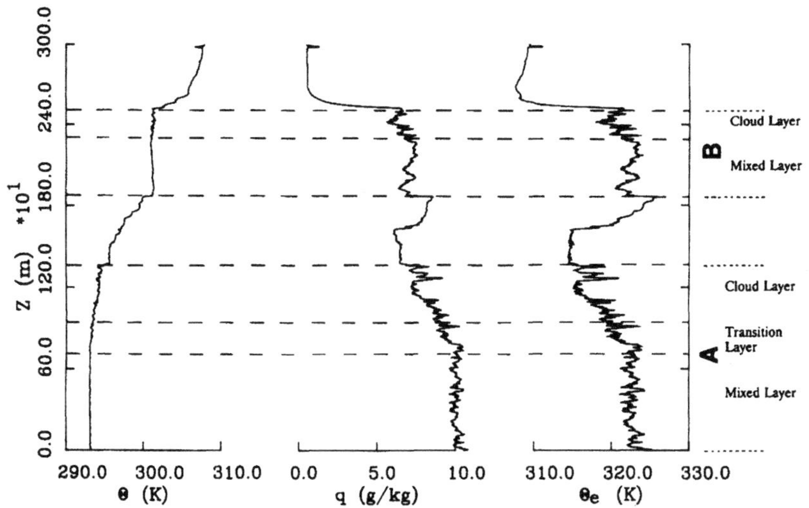

Figure 3.6.7. Aircraft observations during FASINEX showing decoupled cloud layers in the MABL (from Rogers, 1989).

Since the MABL is often associated with clouds, it is more appropriately discussed under cloud-topped atmospheric boundary layers (CTBLs), with appropriate allowances made for distinctions between the ABL over land and that over water. The CTBL is the topic of the next section.

3.7 CLOUD-TOPPED ATMOSPHERIC BOUNDARY LAYER (CTBL)

The presence of clouds introduces additional complexities into both the dynamics and the thermodynamics of the ABL. There are two primary influences. Firstly, the presence of clouds alters the radiative fluxes in the ABL because of the absorption of the SW component by liquid water droplets in the cloud and the change in LW emission characteristics due to the presence of the cloud. Clouds also change the albedo at the top of the ABL. Cloud-top radiative cooling often provides an additional TKE source so that a cloud-topped ABL can be mixed from both top and bottom. Radiative flux divergence associated with clouds constitutes internal sources and sinks of heat that are often more important than the surface fluxes to the dynamics of the ABL. Secondly, the phase conversions involved in the formation and dissipation of clouds in the ABL imply another set of large internal heat sources and sinks in the ABL (due to latent heat of vaporization released during condensation and absorbed during evaporation), and this also has profound effects on ABL dynamics. Overall, the complications introduced are quite immense and often intractable. Yet, given the importance of the CTBL (particularly over the ocean) to issues such as global warming (Ramanathan *et al.,* 1989), it is imperative that these complexities be dissected, studied, understood, and modeled. Because the subject is complex enough to require a dedicated lifetime of study, all we can do here is to borrow heavily from summaries and reviews existing in literature. We depend heavily upon Garratt (1992), Stull (1988), and Sorbjan (1989) for the following material that forms a highly condensed discussion in the context of small scale geophysical processes. For details, we refer the reader to these original references and recent issues of journals such as the *Journal of Atmospheric Sciences,* the *Journal of Applied Meteorology,* and the *Journal of Geophysical Research.*

The clouds near the top of a CTBL can be stratus, stratocumulus, or cumulous. Of these, the stratocumulus clouds generated by convective processes, with a thickness of a few to several hundred meters, are the most important from many considerations. For example, extensive regions of the MABL are covered by stratocumulus clouds and hence are very important to climatic aspects of air–sea exchange. Cumulous clouds are more characteristic of tropics and extratropics and do play a very prominent role in air–sea exchanges there.

When faced with the possibility of phase conversions, it is essential to keep track of the liquid water content q_l in a parcel of air. The virtual potential temperature of an air parcel containing liquid water can be written as

$$\theta_v = \theta\left[1 + 0.61 r - r_l\right] \tag{3.7.1}$$

The equivalent potential temperature, the temperature a parcel of moist air would attain if allowed to ascend until all its water vapor content condensed out and then descended along the dry adiabatic line to sea level, is of considerable importance thermodynamically:

$$\theta_l = \theta + \frac{L}{c_p}\frac{\theta}{T} r \sim \theta + \frac{L}{c_p} r \tag{3.7.2}$$

The liquid water potential temperature is

$$\theta_l = \theta - \frac{L}{c_p}\frac{\theta}{T} \cdot r_l \sim \theta - \frac{L}{c_p} r_l \tag{3.7.3}$$

and the total water mixing ratio is the sum of the water vapor mixing ratio and the liquid water mixing ratio:

$$r_T = r + r_l \tag{3.7.4}$$

Inside a cloud, because of the phase conversions and the associated heat transfer, potential temperature and water vapor mixing ratio are not conserved; instead, the equivalent potential temperature, the liquid water potential temperature, and the total water mixing ratio are conserved quantities. The level at which a rising, unsaturated air parcel attains saturation is known as the local or lifting condensation level (LCL) and defines the cloud base (Figure 3.7.1). If the parcel intrudes above this point, the water vapor in the parcel begins to condense and the resulting latent heat release can make the parcel more buoyant. The level at which the parcel first becomes more buoyant than its surroundings is known as the level of free convection (LFC). The parcel continues to rise until its potential temperature becomes equal to the surrounding air masses. This level is the limit of convection (LOC). The cloud layer extends mostly between the LCL and LOC, although the inertia of air parcels makes them overshoot the LOC and therefore the cloud top is usually above the LOC. The most important energetic quantity for cloud convection is the convective available potential energy (CAPE), which is the buoyancy of a parcel of air integrated between the LFC and the LOC. The CAPE defines a velocity scale that determines the upward vertical velocity of a parcel at the LOC and hence the cloud-top level:

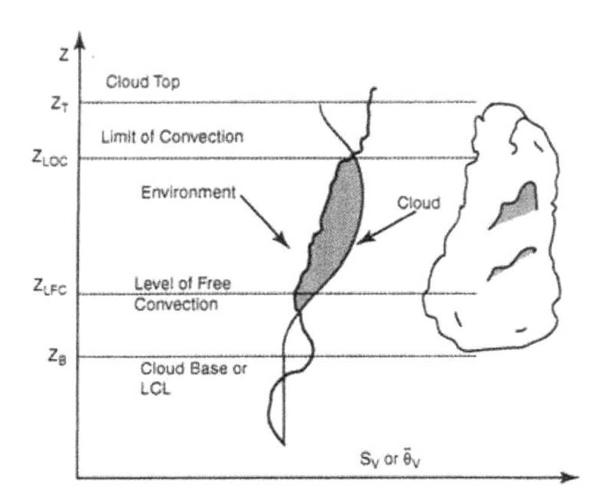

Figure 3.7.1. A sketch illustrating the various levels associated with a cloud layer.

$$\text{CAPE} = \int_{z_{LFC}}^{z_{LOC}} \frac{g}{\theta} \Delta\theta_v(z)\, dz \tag{3.7.5}$$

Evaporative available potential energy (EAPE) applies, on the other hand, to a sinking cloud parcel and is associated with the evaporation of liquid water in the parcel that makes it heavier than the surroundings and sink toward the bottom. This is important for cloud-top entrainment processes.

Clouds affect both SW and LW radiative fluxes through them. Exactly how this comes about is a complex function of the cloud liquid water content, the cloud temperature, the cloud liquid water droplet distribution, and many other factors that are not well understood. Figure 3.7.2 shows the SW and LW radiative fluxes through an idealized stratocumulus cloud layer (Nichols, 1984; Stull, 1988) and the resulting net radiative fluxes and the net flux divergence (the heating rate). It is clear that although the largest SW radiative heating in the cloud is near the top, the entire cloud layer gets heated with the heating rate that decreases exponentially. The extinction scale is typically 50–150 m and is very much dependent on the liquid water path defined as

$$w_p = \int_{z_B}^{z_T} \rho_a\, r_l\, dz \tag{3.7.6}$$

so that the SW radiative flux can be written as

$$I(z) = I_T - \left(I_T^S - I_B^S\right) \frac{1 - \exp\left[\dfrac{-(z_T - z)}{\lambda}\right]}{1 - \exp\left[\dfrac{-(z_T - z_B)}{\lambda}\right]} \tag{3.7.7}$$

where

$$\lambda \sim 15\, w_p^{1/3} \tag{3.7.8}$$

On the other hand, most of the LW radiative flux divergence occurs within a few tens of meters of the cloud top and cloud base and can be parameterized as (Stull, 1988)

$$\Delta S_T^L = \frac{\sigma}{\rho_a c_{p_a}} \left[T_{\text{CLOUD TOP}}^4 - T_{\text{SKY}}^4\right]$$
$$\Delta S_B^L = \frac{\sigma}{\rho_a c_{p_a}} \left[T_{\text{SURFACE}}^4 - T_{\text{CLOUD BASE}}^4\right] \tag{3.7.9}$$

Figure 3.7.3 shows measured profiles of the upward and downward SW fluxes in a stratocumulus-topped MABL. The observations indicate that the cloud reflects 550 W m^{-2}, about 70% of the 800 W m^{-2} of the impinging SW

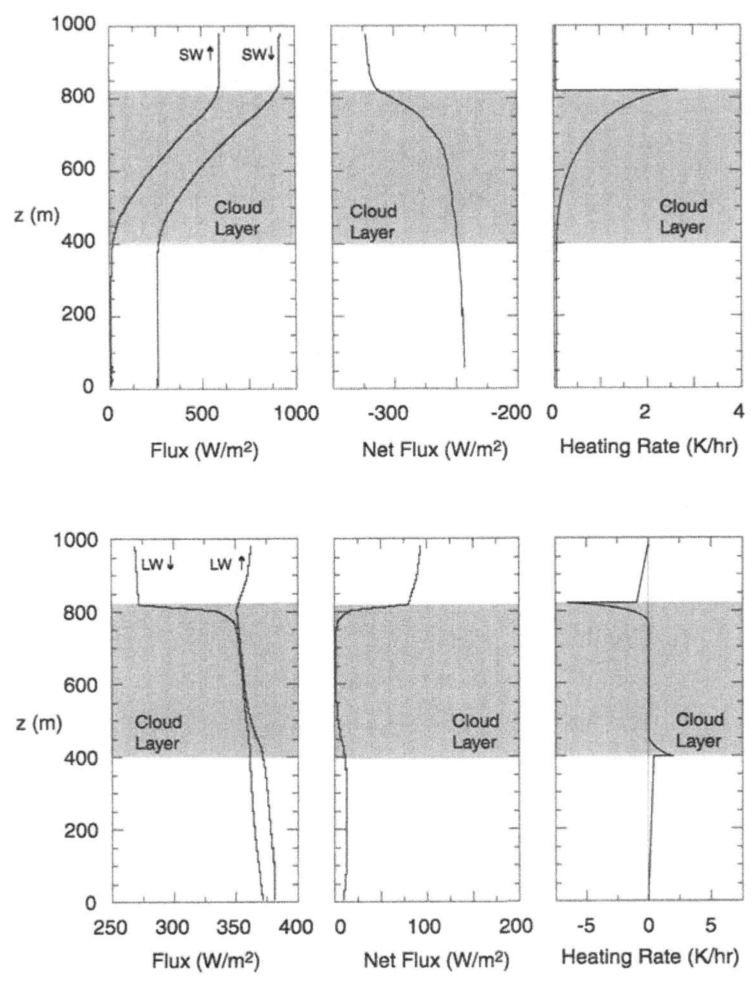

Figure 3.7.2. Modeled distribution of SW (top panels) and LW (bottom panels) radiative fluxes and the resulting heating rates in a CTBL. Left panels show upward and downward fluxes, middle panels show net flux, and right panels shows the heating rate inferred from the net flux profile (from Nicholls, 1984; Hanson, 1987; and Nicholls and Leighton, 1986).

radiation (albedo of a stratocumulus cloud is typically 0.6–0.7), and absorbs about 60 W m^{-2} (in a 400-m thick layer), with about 210 W m^{-2} reaching the sea surface, out of which about 10%, 20 W m^{-2}, is reflected by the sea surface back into the MABL. Clearly, the cloud albedo is the most important influence of the clouds vis-à-vis a cloudless ABL. The net shortwave flux at the sea surface is reduced to 210 W m^{-2}, compared to a cloud-free value of 800 W m^{-2}. Most of

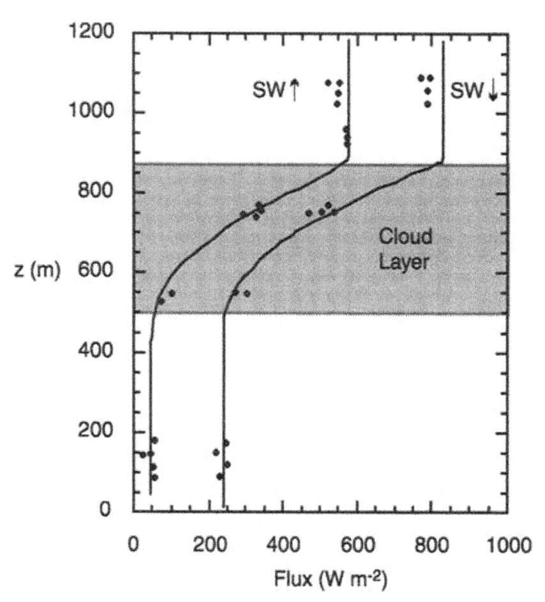

Figure 3.7.3. Profiles of observed (circles) upward and downward SW radiative fluxes and theoretical (dashed lines) in the stratocumulus-topped CTBL (from Slingo *et al.,* 1982a).

this heats the OML. The net radiative heating of the MABL is quite small, about 80 W m^{-2}, mostly inside the cloud layer.

Figure 3.7.4 shows the observed LW radiative fluxes for a strato-cumulus-topped MABL and an ABL over land. Most of the radiative cooling (~60 W m^{-2}) occurs in a layer near the cloud top a few tens of meters thick. This is equivalent to a rate of cooling of 5–10°C hr^{-1} of this layer and can drive cloud-top convective processes.

The CTBL structure is determined by the turbulent motions, radiative fluxes, and cloud physics. Turbulence in the CTBL can be due to convective heating at the bottom as in a dry CABL, the intensity of which is itself determined by the extent and thickness of cloud cover, since the SW radiation that penetrates to the ground (or the sea surface) is determined by the cloud albedo and the absorption in the cloud layer. The turbulence is also generated by shear at the surface as well as near the top as in a dry CABL. An additional mechanism for turbulence generation in a CTBL is cloud-top radiative cooling, which, if sufficiently strong enough to overcome the SW radiative heating near the top, gives rise to convective motions near the top of the CTBL and entrainment of warmer, drier air from above into the cloud and the CTBL. The evaporation of cloud water and the resulting evaporative cooling can make the parcel of entrained air heavier and sink further down into the cloud. This cloud-top entrainment instability can

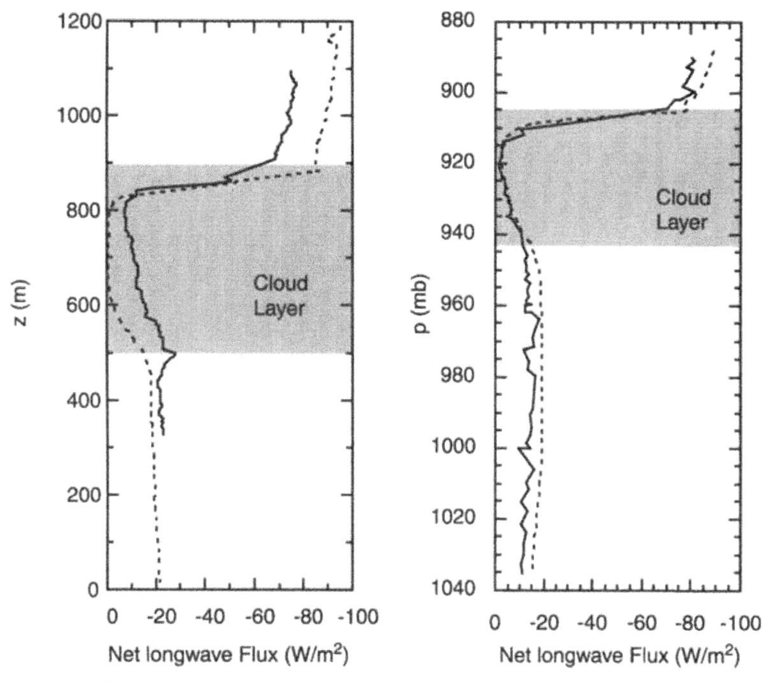

Figure 3.7.4. Observed (solid line) and theoretical (dashed line) LW radiative fluxes for (a) cloud topped marine (from Slingo *et al.*, 1982a) and (b) continental ABL (from Slingo *et al.*, 1982b) based on aircraft observations.

result in vigorous mixing of the drier air aloft and the moist air in the cloud and lead to a rapid dissipation of the cloud layer. The condition for its onset is that the difference between the virtual potential temperature just above the cloud top and that just below be less than a critical value (Deardorff, 1980b; Randall, 1980; Stull, 1988). Figure 3.7.5 shows observations in a stratocumulus-topped ABL driven by cloud-top radiative cooling. The data pertain to both coupled and decoupled cloud layers. Coupled cloud layers occur when the turbulent mixing, driven by either the surface convective fluxes (or shear) or convection due to cloud-top radiative cooling (or both), is energetic enough to couple the cloud layer dynamically with the rest of the ABL. Often, in the absence of strong fluxes or shear driving the mixing in the ABL, strong radiative cooling can constitute a strong enough source of TKE to mix the CTBL from the top all the way to the surface. On the other hand, when the mixing is weak, the cloud layer can be decoupled as in Figure 3.7.6, which shows a decoupled cloud layer ~ 100 m thick underneath a strong inversion, observed during JASIN. Mixing extends several hundred meters below the cloud base, but is not strong enough to be coupled to the 250-m mixed layer near the surface.

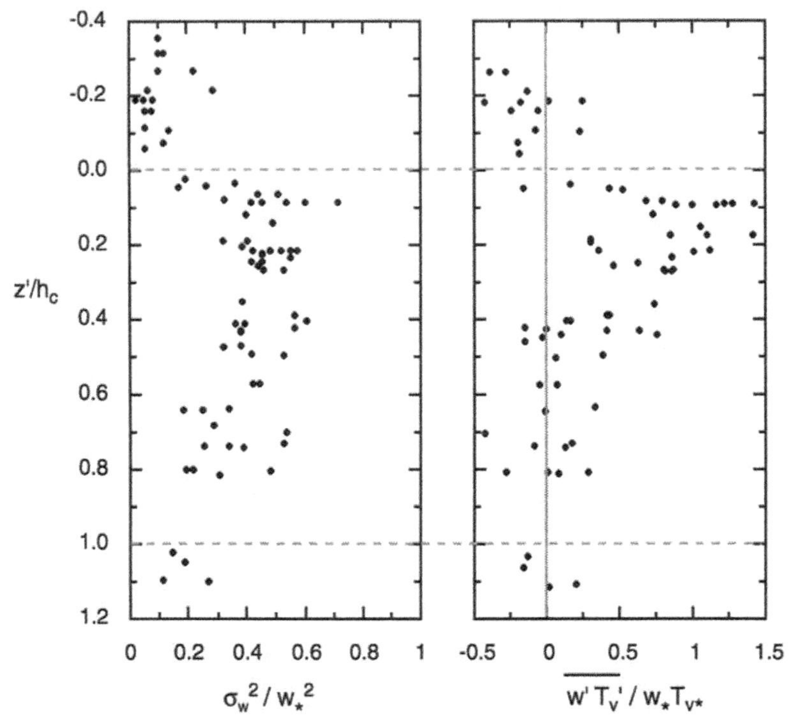

Figure 3.7.5. Normalized profiles of total water flux (left) and buoyancy flux (right) for a CTBL driven by cloud-top convective cooling from aircraft observations (from Nicholls, 1989).

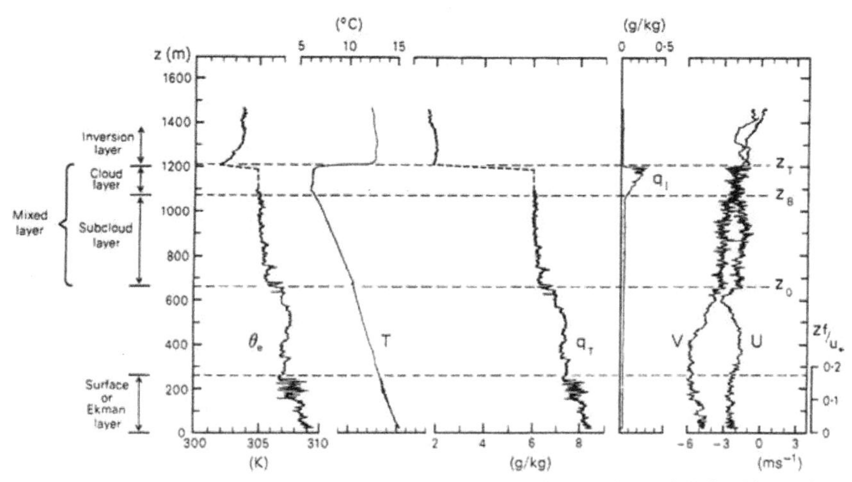

Figure 3.7.6. Properties in a stratocumulus-topped CTBL showing the decoupled cloud layer (from Nicholls and Leighton, 1986).

To sum up, the CTBL is endowed with a richer variety of physical and dynamic processes (compared to the cloudless ABL) that makes it a fascinating as well as a frustrating subject to study and understand. Given the impact of the clouds and their albedo on the climate of Earth, it is inevitable that more attention will be given to the CTBL in the coming decades. Systematic measurements of clouds and their related properties in the past few years (for example, Westphal *et al.*, 1996) under large programs such as FIRE and ASTEX are beginning to provide us with a means to unravel the complexities of processes in a CTBL.

3.8 FLOW OVER TOPOGRAPHIC CHANGES—DOWNSLOPE WINDS

On the 11th of January 1972, Boulder, Colorado, experienced a windstorm (Lilly and Zipser, 1972; Klemp and Lily, 1975) resulting from the generation of severe downslope winds by the atmospheric flow over the Rocky mountain range, which forms a landmark in the study of stratified flow over mountains. This Boulder windstorm, in which the entire tropospheric flow was plunging down and passing over Boulder in a layer a mere two kilometers thick, was the first severe downslope wind event to be well observed by aircraft and other means, and well documented. Figure 3.8.1 shows a cross-section of the flow over the Rockies that day which illustrates dramatically the flow acceleration in the lee of the continental divide. The resulting winds exceeding 40 m s^{-1} caused extensive damage in their wake. Similar wind events occurred in Boulder during 1998 and 1999 as well.

Severe downslope windstorms are quite common at many mountainous sites around the world. The Yugoslavian Bora, a severe northeasterly wind flowing over the Dinaric Alps along the Yugoslavian Adriatic coast, has been extensively studied during the ALPEX program (Smith, 1987). During early 1982, several extended periods of bora enabled extensive aircraft observations of the atmospheric structure to be made. For the first time, extensive turbulence measurements were made possible, and the strong turbulence fields on the downslope side of the Alps were documented (Smith, 1987). Other field programs such as PYREX have fostered a better understanding of flows over mountainous terrain (Bougeault *et al.*, 1990).

Flow of stratified atmospheric column over mountain ranges has always fascinated many due to the spectacular internal lee waves that are formed on the downstream side (Smith, 1976). These are often made visible by condensation of the moisture in the air in the up-flow part of the waves that develops into spectacular cloud formations. Satellite photographs often show extensive trapped lee waves in the vicinity of mountain ranges. Wurtele *et al.* (1996) present a

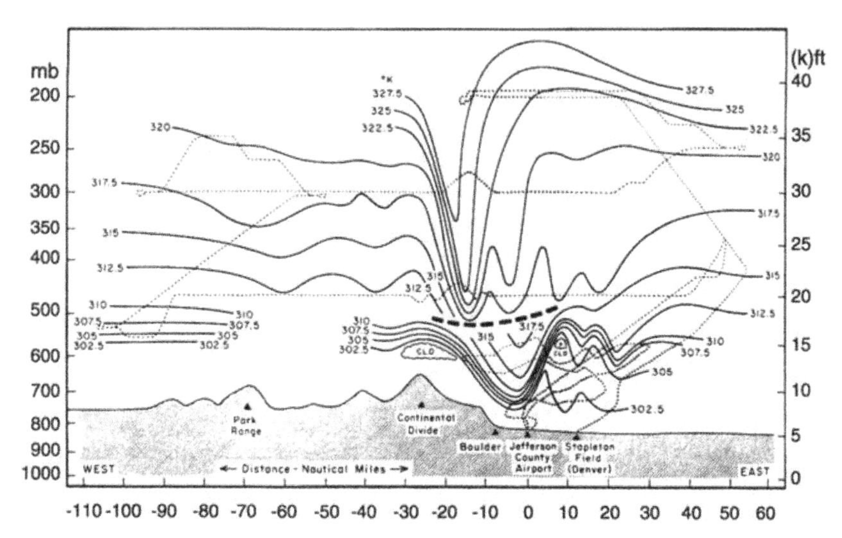

Figure 3.8.1. Potential temperature cross section across the Rockies near Boulder, Colorado, showing the dramatic downslope winds (from Lilly and Zipser, 1972).

satellite image of the cloud pattern over the western United States showing an extensive field of lee waves extending from the coastal mountain ranges to the Rockies, and Hunt *et al.* (1996) show a similar image over Great Britain (Figure 3.8.2). The elegant beauty and practical importance of this natural phenomenon have been catalysts in its investigation since the very first attempts by Lyra (1940) and Queney (1948) in the 1940s to modern times (Kim and Arakawa, 1995). Consequently, the literature on the subject is vast and several reviews exist (for example, Queney *et al.,* 1960; Long, 1972; Nicholls, 1973; Baines, 1987). Smith (1979, 1989) has conducted periodic reviews of the subject since the seventies. Smith (1979) is a particularly comprehensive survey of our knowledge of the influence of mountains on the atmosphere at the time. Smith (1989) updates the survey and details the various gaps in our current understanding of the flow over mountains. More recently, Wurtele *et al.* (1996) have given a particularly succinct review of many aspects of flow over mountains. The reader is referred to these reviews for a good introduction to the subject.

Flow over mountain ranges is of obvious importance to local micro-meteorology. It is clear that severe downslope winds in the lee of the mountains, that are characterized by not only very strong winds but also extreme gustiness, can cause extensive damage to property in places such as Boulder at the foothills of the Rockies and Owens Valley in the lee of the Sierra Nevada range in California. What is not so obvious is the serious westerly bias in operational weather forecasts that can result from neglecting the gravity wave

Figure 3.8.2. Visible satellite image from NOAA 11 showing a mixture of trapped lee waves and orographic cirrus over the UK (courtesy of J. Hunt).

drag due to mountain ranges (McFarlane, 1987; Shutts, 1995; Kim and Arakawa, 1995). Inaccuracies in parameterization of the momentum and dissipation effects of the extensive mountain ranges in midlatitudes on the globe are thought to be partly responsible for the poor skill of the GCMs in predicting the atmospheric state over timescales of a week and beyond in medium-range weather forecasts.

The energy radiated up into the troposphere by the waves generated in the lee of the mountains is extracted from the mean flow. This is accompanied by a downward momentum flux that constitutes an effective drag exerted by the mountain on the atmospheric column and is an efficient mechanism for transfer of momentum (and energy) from near-surface layers to the layers near and beyond the top of the troposphere. These lee waves can therefore constitute a significant source of energy and momentum to the middle atmosphere (for example, Andrews *et al.*, 1987). When these propagating internal waves break or encounter critical layers, they can transfer their momentum to the mean flow. At the ground level, this manifests itself as a pressure difference across the mountain and hence as a form drag, even were the flow to be inviscid. The drag exerted by the mountain clearly retards the atmospheric column, but because of the possibility of wave propagation and breaking aloft, the net effect is that the drag is felt by the atmospheric layers much above the surface. This is the so-called low-drag regime of the flow. Severe downslope flow conditions, on the other hand, constitute a high-drag regime and have an even higher impact. They not only cause strong changes in the flow and strong retardation of the atmospheric column as a whole, but can also lead to clear-air turbulence throughout the troposphere.

Theoretical investigations of flow over mountains started in the 1940s (Lyra, 1940; Queney, 1948; Scorer, 1949) with the use of two-dimensional, linear, inviscid, stratified flow equations under the Boussinesq approximation. Using the hydrostatic approximation, Queney (1948) derived a simple linear solution to the flow at constant velocity and constant stratification over a Witch of Agnesi-shaped mountain of height H and half-width L: $h(x) = H L^2/(L^2+x^2)$. This solution is periodic in z with a wavenumber of N/U, where N is the local buoyancy frequency and U is the flow velocity, but decays in the downstream direction with a scale length of L. The drag exerted by the mountain on the flow in this case is $D_L = (\pi/4)\rho_0 UNH^2$, so that the drag coefficient defined as $C_{DH} = D_L N/\left(\rho_0 U^3\right)$ is $(\pi/4)Fr_I^2$, where $Fr_I = NH/U$ is the inverse Froude number. The drag is independent of the width of the mountain, consistent with the hydrostatic approximation. For nonhydrostatic flow, the half-width L is also important, and the relevant parameter is $S_H = NL/U$. For $S_H < 4$, nonhydrostatic effects must be taken into account (see Wurtele *et al.*, 1996). For small S_H, $C_D = C_{DH}(4S_H^2/3)$. These values for the drag due to a linear mountain are often used to normalize the mountain drag in many numerical simulations involving nonlinear flow cases, with the high-drag flow being defined as one with a normalized drag value of more than 2. The mountain drag is simply the form drag due to higher wind speed and hence lower pressure along the leeward side of the mountain and is calculated simply by $D = \int_{-\infty}^{\infty} p(dh/dx)dx$, where p is the perturbation pressure at the ground level. This expression can also be used to deduce mountain drag from pressure observations (Smith, 1978).

In the fifties, Long (1953, 1954, 1955) undertook a series of theoretical and laboratory investigations that greatly advanced our understanding of stratified flow over obstacles. His laboratory experiments illustrated quite lucidly phenomena such as internal lee waves. Until Long (1955) showed that the linear governing equations for the constant velocity and stratification case considered above are also valid for nonlinear flow disturbances due to a finite-amplitude mountain, most theoretical studies of flow over mountains invoked linear theory. Since then, many investigations have relied upon Long's equations for studying nonlinear flow over finite-amplitude mountains (Huppert and Miles, 1969, for example). The validity of Long's equations in the presence of upstream effects has often been questioned, since the derivation assumes uniform stratification (constant N) and dynamic pressure (constant upstream velocity U) and absence of upstream effects. However, with the advent of modern computers, there is no longer the need to invoke approximations such as this and much of what we now know about flow over mountains comes from numerical simulations (Clark *et al.,* 1994) and of course observations such as ALPEX (Smith, 1987). Clark *et al.* (1994) present 2D and 3D numerical simulations of the 1979 Colorado Front Range windstorm and comparison with observations (Zipser and Bedard, 1982).

Of the many effects of mountains on the atmosphere, the downslope winds (variously called Chinook, Bora, Foehn, or Santa Ana) are the most spectacular. But effects such as the torrential rains that result from the lifting of water-vapor-laden air parcels over their condensation level on the upwind slope in regions such as along the west coast of India during summer monsoons, the upstream blocking of air masses, and cyclogenesis in the lee of mountains are equally important, but have received less attention (Baines, 1987; Pierrehumbert and Wyman, 1985). While studies of flow of stratified fluids over two-dimensional mountains are extensive, three-dimensional effects have not been that well studied. Only in recent years have meaningful simulations of flow over three-dimensional mountains become possible (Eliassen and Thorsteinsson, 1984; Thorsteinsson, 1988; Clark *et al.,* 1994). Three-dimensional effects are particularly important since often a stratified fluid tends to flow around obstacles instead of over them and the resulting blocking and upstream effects are quite important.

During severe downslope events, the drag exerted by the mountain is large, and this high-drag state, which necessarily involves nonlinear effects, has been studied in greater detail in recent years (Clark and Peltier, 1984; Smith, 1985; Bacmeister and Pierrehumbert, 1988). There now exists a better understanding of the behavior of the atmosphere during such events (Smith, 1985). Nevertheless, our understanding of the details of the turbulence generated on the downslope, due to high winds and the hydraulic jump undergone by the supercritical flow, and the resulting flow variability and buffeting is still gappy. Curvature of the streamlines has an important effect on the turbulent stratified flow over mountains and these effects may need to be more accurately accounted for (Kantha and Rosati, 1990; Kaimal and Finnigan, 1994). In this section, we will concentrate on downslope effects. For other equally important effects of mountains, such as the upstream effects, the reader is referred to the references cited above. We will also restrict ourselves to two-dimensional cases.

Consider a two-dimensional mountain. It is a common practice to consider an idealized mountain with a Gaussian or Witch of Agnesi shape and simplified upstream flow conditions to investigate the influence of the mountain on the flow. Given a particular mountain shape, the most important parameters then are the velocity of the flow upstream U (assumed to be uniform in height), the buoyancy frequency of the fluid upstream N (assumed to be uniform in height also), the height H and width L of the mountain, and f, the Coriolis parameter. The ABL thickness upstream, h, is also an important parameter if frictional effects are to be considered. The most important nondimensional number that can be formed out of these is the so-called Froude number, Fr = U/NH. The other parameters are S_H = NL/U, S_R =U/ fL, and h/H.

Parameter S_H is the ratio of the timescale for the flow to pass over the mountain to the buoyancy timescale. For S_H much larger than unity, hydrostatic approximation to the governing equations suffices. Otherwise, a nonhydrostatic model needs to be used. S_R is the Rossby number and should be large for rotational effects to be important. In many studies on flow over mountains, S_R is put to zero. A characteristic length scale can be defined by combining F and S_R: L_R=NH/f. This is the internal Rossby radius of deformation. When topography is steep [as indicated by the parameter $(Fr\ S_R)^{-1}$ =NH/fL being larger than unit], L_R indicates the size of the deceleration zone upstream of the mountain. Parameter L_R/L is an important indicator of the upstream influence. For values of this parameter much less than 1, as in broad continental mountain ranges, the upstream influence is negligible (Pierrehumbert and Wyman, 1985), while when the parameter is comparable to 1 or greater, as in coastal mountain ranges, the upstream influence is significant. Also, the along-shore flow in the latter case can be significantly ageostrophic. The parameter h/H is seldom considered in studies of flow over mountains, since most studies neglect frictional effects of the planetary boundary layer.

The flow pattern depends very much on the value of the inverse Froude number Fr_I = NH/U, which can be looked upon as the ratio of the buoyancy forces to the inertial forces. It is a measure of the relative importance of potential and kinetic energies in stratified flow around and over topography. For Fr_I = 0, neutral stratification, the flow accelerates over the mountain top, and for nongently sloping hills and mountains there is invariably boundary layer separation and turbulence in the lee of the mountain. Even when there is no separation as in gently sloping small hills, the turbulence over the hill is affected by the mean streamline curvature (Zeman and Jensen, 1987). The most salient effects are those associated with the speedup at the top and separation in the lee (Taylor and Teunissen, 1987; Taylor *et al.*, 1987; Kaimal and Finnigan, 1994).

For strong stable stratification ($Fr_I > 1$), the flow is blocked upstream for strictly or nearly two-dimensional flows (Baines, 1987). The blocking phenomenon is observed in the laboratory (Long, 1954, 1955; Baines and Hoinka, 1985) and often in the field. For typical values of N in the atmosphere (10^{-2} to 10^{-1} s^{-1}), heights as low as a few hundred meters can lead to blocking. Such blocking is called cold air damming and is frequently observed along the east coast and west coasts of the United States. For more realistic, three-dimensional mountains, there is some upstream blocking, but the stratified fluid tends to flow around the sides rather than go over the top. It is near $Fr_I \sim 1$ that most interesting processes occur. Standing internal waves are formed in the lee of the mountains and these become quite large under resonance conditions, giving rise to large rotors and spectacular lenticular clouds.

Three-dimensional effects are harder to quantify. If L_A indicates the scale of the flow in the along-mountain direction, then the flow characteristics depend on the ratio L/L_A, assuming the scale of the flow in the cross-mountain direction scales with L. Normally, this ratio is large and the along-mountain flow is pretty much in geostrophic balance, but situations exist where that is not the case. This is true especially for coastal mountain ranges. Gaps in mountain ranges, whether inland or coastal, complicate the flow further, with the nature of the flow through the gap depending very much on the ratio of the gap width to the Rossby radius of deformation. Flow from the Gulf of Mexico through the gaps in the Sierra Madre coastal range in Central America near the Gulf of Tehuantepec and Papagayo cause intense jets with substantial impact on the eastern Pacific in the form of large, wind-driven, anticyclonic eddies spun up by the gap winds (Trasvina *et al.*, 1995).

The ABL height has an important effect on the flow. The upstream flow can then be looked upon as consisting of essentially a nearly neutrally stratified, well-mixed layer capped by an inversion. If the ABL height is much larger than the height of the mountain ($h/H >> 1$), the flow behaves much like a neutrally stratified flow, with some drawdown of the inversion over the mountain top. If $h/H << 1$, the shallow ABL tends to flow around a three-dimensional hill and gives rise to phenomena such as Karman vortex streets and atmospheric ship wakes often made visible by the presence of clouds. See Wurtele *et al.* (1996) for examples.

The presence of vertical shear also has a dramatic effect on the mountain waves. The relevant parameter is then $S_s = (H/U)(dU/dz)$ or the gradient Richardson number $Ri_g = N^{2/}/(dU/dz)^2$. Scorer (1949) was the first to consider this case. For positive dU/dz, that is, flow velocity increasing with height, there can exist a critical level, $z = Ri_g^{1/2}/k_h$, beyond which propagation is not possible. Trapped waves result. For negative dU/dz, the level at which the flow reverses is also the critical layer where the phase speed of the wave is equal to the flow speed (Maslowe, 1986). Critical layers are regions of wave breakdown and clear-air turbulence.

Kaimal and Finnigan (1994) summarize the current knowledge of flow over hills. Belcher and Hunt (1998) review the status of our understanding of neutrally stable turbulent boundary layer flow over hills (and waves). Our understanding of flow over small hills has been greatly improved by the observations made recently during the Askervein Hill project (Taylor and Teunnison, 1987; Taylor *et al.*, 1987; see also Kaimal and Finnigan, 1994). A small, elliptically shaped hill in Scotland (Askervein Hill, 115 m high and 1 and 2 km long in the two horizontal directions), was extensively instrumented by an international team of investigators. Two 50-m towers and an additional fifty 10-m towers enabled unusually extensive coverage of the hill along the major axis and two sections perpendicular to it [see Figure 3.8.3, from Taylor *et al.* (1987)]. The objective was to investigate the flow speedup over the hill and its effect on the flow characteristics. Since $H/L_h < 0.50$, where L_h is the half-width of the hill at half-height, the flow does not separate over the lee of the hill. For low hills such as this, the stratification effects are also not very significant for typical wind speeds ($Fr_I < 0.1$). Rotational effects can also be neglected ($S_R \sim 0.01$–0.1). The flow speeds up at the top of the hill by approximately 1.6 H/L_h (Taylor and Teunissen, 1987; Kaimal and Finnigan 1994), about 80% for Askervein. The flow in the wake of the hill depends very much on whether the flow is separated or not, and even for unseparated flow, the influence of the hill extends many tens of hill heights downstream (Kaimal and Finnigan, 1994). While the flow sufficiently far downstream tends to exhibit a universal wake profile, the wake flow in the immediate vicinity of the hill very much depends on local hill characteristics.

In the absence of density stratification, turbulence characteristics of the flow are affected by the rapid flow acceleration and deceleration over the hill and the streamline curvature. Oncoming turbulence is subjected to additional strain rates, dU/dx and U/R, where x is the streamwise coordinate and R is the radius of curvature of the hill (in addition to the strain rate due to shear dU/dz). It is well known that turbulence is sensitive to any additional strain rates imposed on it (Townsend, 1976) and responds strongly. For example, straining due to strong acceleration and convex curvature can completely suppress turbulence. In general, the behavior of turbulence when subjected to additional strain rates is intermediate between rapid distortion, where the mean flow acceleration is too large for turbulence to adjust to local conditions and hence tends to retain its upstream characteristics, and local equilibrium, where the flow changes are slow enough for it to be in equilibrium with local mean flow conditions. The former implies turbulence memory effects, since the timescales associated with the

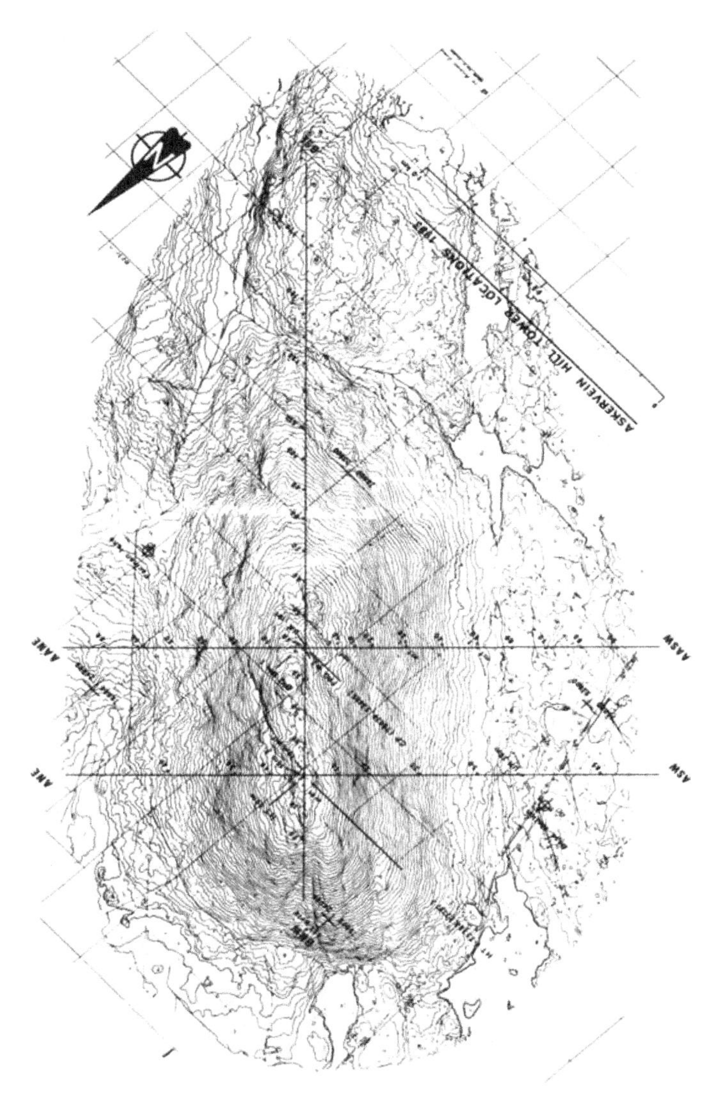

Figure 3.8.3. Askervein contour map showing tower locations used during the 1983 experiment. Contour interval is 2 m (Taylor *et al.*, 1987, with kind permission from Kluwer Academic Publishers).

strain rate are comparable to or smaller than the eddy turnover timescale, the latter being exactly the opposite. These additional strain rates can be taken into account in turbulence models. For example, there is a strong analogy between density stratification and curvature effects and this analogy has long been exploited (Bradshaw, 1969).

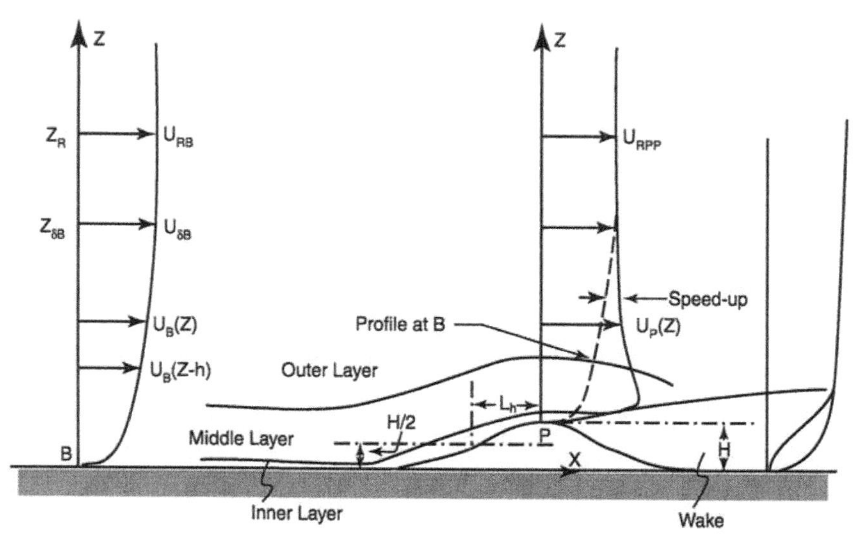

Figure 3.8.4. Flow over Askervin Hill (from Taylor and Teunissen 1987, with kind permission from Kluwer Academic Publishers).

Away from the inner layer (Figure 3.8.4), where the turbulence is roughly in local equilibrium and increases slightly, turbulence intensity decreases substantially in the outer layer, where rapid distortion approximation holds approximately (Teunissen *et al.,* 1987). Zeman and Jensen (1987) have applied a second-moment closure model to successfully simulate the flow characteristics over Askervein Hill.

Both stratification and curvature effects are important and lead to significant influence of streamline curvature on aspects such as Monin–Obukhoff similarity and the critical flux Richardson number for the extinction of turbulence (Kantha and Rosati, 1990). Figure 3.8.5 shows Monin–Obukhoff similarity functions ϕ_M and ϕ_H for various values of the curvature similarity variable $z_c = (\kappa z U)/(R u^*)$, and the variation of the critical flux Richardson number with curvature Richardson number $Ri_c = (U/R)/(dU/dz)$, in the limit of local equilibrium (Kantha and Rosati, 1990). Figure 3.8.6 shows simulations of flow over a two-dimensional mountain with and without curvature effects on turbulence, using second-moment closure (from Kantha and Rosati, 1990). Kantha and Rosati (1990) neglected flow acceleration effects (dU/dx term) on turbulence in their simulations.

For steeper hills, separation occurs over the lee of the hill (Figure 3.8.7). For smooth hills, a lee angle of greater than 18°–20° causes separation, but the critical angle for separation of the flow depends very much on the roughness

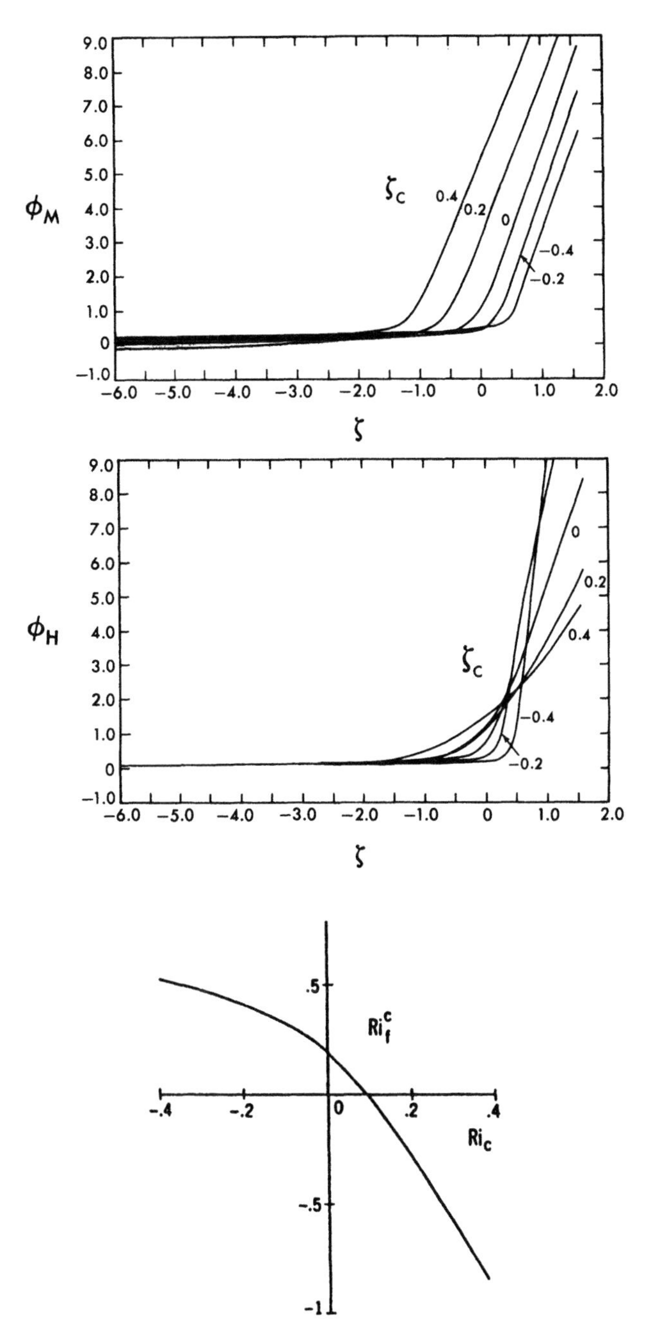

Figure 3.8.5. Monin–Obukhoff similarity functions as a function of the curvature Richardson number (from Kantha and Rosati, 1990).

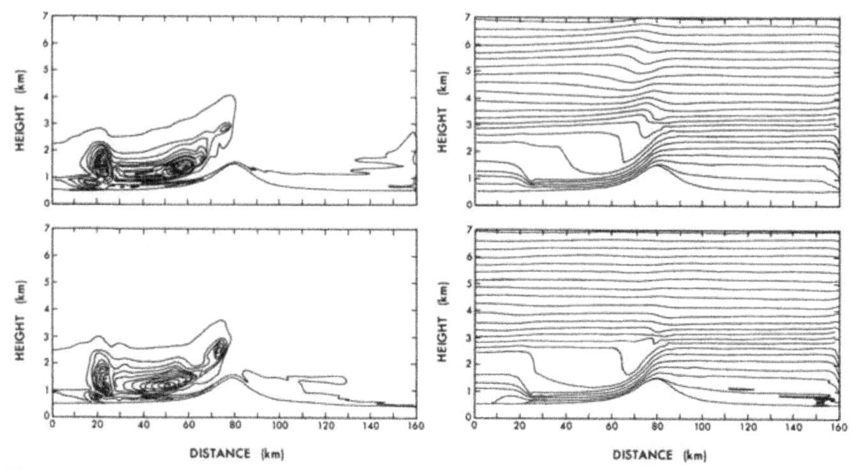

Figure 3.8.6. Model simulations of flow over an idealized mountain showing downslope winds and the associated hydraulic jump. Left panels show the turbulence intensity, while the right panels show the potential temperature. Note the high turbulence levels in the lee of the mountain (from Kantha and Rosati, 1990).

of the underlying surface (Kaimal and Finnigan, 1994). This is not very surprising since the characteristics of the approaching turbulent boundary layer, which determine the location and extent of separation over the lee, depend very much on the underlying surface roughness.

Next we consider flow over mountains, where stratification effects are important. Mountains are typically several kilometers in height and tens to hundreds of kilometers in length. Therefore, rotational effects cannot be neglected ($S_r \sim 0.1$), especially in considering the upstream effects and cyclogenesis in the lee. However, for investigating downslope winds, it is not important to retain rotational terms. For most purposes, it is possible to use hydrostatic models to investigate downslope winds. Nonhydrostatic effects become important only for very steep mountains.

The most important factor that influences the evolution of the flow is the upstream condition, and the most important parameter is the inverse Froude number Fr_I . For small values of Fr_I, linear theory is valid and the most salient aspect of the flow is the generation of internal lee waves, with a nonzero vertical group velocity. These provide the principal means of transferring the mountain-induced drag to the flow. As Fr_I increases further, nonlinear effects become more and more important. Beyond a certain value of Fr_I, the internal waves break and give rise to strong turbulence in the lee of the mountain. It is this kind of turbulence that caused severe damage during the 1978 Boulder windstorm (Zipser and Bedard, 1982).

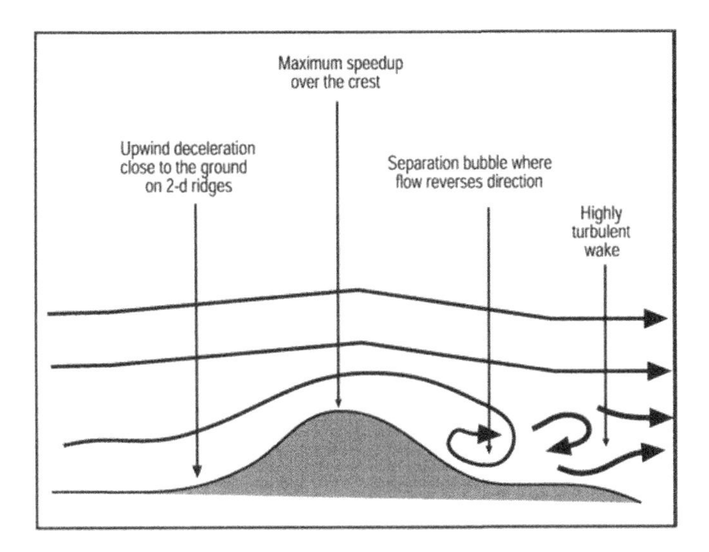

Figure 3.8.7. Flow separation in the lee of a hill (from *Atmospheric Boundary Layer Flows* by J. C. Kaimal and J. J. Finnigan. Copyright 1994 Oxford University Press, Inc. Used by permission of Oxford University Press, Inc.).

Smith (1987) describes aircraft measurements made during the Yugoslavian Bora over the Dinaric Alps. Figures 3.8.8 and 3.8.9 show observations of flow characteristics along a section across the Alps. Large flow accelerations in the lee of the mountain are evident in the potential temperature contours. Figure 3.8.8 shows the large turbulence generated aloft, with vertical velocity variance reaching 10 $m^2\ s^{-2}$. Strong downslope winds occurred in all five bora conditions observed (Smith, 1987).

Smith (1985) developed a simple hydraulic theory to quantify downslope winds resulting from internal wave breaking aloft. He hypothesized that there exists a critical streamline aloft, above which the flow is essentially undisturbed. The flow below the dividing streamline accelerates over the mountain from subcritical to supercritical conditions (Figure 3.8.10). The mean flow between the splitting streamline is weak, but involves strong turbulence and mixing and uniform density. By strict analogy to a single-layer hydraulic flow transitioning from a subcritical to a supercritical state, in which

$$H_u = 1 + \frac{1}{2}F_u^2 - \frac{3}{2}F_u^{2/3} \tag{3.8.2}$$

where $F_u = U/\sqrt{g'H_u}$ is the Froude number and H_u is the layer depth upstream undergoing this acceleration, for continuously stratified fluid upstream,

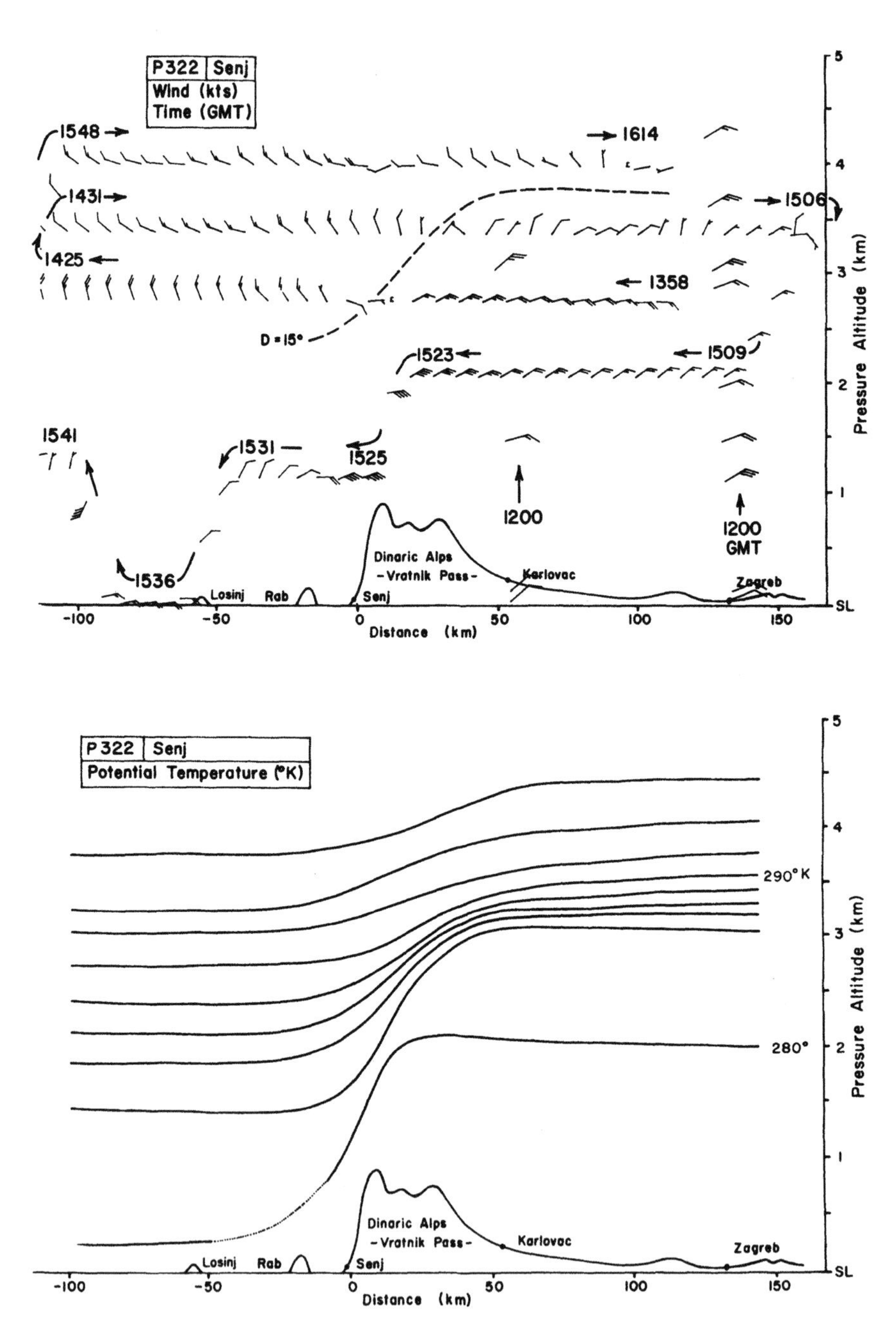

Figure 3.8.8. Wind and potential temperature distributions from aircraft observations during the Yugoslavian Bora, showing the severe downslope winds (from Smith, 1987).

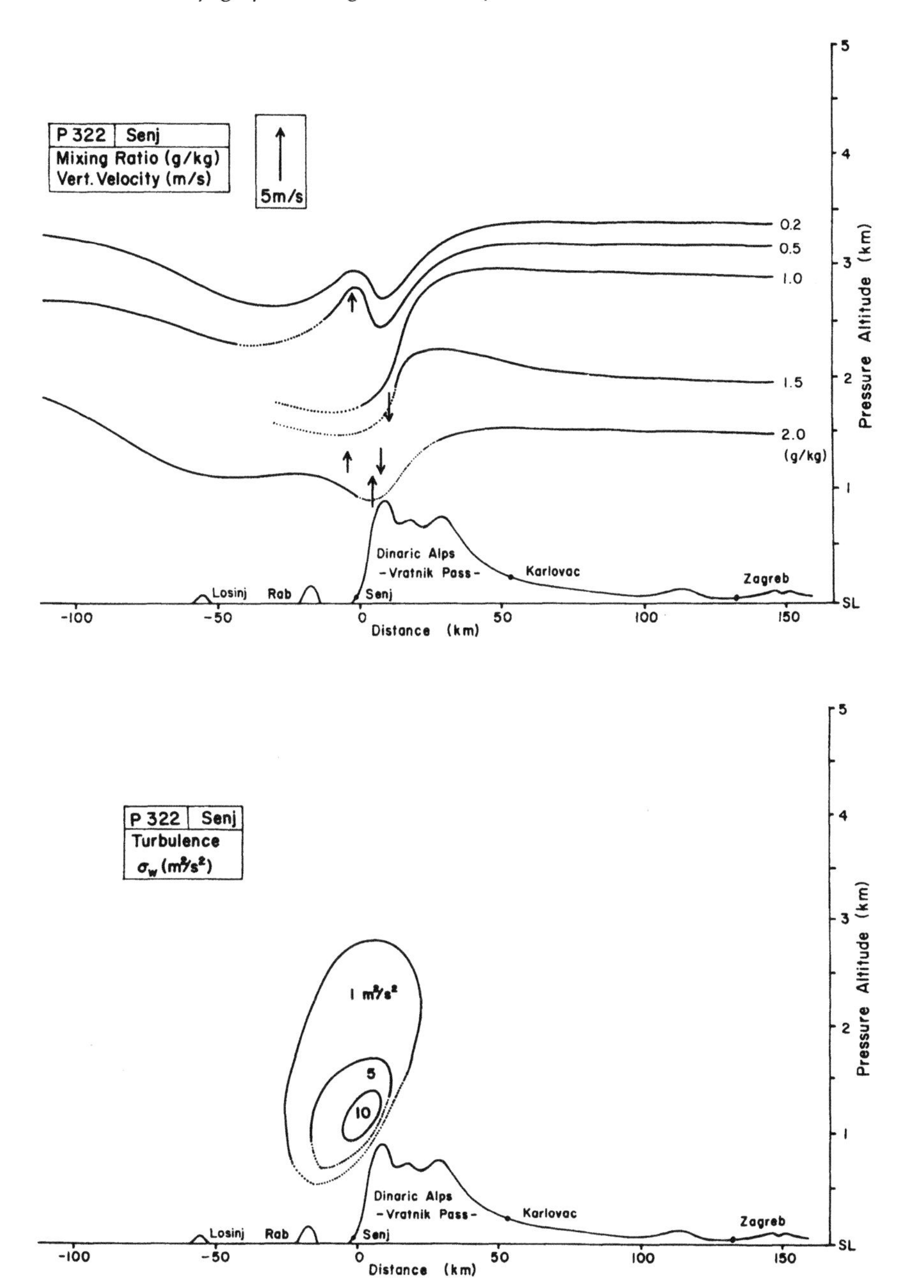

Figure 3.8.9. Water vapor mixing ratio and turbulence intensity distributions from aircraft observations during the Yugoslavian Bora, showing the severe downslope winds and strong turbulence (from Smith, 1987).

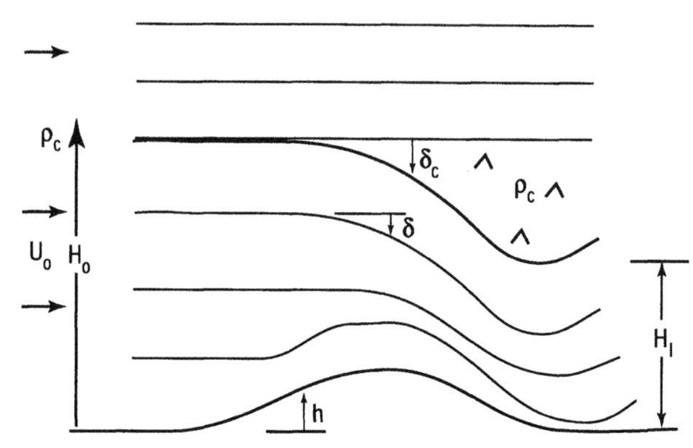

Figure 3.8.10. Idealized model of the downslope winds (from Smith, 1985).

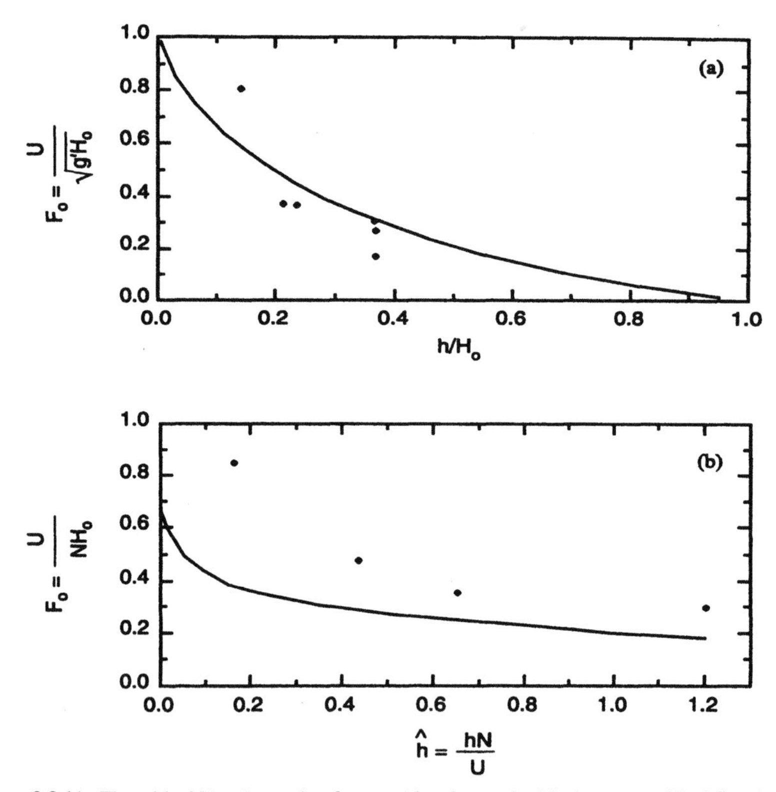

Figure 3.8.11. The critical Froude number for transition from subcritical to supercritical flow in (a) a sharply defined stable layer and (b) uniform stability (from Smith 1987). Curves are from the prediction of hydraulic theory; points are from several bora cases, the Boulder windstorm, and defined the Windy Gap, Wyoming windstorm of 1976.

$$\frac{NH_u}{U} = \frac{NH}{U} - \hat{\delta} + \cos^{-1}\left(\frac{NH/U}{\hat{\delta}}\right) \tag{3.8.3}$$

where

$$\hat{\delta} = \frac{1}{\sqrt{2}}\left\{\left(\frac{NH}{U}\right)^2 + \frac{NH}{U}\left[\left(\frac{NH}{U}\right)^2 + 4\right]^{1/2}\right\}^{1/2} \tag{3.8.4}$$

Smith (1987) plots upstream Froude numbers for both cases (Figure 3.8.11). He also overlays on these the observations of downslope winds for a neutral layer capped by an inversion and those with nearly continuous stratification. The agreement between hydraulic theory and observations is quite reasonable.

Figure 3.8.11 shows the variation of NH_u/U with NH/U. The severe downslope cases correspond to values of $NH_u/U > 1.5$, in good agreement with earlier studies for breaking and high-drag states. A value of $3\pi/2$ is used to determine NH_u/U in gravity wave parameterization schemes (Hunt *et al.*, 1996). With the depth of the splitting streamline upstream (H_u) known, it is possible to

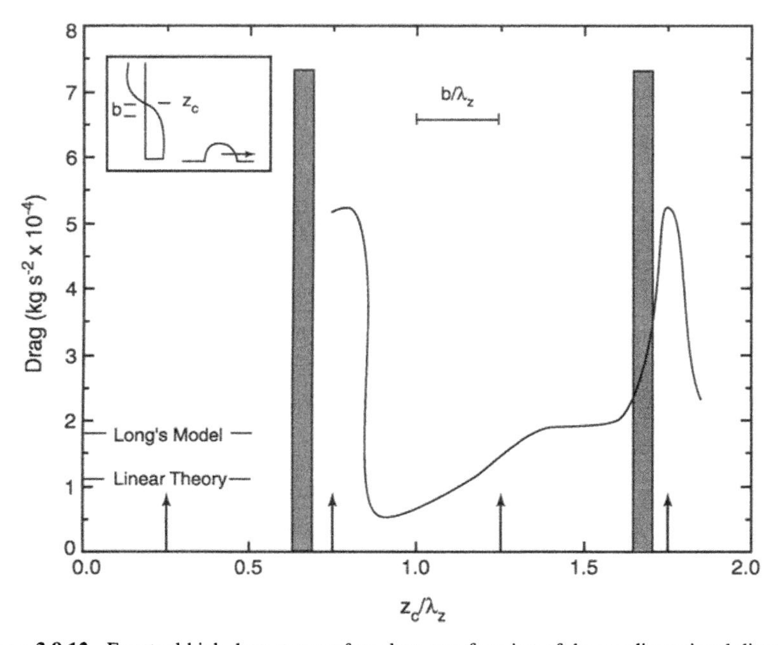

Figure 3.8.12. Eventual high drag state surface drag as a function of the nondimensional distance of the critical level above the ground. Note the resonant peaks centered on 0.75 and 1.75 (from Clark and Peltier, 1984).

obtain solutions to the flow over the mountain. Smith (1985) presents such solutions and compares them to numerical solutions obtained by Clark and Peltier (1984) for cases where wind reversal was used to define and fix the position of the critical layer (the location of the dividing streamline). Figure 3.8.12 shows the high-drag states from Smith's hydraulic theory, which agree reasonably well with the two peaks corresponding to Clark and Peltier's high-drag cases. In the rest of the regime, the drag is low and corresponds to that given by linear theory. Large flow accelerations result in the lee of the mountain and this supercritical flow can decelerate through a hydraulic jump downstream of the mountain, causing large gusts and property damage. Simulations by Kantha and Rosati (1990) clearly show the existence of such a hydraulic jump (Figure 3.8.7).

Numerical simulations of flow over mountains have been increasingly relied upon in recent years. Most use Richardson-number-dependent turbulent diffusivity. The simulation of Kim and Arakawa (1995) is among the first to use second-moment closure (see for example, Kantha and Rosati 1990) to properly account for turbulent mixing in these simulations.

3.9 FLOW OVER PLANT CANOPY

In earlier sections, we considered the ABL over land surfaces with small enough roughness elements that we could essentially ignore their heights above the surface vis-à-vis the observation point. However, vegetation covers a large part of the Earth's surface, either as cultivated crops or as natural vegetation in the form of grasses, brush, and dense forests. The effect of vegetation (especially the tall variety) on the lower ABL is of considerable interest. Also the flow within the plant and forest canopies is important to the micrometeorological conditions there and hence of interest to agriculture and forestry. It is obvious that the vegetation affects the flow within the canopy. Its effect on the ABL itself is less direct, though not less important. In this section we will discuss briefly the effects of vegetation on the lower ABL.

The influence of vegetation on the ABL depends on the characteristics of plants composing the vegetation and their spatial distribution. Dense forest canopy such as the one that covers most of the Amazon and the prairie grasses and crops covering relatively flat terrain in the Midwest are two examples of vegetation that has a large effect on the ABL. This is the kind that we will focus on primarily, although sparsely vegetated, semidry, desert-like environments also influence the ABL characteristics. Vegetation affects exchanges of both momentum and scalar quantities between the atmosphere and the ground surface. Pioneering work was done in this area by Penman and Long (1960), but important contributions have since been made by Finnigan (1979a,b, 1985),

Thom (1971, 1975), and Raupach (1988, 1991). For a good but dated review, see Raupach and Thom (1981). For a more recent summary, see Kaimal and Finnigan (1994).

The fact that vegetation changes the roughness felt by the ABL is obvious. However, the effective roughness depends very much on the character and distribution of individual plants. It is straightforward to see that for a given type of vegetation, as the spatial density of plants grows, the effective roughness increases until a point is reached where the distribution is so dense that the plants form a continuous surface underneath, and the effective roughness decreases. The canopy also effectively displaces the bottom of the ABL upward. This false bottom can be taken into account by a "displacement" in the lower ABL flux relationships. By accounting for the effective roughness and displacement, most of the similarity relationships derived earlier can also be applied to flow over plant canopy.

Vegetation also influences the heat, water vapor, and gas transfer between the atmosphere and the ground surface. Plants tap into groundwater and exert active control over the transfer of CO_2 and water vapor across their leaf surfaces by opening and closing tiny pores in their leaves called stomata. This active control driven by the biological needs and the reaction to the ambient environment in the canopy itself plays an important role in determining the transfer of water vapor and CO_2 from the leaves to the atmosphere. These evapotranspiration effects are important to the determination of sensible and latent heat fluxes over plant canopy.

Similarity relationships in the constant flux layer now depend also on the displacement height d, or more appropriately (z–d),

$$dF/dz \sim f(z-d, u^*, Q_F, Q_b) \tag{3.9.1}$$

where Q_F is the kinematic flux of that property in the surface layer. Equation (3.9.1) can be rewritten in nondimensional form:

$$\phi_F = \frac{\kappa(z-d)}{F_*}\frac{\partial F}{\partial z} = \phi_F\left(\frac{z-d}{L}\right) = \phi_F(\zeta) \tag{3.9.2}$$

$$\zeta = \frac{z-d}{L} \tag{3.9.3}$$

Note that in general the displacement heights for scalars are not the same as that for momentum, just like the equivalent roughnesses are not the same. However, this distinction is often ignored in practice. With this modification, the M-O similarity relationships derived in Section 3.3 should apply here also. For example, under neutral stratification, the profiles of velocity and a scalar quantity such as temperature and humidity become

$$\frac{U}{u_*} = \frac{1}{\kappa} \ln\left(\frac{z-d}{z_0}\right)$$
$$\frac{F-F_s}{F_*} = \frac{Pr_t}{\kappa} \ln\left(\frac{z-d_F}{z_{0F}}\right) \tag{3.9.4}$$

and for stratified situations,

$$U(z) = \frac{u_*}{\kappa}\left[\ln\frac{z-d}{z_0} - \psi_M\left(\frac{z-d}{L}\right)\right] \tag{3.9.5}$$

$$T(z) - T(z_r) = \frac{Pr_t \theta_*}{k}\left[\ln\frac{z\text{-}d_T}{z_r - d_T} - \psi_H\left(\frac{z\text{-}d_T}{L}\right) + \psi_H\left(\frac{z_r\text{-}d_T}{L}\right)\right] \tag{3.9.6}$$

$$q(z) - q(z_r) = \frac{Pr_t q_*}{k}\left[\ln\frac{z\text{-}d_E}{z_r - d_E} - \psi_E\left(\frac{z\text{-}d_E}{L}\right) + \psi_E\left(\frac{z_r\text{-}d_E}{L}\right)\right] \tag{3.9.7}$$

where z_r refers to a reference level in the surface layer and $\Psi_{M,H,E}$ the usual stability functions (see Section 4.2). The disparity between z_0 and z_F is even more exaggerated for flow over canopy, since now the pressure differences in the highly turbulent separated wake flow behind individual plant stalks form the principal momentum transfer mechanism within the canopy. An equivalent pressure mechanism is absent for scalars. Therefore, the scalar roughness lengths can be as little as only 20% of the momentum roughness scale (Thom, 1975), even though the precise value depends on the canopy.

However, all the unknowns are now pushed into d, the displacement height, which falls within the canopy height z_c, but in its top 20–30%. Observations have shown that over a wide range of canopies, $d \sim 0.75\, z_c$ is a very good and consistent approximation (Kaimal and Finnigan, 1994). The roughness scale itself is a function of canopy density and reaches a maximum when the plants are crowded together close enough to prevent each plant from absorbing any more momentum. Kaimal and Finnigan (1994) state that this maximum value is $z_0 \sim 0.2\, z_c$. Typical values for z_0 and d are 1.2 and 6 m over Landes Forest in France (Parlange *et al.,* 1995).

These relations are, of course, valid only sufficiently far away from the roughness elements, in this case, sufficiently away from the canopy top. Figure

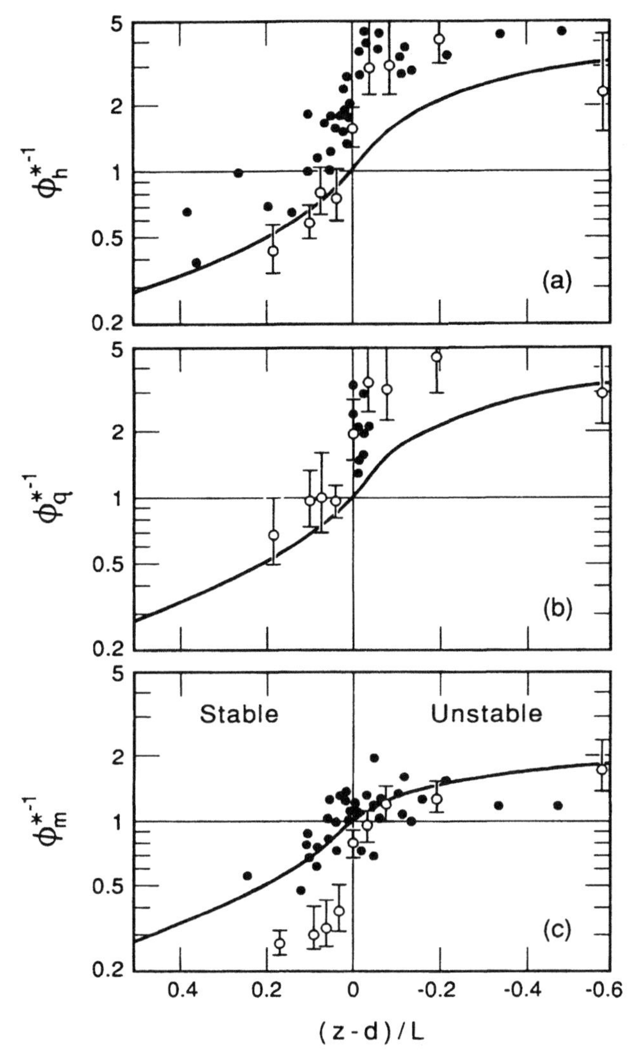

Figure 3.9.1. Monin–Obukoff similarity functions as a function of stability for (a) heat, (b) humidity, and (c) momentum above the plant canopy (from *Atmospheric Boundary Layer Flows* by J. C. Kaimal and J. J. Finnigan. Copyright 1994 Oxford University Press, Inc. Used by permission of Oxford University Press, Inc.).

3.9.1, from Kaimal and Finnigan (1994), shows the velocity and scalar profiles and their departures from the conventional M-O similarity profiles.

The most important aspect of scalar transfer through a plant canopy is the close association between the sensible and latent heat flux. The latter is governed by the evaporation from leaf surfaces and is therefore dependent on whether the stomata are exerting control as in dry conditions or whether the water is plentiful so that plant physiological control is unimportant. Evapotranspiration is one of the most important aspects of hydrological balance and a great deal of effort has been devoted to estimating the partitioning of incident radiant energy into sensible and latent heat fluxes from vegetation-covered surfaces, and hence the evaporation rate [see Parlange *et al.* (1995) for a recent review and an extensive list of past work in the area]. The importance of land surface parameterization in global GCMs has also provided the impetus for advances in this area (Avissar and Verstraete, 1990; Wood, 1991). Based on the earlier work by Thom (1975), Raupach and Finnigan (1988), and others, Kaimal and Finnigan (1994) derive a relationship between the ratio of the latent heat flux to the total heat flux as a function of the heat and mass transfer coefficients. This is called the combination equation,

$$\frac{H_l}{H_t} = \frac{L_E E}{H_s + L_E E} = \frac{1}{1+B_r} = \frac{\varepsilon_s + \dfrac{C_H}{C_A}}{\varepsilon_s + \dfrac{C_H}{C_E} + \dfrac{C_H}{C_C}} \tag{3.9.8}$$

where B_r is the Bowen ratio, the ratio of sensible heat flux H_s to latent heat flux H_l, and the bulk coefficients C_H, $C_{E,}$, C_C, and C_A are given by

$$\begin{aligned} H_s &= \rho c_p \frac{u_*}{C_D} C_H (T_s - T_a) \\ E &= \rho \frac{u_*}{C_D} C_E (q_s - q_a) = \rho \frac{u_*}{C_D} C_C (q_s^{sat} - q_s) \end{aligned} \tag{3.9.9}$$

and

$$H_t = \rho L_E \frac{u_*}{C_D} C_A (q_a^{sat} - q_a) \tag{3.9.10}$$

While C_D, C_H, and C_E have their usual meaning (drag coefficient and Stanton and Dalton numbers), C_C is proportional to the inverse of the stomatal resistance. It is supposed to capture the ease of water vapor transfer from the leaf interior to the surface, with the transfer assumed to depend on the ambient specific humidity q_s and the specific humidity in stomatal air cavities $q_s{}^{sat}$, which is assumed to be saturated at the leaf temperature T_S. It can be called the stomata number. The coefficient C_A is supposed to represent the ambient total heat transfer coefficient with respect to specific saturation deficit at level z. Note that

ε_s is the rate of change of saturated specific humidity with temperature (~2.2 at 20°C):

$$\varepsilon_s = \frac{L_E}{c_p}\frac{dq^{sat}}{dT} \tag{3.9.11}$$

Also

$$H_t = H_s + L_E E = SW_\downarrow + LW_\downarrow - SW_\uparrow - LW_\uparrow - H_G - SW_{ph} \tag{3.9.12}$$

where SW and LW indicate the short- and longwave radiative fluxes (the arrows indicate the downwelling and upwelling aspect), H_G is the heat flux to the ground or to storage, and SW_{Ph} is the amount consumed by photosynthesis of plants (~2% of the net SW and LW flux).

Equations (3.9.8) to (3.9.12) are used for parameterization of evaporation in global models and estimating the evaporation rate over land surfaces. Equation (3.9.8) has very interesting limiting properties (Thom, 1975; Kaimal and Finnigan, 1994):

1. $C_{H,E} \ll C_C$. Here vegetation is wet and stomatal resistance is not an important factor, and the third term in the denominator drops out. This corresponds to the case of "potential evaporation," since evaporation can be as much as physically feasible with no control exerted by the physiology of the plants.

2. $C_{C,A} \gg C_{H,E}$. Here $H_l = \varepsilon_s H_s$. This corresponds to "equilibrium evaporation," where the physiology of plants exerts no control, because the limiting factor is the ability of the water vapor to diffuse away from the leaf surface. The situation corresponds to light winds and humid conditions, such as asymptotic conditions downstream of well-irrigated crops (i.e., crops free from water stress).

3. $C_{C,A} \ll C_{H,E}$. Here stomatal control is dominant and the situation corresponds to dry windy conditions over irrigated crops, "oasis evaporation."

Parlange *et al.* (1995) present a slightly generalized version of Eq. (3.9.8),

$$H_L = L_E E = \beta\left[A\frac{\varepsilon_s H_t}{\varepsilon_s + \frac{C_H}{C_E} + \frac{C_H}{C_C}} + B\frac{\rho L_E \frac{u_*}{C_D} C_H (q_a^{sat} - q_a)}{\varepsilon_s + \frac{C_H}{C_E} + \frac{C_H}{C_C}}\right] \tag{3.9.13}$$

where A and B are constants that take different values in different formulations.

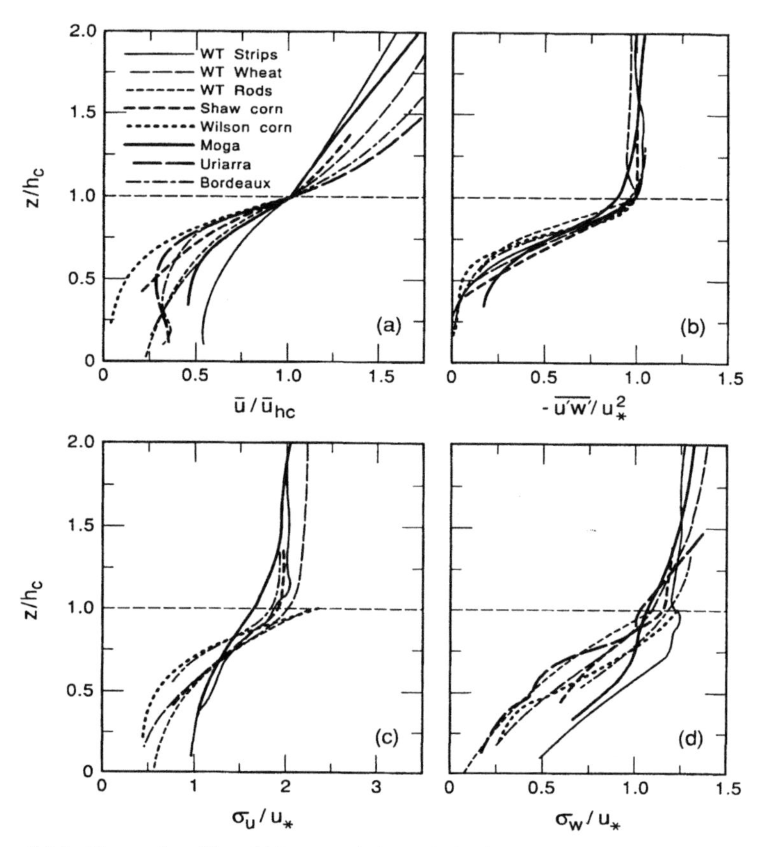

Figure 3.9.2. Observed profiles of (a) mean wind speed, (b) shear stress, (c) standard deviations of u, and (d) standard deviations of w in and above plant canopy (from *Atmospheric Boundary Layer Flows* by J. C. Kaimal and J. J. Finnigan. Copyright 1994 Oxford University Press, Inc. Used by permission of Oxford University Press, Inc.).

For example, for the potential evaporation case, $A = B = 1$; for equilibrium evaporation, $A = 1$, $B = 0$; etc. (see Parlange *et al.,* 1995). The Budyko–Thornwaite–Mather parameter β is taken as some function of surface water availability, put equal to 1.0, but allowed to go to zero as water availability becomes less and less.

Estimating the various numbers related to the evapotranspiration is quite difficult without considering the flow inside the plant canopy. A great deal of empiricism is involved (see Brutsaert, 1982, 1991; Parlange *et al.,* 1995). We cannot go into details here, but refer the reader to the references cited in this chapter, particularly Denmead and Bradley (1987), Finnigan (1979a,b, 1985), Monteith (1975, 1976), Raupach (1988,1991), Raupach and Thom (1981), and

Thom (1971, 1975). Garratt (1992), Kaimal and Finnigan (1994), and Parlange *et al.* (1995) are excellent starting points. Articles in *Boundary Layer Meteorology* are also useful. Briefly, the most important characteristic of the turbulence inside plant canopies is the generation mechanism. In the lower parts, it is principally wake turbulence, turbulence created by separation behind plant stalks. As such it is hard to parameterize and model. In the upper part of the canopy, it is the large eddies that arise due to the strong shear in the crown that are dominant. Figure 3.9.2 shows the profiles of the mean velocity, the shear stress, and the two components of turbulence velocity as a function of height in the canopy, for neutral stratification (from Kaimal and Finnigan, 1994). The data extend over a wide range of canopies, both in the wind tunnel experiments and in the field. The collapse of the velocity profiles when normalized by the velocity at the top of the canopy is noteworthy. Similar high quality data are not readily available for stratified conditions (but see Garratt, 1992; Kaimal and Finnigan, 1994).

The interaction with plants of the large turbulent eddies in the surface layer sweeping along the crown of the canopy of flexible cereal crops has striking visual impact. On a windy day, a ripe standing crop of wheat sustains coherent, wave-like motions called honamis, aesthetically pleasing to the eye because of

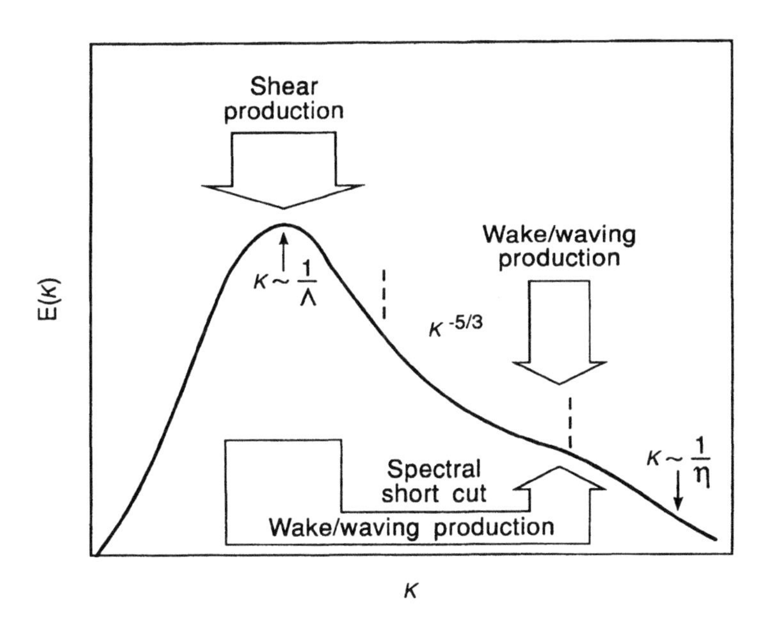

Figure 3.9.3. A schematic showing the short-circuiting of the spectral transfer by plant wake/waving production of TKE (from *Atmospheric Boundary Layer Flows* by J. C. Kaimal and J. J. Finnigan. Copyright 1994 Oxford University Press, Inc. Used by permission of Oxford University Press, Inc.).

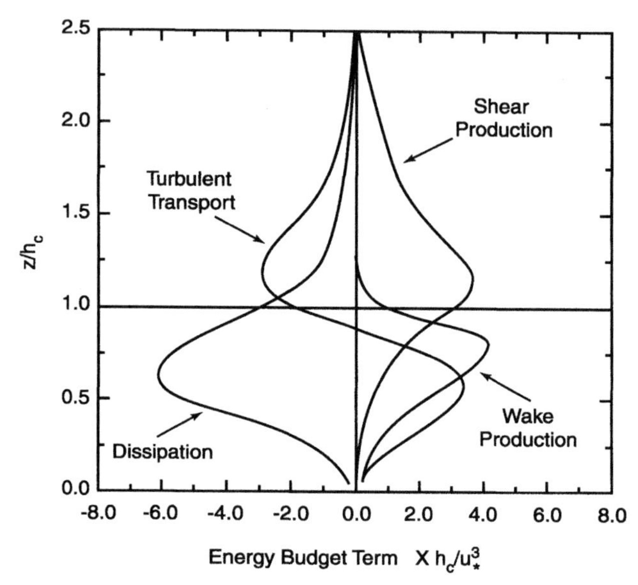

Figure 3.9.4. The TKE budget showing the importance of wake production inside the canopy (from *Atmospheric Boundary Layer Flows* by J. C. Kaimal and J. J. Finnigan. Copyright 1994 Oxford University Press, Inc. Used by permission of Oxford University Press, Inc.).

the waving of the "golden" grain-laden tops in response to these "gusts." The phenomenon occurs because of the resonant coupling between the characteristic frequency of the energy-containing eddies (q/l) and the natural elastic frequency of the stalks. The resulting flow, resembling a "golden carpet" undergoing undulatory motions, testifies to the striking beauty of many natural phenomena.

The waving of plant stalks has practical implications, as well. It constitutes an efficient mechanism for extracting energy from energy-containing large eddies near the crown and transferring them to smaller scales of the order of the plant wake size deep within the canopy. There is, therefore, likelihood of a secondary peak in the turbulence spectrum much removed from the usual spectral peak (Figure 3.9.3). Especially for plants such as cereal crops (rice, wheat, barley), where there can be efficient resonant coupling, this mechanism represents the bypassing of the usual spectral transfer from large eddies down the spectrum. The energy is instead passed onto the strain energy of the plants and then transferred directly to small eddies.

We will close by presenting the TKE budget inside a plant canopy under neutral conditions (Figure 3.9.4). The various terms are scaled by z_c/u_*^3. There occur two extra terms in the TKE budget unique to plant canopies, representing (1) wake production, a source term representing production of TKE due to flow separation behind plant stalks, and (2) a plant waving term, a net sink term representing destruction of TKE, since while part of the energy of plant waving motions is passed on to small eddies quite efficiently, part is also dissipated in plant internal friction; thus, there is no extra production, just an incomplete transfer in spectral space. Note the strong shear production near the crown, but the negligible one deep within. The behavior of the transport term is also noteworthy. It represents a strong sink immediately above the canopy, but a strong source deep within. In combination, this suggests that in lower parts of the canopy, the turbulent transport is important to maintaining turbulence. Turbulence is imported, not locally produced (Kaimal and Finnigan, 1994).

These, then, are some of the fascinating aspects of flow over plant canopies. The field is still evolving and much more remains to be done. Because of its practical importance to agricultural micrometeorology and forestry, it is sure to receive increasing attention in coming years.

3.10 INTERNAL BOUNDARY LAYERS

Finally, we will briefly describe aspects of change in the ABL characteristics due to changes in the underlying surface. In earlier chapters we dealt with ABL over a homogeneous terrain. Most of what we know about the ABL, its structure and behavior, is from observations, theory, and even LES numerical models that pertain to this situation. Traditionally, this has been the favored condition for the study of the ABL, simply because it is easier to understand and parameterize. Investigators went out of their way to locate relatively flat and homogeneous terrain on which to deploy their measurement arrays (Sorbjan, 1989). The 1953 Great Plains, the 1967 Wangara, the 1968 Kansas, and the 1973 Minnesota Experiments, and other ABL studies in the seventies, took place on such sites. These observations and LES's have added immensely to our knowledge of the ABL structure and have solidified the foundations of simple scaling arguments such as the Monin–Obukhoff similarity relations. However, it is being increasingly realized that inhomogeneities of the underlying surface on a variety of spatial scales are ubiquitous and need to be taken into account, especially since the increasing computing capabilities have continuously brought down the scales we can simulate efficiently in our computer simulations of the atmosphere. Satellite remote sensing has reinforced this perception. As we saw in Chapter 3.4, the NABL is affected by even very small terrain slopes. Drainage flows are a characteristic feature of hilly terrain. Even nonhilly terrains have significant topographic changes and the ABL adjusts to this change in terrain via an internal boundary layer, whose development depends very much on how the turbulence readjusts to the changed bottom boundary conditions.

Changes in the bottom boundary conditions as seen by an ABL are essentially of two kinds. The first kind involves changes in the dynamic boundary conditions due to changes in roughness of the underlying surface. The second kind involves changes in the thermodynamic fluxes from the underlying surface. The former is very common over land—for example, flow transitioning from a barren or sparsely vegetated landscape to heavily wooded conditions. The latter is less common over land, but occurs routinely when flow transitions from land to water or vice versa. The effects are quite spectacular during cold air outbreaks east of the continents, when a cold dry ABL encounters warm ocean waters. The result is the transfer of very large quantities of heat from the ocean to the ABL, often as much as 1000 W m^{-2}. Often these large heat transfer rates lead to explosive cyclogenesis. A similar phenomenon occurs when an ABL transitions from a cold ice surface to a relatively warm ocean during off-ice wind conditions in the marginal ice zones of the subpolar regions. Once again, intense convection drives vigorous turbulence in the ABL and leads to spectacular cumulus formations (Figure 3.10.1). Kantha and Mellor (1989a) have modeled the change in ABL during off-ice wind conditions using a two-dimensional model (in the x–z plane) incorporating second-moment closure. Such flow involves changes in both the dynamic and the thermodynamic conditions, since there is an effective change in the roughness of the underlying surface, as well.

Another example is the land–sea interface in a coastal ocean and its effect on land and sea breezes. Because of the increased attention being given to littoral regions of the world, there is a need to understand the response of the lower ABL to inhomogeneities in the underlying surface. Unfortunately, we can provide only a brief discussion of these aspects here. The reader is referred to Garratt (1990, 1992) and Kaimal and Finnigan (1994) for a more thorough discussion of this topic.

Consider first the constant flux layer of the ABL over land surfaces (over the ocean, horizontal homogeneity is normally a good approximation). The relevant vertical scale here is the thickness of the surface layer (z_s). If the scale of variation of the underlying roughness elements and other characteristics (l_r) is such that $l_r >> z_s$, then locally, the various similarity relationships derived in Section 3.3 should still be approximately valid. However, when there is an effective discontinuity in the characteristics of the underlying surface, such as across a land–sea interface, an internal boundary layer begins to grow at the discontinuity and the horizontal gradients cannot be neglected. It is this situation we will address here. Pioneering contributions were made in this area by Bradley (1968), who measured the changes in the velocity profiles and surface stress from changes in the roughness from both smooth to rough and rough to smooth; Dyer and Crawford (1965), who measured the changes as air flowed from hot dry to cool irrigated land surface; and Mulhearn (1977), who applied the self-similarity concepts in aerodynamics (Townsend, 1966, 1976) to these situations to explain the changes. To date, despite the importance of the problem, there has not been a concerted effort (like the Kansas or Minnesota Experiment) to observe and quantify the effect of changing terrain on the ABL structure.

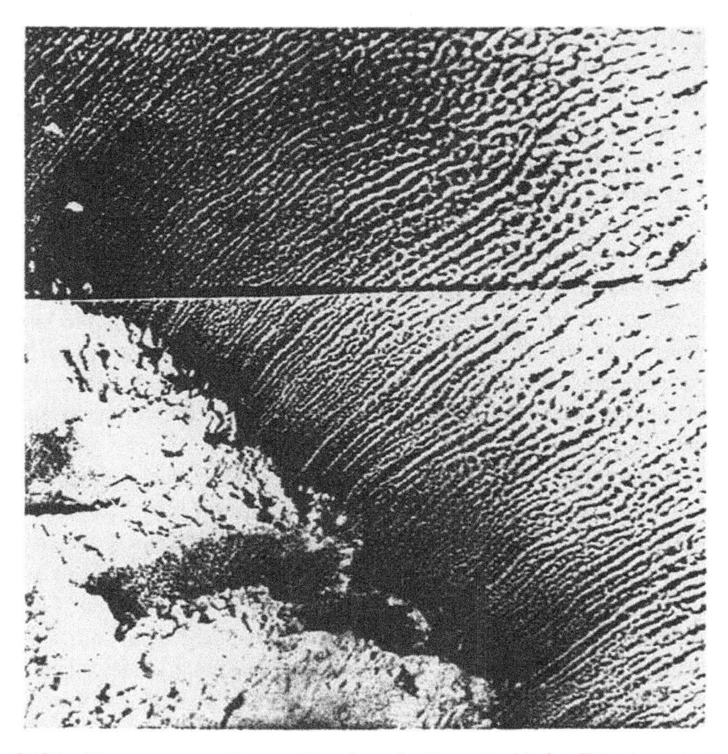

Figure 3.10.1. Vigorous convection resulting from the flow of cold air off the ice onto the warm water in a marginal ice zone (NOAA satellite photo).

For simplicity, consider flow perpendicular to a discontinuity (or, more appropriately, a rapid change) in the characteristics of the underlying surface. There are two possible changes: (1) a change in the effective roughness of the surface—this affects principally the momentum exchange and the velocity structure—and (2) a change in the property such as temperature that affects the scalar fluxes—this can have profound effects on both the momentum and scalar exchanges and the ABL structure. Normally, both effects are present to varying degrees. An example where both effects are strong is the flow over a marginal ice zone (MIZ) as seen in spectacular satellite imagery of off-ice flow. In any case, the flow responds by growing an internal boundary layer (IBL) that starts at the discontinuity and grows, merging eventually with the upstream boundary layer. The response of a boundary layer to abrupt changes in roughness and in heat flux is a classical problem in fluid mechanics and considerable literature exists on the subject (see Schlichting, 1977; Townsend, 1976). Unfortunately, the situation pertains to neutral stratification and the stratification effects are predominant in the ABL. For example, if the abrupt change involves transition of the ABL to a colder surface, the turbulence in the ABL is suppressed. The IBL is confined to a fraction of the upstream ABL and grows very slowly, taking tens, often hundreds, of kilometers to grow to the original thickness. This situation occurs during on-ice flows in the MIZ (Kantha and Mellor, 1989a; Overland *et al.*, 1983), when the ABL over relatively warm water (–1.7°C) flows onto cold ice (–20 to 30°C) and the ABL gives up heat to the ice. The roughness also increases, but this is a relatively minor effect. A land-based inversion develops and the IBL is maintained by the shear in the flow acting against gravitational forces and grows slowly, although the increase in roughness helps. On the other hand, for off-ice flows, the IBL grows quite rapidly because of the large heat flux (several hundreds of W m^{-2}) from the relatively warm water to the cold ABL, even though the roughness decreases. Vigorous convection ensues, and clouds form along rows aligned with the flow, which eventually merge into solid cloud cover (Figure 3.10.1).

IBLs are ubiquitous features in the atmosphere. They occur along MIZs and over any thermal front such as that associated with warm ocean currents, such as the Gulf Stream and Kuroshio, flowing adjacent to cold shelf waters during winter and the land–sea interface at the coast, especially during winter when the land–sea temperature contrast can be a few tens of degrees Celsius. Over land, transition from prairies to wooded lands and flow over and from a lake are causes for the growth of the IBL. Lake effects are particularly spectacular during winter, when the ensuing convection introduces considerable moisture into the ABL, which is precipitated as snow downstream of the lake. But by far, the most profound effects occur in the coastal zone due to land–sea contrasts. Coastal zones are regions of formation of diurnal land–sea breezes, especially during summer, when the differential heating and cooling of land surfaces relative to the oceans give rise to temperature contrasts of as much as 10°C. During the day, the land is warmer and the flow is from the sea onto the land (sea breeze), and during the night, the land is colder and the flow is from land onto the sea (land breeze). This land–sea breeze system (LSBS) is of considerable importance to the local meteorology of coastal regions and regions near the Great Lakes. The fronts formed by the convergence of the LSBS flows with the ambient flows are often visible in satellite imagery, a classic example being the shuttle photo of a land breeze off the coast of Florida. The fronts associated with the LSBS are often regions of cumulus clouds and local thunderstorms. While the general nature of the mesoscale LSBS is well understood, its myriad interactions with the ambient large scale flows are less understood, ill observed, and poorly modeled. For example, not much is known about the transition from land to sea breeze and vice versa, although these are diurnal processes of profound importance to coastal communities.

During winter cold-air outbreaks, the large land–sea temperature contrast (as much as 30°C) gives rise to explosive cyclogenesis along the eastern sides of continents in midlatitudes. These processes were studied during the Genesis of

Atlantic Lows Experiment (GALE) and the Experiment on Rapidly Intensifying Cyclones over the Atlantic (ERICA) field experiments (Doyle and Warner, 1990; Hadlock and Kreitzberg, 1988). The wintertime thermal fronts that develop off the U.S. coast in the Gulf of Mexico due to cold land/shelf water and warm offshore waters are often the source of winter storms over the eastern United States. In all these cases, the structure and growth of the IBL are central to the problem of frontogenesis.

The most important characteristic of an IBL is the departure from equilibrium conditions that prevailed upstream. Consequently, none of the similarity relationships whose very basis is the assumption of local flow equilibrium are valid, at least in the immediate vicinity of the abrupt change. The flow and turbulence in the IBL eventually relaxes to a quasi-equilibrium state, but in the vicinity of the change, pressure gradient, baroclinic forcing, advection, and three-dimensional effects are all important. The turbulence field is also far from equilibrium. Consequently, numerical models with good parameterization of turbulence are essential for dealing with IBLs (for example, Kantha and Mellor, 1989a). LES studies have principally been confined to horizontally homogeneous ABLs and have lagged behind in applications to IBLs. The problem is compounded by a dearth of observational data on the small scales of change in an IBL.

Fortunately, much is known about the IBL due to roughness changes under neutral stratification, mainly due to the observations of Bradley (1968) and the application of the classical boundary layer concept of self-similarity (Townsend, 1966) by Mulhearn (1977). The roughness change accelerates (rough to smooth) or decelerates (smooth to rough) the flow (Figure 3.10.2), and turbulence intensity decreases (increases) adjacent to the surface, and this change is diffused into the IBL. Pressure forces are also important in the immediate vicinity of the change, but are usually ignored. The parameter that characterizes the magnitude of the change encountered by the flow is $M = \ln(z_{0d}/z_{0u})$, where subscripts u and d refer to upstream and downstream conditions. Figure 3.10.2 shows the velocity profiles in the IBL for both smooth to rough and rough to smooth transitions for $|M|=4.8$. The profiles can be described by a modified logarithmic law,

$$\frac{U}{u_{*d}} = \frac{1}{\kappa}\ln\frac{z}{z_{0d}} + f(z/\delta_i) \qquad (3.10.1)$$

$$f(z/\delta_i) = 0 \qquad z/\delta_i << 1$$

$$f(z/\delta_i) = \frac{u_{*u}}{u_{*d}}\frac{1}{\kappa}\ln\left(\frac{z}{z_{0u}}\right) - \frac{1}{\kappa}\ln\left(\frac{z}{z_{0d}}\right) \qquad z/\delta_i > 1 \qquad (3.10.2)$$

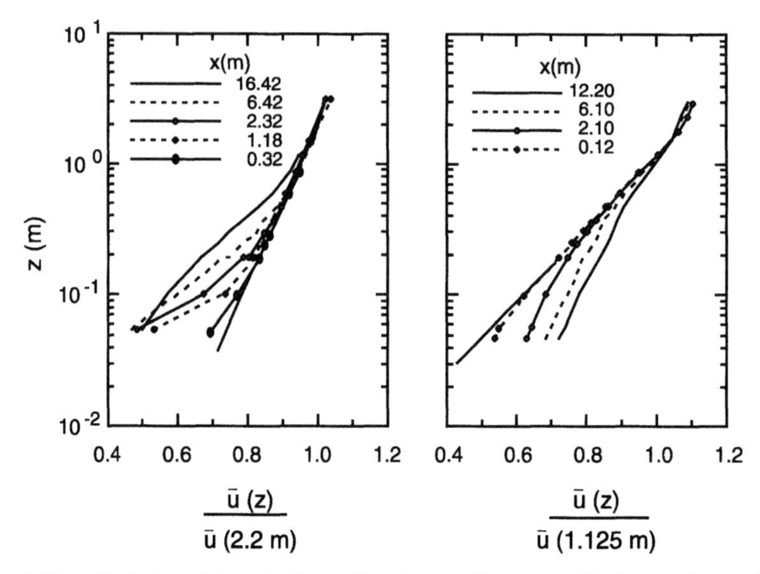

Figure 3.10.2. Evolution of the velocity profile after a sudden change in the roughness of the underlying surface (from Bradley 1968). (Left) smooth to rough, (right) rough to smooth.

where δ_i is the thickness of the IBL. These relationships yield a logarithmic law in the lower IBL, where the turbulent flow is nearly in equilibrium with the new surface, and the upstream profile above the IBL. The thickness of the IBL itself can be described as a function of the downstream distance x by (Kaimal and Finnigan, 1994)

$$\frac{\delta_i}{x}\left[\ln\left(\frac{\delta_i}{z_{0a}}\right)-1\right] \sim \text{constant}; \quad z_{0a}^2 = \left(z_{0u}^2 + z_{0d}^2\right)^{1/2} \tag{3.10.3}$$

The concept of self-similarity in aerodynamics has been applied to the IBL to scale the velocity profiles in the IBL (Mulhearn, 1977):

$$\frac{U_d(z)-U_u(z)}{u_{*d}-u_{*u}} \cong \frac{1}{\kappa}g(\eta); \quad \eta = \frac{z}{\delta_i} \tag{3.10.4}$$

Figure 3.10.3 shows the observed and theoretical forms of g. Figure 3.10.4 shows the surface stress ratio as a function of fetch. The phenomenon of overshoot is noteworthy. This is simply because only the flow in the vicinity of the surface adjusts to the abrupt change and the flow aloft takes a while to

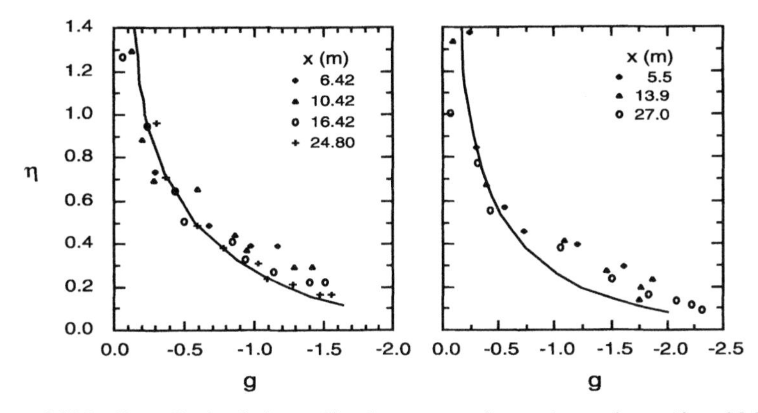

Figure 3.10.3. Normalized velocity profiles downstream of a roughness change (from Mulhearn, 1977).

accelerate or slow down in response. Also, the relaxation to equilibrium conditions is more rapid for smooth to rough transitions than vice versa, simply because the vertical diffusion of turbulence (and hence the changes felt) depends on the downstream friction velocity, which is smaller for the rough to smooth transition case.

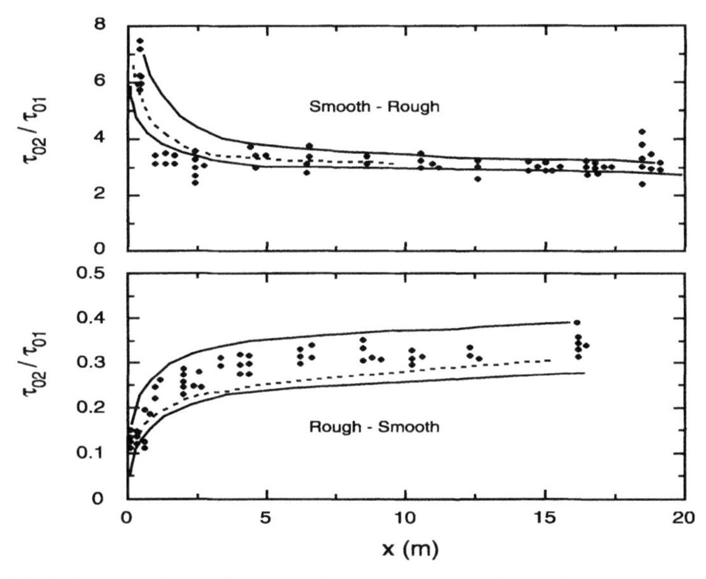

Figure 3.10.4. Relaxation of the surface stress downstream of a roughness change showing various theories as lines and observations as dots (from Atmospheric *Boundary Layer Flows* by J. C. Kaimal and J. J. Finnigan. Copyright 1994 Oxford University Press, Inc. Used by permission of Oxford University Press, Inc.).

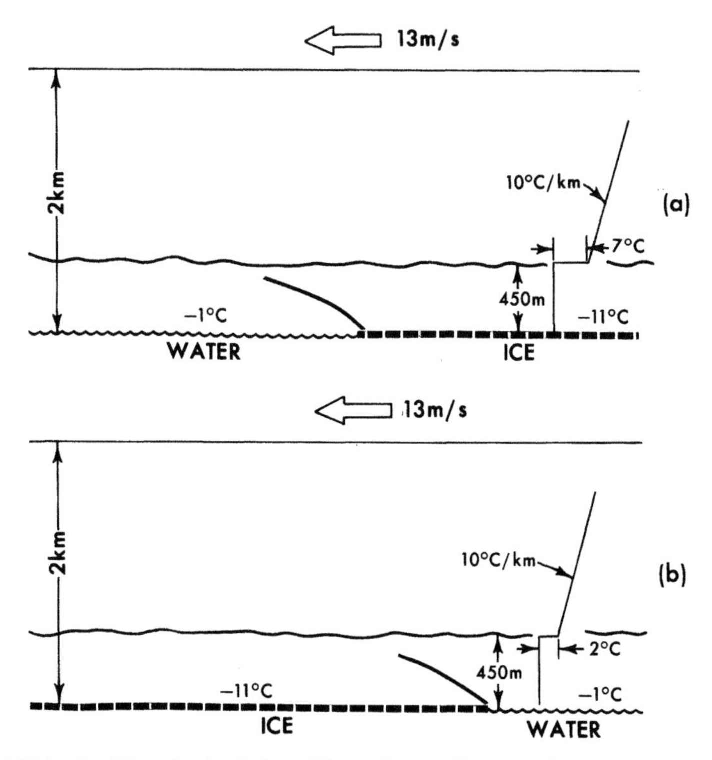

Figure 3.10.5. Conditions for simulating off-ice and on-ice flow conditions in Kantha and Mellor (1989) ABL model.

Overall, the IBL in a neutrally stratified flow is far simpler than that in a stratified flow, since here the roughness changes are usually overwhelmed by stratification effects. Detailed discussion of this situation is, however, beyond the scope of this book and the reader is referred instead to the current literature on the subject in journals like the *Journal of Atmospheric Sciences* and *Boundary Layer Meteorology*. A particularly interesting article is the review by Atkinson and Zhang (1996) on mesoscale shallow convection in the atmosphere, which is manifest in the form of distinctive cloud patterns, both linear cloud streets and hexagonal cellular convection patterns, typically a few to a few tens of kilometers in the horizontal with an IBL of depth f 500 m to 2 km. Cold-air outbreaks off eastern parts of the continental masses during winter and off-ice cold-air flow onto the water in a marginal ice zone are spectacular examples (see Figure 3.10.1). Here, the roll clouds aligned with the wind and spaced 1–2 km apart, and a few tens of kilometers long, break up into cellular patterns and eventually merge to form a solid stratocumulus deck. Cloud streets arise in the

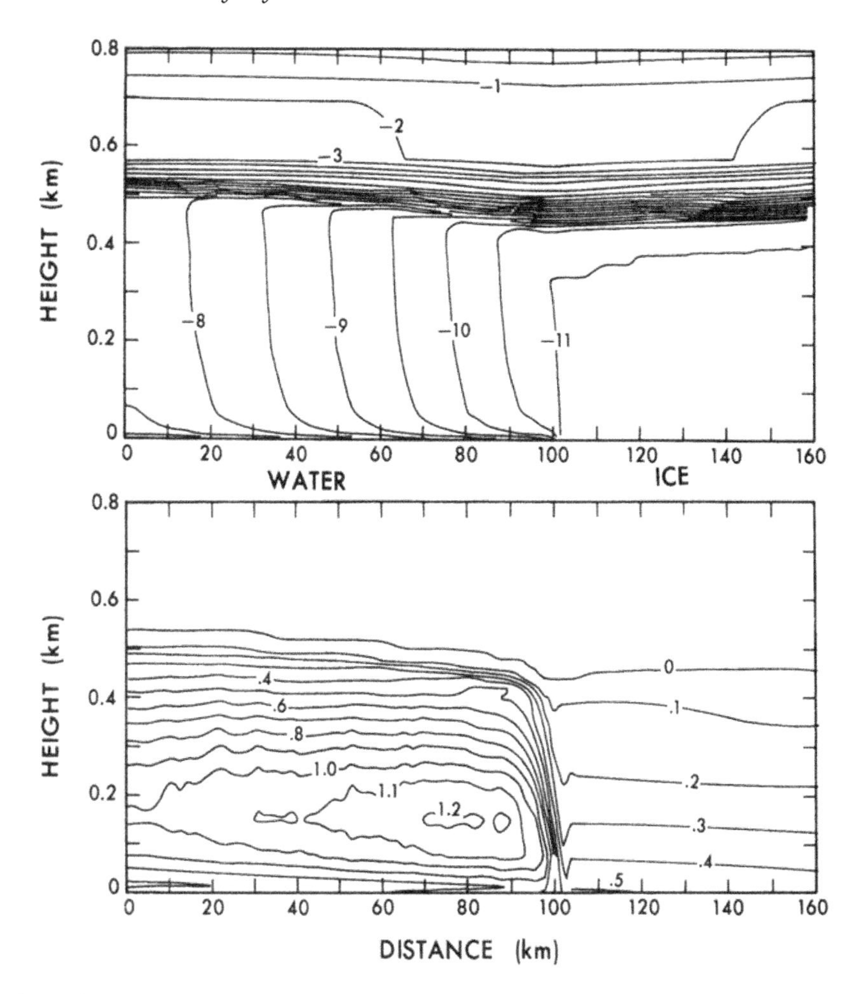

Figure 3.10.6. Temperature and TKE distributions for off-ice flow conditions in the ABL simulations of Kantha and Mellor (1989). Note the strong convective trbulence over water.

rising air between rolls aligned in the wind direction and spanning the IBL in the vertical. Roll-like features exist in ABLs over land surfaces also. Open cell convection patterns are found over the east sides of continents over warm water, driven by convection from below, whereas closed cell convection patterns are found on the west side over cold water, driven perhaps by radiative cooling at the cloud tops of the stratocumulus decks frequent in these regions (Atkinson and Zhang, 1996). A wide range of physical processes are involved, including thermal and dynamical instabilities, and gravity waves and turbulence. The

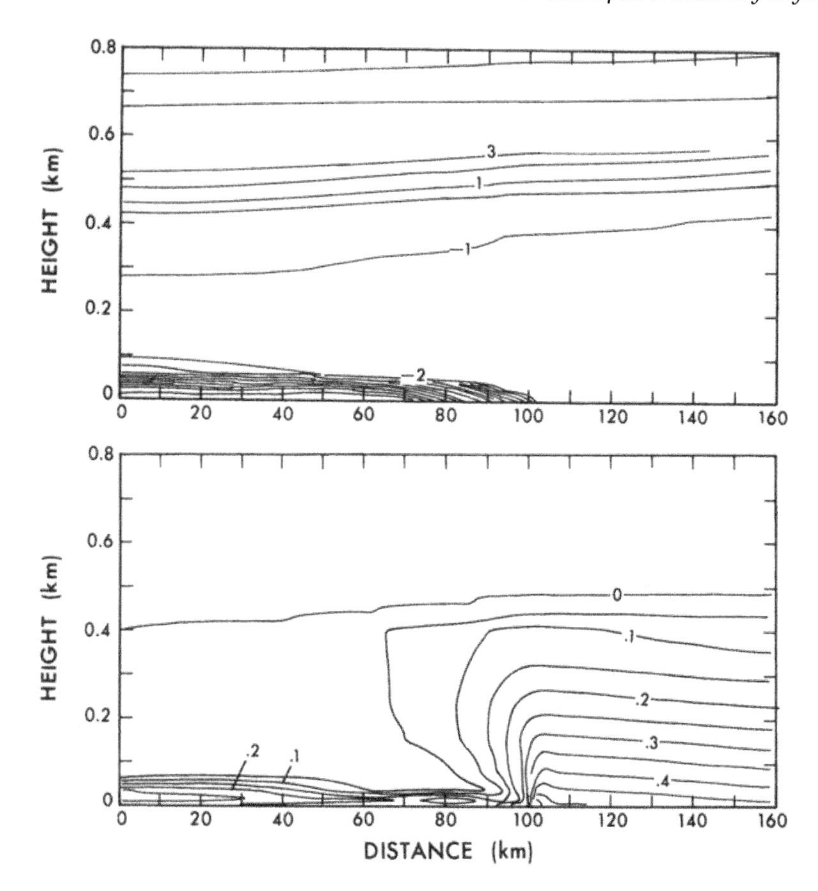

Figure 3.10.7 Temperature and TKE distributions for on-ice flow conditions in the ABL simulations of Kantha and Mellor (1989). Noe the strong damping of turbulence over ice.

reader is referred to Atkinson and Zhang (1996) for a fascinating review of shallow mesoscale convection in the atmosphere.

Examples of numerical solutions to the ABL over a MIZ can be found in Overland *et al.* (1983) and Kantha and Mellor (1989a). Figures 3.10.5 to 3.10.7 show the IBL evolution for off-ice and on-ice flows from simulations with a second-moment turbulence closure model (Kantha and Mellor, 1989a). Note the strong increase in turbulence intensities, and the rapid growth of the IBL in the former and the formation of a shallow IBL in the latter. More details such as the influence of associated roughness changes and overshoot in the surface stress can be found in Kantha and Mellor (1989a).

3.11 MODELING THE ATMOSPHERIC BOUNDARY LAYER

As for the OML, models for the ABL can be classified broadly into two categories (excluding the LES and DNS approaches): (1) slab (single layer) models and (2) diffusion (multilevel) models. Modern large scale atmospheric models have either a slab ABL parameterization included at the bottom of the atmospheric column or incorporate multilevel formulation implicitly, the vertical resolution and complexity depending on the resources that can be spared (Deardorff, 1972b). In cases where an ABL model is used by itself for applications such as air quality monitoring, the boundary conditions at the top of the model domain must be suitably prescribed. A common practice is to use the output of an atmospheric model. Slab models (for example, Deardorff, 1972; Overland *et al.,* 1979; Mass and Dempsey, 1985; Wilczak and Glendening, 1988) are simpler but cruder. Their main advantage is that they are not as computer-resource-intensive so that they can be run at the high horizontal resolutions needed to more accurately model the effects of the surrounding orography and air–land interface than is practicable in even a regional atmospheric model. It is therefore possible to use them sandwiched between regional models of the ocean and the atmosphere for, say, more accurate simulation of ocean surface currents in regions surrounded by orography (Clifford *et al.,* 1997; Horton *et al.,* 1997). Diffusion models based either on K-ε or second-moment closure (see Chapter 1), on the other hand, are best incorporated into the atmospheric model itself (Burke, 1988) at whatever horizontal and vertical resolutions deemed feasible. However, process studies using simple 1D and 2D models can be done with either (for example, Overland *et al.,* 1983; Kantha and Mellor, 1989a). Because of the need for an accurate knowledge of the state of the ABL for pollution, air quality, and many other studies, there exists a voluminous amount of work on modeling and parameterization of the ABL (often called PBL) and the reader is referred to reviews on the topic by, for example, Blackadar (1979), Wyngaard (1982, 1992), and Holt and Raman (1988). Articles on ABL modeling can be found regularly in the journal *Boundary Layer Meteorology.* Here we present a brief overview for completeness. We will discuss only the 1D case; extension to 2D and 3D cases are straightforward, though nontrivial.

3.11.1 SLAB MODELS

Here the governing equations are integrated through the layer, or equivalently, all variables are height-independent (Lavoie, 1972; Overland *et al.,* 1979; Horton *et al.,* 1997). Hydrostatic approximation suffices. The governing equations are

$$\frac{\partial u_k}{\partial x_k}+\frac{\partial w}{\partial z}=0$$

$$\frac{\partial u_j}{\partial t}+\frac{\partial}{\partial x_k}\left(u_k u_j\right)+\frac{\partial}{\partial z}\left(w u_j\right)+\varepsilon_{j3k} f u_k=-\frac{1}{\rho}\frac{\partial p}{\partial x_j}+\frac{\partial}{\partial z}\left(K_M\frac{\partial u_j}{\partial z}\right)+A_M\frac{\partial^2 u_j}{\partial x_k \partial x_k} \tag{3.11.1}$$

$$\frac{\partial p}{\partial z}=-\rho g$$

$$\frac{\partial \theta}{\partial t}+\frac{\partial}{\partial x_k}\left(u_k \theta\right)+\frac{\partial}{\partial z}\left(w \theta\right)=S_\theta+\frac{\partial}{\partial z}\left(K_H\frac{\partial \theta}{\partial z}\right)+A_H\frac{\partial^2 \theta}{\partial x_k \partial x_k} \tag{3.11.2}$$

$$\frac{\partial q}{\partial t}+\frac{\partial}{\partial x_k}\left(u_k q\right)+\frac{\partial}{\partial z}\left(w q\right)=S_q+\frac{\partial}{\partial z}\left(K_H\frac{\partial q}{\partial z}\right)+A_H\frac{\partial^2 q}{\partial x_k \partial x_k} \tag{3.11.3}$$

where f is the Coriolis parameter; g is the gravitational acceleration; u_j is the horizontal velocity; w is the vertical velocity; and z is the vertical direction. θ is the potential temperature and q is the specific humidity. A_M and A_H are the horizontal eddy viscosity and eddy diffusivity. Subscripts j and k indicate the two components of velocity and take values of 1 and 2 only, and repeated indexes imply summation. K_M is vertical diffusivity of momentum and K_H is that of a scalar, temperature and humidity. S_θ denotes source and sink terms—sources arising from condensation of water vapor when values reach above saturation and sinks arising from radiative cooling of the atmospheric column at night. Nocturnal radiative cooling is hard to parameterize since it depends on a variety of factors such as the water vapor content and the cloud cover. The simplest, though not too realistic, is to assume a uniform cooling rate throughout the column, about 1–2 °C hr^{-1}. S_q arises due to condensation and evaporation of water vapor. For simplicity, we will consider the dry case only. Using the perfect gas relation,

$$p=\rho RT \tag{3.11.4}$$

where R is the gas constant and T is the temperature, and noting that the potential temperature is defined by

$$\frac{\theta}{\theta_0}=\frac{T}{T_0}\left(\frac{p_0}{p}\right)^{(\gamma-1)/\gamma} \quad ; \quad R=c_p-c_v,\ \gamma=c_p/c_v=1.4 \qquad (3.11.5)$$

where subscript 0 denotes reference values ($\theta_0 = T_0$) and c_p and c_v are specific heats at constant pressure and constant volume, the pressure gradient terms in the momentum balance can be written as

$$\frac{1}{\rho}\frac{\partial p}{\partial x_j}=\frac{\theta}{\theta_0}\frac{\partial \pi}{\partial x_j}\sim\frac{\partial \pi}{\partial x_j};\quad \pi=c_p\theta_0\left(\frac{p}{p_0}\right)^{(\gamma-1)/\gamma} \qquad (3.11.6)$$

The hydrostatic balance becomes

$$\frac{\partial \pi}{\partial z}=-g\frac{\theta_0}{\theta} \qquad (3.11.7)$$

These then are the equations that need to be integrated across the slab. Doing so, the momentum equation becomes

$$\frac{\partial}{\partial t}\left[(h-s)U_j\right]+\frac{\partial}{\partial x_k}\left[(h-s)\left(U_k-U_k^a\right)\left(U_j-U_j^a\right)\right]+\varepsilon_{j3k}f(h-s)\left(U_k-U_k^a\right)$$
$$=A_M(h-s)\frac{\partial^2 U_j}{\partial x_k\partial x_k}+\frac{g}{\theta_0}\left[\Delta\theta(h-s)\frac{\partial h}{\partial x_j}+\underline{\frac{(h-s)^2}{2}\frac{\partial \Theta}{\partial x_j}}\right]$$
$$-C_D\left(U_kU_k\right)^{1/2}U_j+\underline{W_e\left(U_j^a\right)} \qquad (3.11.8)$$

where U_j is the vertically averaged velocity in the layer; U_j^a is the horizontal velocity above the layer; θ_0 is the reference potential temperature; Θ is the potential temperature in the layer; $\Delta\theta = \Theta - \theta^a$, where θ^a is the temperature above the layer; h is the height of the layer; s is the orographic height; (h – s) is the slab thickness; C_D is the surface drag coefficient; and W_e is the vertical entrainment velocity, positive when the slab is entraining and growing, and zero when it is shrinking. The pressure gradient term in Eq. (3.11.8) has been replaced using the momentum balance aloft,

$$\frac{\partial}{\partial x_k}\left(U_k^a U_j^a\right)+\varepsilon_{j3k} f U_k^a=-\frac{\partial \pi}{\partial x_j} \tag{3.11.9}$$

where $\partial\pi/\partial x_j$ is the horizontal pressure gradient above the layer. The velocity U_j^a is assumed to be provided by a coarser resolution meteorological model. The equation of continuity for the slab is

$$\frac{\partial}{\partial t}(h-s)+\frac{\partial}{\partial x_k}\left[U_k(h-s)\right]=\underline{W_e} \tag{3.11.10}$$

The potential temperature evolves as

$$\frac{\partial}{\partial t}[(h-s)\Theta]+\frac{\partial}{\partial x_k}[(h-s)U_k\Theta]=W_e\theta^a+Q+A_H(h-s)\frac{\partial^2\Theta}{\partial x_k\partial x_k} \tag{3.11.11}$$

Here θ^a is the potential temperature aloft, and Q is the kinematic heat flux from the ground to the layer calculated using

$$Q=C_H\left(U_k U_k\right)^{1/2}\left(\theta_S-\Theta\right) \tag{3.11.12}$$

where C_H is the dimensionless heat exchange coefficient and θ_S is the land or water surface temperature.

Equation (3.11.12) assumes the heat flux at the bottom of the slab to be entirely sensible. However, it is simple enough to carry an additional equation for specific humidity q similar to Eq. (3.11.11) and an exchange equation for water vapor. The equation of state must then be modified to include the specific humidity. This is simple enough to do since the perfect gas law can still be applied (see Appendix B). Including phase changes is, however, difficult, especially cloud physics. It is simplest to cap the relative humidity at 100% and convert the excess water to precipitation when supersaturation occurs. The latent heat released is used to heat the layer. The lapse rate $\partial\theta/\partial z$ above the layer is held fixed so that

$$\theta^a=\hat{\theta}^a+(h-\hat{h})(\partial\theta^a/\partial z) \tag{3.11.17}$$

where $\hat{\theta}^a$ is the initial temperature above the layer and $\hat{h}$ is the initial height of the layer. The entrainment rate W_e can be specified as a function of the

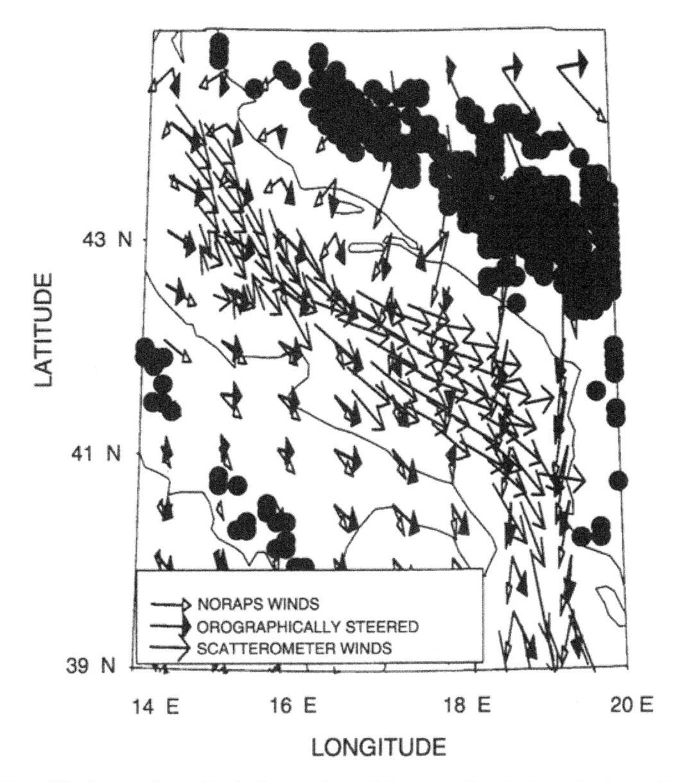

Figure 3.11.1. Winds over the Adriatic Sea, as derived from a regional atmospheric model, from an ABL model with proper orographic steering, and from a scatterometer (from Horton *et al.*, 1997).

Richardson number Ri defined as

$$\mathrm{Ri} = \frac{g(h-s)\Delta\theta}{\theta_0\left[\left(U_k - U_k^a\right)\left(U_k - U_k^a\right)\right]} \tag{3.11.18}$$

The most popular is the $W_e \sim Ri^{-1}$ empirical relationship. Another is to assume dynamical instabilities keep Ri at a constant value and adjust the layer height to achieve that (similar to the dynamical instability slab model in Chapter 2). If the underlined terms in Eqs. (3.11.8) and (3.11.10) are ignored, the model is similar to that used by Eddington *et al.* (1992) that ignores thermodynamics. A_M and A_H are prescribed empirically; the higher the model resolution, the smaller these values.

In limited domain models, boundary conditions for height, speed, and air temperature must be specified along the open lateral boundaries of the model domain. Conditions aloft must be prescribed. Both these can be time-dependent and derived from an atmospheric model forecast. Clifford *et al.* (1997) and Horton *et al.* (1997) used the NORAPS model forecasts to drive a high resolution slab ABL model to capture the influence of surrounding orography on low level winds. The resulting winds were used to drive regional ocean circulation models of the Red Sea and Mediterranean Sea. Orographic steering thus produced improved forecasts of circulation in these seas. Figure 3.11.1 shows an example of the improvement produced by the ABL model.

A possible improvement is to embed a constant thickness (say 30–50 m) surface layer at the bottom of the modeled layer and use well-known Monin–Obukhov constant flux layer relationships derived earlier in this chapter to relate the difference between the layer and the surface properties such as U_j and $(\Theta - \Theta_s)$ to the appropriate fluxes. See Overland *et al.* (1979) for application of such a model to a wide variety of flow situations.

3.11.2 MULTILEVEL MODELS

A fairly recent review can be found in Holt and Raman (1988). Basically these use the primitive equations (3.11.1)–(3.11.3), (3.11.6), and (3.11.7). Note that in these equations, the velocity u_j and θ are functions of the vertical coordinate z. There also arise vertical diffusion terms to contend with.

The solution of this set requires prescription of the heat flux and the momentum flux at the bottom of the ABL and a suitable condition on the mean properties at the top of the model domain. The bottom fluxes are based on Monin–Obukhov similarity, and at the top Newtonian damping to prescribed velocity and temperature aloft is used.

The basic problem is then to specify or calculate K_M and K_H. Since these are the result of turbulent mixing processes, one needs a turbulence model as well. Simplest is to prescribe a K-profile. Numerous such K-profile parameterizations are possible and are listed by Holt and Raman (1988). The main problem is the stability dependence of the K-profile, which has to be prescribed a priori. One possibility is Prandtl's mixing length approach that ties the eddy viscosity to the mean gradient and a length scale, the mixing length, whose value and functional form have to be prescribed suitably. One popular mixing length parameterization due to Blackadar (1962) is $l = \kappa z\,(1+\kappa z/l_0)^{-1}$, which gives $l = \kappa z$ close to the ground, asymptoting to $l = l_0$ toward the edge of the ABL.

However, a better approach is to tie the eddy viscosity and diffusivity to the turbulence velocity scale. Thus K_M, $K_H \sim K^{1/2} l$, where K is the TKE. This then requires a prognostic equation for K to be solved, but stability effects are now included. In addition, l needs to be prescribed as before or computed using either a K-ε or some other closure. Chapter 1 dealt with this topic in some detail. Holt and Raman (1988) compared various formulations in a 1D model and concluded that the use of the TKE equation (either with a prescribed or a calculated l) yields better results for not only the mean quantities but also the turbulence, especially the TKE budget in the ABL.

If one ignores LES's, then second-moment closure (see Chapters 1 and 2) is perhaps the best compromise when accuracy is essential. This explains the popularity of the Mellor and Yamada (1982) model in its various forms in ABL applications. In its simplest form, it reduces to a two-equation turbulence model, with stability-dependent K_M and K_H (see Chapter 2), and has been used for many ABL applications as well as routine inclusion in atmospheric models. See Yamada (1983) for a typical application (see also Mellor and Yamada, 1982). However, these formulations have not allowed for countergradient fluxes important to convective ABLs (Deardorff, 1966, 1972). An ABL model that does so specifically within the context of the K-profile approach is the one by Troen and Mahrt (1986).

The above slab and multilevel formulations can be applied to 2D and 3D ABL models also. For application of a slab model to the ABL over sea ice see Overland *et al.* (1983), and for that of a second-moment closure diffusion model to the same problem, see Kantha and Mellor (1989). Results from the latter were presented in the last section.

LIST OF SYMBOLS

α	Angle between the geostrophic wind and the shear stress
δ	Boundary layer thickness
δ_I	Internal boundary layer thickness
ε	Dissipation rate of TKE
$\phi(n)$	Frequency spectrum
λ	Wavelength
μ	Monin–Kazanski parameter
ν, ν_t	Kinematic viscosity (molecular and turbulent)
θ_*	Friction temperature
θ_* θ_v, θ_l	Potential, virtual potential , and liquid water potential temperatures
ϕ_M, ϕ_H, ...	Monin–Obukhoff similarity functions
ρ	Density
ζ	Monin–Obukhoff similarity variable
κ	Von Karman constant
τ	Shear stress

Ω	Angular rotation rate of the Earth
c_d	Drag coefficient
d, d_T, d_E	Displacement scales for momentum, heat, and water vapor
f	Coriolis frequency
g	Acceleration due to gravity
h	Inversion height
k	Magnitude of the wavenumber vector; also summation index that takes values of 1 to 3
k_3	Vertical component of the wavenumber
l	Integral microscale of turbulence
n	Wave frequency
p	Pressure
q, q_*	specific humidity and friction specific humidity
r, r_l	Water vapor mixing ratio and liquid water mixing ratio
t	Time
w	Vertical component of velocity perturbation
u_a^*, u_*	Friction velocity
u_f^*	Free convection velocity scale
x	Horizontal coordinate
x_3	Vertical coordinate
x_i	Coordinates
z	Vertical coordinate
z_0, z_{0F}	Roughness scale for momentum and scalars
z_{0u}, z_{0d}	Upstream and downstream roughness scales

A_e	Eddy viscosity
A, B, C	Constants in the surface layer
B_r	Bowen ratio
C	Wave phase speed
C_g	Group velocity
C_D, C_H, C_E	Bulk transfer coefficients for momentum, heat, and moisture
C_{DN}, C_{HN}	Neutral bulk transfer coefficients
E	Energy density of the wave
Fr, Fr_I	Froude number and inverse Froude number
G	Geostrophic velocity
H	Height of the mountain
H_0, H_t	Heat flux at the ground and entrainment heat flux at the inversion
K_M, K_H	Turbulent (eddy) mixing coefficients for momentum and heat
L	Monin–Obukhoff length scale; also width of the mountain
L_E	Latent heat of evaporation
L_R	Internal Rossby radius of deformation
N(z)	Buoyancy frequency

Nu	Nusselt number
P	Mean pressure; also potential energy
Pr_t	Turbulent Prandtl number
Q_b	Surface buoyancy flux
Q_0, Q_t	Kinematic heat flux at the ground and kinematic entrainment heat flux at the inversion
R	Radius of curvature
Ra	Rayleigh number
Ri_c	Curvature Richardson number
Ri_f, Ri_g	Flux and gradient Richardson numbers
Ro, Ro^*	Rossby and friction Rossby numbers
S_R	Rossby number of flow over a mountain
T, T_v	Absolute temperature and virtual temperature
T^*	Friction temperature
U, U_{10}	Wind velocity and wind velocity at 10 m
U_u, U_d	Velocity upstream and downstream

Chapter 4

Surface Exchange Processes

In this chapter, we outline some important features of the exchange of momentum, heat, mass, and gases between the oceans, the atmosphere, and land surfaces. Air–sea exchange is central to the determination of the state of the atmosphere over timescales larger than about a few days. It plays a crucial role in processes such as the ENSO (El Niño–Southern Oscillation). The sensible and latent heat exchanges occurring in the warm pool region of the western tropical Pacific are thought to be important to determining the onset, growth, and decay of the ENSO, which has a global impact on weather and precipitation characteristics. The transfer of greenhouse gases such as CO_2 across the air–sea interface has important implications for climate change. Exchanges of heat and moisture between the atmosphere and the soil or vegetation are also part of the global hydrological cycle and must be considered in a global energy balance. Vegetation also exchanges important gases with the atmosphere such as oxygen and carbon dioxide, and these exchanges are modulated by the atmospheric boundary layer. Kaimal and Finnigan (1994), Kraus and Businger (1994), and Kagan (1995) are recent treatises covering the topic of air–surface interactions. An excellent discussion of flux measurement techniques and the associated errors can be found in Dabbert *et al.* (1993) and Smith *et al.* (1996a); the latter is also a recent overview of air–sea fluxes. Recent advances in the field can be found in journals such as the *Journal of Geophysical Research of the American Geophysical Union,* the *Journal of Atmospheric Sciences,* the *Journal of Physical Oceanography,* and the *Journal of Atmospheric and Oceanic Technology of the American Meteorological Society,* and *Boundary Layer*

Meteorology, among others. In this chapter, we will review how the exchange of properties across various surfaces are determined and outline some recent advances in the field.

4.1 SURFACE ENERGY BALANCE

The total heat flux consists of both radiative fluxes and turbulent fluxes. Since the higher temperature of the Sun results in radiation emitted at shorter wavelengths (ultraviolet, visible, and infrared regions) than the cooler Earth system, the usual practice is to divide the radiative fluxes into longwave (LW) and shortwave (SW) components. The longwave radiation is emitted by the atmosphere and the surface, whether the surface consists of vegetation, ice, water, or soil. The turbulent heat fluxes are the latent heat flux, or the exchange of heat due to the process of evaporation, and the sensible heat flux. The turbulence of the atmospheric boundary layer impacts directly on the turbulent heat fluxes but not on the radiative fluxes. In addition, the properties of the surface directly affect the transfer of heat between the atmosphere and the surface.

At the atmosphere–surface interface, there must exist a balance between the incoming and the outgoing components of various fluxes, including the heat flux. If this interface is assumed to be infinitesimally thin, then the net fluxes of all scalars, including heat, must vanish at the surface. In the most general way, this can be written as

$$SW_{\downarrow} + LW_{\downarrow} - SW_{\uparrow} - LW_{\uparrow} - H_s - H_L - H_g = 0 \tag{4.1.1}$$

There is thus a balance between the downwelling shortwave ($SW_{\downarrow}$) and longwave ($LW_{\downarrow}$) radiative fluxes, the upwelling $SW_{\uparrow}$ and $LW_{\uparrow}$ fluxes, and the sensible (H_s) and latent (H_L) heat fluxes. H_g is the heat flux to the ground. All these quantities are in W m^{-2}. The convention here is that a term that adds heat to the interface has a positive sign in front of it (for example, $SW_{\downarrow}$) and the term that takes heat away from the interface has a negative sign in front of it (for example, H_L). The radiative fluxes as written above are all positive quantities. The sensible and latent heat fluxes are normally positive (heat loss for the ocean and a gain to the atmosphere) over the ocean, but can be of either sign over the land and may change sign over a diurnal cycle. During warm air outbreaks over the ocean, these fluxes can change sign as well. The Earth's annual global mean energy budget has been estimated using several different data sets and techniques. A recent survey, performed by Kiehl and Trenberth (1997), produced values shown in Figure 4.1.1.

Considerable effort has been expended in determining these fluxes as a function of the ambient parameters, since it is quite difficult to measure these

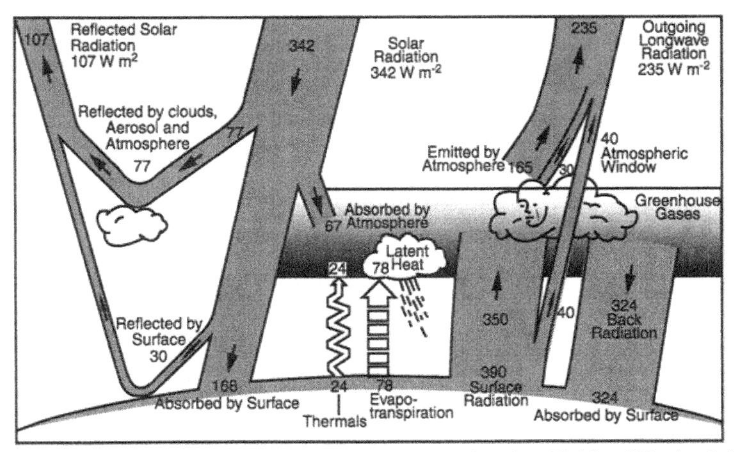

Figure 4.1.1. The Earth's annual global mean energy budget based on Kiehl and Trenberth (1997). Units are W m^{-2}.

fluxes directly. The next section describes radiative flux parameterizations, and the rest of the sections in this chapter deal with the parameterization of the turbulent heat fluxes H_s and H_l.

4.2 RADIATIVE FLUX PARAMETERIZATION

Direct measurements of the radiative fluxes, either SW or LW, at an air–sea, air–ground, or air–ice interface are seldom possible. These fluxes require sensitive instruments and great care in measurements, and, over the sea, can seldom be made except from specially equipped research vessels. Instrumented surface buoys are becoming a viable alternative. Satellite-borne sensors are being increasingly used, but they have difficulty in measuring all the components of the radiation balance at the bottom of the atmospheric column. Also, at present, there is no observational network for radiation measurements, especially over ice and the oceans. Therefore, empirical formulas still form the basis for estimating the radiative and other fluxes at the air–sea, air–ground, or air–ice interfaces. Given the importance of being able to accurately estimate all the components of energy balance at these interfaces (Fairall *et al.*, 1996a), increasing attention is being given to evaluating these formulas (e.g., Schiano, 1996; Key *et al.*, 1996; see also Isemer and Hasse, 1987). This section contains a summary of the current status of radiative flux parameterization, drawing mostly upon an excellent summary by Key *et al.* (1996).

Accurate parameterization of the SW and LW radiative fluxes is important to modeling the evolution of the atmosphere, ocean, and sea ice. They form a substantial component of the energy balance at these various interfaces, and are therefore integral to air–sea exchange and climate-related questions. High quality data sets are very difficult to gather at sea. SW solar radiation is measured by high precision spectral pyranometers, which measure the incident radiation in the 280- to 2800-nm range to an accuracy of perhaps 2%. At sea, the main sources of error are the influence of the ship superstructure and its shadowing effects, clouding of the pyranometer domes by salt spray and rain debris, and departure from the horizon due to ship motion. Observations made during rough seas and heavy rain have to be discarded. Periodic cleaning and calibration are essential to maintaining the accuracy of the measurements whether at sea or over land or ice. Downwelling LW radiative fluxes are measured by pyrgeometers and here, in addition to the usual problems, radiation by heating of the domes is a major source of contamination and great care needs to be exercised in measuring the dome temperature and accounting for its effects to maintain desired accuracies.

The fate of solar radiation incident at the top of the atmosphere is quite complex. Some of it is absorbed and scattered by the intervening atmospheric column. Part is absorbed by the ground or transmitted into the ice or sea; the radiative energy is reemitted in the form of LW (thermal) radiation, which is in turn absorbed, scattered, and partially reradiated by the atmosphere. The principal sources for absorption and scattering in the atmosphere are its constituent gases: ozone, carbon dioxide and oxygen, water vapor, and aerosols. Clouds composed of condensed liquid water play a dominant role in the absorption, reflection, and reemission of solar radiation. High level clouds efficiently reflect the SW solar radiation back into space and therefore their albedo effects are crucial to the overall energy balance of the Earth. In addition, low level clouds cause multiple reflections between the cloud base and the surface. Therefore the downwelling LW flux is a result of scattering and emission from the whole atmospheric column, although the lowest hundred meters or so are the most relevant to the surface fluxes. An estimation of the top of the atmosphere and surface values for the radiative fluxes over various wavelengths are shown in Figures 4.2.1 and 4.2.2.

Since the downwelling SW and LW radiative fluxes at the surface are a complex function of the many properties of the atmospheric column, the best possible estimate is by the use of radiative transfer models. Such models, however, require information on vertical distribution of temperature, moisture, aerosols, and clouds, which is seldom available. For this reason, satellite-sensed data on radiation and clouds are being increasingly used in conjunction with these models to provide a more complete and useful characterization of the radiation (Rossow and Schiffer, 1991). However, there are many applications that require information not readily available from the satellite-sensed data. Instead, empirical parameterization formulas, unfortunately often inadequately substantiated, form the mainstay of most models. These formulas are designed to accept rather minimal input data such as solar zenith angle, cloud cover, and sea

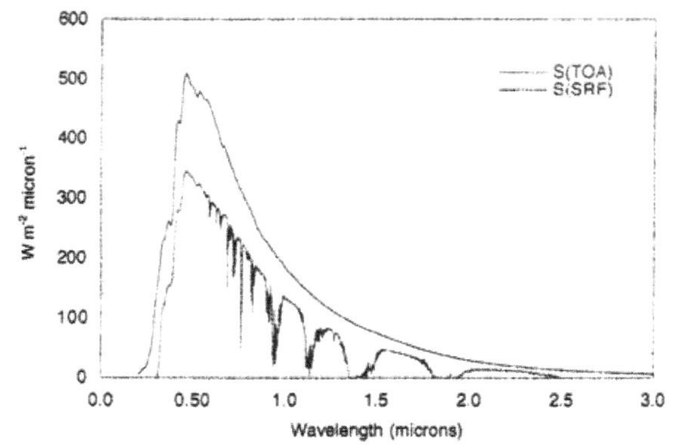

Figure 4.2.1. Downward shortwave flux (W m^{-2} μ m^{-1}) at the top of the atmosphere and at the surface for cloudy conditions (from Kiehl and Trenberth, 1997).

level atmospheric temperature that are relatively easy to obtain or measure. A variety of parameterization schemes exist for both SW (Kimball, 1928; Berliand, 1960; Laevastu, 1960; Lumb, 1964; Tabata, 1964; Budyko, 1974; Reed, 1977; Jacobs, 1978; Bennett, 1982; Lind *et al.*, 1984; Shine, 1984; Dobson and Smith, 1988) and LW (Efimova, 1961; Marshunova, 1966; Idso and Jackson, 1969; Zillman, 1972; Clark *et al.*, 1974; Bunker, 1976; Reed, 1977; Jacobs, 1978; Andreas and Ackley, 1982) radiative fluxes. Some of these are for clear conditions and some are for cloudy skies. Schiano (1996) has done a careful study of the widely used Reed (1977) SW scheme over the western Mediterranean Sea and Key *et al.* (1996; see also Katsaros, 1990) have evaluated both SW and LW schemes specifically for use in sea-ice models.

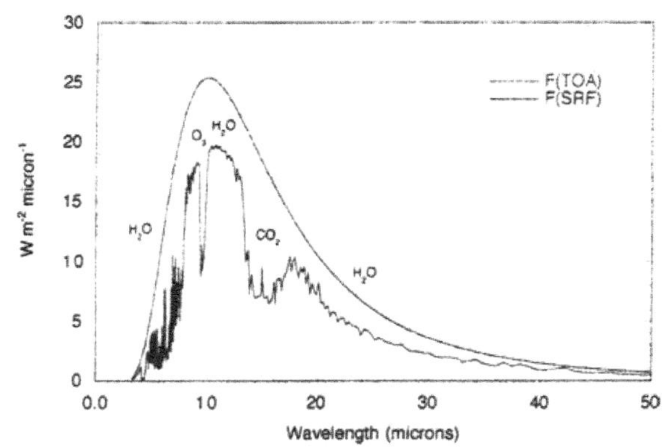

Figure 4.2.2. Surface and top-of-atmosphere upward longwave flux (W m^{-2} μ m^{-1}) for global cloudy conditions. Various gases contributing to the absorption and emission of longwave radiation are denoted (from Kiehl and Trenberth, 1997).

4.2.1 Downwelling Shortwave Radiative Flux

From observations over the mid-Atlantic, Lumb (1964) suggested a formula suitable for hourly, daily, and monthly values for downwelling SW fluxes (in W m^{-2}) for clear skies,

$$SW_{\downarrow} = S_o \cos\theta_z (0.61 + 0.20\cos\theta_z) \tag{4.2.1}$$

where S_o is the solar constant and θ_z is the solar zenith angle (in radians). The zenith angle is given by (Zheng and Anthes, 1982)

$$\sin\theta_z = \sin\phi \, \sin\delta - \cos\phi \, \cos\delta \, \cos\psi \tag{4.2.2}$$

where ϕ is the latitude, δ is the declination, and ψ is the hour angle (all in radians):

$$\delta = 0.409\cos\left[\frac{2\pi(D_y - 173)}{365.25}\right]$$
$$\psi = \left[\left(\frac{\pi T_{UTC}}{12}\right) - \lambda_c\right] \tag{4.2.3}$$

D_y is the day of the year, T_{UTC} is the Coordinated Universal Time in hours, and λ_e is the longitude in radians (positive west).

The atmospheric transmittance is implicitly included in Lumb's formula. However, the transmittance varies between 0.69 and 0.72, and is strongly dependent on the aerosol and water vapor content of the atmospheric column. This formula has been found sensitive to location and does not include the effect of water vapor explicitly. From data over the Indian Ocean, Zillman (1972) derived a formula that includes the effect of the near-surface vapor pressure p_v in bars:

$$SW_{\downarrow} = S_o \cos^2\theta_z \left[1.085\cos\theta_z + (2.7 + \cos\theta_z)p_v + 0.1\right]^{-1} \tag{4.2.4}$$

Shine (1984) suggests a slightly revised version:

$$SW_{\downarrow} = S_o \cos^2\theta_z \left[1.2\cos\theta_z + (1.0 + \cos\theta_z)p_v + 0.0455\right]^{-1} \tag{4.2.5}$$

Cloud cover is an additional parameter that cannot be ignored. The downwelling shortwave radiation after the inclusion of clouds ($SW_{\downarrow}^{c}$) was parameterized by Berliand (1960) as

$$SW_{\downarrow}^{c} = SW_{\downarrow}(1 - \alpha\, C) \tag{4.2.6}$$

where α depends on the latitude (0.45 at 75°). Jacobs (1978) suggests a value of 0.33 for Baffin Bay and Bennett (1982) 0.52 for the Arctic.

Based on measurements at six coastal sites, Reed (1977) suggested the approximation

$$SW_{\downarrow}^{c} = SW_{\downarrow}[1 + 0.0019(90 - \theta_{z}^{n}) - 0.62\, C] \tag{4.2.7}$$

for mean daily insolation based on the noon solar zenith angle in degrees and fractional cloud cover C, where the clear sky value is given by (Seckel and Beaudry, 1973)

$$SW_{\downarrow} = A_0 + A_1 \cos\beta + B_1 \sin\beta + A_2 \cos 2\beta + B_2 \sin 2\beta \tag{4.2.8}$$

where $\beta = (D - 21)(360/365)$, D is time of the year in days, and the coefficients are functions of latitude (see Table 1 from Schiano, 1996). This formula has been checked against a few hundred coastal observations by Reed (1977) and against observations over the oceans by Reed (1982) and Reed and Brainard (1983). Cloud cover is usually measured in oktas (1/8th of the sky) and Eq. (4.2.4) is valid only for $C > 2$ oktas. For lesser values, clear sky values need to be taken. This formula has proved useful although it overestimates the clear sky insolation in the tropics by 2%. It also gave generally good agreement with measurements over the western Mediterranean given the increased aerosol content and hence lower atmospheric transmission coefficient. Tabata (1964) recommends a formula similar to Eq. (4.2.7), but with coefficients of 0.00252 and 0.716 instead. Simpson and Paulson (1979), Isemer and Hasse (1987), and Dobson and Smith (1988) have used Tabata's formulation. Simpson and Paulson (1979) and Frouin *et al.* (1988) have evaluated different insolation formulas and conclude that Reed's formulation is the simplest to use, even if it overestimates by as much as 6%. Garrett *et al.* (1993) and Gilman and Garrett (1994) have also found Reed's formula to overestimate the solar insolation, but this is attributed to incorrect use for very low cloud covers (<6%) and the effect of aerosols. Schiano (1996) also found the effect of atmospheric aerosols to be the principal factor in the error, but overall its daily insolation value agreed well with the measurements over the western Mediterranean, except for a bias of 5 W m^{-2}.

From midlatitude data, Laevastu (1960) suggested

$$SW_{\downarrow}^{c} = SW_{\downarrow}(1 - 0.6\, C^{3}) \tag{4.2.9}$$

which has been used by Parkinson and Washington (1979), but this generally overestimates the SW radiation except for high values of C. Berliand (1960) suggested a quadratic dependence on cloud cover,

$$SW_{\downarrow}^{c} = SW_{\downarrow}(1 - 0.38\,C - \alpha C^{2}) \tag{4.2.10}$$

with the coefficient α dependent on latitude (0.14 at 85°, 0.41 at 55°, and 0.38 at 45°) (Budyko 1974). Bishop and Rossow (1991) consider this to be better than Reed's over land surfaces, although it is considered to be poor over the ocean. Lumb (1964) used an equation of the type like Eq. (4.2.1) for cloudy skies, but with the coefficients dependent on nine categories of cloud types.

By far the most complex is that of Shine (1984), which includes the surface albedo a and cloud optical depth (unitless) d_c,

$$SW_{\downarrow}^{c} = SW_{\downarrow}(1-C) + C\{(53.5 + 1274.5\cos\theta_z)\cos^{0.5}\theta_z\,[1 + 0.139(1 - 0.935a)d_c]^{-1}\} \tag{4.2.11}$$

where the first term is given by Eq. (4.2.5). From comparisons with *in situ* measurements, Key *et al.* (1996) find this to be the most accurate for use at high latitudes and for sea-ice modeling. However, Reed's formula over the ocean and Berliand's over land are much simpler to use.

4.2.2 Downwelling Longwave Radiative Flux

The clear sky downwelling LW flux depends on the near-surface air temperature T_a and vapor pressure p_v and is usually parameterized as

$$LW_{\downarrow} = \sigma T_a^4 (a + b p_v^{0.5}) \tag{4.2.12}$$

where σ is the Stefan–Botzmann constant. a and b are constants assigned different values by different investigators. Brunt (1932) suggests a= 0.526, b=2.372; Berliand and Berliand (1952) use a=0.61, b=1.834; Marshunova (1966) suggests a=0.67, b=1.581 for Arctic drifting stations, and average values of a = 0.65, b=1.866, for coastal Arctic stations. Efimova (1961) suggests instead

$$LW_{\downarrow} = \sigma T_a^4 (0.746 + 6.6 p_v) \tag{4.2.13}$$

for low humidities typical of the Arctic.

From low and midlatitude data Swinbank (1963) suggests

$$LW_{\downarrow} = \sigma T_a^4 (9.365 \cdot 10^{-6} T_a^2) \tag{4.2.14}$$

for the temperature range 275 K $\leq T_a \leq$ 302 K. Zillman (1972) uses a coefficient of 9.2 instead of 9.365 for the southern oceans. Note that this formula depends on only the air temperature.

A more general formula was developed by Idso and Jackson (1969):

$$LW_{\downarrow} = \sigma T_a^4 [1 - 0.261 \exp\{-7.77 x 10^{-4} (273.15 - T_a)^2\}] \tag{4.2.15}$$

There are formulas based on both temperature and vapor pressure

$$LW_{\downarrow} = \sigma T_a^4 [0.70 + 5.95 x 10^{-5} p_v^{1500/T_a}] \tag{4.2.16}$$

by Idso (1981) from measurements in Arizona. However, for Arctic applications Andreas and Ackley (1982) suggest a value of 0.601 for the first constant. Parkinson and Washington (1979) use Eq. (4.2.13) in ice modeling.

For cloudy skies, it is necessary to increase the clear sky value because of LW emission by the cloud base. Jacobs (1978) suggests

$$LW_{\downarrow}^{c} = LW_{\downarrow} (1 + 0.26\ C) \tag{4.2.17}$$

This is similar to Marshunova (1966) but the constant is different: 0.16 for Arctic stations and 0.22 for drifting stations in summer, and 0.31 and 0.30 correspondingly for winter. Budyko (1974) suggests a quadratic dependence on cloud cover. Zillman (1972) suggests

$$LW_{\downarrow}^{c} = LW_{\downarrow} + 0.96\ C\ \sigma T_a^4 (1 - 9.2 x 10^{-6} T_a{}^2) \tag{4.2.18}$$

Maykut and Perovich (1987) suggest for the Arctic

$$LW_{\downarrow}^{c} = \sigma T_a^4 (0.7855 + 0.2232 C^{2.75}) \tag{4.2.19}$$

By far the most complex is the parameterization suggested by Schmetz *et al.* (1986) based on cloud properties such as cloud base temperature,

$$LW_{\downarrow}^{c} = LW_{\downarrow} + (1 - \varepsilon)\varepsilon_c C\ \sigma T_a^4 \exp[(T_b + T_a)/46] \tag{4.2.20}$$

where ε and ε_c are effective sky and cloud emissivities, and T_b is the cloud base temperature. However, there is usually inadequate input for use of this kind of equation. Instead, simpler ones are more widely used. Key *et al.* (1996) find Eqs. (4.2.13) and (4.2.14) to give best results for the Arctic.

Bignami *et al.* (1995) have compared their measurements of upwelling and downwelling LW fluxes in the Mediterranean with many of the above formulations and have concluded that the Efimova-type parameterization (4.2.13) with constants 0.684 and 0.0056 fits their clear sky hourly data quite well with nearly zero bias and a rms of about 10 W m^{-2}. For cloudy skies they find

$$LW_{\downarrow}^{c} = LW_{\downarrow}(1 + 0.1762\, C^2) \qquad (4.2.21)$$

fits their data well with a near zero bias and a rms of 14 W m^{-2}. None of the other formulas that they used for calculating the LW flux had as small a bias and rms error.

Josey *et al.* (1997) have compared measurements of downwelling LW flux in the North Atlantic and the Southern Ocean to estimations based on Clark *et al.* (1974), Bunker (1976), and Bignami *et al.* (1995) formulas, and conclude that the Clark *et al.* (1974) formulation for the net LW flux led to the least bias and recommend its use in midlatitudes. It contains a latitude-dependent cloud cover coefficient.

4.2.3 Upwelling Shortwave and Longwave Radiative Fluxes

Upwelling SW flux results from specular and diffuse reflection of incident SW radiation by the underlying surface, whereas the upwelling LW flux is due not only to reflection of downwelling IR energy, but also emission by the underlying surface. The upwelling SW radiative flux is

$$SW_{\uparrow} = \alpha SW_{\downarrow} \qquad (4.2.22)$$

where α is the albedo of the surface, which depends very much on the nature of the surface [see Stull (1988) and Garratt (1992) for tables of albedo and emissivity of natural surfaces]. Soil albedo can vary from 0.1 for a wet soil to as much as 0.35 for a dry one. Ice albedo is typically 0.6, whereas snow albedo can be as high as 0.9. The albedo of plant canopy varies between 0.1 and 0.2. The albedo of the ocean depends on the sea state and the sun angle; it is typically 0.05 for high sun angles, and varies between 0.1 and 0.5 for low sun angles (Payne, 1972).

The upwelling LW flux is

$$LW_{\uparrow} = \varepsilon\sigma T_s^4 + \alpha_l LW_{\downarrow} \tag{4.2.23}$$

where T_s is the surface temperature, ε is the emissivity of the surface, and α_l is the LW reflectance or albedo of the surface (~0.045 for the ocean). Emissivity typically ranges between 0.95 and 0.98, the typical value being 0.97 for the ocean (see Stull, 1988; Garratt, 1992). Often the upwelling and downwelling LW fluxes are combined in the same formula (for example, Bignami *et al.,* 1995) assuming that T_s and T_a are equal (if only one of these quantities is known, there is no recourse but to assume they are the same).

It is apparent that the best way to characterize the SW and LW radiative fluxes at the bottom of the atmospheric column is either to measure them *in situ* accurately or, if this is not possible, to appeal to radiative models and satellite data [ISCCP, International Satellite Cloud Climatology Project (Rossow and Schiffer, 1991)] with sufficient input data on the state of the atmospheric column. However, simple bulk parameterizations such as the ones discussed above will continue to be useful, especially for modeling the various components of the Earth's systems. It must be kept in mind that these formulas can often involve significant errors, especially due to inaccurate estimation of cloud cover and the atmospheric transmission coefficient due to aerosols.

4.3 FLUX BALANCE AT THE AIR–SEA INTERFACE

From the point of view of both the oceans and the atmosphere, we need to determine the sensible and latent heat fluxes from the ocean to the atmosphere. These fluxes are central to the interaction and coupling between the atmosphere and the ocean. There must also exist a net momentum balance. However, for air–sea exchange purposes, we need to know simply the momentum flux from the atmosphere to the oceans, and therefore the transfer of momentum from winds to surface waves is usually relevant only in so far as it affects the net transfer to the ocean currents. Similarly, the water vapor and gas fluxes from the oceans to the atmosphere, which are usually net losses to the ocean, need to be determined. All these fluxes of heat, mass, momentum, and gases are determined by turbulent processes in the surface layers of the atmosphere and the oceans adjacent to the air–sea interface, with surface waves playing an important role by virtue of their ability to act as sinks of momentum, to determine the "roughness" of the sea surface, and to disrupt the aqueous molecular sublayer responsible for transfer of scalar properties across the interface. At sufficiently high wind speeds, spray and droplets ejected into the atmosphere and air bubbles entrained into the ocean during wave breaking directly affect the water vapor and gas exchange between the two media.

Figure 4.3.1 shows the heat and momentum balance at the air–sea interface; that at the air–ice interface is also shown . If these interfaces are assumed to be infinitesimally thin, then the net fluxes of all scalars, including heat, must vanish at the interface. For the air–sea interface, this means

$$SW_{\downarrow} + LW_{\downarrow} + H_{pr} - SW_{\uparrow} - LW_{\uparrow} - H_s - H_L - SW_{\downarrow}^{P} = 0 \qquad (4.3.1)$$

where $SW_{\uparrow} = \alpha\ SW_{\downarrow}$; α is the albedo of the ocean surface. There is thus a balance between the downwelling shortwave ($SW_{\downarrow}$) and longwave ($LW_{\downarrow}$) radiative fluxes, and the upwelling $SW_{\uparrow}$ and $LW_{\uparrow}$ fluxes, the sensible (H_s) and latent (H_L) fluxes, the heat flux due to any precipitation (H_{pr}), and the solar radiative flux penetrating into the ocean ($SW_{\downarrow}^{P}$). As in Eq. (4.1.1), these quantities are in W m^{-2} and the sign conventions are the same. The radiative fluxes as written above are all positive quantities. The sensible and latent heat fluxes are normally positive (heat loss for the ocean and a gain to the atmosphere) and the precipitation heat flux is negative.

Over the sea, if the diurnal variability of the sea surface temperature is small, there is little deviation in either the latent heat flux or the sensible heat flux from day to night. Also, under conditions when the air temperature over water does not deviate strongly from the sea surface temperature, the sensible heat flux is small compared to the latent heat flux.

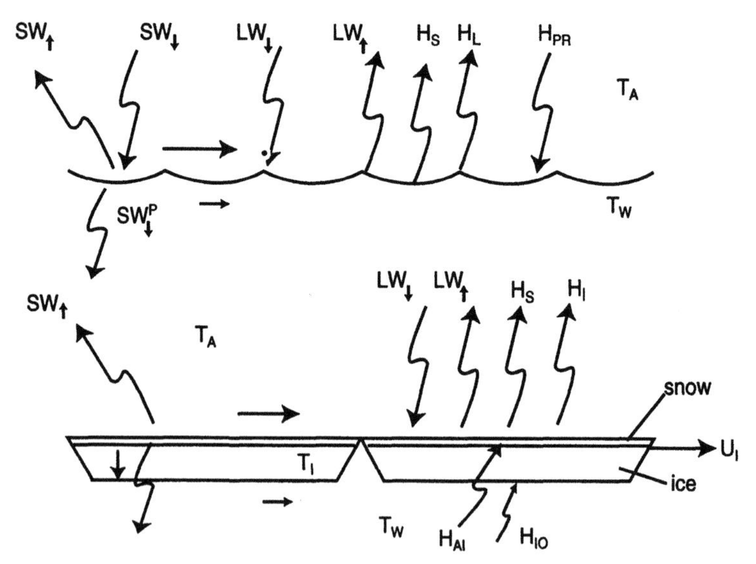

Figure 4.3.1. The flux balance at the air–sea and air–ice interfaces. The fluxes are as given in the text.

There are several additional terms in Eq. (4.3.1) which are particular to the air–sea interface. H_{pr} is the heat flux due to any rain or snow. Since raindrops are usually at the wet bulb temperature T_{wb}, it can be written as

$$H_{pr} = \rho_f c_{pf} \dot{P}_r (T_{wb} - T_s) \tag{4.3.2}$$

where the density and the specific heat pertain to fresh water. $\dot{P}_r$ is the rainfall rate (m s^{-1}); T_s is the skin temperature. In the tropics, T_{wb} is usually less than the ambient air temperature by a few degrees, typically 3°C (see Gosnell *et al.*, 1995) and therefore rainfall normally constitutes a heat loss at the interface (and hence promotes mixing). It is, however, theoretically possible for the ocean to gain heat by precipitation over cold oceans during warm air outbreaks. Using Classius–Clapeyron relationships, it is possible to derive an expression for heat flux in terms of air–sea temperature difference (Gosnell *et al.*, 1995):

$$\begin{aligned} H_{pr} &= \rho_f c_{pf} \dot{P}_r \gamma (T_a - T_s) \\ \gamma &= \varepsilon\left(1 + B_r^{-1}\right) \\ \varepsilon &= \left[1 + \left(\frac{T_a - T_{wb}}{q_s(T_{wb}) - q(T_a)}\right)\frac{dq_s}{dT}\right]^{-1}; B_r = \frac{c_{pa}(T_s - T_a)}{L_E\left(q_s(T_s) - q(T_a)\right)} \end{aligned} \tag{4.3.3}$$

For the western Pacific region that is typical of the tropics, the wet bulb factor $\varepsilon \sim 0.2$ and the Bowen ratio $B_r \sim 0.1$, and for a rainfall rate of 2 cm hr^{-1}, the negative heat flux is as high as 250 W m^{-2}. This is a significant factor in oceanic mixing on timescales comparable to the rain duration. Over longer timescales, the average value of this heat loss is a few W m^{-2} (average rainfall rate ~ 0.04 cm hr^{-1}, with an air–sea temperature difference of 1.5°C) and therefore negligible. The buoyancy effects due to freshwater precipitation overwhelms the heat loss effects and the net effect is that mixing is suppressed in the upper layers.

If the precipitation is in the form of snow, the ocean must give up heat to melt the snow and hence

$$H_{pr} = \rho_f c_{pf} \dot{P}_{sn} (T_{sn} - T_s) - \rho_f \dot{P}_{sn} L_F \tag{4.3.4}$$

where $\dot{P}_{sn}$ is the snow fall rate and L_F is the latent heat of fusion. To a good approximation, the snow temperature T_{sn} can be taken as 0°C. Precipitation is also associated with a mass flux (so is evaporation) and a salt flux; the former is inconsequential (in a Boussinesq fluid) and is usually neglected, while the latter is important to the salt budget and the mixed layer dynamics. The salinity flux due to precipitation is

$$F_{Spr} = \dot{P}_{r,sn}(-S_s) \tag{4.3.5}$$

where S_s is the surface salinity of the ocean and the salinity of precipitation has been safely ignored. The net salinity flux to the ocean is

$$F_{Spr} = (\dot{P}_{r,sn} - \dot{E})(-S_s); \quad \dot{E} = H_l / L_E \tag{4.3.6}$$

where $\dot{P}_{r,sn}$ is the precipitation rate (m s^{-1}), $\dot{E}$ is the evaporation rate, and L_E is the latent heat of evaporation. Precipitation tends to suppress mixing because of the stabilizing effect of fresh water precipitated onto the salty water of the ocean. The net buoyancy flux due to precipitation is normally stabilizing, since the salinity effect overwhelms any thermal effect,

$$F_{bpr} = g(\beta \frac{H_{pr}}{\rho c_p} - \beta_s F_{Spr}) \tag{4.3.7}$$

where the properties now pertain to saline ocean water, β and β_s being the expansion coefficients for heat and salt, respectively. Positive buoyancy flux is stabilizing.

Momentum (and mass) is a conserved quantity. (Mechanical energy is not, if we ignore conversion to heat, which does not play any role in the dynamics or thermodynamics for the low velocities typical of geophysical flows.) Therefore, the momentum flux must also be in balance at the air–sea interface. This just means continuity of stresses at the interface, in particular, tangential stresses,

$$\tau_a + \tau_{pr} = \tau_w + \tau_{wv} \tag{4.3.8}$$

where τ_a is the air-side stress (the shear stress applied by the atmosphere to the ocean), τ_w is the water-side stress (negative of the drag exerted by the ocean on the atmosphere), τ_{wv} is the momentum flux radiated out by propagating surface waves generated by the wind, and τ_{pr} is the momentum flux due to precipitation. Enormous effort has gone into parameterizing τ_a in terms of the atmospheric variables (see the rest of this section) and τ_{wv} (see Chapter 5), since they determine the value of τ_w. The momentum flux to the surface waves is a drag exerted on the atmosphere and is therefore important. It is especially important in the initial stages of development of the wave field (meaning short fetches or small times since the any changes in the wind), since a considerable fraction of the momentum flux from the atmosphere goes then into generating the waves, with the remainder going directly into ocean currents. For a mature wave field, however, equilibrium conditions prevail (see Chapter 5) and most of the momentum flux put into waves is immediately "lost" and transferred to the currents, and τ_{wv} can be safely neglected. It is difficult in practice to compute τ_{wv} without a wave generation model such as WAM, and it is a normal practice to ignore τ_{wv} and put $\tau_w = \tau_a$ in the absence of any precipitation.

The momentum flux due to precipitation arises from the fact that the raindrops (or snow particles) carry horizontal momentum at the time of their impact with the ocean. Normally, a major fraction of the deceleration of air occurs in the bottom few tens of meters and therefore the precipitation traversing this thin layer does not have a chance to equilibrate with the ambient airflow speed and hence hits the water surface with a forward momentum. The velocity at the time of impact is a complicated function of the history of momentum exchange between the falling drops (flakes) and the atmosphere. For practical purposes, one can assume that the precipitation has the same velocity as the atmosphere at some height (normally taken as anemometric height, 10 m), and then one can compute the reduction in its velocity due to the drag of the atmosphere in its further descent to the ocean surface. This reduction is a strong function of the particle size involved, since the settling velocity depends on the size. Settling velocity can be computed by balancing the gravitational force on the drop by the vertical component of the drag. Stokes' law, valid for low Reynolds numbers, can be used even for large drops and, in general, correction to this law due to finite droplet concentration per unit volume of air is negligible. Large raindrops (>1–2 mm in diameter) fall quickly and therefore experience little deceleration. Very small drops, because of their small settling velocities, experience large deceleration. In practice, it is not worthwhile solving for the forward velocity; instead, it is assumed to be a fraction f of the velocity at anemometric height (Caldwell and Elliott, 1971) and therefore

$$\tau_{pr} = m\rho_f \dot{P}_r U_{10} \tag{4.3.9}$$

with m ~ 0.85. The importance of this momentum flux in high precipitation rates can be seen by taking its ratio to τ_a,

$$R = \tau_{pr} / \tau_a \sim (\dot{P}_r U_{10} f) / u_{*w}^2 \tag{4.3.10}$$

neglecting the difference in density of fresh and salty waters. For 8 m s^{-1} winds, u_{*w} is 0.01 m s^{-1} and this ratio is roughly unity for a rainfall rate of ~5 cm hr^{-1}, not unusual during intense squalls and storms. Even for a more reasonable rainfall rate of 1 cm hr^{-1}, this ratio is ~20% and cannot be ignored. This represents a direct transfer of momentum from the upper parts of the atmosphere to the water surface through raindrops.

Both rain-induced buoyancy flux and momentum flux must be taken into account in computing quantities such as the M-O length scale (Section 3.3):

$$L = \frac{u_*^3 \left(1 + \frac{\tau_{pr}}{\tau_a}\right)^{3/2}}{\kappa\left(Q_b + F_{bpr}\right)} \tag{4.3.11}$$

For completeness, we will now deal with the air–ice interface (see Figure 4.3.1). As shown, the air–ice interface is very similar to the air–sea interface except that the albedo is much higher for ice (and snow if the ice is covered by snow, as it invariably is during wintertime) and hence the portion of shortwave flux penetrating into ice is much smaller. Also, there is a conductive heat flux from (or to) the ice/snow cover. During winter, this heat flux can be substantial and hence important to the ice budget, because of the large difference between the air-side (–20 to –30°C) and water-side (–1.7°C) temperatures of the ice/snow cover. This heat flux is however determined by the amount of heat transferred by the ocean to the underside of the ice. During summer, some of the snow and ice at the surface melts and the latent heat of fusion needed for this phase conversion H_m must also be accounted for:

$$SW_{\downarrow} + LW_{\downarrow} + H_{pr} - SW_{\uparrow} - LW_{\uparrow} - H_s - H_L + H_c - H_m = 0 \tag{4.3.12}$$

The momentum balance now involves ice dynamics. Normally, the balance is between the stress exerted by the atmosphere, the drag exerted by the ocean on the ice floe, and the Coriolis acceleration. This is called free drift and is a good approximation in the interior of the ice pack. However, close to shore, the internal ice stresses (τ_I) become large and constitute a significant component of the momentum balance. It is not unusual for the internal ice stresses to nearly balance the applied wind stress, leading to little ice motion and very little drag exerted on the water. In addition, the ice has a different roughness length, and thus the exchange of turbulent heat energy between the ice and the atmosphere will be different than that between the air and the water.

4.4 FLUX BALANCE AT THE AIR–LAND INTERFACE

For an air–ground interface, the heat balance is similar to that shown in Eq. (4.1.1), except that it is also necessary to include the molecular energy transfer into the soil (H_g) and the heat flux due to precipitation as described in the preceding section (H_{pr}). H_g can either be positive or negative (into or out of the ground) depending on whether it is daytime or nighttime and summer or winter (see Garratt, 1992, for a thorough discussion of this aspect). While this flux (a few tens of W m^{-2}) is not too important to the net heat flux at the bottom of the atmospheric column during the day, it is important for the nocturnal boundary layer, and for freezing and thawing of permafrost surfaces in high latitudes. The balance can be written as

$$SW_{\downarrow} + LW_{\downarrow} + H_{pr} - SW_{\uparrow} - LW_{\uparrow} - H_s - H_L - H_g = 0 \qquad (4.4.1)$$

As shown in Figure 4.4.1, the energy balance across a soil surface is considerably simpler than that across a water surface. A soil surface is much less changeable in time than is a water surface, and thus it is much easier to characterize the surface roughness, which directly impacts the transfer of heat between the atmosphere and the surface. Surface roughness lengths for various types of terrain are shown in Figure 3.2.3. Some typical values for these fluxes at various times and differing soil types are shown in Figure 3.4.4. In contrast to the values at the air–sea interface, in general, the latent heat flux is smaller, since the evaporation is less. The sensible heat flux is also much greater and has a stronger diurnal variability, due to the larger change in ground temperature from day to night and the larger difference between the ground and the air temperature than in the air–sea system.

However, once vegetation exists on the surface, there are a number of other parameters which need to be determined, such as the vegetation coverage, the canopy height, and plant reflectivity. A small fraction (~2%) of the SW flux is consumed by energy requirements of photosynthesis (SW_{ph}) if vegetation is present. The evaporative flux, and hence the latent heat flux, is a complicated function of the vegetation covering the surface and the availability of water. Thus the heat flux balance becomes:

$$SW_{\downarrow} + LW_{\downarrow} + H_{pr} - SW_{\uparrow} - LW_{\uparrow} - H_s - H_L - H_g - SW_{ph} = 0 \qquad (4.4.2)$$

The most important aspect of scalar transfer through a plant canopy is the close association between the sensible and the latent heat flux. The latter is governed by the evaporation from leaf surfaces and is therefore dependent on whether the stomata are exerting control as in dry conditions or whether the water is plentiful so that plant physiological control is unimportant. Evapotranspiration is one of the most important aspects of hydrological balance. The ratio of the latent heat flux to the total heat flux is given by combination equations (3.9.8) or (3.9.13). For more details, see Section 3.9.

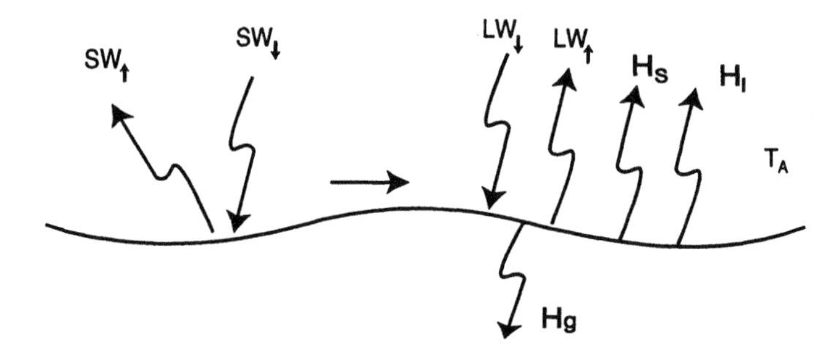

Figure 4.4.1. The flux balance at the air–ground interface. The fluxes are as given in the text.

The momentum flux balance at the air–land interface is less complicated than that for the air–ocean (or air–ice) interface. The stress exerted by the atmosphere is equal to the drag exerted by the ground surface, even if the surface is mobile, such as over sand dunes, since the radiated momentum flux is not important. This balance is shown in Figure 4.4.1. A further description of the impact of flow over rough terrain can be found in Chapter 3.

4.5 BULK EXCHANGE COEFFICIENTS

It is easy to see that air–sea exchange involves complex turbulent processes in both media. The air–ground case is somewhat simpler as it involves only one turbulent layer. Traditionally, the turbulent surface fluxes have been parameterized by bulk transfer laws, with coefficients empirically determined. This pushes all our "ignorance" of details of air–sea interaction into the bulk coefficients, and naturally there has been a considerable effort in determining the dependence of these coefficients on various parameters in the problem, principally the wind speed and stratification, and increasingly the surface wave field as well for use over water. It is important to note that the following treatment is equally applicable to the air–ice interface with appropriate changes.

The momentum, sensible heat, latent heat, water vapor, and gas fluxes across the air–sea interface can be written as

$$\tau = -\rho\,\overline{uw} = \rho\,u_*^2 \tag{4.5.1a}$$

$$H_s = -\rho\,c_p\,\overline{w\theta} = \rho\,c_p\,u_*\theta_* \tag{4.5.1b}$$

$$H_L = -\rho\,L_E\,\overline{wq} = \rho\,L_E\,u_*q_* \tag{4.5.1c}$$

$$E = -\rho\,\overline{wq} = \rho\, u_* q_* = H_L / L_E \quad (4.5.1d)$$

$$G = -\rho\,\overline{wc} = \rho\, u_* c_* \quad (4.5.1e)$$

where the bars denote covariances between the fluctuating vertical velocity and the appropriate fluctuating quantity being transferred across the interface. ρ is the density of air, u_* is the friction velocity, $\theta_*=H_S/(\rho c_p u_*)$ is the surface layer temperature scale, $q_*=H_L/(\rho L_E u_*)=E/(\rho u_*)$ is the surface layer humidity scale, $c_*=G/(\rho u_*)$ is the surface layer concentration scale, τ is the shear stress, H_s is the sensible heat flux, H_L is the latent heat flux, E is the water vapor flux, and G is the gas flux. c_p is the specific heat and L_E is the latent heat of evaporation. The surface layer is the region of the boundary layer near the interface (as discussed in Chapter 3).

It is preferable to deal with kinematic fluxes, which are the covariances,

$$Q_s = -\overline{w\theta} = H_S / \rho c_p \quad (4.5.2a)$$

$$Q_l = -\overline{wq} = H_l / \rho L_E \quad (4.5.2b)$$

$$Q_g = -\overline{wc} = G / \rho \quad (4.5.2c)$$

where Q_s is the kinematic sensible heat flux, Q_l is the kinematic latent heat (evaporative heat) flux, and Q_g is the kinematic gas flux. Note that all the quantities above pertain to air. For example, u_* is the friction velocity on the air side (the value on the ocean side would be smaller by a factor of 32).

The surface turbulent fluxes can also be written using bulk formulas,

$$\tau = \rho C_{Dh} (U_a - U_s)^2 \quad (4.5.3a)$$

$$H_s = \rho c_p (U_a - U_s) C_{Hh} (T_s - T_a) \quad (4.5.3b)$$

$$H_L = \rho L_E (U_a - U_s) C_{Eh} (q_s - q_a) \quad (4.5.3c)$$

$$E = \rho\, (U_a - U_s) C_{Eh} (q_s - q_a) \quad (4.5.3d)$$

$$G = \rho\, (U_a - U_s) C_{Gh} (c_s - c_a) \quad (4.5.3e)$$

where U_a is the wind speed at reference height z_h above the surface (usually 10 m, the standard anemometric height), and U_s is the component of surface velocity along the wind direction (zero for a stationary surface). T is the temperature, q is the specific humidity, and c is the gas concentration. Subscripts a and s denote values pertaining to the atmosphere at height z_h and immediately above the surface, with the exception of c_s, which denotes the surface value of the concentration of the dissolved gas in the ocean. Note that q_s should be the value over salty water (roughly 98% of the pure water value) if measurements are being made over the ocean. In regions of the global oceans where currents are strong (1–2 m s^{-1}) and winds are weak (a few m s^{-1}), such as in the tropics during easterly trade wind regimes, U_s is not negligible compared to U_a, although in most cases, U_s is usually ignored in the above bulk formulas. Also while the above expressions are strictly valid only for potential temperature θ, within the thin surface layer, where the above relations are valid, in most cases, it is possible to use the actual temperature T since adiabatic compression effects are negligible (potential temperature is higher by 0.01 z).

C_{Dh} is the drag coefficient, and C_{Gh} is the gas transfer coefficient. Coefficients of heat and water vapor exchange C_{Hh} and C_{Eh} are also known as Stanton and Dalton numbers. The subscript h in these is to remind ourselves that the values of these coefficients depend on the reference height z_h, although it is traditional to omit it, in which case it is implied that standard anemometric height is used as the reference height. If some other height is used (usually smaller values, for example, 2 m for ocean buoys), as long as it is within the constant flux layer, it is possible to derive the anemometric height values appealing to Monin–Obukhoff similarity theories (Section 3.3). See Donelan (1990) for systematic errors introduced by assuming the fluxes are constant in the constant flux layer.

$$C_{Dh} = \frac{u_*^2}{(U_a - U_s)^2} \tag{4.5.4a}$$

$$C_{Hh} = \frac{H/\rho c_p}{(U_a - U_s)(T_s - T_a)} = \frac{u_* \theta_*}{(U_a - U_s)(T_s - T_a)} \tag{4.5.4b}$$

$$C_{Eh} = \frac{E/\rho}{(U_a - U_s)(q_s - q_a)} = \frac{u_* q_*}{(U_a - U_s)(q_s - q_a)} \tag{4.5.4c}$$

$$C_{Gh} = \frac{G/\rho}{(U_a - U_s)(c_s - c_a)} = \frac{u_* c_*}{(U_a - U_s)(c_s - c_a)} \tag{4.5.4d}$$

The importance of the bulk exchange coefficients C_{Dh}, C_{Hh}, C_{Eh}, and C_{Gh} lies in the fact that the surface turbulent fluxes are seldom measured directly, but are instead inferred from measurements of atmospheric properties such as wind speed, temperature, humidity, and gas concentration at a particular height (the anemometric height) in the atmospheric surface layer using the above bulk formulas. Therefore any errors in these coefficients translate to quite serious errors in the values for the fluxes.

4.5.1 FLUX MEASUREMENTS

Direct accurate measurements of turbulent momentum and heat fluxes, whether over land or sea, require great care given to many details such as the frequency response of sensors, processing of the resulting time series, and the nature and location of the measurement site (Lenschow, 1994; Dabberdt *et al.*, 1993; Kaimal and Finnigan, 1994). Direct measurement of gas exchange is even harder since fast response instruments measuring gas concentrations are not available and a modification of the eddy correlation method (see below) called the eddy accumulation method must be used, where air is collected in two reservoirs, one when the vertical velocity is positive and another when it is negative (Dabberdt *et al.*, 1993; Smith *et al.*, 1996a). Most of our knowledge about air–surface turbulent fluxes comes from measurements in the surface layer over land from well-instrumented towers (for example, Dabberdt *et al.*, 1993). It is assumed that some of this is directly transferable to air–sea exchanges by invoking the Monin–Obukhoff surface layer similarity theory, even though there is a significant difference between the two cases in the form of a mobile interface entertaining surface wave motions that constitute dynamics-determined, ever-changing surface roughness elements in the latter. There have been very few direct measurements of fluxes, especially over the oceans, since they are hard to make, and require sophisticated instrumentation and extreme care in interpreting the results. Direct measurements of fluxes over the sea can be made only from specially equipped oceanographic research ships, aircraft, and stationary masts or moorings, with sensors fast enough to sample the turbulent fluctuations of the relevant quantity in the entire frequency space of interest (see Smith *et al.*, 1996a).

A simpler flux inference technique involves measuring the relevant mean properties at two or more levels in the surface layer and deducing the vertical gradient of the property. Invoking Monin–Obukhoff similarity theory, one can deduce the flux of the property involved. This is the so-called profile method, which while well suited to flux measurements over land, is unsuitable for measurements from ships since even small flow distortions due to the ship superstructure and the ship's heat island effect can lead to large errors in gradients and hence fluxes deduced from them.

A direct flux measurement method involves measurements of the covariances of the fluctuating velocity in the vertical w and the fluctuations of the relevant property pertaining to the flux (the fluctuating velocity component in the direction of the wind, temperature, humidity, and gas concentration), using sensors capable of resolving these fluctuations well, in the frequency space of interest. It is called the eddy correlation method (Smith *et al.,* 1996a) and even under ideal conditions the accuracy of the method depends very much on the averaging interval for the covariances. Clearly, the averaging interval must be long enough to encompass turbulent fluctuations of all frequencies, but only turbulent fluctuations, and hence must exclude secular changes in mean quantities from contaminating the results. More serious is the fact that the sensors are often attached to a nonstationary platform. The method works best when the sensors are immobile. Measurements made from aircraft are subject to contamination by aircraft motions in response to wind gusts and control surface actuations. In measurements over sea by a ship, which executes complex motions in space in response to the wave field, unless these motions are measured and accounted for, the results are likely to be seriously contaminated by ship motions, especially in high seas. Consequently, corrections due to ship motions can be several orders of magnitude higher than the turbulent fluctuations that are being measured (Dupuis *et al.,* 1997). Also any measurements from ships involve flow distortions due to the superstructure of the ship itself, and require great care in ensuring that the sensor is on the upwind side of the ship and far away from the ship to safely neglect ship-induced flow distortions. In contrast, measurements from masts are mostly immune to such errors, but are essentially confined to shallow waters. Even then there are considerable difficulties in making these measurements. At high wind speeds over the ocean, contamination of sensors by salt spray is a perennial problem (see DeCosmo *et al.,* 1996, for a thorough discussion of the analysis procedures and uncertainties of flux measurements). At low wind speeds, sampling problems cause uncertainties (for example, Mahrt *et al.,* 1996). Measurements over land are somewhat easier as the platform (such as a mast or tower) is stationary. However, the cost of the fast-response instruments is still large and distortion of the flow near the sensor still occurs.

The second direct method involves measuring the spectra of velocity and scalar fluctuations in frequency space and relating these measurements to the dissipation rates of TKE and scalar variances. Once the dissipation rates are determined, similarity laws of the atmospheric surface layer can be invoked to extract the needed fluxes from the dissipation rates. This method is called the inertial dissipation method and is the preferred method when making flux measurements from moving platforms, since fluxes can be derived without knowledge of platform motions. The method is, however, indirect (unlike the eddy correlation method) and its accuracy is very much dependent on the validity of underlying implicit assumptions that the turbulence is isotropic and in local equilibrium so that the production of TKE and various variances which are dependent on the corresponding turbulent fluxes are in balance with their dissipation rates. It also requires transformation of the measured spectra from frequency space to wavenumber space, and invoking the existence of the inertial subrange in the wavenumber spectrum, where the Kolmogoroff –5/3 law prevails for both the velocity and the scalar spectrum and relates the spectral density to the dissipation rate. In order for this method to work, a broad inertial subrange must exist and remain uncontaminated by platform motions. These conditions are usually met. The conversion from frequency space to wavenumber space involves appealing to Taylor's frozen turbulence hypothesis, which states that the turbulent fluctuations seen by a sensor when the flow sweeps the turbulent eddies past the sensor appear as if the turbulence is frozen and remains unchanged, which is equivalent to $k = n/U$, where k is the wavenumber, and n the frequency. Using $n = kU$ in the one-dimensional longitudinal velocity spectrum $E_{11}(k)$ (Section 1.5), one can obtain the frequency spectrum $S(n)$,

$$E_{11}(k) = \frac{18}{55} C\varepsilon^{2/3} k^{-5/3} \Rightarrow S(n) = \frac{18}{55} C\varepsilon^{2/3} U^{2/3} n^{-5/3} \qquad (4.5.4e)$$

where the Kolmogoroff constant $C \sim 1.6$. By fitting a –5/3 power law regression line to the inertial subrange in a log–log plot of S vs n, one can determine the dissipation rate ε. For transverse velocity spectra E_{22} or E_{33}, $C = 2.13$ (equal to $4/3 \times 1.6$). The factor 4/3 in the relationship between longitudinal and transverse spectra has been verified by Dupuis *et al.* (1997). Once ε is known, appealing to the TKE equation for a stationary, horizontally homogeneous surface layer,

$$\Phi_\varepsilon(\zeta) = \frac{\kappa z}{u_*^3}\varepsilon = \Phi_M(\zeta) - \zeta - \Phi_T(\zeta) \qquad (4.5.4f)$$

where the last term accounts for local imbalance due to turbulence transport terms in the equation and provides the relationship needed to determine u_*. Large and Pond (1982) and Fairall and Edson (1994) ignored the last term and considered turbulence to be in local equilibrium, but this is valid only under near-neutral conditions and Dupuis *et al.* (1997) suggest that over a range of conditions from neutral to highly unstable, the term should be parameterized as $-0.65\,\zeta$. Thus the universal functional form for Φ_ε is known and knowing the value of the Monin–Obukhoff length scale L enables u_* to be determined. Since L involves u_*, this is usually an iterative process and care must be exercised to ensure convergence. Dupuis *et al.* (1997) suggest that the imbalance term is essential for convergence. Determination of L itself requires knowledge of the buoyancy flux, or equivalently the virtual temperature flux in the atmosphere. If the scalar spectrum is also measured, a similar approach can be used to determine the buoyancy flux. For example, the virtual temperature flux, neglecting the turbulent transport terms in the variance conservation equations, is

$$\overline{wT_v} = \left(\kappa z u_* \varepsilon_{TV} \Phi_{TV}(\zeta)\right)^{1/2} \tag{4.5.4g}$$

where Φ_{TV} is a universal function assumed to be the same as Φ_H. In cases where the scalar spectrum is not measured, one simply uses bulk parameterization to determine the buoyancy flux and hence L and the momentum flux. Still, the free convection limit is a major problem, leading to a nonzero value of u_* for zero wind speed.

Taylor's frozen turbulence hypothesis requires that the turbulence intensity u be small compared to the mean flow speed relative to the sensor U_r (see Chapter 1). This condition is approximately $u < 0.1\ U_r$ but may not always be satisfied, whether the measurement is made from a moving ship or a stationary mast [but see Lumley and Terray (1983), who suggest that the hypothesis is still valid if C and U are suitably redefined], especially for highly unstable conditions and ship speeds less than 1 m s^{-1}. Aircraft measurements, on the other hand, are well suited to this method, simply because irrespective of the wind speed, the sensors are being traversed at great speed across the turbulent fluid and the eddies are effectively "frozen" during sampling by the probe. Nevertheless, the method is somewhat insensitive to motions of the platform itself and as long as the turbulence is being swept past the sensor at a large enough mean speed, it is a viable technique (see Fairall and Larsen, 1986). See Fairall *et al.* (1990, 1996a) and Smith *et al.* (1992) for a comparison of the eddy correlation and inertial dissipation methods. Considerable uncertainty exists still as to which method is better. Janssen (1999) argues that the inertial dissipation involves so many assumptions that do not account for the sea state and the many corrections needed can be so uncertain that eddy correlation technique is preferable. However, the latter suffers from the influence of platform motion and flow distortions, so that both methods appear to have deficiencies.

Use of the bulk formulas is based on the hope that if the bulk coefficients can be determined and their functional dependence on various parameters discerned accurately from those few observations, where surface fluxes and the relevant parameters have been measured accurately, then these formulas can be used with confidence to infer fluxes in situations when only the relevant parameters are known. Though straightforward in principle, the application is beset with problems since the functional relationships are not known unambiguously and not all the relevant parameters are accurately measured during the calibration process. Most current efforts have concentrated on deducing the dependence of bulk exchange coefficients C_{Dh}, C_{Hh}, C_{Eh}, and C_{Gh} on wind speed and the stability of the air column. Over the ocean there is an implicit assumption that the underlying surface wave field is fully developed and hence the sea surface roughness can be inferred without knowing the details of the wind wave spectrum. This assumption is seldom satisfied, as the scatter in the functional relationship is quite large and the resulting accuracies of fluxes compromised to varying degrees. Even though the dependence of bulk coefficients on surface waves is approximately known, most often surface wave field observations are not available, and the best that can be done is to use bulk coefficients that assume a fully developed wave field. Over land, if vegetation exists, the roughness length and bulk transfer coefficients must be adjusted appropriately. The adjustment is based on the types of plants, the average canopy height, and the areal density of the vegetation.

Another problem in the use of these bulk formulas over the ocean is the fact that they are valid only for forced convection conditions of air–sea transfer and are not valid in the limit of zero wind speed (more appropriately, zero $U_a - U_s$), the free convection limit. While the bulk formulas indicate correctly a zero stress at the air–sea interface in this limit, the scalar fluxes are not necessarily zero. While it is possible to account for low wind speeds by large increases in exchange coefficients, the physics of air–sea exchange is fundamentally different when the wind speeds are low, and free convection is the dominant mechanism of transfer. Finally, at large wind speeds at which there is extensive breaking of surface waves, the scalar fluxes are affected by the direct transfer of heat, water vapor, and gases across the air–sea interface by spray and droplet ejection and air bubble entrainment mechanisms. Bulk formulas can be looked upon as linear approximations that retain only the leading linear term in what should appropriately be an infinite series involving nonlinear terms that presumably account for some of these additional physical mechanisms. Once again, suitable adjustments in bulk coefficients are possible, but fail to account for the fundamentally different mechanism of transfer that prevails when the interface is torn apart violently and periodically as compared to when it is basically intact.

In spite of the above-mentioned problems, considerable effort has been expended over the past few decades in determining the bulk transfer coefficients (Garratt, 1977; see also Garratt, 1992). A recent review can be found in Geernaert (1990). Blanc (1985) also provides a summary, and Smith (1988) summarizes evaporation measurements. Smith *et al.* (1996a) catalog the many field experiments (IIOE, BOMEX, ATEX, GATE, AMTEX, JASIN, MARSEN, HEXOS, CODE, FASINEX, TOGA/COARE, RASEX, MBLP, COPE) that have been conducted over the past 30 years that involve measurement and study of air–sea fluxes, and provide a succinct summary of progress in air–sea flux research over that time. Large and Pond (1981, 1982) made the now-classic measurements of air–sea fluxes, but more recent ones can be found in Godfrey *et al.* (1991) and Fairall *et al.* (1996a). The advantage of the latter two stems from the fact that very careful measurements were made in low wind regimes characteristic of the tropical Pacific during TOGA/COARE using several modern techniques, specifically to provide a more accurate means of quantifying the air–sea evaporative fluxes. DeCosmo *et al.* (1996) have measured the heat and moisture transfer coefficients at high wind speeds. Enriquez and Friehe (1997) report on measurements during CODE in 1982 and SMILEX (Shelf MIxed Layer EXperiment) in 1989 obtained by aircraft in coastal waters.

The bulk exchange coefficients over the ocean are a function of height, wind speed at that height, ambient stratification, and the surface wave field. Height dependence can be avoided by considering values at a standard reference height (10 m, anemometric height). The usual practice is to convert measurements at whatever height they were made to the standard anemometric height using well-known profile laws in the atmospheric surface layer. In the absence of ambient stratification, one can expect the bulk coefficients to be functions of only the wind speed and surface wave field, if the wave field is not in equilibrium with prevailing winds. The atmosphere over the ocean is seldom neutrally stratified, although it is often close to neutral stratification (unlike the ABL over land). In that case, the coefficients are converted to equivalent neutral values using the well-known Monin–Obukhoff similarity law for the surface layer (see Chapter 3). Briefly, the profiles of velocity, temperature, humidity, and gas concentration in the marine surface layer (normally a few tens of meters thick) are written as

$$\frac{\kappa z}{u_*}\frac{\partial U}{\partial z} = \phi_M\left(\frac{z}{L}\right) \tag{4.5.5a}$$

$$\frac{\kappa z}{\theta_*}\frac{\partial T}{\partial z} = \mathrm{Pr}_t\,\phi_H\left(\frac{z}{L}\right) \tag{4.5.5b}$$

$$\frac{\kappa z}{q_*}\frac{\partial q}{\partial z} = \mathrm{Sc}_t\phi_E\left(\frac{z}{L}\right) \tag{4.5.5c}$$

$$\frac{\kappa z}{c_*}\frac{\partial c}{\partial z} = \mathrm{Sc}_t\phi_G\left(\frac{z}{L}\right) \tag{4.5.5d}$$

where L is the Monin–Obukhoff length scale and z/L is the Monin–Obukhoff similarity variable ζ (see Chapter 3):

$$L = \frac{-u_*^3 T_v}{\kappa g\, \overline{w\theta_v}} \tag{4.5.6}$$

Note the presence of the turbulent Prandtl and Schmidt numbers on the RHS of

Eqs. (4.5.5b–d). This definition avoids the problem of integrability of stability functions $\phi_{H,E,G}$ (see Appendix A of Enriquez and Friehe, 1997). The constant κ is the von Karman constant (0.40–0.41). T_v is the virtual temperature equal to T (1 + 0.61 q) and $\overline{w\theta_v}$ is the net heat flux at the air–sea interface that accounts for the moisture flux as well:

$$\overline{w\theta_v} = \overline{w\theta}\left(1+0.61\,q\right)+0.61\,T\,\overline{wq} \tag{4.5.7}$$

Note that the temperatures are absolute values and should be in K (°C + 273.15).

The principal contrast to land surfaces is that the influence of humidity and moisture flux cannot be neglected over sea under most conditions. The air is always moist over the sea (and ice) and the evaporative flux significant. The functional form of the Monin–Obukhoff similarity (profile) functions on the right-hand side is well known from extensive measurements over land surfaces. It is often assumed that the same forms are obtained over sea as well. This is only approximately true since the underlying surface wave field is a major factor, especially in heavy seas or in the presence of large swells. Nevertheless, algebraically simple forms deduced from land ABL measurements and conducive to analytical integration in the vertical are used (Paulson, 1970) (see Chapter 3). Henceforth, we assume that the velocities are indeed referred to the moving air–sea interface and put $U_s = 0$. Integration in the vertical (see Chapter 3) gives (note $\zeta = z/L$)

$$U(z) - U_s = \frac{u_*}{\kappa}\left[\ln\frac{z}{z_0} - \psi_M\left(\zeta\right)\right] \tag{4.5.8a}$$

$$T(z) - T_s = \frac{Pr_t\theta_*}{\kappa}\left[\ln\frac{z}{z_{0T}} - \psi_H\left(\zeta\right)\right] \tag{4.5.8b}$$

$$q(z) - q_s = \frac{Pr_t q_*}{\kappa}\left[\ln\frac{z}{z_{0E}} - \psi_E\left(\zeta\right)\right] \tag{4.5.8c}$$

$$c(z) - c_s = \frac{Pr_t c_*}{\kappa}\left[\ln\frac{z}{z_{0G}} - \psi_G\left(\zeta\right)\right] \tag{4.5.8d}$$

where

$$\psi_{M,H,E,G}\left(\zeta\right) = \int_0^{\zeta}\left[1-\phi_{M,H,E,G}\left(\zeta'\right)\right]\frac{d\zeta'}{\zeta'} \tag{4.5.9}$$

and we have substituted Pr_t for Sc_t in Eqs. (4.2.8c) and (4.2.8d). These are all turbulent quantities and should be the same irrespective of which scalar is being transported. Pr_t, the turbulent Prandtl number, is ~0.86. Quantities z_0, z_{0T}, z_{0E}, and z_{0G} are known as roughness length scales for momentum, temperature, humidity, and gas. These roughness scales are directly related to the bulk exchange coefficients:

$$C_{Dh} = \frac{\kappa^2}{\left[\ln\frac{z_h}{z_0} - \psi_M(\zeta_h)\right]^2} \tag{4.5.10}$$

$$C_{Hh,Eh,Gh} = \frac{\kappa^2}{Pr_t\left[\ln\frac{z_h}{z_0} - \psi_M(\zeta_h)\right]\left[\ln\frac{z_h}{z_{0T,0E,0G}} - \psi_{H,E,G}(\zeta_h)\right]} \tag{4.5.11}$$

In other words, knowing the roughness scales is equivalent to knowing the bulk exchange coefficients and vice versa, provided the stratification is also known. The standard practice is to put z_h = 10 m in the above equations. The exchange coefficients often carry a subscript 10 to denote standard anemometric height values (e.g., C_{D10}). Notice that the exchange coefficients are functions of stratification. The exchange coefficients that would be realized under neutral stratification conditions are functions of only the roughness scales, so that roughness scales determine the so-called "neutral" exchange coefficients:

$$C_{DNh} = \frac{\kappa^2}{\left[\ln\frac{z_h}{z_{0N}}\right]^2} \tag{4.5.12}$$

$$\begin{aligned} C_{HNh,ENh,GNh} &= \frac{\kappa^2}{Pr_t\left[\ln\frac{z_h}{z_{0N}}\right]\left[\ln\frac{z_h}{z_{0TN,0EN,0GN}}\right]} \\ &= \frac{C_{DNh}}{Pr_t\left[1 - \frac{1}{\kappa}C_{DNh}^{1/2}\ln\frac{z_{0TN,0EN,0GN}}{z_{0N}}\right]} = \frac{\kappa C_{DNh}^{1/2}}{Pr_t\left[\ln\frac{z_h}{z_{0TN,0EN,0GN}}\right]} \end{aligned} \tag{4.5.13}$$

The utility of the above expressions is that flux measurements under various stratification conditions are converted to C_{DNh}, C_{HNh}, C_{ENh}, and C_{GNh} values for easy comparison between observations. Note that the scalar coefficients are related to the drag coefficient through the corresponding roughness scales. A suffix of 10 is often used to denote standard anemometric values of the neutral exchange coefficients as well (e.g., C_{DN10}). The nonneutral values at any reference height can be related to neutral values at the 10-m reference height (to which observations are often reduced to for easy comparison between different data sets) by

$$C_{Dh} = \frac{C_{DN10}}{\left[1 + \frac{1}{\kappa} C_{DN10} C_{DN10}^{-1/2} \left(\ln \frac{z_h}{10} - \psi_M(\zeta_h) \right) \right]^2} \tag{4.5.14}$$

$$C_{Hh,Eh,Gh} = \frac{C_{HN10,EN10,GN10} \left(\frac{C_{Dh}}{C_{DN10}} \right)^{1/2}}{\left[1 + \frac{1}{\kappa} C_{HN10,EN10,GN10} C_{DN10}^{-1/2} \left(\ln \frac{z_h}{10} - \psi_{H,E,G}(\zeta_h) \right) \right]} \tag{4.5.15}$$

Functions ϕ and ψ depend on the height normalized by the Monin–Obukhoff length scale. The usual practice is to assume algebraically simple expressions for $\phi_{M,H,E,G}$ (for example, Oncley *et al.*, 1990),

$$\phi_M = (1 + \alpha_S \zeta) \tag{4.5.16}$$

$$\begin{gathered} Pr_t \, \phi_{H,E,G} = Pr_t \, (1 + \beta_S \zeta) \\ Pr_t \sim 0.72 - 0.90, \quad \alpha_s \sim 8.1, \quad \beta_s \sim 9.4 \end{gathered} \tag{4.5.17}$$

for stable conditions (ζ>0), and

$$\phi_M = (1 - \alpha_u \zeta)^{-1/4} \tag{4.5.18}$$

$$\begin{gathered} Pr_t \, \phi_{H,E,G} = Pr_t \, (1 - \beta_u \zeta)^{-1/2} \\ \alpha_u \sim 15, \quad \beta_u \sim 9 \end{gathered} \tag{4.5.19}$$

for unstable conditions ($\zeta < 0$), so that analytical expressions are possible for the stability factors (Paulson, 1970),

$$\psi_M = -\alpha_s \zeta, \quad \psi_{H,E,G} = -\beta_s \zeta \tag{4.5.20}$$

for stable conditions and

$$\psi_M = 2\ln\left[\frac{1+\phi_M^{-1}}{2}\right] + \ln\left[\frac{1+\phi_M^{-2}}{2}\right] - 2\tan^{-1}\left[\phi_M^{-1}\right] + \frac{\pi}{2} \tag{4.5.21}$$

$$\psi_{H,E,G} = 2\ln\left[\frac{1+\phi_{H,E,G}^{-1}}{2}\right] \tag{4.5.22}$$

for unstable conditions. These do not obey the free convective similarity scaling under the free convective limit. Note the presence of Pr_t on the LHS of Eqs. (4.5.17) and (4.5.19), consistent with definitions of stability functions $\phi_{H,E,G}$ in Eq. (4.5.5). This is what makes closed form expressions possible for ψ in Eq. (4.5.22) (Enriquez and Friehe, 1997). There are however many other possible profiles that are in use, for reasons of both analytical integrability and numerical well behavedness. Benoit (1977) recommends for unstable conditions

$$\psi_M = \ln\left[\frac{z}{z_0}\right] + \ln\left[\frac{(\chi_0^2+1)(\chi_0+1)^2}{(\chi^2+1)(\chi+1)^2}\right] + 2\left[\tan^{-1}\chi - \tan^{-1}\chi_0\right] \tag{4.5.23}$$

$$\psi_{H,E,G} = \ln\left[\frac{z}{z_0}\right] + 2\ln\left[\frac{\chi_0^2+1}{\chi^2+1}\right], \quad \chi_0 = \left(1-16\frac{z_0}{L}\right)^{1/4} \tag{4.5.24}$$

Beljaars and Holtslag (1991) use for stable conditions

$$\psi_M = a(\zeta) + b\left(\zeta - \frac{c}{d}\right)\exp(-d\zeta) + b\frac{c}{d}$$
$$\psi_{H,E,G} = \left(1+\frac{2}{3}a\zeta\right)^{3/2} + b\left(\zeta - \frac{c}{d}\right)\exp(-d\zeta) + b\frac{c}{d} - 1 \tag{4.5.25}$$

where a, b, c, and d are empirical constants. A scalar profile that satisfies asymptotic $\phi \sim \zeta^{-1/3}$ behavior in the limit of free convection is (Fairall *et al.*, 1996a)

$$\psi_{H,E,G} = 1.5 \ln\left[\frac{\chi^2+\chi+1}{3}\right] - \sqrt{3}\tan^{-1}\left[\frac{2\chi+1}{\sqrt{3}}\right] + \frac{\pi}{\sqrt{3}} \qquad (4.5.26)$$

$$\chi = (1-\beta\zeta)^{1/3}$$

Large *et al.* (1994), on the other hand, use profiles of the form of Eqs. (4.5.14) and (4.5.15) for stable conditions, but for unstable conditions, they splice –1/3 profiles in the free convection limit to profiles of the form of Eqs. (4.5.16) and (4.5.17). Because of the large scatter in the basic observations, especially in the limit of the M-O length $L \to 0$, it is not clear which profile is preferable.

The neutral bulk exchange coefficients are related to nonneutral values:

$$C_{DNh} = \left[C_{Dh}^{-1/2} + \frac{1}{\kappa}\psi_M(\zeta_h) - \frac{1}{\kappa}\ln\left(\frac{z_{0N}}{z_0}\right)\right]^{-2} \qquad (4.5.27)$$

$$C_{HNh,ENh,GNh} = C_{DNh}^{1/2}\left[\frac{C_{Dh}^{-1/2}}{C_{Hh,Eh,Gh}} + \frac{Pr_t}{\kappa}\psi_{Hh,Eh,Gh}(\zeta_h) - \frac{Pr_t}{\kappa}\ln\left(\frac{z_{0TN,0EN,0GN}}{z_{0T,0E,0G}}\right)\right]^{-1} \qquad (4.5.28)$$

While the roughness scales with stratification are different than those that would be attained under neutral conditions, often this distinction is ignored as far as scalar roughness lengths are concerned and the last term on the RHS of Eq. (4.5.28) is put to zero. For z_0, if one invokes Charnock's law, then $z_{0N}/z_0 = C_{DN}/C_D$ and

$$C_{DN} = \left[C_D^{-1/2} + \frac{1}{\kappa}\psi_M - \frac{1}{\kappa}\ln\left(\frac{C_{DN}}{C_D}\right)\right]^{-2} \qquad (4.5.29)$$

4.5.2 Measurements of Air–Sea Bulk Transfer Coefficients

Most measurements for air–sea fluxes ignore the surface wave field and present bulk exchange coefficients as functions only of the wind speed. The usual form is

$$C_{DN,HN,EN,GN} = a_{D,H,E,G} + b_{D,H,E,G}\, U_{10}^{\alpha_{D,H,E,G}} \qquad (4.5.30)$$

Since it is the usual practice to report the exchange coefficients at the standard anemometric height, irrespective of whether they were measured by aircraft flying at heights of 30–50 m, buoys at a height of ~2–4 m, or ships, we will henceforth ignore the subscript 10 on exchange coefficients. Table 4.5.1 (from Geernaert, 1990) shows a compilation of numerous measurements of the drag coefficient C_{DN} over the oceans, and Figure 4.5.1 shows the corresponding variation of C_{DN} with wind speed. Figure 4.5.2 shows C_D values compiled by Said and Druilhet (1991) corresponding to the investigations listed in Table 4.5.2 (from Said and Druilhet, 1991). The scatter in both plots is quite large and some of it at least can be attributed to the fact that the underlying surface wave field has been assumed to be in equilibrium with the wind. This fact is underscored by the outliers in Figure 4.5.3, which shows measurements of C_D by the inertial dissipation method in the western tropical Pacific by Bradley *et al.* (1991) during low wind conditions. The outliers correspond to situations where there was a sudden shift in the winds and presumably the surface wave field did not have time to adjust to the shifting winds. For most open ocean applications one can use the relationship due to Garratt (1977) to within 30% accuracy (Geernaert, 1990):

$$10^3 \cdot C_{DN} = 0.75 + 0.067\, U_{10} \qquad (4.5.31)$$

More recent measurements by Atakturk and Katsaros (1999) over Lake Washington suggests coefficients 0.87 and 0.078 in Eq. (4.5.31). They made careful measurements including the surface wave field spectrum, and were able to reduce the scatter in the above correlation by making explicit use of the root-mean-square deviation of the sea surface in determining the roughness scale and hence C_{DN}.

Measurements of C_{HN}, C_{EN} at sea are much harder to obtain and less frequent than those for C_{DN}. The few measurements that exist are summarized in Table 4.5.3 (from Geernaert, 1990; see also Smith, 1988, 1989) and Table 4.5.4 (from Said and Druilhet, 1992). Most of these use eddy-correlation methods, but some use inertial dissipation techniques. Geernaert (1990) suggests that to within 30% accuracy,

$$10^3 \cdot C_{EN} = 0.48 + 0.083\, U_{10} \qquad (4.5.32)$$

$$10^3 \cdot C_{GN} = 0.55 + 0.083\, U_{10} \qquad (4.5.33)$$

Figure 4.5.4 shows recent measurements of C_H, C_{HN} and C_E, C_{EN} during a 1988 cruise in the western tropical Pacific (Bradley *et al.*, 1991). Figure 4.5.5 shows measurements of $C_E$during a 1990 cruise (Bradley *et al.*,1993). These

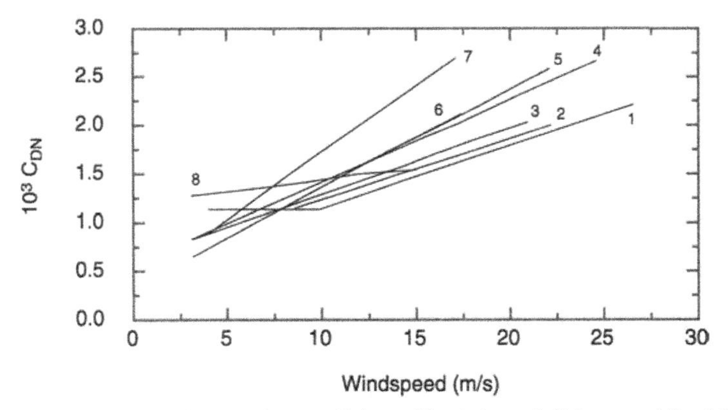

Figure 4.5.1. Dependence of neutral drag coefficient with wind speed: 1) Large and Pond (1981), over deep open ocean; 2) Smith (1980) over deep, coastal ocean; 3) Smithe and Banke (1975) over deep water; 4) Geernaert *et al.* (1987) over North Sea depth of 30 m; 5) Geernaert *et al.* (1986) over North Sea depth of 16 m; 6) Sheppard *et al.* (1972), over Lough Neagh depth of 15 m; 7) Donelan (1982) over Lake Ontario at 10-m depth; and (8) Graf *et al.* (1984) over Lake Geneva at 3 m depth. (From Geernaert, 1990, with kind permission from Kluwer Academic Publishers).

are some of the most recent and the most accurate measurements of evaporative fluxes made using the eddy-correlation method. Figure 4.5.6 shows bulk coefficients estimated from ATLAS buoy observations in the tropical Pacific. Figure 4.5.7 shows a compilation of evaporation measurements of Bradley *et al.* (1991) at low wind speeds plotted along with the high wind speed observations from Large and Pond (1982) by Wu (1992), who recommends

$$\begin{aligned} 10^3 \cdot C_{EN} &= 1.12 - 0.601 \ln U_{10} \quad 0.1 < U_{10} < 0.85\,\mathrm{m\,s^{-1}} \\ &= 1.20 - 0.121 \ln U_{10} \quad 0.85 < U_{10} < 6.5\,\mathrm{m\,s^{-1}} \end{aligned} \tag{4.5.34}$$

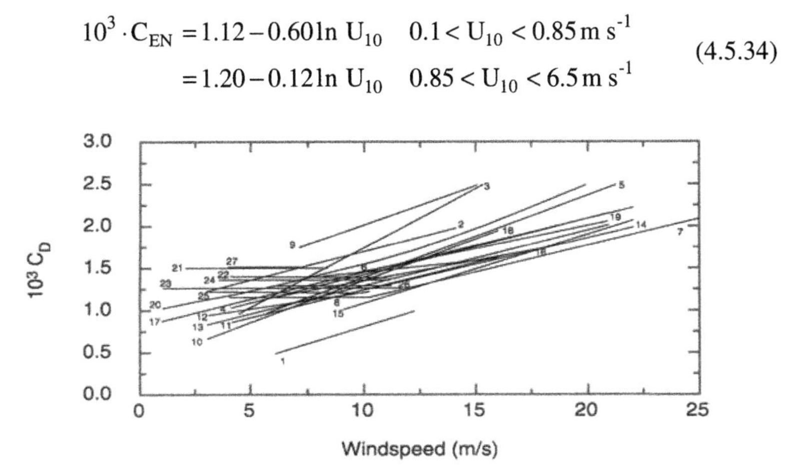

Figure 4.5.2. Neutral drag coefficient as a function of U_{10} (from Said and Druilhet, 1991 with kind permission from Kluwer Academic Publishers). The numbers refer to the authors quoted in Table 4.5.2.

TABLE 4.5.1
Surface Layer Measurements of the Neutral Drag Coefficient (from Geernaert, 1990)

Source	Windspeeds (m/sec)	$10^3 C_{DN}$	Seat (%)	N	[Method]	Platform
Geernaert *et al.* (1987)	5–25	.58 + .085 U	20	116	ec	tower North Sea
Geernaert *et al.* (1986)	5–21	.43 + .097 U	12	186	ec	mast North Sea
Graf *et al.* (1984)	7–17	1.09 + .094 U	—	145	wp	mast Lake Geneva
Donelan (1982)	4–17	.37 + .137 U	28	120	ec	tower Lake Ontario
Large & Pond (1981)	5–19	.46 + .069 U	28	120	ec	tower Atlantic
Large & Pond (1981)	4–10 10–26	1.14 .44 + .063 U	16 16	590 1001	diss diss	tower/ship open ocean
Smith (1980)	6–22	.61 + .063 U	25	120	ec	tower Atlantic
Krugermeyer *et al.* (1978)	3–8	1.30	30	394	wp	buoy North Sea
Khalsa & Businger (1977)	3–12	1.42	22	12	diss	ship open ocean
Smith & Banke (1975)	2.5–21	.63 + .066 U	30	111	ec	mast Atlantic
Hedegåard (1975)	3–14	.64 + .14 U	30	80	ec	mast Kategatt
Kondo (1975)	3–16	1.2 + 0.25 U	15	—	waves	tower Pacific coast
Davidson (1974)	6–11.5	1.44	?	114	ec	FLIP buoy open ocean
Wieringa (1974)	4.5–15	0.6 U 0.86 + .058 U	20	126	ec	tower Lake Flevo
Denman & Miyake (1973)	4–18	1.29 + .03 U	17	70	diss	ship open ocean
Kitaigorodskii *et al.* (1973)	3–11	0.9 to 1.6	>	29	ec	tower Caspian Sea
Hicks (1972)	4–10	$0.5\ U^{.5}$	25	75	ec	tower Bass Strait
Paulson *et al.* (1972)	2–8	1.32	25	19	wp	buoy open ocean
Sheppard *et al.* (1972)	2.5–16	.36 + .1 U 0.86 + .058 U	20	233	wp	tower Lough Neagh
DeLeonibus (1971)	4.5–14	1.14	30	78	ec	Bermuda tower Atlantic Ocean
Pond *et al.* (1971)	4–8	1.52	20	20	ec	FLIP buoy open ocean

(*continues*)

TABLE 4.5.1 (*continued*)

Source	Windspeeds (m/sec)	10^3C_{DN}	Seat (%)	N	[Method]	Platform
Brocks & Krugermeyer (1970)	3–13	1.18+.016 U	15	152	wp	buoy North Sea
Hasse (1970)	3–11	1.21	20	18	ec	buoy North Sea
Miyake *et al.*	4–9	1.09	20	8	ec	UBC site on
(1970)	4–9	1.13	20	8	wp	Spanish Bank
Ruggles (1970)	2.5–10	1.6	50	276	wp	mast Buzzards Bay
Hoeber (1969)	3.5–12	1.23	20	787	wp	buoy open ocean
Weiler &	2–10.5	1.31	30	10	ec	UBC mast on
Burling (1967)	2.5–4.5	0.90	75	6	wp	Spanish Bank
Zubkovskii & Kravchenko (1967)	3–9	0.72+.12 U	15	43	ec	buoy Black Sea

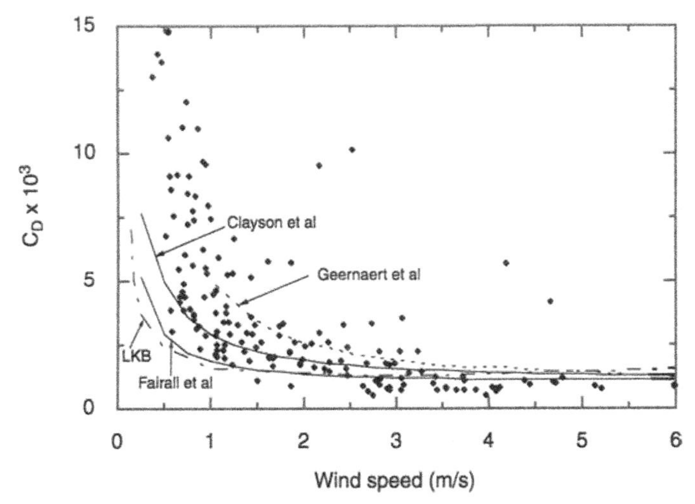

Figure 4.5.3. Drag coefficient measured by inertial dissipation method over the western tropical Pacific (from Bradley *et al.*, 1991). Models are from Liu *et al.* (1979), Geernaert *et al.* (1988), Fairall *et al.* (1996) and Clayson *et al.* (1996).

The measurements of evaporative fluxes from the ocean are particularly important since a major fraction of the heat transfer in the tropics is through evaporation and this drives the coupled atmosphere–ocean system responsible

TABLE 4.5.2
C_D Values Compiled by Said and Druilhet (1992) Corresponding to the Values Shown in Figure 4.5.2

Author	Comp. scheme	Exp. mean	Exp.	$10^3 C_d(10)$	ms	$U(10)$ range	$10^3 C_d$	N^0
Davidson (1974)	Eddy correl.	P	BOMEX Barbads	$-0.015 + 0.085\ U_{10}$		6–12	0.68	1
Deacon and Webb (1962)			COMPIL.	$1 + 0.07\ U_{10}$		3–14	1.7	2
Donelan (1982)	EC	T	Lake Ontario	$0.37 + 0.137\ U_{10}$		4–17	1.74	3
Francey and Garratt (1978)	EC	T	Amtex 75	$0.77 + 0.085\ U_{10}$		5–15	1.6	
Garratt (1977)	EC P	P	COMPIL.	$0.75 + 0.067\ U_{10}$		4–21	1.42	4
Geernaert *et al.* (1986)	EC	P	MARSEN North Sea	$0.43 + 0.097\ U_{10}$	± 0.17	5–22	1.3	5
Kondo (1975)	M		Sagami Bay (Japan)	$1.2 + 0.025\ U_{10}$ $\ldots 0.073\ U_{10}$		8–16 25 50	1.45	6
Large and Pond (1981)	D EC	B–S	Jasin and Bredford	1.14 $0.49 + 0.065\ U_{10}$		10 10–14	1.14	7 8
Merzi and Graf (1985)	P	P	Lake Leman	$1.09 + 0.094\ U_{10}$		7–17	1.99	9
Sheppard *et al.* (1972)	P		Lough Neagh	$0.36 + 0.1\ U_{10}$		3–16	1.36	10
Smith (1974)	EC		Atlantic Ocean	$0.58 + 0.068\ U_{10}$	± 0.24	4–16	1.26	11
			Lake Ontario	$0.82 + 0.0039\ U_{10}$	± 0.20	3–10	1.21	12
Smith and Banke (1975)			COMPIL.	$0.63 \pm 0.006\ U_{10}$	± 0.23	3–21	1.29	13
Smith (1980)	EC	P	Sable Island	$0.61 + 0.063\ U_{10}$		6–22	1.24	14
			(Nova Scotia)	$0.27 + 0.082\ U_{10}$		9–22	1.09	15
Wieringa (1974)	EC P	P	Lake Flevo	$0.87 + 0.048\ U_{10}$		5–15	1.35	16
Wu (1980)	EC P		COMPIL.	$0.8 + 0.065\ U_{10}$		1–22	1.45	17

Garratt (1977)			COMPIL.	$0.51\ U_{10}0.45$		4–21	1.47	18
Wieringa (1974)	EC	P	Lake Flevo	$0.7\ U_{10}^{0.3}$		5–15	1.4	19
	P							
Kitaigorodskyi (1973)	EC	T	Caspian Sea	$1.2\ (\mathrm{Re}^*)^{0.15}$		1–22	1.56	20
and Wu (1980)								
Andreas and	EC	M	Leads and	1.49		2–8	1.49	21
Murphy (1986)	P		Polynyas					
Dunckel *et al.*	EC	B	ATEX	1.39		4.5–11	1.39	22
(1974)	D		Equatorial	1.26			1.26	
	P		Atlantic	1.56			1.56	
Hasse *et al.* (1978)	P	B	GATE-Tropical	1.25		1–12	1.25	23
			Atlantic					
Krügermeyer (1976)	P	B		1.34		3.5–11	1.34	24
Mitsuta and	EC	S	NW Pacific	1.2		3–9	1.2	25
Fujitani (1974)								
Nicholls (1985)	EC	A	JASIN	1.29	±0.08	6–11	1.29	26
Pond and	EC	P	BOMEX	1.5	±0.26	4–7	1.5	27
Phelps (1971)	D	S		1.5	±0.40			
Emmanuel (1975)	EC		Lake Hefner	1.15	±0.2	2.7–8	1.15	
	P			1.34	±0.3			
Greenhut (1981)	EC	A	GATE-Tropical	1.87		5–6		
			Atlantic	$1.21 (Z_0 = aU^*)$				
Greenhut and	EC	A	EPOCS	$2.37\ (Z_0 = au^*)$	±0.56	2.5–15		
Bean (1981)			Equat. Pacif.					
Greenhut (1983)	EC	A	NORPAX	1.9		7–12		
			N Pacific	$1.22\ (Z_0 = aU^*)$				
Hicks and	EC	P	Bass Strait	1.1	±0.1	2–10	1.1	
Dyer (1970)			Australia					
McBean and	EC	A	Ontario Lake	2.6			2.6	
McPherson (1976)								
Pond *et al.* (1974)	P		Arabian Sea	1.49	±0.28	2–8	1.49	
	P		BOMEX	1.48	±0.21	2.5–8	1.48	
Smith and	EC		near Sable	1.6		8–21	1.6	
Banke (1975)			Island					
Tsukamuto	EC		AMTEX	1.32		3–13	1.32	
et al. (1975)			China Sea					

TABLE 4.5.3
Surface Layer Measurements of Stanton and Dalton Numbers (from Geernaert, 1990; see also Smith, 1988, 1989)

Source	10^3 CHN	10^3 CEN	Remarks
Geernaert *et al.* (1987)	0.75	—	Slightly stable only
Large and Pond (1982)	0.69/1.08	—	Stable/unstable
Anderson and Smith (1981)	0.82/1.12	1.27	C_{HN}: stable/unstable
Smith (1980)	0.83/1.10	—	Stable/unstable
Davidson *et al.* (1978)	1.2	1.10	Diss method
Friehe and Schmitt ((1976)	0.91	0.64	
Smith (1974)	1.2	1.41	
Muller-Glewe and Hinzpeter (1974)	1.0	0.77	
Dunckel *et al.* (1974)	1.5	1.40	
Pond *et al.* (1974)	1.5	1.55	

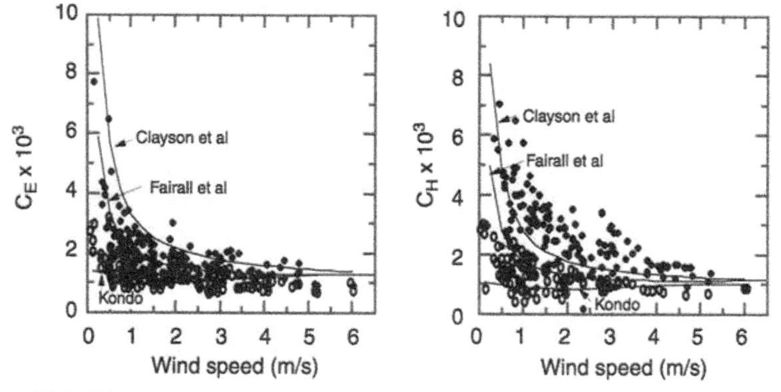

Figure 4.5.4. Eddy correlation measurements of heat and moisture bulk transfer coefficients in the western tropical Pacific in 1988 (from Bradley *et al.*, 1991). Models are from Fairall *et al.* (1996), and Clayson *et al.* (1996) and Kondo (1975) for neutral conditions (1975). Solid circles show measured values; hollow circles show neutral values.

for the ENSO phenomenon. These observations are particularly difficult to make at the low wind speeds characteristic of the tropics. Nevertheless, careful measurements by Fairall *et al.* (1996a) in the western Pacific region during 1992 suggest a near-constant value for C_{EN} of 1.11×10^{-3} with a slight decrease with wind speed. They also find that stress measurements by inertial dissipation techniques involve far less scatter than those by covariance methods, mainly due to the difficulty of correcting for ship motion. Enriquez and Friehe (1997), in

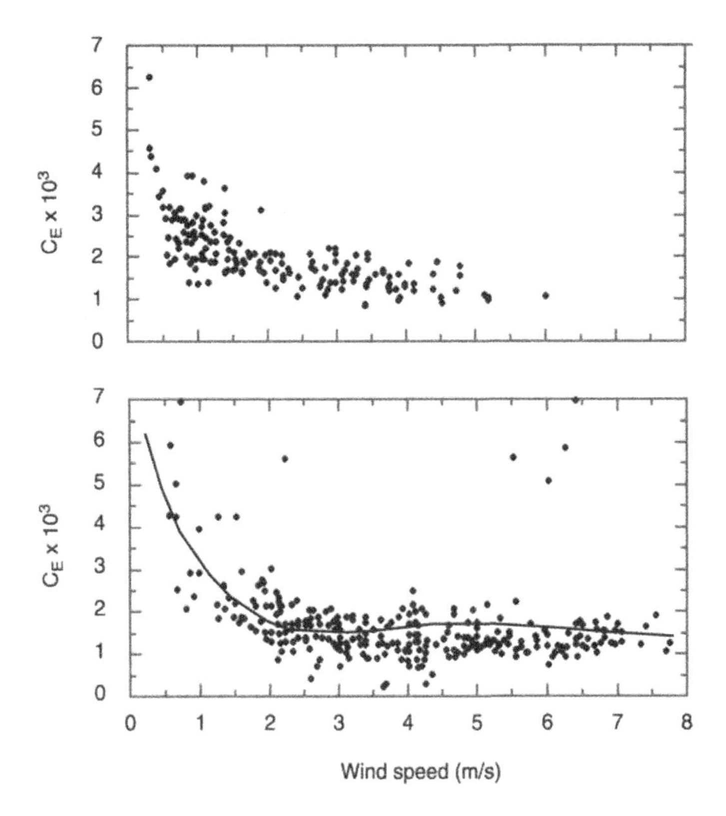

Figure 4.5.5. Measurements of moisture bulk transfer coefficients in the western tropical Pacific (from Bradley *et al.*, 1993). Measurements were all performed during mainly low wind periods of May 1988 and 15 September 1990. Bulk model values of Liu *et al.* (1979) are shown by the line.

measurements over coastal waters during CODE and SMILE, found that, while $C_{DN} \sim (0.509 + 0.065\ U_{10}) \times 10^{-3}$, despite the large scatter, C_{HN} and C_{EN} are roughly constant at $(1.05 \pm 0.39) \times 10^{-3}$. The consensus at present therefore is to regard C_{HN} and C_{EN} as roughly constants with wind speed at a value of about 1.1×10^{-3}.

Very careful measurements at high wind speeds up to 18–23 m s^{-1} of heat and water vapor exchange during the Humidity Exchange Over the Sea (HEXOS) Main Experiment (HEXMAX) from a fixed tower in the North Sea in 1986 also indicate a near-constant value of 1.1×10^{-3} for C_{HN} (Smith *et al.*, 1992). Elaborate precautions were taken during these observations to prevent sensor contamination by salt spray at high wind speeds. The reader is referred to this paper for an idea of the difficulties in making accurate reliable measurements of air–sea transfer at sea, even from fixed platforms. Dupuis *et al.* (1997) used the

TABLE 4.5.4
Measurements of C_{HN}, C_{EN} (from Said and Druilhet, 1992)

Author	Comp. scheme	Exp.	Exp.	$10^3\ C_t(10)$	$10^3\ C_q(10)$	$U(10)$ range
Anderson and Smith (1981)	EC	M	Sable Island (Nova Scotia)	S: $10^3\overline{W'T'} = 0.82\ U\Delta T + 1.69$ I: $10^3\ \overline{W'T'} = 1.12\ U\Delta T + 2.24$	N: 1.27 (±0.26) $0.55 + 0.083\ U$	5.4–11.3
Andreas and	EC	M	Leads &	1.0	$= C_t$	2–8
Murphy (1986)	P	COMPIL.	Polynyas			
Antonia *et al.* (1978)	EC	P	Bass Strait		0.82 (±0.15)	
Coulman (1979)	EC	A	Coral Sea	1.4 (±0.3)	1.34 (±0.3)	2–10
Dunckel (1974)	P	B	ATEX Equat. Atlant.	1.5	1.28	4.5–11
Emmanuel (1975)	EC		Hefner Lake	1.1 (±0.3)	1.34 (±0.3)	2.7–8
Francey and Garratt (1978)	EC	T	AMTEX 75	1.2 (±0.5)	1.5 (±0.3)	5–15
Friehe and	P	P	COMPILATION	1.15 (±0.3)	1.34 (±0.4)	3–7
Schmitt (1976)	EC			$10^3\ \overline{W'T'} = 1.41\ U\Delta T + 1.2$	1.32 (±0.7)	
Fujitani (1981)	EC	S	AMTEX 74–75		1.05	5–15
Garratt and Hyson (1975)	EC	T	AMTEX 74	1.2 (±0.3)	1.6 (±0.3)	4–16
Greenhut and	EC	A	EPOCS	1.4 (±0.19)	1.1 (±0.05)	2.5–15
Bean (1981)	P		tropic. CLM			
Hasse (1978)	P	B	GATE-Tropical Atlantic	1.34	1.15	1–12
Hicks and Dyer (1970)	EC		Bass Strait (Australia)	1.4 (±25%)		2–10

Hicks *et al.* (1974)	EC		Coral reef (Australia)		1.1	3–10
Kitaigorodskyi (1973)	EC	T	Caspian sea	$= C_q(10)$	$1.1(10^{-0.616+0.137U})$	1–22
					1.2 to 10 m/s	
Krügermeyer (1976)	P	B		$1.42(1 - 1.6/U^2)$	$1.2(1 - 1.4/U^2)$	3.5–11
Kruspe (1977)	EC	S	German Bight Baltic Sea		1.36 (±0.25)	9–10
Large and Pond (1982)	D	B-S	Jasin and	S: 0.66	1.15 (±0.22)	4–14
	EC		Bredford	I: 1.13		
Müller-Glewe and Hinzpeter (1974)	EC		Baltic Sea		1.0	8
Nicholls (1985)	EC	A	JASIN		0.9 (±0.1)	6–11
Pond *et al.* (1971)	EC	P	San Diego	1.0		4–7
	EC	P	BOMEX and		1.18(±0.17)	4–7.5
	D	S	San Diego		1.20 (±0.25)	
Pond *et al.* (1974)	P	P	Arabian Sea		1.36 (±0.40)	2–8
	P	B	BOMEX	1.47 (±0.64)	1.41 (±0.18)	2.5–8
Smith (1974)	EC	T	Lake Ontario	1.3 (±0.5)	1.2 (±0.3)	3–10
Smith and Banke (1975)	EC		offshore Sable Island	1.5 (±0.4)		8–21
Smith (1980)	EC	P	Sable Island	S: $10^3\overline{W'T'} = 0.93U\Delta T + 0.1$ I: $10^3\overline{W'T'} = 1.1U\Delta\theta + 3.2$		6–22
Smith and Anderson (1988)	EC	P	HEXOS North Sea		1.2	5–18
Thorpe (1973)	EC	M	Arctic Sea	1.2 (±0.7)	0.55 (±0.23)	
Tsukamoto *et al.* (1975)	EC		AMTEX China Sea	1.40	1.28	3–13

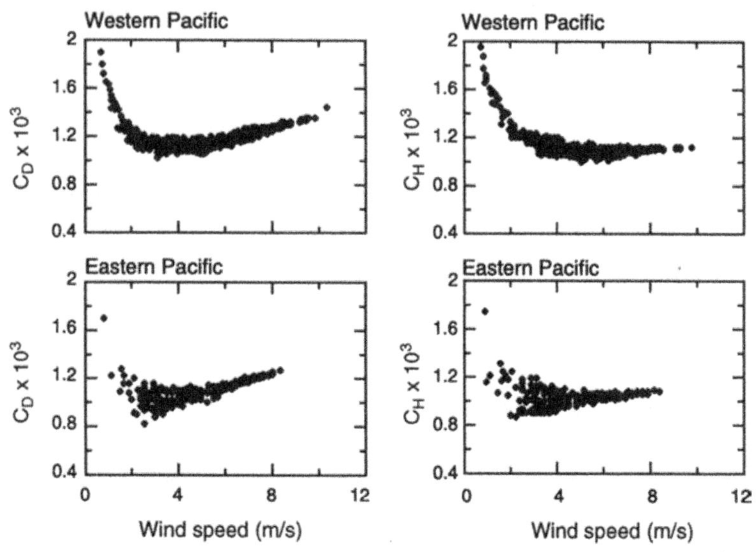

Figure 4.5.6. Measurements of bulk exchange coefficients from buoys in the tropical Pacific (from Liu *et al.*, 1990). (a) and (b) show values from the western and eastern tropical Pacific Ocean, respectively. Values of C_D are shown on the left, and those of C_H are shown on the right.

inertial dissipation method to measure C_{DN} and C_{HN}, C_{EN} at low wind speeds and suggest

$$
\begin{aligned}
10^3 \cdot C_{DN} &= 0.668 + 11.7\, U_{10}^{-1} \quad & U_{10} < 5.5\text{m s}^{-1} \\
10^3 \cdot C_{EN} &= 0.66 + 2.79\, U_{10}^{-1} \quad & U_{10} < 5.2\text{m s}^{-1} \\
&= 1.20 = 10^3 \cdot C_{HN} \quad & U_{10} > 5.2\,\text{m s}^{-1}
\end{aligned}
\tag{4.5.35}
$$

implying a rapid increase in the exchange coefficients as the wind speed decreases within the smooth flow regime.

Taken together with the TOGA/COARE measurements of Fairall *et al.* (1996a), the constancy of the scalar bulk transfer coefficient is hard to explain. The surface area covered by whitecaps from breaking waves increases as the cube of the wind speed. Breaking waves are important to air–sea transfer, because of the entrainment of air bubbles into the water and ejection of droplets by breaking waves into the air, which constitute mechanisms that circumvent the molecular sublayers at the air–sea interface and hence presumably are more efficient in effecting air-sea transfer of water vapor and gas. Therefore, at high wind speeds, the water vapor flux across the air–sea interface is a combination of the usual evaporative flux from the interface and the evaporation from spray droplets ejected into the atmosphere from breaking waves. As the wind speed increases, the spray droplet contribution can be expected to become more significant to the water vapor flux (Andreas, 1992, 1998 for example). Some

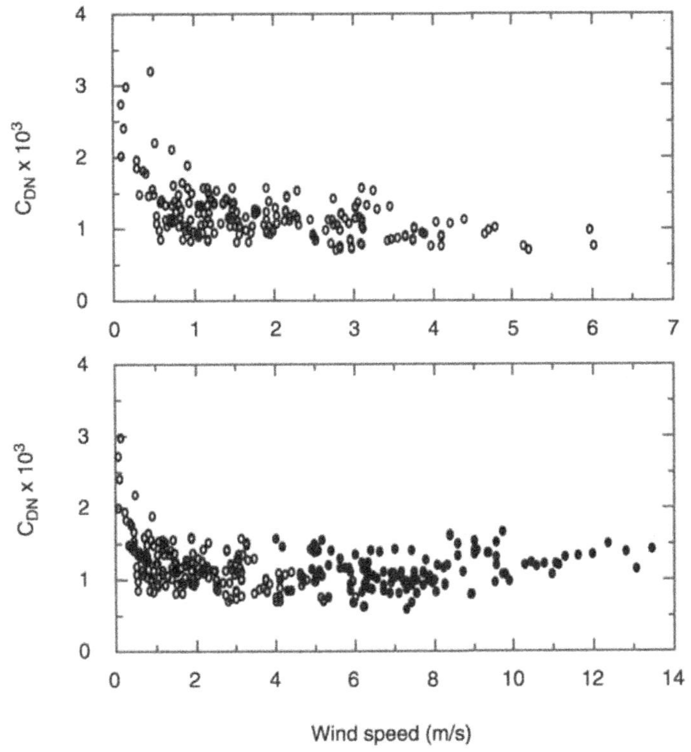

Figure 4.5.7. Bulk transfer coefficient for moisture at low and intermediate wind speeds (from Wu, 1992). Those reported by Bradley *et al.* (1991) are reproduced as open circles and those by Large and Pond (1982) as solid circles.

theoretical studies suggest an effective transfer rate 2–3 times larger than that indicated by conventional bulk transfer across the air–sea interface at wind speeds of 15–25 m s^{-1}. Yet the observations do not indicate this. On the contrary, the measurements are in agreement with the surface renewal theories (for example, Liu *et al.*, 1979), even though these theories do not account for breaking waves and droplet ejection. One possibility is the negative feedback of evaporating droplets, which increase the humidity adjacent to the air–sea interface and hence decrease the evaporative transfer across the interface, which offsets the evaporation from droplets. The droplet evaporation should also lead to a decrease in the air temperature and hence an increase in sensible heat flux from the ocean. It is worth noting that only the direct evaporative flux from the air–sea surface is relevant to the ocean-side heat balance, since the ejection and subsequent evaporation of droplets do not extract any heat from the ocean. The effect on the ocean is only indirect in the sense that the change in the air properties adjacent to the ocean surface can cause changes in the sensible and latent heat fluxes from the ocean. If the droplet evaporation plays an increasing role in water vapor transfer as wind speed increases, the constancy of C_{EN} implies that the ocean is actually losing relatively less heat from evaporation at high wind speeds than it would in the absence of wave breaking, even though the net evaporative transfer is the same! However, no measurements exist of the actual loss of heat by the ocean due to evaporation at these high wind speeds.

When a very moist, warm air mass moves over cold coastal waters, direct condensation of water vapor can occur at the ocean surface, with an equivalent heat transfer to the ocean from the atmosphere of as much as 50 W m^{-2}. This is similar to the process of condensation over land from warm moist air over cooler land surfaces, leading to similar heat transfer rates (Garratt, 1992; Enriquez and Friehe, 1997).

The large scatter in the functional relationships of the exchange coefficients with wind speed is due to surface layer processes that are not properly understood and parameterized (Donelan, 1990), especially those related to the surface wave field, which may not always be in equilibrium with the prevailing wind. It is the wave field that determines the roughness characteristics of the sea surface and hence the fluxes across the air–sea interface. For a fully developed sea, the roughness scale is given by the Charnock relationship,

$$z_0 = c_3^1 \frac{u_*^2}{g} \tag{4.5.36}$$

The value of the Charnock constant in literature ranges from 0.011 to 0.0185 (Garratt, 1992). A value of 0.016 is often used [recent measurements by Fairall *et al.* (1996a) indicate a value of 0.011]. The state of a surface wave field can be characterized very simply by the value of the ratio of the phase speed c to u_* of the peak of the spectrum. Wave age c/u_* is large (close to 20–30) for fully developed wave fields and smaller for developing ones. For a developing wave field the roughness scale should be a function of c/u_* as well. Therefore it is reasonable to expect that the drag coefficient should be a function of c/u_*.

$$C_{DN} \cdot 10^3 = 0.012 \left(\frac{c}{u_*} \right)^{-2/3} \tag{4.5.37}$$

Figure 4.5.8 (from Geernaert, 1990) and Figure 4.5.9 (from Bergstrom and Smedman, 1994) show measurements that indeed support this functional relationship. However, as Bergstrom and Smedman point out, the correlation between C_{DN} and c/u_* could be spurious, since u_* is the normalizing variable on both axes. Other recent measurements disagree with the functional form for C_{DN} presented by Geernaert (1990).

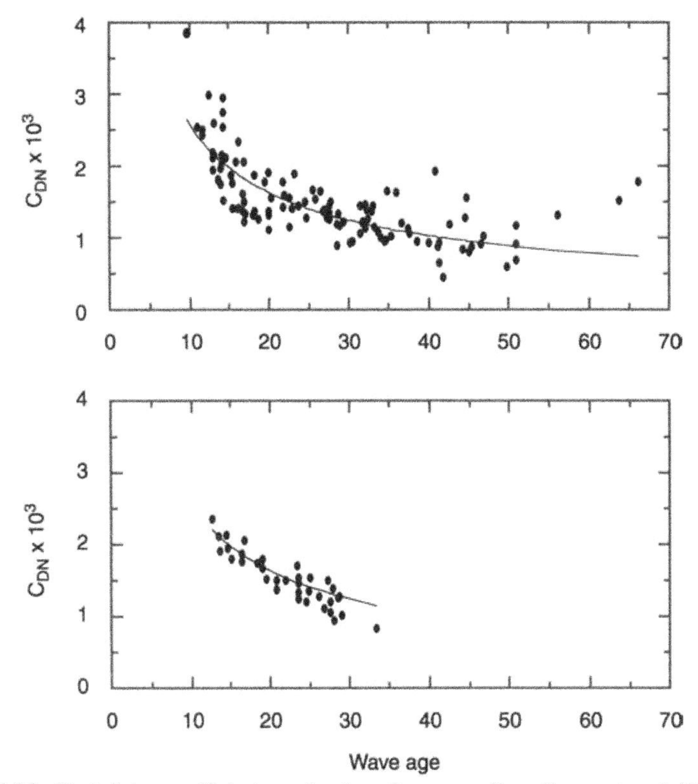

Figure 4.5.8. Neutral drag coefficient as a function of wave age (from Geernaert *et al.* 1987). The upper panel shows data from the North Sea, the lower panel from the MARSEN experiment. The best-fit line is $C_{DN} = 0.012\,(C_o/u_*)^{-2/3}$.

As we saw earlier, C_{DN} and the roughness scale z_0 are simply related to each other by Eq. (4.5.12), but the latter is directly related to the properties of underlying surface wave field. Therefore, it is more logical to derive relations for z_0 instead. Generalizing Charnock's expression for z_0 to make the Charnock constant a function of wave age,

$$g\,z_0/u_*^2 = f\,(c/u_*) \sim c_0\,(c/u_*)^m \qquad (4.5.38)$$

Maat *et al.* (1991), after a careful examination of HEXMAX 1986 field data, conclude that m = -1; $c_0 \sim 0.48$ according to Smith *et al.* (1992).

The value of m, the exponent of (c/u_*) in Eq. (4.5.38), has important physical implications. Toba et al. (1990) suggest that this exponent should be +1, which implies that mature waves are aerodynamically rougher than young waves. This is contrary to the notion that in a mature wave field, a considerable portion of the wave spectrum is moving at speeds close to the wind speed and hence extracting

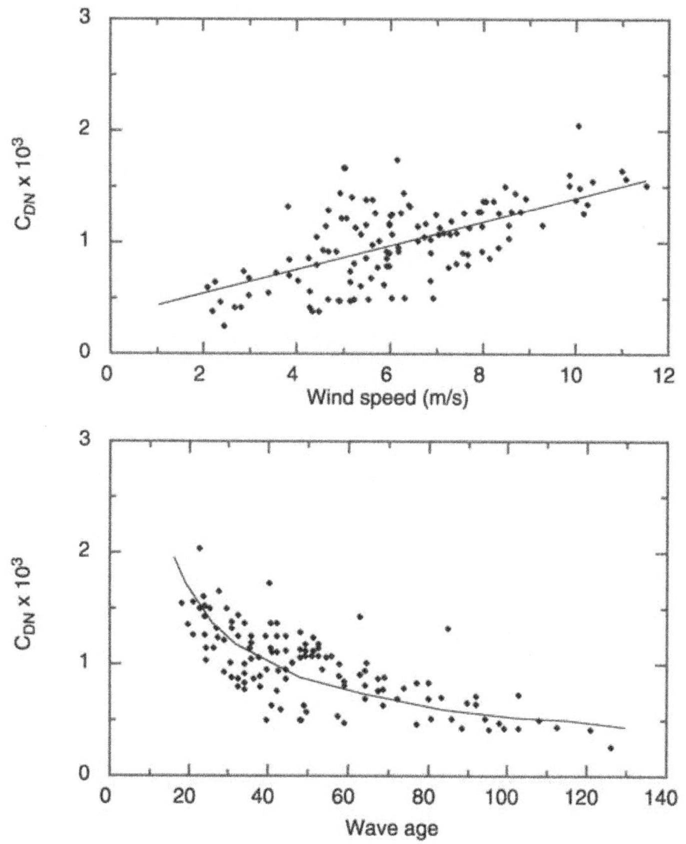

Figure 4.5.9. Neutral drag coefficient at 10 m as a function of wind speed (top panel) and wave age (from Bergstrom and Smedman, 1994 with kind permission from Kluwer Academic Publishers).

little energy from the wind, whereas in early stages of wind wave development, a larger fraction of the spectrum is moving much slower than the wind speed and hence capable of extracting energy from the wind. In other words, younger waves should be more aerodynamically rougher than more mature waves. This calls for m to be negative. Indeed, there is increasing evidence and support for m = -1 (see references cited in Atakturk and Katsaros, 1999). The Charnock formulation implies m = 0.

Charnock's expression for the roughness scale z_0 is valid for fully developed "rough" seas. Often for near-calm wind conditions the sea surface appears "glassy," the boundary layer immediately adjacent to the air–sea interface is viscous and laminar, and the "roughness" scale is the molecular sublayer thickness (ν/u_*). When the winds increase but are still low, the first set of waves generated are capillary gravity waves, and therefore at low wind speeds the roughness is determined by capillary waves, the scale being (γ/u_*^2). An approximate relationship that takes into account these aspects can be written (Wu, 1994; Bourassa *et al.*, 1995) with z_o [remember z_o and C_{DN} are related by Eq. (4.5.12)] as a linear combination of the above three scales:

$$z_0 = c_1 \frac{\nu}{u_*} + c_2 \frac{\gamma}{u_*^2} + c_3 \frac{u_*^2/g}{c/u_*} \tag{4.5.39}$$

$$c_1 \sim 0.11, \quad c_2 \sim 0.016, \quad c_3 \sim 0.8$$

A relationship such as this permits an initially laminar boundary layer, whose thickness (and therefore z_0) decreases with increases in wind speed until capillary waves are generated by laminar instability of the air–sea interface to wind forcing. This occurs at about $u_* = 0.044$ m s^{-1} for a clean surface (0.2 m s^{-1} if the surface is covered by a slick). For typical oceanic conditions, it is possible to use a u_* of 0.07 m s^{-1} ($U_{10} \sim 2$ m s^{-1}) as the condition for onset of capillary waves (note that the waves begin to appear on the surface at u_* value of about 0.02 m s^{-1}, but they are nearly invisible to the naked eye). z_0 increases at this point to values much above the equivalent laminar values. At higher wind speeds, the short gravity waves begin to dominate the drag and z_0 ($u_* \sim 0.15$ m s^{-1}, $U_{10} \sim 4$ m s^{-1}). Beyond this point, it is the gravity waves that are important in determining the sea surface roughness. The composite z_0^t given by the following relationship, where the dependence of z_0 on wave age in the gravity wave limit is ignored and replaced by Charnock's relationship, is shown in Figure 4.5.10:

$$z_0^t = c_1 \frac{\nu}{u_*} + c_2 \frac{\gamma}{u_*^2} + c_3^1 \frac{u_*^2}{g} \tag{4.5.40}$$

$$c_1 \sim 0.11, \quad c_2 \sim 0.016, \quad c_3^1 \sim 0.016$$

Also shown is the effective roughness scale z_0^e, which uses only the appropriate viscous, capillary, and gravity wave roughness scales as the wind speed increases. The comparison between z_0^t and z_0^e shows that, given the uncertainties in parameterizing z_0, the difference between the two is not too significant and either can be used to obtain z_0 as a function of u_*, and by extension in Eq. (4.5.39), which includes the wave age dependence of z_0, as well. Normally, wave age is unknown and therefore Eq. (4.5.40) is more practical.

Using data collected during the Riso Air Sea Experiment (RASEX) from a tower 2 km off the Danish coast but in shallow waters 4 m deep, Mahrt *et al.* (1996) attempted to investigate whether the roughness scale increases due to capillary waves under light wind conditions (Wu, 1994) or if smooth conditions prevail as predicted by Liu *et al.* (1979). The extraordinary difficulties of making stress measurements at weak winds by the eddy correlation techniques resulted

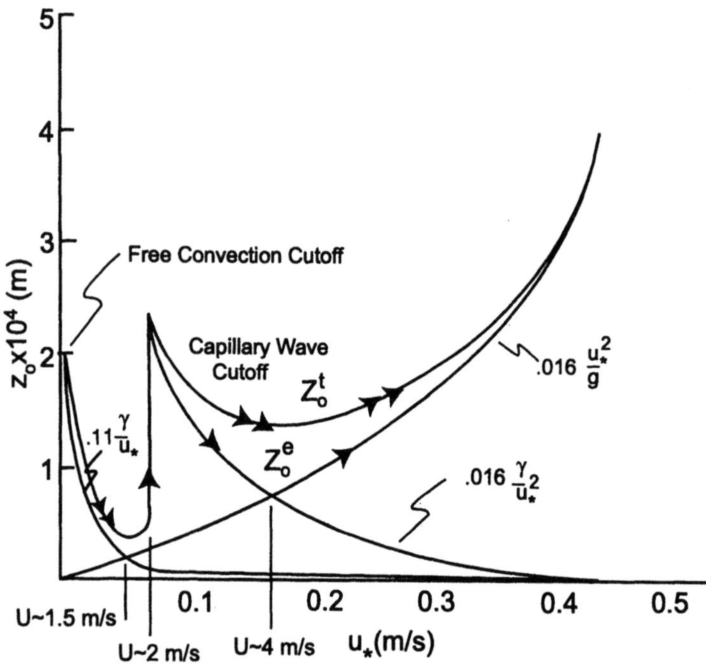

Figure 4.5.10. Roughness scale as a function of friction velocity, showing the influence of capillary waves.

in large uncertainties that precluded any conclusions in this regard. The drag coefficient did however reach a minimum of about 10^{-3} at about 4 m s^{-1}, before increasing to ~2–3 × 10^{-3} at about 1 m s^{-1}. Vickers and Mahrt (1997) also report on these measurements. For a given wind speed, they found larger C_D for younger steeper slow-moving waves representative of short fetches than for longer fetches. They found that their measurements were even more strongly dependent on wave age and the above formula, while fitting their data well, overpredicted C_D but underpredicted the sensitivity to wave age. They seem to conclude that the capillary wave term was not important.

It is now generally accepted that the neutral drag coefficient C_{D10N} depends not only on the wind speed, but also the sea state, with its value increasing with decreasing wave age, because of increased surface roughness associated with younger waves (Donelan, 1990; Smith *et al.*, 1992; Donelan *et al.*, 1993). However, the influence of swell on the drag coefficient is still uncertain. Donelan *et al.* (1997) present measurements that show that the presence of cross and counter swells can result in much higher values for the drag coefficient than the value for the normal sea. They also suggest that inertial dissipation method of measuring the momentum flux may result in significant errors in the presence of swells. Instead, eddy correlation methods, with sensor motion corrected using modern highly sensitive motion sensing technology, may be superior to inertial dissipation techniques, even when the measurements are made from ships and buoys.

4.5.3 MEASUREMENTS OF BULK TRANSFER COEFFICIENTS OVER LAND

The bulk coefficients C_H, $C_{E,}$, C_C, and C_A over land are given by

$$\begin{aligned} H_s &= \rho c_p \frac{u_*}{C_D} C_H (T_s - T_a) \\ E &= \rho \frac{u_*}{C_D} C_E (q_s - q_a) = \rho \frac{u_*}{C_D} C_C (q_s^{sat} - q_s) \end{aligned} \tag{4.5.41}$$

and

$$H_t = \rho L_E \frac{u_*}{C_D} C_A (q_a^{sat} - q_a) \tag{4.5.42}$$

While C_D, C_H, and C_E have their usual meaning (drag coefficient and Stanton and Dalton numbers), C_C is proportional to the inverse of the stomatal resistance. It is supposed to capture the ease of water vapor transfer from the leaf interior to the surface, with the transfer assumed to depend on the ambient specific humidity q_s and the specific humidity in stomatal air cavities q_s^{sat}, which is assumed to be saturated at the leaf temperature T_S. It can be called the stomata number. The coefficient C_A is supposed to represent the ambient total heat transfer coefficient with respect to a specific saturation deficit at level z. Note that ε_s is the rate of change of saturated specific humidity with temperature (~2.2 at 20°C):

$$\varepsilon_s = \frac{L_E}{c_p} \frac{dq^{sat}}{dT} \tag{4.5.43}$$

Also

$$H_t = H_s + L_E E = SW_{\downarrow} + LW_{\downarrow} - SW_{\uparrow} - LW_{\uparrow} - H_G - SW_{ph} \tag{4.5.44}$$

where SW and LW indicate the short- and longwave radiative fluxes (the arrows the downwelling and upwelling aspect), H_G is the heat flux to the ground or to storage, and SW_{Ph} is the amount consumed by photosynthesis of plants (~2% of the net SW and LW flux).

Equations (4.4.2) and (3.9.8), and (4.5.41) and (4.5.42), are used for parameterization of evaporation in global models and estimating the evaporation rate over land surfaces. Equation (3.9.8) has very interesting limiting properties (Thom, 1975; Kaimal and Finnigan, 1994):

1. $C_{H,E} \ll C_C$. Here vegetation is wet and stomatal resistance is not an important factor, and the third term in the denominator drops out. This corresponds to the case of "potential evaporation," since evaporation can be as much as physically feasible with no control exerted by the physiology of the plants.
2. $C_{C,A} \gg C_{H,E}$. Here $H_l = \varepsilon_s H_s$. This corresponds to "equilibrium evaporation," where physiology of plants exerts no control, because the limiting factor is the ability of the water vapor to diffuse away from the leaf surface. The situation corresponds to light winds and humid conditions, such as asymptotic conditions downstream of well-irrigated crops (i.e., crops free from water stress).
3. $C_{C,A} \ll C_{H,E}$. Here stomatal control is dominant and the situation corresponds to dry windy conditions over irrigated crops, "oasis evaporation."

Parlange *et al.* (1995) present a slightly generalized version of Eq. (3.9.8), Eq. (3.9.13), which is repeated here for convenience:

$$H_L = L_E E = \beta \left[A \frac{\varepsilon_s H_t}{\varepsilon_s + \frac{C_H}{C_E} + \frac{C_H}{C_C}} + B \frac{\rho L_E \frac{u_*}{C_D} C_H (q_a^{sat} - q_a)}{\varepsilon_s + \frac{C_H}{C_E} + \frac{C_H}{C_C}} \right] \tag{4.5.45}$$

where A and B are constants that take different values in different formulations. For example, for the potential evaporation case A = B = 1, for equilibrium evaporation A = 1, B = 0, etc. (see Parlange *et al.*, 1995). The Budyko–Thornwaite–Mather parameter β is taken as some function of surface water availability, put equal to 1.0, but allowed to go to zero as water availability becomes less and less.

4.6 SURFACE RENEWAL THEORY

Bulk transfer relationships outlined above are entirely empirical, and invalid in the free convection limit and perhaps at high wind speeds as well. It is therefore preferable to derive functional relationships for air–sea fluxes that are rooted firmly on theoretical arguments based on the physics of air–sea transfer. The surface renewal theory of Brutsaert (1975) is an example of an attempt to bring turbulence theory to bear on the problem. Liu, Katsaros, and Businger (1979) were the very first ones to derive a physics-based model of air–sea transfer. This theory has been employed recently by Clayson *et al.* (1995). In this section, we will examine the basics of these two formulations for examination of air–sea fluxes. We will also discuss the use of surface renewal theory in plant canopies.

If we recognize the fact that the only mechanism for scalar transfer through an air–sea interface at low to moderate wind speeds is the molecular diffusion through thin molecular sublayers that exist on either side of the interface (Figure 4.6.1), then it is natural to attempt to derive an expression for the thickness of these sublayers that in turn determines the magnitude of the fluxes. These layers have temporal and spatial variability, since they are easily disrupted by eddies in the adjacent turbulent regions of the boundary layers (and any wave breaking processes). If t^* is the average time during which the sublayers exist between disruptions, then the average thicknesses of the sublayers are

$$\delta \sim \left(\nu t^*\right)^{1/2}; \quad \delta_{T,E,G} \sim \left(k_{T,E,G}\, t^*\right)^{1/2} \tag{4.6.1}$$

The kinematic fluxes are

$$u_*^2 \sim \nu \frac{\Delta U}{\delta} \sim \nu \frac{\Delta U}{\left(\nu t^*\right)^{1/2}} \tag{4.6.2}$$

$$Q_H \sim k_T \frac{\Delta T}{\delta_T} \sim k_T \frac{\Delta T}{\left(k_T t^*\right)^{1/2}} \tag{4.6.3}$$

$$Q_E \sim k_E \frac{\Delta q}{\delta_E} \sim k_E \frac{\Delta q}{\left(k_E t^*\right)^{1/2}} \tag{4.6.4}$$

$$Q_G \sim k_G \frac{\Delta c}{\delta_G} \sim k_G \frac{\Delta c}{\left(k_G t^*\right)^{1/2}} \tag{4.6.5}$$

where Δ denotes the change across the sublayers. Therefore

$$\frac{\Delta U}{u_*} \sim u_* \left(\frac{t^*}{\nu}\right)^{1/2} \tag{4.6.6}$$

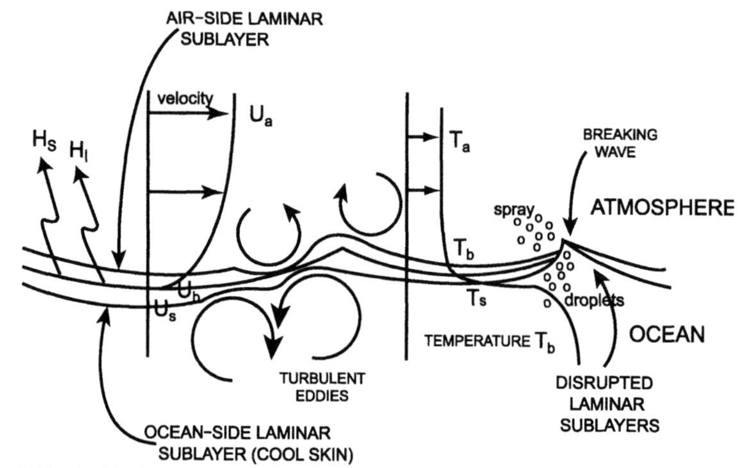

Figure 4.6.1. An idealized version of the air–sea interface showing the laminar sublayers on the air and water side. Also shown are velocity and temperature profiles.

$$\frac{\Delta T}{\theta_*} \sim \frac{\Delta T}{Q_H / u_*} \sim u_* \left(\frac{t^*}{k_T}\right)^{1/2} \tag{4.6.7}$$

$$\frac{\Delta q}{q_*} \sim \frac{\Delta q}{Q_E / u_*} \sim u_* \left(\frac{t^*}{k_E}\right)^{1/2} \tag{4.6.8}$$

$$\frac{\Delta c}{c_*} \sim \frac{\Delta c}{Q_G / u_*} \sim u_* \left(\frac{t^*}{k_G}\right)^{1/2} \tag{4.6.9}$$

The crucial question concerns the timescale t^*. If one assumes that t^* is the timescale of the Kolmogoroff eddies in the turbulent flow and if we take the roughness scale z_0 as the integral length scale (so that dissipation $\varepsilon \sim u*^3/ z_0$), we get

$$t^* = (\nu/\varepsilon)^{1/2} \sim \left(\nu z_0 / u_*^3\right)^{1/2} \tag{4.6.10}$$

$$\frac{\Delta U}{u_*} \sim Re_r^{\ 1/4} \tag{4.6.11}$$

$$\frac{\Delta T}{\theta_*} \sim Re_r^{\ 1/4} Pr^{1/2} \tag{4.6.12}$$

$$\frac{\Delta q}{q_*} \sim \mathrm{Re_r}^{1/4}\mathrm{Sc}^{1/2} \tag{4.6.13}$$

$$\frac{\Delta c}{c_*} \sim \mathrm{Re_r}^{1/4}\mathrm{Le}^{1/2} \tag{4.6.14}$$

where

$$\mathrm{Re_r} = \frac{u_* z_0}{\nu}, \quad \mathrm{Pr} = \frac{\nu}{k_T}, \quad \mathrm{Sc} = \frac{\nu}{k_E}, \quad \mathrm{Le} = \frac{\nu}{k_G} \tag{4.6.15}$$

The corresponding layer thicknesses are

$$\delta = \frac{\nu}{u_*}\frac{\Delta U}{u_*} \tag{4.6.16}$$

$$\delta_T = \frac{k_T}{u_*}\frac{\Delta T}{\theta_*} \tag{4.6.17}$$

$$\delta_E = \frac{k_E}{u_*}\frac{\Delta q}{q_*} \tag{4.6.18}$$

$$\delta_G = \frac{k_G}{u_*}\frac{\Delta c}{c_*} \tag{4.6.19}$$

Note that other plausible scalings exist for t* (see Section 4.7) which would lead to different functional relationships in Eqs. (4.6.11) to (4.6.14). However, the Kolmogoroff timescale appears to be the most popular (Liu *et al.*, 1979; Clayson *et al.*, 1995). The constants in Eqs. (4.6.11)–(4.6.14) can be determined by matching the profiles in the laminar layers to the turbulent profile outside. Following extensive measurements on viscous sublayers in laboratory turbulent boundary layers, which indicate an exponential profile within the viscous layer, Liu *et al.* (1979) assumed

$$U - U_s = \Delta U\left[1 - \exp(-z/\delta)\right] \tag{4.6.20}$$

$$T - T_s = \Delta T\left[1 - \exp(-z/\delta_T)\right] \tag{4.6.21}$$

$$q - q_s = \Delta q\left[1 - \exp(-z/\delta_E)\right] \tag{4.6.22}$$

$$c\text{-}c_s = \Delta c\left[1\text{-}\exp\left(\text{-}z/\delta_G\right)\right] \quad (4.6.23)$$

where

$$\Delta U = U_b - U_s \quad (4.6.24)$$

$$\Delta T = T_b - T_s \quad (4.6.25)$$

$$\Delta q = q_b - q_s \quad (4.6.26)$$

$$\Delta c = c_b - c_s \quad (4.6.27)$$

The subscript s denotes the surface values whereas the subscript b denotes the bulk values. The profiles in the turbulent region outside are given by Eqs (4.5.8a)–(4.5.8d) and (4.5.9). Now one requires that the profiles in the molecular sublayer and the turbulent region outside match smoothly at some value of z. This requires continuity of both the value and the slope of the variables at the matching point. This in principle determines the values of

$$C = \frac{\Delta U}{u_*} = \frac{\delta u_*}{\nu} \quad (4.6.28)$$

$$S = \frac{\Delta T}{\theta_*} = \frac{\delta_T u_*}{k_T} = CPr^{1/2} \quad (4.6.29)$$

$$D = \frac{\Delta q}{q_*} = \frac{\delta_E u_*}{k_E} = CSc^{1/2} \quad (4.6.30)$$

$$D_G = \frac{\Delta c}{c_*} = \frac{\delta_G u_*}{k_G} = CLe^{1/2} \quad (4.6.31)$$

The above arguments are valid when the sea surface appears hydrodynamically rough to the turbulent boundary layer flow. Under these conditions, momentum is transported across the air–sea interface by pressure forces acting on roughness elements. This requires that the Reynolds number based on the roughness scale be large, $Re_r > 1.2$. However, the same relationships hold even when the sea surface is hydrodynamically smooth and the momentum is transported by viscous diffusion across the interface, if z_0 is replaced by the viscous scale ν/u^*. Liu *et al.* (1979) explored the matching conditions for smooth flow conditions and discovered that a value of $z_0 = 0.11\ \nu/u^*$ gives a smooth transition between the laminar and turbulent profiles for the velocity at z = 47 ν/u* and C = 16. The profiles for other properties such as temperature are also found to match smoothly at this value of z and therefore it is possible to find the values of the "roughness" scales for scalars as well:

$$\frac{z_{0T,0E,0G}}{z_0} = \frac{47}{Re_r} \exp\left[-\frac{k}{Pr_t} (S,D,D_G) \right] \tag{4.6.32}$$

Note that $Re_r = 0.11$, and S, D, and D_G are known from Eqs. (4.3.29)–(4.3.31) once C is known. Figures 4.6.2 and 4.6.3 show the excellent agreement of the model with some laboratory data (Liu *et al.*, 1979).

For the hydrodynamically rough regime ($Re_r > 1.2$), Liu *et al.* find that

$$C \sim 9.3\, Re_r^{1/4} \tag{4.6.33}$$

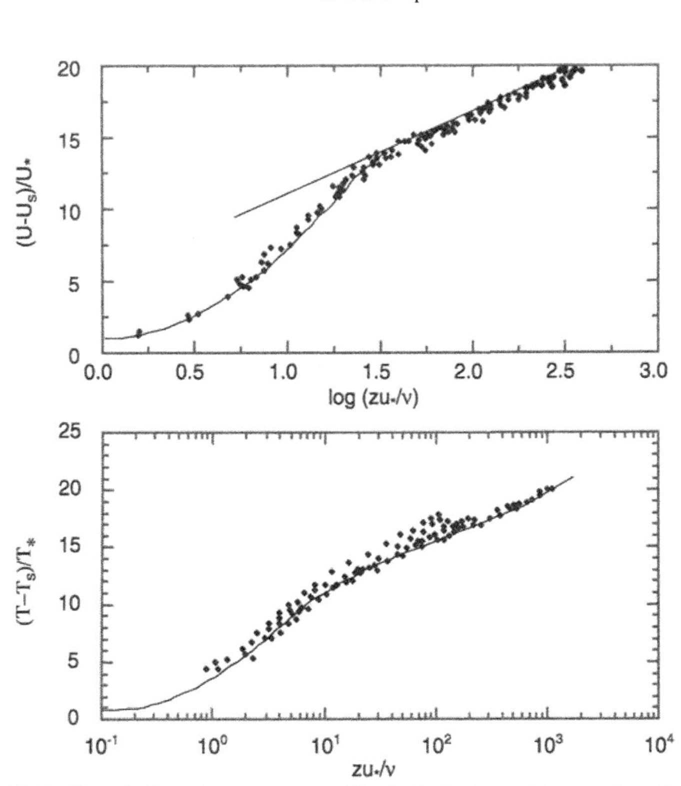

Figure 4.6.2. The velocity and temperature profiles in the laminar sublayer evaluated with the Liu *et al.* model compared with laboratory measurements by Deissler and Eian (1952) (from Liu *et al.*, 1979).

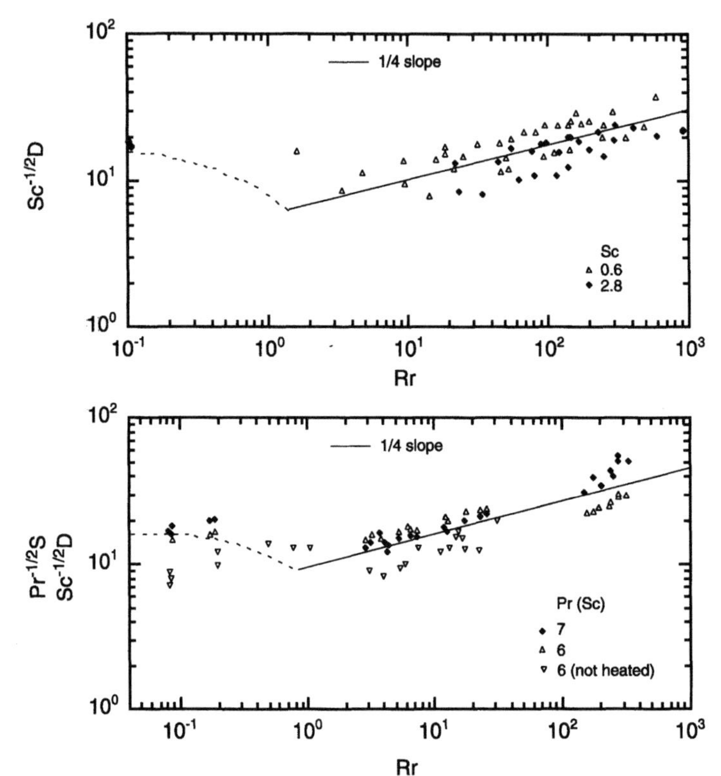

Figure 4.6.3. Normalized temperature and humidity changes across the laminar sublayer as a function of the roughness Reynolds number Re_r (from Liu *et al.*, 1979). Measurements in the top panel are from Chamberlain (1968), and those in the bottom panel from Mangarella *et al.* (1973).

best describes the velocity transition, but were unable to find matching conditions for scalars that would enable us to find expressions for roughness scales for scalars. At this point they abandon the approach detailed above for determining the scalar roughness scales and instead recommend empirical parameterizations of the form

$$\frac{z_{0T}}{z_0} = aRe_r^{b-1}, \quad \frac{z_{0E}}{z_0} = cRe_r^{d-1} \tag{4.6.34}$$

where the values of a, b, c, and d are given in Table 4.3.1. Note that these values are specific to air for which Pr ~ 0.71 and Sc ~ 0.595. They do not present values for gas transfer.

So far we have not commented on the surface values of temperature, humidity, and gas concentration. The air is always assumed to be saturated with water vapor (100% relative humidity) at the sea surface and therefore the value of specific humidity is known if the surface temperature is known. However, in many observations, surface temperature is seldom measured. Instead a bulk temperature at some depth is measured, and because of the existence of molecular sublayers on the water side of the air–sea interface, the surface temperature differs from the bulk temperature (see Section 4.7) and the two need to be related to each other. The difference depends principally on the net heat transfer at the air–sea interface. The ocean loses heat normally by sensible and latent heat fluxes and longwave radiation, but gains heat by longwave radiation from the atmosphere and clouds, and during the day by solar heating. The net heat loss is a complicated function of mixing and heating processes on the water side and the net fluxes on the air side. Nevertheless, it is possible to include the bulk–skin temperature difference in computing fluxes across the air–sea interface. For many practical purposes, it may suffice to reduce the bulk temperatures by 0.3°C to obtain surface or skin temperatures. However, caution is necessary since during the day solar heating can lead to large bulk–skin temperature differences. Then the only recourse is to employ a turbulent mixing model on the water side to determine what the bulk–skin difference should be, and this involves knowing the air–sea fluxes so that the problem is intricately coupled, but nevertheless solvable (Wick, 1995; Wick *et al.*, 1996; Clayson *et al.*, 1996).

The velocity at the air–sea interface, U_s, is seldom zero, even in the absence of strong oceanic currents. Wind-induced drift current studies show that it is a good approximation to take $U_s \sim 0.55\, u^*$ (Wu, 1975). Of course in the presence of strong currents and weak winds, it is essential to add the component of the surface current in the direction of the wind as well. This is particularly important for the tropics, where the easterly trade winds that prevail are weak (a few m s^{-1}) and the surface currents in the equatorial wave guide are strong (1–2 m s^{-1}).

That leaves only the momentum roughness scale z_0 to be determined. The scalar roughness scales can then be parameterized in terms of the momentum roughness scale and the problem reduces to solving Eqs. (4.5.8), (4.5.14)–(4.5.20), and (4.6.28)–(4.6.31) along with the equation for z_0 for u^*, θ^*, and q^*, and hence air–sea fluxes. This is done iteratively starting from neutral values, and if only bulk temperatures are available using a mixed layer model on the water side as well in the iterative process. The roughness scale can be determined from an equation such as Eq. (4.5.33). Since z_0 and the drag coefficient C_D are related [Eq. (4.5.10) or (4.5.12), for example], it is often possible to use an empirical relationship [Eq. (4.5.24) or (4.5.25)] for the drag coefficient to determine z_0. Since these empirical relationships assume zero surface current ($U_s = 0$), it may be necessary to multiply z_0 so determined by a factor $\exp(kU_s/u^*) \sim 1.25$. However, given the large scatter in these empirical relationships, this refinement may be quite inconsequential.

In the limit of zero wind speed, or more appropriately zero ($U_a - U_s$), the above scaling based on u* fails. In this free convection limit, the heat transfer across the air–sea interface is by free convection processes, for which

$$Nu \sim Ra^{1/3} \tag{4.6.35}$$

where

$$Nu = \frac{Q_s d}{k_T \Delta T} \tag{4.6.36}$$

is the Nusselt number and

$$Ra = \frac{g\beta \, \Delta T \, d^3}{\nu k_T} \tag{4.6.37}$$

the Rayleigh number, is a good approximation so that

$$Q_s \sim (g\beta k_T)^{1/3} (\Delta T)^{4/3} Pr^{-1/3} \tag{4.6.38}$$

can be used for the heat transfer. A similar relationship can be used for gas transfer.

Figure 4.6.4 shows calculations of sensible and latent heat fluxes using the Liu *et al.* method compared to the eddy-correlation measurements made during the TOGA/COARE cruise in the tropical western Pacific. The agreement is quite reasonable.

The widely used Liu *et al.* (1979) parameterization for air–sea fluxes invokes a particular form for the surface renewal time and uses a particular form of empirical relationships for scalar roughnesses. Other forms are possible as well (see Section 4.7). There exists an extensive engineering literature on heat and mass transfer across smooth and rough surfaces in a turbulent boundary layer (for example, Yaglom and Kader, 1974; Kader and Yaglom, 1972; Kader, 1981) that can be brought to bear on the air–sea transfer problem. Brutsaert (1975) was one of the very first to point out that the roughness scales for scalars are necessarily different from that for momentum and that ignoring this disparity can lead to serious errors. Yaglom and Kader (1974) pointed out that because the molecular sublayers intervene for scalar transfer even in a hydrodynamically rough flow, the scaling is such that scalar transfer can actually be smaller for rough surfaces compared to smooth ones! We will outline here an alternative approach to deriving the scalar roughnesses.

Consider for simplicity a neutrally stratified flow, a condition applicable to most laboratory investigations of heat and mass transfer in engineering

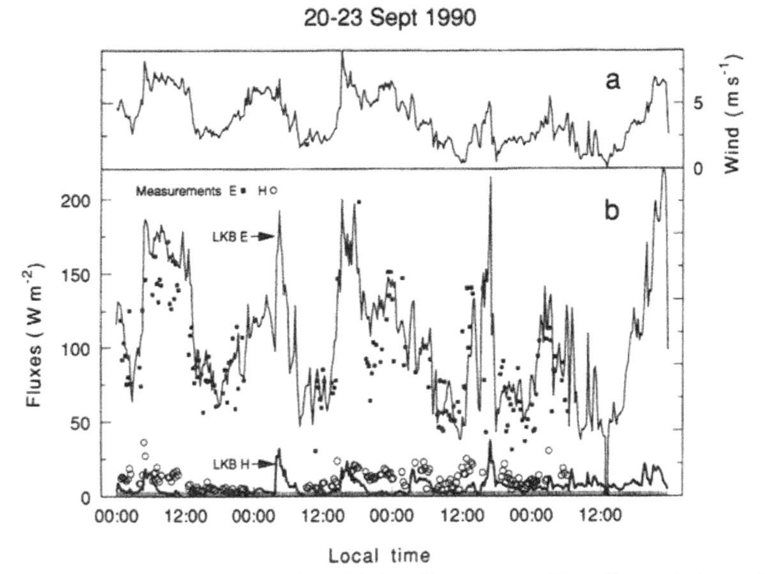

Figure 4.6.4. Comparison of latent and sensible heat fluxes measured by eddy correlation methods with those calculated from Liu *et al.* (1979) bulk formulas (from Bradley *et al.*, 1991).

applications. Then the profiles of velocity and temperature in the turbulent region outside the molecular sublayers can be written as (assuming $U_s = 0$)

$$U/u_* = \frac{1}{\kappa}\ln(z/z_0) \tag{4.6.39}$$

$$\frac{(T-T_b)}{\theta_*} = \frac{Pr_t}{\kappa}\ln\left(\frac{z}{z_0}\right) \tag{4.6.40}$$

where T_b, the bulk temperature, is unknown. Now, the temperature equation can be rewritten as

$$\frac{T-T_s}{\theta_*} = \frac{Pr_t}{\kappa}\ln\left(\frac{z}{z_{0T}}\right) = \frac{Pr_t}{\kappa}\ln\left(\frac{z}{z_0}\right) + \frac{T_b-T_s}{\theta_*} \tag{4.6.41}$$

so that

$$\frac{z_{0T}}{z_0} = \exp\left[-\frac{\kappa}{Pr_t}\cdot\frac{\Delta T}{\theta_*}\right] \tag{4.6.42}$$

This shows that the distinction between the temperature and the momentum roughness scales arises simply because of the existence of a change in temperature across a molecular sublayer. If this change is deduced on a semianalytical basis (Yaglom and Kader, 1974), or simply empirically, it is possible to compute the temperature roughness scale. Brutsaert (1975) tabulates the various observations for both hydrodynamically rough ($Re_r > 2.0$). and smooth ($Re_r = 0.135$) surfaces. For a smooth air–sea interface,

$$\frac{z_{0T}}{z_0} = \exp\left[-\frac{\kappa}{Pr_t}\left(13.6Pr^{2/3} - \frac{13.5}{Pr_t}\right)\right] \quad (4.6.43)$$

(Brutsaert, 1975)

$$= \exp\left[-\frac{\kappa}{Pr_t}\left(12.5Pr^{2/3} - 10.24\right)\right] \quad (4.6.44)$$

(Kader and Yaglom, 1972)

$$= \frac{7.4}{Pr}\exp\left[-\frac{\kappa}{Pr_t}\left(3.85Pr^{1/3} - 1.3\right)^2\right] \quad (4.6.45)$$

(Kader, 1981)

Of these, Kader's (1981) formulation is the most recent and the most preferable. For a rough air–sea interface, Brutsaert (1975) recommends

$$\frac{z_{0T}}{z_0} = \exp\left[-\frac{\kappa}{Pr_t}\left(7.3\,Re_r^{1/4}\,Pr^{1/2} - 5\,Pr_t\right)\right] \quad 0.6 \le Pr \le 6 \quad (4.6.46)$$

while Yaglom and Kader (1974) suggest

$$\frac{z_{0T}}{z_0} = \exp\left[-\frac{\kappa}{Pr_t}\left\{0.55\left(\frac{u_* h_0}{\nu}\right)^{1/2}\left(Pr^{2/3} - 0.2\right) + 9.5 - \frac{Pr_t}{\kappa}\ln\frac{h_0}{z_0}\right\}\right] \quad Pr \le 14$$

$$(4.6.47)$$

where h_0 is proportional to the actual height of the roughness elements, which can be taken as 20 to 30 z_0. For very large values of Pr (or Sc) (irrelevant to the air–sea interface, but important for salt transfer across the ice–ocean interface),

$$\Delta T/\theta^* \sim Re_r^{\,1/3}\,Pr^{\,2/3} \quad (4.6.48)$$

For the transition regime ($0.135 < Re_r < 2.0$), it is necessary to interpolate between the smooth and the rough values.

The Brutsaert formulation is perhaps the simplest to use for the air–sea flux problem and was used by Clayson *et al.* (1996) to compute the sensible and latent heat fluxes in the TOGA/COARE region using surface renewal theory. Their results show some definite improvement over Liu *et al.*'s formulation (Figure 4.6.4), when compared with measurements made using eddy-correlation techniques.

Surface renewal techniques over surfaces with vegetation have a slightly different meaning than when the term is used for air–sea exchange processes. Over a surface with vegetation, the vegetation is usually assumed as acting as a source of the appropriate scalar in question (heat, moisture, etc.). In this case, when a parcel leaves the atmosphere above the canopy and enters the canopy (which is also referred to as sweeping), the parcel is now in contact with a source region in which dissipation effects are important. The parcel is ejected from the canopy into the atmosphere above the canopy when another parcel displaces the original parcel. The original parcel now has altered characteristics based on the sources within the canopy itself. Katul *et al.* (1996) contains a review of the work in this area.

4.6.1 High Wind Speeds

Finally, it is important to remember that the surface renewal theory is only valid for low to moderate wind speeds, where the air–sea interface is more or less intact. At high wind speeds, bubble injection into the water and droplet and spray ejection into the air radically transform the underlying processes of heat and mass transfer across the air–sea interface and hence the fluxes. On the ocean side, wave breaking has even more serious consequences. Recent measurements have shown that in the upper 1–2 m, breaking processes produce highly elevated dissipation rates, and the implications for the molecular layers and air–sea fluxes even without spray ejection and droplet injection are quite substantial. Farmer (1998) provides a succinct discussion of processes occurring on the ocean side of the air–sea interface at high wind speeds.

Air–sea exchange processes are highly nonlinear functions of the wind speed, and one severe oceanic storm can effect exchanges equivalent to long periods of relative calm. Yet it is precisely these high wind conditions that are least understood, principally because of the difficulty of making ship-based *in situ* measurements and the problems associated with contamination of sensors by sea spray and the violent sea surface. Remote sensing is devoid of these difficulties but usually lacks adequate ground truth data essential for calibration. Nevertheless, recent advances such as observations of bubble clouds by underwater acoustic (sonar) sensors are enabling progress to be made (Farmer, 1998). At high wind speeds, wave breaking begins to dominate the air–sea interface (Thorpe, 1995; Melville, 1996). The fraction of the ocean surface covered by breaking waves increases rapidly with wind speeds beyond 10–15 m s^{-1}, and at speeds of ~30 m s^{-1} and beyond, nearly the entire air–sea interface is covered by breaking waves and the associated droplets and spray on the air side and air bubbles on the ocean side. This means that on the air side, the atmosphere senses a layer of air heavily laden with water droplets of various sizes, rather than the air–sea interface directly. On the water side, extensive bubble clouds inhabit the immediate vicinity of the interface. Under these conditions, our conventional notion of an "intact" air–sea interface undergoing sporadic disruptions is no longer valid, and droplets and bubbles mediate air–sea exchanges, especially the water vapor transfer to the atmosphere, and gas transfer to the ocean, respectively.

On the ocean side, bubble clouds due to air entrainment by breaking waves are important to air–sea exchange processes. The clouds penetrate to depths on the order of the amplitude of the breaking waves. While large bubbles rise quickly to the surface, smaller ones are carried deeper by vertical motions due to processes such as Langmuir circulations. Fortunately, air bubbles resonate at natural frequencies that are functions of principally the bubble radius, and it is therefore possible to measure bubble cloud properties such as size distributions either by passive hydrophone arrays or by active sonars. Acoustic energy transmitted by sonars is scattered most efficiently by bubble clouds with natural frequencies close to that of the sonar frequency and this enables various bubble properties to be measured "remotely." Given the importance of dissolution of gases such as CO_2 carried by bubbles in seawater (Farmer *et al.,* 1993), it is likely that increasing emphasis will be placed in the coming years on observing and understanding air–sea exchanges at high wind speeds.

On the air side, droplets created by breaking waves provide an additional pathway for the transfer of water vapor to the atmosphere (Wu, 1974, 1990). Droplet-mediated transfer however involves no latent heat transfer from the ocean. The heat needed to evaporate the droplets comes instead from the layers of the atmosphere adjacent to the air–sea interface. This in turn cools that part of the boundary layer. The droplet evaporation also increases its humidity. The result is that the droplets cause the surface layer of the ABL to be wetter and colder. If all the droplets ejected evaporate fully, the heat flux at the air–sea interface would remain unaltered. However, not all droplets evaporate, and those that fall back into the ocean are colder and this therefore constitutes additional sensible heat loss to the ocean. At high wind speeds, the droplet-mediated sensible heat flux from the ocean to the atmosphere, H_{sd}, can be substantial. As far as the atmosphere is concerned, there exists an additional latent heat flux from the lower layers to the upper layers. As can be expected, the ability of the droplets to transfer water vapor to the atmosphere depends on the difference between the degree of undersaturation. Clearly, if the layers are saturated, the droplets fall back without any evaporation. Thus, the droplet-mediated latent heat flux, H_{ld}, could be parameterized as being proportional to the difference between the saturation humidity q_{sa} and the humidity q_a at the reference height.

Let F_d be the flux of droplets near the surface. Then the potential evaporative heat flux, if all the droplets evaporated, is $H_{lp} = L_e F_d$. However, according to

Fairall *et al.* (1998), who have attempted a systematic look at the problem of parameterizing H_{sd} and H_{ld}, the fraction of the droplets evaporated is usually less than 0.5. They define a droplet-affected layer of thickness z_d (which should scale like U^2/g) and parameterize H_{ld} as proportional to $(q_{sdw} - q_d)$, where d denotes conditions at $z = z_d$. The subscript w denotes the value at the wet bulb temperature. Since $T_w = T - (L_e/c_{pa})(d_v/d_t)\ (q_{sw} - q)$, where d_v and d_T denote diffusivities of water vapor and temperature, and since the Classius–Clapeyron condition relates q_{sw} to q_s by $q_{sw} = q_s + (\varepsilon/R)(L_e/T^2)(T_w - T)$, the quantity $(q_{sw} - q_s) = \beta = [1 + (\varepsilon L_e^2)/(Rc_{pa}T^2)]^{-1}$. The value of β ranges from 0.21 at 303 K to 0.59 at 273 K. The latent heat flux $H_{ld} \sim \beta_d\ (q_{sd} - q_d) = \gamma\beta_a\ (q_{sa} - q_a)$, where γ has a value around unity [0.9 at 15 m s^{-1} to 1.08 at 40 m s^{-1} according to Fairall *et al.* (1998)]. The droplet-mediated fluxes can therefore be written as

$$H_{ld} = \rho_a\, L_e\, \beta_a\, \gamma\, C_{Ed}\, (q_{sa} - q_a)$$

$$H_{sd} = \rho_a\, c_{pa}\, \gamma\, C_{Hd}\, (T_s - T_a) \tag{4.6.49}$$

Indications are that the coefficients C_{Hd} and C_{Ed} are strong functions of the wind speed, being proportional to U^n and U^m, respectively, the area fraction of the whitecaps A_{wc} being an important parameter of dependence. Fairall *et al.* (1998; see also Andreas, 1992) indicate that $C_{Ed} \sim 3.6 \times 10^{-9}\ U_{10}^{5.4}$ and $C_{Hd} \sim 3.2 \times 10^{-8}\ U_{10}^{3.4}$. Ling (1993) suggests instead that $C_{Ed} = 0.88 \times 10^{-3}\ U_{10}^{2}$. While careful observations do indicate that the surface layers cool appreciably (by several K) during tropical cyclones, measurements needed to accurately parameterize droplet-mediated fluxes at high wind speeds are very sparse. According to the Fairall *et al.* (1998) parameterization, the droplet-mediated latent heat flux becomes comparable to the regular latent heat flux at a speed of about 28 m s^{-1}. At 40 m s^{-1}, the droplet-mediated sensible heat flux becomes comparable to the regular sensible heat flux. This is the same speed at which the whitecap area fraction reaches unity. Since $F_d \sim 5.0 \times 10^{-6}\ \rho_w\ A_{wc}$, the absolute upper bound to the latent heat flux is $H_{lp} \sim 12{,}000$ W m^{-2}.

On the other hand, there have been suggestions (for example, Smith *et al.*, 1993; Wu, 1998) that the contribution of sea spray to the moisture flux is likely to be modest even under high wind conditions. It is important to remember that the moisture flux is not only a function of the wind speed, but also of the undersaturation of the surface layer. Relevant observational data are sparse and the influence of droplet evaporation on the heat budget in the marine boundary layer is still an open question. More recently, Andreas (1998) has examined sea spray generation up to wind speeds of 32 m s^{-1}.

4.7 COOL SKIN OF THE OCEAN

The skin of the ocean is found to be invariably a few tenths of degrees cooler than the water a few millimeters below the surface (Ewing and McAlister, 1960; Saunders, 1967; McAlister and McLeish, 1969; Paulson and Parker, 1972; Grassl, 1976; Katsaros, 1980; Paulson and Simpson, 1981; Robinson *et al.*, 1984; Schluessel *et al.*, 1990). Much larger temperature differences can be found between the skin and the waters a few meters below, especially during strong insolation and weak winds (Coppin *et al.*, 1991). This is called the bulk–skin temperature difference and is of importance to considerations of heat storage in the upper layers of the ocean, and heat and mass exchanges across the air–sea interface. In air–sea exchange, it is the SST or the temperature of the skin that is important. It is also the skin temperature that satellite-orbited radiometers such as the AVHRR sense. However, the heat storage capacity of the skin layer, which is only a millimeter thick, is negligible, and it is the bulk temperature in the OML that determines the heat storage in the upper layers of importance to long-term air–sea interactions.

The existence of the cool skin of the ocean (Figure 4.7.1) is simply due to the fact that right at the surface, the net heat balance even during strong solar insolation and weak winds is from the ocean to the atmosphere. This is because normally the sensible and latent heat fluxes at the air–sea interface are net losses from the ocean. Added to this sensible and evaporative cooling of the surface is the net longwave emission at the surface (the difference between the outgoing LW radiation emitted by the ocean surface and the incoming LW radiation emitted from the atmosphere and the clouds), which is also normally a heat loss. The incoming shortwave solar visible, infrared, and ultraviolet radiation is absorbed by the upper layers to differing degrees, with the infrared and near-infrared absorbed within the upper meter of the water column (the ultraviolet part is absorbed in the upper 3–5 m), but the visible part penetrates up to 100 m depending on the turbidity of the water. Consequently, the SW solar absorption in the millimeter skin of the ocean is usually small and the ocean surface normally loses heat even during a calm, cloudless summer day. This heat loss at the surface requires a flux of heat from the interior. However, the only mechanism that can transfer this heat from water to air across an intact air–sea interface is molecular conduction; turbulence is damped close to the surface. In order to accommodate the large heat losses at the surface by conduction right below, the temperature gradient has to be large enough. This causes the skin temperature to drop such that the resulting temperature gradient can accommodate the heat flux from the interior. The result is a cool skin that is normally a millimeter or so thin with a skin temperature 0.1–0.5°C below the bulk temperature a millimeter or so below the surface.

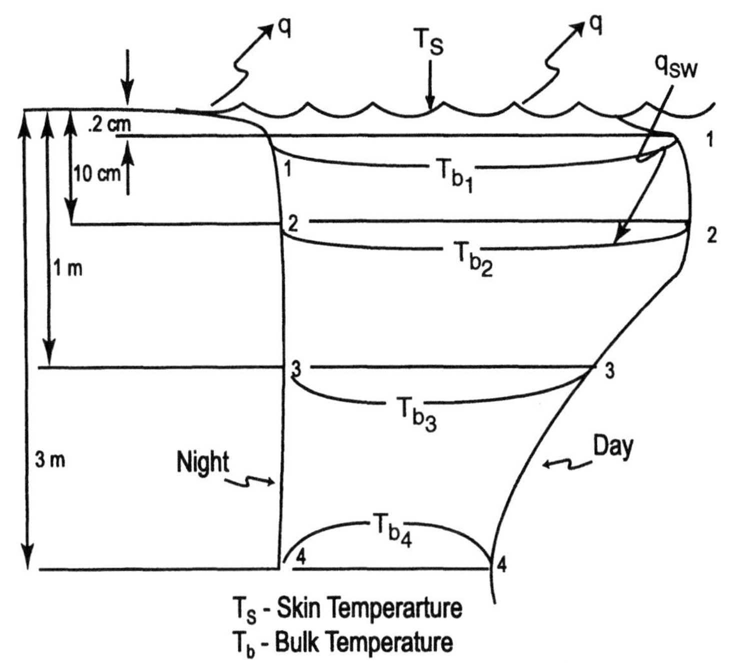

Figure 4.7.1. An idealized rendering of the upper ocean showing the skin and various bulk temperatures during day and night.

A similar effect occurs also for gas transfer at the air–sea interface. In fact in many ways, the air–sea transfer of dissolved gases is similar to the air–sea transfer of heat and water vapor. Unlike the momentum exchange between the ocean and the air, which can proceed by mediation of pressure forces, air–sea transfer of scalar properties can only take place through the molecular sublayers by molecular diffusion. There are no terms analogous to pressure gradient terms in the momentum equation in the scalar conservation equations. This has some interesting consequences. Normally, laws of turbulent exchange do not involve molecular properties of flow, and hence are independent of molecular properties of the fluid in the asymptotic limit of large enough Reynolds number. For example, the roughness length scale for momentum exchange for a fully turbulent boundary layer adjacent to the air–sea interface is independent of molecular viscosity, and depends only on the size of the roughness elements. However, the roughness length scales for scalar transfer, such as heat, water vapor, and dissolved gases, in such a fully turbulent boundary layer are functions of molecular diffusion coefficients, since ultimately the exchange between air and sea right at the interface is by molecular diffusion (Yaglom and Kader, 1974).

Temperature measurements from *in situ* sensors on buoys and moorings, and measurements using buckets or engine intakes on ships, measure the bulk temperature a few tens of centimeters to a few meters below the surface. However, satellite remote sensing of the ocean SST has the best coverage of the global oceans, both temporally and spatially. In order for this to be useful for applications to heat storage considerations, it is essential to relate the skin temperature sensed to the bulk temperatures representative of the OML. Traditionally, algorithms such as MCSST (MultiChannel SST) and CPSST (Cross Product SST) have been used to regress the radiance sensed by the radiometer to buoy-measured temperatures. These algorithms have a bias of about 0.1–0.2°C and an rms error of 0.7–1.2°C (Wick *et al.,* 1992). The principal problem is simply the wide variation in bulk temperatures depending on the intensity of solar insolation and the wind stress at the sea surface. Figure 4.7.1 also shows typical nighttime and daytime temperature profiles in the upper few meters of the ocean. Due to the absence of solar insolation, the ocean normally cools during the night and the resulting convection essentially homogenizes the upper layers so that the bulk temperature is roughly the same no matter what depth it is referred to as long as it is a few millimeters below the surface and within the mixed layer. The skin–bulk temperature difference is always negative. However, during the day, because of solar volumetric heating in the bulk, the bulk temperature depends very much on the depth of the measurements, and the skin–bulk temperature differences can reach values as high as 2–3°C. More seriously, neither the bulk nor the skin temperature is representative of the heat storage in the mixed layer under these conditions. Also, the degree to which the surface layers can be heated depends also on the salinity stratification. If there exists a brackish layer at the surface due to a rainstorm or runoff, then this tends to suppress mixing and lead to larger temperatures close to the surface.

It is therefore important to remember that a nighttime bulk temperature is more representative of the heat storage in the mixed layer, especially after convection has had a chance to mix the heat deposited in near-surface layers through the bulk of the mixed layer, but has not gone on long enough to have caused significant cooling of the mixed layer itself. It is also interesting to note that the nighttime bulk–skin temperature difference is a direct indicator of the net heat loss at the sea surface and hence it may be possible to infer the hard-to-measure turbulent sensible and latent heat exchanges from this difference. This idea has been tested with an *in situ* sensor designed to determine the bulk–skin temperature difference and thus to calculate the surface heat flux (Suomi *et al.,* 1996). McKeown *et al.* (1995) have used a radiometer on an aircraft to measure the bulk–skin temperature difference by making dual-channel measurements. With this information and knowledge of the wind stress from an active microwave device and plausible corrections for net LW loss at the sea surface, it may in principle be possible to infer the combined sensible and latent turbulent heat fluxes at the sea surface remotely, provided the various quantities can be measured with adequate precision.

The bulk-skin temperature differences on diurnal timescales are shown in Figures 4.7.2 and 4.7.3 from observations by Coppin *et al.* (1991) in the western tropical Pacific. Figure 4.7.4 shows a histogram of bulk–skin temperature differences with a mean value of around 0.3°C.

The cool skin of the ocean is continuous neither in space nor temporally. The skin is readily disrupted by breaking waves and occasional sweep-past of eddies from the adjacent turbulent region. However, because the timescales involved are molecular and the thickness rather small, the layer reestablishes itself quite readily, within a matter of several seconds after disruption by eddies or wave breaking. It is this average timescale during which the molecular layer exists that determines its average thickness and hence the bulk–skin difference. Various parameterizations proposed for the skin layer can be looked upon as essentially proposing different residence timescales for the molecular layer. If t^* is the skin timescale, then the thickness of the molecular sublayer is

$$\delta_T \sim \left(k t^*\right)^{1/2} \tag{4.7.1}$$

Then the kinematic heat flux through the sublayer and hence through the air–sea interface is

$$Q_H \sim k \frac{\Delta T}{\delta_T} \sim \frac{k \Delta T}{\left(k t^*\right)^{1/2}} \tag{4.7.2}$$

where ΔT is the temperature drop across the molecular sublayer. This can be rearranged to write

$$\frac{\Delta T}{T_*} = \frac{\Delta T}{(Q_H / u_*)} \sim u_* \left(\frac{t^*}{k}\right)^{1/2} \tag{4.7.3}$$

where T_* is the friction temperature. Note that once we know what t^* is, we know what the bulk–skin temperature difference is, given the heat flux across the air–sea interface. Here Q_H is a heat loss (kinematic) and the skin is a cool skin. If for some reason Q_H is a heat gain, then we have a warm skin. The skin is seldom warm and is always cool during the night because of the absence of SW heating of the skin. This is often a powerful check against instrumental and measurement errors.

The first model for the cool skin was proposed by Saunders (1967). His model is equivalent to a skin timescale $t^* = t^*_1 \sim (\nu / u^{*2})$. Then $\delta_T \sim \nu / u_*$ and

$$\frac{\Delta T}{T_*} = Pr^{1/2} \tag{4.7.4}$$

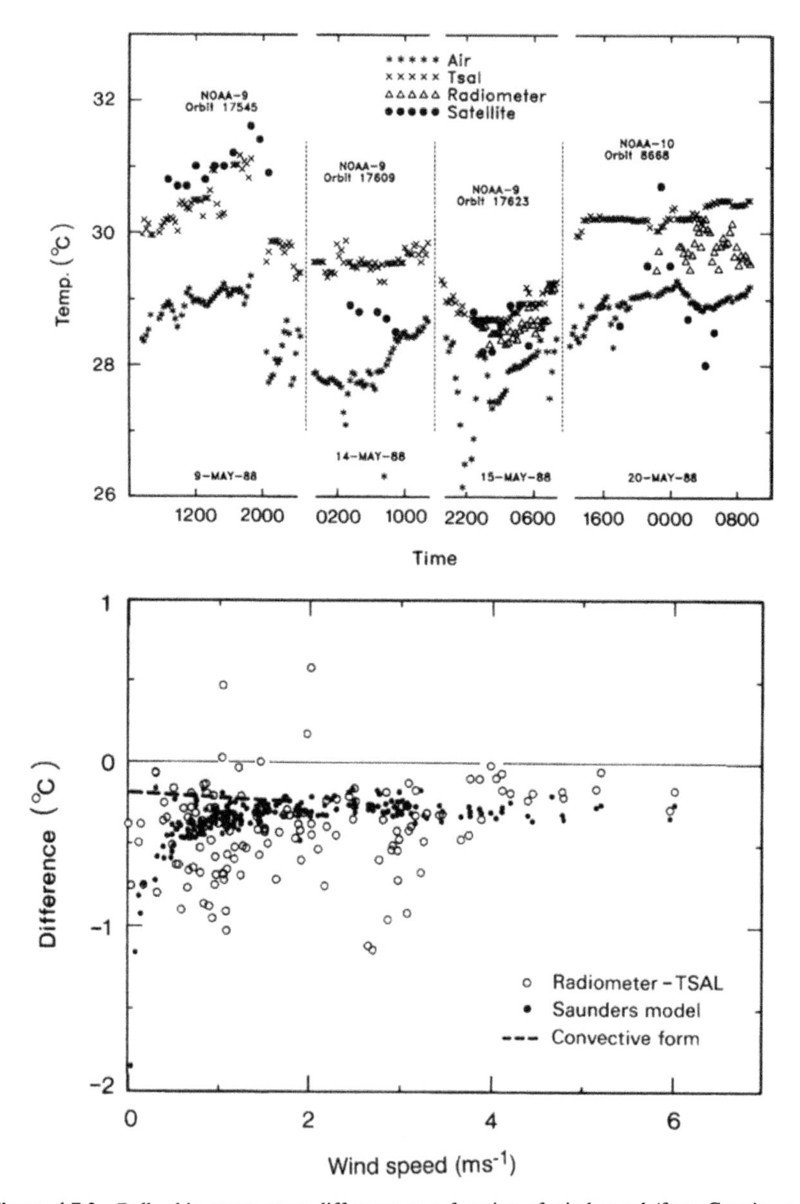

Figure 4.7.2. Bulk–skin temperature difference as a function of wind speed (from Coppin *et al.*, 1991). The upper panel shows the air temperature, infrared radiometer SST, bulk sea temperature, and satellite-derived SST during four periods. The lower panel shows differences between the bulk sea temperature and the sea surface temperature in addition to model predictions.

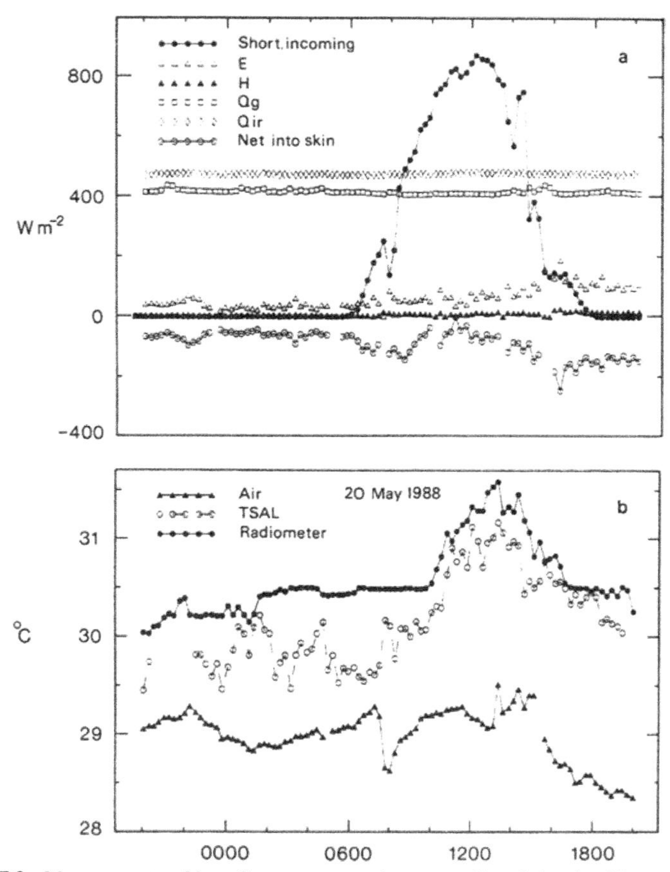

Figure 4.7.3. Measurements of heat flux parameters (upper panel) and the air, skin, and bulk sea temperatures (lower panel) in the western tropical Pacific (from Coppin *et al.*, 1991).

This has been observed to be roughly valid at moderate wind speeds. At low wind speeds, free convection dominates and if one uses the high Rayleigh number limit of the Nusselt number Nu and the Rayleigh number relationship in free convection, namely the 1/3 power law (there are indications that this power law may not be strictly correct, but it is a good approximation for most purposes),

$$\mathrm{Nu} = \frac{Q_H}{\left(\dfrac{k\Delta T}{\delta_T}\right)} \sim \left(\frac{g\beta\Delta T\delta_T^3}{k\nu}\right) = \mathrm{Ra}^{1/3} \tag{4.7.5}$$

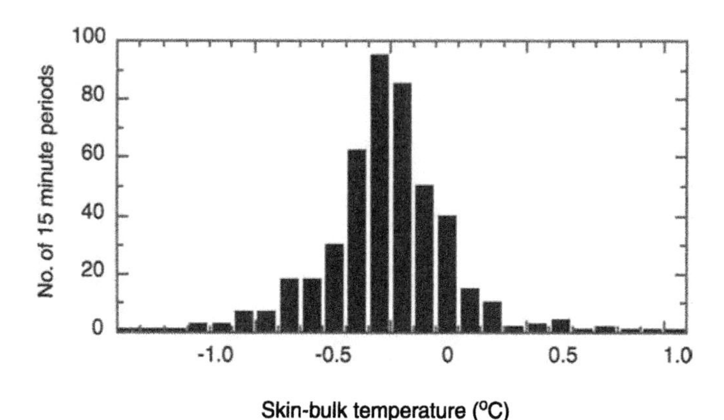

Figure 4.7.4. Histogram of bulk–skin temperature differences (from Coppin *et al.*, 1991).

Rearranging, one gets (Katsaros, 1976)

$$\frac{\Delta T}{T_*} \sim \Pr^{1/2}\left(\frac{u_*^4}{g\beta Q_H \nu}\right)^{1/4} = \Pr^{1/2} Ri_{cs}^{-1/4} \tag{4.7.6}$$

where Pr is the Prandtl number, Ra is the Rayleigh number, and Ri_{cs} is the Richardson number,

$$Ri_{cs} = \frac{g\beta Q_H \nu}{u_*^4} \tag{4.7.7}$$

The residence timescale is $t^* = t^*_2 = (\nu / g\beta Q_H)^{1/2}$.

For large values of wind speeds, intermittent disruption of the sublayer by wave breaking becomes important (Wu, 1985; Csanady, 1990). Since wave breaking requires vertical accelerations at the wave crests to be comparable to gravitational acceleration, the relevant timescale is u^*/g, and if we choose $t^* = t^*_3 \sim u^*/g$, then

$$\frac{\Delta T}{T_*} \sim \left(\frac{u_*^3}{gk}\right)^{1/2} = \Pr^{1/2}\left(\frac{u_*^3}{g\nu}\right)^{1/2} = \Pr^{1/2} Ke^{1/2} \tag{4.7.8}$$

where Ke is the Kuelegan number ($u^{*3}/\nu\, g$). This is the wave breaking limit quoted by Soloviev and Schluessel (1994). If one assumes that the surface wave field is fully developed, then Charnock's relationship relates the roughness scale to u^* and g: $z_0 \sim u^{*2}/g$, and therefore the timescale is $t^* = t^*_3 \sim z_0/u^*$.

There are systematic departures from Saunders' law for the cool skin and the proportionality constant is a function of wind speed (Schluessel *et al.,* 1990), and an alternative timescale can be proposed for t* based on the surface renewal hypothesis of Brutsaert (1975). Here it is assumed that the turbulent eddies in the adjacent turbulent boundary layer periodically sweep the molecular layer away. The smallest eddies possible in any turbulent flow are the Kolmogoroff eddies. If one assumes then that the residence timescale is this Kolmogoroff timescale, $t^* = t^*_4 \sim (\nu/\varepsilon)^{1/2}$. If we take the integral scale to be z_0, the roughness scale, then $\varepsilon = u_*^3/z_0$ and $t^*_4 \sim (\nu z_0/u^{*3})^{1/2}$. If one assumes fully developed seas and Charnock's law, $t^*_5 \sim (\nu/u^* g)^{1/2}$ and

$$\frac{\Delta T}{T_*} \sim \Pr^{1/2}\left(\frac{u_*^3}{g\nu}\right)^{1/4} = \Pr^{1/2}\mathrm{Ke}^{1/4} \qquad (4.7.9)$$

Note that t^*_4 is the geometric mean of t^*_1 and t^*_3; in other words, the timescale for the skin under the surface renewal hypothesis is the geometric mean of those under Saunder's and wave breaking limits: $t^*_4 \sim (t^*_1\, t^*_3)^{1/2}$. Extensive and very careful analyses of bulk–skin observational data by Wick (1995) suggests that at low to moderate wind speeds, it is possible to approximate the bulk–skin difference by a formula that transitions smoothly between the free convection and surface renewal hypothesis limits,

$$t^* = a_4 t_4^* + (a_2 t_2^* - a_4 t_4^*)\,\exp\left(\mathrm{Ri}_{cs}^{c}/\mathrm{Ri}_{cs}\right)$$

$$\frac{\Delta T}{T_*} \sim \Pr^{1/2}\left[\frac{u_*^2 t^*}{\nu}\right]^{1/2} \qquad (4.7.10)$$

where a_2, a_4, and $\mathrm{Ri}_{cs}{}^{c}$ are empirically determined. Soloviev and Schluessel (1994) suggest

$$\frac{\Delta T}{T_*} \sim \Pr^{1/2}\left[1+\mathrm{Ri}_{cs}/\mathrm{Ri}_{cs}^{C}\right]^{-1/4}\left[1+\mathrm{Ke}/\mathrm{Ke}^{C}\right]^{-1/2} \qquad (4.7.11)$$

$$\mathrm{Ri}_{cs}^{C} \sim 1{\cdot}5\;10^{-4}, \quad \mathrm{Ke}^{C} \sim 0{\cdot}18$$

The data at wind speeds above 8 m s^{-1} are unreliable because of the various errors in radiometer and bulk temperature measurements, but the wave breaking limit should be valid at high wind speeds. Figure 4.7.5, from Wick *et al.* (1996), shows the variation of ΔT with wind speed and heat flux. Note the roughly linear increase with Q_H and u* and then the leveling off at high wind speeds.

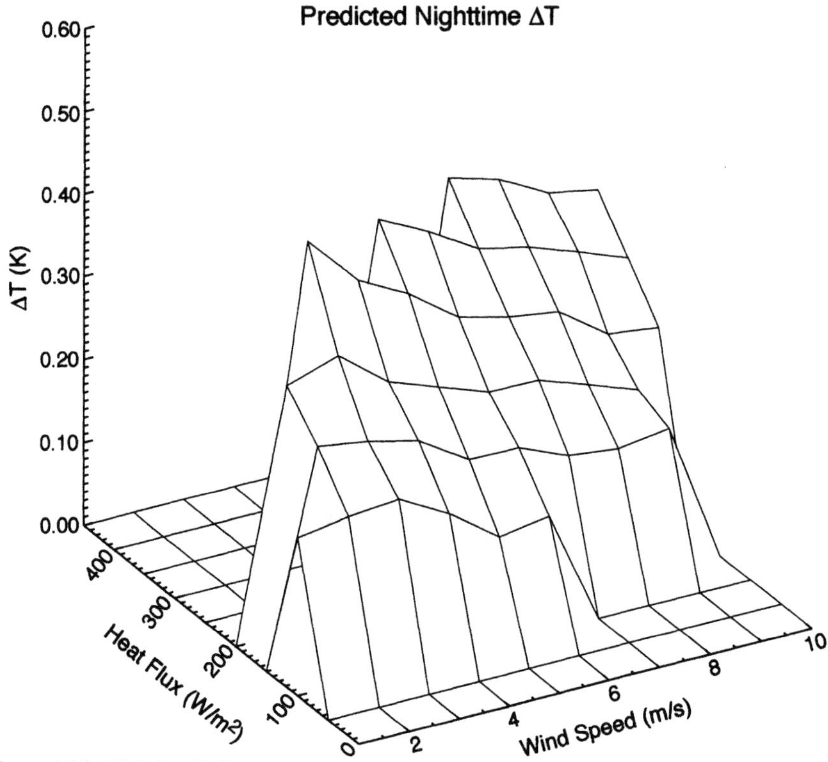

Figure 4.7.5. Nighttime bulk–skin temperature difference as a function of wind speed and heat flux (from Wick, 1995).

Wick (1995) restricted his analysis to nighttime data, when it is assured that the skin is cool and there are no complications due to the solar heating of the upper layers. His results (Figure 4.7.5) indicate a cool skin difference of 0.1–0.4°C, consistent with all carefully made observations of the cool skin. Even this seemingly small difference can amount to a large error in deducing the sensible and latent heat fluxes across the air–sea interface. For example, if the air–sea temperature difference based on bulk temperature is 1.5°C, a 0.3°C cooler skin, if not taken into account in computing heat fluxes, amounts to an overestimate in the computed sensible heat flux of 25% and one in latent heat flux of 10%. In the region of the western tropical Pacific, where the latent heat fluxes are of O (100 W m^{-2}), the error is 10 W m^{-2} (sensible heat flux is smaller and less important), a serious error if one remembers the net heat flux into the ocean in this region is of the same order. It is often said that for accurate modeling of the atmosphere over the tropics, an accuracy in SST of about 0.1°C is desirable, which can be less than the bulk–skin temperature difference.

Fairall *et al.* (1996) follow Saunders' (1967) approach, but incorporate both free convective and shear-forced regimes to arrive at an expression for the constant in Saunders' equation that depends on the wind speed up to 3 m s^{-1}, but remains constant at 6.0 thereafter:

$$\lambda = 6\left\{1+\left[16\frac{g\alpha\rho c_p v^3}{u_*^4 k^2}\left(Q+\frac{\beta\rho c_p}{\alpha L}Q_L\right)\right]^{3/4}\right\}^{-1/3} \tag{4.7.12}$$

This parameterization includes the effect of evaporation by including salinity S and the salinity expansion coefficient β.

Donlon and Robinson (1997) report on an extensive set of bulk–skin temperature difference (ΔT) measurements along a series of transects in the Atlantic from 20° S to 52° N in 1992. They found that ΔT decreases as the wind speed increases, asymptoting to about 0.1°C at wind speeds larger than 10 m s^{-1}. They also compare measured ΔT values to those inferred from various parameterizations, including those of Saunders (1967), Soloviev and Schluessel (1994), Fairall *et al.* (1996), and Wick (1995; see also Wick *et al.,* 1996), and conclude that while none of these adequately reproduce the measurements, Wick *et al.* (1996) appears to be the best of the lot. In view of the dependence of ΔT on wind speed, they suggest that it is inappropriate to assume a constant value of 0.3°C in calibrating satellite IR measurements.

Wick and Jessup (1998), after careful examination of the modulation of skin–bulk temperature difference by swell, conclude that the major effect is due to preferential wave breaking on the forward face of the swell. This highlights the importance of taking into account the extent of wave breaking in modeling the skin–bulk temperature difference. Current models (Soloviev and Schlussel, 1994; Wick *et al.,* 1996) assume sporadic disruption of the skin layer, but do not explicitly account for wave breaking in deriving the mean time for reestablishment of the skin layer.

Solar heating during the day complicates matters considerably. For calculating heat transfer across the air–sea interface, it is the skin temperature that needs to be used. Most often, it is some sort of bulk temperature that is available at some depth and clearly the resulting error can be substantial (see Figure 4.7.1 above). For low wind conditions in the tropics, Fairall *et al.* (1996) saw afternoon peak differences between the skin and the 5-cm temperature approaching 4°C. Deducing the skin temperature for use in heat flux calculations from the bulk temperature measured at some depth clearly requires the mixing and the solar extinction in the upper layers to be modeled correctly and is therefore a nontrivial endeavor. Wick (1995) has used a mixed layer model based on second-moment closure (Kantha and Clayson, 1994) to compute the skin–bulk temperature differences for various sets of data, including the Franklin data set taken during the TOGA/COARE program. Figure 4.7.6 shows a comparison between modeled and observed temperatures; the agreement is quite good given the various uncertainties such as the water clarity, horizontal inhomogeneity, and imprecisions in measurements of various parameters related to the air–sea heat exchange and the skin temperature itself. The heat flux itself is computed using the surface renewal hypothesis and not bulk formulas. Solar extinction in both the infrared and the visible portions of the spectrum is given careful treatment in Wick's model (see Wick, 1995, for details).

There are many similarities between gas transfer and heat transfer across an air–sea interface (Brutsaert and Jerka, 1984; Soloviev and Schluessel, 1994). For one, both heat and gases have to be ultimately transferred across a molecular sublayer, and hence analyses similar to those for heat transfer apply to gas transfer. There has been a considerable amount of work done on this subject in recent years (see, for example, Kitaigorodskii, 1983, 1984; Brutsaert and Jerka, 1984), and the topic has assumed much greater importance in view of the importance of greenhouse gases such as CO_2 and photochemically produced gases such as carbonyl sulfide, which have high residence times in the stratosphere and therefore an influence on the ozone budget (Kantha and Clayson, 1994). Air–sea exchange of these gases has important climatic implications.

The gas transfer rate across the air–sea interface can be written in terms of a gas transfer coefficient C_g, which has units of velocity and is often called the piston velocity (Peng *et al.*, 1979; Broeker *et al.*, 1986):

$$c_g \sim \left(\frac{k_g}{t^*} \right)^{1/2} \tag{4.7.12}$$

C_g can be written as $C_g \sim (k_g / t^*)^{1/2}$, where k_g is the gas diffusion coefficient. Hence

$$\frac{c_g}{u_*} \sim \frac{1}{u_*} \left(\frac{k_g}{t^*} \right)^{1/2} \tag{4.7.13}$$

Using the various timescales deduced for the cool skin, we get in the convection limit of low wind speeds

$$t^* = t_2^* = \left(\frac{\nu}{g\beta_g Q_g} \right)^{1/2}$$
$$\frac{c_g}{u_*} \sim Sc^{-1/2} \left(\frac{u_*^4}{g\beta_g Q_g \nu} \right)^{-1/4} \sim Sc^{-1/2} Ri^{-1/4} \tag{4.7.14}$$

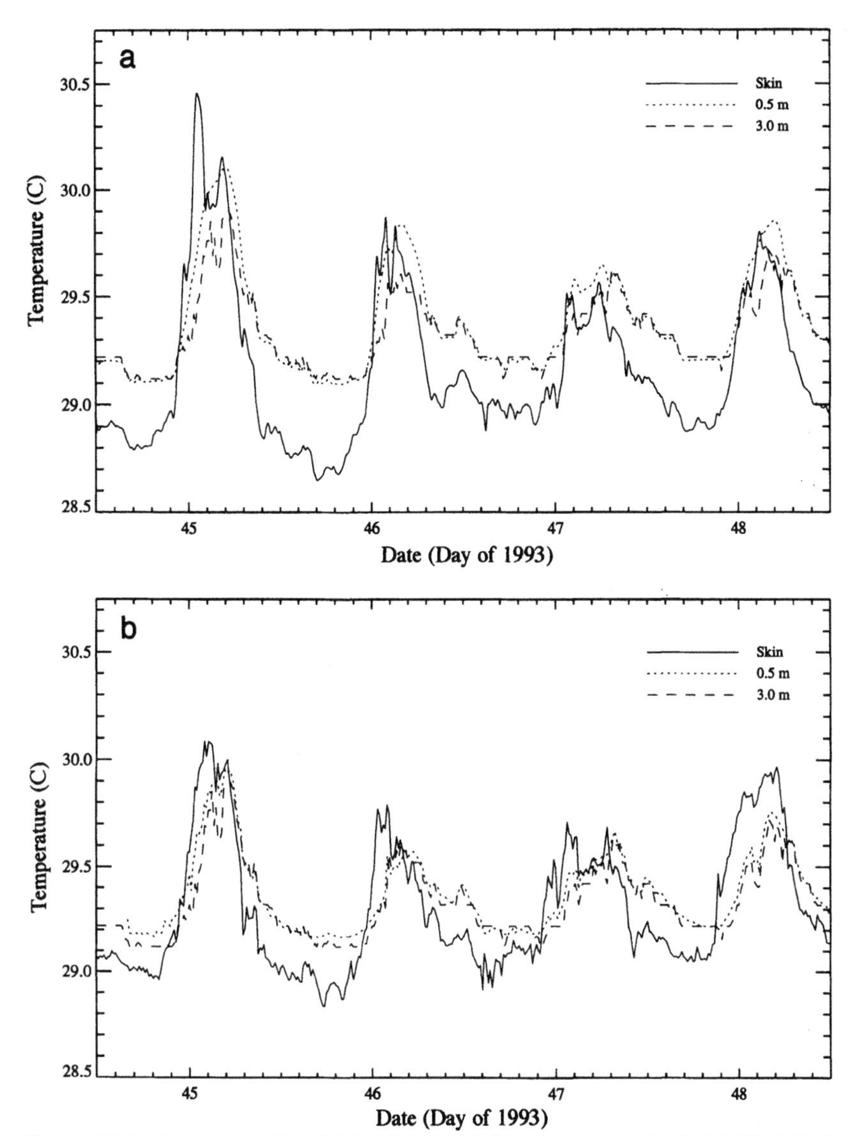

Figure 4.7.6. Comparison of modeled and observed skin temperatures for TOGA/COARE (from Wick, 1995).

where Q_g is the kinematic flux of gas across the interface (with units of concentration times velocity), and β_g is the coefficient of expansion due to gas dissolution. $Sc = (\nu / k_g)$ is the Schmidt number, the ratio of kinematic viscosity to gas diffusivity in water. Note that Ri is based on β_g. Saunders' formulation for gas transfer becomes

$$t^* = t_1^* = \frac{\nu}{u_*^2}$$
$$\frac{c_g}{u_*} \sim Sc^{-1/2} \tag{4.7.15}$$

The wave breaking limit yields

$$t^* = t_3^* = \frac{u_*}{g}$$
$$\frac{c_g}{u_*} \sim Sc^{-1/2}\left(\frac{u_*^3}{g\nu}\right)^{-1/2} = Sc^{-1/2}Ke^{-1/2} \tag{4.7.16}$$

and the surface renewal hypothesis gives

$$t^* = t_4^* = \left(\frac{\nu}{u_* g}\right)^{1/2}$$
$$\frac{c_g}{u_*} \sim Sc^{-1/2}\left(\frac{u_*^3}{g\nu}\right)^{-1/4} = Sc^{-1/2}Ke^{-1/4} \tag{4.7.17}$$

Figure 4.7.7 (from Soloviev and Schluessel, 1994) shows measured gas transfer rates plotted against friction velocity.

While we have referred to the skin as a cool skin, which it invariably is, it does not have to be. One can imagine situations where the ocean is a net gainer of heat. Take, for example, the situation when a warm continental air mass flows over a colder ocean and imagine a cloudy ABL. Under these conditions, the net heat flux is into the ocean and the skin will be warm. But such conditions are rather exceptional.

The above models have concentrated on the skin under nonprecipitating conditions. However, rain can have a large effect on the skin and the skin–bulk temperature difference. Schluessel *et al.* (1997) have investigated the effects, which include an alteration in the surface renewal theory due to increased mixing by raindrops, and additional cooling of the skin due to the cooler rain. The combined effects of these results on the bulk–skin temperature difference are shown in Figure 4.7.8. In addition, a haline molecular diffusion layer can be created which causes salinities in the skin to be less than the bulk by up to 4 psu (Schluessel *et al.,* 1997).

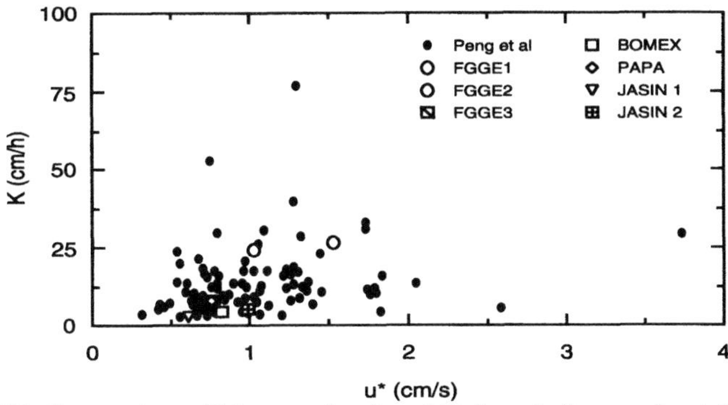

Figure 4.7.7. Gas transfer coefficient as a function of friction velocity according to Broecker and Peng (1974), Peng *et al.* (1979), and Broecker *et al.* (1986) (from Soloviev and Schluessel, 1994).

The phenomenon of cool skin may provide an indirect and remote means of inferring the near-surface turbulence properties. Since the recovery time for the cool skin disrupted by breaking waves depends on the turbulence underneath, infrared probing of the sea surface by highly sensitive multichannel infrared sensors (Jessup *et al.,* 1997a) might be helpful in not only inferring the cool skin characteristics at high wind speeds, but also the small scale turbulence near the air–sea interface.

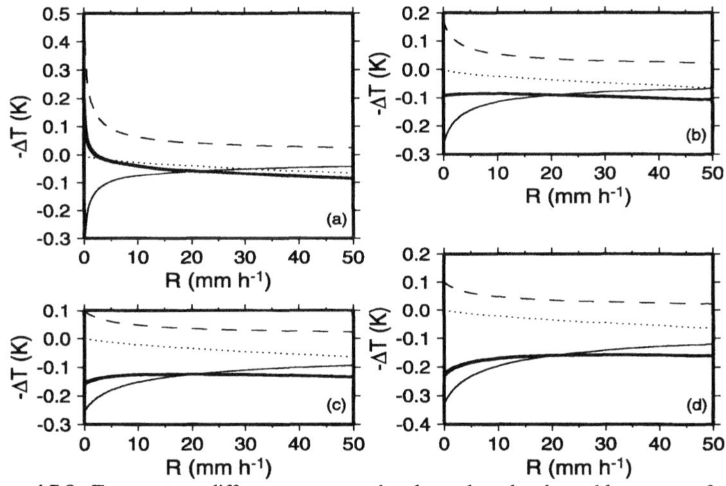

Figure 4.7.8. Temperature differences across the thermal molecular sublayer as a function of rainrate for (a) u=1 m s-1, (b) u=5 m s-1, (c) u=10 m s-1, and (d) u=15 m s-1 (from Schluessel *et al.* 1997, with kind permission from Kluwer Academic Publishers). The curves correspond to differences due to cooling by turbulent and longwave fluxes (thin solid), warming by solar ratiation (dashed), rain-induced cooling (dotted) and the combined effect (thick solid).

4.8 SATELLITE-MEASURED FLUXES

It is a difficult and costly endeavor to measure surface fluxes directly by appropriate *in situ* instrumentation over the global oceans and land system. Air–sea fluxes are only measured directly occasionally as during the TOGA/COARE program (Webster and Lukas, 1992; Fairall *et al.*, 1996). Even then, some piece of the puzzle is often missing and the missing information has to be deduced indirectly and often less accurately. Nevertheless careful point measurements such as by Young *et al.* (1992) and Fairall *et al.* (1996) have greatly increased the air–sea flux data base on which to base our ideas of air–sea exchange and derive more accurate parameterizations of such exchanges. It is less difficult and costly to measure the fluxes directly over land, and most of our information about the ABL and its structure comes from land programs such as BLX96 (Stull *et al.*, 1997) and Kansas (Izumi, 1971).

A more likely scenario that prevails most often is that only certain parameters central to the global exchange of heat, moisture, momentum, and mass are available from either observations or as products of NWP (numerical weather prediction) models from various regional centers. *In situ* observational data are quite sparse temporally and spatially, and cannot be relied upon on a global basis. The NWP products are inevitably affected by the skill of the model used, and even with the best model, the model-produced flux parameters are of uneven validity on all time and space scales of interest, especially parameters such as cloud cover and precipitation. These are the primary reasons for the attractiveness of satellite-derived fluxes. They have global coverage and often good temporal sampling as well. If parameters relevant to surface exchange can be obtained from satellite-borne sensors and the fluxes deduced, the vastly improved temporal and spatial coverage would be useful for routine monitoring of the exchanges as well as input into models of the land/atmosphere/oceans, that could in turn increase our understanding and ability to predict the behavior of the coupled system.

The principal problem in using remote sensing is that the remotely sensed electromagnetic radiation as measured by the sensor orbiting several hundred to thousands of kilometers above the Earth's surface must somehow be related to the geophysical parameters of interest. For remote sensing of properties at or near the surface, the intervening atmosphere is a "nuisance" that often corrupts the signal emitted by the surface as seen by the sensor aloft. For an altimeter, which essentially measures the time for a microwave signal emitted by it to be reflected by the sea surface back to the device, propagation delays induced by the intervening troposphere and ionosphere are effects that need to be corrected before the distance between the sensor and the sea surface can be deduced to the precision required, usually an rms value of a few centimeters. The fact that modern altimeters such as TOPEX/Poseidon can do so is a real testimony to the advanced science and technology that can be brought to bear on the problem. For a sensor such as SeaWIFS that measures the ocean color and hence indirectly the oceanic primary productivity, the majority of the signal at the sensor is that from the intervening atmosphere and considerable ingenuity is needed to detect the small shift in the spectral intensity of upwelling radiation from beneath the ocean surface from the predominantly blue part of the spectrum to blue-green due to the chlorophyll concentration in the upper layers.

Deducing air–sea fluxes remotely requires measurements of the parameters governing air–sea heat balance (see Section 4.1). This includes the shortwave and longwave radiation impinging on the ocean surface and the longwave emission from the sea surface, in other words, components of the radiation balance at the air–sea interface. It is in principle possible to measure these quantities by radiometers (Katsaros, 1990). On the other hand, measurement of turbulent air–sea fluxes cannot be done directly and these quantities need to be inferred (Liu, 1990). In order to determine the surface turbulent fluxes using satellite-based data, it is necessary to use the bulk equations (4.5.3) and a formulation for determining the turbulent fluxes from these bulk parameters. This then requires the knowledge of the surface and near-surface air temperature (or the difference between the two), the surface and near-surface specific humidity (or the difference between the two), and the near-surface wind speed. If the fluxes are being calculated over land, further information is needed, such as the type and coverage of vegetation. Because of the advantages of the satellite-derived fluxes, this is an active area of research, and thus what follows is only an overview of various methods being used to determine the fluxes from satellites.

For remote sensing of air–sea fluxes, the principal oceanic quantity of interest is the sea surface temperature. Infrared and microwave emissions from the ocean surface can be used to deduce the SST, but careful considerations have to be given to the absorption characteristics due to gases and aerosols in the intervening atmosphere. Principally, strong absorption bands have to be avoided and only those spectral ranges where the atmosphere is sufficiently transparent to upgoing longwave infrared and microwave radiation have to be measured. Infrared sensors such as AVHRR employ different spectral bands so that radiances from these bands can be combined optimally to obtain the SST more accurately. The technique is useful only in cloud-free regions, which severely restricts its usefulness in many regions of the world where cloud cover is a rule rather than an exception. Even then, accounting for water vapor in the intervening atmospheric column is a difficult task. Nevertheless, it is now possible to sense the SST in cloud-free regions by AVHRR with a bias of a few tenths of degrees and an rms of less than 1°C (Wick *et al.,* 1992). The fact that most of the infrared and near-infrared radiation sensed comes from the skin of the ocean is quite useful to air–sea exchange since it is the skin temperature that governs the air–sea transfer. Several methods are being evaluated in order to fill in the gaps left by cloud coverage [see Figure 4.8.1 and Clayson *et al.* (1996)].

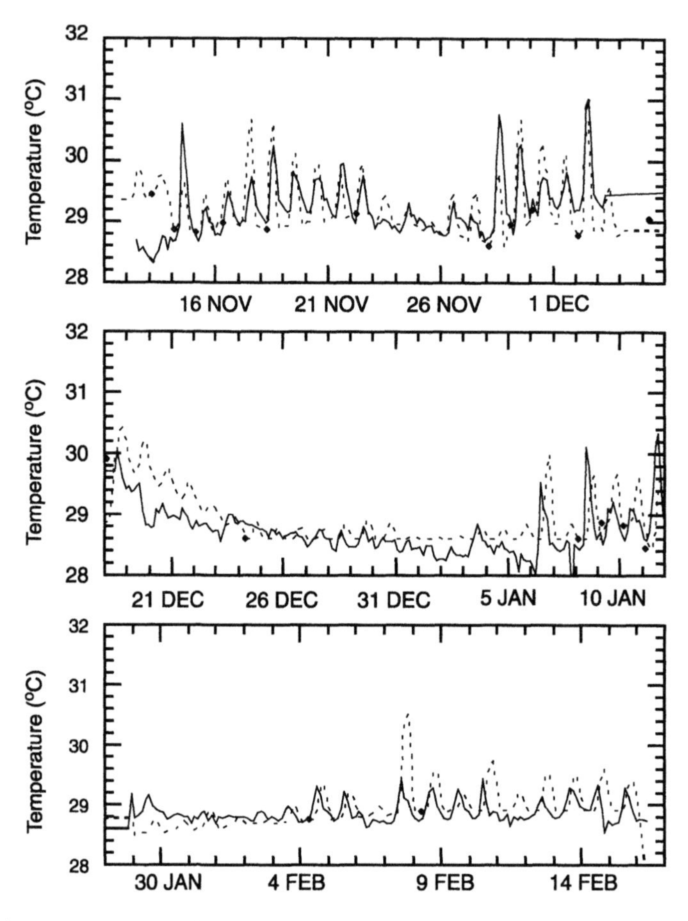

Figure 4.8.1. Time series of sea surface temperature for three different time periods during TOGA/COARE. Ship measurements at a 5-cm depth are shown by the solid line, values of skin SST determined from AVHRR data are shown by the diamonds, and the satellite-derived values of skin SST using the Clayson *et al.* (1996) parameterization for diurnal amplitude is shown by the dotted line (from Clayson *et al.*, 1996).

Passive microwave sensors primarily detect the changes in the radiance brought on by changes in roughness of the sea surface due to the action of the prevailing winds, and hence the wind stress (and wind speed) acting at the sea surface (Swift, 1990, for example). They do contain information on the temperature of the upper few centimeters (depending on the wavelength) and hence possibly information on bulk ocean temperatures. But the frequency bands of currently orbited microwave sensors such as SSMI are ill suited to deriving the bulk temperatures and are optimized primarily for detecting the increased microwave brightness from a rough sea surface. If it ever became possible to remotely measure the bulk and skin temperatures accurately, the difference contains information vital to deducing the net air–sea heat transfer rate directly without having to measure the ABL properties remotely and appeal to bulk formulas.

Determination of the temperature over a land surface contains some similarities to methods used over the ocean. As with sea surface temperature measurements, ground temperature measurements can be deduced from infrared and microwave emissions, with appropriate consideration to absorption characteristics in the atmosphere. Thus the AVHRR is a useful tool for the ground temperature as well. However, the existence of a vegetation canopy over many land surfaces greatly complicates the estimation of the ground temperature, particularly in regions in which there is a mixture of vegetation and bare soil. A satellite-based estimation of the vegetation density, provided by the NOAA AVHRR instruments, is used for determination of the vegetation type, density, etc. These data, called the Normalized Difference Vegetation Index (NDVI), are indicative of the level of photosynthetic activity in the vegetation. Various schemes for determination of the appropriate combined ground/canopy temperature to use for the sensible heat flux from the radiometer data exist. Dual-source models, or models using the radiometric temperature and other surface or satellite-based data, exist for relating the radiometric surface temperature to the true surface temperature, the vegetation temperature, and the temperature within the canopy itself. Zhan *et al.* (1996) contains a description of various methods used for estimating the appropriate temperature for use in sensible heat flux calculations using satellite-based techniques.

Another parameter needed for determining the surface sensible heat flux is the near-surface air temperature, which is difficult to measure directly. Current atmospheric profilers such as TOVS are not capable of providing accurate measurements of temperature (and humidity) in the lower parts of the ABL, and the surface values are at present beyond their capability because of the crude vertical resolution. One approach to determining T_a over the ocean from satellites is to use the satellite-derived values of q_a with an assumed relative humidity in order to determine T_a (e.g., Liu, 1986). The use of relative humidity requires an accurate value of q_a and a good assumption of the relative humidity. Jourdan and Gautier (1995) determined T_a from a relationship between T_a and precipitable water. Figure 4.8.2, from Liu (1990), demonstrates that there often exists an excellent correlation at high and midlatitudes between the saturation value and the humidity, suggesting that it is a good approximation to regard the relative humidity to be a constant. This is not always true, however, especially in the tropics. Fortunately, in the tropics, latent heat flux dominates sensible flux in air–sea exchange by a factor of more than four most often, and errors in retrieval

Figure 4.8.2. Time series showing excellent correlation between the surface level humidity and the

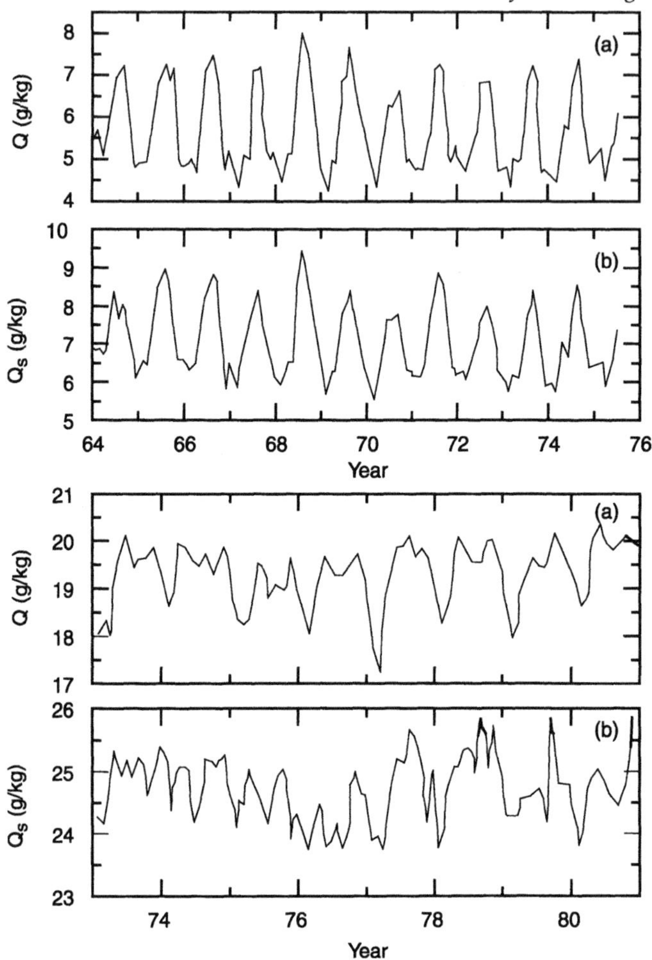

saturation value (from Liu 1990). Top two panels are from OWS I (59° N, 19° W); bottom panels are from Truk atol (7° N, 150° E). (a) shows the surface-level mixing ratio and (b) the saturation mixing ratio at surface-level temperature.

of sensible heat fluxes are less serious, at least in an overall sense. Clayson *et al.* (1996; see also Clayson, 1995) have used a different method to estimate (T_a–T_s) in the tropical western Pacific directly from observations on clouds, based on the hypothesis that the type of clouds present is a reflection of the prevailing static stability in the atmospheric column. This hypothesis, though quite crude, might be better than assuming a constant relative humidity to obtain T_a from q_a in the tropics.

Determination of the near-surface air temperature over land surfaces makes use of essentially the same information as over the ocean. TOVS data can be used for this purpose; if used in conjunction with other data such as model data or *in situ* data it is possible to retrieve fairly accurate near-surface air temperatures, especially at monthly timescales (e.g., Choudhury, 1997).

Microwave radiometers on satellites can measure the precipitable water (W) in the atmospheric column to approximately the same accuracy as *in situ* sensors such as radiosondes, 0.2 g kg^{-1} (Liu, 1990). W is the total (column-integrated) water vapor content. However, this is not the quantity of interest to deducing the latent heat flux at the air–sea interface. Instead one needs to know either the surface layer humidity or the difference between the ocean surface and surface layer humidities. Fortunately, there appears to be a very good correlation empirically between W and the surface layer mixing ratio. Figure 4.8.3, from Liu (1990), demonstrates this quite well (with a rms of 0.73 g kg^{-1} and a correlation coefficient of 0.99!). Liu and Niiler (1984) demonstrated the feasibility of deducing surface level mixing ratios from microwave radiometer measurements by comparisons with measurements at 27 stations in the tropical oceans over nearly 4 years, with excellent results (see Liu, 1990). Other algorithms use combinations of W and SST in order to determine q_a (Miller and Katsaros, 1992; Clayson *et al.,* 1996).

The precipitable water vapor can also be used for the determination of the surface air specific humidity over land surfaces . For these cases, W can be obtained from TOVS data. Semiempirical equations deriving the mean vapor pressure from W have been used (e.g., Choudhury, 1997) in order to determine the near-surface air specific humidity.

The sea surface mixing ratio can be determined from the SST under the assumption that the air immediately adjacent to the air–sea interface is saturated and by appealing to the Clausius–Clapeyron equation (see Appendix B). However, accurate retrievals of SST are essential to accurate deduction of surface humidity. An error of 0.5°C in SST corresponds to an error of 0.6 g kg^{-1} in the surface mixing ratio.

Over land, a determination of the appropriate surface humidity to use is much more complicated than that over the ocean. The complications arise in part because of the variable humidity characteristics of vegetation canopies and of bare surfaces, and because the humidity characteristics of each of these surfaces vary depending upon ambient conditions. Over bare soil, it is necessary to know both the ground temperature and the soil moisture content. The moisture availability of a vegetated surface depends upon the bulk stomatal resistance, which in turn depends upon a number of features. Estimates of these parameters using information from both ground temperature and NDVI data have been examined (Vukovich *et al.,* 1997).

Wind speed is a crucial feature to determine, as it affects both the latent heat flux and the sensible heat flux. Current methods for determining the surface wind

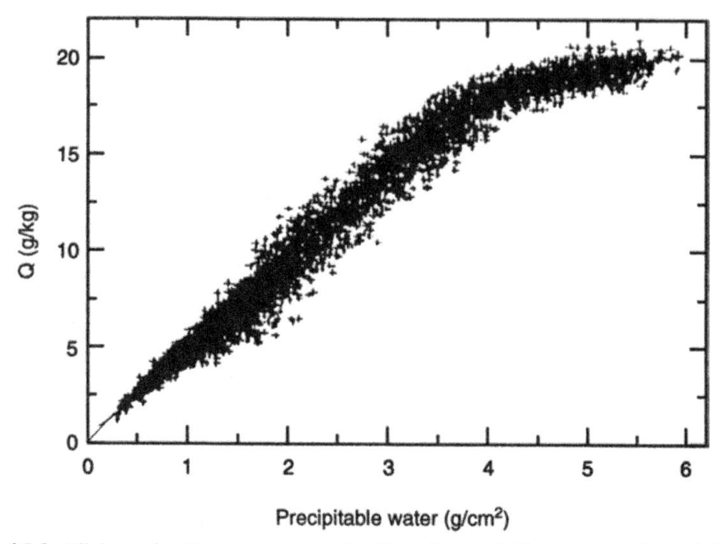

Figure 4.8.3. Mixing ratio of water vapor as a function of precipitable water over the global oceans over a period of 17 years. Each point is a monthly average. The solid line represents the global relation by Liu (1986) (from Liu, 1990).

speed over the ocean come from both passive and active microwave scanners (e.g., Wentz, 1992; Freilich and Dunbar, 1993), and infer the wind speed at a given height above the ocean from the roughness characteristics of the sea surface (Figure 4.8.4 shows a comparison and approximate satellite coverage). No comparable method has been found to use satellite-based data for estimating the wind speed over land.

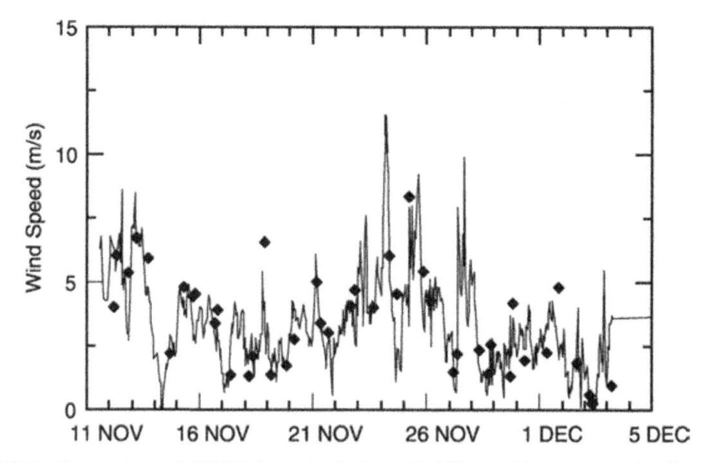

Figure 4.8.4. Comparison of SSM/I derived wind speeds (diamonds) to measured values (lines) (from Clayson and Curry 1996).

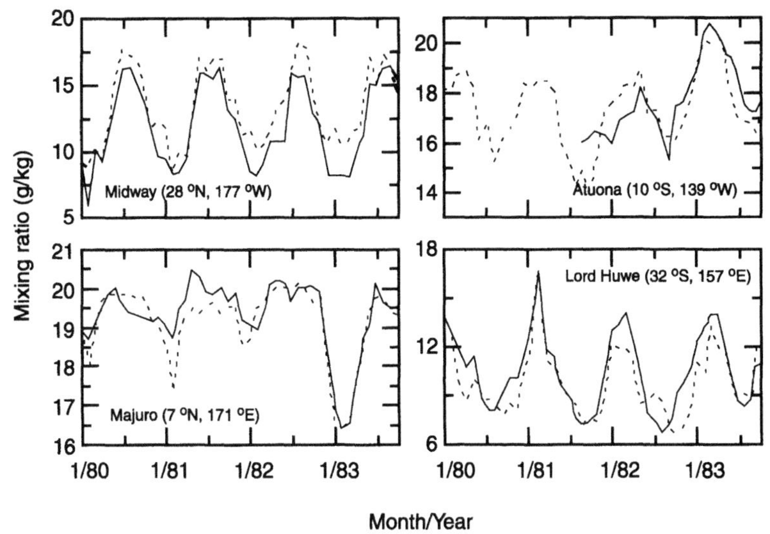

Figure 4.8.5. Comparison of satellite-derived surface level mixing ratio derived from Nimbus/SMMR (dashed line) with *in situ* measurements (solid line) in the Pacific (from Liu 1990).

With these parameters measured from satellite-borne sensors, it is possible to deduce the latent and sensible heat flux at the air–sea interface from bulk formulas (see Section 4.2 and 4.3). An example of such a calculation from an earlier sensor (SMMR on Seasat) compared to ship measurements is shown in Figure 4.8.5, from Liu (1990). Satellite retrievals since then have enabled temporal and spatial variability of latent heat fluxes over important regions such as the Indian Ocean to be monitored and analyzed. Figures 4.8.6 and 4.8.7 show a global monthly average latent heat flux, and Figure 4.8.8, from Clayson and Curry (1996), shows comparisons of three-hourly values of latent and sensible heat flux in the western Pacific.

Since a microwave radiometer measures radiances at various frequencies (10 on SMMR, 4 on SSM/I), efforts have been made to relate the radiances directly to geophysically important quantities such as the latent heat flux, without the intervening step of deducing the mixing ratios and appealing to bulk formulas. Liu *et al.* (1990), shows the feasibility of such direct retrievals at least on certain spatial and temporal scales.

Due to the fewer parameters available from satellite-based data over the land, generally either model or *in situ* observations are blended with the satellite data to derive the heat fluxes. Alternative methods to the bulk equations, such as the use of the Thornthwaite (1948) concept of potential evaporation, or the Penman–Monteith equations for determination of the evaporation from a vegetated surface, have been explored. A summary of these and other similar schemes can be found in Jensen *et al.* (1990).

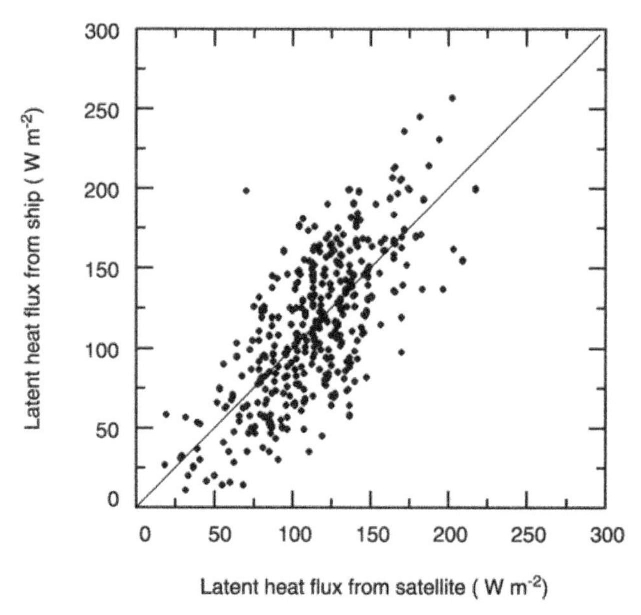

Figure 4.8.6. Comparison of satellite-derived latent heat fluxes with those measured by ships. Each point is a 2° latitude by 2° longitude monthly mean. Only those bins between 40° N and 40° S and 600 km from land are included (from Liu 1990).

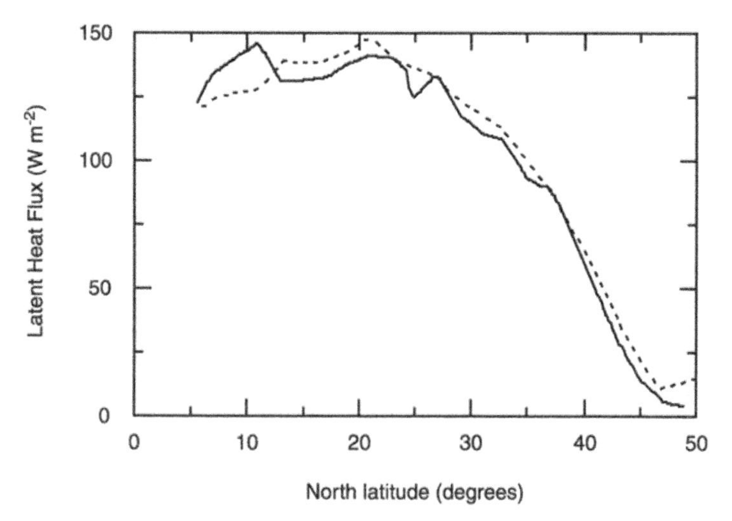

Figure 4.8.7. Comparison of zonal mean latent heat fluxes from ships (dashed line) and satellites (solid line) from 1982 (from Liu 1990).

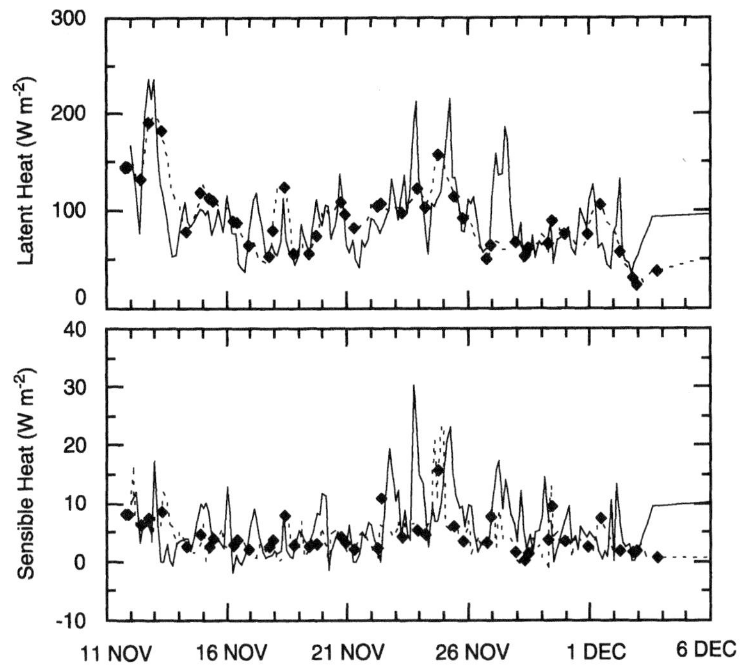

Figure 4.8.8. Comparison of measured latent (top panel) and sensible (bottom panel) heat fluxes with satellite-retrieved ones during TOGA/COARE (from Clayson and Curry 1996).

The surface radiation budget at the surface requires measurement of the net shortwave radiation into, and the net longwave radiation out of, the ocean and land surfaces. ERBE (the Earth Radiation Budget Experiment) was highly successful in monitoring the global radiative budget as measured at a level much above the atmosphere from several sensors. Consequently, there has been a considerable increase in our knowledge, for example, of the overall albedo of the Earth and the meridional transport of heat by the atmosphere from the tropics to the high latitudes. However, measuring the surface radiation budget is much harder, since it requires knowledge of and or measurements of the radiative transfer through the atmospheric column. Heavy reliance has to be placed on an optimum combination of radiative transfer models of the atmosphere and satellite measurements. The transmissivity of the atmosphere is around 0.7, but can vary over quite a wide range depending on the state of the air column.

While the SW radiative flux at the top of the atmosphere is well known, the value at the sea surface is a function of principally the absorption by water vapor and ozone, and scattering by aerosols in a cloud-free atmosphere, and the cloud cover. The water vapor and cloud cover are the most important to SW radiative flux at the surface. The LW radiative flux incident on the surface from the atmosphere itself depends on principally the CO_2, ozone, and water vapor and temperature in a cloud-free atmosphere, and radiation by the cloud bottoms depends on the type and extent of cloud cover. It is therefore essential to measure the temperature, water vapor, and cloud cover (especially low-level clouds for LW) distributions in the atmospheric column remotely to deduce these quantities.

Katsaros (1990) indicates that Lumb's parameterization of hourly insolation in terms of cloud cover (Lumb, 1964) is perhaps the most accurate, while Reed's formula for daily averages (Reed, 1977) is roughly similar. It appears (Katsaros, 1990) that surface SW irradiance can be obtained from satellite measurements with about a 10% error for daily averages, while hourly values can be in error by as much as 20% (compare this to the 5% accuracy of *in situ* measurements by pyranometers). Monthly averages can be estimated even more accurately. However, hourly values are needed for many applications. Since the SW irradiance at the ocean surface is the biggest component of the surface radiation budget, even small percentage errors are quite consequential. Nevertheless as Figure 4.8.9 from Gautier *et al.* (1980), shows satellite retrieval of surface SW irradiance is quite reliable at least on certain timescales. Use of satellite-retrieved profiles in combination with radiative transfer models has the potential for increased accuracies. The most widely used surface irradiance values (for example, Curry *et al.*, 1993) use the values of cloud properties from the International Satellite Cloud Climatology Project (ISCCP) (Rossow and Schiffer, 1991) and temperature and humidity profiles from an atmospheric sounder such as TOVS (Tiros Operational Vertical Sounder). Measurement of SW flux at the top of the atmosphere, combined with estimates of cloud properties, water vapor,

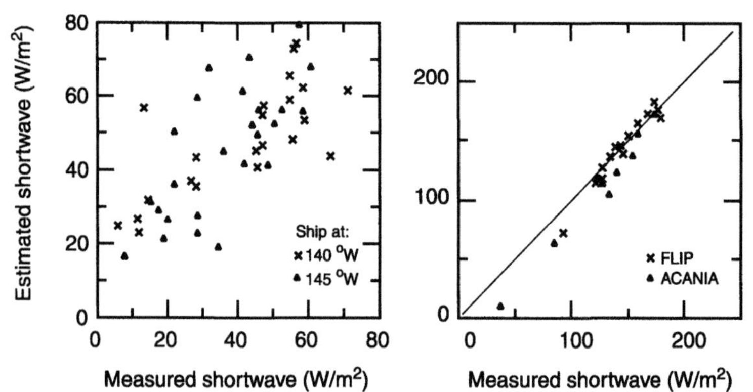

Figure 4.8.9. Satellite retrieval of daily-averaged SW radiative fluxes compared to measured values (from Katsaros 1990). Left panel shows comparisons in the STREX experiment (after Gautier and Katsaros, 1984) at 50° N and right panel shows comparisons in the MILDEX experiment (after Frouin *et al.* 1988).

and aerosol content in the atmospheric column, has been used to derive surface irradiance from radiative transfer models routinely used in numerical models of the atmosphere. Atmospheric temperature and humidity profiles are retrieved from TOVS, and cloud properties such as optical thickness and cloud-top temperature are obtained from ISCCP data sets. Cloud base temperatures needed for calculating downwelling LW radiation cannot be measured from satellites and have to be supplied from the climatology of cloud thicknesses. Aerosol distributions also have to be estimated and appear to be the most uncertain for a cloud-free atmosphere. Typical results from such a "satellite" retrieval are shown in Figure 4.8.10 (from Curry *et al.*, 1993), compared to *in situ* measurements. In general, there is a reasonable agreement between the satellite-retrieved radiative fluxes and *in situ* measurements.

The outgoing SW radiation from the surface consists of the reflected part and the part that upwells from the interior. Over the ocean, it is essentially determined by the albedo of the ocean surface, which can vary between a low value of 0.06 when the Sun is overhead to as high as 0.40 at low incidence as, and ingles dependent on the diffuseness of incident radiation (Payne, 1972). Roughness of the sea surface also plays an important role. In general, these dependences are known, and albedo and hence outgoing SW radiation can be determined quite readily and quite accurately (Katsaros, 1990). The albedo over land or ocean surfaces can be determined from satellite data that accurately describe the surface characteristics (such as the NDVI for vegetation surfaces).

vapor profiles, atmospheric gases, and most importantly the downward radiation from cloud bases. The Lind and Katsaros (1982) parameterization is typical of the generally high accuracies possible. Satellite-based retrieval schemes however depend on the use of satellite-derived (LW irradiance at the ocean surface is a function of the temperature and water such as TOVS) vertical profiles of temperature and water vapor content, and cloud properties, combined with a radiative transfer model. But cloud cover and cloud base temperature, or equivalently, cloud thicknesses, are needed and are hard to measure or estimate. In general, the accuracies are just about the same as empirical parameterizations. At low latitudes, the variability of LW radiation is quite small, because of the relatively constant humidities and air temperatures. It is the high latitudes that exhibit larger variability (Katsaros, 1990). Nevertheless, attractiveness of global coverage makes satellite retrieval of downwelling LW radiation at the sea surface quite useful, since errors appear, despite all the uncertainties, rather small, relative to those associated with the SW radiation (typically 15 W m^{-2}).

Outgoing LW radiation from the sea surface can be readily estimated if the SST is known. The errors resulting from errors in the measurement of SST from satellites appear to be quite inconsequential compared to the other components of the radiation budget (~5 W m^{-2}).

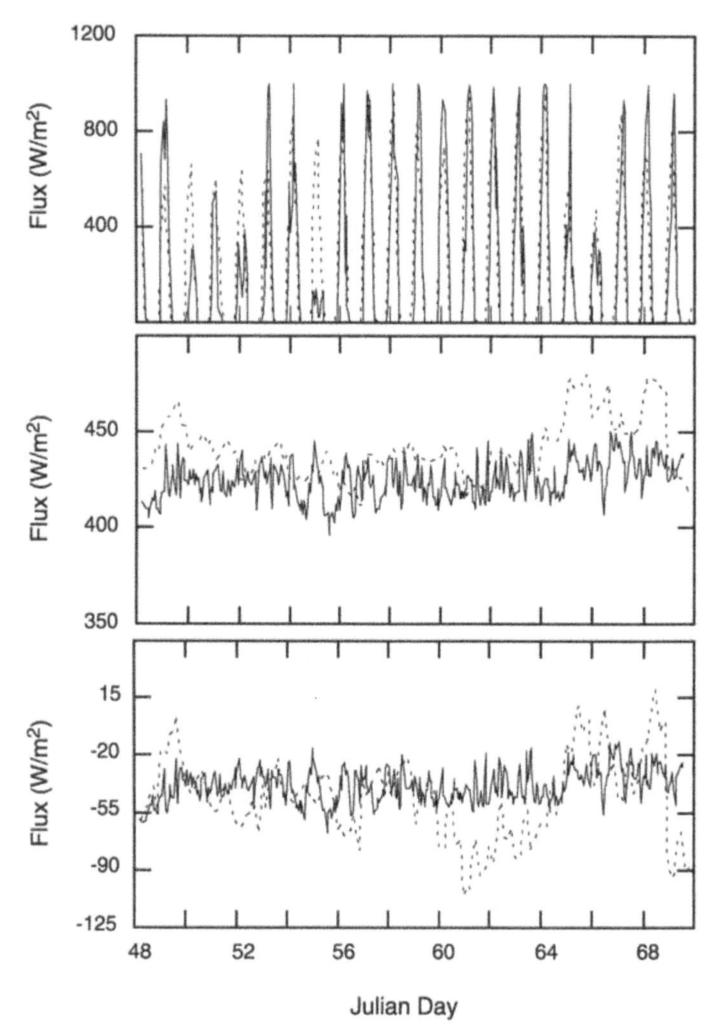

Figure 4.8.10. Satellite retrieved (dashed line) downwelling SW (top panel), downwelling LW (middle panel), and net LW (bottom panel) radiative fluxes compared to measured *in-situ* values (solid line) during TOGA (from Curry *et al.* 1993).

Net LW radiation over land surfaces requires similar input and techniques as over the ocean. Comparison of satellite-derived values for the incident longwave flux with surface observations provides an estimate of the rms error of 15 W m^{-2} and a bias of less than 15 W m^{-2} (Rossow and Zhang, 1995).

To conclude, given the promise of global coverage and reasonable temporal sampling, satellite-derived fluxes are bound to be important components of any future monitoring of the surface fluxes and important input to climate and air–sea coupled models. The desired accuracies of 5–10 W m^{-2} needed for global climate change applications are still elusive, but the situation is improving steadily.

LIST OF SYMBOLS

δ, $\delta_{T,E,G}$ — Laminar sublayer thicknesses for momentum, heat, water vapor, and gas

ε — Dissipation rate of TKE

γ — Kinematic surface tension coefficient

λ — Wavelength

ν, ν_t — Kinematic viscosity (molecular and turbulent)

θ_* — Friction temperature

θ_* θ_v, θ_l — Potential, virtual potential, and liquid water potential temperatures

ϕ_M, ϕ_H, ... — Monin–Obukhoff similarity functions

ψ_M, ψ_H, ... — Stability profile functions

ρ — Density of air

ζ — Monin–Obukhoff similarity variable

κ — Von Karman constant

τ_A, τ_w — Shear stress on the atmospheric side and the water side of the interface

τ_{wv}, τ_{pr}, τ_I — Momentum flux to waves, momentum flux from precipitation, and ice stresses

ΔU — Velocity change across the sublayer

ΔT, Δq, Δc — Temperature, humidity, and concentration changes across the laminar sublayers

c_p — Specific heat

c, c_a, c_s — Concentration, concentration at anemometric height, and surface concentration

c_* — Friction concentration

f — Coriolis frequency; also a fraction

g — Acceleration due to gravity

h_0 — Height of roughness elements

k — Magnitude of the wavenumber vector

k_T, k_E, k_G — Kinematic heat, water vapor, and gas diffusivities

l — Integral microscale of turbulence

q, q_*, q_s — Specific humidity, friction specific humidity, and saturation specific humidity

t	Time
t^*	Surface renewal timescale
u_a^*, u_*	Friction velocity
u_f^*	Free convection velocity scale
x_3	Vertical coordinate
x_i	Coordinates
z	Vertical coordinate
z_a	Anemometric height
z_0, z_{0T}, z_{0E}	Roughness scale for momentum, heat, and water vapor
A_e	Eddy viscosity
A, B, C	Constants in the surface layer
B_r	Bowen ratio
BF	Buoyancy flux
C_p	Wave phase speed at spectral peak
C_g	Group velocity
C_d	Drag coefficient
C_H, C_E, C_G	Bulk transfer coefficients for sensible heat (Stanton number), evaporation (Dalton number), and gas
C_{DN}, C_{HN}	Neutral bulk transfer coefficients
E	Evaporative flux
$\dot{E}$	Evaporation rate
Fr	Froude number
G	Geostrophic velocity; also gas flux
K_M, K_H	Turbulent (eddy) mixing coefficients for momentum and heat
K_e	Kuelegan number
H_s, H_l, H_{pr}	Sensible, latent, and precipitation heat fluxes
H_g	Heat flux to the ground
H_{AI}, H_{IO}	Heat flux at the air–ice and ice–ocean interfaces
L	Monin–Obukhoff length scale
L_E, L_F	Latent heat of evaporation and latent heat of fusion
Le	Lewis number
LW	Longwave radiative flux
Nu	Nusselt number
Pr	Prandtl number
Pr_t	Turbulent Prandtl number
$\dot{P}_r$, $\dot{P}_{sn}$	Precipitation rate of rain and snow
Q_b	Surface buoyancy flux
Q_s, Q_l, Q_g	Kinematic fluxes of sensible and latent heat, and of gas
Ra	Rayleigh number
Re_r	Roughness Reynolds number
Ri_{cs}	Cool skin Richardson number

Ri_f, Ri_g	Flux and gradient Richardson numbers
Ro, Ro^*	Rossby and friction Rossby numbers
S, S_s	Salinity and surface salinity
Sc	Schmidt number
SF	Salinity flux
SW	Shortwave radiative flux
Sw_p, Sw^p	Shortwave flux to photosynthesis and penetrative SW radiation
T, T_v, T_{wb}	Absolute temperature, virtual temperature, and wet bulb temperature
T_r, T_{sn}	Temperature of rain and snow
T_s, T_a	Skin temperature and air temperature
T^*	Friction temperature
U, U_{10}, U_a	Wind speed, wind speed at 10 m, and wind speed at anemometric height
U_s	Ocean surface speed

Chapter 5

Surface Waves

In this chapter, we will discuss the important characteristics of the oceanic surface waves and their impact on the OML. We will address their generation, propagation, dissipation, and interactions. Surface gravity waves have been studied extensively for nearly two centuries and the resulting literature is both vast and varied. It is impossible to do justice to this vital topic in a brief survey such as this. All we can do is touch upon those aspects that are fundamental to the problem of the OML and air–sea exchange in the hope of providing some basic theoretical underpinnings and refer the reader to appropriate literature and many excellent monographs on the subject. Despite its age, Phillips' monograph (Phillips, 1977) is still an authoritative treatment of the subject and much of the elementary material here is derived from this book. A fascinating description of surface waves from a seaman's perspective can be found in Kinsman (1984). LeBlond and Mysak (1978) is another valuable reference. But by far the most useful is the recent summary of the dynamics and modeling of ocean surface waves by Komen *et al.* (1994), which is highly recommended as a reference source for students and experts alike. There exists of course a vast literature, literally thousands of articles on one or another aspect of surface gravity waves scattered around in myriad scientific journals, the most pertinent to our readers being the *Journal of Fluid Mechanics,* the *Journal of Physical Oceanography,* and the *Journal of Geophysical Research (Oceans).* Here we will attempt to provide a modern but succinct review of this fascinating topic. As we shall see, while much is known about surface gravity waves, despite two centuries of effort by many brilliant minds, many uncertainties remain and much work lies ahead, especially in the area of breaking waves and the layers immediately adjacent to the dynamic, ever-changing air–sea interface.

5.1 SALIENT CHARACTERISTICS

Gravity waves on an air–sea interface are a fascinating as well as an important component of air–sea exchange. Surface waves mediate the exchange of principally the momentum between the winds and the upper ocean; however, they are important to heat, mass, and gas exchange as well. Breaking waves disrupt the pervasive molecular sublayer which exists adjacent to the air–sea interface and which regulates the transfer of heat and dissolved gases to the atmosphere. Air bubbles entrained into the water and the spray of tiny droplets entrained into the air during wave breaking are important mechanisms of mass, heat, and gas transfer that are also probably the least understood and not easily quantified. Most of the attention in the past has been focused on momentum exchange issues involved with surface waves, principally those that do not involve extensive breaking or "whitecapping." Only recently has wave breaking been given the attention it deserves in air–sea exchange (Banner and Grimshaw, 1992; Banner and Peregrine, 1993; Melville, 1994, 1996).

Surface waves are the inevitable consequence of wind action over the air–sea interface. While the causal relationship between the two is quite apparent even to a most casual observer, the details of the interaction are extremely complex, and despite decades of research, it is fair to say that an accurate theoretical model does not exist. This is not to say that we cannot model wave growth in the oceans. As a result of international collaboration among experts in the field, an empirical wave model WAM has been constructed and refined over the past decade (WAMDI Group, 1988; Komen *et al.,* 1994) that provides a reasonably good characterization of the oceanic surface wave field useful for many operations at sea. However, the model is empirical and not perfect in its depiction of wave-related processes, and the details of wind-wave generation, dissipation, and transfer of energy across the spectrum are still too poorly understood to be accurately quantified.

Surface waves, like oceanic tides, were one of the very first topics that became amenable to theoretical analysis. Because it is possible to treat them using linear, inviscid governing equations (potential theory), and because the phenomenon is so fascinating to many, excellent theoretical treatments became available fairly early in the history of surface waves. While these treatments enabled a better understanding of surface waves, they were of little use in practical applications to understanding and predicting the oceanic sea state. The principal problem is the stochastic nature of the surface wave field in the oceans and the difficulty of quantifying the source, sink, and transfer terms in spectral space. Not until the 1950s were plausible generation mechanisms postulated and quantified (Phillips, 1957; Miles, 1957) and spectral descriptions offered. Phillips' celebrated –5 power law for the frequency spectrum in the saturated range of the wind-wave spectrum (Phillips, 1958) was a landmark achievement akin to the Kolmogoroff universal range spectrum in turbulence. But unlike the Kolmogoroff spectrum, it has not stood the test of time and had to be revised when more accurate field observations displayed systematic departures of the wind-wave spectrum from this spectral shape (Kitaigorodskii, 1983; Phillips, 1985). Not until the eighties did it become possible to construct an empirical model for the evolution of the wind-wave spectrum (SWAMP, 1985; WAMDI, 1988; Komen *et al.*, 1994), based partly on advances of the understanding gained by the work of Hasselmann *et al.* (1985) on resonant wave–wave interactions (see also Phillips, 1977) and the transfer of wave energy in spectral space. It now appears that such interactions are key to understanding the resulting spectral shape of the wind-generated surface waves.

Air–sea coupling involves surface waves at the air–sea interface. The details of the momentum transfer from winds to the ocean necessarily include surface waves. They figure prominently in the determination of the bulk transfer coefficients. As far as the OML is concerned, the direct influence is through the process of wave breaking. It is this process that alters mixing in the upper few meters of the ocean, disrupts the molecular sublayer that mediates heat and gas exchange with the atmosphere, and, if violent enough, causes extensive injection of droplets and spray into the air and air bubbles into the water. However, wave breaking is a hard process to quantify in a random wave field and herein lies one of the principal difficulties and the focus of current research (Jessup, 1995; Melville, 1996).

5.2 LINEAR WAVES FROM POTENTIAL THEORY

Gravity waves on an air–water interface were one of the early triumphs in application of Newton's laws of motion to practical fluid dynamical problems. The reason for this success is simply the fact that surface waves can be treated using linear, incompressible, inviscid, irrotational equations of motion, in other words, potential theory, to a fairly good degree of approximation. This is partly because in an otherwise quiescent fluid sustaining surface wave motions, the influence of molecular viscosity is confined to a very thin layer near the free surface, and the bottom (if the water is shallow). From simple dimensional analysis, the thickness δ of these oscillating boundary layers can be shown to be O , where ν is $(\nu / n)^{1/2}$ the viscosity and n is the frequency. The scale of the wave is given by its wavelength λ, the characteristic velocity is the orbital velocity of a fluid particle under the action of the wave ($u \sim n\,\lambda$), and therefore the relevant Reynolds number is $R_w \sim (\lambda^2 n/\nu) \sim (n/k^2\nu)$, where k is the magnitude of the wavenumber vector. The ratio $\delta/\lambda \sim R_w^{-1/2}$, and this is a very small quantity. The rate of strain in these layers is also small and therefore their contribution to wave dissipation is small as well. Thus in idealized conditions at least, vorticity is confined to these very thin layers and therefore the bulk of the water column under wave motion is irrotational. This may be one reason why potential theory works so well for surface waves. However, in the real ocean, the layers adjacent to the air–sea interface are invariably turbulent and highly vortical. They can interact with the nearly irrotational wave motions and exchange energy with them. Nevertheless, the astonishing fact is that low frequency swell is known to propagate across entire ocean basins with little attenuation (Munk *et al.*, 1963) and potential theory results are generally valid. This must mean that at least at some space/time scales (ignoring small capillaries where coupling appears to be strong), the coupling between the irrotational surface wave motions and highly vortical turbulent motions must be small.

In an irrotational flow the vorticity, the curl of the velocity, is zero, so that a velocity potential ϕ can be defined, $u_i = \partial\phi/\partial x_i$, that automatically satisfies this condition. By virtue of the continuity equation for incompressible flows, $\partial u_i/\partial x_i = 0$, we obtain Laplace's equation for the velocity potential:

$$\frac{\partial^2\phi}{\partial x_i \partial x_i} = 0 \qquad (5.2.1)$$

Of course, this needs to be solved subject to boundary conditions at the bottom and at the free surface. The bottom condition is simply that it be impermeable. Let d be the depth of the fluid column. For a flat bottom, $u_3=0$,

$$\left.\frac{\partial\phi}{\partial x_3}\right|_{x_3=-d} = 0 \qquad (5.2.2)$$

It is the top boundary condition that is quite difficult to treat, because the condition applies at the free surface deformed by the wave motion, whose shape is of course unknown a priori, and herein lies the difficulty of the subject. Even though the governing equations are linear, nonlinearity is introduced through this free surface boundary condition. The traditional simplification, due to Stokes, is to apply the boundary condition at the undeformed interface ($x_3 = 0$) and invoke series expansions about it in terms of a small parameter, in this case the wave slope $\varepsilon = ka$, where a is the wave amplitude. As long as the wave slope is small (it never exceeds a value of about 0.45, the Stokes limit for finite-amplitude waves), this procedure is valid, although convergence has never been proved formally for the general case of unsteady waves (Phillips, 1977).

At the free surface, both a kinematic and a dynamic condition need to be satisfied. If $x_3 = \zeta\,(x_\alpha, t)$ ($\alpha = 1,2$) denotes the free surface, then right at the free surface, the kinematic condition is

$$\frac{\partial\phi}{\partial x_3} = u_3^\zeta = \left(\frac{\partial}{\partial t} + u_\alpha \frac{\partial}{\partial x_\alpha}\right)\zeta = \frac{\partial\zeta}{\partial t} + \frac{\partial\phi}{\partial x_\alpha}\frac{\partial\zeta}{\partial x_\alpha} \tag{5.2.3}$$

This condition requires that a fluid particle at the free surface remain there, in other words, the free surface is a material surface. The dynamical condition relates the pressure difference across the interface to the surface tension force. The momentum equation for an inviscid potential flow is

$$\frac{\partial u_i}{\partial t} + \frac{\partial}{\partial x_i}\left(p + \frac{1}{2}u_k u_k + g\,x_3\right) = 0 \tag{5.2.4}$$

where p is the pressure. The convention followed here is that when Greek symbol α is used as a subscript, the summation take place over $\alpha = 1$ and 2 only, since quantities pertaining only to the horizontal direction are involved. This restriction does not apply to summation involving other indexes such as i, j, and k.

For a potential flow, this can be integrated to yield the Bernoulli equation

$$\frac{\partial\phi}{\partial t} + p + \frac{1}{2}\frac{\partial\phi}{\partial x_k}\frac{\partial\phi}{\partial x_k} + g\,x_3 = 0 \tag{5.2.5}$$

after incorporating the arbitrary integration constant (actually a function of time) into the velocity potential itself. Taking the total derivative of this equation,

$$\frac{dp}{dt} + \left(\frac{\partial^2\phi}{\partial t^2} + g\frac{\partial\phi}{\partial x_3}\right) + \frac{\partial}{\partial t}\left(\frac{\partial\phi}{\partial x_k}\frac{\partial\phi}{\partial x_k}\right) + \frac{1}{2}\left[\frac{\partial\phi}{\partial x_i}\frac{\partial}{\partial x_i}\left(\frac{\partial\phi}{\partial x_k}\frac{\partial\phi}{\partial x_k}\right)\right] = 0 \tag{5.2.6}$$

The dynamical condition involves the pressure at the free surface and is

$$p = p_a - \gamma \left\{ \frac{\partial^2 \zeta}{\partial x_1^2} \left[1 + \left(\frac{\partial \zeta}{\partial x_2} \right)^2 \right] + \frac{\partial^2 \zeta}{\partial x_2^2} \left[1 + \left(\frac{\partial \zeta}{\partial x_1} \right)^2 \right] - 2 \frac{\partial^2 \zeta}{\partial x_1 \partial x_2} \frac{\partial \zeta}{\partial x_1} \frac{\partial \zeta}{\partial x_2} \right\} \left[1 + \frac{\partial \zeta}{\partial x_\alpha} \frac{\partial \zeta}{\partial x_\alpha} \right]^{-3/2} \quad (5.2.7)$$

where γ is the kinematic surface tension, surface tension divided by density, and p_a is the kinematic atmospheric pressure. The surface tension term involves the sum of the inverse principal radii of curvature of the free surface. It is traditional to align the x_1-axis in the direction of wave propagation and hence retain only the terms involving x_1 in one-dimensional wave propagation problems, and then Eq. (5.2.7) simplifies to a more familiar form in, for example, Phillips (1977). Note that both the kinematic and the dynamic boundary conditions have to be applied at $x_3 = \zeta$, and the pressures as defined here are kinematic, that is, they have been divided by density. Note also the presence of nonlinear terms in the Bernoulli equation. The atmospheric pressure cannot usually be regarded as a constant and in fact must be obtained as part of the solution for problems involving wave generation, since the differential pressure between the forward and the rearward faces of the wave input energy into wave motion. However, for freely propagating waves, atmospheric pressure perturbations can be ignored and p_a assumed to be constant (swell freely propagating against the wind can transfer its energy to the wind and this approximation must be relaxed if its decay is to be quantified). Now seek solutions in terms of perturbation expansions involving the wave slope as a small parameter ($\varepsilon = ka$):

$$\phi = \sum_{n=0}^{\infty} \varepsilon^n \phi_n \quad (5.2.8)$$

Substitution in the governing equation and the boundary conditions gives

$$\left. \begin{aligned} \frac{\partial^2 \phi_n}{\partial x_\alpha \partial x_\alpha} = \nabla^2 \phi_n &= 0 \\ (\partial \phi_n / \partial x_3)_{x_3 = -d} &= 0 \end{aligned} \right\} \quad n = 0, 1, \cdots \quad (5.2.9)$$

$$\left. \begin{aligned} \frac{\partial \zeta_0}{\partial t} &= \frac{\partial \phi_0}{\partial x_3} \\ \frac{\partial^2 \phi_0}{\partial t^2} + g \frac{\partial \phi_0}{\partial x_3} &= \gamma \frac{\partial^2}{\partial x_\alpha \partial x_\alpha} \left(\frac{\partial \phi_0}{\partial x_3} \right) \end{aligned} \right\} \quad \text{at } x_3 = 0 \quad (5.2.10)$$

$$\left.\begin{aligned} \frac{\partial\zeta_1}{\partial t} &= \frac{\partial\phi_1}{\partial x_3} + \zeta_0 \frac{\partial^2\phi_0}{\partial x_3^2} - \frac{\partial\phi_0}{\partial x_\alpha}\frac{\partial\zeta_0}{\partial x_\alpha} \\ \frac{\partial^2\phi_1}{\partial t^2} + g\frac{\partial\phi_1}{\partial x_3} &= -\zeta_0 \frac{\partial}{\partial x_3}\left(\frac{\partial^2\phi_0}{\partial t^2} + g\frac{\partial\phi_0}{\partial x_3}\right) - \frac{\partial}{\partial t}\left(\frac{\partial\phi_0}{\partial x_k}\frac{\partial\phi_0}{\partial x_k}\right) \end{aligned}\right\} \text{ at } x_3 = 0 \qquad (5.2.11)$$

and so on.

Equations (5.2.9) to (5.2.11) can be solved readily to obtain solutions for surface wave motions. For simplicity, we can orient the x_1-axis in the direction of wave motion and consider the flow to be two-dimensional in the x_1–x_3 space. In that case, it is convenient to revert to the conventional notation and put $x_1 = x$, and $x_3 = z$. Dropping suffix 0 and letting

$$\zeta = a\cos(kx - nt) \qquad (5.2.12)$$

the velocity potential to zeroth order is

$$\phi = \frac{na \cosh k(z+d)}{k \sinh kd}\sin(kx - nt) \qquad (5.2.13)$$

For the general case, the wavenumber $k = (k_\alpha\, k_\alpha)^{1/2}$ and kx above has to be replaced by $k_\alpha x_\alpha$. Because of the dynamical condition at the free surface, the frequency n is related to the wavenumber k by the dispersion relation (see Figure 5.2.1):

$$n^2 = gk\left(1 + \gamma\frac{k^2}{g}\right)\tanh kd \qquad (5.2.14)$$

This can be obtained by substituting Eq. (5.2.13) into Eq. (5.2.10). Note that the dispersion relation involves only the magnitude of the wavenumber vector, not its direction (contrast this to small scale internal waves, whose dispersion relation is independent of the magnitude of the wave vector, but dependent on the direction of propagation with respect to the vertical; see Chapter 6). Therefore surface gravity waves are isotropic, in the sense that the direction of propagation does not matter in the dispersion relation. Note, however, that the directional distribution of surface waves is almost always nonisotropic. The phase speed $c = (n/k)$. The immediate consequence of the dispersion relationship (5.2.14) is that the group velocity $c_{g\alpha} = \partial n/\partial k_\alpha$ is in the same direction as the

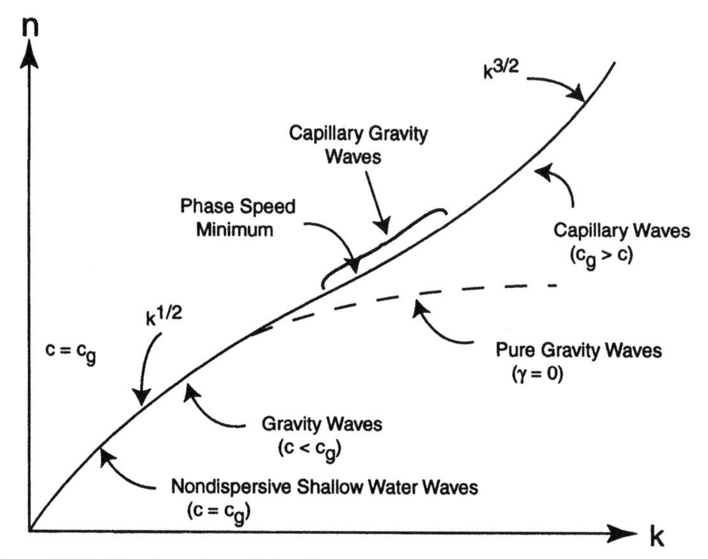

Figure 5.2.1. The dispersion relation for gravity waves, delineating the different regimes.

phase velocity, but in general, its magnitude is not the same as c. Since the dispersion relation is isotropic, we will use c and c_g to denote the magnitudes of the phase velocity and group velocity henceforth, with the understanding that since these velocities are in the direction of the wavenumber vector, one needs to multiply these magnitudes by the unit vector in the direction of the wavenumber vector to get the velocities.

When kd << 1, the waves are called shallow water waves, and ignoring surface tension, the dispersion relation becomes

$$n^2 = gdk^2, \quad c = c_g = (gd)^{1/2} \tag{5.2.15}$$

Shallow water waves are nondispersive ($n/k = \partial n/\partial k$), which means if one creates an arbitrary deflection of the free surface composed of many different wavenumbers, all of them travel with the same phase speed and hence the disturbance travels without change of shape, at least in the limit of linear infinitesimal waves.

When kd >> 1, the waves are called deep water waves. In practice this condition is unnecessarily restrictive and waves are essentially deep water waves if kd > π or the water is deeper than half the wavelength! The dispersion relation is then independent of the depth d of the water column:

$$n^2 = g\,k\left(1+\frac{\gamma k^2}{g}\right)$$
$$c^2 = \frac{g}{k}\left(1+\frac{\gamma k^2}{g}\right) \qquad (5.2.16)$$
$$c_g = \frac{c}{2}\left(1+3\frac{\gamma k^2}{g}\right)\left(1+\frac{\gamma k^2}{g}\right)^{-1}$$

For deep water waves, $\phi \sim e^{kz}$, that is, the wave motion decays exponentially with depth, the scale of this decay being the wavelength. The phase velocity c has a minimum when $k = (g/\gamma)^{1/2}$. For wavenumbers higher than this wavenumber, the restoring forces are due primarily to surface tension and the waves are called capillary waves, and for those less than this, the restoring forces are due to gravity and these are the deep water gravity waves. There is little energy in the capillary wave part of the amplitude spectrum of oceanic surface waves, but capillary waves do contribute significantly to the surface slope spectrum (Phillips, 1977). The wavelength associated with this minimum phase speed is ~1.7 cm and the corresponding phase and group speeds are ~23 and 18 cm s^{-1} for pure water ($\gamma \sim 7.4 \times 10^{-5}\ m^3\ s^{-2}$). The timescales for the generation of capillary/capillary–gravity waves is small and hence they quickly adjust to changing wind conditions. Microwave sensing of the sea surface involves principally these waves, since the scattered radiation is due to Bragg scattering and principally from waves with wavelengths twice the wavelength of the incident radiation, which is usually in the centimeter range. As such, it is thought that the scattered microwave energy has information about the "instantaneous" wind stress acting at the ocean surface. This is indeed extremely valuable for many applications that need information about the wind stress at the ocean surface.

For pure capillary waves, g drops out of the dispersion relation:

$$n^2 = \gamma k^3, \quad c=(\gamma k)^{1/2}, \quad c_g = 3c/2 \qquad (5.2.17)$$

Note that the group velocity of capillary waves is larger than the phase velocity! This is the reason that for a group of propagating capillary waves, wave crests appear to be destroyed at the front of the group and created at the rear, exactly the opposite of a group of gravity waves, whose group velocity is half their phase velocity, so that the crests appear to be created continuously at the front and destroyed at the rear of the group.

For pure gravity waves, the second term involving surface tension is small and the dispersion relation becomes

$$n^2 = gk, \quad c = (g/k)^{1/2}, \quad c_g = c/2 \tag{5.2.18}$$

Thus, for gravity waves, the group velocity lies in the range $(c/2) < c_g < c$, while for capillary waves, the range is $c < c_g < (3c/2)$. There are several additional quantities of interest. The expressions for these correct to first order are given here without derivation (see Phillips, 1977, for details). Equipartition between kinetic energy and potential energy holds in a conservative system in the linear limit, and the total energy density (averaged over the wave period) for capillary–gravity waves can thus be written as

$$E = \frac{\rho}{2} \cdot \frac{n^2 a^2}{k} \coth kd \tag{5.2.19}$$

The dependence on surface tension enters through the dispersion relation. For pure capillary waves,

$$E = \frac{\rho}{2} \gamma k^2 a^2 = \rho \gamma \overline{\frac{\partial \zeta}{\partial x_j} \frac{\partial \zeta}{\partial x_j}} \tag{5.2.20}$$

For pure gravity waves,

$$E = \frac{\rho}{2} g a^2 = \rho g \overline{\zeta^2} \tag{5.2.21}$$

These expressions can be used for an arbitrary spectrum of surface waves. Note the interesting fact that the energy density involves wave slopes for pure capillary waves and wave amplitudes for pure gravity waves. As we said earlier, the major contribution to the energy spectrum is from the gravity waves, and to the slope spectrum it is from the capillary range. If indeed the details of transfer of energy from wind to waves involve wave slope considerations at all spatial scales, capillary waves will figure prominently in air–sea exchange. At least for this reason they are important to study. The magnitude of the mean momentum per unit area (which can be interpreted as the mass flux per unit width defined in the most general case as $M_\alpha = \rho \overline{\int_{-\infty}^{\zeta} u_\alpha dz} = -\rho \overline{\phi(\zeta) \nabla_\alpha \zeta}$, where the overbar indicates average over a cycle and u_α is the velocity component) is

$$M = \frac{E}{c} = \frac{\rho}{2} n a^2 \coth kd \tag{5.2.22}$$

This is a general result that holds for two-dimensional waves and to all orders. The magnitude of the energy flux associated with these waves is given by $E_f = Ec_g$, while that of the momentum flux is $M_f = E_f/c = Ec_g/c$. These fluxes are of course in the direction of the propagation of waves.

An additional quantity of great interest is the action density, which is defined as $A = E/n$. Irrespective of whether the waves are swiftly moving tsunamis, slower moving swell, or an entire spectrum of gravity waves, the wave propagation is governed by laws of conservation. In the presence of background ocean currents, the energy density of the waves is not a conserved quantity. Instead the action density is conserved following the wave group, if dissipation effects can be ignored (see Whitham, 1974, for a lucid discussion of these aspects). Thus, conservation of action is a useful alternative to conservation of energy in wave propagation problems (see Chapter 1 of Komen *et al.*, 1994).

The deviation of kinematic pressure from hydrostatic balance due to wave motion (to zeroth order, $p = -\partial\phi/\partial t$) is

$$p = ga \frac{\cosh k(z+d)}{\cosh kd} \cos(kx - nt) \tag{5.2.23}$$

and therefore the average kinematic pressure is

$$\bar{p} = -\frac{1}{2} n^2 a^2 \sinh^2 k(z+d) / \sinh^2 kd \tag{5.2.24}$$

From the Eulerian point of view, the fluid particles under the action of a propagating surface wave execute, to zeroth order, perfect closed elliptic orbits with major and minor axes,

$$a \frac{\cosh k(z+d)}{\sinh kd}, \quad a \frac{\sinh k(z+d)}{\sinh kd} \tag{5.2.25}$$

with the major axis aligned with the horizontal in the direction of wave motion. In deep water, these are perfect circles, with the diameter largest at the surface but decreasing exponentially with depth (Figure 5.2.2). For many considerations, the details of these particle motions themselves are of great interest. Transport of

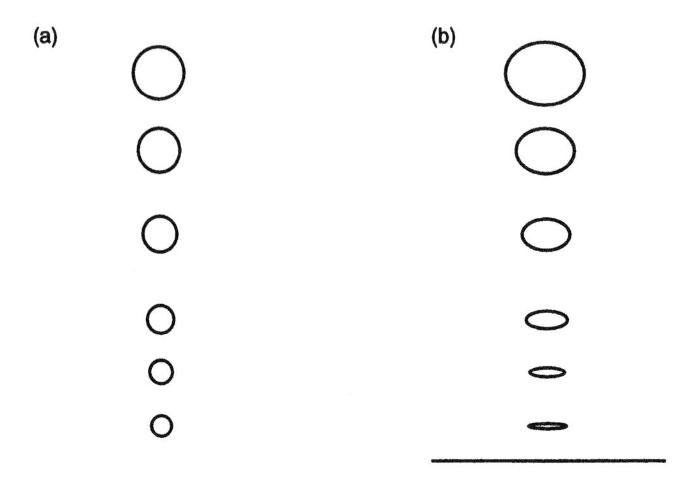

Figure 5.2.2. Orbital velocities under the action of gravity waves for (a) deep water waves and (b) shallow water waves.

pollutants, such as spilled oil, and floating debris is observed to have a nonzero mean velocity due to wave motion. Thus a Lagrangian point of view is more appropriate for transport considerations. Phillips (1977) shows that the Lagrangian velocity of a fluid particle initially at the position $x_\alpha = x_\alpha^0$ is, correct to first order,

$$u_\alpha^l\left(x_\alpha^o,t\right) = u_\alpha\left(x_\alpha^o,t\right) + \left[\int_0^t u_\alpha\left(x_\alpha^o,t'\right)dt'\right]\frac{\partial}{\partial x_\alpha^o}u_\alpha\left(x_\alpha^o,t\right) \qquad (5.2.26)$$

Thus to zeroth order, the Eulerian and Lagrangian velocity fields are identical. Also there is no mass flux below the troughs in the Eulerian frame of reference (Starr, 1945). The difference shows up when the first-order term is evaluated. This has a *nonzero* mean value in the horizontal direction,

$$\bar{u}^l = \frac{c}{2}(ka)^2\,\frac{\cosh 2k\left(z+d\right)}{\sinh^2 kd} = V_S \qquad (5.2.27)$$

and $\overline{u_3^l} = 0$. The particle orbits are not closed to this order and there is a slow drift in the direction of the wave motion on the order of $(ka)^2$. This was pointed out by Stokes himself and is known as the Stokes drift. The Stokes drift is c $(ka)^2$coth (kd) at the surface. For deep water waves (kd >>1), the Stokes drift is equal to c $(ka)^2$ exp (2kz). This is extremely important in the oceans as it causes a steady drift of floating objects due to a propagating wave field even if there are no mean currents. The presence of the Stokes drift of the ocean surface wave field is also responsible for creating cellular motions in the OML known as Langmuir circulation (Leibovich, 1983). These make themselves visible by accumulating debris at regions of the surface convergences between counter-rotating horizontal cells roughly aligned with the wind. They are also known as windrows (see Section 2.4). In his own wonderful, inimitable style, Longuet-Higgins (1986) discusses the Eulerian and Lagrangian aspects of surface waves, including topics such as mass transport and particle trajectories in steep nonlinear waves.

We mentioned briefly that there are very thin oscillatory vortical (boundary) layers at the surface and the bottom due to wave motions. The surface layer does not contribute much to dissipation of waves. In deeper water, the dissipation of wave motion is primarily due to viscosity, and by simple dimensional considerations, it should be evident that the decay timescale should depend only on νk^2. Phillips (1977) shows that

$$\frac{1}{E}\frac{\partial E}{\partial t} = -4\nu k^2 \tag{5.2.28}$$

In shallow water, it is the bottom boundary layer that attenuates the wave and the attenuation rate is

$$\frac{1}{E}\frac{\partial E}{\partial t} = -\left(\frac{k\nu}{\delta}\right)\operatorname{cosech}(2kd) \tag{5.2.29}$$

where $\delta = (2\nu/n)^{1/2}$ is the bottom boundary layer thickness.

The surface boundary layer is sensitive to any contamination at the surface by a film of viscous oil (Phillips, 1977). Oil on the water has been known to very effectively damp short waves, which is responsible for the glassy appearance of the sea surface with an oil slick on it. For a densely packed surface film, Phillips (1977) shows that the rate of attenuation of wave energy is

$$(k\nu/\delta)\coth kd \tag{5.2.30}$$

which is a factor of $(R_w)^{1/2}$ greater than that for a clean surface.

The presence of the bottom boundary layer in water of finite depth also leads to small first-order mean velocity near the bottom in the direction of wave motion, called the streaming velocity (Phillips, 1977). This was shown by Longuet-Higgins to be

$$\overline{u}^b = \frac{3}{4}\frac{c\,(ka)^2}{\sinh^2 kd} \tag{5.2.31}$$

Note that it is independent of viscosity even though nonzero viscosity is essential for its existence. It is also independent of the boundary layer thickness. This needs to be added to the Stokes drift,

$$\overline{u}^s = \frac{1}{2}\frac{c\,(ka)^2}{\sinh^2 kd} \tag{5.2.32}$$

to obtain the total mass transport velocity near the bottom. This bottom streaming is of great importance to transport of sand and sediments by waves. When there is a surface slick, Phillips (1977) shows that the oil slick streams ahead of the fluid particles at the surface by a velocity excess of

$$\Delta\overline{u} = \frac{3}{4} c\,(ka)^2 \coth^2(kd) \tag{5.2.33}$$

derived by assuming that the mean vorticity below the surface boundary layer vanishes to $O(ka)^2$ and invoking then an analogy with the bottom boundary layer to derive the velocity jump across the surface boundary layer. The result is independent of viscosity and the boundary layer thickness. The shear at the surface is also of interest and is $O(\Delta\overline{u}/\delta)$. However, the vorticity below the surface boundary layer is not negligible and Craik (1982) uses the argument that to prevent the waves from decaying, a vertical momentum flux equal to $(1/c)(\partial E/\partial t) = -\gamma E/c$ must exist. Since this is the viscous stress at the surface, $\rho\nu(\partial\overline{u}/\partial z)_{z=0}$, from Eqs. (5.2.19) and (5.2.30), the shear at the surface is $2\Delta\overline{u}/(3\delta)$, with $\Delta\overline{u}$ given by Eq. (5.2.33). This is $O(k\delta)^{-1}$ times larger than the value for a clean surface. In addition, this surface layer is susceptible to spanwise perturbations that can produce streamwise rolls (Craik, 1982).

One cannot but be amazed at how many important effects of surface waves can be explained simply by using linear wave theory with suitable modifications to account for viscous effects (or finite amplitude) effects! It is therefore no wonder surface waves are a shining example of success in applying the laws of nature to physical processes. However, a word of caution is also in order. The above results have been derived assuming the flow is laminar. Seldom does one invariably turbulent, and then even if one assumes that the molecular viscosity find surface waves propagating over a laminar ocean. The upper mixed layer is invariably turbulent, and then even if one assumes that the molecular viscosity can be conveniently replaced by an effective turbulent viscosity for those quantities that depend explicitly on viscosity, it is not clear what values, if any, one should use. There has been additional work done since Phillips (1977) on the surface-wave-induced bottom boundary layer (for example, Jacobs, 1984), but little is known about the boundary layer at the surface. Wave motions at the air–sea interface acted upon by turbulence in the adjoining ABL and OML are one of the hardest flow problems to tackle, and despite valiant attempts (for example, see Komen *et al.*, 1994), much work remains.

5.3 FINITE-AMPLITUDE EFFECTS

Infinitesimal surface waves are sinusoidal and symmetrical w.r.t. the air–sea interface. Finite-amplitude waves, however, tend to have sharper crests and flatter troughs. In other words, they have harmonics bound to the primary sinusoidal wave. A wave of permanent but general form can be Fourier-decomposed into an infinite sum of harmonic components, all of which must travel at the same phase speed in order that the waveshape be maintained without change. Thus the mth harmonic will be of the form cos [m(kx–nt)]. An infinite finite-amplitude wave train of single wavenumber component and permanent shape is called a Stokes wave. It is a nondispersive wave and is ideally supposed to propagate unchanged at its phase speed c. The existence of the Stokes wave was doubted ever since Stokes postulated it in 1840s, but Levi-Civita proved its existence in the 1920s. The limiting Stokes wave was shown to exist in as late as the 1980s. However, it turns out to be unstable to sideband instability and tends to degenerate into a group of waves. Nevertheless, a Stokes wave has properties germane to wind-generated waves.

Calculation of the shape and propagation characteristics of a Stokes wave is difficult because of the nonlinearity of the free surface boundary condition, but perturbation expansions using the wave slope as a small parameter make it possible. Stokes computed the first five coefficients in the expansion, Kinsman (1984) presents expansions to fourth order, and Drennan *et al.* (1988) over 150! If we look at higher order solutions to account for nonnegligible wave slope (ka), we find to order eight (Drennan *et al.*, 1988; Donelan and Hui, 1990)

$$\begin{aligned} c^2 &= \frac{g}{k}\left[1+(ka)^2+\frac{(ka)^4}{2}+\frac{(ka)^6}{4}-\frac{22(ka)^8}{45}\right] \\ \zeta &= \frac{1}{k}\sum_{m=1}^{8}\beta_m \cos m(kx-nt) \end{aligned} \qquad (5.3.1)$$

$$
\begin{aligned}
\beta_1 &= (ka) - \frac{3}{8}(ka)^3 - \frac{211}{192}(ka)^5 - \frac{14411}{5120}(ka)^7 \\
\beta_2 &= \frac{(ka)^2}{2} + \frac{(ka)^4}{3} - \frac{13}{48}(ka)^6 - \frac{231}{160}(ka)^8 \\
\beta_3 &= \frac{3}{8}(ka)^3 + \frac{99}{128}(ka)^5 + \frac{3783}{5120}(ka)^7 \\
\beta_4 &= \frac{1}{3}(ka)^4 + \frac{217}{180}(ka)^6 + \frac{4987}{2160}(ka)^8 \\
\\
\beta_5 &= \frac{125}{384}(ka)^5 + \frac{15769}{9216}(ka)^7 \\
\beta_6 &= \frac{27}{80}(ka)^6 + \frac{13131}{5600}(ka)^8 \\
\beta_7 &= \frac{16807}{46080}(ka)^7 \\
\beta_8 &= \frac{128}{315}(ka)^8
\end{aligned}
\qquad (5.3.1)
$$

where ka is the wave slope and a is the wave amplitude, half the crest to trough distance.

The waveshape is shown for various wave slopes in Figure 5.3.1. The phase speed is now amplitude-dependent (see Figure 5.3.2) and reaches its maximum at ka ~ 0.437, well before ka ~ 0.4432, the theoretical maximum slope corresponding to the Stokes limiting wave. This may not be of much practical importance, since in practice, the waves break at ka values below 0.4, well before the theoretical maximum in phase velocity can be attained. Nevertheless, there is a weak amplitude dispersion in surface waves, although the magnitude is not large enough to be important in most cases. Figure 5.3.2 also shows the phase and group velocities as functions of wave slope. It is interesting that the velocity at the wave crest increases with wave slope and approaches 90% of the phase speed for ka ~ 0.4.

Stokes himself derived the limiting form for steady waves and showed that the limiting form for a finite-amplitude wave is a sharp crest subtending an angle of 120° (see Lamb, 1945). This is obtained by imposing the condition that the forward fluid particle velocity be equal to the phase speed of the surface wave in this limit (Kinsman, 1984). It is easily shown that the vertical acceleration at the crest of the fluid particle in its orbital motion reaches a value equal to g/2 at the limiting condition. The corresponding wave slope (ka) is ~0.4432 and the limiting phase speed is over 9% larger than that of the linear wave. Beyond this

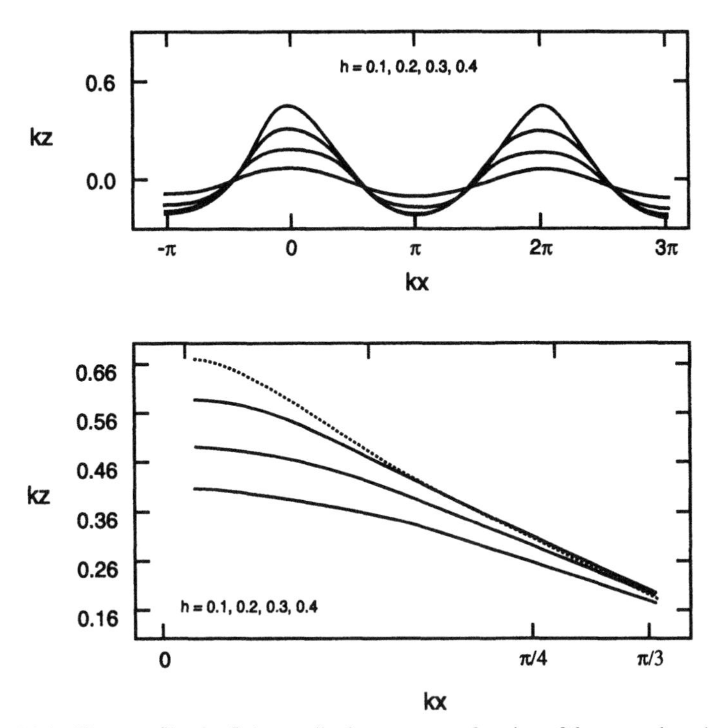

Figure 5.3.1. Wave profiles for finite-amplitude waves, as a function of the wave slope (top). The bottom chart shows details near the crest for waves approaching maximum steepness (from Drennan *et al.*, 1988).

slope, one can expect the particle velocity to exceed the phase speed and hence the wave to "break." The top of the crests is essentially torn off and "spilt," engulfing air and hence appearing as whitecaps to the naked eye. In shallow water, breaking also takes the form of plunging breakers in addition to spilling ones.

There is, however, no observational evidence for the existence of a limiting Stokes wave, although Kinsman (1984) suggests that this limiting crest angle is indeed attained (even in shallow water, even though symmetry is lost). Only shorter waves in the deep water gravity wave spectrum ever approach this wave slope and this limiting angle. Also, this criterion applies only to a steady wave train. Most wave trains become unstable and break at far lesser values of the slope (Phillips, 1985). Around wave slopes of about 0.3, waves become susceptible to sideband Benjamin–Feir instabilities and at higher slopes to three-dimensional instabilities, both of which eventually lead to breaking (Melville, 1996). The fastest growing Benjamin–Feir instability occurs at ka ~ 0.38, much less than the theoretical value, leading to wave breaking and whitecapping

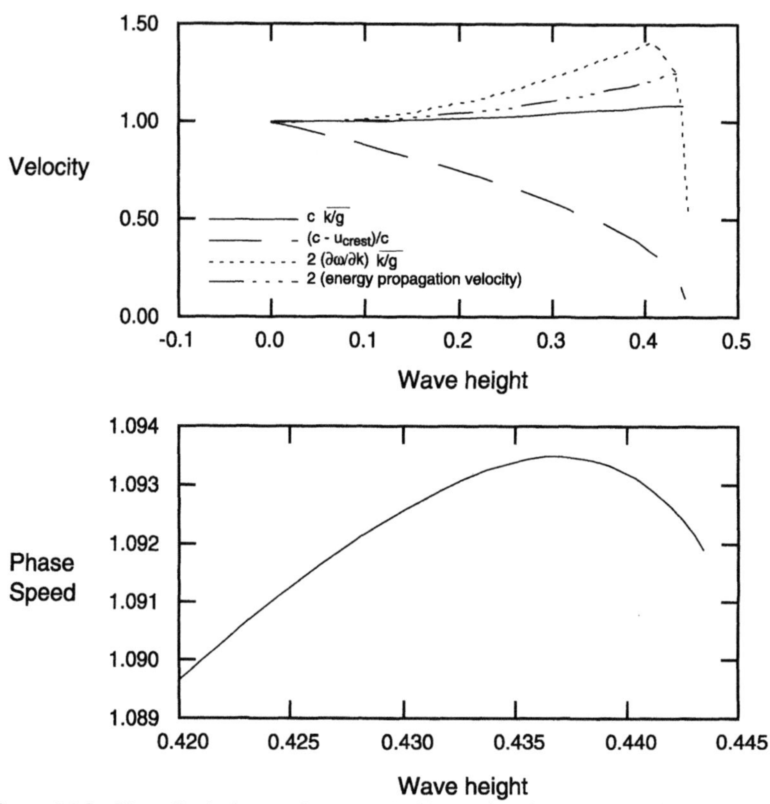

Figure 5.3.2. Normalized phase and group velocities as functions of wave slope. Note the very small effect of finite amplitude on wave propagation (figure courtesy of W. Drennan).

beyond this amplitude. Longuet-Higgins and Dommermuth (1997) have shown that the flow in the vicinity of the crests of steep waves is unstable, leading to overturning at values of ka much less than 0.4432. Such overturning is purely an irrotational phenomenon, and neither surface tension nor viscosity enter the picture (Longuet-Higgins and Cokelet, 1976). Longuet-Higgins (1985a) discusses the distinction between the real, Lagrangian and apparent Eulerian accelerations in a Stokes wave, and also shows that unsteady waves can reach vertical accelerations that are larger than those in steady Stokes wave of the same steepness, and hence become more susceptible to breaking. Longuet-Higgins (1972) shows that the profile of the limiting Stokes wave can be represented accurately by a simple expression of the form

$$d\zeta / dx = \tan x, \quad |x| < \pi/6 \tag{5.3.2}$$

by exploiting the analogy between a water wave and a grandfather clock (Longuet-Higgins, 1979)! Finally, Longuet-Higgins (1985b; see also Balk, 1996) has discovered quadratic Stokes identities that permit rapid and economical computations of steep Stokes waves and properties such as the mass transports and particle trajectories associated with them. These aspects are beyond the scope of this brief review, but suffice it to say that despite two centuries of theoretical work on water waves by powerful mathematicians and fluid dynamicists, there are still many surprises.

5.4 RESONANT WAVE–WAVE INTERACTIONS

The mere fact that the infinitesimal wave theory works well and can explain many salient features of surface waves is proof enough that the nonlinear effects are rather small in most cases. Otherwise infinitesimal surface wave theory would not have been such an outstanding example of success in the application of natural laws to an important physical process so early in the history of quantitative science that Sir Isaac Newton and others launched. Nevertheless, these small nonlinear effects happen to play a very significant role in the transfer of energy in spectral space and current research indicates that they play a dominant and not a secondary role in determining the shape of the wind-wave spectrum. In this section, we will explore how this rather paradoxical situation arises.

Take a simple linear oscillator with a natural frequency n (such as a simple pendulum). If this oscillator is forced externally by a very small periodic force of frequency ω and amplitude ε, then the governing equation for the amplitude of the oscillation is

$$\frac{d^2a}{dt^2}+n^2a=\varepsilon e^{i\omega t} \tag{5.4.1}$$

keeping in mind that by convention, only the real part is retained in the solution. The equilibrium solution and the exact solution at resonance ($n = \omega$) are

$$a_{eq}(t)=\frac{\varepsilon e^{i\omega t}}{\left(n^2-\omega^2\right)};\ a_{ex}(t)=\left[a_0-i\left(\frac{\varepsilon}{2n}\right)t\right]e^{int} \tag{5.4.2}$$

which illustrates the fact that unless the excitation frequency ω is close to the natural frequency n, the response of the system is also very small (Figure 5.4.1). Only resonant forcing can build up sufficient amplitudes, and the growth rate will be initially linear, as can be seen from Eq. (5.4.2).

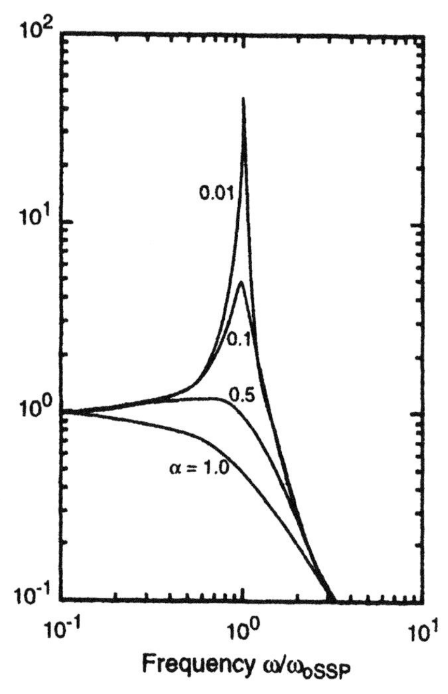

Figure 5.4.1. Response of a linear system to periodic forcing for different frictional damping values. Note the large response around the resonant frequency, that would be infinite in the absence of frictional damping (from Dietrich, G., K. Kalle, W. Krauss, and G. Siedler, *General Oceanography,* Copyright 1980, John Wiley. Reprinted by permission of John Wiley & Sons, Inc.).

Equations governing infinitesimal wave propagation are linear approximations obtained from the generally nonlinear governing equations by retaining only the zeroth-order terms in a perturbation expansion involving a small parameter ε such as the wave slope (see Section 5.2) and are best written as

$$L(\zeta) = \varepsilon N(\zeta) = \text{(Quadratic terms)} + \text{(Cubic terms)} + \ldots \qquad (5.4.3)$$

where L is the linear operator and N pertains to the neglected nonlinear terms involving products of the dependent variable ζ. Now consider two infinitesimal waves that satisfy the linear equation $L(\zeta) = 0$:

$$\zeta_m = a_m \exp i\left[\mathbf{k}_m \cdot \mathbf{x} - n_m t\right] \qquad (m=1,2) \qquad (5.4.4)$$

A note on notation is in order here. In Eq. (5.4.4) and below, boldface characters will be used to indicate that the quantities are two-dimensional vectors and any subscripts on these quantities refer to different wavenumbers, not components of a single wavenumber (unfortunately, retaining tensorial notation would lead to awkward notational complexities). The frequency and wavenumber of the waves must necessarily satisfy a dispersion relationship of the form $n_m = n_m(\mathbf{k}_m)$. In the linear theory, the nonlinear terms, of which the leading terms are of the form

$$a_1 a_2 \exp i\left[(k_1 \pm k_2)\cdot x - (n_1 \pm n_2)t\right] \tag{5.4.5}$$

that arise due to the interaction of these two infinitesimal waves are neglected and these modes propagate unchanged. However, to first order, the above terms constitute a small forcing of the linear system at wavenumbers $(\mathbf{k}_1 \pm \mathbf{k}_2)$ and frequency $(n_1 \pm n_2)$. The terms are small (of the order ε) since they are quadratic in amplitude of the infinitesimal waves. Now the response of the system will also be small, of the order ε, *unless* there exists a natural mode at this wavenumber and frequency,

$$k_3 = \mp(k_1 \pm k_2); \quad n_3 = \mp(n_1 \pm n_2) \tag{5.4.6}$$

in which case, the response is vigorous and the natural mode grows (initially linearly) at the expense of the interacting forcing modes. When the forced mode reaches its maximum amplitude, then the depleted forcing mode(s) can now extract energy by the same mechanism from the forced mode. The two forcing modes and the forced mode therefore constitute a triad of waves interacting with each other resonantly and exchanging energy among themselves. The total energy is conserved, but continuously exchanged among one another. In spectral space, this resonant interaction constitutes redistribution of energy in wavenumber–frequency space. Thus quadratic resonant interactions occur among a triad of waves whose wavenumbers and frequencies satisfy

$$\begin{aligned} k_1 \pm k_2 \pm k_3 &= 0 \\ n_1 \pm n_2 \pm n_3 &= 0 \\ n_i &= n_i(k) \qquad i = 1,2,3 \end{aligned} \tag{5.4.7}$$

We considered only the leading order nonlinear terms. The next higher order nonlinear terms can give rise to higher order interactions. For example, cubic resonant interactions are possible among a tetrad of waves whose wavenumbers and frequencies satisfy

$$\begin{aligned} k_1 \pm k_2 \pm k_3 \pm k_4 &= 0 \\ n_1 \pm n_2 \pm n_3 \pm n_4 &= 0 \\ n_i &= n_i(k) \qquad (i = 1,2,3,4) \end{aligned} \tag{5.4.8}$$

and so on. The question of whether quadratic or cubic or even higher order resonant interactions occur among waves depends critically on whether the dispersion relationship for these waves permits such solutions. For pure gravity waves, triad resonant interactions are not possible, but tetrad resonant interactions are. For capillary–gravity waves (and internal waves—see Chapter 6), triad interactions are possible. This can be shown by a simple geometric construction (Figure 5.4.2) due to McGoldrick (1965). The dispersion relation ship is such that the curve in frequency–wavenumber space is shaped like a trumpet with the lips at the outflow end bent back, of which Figure 5.4.2 shows

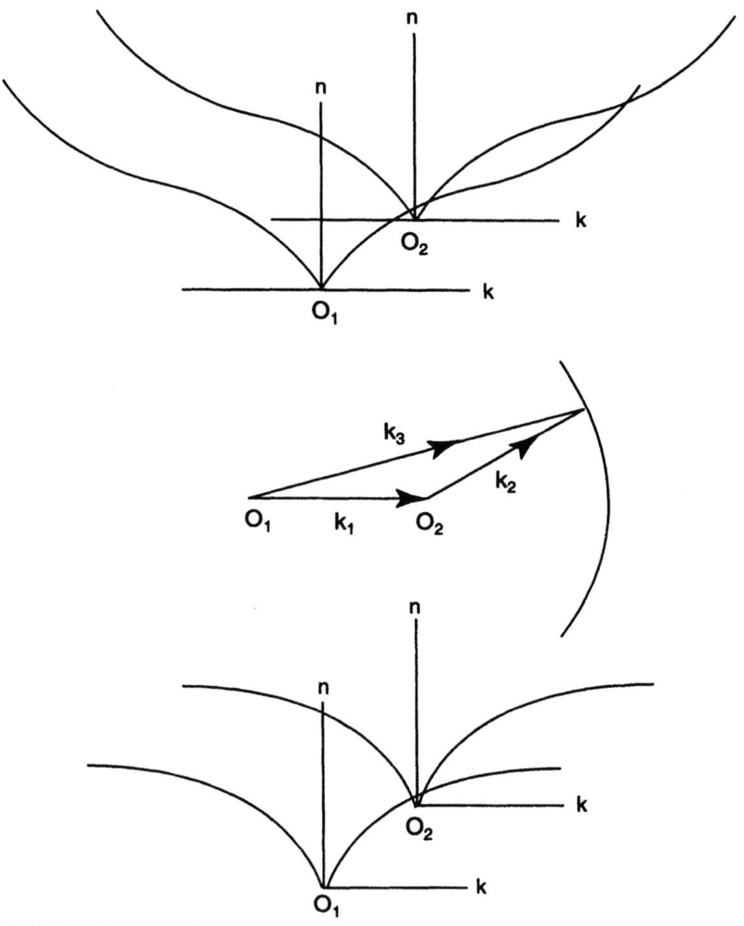

Figure 5.4.2. Triad resonant interactions among capillary waves (top) and the plan view of the triad that undergoes interactions (middle). The bottom figure shows why resonant triad interactions are not possible for pure gravity waves.

only a cross section. The projection onto the horizontal plane of the intersection in space of the two distorted-trumpet shapes defines the locus of wavenumbers that can interact resonantly with one another. In this case,

$$k_1 \pm k_2 = k_3 ; \qquad n_1 \pm n_2 = n_3 \tag{5.4.9}$$

If there is no intersection of the two, as, for example, for pure gravity waves (see Figure 5.4.2), resonant interactions are not possible, since the dispersion relationship does not permit them. When interactions are permitted, the dispersion relationship sees to it that the interactions are only among selected modes (that satisfy the dispersion relationship). This is in marked contrast to turbulence where there is no such restriction on modes interacting with each other. Any turbulence component in the wavenumber–frequency space can in principle interact with any other, although in practice the strongest interactions apparently occur among neighboring components, not distant ones (very much similar in this sense to interacting waves).

Because of the smallness of the nonlinear terms, resonant nonlinear interactions among waves are *weak* and hence very slow, and the timescale needed for significant energy exchange among modes is on the order of the wave period divided by the slope of the wave, ~$(n\,k\,a)^{-1}$. This is in marked contrast to turbulence where eddies (or modes) transfer a significant fraction of their energy through *strong* nonlinear interactions to other modes within a very short timescale on the order of their turn over the timescale or the period. The wave modes, on the other hand, lose a significant fraction of their energy only over many, many wave periods. This is one of the reasons that infinitesimal wave theories are so successful in describing many basic aspects of wave motions.

For capillary waves, as for all surface waves, the nonlinearity comes from the surface boundary condition, which is to first order (see Section 5.2).

$$\begin{aligned} &\frac{\partial^2 \phi}{\partial t^2} + g\frac{\partial \phi}{\partial x_3} - \gamma \frac{\partial^2}{\partial x_j \partial x_j}\cdot\frac{\partial \phi}{\partial x_3} = -\zeta \frac{\partial}{\partial x_3}\left(\frac{\partial^2 \phi}{\partial t^2} + g\frac{\partial \phi}{\partial x_3}\right) + \cdots \\ &\frac{\partial \zeta}{\partial t} - \frac{\partial \phi}{\partial x_3} = \zeta \frac{\partial^2 \phi}{\partial x_3^2} + \cdots \qquad \text{at } x_3 = 0 \end{aligned} \tag{5.4.10}$$

Assume solutions of the form

$$\begin{aligned} \phi &= \sum_{i=1}^{3} a_i(t)\exp i\left[k_i \cdot x - n_i t\right] + CC \\ \zeta &= \sum_{i=1}^{3} \frac{ik_i}{n_i} a_i(t)\exp i\left[k_i \cdot x - n_i t\right] + CC \end{aligned} \tag{5.4.11}$$

where the amplitudes a_i are slowly varying functions of time, and the wavenumber and frequency of each component satisfy the dispersion relationship. Note that k_i is the magnitude of the wavenumber $\mathbf{k}_i$. CC stands for the complex conjugate so that the terms on the LHS are real. It is then possible to derive equations for rate of change of amplitudes of each of the three interacting waves in the form (see Simmons, 1969; Phillips, 1977)

$$\frac{da_1}{dt} = ihn_1 a_2 a_3$$
$$\frac{da_2}{dt} = ihn_2 a_3^* a_1 \quad (5.4.12)$$
$$\frac{da_3}{dt} = ihn_3 a_1 a_2^*$$

The triad satisfies the resonance condition

$$k_1 = k_2 + k_3 ; \qquad n_1 = n_2 + n_3 \quad (5.4.13)$$

Coefficient h is a function of the wavenumbers involved and their geometrical configuration, whose calculation involves very tedious algebra which cannot be repeated here. Certain properties are worth pointing out. The rate of change of total energy among the triad is

$$\frac{d}{dt}(E_1 + E_2 + E_3) \cong \frac{d}{dt}\left(a_1 a_1^* + a_2 a_2^* + a_3 a_3^*\right)$$
$$= ih\left(n_1 - n_2 - n_3\right)\left(a_1^* a_2 a_3 - a_1 a_2^* a_3\right) = 0 \quad (5.4.14)$$

by virtue of the resonance condition. In other words, total energy is conserved and energy is simply exchanged among the waves.

Since the action density A = 2 (a a*/ n), it is easy to show that

$$\frac{d}{dt}(A_1 + A_2) = \frac{d}{dt}(A_1 + A_3) = 0 \quad (5.4.15)$$

Note that total action density is *not* conserved. If initially one component is zero ($a_1 = 0$, say), then Eq. (5.4.12) shows that it grows at an initial growth rate that is linear:

$$a_1 = ihn_1 a_2 a_3 \cdot t \quad (5.4.16)$$

Figure 5.4.3 shows both the energy and the action densities in each component as varying functions of time. Note that time T_q is many, many wave periods,

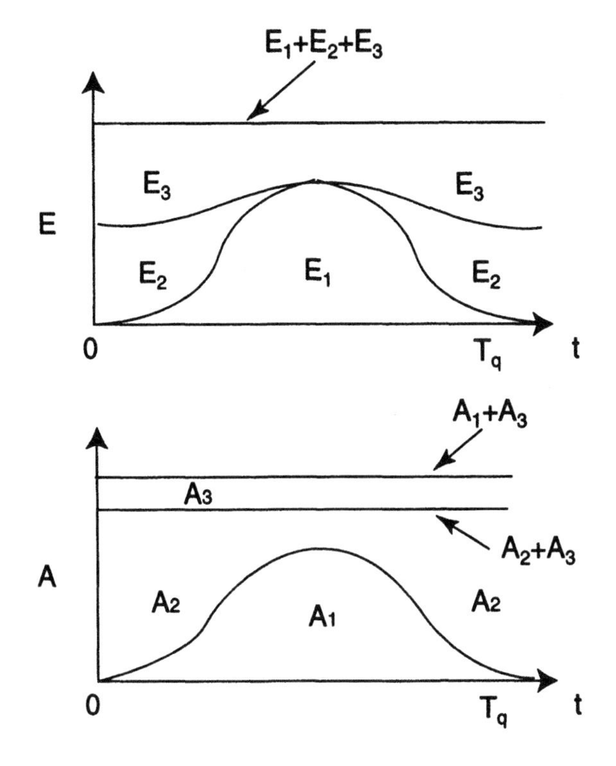

Figure 5.4.3. Interchange of energy density and action among the three waves during triad resonant interactions. Note that the energy density is conserved, but not the wave action.

being on the order of the characteristic wave period divided by the characteristic initial wave slope of, say, the components containing the energy initially.

Another consequence of such triad interactions is that under certain conditions, exponential growth is possible for an initially small wave component. If, for example, a_2 and a_3 are small initially, then

$$\frac{da_1}{dt} \sim 0$$
$$\frac{d^2 a_2}{dt^2} = \frac{d}{dt} \cdot \frac{da_2}{dt} \sim ihn_2 a_1 \frac{da_3^*}{dt} = h^2 n_2 n_3 a_1 a_1^* a_2 \tag{5.4.17}$$

so that

$$a_2 \sim \exp\left[\left(n_2 n_3\right)^{1/2} h a_1 t\right]$$

is a possible solution indicating possible exponential growth of a_2. A similar solution is obtained for a_3. If $k_2 = k_3 = k_1/2$, this is known as subharmonic parametric instability, in which a wave mode interacts with its second harmonic and decays rapidly, while the initially infinitesimal harmonic perturbation grows exponentially. This is one of the ways in which a wave train can become unstable to any perturbation at a frequency equal to its harmonic frequency, provided such resonant interactions are possible. A uniform wave train is degenerated into a triad, each which in turn can become unstable, so that ultimately a group of wave modes results. This "sideband instability" occurs for higher order interactions as well.

Capillary waves have high curvatures and slopes, and large wavenumbers and short periods. Consequently, the resonant interaction time is rather small. However, they dissipate rapidly as well so that it is the competing dissipation and resonant generation mechanisms that determine their configuration. One important mode of interaction among capillary waves is the interaction of two capillary modes with a gravity mode. Another involves a capillary–gravity mode (of about 2.4-cm wavelength) producing another of twice its own wavenumber (or half its wavelength) by resonant self-interaction (Simmons, 1969; see Phillips, 1974). The energy in the primary mode is then transferred to its second harmonic and the primary wave gets increasingly distorted. This is different from the sideband instability, since the primary wave interacts with itself to produce a harmonic, and not with two initially small harmonics. Such harmonic resonance is also possible at the next higher order, where a primary mode interacts with two of itself to produce a harmonic with thrice its own wavenumber. Such harmonic resonances are responsible for generating an entire ripple train over a longer gravity–capillary wave. For steep primary waves, the high curvature at the crest spawns capillary waves that are stationary with respect to the crest, but propagate energy forward of the crest (because their group velocity is higher than their phase velocity), thus providing an efficient mechanism for dissipation of the short gravity wave. These "parasitic" capillaries might be quite important for limiting the amplitude of such waves, since the energy input from the wind can be efficiently dissipated by capillary waves, preventing further growth (Longuet-Higgins, 1995). See also Longuet-Higgins (1992) for a discussion of the damping of short gravity waves by capillary waves.

As we have seen earlier, for pure gravity waves, resonant quadratic interactions among a triad are not permitted by the dispersion relation. However, cubic resonant interactions, which are much weaker than quadratic interactions,

$$\mathrm{L}(\zeta) = (\text{Cubic terms}) + \cdots \tag{5.4.18}$$

are possible among a tetrad of waves

$$\begin{aligned} k_1 \pm k_2 \pm k_3 \pm k_4 &= 0 \\ n_1 \pm n_2 \pm n_3 \pm n_4 &= 0 \end{aligned} \tag{5.4.19}$$

that satisfy the dispersion relation. This happens because the RHS forcing term now consists of triple sums and differences of the wavenumbers and frequencies of the three interacting primary waves, and this can couple efficiently to a free mode. The existence of tetrad interactions in pure gravity waves can be shown once again by geometrical construction (see Figure 5.4.4). Now the dispersion relation in the three-dimensional frequency–wavenumber space is described by a trumpet-shaped surface; the horizontal projection of the locus of its intersection with another but inverted trumpet with an origin not lying on its surface describes the wavenumbers that are permitted to undergo resonant interactions. The trajectories of the interacting tetrad of wavenumbers are shown in Figure 5.4.4. Any two points on a particular curve define a set of four wavenumbers that can undergo resonant interactions, exchanging energy and action density with one another:

$$\begin{aligned} k_1 + k_2 &= k_3 + k_4 \\ n_1 + n_2 &= n_3 + n_4 \\ n_i &= (gk_i)^{1/2} \qquad (i = 1,2,3,4) \end{aligned} \tag{5.4.20}$$

The existence of such resonant interactions was first shown by Phillips (1960) and Hasselmann (1962). These interactions are very weak, much weaker than triad interactions, simply because the nonlinear terms are cubic in amplitudes of interacting infinitesimal waves. The interaction time for significant energy exchange among components of the tetrad is even larger than that for a triad, it now being the characteristic wave period divided by the square of the wave slope. Cubic interactions are possible for capillary–gravity waves as well (see Phillips, 1974).

The algebra involved in computing the rate of change of amplitude of various components is even more tedious than that for the triad interactions and will not be repeated here (see Phillips, 1977, for example). The growth rates can be summarized as (Phillips, 1974)

$$\begin{aligned} n_1 \frac{da_1}{dt} &= ia_1\left(g_{11}a_1a_1^* + g_{12}a_2a_2^* + g_{13}a_3a_3^* + g_{14}a_4a_4^*\right) + iha_2^*a_3a_4 \\ n_2 \frac{da_2}{dt} &= ia_2\left(g_{21}a_1a_1^* + \cdots\right) + iha_1^*a_3a_4 \\ n_3 \frac{da_3}{dt} &= ia_3\left(g_{31}a_1a_1^* + \cdots\right) + iha_1a_2a_4^* \\ n_4 \frac{da_4}{dt} &= ia_4\left(g_{41}a_1a_1^* + \cdots\right) + iha_1a_2a_3^* \end{aligned} \tag{5.4.21}$$

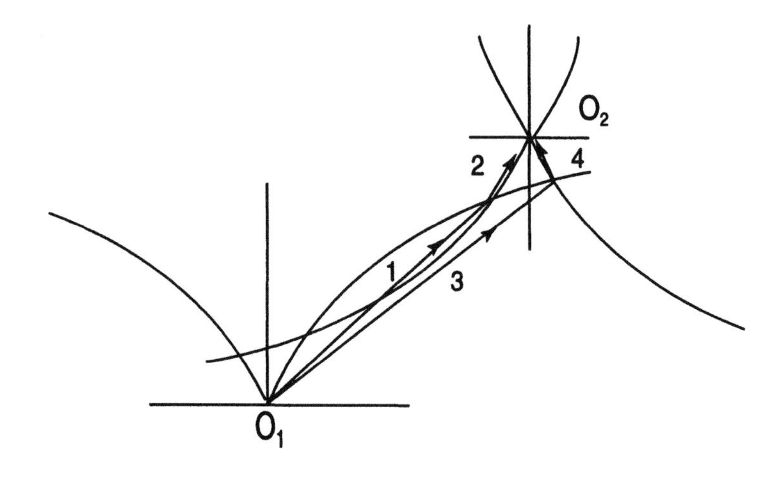

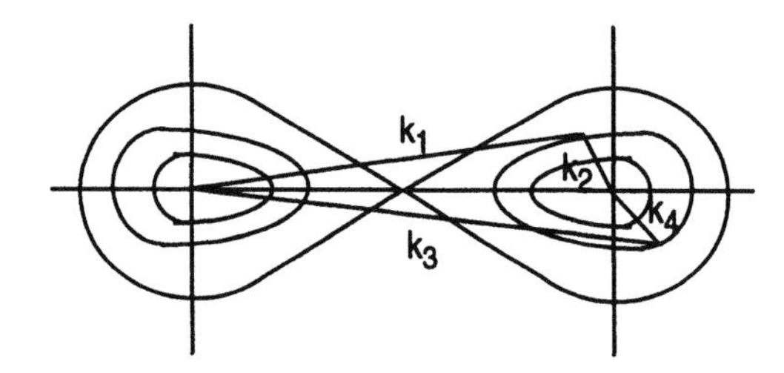

Figure 5.4.4. Tetrad resonant interactions among gravity waves (top) and the corresponding plan view of the wavenumber tetrad that describe a figure of eight loop.

where g_{ij} are complicated but real functions of the wavenumber configuration of the interacting tetrad. Important consequences of tetrad interactions can be deduced from these equations. For example, the interaction is such that the total energy *and* action density of the tetrad are conserved:

$$\frac{d}{dt}(E_1 + E_2 + E_3 + E_4) = \frac{d}{dt}(A_1 + A_2 + A_3 + A_4) = 0 \qquad (5.4.22)$$

$$\frac{d}{dt}(A_1 + A_3) = \frac{d}{dt}(A_2 + A_4); \frac{d}{dt} A_1 - A_2 = 0$$

At any point in time, two of the components have decreasing energy densities and their loss is the gain of the remaining two, which have increasing energy densities. The action density of each pair composed of one with increasing and one with decreasing energy density is conserved. Figure 5.4.5 shows the change

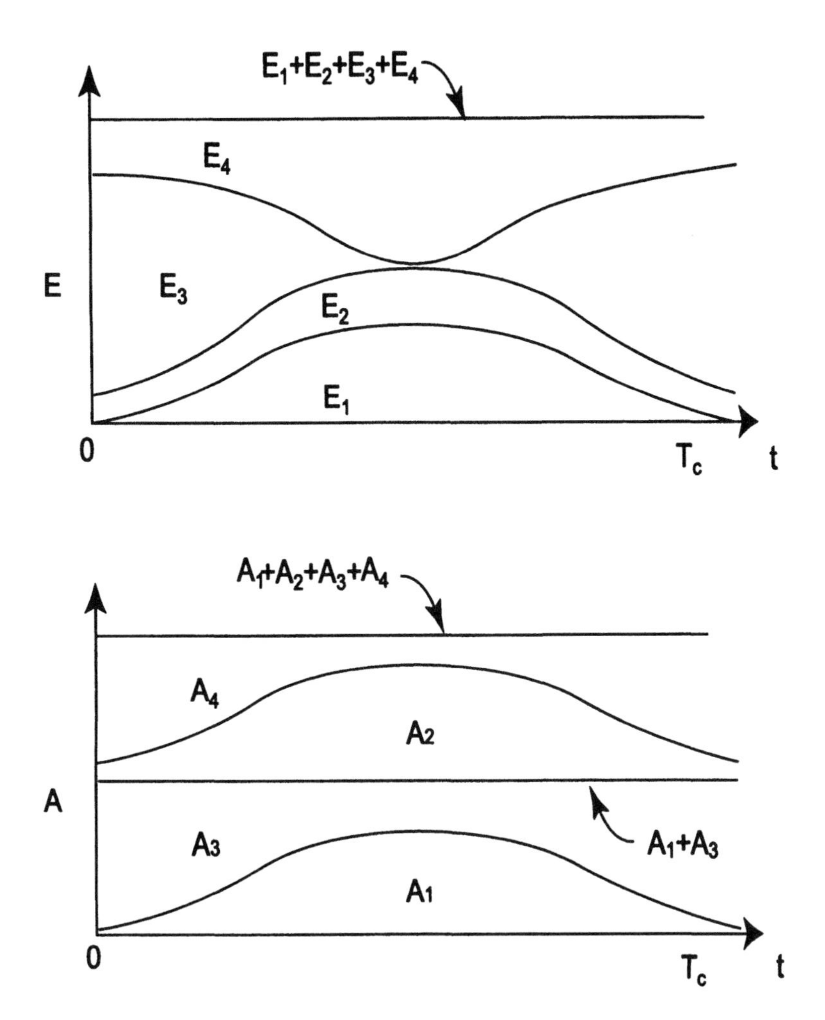

Figure 5.4.5. Interchange of energy density and action among the four waves during tetrad resonant interactions. Note that both the energy density and the wave action are conserved.

in action and energy densities of various components. Note that the cubic interaction time T_c is even larger than the quadratic interaction time T_q ($T_c >> T_q >> 1/n$). This time, during which there is a substantial change in the energy density of two of the components compared to the other two, is proportional to the characteristic wave period divided by the square of the characteristic wave slope [$T_c \sim (n\ a^2\ k^2)^{-1}$]. Therefore tetrad interactions among gravity waves are very weak and very selective. This does not mean they are unimportant. In fact, they play a dominant role in shaping the wind-wave spectrum (see Section 5.5). Even more important than the timescale T_c associated with energy (and action) transfer in the spectrum is the timescale associated with irreversible transfer of energy in a homogeneous wave field (Hasselmann, 1962) due to the fact that the waves are mostly weakly nonlinear so that their statistics are nearly Gaussian: $T_r \sim (n\, a^4\, k^4)^{-1}$. While T_c is on the order of a few minutes, T_r is on the order of an hour or so, and can be a non-insignificant fraction of the inertial period and hence important to the development of the energy-containing part of the wind-wave spectrum near its spectral peak, where resonant interactions are a dominant influence.

A particular case of interest is when two wavenumbers k_1 and k_2 are orthogonal. This situation is shown in Figure 5.4.6 and is well suited to simulation in a laboratory with two wave generators perpendicular to each other. The interaction condition is then

$$\begin{aligned} k_1 &= k_3, \quad k_2 + k_3 = 2k_1 \\ n_1 &= n_3, \quad n_2 + n_3 = 2n_1 \end{aligned} \tag{5.4.23}$$

Since initially the amplitudes a_1 and a_2 of the generated modes k_1 and k_2 are roughly constant,

$$\begin{aligned} n_3 \frac{da_3}{dt} &= \frac{i}{2} h a_1^2 a_2^* \\ a_3 &= \left(i h a_1^2 a_2^* / 2 n_3 \right) t \end{aligned} \tag{5.4.24}$$

This gives rise to a linear initial growth rate of the forced mode k_3. This experiment was used by McGoldrick *et al.* (1966) to confirm the occurrence and growth rate of tetrad resonant interactions. The wavenumber component k_3, initially zero, grew as a result of the interaction, and its growth rate was verified to be that given by theory (see Phillips, 1977).

By far, the most important case of cubic resonant interactions is that corresponding to the instability of a Stokes wave, discovered by Benjamin and Feir (1967). A perturbation to this wave at an adjacent wavenumber grows and the Stokes wave degenerates into a group of waves, an effect well known to laboratory modelers of surface waves. When both a_1 and a_2 are small compared to a_3, and thus are small perturbations to a_3, then both a_1 and a_2 can grow at the expense of a_3. If in addition, there is a slight mismatch from resonant conditions to compensate for the effects of amplitude dispersion, the disturbance amplitudes grow exponentially. This combination of near-resonance and amplitude dispersion is thought to be responsible for the Benjamin–Feir instability of a Stokes wave (see Phillips, 1977). Therefore a Stokes wave is unstable to

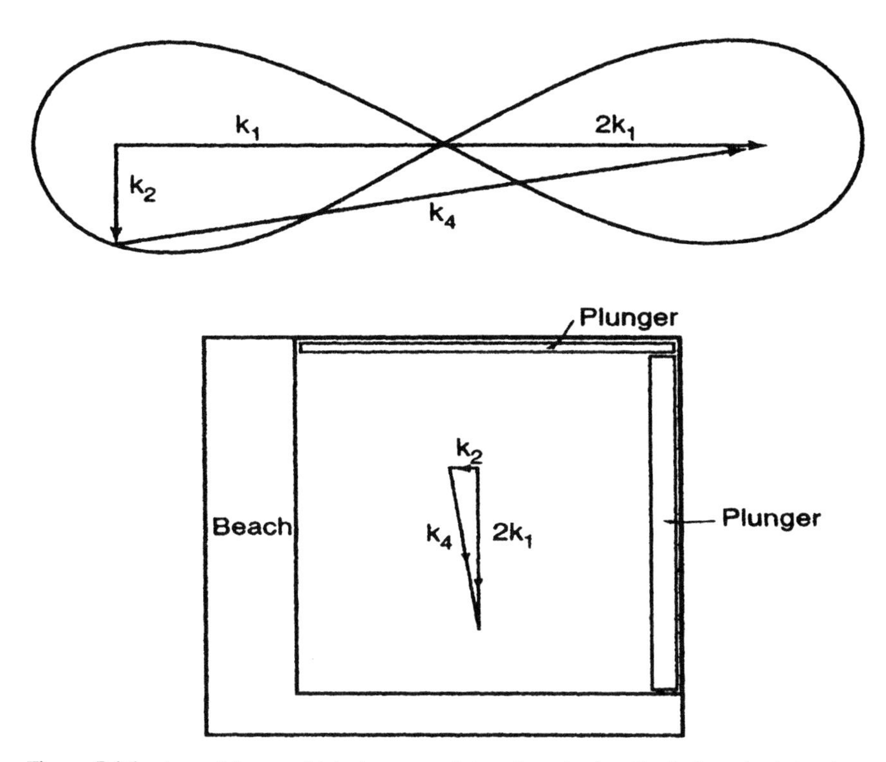

Figure 5.4.6. A special case of tetrad resonant interactions (top) well-suited to simulation in a laboratory wave tank (bottom).

perturbations at the wavenumbers just inside the figure-eight loop as shown in Figure 5.4.7, with the growth rate proportional to the square of the slope of the primary wave. The steeper the primary wave, the more rapid its degeneration into a series of groups. Thus as Phillips points out, a Stokes wave, proven to exist early this century, after lingering doubts about its existence for a while, has now been proven to be unstable to sideband perturbations (see Phillips, 1977, for a more detailed discussion)!

Resonant interactions among surface gravity waves are both selective and weak, yet they have very important consequences to transfer of energy in spectral space. These aspects are discussed further in Section 5.5.

5.5 WIND-WAVE SPECTRUM

Even to the most casual observer, it is apparent that wind waves on the ocean surface are random. They are simply the superposition of random waves with a

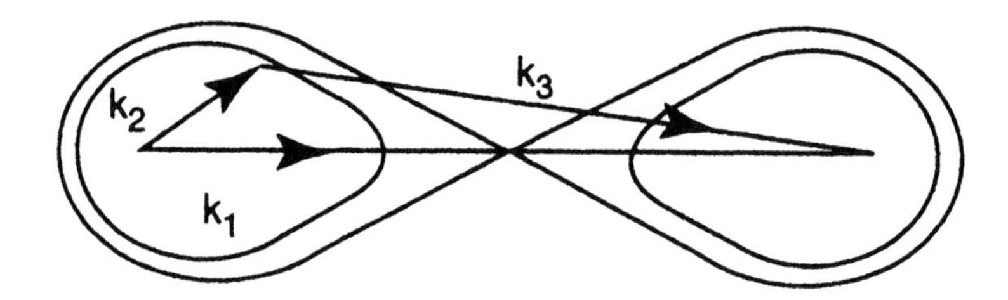

Figure 5.4.7. The wavenumbers that render the Stokes wave unstable.

wide variety of wavenumbers and frequencies with directions confined mostly to a small angle on either side of the wind direction. Because of their stochastic nature, a statistical description is unavoidable.

Let ζ (**x**,t) and ζ (**x**+**r**, t+τ) be the sea surface displacements at point **x** at time t and point (**x**+**r**) at time (t+τ). Their covariance is

$$R(x,t,r,\tau)=\overline{\zeta(x,t)\zeta(x+r,t+\tau)} \tag{5.5.1}$$

For a homogeneous and stationary wave field, R is an even function of **r** and τ. The wavenumber–frequency spectrum F(**k**, n) is the Fourier transform of the covariance R:

$$F(k,n)=(2\pi)^{-3}\iint R(r,\tau)\exp\left[-i(k\cdot r-n\tau)\right]dr\,d\tau \tag{5.5.2}$$

$$\overline{\zeta^2}=\iint F(k,n)\,dk\,dn \tag{5.5.3}$$

The variance of the sea surface displacement is the integral of F(**k**, n) over the entire wavenumber–frequency space. For a homogeneous, stationary wave field, F is real and positive definite for all **k** and n, and contains the most complete information on the wave field. The wavenumber spectrum ψ (**k**) can be obtained thusly,

$$\psi(k)=\int_{-\infty}^{\infty}F(k,n)\,dn \tag{5.5.4}$$

Also

$$\psi(k)=(2\pi)^{-2}\int\overline{\zeta(x,t)\zeta(x+r,t)}\exp(-ik\cdot r)\,dr \tag{5.5.5}$$

Therefore

$$\overline{\zeta^2} = \int_{-\infty}^{\infty}\int_{-\infty}^{\infty} \psi(k_1,k_2)\, dk_1 dk_2 = \int_{0}^{\infty}\int_{-\pi}^{\pi} \psi(k,\theta)\, k dk\, d\theta = \int_{0}^{\infty} \psi(k) dk \tag{5.5.6}$$

where $\psi(k_1,k_2)$ or equivalently $\psi(\mathbf{k})$, $\psi(k,\theta)$ are the directional wavenumber spectrum in Cartesian and polar coordinates, and $\psi(k)$ is the omnidirectional wavenumber spectrum given by $\psi(k) = \int_{-\pi}^{\pi} \psi(k,\theta)\, dk\, d\theta$. Note that we have used the same symbol to describe all three. This is not a problem since the arguments usually make it clear which of the three we are dealing with. Similarly the frequency spectrum can be obtained from $F(\mathbf{k}, n)$

$$\Phi(n) = 2\int F(k, n)\, dk \tag{5.5.7}$$

Note that the function is nonzero only for positive n values by this definition. Also

$$\Phi(n) = \frac{2}{\pi}\int_{0}^{\infty} \overline{\zeta(x, t)\zeta(x, t+\tau)}\, \cos(n\tau)\, d\tau \tag{5.5.8}$$

and

$$\overline{\zeta^2} = \int_{0}^{\infty} \Phi(n) dn \tag{5.5.9}$$

Now, since the wave components satisfy the dispersion relationship (in deep water $n^2 = gk$), then the various spectra are related. This requires that the short waves be freely traveling modes and there be no Doppler shift of short gravity waves by long waves near the peak of the spectrum due to advection of short waves by long ones. In particular, the wavenumber spectrum

$$\psi(k) = \psi(k,\theta) \tag{5.5.10}$$

is then related to the directional frequency spectrum

$$\Phi(n,\theta)=\left[k\psi(k)/\left(\frac{dn}{dk}\right)\right]_{k=n^2/g} \tag{5.5.11}$$

since

$$\overline{\zeta^2}=\int_0^{2\pi}\int_0^{\infty}\psi(k,\theta)\,k\,dk\,d\theta \tag{5.5.12}$$

$$=\int_0^{2\pi}\int_0^{\infty}\Phi(n,\theta)dn\,d\theta \tag{5.5.13}$$

It is also useful to define the wavenumber spectrum of the sea surface slope in the x_j direction:

$$S_\alpha(k)=(2\pi)^{-2}\int\overline{\frac{\partial\zeta(x)}{\partial x_\alpha}\frac{\partial\zeta(x+r)}{\partial x_\alpha}}\exp(-ik\cdot r)\,dr \tag{5.5.14}$$

$$=k_\alpha^2\,\psi(k) \tag{5.5.15}$$

The spectrum of total slope is then

$$S(k)=\sum_{\alpha=1}^{2}S_\alpha(k) \tag{5.5.16}$$

and therefore the total mean-square slope (mss) of the sea surface is related to the slope spectrum by

$$\begin{aligned}
\text{mss}&=\overline{\frac{\partial\zeta}{\partial x_\alpha}\frac{\partial\zeta}{\partial x_\alpha}}=\int S(k)\,dk\\
&=\int_{-\infty}^{\infty}\int_{-\infty}^{\infty}\left(k_1^2+k_2^2\right)\psi(k_1,k_2)\,dk_1dk_2\\
&=\int_0^{\infty}\int_{-\pi}^{\pi}k^2\psi(k,\theta)\,kdk\,d\theta=\int_0^{\infty}k^2\psi(k)dk\\
&=\int_0^{\infty}\int_{-\pi}^{\pi}k^2\cos^2\theta\,\psi(k,\theta)\,kdk\,d\theta+\int_0^{\infty}\int_{-\pi}^{\pi}k^2\sin^2\theta\,\psi(k,\theta)\,kdk\,d\theta
\end{aligned} \tag{5.5.17}$$

where the last equation simply sums up the contribution from the x_1 and x_2 directions. The spectrum of instantaneous traverse in the x_1 and x_2 directions is the Fourier transform of $\overline{\zeta(x)\zeta(x+r)}$, where the overbar indicates a spatial average, is

$$\chi(k_{1,2}) = \int_{-\infty}^{\infty} \psi(k_1, k_2)\, dk_{2,1} \tag{5.5.18}$$

It is normally the omnidirectional or 1D spectra that one deals with mostly, although for many applications such as microwave sensing of the ocean surface the directional spectra are very important. Similarly, since backscattered microwave radiation depends very much on the high wavenumber portion of the spectrum, in addition to the frequency and wavenumber spectra of heights, that of slope becomes important. Note that the omnidirectional slope spectrum S(k) is related simply to the omnidirectional height spectrum $\psi(k)$ by $S(k) = k^2\psi(k)$. A curvature (or so-called saturation) spectrum can also be defined, the omnidirectional form of which is $C(k) = k^3\psi(k)$.

It is also useful to consider action conservation in spectral space when dealing with linear waves and the action spectral density $N(\mathbf{k}_\alpha)$ can be defined as

$$N(k) = \frac{g}{n}\psi(k) = \left(\frac{g}{k}\right)^{1/2} \psi(k) \tag{5.5.19}$$

Action spectral density obeys the conservation law

$$\frac{d}{dt}N(k) = \frac{\partial}{\partial t}N(k) + (Cg_\alpha + U_\alpha)\frac{\partial}{\partial x_\alpha}N(k) = Y(k) + S_w(k) - D(k) \tag{5.5.20}$$

$$Y(k) = -\partial\left[T_\alpha(k)\right]/\partial k_\alpha \tag{5.5.21}$$

where $Y(\mathbf{k})$ represents the divergence of the spectral flux of action through wavenumber **k** by resonant wave–wave interactions. The term denotes redistribution in spectral space of action density, and its integral over the entire wavenumber space should vanish. S_w is the input of action by wind and D is the dissipation at wavenumber **k**. U_α is the background current. Equation (5.5.20) is the centerpiece of wind-wave development research and modeling (an alternative is to deal with a similar equation for energy density itself). It is a statement of the rate of change of action density of a wave packet of wavenumber **k** being advected in the fluid with an effective velocity that is the sum of the group velocity and the velocity of the background medium. If the RHS were zero, the action density of the wave packet would be constant. The source and sink terms on the RHS determine the effective shape of the wave spectrum. Unfortunately, these terms are hard to quantify, since even the basic physical mechanisms are often uncertain and the algebra involved in the derivation of Y(**k**) incredibly complex. However, considerable progress has been made in the past three decades; for example, Figure 5.5.1, from Komen *et al.* (1994), shows the relative magnitude of these sources and sinks for young and old wind seas.

Resonant interactions among surface gravity waves are both selective and weak, yet they have very important consequences for transfer of energy in spectral space. Unfortunately, the derivation of the rate of increase of the energy (or the action) density at a particular point in the spectral space from Eq. (5.4.21) is nontrivial. We will therefore state the results instead and refer the readers to Komen *et al.* (1994, Section II.3) instead. The rate of increase of spectral density of wave action for a wave of wavenumber $\mathbf{k}_1$ can be expressed in the form (Phillips, 1977; Komen *et al.*, 1994)

$$\frac{dN_1}{dt} = \iiint \{(N_1 + N_2)N_3N_4 - (N_3 + N_4)N_1N_2\} Q(k_1, k_2, k_3, k_4)$$
$$\delta(n_1 + n_2 - n_3 - n_4)\,\delta(k_1 + k_2 - k_3 - k_4)\,dk_2\,dk_3\,dk_4 \qquad (5.5.22)$$

where

$$N_i = N(k_i) = \frac{g}{n}\,\psi\,(k_i)$$
$$n_i = (gk_i)^{1/2} \qquad (5.5.23)$$

and Q is a function of the four wavenumbers. Dirac delta functions ensure that resonance conditions are satisfied. The integrand is cubic because of the cubic interactions. Note that the RHS of Eq. (5.5.22) is the term Y(k) in Eq. (5.5.20). Evaluation of Q is extremely tedious even by numerical means (see, for example, Hasselmann, 1962). However, numerical solutions have demonstrated the existence of asymptotically stationary solutions (Komen *et al.*, 1984). For a given spectral shape, it is possible to evaluate numerically the energy flux through the spectral space using the above equations. They show that the energy transfer in spectral space occurs primarily among groups of nearly identical wavenumbers. Also since the energy transfer is proportional to the fourth power of the average wave slope, the resonant interactions can be expected to be strong

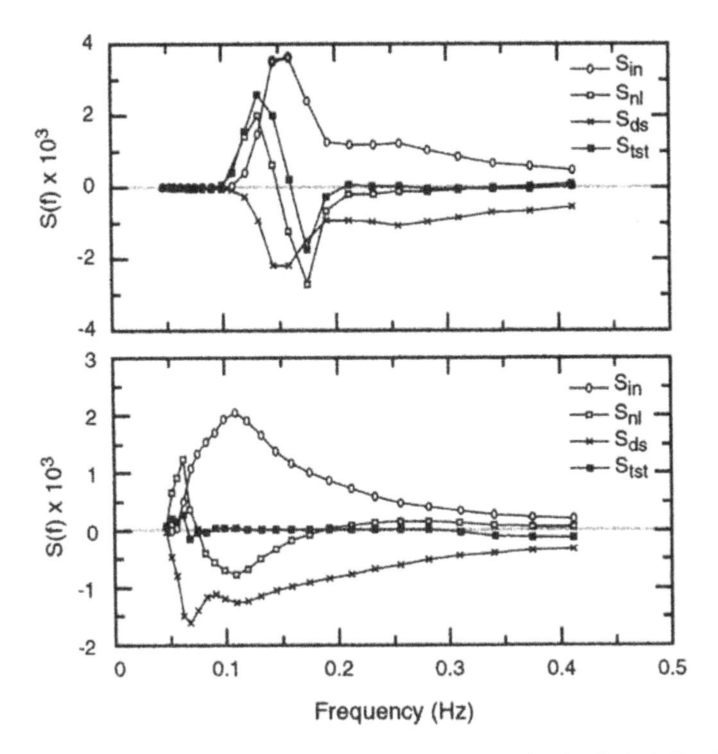

Figure 5.5.1. Energy balance in spectral space for (top) young duration-limited (T = 3 hr) and (bottom) old wind sea (T = 96 hr) (from Komen *et al.*, 1994). S_{in} is the input of energy by the wind, S_{nl} is the nonlinear transfer, S_{ds} is the dissipation of energy, and S_{tot} is the sum total of all three.

near the peak of the spectrum. When the four wavenumbers are nearly identical, then

$$Q=4\pi k_p^6 \tag{5.5.24}$$

and this leads to considerable simplifications in algebra. The principal result (see Figure 5.5.2) is that the energy flow is outward along the figure-eight lines as shown (Phillips, 1977).

Computations of resonant interactions necessarily involve starting from a spectral shape, since it is the primary factor controlling the intensity of transfer in spectral space. In general, any departure from "equilibrium" spectral shape results in a large increase in energy transfer by resonant interactions that tends to restore the spectral shape. Calculations have been done both for a narrow spectrum (Longuet-Higgins, 1976) and a JONSWAP (Joint North Sea Wave Project) spectrum (Fox, 1976). Both show energy transfer from a narrow spectral

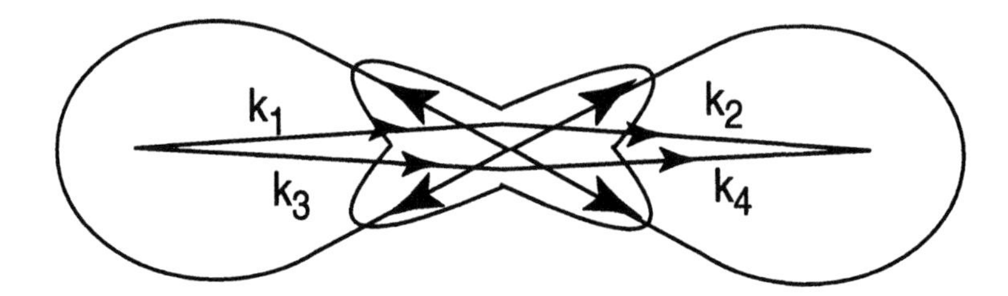

Figure 5.5.2. The energy flow during interactions between nearly identical wavenumbers.

peak to *both* lower and higher wavenumbers (Phillips, 1977), thus broadening the peak. The transfer of energy to the forward face of the spectrum by resonant interactions is very important to the growth of the lowest wavenumbers of the wind-wave spectrum, since the wind may not be very efficient in putting energy into these long wavelengths. The nonlinear interactions also tend to sharpen a broad peak, making it narrower. In either case, the peak wavenumber is lowered and the forward face of the spectrum grows (Donelan and Hui, 1990). This shift of the peak depends on the wave age; it is more significant for young seas. For old seas, it is small, since then the waves at the peak are running close to the wind speed and hence receiving little energy from winds that they can hand over to other parts of the spectrum.

Resonant wave–wave interactions play an important role in determining the shape of the wind-wave spectrum. Current theories for the equilibrium spectrum are based on postulating that the flux divergence in spectral space at a given wavenumber due to wave–wave interactions is at least as important as the input by wind, and dissipation at that wavenumber (Phillips, 1985). This leads to the so-called Toba –4 power law for the frequency spectrum of wind waves immediately beyond the spectral peak (see Section 5.5). This power law was first established theoretically by Zakharov and Filonenko (1966).

McLean (1982) showed that finite-amplitude waves in deep water are subject to instability as a consequence of five-wave interactions. He also found that this instability dominates resonant four-wave interactions in shallow water. Lin and Perrie (1997) have shown that in deep water, nonlinear transfer due to resonant tetrad interactions is two orders of magnitude larger than that due to five-wave interactions, but in shallow water it is comparable. When the water is less than 10 m deep and the waves are steep, as in near-shore regions of the coast, the five-wave interactions overwhelm the tetrad interactions, and become the dominant mechanism in wave evolution. Observationally, the process of transfer of momentum and energy from winds to waves is such that the spectrum appears to "saturate" at higher wavenumbers (frequencies), while continuing to grow at lower ones. While the wind can be expected to continue transferring energy to the "saturated" waves, the energy appears to be "lost" immediately by some process, thus limiting the spectral density at wavenumbers sufficiently far removed from the spectral peak.

Figure 5.5.3 shows the development of frequency spectra with increasing fetch as measured during the landmark JONSWAP experiments, displayed in a variance-preserving plot (Hasselmann *et al.,* 1973). It shows the typical growth of the spectral peak and its shift toward lower and lower wavenumbers with increasing fetch. It also shows that higher frequencies beyond the spectral peak appear to reach a "limit" and not grow beyond this limit irrespective of any further increase in fetch. This suggests that the details of the spectrum may not depend on the wind input under these "saturated" conditions, but more on an inherent property of the wave that causes breaking. Then external parameters such as u_* that characterize wind input drop out of the picture. If one assumes that the vertical accelerations have an upper limit which is a fraction of g, beyond which waves "break," then g becomes the only important external parameter in the problem. Then internal quantities such as the spectral density of the frequency spectrum (or the wavenumber spectrum) should depend only on g and the local spectral parameter n (or k). Simple dimensional arguments then yield the celebrated –5 power law of Phillips (1958) for the frequency spectrum,

$$\Phi(n) = \alpha g^2 n^{-5} \tag{5.5.25}$$

where α is the Phillips constant (with a value of around 0.015). Equivalently, since only length scales are involved, the wavenumber spectrum can depend only on k (not g) and hence

$$\psi(k_j) = \psi(k,\theta) = f(\theta) k^{-4} \tag{5.5.26}$$

The derivation of these spectral shapes assumed implicitly that the saturation spectral density does *not* depend on the wind speed or wind stress. Only the limiting configuration of waves would then be important in determining the wavenumber spectrum. Since the wavenumber and frequency spectra are related by the dispersion relation, the shape of the frequency spectrum ensues as well.

Early observations showed that while individual spectra often deviated from the –5 power law, the ensemble of a variety of data followed the –5 power law reasonably well (Figure 5.5.4). For 25 years, this spectral relationship was the basis of most analyses of wind waves and in fact was the basis of the empirically derived spectrum from the JONSWAP (Hasselmann *et al.,* 1973) experiments.

Kitaigorodskii *et al.* (1975) suggested that the existence of a saturated range can only be postulated for spatial statistical characteristics and therefore the

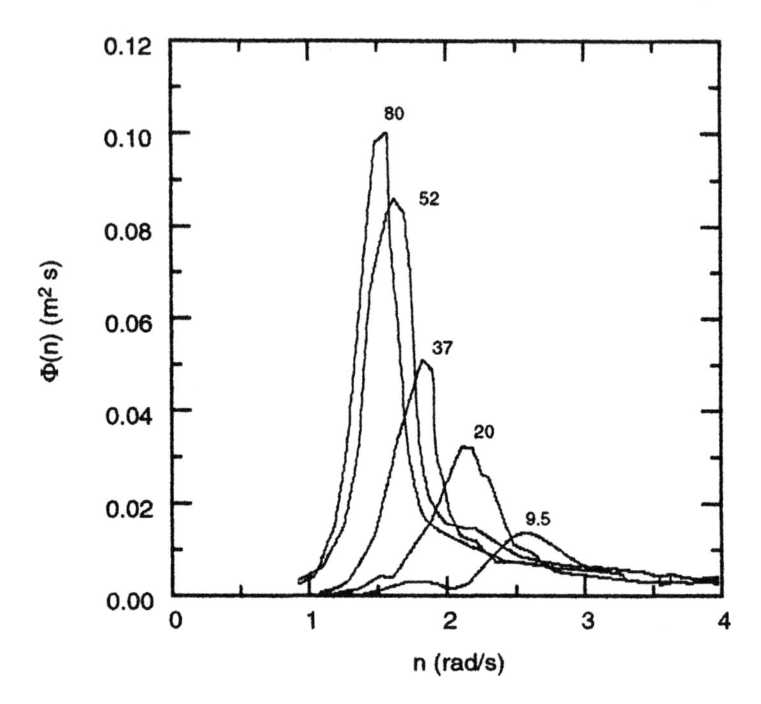

Figure 5.5.3. The development of the wave frequency spectrum with fetch, as measured during JONSWAP. Note the shift of the peak toward lower and lower wavenumbers with increasing fetch (in km) (from Hasselmann *et al.*, 1973).

wavenumber spectrum (thus yielding the k^{-4} power law for the wavenumber spectrum), while the frequency spectrum should be derived from the wave spectrum using the dispersion relation for gravity waves and the relationship between the wavenumber spectrum and the frequency spectrum [Eq. (5.6.11)]. For deep water waves this leads to the –5 power law for the frequency spectrum. However, Kitaigorodskii *et al.* (1975) showed that for shallow water waves, a –3 power law results:

$$\Phi(n) = \frac{\alpha}{2} g d n^{-3} \tag{5.5.27}$$

Experimental data in shallow water confirmed the possibility of deriving separate power laws for deep and shallow water frequency spectra from the sameuniversal wavenumber spectrum. However, determination of an accurate value for the Phillips constant proved elusive and Hasselmann *et al.* (1973) summarized data that well illustrated the variability of this constant (Figure 5.5.5). Attempts to explain the variability by accounting for the Doppler shift of short waves due to advection by ong waves and intermittency of short waves

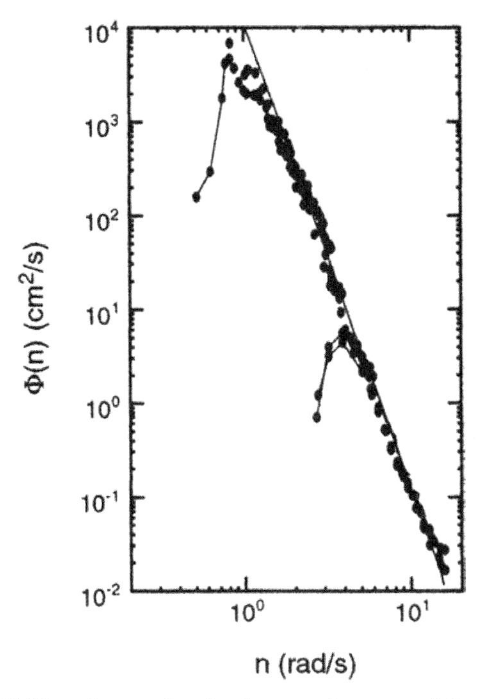

Figure 5.5.4. The Phillips –5 power law for the saturation range of the wind-wave spectrum, compared with field observations. Note the systematic departure of individual spectra from the –5 power law (from Phillips 1977, reprinted with the permission of Cambridge University Press). The shape of the spectral peak is included in only three cases; otherwise only the saturated part of each spectrum is shown.

proved fruitless. Also more careful measurements in the eighties began to cast doubts on the form of the frequency spectrum and hence the saturation range hypothesis itself. Toba (1973) found empirically that the frequency spectrum was better represented by a –4 power law,

$$\Phi(n) = \left(\frac{u_*}{c}\right)\beta g^2 n^{-5} = \beta u_* g n^{-4} \qquad (5.5.28)$$

where β is the so-called Toba constant (roughly 0.11). This is just u_*/c times the Phillips spectrum. Extensive observations by Kawai *et al.* (1977), and Forristall (1981), Kahma (1981a,b), and Donelan *et al.* (1982) in the eighties, confirmed that the frequency spectra beyond the spectral peak obeyed the –4 law much better and also showed a dependence on the wind (U or u_*). Figures 5.5.6 and 5.5.7 present data from Donelan *et al.* (1982) and Forristall (1981) that clearly show the existence of a –4 range in the frequency spectrum beyond the spectral

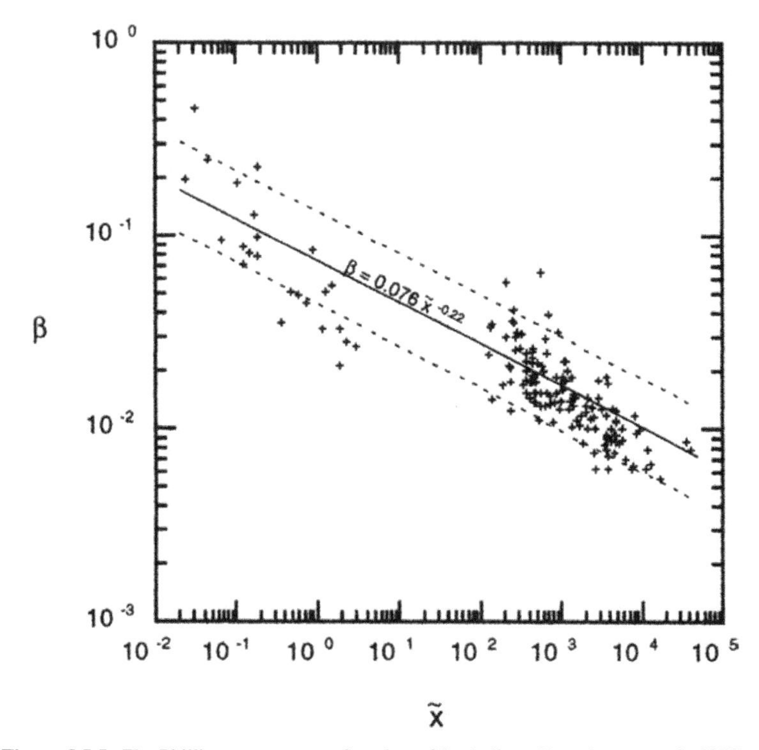

Figure 5.5.5. The Phillips constant as a function of fetch (from Hasselmann *et al.*, 1973).

peak. The idea of a "saturation" limit to spectral density had to be abandoned as untenable (Phillips, 1985).

Borrowing from Kolmogoroff's celebrated equilibrium range in turbulence theory, Kitaigorodskii (1983) took the lead in suggesting that there exists instead an "equilibrium" range beyond the spectral peak, where energy input from wind, dissipation, and spectral flux divergence all vanish. Wind is assumed to put in energy to large scales and dissipation is supposed to occur primarily at small scales, very much similar to turbulence, and it is supposed that there is just a constant spectral flux (directionally integrated) across the equilibrium range. If one takes this flux as a cubic in directionally averaged wavenumber spectrum density and then supposes that the flux ~ $u*^3$, dimensional analysis yields

$$\psi(k) = u_* g^{-1/2} k^{-7/2} = \frac{u_*}{c} k^{-4} \tag{5.5.29}$$

and the corresponding frequency spectrum (by virtue of the dispersion relation)

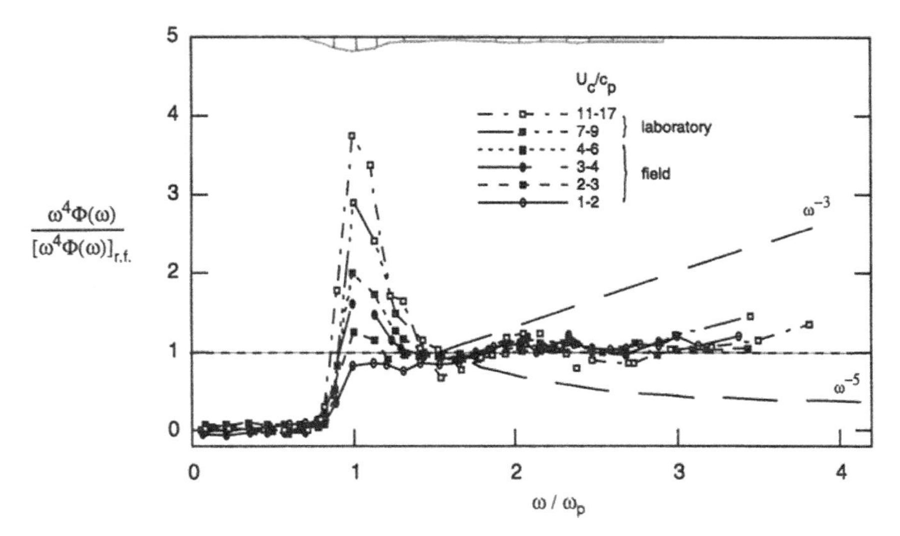

Figure 5.5.6. Compensated frequency spectra from observations showing a broad –4 power law region beyond the spectral peak (from Donelan *et al.,* 1985). The vertical bars at the top of the figure are an estimate of the 90% confidence limits based on the standard error of the mean.

$$\Phi(n) = \beta \frac{u_*}{c} g^2 n^{-5} \tag{5.5.30}$$

which is Toba's empirical spectrum.

A historical note is in order here. The ideas related to scaling the wind-wave spectrum in analogy with the Kolmogoroff inertial cascade approach to scaling the turbulence spectrum can be found in Russian literature far before Kitaigorodskii's (1983) work. Zakharov and his co-workers in Russia have been particularly active since the 1960s in the study of wave–wave interactions in a random wave field (Zakharov, 1968; see Komen *et al.,* 1994, for a discussion of these aspects) and the application of "weak turbulence" theory to the wind-wave spectrum (see references cited in Zakharov, 1992).

Phillips (1985) argued that Kitaigorodskii's hypothesis of wind input at low wavenumbers and dissipation at high wavenumbers, followed by energy transfer across the spectrum from large waves to small ones through the equilibrium range, is flawed, and instead proposed an alternative statistical equilibrium spectral model where the flux divergence, energy input, and dissipation are all important and nonvanishing and in balance with another in the equilibrium range [the LHS of Eq. (5.6.20) is zero]. He argued that action input from wind does not iscease at small scales (it is actually larger) and breaking occurs over a wide range of the spectrum as well. He therefore suggests that all three terms are proportional to each other in the equilibrium range away from the spectral peak

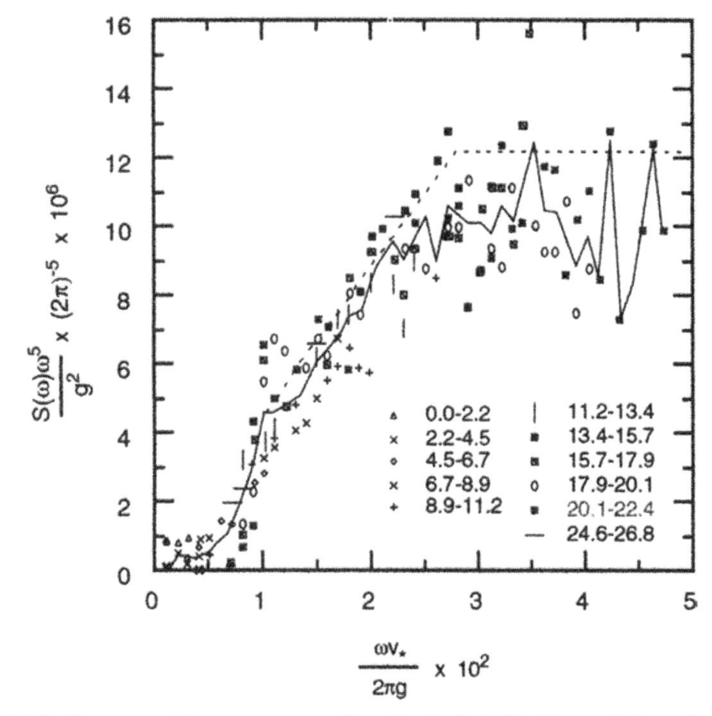

Figure 5.5.7. Compensated frequency spectra (from Forristall, 1981). The symbols are for various wind speeds in m/s.

and away from very small scales. He starts with the divergence of the spectral flux $T(k_j)$, and shows that this term scales like $B^3(k_j)$, where $B(k_j) = k^4\ \psi\ (k_j)$. He then argues that the wind input term

$$S_w\left(k_j\right) \sim g\,k^{-4}\left(\frac{u_*}{c}\right)^2 B\left(k_j\right)\cos^{2p}\theta \tag{5.5.31}$$

Requiring balance between these terms provides a value for $B\left(k_j\right)$,

$$B\left(k_j\right) \sim \frac{u_*}{c}\cos^p\theta \tag{5.5.32}$$

and therefore the wavenumber spectrum becomes

$$\psi\left(k_j\right) \sim \frac{u_*}{c}k^{-4}\cos^p\theta \tag{5.5.33}$$

Using the relationship between wavenumber spectrum and frequency spectrum [Eq. (5.6.11)], Toba's empirical spectrum results. Phillips goes on to derive many useful properties from the resulting spectrum. For example, the slope frequency spectrum is

$$S(n) \sim \beta u_* / g \tag{5.5.34}$$

Figure 5.5.8 shows the excellent agreement of the above spectral form with observational data. The normalized frequency spectrum has the form

$$\frac{\Phi(n) g^3}{u_*^5} \sim \left(\frac{n u_*}{g} \right)^{-4} \tag{5.5.35}$$

in Forristall's data, although for $(n u_*/g) > 0.17$, the behavior is more like $(nu_*/g)^{-5}$. This roll-off is evident for other data as well but at higher values of (nu_*/g). Phillips (1985) cites some other data that support the old saturation hypothesis. This suggested that while observations indicate that there exists an

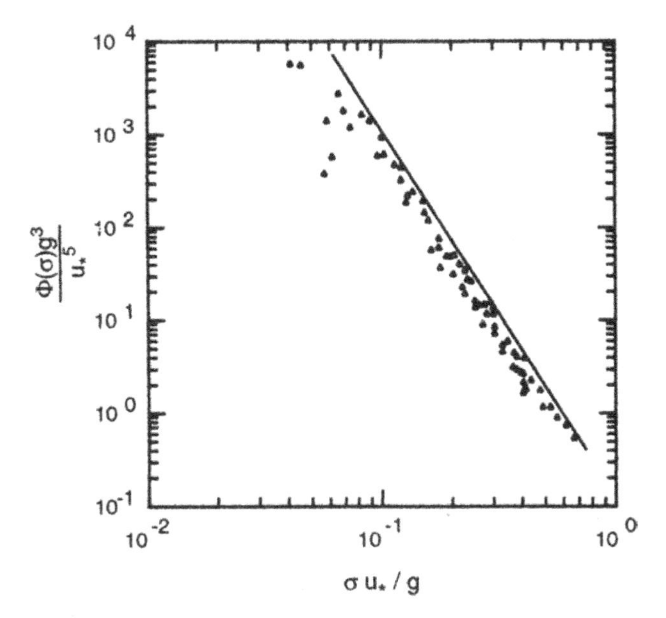

Figure 5.5.8. A dimensionless plot of the frequency spectrum in the equilibrium range. The wind wave spectrum shows the –4 power law (from Phillips, 1985).

equilibrium subrange in the wind-wave frequency spectrum beyond the spectral peak, there is no unequivocal support for the existence of the equilibrium range over the *entire* range of high wavenumbers. However, Phillips (1985) suggested that this equilibrium range exists up to wavenumbers characterizing significant advection effects by long waves, in other words, for all freely traveling waves.

Kitaigorodskii (1986) suggested instead that the wind-wave frequency spectrum consists of both the Kolmogoroff-type equilibrium and Phillips' saturation ranges. He cited evidence from Forristall (1981) (Figure 5.5.8) and Kahma (1981a,b) (Figure 5.5.9) that shows a sharp transition from the "inertial subrange" –4 law to the Phillips subrange –5 law at a frequency corresponding to $(n_t u_*/g) \sim 0.15–0.17$. This transitional frequency can be obtained by equating the spectral densities as given by the Phillips and Toba laws,

$$\Phi(n) = \alpha g^2 n^{-5} = \beta u_* g n \tag{5.5.36}$$

so that $(n_t u_*/g) = \alpha/\beta$, the ratio of the Phillips constant to Toba's constant! Carefully reevaluating the Phillips constant using data valid only in the

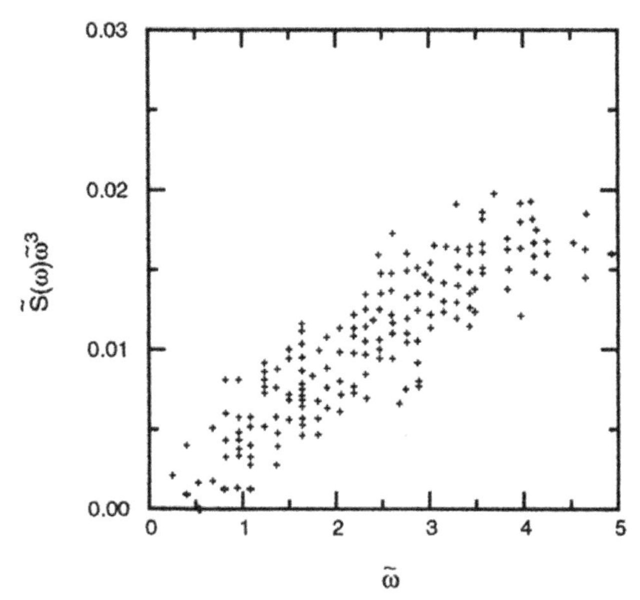

Figure 5.5.9. Compensated wind wave spectrum from Kitaigorodskii (1986) showing support for the saturation range.

saturation range of the spectrum, Kitaigorodskii (1986) derived a value of 0.015. Toba's constant is known to be ~0.11 (Forristall, 1981; Kahma, 1981a,b) so that this transitional value is ~0.14, quite close to the observed value of 0.17 reported by Forristall (1981).

This indeed appears to have been confirmed by more and more observations. Kahma and Calkoen (1992) collected frequency spectra from various field experiments including JONSWAP, the Bothnian Sea, and Lake Ontario, and after a very careful reanalysis, derived a "grand average" spectrum that indeed shows (Figure 5.5.10) a well-defined transition from Toba's equilibrium range spectrum to Phillips saturation range spectrum at $(nU_{10}/g) \sim 5$ or, equivalently, $(nu_*/g) \sim 0.17$ as originally suggested by Forristall (1981) on empirical, and Kitaigorodskii (1986) on theoretical, grounds. Thus as of present, after nearly 35 years since Phillips's discovery of the saturation range, the spectral shape at frequencies beyond the spectral peak might once again be on a solid footing! Since the n^{-4} (or $k^{-3.5}$) spectral shape continued to high frequencies the energy-(wavenumbers) would lead to excessive spectral densities, and since mean-square wave slopes would exceed those estimated from radar backscattering, the break to the n^{-5} (or k^{-4}) spectral shape at a frequency of $n \sim 3\ n_p$, that is beyond the energy-containing n_p to $3\ n_p$ range of frequencies, appears reasonable and supported by observations (Donelan and Hui, 1990). In the capillary–gravity range of the spectrum, microwave scatter measurements indicate high wind dependence and the capillary range is heavily damped by viscous forces.

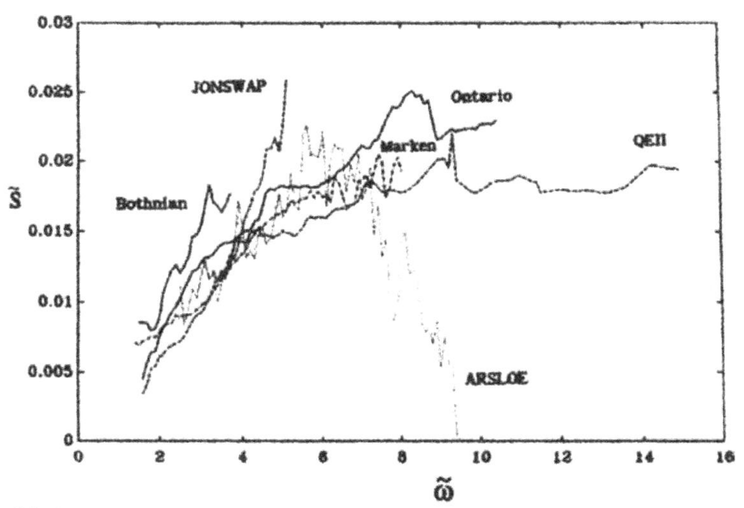

Figure 5.5.10. Grand average compensated spectrum from various observations showing transition from the Toba form to the Phillips form (from Kahma and Calkoen, 1992).

However, since the energy contained in these parts of the spectrum is usually a small fraction of the total, for many purposes (not microwave sensing), it is permissible to ignore these portions of the spectrum and consider the spectrum to consist of only the n^{-4} and n^{-5} segments beyond the spectral peak.

With the spectral shape known with reasonable certainty, one can calculate a variety of properties associated with wind waves (see Phillips, 1985, for example). If one assumes that the wind-wave spectrum is truncated at the low wavenumber end at the spectral peak n_p, and consists of the equilibrium range (5.5.28) from n_p to n_t and the saturation range (5.5.25) from n_t to ∞, integration of the spectrum readily leads to the energy density

$$E_d = \rho g \overline{\zeta^2}; \quad \frac{g^2 \overline{\zeta^2}}{u_*^4} = \frac{2\beta^4}{3\alpha^3} + \frac{\beta}{3}\left(\frac{c_p}{u_*}\right)^3 \tag{5.5.37}$$

where $\overline{\zeta^2}$ is the wave amplitude, and c_p is the phase speed of the spectral peak. The quantity c_p/u_* is the wave age; the larger its value, the more developed the wave field. The energy flux from breaking waves is proportional to $n_p E_d$ and can be written as

$$\frac{E_f}{\rho u_*^3} = \gamma_1 \left[\frac{2\beta^4}{3\alpha^3}\frac{u_*}{c_p} + \frac{\beta}{3}\left(\frac{c_p}{u_*}\right)^2\right] \tag{5.5.38}$$

where γ_1 is an unknown constant. Observations (for example, Phillips, 1977) suggest that

$$\frac{g^2 \overline{\zeta^2}}{u_*^4} \sim 0.0033\left(\frac{c_p}{u_*}\right)^4; \quad \frac{c_p}{u_*} \sim 2.2\left(\frac{gx}{u_*^2}\right)^{1/4}; \quad k_p^2 \overline{\zeta^2} \sim 0.0033 \tag{5.5.39}$$

with an upper bound on c_p roughly equal to U_{10}, the wind speed at anemometric height, so that the upper bound on c_p/u_* is $(c_d)^{-1/2}$, a value of roughly 26. This corresponds to the fully developed wave field.

$$\frac{E_f}{\rho u_\#^3} = \gamma_2 = 3.787\left[28.9\frac{u_*}{c_p} + 0.037\left(\frac{c_p}{u_*}\right)^2\right] \tag{5.5.40}$$

where $u_\#$ is the water-side friction velocity. $\gamma_2 \sim 100$ for a fully developed wave field. When waves break, they give up some of their energy to turbulence, which is eventually dissipated. However, momentum cannot be destroyed but is transmitted to currents instead. The associated momentum flux is E_f/c_p and can be written as

$$\frac{M_f}{\rho u_\#^2} = 0.133\left[28.9\left(\frac{u_*}{c_p}\right)^2 + 0.037\left(\frac{c_p}{u_*}\right)\right] \tag{5.5.41}$$

Note that we have used a composite Toba–Phillips spectrum. Use of the Toba spectrum alone (5.5.28) and the Phillips spectrum alone (5.5.25) is possible and leads to the following expressions for the normalized squared amplitude and energy flux:

$$\frac{g^2\overline{\zeta^2}^P}{u_*^4} = \frac{\alpha}{4}\left(\frac{c_p}{u_*}\right)^4; \quad \frac{E_f^P}{\rho u_\#^3} = \gamma_3 = 0.0057\left(\frac{c_p}{u_*}\right)^3$$

$$\frac{g^2\overline{\zeta^2}^T}{u_*^4} = \frac{\beta}{3}\left(\frac{c_p}{u_*}\right)^3; \quad \frac{E_f^T}{\rho u_\#^3} = \gamma_4 = 0.148\left(\frac{c_p}{u_*}\right)^2 \tag{5.5.42}$$

The values of these normalized energy fluxes are also shown in Table 5.5.1 in parentheses. While it is clear that what spectral shape one uses influences the TKE flux from breaking waves, the composite spectrum is at present the most defensible of the three.

We have so far looked at the wind-wave spectrum from the perspective of its frequency content. Such a focus on the frequency spectrum ignores the directional properties of the wind-wave field. An alternative approach, thought by some to be more fundamental (for example, Kitaigorodskii *et al.*, 1975; Phillips, 1977) involves dealing with the wavenumber spectrum of the wind

TABLE 5.5.1
Values for Normalized Energy and Momentum Fluxes as Functions of the Normalized Phase Speed and Fetch

C_p/u_*	gx/u_*^2	$E_f(E_f^P, E_f^T)/\rho u^3$	$M_f/\rho u^2$
5.0	1.5×10^4	0.8(0.7, 3.7)	0.005
12.0	4.6×10^5	20.2(9.8, 21.3)	0.059
26.0	1.1×10^7	100(100,100)	0.134

waves, including their directional spreading, which is a function of the frequency. Basically, the argument that the spectral shape of the wavenumber spectrum follows k^{-4} law underpins this approach. Assuming that this certainly holds in the mean wave propagation direction [with some supporting evidence, for example, Donelan et al. (1996)], Banner (1990a) has shown that the frequency spectrum is also a function of the directional spreading of wind waves. While the postulates concerning the shape of the frequency spectrum are appealing, especially in view of their simplicity, the situation is more complex and it is therefore prudent to regard the postulated shape of the frequency spectrum as an approximation to reality. Because of the incredibly complex processes that govern the details of the distribution of energy in the wavenumber–frequency spectrum of wind waves, much work lies ahead.

In microwave remote sensing of the sea surface, precise knowledge of the high wavenumber part of the wind-wave spectrum (short gravity and gravity–capillary waves) that contributes heavily to the sea surface roughness (and transfer of energy from the wind to the ocean) is essential. Since the small scale capillary and gravity–capillary waves are affected by long waves (mainly swell), the low wavenumber part is also important. Directional information is also needed. Also, an analytical form for this directional wavenumber spectrum is helpful. With these in mind, Elfouhaily *et al.* (1997) have derived an analytical form for the directional wavenumber spectrum that satisfies Cox and Munk's (1954) results for the mean-square slope and laboratory measurements of the gravity–capillary wave curvatures. The reader is referred to Elfouhaily *et al.* (1997) for a discussion of the various forms for the directional wavenumber spectra proposed in the past.

Quite recently, Makin and Kudryavtsev (1999) and Kudryavtsev *et al.* (1999) have examined the theoretical basis for deriving the spectrum of short wind waves (few millimeters to few meters) and their implications to air-sea coupling. Their model assumes that the capillary range of the spectrum is determined by parasitic capillaries, which drain energy from short gravity waves through nonlinear cascade of energy (which is made up by input to the short gravity waves), which is in turn dissipated by molecular viscosity. This balance between energy input from very short gravity waves and viscous dissipation determines the shape of the capillary range of the spectrum as proportional to $u^3_* \nu^{-1} \gamma^{-1} k^{-2}$. The k^{-2} shape of the spectrum in this range agrees with measurements of Jahne and Riemer (1990) and Zhang (1998). The reader is referred to the above references for an idea of the approach needed to model the sources and sinks of wave energy in spectral space and derive the shape of the spectrum, which in turn can be used to parameterize the fraction of momentum deposited by wind into wind waves and hence unavailable to drive the upper ocean directly (and a list of recent references on the topic).

5.6 SIMILARITY THEORY FOR GROWTH OF WIND WAVES

Winds over the ocean are seldom steady or unidirectional over long periods of time. The spatial and temporal scales of wind variability are quite large and complex, and this is responsible for one of the difficulties in quantifying the oceanic wave field for operational purposes. Since the wave field at a given location and time is a function of the history of the wind forcing, not just at that location but over a broad region encompassing the point, numerical models of wind-wave growth driven by surface winds, with accurate temporal and spatial structures deduced by some means, are indispensable for reliable hindcasts and forecasts of the wave field. We will consider the state of wind-wave prediction the next section. Here we will attempt to quantify the growth characteristics of wind-generated waves from similarity theory and field observations.

Consider the case of a steady wind that has been blowing offshore in a direction perpendicular to the shoreline of a long straight coast. Assume that thewind has been blowing long enough for the wave field at any point to have reached its asymptotic growth limit. This is one of the best conditions to study wind-wave growth under, since the wave growth depends principally on the fetch, or the distance from the shoreline, and time drops out of the picture. Such waves are called fetch-limited (as compared to duration-limited ones where the fetch is unlimited, but the duration for which wind has been blowing over that fetch has been short), and afford one of the best chances to study, understand, and quantify wind-wave growth. JONSWAP measurements (Hasselmann *et al.*, 1973) off the German coast were made under these conditions. The only external parameters in the problem then are the wave fetch x, the wind stress u_* (or equivalently wind speed U_{10} at the anemometric height of 10 m), and the gravitational acceleration g. Simple similarity theory (scaling argument) can then be used to relate the internal parameters of wave growth such as the energy density of the waves E and the location of the spectral peak n_p (or equivalently the phase speed of waves at the spectral peak c_p) to these external parameters (Kitaigorodskii, 1962). Then the nondimensional energy density $\tilde{E}$ (or $\tilde{E}^*$), the peak frequency $\tilde{n}_p$ (or $\tilde{n}_p^*$), and the frequency spectrum $\tilde{\Phi}$ (or $\tilde{\Phi}^*$) depend only on the nondimensional fetch $\tilde{x}$ (or $\tilde{x}^*$):

$$\tilde{E} = \frac{g^2}{U_{10}^4}\left(\frac{E}{\rho g}\right) = f(\tilde{x}) \sim P\tilde{x}^p$$

$$\tilde{n}_p = \frac{n_p U_{10}}{g} = g(\tilde{x}) \sim Q\tilde{x}^q$$

$$\tilde{\Phi}=\frac{g^3\Phi}{U_{10}^5}=h(\tilde{x},\tilde{n})$$

$$\tilde{x}=\frac{gx}{U_{10}^2};\ \tilde{n}=\frac{nU_{10}}{g} \tag{5.6.1}$$

Quantities with asterisks indicate normalization by u_* instead of U_{10} and similar relations would hold. Note that $E\sim\int_0^\infty \Phi(n)dn$. For duration-limited wave growth, the independent variable is $\tilde{t} = gt/U_{10}$. For a fully developed wave field, when a uniform, steady wind has been blowing long enough over an unlimited fetch, these functions should asymptote to constant values.

Both U_{10} (it should be the value at a distance sufficiently far away from the wave influence, strictly speaking, for similarity arguments to hold) and u_* have been used for scaling interchangeably in the past. Observationally, it is easy to measure U_{10}. Invariably u_* has to be deduced from U_{10}, using a drag law. While U_{10} and u_* are related by the drag law, the relationship is affected by ambient stratification, and since this is often not measured, it complicates the task of deriving reliable estimates for wave growth (Kahma and Calkoen, 1992) and contributes to the scatter. When u_* is actually measured, it is still the best to use. In other cases, it is still not clear what is the best and most accurate variable for practical use: U_{10}, u_*, or some other measure of the wind, such as the value at a height of half the dominant wavelength, as proposed by some. Use of an internal parameter such as the phase speed (or equivalently the frequency) at the spectral peak as the normalizing variable, as some Japanese authors do, is possible and avoids this problem. Figure 5.6.1 shows various wind-wave growth data that exhibit a power law relationship between $\tilde{E}$ and $\tilde{x}$ over a wide range of fetches, but the constant of proportionality from Bothnian Sea experiments differs from that from JONSWAP. This prompted Kahma and Calkoen (1992) to carefully reexamine and reanalyze several data sets:

1. Subset of JONSWAP (Hasselmann *et al.,* 1973; Muller, 1976)
2. Bothnian Sea (Kahma, 1981a,b)
3. Lake Ontario (Donelan *et al.,* 1985)
4. QE II reservoir (Birch and Ewing, 1986)
5. ARSLOE (Rottier and Vincent, 1982)
6. Lake Marken (Bouws, 1986)

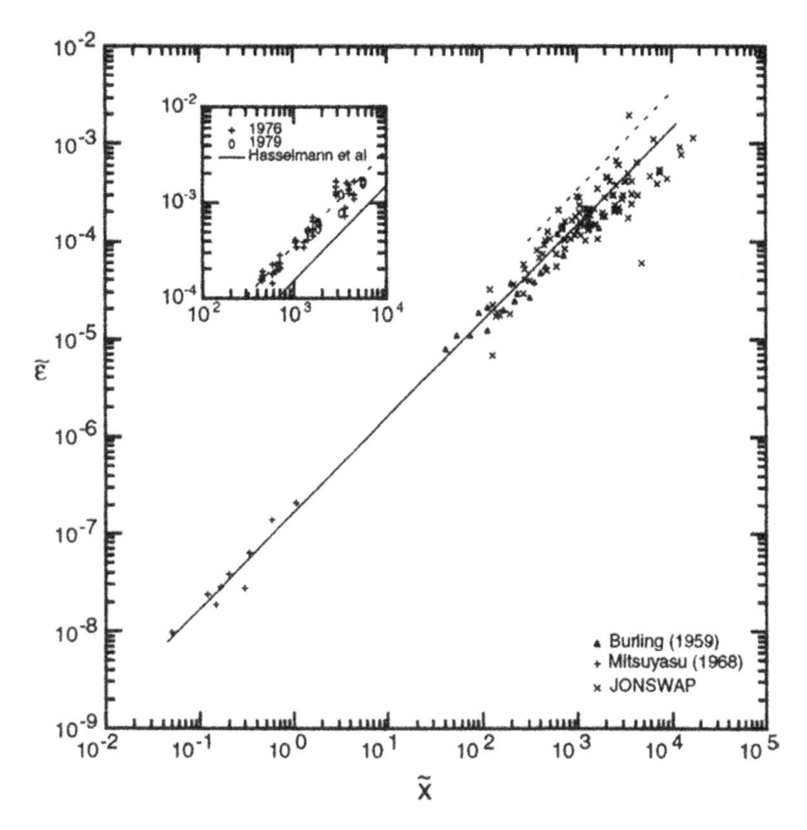

Figure 5.6.1. The dependence of normalized spectral energy on normalized fetch showing discrepancies in the proportionality coefficient (from Kahma and Calkoen, 1992).

Now, what the ocean feels is the stress exerted by the winds at its surface. So, it is not unreasonable to expect that the friction velocity, when measured accurately or deduced by drag laws that properly account for stability effects in the atmosphere, should be the appropriate scaling parameter irrespective of the stratification. Scaling based on wind velocity would be more affected by stratification. However, in many cases, u_* and all the parameters needed to deduce the stratification effects were never measured. Kahma and Calkoen (1992) found that when u_* was deduced by a consistent drag law accounting for stratification effects, the disagreement between observations taken under stable and unstable stratification conditions could be reduced by the use of u_*, instead of U_{10}. Nevertheless, it was not possible to remove all the discrepancies and they recommend instead the following for stably stratified, unstably stratified, and composite data sets.

$$
\begin{array}{llll}
P=9.3\times10^{-7}, & p=0.76; & Q=12.0, & q=-0.24 \text{ (stable)} \\
=5.4\times10^{-7}, & =0.94; & =14.2, & =-0.28 \text{ (unstable)} \\
=5.2\times10^{-7}, & =0.90; & =13.7, & =-0.27 \text{ (composite)}
\end{array}
\tag{5.6.2}
$$

A note of caution is in order here. The results in Eq. (5.6.2) are based on failure of u_* as the scaling variable, which is difficult to understand, and hence should be regarded as tentative. Figure 5.6.2 shows the normalized energy and frequency peak as a function of the nondimensional fetch. These relations are

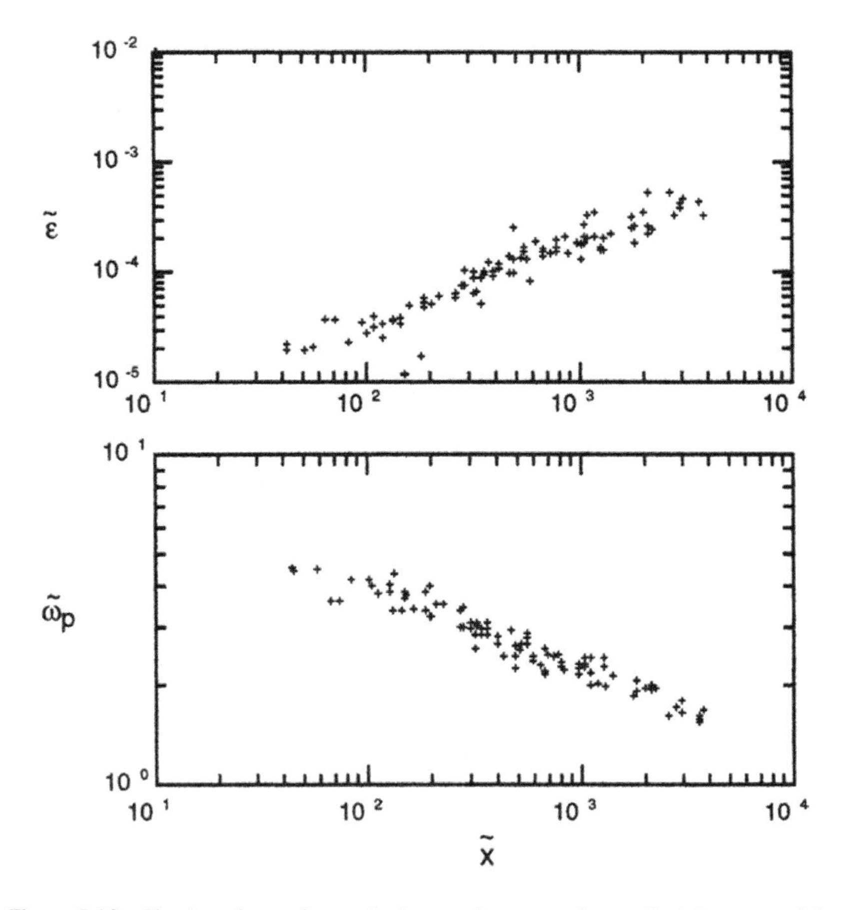

Figure 5.6.2. The dependence of normalized spectral energy and normalized frequency of the spectral peak on normalized fetch (from Kahma and Calkoen, 1992).

only valid for moderate fetch ($\tilde{x}$ < 8000). When the wind-wave field is close to full development, the growth rate slows down dramatically and both E and n_p tend to reach asymptotic values ($c_p \rightarrow U_{10}$).

Observations also show an overshoot in the spectral density at any particular frequency (Figure 5.6.3). Initially the growth at a frequency is steep; the spectral density overshoots its equilibrium value before settling down to the equilibrium value. This is a particularly important feature of the wind-wave spectrum. Figure 5.5.6 shows that the peak enhancement depends on the wave age. This is an important aspect of empirical spectra such as the JONSWAP spectrum (see Section 5.8).

Kitaigorodskii's similarity scaling laws have been helpful in systematic analyses of wind-wave observations. As discussed above, they were the scaling basis for experiments such as JONSWAP. Nevertheless, it is important to remember some of the limitations. Wind conditions are never steady in the field and the gustiness and the directional changes affect the wave field. Any prevailing currents would also affect the spectrum. While these effects could be included in the scaling arguments, there is usually a dearth of observational data to infer the precise form of the functional relationships.

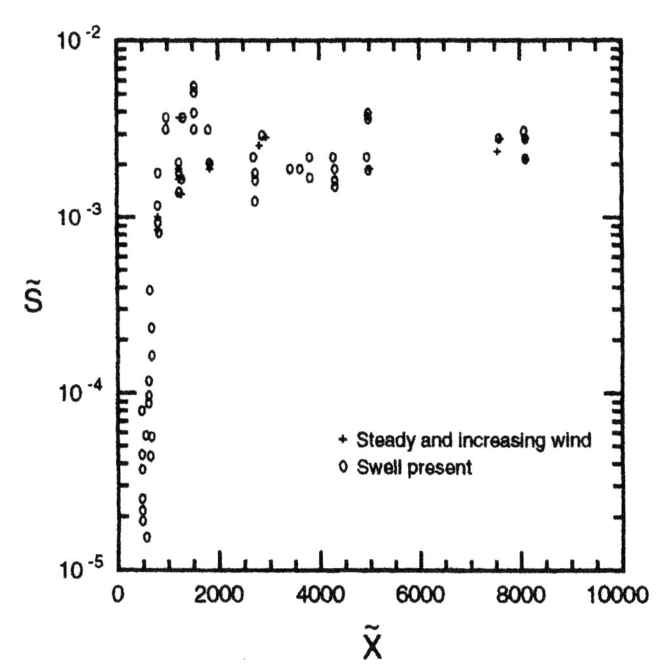

Figure 5.6.3. The overshoot in spectral energy density at a fixed normalized frequency (from Kahma and Calkoen, 1992).

5.7 WIND-WAVE GENERATION, DISSIPATION, AND PROPAGATION

5.7.1 GENERATION

While it is clear to even the most casual observer that the wind generates waves, the details on how the wind transfers momentum to waves are still pretty much sketchy. This is simply because this transfer involves an extremely complex interaction between the turbulent wind field over the mobile waves, with significant feedback from the waves on the wind field itself. In other words, the problem is coupled. Jeffreys (1924) was the first one to propose a mechanism for wind-wave generation. He suggested that the air flow over a wave creates a sheltering effect on the downwind face so that pressure acting on the upwind face can transfer momentum to the wave. The mechanism accounts for both wave growth by wind as well as its decay in adverse wind conditions. The problem is, however, transferred to calculating the sheltering coefficient. The mechanism itself was disputed and not much progress was made until the late fifties, at which time Phillips (1957) proposed a resonant mechanism for generation. The pressure fluctuations due to advection of turbulent eddies by the wind past the air–sea interface can efficiently couple with a free mode if the relevant wavenumber and frequency satisfy the dispersion relationship. Thus resonant excitation of wind waves is possible simply due to turbulent pressure fluctuations. The growth rate is linear. The mechanism appears to prevail at the very initial stages of wind-wave generation, when the amplitudes are quite small (Phillips, 1977; Kahma and Donelan, 1988). However, this mechanism does not explain the exponential rate observed in wind-wave growth immediately thereafter. This problem was solved to some extent by Miles (1957) who proposed a mechanism where the shear flow over the waves transmits energy to waves via critical layer processes.

Miles boldly ignored the turbulent aspects of the wind and considered only the laminar shear flow, which because of the gradient of shear, transfers energy from the winds to the wave through a critical layer that exists when the frame of reference is fixed to the moving wave (Figure 5.7.1). Because the mechanism is of the nature of an instability, the growth rate is exponential. Miles' mechanism presupposes existence of the wave, to which wind then puts in more energy to make it grow. In a way, the Phillips and Miles mechanisms are complementary and sequential. However, the growth rates were originally thought to be underestimated by a factor of 2 to 3 by Miles' mechanism (but see below). Also since it depends on the existence of a critical layer (layer at which the wind velocity equals the phase speed of the wave), the mechanism is inapplicable to the observed damping of wind waves by adverse wind.

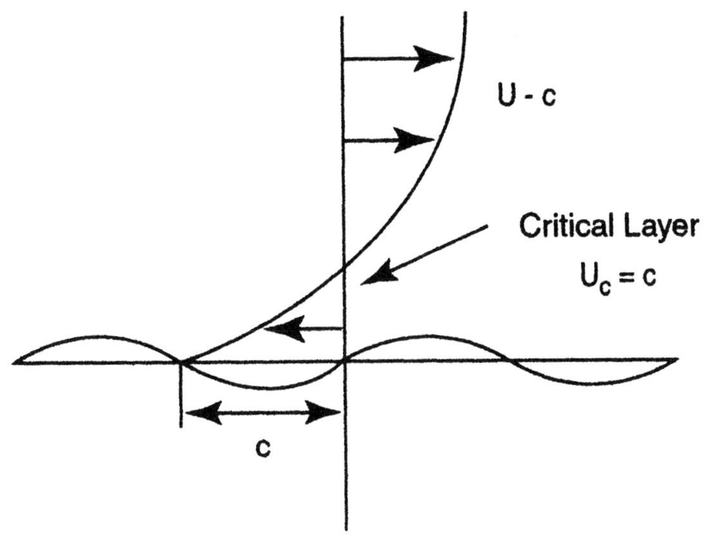

Figure 5.7.1. Laminar shear flow which, because of the gradient of shear, transfers energy from the winds to the wave through a critical layer that exists when the frame of reference is fixed to the moving wave.

The critical layer (where the wind speed equals the wave speed) is a region of high vorticity and it is in this region where interactions between the fluctuating wave-induced motion and the mean flow transfer energy from the mean flow to wave-correlated motions in the air take place. This energy is then transferred to the wavy air-sea interface by processes that are intimately tied to turbulent motions in the layer adjacent to the interface. Miles did not tackle the problem of how this transfer actually takes place, which requires modeling the motions in this turbulent surface layer. The energy flux to the wave is proportional to the product of the variance of the wave-induced vertical velocity fluctuations at this level and the ratio of the second derivative of the flow velocity to its first derivative at this level. However, Phillips points out that below c/u^* of about 10, the matched layer is within the viscous sublayer and the curvature of the wind profile is not very large, and therefore the Miles mechanism, in which the energy flux to wave motions is proportional to the second derivative of the flow velocity, is quite inefficient. At values of c/u^* much larger than 10, the critical layer is much higher above the wavy surface, and since the wave-induced fluctuations decrease rapidly with increasing distance above the ocean surface and since the energy flux is proportional to their variance, the Miles mechanism becomes quite inefficient once again. Thus only in a narrow range of c/u^* (10 to 15) does the Miles mechanism work efficiently to transfer energy from wind to waves (Phillips, 1977).

Miles ignored turbulent fluctuations in the wind field. It appears that the details of the turbulent stress and pressure field in the air flow in the immediate vicinity of the wave might be important (but see below) to the generation process. The momentum flux to the waves depends on the normal stress component of the turbulent flow in phase with the wave slope and with the tangential stress component in phase with the wave elevation (the energy flux, the rate of work done per unit area, depends on the correlation between the pressure at the surface and the vertical velocity there). There are indications however that the contribution of the latter to wave growth is generally much smaller at least for short gravity waves [it may however be important for generation of spray at high wind speeds—see Phillips (1977)]. It is therefore necessary and sufficient to know the magnitude of the covariance between pressure fluctuations and the wave slope. This however requires calculation of the deformed turbulent flow over the wave, often separated, and herein lies the difficulty in computing and predicting wind-wave growth.

Also observations of relevant flow properties are very hard to make in either the field or a laboratory because of the difficulty of making measurements in the immediate vicinity of a mobile interface, and therefore it is difficult to validate wave generation theories. Quantification of the energy input from the wind to waves involves measuring (or computing) the wave-induced pressure perturbation in phase with the wave slope. This is hard to do since measurement of a small quantity very close to a moving interface is essential, especially at the shorter wavelengths of the spectrum, at which the wind is most efficient at transferring energy. Also, in the field, the problem is compounded by an unknown dissipation rate at that wavenumber, and spectral energy transfer among the other wave components that exist. In numerical calculations, the pressure–wave slope covariance is extremely sensitive to the details of the coupled turbulent flow over the wave, and how exactly turbulence closure is effected appears to make a difference. In potential flow, the high pressure is at the trough of the wave, whereas a small shift of the high pressure up the downwind face by only a few tens of degrees or so is enough to cause exponential growth (Donelan and Hui, 1990), and clearly, the degree of this phase shift is important and hard to calculate accurately, especially when the waves are breaking and the flow separation in the lee of the breaking waves tends to enhance the energy transfer. Figure 5.7.2 (from Phillips, 1977) shows measurements of this phase shift and it is evident that the phase shift is maximum around c/u_* of 10 and decreases rapidly above and below this value. It is also evident from the figure that the Miles mechanism underestimates the phase shift, and since the critical layer contribution is negligible outside c/u_* values of 10–15, the figure also demonstrates the importance of turbulent normal stress fluctuations at almost all values of c/u_* to the phase shift and hence the wave generation process.

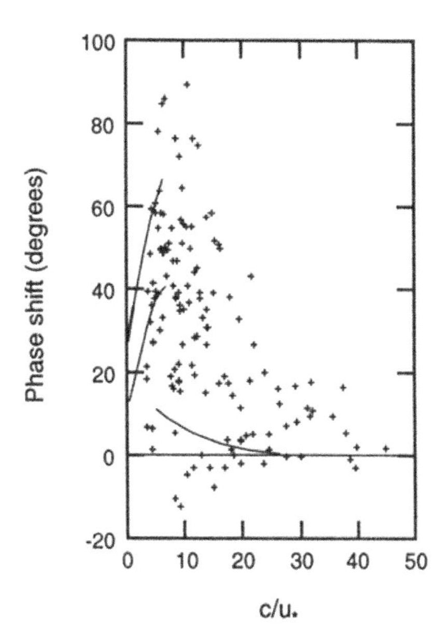

Figure 5.7.2. The phase shift of the pressure from 180° (from Phillips 1977, reprinted with the permission of Cambridge University Press).

Miles' (1957) work was a milestone in wind-wave studies. However, it is not the entire story. It does not work well for the multiscale problem, which is invariably the case for a wind-driven sea. It is not valid for steep breaking waves. In this case, there is airflow separation in the lee of the wave crests, with a consequent enhancement of momentum flux by wind to the waves (Banner, 1990b), the classic "sheltering mechanism" of Jeffreys (1924). Neither Miles' theory nor the later turbulent surface layer models (for example, Al-Zanaidi and Hui, 1984) account for flow separation. More importantly, for waves encountering adverse winds, the theory does not apply, since there is no critical layer. Thus it cannot account for damping of waves by adverse winds. Even for favorable winds, it does not work well if the critical layer is too close to or too far from the interface. Wave models which depend principally on this theory or its modification have difficulty predicting the low frequency components of the wind-wave spectrum correctly. Despite its documented successes, WAM, a heavily used wave model (see below), has difficulty predicting swell accurately.

Al-Zanaidi and Hui (1984) numerically computed the turbulent flow over a wave using a two-equation model of turbulence and, assuming no separation, showed that the fractional growth rate of wave amplitude β / n per radian can be written as

$$\frac{\beta}{n} = b\frac{\rho_a}{\rho_w}\left[\frac{U_w}{c} - 1\right]^2 \tag{5.7.1}$$

The fractional growth rate is related to the rate of work done by the wind as given by the pressure–vertical velocity correlation (see Plant, 1982). Here b plays the role of the coefficient in Jeffreys' sheltering mechanism. U_w is the wind speed at the height of one wavelength. The coefficient b equals 0.04 for smooth (and transitional) flow, 0.06 for rough flow, –0.04 for rough adverse flow, and –0.024 for smooth adverse flow. Thus their calculations apply to both growth and decay of wind waves by the action of the wind. Figure 5.7.3 shows the observed growth rates compared with calculated ones. The agreement, especially the dependence on wave age c/u_*, is quite good. Thus Jeffreys' sheltering mechanism appears to be valid.

Field measurements show that the growth rate near the peak of the spectrum is linear in spectral density but proportional to the wind stress (Plant, 1982),

$$\frac{\beta}{n} = 0.04\left(\frac{u_*}{c}\right)^2 \cos\theta \tag{5.7.2}$$

where θ is the angle between the wind and waves. Generalization of Eq. (5.7.1) for an arbitrary angle between the wind and the waves gives (Donelan and Pierson, 1987)

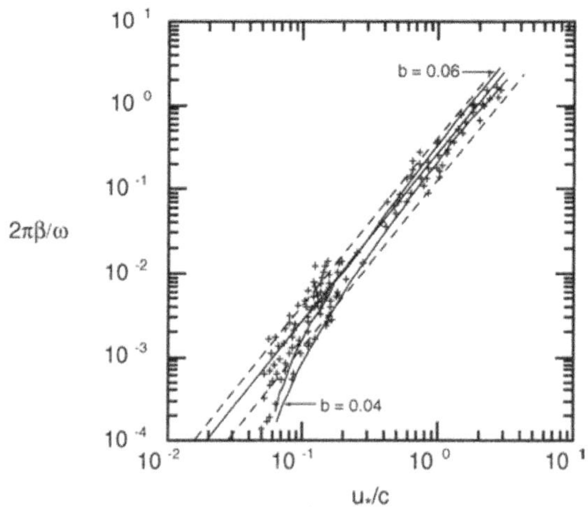

Figure 5.7.3. Fractional growth rate as a function of the inverse of the normalized phase speed (from Al-Zanaidi and Hui, 1984).

$$\frac{\beta}{n} = b\frac{\rho_a}{\rho_w}\left[\frac{U_w}{c}\cos\theta - 1\right]\left|\frac{U_w}{c}\cos\theta - 1\right| \tag{5.7.3}$$

where U_w is referred to a height π/k, half the wavelength ($b \sim 0.194$).

The details of wind-wave generation at low wind speeds are quite germane to aspects involving sea surface roughness. The air–sea interface is quite stable and at wind speeds below about 1.5 m s^{-1}, the surface is smooth and the salient length scale is the viscous length scale (ν / u_*). This is the effective "roughness" scale of the sea surface. Above this speed, however, the interface becomes unstable and capillary waves are generated. This leads to a dramatic increase in the sea surface roughness. The exact wind speed at which this occurs depends on the condition of the sea surface. When the sea is covered by an oil slick, the increased surface tension requires a higher wind speed (5.5 m s^{-1}) for the capillaries to be initiated. In either case, the capillary waves have a wavelength of about 3.5 to 4 cm. Phillips' mechanism of resonant wave generation by turbulent pressure fluctuations is apparently most efficient (Phillips, 1977) at wavelengths corresponding to the minimum phase speed, 1.7 cm. The mechanism apparently produces two wave trains at an angle to each other [$\theta = 2 \cos^{-1} (c_m/U)$, where U is the advection speed of the pressure patterns], giving rise to a characteristic dimpled rhomboid pattern of capillaries called cat's paws. At higher wind speeds, gravity waves are excited.

In early 1980s, Klaus Hasselmann at Hamburg gathered a group of wave modelers together and set in motion a series of collaborations, the ultimate result of which has been the third-generation wave model called WAM (WAve Model) that is now extensively used around the world for both research and operational applications. The WAM group also produced a summary report in the form of a book (Komen *et al.*, 1994). This book constitutes an up-to-date summary of the state of our knowledge base about surface wave processes that went into constructing the WAM model. We will report on some of their findings on generation mechanisms briefly here; for details, the reader is referred to Komen *et al.* (1994), who also report extensively on numerical modeling of waves, wave hindcasting/forecasting, and comparisons with observations from buoys and satellites.

Miles' mechanism of wind-wave generation ignored turbulence effects. There has been considerable controversy in the intervening years about the importance of turbulence and the influence of waves on the wind itself. Since field observations are very difficult to make and the results often ambiguous, and controlled laboratory investigations on wave growth, interactions, and decay are often in a parameter range inappropriate for application to the field, it has been extremely difficult to unambiguously ascertain the correctness of a particular hypothesis and approach to characterizing the growth, decay, and interactions.

The WAM approach is semiempirical, based on a judicious combination of theory and observations, and has been remarkably successful in practical applications. The source term for waves in the WAM model is based on Miles' mechanism, modified for the effect of gustiness (turbulence) at low frequencies and the feedback effect of wave field on the wind profile via a quasi-linear theory.

Miles' mechanism is simply the classical instability theory of a laminar shear flow applied to a mobile air–water interface in the form of a propagating surface wave. The governing Boussinesq equations can be reduced using small perturbation theory applied to the classical Taylor–Goldstein equation (see Komen *et al.*, 1994, Chapter II, for details) for the normalized wave-induced vertical velocity perturbation $\chi = w / w_0$,

$$\begin{aligned} &\frac{1}{\chi}\frac{d^2\chi}{dz^2} = k^2 + \frac{1}{U}\frac{d^2U}{dz^2} + \frac{g}{U^2}\frac{1}{\rho_a}\frac{d\rho_a}{dz} \\ &\chi = 1 \text{ for } z = 0;\ \chi \to 0 \text{ as } z \to \infty \\ &U(z) = \bar{U}(z) - c \end{aligned} \tag{5.7.4}$$

where U(z) is the wind speed relative to the propagating wave (c is the phase velocity), k is the wavenumber, and ρ_a is the air density. A subscript of 0 indicates values at z=0. This equation has been obtained by assuming the wave dispersion relation to be of the form $\hat{c} = c + (\rho_a / \rho_w) c'$, where the imaginary part of c' gives the growth rate β of wave amplitude (half the growth rate of wave energy),

$$\beta = \frac{-i}{4} c \frac{\rho_a}{\rho_w} W_0 = \frac{-i}{4} c \frac{\rho_a}{\rho_w} \left(\chi^* \frac{d\chi}{dz} - \chi \frac{d\chi^*}{dz} \right)_0 \tag{5.7.5}$$

where W is the Wronskian. Miles (1957; see also Miles, 1993) solved Eq. (5.7.4) ignoring the stratification terms, which then becomes the Rayleigh equation. The growth rate depends on the Wronskian, which for the Rayleigh equation is a constant except at the critical layer z_c (where U=0). The value of the Wronskian determines the growth rate:

$$\beta = -\frac{\pi}{2} c \frac{\rho_a}{\rho_w} \left(\frac{\dfrac{d^2U}{dz^2}}{\left|\dfrac{dU}{dz}\right|} |\chi|^2 \right)_{z=z_c} \tag{5.7.6}$$

This is the Miles classical result for wind-wave growth. Negative curvature of the wind profile at the critical height is essential for growth. A logarithmic profile gives that. The physical mechanism is the induced vortex force (which is proportional to the gradient of the wave-induced stress τ_w, and hence is a delta function at the critical height) acting to slow down the mean flow and transfer the momentum to the wave (see Lighthill, 1962, for a lucid discussion of the Miles mechanism). The result is easily extended to stratified flows (see Komen *et al.*, 1994).

Equation (5.7.4) can be solved numerically for any arbitrary wind profile. Monin–Obukhoff similarity theory (see Chapter 3) can be applied to infer the velocity and density profiles. The neutral stratification profile is given by log law,

$$\bar{U}(z) = \frac{u_*}{\kappa} \ln\,(z/z_0);\; z_0 = \alpha u_*^2/g;\; \alpha \sim 0.0144 \tag{5.7.7}$$

where the Charnock relation has been used to infer the roughness of the sea surface, with the coefficient of proportionality assumed to be constant, although it is a function of wave age. For stratified cases, the profile also depends on the normalized Monin–Obukhoff length scale $L_* = gL/u_*^2$. The growth rate therefore depends only on L_* and c/u_*. Figure 5.7.4 shows the growth rate compared with field observations of Snyder *et al.* (1981) and some laboratory results. The excellent agreement is noteworthy, although it depends on the exact value of Charnock's constant used, since the growth rate normalized by wave frequency is inversely proportional to this constant. There is considerable uncertainty in this value, being anywhere from 0.011 to 0.018, with the most recent measurements during TOGA/COARE in the western Pacific showing a value of 0.011 (Fairall *et al.*, 1996a). The higher value of 0.018 may be more appropriate to shallow waters, while the smaller value is the best for open ocean conditions. It is a very sensitive function of the drag coefficient (say, referred to the 10-m anemometric height),

$$\alpha = \frac{10g}{u_*^2} \exp\left(-\kappa C_{d10}^{-1/2}\right) \tag{5.7.8}$$

with an error of few tens of percent in the drag coefficient causing a several-fold error in Charnock's constant.

Gustiness, wind variability at frequencies lower than the wave frequency, may play an important role in wave growth. It is accounted for by Komen *et al.* (1994) by fitting an empirical curve to the growth rate

$$\beta = n\ \max\left(0.2\frac{\rho_a}{\rho_w}\left(28\frac{u_*}{c} - 1\right)^2, 0\right) \tag{5.7.9}$$

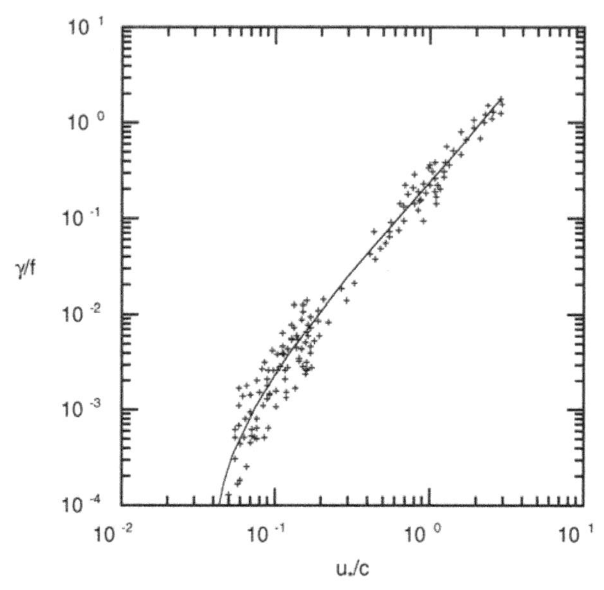

Figure 5.7.4. Fractional growth rate used in WAM compared to observations (from Komen *et al.*, 1994).

where n is the wave frequency and u_* is the mean friction velocity, whose probability distribution is considered to be a Gaussian. The influence is important for low frequency waves (low values of u_*/c) and hence the later stages of wind-wave growth.

The effect of small scale turbulence on wave growth is still rather uncertain, although Komen *et al.* (1994), based on the eddy viscosity approach to modeling small scale turbulence and numerical modeling results, suggest that the growth rate can be written as

$$\beta = 2\kappa\, n \frac{\rho_a}{\rho_w} \left(C_d^{-1/2}(k) \left(\frac{u_*}{c} \right)^2 \cos\theta - \frac{u_*}{c} \right) \tag{5.7.10}$$

where the drag coefficient is referred to height k^{-1}, and θ is the angle between wind and waves. This relationship causes damping of low frequency waves and damping of waves propagating faster than the wind and against the wind.

So far, the effect of waves on the wind, especially the wind profile, the details of which are critical to wave generation, has been ignored. However, this influence is substantial. Observations indicate that the drag coefficient (and the roughness) depends on the wave age $A_w = c_p/u_*$, with subscript p referring to the peak of the spectrum, young seas corresponding to $A_w < 5$–10, and mature seas to $A_w > 20$.

The partitioning of the air–sea momentum flux τ from the wind into τ_w, the fraction that goes first into waves, and τ_c, the part that goes directly into currents, is of considerable interest. Mitsuyasu (1985) found that $\tau_w/\tau \sim 22\,(ka)^2$, and laboratory and field measurements show that this fraction is usually in the range 0.4–0.6 (Melville, 1996). However only ~5% of the momentum flux from the wind to waves (0.05 τ) is carried out of the generation region by waves, with the rest being lost to currents in the generation region itself. The wave-induced stress can be derived from empirical expressions for the wind-wave spectrum since

$$\tau_w = \int dk \frac{\partial}{\partial t}\left(\frac{\psi(k)}{c}\right) \qquad (5.7.11)$$

where $\psi(\mathbf{k})$ is the wavenumber spectrum. This result is a generalization of the result for monochromatic waves that momentum flux is equal to $c^{-1}(\partial E/\partial t)$. Komen *et al.* (1994) show, using a quasi-linear theory and a series of assumptions based on observations, that the wave-induced stress as a fraction of the total τ_w / τ is a strongly decreasing function of the wave age (Figure 5.7.5). For old seas, most of the wind momentum flux goes into the turbulent part, whereas for young seas, wave-induced stress is a considerable fraction. The wave growth rate also depends on the wave age (Figure 5.7.6), with the normalized growth rate larger for older seas. Thus for young seas, there is a strong two-way interaction between waves and wind, with the wind profile strongly modified and the sea rougher than indicated by the Charnock relation, whereas for an old sea, this interaction is weak.

The question of the dependence of the roughness length, or equivalently the drag coefficient, on the maturity of the surface wave field is a particularly important question. As indicated in Chapter 4, Geernaert *et al.* (1987) proposed an empirical relationship for the drag coefficient as a function of wave age, but this was not based on actual measurements of wind-wave spectra. Based on independent wind stress and wave measurements during the North Sea Humidity Exchange over the Sea (HEXOS) (DeCosmo *et al.*, 1996) program, Maat *et al.* (1991) and Smith *et al.* (1992) concluded that the normalized roughness scale, the Charnock constant gz_0/u_*^2, is inversely proportional to the wave age:

$$\alpha = \frac{gz_0}{u_*^2} = 0.48\left(\frac{c_p}{u_*}\right)^{-1} \qquad (5.7.12)$$

Figure 5.7.7 shows a comparison of Eq. (5.7.12) with carefully selected laboratory tank and field data, by Donelan *et al.* (1993), who argue that for a

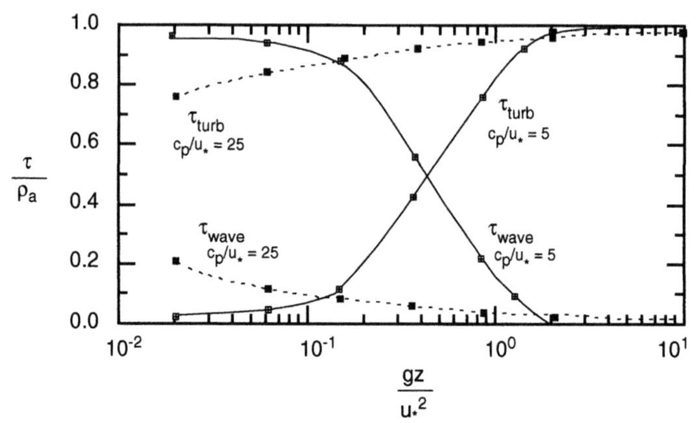

Figure 5.7.5. Turbulent and wave portions of the shear stress as functions of the height above the free surface for young and mature seas (from Komen *et al.*, 1994).

variety of reasons, the laboratory waves tend to be smoother and hence should be disregarded. If laboratory data are discarded, then there is very good agreement between various field data and the above relationship, which is consistent with

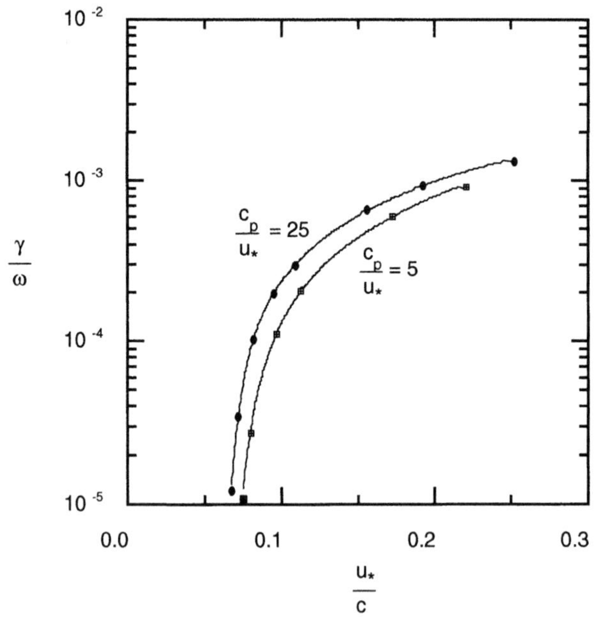

Figure 5.7.6. Normalized growth rate as a function of normalized phase speed for young and old seas (from Komen *et al.*, 1994).

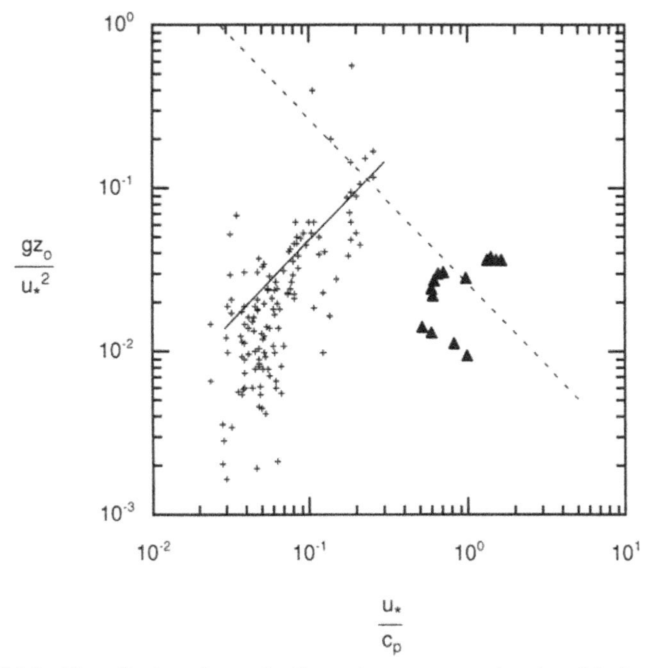

Figure 5.7.7. Normalized roughness, the Charnock constant, as a function of the inverse wave age for field and laboratory observations (from Donelan *et al.*, 1993). Triangles denote laboratory data.

the notion that young waves are rougher than old waves. Jones and Toba (1995; see also Toba and Jones, 1992), in an effort to fit both laboratory and field data, proposed a 1/2 power law in Eq. (5.7.11) that makes the normalized sea surface roughness increase with wave age. Figure 5.7.8 shows a comparison of laboratory data and data collected in the Bass Strait with the Jones and Toba (1995) law. Donelan *et al.* (1995) argue that sea surface roughness should decrease with age, and the laboratory and field data taken separately do indeed behave that way. They also suggest that for a constant wind speed, the drag coefficient or equivalently the roughness should decrease with wave age (Figure 5.7.9), whereas the Jones and Toba (1995) law would be exactly the opposite. While we believe that laboratory waves tend to be inherently different from waves in the field, and therefore should be excluded, and hence the sea surface roughness should decrease with wave age, at this point in time, the controversy remains unresolved. Even though the wave age is the most important parameter indicative of the degree of development of a wind-wave field, it may not be the sole parameter. It is for this reason that it is unlikely that the drag coefficient (and the roughness length) can be parameterized as solely a function of wave age

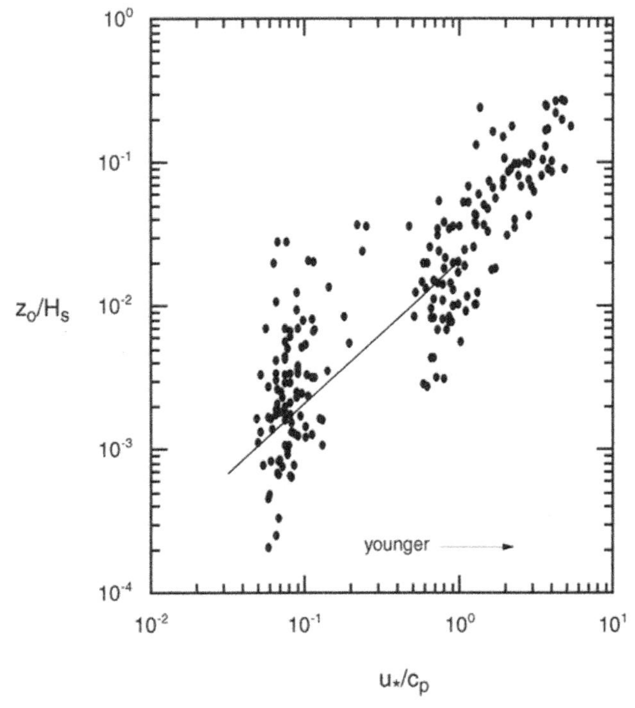

Figure 5.7.8. Roughness length normalized by the significant wave height as a function of the inverse wave age (from Jones and Toba, 1995).

(Donelan, 1990) as is done in Eq. (5.7.12), although it may be considered as approximately valid for practical applications.

Johnson *et al.* (1998), in a recent analysis of data collected during the Riso Air–Sea Exchange (RASEX) experiment in shallow (3–4 m) fetch-limited waters off Denmark, found it difficult to estimate the wave age dependence of roughness, because of the errors in measurement of friction velocity and the large scatter resulting therefrom. The large scatter is typical of such measurements, and therefore while it is certain that the Charnock parameter decreases with wave age, it is difficult to derive with great certainty the rate of decrease. Combining RASEX data and data from earlier studies in 30-m water of the German Bight (Geernaert *et al.*, 1987), HEXOS studies in 18-m water off the Dutch coast (Maat *et al.*, 1991; Smith *et al.*, 1992), and 12-m water in Lake Ontario (Donelan *et al.*, 1993), Johnson *et al.* (1998) conclude that the best fit to the mean values derived from each data set of the dimensionless Charnock constant is given by Eq. (5.7.12), but with a coefficient of 1.89 (instead of 0.4),

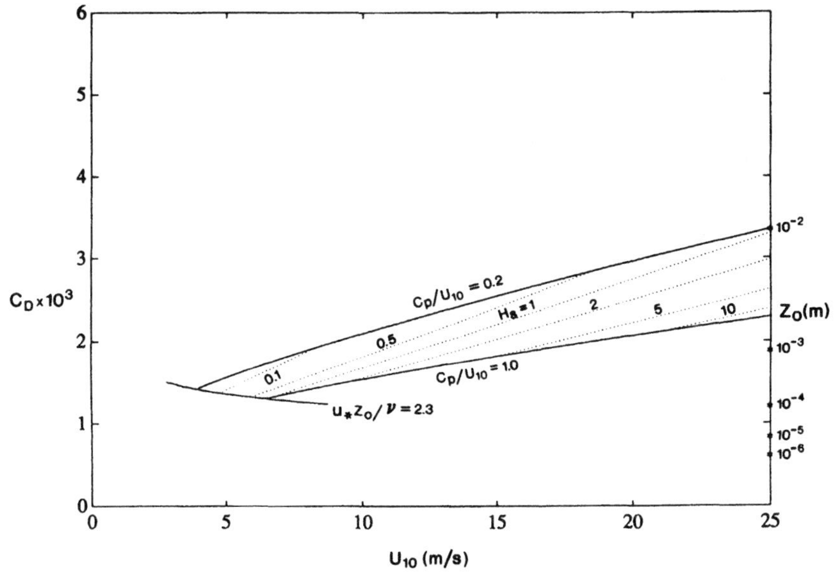

Figure 5.7.9. Drag coefficient as a function of wind speed for various significant wave heights (from Donelan *et al.*, 1995).

and an exponent of −1.59 (instead of −1.0). They also found that a constant value of $\alpha \sim 0.018$ adequately reproduces friction velocities measured during RASEX.

A note of caution is in order here. Relations such as Eq. (5.7.12) may significantly overestimate the drag coefficient (Yelland *et al.*, 1998) by as much as 15–25% when compared to typical open ocean values (for example, Large and Pond, 1981). Many of the measurements, on which such relations such as Eq. (5.7.12) are based, have been carried out from towers in shallow waters. These platforms are stable and the flow distortion induced by the tower itself small. Open ocean measurements, on the other hand, necessarily involve platforms that are subject to the action of the waves themselves, and extreme care is necessary to prevent contamination of wind stress measurements by platform motion and distortion of airflow around the ship. However, the wind-wave environment is likely to be different in coastal waters and the open ocean, with swell contributing prominently to the wind-wave spectrum in open ocean. For winds blowing offshore, the coastal wave spectrum is single-peaked due to the absence of swell. For waves propagating toward the shore, the long waves tend to shoal and steepen on encountering shallow bottom depths. These differences might be responsible for the fact that the Charnock constant in open ocean measurements is ~0.011, while that in shallow water is much higher,

~0.018. Smith *et al.* (1996b) suggest that the wave age dependence of wind stress or the Charnock constant may be valid for only single-peaked spectra. If so, it would be of questionable validity in the open ocean.

Thus while there are indications that the normalized roughness scale (Charnock parameter) might very well be a decreasing (perhaps weakly) function of wave age, the exact relationship is to be regarded as still rather uncertain. To compound these difficulties, recent measurements by Yelland *et al.* (1998) from a research vessel in the open ocean have failed to detect any wave age dependence of wind stress. From careful measurements of the wind stress and accounting for the distortions induced by the ship superstructure via a CFD model of the airflow around the research ship, they suggest that most of the scatter in the data can be attributed to experimental errors, and the drag coefficient is simply a function of the wind speed governed by a relationship of the Large and Pond (1981) form: $1000\ C_{D10n} = 0.50 + 0.071\ U_{10n}$ ($6 < U_{10n} < 26$ m s^{-1}). The wave age dependence, if any, was insignificant. Given the fact that the momentum flux from the wind to the waves is a function of the correlation between surface pressure and the wave slope, it is quite possible that it is the wind-wave slope spectrum, with major contribution from the shorter components of the spectrum (which are in turn affected by the longer waves), that is important to this issue. Clearly, even as we learn more and more about wind waves, the uncertainties associated with many of its aspects continue to flourish.

5.7.2 DISSIPATION

Dissipation mechanisms of surface waves are even more poorly understood. The principal mechanism for gravity waves is wave breaking. However, generation of parasitic capillary waves can be another sink. For capillary gravity waves, viscous decay is quite important. The fractional decay rate of the wave amplitude is (Donelan and Hui, 1990)

$$\frac{1}{a}\frac{da}{dt} \sim -2\nu\, k^2 \qquad (5.7.13)$$

[similar to Eq. (5.2.28) for energy]. The e-folding timescale is $(2\ k^2\ \nu)^{-1}$. Because of the quadratic dependence on wavenumber, high wavenumbers (capillary and gravity–capillary waves) are rapidly damped and viscous extinction is the principal mechanism. As anyone watching the surface of a pond on a blustery day knows, the capillary waves excited by a wind gust do not last long after the gust is gone. Thus at the high wavenumber capillary range of the wind-wave spectrum, viscous dissipation dominates and this is quite important to microwave remote sensing. The mechanism is not of much consequence to ordinary gravity waves and to the overall energy balance.

The second mechanism for loss of energy by gravity waves is generation of parasitic capillaries near the crests on short gravity waves. When the waves become steep, the crests become sharper with increases in wave amplitude; the curvature near the crest increases as well and this gives rise to capillary waves a few centimeters in length on the forward face of the gravity wave. These capillaries draw energy from the primary gravity wave, which is quickly lost to viscous damping and therefore constitute an energy sink for the gravity waves. This process is hard to parameterize and none of the current wave models including WAM include this mode of dissipation. Longuet-Higgins (1992) discusses this process and its impact.

The third mechanism, wave breaking, is the most important mechanism for dissipation of gravity waves. Wave breaking can be brief and violent, as in plunging breakers, where the wave steepens abruptly and breaks, and a cylinder of air is encompassed. The energy loss is rapid and large and therefore another such event in rapid succession is rare. Wave breaking can also be gentle and prolonged, as in spilling breakers. In this case, energy drain is slow. The wave crest spills over, entraining air and dissipating part of its energy. However, these processes are extremely difficult to characterize, and our knowledge of wave breaking remains essentially empirical, and imperfect even at that. Nevertheless, the wave dissipation processes have to be modeled for numerical wave prediction purposes. Given the importance of wave breaking processes to air–sea transfer of heat and gases, this phenomenon has been getting increased attention in recent years (see Section 5.9).

There are few observations of wave breaking processes in the laboratory. Duncan (1981) and Rapp and Melville (1990) have looked at single breaking events. Duncan *et al.* (1994) and Bonmarin (1989) have also studied the mechanics of unsteady group breaking events in the laboratory. These are discussed in Section 5.9. Theoretical approaches to quantifying wave dissipation by breaking are also very few. There are basically three approaches (Komen *et al.*, 1994). One approach is to consider a physical model of breaking and whitecapping and estimate the drag exerted on the wave. The second approach, originally due to Longuet-Higgins (1969), is to make use of the limiting steepness condition and the probability density function of wave heights. The limit on amplitude of a wave, beyond which it will break, is given by the Stokes wave, where the downward acceleration at the crest reaches the limiting value of g/2. Any wave above that amplitude breaks until the amplitude is reduced to the limiting value. Then the energy loss per wave cycle is

$$\Delta E = \frac{\rho g}{2} \int_{a_S}^{\infty} (a^2 - a_S^2) P(a) da; \quad a_S = \frac{g}{2n^2} \qquad (5.7.14)$$

where a_S is the Stokes limiting wave amplitude and P(a) is the probability density function of wave amplitudes. Assuming a narrow-banded Gaussian sea for which the wave heights are Rayleigh distributed, Longuet-Higgins derived the fractional energy loss per wave cycle. This approach can be extended to relax the narrow-banded sea assumption (see Komen *et al.*, 1994). Presumably, this breaking mechanism is applicable in the "saturation" range of the spectrum. It is important to note however that there are shortcomings to this approach. First of all, waves become unstable and tend to break long before the Stokes wave limit is reached, as discussed earlier. The expression for the limiting amplitude involves a series of unduly restrictive assumptions, and the Gaussian statistics are not entirely correct (Srokosz, 1990).

The third approach is that due to Phillips (1985), in which he postulated that in the equilibrium range of the wave spectrum, all three terms, the wind input, nonlinear transfer, and wave dissipation, must be proportional to one another. Then, from our better knowledge of wind input and nonlinear transfer processes, it is possible to parameterize dissipation (see Section 5.5). This approach, however, does not specify any dynamical mechanism.

None of these methods have proved entirely satisfactory. Modeling the spectral wave dissipation term is the principal weak point of wind-wave models. For lack of a better alternative, wave models such as WAM use a plausible functional form and tune the constants to force agreement of the resulting wind-wave spectrum in simple well-known cases such as the fetch-limited wave growth case, for which accurate observational data are available.

In shallow water, other complications arise. The waves feel the bottom and hence bottom friction is a major dissipative mechanism. In addition, the bottom roughnesses act as scatterers. In regions where the bottom is muddy, waves can be excited in the porous mobile medium, which represents a drain on the surface wave energy. These effects need to be parameterized suitably, at least the bottom friction effects. Bottom friction is especially important for swells, which due to their large wavelengths begin to feel the bottom on the continental shelf long before they break near the shore.

There is an enormous amount of energy in wind waves over the ocean. Rough estimates indicate that as much as half of the kinetic energy in the upper ocean might reside in its surface wave field. If so, this constitutes an incredibly large potential source of energy for upper ocean processes. Most of this energy is dissipated in shallow water along the ocean margins by spilling breakers in the surf zone, but a significant fraction is dissipated in the deeper regions as well by deep water breaking processes and contributes to mixing in the upper ocean. Wave breaking transfers momentum in the waves to ocean currents, but most of the energy is dissipated by the resulting turbulence. Nevertheless, wave breaking processes are important to mixing in the upper few meters of the ocean everywhere around the globe and are certainly central to the air–sea exchange of heat and gases.

Since waves are primarily generated by winds over a certain part of the ocean basin and then propagate to other parts, they constitute an efficient mechanism for transfer of energy and momentum originally resident in the winds from one region of the basin to another, without the intervention of oceanic currents or other oceanic wave motions. They also constitute a nonlocal dissipation mechanism for the kinetic energy resident in the ABL winds. Because of the steady Stokes drift currents they induce, they constitute an important mechanism for transport of anything at or near the ocean surface (especially when the underlying ocean currents are weak). As a result, pollutants such as spilled oil can travel long distances even in the absence of underlying currents and often end up on beaches and wildlife habitats.

Waves can also transfer energy to the wind if they are propagating faster than the prevailing wind, when, for example, the wind dies suddenly after the wave field is generated. Figure 5.7.10 shows wind profiles measured over Lake Ontario that show the formation and decay of wave-driven wind. The physical mechanisms involved are extremely complex and overall there has not been much progress in deciphering the details of the energy transfer from the waves to the wind.

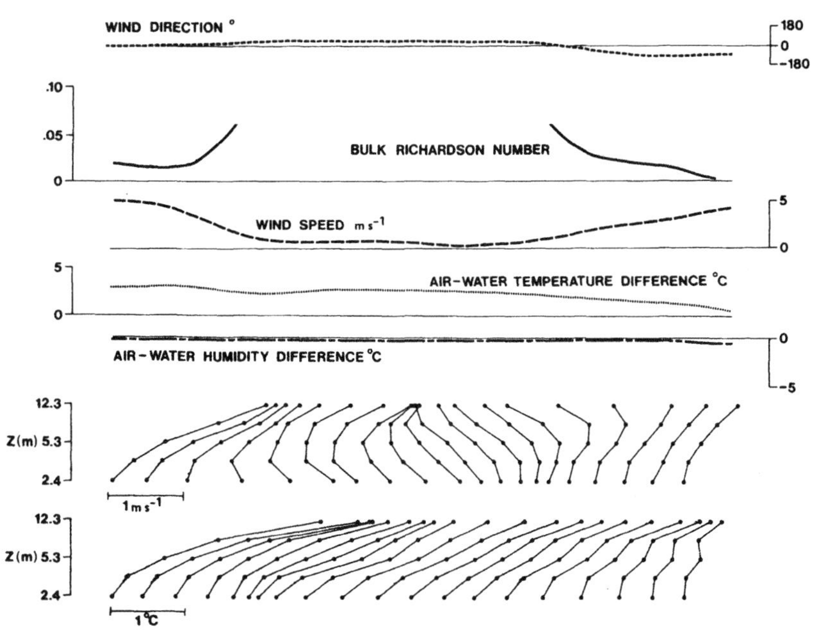

Figure 5.7.10. Wind and temperature profiles over Lake Ontario, showing the influence of waves running against the wind (from Donelan *et al.* 1974).

5.7.3 PROPAGATION

When wind generates and sets in motion a group of surface waves, once the wind input ceases, the higher frequency components of the spectrum decay more rapidly than their lower frequency counterparts. The low frequency components called swells propagate over vast distances with little change, unless they encounter severe adverse winds along the way. This is simply because the wave slopes are small and hence they are not dissipated by wave breaking, and because of the low wavenumbers, viscous damping is negligible. Locality of resonant wave–wave interactions ensures that there is little energy lost by this mechanism as well. By the same token, the inefficiency of wind input at low wavenumbers ensures they do not gain much energy or lose much in adverse winds. Consequently once generated and set in motion, they travel essentially unchanged. They more or less follow the great circle paths although the ellipticity of the Earth and the rotational effects cause small deviations from this path (Munk *et al.,* 1963; Snodgrass *et al.,* 1966). In fact swells created on the other side of the ocean basins have been detected along many coasts. By examining the records from neighboring wave gauge stations for phase differences, it is possible to identify the source region of the swell and the time of its creation.

An extreme case of such long distance propagation by surface waves, in this case those generated by submarine quakes and crustal readjustments along subduction zones and fault lines, occurs principally in the Pacific ocean, with its western rim characterized by the subduction of the Pacific plate under the Asian one due to tectonic forces. Consequently, long wavelength gravity waves called tsunamis are generated and set in motion. These nondispersive waves, which because of their long wavelength can feel the deep ocean bottom and propagate at the speed of shallow water waves [$c = (gd)^{1/2}$], travel swiftly (at speeds ~ 200 m s^{-1}) clear across the Pacific basin in less than half a day and impact the coast. Their amplitudes when they are in the open ocean are on the order of a few tens of centimeters and therefore their passage is hardly noticeable. However, they carry enormous energies and grow to incredible amplitudes when they reach shallow water. A recent tsunami in the Sea of Japan wiped out everything up to a height of 30 m, creating incredible devastation and loss of life on a small Japanese island. Because of the potential for catastrophic impact of tsunamis, the Pacific basin is surrounded by seismic sensors that can quickly localize the source region of tsunamis when an underocean disturbance occurs and can often issue advance warnings to shore communities likely to be impacted. However, when propagation distances involved are small, as in the above incident, there is usually not enough time to issue warnings that can avert a disaster.

Irrespective of whether the waves are swiftly moving tsunamis, slower moving swell, or an entire spectrum of gravity waves, the propagation is governed by laws of conservation. Equation (5.5.20) is the most general form (Phillips, 1977; Komen *et al.*, 1994) governing action density in spectral space. The wave group follows the ray path given by

$$\frac{\partial x_\alpha}{\partial t} = U_\alpha + c_{g\alpha} \tag{5.7.15}$$

Equation (5.5.20) can be used to predict surface wave propagation. The major difficulty is that the terms on the RHS are not known with certainty as discussed earlier. The equations apply quite well to propagation of swell since the RHS is very nearly zero. Note that the ray approximation in Eq. (5.7.15) excludes diffraction effects.

In the absence of underlying currents, action density conservation is equivalent to conservation of energy density along ray paths. However, in the presence of currents, there is energy exchange between the mean flow and the waves, and energy density of the wave is not conserved. If an equation is written for the energy density, there occur additional terms indicating the interaction between the mean flow and wave motions called radiation stress terms [see Phillips (1977) for detailed derivation and Longuet-Higgins and Stewart (1964) for a physical description]:

$$\frac{\partial E(k)}{\partial t} + \frac{\partial}{\partial x_\alpha}\left[E\left(c_{g\alpha} + U_\alpha\right)\right] + S_{k\alpha}\frac{\partial U_k}{\partial x_\alpha} = 0 \tag{5.7.16}$$

$S_{\alpha k}$ represents the excess momentum flux to the mean motion due to superimposed wave motions. They work on the horizontal shear of the mean flow to transfer energy from wave motions to the mean flow or vice versa. They are somewhat analogous to Reynolds stresses in a turbulent flow which interact with the mean shear to transfer energy from mean flow to turbulence, except that in turbulent flow the energy flow is one way, from the mean to turbulence, whereas with wave motions it can be either way. An approximate form for S_{jk} for pure gravity waves (Phillips, 1977) is

$$S_{jk} = E\frac{c_g}{c} l_j l_k + \frac{E}{2}\left(2\frac{c_g}{c} - 1\right)\delta_{jk} \tag{5.7.17}$$

where $l_j = k_j / |k|$ is the unit vector. In deep water, with the x_1-axis aligned in the direction of wave propagation,

$$\begin{aligned} S_{11} &= \left(2\frac{c_g}{c} - \frac{1}{2}\right) \sim \frac{E}{2} \\ S_{12} &= S_{21} = 0; \quad S_{22} = E\left(\frac{c_g}{c} - \frac{1}{2}\right) \sim 0 \end{aligned} \tag{5.7.18}$$

Radiation stresses play an important role in many aspects of wave propagation (Longuet-Higgins and Stewart, 1964). For waves propagating over decreasing water depths onto a beach, they cause the mean water level to actually decrease in the region beyond the breakers in the surf zone, because of the increase in radiation stress. This is the so-called setdown, proportional to the product of the amplitude and the slope ($a \cdot ka$) of the approaching waves. The waves break when their amplitude becomes comparable to local water depth, and inward of the breaker zone, there is an increase in the mean water level, called setup, because of a decrease in the radiation stress (Phillips, 1977). This wave setup often contributes to flooding along the coastline.

One of the most dramatic impacts of radiation stresses is when a wave train runs into an adverse current. For steady background current, the kinematic conservation of wave crests (see Appendix D for a more general form) gives

$$\frac{\partial}{\partial x}(n + kU) = 0 \tag{5.7.19}$$

where the x-axis is aligned with the direction of wave propagation and U is the component of the mean current in the x direction. If subscript 0 corresponds to wave conditions in the region of zero currents,

$$k(U + c) = k_0 c_0 \tag{5.7.20}$$

then using the deep water wave dispersion relationship to substitute for the wavenumbers,

$$\frac{c}{c_0} = \frac{1}{2}\left[1 + \left(1 + \frac{4U}{c_0}\right)^{1/2}\right] \tag{5.7.21}$$

This equation shows that when the currents are adverse, there exists a value of $U_c = -c_0/4$ at which $c = c_0/2 = -2\,U_c$ and the group velocity is exactly equal to the current velocity, and the wave energy cannot propagate beyond this point. Action conservation then gives

$$\frac{\partial}{\partial x_1}\left[\frac{E}{n}(U + c_g)\right] = 0 \tag{5.7.22}$$

so that

$$E\left(U + \frac{c}{2}\right)c = \frac{1}{2} E_0 c_0^2$$

$$\frac{a}{a_0} = \frac{c_0}{\left[2c\left(U + \frac{c}{2}\right)\right]^{1/2}} \tag{5.7.23}$$

The amplitude of the wave increases in adverse currents until it becomes very large at the location of the critical current. The energy density increases from two causes: first, the wavenumber increases in adverse currents and the waves become shorter, and second, the action of the radiation stress transfers energy from mean flow to wave motions. Far before the energy density becomes very large, the wave breaks and loses its excess energy.

The impact of adverse currents on a wave field is dramatically evident near strong tidal currents and oceanic currents such as the Gulf Stream and the Kuroshio. The sea becomes very rough as the waves run into increasing adverse currents and there is extensive breaking. Once beyond the zone of increasing currents, and in the zone of decreasing adverse currents, the radiation stress transfers energy from the waves to the mean motion and the sea becomes calm once again. Waves traveling against an ebb tide at the mouth of an estuary and against rip currents along a shoreline are dramatic examples.

Another dramatic example of such wave–current interaction occurs in regions like the Andaman and Sulu seas, where large internal wave solitons are generated by tidal currents. Osborne and Burch (1980) show an examples for the Andaman Sea. The current patterns setup by these internal solitons is such that at the ocean surface there exists a convergent current field at the forward face of these solitons and a divergent field on the rearward face. Thus as these soliton packets propagate through the ocean, the existing surface wave field encounters increasingly adverse currents on the forward face of each soliton and extensive breaking takes place, with the sea becoming quite rough in a 500–800-m wide strip extending from horizon to horizon. With the passage of the soliton, the sea becomes calm once again. The effect is dramatic since the breaking process and the choppy sea is accompanied by a distinct roar. The alternating periods of intense roar and succeeding calm are the only manifestations of a subsurface process hidden from view, and to the uninitiated can be quite an unnerving experience, especially since no causal mechanism is apparent. The above expressions derived for steady currents apply to this situation as well if U is replaced by the current induced at the surface by the internal wave motions. When the induced velocity reaches the group velocity of the surface waves, the surface waves become very large. Far before then, they break and shed their excess energy.

Finally, we will mention one aspect of short gravity waves, namely their coexistence with longer ones in the wind-wave spectrum and with a drift layer with significant wind-induced drift velocities. These are shown schematically in

Figure 5.7.11. The existence of a very thin molecular drift layer at the interface a few millimeters thick under most conditions is well known and Figure 5.7.12 shows the velocity in the drift layer obtained in laboratory experiments on very young waves. The typical surface drift velocity is $U_d \sim 0.6\ u_*$. As noted by Wu (1975), for young waves, the fraction of the momentum flux from wind to surface currents via viscous stresses is large and therefore the near-surface current shear is also large. For more mature waves in the field, breaking transfers a momentum flux comparable to the wind stress from wind to the water (Melville and Rapp, 1985). So the near-surface shear and hence the drift current is likely to differ substantially from that in Figure 5.7.12, except at short fetches and durations. Unfortunately, observations of near-surface current for mature well-developed wave fields are not yet reliable.

The presence of the thin vortical drift layer can lead to breaking of short gravity waves (of wavelengths on the order of 10 cm or less) at amplitudes less than those in the absence of the current (Phillips and Banner, 1974). In the frame of reference moving at c, the phase speed of the gravity wave, the Bernoulli equation at the surface is essentially a statement of conservation of energy,

$$\frac{u^2}{2} + g\zeta = \text{constant} \tag{5.7.24}$$

where u is the net horizontal velocity and ζ is the elevation. Applying Eq. (5.7.24) to the point of incipient breaking where $\zeta = \zeta_m$ and $u = 0$, and at the

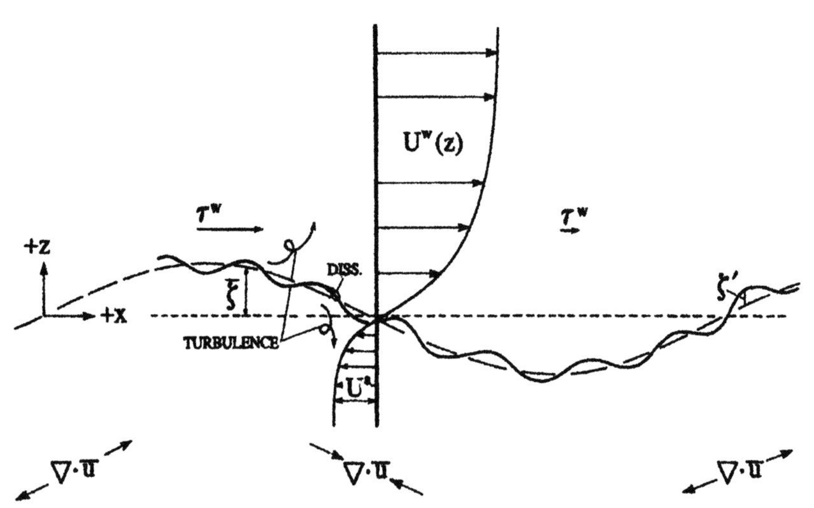

Figure 5.7.11. Small waves on long surface waves (from Smith, J. A., Modulation of short wind waves by long waves, in *Surface Waves and Fluxes*, Vol. 1, eds. G. L. Geeranert and W. J. Plant, Kluwer Academic, 247–284, 1990, with kind permission from Kluwer Academic Publishers).

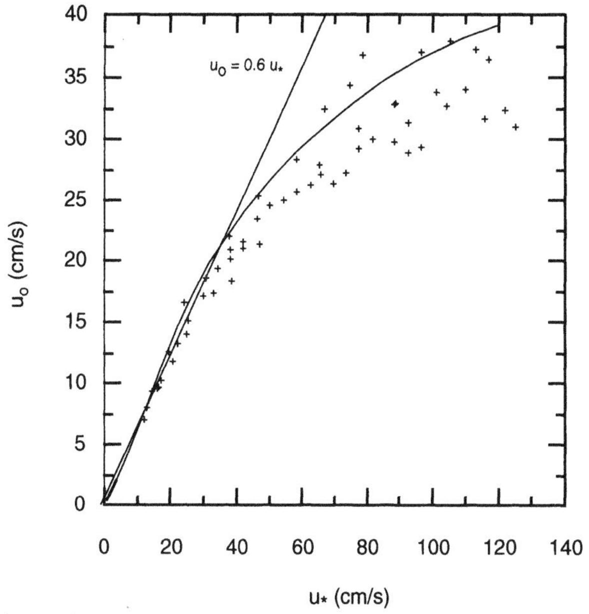

Figure 5.7.12. Surface drift velocity as a function of friction velocity (From Smith, J. A., Modulation of short wind waves by long waves, in *Surface Waves and Fluxes*, Vol. 1, eds. G. L. Geeranert and W. J. Plant, Kluwer Academic, 247–284, 1990, with kind permission from Kluwer Academic Publishers).

zero-crossing point $\zeta=0$, and $u=-c+U_d$, the amplitude for incipient breaking is

$$\zeta_m = \frac{(U_d - c)^2}{2g} \tag{5.7.25}$$

The ratio of ζ_m to the Stokes limit in deep water $a_s = g/(2n^2) = c^2/(2g)$ is

$$\frac{\zeta_m}{c^2/2g} = \left(1 - \frac{U_d}{c}\right)^2 \tag{5.7.26}$$

If in addition, long waves are present, the short waves are advected around by the orbital velocities induced by long waves. This leads to Doppler shifting of the shortwave frequencies. More importantly, short waves encounter increasing and decreasing horizontal currents as the long wave passes by. When the short waves feel an increasingly adverse current, they steepen and break, losing excess energy; when they see a favorable current, they regain some of their energy. Added to this is the effect of the drift layer superimposed on the long wave orbital velocities that has an effect on short wave breaking (Phillips, 1977). To complicate matters further, the wind action itself appears to enhance short gravity waves, compensating somehow for the drift enhancement of dissipation (Smith, 1990). This in turn is compounded by the modulation of the surface roughness by the short waves riding on long waves, leading to the modulation of stress along the long wave, and this appears to be a rather important factor. Overall, short waves riding on long waves appear to be pretty much in equilibrium with the wind, responding quite quickly to wind changes. But the overall picture is quite complex and hard to describe theoretically, but important enough for microwave remote sensing to merit continued study.

5.8 WIND-WAVE PREDICTION

Accurate wind-wave forecasts are important for marine operations. Naval forces, commercial shipping operations, and offshore exploration and drilling activities are all potential consumers of such information. Such forecasts require two essential ingredients: (1) accurate models of wave growth, and (2) accurate forecasts of marine surface wind fields. There are several possible methods, but given the complexities of the problem, the most useful method relies on numerical models of the wave growth in spectral space. Given the fact that our knowledge of the source, sink, and transfer processes are incomplete, especially in spectral space, and given the complex spatial and temporal variability of forcing, the accuracies achieved by currently widely used models of wind wave growth are truly amazing. Here we will provide a brief summary. The reader is referred to Sobey (1986) and Komen *et al.* (1994) for recent reviews.

Empirical wind-wave prediction becomes possible if one makes use of the observed properties of wind-wave growth with fetch (and duration) suitably nondimensionalized as described above. The very first attempt at wind-wave forecasting (Sverdrup and Munk, 1947) used this empirical approach, where the spectral shape was assumed to be that of the fetch (or time)-limited equilibrium spectrum of some assumed form. It was then a matter of computing the parameters of the normalized spectrum for a given fetch or time from known observational dependence on nondimensionalized fetch or time. Most frequently used spectral shape was the Pierson–Moskowitz form spectrum. The directional distribution was also an empirical equilibrium distribution. This approach is still useful for rough and quick estimates of the wave field. However, it does not depict the wind-wave evolution properly, since even with equilibrium rather than saturation theory for the waves beyond the peak of the spectrum, the wave components are assumed to stop growing when they reach the equilibrium level, whereas observations clearly show that the spectral densities overshoot the equilibrium value before settling down. The method is equivalent to predicting only the location and magnitude of the spectral peak, with the rest of the spectrum assumed to be fully evolved. The Pierson–Moskowitz spectrum

(Pierson and Moskowitz, 1964) was widely used in this approach:

$$\phi(n) = \alpha g^2 n^{-5} \exp\left[-\beta\left(\frac{n}{n_p}\right)^{-4}\right] \qquad (5.8.1)$$
$$\alpha \sim 8.1\times10^{-3}, \quad \beta \sim 1.25$$

Thus knowing the frequency of the spectral peak n_p, which is a function of the nondimensional fetch, determines the spectrum and the energy content of the wave field. However, the narrowness of the spectrum around the peak is an additional parameter that needs to be considered often. The empirical fetch-limited JONSWAP spectrum that does so can be written as

$$\phi(n) = \frac{\alpha g^2}{(2\pi)^4} n^{-5} \exp\left[-\frac{5}{4}\left(\frac{n}{n_p}\right)^{-4}\right]\gamma^{\Gamma}; \; \Gamma = \exp\left[-\left(n - n_p\right)^2 / 2\sigma_{f,b}^2 n_p^2\right] \qquad (5.8.2)$$

There are five parameters: α is the Phillips constant, n_p the peak frequency, γ the peak enhancement factor, and σ_f and σ_b the spectral widths at the forward ($n < n_p$) and backward ($n > n_p$) faces of the peak. These parameters are functions of the nondimensional fetch:

$$\frac{n_p U_{10}}{g} \sim 3.5\,\tilde{x}^{-0.33}, \; \alpha \sim 0.076\,\tilde{x}^{-0.22} \qquad (5.8.3)$$
$$\gamma = 3.3, \; \sigma_f = 0.07, \; \sigma_b = 0.09$$

Alternatively, a more recent form [Eqs. (5.6.1) and (5.6.2)] can be used for the dependence of n_p on fetch. This spectrum was based on the n^{-5} spectral law for the saturation range. Based on Donelan *et al.* (1985), Donelan and Hui (1990; see also Komen *et al.*, 1994) suggest a form different from that of the JONSWAP spectrum (one that is based on the n^{-4} spectral shape instead of n^{-5}, with a different form for the peak enhancement factor and different directional spreading),

$$\Phi(n) = \alpha g^2 n^{-5}\left(\frac{n}{n_p}\right)\exp\left[-\left(\frac{n}{n_p}\right)^{-4}\right]\gamma^{\Gamma}$$

$$\Gamma = \exp\left[-\frac{\left(n - n_p\right)^2}{2\sigma^2 n_p^2}\right]$$

$$\alpha = 0.006\left(U_{10}/c\right)^{0.55} \qquad 0.83 < \frac{U_{10}}{c} < 5$$

$$\sigma = 0.08\left[1 + \frac{4}{\left(U_{10}/c\right)^3}\right] \qquad 0.83 < \frac{U_{10}}{c} < 5 \tag{5.8.4}$$

$$\gamma = \begin{cases} 1.7 & 0.83 < \frac{U_{10}}{c} < 1 \\ 1.7 + 6.0\log\frac{U_{10}}{c} & 1 \le \frac{U_{10}}{c} < 5 \end{cases}$$

where U_{10} is the component of wind in the mean wave direction. They recommend a directional frequency spectrum (an example of the observed directional spectrum is shown in Figure 5.8.1) of the form

$$\Phi(n,\theta) = \frac{1}{2}\Phi(n)\beta\,\mathrm{sech}^2\beta\left[\theta - \bar{\theta}(n)\right]$$

$$\begin{aligned} \beta &= 2.61\left(\frac{n}{n_p}\right)^{1.3} && 0.56 < \frac{n}{n_p} < 0.95 \\ &= 2.28\left(\frac{n}{n_p}\right)^{-1.3} && 0.95 < \frac{n}{n_p} < 1.6 \\ &= 1.24 \quad \text{otherwise} \end{aligned} \tag{5.8.5}$$

where $\bar{\theta}$ is the mean wave direction, and c is the phase speed.

The first-generation models of the 1960s and 1970s did not model the energy balance in spectral space. Also, they were based on concepts that extensive measurements in recent years have proven inaccurate. The most important deficiency was that the nonlinear transfer of energy among wave components was grossly underestimated and wind input overestimated. Observations in the 1970s and 1980s effected a fundamental change in our concepts of energy balance in spectral space and demonstrated the importance of nonlinear transfer in spectral space in addition to wind input and dissipation. This was reinforced by the painstakingly careful theoretical work by Hasselmann's group on resonant wave–wave interactions. This led to the development of second-generation

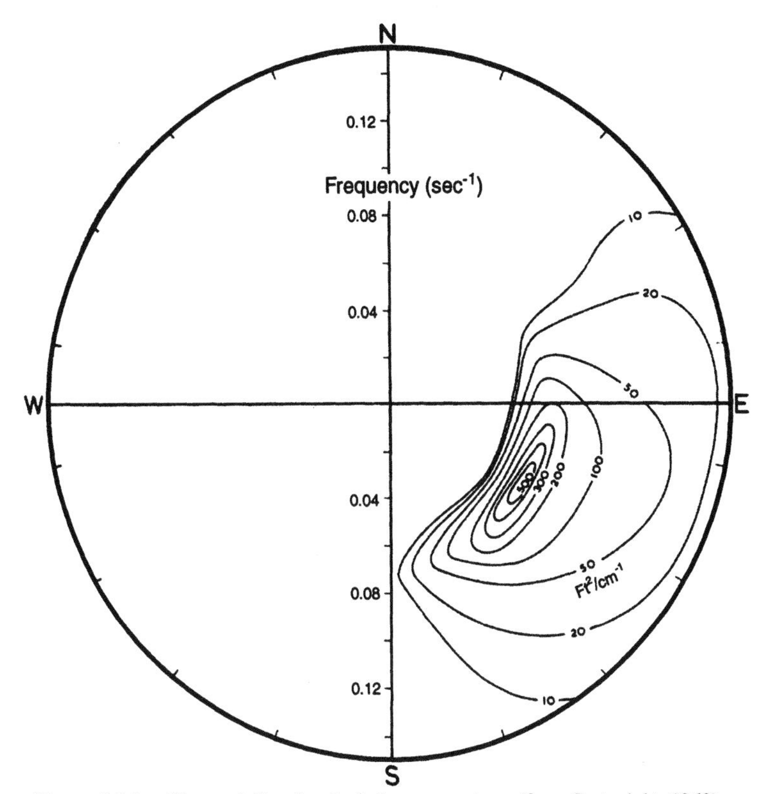

Figure 5.8.1. Observed directional wind-wave spectrum (from Cartwright, 1963).

models based on the spectral energy conservation equation, but for numerical reasons, the nonlinear transfer terms were simplified to the extent that ad hoc conformity to an assumed spectral shape beyond the peak was forced upon the model at one stage or another. While this worked for simple wind cases, these models failed to reproduce the wave field under complex, shifting wind fields accurately. SWAMP Group (1985) details the many models existing at the time and presents detailed comparison of each with data for conditions ranging from simple fetch- and time-limited growth to complex conditions corresponding to hurricanes and fronts. It is a fascinating summary and careful evaluation of the state-of-the-art as of the early eighties in numerical wind-wave prediction, which proved to be less than satisfactory. Figures 5.8.2 and 5.8.3 show the performance of various second-generation models for fetch- and duration-limited test cases.

Both the second- and third-generation wave models are based upon the conservation equation for the spectral energy density of the two-dimensional

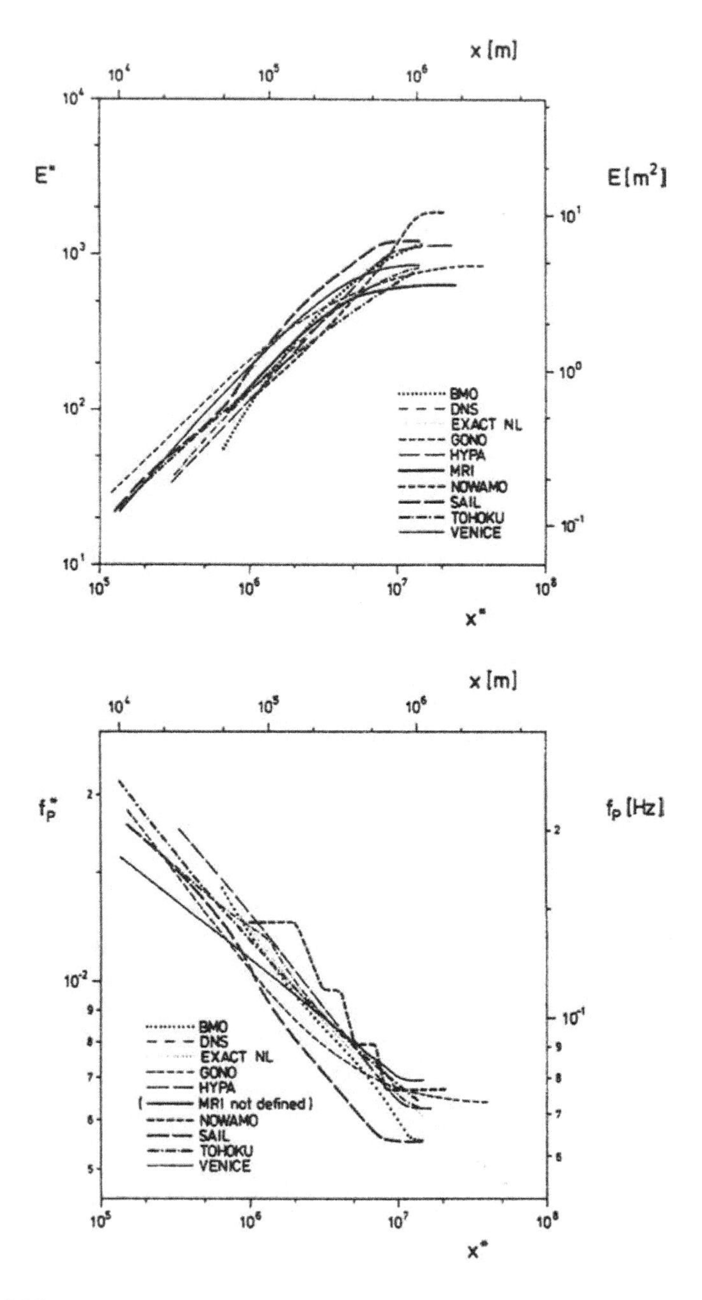

Figure 5.8.2. Normalized energy (top) and peak frequency (bottom) as a function of normalized fetch from various second-generation wave models for the fetch-limited case (from WAMDI, 1988).

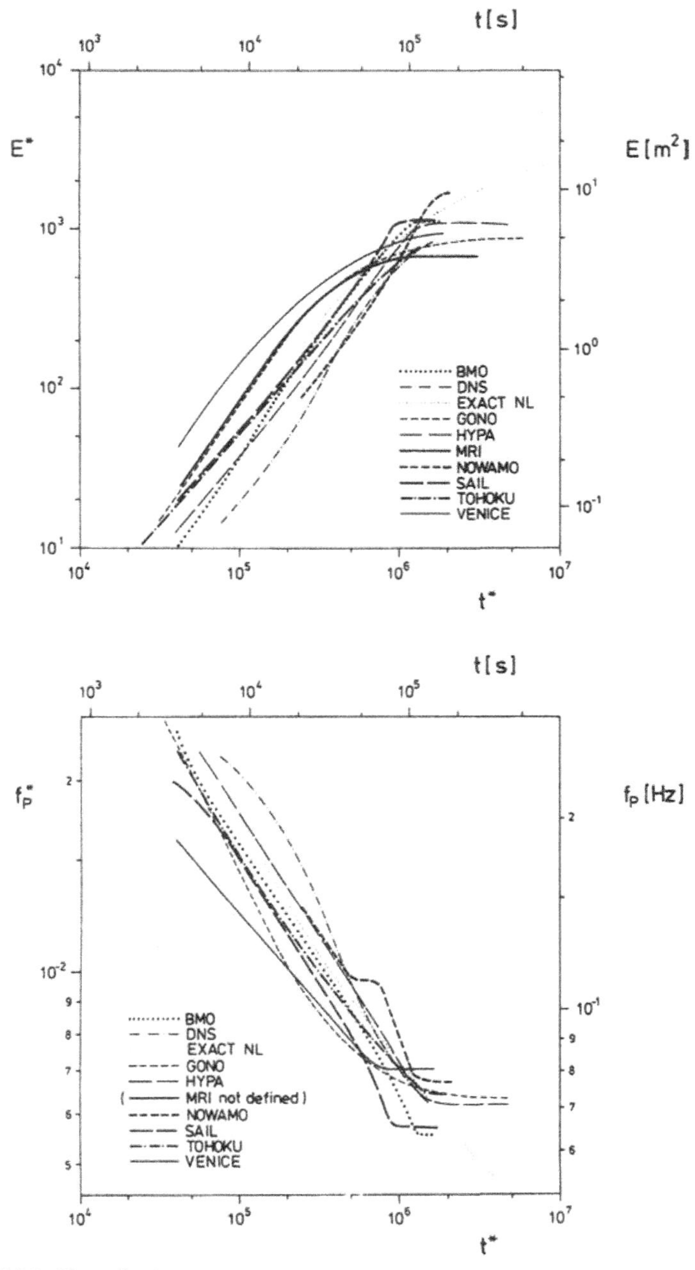

Figure 5.8.3. Normalized energy (top) and peak frequency (bottom) as a function of normalized fetch from various second-generation wave models for the duration-limited case (from WAMDI, 1988).

wave spectrum. The principal difference between the second generation (SWAMP Group, 1985) and the third generation (WAMDI Group, 1988; Komen *et al.,* 1994) is that the latter computes the spectrum from first principles. Both employ the same basic equation but differ in how the source, sink, and spectral transfer terms on the RHS are modeled. Since it is still numerically intensive to explicitly calculate the full nonlinear transfer integral, a parameterization with the same number of degrees of freedom as the spectrum itself is employed using the discrete interaction approximation of Hasselmann *et al.* (1985). The source function is adopted from empirical data of Snyder *et al.* (1981), but rescaled by u* instead of U_{10}, following Komen *et al.* (1984). The dissipation function is assumed to be a modified version of the form proposed originally by Komen *et al.* (1984). The third-generation wave prediction model WAM is therefore tuned to essentially reproduce the fetch-limited uniform wind case for which extensive observations are available (WAMDI Group, 1988). However, when applied to more complex situations, such as the hurricanes in the Gulf of Mexico or storms in the North Sea, WAM has demonstrated considerable skill in hindcasting the resulting wave field. Consequently, it is now routinely used for forecasting surface waves at many weather prediction centers around the world, which now routinely put out wave forecasts in addition to weather forecasts. In forecasting applications, however, the skill of even the most accurate wave model is dependent on the accuracy of the forecast wind fields and therefore the skill deteriorates rapidly for longer forecast ranges.

The model can be applied regionally or globally and for deep or shallow water. The details are presented by WAMDI Group (1988) and Komen *et al.* (1994). Here we will present a brief summary and a few results. The conservation equation for the density of the two-dimensional frequency spectrum $\Phi\ (n, \theta; \phi, \lambda; t)$ is

$$
\begin{aligned}
&\frac{\partial \Phi}{\partial t}+\frac{1}{\cos\phi}\frac{\partial}{\partial \phi}\left(\frac{d\phi}{dt}\cdot\cos\phi\cdot\Phi\right)+\frac{\partial}{\partial\lambda}\left(\frac{d\lambda}{dt}\Phi\right)+\frac{\partial}{\partial\theta}\left(\frac{d\theta}{dt}\Phi\right)\\
&\qquad = S_{in}+S_{ds}+S_{nl}+S_{bf}\\
&\frac{d\phi}{dt}=\frac{c_g}{R}\cos\theta \qquad\qquad (5.8.6)\\
&\frac{d\lambda}{dt}=\frac{c_g}{R}\frac{\sin\theta}{\cos\phi}\\
&\frac{d\theta}{dt}=\frac{c_g}{R}\sin\theta\tan\phi+\frac{1}{kR}\left(\sin\theta\frac{\partial d}{\partial\phi}-\frac{\cos\theta}{\cos\phi}\frac{\partial d}{\partial\lambda}\right)\frac{\partial\omega}{\partial d}
\end{aligned}
$$

where λ and ϕ are the longitude and latitude, and θ is the direction measured clockwise from north. The last term denotes the great circle refraction augmented by refractions due to changes in depth d of shallow water. The source functions on the RHS correspond to wind input, dissipation, nonlinear transfer, and bottom friction (for shallow water):

$$
\begin{aligned}
S_{in} &= \max\left[0,\, 0.25\frac{\rho_a}{\rho_w}\left(28\frac{u_*}{c}\cos\theta - 1\right)\right] n\Phi \\
S_{ds} &= -2.33\cdot 10^{-5}\,\bar{n}\left(\frac{k}{\bar{k}}\right)\left(\frac{\bar{\alpha}}{\bar{\alpha}_{PM}}\right)^2 \Phi \\
\bar{n} &= \frac{E}{F} = \frac{\iint \Phi(n,\theta)\,dn\,d\theta}{\iint \Phi(n,\theta)\frac{1}{n}\,dn\,d\theta} \\
\bar{k} &= \frac{E^2}{\left[\int k^{-1/2}\psi(k)\,dk\right]^2} \\
\bar{\alpha} &= E\bar{k}^2 \rightarrow E\bar{n}^4 g^{-2} \quad \text{as} \quad \bar{k}d \rightarrow \infty \\
\bar{\alpha}_{PM} &= 3.02\times 10^{-3} \\
S_{bf} &= -\frac{0.038}{g^2}\cdot\frac{n^2}{\sinh^2 kd}\Phi \\
c &= \left(\frac{g}{k}\tanh kd\right)^{1/2}, \quad c_g = \frac{1}{2}\left(\frac{g}{k}\tanh kd\right)^{1/2}\left[1 + \frac{2kd}{\sinh 2kd}\right]
\end{aligned}
\tag{5.8.7}
$$

The nonlinear source terms use two-dimensional discrete interaction approximation to the full five-dimensional one of all quadruplets by considering only a mirror symmetrical pair of discrete interaction configurations, an approximation that apparently works satisfactorily for fetch- and duration-limited wave growth (WAMDI Group, 1988). The spectrum is predicted at 25 frequency and 12 directional bands (for example, 0.042–0.41 Hz), and beyond the high frequency limit, a diagnostic n^{-4} tail is added (n^{-5} tail gives similar results, since the energy content is negligible in the tail):

$$
\Phi(n,\theta) = \Phi(n_1,\theta)\cdot\frac{c_{g1}}{c_g}\left(\frac{k}{k_1}\right)^{-2.5} \qquad (n > n_1) \tag{5.8.8}
$$

The details of the numerical implementation, as well as the justification for the carefully selected forms for the source and sinks, can be found in WAMDI

Group (1988) and Komen *et al.* (1994). The software package includes pre- and postprocessing routines as well as simple test cases to assure proper implementation (see Komen *et al.*, 1994, for details).

Figure 5.8.4 shows the evolution of the spectrum for the fetch-limited case. Figure 5.8.5 shows comparisons of measured and calculated spectrum for Hurricane Camille (from WAMDI Group, 1988). To provide another measure of the WAM skill, we present Figures 5.8.6 and 5.8.7, from Komen *et al.* (1994). These figures compare the winds and the waves measured by buoys deployed off the east coast of the United States during the 1990–1991 Surface Wave Dynamics Experiment (SWADE). Figure 5.8.6 compares buoy-measured winds at two locations with those from two different wind products, one an operational product from ECMWF and the other manually, carefully reanalyzed wind fields. Figure 5.8.7 compares the buoy-measured significant wave heights with those from WAM using these two wind products. The results emphasize the crucial nature of accurate wind input for accurate numerical wave predictions. While the model predictions are reasonable with operational wind input, its skill is remarkable when a more accurate wind field is available. Figure 5.8.8 shows the spatial distribution of the wave field during the SWADE storm for the two wind

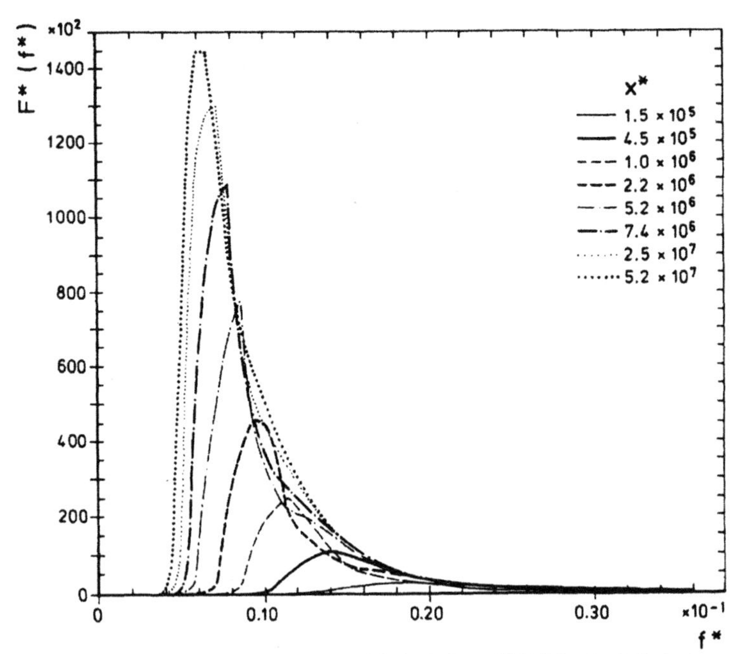

Figure 5.8.4. The evolution of spectrum with fetch from WAM for fetch-limited wave growth (from WAMDI, 1988).

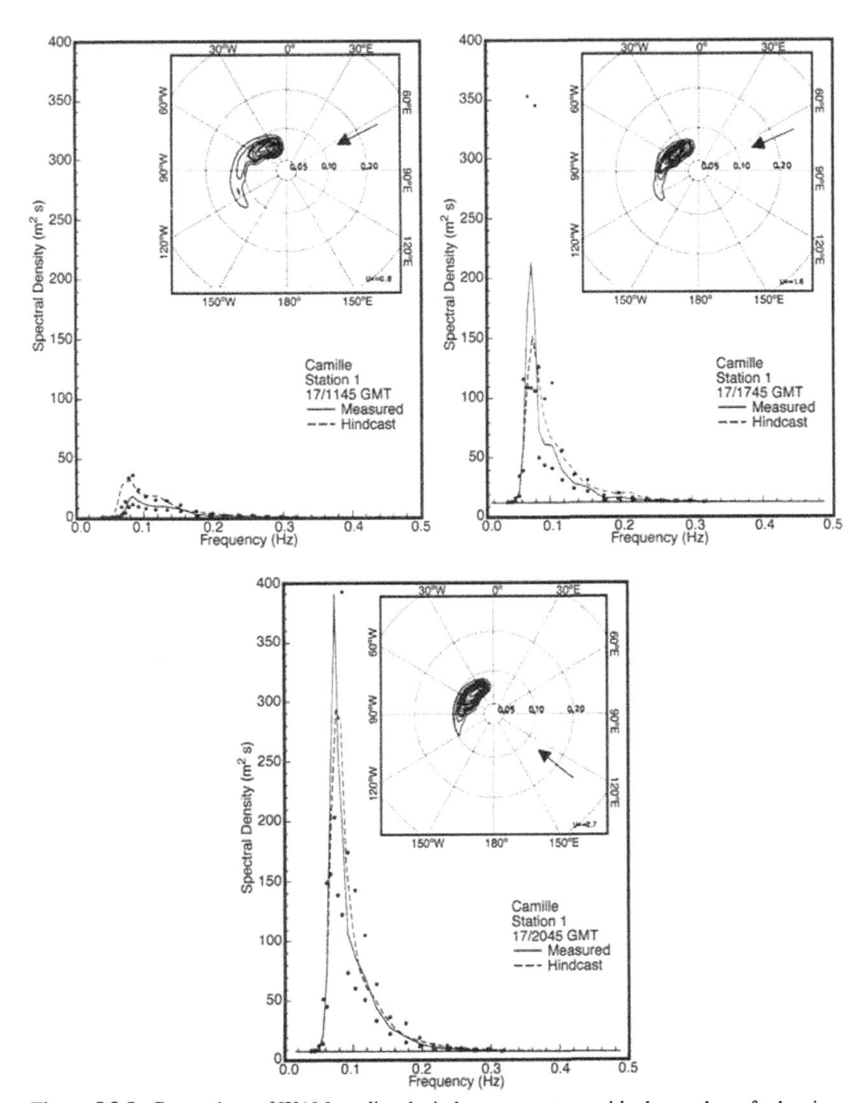

Figure 5.8.5. Comparison of WAM-predicted wind-wave spectrum with observed one for hurricane Camille in the Gulf of Mexico for three separate times (from WAMDI, 1988).

cases. This comparison highlights the prevailing problem in many oceanic predictions involving winds as input. It is often the quality of the wind forcing prescribed that determines the skill of the model. This is especially true in coastal oceans and semienclosed seas, where orographic effects and diurnal

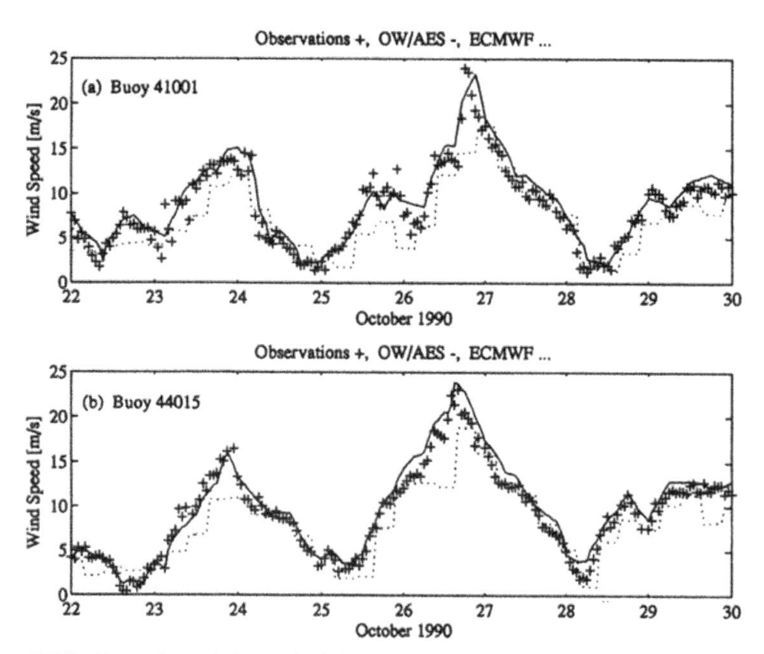

Figure 5.8.6. Comparison of observed winds at two buoys off the U.S. coast and wind analyses during SWADE (from Komen *et al.*, 1994).

variability in winds need to be faithfully reproduced before the winds can be used with confidence in deriving oceanic circulation and wave fields, etc., from numerical models.

While it is clear that the empirically derived source functions in the model could be refined in due time as dictated by more accurate observations, and the calculation of the nonlinear transfer term improved, the model has already shown useful skill in hindcasting waves. The model is widely available and is being widely used around the world by operational centers. However, more work is still needed. Heimbach *et al.* (1998) have compared WAM-derived monthly mean significant wave heights in various ocean basins to ERS-1 SAR-retrieved values. The comparison shows a systematic underestimation by WAM, mostly due to underestimation of swell by 20–30%, possibly due to strong damping in WAM at low frequencies. Simulation of the swell portion of the wind-wave spectrum by WAM has been particularly disappointing.

For practical applications, it is often necessary to know not the total energy density of the wave field or equivalently the rms amplitude, but the significant wave height, which is the average of the heights of one-third of the highest

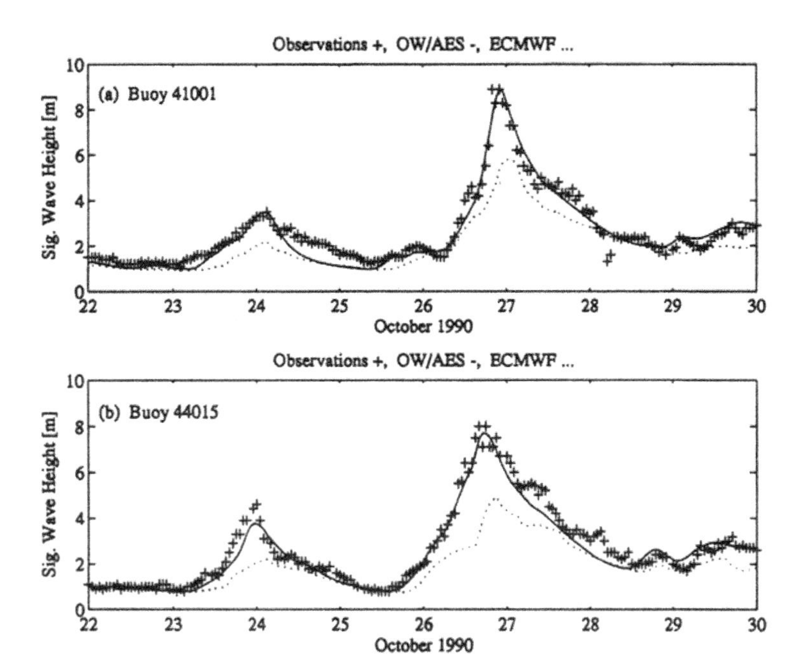

Figure 5.8.7. Comparison of observed and WAM-predicted waves at two buoys off the U.S. coast during SWADE. The plot shows the paramount importance of wind forcing on the accuracy of waves predicted (from Komen *et al.*, 1994)

waves (crest to trough). For a given probability distribution of wave heights, the significant wave height is related to the rms wave height and hence can be easily computed once the parameters of the spectrum are determined. Although the probability distribution of sea surface height appears to be fairly close to a Gaussian (Figure 5.8.9), it is misleading because the presence of short gravity–capillary waves located preferentially on the forward faces of steep gravity waves causes considerable skewness. Sea surface slope distribution is even more skewed since the windward faces of wind waves tend to be steeper than the leeward ones. Nevertheless, if the wave field consists of a superposition of many waves generated by wind in different regions, the phases are random and one can treat the sum as governed by the central limit theorem (which predicts a normal distribution) and so approximate the sea surface displacement by a Gaussian:

$$P(\zeta) = \left(2\pi\overline{\zeta}^2\right)^{-1/2} \exp\left[-\frac{\zeta^2}{2\overline{\zeta}^2}\right] \tag{5.8.9}$$

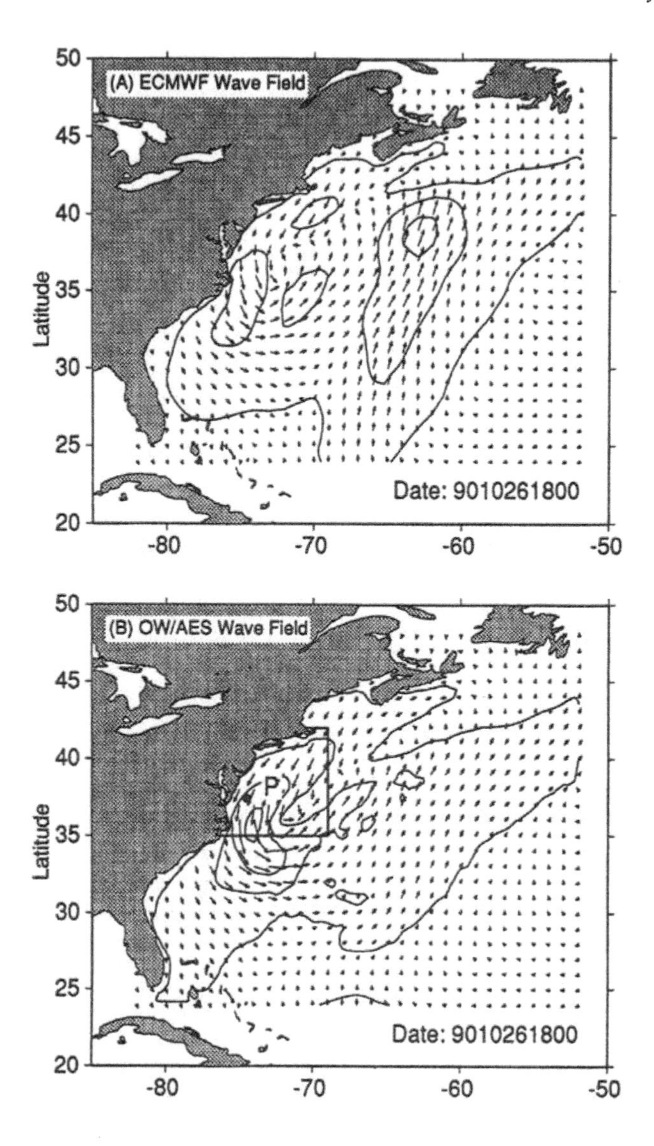

Figure 5.8.8. Wave field predicted off the U.S. coast during SWADE using the two wind analyses ECMWF (top) and OW/AES (bottom) (from Komen *et al.*, 1994).

The probability distribution of the wave height maxima depends on the width of the spectrum. For a very broad spectrum it is once again a Gaussian. However, if the spectrum is narrow (as, for example, when only swell is present), the

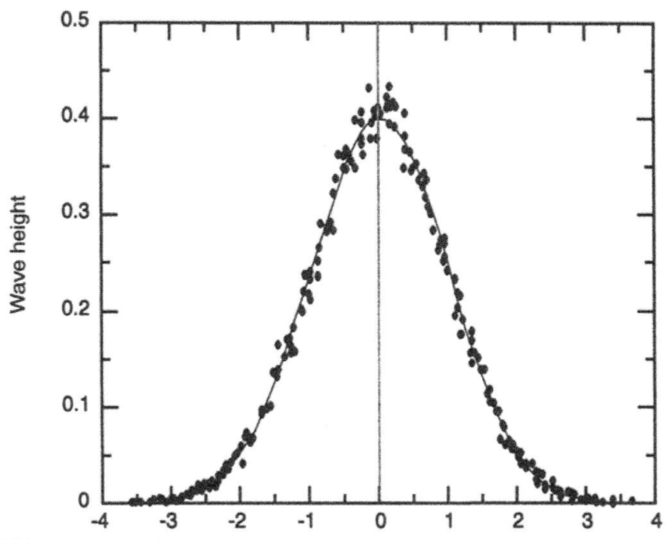

Figure 5.8.9. Probability distribution of wave heights. Even though the departures from the Gaussian are small, they are nevertheless important (from Carlson *et al.*, 1967).

distribution is best approximated by a Rayleigh distribution:

$$P(\zeta_m) = \left[\frac{2\zeta_m}{\left(\overline{\zeta^2}\right)^{1/2}} \exp\left(-\frac{\zeta_m^2}{2\overline{\zeta^2}} \right),\ 0 \right] \quad [\zeta_m > 0, \zeta_m < 0] \tag{5.8.10}$$

The wind-wave spectrum is usually quite narrow and therefore the distribution of maxima is closer to the Rayleigh distribution. For this distribution, the significant wave height is given by

$$H_{1/3} = 4\left(\overline{\zeta^2}\right)^{1/2} = 4\left[\int_0^{2\pi}\int_0^{\infty} \Phi(n,\theta)\, dn\, d\theta \right]^{1/2} \tag{5.8.11}$$

5.9 BREAKING WAVES

The air–sea interface acts as a significant barrier to the exchange of heat and gases between the ocean and the atmosphere. This is simply due to the fact that the transfer of these scalar quantities is mediated by a molecular sublayer at the air–sea interface and the associated molecular diffusion. However, the sea surface is invariably covered with surface waves, and under certain conditions, these waves break. When they do, additional transfer mechanisms come into play that are more efficient in effecting the transfer across the interface. For example, breaking waves entrain air in the form of small bubbles that are propelled into the water to depths on the order of the depth of breaking and therefore more efficiently transfer properties to the water. Similarly, spray and droplets ejected into the air during breaking are an efficient mechanism for transferring water vapor, heat, and dissolved gases from the ocean to the atmosphere. Breaking waves also create additional turbulence and mixing. These are the reasons why wave breaking is quite important to air–sea exchange processes. Wallace and Wirick (1992) have shown that the air–sea gas fluxes increase with increased surface wave activity. Kitaigorodskii (1984) has proposed that the gas transfer velocity is proportional to $Sc^{-1/2}(\nu\varepsilon_S)^{1/4}$, where ε_s is the dissipation rate at the surface, Sc is the Schmitt number, and therefore an increase in the dissipation rate by, say, a factor of 50, as recent observations suggest, leads to an increase in gas transfer by nearly a factor of 2.5. This is quite significant to air–sea interactions and their impact on climate.

Wave breaking disrupts the cool skin of the ocean, and the appearance of warmer water at the surface during a breaking event [see the dramatic IR images presented by Jessup (1995) and Jessup *et al.* (1997a,b)] is important to remote sensing of SST. It is also important dynamically. Surface waves mediate the transfer of momentum between the atmosphere and the ocean. Estimates (Mitsuyasu, 1985; Rapp and Melville, 1990) are that a large fraction of the momentum flux from the atmosphere to the ocean is transferred initially to surface waves, yet only a small fraction (a few percent) of the momentum flux is carried out of the generation region by waves (Melville, 1994). This is simply due to the fact that wave breaking transfers most of the momentum in the surface wave field to the currents. In the process, it also constitutes, a significant additional source of TKE for mixing in the upper ocean. The resulting increase in mixing near the air–sea interface also effects increased transfer across the interface of heat and gases, even in the absence of spray production and bubble injection (Dahl and Jessup, 1995). Extensive tower-based measurements during the Water Air Vertical Exchange Studies (WAVES) project in the second half of the 1980s (Agarwal *et al.*, 1992; Drennan *et al.*, 1996; Terray *et al.*, 1996) and other observations using vertical profilers (Anis and Moum, 1992, 1995) and from a submarine (Osborn *et al.*, 1992) have demonstrated the existence of a region of very high dissipation rate in layers adjacent to the air–sea interface that is of profound importance to matters related to mixing in the upper layers and air–sea exchanges.

Wave breaking is one well-known mechanism for dissipation of surface wave energy. However, it has been difficult to quantify—even more difficult than characterization of the generation mechanisms for surface waves. Consequently, there exists a significant gap in our knowledge and ability to model surface waves for various marine applications.

Because of their importance in a variety of applications, breaking waves have been getting increased attention in recent years (Banner and Grimshaw, 1992; Banner and Peregrine, 1993; Gemmrich *et al.*, 1994; Melville, 1996). Breaking of surface waves that involve air entrainment or whitecaps is invariably due to spilling on the forward face of the wave. The sea surface area covered by whitecaps is a strong function of the wind speed. Whitecaps begin to appear on the sea surface at a wind speed of about 3 m s^{-1}. Phillips (1985) and Wu (1988) find that the fraction of the sea surface covered by whitecaps is proportional to the cube of the friction velocity,

$$f_w \sim 0.2 u_*^3 \tag{5.9.1}$$

where u_* is the air-side friction velocity in m s^{-1}. Therefore the fractional area covered by whitecaps is never larger than a few percent (Wu, 1995) even at high wind speeds (2% at 12 m s^{-1}). Yet, they are important to air–sea transfer, because of the entrainment of air bubbles into the water and ejection of droplets by breaking waves into the air. These constitute mechanisms that circumvent the molecular sublayers at the air–sea interface and hence presumably are more efficient in effecting the air–sea transfer of water vapor and gas. Unfortunately, very little is reliably known about wave breaking and even the extent of it. For example, Ding and Farmer (1994) find a much weaker dependence of the fraction of whitecapping, $f_w \sim U_{10}$.

A dearth of measurements close to the air–sea interface, especially at high wind speeds, and reliance on observations at deeper levels led to the traditional view that the law of the wall prevails near the interface. However, recent measurements of dissipation rate in the upper layers of the ocean immediately adjacent to the air–sea interface at high wind speeds (Gargett, 1989; Drennan *et al.*, 1992, 1996; Agarwal *et al.*, 1992; Anis and Moum, 1992, 1995; Osborn *et al.*, 1992) have demonstrated quite conclusively that the dissipation rate in these upper layers is one to two orders of magnitude larger than expected from the law of the wall scaling arguments. This is definitely due to the presence of breaking waves and their influence on a shallow layer with a depth on the order of the wave height. Measurements of dissipation rate in the northeast Pacific show marked elevations in the surface layer and with a z^{-4} behavior much different from the classical z^{-1} wall layer dependence (Denman and Gargett, 1995). Figure 2.2.4 shows dissipation rates near the ocean surface scaled as

$$\frac{\varepsilon z}{u_\#^3} = \frac{1}{\kappa} f\left(\frac{gz}{u_\#^2}\right) \tag{5.9.2}$$

Equation (5.9.2) is equivalent to assuming that the wave field is fully developed. The WAVES observations were mostly for immature wave fields.

Laboratory measurements of Rapp and Melville (1990) also show enhanced dissipation levels due to surface wave breaking. These observations (Melville, 1994) show that more than 90% of the energy lost due to breaking is lost quite quickly, within about four wave periods, with the breaking event itself lasting approximately a wave period, and more than 50% of the energy loss is expended in entraining air bubbles against the action of gravitational forces. The bulk of the dissipation takes place in a layer of a depth on the order of the height of the breaking wave and not its wavelength. After four wave periods, the dissipative layer is 1–2 wave heights thick, but even after 100 wave periods, its depth is still on the order of the wave height. This suggests that the influence of wave breaking in elevating the dissipation rate in the upper ocean is normally confined to a layer several meters thick near the air–sea interface.

Field observations of wave breaking processes are inherently difficult to make. In addition to conventional techniques, active microwave and acoustic methods are being used (Melville, 1996). Acoustic methods appear to be especially promising. Breaking waves lead to bubble entrainment, and the bubble cloud in turn causes generation and radiation of acoustic energy that is useful for tracking and quantifying wave breaking. At frequencies greater than 1/2 kHz, the radiated acoustic energy appears to correlate well with the energy dissipated (Melville, 1996). Ding and Farmer (1994) have been able to use this radiated acoustic energy to infer the characteristics of the breaking events.

Gargett (1989) found that the dissipation rate in the upper layers of the ocean decayed as z^{-4}. This decay law is similar to that due to turbulence created by a stirring grid (Hopfinger and Toly, 1976), which can be characterized by a turbulence velocity scale $q \sim z^{-1}$, and a turbulence length scale $\ell \sim z$, where z is the distance from the grid, so that the dissipation rate $\varepsilon \sim q^3/\ell \sim z^{-4}$ and the eddy viscosity ($\sim q\ell$) is constant. Thus there is a rapid decay of TKE ($\sim z^{-2}$) in this layer and therefore its contribution to mixing in the bulk of the OML and hence to its deepening may not be very significant, unless the OML is rather shallow. The presence of this layer of elevated turbulence and dissipation rate is however quite important to air–sea exchanges.

Terray *et al.* (1996) present a comprehensive review of the enhanced dissipation in the near-surface layers. They argue that since the depth of the region of wave breaking scales as the wave height for a monochromatic wave (Melville, 1994, 1996), for a spectrum of waves, the relevant scale should be the significant wave height H_s. Using Eq. (5.7.1) to estimate the growth rate and hence the flux of energy to growing waves from the expression,

$$\frac{F}{\rho_w} = \frac{\tau_a \bar{c}}{\rho_w} = u_{*a}^2 \bar{c} \frac{\rho_a}{\rho_w} = g \int\int \frac{\partial \phi(n,\theta)}{\partial t} dn d\theta = g \int\int \beta\, \phi(n,\theta) dn d\theta \qquad (5.9.3)$$

they show that $\bar{c}/c_p$ increases linearly with u_*/c_p up to u_*/c_p values of 0.075, beyond which it remains constant at approximately 0.5 (see Figure 5.9.1). Since dimensional arguments suggest that

$$\frac{\varepsilon H_s}{F/\rho_w} = f\left(\frac{z}{H_s}, \frac{c_p}{u_*}\right) \tag{5.9.4}$$

they argue that close to the surface, the second term is unimportant, thereby deriving a power law dependence of normalized dissipation rate on z/H_s,

$$\frac{\varepsilon H_s}{F/\rho_0} = 0.3\left(\frac{z}{H_s}\right)^{-2} \tag{5.9.5}$$

in the range of wave age c_p/u_* between 4.3 and 7.4 (the exponent is derived from a fit to data). This contrasts with the scaling used in Figure 2.2.4, based on Eq. (5.9.2), which displays considerable scatter and fails to collapse the WAVES data.

Craig and Banner (1994) used second-moment closure to simulate turbulence generated by breaking waves and found a $z^{-3.4}$ power law dependence for dissipation rate. Anis and Moum (1992) found a z^{-3} dependence in their microstructure measurements. Drennan *et al.* (1996) also found a z^{-2} dependence in ship observations during SWADE consistent with the WAVES tower observations of Terray *et al.* (1996). Thus there is considerable disagreement as to the exact value for the exponent of the power law and it is not clear at this point what the decay rate should be.

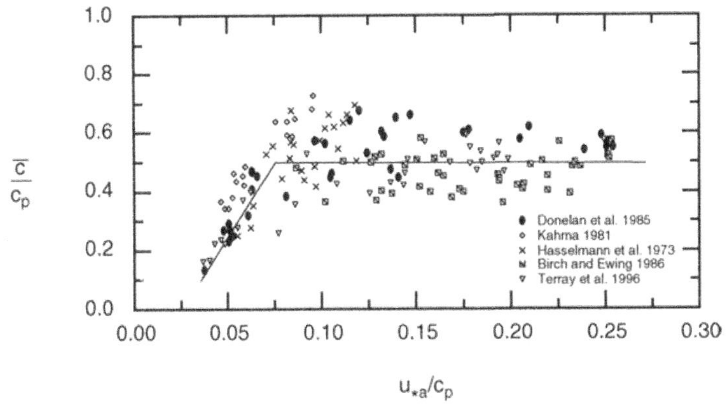

Figure 5.9.1. Dependence of the ratio of the effective velocity in the wave energy flux to the phase speed at the peak of the spectrum with inverse wave age. Note the nearly constant value for young waves (from Terray *et al.*, 1996).

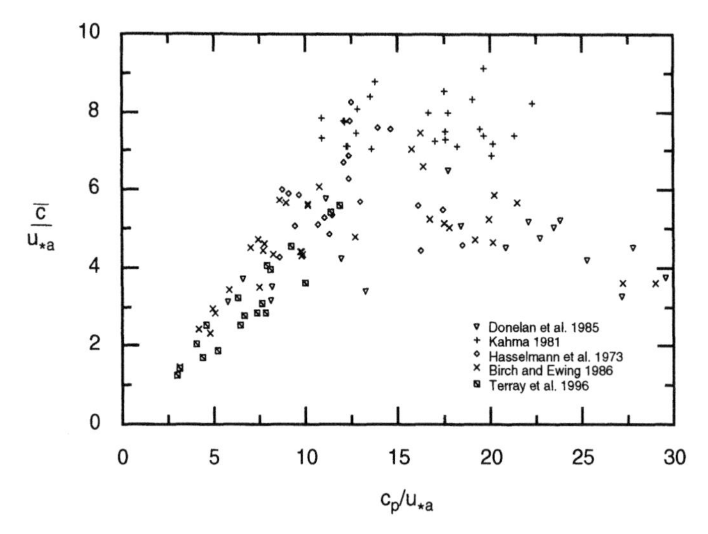

Figure 5.9.2. Dependence of the effective velocity in the wave energy flux normalized by friction velocity on wave age (from Terray *et al.*, 1996).

Sufficiently far away from the wall (at $z \sim 3.6 H_s \overline{c} / u_*$), one recovers the law of the wall scaling:

$$\frac{\varepsilon z}{u_{\#}^3} = \frac{1}{\kappa} \tag{5.9.6}$$

This depth at which the law of the wall is attained can be as large as 8–10 m for a very mature sea (wave age ~ 30), consistent with the observations of Anis and Moum (1992) and Osborn *et al.* (1992). Immediately adjacent to the wall, for values of z less than about 0.6 H_s, it is possible that the dissipation rate is approximately constant and equal to the value given by Eq. (5.9.5) at $z = 0.6\ H_s$, although experimental evidence for this is still lacking.

Note that the ratio of the energy flux from the wind to the waves to that directly to currents is $\overline{c} / U_s \sim 2\overline{c} / u_*$, where based on Jin Wu's estimate (Wu, 1975) the drift current U_s has been put equal to $u_*/2$. Figure 5.9.2, from Terray *et al.* (1996) and showing the dependence of $\overline{c} / u_*$ on wave age, illustrates the gross underestimate in the energy flux to the ocean if the pathway through breaking waves is ignored. Models of mixing in the OML must account for this energy flux, especially for shallow diurnal OMLs.

Thorpe (1992b, 1995) has reviewed the dynamics of the surface layer, including the bubble-cloud-affected surface layer. Thorpe (1993) suggests that the energy loss per unit surface area due to wave breaking can be parameterized as

$$\frac{E}{\rho_0} \sim 3.0\text{x}10^{-5}\,U_{10}^{3}\left(\frac{c_b}{c_p}\right)^{5} \tag{5.9.7}$$

where c_b is the characteristic speed of the breaking waves. The value of c_b/c_p varies from 0.4 to 0.75 (Melville, 1994; Ding and Farmer, 1994). The phase speeds c_b is approximately equal to $\overline{c}$, since waves much beyond the spectral peak are saturated.

The enhanced dissipation rate and entrainment of air bubbles by breaking waves both enhance gas transfer across the air–sea interface. Waves begin to break around 3–5 m s^{-1} and the enhanced dissipation rate in the surface layer leads to enhanced gas transfer rates (Kitaigorodskii, 1984) even without the entrainment of air bubbles. Bubble entrainment however increases rapidly with the increase of wind speed. Above a wind speed of 12 m s^{-1}, bubbles may therefore dominate the gas flux across the interface. See Thorpe (1992b) for a recent review of the subject of bubble clouds formed by wave breaking and their role in air–sea gas transfer.

Observations are continuously adding to our knowledge of wave breaking processes. But much remains to be done, especially in modeling their effects on air–sea exchange. Melville (1996) points out that strongly wind-forced short wavelength (10–100 cm) surface waves which have steep forward faces and which break almost continuously make their appearance at very high wind speeds but have not at all been studied. Studies of breaking waves and their impact are still in their infancy.

5.10 SATELLITE MEASUREMENTS OF OCEAN WAVES

Traditionally waves have been measured using *in situ* sensors. In addition to special observational programs such as HEXOS and RASEX, autonomous buoys deployed in the coastal oceans of the United States and elsewhere have carried wind and wave sensors that have increased our knowledge of the wave field around the globe. However, these traditional methods have grossly inadequate spatial and temporal coverage. But the advent of satellite-borne microwave sensors has enabled wave fields to be measured and monitored on a global basis. In addition to the sea surface height, satellite radar altimeters provide information on the significant wave height. The distortion of the radar pulse by the sea surface contains information on the sea state, and the significant wave height can be estimated from the slope of the leading edge of the altimeter return pulse, if one assumes that the spatial statistics of the sea surface are linear and Gaussian

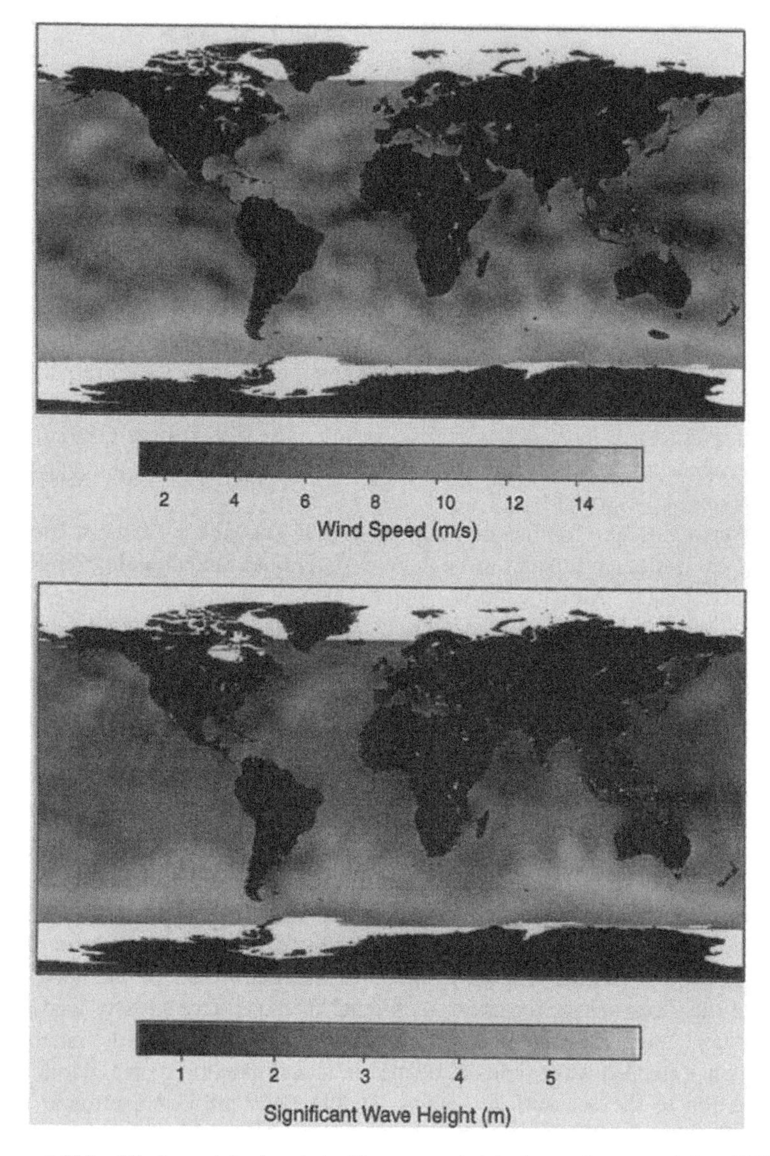

Figure 5.10.1. Wind speed (top) and significant wave height (bottom) measured from TOPEX altimeter (courtesy of NASA).

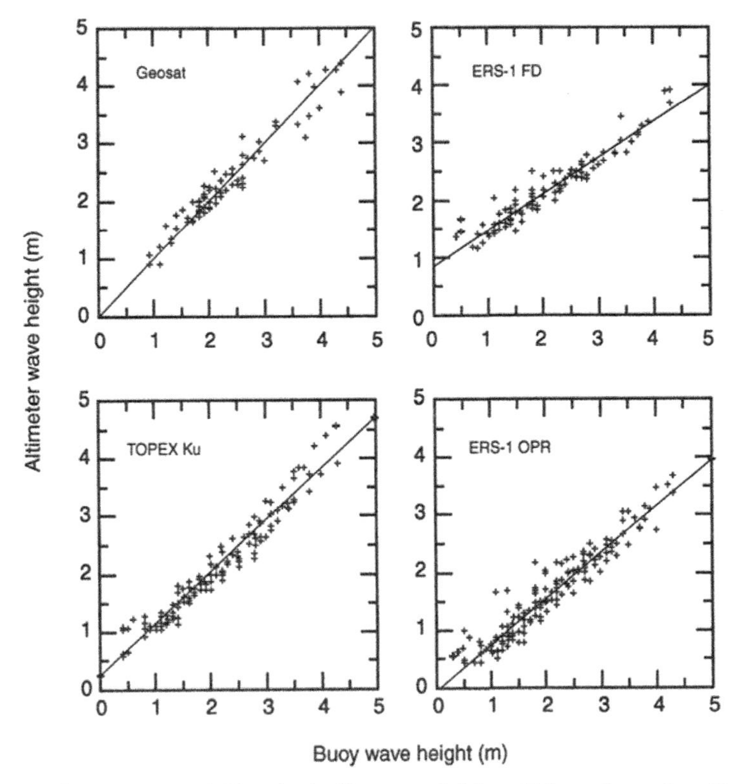

Figure 5.10.2. Comparison of altimetric significant wave heights with buoy observations. The two are very well correlated (from Cotton and Carter, 1994).

(Cotton and Carter, 1994). Past altimetric missions such as SEASAT and GEOSAT, current ones such as ERS-1/2 and TOPEX/Poseidon, and future ones such as Jason1 promise to add considerably to our knowledge of wind-wave climate in the global oceans. Figure 5.10.1 is a typical example from the TOPEX altimeter, which shows significant wave heights during a 10-day cycle. The large wave heights in the southwest Atlantic are noteworthy. Figure 5.10.2, from Cotton and Carter (1994), shows comparison of altimetric monthly mean significant wave heights derived from GEOSAT, ERS-1, and TOPEX altimeters with observed values from 24 NDBC open-ocean buoys deployed along the east and west coasts of the United States and off Hawaii. For ERS-1, both fast delivery (FD) and off-line products (OPR) are compared. The agreement is remarkably good. Figure 5.10.3 shows similar comparisons of wind speed and wave height with buoy-measured values from Barstow (1996), showing once again the excellent correlation between the two.

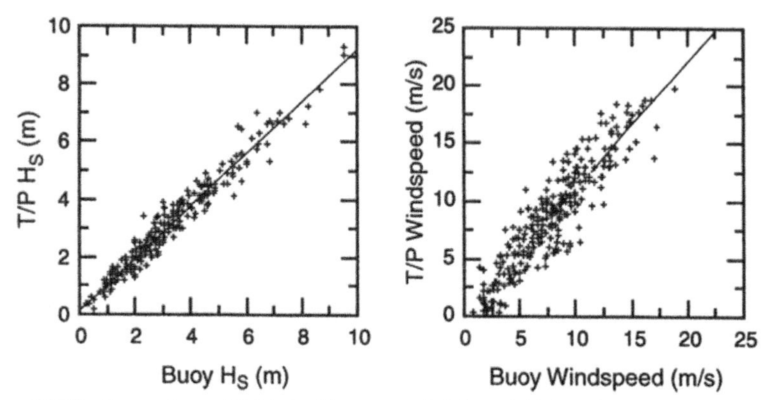

Figure 5.10.3. Comparison of (left) significant wave height and (right) wind speed from TOPEX altimetry compared with buoy observations (from Barstow, 1996).

As a result, we now have global wave climatology from GEOSAT, ERS-1, and TOPEX sources, starting from the mid-1980s to present, with some gaps. Efforts such as that of Young and Holland (1996), who have put together monthly wave climatology from GEOSAT altimeter data on a CD-ROM, are underway to make these data widely available.

Microwave energy, Bragg-backscattered from capillary/gravity–capillary waves, enables a synthetic aperture radar (SAR) to provide information on the wave field. While the extraction of the wave field information is complicated by the modulation of these short waves by longer ones, the added advantage of SAR is the possibility of deriving the two-dimensional wave spectrum. Komen *et al.* (1994) discuss the use of SAR on board ERS-1 for measuring the significant wave height and the mean direction of waves globally.

One of the reasons given for the development of the advanced third-generation numerical wave model WAM is its potential use in conjunction with satellite observations for wave nowcasting and forecasting applications around the globe (Komen *et al.*, 1994). Tremendous strides have been made toward this goal as can be seen by the examples of operational applications provided by Komen *et al.* (1994). Satellite wave field observations have been used not only for routine monitoring of the wave field and verification of wave model results, but also for assimilation into wave models for a more realistic depiction of the wave fields by the model. Assimilation of satellite observations into the WAM model is described in great detail by Komen *et al.* (1994).

Remote sensing of sea state by microwave sensors is an important and steadily and rapidly growing field. It requires study of the details of the interactions between high frequency (low wavelength) electromagnetic radiation and small scale capillary–gravity waves in the wind-wave spectrum, and of the reflection of electromagnetic wavefronts by a sea surface roughened by surface gravity waves. These topics are however beyond the scope of this book, and the reader is referred instead to books by Apel (1987) and Komen *et al.* (1994).

To sum up, this fascinating centuries-old topic of surface gravity waves is by no means fully deciphered at present. It still holds many secrets and surprises, and much work lies ahead.

LIST OF SYMBOLS

α	Phillips constant; also Charnock constant
β	Toba constant; also wave growth rate
δ	Wave boundary layer thickness
γ	Kinematic surface tension coefficient
ε	Dissipation rate
φ	Velocity potential; also latitude
$\phi(n)$	Frequency spectrum
$\phi(n,\theta)$	Directional frequency spectrum
κ	Von Karman constant
λ	Wavelength; also longitude
χ	Normalized wave-induced velocity perturbation
ν, ν_t	Kinematic viscosity (molecular and turbulent)
θ	Angle of the waves to wind direction
ρ_w, ρ_a	Density of water and air
ζ	Sea surface height
τ_O, τ_w	Shear stress and wave-induced shear stress
$\Psi(k_j)$	Wavenumber spectrum
a	Wave amplitude
c, c_g	Phase velocity and group velocity
$_d$	Drag coefficient
$_p$	Phase speed at the spectral peak
d	Depth of the water column
f	Coriolis parameter
g	Acceleration due to gravity
g_{ij}	Coefficients in a resonant wave–wave interaction matrix
j	Summation index (takes only values of 1 and 2)
k	Wavenumber; also summation index that takes values of 1 to 3
k_T, k_S, k_c	Diffusivity of heat and salt, and a passive scalar (kinematic)
l	Integral microscale of turbulence
n	Wave frequency
p, p_a	Pressure and atmospheric pressure

t	Time
w	Vertical component of wave-induced velocity
u_*	Friction velocity corresponding to the air side
$u_\#$	Friction velocity corresponding to the water side
x	Horizontal coordinate; also fetch
x_i	Coordinates
x_3	Vertical coordinate
z	Vertical coordinate
z_0	Roughness scale
z_c	Critical layer height
A	Wave action
C	Wave phase speed
C_g	Group velocity
E	Energy density of the wave
$E(k_i)$	Wavenumber spectrum
F	Energy flux from wind to waves
$F(k_j,n)$	Wavenumber–frequency spectrum
$H_{1/3}$, H_s	Significant wave height
L	Monin–Obukhoff length scale
$N(k_j)$	Spectral density of wave action
P	Mean pressure; also probability
R	Covariance
R_w	Wave Reynolds number
$S(k_j)$	Spectrum of sea surface slope
S_{ij}	Radiation stress
u_i	Fluid velocity
U_{10}	Wind velocity at anemometric height
V_S	Stokes drift velocity due to surface waves

Chapter 6

Internal Waves

In this chapter, we will discuss the salient features of oceanic internal waves and their generation, dissipation, propagation, and mutual interactions. Tidally generated internal wave solitons are also described. The role of deep-sea internal waves in abyssal mixing is discussed. This review relies on and borrows heavily from the articles by Olbers (1983) and Munk (1981), and the monograph by Phillips (1977). In the interests of brevity, many interesting details are omitted; the reader is referred to these sources, which have a more thorough treatment of the subject, especially the article by Olbers (1983). Proceedings of a workshop on oceanic internal gravity waves held in Hawaii (Muller and Henderson, 1991) is also useful. There exists of course considerable literature on one or another aspect of internal gravity waves scattered around in scientific journals, the most pertinent to our readers being the *Journal of Fluid Mechanics,* the *Journal of Physical Oceanography,* and the *Journal of Geophysical Research.* Though virtually invisible to the naked eye, internal waves, both in the atmosphere and the oceans, are a fascinating and important part of small scale processes in geophysical flows. Most of the treatment for oceanic internal waves applies equally well to atmospheric internal waves also, but we will not deal with internal waves in the atmosphere explicitly here. Their one important distinguishing characteristic is that, as they propagate up into the upper atmosphere (stratosphere, mesosphere) away from their source regions in the lower troposphere, their amplitude increases rapidly commensurate with the decrease in the ambient density, and therefore they are quite ubiquitous in remotely sensed data of the upper atmosphere.

6.1 SALIENT CHARACTERISTICS

Internal waves occur in stably stratified fluids through the restoring action of the buoyancy forces on water parcels displaced from their equilibrium position. A good example is internal (or interfacial) waves on the buoyancy (or density) interface between two layers of a stably stratified fluid. Akin to the air–sea interface, when this interface is disturbed, waves are radiated away horizontally along the interface. In the interior of a continuously stratified rotating fluid, with characteristic frequencies of N and f, where N is the buoyancy or Brunt–Vaisala frequency and f is the inertial frequency, free internal waves are radiated at an angle to the vertical, but are confined to frequencies between f and N. In the mid/high latitude upper ocean, N is typically one or two orders of magnitude larger than f (N ~ 10^{-3} to 10^{-2} s^{-1}; f ~ 10^{-4} s^{-1}). Therefore at low frequencies, close to f, rotational effects are important, and the waves are called inertial-internal waves. At high frequencies, close to N and far from f, rotational effects are negligible and the waves are called simply internal waves. In the following, we will not always use the term inertial-internal waves, but it is understood that internal waves mean inertial-internal waves when their frequency is comparable to the inertial frequency. In the mid/high latitude deep oceans though, f and N (typically 10^{-4} s^{-1}) are comparable. It is worth noting that f goes to zero as the equator is approached and therefore the rotational effect becomes smaller (f_y, the horizontal component of rotation, becomes important), and this has an important effect on the internal wave (IW) structure and dynamics. Most of our knowledge about internal waves comes from midlatitude oceans, although more recent observations are adding considerably to internal wave processes in the equatorial waveguide (for example, Moum *et al.,* 1992a) and in high latitude oceans (for example, Levine *et al.,* 1986, 1997). The earliest work on the subject of internal waves is due to Stokes (1847) for two-layer fluids and Rayleigh (1883) for continuously stratified ones.

Internal waves are ubiquitous in the oceans. They show up in temperature, salinity, and current measurements almost anywhere in the ocean and have often been regarded as a mere nuisance, unwanted, and to be removed from oceanic measurements. They produce "scintillation" in acoustic propagation through the oceans and therefore are important to underwater acoustics, acoustic tomography, and detection of underwater vehicles. But their importance to mixing in the deep ocean and hence the dynamics of ocean circulation has been recognized only in recent years. They are part of the cascade of energy from large scales to dissipation scales in the oceans. The mixing they produce disperses pollutants in the deep ocean (and so do internal waves in the atmosphere). About 3.5 TW of energy loss in the Earth–Moon–Sun gravitational system occurs by generation and dissipation of oceanic tides (Kantha *et al.,* 1995); it is generally thought that internal tides account for 10–15% of this tidal dissipation (Wunsch, 1975; Kantha, 1998; Kantha and Tierney, 1997; Munk and Wunsch, 1998). Their dissipation more than accounts for the 10^{-5} m^2 s^{-1} vertical mixing observed in the deep oceans by modern microstructure measurements (Kunze and Sanford, 1996). The energy density in the deep-sea internal wave field, an average of 3800 J m^{-2} (the canonical Garrett–Munk Spectrum), is larger than that in the barotropic tides (an average of 1800 J m^{-2}). Unlike the sea surface, which can be calm at times, the ocean interior is never calm because of the ever-present internal waves.

The typical velocity associated with internal waves is 5 cm s^{-1}, typical vertical length scales range from a few meters to about one kilometer, typical amplitudes from meters to tens of meters, typical periods from minutes to hours, and typical horizontal scales from a few meters to a few tens of kilometers (Olbers, 1983; Muller *et al.*, 1986). Internal tides have wavelengths of 100 (semidiurnal) to 200 (diurnal) km, and amplitudes often as high as 100 m at some places during spring tides. These scales overlap the scales of turbulent motions in the ocean and even though these time and space scales can be measured easily, it is often difficult to separate the two. Extracting internal waves from the background "noise" and potential contamination by fine structure and Doppler shift due to prevailing ocean currents involves a series of difficult steps (Muller *et al.*, 1986). Herein lies one difficulty in studying internal waves.

The very first observation of internal waves was made by Nansen a century ago, when he observed that his sailing ship slowed down considerably in the Barents Sea when it encountered a layer of brackish water overlaying saltier water. This "dead water" phenomenon was explained by Ekman as being due to the energy expended by the ship in generating internal waves on the interface between the two layers and the consequent drag on the ship. Lord Rayleigh was the first to investigate internal waves in continuously stratified fluids. In recent times, the development of dropped fine structure and microstructure measuring instruments, moored arrays, and towed sensors has enabled internal waves in the oceans to be measured and quantified. These measurements have shown that on the average, a remarkable universality exists in the shape of the spectrum of midlatitude deep-sea internal waves (see Section 6.6).

Examples of internal waves are the now-classic observations of the vertical displacement frequency spectrum of an isotherm off California by Cairns (1975) and towed vertical displacement spectra by Katz (1975). Figure 6.1.1 shows observations of the frequency spectrum of vertical displacement of an isotherm due to IWs at a depth of 350 m, 800 km off southern California, measured by Cairns (1975) using a device attached to a mooring that oscillated in the vertical about an isotherm. It shows a –2 power law decay and a sharp peak near the buoyancy frequency before an abrupt drop-off at higher frequencies. Figure 6.1.2 shows the wavenumber spectra of vertical displacements of the 12°C isotherm in the Sargasso Sea measured by Katz (1975) using a towed device that also oscillated around an isotherm. It also shows a –2 power law for the spectrum over a broad range of wavenumbers. Figure 6.1.3 shows the frequency spectrum of currents from a subsurface mooring at a 600-m depth in the western

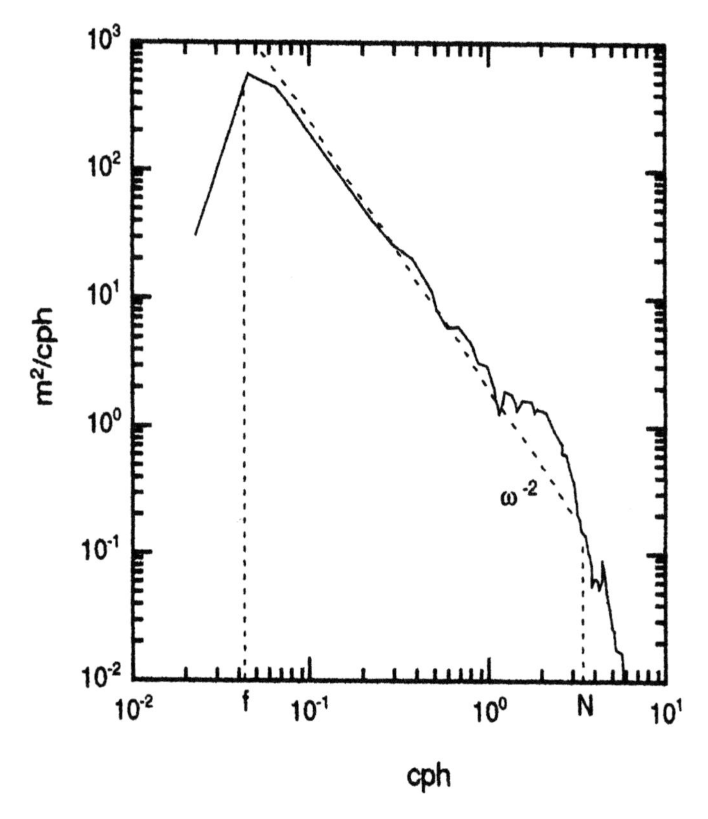

Figure 6.1.1. Frequency spectrum of vertical displacement of isotherms due to internal waves (from Cairns and Williams, 1976).

North Atlantic (Fu, 1981). Prominent peaks at tidal and inertial frequencies can be seen. Finally, Figure 6.1.4 shows the displacement and horizontal current spectra from careful measurements conducted during IWEX (Internal Wave Experiment) (Muller *et al.*, 1978). Modern remote sensing instruments such as synthetic aperture radar (SAR) have indicated the presence of internal waves throughout the global oceans; this they do by detection of the modulation the internal waves produced in sea surface roughness. Large internal wave solitons generated by the passage of tidal currents through narrow straits and passages have been detected in the Sulu and Andaman seas. Figure 6.1.5 shows large internal wave soliton packets generated and propagating in the Sulu Sea (Apel *et al.*, 1985). Both internal tides and internal wave solitons are often detectable in precision altimetry, even though their surface manifestation is a modulation of the sea level by a few centimeters. Figure 6.1.6 shows M_2 internal tides observed

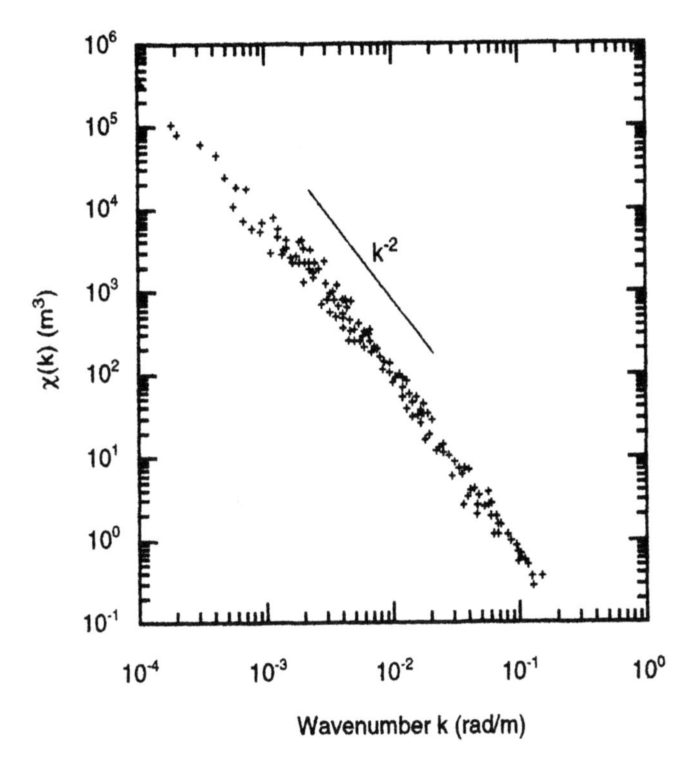

Figure 6.1.2. Wavenumber spectrum of vertical displacement of isotherms due to internal waves (from Katz, 1975).

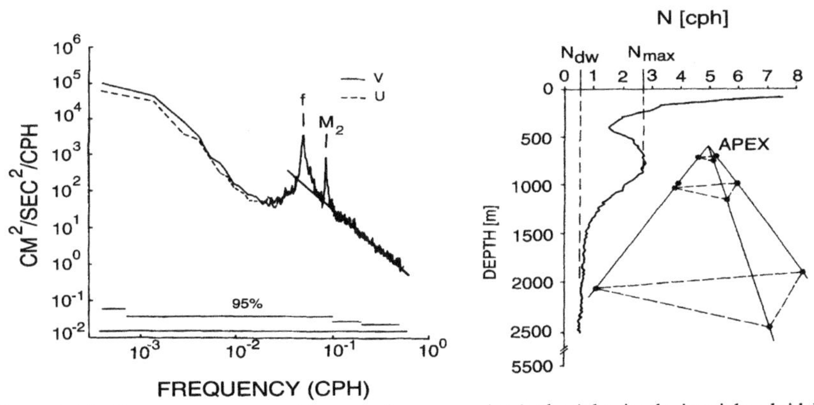

Figure 6.1.3. Frequency spectrum of currents from a mooring in the Atlantic; the inertial and tidal frequencies are shown (left from Fu, 1981) (Right) Profile of buoyancy frequency at the IWEX site; the geometry of the IWEX array is schematically indicated (from Muller *et al.* 1978).

by the NASA/CNES TOPEX/Poseidon precision altimeter (Kantha and Tierney, 1997), and Figure 6.1.7 shows the IW solitons observed by the same altimeter in the Sulu Sea.

It is possible to divide internal waves into three components (Levine *et al.*, 1997) based on frequency: (1) near-inertial waves that have a frequency close to the local inertial frequency and hence are strongly influenced by rotation, (2) internal tides at both semidiurnal and diurnal frequencies generated by barotropic tidal currents flowing over topographic changes, and (3) the internal wave "continuum" between frequencies N and f. Most of the work on internal waves (see Munk, 1981) is on the third, although the second is getting increased attention. Inertial-internal waves are principally generated in the upper ocean by storms and propagate into the interior, and therefore exhibit considerable variability in time and space. Kunze (1985) derives a dispersion relation for these waves in the presence of geostrophic shear (such as due to mesoscale eddies) that suggests that in a warm core (anticyclonic) eddy, near-inertial energy generated by winds propagates downward and becomes trapped in the core of the eddy, whereas in a cold core (cyclonic) eddy, the energy propagates horizontally away from the eddy. This inertial chimney (Lee and Niiler, 1998) is

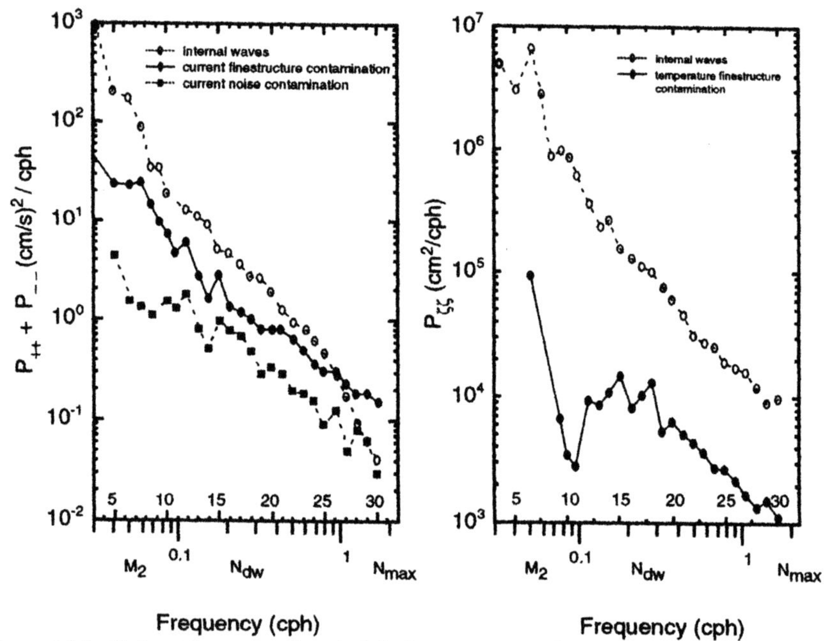

Figure 6.1.4. Horizontal current and vertical displacement spectra from IWEX; the distribution of buoyancy frequency with depth is also shown (from Muller *et al.*, 1978).

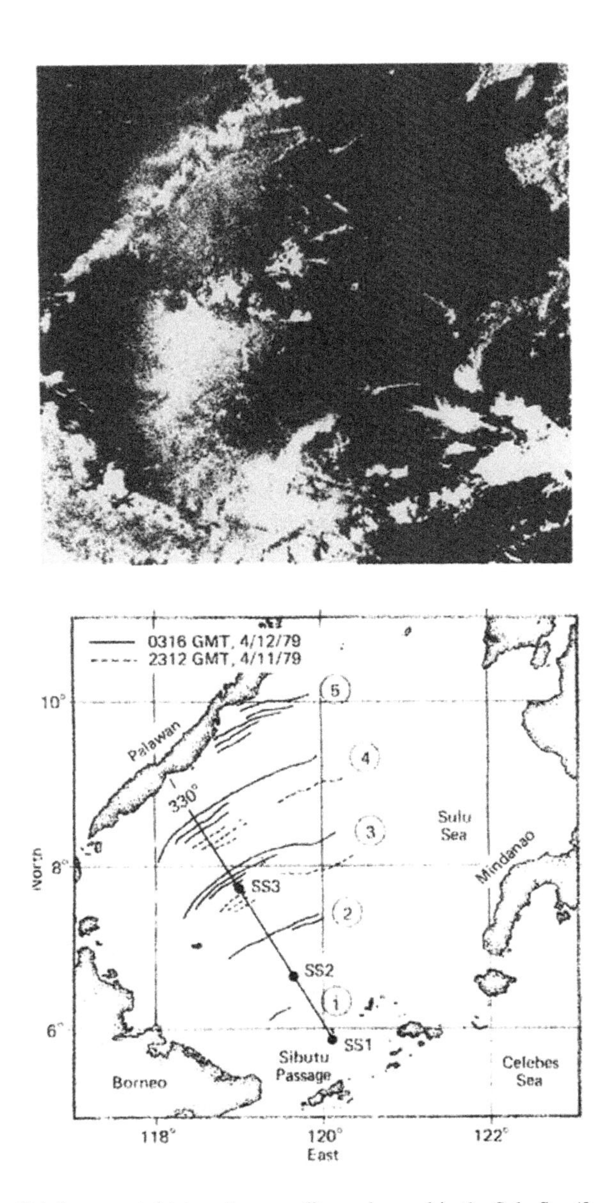

Figure 6.1.5. Tidally generated internal wave solitons observed in the Sulu Sea (from Apel *et al.*, 1985).

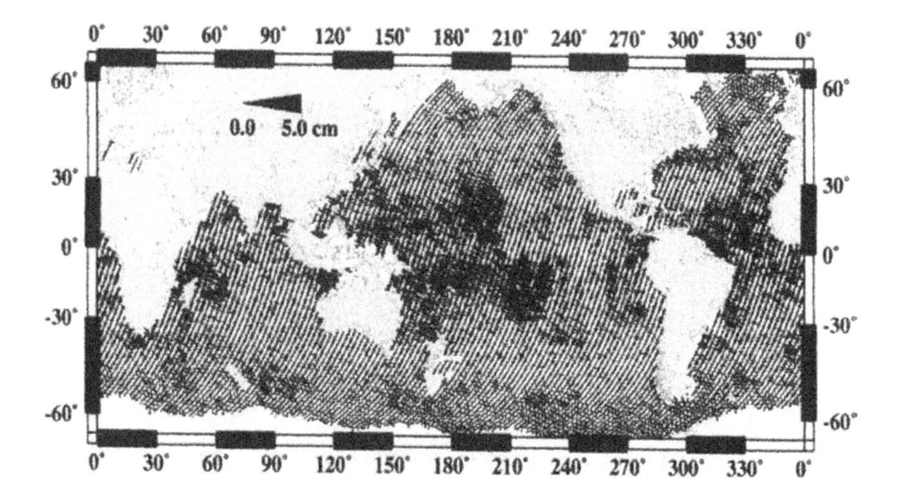

Figure 6.1.6 Internal tides observed by the NASA TOPEX precision altimeter in the global oceans. Note the large amplitudes near midocean ridges, islands, and other topographic changes.

an efficient mechanism for propagation of wind-generated near-intertial energy into deeper layers. Internal tides are also site-specific. Consequently, unlike the continuum, which is remarkably steady in time and homogeneous in space, and hence describable by a universal spectrum (Garrett and Munk, 1972, 1979; Munk, 1981), no such description exists for these two.

The internal wave field in the ocean consists of a superposition of many waves with different frequencies, wavenumbers, and amplitudes. Waves are generated at different locations by whatever mechanisms that prevail at each, propagate, and fill the ocean interior, undergoing strong, rapid nonlinear transfer of energy between frequencies in the process, before they dissipate and contribute to internal mixing in the oceans. The result is a fairly random and nearly universal internal wave continuum. Whatever small differences exist help identify the sources and sinks (Wunsch, 1975). Because of its stochastic nature, this field is best described statistically. Unlike surface waves but not unlike turbulence, three-dimensional space and time are involved in such a description. One needs to characterize internal waves by a wavenumber–frequency spectrum. The forcing of internal waves is broadband, the generation mechanisms many, and the nonlinear transfer in spectral space quite strong. In general, the spectrum can be represented conveniently as a superposition of linear vertically propagating waves with random amplitudes and phases, subject to constraints dictated by the dispersion relation. However, because of the strong nonlinear interactions, it is difficult to derive the spectrum theoretically (a situation not

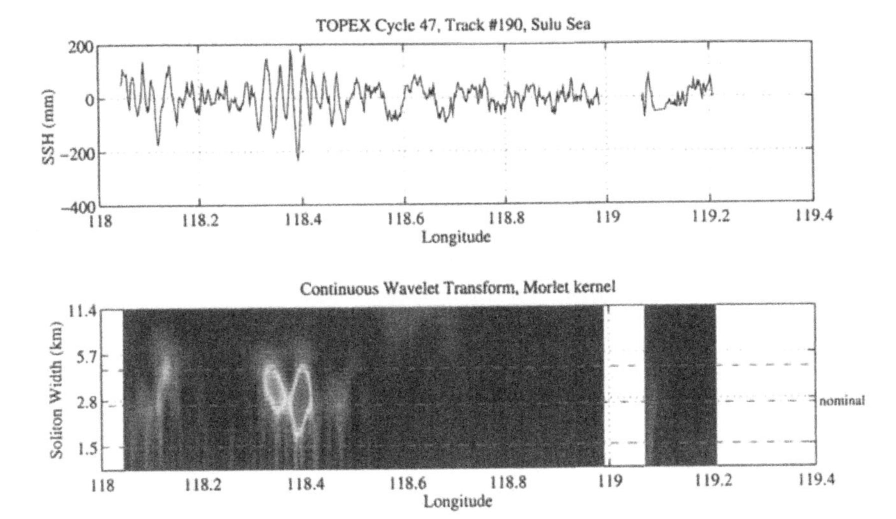

Figure 6.1.7 Internal wave solitons observed in 10-Hz TOPEX altimetric records in the Sulu Sea. The top panel shows the along-track signal, while the bottom panel shows its wavelet transform showing the highlighted soliton packet in alongtrack distance–period space.

unlike that in turbulence), despite many attempts (McComas and Muller, 1981; Henyey *et al.,* 1986) (for a survey, see Olbers, 1983; Muller *et al.,* 1986). Empirical construction has been possible and a notable success in this direction is the Garrett–Munk (GM) empirical spectrum of the midlatitude deep-ocean internal wave field (Munk, 1981). Observations indicate that while there are deviations from the universal GM spectrum, overall the internal wave field in the deep ocean, away from sources and sinks of internal wave energy such as seamounts and canyons, has a remarkably universal shape and level (typical value is 3.8×10^3 J m^{-2} within a factor of two or so). The principal reason appears to be the strong and rapid resonant interactions among internal wave modes that tend to rapidly and efficiently restore the spectrum to a universal equilibrium form.

For most purposes in dealing with inertial-internal waves, those internal waves that have frequencies low enough to be affected by rotation, it is adequate to regard the rotation rate as constant (f-plane approximation). For those waves that traverse large distances in the meridional direction, latitudinal variations of the planetary rotation, for the most part, can be treated using WKB approximations. Horizontal variations in the ambient currents and buoyancy frequency can also be dealt with using WKB methods, as long as the scales of the variability in the propagating medium are much larger than the length and timescales associated with these waves. Most often it is the vertical gradients of currents and density that are important to the propagation of internal waves and it is permissible to assume horizontal homogeneity of the medium. Therefore, the most important parameters for the internal wave problem are the buoyancy frequency N(z) and the current U(z). N has a maximum immediately below the mixed layer in the seasonal thermocline (typically 10^{-2} s^{-1} or a period of 10 min) and another in the permanent thermocline (roughly 10^{-3} s^{-1}, a period of 1.5 hr). In the deep ocean N can be as low as 10^{-4} s^{-1} (a period of 16 hr).

Internal wave motions are also important in the atmosphere, not only because they can transfer momentum and energy from near the ground level up into the atmospheric column, but also from considerations such as clear-air turbulence resulting from breaking internal waves. Their dynamics are very much similar to oceanic internal waves, but the sources are somewhat different. Orography such as mountain ranges are primary sources of internal waves in the atmosphere. Spectacular cloud formations showing internal waves in the lee of mountain ranges such as the Rocky Mountains and Sierra Madre are routinely visible in satellite imagery (see Chapter 3). These internal waves radiate energy and momentum upward well into the troposphere and the upper atmosphere. The rapid decrease of atmospheric density with height causes the wave amplitudes to also increase rapidly with height and this is why internal wave motions are very prominent in the upper atmosphere. Critical layer processes and exchange of momentum between internal gravity waves and the mean flow are also important to the dynamics of the lower atmosphere. Atmospheric tides (see Chapter 6 of Kantha and Clayson, 1999) are essentially internal waves with semidiurnal and diurnal frequencies generated by solar radiation absorption in the troposphere and the middle atmosphere. Vertically propagating long-period internal waves generated near the equator are thought to be responsible for the quasi-biennial oscillation with a period of 26 months observed in the tropical stratosphere (Lindzen, 1990). Large eddies in the daytime CABL also generate internal waves (which drain energy from these eddies) at the top of the inversion capping the CABL, and intermittent breaking of internal waves constitutes an important source of mixing in the stably stratified NABL (see Chapter 3). Internal waves in the vicinity of the capping inversion are evident in remote probing of the atmospheric boundary layer by lidars and radars. Typical buoyancy period in the atmosphere is about 5 minutes (Lindzen, 1990), comparable to the values in the oceanic seasonal thermocline.

6.2 GOVERNING EQUATIONS

Internal wave motions can be modeled using incompressible Navier–Stokes equations under Boussinesq approximation. Neglecting viscous stresses,

$$\frac{\partial \tilde{u}_k}{\partial x_k} = 0 \tag{6.2.1}$$

$$\frac{\partial \tilde{u}_j}{\partial t} + \frac{\partial}{\partial x_k}\left(\tilde{u}_j \tilde{u}_k\right) + \varepsilon_{jkl} f_k \tilde{u}_l = -\frac{\partial \tilde{p}}{\partial x_j} + \tilde{b}_j \delta_{j3} \quad (j=1,2,3) \tag{6.2.2}$$

$$\frac{\partial \tilde{\rho}}{\partial t} + \frac{\partial}{\partial x_k}\left(\tilde{u}_k \tilde{\rho}\right) = 0 \tag{6.2.3}$$

where $\tilde{\rho}$, $\tilde{p}$, $\tilde{u}_i$ are the density, pressure divided by density, and velocity, and $\tilde{b}_j$ is the buoyancy $(= g\tilde{\rho}/\rho_0)$, where g is the gravitational acceleration. Now consider a small departure from equilibrium with the fluid stably stratified in the vertical and in hydrostatic equilibrium:

$$\begin{aligned} \tilde{u}_j &= u_j \\ \tilde{\rho} &= \overline{\rho}(z) + \rho \\ \tilde{p} &= P + p \end{aligned} \tag{6.2.4}$$

The equations for perturbation quantities ρ, u_j, and p become

$$\frac{\partial u_j}{\partial x_j} + \frac{\partial w}{\partial z} = 0 \tag{6.2.5}$$

$$\frac{\partial u_j}{\partial t} + u_k \frac{\partial u_j}{\partial x_k} + \varepsilon_{jkl} f_k u_l = -\frac{\partial p}{\partial x_j} \tag{6.2.6}$$

$$\frac{\partial w}{\partial t} + u_k \frac{\partial w}{\partial x_k} = -\frac{\partial p}{\partial z} + b \tag{6.2.7}$$

$$\frac{\partial \rho}{\partial t} + u_k \frac{\partial \rho}{\partial x_k} + w \frac{d\overline{\rho}}{dz} = 0 \tag{6.2.8}$$

with $j = 1, 2$ (indexes k and l assume values of 1 to 3), and we have separated the horizontal component equation from the vertical component of velocity w. Eliminating the p terms in Eq. (6.2.6) and (6.2.7) by cross differentiation,

$$\frac{\partial}{\partial t}\left(\frac{\partial u_j}{\partial z}-\frac{\partial w}{\partial x_j}\right)-\frac{\partial b}{\partial x_j}+\varepsilon_{jkl}f_k\frac{\partial u_l}{\partial z}=\frac{\partial}{\partial x_j}\left(u_k\frac{\partial w}{\partial x_k}\right)-\frac{\partial}{\partial z}\left(u_k\frac{\partial u_j}{\partial x_k}\right) \tag{6.2.9}$$

Equation (6.2.8) can be written as

$$\frac{\partial b}{\partial t}=N^2(z)w\text{-}u_k\frac{\partial b}{\partial x_k} \tag{6.2.10}$$

where $N^2(z)=-\frac{g}{\rho_0}\frac{\partial\bar{\rho}}{\partial z}$ is the square of the buoyancy (Brunt–Vaisala) frequency. Taking $\partial/\partial t$ of Eq. (6.2.9) and using Eq. (6.2.10),

$$\begin{aligned}\frac{\partial^2}{\partial t^2}\left(\frac{\partial u_j}{\partial z}-\frac{\partial w}{\partial x_j}\right)-N^2(z)\frac{\partial w}{\partial x_j}+\varepsilon_{jkl}f_k\frac{\partial^2 u_l}{\partial t\partial z}\\ =\frac{\partial}{\partial t}\left\{\frac{\partial}{\partial x_j}\left(u_k\frac{\partial w}{\partial x_k}\right)-\frac{\partial}{\partial z}\left(u_k\frac{\partial u_j}{\partial x_k}\right)\right\}+\frac{\partial}{\partial x_j}\left(u_k\frac{\partial b}{\partial x_k}\right)\end{aligned} \tag{6.2.11}$$

Now taking the horizontal divergence $\partial/\partial x_j$ and using Eq. (6.2.5),

$$\begin{aligned}\frac{\partial^2}{\partial t^2}\left(-\frac{\partial^2 w}{\partial z^2}-\frac{\partial^2 w}{\partial x_j\partial x_j}\right)-N^2(z)\frac{\partial^2 w}{\partial x_j\partial x_j}+\frac{\partial}{\partial t\partial z}\left(f_k\varepsilon_{jkl}\frac{\partial u_l}{\partial x_j}\right)+\varepsilon_{jkl}\frac{\partial f_k}{\partial x_j}\frac{\partial^2 u_l}{\partial t\partial z}\\ =\frac{\partial^3}{\partial t\partial x_j\partial z}\left(-u_k\frac{\partial u_j}{\partial x_k}\right)+\frac{\partial^2}{\partial x_j\partial x_j}\left[u_k\frac{\partial b}{\partial x_k}+\frac{\partial}{\partial t}\left(u_k\frac{\partial w}{\partial x_k}\right)\right]\end{aligned} \tag{6.2.12}$$

Note that in Eqs. (6.2.9)–(6.2.12), j can take values of 1 and 2 only, while k can assume values of 1, 2, and 3. If we recognize

$$\omega_k=\varepsilon_{jkl}\frac{\partial u_l}{\partial x_j} \tag{6.2.13}$$

is the vorticity vector, and invoking f-plane approximation $\frac{\partial f_k}{\partial x_j} = 0$,

$$\frac{\partial^2}{\partial t^2}\left(\frac{\partial^2 w}{\partial x_j \partial x_j} + \frac{\partial^2 w}{\partial z^2}\right) + N^2(z)\frac{\partial^2 w}{\partial x_j \partial x_j} + \frac{\partial}{\partial z}\left(f_k \frac{\partial \omega_k}{\partial t}\right) = R\left(x_j, z, t\right) \qquad (6.2.14)$$

where

$$R\left(x_j, z, t\right) = \frac{\partial^3}{\partial t \partial x_j \partial z}\left(u_k \frac{\partial u_j}{\partial x_k}\right) - \frac{\partial^2}{\partial x_j \partial x_j}\left[u_k \frac{\partial b}{\partial x_k} + \frac{\partial}{\partial t}\left(u_k \frac{\partial w}{\partial x_k}\right)\right] \qquad (6.2.15)$$

An equation for vorticity can be written,

$$\frac{\partial \omega_k}{\partial t} + \frac{\partial}{\partial x_j}\left(u_j \omega_k\right) + \frac{\partial}{\partial z}\left(w \omega_k\right) = -\frac{\partial}{\partial x_l}\left(f_k u_l\right) \qquad (6.2.16)$$

and therefore the following expression can be substituted in Eq. (6.2.14):

$$f_k \frac{\partial \omega_k}{\partial t} = -f_k \frac{\partial}{\partial x_l}\left(f_k u_l\right) + f_k\left[-\frac{\partial}{\partial x_j}\left(u_j \omega_k\right) - \frac{\partial}{\partial z}\left(w \omega_k\right)\right] \qquad (6.2.17)$$

We then obtain, recognizing that $\left(f_1, f_2, f_3\right) \equiv \left(0, 0, f\right)$ and therefore that only the vertical component ω_3 need be accounted for in the vorticity equation,

$$\frac{\partial^2}{\partial t^2}\left(\frac{\partial^2 w}{\partial x_j \partial x_j} + \frac{\partial^2 w}{\partial z^2}\right) + N^2(z)\frac{\partial^2 w}{\partial x_j \partial x_j} + f^2 \frac{\partial^2 w}{\partial z^2} = \bar{R}\left(x_j, z, t\right) \qquad (6.2.18)$$

where

$$\bar{R}\left(x_j, z, t\right) = R\left(x_j, z, t\right) - f_k\left[\omega_l \frac{\partial u_k}{\partial x_l} - \frac{\partial}{\partial x_j}\left(u_j \omega_k\right) - \frac{\partial}{\partial z}\left(w \omega_k\right)\right] \qquad (6.2.19)$$

We have therefore managed to obtain an equation for the vertical velocity perturbations in a stably stratified fluid on a rotating plane. $\bar{R}$ contains all the nonlinear terms in the equation. If we now consider only infinitesimal disturbances and assume zero mean shear, $\partial u_j / \partial z = 0$, then we get $\bar{R} = 0$ and the governing equation for infinitesimal internal waves in a stably stratified fluid becomes

$$\frac{\partial^2}{\partial t^2}\left(\frac{\partial^2 w}{\partial x_k \partial x_k}\right) + N^2(z)\frac{\partial^2 w}{\partial x_j \partial x_j} + f^2 \frac{\partial^2 w}{\partial z^2} = 0 \qquad (6.2.20)$$

We can seek wave-like solutions to this equation,

$$w(x_j, z, t) = W(z)\exp\left[i(k_j x_j - nt)\right] \tag{6.2.21}$$

where k_j is the horizontal component of the wavenumber (k_3 is the vertical wavenumber) and n is the intrinsic frequency of the internal wave. Equation (6.2.20) then reduces to

$$\frac{d^2W}{dz^2} + \left(\frac{N^2(z) - n^2}{n^2 - f^2}\right) k_h^2 W = 0 \tag{6.2.22}$$

where $k_h = \left(k_1^2 + k_2^2\right)^{1/2}$ is the magnitude of the horizontal wave vector.

6.3 VERTICALLY PROPAGATING SMALL SCALE INTERNAL WAVES

When the vertical scale of the ambient stratification is large compared to the scale of the internal waves, N(z) can be regarded as a slowly varying function of z, and WKB methods can be used to obtain solutions to Eq. (6.2.20). However, it is easier to regard N(z) as constant in Eq. (6.2.22) and seek solutions. The two methods are equivalent. In the former, N(z) is regarded as a constant locally, whereas in the latter, it is assumed *explicitly* to be a constant. If we now seek solutions of the form

$$W(z) = A\exp(ik_3 z) \tag{6.3.1}$$

so that we are seeking solutions of the form

$$w(x_j, z, t) = A\exp\left[i(k_j x_j + k_3 z - nt)\right] \tag{6.3.2}$$

substitution of Eq. (6.3.1) in Eq. (6.2.22) or Eq. (6.3.2) in Eq. (6.2.20) gives the dispersion relation for internal waves,

$$\left(\frac{k_3}{k_h}\right)^2 = \frac{N^2 - n^2}{n^2 - f^2} = \tan^2\theta \tag{6.3.3}$$

where θ is the inclination of the *total* wavenumber vector to the horizontal plane.

This equation can be rearranged to yield

$$n^2 = N^2 \cos^2 \theta + f^2 \sin^2 \theta \tag{6.3.4}$$

Note that

$$\frac{k_3}{k} = \sin\theta, \quad \frac{k_h}{k} = \cos\theta, \tag{6.3.5}$$

where k is the magnitude of the wavenumber vector,

$$k^2 = k_h^2 + k_3^2 = k_j k_j + k_3^2 \tag{6.3.6}$$

There are two important consequences of the dispersion relation for internal waves [Eqs. (6.3.3) or (6.3.4)]:

1. The intrinsic frequency n is independent of the magnitude of the wave vector k and depends *only* on the inclination of the wave vector to the horizontal, angle θ:

$$n = n(\theta) \tag{6.3.7}$$

2. The intrinsic frequency is bounded on both sides:

$$f \le n \le N \tag{6.3.8}$$

This means that internal waves with frequencies greater than the Brunt–Vaisala frequency or with frequencies less than inertial frequency cannot exist. In other words, the period of wave motions must be less than the inertial period and greater than the buoyancy period. Internal waves that satisfy the dispersion relations [Eqs. (6.3.3) and (6.3.4)] are also called inertial-internal waves. When $f = 0$, or when $n \gg f$, it is possible to ignore f in all considerations so that the dispersion relation becomes

$$n = N \cos\theta \tag{6.3.9}$$

Such waves are simply called internal waves.

The independence of the intrinsic frequency from the magnitude of the wave vector is a very unusual property peculiar to internal waves. Most other wave motions have n dependent on k (for example, surface gravity waves and planetary Rossby waves). Internal waves are anisotropic and their frequency depends on the direction of their propagation with respect to the vertical.

For the case of no rotation (f = 0), a simple dynamical explanation for Eq. (6.3.9) was given by Phillips (1967). Consider a fluid particle displaced upward along a plane inclined at an angle θ to the vertical, by a distance x parallel to the plane (see Figure 6.3.2). Its excess density relative to its surroundings is the background density gradient times its vertical displacement $(x\cos\theta)$. This gives rise to a buoyancy force that acts vertically downward. The component of this along the plane of motion is $\left(\frac{\partial\bar{\rho}}{\partial z}x\cos\theta\right)\cdot g\cos\theta$. This must be balanced by the acceleration downward of the fluid parcel along the inclined plane:

$$\begin{aligned} \rho_0\frac{d^2x}{dt^2} &= -g\frac{\partial\bar{\rho}}{\partial z}\cdot x\cos^2\theta \\ \frac{d^2x}{dt^2}-\left(N^2\cos^2\theta\right)x &= 0 \end{aligned} \tag{6.3.10}$$

This is a simple harmonic oscillator with frequency $n=N\cos\theta$. Therefore the fluid particle oscillates in a plane inclined at an angle to the vertical with frequency $N\cos\theta$. When $\theta = 0$, n = N; a fluid particle displaced in the vertical in a stably stratified fluid oscillates in the vertical plane with a frequency equal to the buoyancy frequency of the fluid. This is the maximum frequency possible for wave motions. If one remembers that because of the incompressibility of the fluid, particle motions are constrained to planes normal to the wave vector k_i, then θ is also the inclination of the wave vector to the horizontal.

For low frequency inertial-internal waves (n close to f), the particle motion is primarily in a plane very close to the horizontal plane (Figure 6.3.1). If N = 0, the resulting fluid motions are called inertial oscillations, and Eq. (6.3.4)

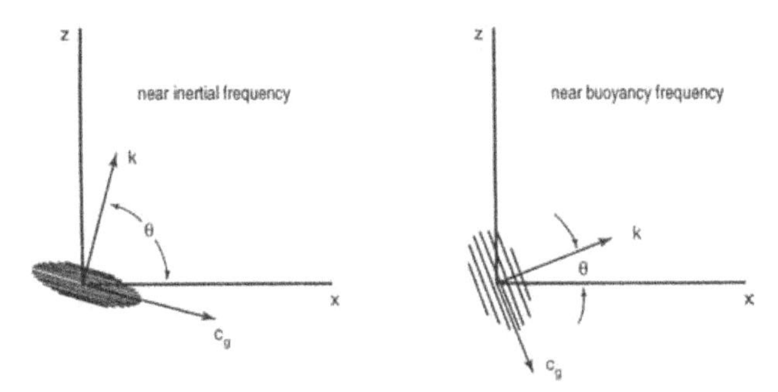

Figure 6.3.1. Relationship of the wavenumber vector and the group velocity vector to the horizontal for a nearly inertial and nearly internal wave (from Munk, 1982).

becomes

$$n = f \sin\theta \tag{6.3.11}$$

These oscillations can be explained dynamically as follows. If a fluid particle is once again displaced in a plane inclined at angle θ, there is a Coriolis force that acts perpendicular to the displacement velocity and its magnitude is proportional to the velocity dx/dt and the component of f perpendicular to the plane of motion $f \sin\theta$. The acceleration of the particle is therefore

$$\frac{d^2x}{dt^2} + i\frac{dx}{dt} \cdot f \sin\theta = 0 \tag{6.3.12}$$

where $x = x_1 + ix_2$, a complex quantity needed to account for the non-unidirectional motion caused by the fact that the Coriolis force acts perpendicular to the fluid motion. The solution is $\frac{dx}{dt} \sim \exp(-i \cdot f \sin\theta \cdot t)$, which describes inertial oscillations with frequency $f \sin\theta$. The particle traces a circle in the inclined plane once every inertial period.

For inertial-internal waves $f \le n \le N$, the motion is influenced by restoring forces due to both buoyancy and the action of Coriolis acceleration. Consequently, the particle motion is elliptic on a plane inclined at an angle θ to the vertical, with the minor axis horizontal. For $n \to N$, the displacement is primarily vertical, and restoring forces due to buoyancy dominate. For $n \to f$, the displacement is primarily horizontal and the Coriolis force produces an inertial oscillation (see Figure 6.3.1)

The particle velocities can be written as (Olbers, 1983)

$$w(x_j, z, t) = -r\left(\frac{n^2 - f^2}{N^2 - n^2} nk_3\right) \exp\left[i\left(k_j x_j + k_3 z - nt\right)\right]$$

$$u_1(x_j, z, t) = r(nk_1 + ifk_2) \exp\left[i\left(k_j x_j + k_3 z - nt\right)\right] \tag{6.3.13}$$

$$u_2(x_j, z, t) = r(nk_2 - ifk_1) \exp\left[i\left(k_j x_j + k_3 z - nt\right)\right]$$

It is implied that only the real part is taken in Eqs. (6.3.11)–(6.3.13). These velocities indicate elliptic polarization of particle motion that takes place only in the plane orthogonal to wavenumber vector $\vec{k}$ because of the constraints imposed by incompressibility. For near-inertial frequencies ($n \to f$), it is easy to see that $w \to 0$ and the particle motion is almost horizontal $(k_h >> k_3)$ and circular. As n increases, the ellipse is more and more inclined toward the vertical and its eccentricity increases. As $n \to N$, the motion becomes primarily vertical $(k_3 >> k_h)$. If r is chosen to be

$$r = \frac{1}{nk_h}\left(\frac{N^2 - n^2}{N^2 - f^2}\right)^{1/2} a \tag{6.3.14}$$

the wave density, averaged over a wave period, can be written as

$$E = \frac{1}{2}\left(\overline{u_k u_k^*} + N^2 \overline{\zeta\zeta^*}\right) = 2a^2 \tag{6.3.15}$$

where a is the current amplitude, ζ is the vertical displacement, and * indicates complex conjugates:

$$\zeta(x_j, z, t) = -\frac{ia}{n}\left(\frac{n^2 - f^2}{N^2 - f^2}\right)^{1/2} \exp\left[i\left(k_j x_j + k_3 z - nt\right)\right] \tag{6.3.16}$$

The pressure is given by

$$p(x_j, z, t) = c\left(n^2 - f^2\right)\exp\left[i\left(k_j x_j + k_3 z - nt\right)\right] \tag{6.3.17}$$

and

$$w(x_j, z, t) = a\left(\frac{n^2 - f^2}{N^2 - f^2}\right)^{1/2} \exp\left[i\left(k_j x_j + k_3 z - nt\right)\right] \tag{6.3.18}$$

The energy flux vector is

$$F_i = \overline{pu_i} = 2a^2 c_{gi} = E c_{gi} \tag{6.3.19}$$

This shows that the wave energy flux is the product of its energy density E and its group velocity c_{gi}. This expression holds for many types of wave motions. The group velocity c_{gi} is given by

$$c_{gi} = \frac{\partial n}{\partial k_i} \tag{6.3.20}$$

An expression for this can be derived using the dispersion relation (6.3.3),

$$c_{g_1}, c_{g_2}, c_{g_3} = \frac{N^2 - n^2}{n k_h^2} \frac{n^2 - f^2}{N^2 - f^2} \left(k_1, k_2, -\frac{n^2 - f^2}{N^2 - n^2} k_3 \right) \tag{6.3.21}$$

The phase velocity is, of course,

$$c_i = \frac{n}{k} k_i \tag{6.3.22}$$

It is easy to see that $c_i c_{gi} = 0$, i.e., the phase propagation and group propagation are orthogonal to each other. Energy flux is therefore parallel to the wave crests (Figure 6.3.1). This was shown in beautiful experiments in a laboratory tank by Mowbray and Rarity (1967) for pure internal waves (Figure 6.3.2). This is once again just a consequence of the transverse nature of wave motions in incompressible fluids. The fluid velocity is perpendicular to the wave vector, and hence to the direction of phase propagation, so that the energy flux vector $\overline{pu_i}$ is perpendicular as well and so is the group velocity.

The kinetic energy density is given by

$$K = \frac{\rho_0}{2} \left(\overline{u_1 u_1^* + u_2 u_2^* + w^2} \right) \tag{6.3.23}$$

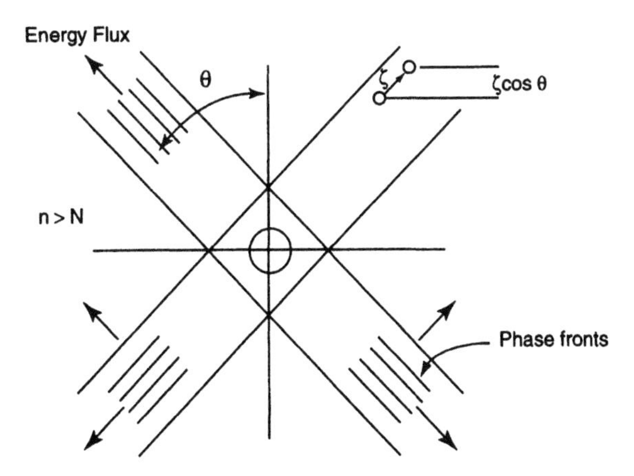

Figure 6.3.2. Sketch showing propagation of internal waves away from the oscillating source in a linearly stratified fluid. Note the characteristic St. Andrews cross pattern and the orthogonality of phase and group velocity vectors (adapted from Mowbray and Rarity, 1967).

and therefore

$$K=\frac{\rho_0}{2}a^2\left[\frac{n^2-f^2}{N^2-f^2}+\frac{n^2+f^2}{n^2}\cdot\frac{N^2-n^2}{N^2-f^2}\right] \tag{6.3.24}$$

The potential energy density is

$$P=\frac{\rho_0}{2}N^2\overline{\zeta^2}=\frac{\rho_0}{2}a^2\frac{N^2}{n^2}\frac{n^2-f^2}{N^2-f^2} \tag{6.3.25}$$

There is no equipartitioning of energy between potential and kinetic energy for inertial-internal waves. As $n \to f$, $P \to 0$ and the energy is totally in horizontal inertial oscillations. $c_g \to 0$ for $n \to f$ (and $n \to N$).

It is instructive to consider pure internal waves ($f = 0$). Then we can write

$$\begin{aligned}
\zeta(x_j, z, t) &= -\frac{ia}{N}\exp\left[i\left(k_j x_j + k_3 z - nt\right)\right]\\
w(x_j, z, t) &= (a\cos\theta)\exp\left[i\left(k_j x_j + k_3 z - nt\right)\right]\\
u_j(x_j, z, t) &= \frac{k_j}{k_h}(a\sin\theta)\exp\left[i\left(k_j x_j + k_3 z - nt\right)\right]\\
p(x_j, z, t) &= -\frac{n^2}{Nk_h}(a\tan\theta)\exp\left[i\left(k_j x_j + k_3 z - nt\right)\right]\\
K = P &= \frac{\rho_0}{2}a^2
\end{aligned} \tag{6.3.26}$$

There is therefore equipartitioning of energy density between kinetic and potential energy for pure internal waves.

Figure 6.3.1 shows the relationship between the wavenumber vector $\vec{k}$ (and phase velocity $\vec{c}$) to the group velocity $\vec{c}_g$ (and fluid velocity $\vec{u}$) for predominantly inertial and predominantly internal wave packets with crests and troughs normal to the paper. The packet slides sideways along the crests.

The ansiotropic nature of internal waves is responsible for its unusual properties of reflection at a sloping boundary. Since the frequency of the incident and reflected waves must be the same, their inclination to the vertical must be the same, no matter what the inclination of the sloping boundary might be. The angles of incidence and reflection are therefore *not* the same, except for a horizontal boundary. This often leads to reduction or expansion of the reflected beam, leading to transfer of energy in wavenumber space. At the same time, the resulting velocity due to the combination of incident and reflected waves must be parallel to the reflecting boundary. For a given frequency, there is a value of angle α of the boundary for which the reflection is along the boundary

$$\alpha_c = 90 - \theta = \tan^{-1}\left(\frac{n^2 - f^2}{N^2 - n^2}\right)^{1/2} \tag{6.3.27}$$

For angles less than α_c, the wave energy is forward reflected. For $\alpha > \alpha_c$, the reflection is backward, back toward the direction in which the wave is incident (Figure 6.3.3). Wunsch (1972) suggested that the peak in the spectrum

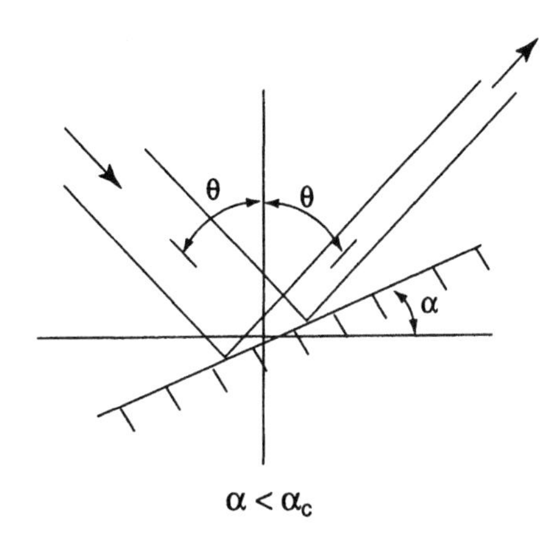

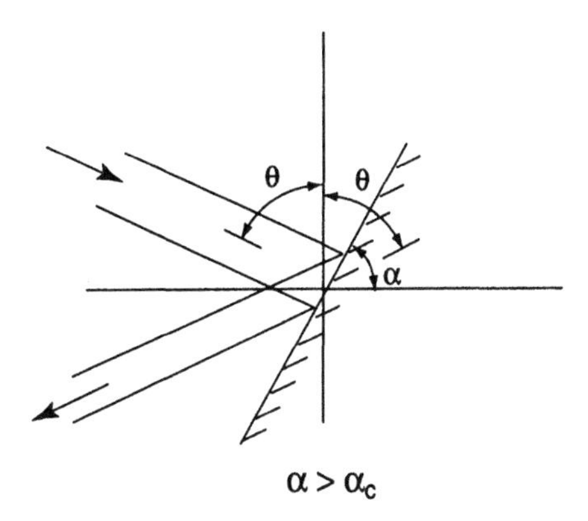

Figure 6.3.3. Reflection of internal waves at a sloping boundary for (top) incidence angle less than critical and (bottom) greater than critical.

of temperature fluctuations off Bermuda is due to the accumulation of wave energy at frequency n determined by Eq. (6.3.27). Off Bermuda, $\alpha \sim 13°$ and N = 2.6 cph, yielding the peak at n ~ 0.6 cph, close to the observed peak at 0.5 cph (Munk, 1981).

Pure inertial waves (N = 0, n > f) have similar reflection properties since they are also ansiotropic waves with only the inclination to the horizontal appearing in the dispersion relation.

An alternative way of looking at reflection properties is to consider which frequencies are forward reflected (or transmitted) and which ones are backward reflected (reflected) for a given slope.

$$n_c^2 = N^2 \sin^2 \alpha + f^2 \cos^2 \alpha \tag{6.3.28}$$

gives the critical frequency n_c for a given α. For $n \to n_c$, the group velocity of incident and reflected waves are in the same direction, and for $n < n_c$ in the opposite directions. Thus, low frequency components ($n < n_c$) of an incident wave field are reflected back, impinging on a sloping bottom, and high frequency components proceed upslope (Olbers, 1983). The wavenumbers of the reflected and incident waves are related thusly (Olbers, 1983)

$$\begin{aligned} k_1^r &= \frac{\left(1+\tan^2\alpha \sin^2\theta\right)k_1^i + 2k_z^i \tan\alpha}{\left(1-\tan^2\alpha\sin^2\theta\right)} \\ k_2^r &= k_2^i \\ k_3^r &= -\frac{\left(1+\tan^2\alpha\sin^2\theta\right)k_3^i + 2k_1^i \tan\alpha\sin^2\theta}{\left(1-\tan^2\alpha\sin^2\theta\right)} \end{aligned} \tag{6.3.29}$$

Near the critical angle α_c given by $\tan^2\alpha_c \sin^2\theta = 1$, or near critical frequency n_c, there can be an accumulation of energy near a sloping bottom as Eriksen (1982) showed.

$$E_r = E_i \frac{k_3^r}{k_3^i} \sim E_i / \left(n - n_c\right)^2 \tag{6.3.30}$$

so that as $n \to n_c$, E_r, the energy density, increases rapidly.

6.4 VERTICAL STANDING MODES

Equation (6.2.20) can be used to obtain solutions for a wave propagating horizontally with a large enough wavelength to feel the bottom. The solution depends on the functional form of N(z). Closed form solutions are possible only for specific distributions of the Brunt–Vaisala frequency. One of the simplest cases is for N = constant. The appropriate boundary conditions are

$$W = 0 \text{ at } z = -H \tag{6.4.1}$$

and

$$n^2 \frac{dW}{dz} - g k^2 W = 0 \text{ at } z = 0 \tag{6.4.2}$$

The latter is derived from constancy of pressure at the free surface. Equation (6.4.2) can be simplified further. Near the free surface, substituting

$$W(z) \sim \exp\left[i \left(\frac{N^2 - n^2}{n^2 - f^2} \right)^{1/2} kz \right] \tag{6.4.3}$$

the ratio of the first term to the second term is

$$\frac{n^2}{gk} \left(\frac{N^2 - n^2}{n^2 - f^2} \right)^{1/2} \sim \frac{n^2}{n_s^2} \sim 0\left(10^{-2}\right) \tag{6.4.4}$$

where n_s is the frequency of the surface wave with the same wavenumber k. Therefore, to a good accuracy, a rigid lid approximation holds for internal waves (for $n \gg f$), which produce very small vertical displacements at the free surface. These small surface displacements are nevertheless important for remote sensing of internal waves such as solitons in altimetric data (see Section 6.7). Thus, Eq. (6.4.2) can be replaced by

$$W = 0 \text{ at } z = 0 \tag{6.4.5}$$

It can be verified readily that

$$W = a \sin mz \tag{6.4.6}$$

where

$$m^2 = k^2 \frac{N^2 - n^2}{n^2 - f^2} \tag{6.4.7}$$

satisfies Eq. (6.2.22). Thus there exist an infinite number of discrete internal wave modes satisfying the relationship

$$n_n^2 = \frac{k_h^2 N^2 + m_n^2 f^2}{m_n^2 + k_h^2}, \qquad m_n H = n\pi \qquad n = 1, 2 \cdots \tag{6.4.8}$$

For a general N(z) distribution, Eqs. (6.2.22), (6.4.1), and (6.4.2) constitute an eigenvalue problem, the solution of which yields the modal structure and the dispersion relation. Analytical solutions are, however, not possible. In general, in the range of depths were $n < N(z)$, oscillatory solutions exist. The first mode has a single maximum for w and corresponds to up and down motions of this region, the pycnocline. The second mode has a zero crossing and corresponds to pulsatory motions of the pycnocline.

It is found that the lowest mode is usually the most energetic, especially when the pycnocline is sharp. Then it is very much similar to a surface wave on the air–sea interface. For the most simple case of $f = 0$, and $N = 0$ outside the sharp pycnocline, it is possible to obtain solutions by simple matching of the solutions in the upper and lower layers across the buoyancy interface (Phillips, 1977; Munk, 1981). In general, to derive dispersion relations for modes that leak their energy into waves propagating into the interior, we will follow a slightly different approach.

Consider the buoyancy distribution in Figure 6.4.1. Locations $z = -d_+$ and $z = -d_-$ denote the upper and lower edges of a diffuse interface $(\varepsilon = d_- - d_t)$, where the buoyancy frequency N(z) is prescribed. Let $N = N_0$ in the lower layer and $N = 0$ in the upper layer. The solution in the upper layer satisfying the boundary condition (6.4.5) is

$$W = A\sinh(k_h z) \qquad (0 \le z \le -d_+) \tag{6.4.9}$$

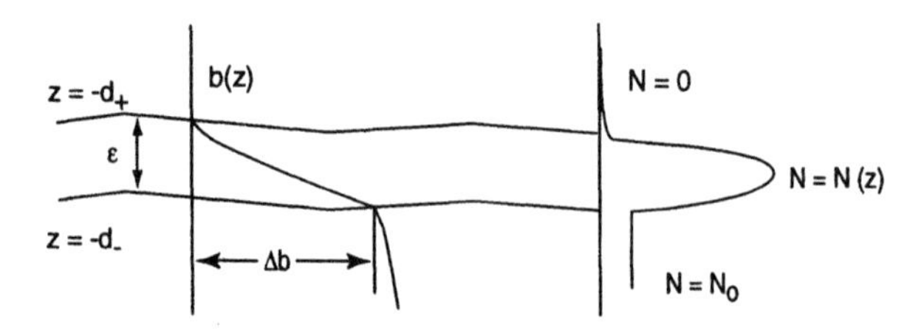

Figure 6.4.1. Idealized buoyancy and buoyancy frequency distributions across a diffuse pycnocline.

The solution in the lower layer is

$$W = B\exp(ik_z z) \qquad (d_- \le z) \tag{6.4.10}$$

where

$$\frac{k_z}{k_h} = \left(\frac{N_0}{n}\right)^2 - 1 \tag{6.4.11}$$

To allow for internal waves radiating energy down away from the pycnocline, n and k_z must be regarded as complex values.

In the pycnocline itself, analytical solutions are not possible for a general distribution of N(z). But for a diffuse thermocline, perturbation solutions in terms of parameter $k_h \varepsilon$ $(k_h\varepsilon << 1)$ are possible. The idea is to obtain W and dW/dz in the thermocline correct to $O(k_h\varepsilon)^2$ and match these to the solutions in the upper and lower layers. The zeroth-order solution yields the dispersion relation for a sharp pycnocline, which can only sustain the lowest mode. For the lowest mode,

$$W(-d_+) = W(-d_-) \tag{6.4.12}$$

$$\frac{dW}{dz}(-d_+) = -\frac{dW}{dz}(-d_-) = \int_{-d_-}^{-d_+} \frac{d^2W}{dz^2} dz \tag{6.4.13}$$

Substituting for d^2W/dz^2 from Eq. (6.2.22),

$$\begin{aligned} \text{L.H.S} &= \underset{\varepsilon \to 0}{\text{Lt}} \int_{\varepsilon} k_h^2 W \left[1 - \frac{1}{n^2}\frac{d\Delta b}{dz}\right] dz \\ &= -k_h^2 W(-d) \frac{\Delta b_0}{n^2} \end{aligned} \tag{6.4.14}$$

Using Eqs. (6.4.9) and (6.4.10) in Eqs. (6.4.12) and (6.4.14) gives

$$n^2 = k_h \Delta b \left[\coth(k_h d) + i\frac{k_z}{k_h}\right]^{-1} \tag{6.4.15}$$

This is correct to $O(k_h\varepsilon)$. Replacing the second term by coth $k_h(D\text{-}d)$, where D is the depth of the ocean, gives the classical dispersion relation for the lowest mode internal wave on a sharp pycnocline,

$$n^2 = k_h\Delta b\left\{\coth(k_h d) + \coth\left[k_h(D\text{-}d)\right]\right\}^{-1} \tag{6.4.16}$$

Expanding W(z) in the pycnocline in terms of $k_h\varepsilon$, it can be shown

$$W(-d_-) = W_0\left\{1 + k_h\varepsilon\left[\coth(k_h d_+) - \beta(1+I_1) + 0(k_h\varepsilon)^2\right]\right\} \tag{6.4.17}$$

$$\frac{dW}{dz}(-d_-) = W_0 k_h \left\{\begin{matrix}\left[\beta - \coth(k_h d_+) - (k_h\varepsilon)\right] \\ \left[1 + \beta I_1(\coth k_h d_+ - \beta) + \beta^2 I_2\right] + 0(k_h\varepsilon)^2\end{matrix}\right\} \tag{6.4.18}$$

where

$$\beta = \frac{k_h\Delta b}{n^2} \tag{6.4.19}$$

$$I_1 = \frac{1}{\varepsilon}\int_{-d_+}^{-d_-}\frac{\Delta b(z)}{\Delta b_0}dz \tag{6.4.20}$$

$$I_2 = \frac{1}{\varepsilon}\int_{-d_+}^{-d_-}\left[\frac{\Delta b(z)}{\Delta b_0}\right]^2 dz \tag{6.4.21}$$

Note that $0 < I_2 < I_1 < 1$. For linear stratification, $I_1 = -1/2$, $I_2 = -1/3$. Matching Eqs. (6.4.17) and (6.4.18) to those given by Eq. (6.4.10) gives

$$\frac{n^2}{n_0^2} = 1 - k_h\varepsilon\left\{I_2\coth(k_h d) + \frac{ik_z}{k_h}(I_2 - 2I_1 - 1) + \frac{1}{2}\left[1 + 2i\frac{k_z}{k_h}\coth(k_h d) + \coth^2(k_h d)\right] / \left[\coth(k_h d) + \frac{ik_z}{k_h}\right]\right\} \tag{6.4.22}$$

where

$$n_o^2 = \frac{k_h \Delta b_0}{\coth\left(k_h d + \frac{ik_z}{k_h}\right)} \qquad (6.4.23)$$

This is the dispersion relation for the lowest mode on a diffuse pycnocline. In general, n, n_0, and k_z are complex. For a standing vertical mode, ik_z / k_h can be replaced by coth [k_h (D-d)].

Equation (6.4.16) is simpler to deal with. Some properties of the lowest mode can be investigated using this simpler dispersion relation (Phillips, 1977; Munk, 1981). In the deep oceans, $k_h(D\text{-}d) \gg 1$ and therefore

$$n^2 = k_h \Delta b_0 / \left[1 + \coth(k_h d)\right] \qquad (6.4.24)$$

Let

$$\zeta = a \exp\left[i(k x - n t)\right] \qquad (6.4.25)$$

be the amplitude of the pycnocline displacement due to a lowest mode internal wave propagating in the x direction (note we have put $k_h = k$). Following Phillips (1977), various properties of the wave can be written down. The flow outside the pycnocline is irrotational. The vertical velocity of the pycnocline is

$$w = -ina \exp\left[i(kx - nt)\right] \qquad (6.4.26)$$

The wave-induced change of horizontal velocity across the pycnocline is

$$\Delta u = na\left[\coth(kd) + \coth k\,(D\text{-}d)\right]\exp\left[i(kx - nt)\right] \qquad (6.4.27)$$

This change is quite large and gives rise to a large shear at the pycnocline. From the Miles–Howard theorem (Miles, 1961, 1963; Howard, 1961), it is possible to ascertain if the flow is unstable. This requires a condition on the gradient Richardson number. For a shear flow, $Ri_g < 1/4$ is a necessary but not sufficient condition for instability. Equivalently,

$$\left|\frac{1}{N}\frac{du}{dz}\right| > 2 \qquad (6.4.28)$$

From the incompressibility condition,

$$iku(z) = \frac{dw(z)}{dz} \tag{6.4.29}$$

so that

$$\frac{1}{N}\frac{du}{dz} = \frac{1}{ikN}\frac{d^2w}{dz^2} = ka\left(\frac{N}{n} - \frac{n}{N}\right)\exp\left[i(kx\text{-}nt)\right] \tag{6.4.30}$$

making use of Eq. (6.4.22). For $n \ll N$,

$$\left|\frac{1}{N}\frac{du}{dz}\right| \sim ka\frac{N}{n} \tag{6.4.31}$$

The remarkable aspect of this relationship is that the dynamical stability is the *lowest* at the point in the pycnocline where the density stratification is most stable, i.e., the location of maximum N. Equation (6.4.31) is useful for investigating how energetic low modes on a sharp pycnocline decay. It shows that when the wave slope exceeds a certain value given by

$$ka > 2\frac{N}{n} \tag{6.4.32}$$

then a local instability may develop leading to a patch of turbulence near the crest or trough of the wave. This mixes up the fluid, decreases N, and restores stability. Therefore any such generation of mixing in the pycnocline is likely to be sporadic (intermittent) in space and time. In the presence of background mean shear, breaking is likely to occur at the crest or trough depending on whether the shear augments or decreases the wave-induced shear.

From the momentum equation at the surface (z = 0), it can be shown (Phillips, 1977) that the free surface displacement due to a first mode internal wave is

$$\eta_0 = -\frac{an^2}{gk\sin kd}\exp i(kx\text{-}nt) \tag{6.4.33}$$

since $n^2 \sim gk\frac{\Delta\rho}{\rho_0}$, $|\eta_0| \sim -a\cdot\frac{\Delta\rho}{\rho_0}$. The free surface displacement due to the internal wave is smaller by a factor $\Delta\rho/\rho_o\left(\sim 10^{-3}\right)$ and is in antiphase to thermocline displacement. The phase speed of the internal wave is

$$c=\frac{n}{k}=\left(\frac{\Delta b}{k}\right)^{1/2}\left[\coth(kd)+\coth k(D\text{-}d)\right]^{-1/2} \tag{6.4.34}$$

and its group velocity is

$$\frac{c_g}{c}=\frac{1}{2}+\frac{kd\sinh^2 k(D\text{-}d)+k(D\text{-}d)\sinh^2 kd}{2\sinh(kd)\sinh(kD)\sinh k(D\text{-}d)} \tag{6.4.35}$$

6.5 GENERATION, DISSIPATION, PROPAGATION, AND INTERACTION

6.5.1 GENERATION

Thorpe (1975) surveyed the state of our knowledge on generation, dissipation, and interaction of internal waves and pointed out our inability at the time to quantify many of these. Figure 6.5.1, from Thorpe (1975), is a succinct summary of various generation, dissipation, and interaction mechanisms for internal waves in the ocean. The problem has always been not what processes could generate internal waves, but rather their relative importance. Even with decades of research on internal waves, it is still not possible to quantify these sources accurately. Some progress has been made (Olbers, 1983; Muller *et al.*, 1986) but overall it is still difficult to assign values to each of the sources and sinks, and ascertain their relative importance to internal waves. Thorpe (1975) however did not attach much importance to internal tides as a possible source of internal waves. Various estimates suggest that internal tides could dissipate as much as 10–15% of the power input by the lunisolar system into oceanic tides (Wunsch, 1975, for example), and if this is true, internal tides may constitute a major source of oceanic internal waves (Munk, 1997; Kantha and Tierney, 1997). The discussion that follows below is based on Thorpe's review.

The oceans are principally forced at the top by atmospheric pressure, wind stress, and buoyancy fluxes. Therefore, the energy density is the highest in the upper layers of the water column. Consequently one has to look at the atmospheric forcing and the oceanic mixed layer (OML) for a large fraction of the source of internal wave energy in the deep ocean (ignoring internal tides generated at midocean ridges and seamounts). Any disturbance that displaces water parcels in the vertical away from their positions of equilibrium is a likely candidate for generation of internal waves. It may not however be an effective candidate if the forcing is not near the intrinsic frequencies of the internal waves for the prevailing ambient stratification and the wavenumbers (or scales) being

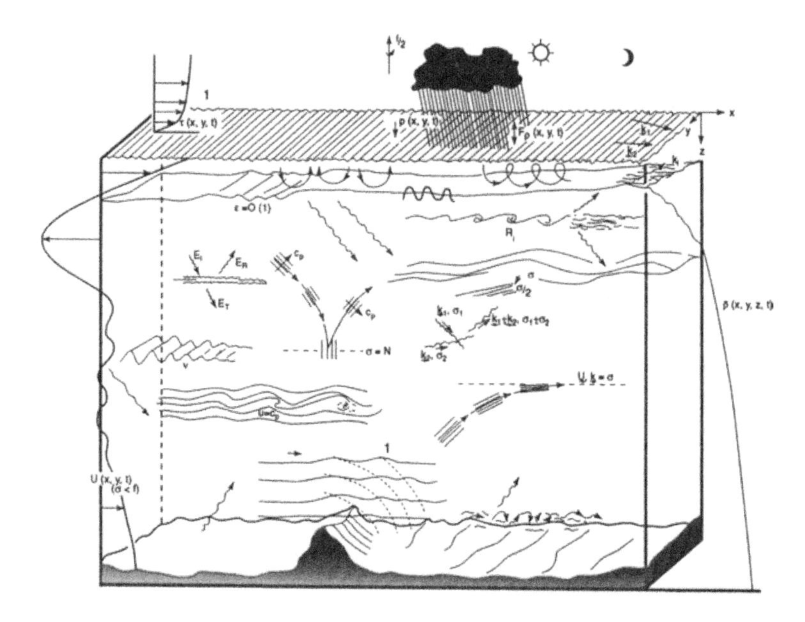

Figure 6.5.1. A schematic of the processes affecting internal waves in the ocean (from Thorpe, 1975). The Moon and the resulting internal tides have been added.

forced. Normally, it is the resonant coupling between the forcing being applied and the waves being generated that leads to efficient generation. Secondly, the forcing needs to be applied over a sufficiently long period of time to build up significant amplitudes. These requirements determine how effective a particular source is, in the generation of a response in any medium, and this is true of internal waves as well.

As Olbers (1983) indicated, it is preferable to consider the source and sink terms in spectral space rather than physical space, because of the stochastic nature of the internal wave field and the broadband characteristic of most internal wave forcing. Also, the strong nonlinear interactions among internal wave modes tend to rapidly redistribute energy across the spectrum. The governing equation is an equation for conservation of action density in wavenumber space,

$$\left[\frac{\partial}{\partial t}+\left(Cg_i+U_i\right)\frac{\partial}{\partial x_i}\right]A\left(k_i,x_i,t\right)=S\left(k_i,x_i,t\right) \tag{6.5.1}$$

where S denotes the sources in spectral space that generate or augment the mode, sinks that extract energy away from the mode, and wave–wave interactions that redistribute energy in spectral space. While characterization of the interaction among wave modes, the classic closure problem, is much simpler for a random weakly nonlinear dispersive wave field (than for turbulence), which can be considered to be close to a Gaussian state (Olbers, 1983), the approximation of weakly nonlinear interacting modes may not be adequate to describe internal waves in the ocean (Holloway, 1980; Muller *et al.*, 1986). The mathematical details of even the weak interaction theory are too complex to be dealt with here. Characterization of sources and sinks is usually done in physical space for a deterministic narrowband forcing and Thorpe (1975) provides an excellent review. But for practical purposes, it is essential to have them characterized in spectral space, since as Olbers (1983) points out the growth rates in spectral space are inherently smaller. Unfortunately, this is harder to do.

There are principally five potential sources of internal wave energy: (1) the OML, (2) the atmospheric forcing at the oceanic surface, (3) the surface waves, (4) the mean flow, and (5) the baroclinic tides. Theoretical studies of internal wave generation, propagation, etc., involve idealizations of the ambient stratification. For internal waves on the seasonal thermocline, it is often adequate to consider two-layer stratification, whereas for deep-ocean internal waves, a continuous but constant N stratification is simpler to treat.

First let us look at the OML as the source of internal waves on the pycnocline and in the deep ocean, since this is likely to be a major source. In this case, one has to include both types of stratification. Turbulent fluctuations in the OML create vertical displacements of the bounding pycnocline at the bottom, thus generating and propagating internal waves to the stably stratified interior. Pioneering work was done by Townsend (1976), who considered pressure fluctuations in a boundary layer adjacent to a stratified medium, without the bounding pycnocline. However, it is easy to see that the magnitude of the undulations produced in the stratified interior, for a given pressure forcing, depends on the stability of the pycnocline itself. For a strong pycnocline, the interface does not deform as much and hence the energy radiated out into the interior will be smaller. Kantha (1977, 1979a,b) has examined internal wave generation for realistic stratifications. Laboratory experiments such as those of Willis and Deardorff (1974) under convective conditions and Kantha (1980b) have amply demonstrated the existence of internal wave motions in the stratified medium adjoining the mixed layer, no matter whether the mixing is caused by convection or shear. Quantification has, however, been rather difficult since the spectrum of pressure fluctuations at the base of the OML has never been measured. Perhaps large eddy simulations (LES's) of the OML might be helpful in this regard in the near future.

Bell (1978) also treated the problem of internal wave generation by turbulence in the OML generated by wind mixing and advected by low frequency inertial currents. He estimates the transfer rate to be

$$S_{OML} \sim \frac{\rho}{2} N_0^3 l \overline{\zeta^2} + 0\left(\frac{N_0 l}{U_0}\right) \tag{6.5.2}$$

where l is the integral length scale, N_0 the buoyancy frequency, U_0 the inertial current, and $\overline{\zeta^2}$ the rms displacement at the base of the mixed layer. A typical value for this flux is 10^{-3} W m^{-2}. The waves excited are of high frequency and therefore this mechanism is important for the upper ocean internal waves.

One atmospheric forcing mode consists of moving pressure disturbances on a variety of spatial and temporal scales. A traveling pressure field can generate wave motions, surface waves generated on the air–sea interface by a ship being a classic example. Since the speed of internal waves is very small (compared to those of surface waves of comparable wavelength), it is a question of what the ratio of the speed of the disturbance is to the speed of the resonant internal wave. If the internal wave cannot keep up with the pressure field, there will be no waves traveling in the direction of the motion of the pressure field, only divergent waves spreading laterally on either side of the pressure point. Now the atmospheric pressure spectrum is broadband and only those frequencies that satisfy the relationship $n = k_i u_i$, where u_i is the advection velocity of the disturbance and k_i is the wavenumber and n the frequency, will be resonantly excited. This is the classic resonant generation mechanism that was first applied to surface waves by Phillips (1957). Leonov and Miropolskiy (1973) have applied such a theory to study the resonant excitation of internal waves for both a two-layer and continuous stratifications (for details, see Thorpe, 1975; Olbers, 1983).

Vertical motions caused by a wind stress field or the pressure field from a traveling buoyancy flux at the surface can also generate internal waves. However, even less is known about the spectrum of these forcings and it is hard to arrive at quantitative estimates for the importance of these mechanisms. Nevertheless, rough estimates indicate that the wind stress forcing is by far the most dominant (~10^{-3} W m^{-2}), and the buoyancy flux forcing the less important (Olbers, 1983).

Resonant interaction among a pair of surface waves and an internal wave is a possible mechanism for transfer of energy from surface waves into internal waves. There is a tremendous amount of energy in surface waves on the air–sea interface around the globe, and even if a small fraction of this energy can be transferred by this mechanism to internal waves, it could constitute a major source of energy for deep-sea internal waves. Also, the surface wave spectrum is fairly steady with continuous input of energy from the wind and therefore can provide a forcing long enough to build up significant amplitudes. The fact that the deep-sea internal wave spectrum observed below sea-ice cover is an order of magnitude weaker than that in midlatitudes might be related to the absence of surface wave motions in the presence of near-continuous ice cover. Ice does transfer the momentum from the winds to the OML through its own motion in response to forcing by the wind and therefore wind stirring is still operative in the Arctic OML, but surface waves are largely absent.

The triad of surface waves and the internal wave participating in such resonant interactions is given by

$$k_1 - k_2 = k_i \text{ and } n_1 - n_2 = n_i \tag{6.5.3}$$

Because of the higher frequencies of surface waves compared to the internal wave, the surface waves have to be of nearly the same frequency. The fastest growing internal wave is almost at right angles to that of the two surface waves, which are such that they propagate in nearly the same direction parallel to each other. The growth rate of the internal wave for a continuously stratified ocean is

$$dA_i/dt \sim 0.3\, A_1\, A_2\, k_2\, N \tag{6.5.4}$$

where A_1 and A_2 are amplitudes of the surface waves. For typical values of $A_1 \sim 1$ m, $A_2k_2 \sim 0.1$, and $N \sim 10^{-3}\ s^{-1}$, the growth rate is 2.5 m day^{-1}. However, Olbers (1983) indicates that while this mechanism is very efficient for generating high frequency internal waves in the upper ocean, it may not be that important for deep-ocean internal waves.

Generation of internal waves by flow over topography is another possible mechanism (for example, Bell, 1975). This mode of generation operates very effectively in the atmosphere, and flow over mountains is a well-studied topic in the atmosphere. The importance of this mechanism in the ocean is not well known. It is possible that in regions where there are strong benthic currents such as in the region of western boundary undercurrents, the mechanism might be quite important. But in general, because steady benthic currents are quite weak in the global oceans, it is not clear how strong this source is, although Olbers (1983) indicates a value of 10^{-3} W m^{-2}, roughly the same as other sources. Weak tidal currents flowing over topographic irregularities in the deep ocean could be an important source of internal tides in the deep ocean, which in turn could spread their energy over the internal wave spectrum by weak nonlinear interactions. The flux into deep ocean internal tides also appears to be on the order of 10^{-3} W m^{-2}.

There is a considerable energy imparted to barotropic ocean tides by the Earth–Sun–Moon system. These tidal motions are dissipated predominantly in shallow water, but a prominent sink term for these motions is baroclinic tides. In regions of strong topographic variations, such as near seamounts, midocean ridges, and most of the continental shelf break, significant baroclinic tides are excited by barotropic tidal motions. SAR imagery from orbiting satellites shows such baroclinic tidal motions all around the world, principally near shelf breaks and narrow straits. It is estimated that the transfer of energy from barotropic to baroclinic tides may be as much as 10–15%. Since the rate of energy input into barotropic tides is nearly 3.5 TW, this constitutes a significant source for internal tides. Such motions often take the form of an internal wave soliton train radiated toward the coast. Each tidal cycle generates a train of 3 to 7 solitons that propagate into shallow water and dissipate. Excellent examples can be found in the Sulu (Figure 6.1.5) and Andaman seas (Osborne and Birch, 1980; Apel *et al.*, 1985), off west Africa, and in the tropical western Pacific (see Section 6.7).

6.5.2 DISSIPATION

If sources of internal waves cannot be well quantified, the situation is worse with their sinks. The dissipation mechanisms for internal waves are even more poorly known and quantified. The concept of a universal internal wave spectrum implicitly assumes a balance between energy input by sources and energy dissipation by sinks, but does not, however, specify the source and sink mechanisms. It is assumed that the waves are "saturated" and any additional input of energy is lost immediately by "breaking" or some other unspecified process.

There are two principal mechanisms that could lead to breaking of internal waves. One is the shear (Kelvin–Helmholtz, or K-H) instability, which takes place if the total local shear (due to both background currents and self-induced shear) exceeds a certain value so that the Miles–Howard criterion $Ri_g > 1/4$ for stability [valid strictly for only parallel shear flows, but see Weissman (1980), who showed that it is reasonably accurate for long internal waves] is violated. Dynamical instability ensues, a patch of turbulence is created mostly at the crest of the internal wave, and the stratification is locally destroyed and static stability restored, until the stage is set for the next course of such events. However, advective gravitational instability, a second kind of instability that ensues when the particle speed at the wave crest exceeds the wave speed and the wave just "tumbles over," could be more prevalent (Thorpe, 1979). In waves with steep isopycnal slopes, particles at the crest are advected forward of the crest, creating a local density inversion—Rayleigh–Taylor instability ensues and breaking results.

Such advective instability can occur in the absence of mean shear. Shear instability often occurs in the presence of mean shear modulated by the shear from internal waves. Both types of instability were demonstrated by beautiful experiments in a tilting tube facility (see Figure 6.5.2) by Thorpe (1979), where a long rectangular tube is filled with stratified fluid and an internal wave is generated by a wave maker at one end. Before the waves reach the other end, the tube is tilted to induce a mean shear in the flow. For steep waves with sharp crests, advective instability ensues, creating a patch of turbulence which spreads out horizontally. In later stages, K-H billows also form from shear instability. From these experiments, Thorpe was able to demonstrate that internal waves on a density interface become advectively unstable if the wave slope exceeds a value of about 0.34, and shear unstable if the ambient shear exceeds 2N. It is not clear which mechanism prevails under the most general condition of a nonzero ambient mean shear (see Munk, 1981).

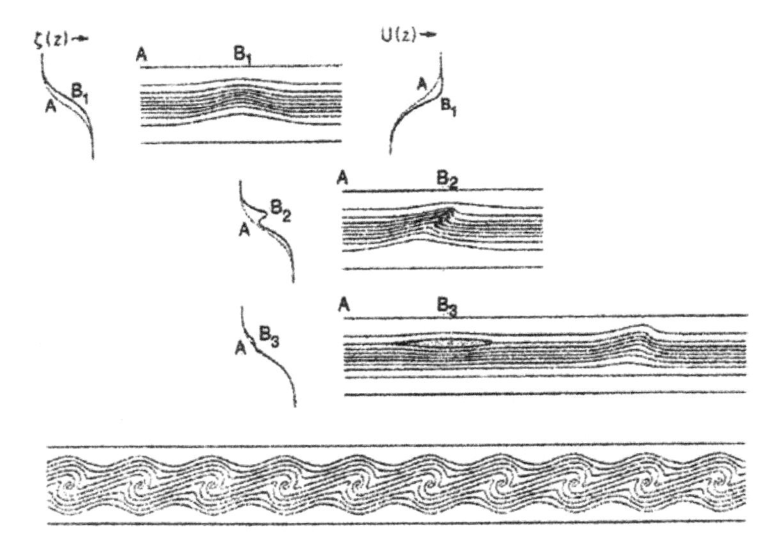

Figure 6.5.2. Experiments in a tilting tube by Thorpe illustrating the dissipation mechanisms for internal waves (from Munk, W., *in Evolution of Physical Oceanography,* Copyright M.I.T. Press, 1981).

An internal wave field in the ocean consists of a random superposition of various modes. Breaking occurs in the presence of other modes and there is randomness superimposed on the whole process. Therefore it is difficult to identify breaking of a particular mode as such. Interaction of internal waves can lead to localized intensification of the density gradient and shear leading to turbulent mixing (McEwan, 1973). McEwan finds that less than one-fourth of the energy goes into increasing the potential energy of the system. The rest is dissipated. Dissipation in an internal wave field is a highly random, sporadic event that is difficult to predict with certainty. It might very well be that this sporadic dissipation of internal wave energy is enough to keep the spectrum close to a universal spectrum. It has been observed that the internal wave energies increase rapidly after a storm, but relaxation to a universal spectrum level is also quite rapid. Similarly, the relaxation of the spectrum to a universal spectrum as one moves away from a source such as seamounts is also quite rapid. This suggests that whatever the dissipation and redistribution mechanisms that prevail in a random field of deep-sea internal waves, they are efficient at redistributing and disposing of excess energy. The precise manner in which this occurs is not well known.

Absorption at critical layers is quite important to internal wave dissipation. This mechanism will be dealt with later. However, as Phillips (1977) points out, this mechanism is a unique way of transferring momentum from bottom currents via internal waves to the mean flow around the critical layer. The same holds for internal waves generated in the upper layers of the ocean and traveling into the deep. Thus critical layer processes are efficient means of redistributing momentum in the water column. They are of course very efficient dissipaters of internal waves, and therefore indirectly of the mean flow energy in the ocean.

Internal waves might be responsible for a large fraction of mixing in the deep ocean and that is why accurate estimates of internal wave breaking is so important. However, the current estimates of breaking are too low (by an order of magnitude) to provide the estimated mixing rates in the deep ocean (Olbers, 1983; Gargett, 1989), if internal tides are ignored (but see Section 6.7). Microstructure measurements are filling in the gaps of our knowledge of dissipation in general in the global oceans (Gregg, 1989).

There exists yet another mechanism for decay of internal waves, this one involving interaction between turbulence and internal waves, germane to internal waves below the OML. Laboratory experiments (Barenblatt, 1978; Phillips, 1977; Kantha, 1980c) have shown that turbulence in a mixed layer can extract energy from an internal wave propagating into the region. Phillips (1977) postulated that this occurs due to the presence of Reynolds stress-like terms in the phase-averaged equations for the wave motions that transfer energy from wave motions to turbulence. More recently Ostrovsky *et al.* (1996) and Ostrovsky and Zaborskikh (1996) have investigated this process both theoretically and experimentally. They find that the damping rate is a strong function of the wavenumber and this mechanism could be an important sink for internal waves in the upper layers of the ocean.

6.5.3 Propagation

Unlike surface waves, internal waves travel at low velocities and therefore even small background velocities can affect their properties significantly. Propagation of internal waves in the stratified interior away from its source can be treated using the WKB approximation, as long as the scales (temporal and spatial) of variation of the properties such as buoyancy frequency N and mean velocity U are larger than the scales of the waves, so that the medium can be considered slowly changing in the frame of reference fixed to the moving wave group. A propagating wave packet in a moving medium does not conserve its energy since there can be an interchange of energy between the medium and the wave group. The action density is however conserved in the absence of dissipation and this principle applies to internal wave packets as well. It is always useful to think in terms of action conservation and the WKB approach, when dealing with propagation of linear waves through a medium with slowly changing properties.

Consider a wave train propagating along a trajectory. A local dispersion relation relates the waves' frequency to the wavenumber vector:

$$\omega = \omega\,(x_i, t, k_i) \qquad (6.5.5)$$

ω is Doppler shifted due to any ambient mean current U_i so that

$$\omega = n + k_i U_i \tag{6.5.6}$$

where n is the intrinsic frequency and k_i is the wavenumber vector. Along the ray path, the waves travel with the group velocity

$$c_{gi} = d\omega / dk_i \tag{6.5.7}$$

and the wavenumber and frequency of the wave are given by the kinematical conservation equation for wave crests:

$$dk_i / dt + d\omega / dx_i = 0 \tag{6.5.8}$$

This is a kinematic relationship between the wavenumber and the frequency. Note that the dispersion relationship relates the two dynamically. The above relations are called ray equations. The energetics of wave motion are described, however, by action conservation. If E is the energy density of the wave packet and A is the wave action density,

$$A = E/n = E/(\omega - U_i k_i), \tag{6.5.9}$$

then action conservation can be written as (Whitham, 1970)

$$dA / dt + d(c_{gi} A) / dx_i = S \tag{6.5.10}$$

where S is the source–sink term, which includes the energy exchange through resonant interactions with other waves, generation, and dissipation. For progressive internal waves, the ray equations become

$$\begin{aligned} \frac{dx_i}{dt} &= U_i + Cg_i \\ \frac{dk_i}{dt} &= -\frac{N}{n} \cdot \frac{N^2 - n^2}{N^2 - f^2} \frac{\partial N}{\partial x_i} - k_j \frac{\partial U_j}{\partial x_i} \\ \omega &= \text{const.} \end{aligned} \tag{6.5.11}$$

where C_{gi}, the intrinsic group velocity, is given by Eq. (6.3.21). For a packet of monochromatic internal waves, action conservation requires

$$d[(c_{gi} + U_i)A] / dx_i = d[(c_{gi} + U_i)(E/n)] / dx_i = 0 \tag{6.5.12}$$

at any position along the ray. Now consider a horizontally homogeneous medium with nonconstant N(z) and U(z). Action conservation gives

$$d(c_{g3}E/n)/dx_3 = 0 \tag{6.5.13}$$

Horizontal wavenumber vector k_h remains constant, but the intrinsic frequency n, the vertical wavenumber vector k_z, and the energy density E change according to

$$k_z(z) \sim \left[N^2(z) - n^2\right]^{1/2}$$

$$E(z) \sim (Cg_z)^{-1} \sim \frac{N^2(z) - f^2}{\left[N^2(z) - n^2\right]^{1/2}} \tag{6.5.14}$$

$$n\text{-}k_h U(z) = \text{const.}$$

$$k_h = \text{const.}$$

For a constant U(z) (no mean shear) and decreasing N(z), the wave might eventually reach a depth z_t where the frequency of the wave $n = N(z_t)$. The wave packet cannot propagate further (because n cannot be larger than N) and is totally reflected (if dissipation and other effects are ignored). This depth is called the turning depth (Figure 6.5.3). The group velocity here is purely vertical. WKB solutions are invalid near the turning depth since vertical scale of the wave becomes large (implying that it is no longer small compared to the scale of background variations), and have to be replaced by solutions involving Airy functions, which are oscillatory on one side of the turning depth and

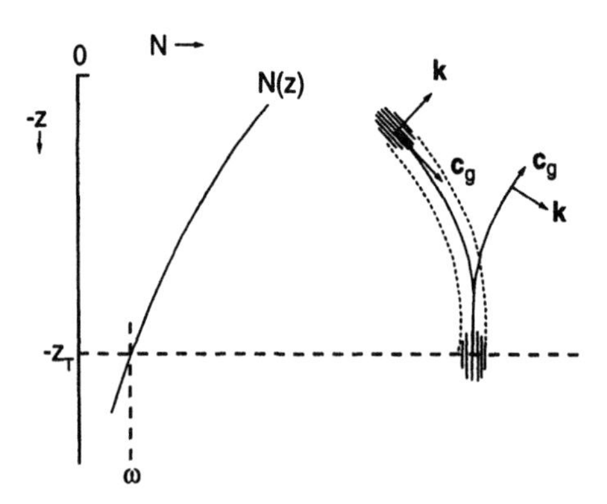

Figure 6.5.3. Reflection of internal waves at the level where n reaches N (from Munk, W., *in Evolution of Physical Oceanography,* Copyright M.I.T. Press, 1981).

damped on the other. Airy functions are solutions to equations of the form

$$d^2y/dz^2 + (z - z_t)y = 0 \tag{6.5.15}$$

The Airy solutions show that E has a finite maximum at the turning depth, and not a singularity. The wave amplitudes are a bit higher at the turning depth z_t and the group velocity is along the vertical, but tends to zero. Since the waves are also reflected at the free surface, the resulting modal shape, which is the consequence of superposition of upward and downward propagating waves of equal energy density, is also shown in Figure 6.5.4. The potential energy spectrum of internal waves is observed to be enhanced at the buoyancy frequency (Figure 6.5.5).

A consequence of this "trapping" is that for short internal waves in a diffuse pycnocline, with a N(z) distribution with a maximum N_m, waves with $n < N_m$ are trapped in a waveguide delineated by upper and lower turning depths on either side of the location of the maximum and are repeatedly reflected back into the waveguide. Therefore the waves are "trapped" in the waveguide and propagate in the horizontal direction along the waveguide (see Figure 6.5.6).

For a nonconstant U(z), if U(z) increases in the direction of propagation, N(z)/U(z) can approach k_h, so that k_z can go to zero. This is the same as the trapping considered above and the wave is reflected.

A similar process of wave reflection occurs with change in latitude for an inertial-internal wave packet propagating poleward; the wave is reflected at the turning latitude θ_t given by

$$n = f = 2\Omega \sin\theta_t \tag{6.5.16}$$

The motion is purely horizontal at the turning latitude and the kinetic energy spectrum is observed to have a peak at n = f (Figure 6.5.5).

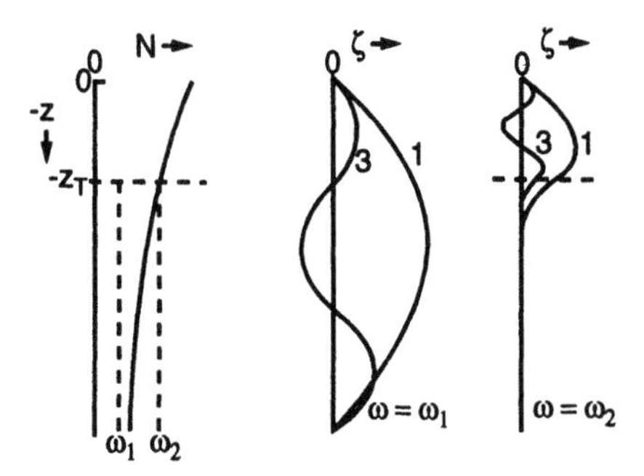

Figure 6.5.4. Propagation of internal waves for different modes and frequencies (from Munk, W., *in Evolution of Physical Oceanography,* Copyright M.I.T. Press, 1981).

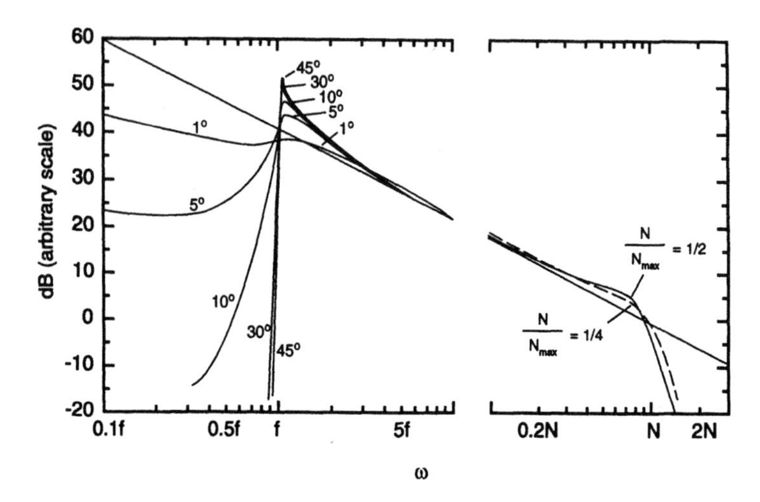

Figure 6.5.5. Kinetic and potential energy spectra showing enhancement of KE at the inertial and PE at the buoyancy frequency (from Munk, W., *in Evolution of Physical Oceanography,* Copyright M.I.T. Press, 1981).

Background vertical mean shear is of more profound consequence to internal waves. The intrinsic frequency now changes with depth and can become zero (or more appropriately equal to f, but we will ignore f in the following). Then,

$$\begin{aligned} &n_1 + k_h U_1 = n_2 + k_h U_2 \rightarrow k_h U_2; \\ &k_{z2} \rightarrow \infty;\ c_{gz2} \rightarrow 0; E_2 \rightarrow \infty \end{aligned} \tag{6.5.17}$$

where subscript 2 denotes the conditions at the depth where the intrinsic frequency goes to zero. Within the ray theory approximation, both the vertical wavenumber and the energy density become singular, although the vertical group velocity and the vertical energy flux go to zero. This is the so-called critical layer absorption.

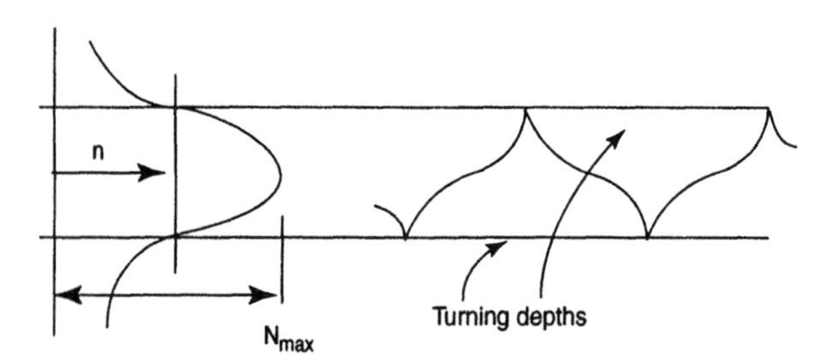

Figure 6.5.6. Trapping of internal waves in the diffuse pycnocline.

Waves do not transfer momentum to the mean flow except when they break or encounter a critical layer where their Doppler-shifted frequency vanishes (neglecting f) and their phase speed equals the current speed (U = C) (Figure 6.5.7). Critical layers are associated with clear-air turbulence in the atmosphere and the phenomenon is similar in the oceans. Bretherton (1966) investigated this process using WKB methods for an ocean of constant N(z) and slowly changing U(z). Near the critical layer, Booker and Bretherton (1967) showed that the frequency ω, vertical wavenumber k_z, the vertical displacement ζ, the vertical velocity w, and the horizontal velocity u vary near the critical layer ($z=z_c$) as

$$\begin{aligned} &\omega \sim |z-z_c|,\ k_z \sim |z-z_c|^{-1},\ \zeta \sim |z-z_c|^{-1/2}, \\ &u \sim |z-z_c|^{-1/2}, w \sim |z-z_c|^{1/2} \\ &n = N\cos\theta = k_h / k \end{aligned} \tag{6.5.18}$$

The intrinsic frequency n vanishes as the wave packet approaches the critical layer. The vertical wavenumber increases and the wave is refracted (Figure 6.5.7). Large vertical displacements, large vertical shears of horizontal velocities, develop which could lead to shear instability, and wave breaking and dissipation. What actually happens depends very much on the value of Ri_g. If $Ri_g > 1/4$, the wave energy flux is attenuated by the critical layer (Booker and Bretherton, 1967) by a factor of $\exp[-2\pi(Ri_g - 1/4)^{1/2}]$.

For large Ri_g, the critical layer absorption is nearly complete, as given by WKB theory (Bretherton, 1966), which is valid under these conditions. This indicates that critical layers in the ocean tend to absorb internal waves. But for $0 < Ri_g < 1/4$, most of the energy is reflected (Jones, 1968); thus a critical layer becomes a reflector, but the reflected wave amplitude is generally less than the incident wave amplitude. Below a particular value of Ri_g (for a given frequency and wavenumber), Jones (1968) found that the reflected wave amplitude can be

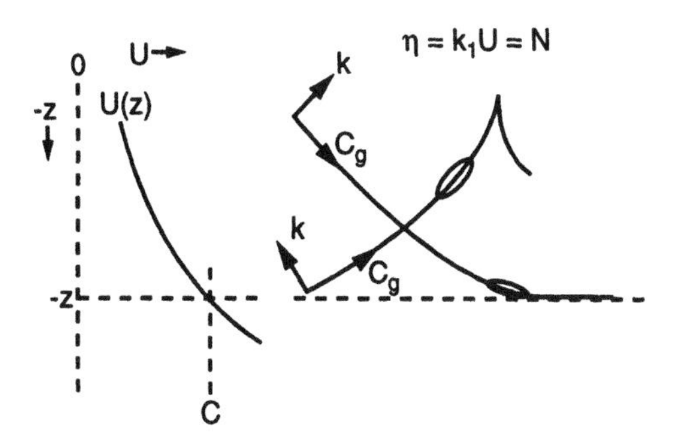

Figure 6.5.7. Refraction of IWs at the critical layer.

larger; this occurs at the expense of the mean flow. The waves extract energy from the mean flow and are called overreflected. Figure 6.5.8, from Munk (1981), shows an analysis of the critical layer process for a tanh profile for U(z).

6.5.4 INTERACTION

Internal waves can also interact with one another and exchange energy among themselves. This is possible because of the nonlinear terms in the governing equations (Phillips, 1977). A triad of internal waves can interact resonantly among themselves if

$$k_3 = k_1 \pm k_2, \quad n_3 = n_1 \pm n_2 \tag{6.5.19}$$

The condition on frequencies reduces to a geometric condition on the inclination of the three wave vectors:

$$\cos\theta_3 = \cos\theta_1 \pm \cos\theta_2 \tag{6.5.20}$$

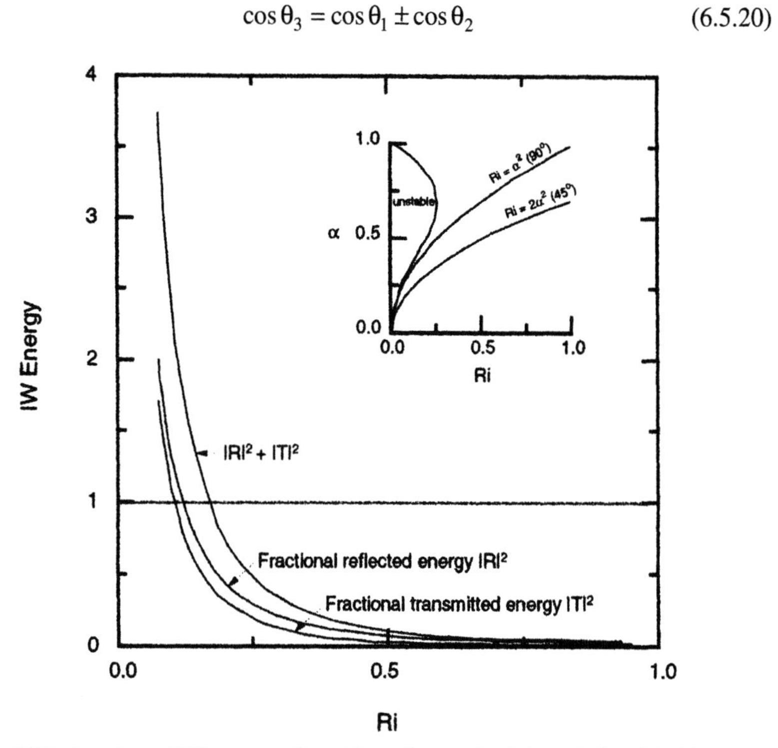

Figure 6.5.8. Fractions of IW energy reflected by and transmitted through the critical layer as a function of the Richardson number (from Munk, W., *in Evolution of Physical Oceanography,* Copyright M.I.T. Press, 1981).

This triad interaction does not increase or decrease the total energy of the triad, but instead causes simply an interchange of energy among the participants. For example, one mode can grow at the expense of the others. McComas and Bretherton (1977) identified three important classes of resonant triad interactions shown in Figure 6.5.9:

1. Induced diffusion. Here a high wavenumber–high frequency mode k_1 interacts with a mode of much lower wavenumber–frequency k_2 to generate another high wavenumber–high frequency mode k_3. Thus the low frequency mode k_2 is induced to diffuse energy to high frequency mode k_3. This tends to fill in any sharp high wavenumber cutoffs in the spectrum (Munk, 1981).

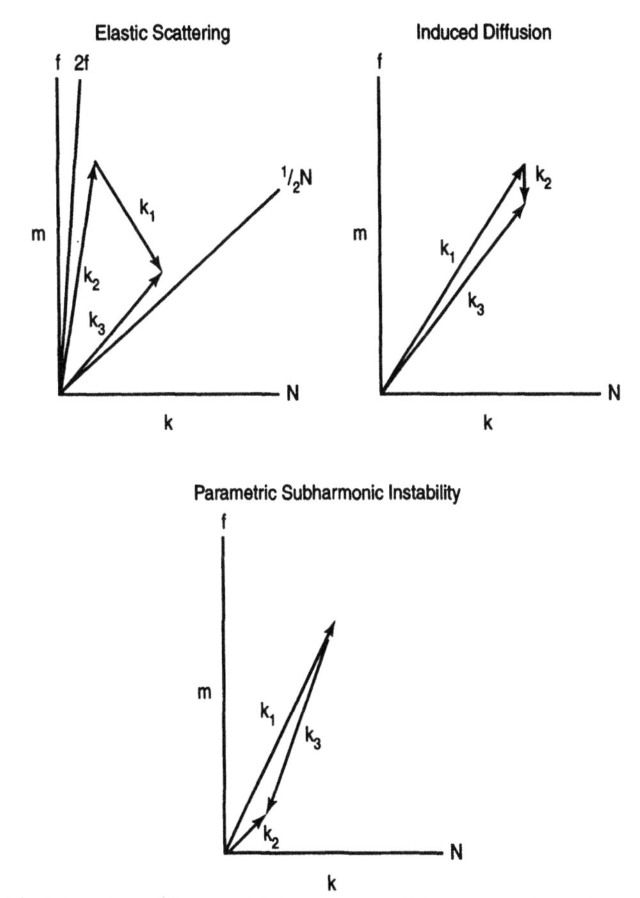

Figure 6.5.9. Three types of resonant interactions possible among an internal wave triad (from Munk, W., *in Evolution of Physical Oceanography,* Copyright M.I.T. Press, 1981).

2. Elastic scattering. Here energy from a high frequency mode k_3 is scattered into another mode $k_1 \sim -k_3$ by interaction with a low frequency mode $k_2 \sim 2\,k_3$. This Bragg scattering tends to equalize upward and downward energy fluxes(for both bottom-generated, upward-propagating internal waves and surface-generated, downward-propagating internal waves) except at very low frequencies. Thus a vertically symmetric spectrum tends to result, except at very low near-inertial frequencies.

3. Parametric subharmonic instability. Here a low vertical wavenumber mode k_2 decays into two high vertical wavenumber modes k_1 and $-k_3$ of about half the frequency

The interaction (relaxation) time, defined as the ratio of energy density to the net energy flux into or out of a mode, determines how fast the mechanism acts. Relaxation times for induced diffusion and elastic scattering are very short, less than about a wave period. Therefore these resonant interactions impose strong constraints on the possible shape of the internal wave spectrum (Munk, 1981). They are responsible for the rapid relaxation of a perturbed spectrum to the equilibrium form with a universal shape. In the GM spectral model, induced diffusion and elastic scattering are in approximate equilibrium, while parametric subharmonic instability controls the energy flow to the high wavenumber–low frequency part of the spectrum at the rate of 6×10^{-4} W m^{-2} (Olbers, 1983).

6.6 GARRETT AND MUNK SPECTRUM

In the early seventies, Garrett and Munk pooled all the internal wave observations available at the time to deduce an empirical model for the wavenumber–frequency spectrum of oceanic internal waves (Garrett and Munk, 1972). Since then, the model has been revised (Garrett and Munk, 1975; Munk, 1981) somewhat, but has withstood the test of time rather well. The internal wave spectrum, in general, appears to have a universal shape and energy level throughout the global oceans, except near seamounts and canyons and at the equator and close to the surface. The rapid and strong nonlinear interactions among internal waves are thought to be responsible for the universal shape. These nonlinear interactions which occur among wave triads appear to redistribute energy and momentum quite efficiently among different wavenumbers so much so that a distorted spectrum, for example, near a seamount, relaxes rapidly toward the universal shape. The GM (Garrett–Munk) spectrum appears to be a useful approximation to the energy distribution at large energy-containing scales of the internal wave spectrum in the area, but Muller *et al.* (1986) questions how well it models the smaller scales where most of the oceanic shear is concentrated (Gargett *et al.*, 1981).

The GM spectrum starts with the assumption that the internal wave energy distribution is isotropic in the horizontal direction so that only a scalar k_h is needed to characterize the waves. It is formulated in terms of discrete vertical modes for an exponentially stratified ocean,

$$N(z) = N_0 e^{z/b} \tag{6.6.1}$$

where $N_0 = 5.2\times10^{-3}\,s^{-1}\,(3\,cph)$ and b ~ 1.3 km, the e-folding scale of N (z).

In other words, upward and downward energy fluxes are assumed to be equal. The energy spectrum per unit mass is written as a function of frequency n and mode number j in separable form,

$$E(n,j) = b^2 N_0 N E_0 B(n) H(j) \tag{6.6.2}$$

where

$$H(j) = \frac{\left(j^2 + j_*^2\right)^{-1}}{\sum_{j=1}^{\infty}\left(j^2 + j_*^2\right)^{-1}}, \qquad \sum_{j=1}^{\infty} H(j) = 1 \tag{6.6.3}$$

$$B(n) = \frac{2}{\pi}\frac{f}{n}\left(n^2 - f^2\right)^{-1/2}; \qquad \int_f^N B(n)\,dn = 1 \tag{6.6.4}$$

where $j_* = 3$, $E_0 = 6\cdot3\times10^{-5}$. The factor $\left(n^2 - f^2\right)^{-1/2}$ allows for a peak at the inertial turning frequency. Away from n = f, the frequency spectrum decays with a –2 slope. Conversion from (n, j), the frequency–mode number space, to (n, k), the frequency–wavenumber space, is done by setting

$$k_z = k_h\left(\frac{N^2 - n^2}{n^2 - f^2}\right)^{1/2} = \frac{\pi}{b}\left(\frac{N^2 - n^2}{N_0^2 - n^2}\right)^{1/2} j \tag{6.6.5}$$

involving a slowly varying N(z), WKB approximation. The spectra as vertical displacements and horizontal velocity become

$$E_\xi(n,j) = \frac{1}{N^2}\left(\frac{n^2 - f^2}{n^2}\right) E(n,j) \tag{6.6.6}$$

$$E_u(n,j) = E_{u_1}(n,j) + E_{u_2}(n,j) = \left(\frac{N^2 - n^2}{N^2}\right)\cdot\left(\frac{n^2 + f^2}{n^2}\right) E(n,j) \tag{6.6.7}$$

The total energy contained in the spectrum is

$$\begin{aligned}\overline{\xi^2} &= \frac{1}{2} b^2 E_0 \frac{N_0}{N} = 0.0053 \frac{N_0}{N} m^2 \\ \overline{u^2} &= \frac{3}{2} b^2 E_0 N_0 N = 0.0044 \frac{N}{N_0} m^2 s^{-2} \\ E_e &= b^2 E_0 N_0 N = 0.003 \frac{N}{N_0} m^2 s^{-2}\end{aligned} \tag{6.6.8}$$

Note that $E_e = \frac{1}{2}\left(E_u + N^2 E_\xi\right)$. The spectral energy level is equivalent to a 7-m rms vertical displacement and a 7 cm s^{-1} rms current. The total KE is three times the PE in the GM spectrum. The energy per unit area of the ocean is

$$\int \rho E_e \, dz \sim \rho b^3 N_0^2 E \sim 3800 \, J m^{-2} \tag{6.6.9}$$

To keep shear finite, a high wavenumber cutoff at about 0.1 cpm is used (Munk, 1981), where the Richardson number and strain rate are of order 1.

This spectrum has turned out to be remarkably universal. Exceptions are found near sources and sinks of internal waves, such as seamounts, and near the equator and in regions of high shear (Munk, 1981). Introducing a continuum of upward and downward propagating modes with vertical wavenumber,

$$k_z = \pm k_h \left(\frac{N^2 - n^2}{n^2 - f^2} \right)^{1/2} \tag{6.6.10}$$

the discrete spectrum corresponds to the continuous spectrum

$$E\left(n, k_z\right) dn \, dk_z = \frac{1}{2} E\left(n, j\right) dn \, dj \tag{6.6.11}$$

with j and k_h $\left(\& \ k_z\right)$ related as in Eq. (6.6.5).

Note that the GM spectrum exhibits four basic features: (1) horizontal isotropy, (2) vertical symmetry, (3) a n^{-2} frequency spectrum with a cusp and increase at the inertial frequency, and (4) a $k^{-2.5}$ slope in the wavenumber spectrum. Due to horizontal isotropy and vertical symmetry, there are only two independent (n, k_z or n, k_h or k_h, k_z) parameters. Figure 6.6.1 shows the shape of the GM spectrum. Olbers (1983) describes how this empirical spectrum was constructed from observations: spectral shape in the frequency domain from

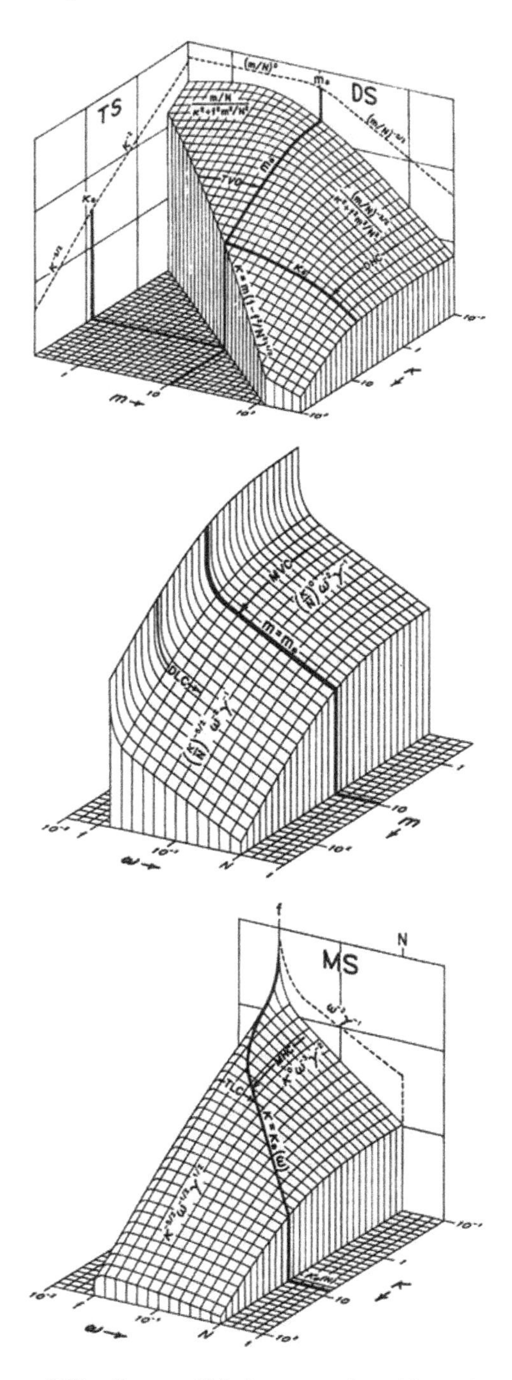

Figure 6.6.1. Garrett and Munk spectrum (from Olbers, 1983).

moored energy spectrum, and that in the wavenumber space from towed and dropped spectra of vertical displacements. The GM spectrum has been extensively used to characterize the deep-ocean internal wave field.

Since the GM spectrum was originally derived from empirical data and refinements made using empirical data, it is not surprising that the internal wave observations, on the average, are consistent with it. However, as IWEX in the Sargasso Sea in 1973 showed, the individual spectra show a great deal of variability about the GM mean. However, given the potential for contamination of measurements by the presence of fine layered structures (Phillips, 1971) and by other noise, this is not surprising. IWEX measurements essentially confirm the universality of internal wave spectrum in midlatitude oceans, away from the tidal and inertial peaks (Olbers, 1983; Muller *et al.,* 1986).

Assuming the spectral shape as given, two independent parameters, A and B, that describe the energy and the wavenumber bandwidth are adequate to describe the GM spectrum:

$$A = E_0 b^2 N_o \ (0.56\ m^2 s^{-1}), B = \pi j_* /(bN_0)\ (1.39\ m^{-1} s) \tag{6.6.12}$$

The frequency and wavenumber spectra all scale with A, and the total energy ($J\ kg^{-1}$) in the entire frequency–wavenumber spectrum is AN. The parameter B defines the wavenumber content of the internal wave field and hence the coherence. Lower coherence is to be expected if at a given frequency the wavenumber bandwidth is higher, meaning that more wavenumbers are present. For ease of comparison of different spectral shapes, it is convenient to define an equivalent vertical wavenumber bandwidth k_e [equal to $\pi BN(z)$ for the GM spectrum], which is the bandwidth of an equivalent rectangular distribution with the same ratio of variance to squared mean (Levine *et al.,* 1997). B_e is related to a horizontal wavenumber bandwidth B_{he} by $B_{he} = k_e[(n^2-f^2)/(N^2-f^2)]^{1/2}$. The GM spectra of the vertical displacement E_ζ and the vertical velocity E_w, and the rotary spectra of clockwise (E_{u-}) and counterclockwise (E_{u+}) horizontal velocity components, can be written as (Levine *et al.,* 1997)

$$\begin{aligned} E_\xi(n) &= A \frac{2f}{\pi N(z)} \frac{\left(n^2 - f^2\right)^{1/2}}{n^3} \quad m^2 s^{-1} \\ E_w(n) &= n^2 E_\xi(n) \quad m^2 s^{-3} \\ E_{u\pm}(n) &= A \frac{fN(z)}{\pi} \frac{\left(n \mp f\right)^2}{n^3 \left(n^2 - f^2\right)^{1/2}} \quad m^2 s^{-3} \end{aligned} \tag{6.6.13}$$

The wavenumber spectrum of vertical displacement is then (Levine, 1990)

$$\tilde{E}_\xi(k) = \int_f^N \left[\frac{2}{\pi} \frac{k_*}{\left(k_*^2 + k^2\right)} \right] E_\xi(n)dn \approx \frac{AB}{\pi\left(k_*^2 + k^2\right)}$$

$$\text{where } k_* = BN(z) \text{ and } \int \left[\frac{2}{\pi} \frac{k_*}{\left(k_*^2 + k^2\right)} \right] dk = 1 \tag{6.6.14}$$

The vertical coherence depends only on B:

$$C_\zeta(\Delta z) = \exp(-BN\Delta z) \tag{6.6.15}$$

The dissipation rate due to the internal wave field was derived by Gregg (1989) as

$$\varepsilon_{1,2} = \left[\left(1.67/\pi^3\right)\cosh^{-1}(N/f), \left(1 + 0.27\pi\right) \right] N^2 f A^2 B^2 \tag{6.6.16}$$

where ε_1 is from the weak interaction theory of McComas and Muller (1981) and ε_2 is from the strong interaction theory of Henyey *et al.* (1986; see also Henyey, 1991). The two differ only in terms of their relative magnitude, with ε_1 being generally several times greater than ε_2. The dissipation rate can be related to the diapycnal diffusivity and hence to vertical heat flux (see Section 6.8). Levine *et al.* (1997) find that ε_1 gives a more realistic estimation of the vertical heat flux (1 W m^{-2}) than ε_2 (7 W m^{-2}) in the Weddell Sea.

If the spectral slope is to be treated as a free parameter, as may be necessary in evaluating observed spectra, these can be modified to (Levine *et al.*, 1986)

$$\hat{E}_\xi(n) = A \frac{f^{1/2}}{JN(z)} \frac{\left(n^2 - f^2\right)^{1/2}}{n^{5/2}} \quad m^2 s^{-1}$$

$$\hat{E}_w(n) = n^2 \hat{E}_\xi(n) \quad m^2 s^{-3}$$

$$\hat{E}_{u\pm}(n) = A \frac{f^{1/2} N(z)}{J} \frac{(n \mp f)^2}{n^{5/2}\left(n^2 - f^2\right)^{1/2}} \quad m^2 s^{-3} \tag{6.6.17}$$

$$J = \int_1^{N/f} \left[\hat{n}\left(\hat{n}^2 - 1\right) \right]^{-1/2} d\hat{n}; \quad \hat{n} = n/f$$

where J is a nondimensional normalization constant. The wavenumber

TABLE 6.6.1
Summarization of Internal Wave Characteristics Observed during Various Field Experiments (from Levine *et al.*, 1997)

Experiment	Location/ Reference	Energy	Bandwidth	N, $10^{-3}s^{-1}$	f, $10^{-4}s^{-1}$	Slope	ϵ 10^{-9} W kg^{-1}	K_d, $10^{-6}m^2s^{-1}$
GM79	[*Munk*, 1981]	1	1	5.24*	1*	−2	0.42*	3.1*
ISW	western Weddell Sea							
Segment 1		0.4–0.6	5	1.6–1.9	1.4	−1.5	...	...
Segments 2 and 3		0.2–0.4	4–5	1.6–1.9	1.4	to	0.021–0.38	1.8–21
Segment 4		1.3–1.6	2	1.6–1.9	1.4	−2	0.023–0.05	19–30
Site C								
Segments 1, 2, and 3		1.0–1.3	...	1.0–1.1	1.4		...	...
IWEX	NW Atlantic [*Müller et al.*, 1978]	0.25	0.9–1.6	0.63–4.7	0.68		0.00013–0.036	0.064–0.33
MATE	NE Pacific [*Levine et al.*, 1986]	0.9	2	1.7–2.3	1.0	−1.7	0.1–0.2	7.4–8.0
AIWEX	western Arctic Ocean *Levine et al.*, 1987; *Levine*, 1990]	0.03–0.07	10	8.7	1.4	−1.2	0.1–0.8	0.4–2.1
CEAREX	eastern Arctic Ocean							
Period 1	[*Wijesekera et al.*,	0.26–0.3	1–2	2.8–6.8	1.4		0.01–0.35	0.23–1.5
Period 2	1993]	0.6–1.3	2–3.3	2.8–6.8	1.4		0.93–3.9	6.0–65
Period 3		1.6–2.3	1–1.3	2.8–6.8	1.4		0.73–5.0	9.8–32

Note. The energy A and bandwidth B are relative to the GM spectrum (Munk, 1981). Estimates of dissipation rate ϵ are computed from the Henyey *et al.* (1986) model. The dissipation rate and diapycnal diffusivity K_d for the GM spectrum is for typical midlatitude N and f values.

spectrum is similar in form to GM form and is approximately equal to 1.24 $AB[\pi\,(k_*^2+k^2)]^{-1}$. Table 6.6.1 and Figure 6.6.2, from Levine *et al.* (1997), summarize existing observations of A and B (and the spectral slope, dissipation rate, and diapycnal diffusivity) relative to GM values. There are hints that the product AB might be roughly constant to within a factor of about 2. A Richardson number $Ri_{IW} = \sigma_s^2/N^2$ can be defined, where σ_s^2 is the variance of vertical shear due to internal waves, and is proportional to Abk_c, where k_c is a cutoff wavenumber. Munk (1981) suggests that there exists a tendency in an internal wave field for Ri_{IW} to be close to the critical value, and if the cutoff wavenumber were also to remain relatively unchanged, the product AB would tend to be constant as well. If true, this would mean that the dissipation rate and hence the vertical heat flux in the deep ocean is a function only of N and f. Levine *et al.* (1997) suggest that more observations and numerical simulations could help ascertain this.

Needless to say, the GM spectrum is not obtained close to the surface. An example (Figure 6.6.3) is the spectrum measured during GATE (Global Atlantic Tropical Experiment), which clearly shows a prominent tidal peak at the M_2 frequency (Kase and Siedler, 1980), a peak in the near-inertial band, and a peak near 3 cph. The latter, the 30-min waves, are waves trapped in the sharp pycnocline, with a local buoyancy frequency of 10 cph.

Observations made in the Beaufort Sea in the Arctic from a drifting ice camp

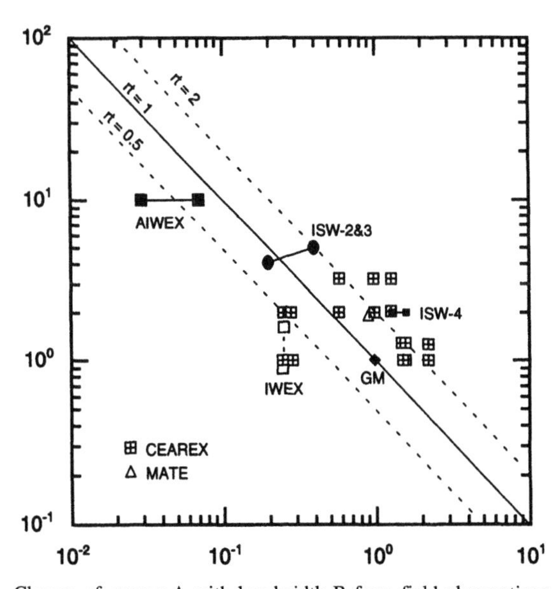

Figure 6.6.2. Change of energy A with bandwidth B from field observations tabulated in Table 6.6.1 (from Levine *et al.*, 1997). Values are relative to the GM spectrum (Munk, 1981). Note that the product AB tends to be a constant.

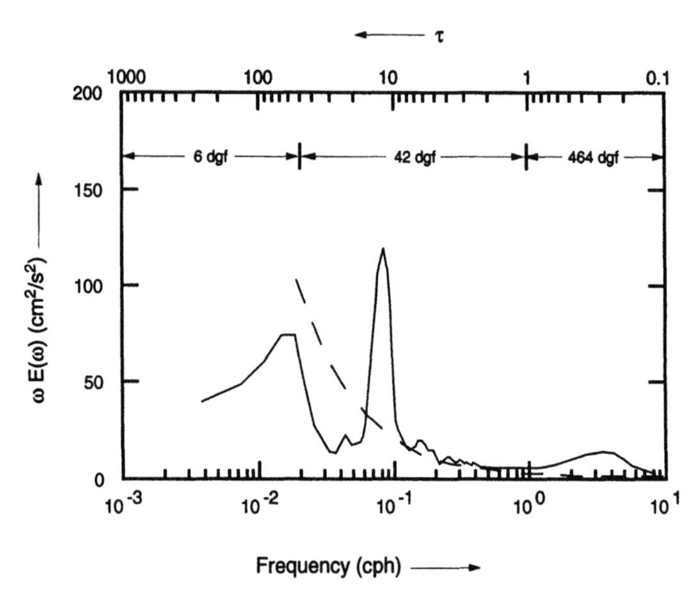

Figure 6.6.3. IW spectrum close to the ocean surface (Olbers, 1983).

during the Arctic Internal Wave Experiment (AIWEX) (Levine, 1990; Levine *et al.,* 1985) and Coordinated Eastern Arctic Experiment (CEAREX) (Wijesekera *et al.,* 1993), and in the Antarctic (Levine *et al.,* 1997), have shown that the IW spectrum beneath the ice pack also departs significantly from the GM spectrum. The sea-ice cover acts as a rigid lid and, by mediating the transfer of momentum by the wind to the ocean, prevents excitation of surface gravity waves. Wind and waves are two important potential sources of IW energy. The ice cover alters drastically the process of generation of internal waves by wind stress and very nearly eliminates their generation by surface waves. The remaining principal mechanism is generation by currents near sharp topographic changes. So it is not surprising that in ice-covered seas, away from topographic features, the internal wave energy densities are low. Figure 6.6.4 shows the AIWEX buoyancy frequency distribution in the vertical as well as the displacement spectrum at a 250-m depth, along with the GM spectrum. The energy densities are lower, 0.03–0.07 of the GM value (Levine *et al.,* 1985), and the slope of the frequency spectrum is lower as well (–1.1 compared to the –2 slope of the GM spectrum). Tides, especially internal tides, are quite weak in the Arctic, and it is possible that this is a contributory factor to the exceptionally low energy levels. However, energy levels appear to be much higher in some regions. Drifting ice camp observations by Padman and Dillon (1991) during CEAREX, and the Arctic Environmental Drifting Buoy Observations by Plueddemann (1992), showed sharp changes in the internal wave energy levels during the drift. Plueddemann (1992) attributes the increase from low levels in the Nansen Basin

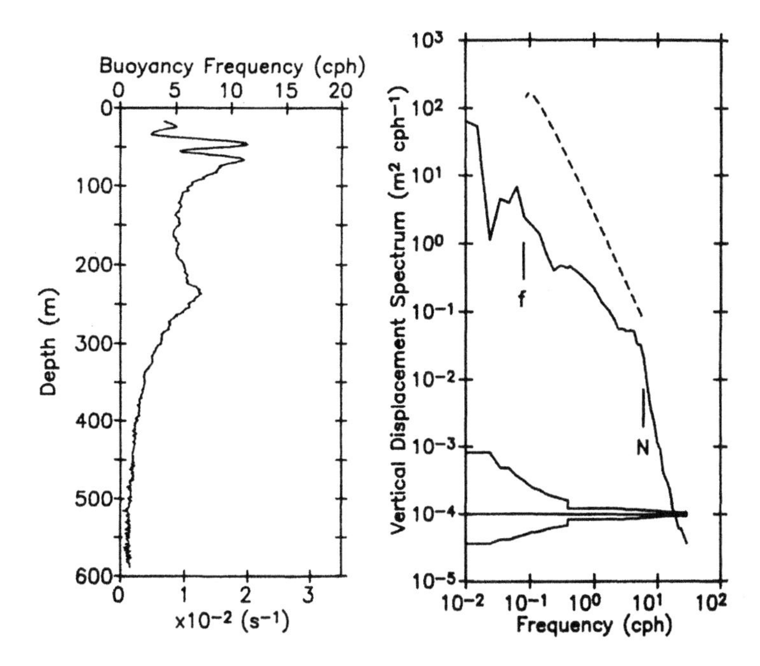

Figure 6.6.4. IW spectrum measured under the ice pack. Note the significantly smaller energy levels. The vertical profile of buoyancy frequency is also shown (from Levine *et al.*, 1987).

to near GM spectrum levels on Yermak Plateau to the presence of barotropic diurnal tidal currents. Levine *et al.* (1997) also observed energy levels lower than the GM value (0.2–0.6 times) in deep water in the upper permanent thermocline (200- to 300-m depth) of the Weddell Sea in the Antarctic, but near or above GM values closer to topographic features where the barotropic tidal current is larger. The slope of the vertical displacement spectra was closer to –1.5. Slope values less than the GM canonical value of –2 are also observed often at lower latitudes (Levine *et al.*, 1983). Figures 6.6.5 shows the vertical displacement frequency spectra measured at two sites, ISW and C, in the Weddell Sea from Levine *et al.* (1997). Figure 6.6.6 shows the vertical displacement wavenumber spectra at ISW that display a k^{-2} slope. The energy level is proportional to AB, but is higher than the GM level, probably due to contamination by the fine structure. The frequency spectra are relatively free from this contamination.

To conclude, while empirical knowledge of internal wave processes and their spectra is now reasonably complete, the dynamical underpinnings are still uncertain. Since other processes such as shear mixing and double diffusion that contribute to the fine structure and variability in the deep oceans operate at similar spatial and temporal scales as internal waves, there exists a considerable overlap between the two and this makes unambiguous interpretation of field

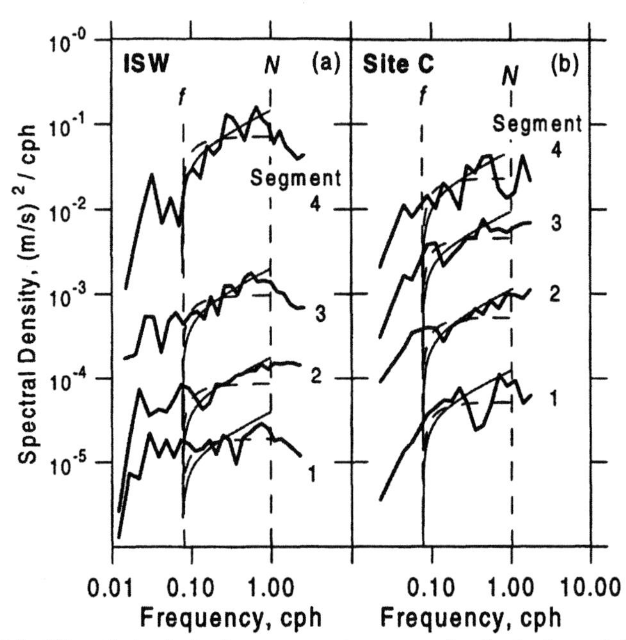

Figure 6.6.5. IW vertical velocity frequency spectra measured under the ice pack in the Weddell Sea at a 250-m depth at site ISW and a 200-m depth at site C (from Levine *et al.,* 1987). The solid line is the modified GM spectrum (Levine *et al.,* 1986) and the dashed line indicates the –1.5 slope. Each successive curve is offset vertically by a factor of 10.

observations and development of insight into dynamics of internal waves particularly difficult. A convincing dynamical approach to deriving the internal wave spectrum is still elusive.

6.7 INTERNAL WAVE SOLITONS

Thus far we have ignored finite-amplitude effects. Because of the small stabilities involved ($\Delta\rho/\rho \sim 10^{-3}$), the vertical displacements involved in internal wave motions can routinely reach several meters in amplitude. In some isolated cases, the displacements can be several tens of meters. Here we will discuss one such case—internal wave solitons generated mostly by tidal action in regions of sharp topographic changes such as narrow straits and canyons near the shelf break. The following is from Osborne and Birch (1980).

Solitons, solitary waves that retain their shape and speed after collisions with each other, have been found in many branches of physics. They are finite-amplitude waves that result from an exact balance between the nonlinear steepening of the waveform and the tendency toward dispersion of the wave in

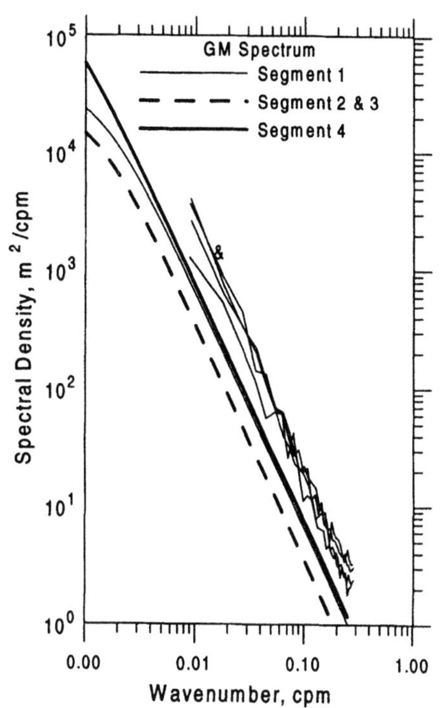

Figure 6.6.6. IW vertical velocity wavenumber spectra measured under the ice pack in the Weddell Sea at site ISW (from Levine *et al.,* 1987). Equivalent GM spectra (Levine *et al.,* 1986) are also shown. The slope of the spectra agrees with the GM value, but the energy levels are higher due to fine structure contamination.

the governing equations, so that their shape and speed remain invariant as they propagate in the medium. Surface solitary waves are governed by the Korteweg–de Vries (K-dV) equation,

$$\frac{\partial \eta}{\partial t} + c\frac{\partial \eta}{\partial x} + \alpha\eta\frac{\partial \eta}{\partial x} + \gamma\frac{\partial^3 \eta}{\partial x^3} = 0$$
$$c=\sqrt{gH},\ \alpha = 3c/2H,\ \gamma=cH^2/6 \tag{6.7.1}$$

where η (x,t) describes the shape of the solitary wave, H is the water depth, and c is the phase speed of the infinitesimal shallow water waves. This equation has the solution

$$\eta(x,t) = A\,\mathrm{sech}^2\left[2(x - C\,t)/\lambda\right] \tag{6.7.2}$$

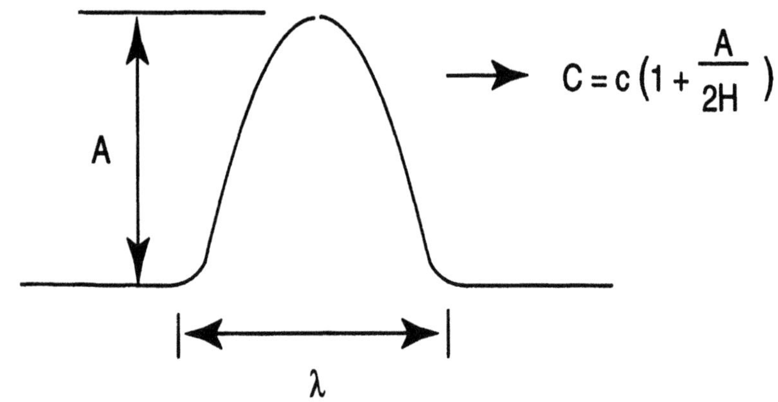

Figure 6.7.1. Internal wave soliton shape.

The term A is the amplitude of the solitary wave, $\lambda = 4(H^3/3A)^{1/2}$ is the characteristic "wavelength," and $C = c\,(1 + A/2H)$ is its phase speed. Its shape is shown in Figure 6.7.1. The permanence of its shape results from an exact balance between the nonlinear term $\alpha\eta\partial\eta/\partial x$ and the dispersion term $\gamma\partial^3\eta/\partial x^3$ in the equation.

Gardner *et al.* (1967) presented analytical solutions to the evolution of an initial localized hump of fluid into a packet of solitons accompanied by a dispersive linear wave train or tail (Figure 6.7.2). The soliton packet consists of n solitons ordered according to their amplitudes $A_m = [(n-m)/(n-1)]^2$ with the largest one propagating at the front of the train, since the larger the amplitude, the higher the phase speed. The number n is determined by number of zeros of the solution to a Schrodinger equation (Osborne and Birch, 1980). At least one soliton emerges for a positive initial waveform and none for a negative one. The separation distances between the individual solitons in the packet increase with time because of the dependence of phase speed on amplitude.

Solitary waves at the interface of a two-layer fluid are governed by a similar

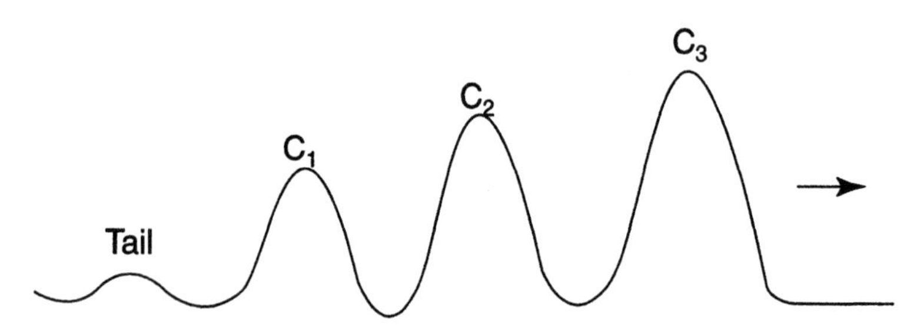

Figure 6.7.2. Soliton train accompanied by a linear wave train at the tail end.

equation, except that η is the interfacial displacement and

$$c=\left[g\frac{\Delta\rho}{\rho}H_1(1+r)\right]^{1/2}, \quad \alpha=-\frac{3c}{2}\left[(1-r)/H_1\right] \tag{6.7.3}$$

$$\gamma=cH_1H_2/6, \quad \Delta\rho=\rho_2-\rho_1, \quad r=H_1/H_2$$

If the upper layer is thinner than the lower one, as is usually the case, the internal soliton has a downward displacement of the form

$$\eta(x,t)=-A\,\mathrm{sech}^2\left[2(x-C\,t)/\lambda\right] \tag{6.7.4}$$

Its phase speed is

$$C=c\,(1-A\alpha/3c) \tag{6.7.5}$$

and its "wavelength" is

$$\lambda=4(-3\gamma/\alpha A)^{1/2} \tag{6.7.6}$$

The velocities in the upper and lower layers are

$$\begin{aligned} u_u(x,t)&=(cA/H_1)\mathrm{sech}^2\left[2(x\text{-}Ct)/\lambda\right] \\ u_l(x,t)&=(cA/H_2)\mathrm{sech}^2\left[2(x\text{-}Ct)/\lambda\right] \end{aligned} \tag{6.7.7}$$

These velocities are in opposite directions but are constant within each layer. Note that since the interface has a downward displacement, the free surface will have an upward displacement of the same form as η but smaller by a factor of $\Delta\rho/\rho$ (Figure 6.7.3). The total energy per unit length is

$$E=\frac{2}{3}\Delta\rho\,g\,A^2\lambda \tag{6.7.8}$$

Note that the sign of α depends on the ratio $r = H_1/H_2$. For $H_1 < H_2$, the upper layer is shallower than the lower one, the internal soliton is a wave of depression, and only an initial waveform that is a depression can generate internal wave solitons. Also by extension, a soliton propagating from deep water onto a shallow shelf can "fission" into a train of rank-ordered solitons. If $H_1 > H_2$, the upper mixed layer is thicker than the bottom layer, and the solitary waves will be waves of elevation, with displacement of the interface upward. When a soliton train in deep water consisting of depression waves propagates

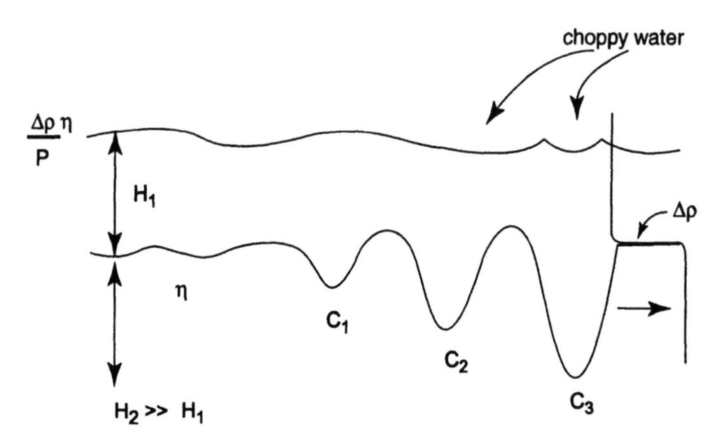

Figure 6.7.3. The IW soliton.

into shallow water, the solitons first disintegrate into dispersive wave trains and then reorganize themselves as a packet of nonlinear elevation waves in shallow water after they pass through a turning point where the upper and lower depths are approximately equal. This fascinating behavior has been observed in SAR images (Liu *et al.*, 1998) as well as simulated by a numerical model consisting of the K-dV equation of the type discussed above (for example, Liu, 1988). Figure 6.7.4, from Liu *et al.* (1998), shows a schematic of the depression solitary wave train evolving into elevation waves. A train of several rank-ordered elevation solitons can emerge from the disintegration of a single depression soliton as it passes through the turning depth. Liu *et al.* (1998) describe observations of nonlinear internal waves off Taiwan in the East China Sea and off Hainan in the South China Sea in ERS-1 SAR imagery.

The amplitudes of solitons in the train can be estimated by (Sandstrom and Oakey, 1995)

$$A_n = A_1\,[1-(n-1)/M] \quad n = 1,2,\ldots, M \tag{6.7.9}$$

with a total energy in the wave train proportional approximately to $A_1^{3/2}(0.4\,M + 0.5)$, where $A_1 \sim 2\,A$, that is, twice the amplitude of the internal tide (Sandstrom and Oakey, 1995). If the energy in the internal tide is also known, N can be computed. The primary dissipation mechanisms for these solitons as they propagate into shallow water are localized shear instability and breaking into turbulence, bottom friction, and scattering into other modes. Vertical shear from wave motion reduces the Richardson number to values less than 0.25 in a layer near maximum stratification (N^2) and breaking occurs preferentially at the solitary wave troughs until the wave amplitude is reduced to the critical value.

The minimum Ri is given by

$$\mathrm{Ri_m} = \frac{C^2}{N^2A^2}\left[1+\frac{2C^2}{N^2L^2}\right]^{-1} \tag{6.7.10}$$

where C is the propagation speed, A is the amplitude, and L is the halfwidth of the wave. The dissipative timescale for the solitons varies from a few tens of hours in deeper waters to a few hours in shallow water. Solitons generated by semidiurnal tides have their energy dissipated within the semidiurnal period. Despite their spectacular nature and large power densities (30,000W m^{-1}), Munk (1997) suggests that their contribution to tidal dissipation, an upper bound of about 30 GW, is relatively insignificant to the tidal energy balance of the global oceans.

Internal wave solitons can reach amplitudes of as much as 90 m with an associated surface soliton of about 9 cm in amplitude. Since the convergences and divergences produced by the currents associated with them modify any existing surface wave field, they become visible both to the naked eye and to instruments such as SAR, because of the changes in the sea surface roughness they produce. They have been observed in a variety of places, including the Andaman Sea, Sulu Sea, Indian Ocean, and Bay of Bengal. They are known as current rips or tide rips, and were first reported in 1861 in Malacca Straits.

Osborne and Birch (1980) present fascinating observations of large solitons in the Andaman Sea. Figure 6.7.5 shows the observed isotherm depression during the passage of one of the solitons (propagation is to the left). The measurements are in substantial agreement with the two-layer model. The packets appeared every semidiurnal tidal cycle, indicating that tidal currents in the straits between Sumatra and the Nicobar Islands may be their source. Crest lengths of 150 km and separation distances between individual solitons of as much as 15 km were observed. The individual solitons were accompanied at their leading edge by choppy water from surface waves breaking due to convergence of currents produced at the surface by the internal soliton.

Since then, internal solitons have been observed in many satellite photographs as well as SAR images. An intensive 2-week experiment was conducted in the spring of 1980 in the Sulu Sea to observe and measure the large solitons that occur there (Liu *et al.*, 1985; Apel *et al.*, 1985). Seventeen soliton packets were observed consisting of solitons with amplitudes as large as 90 m and wavelengths of up to 16 km, propagating at speeds of 2.5 m s^{-1} from the Pearl Bank sill, where they are generated by the ebbing of tidal currents through the narrow straits there, across the Sulu Sea to the shallows near Palawan Island with lifetimes of 2.5 days. The packets show a fortnightly modulation due to the mixed nature of tides in the region. Figure 6.7.6 shows temperature and current

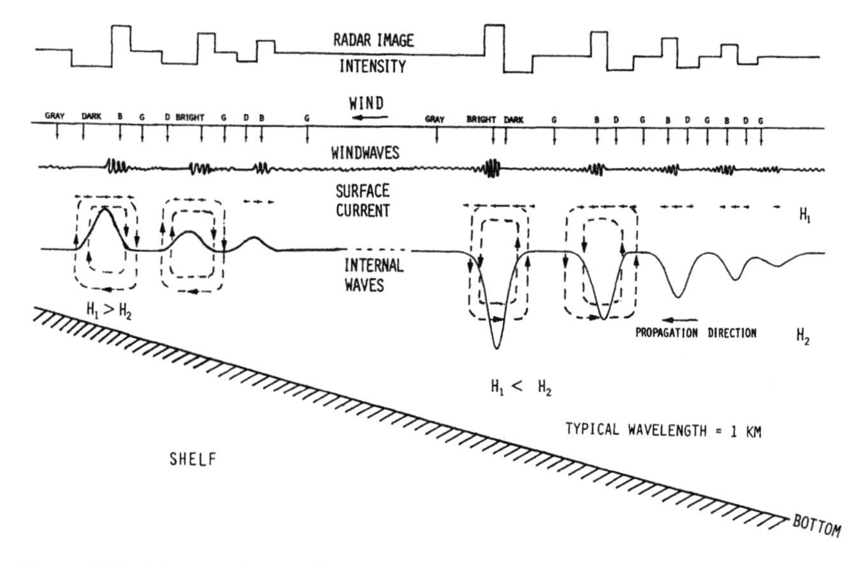

Figure 6.7.4. Schematic diagram of internal waves, surface currents, surface waves, and SAR image intensity variation when depression solitary waves move into shallower water (from Liu *et al.*, 1998).

measurements of one of the packets that clearly displays the rank-ordered solitons in the packet.

Figure 6.1.7 shows internal wave solitons in the Sulu Sea detectable in TOPEX/Poseidon altimetric records. The upper panel shows the along-track SSH anomalies, while the bottom panel shows the corresponding wavelet transforms (see Appendix B of Kantha and Clayson, 2000). The soliton packets are readily seen. They are detectable because of the high sampling rate represented by the 10-Hz data (0.7-km along-track resolution) and the high precision (2–3 cm) of the TOPEX altimeter. Therefore an IW soliton propagating underneath the altimeter is detectable, as long as its free surface displacement is a few centimeters. The large solitons in the Sulu and Andaman seas fall into this category.

Numerous SAR and shuttle observations suggest that internal wave solitons generated by tidal action are ubiquitous along the margins of ocean basins and around islands, and because of their energy content, might constitute a dissipation mechanism for global tides as well as the source of internal waves in marginal seas.

Henyey and Hoering (1997) suggest that these tidally generated internal wave trains are too nonlinear and too highly dissipative to be treated by dissipationless K-dV equations, or even the Joseph equation (see Ostrovsky and Stepanyants, 1989) that can be used when the lower layer is too deep and the shallow water approximation breaks down. The wave amplitude is often several times the upper layer thickness. The passage of the wave train also changes the ambient

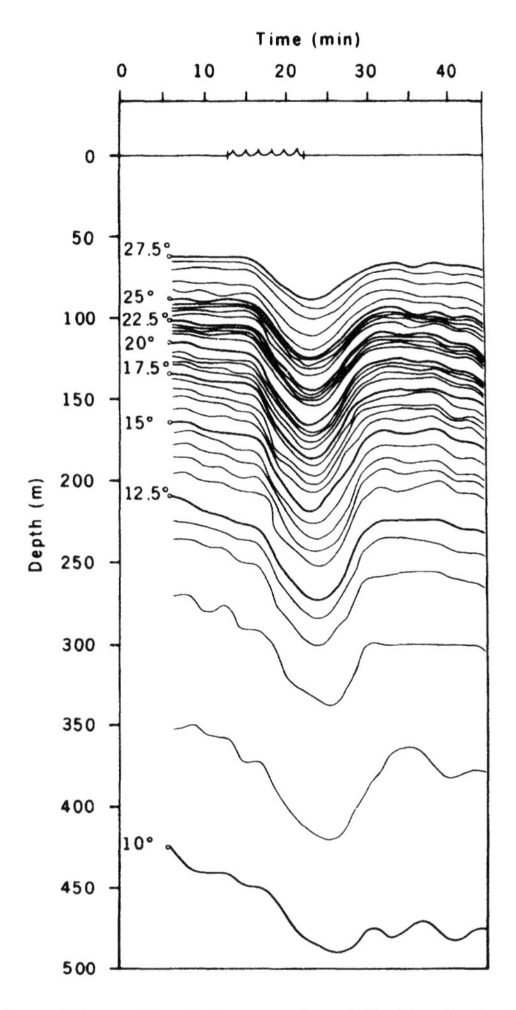

Figure 6.7.5. Observed thermocline displacement from IW soliton in the Andaman Sea (from Osborne and Birch, 1980).

stratification. For these reasons, they suggest that these wave trains are better treated by a two-layer internal hydraulic jump theory. Extensive literature exists on this topic (see Lawrence, 1993), starting from the pioneering investigations of Long (1970) in a non-Boussinesq fluid and numerical solutions by Cummins (1995), as Cummins and Li (1998) point out. Henyey and Hoering (1997) derive a jump relationship governing the energy flux into the internal bore, which for a steadily propagating bore is balanced by dissipation, that is similar to the one derived by Long (1970).

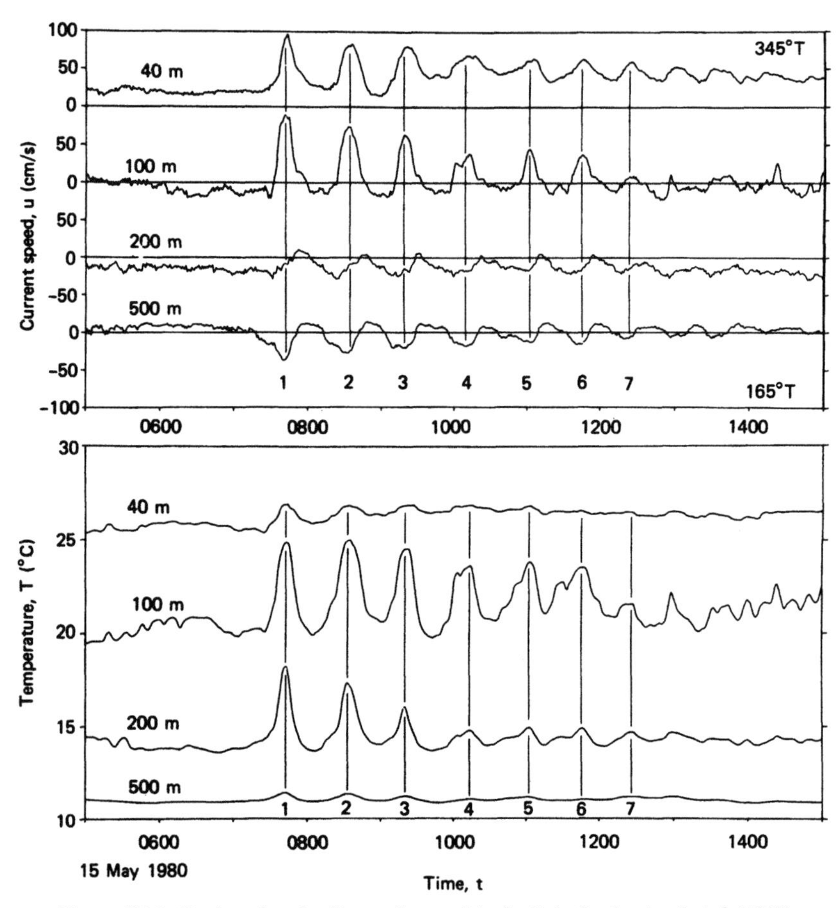

Figure 6.7.6. Rank-ordered solitons observed in the Sulu Sea by Apel *et al.* (1985).

6.8 MIXING IN THE DEEP OCEAN (ABYSSAL MIXING)

Microstructure measurements in the main thermocline and the deep ocean (for example, Moum and Osborn, 1986; Gregg, 1989; Peters *et al.*, 1988) have revolutionized our thinking on mixing in the deep ocean (Kunze and Sanford, 1996) (see also a comprehensive but dated review by Gregg, 1987). Contrary totheoretical expectations from thermocline theory of about 10^{-4} m^2 s^{-1} (Munk, 1966, 1981), these observations have found the vertical mixing coefficient in the abyssal oceans, away from rough topography and boundary currents, to be consistently an order of magnitude less, at about 10^{-5} m^2 s^{-1}, only 10 times the molecular value for the diffusivity of momentum, ~100 times the molecular

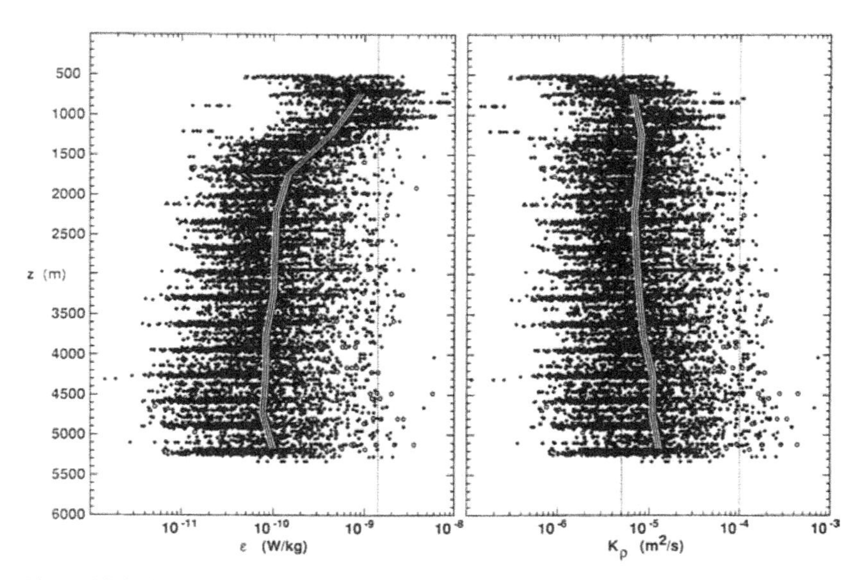

Figure 6.8.1. Vertical profiles of dissipation rate and mixing coefficient in the deep ocean (from Kunze and Sanford, 1996). Note the near-constant values of diffusivity at about 10^{-5} m^2 s^{-1} below a 1000-m depth.

diffusivity of heat and ~10000 times the molecular diffusivity of salt. Tracer release experiments (Ledwell *et al.*, 1993) and numerous full water column measurements in recent years (Toole *et al.*, 1994; Kunze and Sanford, 1996) have confirmed this result. Toole *et al.* (1994) find a depth-independent value for diapycnal eddy coefficient K_ρ of ~10^{-5} m^2 s^{-1} over smooth abyssal plains in the eastern North Atlantic and eastern North Pacific. So do Kunze and Sanford (1996) for the Sargasso Sea, who also found the abyssal diffusivities to be in the range (0.1–0.2) $\times 10^{-4}$ m^2 s^{-1}, very nearly constant over nearly the entire water column from 1000 to 5000 m (Figure 6.8.1) and independent of the bottom slope. The probability distribution is nearly log normal (Figure 6.8.2). Measurements of Ledwell *et al.* (1993) indicated a value of (0.05–0.15) $\times 10^{-4}$ m^2 s^{-1} in the upper kilometer of the water column. Only near strong topographic changes does one find elevated values of mixing. For example, values as high as 3.0×10^{-4} m^2 s^{-1} are found near a seamount (Toole *et al.*, 1994) and as high as 10^{-3} m^2 s^{-1} on top (Toole *et al.*, 1996). In measurements near the Romanche and Chain Fracture Zones in the equatorial Atlantic, Polzin *et al.* (1996) found values as high as 10^{-2} m^2 s^{-1} above the sills and 10^{-3} m^2 s^{-1} over rough topographic features. These observations suggest that the mixing in the abyssal oceans is rather weak except in localized regions near rough topography and sills, and perhaps near lateral ocean boundaries. This is an important finding in oceanography.

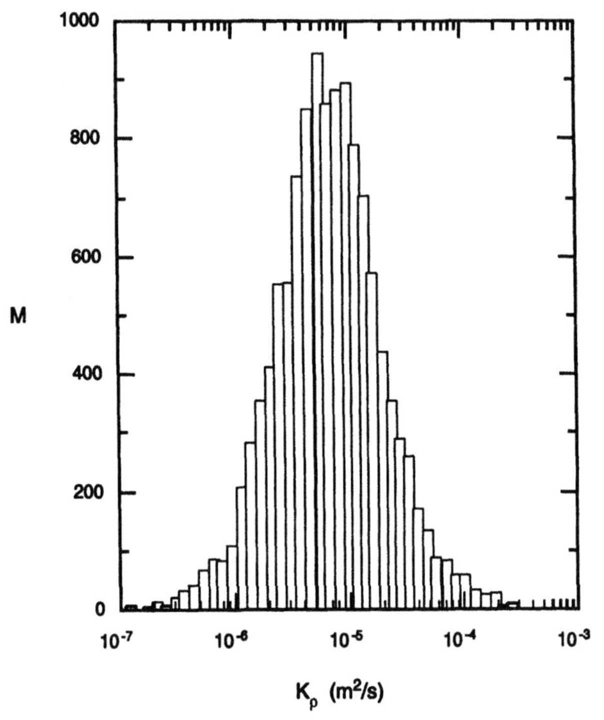

Figure 6.8.2. Probability distribution of mixing coefficients in the deep ocean (from Kunze and Sanford, 1996). Note the log-normal shape.

Long-term stability of the depth of the main thermocline requires that the slow upward transport of water masses in the global oceans needed to compensate for the deep-water formation in subpolar seas must be balanced by the vertical diffusion of density across the pycnocline. This vertical advection–diapycnal diffusion balance requires (Munk, 1966)

$$w \frac{d\rho}{dz} = K_\rho \frac{d}{dz}\left(\frac{d\rho}{dz}\right) \tag{6.8.1}$$

Assuming the eddy diffusivity to be roughly constant, one gets

$$K_\rho = w \frac{d\rho}{dz}\left[\frac{d}{dz}\left(\frac{d\rho}{dz}\right)\right]^{-1} = w\ \ell_N \tag{6.8.2}$$

where ℓ_N is the buoyancy length scale, which is roughly constant in the deep ocean at a canonical value of about 1300 m (see, for example, Kunze and Sanford, 1996). Observations of deep-water formation rates (Carmack and Foster, 1975, for example) suggest a value of about 8×10^{-8} m s^{-1} (~0.7 cm day^{-1}) for w, so that the global average value for the vertical diffusivity assumes a canonical value of 10^{-4} m^2 s^{-1}. The vertical diffusion is of course spatially nonuniform and this global average value can be accounted for by elevated values over a few local "hot spots" (Kunze and Sanford, 1996). In fact, if we assume abyssal values of only 10^{-5} m^2 s^{-1} over the bulk of the global oceans as observations seem to suggest, it is necessary to find only $9 \times 10^{m-3}$% of the global oceans of 10^{-m} m^2 s^{-1} diffusivity. If we assume m ~ 2, the value over the Romanche Fracture Zone (Polzin *et al.,* 1996), only 1% of such areas could account fully for the global budget values needed. Alternatively, strong vertical mixing could be confined to the vicinity of lateral boundaries of the oceans (Munk, 1966), consistent with the assumptions underlying ideal thermocline theory (Luyten *et al.,* 1983) and the finding that lateral advection is important in the thermocline (Jenkins, 1980) and the deep ocean (Sarmiento *et al.,* 1976), with the rest of the global oceans contributing a small fraction. In either case, the low values of diffusivity found over abyssal oceans is not in conflict with the requirements of thermocline theory.

The source of such deep mixing is the deep-sea internal wave field. The source of these deep sea internal waves is still rather uncertain. There are four possible sources, the winds, the surface waves, the bottom roughness, and baroclinic tides. The winds can generate strong inertial currents which can in turn generate internal waves radiating into the deep ocean. Turbulent eddies in the upper mixed layer generated by winds can also leak some of their energy into internal waves (Kantha, 1979a). Resonant interactions between two surface waves of nearly identical wavenumber can generate leaky internal waves (Kantha, 1979b). Strong currents flowing over rough topography at the ocean bottom can generate internal waves. None of these mechanisms have been proven to be adequate to provide the level of the internal wave field and the associated mixing and dissipation. While none of these can be ignored, it is becoming increasingly likely that baroclinic tides generated by barotropic currents flowing over midocean ridges and seamounts are an important source of most if not all of the deep-sea internal waves. Of the 3.5 TW of work done by the Sun and Moon system in generating barotropic tides, nearly 17%, about ~600 GW, is thought to be converted into baroclinic tides [the values are 2.5 TW and ~400 GW for the M_2 component alone (Kantha, 1998; Kantha and Tierney, 1997)]. If all of this energy percolates down the spectrum through resonant wave–wave interactions, it can be shown that it accounts for the level of mixing observed in the abyssal oceans (Munk, 1996). The argument goes as follows:

Observations indicate that the gradient Richardson number in a deep-sea internal wave field is near values indicative of shear instability. Shear instability leads to conversion of a portion of the TKE into potential energy through vertical buoyancy flux. The efficiency of this conversion is given by the mixing efficiency, γ_m, the ratio of the buoyancy flux to the dissipation rate (related to the flux Richardson number Ri_f), which varies widely (0.2–0.4) in stably stratified flows. Recent measurements by Moum (1996a) indicate a value of 0.25–0.33. We will assume a value of 0.25. Thus knowing the dissipation rate, the buoyancy flux can be computed, and knowing the stratification in the water column provides then the diapycnal diffusivity.

$$\frac{g}{\rho_0}\overline{w\rho} \sim \gamma_m \varepsilon \Rightarrow \varepsilon \sim \frac{K_\rho}{\gamma_m}\frac{g}{\rho_0}\frac{d\rho}{dz} \Rightarrow \varepsilon(z) \sim \frac{K_\rho}{\gamma_m}N^2(z) \tag{6.8.3}$$

Assuming canonical values of stratification corresponding to the Garrett–Munk internal wave spectrum,

$$N(z) \sim N_0 \exp(z/\ell);\ \ell \sim 1000\text{ m},\ N_0 \sim 5.2\text{x}10^{-3}\text{s}^{-1}(3\text{ cph}) \tag{6.8.4}$$

and integrating from 0 to $-\infty$, one gets for the dissipation per unit area an upper bound of

$$\overline{\varepsilon} \sim \frac{1}{2}\rho_0 N_0^2\, \ell\, \frac{K_\rho}{\gamma_m} \sim 55\, K_\rho\ \text{W m}^{-2} \tag{6.8.5}$$

In fact, Equations (6.8.2) and (6.8.4) can be rewritten in terms of total dissipation rate (W) in the deep global ocean,

$$\varepsilon_t \sim \frac{Q}{2\gamma_m}\rho_0 N_0^2\, \ell\, \ell_N \tag{6.8.6}$$

where Q is the deep-water formation rate ($m^3\ s^{-1}$), or assuming an efficiency of 0.25, $\varepsilon_t \sim 0.072\ Q$ TW, when Q is in Sverdrups. Thus a 20-Sv deep-water formation rate requires 1.45 TW of total dissipation. Now, a 600-GW tidal work conversion rate to baroclinic tides implies an average dissipation rate of 2.0×10^{-3} W m^{-2}, assuming all the baroclinic tidal energy eventually percolates down the spectrum into the internal wave field and is dissipated in the deep ocean (area ~ 3.05×10^{14} m^2). Thus the upper bound on diapycnal diffusivity, if the mixing is due to baroclinic tides, is 3.5×10^{-5} $m^2\ s^{-1}$ (600 GW from all constituents), and 2.3×10^{-5} $m^2\ s^{-1}$ from M_2 alone. Near midocean ridges and seamounts, the generation regions of internal tides, the conversion rates can be two to three orders of magnitude higher, and hence the isopycnal diffusivity rate also two to three orders of magnitude higher than this value. Abyssal values are a fraction of this value, still consistent with deep-sea dissipation and diffusivity measurements.

Resonant triad wave–wave interactions cause energy to percolate down the spectrum to high wavenumbers at a rate that scales with the buoyancy frequency N and the Garrett–Munk spectral energy level E_0 so that the dissipation rate is

$$\varepsilon(z) \sim \rho_0 N^2(z)\, E_0^2 \tag{6.8.7}$$

Recent work by Gregg (1989), Henyey *et al.* (1986), and Polzin *et al.* (1995) has shown that the dissipation rates observed in the deep ocean by microstructure instruments can be collapsed quite well (within a factor of 2) by scaling it as (see Kunze and Sanford, 1996)

$$\varepsilon(z) = \varepsilon_0 \frac{N^2}{N_0^2} \frac{\left(\overline{U_z^2}\right)^2}{\left(\overline{U_z^2}\right)_{GM}^2} f(R)$$

$$f(R) = \left(\frac{R_{GM}(R+1)}{R(R_{GM}+1)}\right)\left(\frac{R_{GM}-1}{R-1}\right)^{1/2} \tag{6.8.8}$$

where R is the ratio of the shear variance $\overline{U_z^2}$ to the strain rate variance $N^2\overline{\xi_z^2}$ (Kunze and Sanford, 1996). R ~ 3 for the GM spectrum, $\varepsilon_0 \sim 7.8 \times 10^{-10}$ W Kg^{-1}, and $N_0 \sim 5.2 \times 10^{-2}$ s^{-1}. The function f (R), indicative of the rms aspect ratio of the internal wave field $\overline{k_H / k_z}$, enables the scaling to be valid in regions such as the equatorial oceans, where the GM spectrum is not expected to be valid because of the strong vertical shear and near-zero Coriolis parameter. The diapycnal diffusivity is then given by

$$K_\rho = \frac{\varepsilon_0}{N_0^2} \frac{\left(\overline{U_z^2}\right)^2}{\left(\overline{U_z^2}\right)_{GM}^2} f(R) \tag{6.8.9}$$

For the canonical GM spectrum, $K_\rho \sim 0.6 \times 10^{-5}$ m^2 s^{-1}. Kunze and Sanford (1996) have used the above scaling relations to arrive at the values of the dissipation rate and isopycnal diffusivity (Figure 6.8.1) in the Sargasso Sea as described above.

Suffice it to say that there still remains much to be learnt about internal wave processes in the global oceans.

LIST OF SYMBOLS

α	Angle between the wave and the boundary
δ	Wave boundary layer thickness
ε	Dissipation rate
$\phi(n)$	Frequency spectrum
$\phi(n,\theta)$	Directional frequency spectrum
λ	Wavelength
ν, ν_t	Kinematic viscosity (molecular and turbulent)
θ	Inclination of the wavenumber vector to the horizontal
ρ	Density perturbation
ρ_0, $\overline{\rho}$	Reference density and mean density
ζ	Vertical displacement due to internal wave
η	Displacement of the free surface
τ_O, τ_w	Shear stress and wave-induced shear stress
A, B	Constants in the internal wave spectrum
a	Wave amplitude
b	Buoyancy
c, c_g	Phase velocity and group velocity
c_d	Drag coefficient
d, D	Depth of the upper layer and the depth of the water column
f	Coriolis frequency
g	Acceleration due to gravity
j	Summation index (takes values of 1 and 2)
k	Magnitude of the wavenumber vector; also summation index that takes values of 1 to 3.
k_h, k_z	Horizontal and vertical components of the wave vector
k_3	Vertical component of the wavenumber
l	Integral microscale of turbulence; also length scale of stratification
l_N	Buoyancy length scale
n	Wave frequency
p	Perturbation pressure
t	Time
w	Vertical component of velocity perturbation; also vertical advective velocity
u_a^*, u_*	Friction velocity
u_i	Fluid velocity
x	Horizontal coordinate; also fetch
x_3	Vertical coordinate
x_i	Coordinates
z	Vertical coordinate

z_c	Critical layer height
A	Wave action
C	Wave phase speed
C_g	Group velocity
Cg_g, Cg_3	Vertical component of group velocity
E	Energy density of the wave
$F(k_j,n)$	Wavenumber–frequency spectrum
K	Kinetic energy
K_ρ	Diapycnal diffusivity
N(z)	Buoyancy frequency
P	Mean pressure; also potential energy
R	Covariance; also ratio of shear variance to the strain-rate variance
R_{GM}	Ratio of shear variance to the strain-rate variance
Ri	Bulk Richardson number
$S(k_j)$	Spectrum of sea surface slope
U_z	Velocity shear

Chapter 7

Double-Diffusive Processes

In this chapter, we will discuss the salient characteristics of double-diffusive processes. These processes are thought to be important to mixing in the deep ocean, especially the permanent thermocline and hence the thermohaline circulation in the oceans. We will try to summarize our present understanding of these processes. There have been periodic reviews of this topic over the past three decades (Turner, 1974, 1985, 1996; Schmitt, 1994) and this summary borrows heavily from some of them. Workshops and meetings have also been held periodically on the topic and the reader is referred to the corresponding reports (Chen and Johnson, 1984; Schmitt, 1991; Brandt and Fernando, 1996; see also Ruddick, 1998). Once again, the *Journal of Fluid Mechanics,* the *Journal of Physical Oceanography,* and the *Journal of Geophysical Research* are excellent sources of latest advances in the field.

7.1 SALIENT CHARACTERISTICS

When a stratified fluid is statically stable, for example, when a warm layer of water overlies a colder layer, conventional wisdom dictates that there be no ensuing fluid motions (in the absence of shear). There is no mechanism present in this situation to supply the needed energy to work against buoyancy forces. This is true except when the stratification involves two (or more) components with differing molecular diffusivities. Then, instabilities generated simply by virtue of the fact that one component can diffuse away faster than the other can give rise to vigorous fluid motions that lead to efficient mixing in the vertical. These processes are known as double-diffusive processes. They were discovered by Henry Stommel (Stommel et al., 1956) and Melvin Stern (Stern, 1960) and confirmed in the laboratory and the oceans soon after. Unlike turbulent mixing, which tends to increase the potential energy of the system and dissipates the kinetic energy, double-diffusive motions decrease the potential energy of the system. The diffusion coefficient for density is negative (see Ruddick, 1998, for a succinct description). Like many other processes, they have been found to occur in many diverse areas including metallurgy, geology, and stellar physics. The phenomenon is not restricted to two components or to scalar diffusion alone, and is now more appropriately called multicomponent convection (Turner, 1985), but here we will restrict ourselves to the conventional terminology, double diffusion or double-diffusive convection. It can occur in the presence of more than two constituents and due to cross-diffusion of one property due to the gradient of the other.

7.1.1 SALT FINGERS

Consider heat and salt, the two components of stratification in the oceans that diffuse in water at vastly differing rates (typical value for the kinematic diffusivity of heat ~ 1.4×10^{-7}, and that for salt ~ 1.1×10^{-9} m^2 s^{-1}). In the upper kilometer of vast regions in the tropics and subtropics, the excess evaporation over precipitation produces a salinity profile with the salinity decreasing while the temperature also decreases, with depth. This situation is conducive to the formation of salt fingers. Schmitt (1990) reports that 90% of the upper kilometer in the Atlantic at 24° N is salt fingering favorable. Another situation that occurs quite widely in the global oceans is when warmer and more saline water formed in evaporative semienclosed seas flows out into the open ocean. For example, the waters of the Mediterranean Sea are warm and salty due to excess evaporation over precipitation, and this water flows out along the bottom through the Straits of Gibraltar into the Atlantic Ocean and spreads out horizontally at a depth of about 1000 m. Extensive salt finger fields have been found in the Atlantic in this Mediterranean outflow region.

Now imagine that a layer of warm salty water is introduced on top of a colder fresher (but heavier) layer of water. The heat and salt concentrations in the two layers are such that the upper warm layer is lighter than the colder layer below. In other words, the stable temperature stratification overcompensates for the unstable salinity stratification and the fluid system is statically stable. Now imagine that a parcel of fluid is displaced downward from the top layer into the bottom layer. Normally, it would return to the top layer after executing a few oscillations in the vertical. However, because heat diffuses away faster than salt, the parcel loses more heat to the surroundings than salt and consequently becomes heavier. If it loses enough heat, it becomes denser than its surroundings and keeps moving down. By a similar process, a parcel displaced upward tends to keep moving upward. The resulting motion takes the form of fine vertical tube-like structures, the salt fingers, that span across a thick interface separating the two layers and is very efficient at transferring properties between the two layers, even though the system is gravitationally statically stable. The vertical buoyancy flux due to the destabilizing component, in this case salt, is larger than the stabilizing component, in this case heat. The slower-diffusing component, in this case salinity, that is heavy at the top, provides the needed potential energy release to drive the fluid motions. The process drives the system toward a final state that has less potential energy than the initial state, and the stratification increases as well (Figure 7.1.1). Herein lies the uniqueness of double diffusion; unlike turbulent mixing processes at a density interface, which always increase the potential energy at the expense of TKE and erode the interface, weakening and eventually destroying the stratification, double-diffusive processes decrease the potential energy. Thus the eddy viscosity for density is negative. Once the process slows down or terminates, homogenization can occur but only through molecular or turbulent mixing of the two layers. Also, unlike turbulent mixing, which requires an external energy source, such as background mean shear or internal waves, double diffusion is self-driven.

Salt fingers can also occur when the density stratification is continuous, as is the usual situation in the oceans (see Figure 7.1.2), as long as the slower-diffusing component, salt, is heavier at the top. This is the case in many oceanic

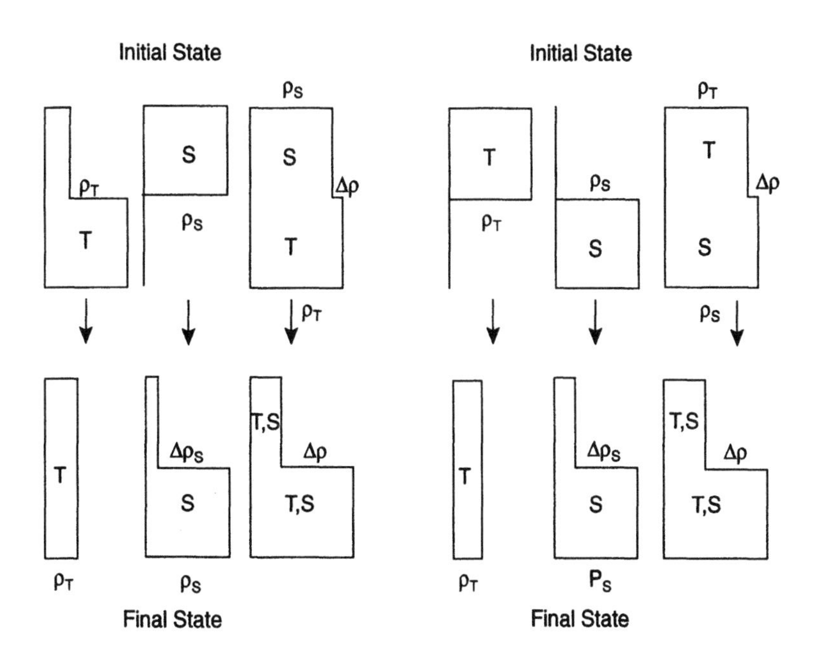

Figure 7.1.1. The initial and final states resulting from salt fingers (left) and double-diffusive convection (right) starting from a stable two-layer stratification, with the warm salty layer on the top (left) and on the bottom (right). The resulting density distribution in each layer is shown at top, and the final rundown state is shown at the bottom.

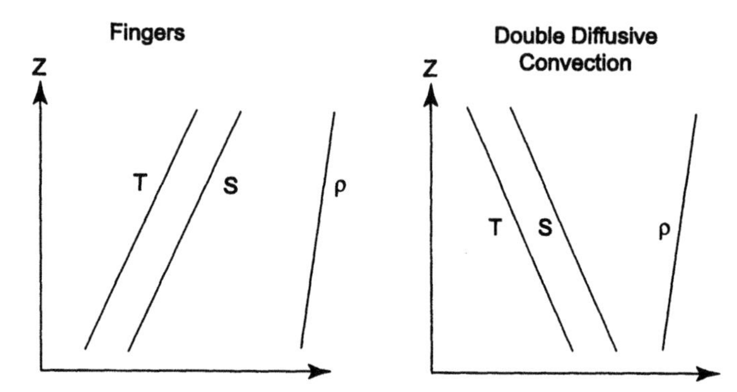

Figure 7.1.2. Continuous stratification conducive to salt fingers (left) and double-diffusive convection (right).

regions, where excess evaporation concentrates the salt in the upper layers. Salt fingers also occur for a different set of constituents, for example, salt and sugar, as long as there is considerable disparity in diffusivities.

7.1.2 Diffusive Convection

The opposite case of diffusive convection occurs when a cold fresh layer is introduced on top of a warmer saltier (but heavier) layer (Figure 7.1.1). In this case, a fluid parcel undergoes growing oscillatory motions because of the differing diffusivities (Stern, 1960; Shirtcliffe, 1967). The final result is a very sharp diffusive interface across which heat and salt are transported quite efficiently, driving a vigorous convection in the two layers that leads to transfer of properties between the two layers in the face of gravitationally statically stable stratification (Turner, 1973). Here the faster-diffusing component, in this case heat, that is heavy at the top, provides the necessary release of potential energy to drive the motion. There is a decrease of potential energy and an increase in stratification of the system. There is a larger buoyancy flux due to heat than due to salt across a diffusive interface. Once again, the density contrast between the layers increases and the eddy diffusivity for density is negative. The process also occurs in continuously stratified fluid as long as the faster-diffusing component is heavier at the top (Figure 7.1.2). Diffusive convection is quite common in high-latitude oceans such as the Arctic (Neal et al., 1969; Neshyba et al., 1971; Kelley, 1984; Padman and Dillon, 1987, 1988) and the Antarctic. Muench et al. (1990) estimate vertical heat fluxes of up to 15 W m^{-2} in these double-diffusive "staircases" that may help keep the ice cover thin. Strong surface cooling and excess precipitation and ice melt cause the surface waters in the polar and subpolar regions to be cooler and fresher than the subsurface water and this is conducive to diffusive convection. It has also been found to occur in natural water bodies stably stratified due to dissolved salts, but with a heat flux from the bottom. Extensive double-diffusive layers have been found at the bottom of Lake Kivu in Africa (Newman, 1976).

Layering also results when a fluid stably and smoothly stratified by one component has a destabilizing gradient or flux of the second component imposed upon it. A classical example is when a fluid with linear stable salinity gradient is heated from below (Turner, 1968). Here, the faster-diffusing component is destabilizing. Convection ensues in a layer immediately adjacent to the bottom and the depth of this layer increases as $t^{1/2}$, where t is time. The temperature and salinity contrasts in the gradient region at the top of the convecting layer also increase, but the net density contrast is, however, small. As the convective layer grows, the gradient region above, which is diffusive-convection-favorable, eventually reaches a critical Rayleigh number and becomes unstable either to Rayleigh convection or overstable oscillations and breaks down. The convective layer at the bottom stops growing and a new layer grows on top of it, and the molecular diffusion of heat above this layer starts this process all over again [Figure 7.1.3 (from Huppert and Linden, 1979)]. The consequence is sequential formation of new convective layers at the top with time, leading eventually to a series of convective layers separated by sharpinterfaces called a thermohaline staircase (Ruddick, 1998; Huppert and Linden, 1979; see also Fernando, 1987; Fernando and Ching, 1991). Turbulent convection carries the heat and salt across the layers, but a major part of their transport through the interfaces is thought to be by molecular diffusion. The

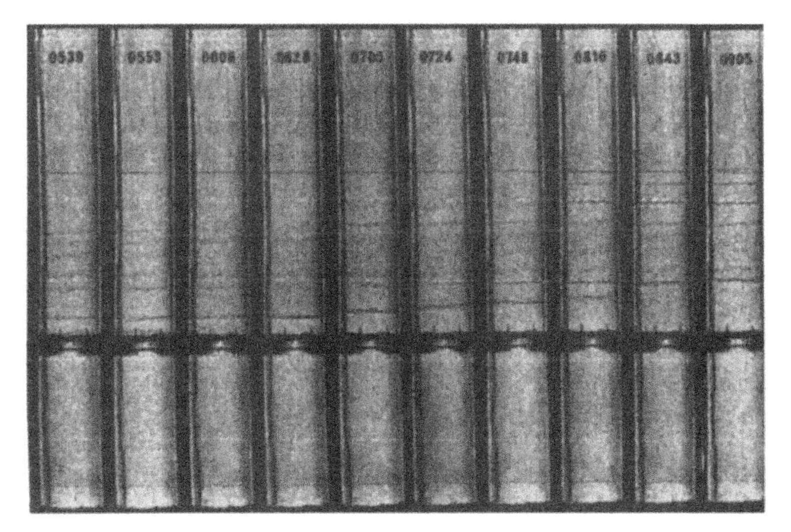

Figure 7.1.3. Diffusive convection resulting from heating from below of a linearly stratified salt solution (from Huppert and Linden, 1979, reprinted with the permission of Cambridge University Press).

layers near the bottom often merge with each other, with the interface between them migrating either up or down. Kelley (1987) provides a thorough discussion of the physics of layer merging and interface thickening. A beautiful example of layered formations in nature is that of Lake Kivu, a bottom-heated saline lake [Figure 7.1.4 (from Newman, 1976)].

Stern and Turner (1969) describe the formation of successive convective layers through the salt finger process when a sugar flux (equivalent S flux) is imposed on top of a salt gradient (equivalent T gradient). Here the destabilizing component has slower diffusivity. Nevertheless, there is a great similarity between the previous case and this in terms of the formation of convective layers. The difference is that fingers form initially in each convective layer, become unstable to collective instability in the form of an internal wave (probably due to superimposed shear), break up, and result in a convective layer that deepens bounded by a thin finger interface ahead of it, whose fingers in turn elongate and break down, producing the next convecting layer. Although still

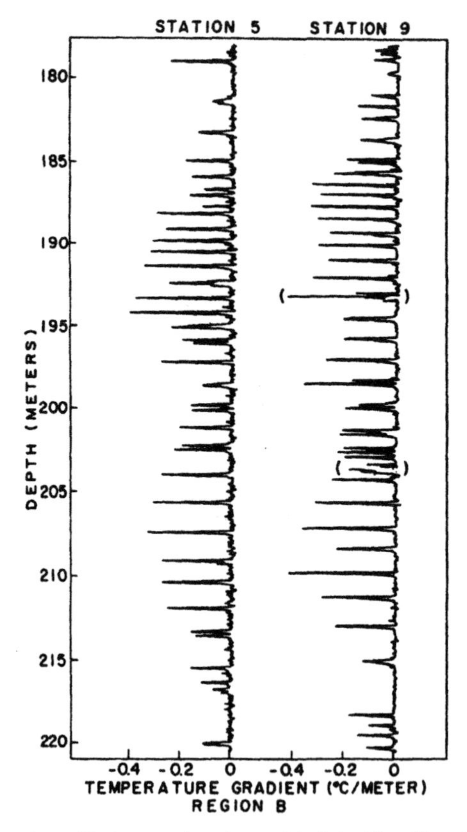

Figure 7.1.4. Double-diffusive layering observed in Lake Kivu (from Newman, 1976).

unsubstantiated, this presumably mimics the mode in which thermohaline staircases form in finger-favorable regions of the permanent thermocline.

It is quite simple to demonstrate the process of salt fingers and diffusive convection using a two-layer system consisting of either sugar and salt or heat and salt. The latter does not last as long as the former because of the difficulty of insulating against heat loss through the walls. The fingers and the sharp diffusive interface are set up in a matter of minutes, but it takes hours (to days, especially for very viscous fluids) for the system to eventually run down. All one needs is a glass beaker (preferably a rectangular tank) and a specific gravity meter to ensure that the appropriate layer is heavier (a specific gravity difference of 0.01 to 0.02 suffices). Then salt and sugar (or hot salty and cold fresh) solutions with the required density difference can then be prepared. The beaker can be half-filled with the lighter layer first and the heavier layer can then be poured slowly down a straw that penetrates to the bottom so that it lifts the lighter one without too much mixing at the interface. If the heavier salty (or colder fresher) layer is introduced at the bottom of a lighter sugar (or warmer saltier) layer, fingers ensue in a matter of minutes and persist for many hours. If the heavier sugary (or warmer saltier) layer is introduced at the bottom of a lighter salty (or colder fresher) layer, a sharp double-diffusive interface results. In either case, vigorous convection can be seen in both layers driven by the destabilizing buoyancy flux through the fingering or double-diffusive interface. The fingering and convective motions can be easily visualized if the beaker is held against a light source, since the refractive index fluctuations cast a shadowgraph that makes the structures visible (or one layer can be dyed). Figure 7.1.5 shows salt fingers and diffusive convection set up in the laboratory (Turner, 1985) by using salt and sugar dissolved in water (sugar diffuses roughly three times slower than salt in water). Note the thick finger interface and the very sharp diffusive interface.

7.1.3 STAIRCASES IN THE OCEAN

In the oceans, double-diffusive processes can be identified by their characteristic stepped or staircase structures in the vertical temperature and salinity profiles. Note, however, that unlike diffusive convection, salt fingers do not necessarily produce staircase structures, and salt finger fluxes have been observed in the North Atlantic without any accompanying staircases. Since their discovery in 1960, many experiments have been conducted to confirm their existence in the oceans and assess their importance to deep-ocean mixing. The most recent such experiment is the Caribbean Sheets and Layers Transect (C-SALT) experiment conducted east of Barbados in 1985 (Schmitt, 1991, 1994). Researchers found an area more than a million square kilometers in extent where thermohaline staircase structures were quite vivid (Figure 7.1.6). Analysis of the data collected shows, however, that while the processes are analogous to those

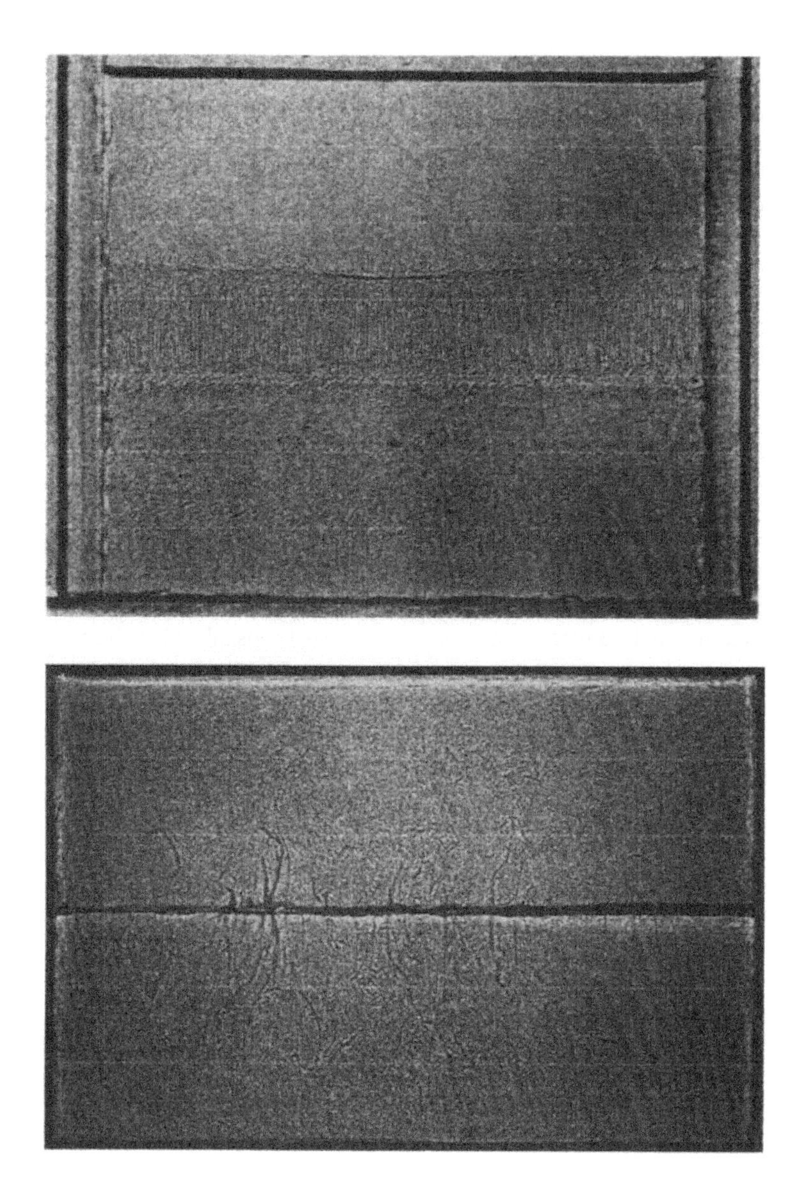

Figure 7.1.5. Salt fingers in the laboratory, created by placing sucrose solution on top of denser NaCl solution and leaving for serveral minutes (top), and double diffusive convection in the laboratory, created by pouring a layer of NaCl solution on top of a layer of denser sucrose solution (bottom). Flow is visualized by using the shadowgraph technique (from Turner, 1995).

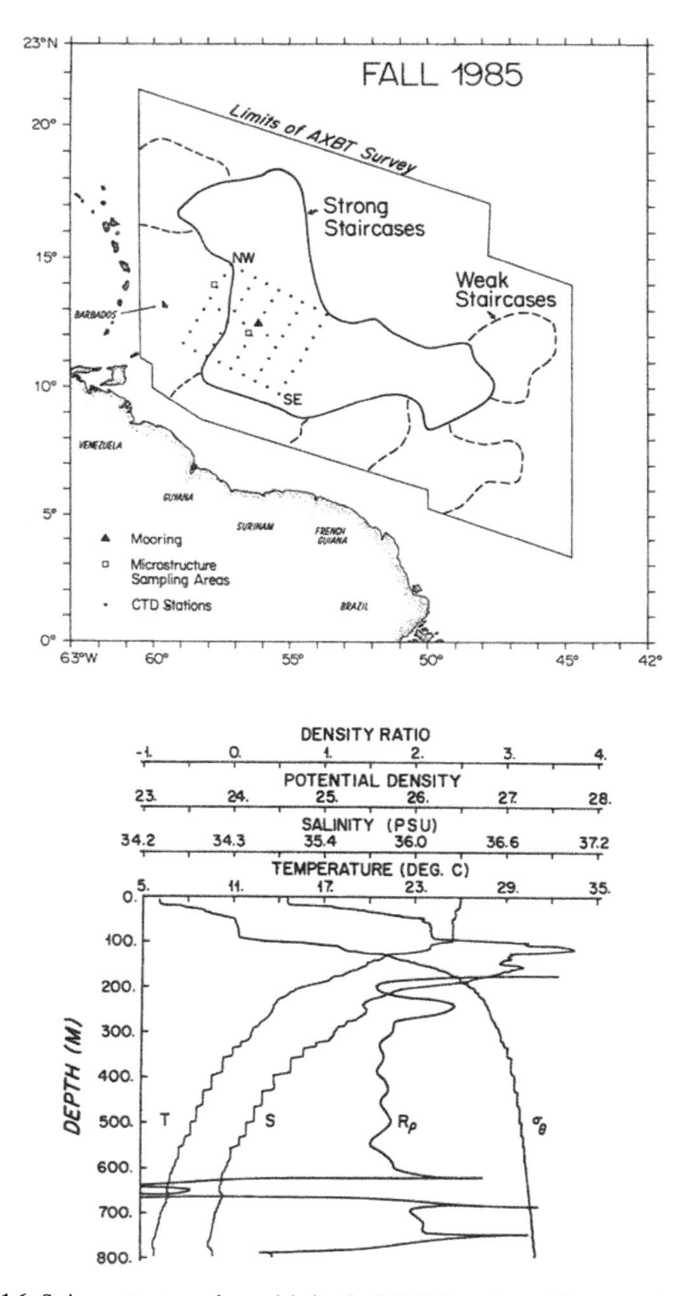

Figure 7.1.6. Staircase structures observed during the C-SALT experiment. The top panel shows the region where such structures were found and studied. The bottom panel shows typical staircase structures found in temperature and salinity (from Schmitt, 1994 with permission from the *Annual Review of Fluid Mechanics,* Volume 26, Copyright 1994, by Annual Reviews).

observed in the laboratory, there are detailed differences as well. For one, the finger interfaces appear to be much thicker than what the laboratory observations indicate they should be (2–5 m typically as opposed to 30 cm), and therefore drive a flux much weaker than those suggested by laboratory measurements. The second important difference is that unlike the fingers observed in the regions like the Mediterranean outflow (Williams, 1975), the finger structures were aligned more toward the horizontal direction than the vertical. There is considerable vertical shear in the deep oceans due to inertial/internal waves and other motions, and it is not always clear what this shear does to salt finger processes—whether it merely realigns them or destroys them completely and periodically.

The C-SALT program found extensive thermohaline staircases east of Barbados with typically 10 layers of 5–40 m thickness between 150- and 600-m depths, with sharp gradient regions 0.5- to 5-m thick, 0.5–1°C temperature and about 0.1 psu salinity contrasts, and coherent over 300–400 km. The most striking aspect of these layers (Figure 7.1.7) is the constancy of the horizontal density ratio at 0.85. This ratio is equivalent to the density (or equivalently buoyancy) flux ratio $R_F=(\beta_T F_T)/(\beta_S F_S)$, where F_T and F_S are heat and salt fluxes, since the principal balance is between horizontal advection of

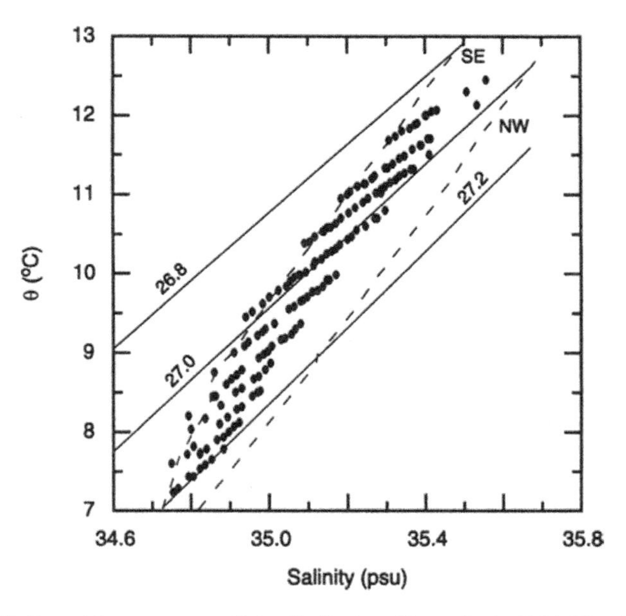

Figure 7.1.7. Potential temperature–salinity distribution of layers in the C-SALT region, showing the crossing of isopycnals by layers with a heat/salt density flux ratio that is remarkably constant at around 0.85 (from Schmitt, 1994 with permission from the *Annual Review of Fluid Mechanics,* Volume 26, Copyright 1994, by Annual Reviews).

temperature and salinity and divergence of vertical heat and salt fluxes (see Schmitt, 1994). The vertical density ratio $R_\rho=(\beta_T\Delta T)/(\beta_S\Delta S)$, the ratio of stratification due to salt to that due to heat (symbols β_T and β_S denote the coefficients of thermal and saline expansion, and ΔT and ΔS the changes in temperature and salinity) is 1.6. For salt fingers the flux ratio is 0.5 to 0.8, somewhat lower than the observed value, but nevertheless indicating that salt finger processes were active. For isopycnal mixing this ratio is 1 and for vertical mixing it should be 1.6 (Schmitt, 1991). Schmitt explained the higher observed value by invoking conductive correction to the fluxes. Using the observed R_F value of 0.72 for a density ratio of 1.6, and adding a correction of 0.1 due to conduction, brings it close to the observed values. The higher value may also be due to intermittent turbulence or migration of interfaces between layers due to nonlinearity of the equation of state (Schmitt, 1994). Kunze (1991) estimates that the salt eddy diffusivity is around $1\text{–}2 \times 10^{-4}$ m^2 s^{-1} in the staircases in the C-SALT region, an order of magnitude larger than the canonical 10^{-5} m^2 s^{-1} value found in the abyssal oceans.

Another aspect of the C-SALT observations was the nearly horizontal orientation of the laminae, unlike observations in other regions where the orientation is predominantly vertical (Schmitt and Georgi, 1982). Kunze (1991) has hypothesized that this tilting is due to inertial wave shear, with a balance between sheet growth and disruption by the prevailing shear. The interfaces were also thicker than predicted by laboratory simulations with correspondingly weak fluxes. There is still no satisfactory explanation for this discrepancy (Ruddick, 1998), although attempts have been made (Kunze, 1990, 1994).

Persistent thermohaline staircases have been found in the Tyrrhenian Sea in the Mediterranean. In the western tropical Atlantic, they have been observed for over 30 years (Schmitt, 1994). These long persistence times suggest that double-diffusive processes might be a quasi-permanent feature of many regions in the global oceans. However, while a majority of the permanent thermocline region of the world is stratified salt-finger-favorable, staircases have not been observed in all these regions. There are competing, often more vigorous processes operative, such as internal waves and turbulent mixing, and the shear and intense mixing associated with them have a tendency to destroy salt fingers or at least prevent them from creating permanent staircases. Although the data are sparse, there appears to be a systematic variation of the vertical density ratio across the thermohaline steps with latitude (Schmitt, 1994), with values of 1.2, 1.3, and 1.6 for the Tyrrhenian Sea (40° N), Mediterranean Sea outflow (35° N), and Caribbean Sea (12° N). Ingham (1966) finds that in 90% of the main thermocline of the Atlantic, the vertical density ratio is less than 2.3. Fingers appear to be most intense when the density ratio is less than ~2 (see Zhang et al., 1998).

Diffusive convection staircases are usually found in high latitudes, such as in the Arctic and Antarctic waters, where cold fresh water formed from ice melt lies above saltier warmer waters (Padman and Dillon, 1989). They are quite ubiquitous in the Arctic at 200- to 400-m depths. The scale of these layers is ~C $(k_T/N)^{1/2}$, where k_T is the thermal diffusivity and N the buoyancy frequency; C is a function of the density ratio R_ρ (Kelley, 1984).

Vertically profiling microstructure probes measure the dissipation rate and scalar variances, and along with information on background gradients, can be used to infer eddy diffusivities, a measure of vertical mixing. They are very good at measuring mixing due to conventional mixing mechanisms, such as turbulence from internal wave breaking. They are at present, however, less able to fully resolve the very fine scale structure (less than 2 cm) involved in finger processes, especially the salinity variances. Because salt finger structures are likely to be vertically oriented, towed instruments are needed to measure the associated microstructure. Nevertheless, the small dissipation rates measured do not imply low mixing rates in thermohaline staircases, because of the different constant in the relationship between diffusivity and dissipation rate. Typical values of $2\text{–}5 \times 10^{-10}$ W kg^{-1} are two orders of magnitude smaller than those measured elsewhere in the deep ocean, but amount to a diffusivity of $2\text{–}20\times10^{-5}$ m^2 s^{-1}, whereas deep-ocean dissipation rate measurements indicate low diffusivities on the order of 10^{-5} m^2 s^{-1} due to internal-wave-induced mixing processes!

Figure 7.1.8 shows towed spectra obtained in the Pacific by Gargett and

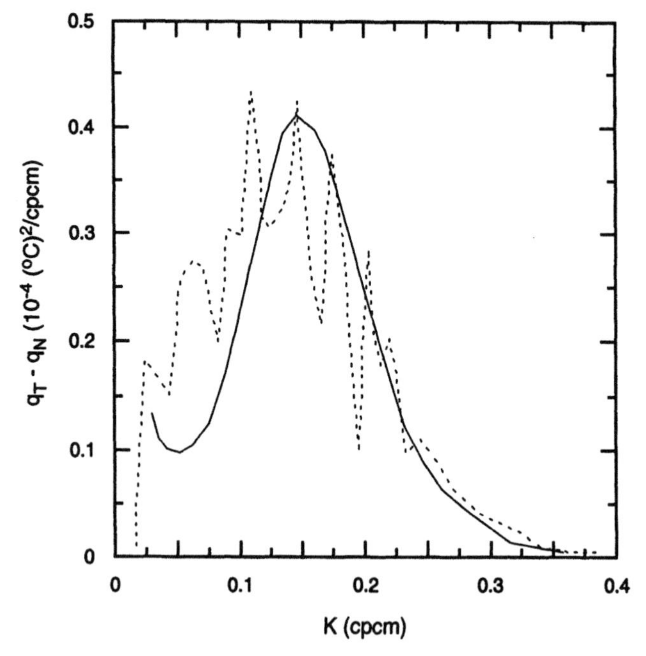

Figure 7.1.8. Observed salt finger spectrum from towed sensors (from Gargett and Schmitt, 1982).

Schmitt (1982) from a tow through salt fingering layers and its comparison with the theoretical +2 power law spectrum from Schmitt (1981). The prominent peak is noteworthy, since it is suggestive of the dominance of a particular scale of fingers. Also the narrowband-limited amplitude nature of salt fingers results in a low Kurtosis value in such records compared to the high values for turbulence, and this has often been used to discriminate between salt fingers and conventional turbulence in towed measurements (Mack and Schoeberlein, 1993). Microstructure observations of staircases have also been made by Mack (1985, 1989) and Marmorino et al. (1987) in the Sargasso Sea, and by Lueck (1987) and Fleury and Lueck (1991) among others.

7.1.4 DENSITY INTRUSIONS

Density intrusions are frequent in the oceans, a classic example being the Mediterranean outflow intruding as a density current into the deep Atlantic. Carmack et al. (1995) report spectacular intrusions in the Arctic, whose T-S properties remain coherent over several thousand kilometers. If, as in this case, the water masses entertain salinity and temperature contrasts, double-diffusive motions can ensue. This has been demonstrated by beautiful laboratory experiments (Figure 7.1.9) using a salt–sugar system by Turner (1978; see also Ruddick and Turner, 1979). The intruding layer has both salt finger and diffusive convection processes occurring, one on the top side and the other on the bottom (which one occurs on which side depending on whether the layer has a slower diffusing component or a faster one). The layer also tilts slightly in the vertical as it intrudes (unlike ordinary intrusions which maintain the same level), depending on whether the net density flux to the layer, which is the sum of the fluxes through the finger and diffusive interfaces into the layer, is negative (layer becomes lighter and rises) or positive (the layer becomes heavier and sinks). For example, in Figure 7.1.10, from Turner (1981), a sugar layer is intruding into a salt-stratified system, with fingers below and a diffusive interface at the top. The layer rises as it intrudes since the finger fluxes dominate and reduce the layer density. Contrast this to the behavior of an intrusion of a simple salt layer into the same solution in Figure 7.1.10. As can be seen, double-diffusive intrusions have a vastly different behavior.

If two identically stably stratified systems containing salt and sugar solutions are set up in a two-chambered tank separated by a partition, and the partition is carefully removed, double-diffusive instabilities cause intrusions to occur in the two fluids, leading to a staircase structure involving finger structures separated by sharp double-diffusive interfaces. Figure 7.1.11 (from Ruddick, 1992) shows layered intrusions that occur when the partition separates the initially homogeneous sugar and salt solutions of identical density. Inversions in both temperature and salinity develop, leading to finger and diffusive convection occurring alternately in the vertical direction. These

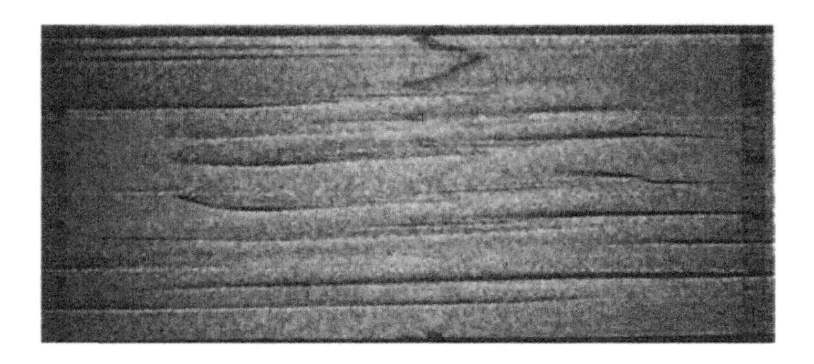

Figure 7.1.9. Double-diffusive intrusions produced by removing a vertical barrier dividing salty and sugary solutions with the same density distribution, with the warm sugary layer on the left and cold salty layer on the right; salt finger fluxes dominate in the top panel (from Ruddick and Turner, 1979, reprinted with permission from Elsevier Science) and double-diffusive convection in the bottom (from Ruddick, 1992).

intrusions can slope either upward or downward depending on whether the finger fluxes or diffusive fluxes dominate (Ruddick, 1992). Figure 7.1.11 explains this aspect more clearly (see also Ruddick and Walsh, 1996; May and Kelley, 1997). A warm salty intrusion can have fingers below and a diffusive interface above so that a fluid parcel moving from A to B loses both heat and salt to both. If the net density flux to fingers below is larger than that to diffusive convection above, the parcel becomes lighter and rises, and the intrusion slopes upward. If the net density flux to fingers below is smaller than that due to diffusive convection above, the parcel becomes heavier and sinks, and the intrusion slopes downward. Thus, the slope of the intrusions indicates whether the diffusive convective flux or the finger flux is more dominant. The intrusions reported by Carmack et al. (1995) appear to be dominated by diffusive convection rather than by salt fingers. Figure 7.1.12 (from Turner, 1981) shows the intrusive layering that results when the partition separates initially identically

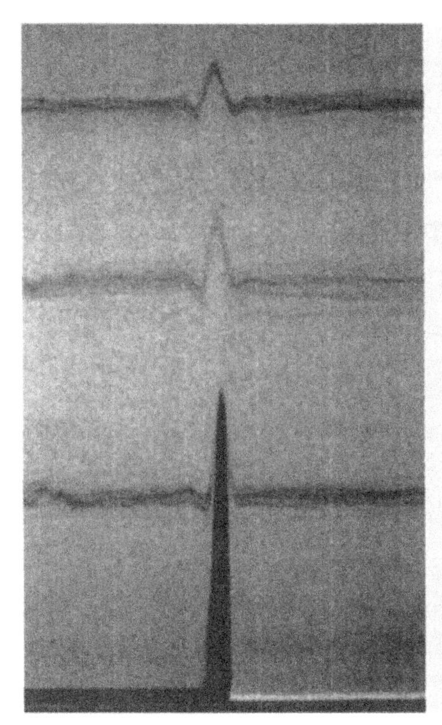
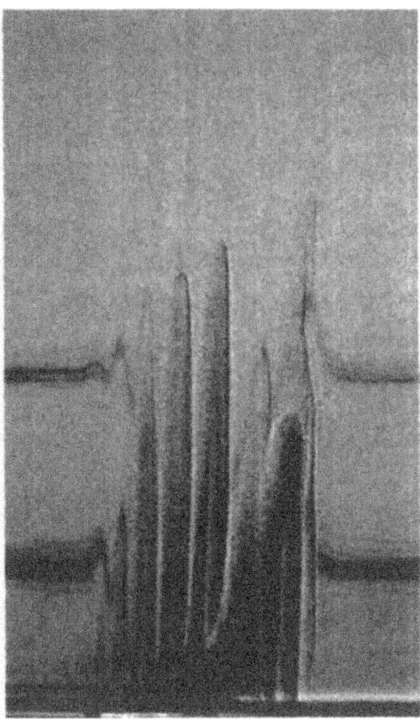

Figure 7.1.10. Shadowgraphs illustrating the difference between a regular density intrusion of a salty layer into a stably-stratified salty fluid (left) and a double-diffusive intrusion of a sugary layer into the same solution (right) (from Turner, J. S., *in Evolution of Physical Oceanography,* Copyright M.I.T. Press, 1981).

linearly stratified salt–sugar systems. The salt–sugar contrast increases with depth and therefore the layers increase in thickness from the top to the bottom. Since the speed of advance of the layers is proportional to their thickness, one gets the "Christmas tree" effect.

7.1.5 MEDDIES

The spreading of the Mediterranean outflow in the Atlantic involves the spawning of huge lenses of warm salty water called Meddies, roughly 100 km in diameter and 200–300 m in thickness initially, moving roughly at 1–2 cm s^{-1} through the surrounding water at a depth of about 1100 m (Armi et al., 1989; Richardson et al., 1989). The salinity and temperature contrasts in the Meddy compared to that in the North Atlantic Central Water are about 0.2–1.0 psu and

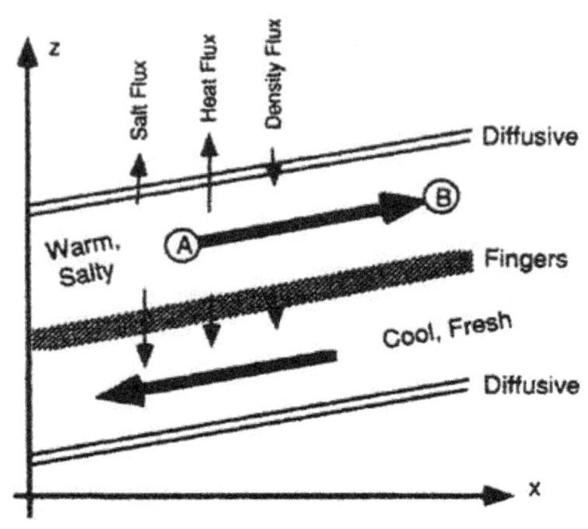

Figure 7.1.11. A sketch of double-diffusive intrusions showing alternating salt fingers and double diffusion. The intrusive layers slope upward to the right, since a parcel A becomes cooler and fresher as it moves to B and becomes lighter when finger fluxes dominate. The sloping will be downward if double-diffusive flux dominates (from Ruddick, 1992).

2 to 3°C. The Meddies have a thickness of about 800 m and rotate anticyclonically with a typical period of 4–6 days. They last typically two to four years unless they run into topography, and during this time travel nearly 500–1200 km more or less intact, but with a gradual decrease in diameter and salinity contrast (Figure 7.1.13). The principal mode of decay of these Meddies appears to be intense mixing at their boundaries with the ambient waters (Ruddick and Hebert, 1988; Hebert et al., 1990; Ruddick, 1992). This involves double-diffusive processes occurring at the edges of the lens through lateral intrusions (Figure 7.1.14). Enhanced dissipation rates are found in micro structure measurements at the top, bottom, and sides of the Meddy, although the

Figure 7.1.12. Double-diffusive interleaving obtained when a barrier separating linearly stratified sugar on the left and salty solution on the right is removed (reprinted from Deep-Sea Research, 26A, Ruddick and Turner, The vertical length scale of double-diffusive intrusions, 903–931, copyright 1979, with permission from Elsevier Science).

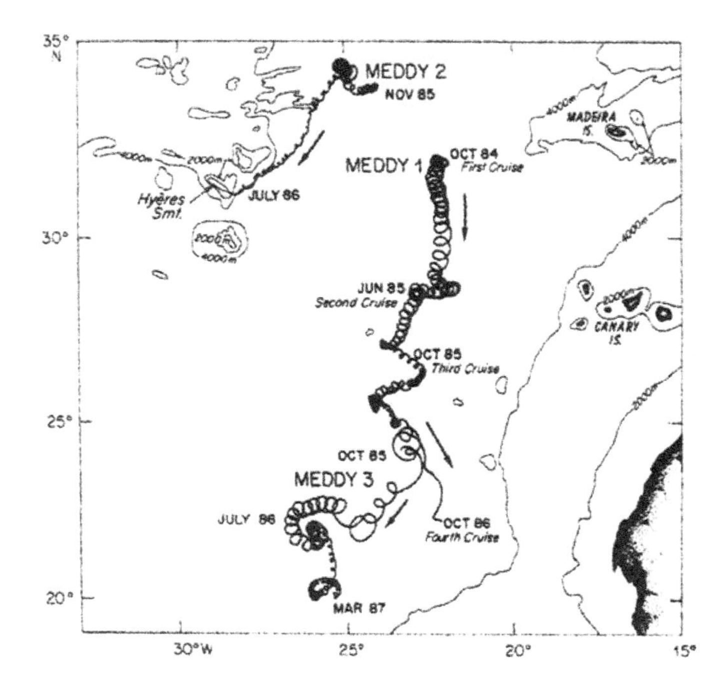

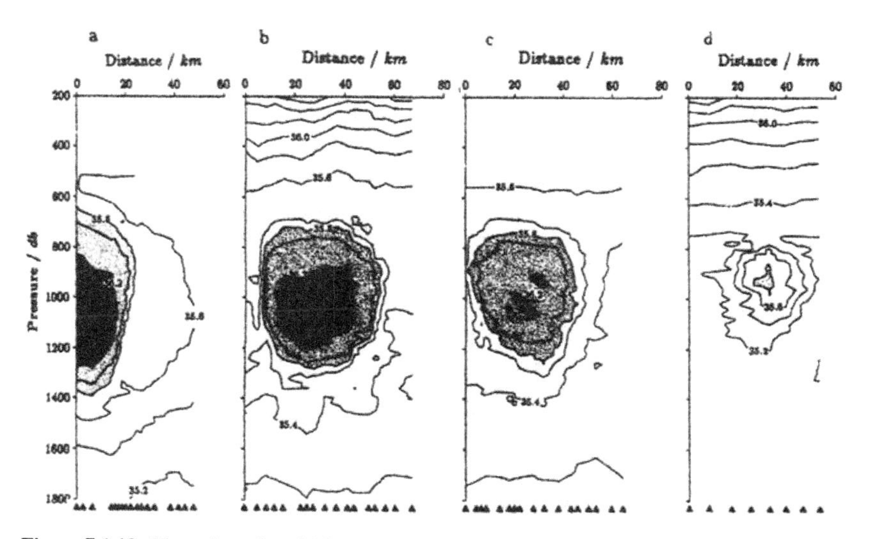

Figure 7.1.13. The trajectories of SOFAR floats emplaced in three of the Meddies generated by the outflow of warmer, saltier Mediterranean waters into the Atlantic through the Straits of Gibraltar (top panel, from Richardson *et al.* 1989). The longevity of Meddy 1 is noteworthy. Meddy 2 ran into a seamount and dissipated. Salinity cross sections of Meddy 1 at four points during its lifetime, October 1984, June 1985, October 1985, and October 1986 (bottom panel, from Armi *et al.* 1989). Note the gradual shrinking of its core of saltier Mediterranean waters.

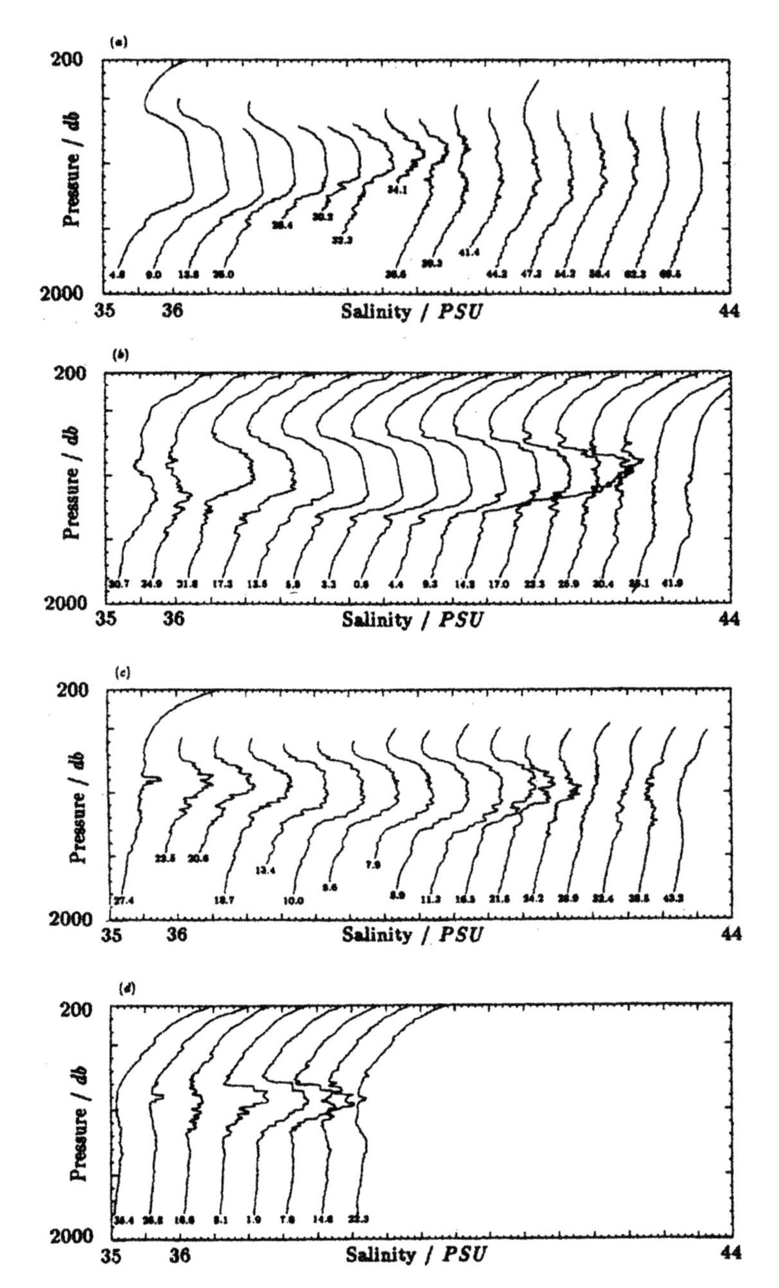

Figure 7.1.14. Salinity profiles through the Meddy at the four points during its lifetime (see Figure 7.1.13). Successive traces have been offset to the right by 0.5 psu. Staircase structures can be seen both on the top (due to double-diffusive convection) and on the bottom (due to salt fingers) (from Armi *et al.*, 1989).

core is stable and has little microstructure. Double-diffusive convective fluxes at the top produce staircase structures with strong shear, especially near the center. In the Meddy Sharon, probed extensively as it drifted slowly southward in the Atlantic, the lower surface also had stepped structures which resulted from salt fingers. The diffusive layers at the top edge had vertical scales of 10–15 m, and the finger layers at the bottom edge had vertical scales of roughly twice those at the top (~25 m). Lateral intrusions at the sides are apparently responsible for most of the mixing (Ruddick and Hebert, 1988). They mix heat, salt, and momentum anomalies outward radially, which are then dispersed into surrounding waters. Hebert et al. (1990) find that the Meddy decay is equivalent to a radial diffusivity of 2 $m^2\ s^{-1}$. These intrusions appear to be dominated by salt finger fluxes (Ruddick, 1992; Ruddick and Walsh, 1996) and eventually reach the center and alter the structure of the lens itself. Ruddick (1992) has made a careful analysis of these measurements to show that it is the double diffusion that is responsible for the observed structure and decay. The isopycnal slopes are consistent with what would be expected for double-diffusive intrusions.

7.1.6 Importance of Double Diffusion

The most salient aspect of double diffusion is its inherent efficiency in effecting vertical transfer of properties, in other words, vertical mixing. Compared to mechanical turbulence, the dissipation rates are low in double-diffusive processes and this translates to greater efficiency of transferring properties in the vertical. McDougall (1988) and Schmitt (1991) point out that the effective vertical eddy diffusivity of salt in salt fingers can be written as

$$K_s^e = \Gamma_f \frac{\varepsilon}{N^2}; \qquad \Gamma_f = \frac{R_\rho - 1}{1 - R_F} \tag{7.1.1}$$

For $R_\rho \sim 1.6$ and $R_F \sim 0.6$, $\Gamma_f \sim 1.5$. In general, Γ_f can be 1 to 4. In contrast, the vertical eddy diffusivity for mechanical turbulence is

$$K_s^e = \Gamma_t \frac{\varepsilon}{N^2}; \qquad \Gamma_t = \frac{Ri_f}{1 - Ri_f} \tag{7.1.2}$$

where Ri_f is the flux Richardson number, which indicates the efficiency of conversion of TKE into potential energy. Ri_f has a value of about 0.15–0.20, thus giving a value of 0.18–0.25 for Γ_t, an order of magnitude smaller than that for salt fingers. The contrast between Γ_f and Γ_t is not surprising, since only a small fraction of the flow energy in turbulence gets converted to potential energy. A major fraction is dissipated. In contrast, the dissipation rate (proportional to the buoyancy flux) in salt fingers is much smaller. Salt fingers are therefore very efficient at converting haline potential energy to thermal potential energy. However, Γ_f values are very sensitive to the value of flux ratio R_F and are very difficult to estimate accurately (Schmitt, 1994).

The fact that differential diffusion of heat and salt can be important to interior water mass structure and meridional thermohaline circulation in the global oceans has prompted several numerical studies. Gargett and Holloway (1992) demonstrated its significant potential impact through simple simulations using the Geophysical Fluid Dynamics Laboratory (GFDL) Modular Ocean Model. The large scale meridional circulation simulated by this model was altered considerably when the ratio of the heat and salt vertical diffusivities was changed to nonunity values. Gargett and Ferron (1996) have shown that even in simple multibox models of thermohaline circulation, differential diffusion of heat and salt can produce a different response to freshwater forcing changes. More recently, Zhang et al. (1998) have rerun these simulations but with proper parameterization of the double-diffusion-driven vertical diffusivities of heat and salt as a function of the local density ratio. They show that even with rather conservative parameterization of double-diffusive effects, the meridional overturning is reduced by 22% compared to the control run with the same diapycnal diffusivities for heat and salt, and the maximum poleward heat transport is reduced by 8%, these values being about twice the values that the Gargett and Holloway (1992) parameterization indicates. The prevailing upwelling–diffusion balance in the vertical in these models is upset by the negative (upgradient) density diffusivity. Quite simply, the downward density flux tends to hinder the upwelling of dense water from the deep by making it even denser.

The contrast between the weakly stratified oceanic mixed layer and the strongly stratified thermocline below is striking. Clearly, the vertical mixing of scalar quantities is generally weak in the upper thermocline (10^{-5} m^2 s^{-1} or less), but strong in the mixed layer (10^{-4} m^2 s^{-1}or more). More interesting is the fact that in the mid-latitude mixed layers, the density ratio $R = \alpha\Delta\theta\ (\beta\Delta S)^{-1}$, the ratio of temperature contribution to the density change over a certain spatial scale to that of salinity, is very nearly 1, whereas in the thermocline below, it asymptotes to a value of very nearly 2 (Schmitt 1999). Observations by Rudnick and Ferrari (1999) in the North Pacific show that over horizontal distances of 20 meters to 10 kilometers, the temperature and salinity gradients in the mixed layer tend to compensate each other so that the horizontal density changes are very small near the surface. Any horizontal density gradients imposed by atmospheric forcing tend to get dissipated by slumping of the heavier fluid beneath the adjacent lighter fluid (so that the isopycnals change from vertical to horizontal), so that only those temperature and salinity changes imposed by atmospheric forcing in which the density ratio is 1 (and hence the density change is negligible) tend to persist (Young 1994). This process can be expected to be active only at spatial scales below the Rossby radius of deformation, since otherwise rotational effects can maintain horizontal density gradients, without the necessity of such slumping. Moreover, the strong mixing inhibits double-diffusive processes, even though the density ratio is near unity and hence favorable to them. However, immediately below the mixed layer, the vertical mixing weakens, permitting double-diffusion, and the density ratio begins to increase towards a value of 2. Salt finger processes are especially strong near density ratio of 1, but weaken considerably as the density ratio approaches 2. It is likely that in the weakly mixed regions below the mixed layer, salt finger processes dominate mixing and since there is a preferential transport of salt in these processes, this could explain why the density ratio increases above 1 and approaches the value of 2 in the mid-latitude thermocline (Schmitt 1999). Thus lateral mixing in the surface mixed layer subject to random atmospheric forcing may lead to density-compensated temperature and salinity changes there, whereas salt fingers may tend to drive the density ratio above 1 in the thermocline below. If substantiated, this would have considerable implications to modeling small scale mixing processes in the upper layers in mid-latitude oceans: the horizontal diffusivity in the mixed layer could be a function of the density gradient and vertical mixing in the thermocline must account for salt finger processes.

To sum up, the fact that self-driven vertical transfers of heat and salt could occur in some statically stable interior regions of the oceans has potentially profound consequences to the water mass structure and ocean mixing in the interior, which in turn affects the meridional overturning and thermohaline convection in the oceans. This has climatic implications. This must however be counterbalanced by the fact that double-diffusive convection is rather fragile and easily modified by background shear and internal wave motions and mixing. While numerical models have shown that the incorporation of differential diffusion of heat and salt in the interior has a significant impact on the meridional circulation and poleward heat transport simulated by ocean models, the importance of double diffusion relative to internal wave and boundary mixing is still largely an unresolved issue.

There now exists a vast literature on multicomponent convection in other fields such as chemistry, geology, metallurgy, and astrophysics. The reader is referred to Huppert (1986), Turner (1985, 1996), and Brandt and Fernando (1996) for excellent reviews. We would like to point out here that double-diffusive processes have come to be recognized as pivotal in explaining and understanding mechanisms such as ancient sulfide ore deposits and mixing in magma chambers. The salient distinction in these cases from double diffusion in oceanography is the presence and impact of phase changes, principally crystallization. Cooling of molten material is almost always accompanied by crystallization of some mineral components, so that the remnant fluid has its composition altered radically, and this is often conducive to double diffusion. In cooling magmas and other fluids involving crystallization, it is the compositional gradients brought about by crystallization that drive diffusive convection, and not so much the thermal contrasts. The different minerals in the solution diffuse at different rates, and crystallization of a component alters the density (when a denser mineral component settles out, the colder remnant fluid becomes lighter). These effects can be reproduced in the laboratory by the use of aqueous solutions such as Na_2CO_3. For example, cooling a Na_2CO_3 solution from the top results in crystallization from the top, with what Turner (1996) calls compositional convection occurring at the leading edge of the advancing solidification front forming a series of convecting layers [Figure 7.1.15 (from Turner, 1996)].

Kantha (1981) offered an alternative mechanism for the formation of elongated basalt columnar structures observed in many ancient lava flows around the world. He suggests that columnar structures might result from "basalt" finger-like process in the cooling magmatic layer. The conventional explanation entrenched in geology invokes analogy to mud cracks and therefore cracking during cooling and solidification. The complex frozen "fluid-like" structures of columnar formations (Devil's Post Pile, for example) and the sharp "turbulent" transition zones between columnar structures (Columbia River Basalt formations) are very difficult to explain without invoking a fluid-dynamical process operative in the molten magmatic layer during the cooling phase. Salt-finger-like processes driven by compositional convection in the cooling magma are most likely to produce such structures that are then frozen, leading to preferential cracking along the junctions during solidification. The

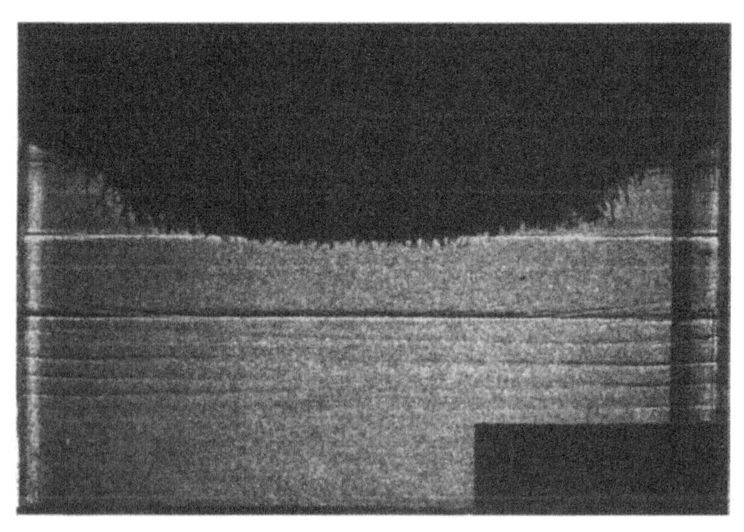

Figure 7.1.15. Compositional convection in Na_2CO_3 solution cooled and crystallized from the top (from Turner, 1996).

exquisitely regular pattern of resulting columns is far from what one would expect if they were formed from cracking in a fluid with unordered, random but presumably homogeneously distributed inclusions and compositions.

These ideas are however still quite controversial in the geological community, as is the idea that layering seen in some ancient igneous rock formations might have been due to diffusive convection. Muller (1998) describes columnar formations very similar to basalt columns formed in desiccating starch–water mixtures. These have similar polygonal cross sections and large length to width ratios typical of basalt column structures. Muller therefore asserts that convective processes have little to do with formation of this kind of structure in basaltic magma, although he acknowledges that his experiments cannot explain the column/entablature nature of many basaltic column formations, which tends to suggest convective layers and basalt finger structures overlying each other, typical of double diffusion. His experiments also used heat applied at the top to promote evaporation and this once again introduces heat and concentration as two diffusive components across the crack front. Clearly, the controversy is yet to be resolved.

7.2 CONDITIONS FOR THE ONSET OF DOUBLE DIFFUSION

Derivation of conditions for the onset of double-diffusive motions involves stability considerations of the fluid. It is a simple extension of the classical Rayleigh approach to the convection problem to double diffusion (Nield, 1967) and was first considered by Stern (1960). Consider a linearly and stably stratified system with linear stratification in temperature and salinity as in Figure 7.1.2. Let d be the thickness of the layer and assume that the layer has free boundaries with temperature and salinity held fixed at these boundaries. The relevant parameters in the problem are the Rayleigh number Ra, the Prandtl number Pr, the density ratio R_ρ, and the ratio of diffusivities τ:

$$Ra=\frac{g\beta_T \Delta T d^3}{\nu k_T}, \quad Pr=\frac{\nu}{k_T}, \quad R_\rho=\frac{\beta_T \Delta T}{\beta_S \Delta S}=\frac{\beta_T \dfrac{\partial T}{\partial z}}{\beta_S \dfrac{\partial S}{\partial z}}, \quad \tau=\frac{k_s}{k_T} \tag{7.2.1}$$

One can also define

$$Ra_S=Ra/R_\rho \tag{7.2.2}$$

Classical linear stability analysis using infinitesimal disturbances to the stratified system shows that for the finger situation ($Ra > 0$, $Ra_s < 0$), the fluid becomes dynamically unstable when

$$\mathrm{Ra} - \frac{\mathrm{Ra_S}}{\tau} > \frac{27}{4}\pi^4 \tag{7.2.3}$$

When τ is small, motion ensues even when $\mathrm{Ra_s} << -\mathrm{Ra}$, that is, when the fluid is still statically stable. The fastest growing motions have a horizontal scale given by wavenumber k:

$$k^4 \sim \frac{\nu k_T d}{g\beta_T \Delta T} \tag{7.2.4}$$

This result holds for even finite-amplitude disturbances and has in fact been employed successfully in many analytical models of salt finger motions. Figure 7.2.1 shows the stability boundaries for salt fingers and diffusive convection. Ruddick (1983) defines a parameter named the Turner angle:

$$\mathrm{Tu} = \tan^{-1}\left(\frac{R_\rho + 1}{R_\rho - 1}\right) \tag{7.2.5}$$

Thus salt fingers occur when 45° < Tu < 90° (strongest for Tu close to 90°) and diffusive convection occurs when –45° > Tu > –90° (strongest for Tu

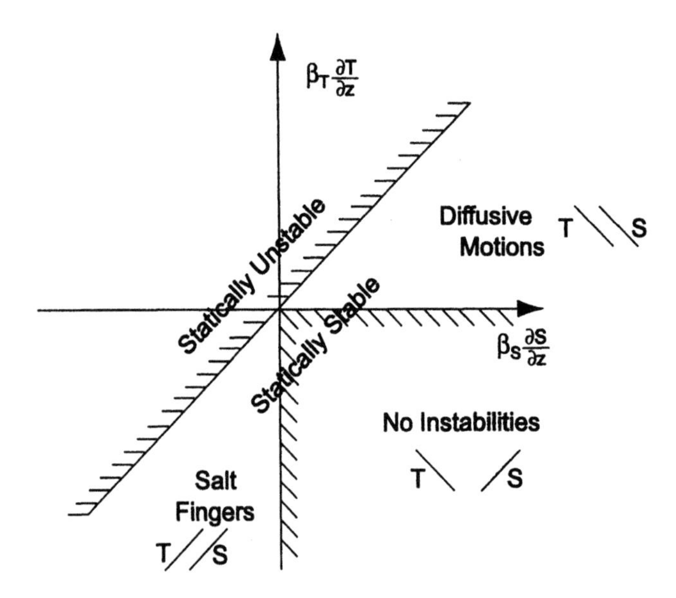

Figure 7.2.1. Stability boundaries for double-diffusive processes, with diffusive motions in the ENE quadrant and fingers in the SSW quadrant.

close to –90°). For 45° > Tu > –45° double diffusion is not possible. For the remainder of the Tu values, the fluid is statically unstable.

Schmitt (1983) has extended the analysis to arbitrary values of Pr and τ, and computed the maximum growth rates and flux ratios (R_F):

$$R_F = \frac{\beta_T F_T}{\beta_S F_S} \tag{7.2.6}$$

Figure 7.2.2 shows the flux ratio for $R_\rho = 2$ from Schmitt (1983). The results are in good agreement with laboratory results for the heat–salt and the sugar–salt systems. A salient conclusion from this analysis is that fingers can be expected to be sluggish for magmas and vigorous for liquid metals.

A similar stability analysis can be done for diffusive interfaces ($R < 0$, $R_s > 0$). A double-diffusive interface can become dynamically unstable (Huppert and Manins, 1973) if

$$R_\rho < \tau^{-3/2} \tag{7.2.7}$$

However, unlike the finger case, the results from nonlinear theories are significantly different in this case. Therefore, numerical approaches have been used to determine if a steady nonlinear convection can ensue. The reader is referred to Turner (1985) and the references cited therein for more details.

In water, cross-diffusion terms (see below) are negligible and only self-diffusion needs to be considered in determining if and when double-diffusive

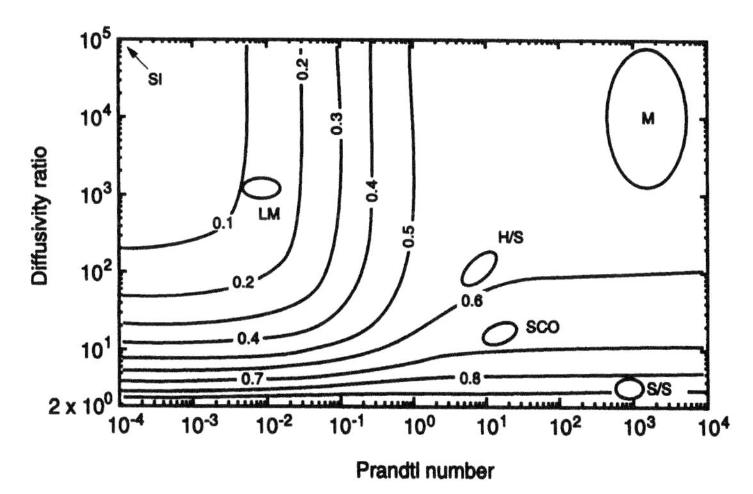

Figure 7.2.2. Flux ratio R_F for a density ratio R_ρ of 2, for various values of diffusivity ratio and Prandtl number. LM denotes liquid metal, M, magmas, H/S, the heat–salt, S/S, the sugar–salt, and SCO, the semiconductor oxide regimes (from Schmitt, 1983).

motions ensue. An interesting situation in many other fields is one where a gradient of one component can produce diffusion of the other through cross-diffusion terms. For example, if heat and salt fluxes F_T and F_S can be written as

$$\begin{aligned} F_T &= -\left(k_T \nabla T + k_{TS} \nabla S\right) \\ F_S &= -\left(k_{ST} \nabla T + k_S \nabla S\right) \end{aligned} \tag{7.2.8}$$

with

$$\rho = \rho_0 \left[1 + \beta_T T + \beta_S S\right] \tag{7.2.9}$$

k_{ST} is proportional to the Soret coefficient and k_{TS} is proportional to the Dufour coefficient. The Soret effect is an example of cross-diffusion where a salt flux results from a temperature gradient. The Dufour effect results in heat flux in the presence of a salt gradient. The Soret effect is more important in liquids. The beautiful photographs of fingers in polymer solutions in Preston et al. (1980) illustrate the importance of cross-diffusion in such systems. McDougall (1983) extended the stability analysis to cross-diffusion for both finger and diffusive cases and sketched the stability boundaries. He showed that for finger instability, the stability condition for nonzero cross-diffusion coefficients is

$$\left(\frac{\beta_S k_{ST}}{\beta_T k_S} - 1\right) + \frac{1}{R_\rho \tau}\left(\frac{\beta_T k_{TS}}{\beta_S k_T} - 1\right) > \frac{27}{4}\frac{\pi^4}{Ra}\left(1 - \frac{k_{TS} k_{ST}}{k_T k_S}\right) \tag{7.2.10}$$

One salient aspect of the presence of cross-diffusion terms is that instability can ensue even when both components are stably stratified. The diffusion of temperature in fingers is unaffected by cross diffusion, but salt diffusion can be effected by temperature gradient by cross-diffusion and can cause instability (see Turner, 1985).

7.3 EXPERIMENTAL RESULTS

Double diffusion has been explored quite extensively in the laboratory (for thorough reviews of laboratory experiments see Turner, 1974, 1981, 1996), simply because the experiments are quite easy to set up and the various fluxes easily measured whereas definitive observations and measurements of quantities like heat and salt fluxes are very difficult to carry out in the oceans. The important quantities that are measured in laboratory experiments are the flux ratio R_F and the heat flux F_T as functions of the density ratio R_ρ and the ratio of diffusivities τ.

Turner (1965), Marmorino and Caldwell (1976), Takao and Narusawa (1980), and Taylor (1988) have conducted careful measurements of the heat and salt fluxes and the flux ratio in diffusive convection. A remarkable aspect is that the flux ratio R^*_F [defined as $(\beta_S F_S)/(\beta_T F_T)$ in diffusive convection literature, the inverse of R_F] is nearly constant and equal to the square root of the ratio of diffusivities τ^* (defined as k_T/k_S, the inverse of τ) at values of R^*_ρ [defined as $(\beta_S \Delta S)/(\beta_T \Delta T)$, the inverse of R_ρ] beyond about 2. Figure 7.3.1 (taken from Takao and Narusawa, 1980) shows the variation of F_T and R^*_F with R^*_ρ. The experimental values for F_T can be fitted very well by (Marmorino and Caldwell, 1976)

$$F_T = F_T^0 A \exp\left\{4.6 \exp\left[-0.54\left(R_\rho^* - 1\right)\right]\right\} \tag{7.3.1}$$

where

$$A = 0.0044\left(\tau^*\right)^{-3/4} \tag{7.3.2}$$

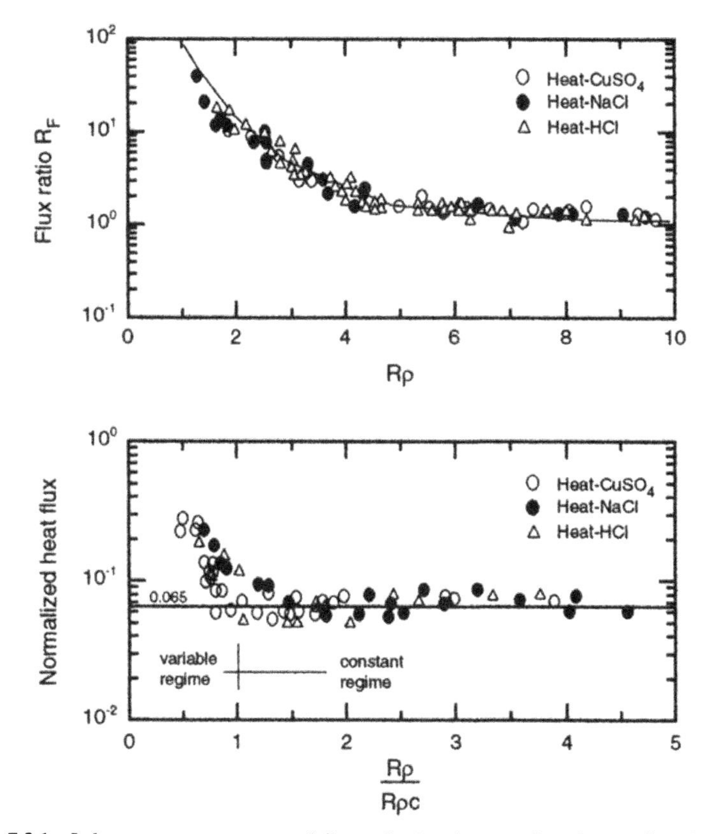

Figure 7.3.1. Laboratory measurements of flux ratio R_F (top panel) and normalized heat flux (bottom panel) across a double-diffusive interface as functions of the density ratio (Reprinted from Takao, S. and U. Narusawa, *Int. J. Mass Heat Transfer,* Copyright 1980, with permission from Elsevier Science).

and

$$F_T^0 = 0.085\left(\frac{k_T}{\nu}\right)^{1/3}\left(g\beta_T k_T\right)^{1/3}\Delta T^{4/3} \qquad (7.3.3)$$

is the so-called flat plate flux value, the flux obtained by putting a fictitious conducting flat plate at the interface. Note the classical 4/3 law normalization used in this and other diffusive convection experiments. Marmorino and Caldwell (1976) suggest a value of 0.101 for the constant A for the heat–salt case. For large values of R^*_ρ, R^*_F can be approximated as

$$R_F^* = 0.034\left(\tau^*\right)^{-1/3} \qquad (7.3.4)$$

provided

$$R_\rho^* > 0.46\left(\tau^*\right)^{-1/3} \qquad (7.3.5)$$

For the oceans, $\tau^* \sim 0.008$, so that $R^*_F \sim 0.17$ and $R^*_\rho \sim 2.3$. Huppert (1971) suggests for the heat–salt case

$$\begin{aligned} R_F^* &= 1.85 - 0.85 R_\rho^* \qquad 1 < R_\rho^* < 2 \\ &= 0.15 \qquad\qquad\qquad R_\rho^* > 2 \end{aligned} \qquad (7.3.6)$$

Traditionally R^*_F has been regarded as a function only of R^*_ρ. Recent experiments (Taylor and Veronis, 1996) in a two-layer salt–sugar system, however, suggest that there is some dependence on Ra_S as well, and this is confirmed by the numerical experiments of Shen and Veronis (1997).

Kelley (1984) has proposed an empirical scaling for the thickness of the convective layers adjoining double-diffusive interfaces. His scaling appears to apply well for both laboratory and oceanic thermohaline staircases. His results are equivalent to assuming that the Rayleigh number in the layers is approximately proportional to the density ratio,

$$Ra = \frac{g\beta_T \Delta T H^3}{\nu k_T} = B R_\rho^*; \qquad B \sim 0.26\times 10^9 \qquad (7.3.7)$$

where H is the layer thickness that scales quite well as

$$\begin{aligned} &H = G\left(k_T / N\right)^{1/2} \\ &G^4 = B R_\rho^*\left(R_\rho^* - 1\right)\frac{\nu}{k_T}, \quad N^2 = -g\beta_T\left(R_\rho^* - 1\right)\Delta T / H \end{aligned} \qquad (7.3.8)$$

where N is the smoothed buoyancy frequency determined from the thickness of the layer and the temperature jump across it. This is unlike in Huppert and Linden (1979), who used the value of N prior to the onset of the staircase structure, even though the two might be close to each other. Kelley's scaling appears to be well substantiated by data from diffusive staircases from both high and low latitude regions as well as saline lakes (Kelley, 1984, 1990; Padman and Dillon, 1987, 1989). Figure 7.3.2 shows the effective heat (K_T) and salt (K_S) diffusivities from Kelley (1984). Note that the ratios of these to molecular diffusivities, $D_{DT} = K_T/k_T$ and $D_{DS} = K_S/k_S$, denote the "effectiveness" of double diffusion in the vertical transfer of heat and salt. They both decrease as the density ratio increases (Figure 7.3.2); D_{DT} drops from ~500 near a R^*_ρ of 1 to ~10 near a R^*_ρ of 10, whereas D_{DS} drops from ~2500 near a R^*_ρ of 1 to ~2.5 near a R^*_ρ of 10.

Turner (1967), Linden (1973), Schmitt (1979a), McDougall and Taylor (1984), and Taylor and Bucens (1989) have measured the salt flux F_S and the flux ratio R_F (F_T/F_S here) in the laboratory for salt fingers in the heat–salt system, and Griffiths and Ruddick (1980) have done so for the sugar–salt system. Figure 7.3.3 shows R_F as function of R_ρ for the heat–salt

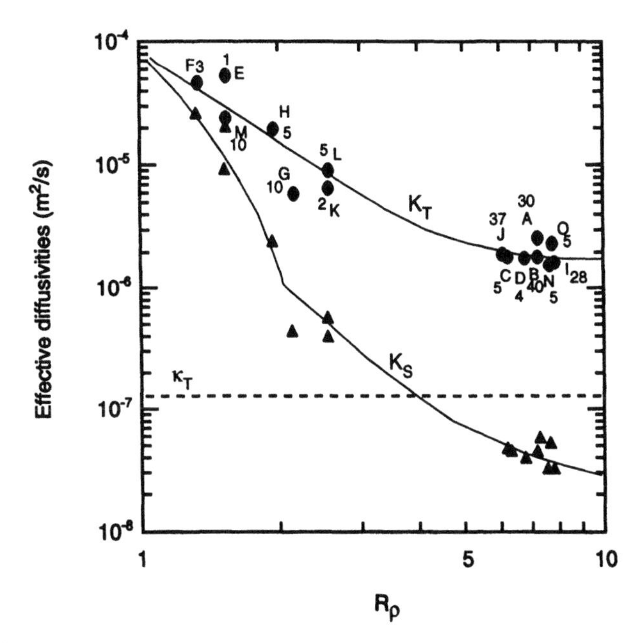

Figure 7.3.2. Effective diffusivities for salt K_S (triangles) and heat K_T (circles) from Kelly (1984). The letters denote data sources listed in Table 1 of Kelley (1984) and the numbers 1000 times the buoyancy frequency in s^{-1}. The molecular value for heat diffusivity is also shown. The solid curves are proposed semiempirical formulations.

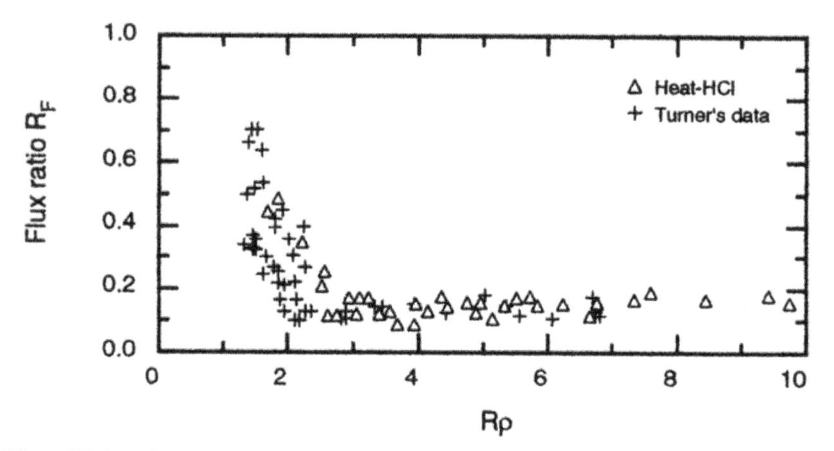

Figure 7.3.3. Laboratory measurements of flux ratio R_F across a salt finger interface as a function of the density ratio (Reprinted from Takao, S. and U. Narusawa, *Int. J. Mass Transfer,* Copyright 1980, with permission from Elsevier Science).

system. The flux ratio once again tends to approach a constant value for large values of R_ρ, and the heat and salt fluxes follow the classical 4/3 law for turbulent convection.

Numerical simulations have also been carried out for salt fingers using two-dimensional models (Shen, 1991, 1993; Shen and Veronis, 1997). While the assumption of two-dimensionality may become unnecessary as computing capability increases, the results should be quite valid for the actual three-dimensional fingers. Figure 7.3.4 (from Shen, 1991) shows the numerical value for the constant in the 4/3 power law for the heat flux, the flux ratio R_F, and the Stern number S_N [see Eq. (7.4.24)] as functions of R_ρ. There is a good agreement between these numerical results and the experimental results of Taylor and Bucens (1989). The temperature and salinity spectra from the model show a rapid decay of variance beyond the wavenumber corresponding to that of the fastest growing mode. The spectrum is, however, not peaked at this value (the buoyancy flux spectrum is peaked at this value), but flatter toward lower wavenumbers. Shen and Veronis (1997) studied the evolution of salt fingers in a two-layer heat–salt system ($\tau \sim 1/80$) for R_ρ of 1.5 to 3.0 and obtained a flux ratio R_F between 0.74 and 0.17 for this range. The most interesting aspect of their numerical simulations is that at high flux ratios, adjacent fingers coalesce to form eddies, leading to horizontally rather than vertically oriented structures of the type observed during C-SALT. They suggest that a horizontal shear may not therefore be essential to producing nearly horizontal laminae as has been presumed so far (for example, Kunze, 1990).

Ruddick and Turner (1979) found that $\beta_S \Delta S \sim \beta_T \Delta T$ across the isopycnal front of the intrusion, and the vertical length scale of the cross-frontal intrusions

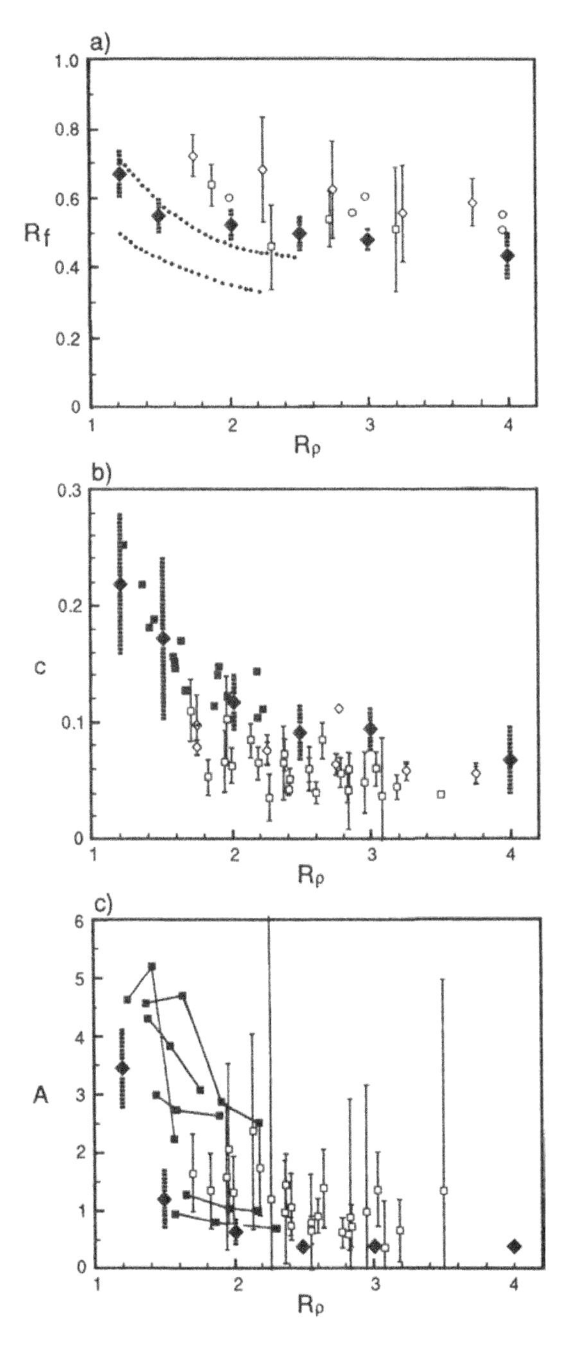

Figure 7.3.4. The flux ratio and the normalized salt flux and Stern number as functions of the density ratio. Numerical model results from Shen (1991) are also shown (filled symbols).

can be scaled to within a factor of two by

$$d \sim g\beta_S \Delta S / N^2 \tag{7.3.9}$$

where N is the buoyancy frequency of the two initially linearly stratified fluids stratified by heat and salt interleaving into each other. The constant of proportionality is different for different constituents. Experimental results on double-diffusive intrusions have been summarized by Turner (1981).

7.4 ANALYTICAL MODELS

Most of the focus in theoretical work on double diffusion has been on salt fingers, with some notable exceptions that deal with diffusive convection. Stern (1969) offered the very first analysis of salt fingers, their structure, and the fluxes of heat and salt through a finger interface, based on a collective instability argument. He followed it up with a variational argument that maximizes the buoyancy flux (Stern, 1976), which has also been adopted by Schmitt (1979b). Kunze (1987) has suggested a Richardson/Froude number criterion instead that appears to explain some of the features of salt fingers. He has also extended it to the case of fingers in background shear (Kunze, 1994, 1996). The basic problem in any analytical model of salt fingers lies in scaling their size and length. Auxiliary conditions are unavoidable, and one or another plausible argument is essential, the only verification being the consistency of results with laboratory and field observations. The reader is referred to Kunze (1987, 1996) for a more comprehensive review. Here we summarize the non-steady-state model. The following comes mainly from Kunze (1987, 1996).

7.4.1 NON-STEADY-STATE MODEL

Consider the conceptual description of salt fingers in a quiescent fluid as represented in Figure 7.4.1, taken from Kunze (1987). A warm, salty layer lies on top of a colder, fresher one. The fingers consist of fluid columns of, say, size d (or equivalently wavenumber k) in which the fluid motion is opposite to that of its immediate neighbors. Thus an alternating pattern of up and down flow prevails in the finger interface transporting heat and salt both vertically up and vertically down through the diffuse finger interface of thickness l. In Kunze's model, the finger length h is not the same as the thickness of the finger interface and this distinction is important. The figure also shows the temperature and salinity gradients in the adjacent upgoing and downgoing fingers. The background stratification in the finger interface is also shown (dotted lines). One can look upon salt fingers as the mechanism the fluid sets up to sharpen the

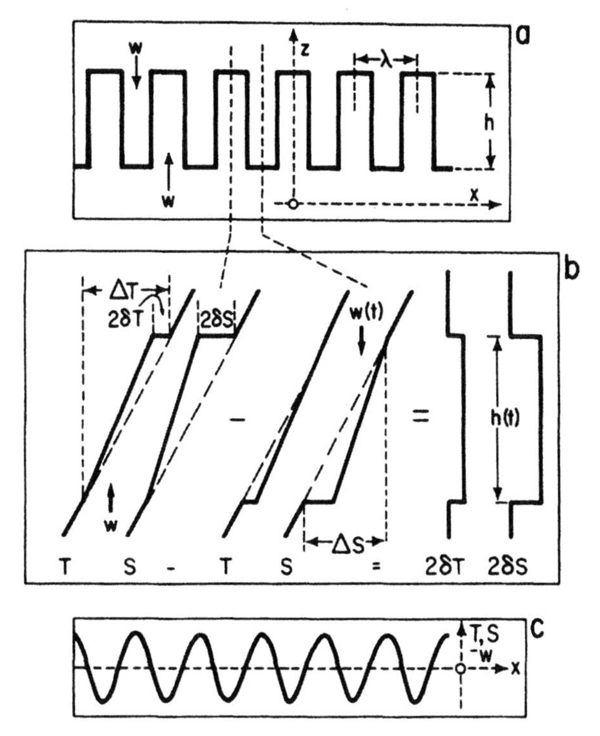

Figure 7.4.1. A sketch of the idealized salt finger model in Kunze's non-steady-state model (from Kunze, 1987).

gradients across the fingers, thus creating vigorous buoyancy-driven motions through the fingers that drain the excess salt in the system.

Notice that the temperature and salinity are continuous at the entry tip of the fingers but that there are strong "jumps" in properties (δT and δS) at the exit tips of the fingers. It is these jumps that drive vigorous convection in layers adjacent to the fingers. It is assumed that the mechanism in the interior of the fingers is what governs the heat and salt fluxes through the fingers and hence this convection. Thus the convection in the adjacent layers is entirely passive and does not influence the fluxes driven through the fingers. In fact, they are not at all essential to the model. The motions are initiated by instabilities due to double diffusion (See Section 7.2) and the fingers lengthen and the horizontal gradients in them sharpen until certain heat and salt fluxes are achieved and maintained. The fluid motion determines the value of the size d and length h of the fingers, the size of the jumps δT and δS, and the intensity of vertical motions (vertical velocity w). For simplicity we have assumed a square structure for fingers that ensures similarity in the two horizontal directions; this allows only one horizontal direction to be retained in the problem (say, x) but without any loss of generality. Extension to two dimensions involves multiplication by geometric factors resulting from the assumed planform of the fingers (Stern, 1976; Kunze, 1986). Schmitt (1994) explores the various forms that can arise from a combination of square and sheet planforms.

With this conceptual model, the flow in the fingers is assumed to be one-dimensional. This is a reasonable assumption since the aspect ratio of fingers is quite large (>50). Only one finger need be considered. The governing unsteady, incompressible equations for one-dimensional motion ($u = 0$) in the fingers were first given by Stern (Stern, 1960; see also Kunze, 1986),

$$\frac{\partial w}{\partial z} = 0 \tag{7.4.1}$$

$$\frac{\partial w}{\partial t} - \nu \frac{\partial^2 w}{\partial x_j \partial x_j} = b = g\left(\beta_T \delta T \text{-} \beta_S \delta S\right) \tag{7.4.2}$$

$$\frac{\partial}{\partial t}(\delta T) - k_T \frac{\partial^2 (\delta T)}{\partial x_j \partial x_j} + w \left.\frac{\partial T}{\partial z}\right|_f = 0 \tag{7.4.3}$$

$$\frac{\partial}{\partial t}(\delta S) - k_S \frac{\partial^2 (\delta S)}{\partial x_j \partial x_j} + w \left.\frac{\partial S}{\partial z}\right|_f = 0 \tag{7.4.4}$$

where w is the vertical velocity, ν is the kinematic viscosity, k is the kinematic molecular diffusivity, β is the volumetric coefficient of expansion, b is the buoyancy, and g is the gravitational constant. Index $j = 1, 2$. Note that the buoyancy force driving motions in each finger is determined by the contrasts δT and δS between itself and adjacent fingers. The subscript f denotes the interior of the finger and subscripts T and S refer to heat and salt, respectively. Note that $\nu \gg k_T \gg k_S$. The density ratio R_ρ (>1) is defined as

$$R_\rho = \beta_T \left.\frac{\partial T}{\partial z}\right|_b / \beta_S \left.\frac{\partial S}{\partial z}\right|_b = \frac{\beta_T \Delta T}{\beta_S \Delta S} \tag{7.4.5}$$

where the subscript b denotes undisturbed (background) values. The gradients in the finger are

$$\begin{aligned} \left.\frac{\partial T}{\partial z}\right|_f &= \left.\frac{\partial T}{\partial z}\right|_b - \frac{dT}{h} \\ \left.\frac{\partial S}{\partial z}\right|_f &= \left.\frac{\partial S}{\partial z}\right|_b - \frac{dS}{h} \end{aligned} \tag{7.4.6}$$

Typical oceanic values of various parameters are (from Kunze, 1986) are as follows: $\beta_T \sim 2 \times 10^{-4}\ °C^{-1}$, $\beta_S \sim 7.5 \times 10^{-4}$ psu^{-1}, $\Delta T \sim 0.6°C$, $\Delta S \sim 0.1$ psu m^{-1}, $R_\rho \sim 1.6$, $\partial T/\partial z|_b \sim 0.3°C$ m^{-1}, $\partial S/\partial z|_b \sim 0.05$ psu m^{-1}, and $h \sim 2$ m. It is possible to define two ratios indicating the relative strengths of jumps at the finger tips:

$$\delta_T = \frac{\delta T}{h \left.\dfrac{\partial T}{\partial z}\right|_b}, \qquad \delta_S = \frac{\delta S}{h \left.\dfrac{\partial S}{\partial z}\right|_b} \tag{7.4.7}$$

Define

$$N_S^2 = g\beta_S \left.\frac{\partial S}{\partial z}\right|_b, \qquad N^2 = g\beta_S \left.\frac{\partial S}{\partial z}\right|_b - g\beta_T \left.\frac{\partial T}{\partial z}\right|_b \tag{7.4.8}$$

where N denotes the buoyancy frequency and N_S the contribution from salt stratification. Consider steadily growing fingers. This is the scenario most theoretical models assume. Kunze (1986) relates w and h thusly,

$$w = \frac{1}{2}\frac{\partial h}{\partial t} \sim \sigma \frac{h}{2} \tag{7.4.9}$$

Note that the continuity equation does not involve time and Eqs. (7.4.2) to (7.4.4) can be written as

$$\sigma\left(\sigma + \nu k^2\right) + N_S^2\left(R_\rho \delta_T - \delta_S\right) = 0 \tag{7.4.10}$$

$$\delta_T = \sigma / \left(2\sigma + k_T k^2\right) \tag{7.4.11}$$

$$\delta_S = \sigma / \left(2\sigma + k_S k^2\right) \tag{7.4.12}$$

where

$$\frac{\partial}{\partial t} \equiv \sigma, \qquad \frac{\partial}{\partial x_j \partial x_j} \equiv k^2 \tag{7.4.13}$$

k is the wavenumber. Because $k_T >> k_S$, the relative temperature contrast δT is smaller than the relative salinity contrast δS. Because of Eq. (7.4.9), Eqs. (7.4.10)–(7.4.12) now involve only four unknowns: σ, k, δT, and δS. A single equation relating σ and k can be derived (Stern, 1960):

$$\left(\sigma+\nu k^2\right)\left(2\sigma+k_T k^2\right)\left(2\sigma+k_S k^2\right)+N_S^2\left[2\left(R_\rho-1\right)\sigma-\left(k_T\text{-}R_\rho k_S\right)k^2\right]=0 \tag{7.4.14}$$

It is reasonable to assume that among all the initial perturbations pertaining to a wide wavenumber spectrum, the fastest growing ones prevail and determine the scale of the fingers. This restriction will relate k and σ in Eq. (7.4.14). To determine the maximum growth rate σ_{max}, note is made of the fact $|\sigma| << \nu\ k^2$, and the temporal term in the vertical momentum equation (7.4.2) [or Eq. (7.4.10)] is neglected. Assuming $R_\rho < k_T / k_S$, a reasonable assumption in the oceans, one can show (Kunze, 1986) that the maximum growth rate corresponds to the wavenumber:

$$k_M^4 \sim \frac{N_S^2\left(R_\rho-1\right)}{\nu k_T}=\frac{N^2}{\nu k_T} \tag{7.4.15}$$

This expression is the same as those derived by Stern (1960, 1975) and Schmitt (1979), and corresponds to a finger size of about 3 cm for staircases east of Barbados (Kunze, 1986). The maximum growth rate is therefore

$$\sigma_M \sim A_\rho\left[\frac{k_T}{\nu}N_S^2\left(1-R_\rho\frac{k_S}{k_T}\right)\right]^{1/2} \tag{7.4.16}$$

where

$$A_\rho=R_\rho{}^{1/2}-\left(R_\rho-1\right)^{1/2} \tag{7.4.17}$$

Substitution into Eqs. (7.4.11) and (7.4.12) gives (for $R_\rho << k_T/k_S$, excluding the sugar–salt case)

$$\delta_T \cong \frac{1}{2}\frac{A_\rho}{B_\rho}, \qquad \delta_S \cong \frac{1}{2}\frac{A_\rho}{C_\rho}\sim\frac{1}{2} \tag{7.4.18}$$

where

$$B_\rho=R_\rho^{1/2}, \quad C_\rho=R_\rho^{1/2}-\left(1-\frac{k_S}{k_T}\right)\left(R_\rho-1\right)^{12}\sim A_\rho \tag{7.4.19}$$

$$\delta_T = \frac{\sqrt{R_\rho} - \sqrt{R_\rho - 1}}{\sqrt{R_\rho}} \tag{7.4.20}$$

$$\delta_S = \frac{\sqrt{R_\rho} - \sqrt{R_\rho - 1}}{\sqrt{R_\rho} - \left(1 - \frac{k_S}{k_T}\right)\sqrt{R_\rho - 1}} \sim 1 \tag{7.4.21}$$

The ratio of heat flux to salt flux R_F, an important quantity extensively measured in the laboratory, can be found without knowing w:

$$R_F = R_\rho \frac{\delta T}{\delta S} \cong A_\rho B_\rho \sim \left(\sqrt{R_\rho} - \sqrt{R_\rho - 1}\right) \tag{7.4.22}$$

This expression is in good agreement with laboratory experimental results (Kunze, 1986). It was given first by Stern (1975). A more general expression valid for $k_S \sim k_T$ can be found in Schmitt (1979), and comparisons with the sugar–salt system in Griffiths and Ruddick (1980).

While the flux ratio is independent of vertical velocity w (and in the growing fingers scenario, h), determination of individual heat and salt fluxes involves w and it is here that one more closure approximation is needed. Kunze's hypothesis of a constant finger Richardson number (or equivalently Froude number) has been popular. He postulates that the salt finger length adjusts such that the gradient Richardson number Ri_g, defined as

$$Ri_g = \frac{N^2}{w^2 k^2} \tag{7.4.23}$$

is maintained at a constant value ~ 0.25. Invoking an analogy to uniform vertically sheared stably stratified flow, where the Miles–Howard theorem indicates a value of 0.25 as the limiting condition for stability of the system, Kunze (1986) argues that Ri_g should attain this value for salt fingers. He points out that the criterion is the same as the collective instability criterion of Stern (1969), who argued that the S_N,

$$S_N = F_b / \nu N^2 \tag{7.4.24}$$

should be a constant. These two are equivalent since $F_b \sim w\Delta b \sim \nu w^2 k^2$. The Stern number S_N can be looked upon as the ratio of effective diffusivity to kinematic viscosity. S_N can also be written as

$$S_N = \frac{1}{2} \cdot \frac{wh}{\nu} \left(\frac{\delta_S - R_\rho \delta_T}{R_\rho - 1} \right) \quad (7.4.25)$$

and hence defines the effective Reynolds number of the flow. There is some limited support for a constant Stern number (McDougall and Taylor, 1984; Schmitt, 1979), although Shen's (1991) numerical simulations suggest that it depends on the density ratio. This closure hypothesis enables determination of the finger length,

$$h \sim D_\rho^{3/2} \left(\frac{\nu^3}{k_T N_S^2} \right)^{1/4} Ri_g^{-1/2} \quad (7.4.26)$$

where

$$D_\rho = R_\rho^{1/2} + \left(R_\rho - 1\right)^{1/2} \quad (7.4.27)$$

since

$$Ri_g^{-1} \sim \left(\frac{k_T}{\nu} \right)^{1/2} \frac{N_S h^2}{\nu} A_\rho^3 \quad (7.4.28)$$

Also

$$w \sim \left(\nu k_T N_S^2\right)^{1/4} D_\rho^{1/2} Ri_g^{-1/2} \quad (7.4.29)$$

This relationship gives typical values for h and w for conditions in staircases east of Barbados of 30 cm and 10^{-3} ms^{-1}. Actual staircases are found to be 2–5 m thick. The heat and salt fluxes at conditions of maximum possible h are given by

$$g\beta_T F_T \cong g\beta_T w \delta T \sim \nu N_s^2 Ri_g^{-1} B_\rho D_\rho \quad (7.4.30)$$

$$g\beta_S F_S \cong g\beta_S w \delta S \sim \nu N_s^2 Ri_g^{-1} D_\rho^2 \quad (7.4.31)$$

Laboratory experiments do indicate a linear dependence of fluxes on the salinity gradient N_S^2. Kunze (1986) derives average fluxes "during" salt finger growth from $h_o \sim 1/k$ to h. These are equal to the above values divided by the factor

$$\ln\left[\left(\frac{\nu}{k_T Ri_g}\right)^{1/2}\left(R_\rho - 1\right)^{1/4} D_\rho^{3/2}\right] \tag{7.4.32}$$

If one further assumes that the finger length is the same as the thickness of the salt finger interface,

$$N_S^2 h = g\beta_S \Delta S \tag{7.4.33}$$

and one recovers the 4/3 law for the fluxes measured in the laboratory,

$$g\beta_T F_T \sim \left(g\beta_S \Delta S\right)^{4/3} k_T^{1/3} Ri_g^{-1/3} A_\rho B_\rho \tag{7.4.34}$$

$$g\beta_S F_S \sim \left(g\beta_S \Delta S\right)^{4/3} k_T^{1/3} Ri_g^{-1/3} \tag{7.4.35}$$

However, in this unsteady-state model, it is necessary to average the maximum flux values over an assumed finger timescale to derive the average values for fluxes and other quantities such as the effective salt diffusivity. This results in complex expressions involving the averaging timescale and an empirical constant related to the value of the critical Richardson/Froude number that may have to be tuned to obtain agreement with observations. We will not present these expressions, but refer the reader instead to Kunze (1987, 1994, 1996). Nevertheless, the finger model of maximum growth rate and a critical Richardson number criterion leads to agreement with laboratory observations. The principal problem is the disagreement with field observations during C-SALT. Here the interfaces were found to be far more diffuse compared to the above theory, and consequently the buoyancy fluxes inferred indirectly through microstructure dissipation rate measurements (buoyancy flux ~ dissipation rate) were a factor of 30 weaker. Kunze (1987) suggests that the thick interfaces in the oceans are actually composed of a series of smaller finger interfaces of smaller thickness over which the salinity contrast would be correspondingly lower and hence the fluxes are smaller. This is quite possible as shown by Linden (1978). He found that an initially thick interface either evolved into a thin interface or split into several thin interfaces separated by convective layers. Schmitt and Georgi (1982) report high optical gradient vertically banded regions 10- to 15-cm thick separated by low gradient ones apparently containing more isotropic structures, whereas Gargett and Schmitt (1982) observed 2- to 5-m thick finger-favorable staircases uniformly filled with narrow bandwidth signals. This question is still wide open.

To conclude, Kunze's model appears to provide reasonable agreement with laboratory flux laws and indicates that the fastest growing fingers prevail. This is also confirmed by the wavenumber and bandwidth of the towed spectra measured by Gargett and Schmitt (1982) in the North Pacific.

The weakest link in Kunze's model is the invocation of a critical gradient Richardson/Froude number for the fingers. Kunze exploited the analogy with vertically sheared stably stratified flow with the velocity vector perpendicular to the gravitational vector. In this situation, the mean kinetic energy of the flow can be used to supply the energy needed to work against buoyancy forces and produce the classic shear instability and Kelvin–Helmholtz billows. But the shear in the fingers is horizontal and the flow is parallel to gravitational direction and already unstable with finger motions ensuing. It is therefore not clear if this is the right mechanism to invoke to determine the finger structure.

7.4.2 Quasi-Steady-State Model

If the phenomenon can be considered to be quasi-steady (for example, Howard and Veronis, 1987; Shen, 1993), the $\partial/\partial t$ terms in Eqs. (7.4.2)–(7.4.4) should be negligibly small,

$$w\left(1-\delta_S\right)-hk_s k^2\delta_S=0 \tag{7.4.36}$$

$$w\left(1-\delta_T\right)-hk_T k^2\delta_T=0 \tag{7.4.37}$$

$$\nu w k^2+N_S^2 h\left(R_\rho\delta_T-\delta_S\right)=0 \tag{7.4.38}$$

where h is the finger length. Howard and Veronis (1987) and Shen (1993) ignore Eq. (7.4.4), equivalent to ignoring the salt diffusivity completely. Shen and Veronis (1997) show that the resulting equations, Eqs. (7.4.2) and (7.4.3), without the $\partial/\partial t$ terms are the same as Prandtl's steady buoyancy layer in a stably stratified fluid adjacent to a heated boundary (Prandtl, 1952). The solution is similar to that of an Ekman layer; the velocity parallel to the boundary and temperature oscillate and decay with distance from the boundary with the boundary layer scale being ~ $O(Ra^{-1/4})$. For a weakly stably stratified case, fluid rises (or sinks) near the boundary with a small horizontal flow connecting the interior to the boundary layer. This horizontal flow is similar to Ekman pumping in rotating Ekman layers with wind stress curl imposed. For salt fingers though, the flow is more akin to the buoyancy-induced flow in a narrow tube that can be built up from this elementary solution (Howard and Veronis, 1987).

These are three equations and five unknowns, w, k, h, δT, and δS, and so two closure assumptions are needed. If we assume as before that the fingers that prevail are the ones that have maximum growth rates and assume then that the wavenumber in the steady finger regime is determined by Eq. (7.4.15), then k is determined. As before, this is adequate to determine the flux ratio (not the fluxes themselves, which need knowledge of w). From Eqs. (7.4.36) and (7.4.37),

$$\frac{\delta_S}{1-\delta_S} = \frac{k_T}{k_S} \cdot \frac{\delta_T}{1-\delta_T} \tag{7.4.39}$$

From Eqs. (7.4.36), (7.4.37), and (7.4.15), an expression can be derived for δ_T:

$$\frac{1-\delta_T}{\delta_T} = \frac{R_\rho - 1}{\delta_S - R_\rho \delta_T} \tag{7.4.40}$$

Equations (7.4.39) and (7.4.40) provide solutions for δ_T and δ_S. An approximate solution valid for small τ can be obtained by assuming $\delta_S \sim 1$ in Eq. (7.4.40):

$$\delta_T = \frac{A_\rho}{B_\rho} = \frac{\sqrt{R_\rho} - \sqrt{R_\rho - 1}}{\sqrt{R_\rho}} \tag{7.4.41}$$

This is similar to Kunze's expression. The flux ratio is the same as Kunze's:

$$R_F = A_\rho B_\rho = \sqrt{R_\rho}\left[\sqrt{R_\rho} - \sqrt{R_\rho - 1}\right] \tag{7.4.42}$$

One more constraint is needed to close the problem and derive the individual fluxes and effective salt diffusivity. Kunze's finger interface shear instability criterion (7.4.21) and Stern's collective instability condition (7.4.22) or Howard and Veronis' (1987) maximum buoyancy flux condition can be used. If we however postulate that the Kolmogoroff wavenumber in the adjacent turbulent layer,

$$k^4 = \varepsilon / \nu^3 \tag{7.4.43}$$

is proportional to the finger wavenumber, this provides the needed closure constraint:

$$k^4 = \frac{\varepsilon}{\nu^3} = \frac{F_b}{\nu^3} = \frac{w^2 k^2}{\nu^2} \tag{7.4.44}$$

Equation (7.4.36) gives

$$h = \frac{w}{k^2} \cdot \frac{1}{k_T} \cdot \frac{1-\delta_T}{\delta_T} \tag{7.4.45}$$

where

$$\frac{1-\delta_T}{\delta_T} = \frac{\sqrt{R_\rho - 1}}{\sqrt{R_\rho} - \sqrt{R_\rho - 1}} \tag{7.4.46}$$

and the wavenumber for fingers that prevail is that for those that had the maximum growth rate during the growth phase:

$$k^4 = \frac{N_s^2 \left(R_\rho - 1\right)}{\nu k_T} \tag{7.4.47}$$

Equations (7.4.44)–(7.4.47) provide the needed solution. In particular,

$$h = \frac{Pr}{k}\frac{1-\delta_T}{\delta_T} = Pr\frac{1-\delta_T}{\delta_T}\left[\frac{N_S^2 \left(R_\rho - 1\right)}{\nu k_T}\right] \tag{7.4.48}$$

$$w = k\nu = \nu\left[\frac{N_S^2 \left(R_\rho - 1\right)}{\nu k_T}\right]^{1/4} \tag{7.4.49}$$

$$wh = \nu \cdot Pr\frac{1-\delta_T}{\delta_T} \tag{7.4.50}$$

The Stern number is a constant consistent with the postulates of Stern (1969) and Kunze (1987),

$$S_N = \frac{wh}{\nu} \cdot \frac{1 - R_\rho \delta_T}{R_\rho - 1} = Pr \tag{7.4.51}$$

and equal to the Prandtl number. The Cox number for temperature is also a constant for a given R_ρ:

$$Co_T = \frac{\left\langle \frac{\partial T}{\partial x_j}\frac{\partial T}{\partial x_j} \right\rangle}{\left(\left.\frac{\partial T}{\partial z}\right|_b\right)^2} = k^2 h^2 \delta_T^2 = Pr^2 \left(1 - \delta_T\right)^2 = Pr^2 \frac{R_\rho - 1}{R_\rho} \tag{7.4.52}$$

Note also that the effective salt diffusivity is the same as that of Kunze's model:

$$K_S^e = \frac{F_S}{\partial \bar{S} / \partial z} = \frac{w\,\delta S}{\Delta S/h} = w h \delta_S \sim \nu \cdot Pr \cdot \frac{\sqrt{R_\rho - 1}}{\sqrt{R_\rho} - \sqrt{R_\rho - 1}} \tag{7.4.53}$$

The value of effective salt diffusivity is consistent with that observed in the pycnocline [5.0×10^{-6} to 1.5×10^{-5} $m^2\ s^{-1}$; see Kunze (1996)]. For $R_\rho \sim 1.6$, $K_s^e \sim 1.1\times 10^{-5}$ $m^2\ s^{-1}$.

A salt finger number can be defined as the ratio of the effective salt diffusivity to molecular diffusivity of salt (the same can be done for diffusive convection using temperature and effective heat flux):

$$D_{DS} = \frac{K_S^e}{k_s} = \frac{F_S}{k_s \partial \bar{S} / \partial z} \sim Sc \cdot Pr \cdot \frac{\sqrt{R_\rho - 1}}{\sqrt{R_\rho} - \sqrt{R_\rho - 1}} \tag{7.4.54}$$

This number denotes the efficiency or efficacy of the salt finger process in transporting salt vertically. Its value is typically ~O (10^4). An expression can be derived for the buoyancy flux of salt, which depends on the salt finger length by assuming $h \sim l_i$, the interfacial thickness:

$$h \sim Pr^{4/3} \left(\frac{g\beta_S \Delta S}{\nu k_T} \right)^{-1/3} \frac{\left(R_\rho - 1\right)^{1/3}}{\left(\sqrt{R_\rho} - \sqrt{R_\rho - 1}\right)^{4/3}} \tag{7.4.55}$$

This expression gives the 4/3 law observed in the laboratory. The corresponding expression for the salt flux is

$$g\beta_S F_S \sim \left(g\beta_S \Delta S\right)^{4/3} k_T^{\ 1/3}\, Pr^{1/3} \left[\sqrt{R_\rho - 1}\left(\sqrt{R_\rho} - \sqrt{R_\rho - 1}\right)\right]^{1/3} \tag{7.4.56}$$

However, there does not appear to be much support in the double-diffusive community for quasi-steady-state models of salt fingers and therefore the above should be considered controversial.

Note that F_S goes to zero, contrary to observations which find higher fluxes, as R_ρ approaches unity. This is a failing of Kunze's unsteady model as well. Thus, these analytical models are of doubtful validity for R_ρ close to unity. Numerical simulations are likely to help resolve some of these problems. Another complication is that the background state in the ocean is never quiescent. Shear instabilities and internal wave breaking tend to generate turbulence patches every dozen or so buoyancy periods and these could disrupt the finger formation process (Linden, 1971; Kunze, 1996).

7.4.3 Salt Fingers in Shear

Salt fingers are often observed in the presence of a weak background shear. Observations show that the fingers in such cases are often aligned with shear and look more like near-horizontal laminae than the vertical finger-like structures observed in classical laboratory experiments. It is not clear what exactly shear does to salt finger fluxes, but at least in C-SALT, the fluxes were quite weak. There can not be any doubt that if the shear across the finger interface is strong, shear instability ensues, leading to vigorous turbulent mixing. Turbulence is known to disrupt fingers in the laboratory. However, if the shear is below the threshold indicated by the Miles–Howard criterion (dU/dz > 2 N), the fingers are likely to remain more or less intact, even if altered in structure. To account for the inconsistencies between the model and the observations of the Cox number, Kunze (1994, 1996) extended his Froude number criterion ($F_r \sim Ri^{-1/2}$, where Ri is the Richardson number he originally defined in 1987) to account for shear by taking the effective Froude number to be the geometric mean of the finger Froude number and the Froude number based on mean shear:

$$\left[\frac{dU/dz}{N}\right]\left[\frac{wk}{N}\right] = Fr_s Fr_f < 1.0 \tag{7.4.57}$$

Using this criterion, and assuming a velocity change across the interface independent of the interface thickness and hence a background shear inversely proportional to interface thickness l_i, he shows that the effective salt diffusivity is

$$K_S^e = \frac{1}{Fr_S^2}\left(\frac{l_0 + l_i}{l_i}\right)\left(\frac{1-\delta_T}{\delta_T}\right)\cdot \nu \tag{7.4.58}$$

For typical values of $l_0 / l_i \sim 10$, $Fr_s \sim 0.4$, he gets $K_s^e \sim 10^{-4}$ m^2 s^{-1}. He also gets a Cox number proportional to l_i, as observed by Marmorino (1989) and Fleury and Lueck (1991). While this extended Froude number constraint can be made to yield the salt finger fluxes observed in the presence of shear by a suitable choice of the parameters, its validity is uncertain and direct observational support not yet available. It is also difficult to separate in the field the conditions when strong shear is periodically disrupting the fingers from those when the shear is weak and just modifying the finger fluxes. Hence modeling salt fingers in the presence of a background shear should be regarded as still an open question. Also Kunze (1994) points out that the criterion becomes singular as shear goes to zero.

7.5 PARAMETERIZATION OF DOUBLE-DIFFUSIVE MIXING

To conclude, analytical models of double-diffusive processes are still in their infancy. Indeed, Kunze's model is not very satisfactory, since at low density ratios, the behavior of the flux ratio is contrary to observations. Thus, we should regard analytical models of salt finger processes as being in a truly tenuous state. Kelley's parameterization of diffusive convection is however a little more successful. Nevertheless, analytical models necessarily imply simplifications and assumptions whose veracity can only be tested by comparison with experiments and oceanic observations. The very small scales associated with salt fingers make it difficult to sample them correctly in the oceans. Vertical fluxes associated with double diffusion are hard to infer. Therefore an attractive alternative approach that has been emerging in recent years is the use of numerical models to study double-diffusive processes (Shen, 1991; Shen and Veronis, 1997), as described earlier. Great strides are being made and this approach should be able to improve our understanding of these processes in the near future.

In the meantime, there is a need to parameterize the vertical transfer of heat and salt by double diffusion in numerical ocean models as best as we can. Zhang et al. (1998) is one of the first (see also Large et al., 1994) to attempt this. They parameterize the effective diapycnal salt and heat diffusivities in the salt fingering regime [$(\partial T / \partial z) > 0, (\partial S / \partial z) > 0; 1 < R_\rho < (k_T / k_S)$] as

$$K_S = \frac{K_*}{\left[1 + \left(R_\rho / R_c\right)^n\right]} + K_b; K_T = \frac{0.7 K_*}{R_\rho\left[1 + \left(R_\rho / R_c\right)^n\right]} + K_b \qquad (7.5.1)$$

where R_c is the critical value of density ratio $R_\rho = (\beta_T \partial T/\partial z)/(\beta_S \partial S/\partial z)$ beyond which the finger transfer of heat and salt falls off rapidly due to the absence of staircases. K_* is the maximum value for finger diffusivity. K_b is the diapycnal eddy diffusivity due to processes other than double diffusion and is independent of R_ρ. The index n controls how fast the diffusivities decay with R_ρ. Zhang et al. (1998) choose $R_c \sim 1.6$, $K_* \sim 10^{-4}$ m^2 s^{-1}, $K_b \sim 3 \times 10^{-5}$ m^2 s^{-1}, and $n \sim 6$.

Zhang et al. (1998) use an extension of Kelley's (1990) in the diffusive convection regime [$(\partial T/\partial z) < 0$, $(\partial S/\partial z) < 0$; $1 > R_\rho > (k_T/k_S)$]:

$$K_T = C\, Ra^{1/3} k_T + K_b;\, K_S = R_\rho R_F \left(K_T - K_b\right) + K_b$$

$$C = 0.0032 \exp\left(4.8 R_\rho^{0.72}\right)$$

$$Ra = 0.25 \times 10^9 R_\rho^{-1.1} \tag{7.5.2}$$

$$R_F = \frac{\beta_S F_S}{\beta_T F_T} = \frac{1/R_\rho + 1.4\left(1/R_\rho - 1\right)^{3/2}}{1 + 14\left(1/R_\rho - 1\right)^{3/2}}$$

Kelley (1984) suggested a slightly different form for C and R_F based on Huppert (1971). It is important to note that in diffusive convection literature R_ρ is often defined as the inverse of that in Eqs. (7.5.1) and (7.5.2).

In an effort to limit double-diffusive mixing to the region of the permanent thermocline, where the temperature and salinity gradients can be expected to be strong enough to drive double-diffusive convection, Zhang et al. (1998) choose to limit the application of Eqs. (7.5.1) and (7.5.2) to that part of the water column where the magnitude of the vertical temperature gradient exceeds a certain value, 2.5×10^{-4} °C m^{-1}. In non-double-diffusion regions, $K_T = K_S = K_b$. The effective diapycnal diffusivity of density is given by

$$K_\rho = \frac{R_\rho K_T - K_S}{R_\rho - 1} \tag{7.5.3}$$

Note that at low and high values of R_ρ, double diffusion weakens considerably and $K_\rho \to K_b$ ($= K_T = K_S$), its upper limit given by non-double-diffusive mixing processes. When the double diffusion is moderately strong, the value of K_ρ is reduced from this value. For strong double-diffusive activity, K_ρ becomes negative. Figure 7.5.1 shows the change of K_T, K_S, and K_ρ with R_ρ.

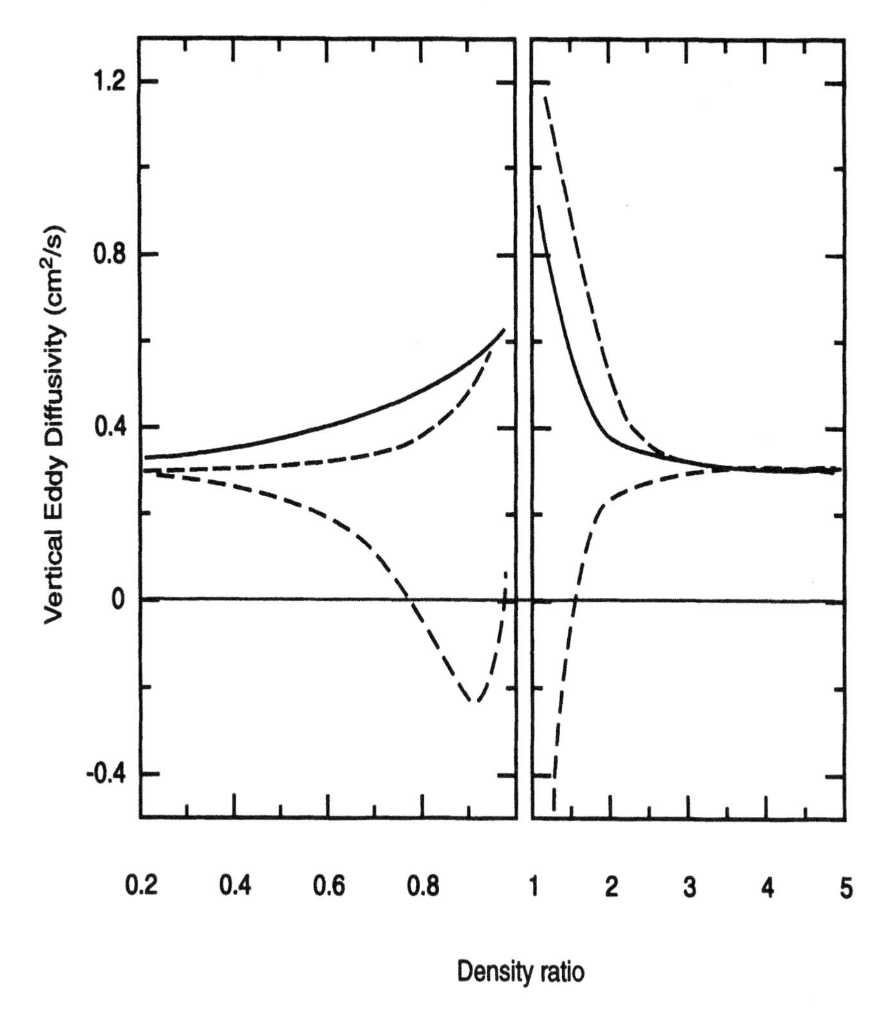

Figure 7.5.1. Dependence of diapycnal eddy diffusivity for temperature (solid line), salinity (broken line), and density (dotted line) on the density ratio R_ρ that Zhang *et al.* (1998) used. The left panel corresponds to diffusive layering and the right panel to salt fingers. The two curves are separated by an unstable region at $R_\rho = 1$, where $R_\rho < 0$, even though the curves appear continuous across $R_\rho = 1$.

To sum up, it is increasingly being realized that double-diffusive processes are too important to ignore in the study of mixing in the global ocean and in numerical models of the global ocean. It may be essential to take into account these along with internal tides [and possibly geothermal heat fluxes at the ocean bottom (Huang 1999)] to model correctly the oceanic thermohaline circulation, of great importance to the global climate and its variability.

LIST OF SYMBOLS

β_T, β_S	Coefficients of volumetric expansion due to heat and salt
δ_T, δ_S	Normalized change of temperature and salinity in the salt finger model
ε	Dissipation rate of TKE
λ	Wavelength
ν, ν_t	Kinematic viscosity (molecular and turbulent)
ρ	Density of water
κ	Von Karman constant
τ	Ratio of molecular kinematic diffusivities of salt and heat (k_S/k_T)
ΔU	Velocity change
ΔT, ΔS	Temperature and salinity changes
b	Buoyancy
c_p	Specific heat
d	Layer thickness
g	Acceleration due to gravity
h	Length of salt fingers
k	Wavenumber
k_M	Wavenumber corresponding to maximum growth rate
k_T, k_S	Kinematic molecular diffusivities of heat and salt
k_{TS}, k_{ST}	Kinematic molecular cross diffusivities (Dufour and Soret diffusion coefficients)
l	Integral microscale of turbulence
l_0, l_i	Thickness of the mixed layer and the interfacial layer
q	Turbulence velocity scale
t	Time
u_f^*	Free convection velocity scale
w	Vertical velocity
x_3	Vertical coordinate
x_i	Coordinates
z	Vertical coordinate
C_T	Cox number for temperature
D_{DT}, D_{DS}	Ratio of effective diffusivity of heat and salt to corresponding molecular diffusivities in double-diffusive processes
F_T, F_S, F_b	Heat, salt, and buoyancy fluxes
Fr	Froude number
H	Layer thickness
K_s, K_T, K_ρ	Effective salt, heat, and density diffusivity
L	Monin–Obukhoff length scale

N, N_s	Buoyancy frequency and buoyancy frequency due to salt stratification
Nu	Nusselt number
Pr	Prandtl number
R_F	Flux ratio, the ratio of the buoyancy flux due to salt to that due to heat in double diffusion and vice versa in salt fingers
R_ρ	Density ratio, the ratio of buoyancy change due to salt to that due to heat
Ra, Ra_S	Rayleigh number and Rayleigh number due to salinity changes
Ri_f, Ri_g	Flux and gradient Richardson numbers
S	Salinity
S_N	Stern number
Sc	Schmidt number
T	Temperature
Tu	Turner angle
U	Horizontal velocity

Chapter 8

Lakes and Reservoirs

In this chapter, we will discuss the salient characteristics of fresh water lakes and reservoirs and the mixing processes that determine their vertical structure. Freshwater lakes are especially important for societal needs. They also exhibit rather unique mixing characteristics, especially the deep freshwater lakes. The contrast with the oceans is quite striking and this is the primary reason for including them in this book. However, there are many similarities in mixing and dynamical mechanisms as well. We will concentrate primarily on the seasonal evolution of the thermal structure of lakes and the factors governing it, not on the water quality aspects. Some of the introductory material is derived from Henderson-Sellers (1984). Two of the most relevant periodicals on the subject are *Limnology and Oceanography* and *Water Resources Research.*

8.1 SALIENT CHARACTERISTICS

Lakes are natural bodies of water, created by volcanic, tectonic, or glacial activity, whereas reservoirs are artificial, man-made impoundments. Lake Baikal (54° N, 108° E) in Russia and Lake Tanganyika in Africa (7° S, 30° E) are examples of lakes created by the tectonic rifting process. Such lakes tend to be narrow and deep. Volcanic craters are often filled with water and form lakes. Crater Lake in Oregon and Lake Nyos in Cameroon, Africa, are good examples of crater/caldera lakes. These are usually small in size, one to a few hundred meters deep, often with a thick layer of sediments at the bottom through which warmer water with dissolved gases and chemicals might diffuse into the bottom waters. Lakes formed by glacial activity, gouging by advancing glaciers, tend to be large and not particularly deep. The Great Lakes in North America and the Canadian and Scandinavian lakes are excellent examples of lakes formed by filling-up of depressions, once the ice sheets covering much of the northern part of the northern hemisphere retreated. Of these, Lake Superior is the largest (by surface area) in the world.

Reservoirs are bodies of fresh water formed by artificial damming of rivers in natural valleys for providing water for irrigation, power generation, and drinking. They tend to be small in size, comparable to a small natural lake, but no less important to the welfare of the community dependent on them for electric power, flood control, agriculture, and drinking water needs. Maintenance of water quality in reservoirs is a nontrivial task, in view of the many sources of potential pollution such as sewage discharges and leaching of fertilizers from surrounding farmland. The largest so far is Lake Nasser in Egypt formed by damming of the Nile River by the Aswan Dam. The depth of a man-made reservoir is typically a few tens of meters, the deepest being less than 200 m deep (the reservoir created by the Three Gorges Dam across the Yangtze River in China is the deepest at about 175 m). The depth is of considerable importance to mixing processes, as we shall see shortly.

Fresh water is the most important characteristic of both lakes and reservoirs. This is what makes them overwhelmingly important to human welfare. For dynamical purposes, it is the vertical temperature stratification produced by the combined action of solar heating, convective and radiative cooling at the surface, and wind mixing that is the most important property of lakes and reservoirs. Evaporative cooling and precipitation also play a role. The circulation itself is mostly driven by winds, although in large lakes, formation of thermal bars in shallow regions around the periphery can drive a baroclinic component of circulation. Thermal bars are thermal fronts separating water masses with temperatures above and below the temperature of density maximum and are common features in many dimictic lakes during spring heating, because of the differential heating of shallow near-shore and deep offshore waters. Nevertheless, the resulting currents, generally speaking, are weaker compared to the flow velocities in rivers, which often feed and drain these reservoirs and lakes. In most lakes, the influence of the river inflows on the thermal structure and circulation is small (except locally), even though the impact may be large from pollution and water quality points of view, since discharges from industries situated on the banks of the rivers often constitute an important source of pollution. There are exceptions. A classic example is the hypersaline Dead Sea, with a salinity of ~275 psu. The stratification in the Dead Sea was maintained principally by freshwater inflow from the River Jordan. With increased use of Jordan River water for irrigation since the 1960s, the lake level dropped by ~4 m and the lake overturned in 1979 (Steinhorn, 1985). Many lakes like Lake Baikal may have both inflowing and outflowing rivers. The residence time of throughflow of fresh water, defined as the volume of the lake divided by the volume flow rate, is an important parameter in lake circulation and thermal (water mass) structure (Carmack et al., 1986). Lakes with low residence times are dominated by river flows, whereas those with large ones are not.

Both thermal stratification and circulation are of importance in the study of lakes. However, it is their combined effect on chemical and biological properties of the lake that is often even more important. In reservoirs, water quality is the overwhelmingly important characteristic, when the water is used for drinking purposes. During summer, the solar heating of the upper layers forms a strong thermocline which isolates the deeper waters from the supply of oxygen from the atmosphere and can lead to hypoxia and a deterioration of drinking water quality. When used for irrigation of rice, the level of water withdrawal becomes very important, because the colder temperatures of deeper waters could be harmful to rice seedlings. Overall, it is the water quality, and hence the indirect effect of thermal structure and circulation on chemical and biological characteristics, that makes lakes and reservoirs important in the current context of small scale processes in geophysics. Eutrophication of initially oligotrophic (containing very small biomass) lakes by natural and anthropogenic input of nutrients and consequent growth in biomass is quite important to water quality considerations. In addition, growth of blue-green algae and toxin-producing phytoplankton is a source of concern as well in reservoirs and small lakes used for drinking water and biomass growth in water bodies used for recreational activities. Such biochemical aspects and water quality are beyond the scope of this book and the reader is referred to existing monographs (for example, Henderson-Sellers, 1984) and journals on the subject (for example, *Limnology and Oceanography and Water Resources Research*). Here we will concentrate on mixing and its effect on the thermal structure of lakes and reservoirs.

The distinguishing property of volcanic crater/caldera lakes is their rich chemical structure due to dissolved chemicals and dissolved gases seeping in from the bottom into the waters. They thus form a window to magmatic processes below. Crater Lake in Oregon is an example of such a lake (McManus et al., 1993). Some crater/caldera lakes are of importance partly because of the potential for natural disaster they constitute. Lake Nyos, a very small lake in the hills of Cameroon, Central Africa, is one such example. The lake is only 1.4 km^2 in area and roughly 210 m deep, with about 100-m-thick sediments at the lake bottom through which warm CO_2- laden waters seep into the bottom layers. Consequently, the lake has large quantities of dissolved CO_2 in waters below the upper mixed layer that could be released into the atmosphere in certain cir-circumstances. Such an exsolution occurred in the August of 1986, when massive quantities (100 million m^3 at STP) of CO_2 were released by the lake to flow down the valley below as a density current, killing 1700 people and much animal life as far as 20 km from the lake (Freeth, 1992; Kantha and Freeth, 1996).

8.2 THERMAL STRUCTURE AND SEASONAL MODULATION

Many of the small scale processes we discussed in the context of the turbulent oceanic and atmospheric mixed layers in earlier chapters have relevance also to lakes and reservoirs. Below the surface layers, mixing is intermittent and weak, driven by internal wave breaking and boundary mixing, and in cases where the water is not completely fresh, possibly by double diffusion (Imberger and Hamblin, 1982; Imberger and Patterson, 1990). The dynamics of the upper layers are determined by winds and buoyancy fluxes. In addition, differential deepening and differential heating/cooling can lead to horizontal gradients and circulation in the water column. In shallow lakes, the wind stress is balanced principally by the bottom stress, and gyre-like circulations are set up by the wind. See Csanady (1985) for a discussion of the dynamics of large lakes.

The dynamics of the surface layer in a lake are governed by the nondimensional number known as the Wedderburn number,

$$We = \frac{c^2}{u_*^2}\frac{h}{L} = Ri\frac{h}{L} \tag{8.1.1}$$

where h is the thickness of the surface layer, $c = (g'h)^{1/2}$ is the relevant internal gravity wave speed, g' being the reduced gravity based on the density change across the base of the layer, u_* is the friction velocity, and Ri is the bulk Richardson number. L is the equivalent fetch, equal to the length of the lake. W_e is the ratio of the maximum baroclinic pressure force (before the upwind surfacing of the thermocline) and the surface wind force. The magnitude of W_e determines the deepening regime of the surface layer. For $W_e >> 1$, corresponding to strong stratification and weak winds, the isotherm tilt due to the applied wind stress is small, horizontal variations are negligible, and the deepening of the mixed layer is essentially one-dimensional and driven by turbulence generated by surface stirring. For $W_e << 1$, the interface tilts strongly with upwelling on the upwind end of the lake, and strong internal shear production occurs, but on a short timescale. In addition, seiches become important to dynamics and mixing. In either case, 1D mixing models may suffice. Only when $W_e \sim 1$ do upwelling and horizontal mixing become important (Spigel et al., 1986).

The type of motions induced by wind fluctuations itself depends on the ratio of the width of the lake b to the internal Rossby radius of deformation, a = c/f. For large b/a, Kelvin waves are generated and go around the lake along the boundaries, whereas for small values of b/a, internal waves propagate across. See Imberger and Hamblin (1982) and Imberger and Patterson (1990) for a detailed discussion of these aspects. For Lake Geneva, b/a ~ 5, and the current fluctuations induced by fluctuating winds are dominated by cyclonic-propagating signals and clockwise-turning currents (Bohle-Carbonell, 1986). For a narrow lake, the internal seiche period is $2(L/c)\,[(H-h)/H]^{1/2}$. Lemmin and Mortimer (1986) describe a procedure to compute internal seiche periods for lakes, including large lakes such as Lake Geneva for which Earth's rotation cannot be ignored. Wind-induced internal seiches in Lake Zurich have been measured and modeled by Horn et al. (1986).

An important difference between oceans and lakes is the difference in their physical size. The average depth of the ocean basins is about 4000–5000 m, whereas the deepest lakes (Lake Baikal in Russia being one of the exceptions with a 1632-m depth) are less than 800 m deep. Most lakes are less than 150 m deep. The horizontal extents are also much different. The largest lake, Lake Superior, is comparable in size to a small semienclosed sea (such as the Persian Gulf). For small lakes and reservoirs, the size is small enough to permit an area-averaged one-dimensional model to be used quite effectively to model physical and biochemical processes. Plots of the horizontal area at each depth and the volume below are known as hypsographic curves.

But by far the most important difference dynamically compared to the oceans is the absence or near-absence of salinity effects on the density structure and the enclosed nature of the water body. While wind forcing, surface cooling, and solar insolation bear much resemblance to the oceanic case, more often than not, the weak stabilizing effect of salinity gradients in combination with the often shallow water depths makes the entire water column susceptible to mixing.

In addition, a unique property of fresh water, namely the density maximum that occurs at ~4°C at atmospheric pressure, has interesting and unique dynamical consequences in dimictic lakes, lakes whose near-surface temperature passes through the 4°C mark during its seasonal evolution. In shallow lakes, this passage implies a tendency for complete overturning and homogenization of the entire water column not once but twice over the year—once during the heating and once during the cooling season. It also implies that the water column gets stably stratified and a shallow mixed layer forms not just during spring–summer heating but also during strong winter cooling when temperatures fall below 4°C. In deep lakes, the situation is further complicated by the fact that the temperature at which the density maximum occurs (T_{md}) decreases with increasing pressure so that for every 100-m increase in depth, the T_{md} decreases by 0.2°C. This prevents the entire water column from being homogenized, only the upper 100–200 m or so. However, deep mixing does occur episodically due to thermobaric instabilities, and instabilities associated with thermal bars that form frequently during spring heating. These episodic mixing events are of great importance to the renewal of oxygen in deep waters and replenishment of nutrients in the euphotic zone of the upper layers by injection from the bottom layers during deep convection events.

The vertical thermal structure is the most important property of lakes and impacts their biochemical characteristics. This is simply because the extent of mixing driven by processes at the lake surface depends on the vertical density structure, which in this case is determined predominantly by the vertical distribution of temperature. This in turn determines the vertical structure of dissolved oxygen and other parameters relevant to water quality considerations. The seasonal evolution of the lake thermal structure is therefore of great interest to limnologists and engineers.

The contrast in temperature between the lake surface and the bottom can be as much as 26°C in deep lakes during the peak of summer, when the deep waters are still cold at around 4°C, whereas the surface waters may have been heated by the Sun to nearly 30°C or more. In contrast, the difference during winter can be quite small, and in fact negligible during overturning when the entire water column is at a uniform temperature of a few degrees Celsius. Figure 8.2.1 shows the typical seasonal evolution of the lake surface temperature (LST) during the year for lakes at various latitudes. The strength of the seasonal modulation of LST is a strong function of the latitude, with the maximum range in the midlatitudes of around 20°C (ranging between 0 and 20°C). The temperature range in tropical and subtropical latitudes is comparatively small, less than 5°C, with the LST varying typically between 20 and 25°C over the year. In high latitudes, some lakes cool down to 0°C often and lake ice forms at the surface during winter. Lake ice may be an important factor to shipping in some large lakes.

Although LST is a function of the local meteorological conditions and the temperature structure of the lake itself, and hence its evolution over any particular year being dependent on the conditions during that year, it is possible to represent the seasonal variation in LST approximately as a function of latitude and time of the year,

$$T(t,\phi) = A_0 + A_1 \sin\left[\pi(t+\phi)/180\right] \tag{8.1.3}$$

where ϕ is 60° for the southern and 240° for the northern hemisphere, and t is the Julian day. Data from over 50 lakes between 26° S and 74° N (Straskraba, 1980) obey this relationship quite well, as can be seen from Figure 8.2.1. For medium-sized lakes, the amplitudes A_0 and A_1 are given by

$$\begin{aligned} A_0 &= 28.12 - 0.34(\theta - 3.4) \pm 2.41 \\ A_1 &= 0.5397 - 4.501\text{x}10^{-2}(\theta - 3.4) \\ &+1.463\text{x}10^{-2}(\theta - 3.4)^2 - 1.965\text{x}10^{-4}(\theta - 3.4)^3 \end{aligned} \tag{8.1.4}$$

where θ is the latitude in degrees.

It is often permissible to neglect the horizontal gradients of temperature and use area-averaged one-dimensional models to study mixing in small lakes and reservoirs. In large lakes such as the Great Lakes the horizontal gradients are important to the lake thermal structure and circulation and therefore a fully three-dimensional model is needed. Figure 8.2.2 shows the formation of thermal bars and the establishment of the horizontal thermal structure in Lake Ontario.

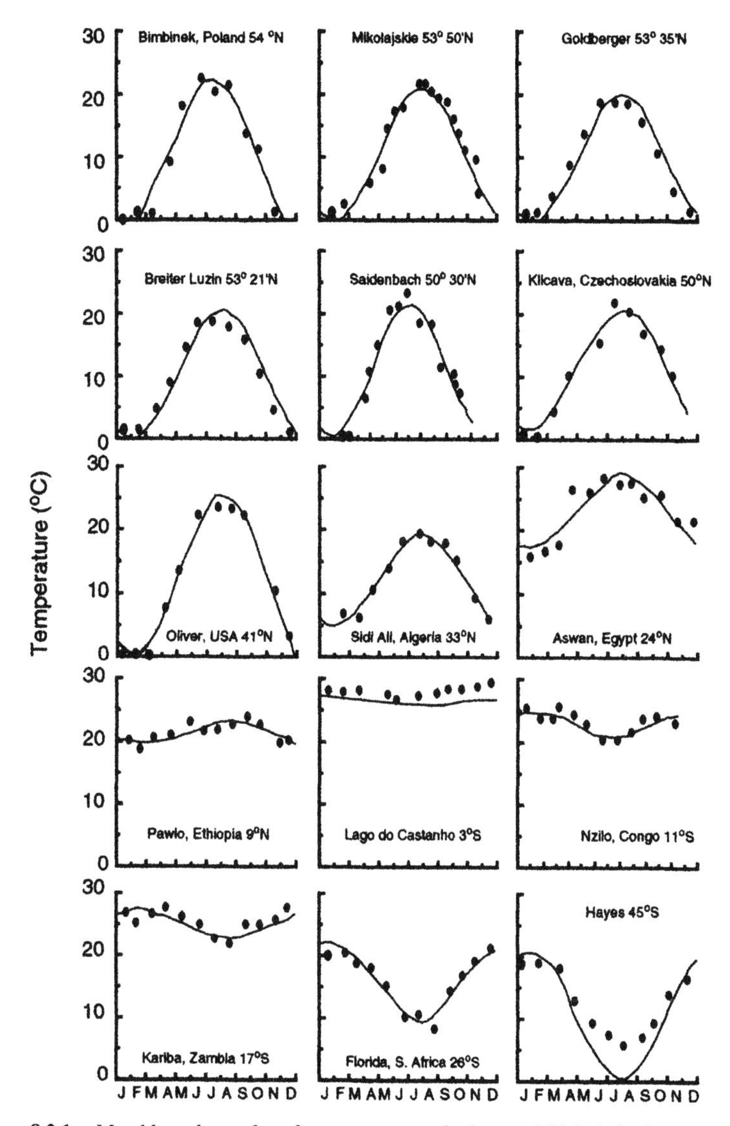

Figure 8.2.1. Monthly values of surface temperature in low and high latitude lakes showing increased seasonal variability with increasing latitude (from Henderson-Sellers, Engineering Limnology, Copyright Pearson Education, 1984; data from Straskraba 1980).

Figure 8.2.3 shows the dependence of density of fresh water on temperature. The maximum at 4°C (277 K) is noteworthy and has interesting dynamical consequences. Imagine a relatively shallow lake (less than 50-m deep) with a uniform temperature of just above 4°C in the entire water column at the onset of

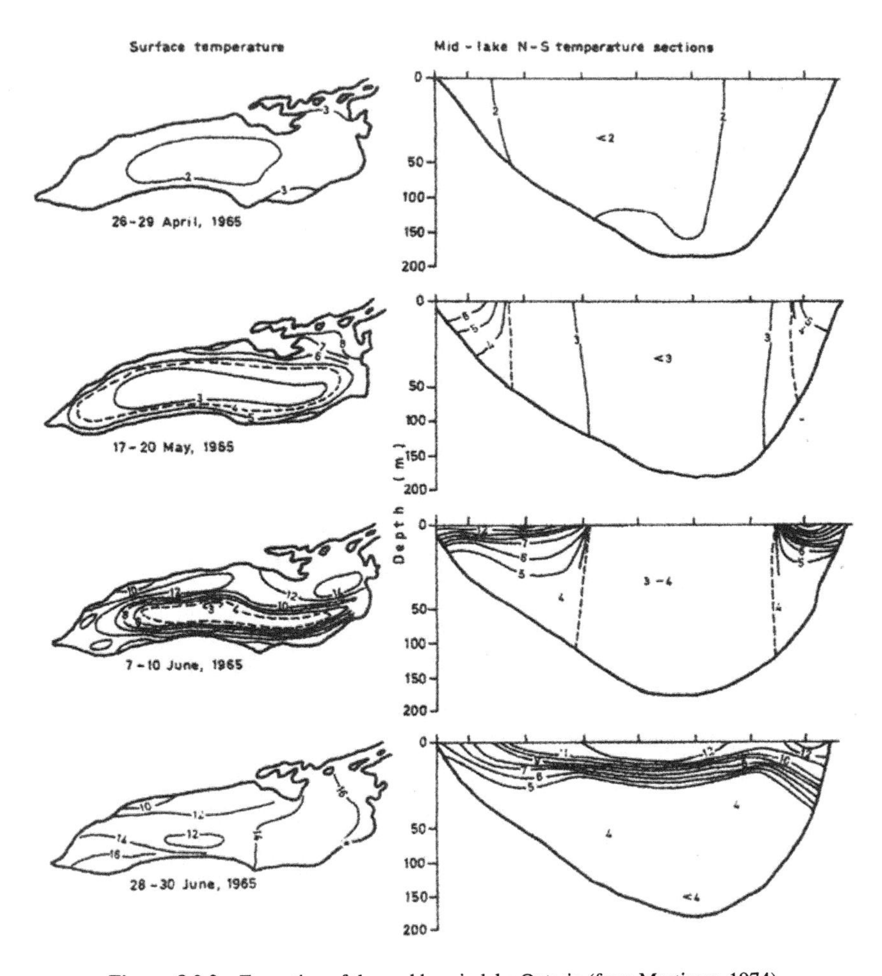

Figure 8.2.2. Formation of thermal bars in lake Ontario (from Mortimer, 1974).

spring. Spring heating causes net heating of the lake, and the extent and degree of such heating are a function of wind mixing, the intensity of solar insolation, and the turbidity of the water. The mixed layer does not extend to the bottom in most lakes that are not shallow. This layer is called the epilimnion. If a strong thermocline forms, it isolates the deeper layers from the upper ones, and the deeper layers (the hypolimnion) tend to stay pretty much at the cold winter values, with a sharp temperature change occurring in the intermediate layers (the metalimnion). The epilimnion might reach temperatures of 15–28°C, depending on the latitude, by the end of summer, while the hypolimnion might undergo a change in temperature of only a few degrees Celsius. The stratification is quite strong during summer, preventing renewal of oxygen in the deeper waters of the

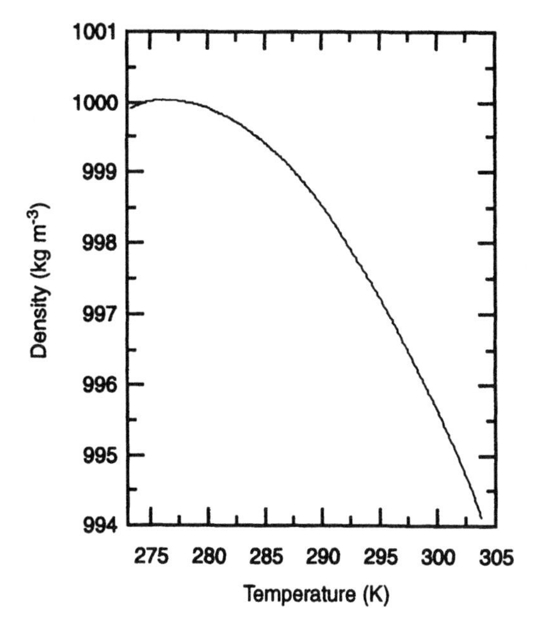

Figure 8.2.3. Density of fresh water as a function of temperature. Note the maximum at 277 K.

hypolimnion. Then fall–winter cooling begins. Since the net heat flux is now out of the water, this causes convection in the water column, which, aided by wind mixing, reduces the change in temperature across the metalimnion. In lakes that are not too deep, the entire water column in the lake can overturn and the water temperature can become uniform from top to bottom. Any further cooling reduces the temperature of the entire water column until the water temperature reaches 4°C. Continued cooling then reduces the density of the upper layers, so that the lake becomes stratified once again. This cooler, less dense layer is much shallower than the upper mixed layer during spring. If cooling is such that the freezing point is reached (0°C), then any further cooling can result in ice formation (for example, in the Great Lakes and Lake Calhoun in Minnesota at 44.9° N, 93.2° W—see Figure 8.2.4).

Ice can form at the surface as congelation ice during calm conditions and as frazil ice in the water column when the wind is strong enough to keep the water well mixed. Eventually, all the ice formed accumulates at the surface. During severe winters, the lake ice can grow to thicknesses of up to a meter, posing a hazard to shipping and other operations. Spring heating begins to heat the epilimnion even before the ice melts, because of penetrative solar heating. It also leads to deepening of the upper mixed layer since solar heating is destabilizing until temperatures reach 4°C. Eventually the ice disappears and the upper layers increase in temperature until the lake becomes isothermal. It remains isothermal

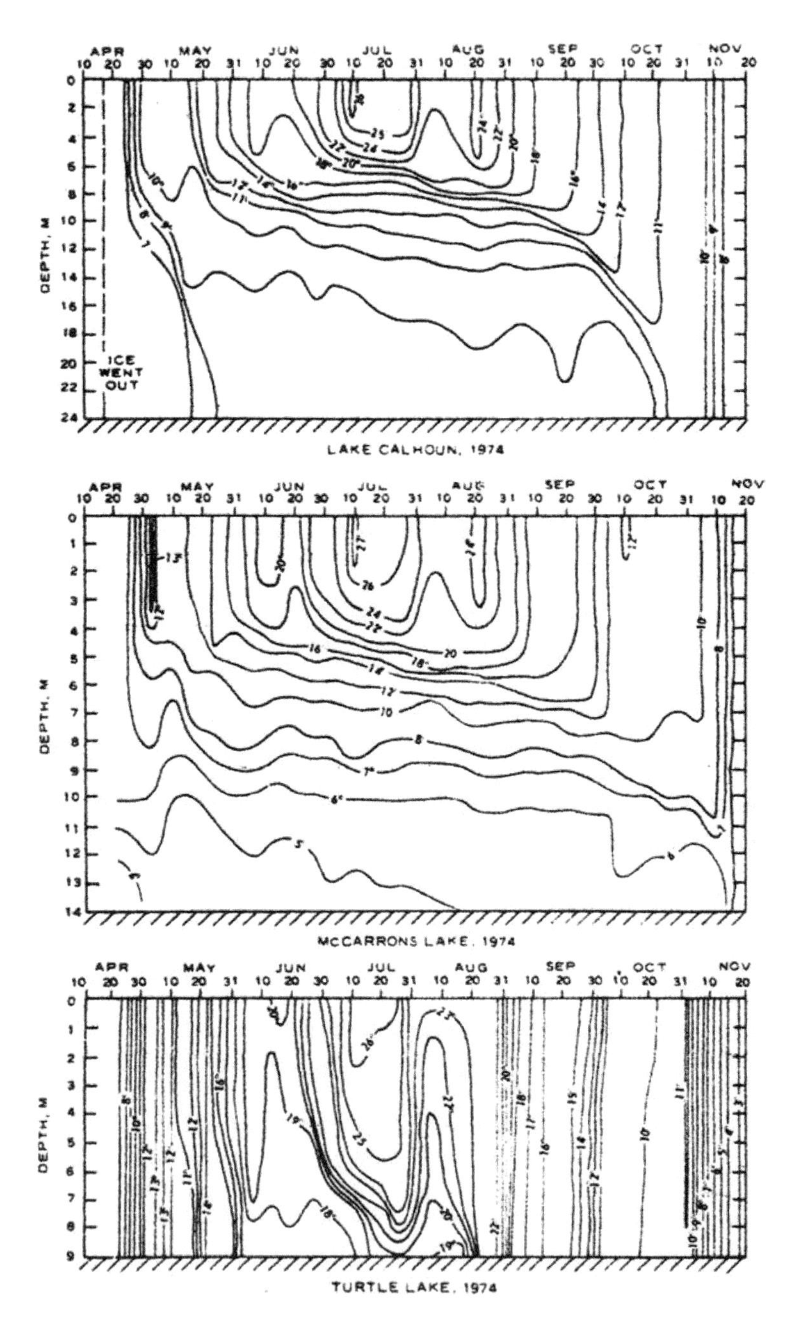

Figure 8.2.4. Seasonal cycle of lake thermal structure from Lake Calhoun, Lake McCarrons, and Turtle Lake. Note the spring heating resulting in large LSTs and shallow thermoclines (from Ford and Stefan, 1980, with permission from the American Water Resources Association).

until the entire water column reaches 4°C and the seasonal cycle of stratification and overturning repeats all over again.

Thus at temperate midlatitudes, there are two overturn periods, one during the autumn cooling and one during the spring heating. Such lakes are called dimictic and require cooling below 4°C. If the minimum temperature remains above 4°C, only the autumn overturn is possible and such lakes are called warm monomictic lakes (Henderson-Sellers, 1984). Figure 8.2.5 shows the seasonal cycle of temperature in a dimictic and a monomictic lake. It shows the temperature

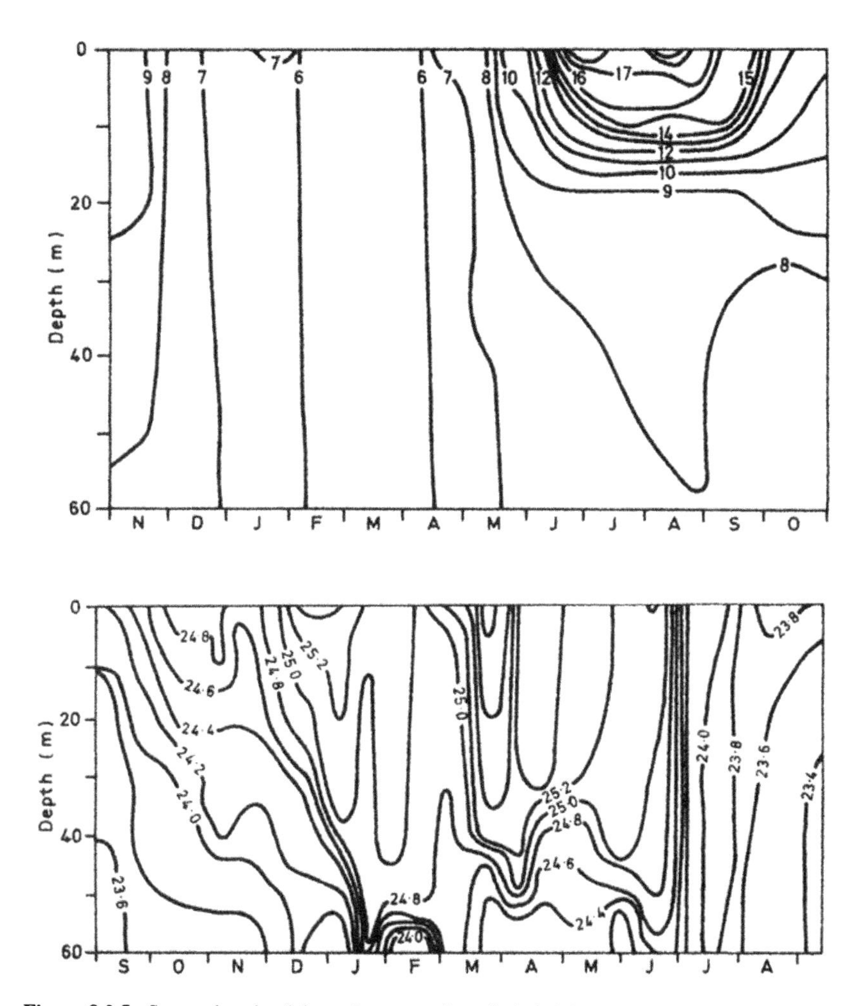

Figure 8.2.5. Seasonal cycle of thermal structure in a dimictic lake, Lake Windermere (top), and a monomictic lake, Lake Victoria (bottom) (from Henderson-Sellers, *Engineering Limnology*, Copyright Pearson Education, 1984).

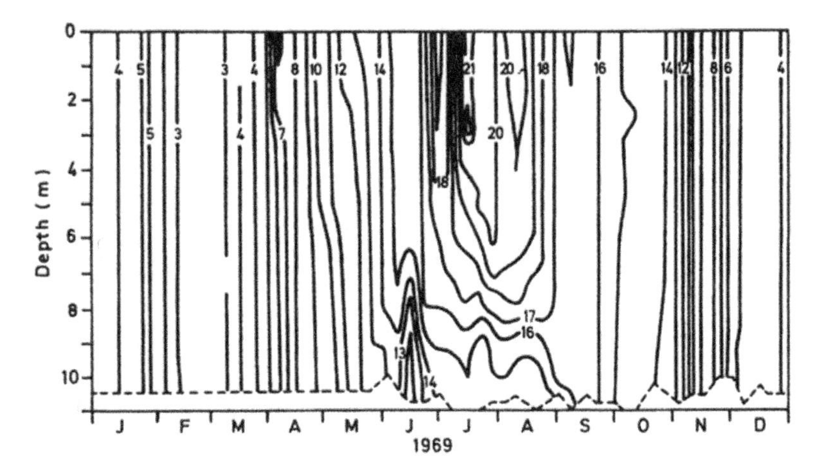

Figure 8.2.6. Seasonal cycle in the shallow Farmoor reservoir in UK (from Henderson-Sellers, Engineering Limnology, Copyright Pearson Education, 1984).

changes in the water column of Windermere, a temperate warm monomictic lake. Tropical and subtropical lakes are also typically warm monomictic lakes. If the maximum temperature in the lake is below 4°C, as in the cold polar regions, only spring overturn is possible and these lakes are called cold monomictic lakes. Figure 8.2.5 also shows an example from such a lake. Amictic lakes have permanent ice cover and therefore do not mix. In the tropics are also found oligomictic lakes with rare periods of continuous circulation and polymictic lakes with frequent or continuous circulation in the water column due to high wind mixing and small seasonal temperature modulation.

Shallow lakes and reservoirs are usually well mixed except for a brief period during the heating season. Wind mixing is usually strong enough to mix the water column to a depth of about 8–10 m, except during calm, high insolation conditions. Figure 8.2.6 shows the seasonal thermal structure of the shallow Farmoor reservoir in the United Kingdom that illustrates this.

Figure 8.2.7, from Wetzel (1995), shows the different lake types at various latitudes and altitudes.

Solar heating is the primary source of stabilization of the upper layers during the spring–summer heating season. The degree of penetration of solar radiative flux in the water column depends very much on the turbidity of the water, to which suspended sediments, dissolved organic matter, and phytoplankton concentration contribute (see Section 3.5). The IR and near-IR components are absorbed in the upper 1 m of the water column and it is the visible part of the spectrum that penetrates into the water column and constitutes internal sources of heat in the lake. For shallow lakes with clear waters, a substantial part of the SW solar radiation reaches the bottom and is partly reflected back up, the amount depending on the albedo of the bottom surface. In cold latitudes, during

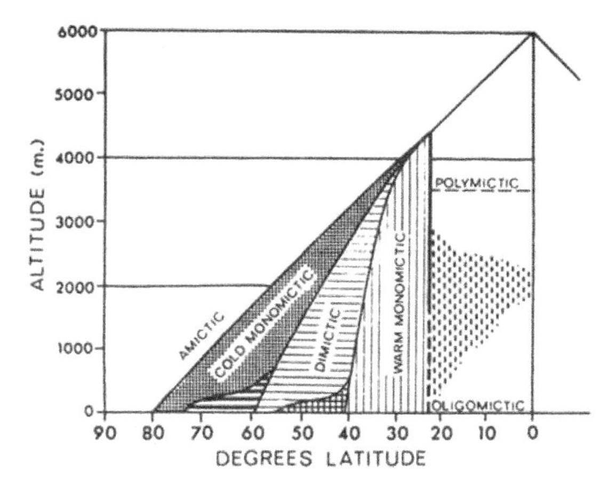

Figure 8.2.7. Lake types at various latitudes and altitudes (from Wetzel, 1975).

the initial stages of spring heating, the amount of SW solar radiation penetrating into the water column depends also on the snow and ice cover over the lake. The albedo of snow is typically 0.7–0.8, depending on its condition, and this can influence the amount of solar radiative heating of the lake. It is therefore

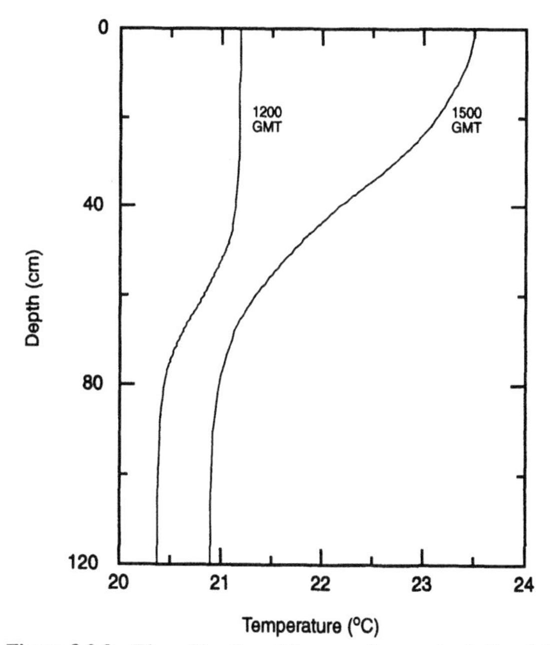

Figure 8.2.8. Diurnal heating of the upper layers of a shallow lake.

important to account for all these influences in modeling the spring heating of the lake waters. See Imberger (1985) and Spigel et al. (1986) for discussion of diurnal mixed layers in lakes and a 1D model of the diurnal mixed layer in Lake Wellington in Australia.

Just as there are short-term changes in the ocean beneath the upper mixed layer, there are temperature changes in the epilimnion on shorter timescales also. Diurnal changes in near-surface layers 1- to 2-m thick can reach the 3–4°C range (see Figure 8.2.8), with the larger mid-afternoon LSTs reached on calm windless days with strong solar insolation. The excess heat is mixed down into the water column by high winds and nocturnal cooling. This kind of intermittent mixing appears to be important in elevating the hypolimnitic temperature much above winter values in temperate lakes (Henderson-Sellers, 1984).

It is relatively straightforward to model the temporal evolution of the thermal structure of small lakes and reservoirs using area-averaged one-dimensional models of turbulent mixing described in Chapter 2, provided the meteorological parameters are known. Parameters such as the air temperature, humidity, wind speed, and SW and LW heat balances at the surface must be known for accurate modeling of the lake thermal structure. Unfortunately, measurements are often sparse, and under these conditions, modeling is based on a series of assumptions and simplifications which have an influence on the accuracy of the results. Figure 8.2.9 shows a successful simulation of the seasonal cycle in Lake Ohrid

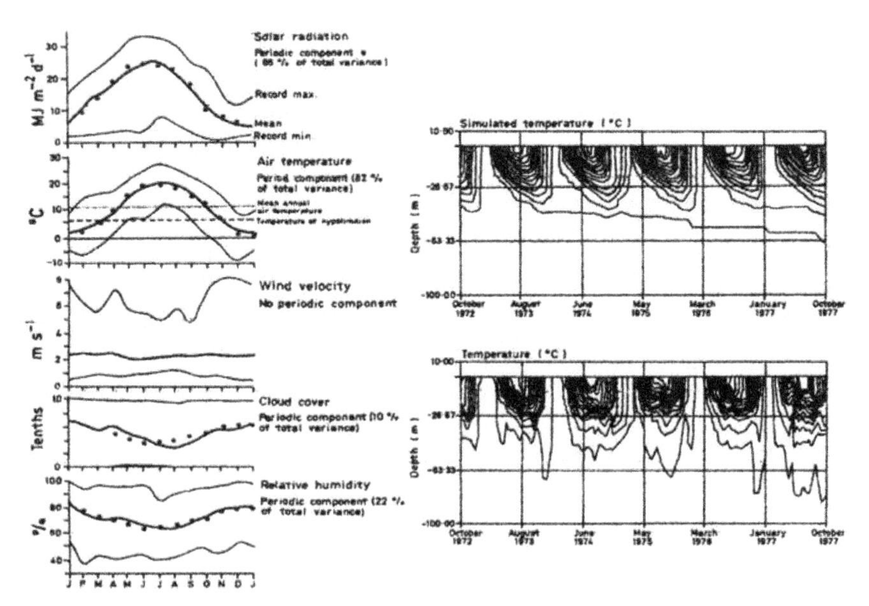

Figure 8.2.9. Simulated and observed seasonal thermal structure in 1977 in Lake Ohrid. Monthly values of meteorological parameters are also shown (from Henderson-Sellers, Engineering Limnology, Copyright Pearson Education, 1984).

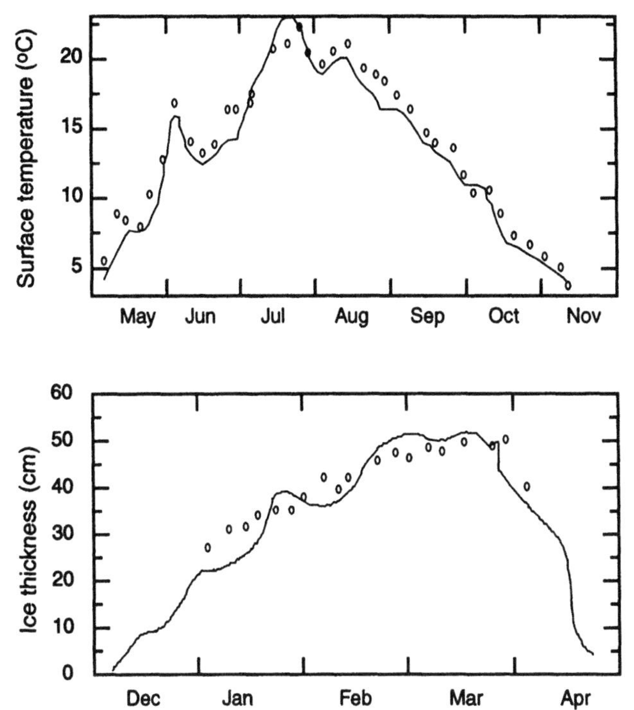

Figure 8.2.10. Simulation of the seasonal LST cycle and the ice thickness in Lake Punnus (from *Turbulent Penetrative Convection,* Zilitinkevich, 1991, Avebury Technical).

from 1972 to 1977 (from Henderson-Sellers, 1984). The success of the model is in large part due to the availability of accurate meteorological data. Figure 8.2.10 shows a successful model simulation of the seasonal cycle (Zilitinkevich, 1990) of Lake Punnus in Russia, including the thermal structure and the ice formed.

DYRESM (Imberger and Patterson, 1981) is a lake mixing model that is used often by limnologists. A recent extension and application to an ice-covered midlatitude lake, Harmon Lake in Canada, can be found in Stefan et al. (1995). See Imberger and Patterson (1981), Henderson-Sellers (1984), and Henderson-Sellers and Davies (1989) for a discussion of the models of the thermal structure of lakes.

Results from a lake mixing model can be used to model and examine the dissolved oxygen (DO) concentration, which is a good indicator of the water quality. High DO concentrations promote aerobic processes, whereas low ones promote anaerobic processes that produce gases such as methane, hydrogen sulfide, and ammonia. These gases impart a bad taste to drinking water and in large concentrations can be toxic. Since the major source of DO is through gas

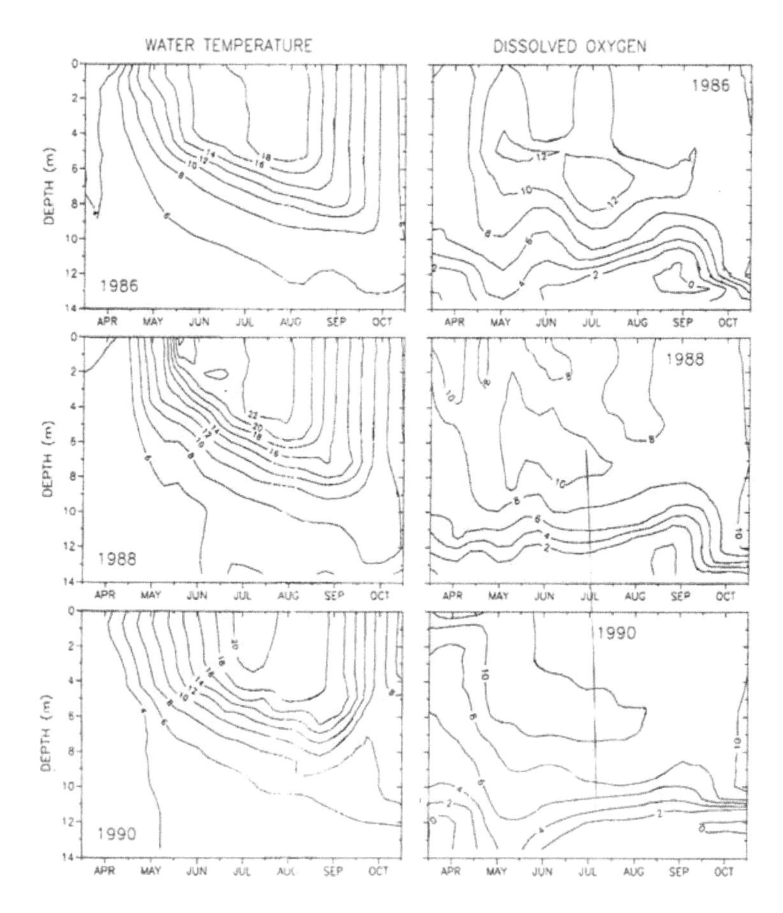

Figure 8.2.11. Seasonal isotherms and D.O. isopleths interpolated from field data from 1986, 1988, and 1990. Isotherms are in increments of 2 °C and isopleths in increments of 2 mg liter^{-1} (from Stefan *et al.* 1995).

transfer from air to the water (photosynthetic processes in the water column are another source of DO), and bacterial action is the primary sink, water that is well mixed and in contact with the lake surface tends to be higher in DO. Reduction or cessation of mixing over long periods of time, such as in the hypolimnion during summer, depletes DO and promotes hypoxic conditions. The situation is very much similar to that in the coastal oceans near the mouths of nutrient-laden river waters such as the Mississippi. Figure 8.2.11 shows the annual thermal and oxygen cycle in Thrush Lake (Stefan et al., 1995); the summer hypoxic conditions in the hypolimnion are quite typical of some freshwater reservoirs and often call for active intervention in the form of artificial mixing by air bubbling devices and such to maintain water quality.

It is not our intention here to present and discuss the many limnological aspects of lakes and reservoirs in their entirety. The subject is too vast for that. Instead, we will concentrate on mixing in deep lakes and reservoirs, and point out the interesting complications they bring about in understanding and modeling the evolution of thermal structure in such lakes. We will provide three examples. One of them is Lake Nyos in Cameroon, a tropical lake with a heavy concentration of dissolved CO_2 which has an important effect on mixing. The others are Crater Lake in Oregon and Lake Baikal in Russia, both of which are deep and greatly influenced by the influence of thermobaric instability on mixing.

8.3 MIXING MECHANISMS

The dynamical mechanisms responsible for mixing in lakes and reservoirs are very much similar to those in the oceans and the atmosphere, with a few exceptions as discussed below. In the upper layers, winds and the vertical shear associated with currents promote mixing. In contrast, cooling, including evaporative effects, promotes mixing only when the temperature is above 4°C, and suppresses mixing if it falls below that value. If the concentration of dissolved solids is small, as it is in most cases, the dynamical effect due to the concentration of dissolved solids by evaporation at the surface is comparatively small. Then only the cooling effect due to latent heat transfer is important to mixing. The same is true of precipitation. This is an important difference from the oceanic case. Solar heating, including penetrative solar heating, tends to suppress mixing, except when the temperature is below 4°C, when it tends to enhance it. If there is a significant heat flux at the bottom, unless there is also a compensating flux of dissolved solids, the bottom layers are mixed by free convection. In Lake Kivu, for example, the heat flux at the bottom leads to spectacular staircase structures. In Lake Nyos, the bottom heat flux is more than compensated for by the flux of dissolved CO_2 so that there is little mixing and the bottom layers heat up. If the water column is shallow, and the water is clear enough, solar insolation penetrates to the bottom and, depending on the nature of the bottom, tends to heat the bottom and cause convective heating of the water column from below, in addition to penetrative heating from the top. In fact, in many lakes, the more rapid springtime heating of the shallow near-shore waters vis-à-vis the deeper offshore waters promotes the formation of thermal bars, which are regions across which the temperature changes from below 4°C to above 4°C. Thermal bars are regions of strong circulation, due to density gradient effects, and also of enhanced mixing as described below. Below the seasonal thermocline and above the benthic boundary layer, the mixing is small and episodic, and due mostly to internal waves.

The fact that fresh water has a maximum density at T_{md} at which the coefficient of thermal expansion changes sign requires care in assessing the stability of the water column. This is because isopycnal surfaces (isosurfaces of potential density) may not always be appropriate from mixing considerations. Neutral surfaces (surfaces along which when a water mass is transported isentropically without any exchange of heat or salinity with its surroundings, it experiences no buoyancy forces) are not in general parallel to isopycnal surfaces (McDougall, 1987). The use of potential density, the density a water mass would acquire if brought isentropically from its depth z to a common reference depth z_r, can lead to errors in computation of the local stability of the water column. This point is of particular importance in deep freshwater lakes like Lake Baikal, where the formation of deep water and ventilation of deep waters depend crucially on the stability in the water column. Peeters et al. (1996) show that the potential density referenced to the surface shows that the water column over 150- to 850-m depths in Lake Baikal is nearly unstable, whereas in reality the water column is quite stable except for a small locally unstable patch at 700 m.

The most distinguishing feature of mixing in freshwater lakes is the possibility of two other mixing mechanisms associated with the unique property of fresh water related to T_{md} (Farmer and Carmack, 1981). Both arise from nonlinear effects (see also Chapter 2). The first one arises from the quadratic dependence of density on temperature near T_{md} and is called the thermostoltic effect. It is caused by the contraction on mixing of waters of different temperatures resulting in a water mass denser than that of either of the parent masses and is thermodynamically irreversible. The resulting instability is called cabbeling instability (Foster, 1972; McDougall, 1987). Such instability can arise near springtime thermal bars, when mixing of waters above and below 4°C across the thermal bar creates a heavier water mass and the resulting cabbeling instability promotes mixing. Shimarev et al. (1993) describe observations of such mixing processes in Lake Baikal near the thermal bars produced near the outflow of the river Salenga in the spring of 1991 (Figure 8.3.1). Cabbeling instability (CI) can occur in both shallow and deep freshwater lakes. It is also important to the deep oceans, where mixing of two water masses of different T and S characteristics can actually lead to a water mass of slightly higher density (McDougall, 1987).

The second mechanism is particular to deep dimictic lakes and arises from the differential compressibility of water. It is caused by the dependence of the temperature of the density maximum (T_{md}) on pressure; T_{md} decreases by 0.21°C for every 100-m increase in depth. This gives rise to the possibility of thermobaric instability (TBI) (Carmack and Farmer, 1982; Carmack and Weiss, 1991) in the water column. Because of the decrease of T_{md} with depth, when a freshwater lake is reversely stratified, the water column becomes conditionally unstable when the base of the mixed layer is pushed by mixing beyond the depth, called the compensation depth, where its temperature matches T_{md} (Carmack and Weiss, 1991). If the water below is nearly neutrally stratified, this initiates convection that could mix the entire water column (Figure 8.3.2). It is

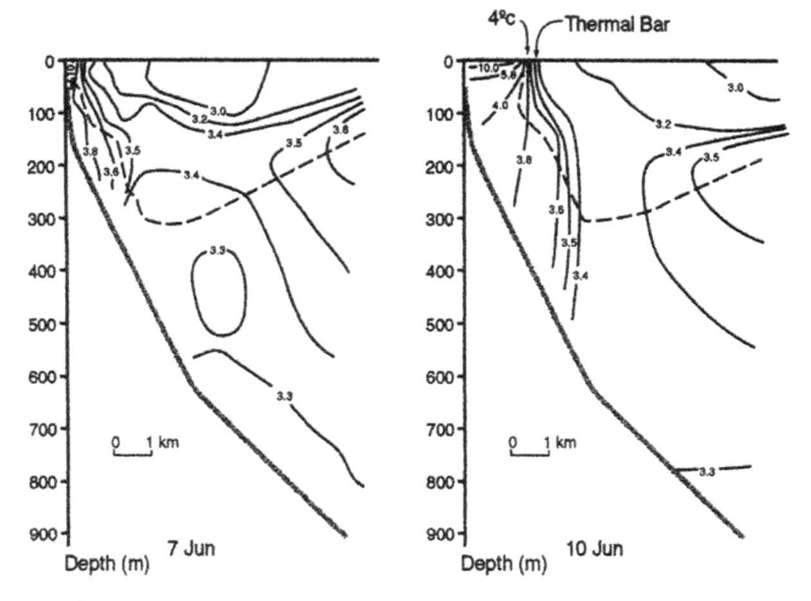

Figure 8.3.1. Thermostaltic (cabbeling) instability in Lake Baikal observed by Shimarev et al. (1993) in the spring of 1991.

thought that TBI is quite important to mixing in deep dimictic lakes such as Lake Baikal, which has nearly zero dissolved solids content (Carmack and Weiss, 1991; but see Shimarev et al., 1993), and could be important, if not central, to mixing in Crater Lake, Oregon, where there is some stable salinity stratification in the water column.

In view of the CI and TBI, it is important to pay special attention to computing the local stability of the water column in deep bodies of water, including fresh water lakes. Traditionally, potential density inthe global oceans, ρ_{pot} $(z,z_r) = \rho$ $[\theta(z,z_r)$, $S(zte)$, $p(z_r)]$, with pontial temperature computed at a single reference level z_r, has been used to compute the buoyancy frequency from $N^2 = -(g/\rho_{pot})d(\rho_{pot})/dz$. This is equivalent to using the thermal and salinity expansion coefficients at the reference depth z_r (which is usually the surface). But this is not necessarily indicative of the local stability of the water column. What is needed in stability and mixing considerations is an assessment of whether a fluid particle when displaced isentropically by an infinitesimal distance in the vertical experiences any restoring buoyancy forces. If the density change due to an infinitesimal isentropic displacement Δz exceeds the existing density change in the water column over the same distance, the water column is locally stable. This condition is equivalent to using thermal and salinity expansion coefficients computed at the local depth. The same holds for deep fresh water lakes.

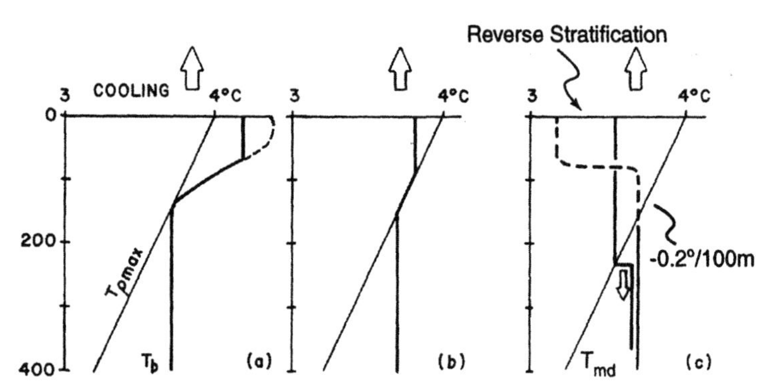

Figure 8.3.2. Thermobaric instability in a deep lake such as Lake Baikal during autumn cooling (Carmack and Weiss, 1991). When the lake is reverse-stratified as in (c), any mixing event which pushes the thermocline past the compensation depth causes deep mixing.

In a numerical model (with discretization in the vertical), one way to assess whether the water column is stable or unstable, from a mixing point of view, is to compute in situ density, $\rho_{in\ situ}(z) = \rho\ [T(z), S(z), p\ (z)]$, at each level. Then, to assess if a water column between levels z_1 and z_2 is stable or not, the densities ρ_1 and ρ_2 of water parcels from depths z_1 and z_2 would attain if moved isentropically to midlevel $(z_1 + z_2)/2$ can be computed. These values can then be used to compute the local value of N. This is preferable to using $\rho_{pot}\ (z_1,z_r)$ and $\rho_{pot}\ (z_2,z_r)$, as is the traditional practice. The transport of water masses is still through the use of conservation equations for potential temperature and salinity, albeit only the former for fresh water lakes.

8.4 LAKE NYOS—THERMAL AND CO_2 STRUCTURE

Some volcanic crater/caldera lakes are important because they pose a potential hazard. They are also interesting because of their interaction with the geological formations immediately beneath the crater. These lakes are freshwater-filled volcanic craters, often with a thick layer of sediments underneath through which dissolved gases and chemicals may seep into the lake waters. The walls are usually steep and the water level is usually much below the level of the rim. The depth of the water column is typically one to a few hundred meters deep. The area of the lake is usually a few to a few tens of square kilometers. Dissolved chemicals and gases make the water column density deviate considerably from that of fresh water. Often the dissolved gases stay dissolved because of the pressure of the overlaying layers of water. Exsolution of these gases is possible if this pressure is somehow released. It is only recently that volcanologists have begun to pay careful attention to mixing processes in caldera/crater lakes (see McManus et al., 1993).

A catastrophic release (Freeth and Kay, 1987; Kling et al., 1987) of an enormous volume (as much as 100 million m^3) of carbon dioxide on August 26, 1986, from Lake Nyos in Cameroon, West Africa (Figure 8.4.1), killed an estimated 1700 people as well as much livestock (Freeth, 1992). The gas is believed to have been released from the CO_2-rich waters of the lake, and after overflowing the rim at the northern end of the lake, the gas is believed to have

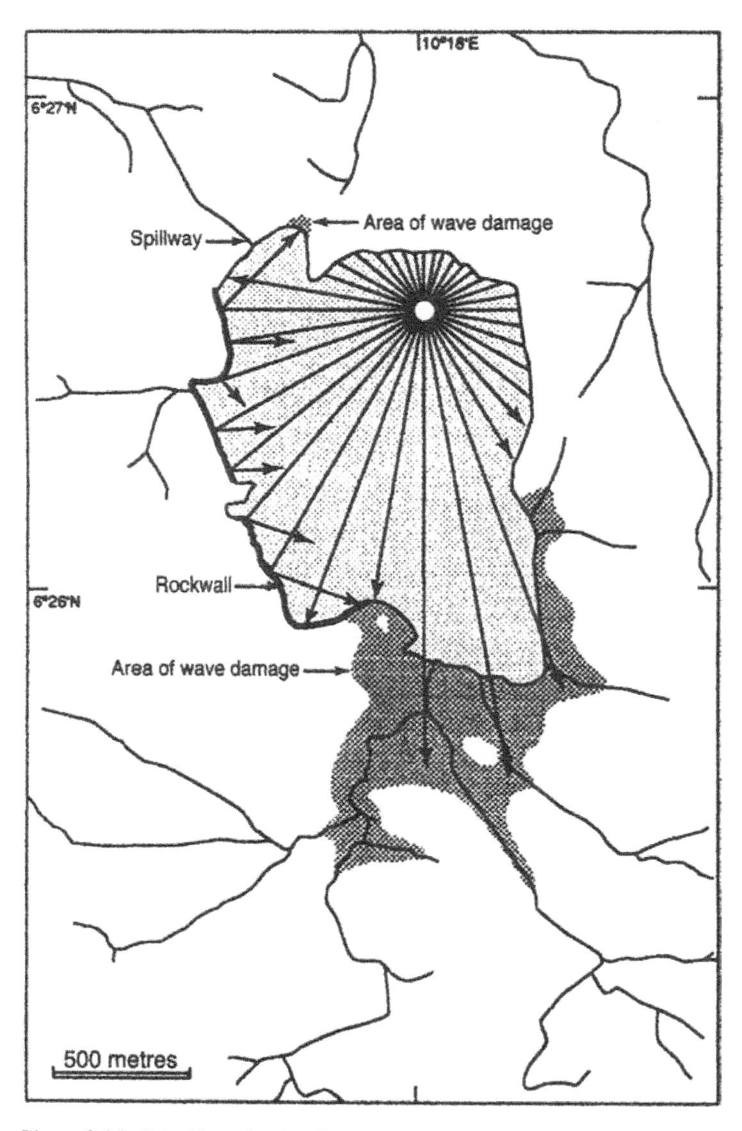

Figure 8.4.1. Lake Nyos, showing the area of damage (from Freeth and Kay, 1991).

flowed as a density current through the low-lying valleys extinguishing most animal life up to 20 km from the lake (Figure 8.4.2). There is now a general consensus that CO_2 was released when circulation within the lake somehow brought gas-rich bottom waters toward the surface (Freeth et al., 1990; Kling et al., 1987; Giggenbach, 1990; Tietze, 1992). The resulting reduction in pressure

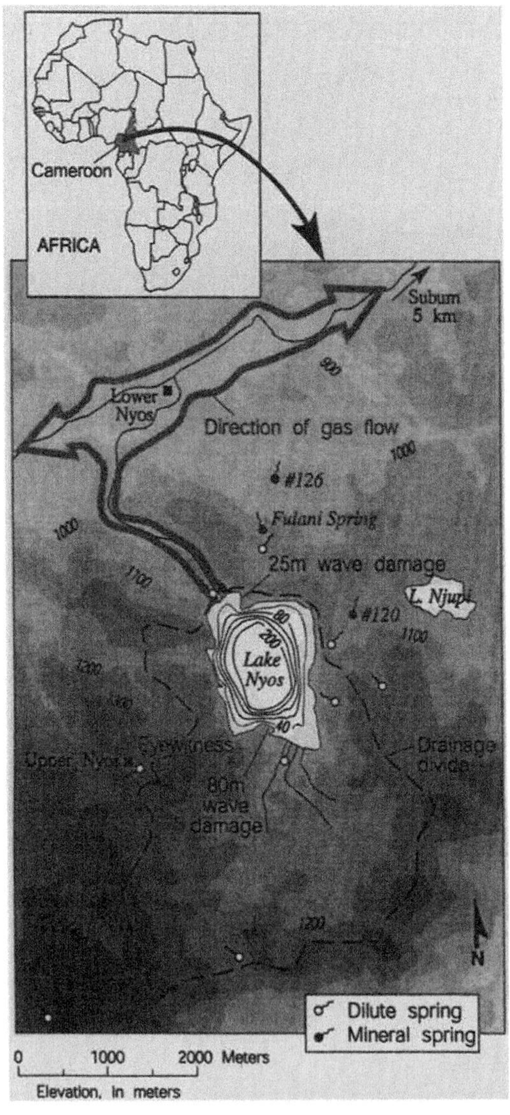

Figure 8.4.2. The route of gas flow in the valley below Lake Nyos (from Evans *et al.*, 1994).

would have caused massive exsolution of the dissolved gas and led to runaway outgassing of massive quantities of gas in a very short time (Tietze, 1992; Evans et al., 1994). But what triggered the release is still not known with certainty.

Whatever the trigger that caused the massive outgassing on that fateful day in 1986, the vertical density structure must have been so as to permit substantial vertical transport of gas-rich lake waters from the depths to the surface. Lake Nyos is now stratified (Evans et al., 1994) with warm, but gas-rich and therefore dense, water overlain by water which is cooler and contains less gas, and it was presumably similarly stratified prior to the 1986 disaster. Postdisaster observations show that CO_2 concentration near the bottom has been increasing steadily (Freeth et al., 1990).

Lake Nyos is roughly rectangular in shape, with a surface area of 1.4 km^2 and a maximum depth of 208 m (Figure 8.4.2). Figure 8.4.3 shows the vertical distribution of temperature and CO_2 and dissolved solid concentrations measured in the lake at various times since the disaster. The temperature and CO_2 and dissolved solids concentrations in the lake are nearly linear except in the bottom 50 m, where there is a steep increase, and in the top 50 m, where the concentrations are much reduced because of the interaction with the atmosphere and the temperature exhibits the influence of the seasonal cycle and river runoff (Figure 8.4.3).

Observations made immediately after the disaster show nearly linear temperature and dissolved gas and solids concentration distributions in the lake (shaded profiles in Figure 8.4.3). Seepage into the lake bottom layers through the sediments caused a rapid increase in CO_2 and dissolved solids concentrations in the bottom layers in the years immediately after the disaster. There has been a decrease in the rate of this increase in recent years, with values equilibrating with the values in the sediments immediately below. If the conditions in the lake before the disaster were similar to what they are now, with the bottom 50 m roughly in equilibrium with the sediments below, it is likely that most of the gas came from these bottom layers. The difference in the CO_2 content in the bottom 50 m at the present conditions and that in the immediate aftermath of outgassing (more than 10^8 kg of CO_2) is enough to account for the amount of gas released during the disaster (100 million m^3 at STP).

It is possible to model the vertical structure of the lake with the help of a numerical model. The model would be based on the hydrodynamics of the lake waters driven at its upper surface by wind, precipitation, and solar heating during the day and radiative cooling during the night, and by the input of heat, dissolved solids, and CO_2 at the bottom of the lake. Provided these forcings are known reasonably accurately or some reasonable scenarios can be postulated, and the initial structure of the lake is also known, it is possible to model the evolution of the vertical structure in the lake using a simple one-dimensional model of the lake.

Kantha and Freeth (1996) applied a comprehensive, one-dimensional numerical model of turbulent mixing (Kantha and Clayson, 1994) to describe the

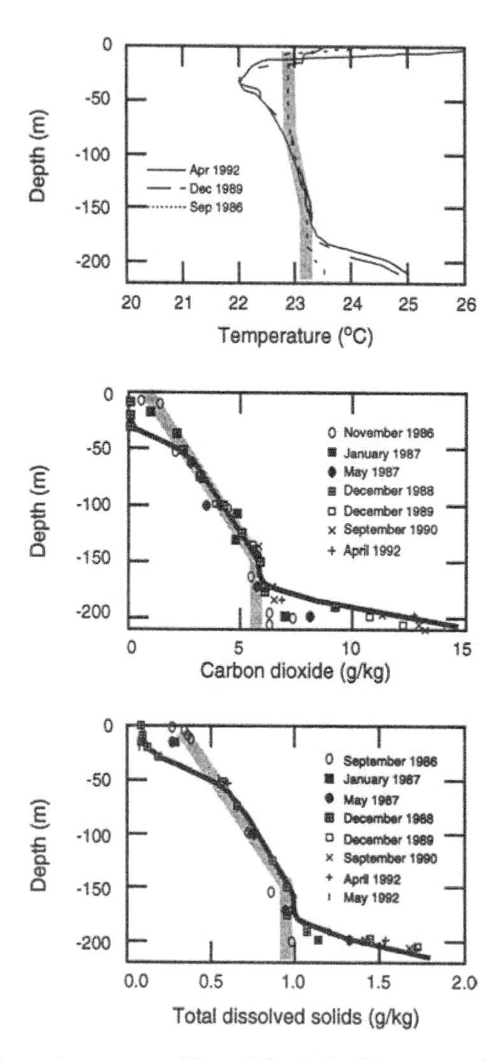

Figure 8.4.3. Observed temperature, CO_2, and dissolved solids concentration at various times since the 1986 disaster (from Kantha and Freeth, 1996).

evolution of the vertical structure of temperature and dissolved CO_2 in Lake Nyos since the outgassing event. A detailed description of the 1D mixing model can be found in Kantha and Clayson (1994) (see also Galperin et al., 1988). Briefly, the model involves solving the Reynolds averaged governing equations for momentum, and conservation of heat and salinity. Vertical mixing is modeled as the sum of background molecular and turbulent diffusion. The turbulence closure problem is effected at the second-moment level, and

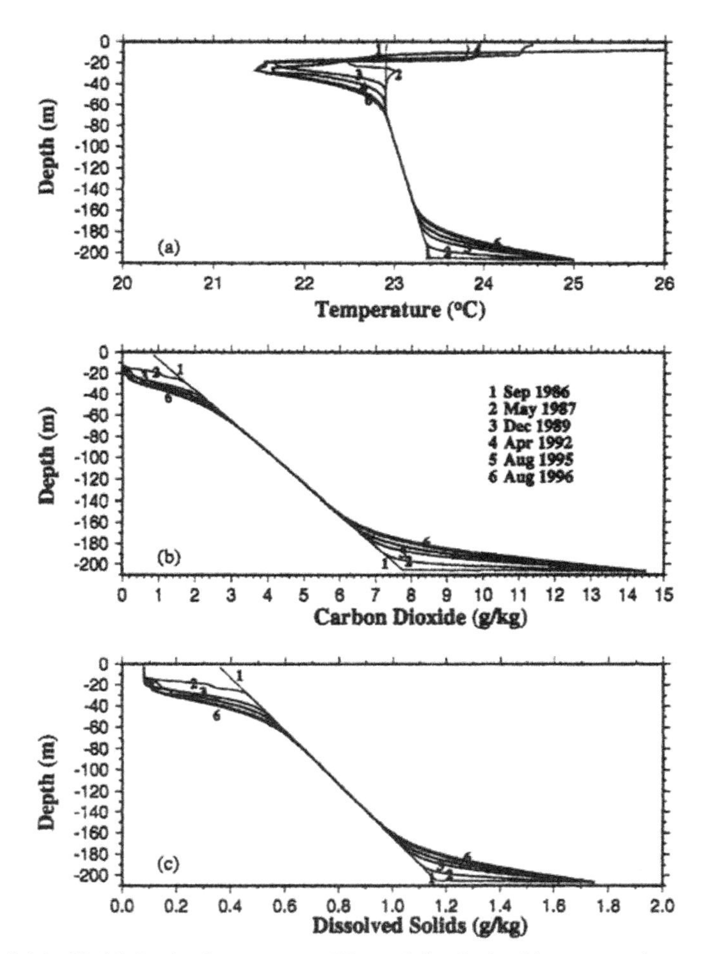

Figure 8.4.4. Model simulated temperature, CO_2, and dissolved solids concentrations at various times since the 1986 disaster (from Kantha and Freeth, 1996).

turbulent diffusion, which depends on the turbulence velocity and length scales, is computed using a two-equation model for these quantities. The influence of stable and unstable stratification on vertical mixing in the water column is properly modeled (see Chapter 2).

The model was initialized from observations made immediately after the disaster, integrated forward for 10 years under plausible atmospheric forcing conditions, and compared with observations made since then. The various scenarios they investigated suggest that it is unlikely that surface cooling might have triggered homogenization and outgassing, but rather some event that led to the destabilization and homogenization of the bottom layers. Thus the trigger for the gas release might have been related to some event near the lake bottom.

Figure 8.4.4 shows the temperature, the dissolved CO_2 concentration, and the dissolved solids concentration profiles in the water column at various times since the disaster as simulated by the model. The agreement with the observed profiles is quite reasonable.

8.5 CRATER LAKE, OREGON

Crater Lake in Oregon is a deep dimictic caldera lake, and is the deepest and one of the clearest lakes in the United States. Its natural beauty and the sensitivity of its level to climatic fluctuations, and therefore the possibility of its use as an indicator of climatic change, have made it the focus of intensive study in recent years (Drake et al., 1990). Consequently, there exists more observational data on mixing processes here than in any other deep dimictic lake in the world.

Crater Lake is situated in the Cascade Range (Bacon and Lanphere, 1990). The lake level is at an elevation of 1882 m above sea level. It was formed by the collapse of the caldera following the explosive eruption of Mt. Mazama 6950 years ago. It is nearly circular with an area of 53.2 km^2 (McManus et al., 1993). The lake consists of two basins, the North Basin with a maximum depth of 589 m in the northeast and the South Basin with a maximum depth of 485 m in the southeast, separated by a 425-m deep sill. The communication between the two basins may be important to the deep circulation in the lake. The lake circulation is also strongly influenced by winds blowing over the caldera, which has steep walls rising to an average elevation of 2100 m above sea level (Figure 8.5.1).

The catchment area is only 25% larger than the lake surface area itself and it therefore acts like a large rain gauge (Redmond, 1990). While observations of the lake level have been made since 1878, 25 years after its discovery, a consistent climatic record apparently exists only since 1931, according to Redmond. Since the lake is also an excellent evaporation pan, its level is a near-perfect indicator of "aridity," if proper allowance is made for seepage. While the lake level fluctuates by about 0.5 m between the spring maximum and the autumn minimum, the net change over a year depends very much on the precipitation, and can be an increase or decrease of close to a meter (Redmond, 1990). The records indicate that while the lake level at present is close to what it was at the beginning of the century, and has been nearly constant over the past 50 years, it dipped to as much as 4 m below the present level in the dust bowl years of the 1930s (Redmond, 1990).

Another interesting aspect of Crater Lake is that in spite of its location, it seldom freezes over during winter, although a sustained ice cover was observed once this century during the winter of 1949 (Kibby et al., 1968). This has very much to do with how the heat stored in the lake water column during the heating season is released during winter cooling by mixing processes in the upper layers.

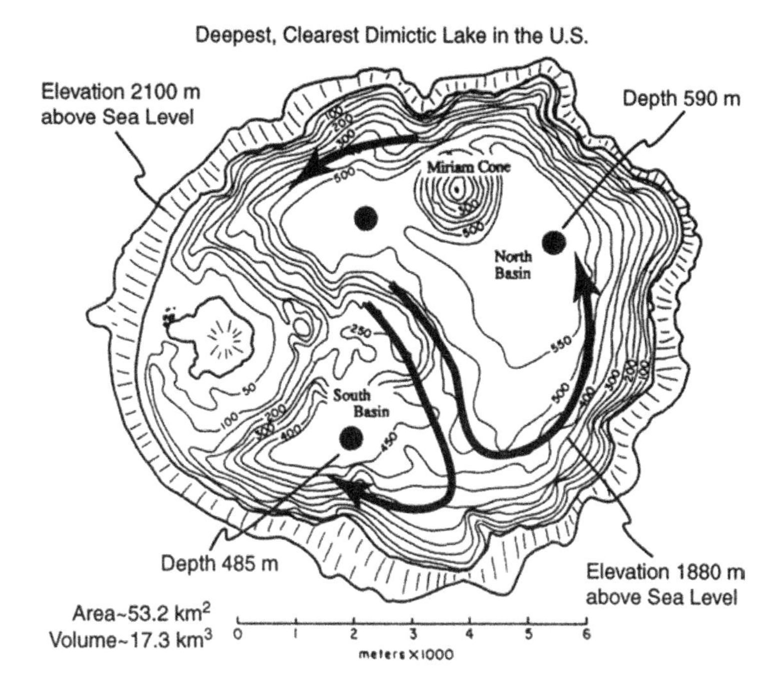

Figure 8.5.1. A sketch of Crater Lake, showing the bathymetry (from McManus et al., 1993, bathymetry from Byrne, 1965).

Normally, winter mixing processes tend to make available the large heat storage in the deep mixed layer and hence prevent the surface from cooling down to 0°C. Nearly continuous observations of the lake's thermal structure have been made in the North Basin since 1990 (McManus et al., 1993; Crawford and Collier, 1997) and these measurements, albeit limited, suggest that the surface seldom cools below about 2°C.

Wintertime deep mixing events are observed regularly by the mooring deployed in the North Basin (Crawford and Collier, 1997), and the dynamical mechanisms responsible for this deep ventilation are still being sorted out. Figure 8.5.2 shows the evolution of the temperature in the water column at a mooring in the North Basin from 1992 to 1994 (Crawford and Collier, 1998). Deep mixing events during each winter can be seen in this record. Thermobaric instabilities (Farmer and Carmack, 1981) are thought to play a central role in deep-water renewal and ventilation in deep dimictic lakes such as these, Lake Baikal being a classic example (Carmack and Weiss, 1991). However, the role of dissolved solids, that give rise to a stable stratification in the water column (McManus et al., 1993) against which any thermobaric instability has to work to mix the water column, makes straightforward interpretation of deep mixing

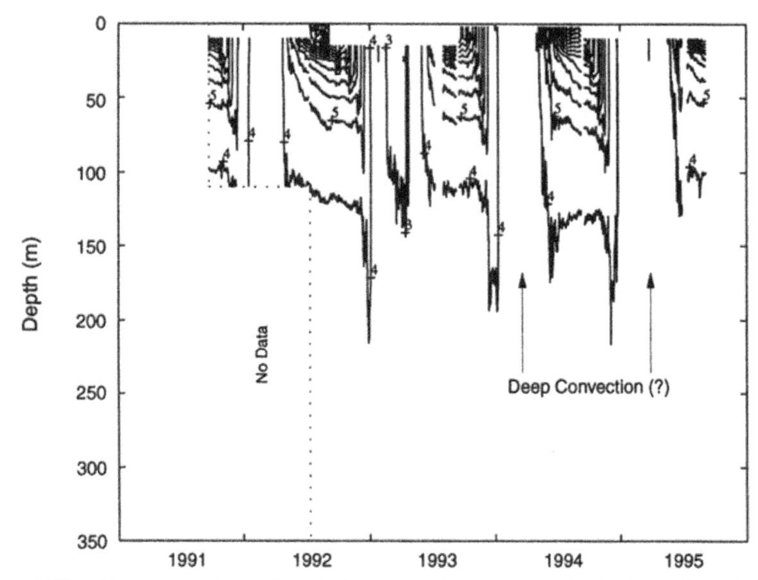

Figure 8.5.2. Temperature observations in the water column at a mooring in the north basin from 1992 to 1994. Note the deep convection events during each winter. Only temperature contours bertween 2 °C and 18 °C are plotted; contour interval is 1.0 °C (from Crawford *et al.*, 1999).

events as being solely due to thermobaric convection somewhat tenuous. In fact, the role of dissolved solids in stably stratifying the lake (Figure 8.5.3), and hence inhibiting mixing processes below the upper mixed layer, was pointed out only recently (McManus et al., 1993). However, salinity measurements in the lake are quite infrequent (except during summer), and inferences regarding the precipitation and evaporation/seepage in the lake are subject to large uncertainties, so that it is difficult to decipher the role of dissolved solids in lake mixing processes by observations alone.

Wintertime deep mixing events are important to the limnology of Crater Lake, especially the biological cycling of nutrients (Dymond et al., 1996). The remarkable clarity of the lake waters (Larson, 1972; Smith et al., 1973) attests to the fact that the major input of nutrients into the euphotic zone is from the deep waters of the lake. Radioisotope measurements (for example, Simpson, 1970) suggest a very short 1- to 2-year renewal timescale for the deep waters of the lake. More recent work on the oxygen budget in the lake (McManus et al., 1996) concludes that the mean residence time for deep water is 2–4 years and the deep lake is partially mixed with surface waters each year. Dymond et al. (1996) indicate that the upward flux of deep-water nitrate accounts for more than 85% of the new nitrogen input into the euphotic zone and the nutrients are recycled many times in the euphotic zone before settling to the bottom. This underscores the importance of deep mixing events, which not only ventilate and oxygenate

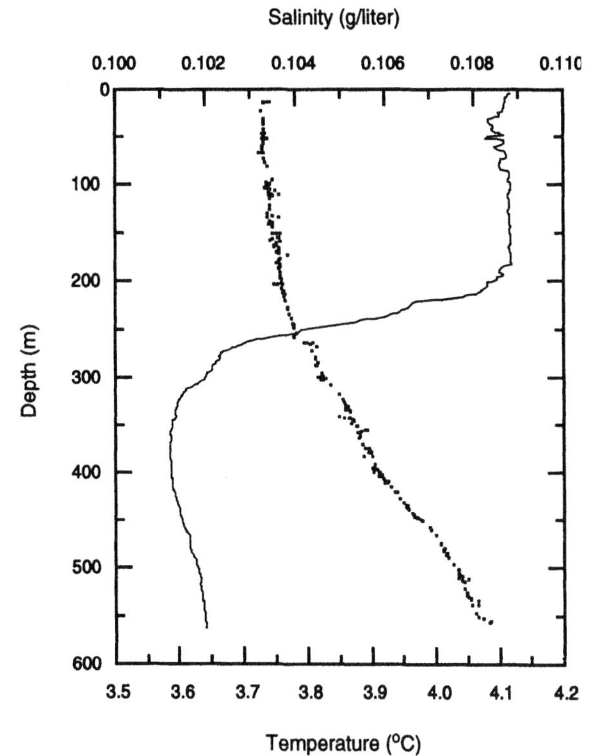

Figure 8.5.3. Typical vertical temperature and salinity profiles during winter (from McManus *et al.*, 1993). Note the stable stratification due to salinity below 200- m depths that might inhibit mixing.

deep lake waters, but also import nutrients critical to primary productivity into the euphotic zone (Crawford and Collier, 1997).

It is thought that TBI is quite important to mixing in dimictic lakes such as Lake Baikal, which has nearly zero salinity (Carmack and Weiss, 1991; but see Shimarev et al., 1993), and could be important if not central to mixing in Crater Lake, where there is significant stable salinity stratification in the water column. Accurate computation of local stability is therefore crucial to simulating TBI in the water column. Kantha et al. (1996) therefore modified the Kantha and Clayson (1994) 1D mixing model to more accurately compute local stability, based on Chen and Millero's (1986) equation of state. This modified model has been applied to simulate mixing in the North Basin, where the lake's thermal structure and meteorological conditions have been monitored since 1991. The evolution of the temperature and salinity in the water column has been investigated for a typical climatological year as well as specifically for the 1993/1994 and 1994/1995 autumn/winter/spring conditions (Kantha et al., 1996).

The model is driven by the wind stress derived from wind measurements over the North Basin, observed solar heating during the day, and the net cooling at the surface inferred from measurements near the crater rim (Crawford and Collier, 1996). Simulations were made for 1993/1994 starting from September 15, when salinity measurements are also available for model initialization, for a period of 240 days.

The fluxes of heat and dissolved solids at the lake bottom have been inferred previously by Williams and Von Herzen (1983) and more recently by McManus et al. (1993). McManus et al. infer a heat flux of about 1 W m^{-2} and a salt flux of about 5 $\mu g\ m^{-2}\ s^{-1}$ at the lake bottom from various measurements in the two basins. However, instead of using these fluxes to prescribe the conditions at the bottom of the lake, we prefer to prescribe Dirichlet conditions on the temperature and dissolved solids concentration at the water–sediment interface similar to what Kantha and Freeth (1996) did for Lake Nyos. Observations indicate that the temperature and salinity values at the bottom of the water column are fairly accurately prescribed as 3.66°C and 0.1085 psu.

The mixed layer model described above, as well as a fully 3D circulation model, has been applied to Crater Lake (Kantha et al., 1996). The most significant difference between the mixing model for the upper ocean and that used here is found in the equation of state. The equation of state used in this study is that due to Chen and Millero (1986), specifically designed for limnological applications (see Appendix B). An additional complication in these deep dimictic lakes is the possibility of thermobaric instability (Carmack and Farmer, 1982; Carmack and Weiss, 1991) in the water column.

Figure 8.5.4 shows the 1D model simulated temperatures in the North Basin during the winter of 1993/1994. Unlike the observed temperature records, no deep convection event is evident. This is possibly due to the stabilizing influence of the dissolved salts as well as the lack of the influence of wind-driven basinwide circulation in a 1D model. The profiles also lack the small scale variability evident in the observations. Figure 8.5.5 shows the temporal evolution of the lake temperatures at the surface and at 50 m during the winter of 1993/1994, as well as the mixed layer depth during the same period. The magnitude of the observed wind stress and solar radiation is also shown. The mixed layer deepens to about 200 m, but no overturning results.

The 3D model used for simulation of the lake thermal structure and its circulation is the University of Colorado version of the sigma-coordinate hydrostatic model developed originally at Princeton by George Mellor's group (see Chapter 9 of Kantha and Clayson, 1999). The core model has been previously documented by Blumberg and Mellor (1987) and Mellor (1991), and the curvilinear version by Kantha and Piacsek (1993, 1996). Several modifications have been made to the core model, including an improved mixed layer model (Kantha and Clayson, 1994) and a data assimilation module capable of assimilating both in situ XBT/CTD data and remotely sensed data such as MCSST (Horton et al., 1997), and altimetric SSH anomalies (Choi et al.,1995;

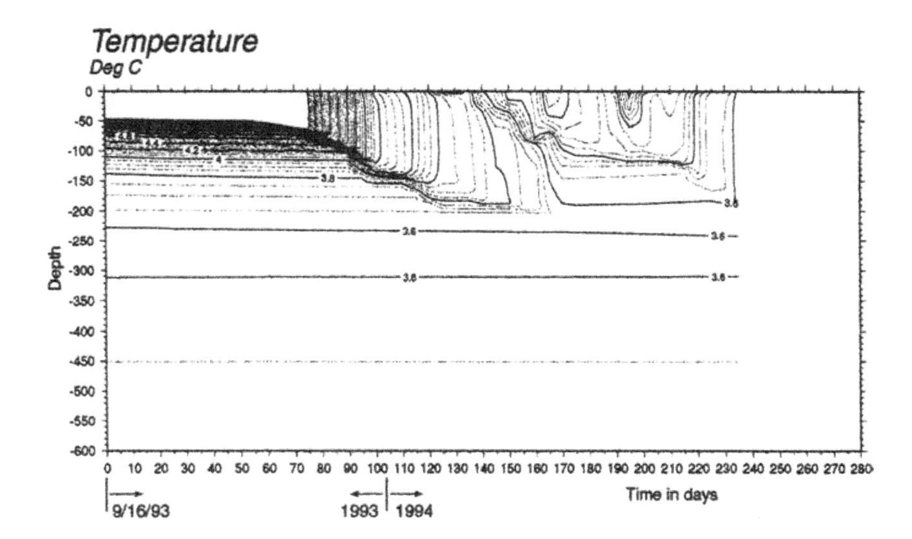

Figure 8.5.4. Temporal evolution of temperature in the water column in the North Basin from a 1-D mixing model (from Kantha and Crawford, 1999a) during the winter of 1993–1994. The mixed layer deepens to about 200 m, but no deep convection results.

Bang et al., 1996). Brief reviews of ocean modeling can be found in Kantha and Piacsek (1993, 1997), and a more extensive one in Kantha and Clayson (1999).

The bottom topography used in the 3D model has been subjected to a nonlinear filter to reduce the topographic gradients that might otherwise cause spurious along-slope currents in a sigma-coordinate model (Haney, 1991). Since spatial distributions of temperature and salinity have never been measured in the lake, we initialized the model with the profiles used for 1D simulations, assuming horizontal homogeneity at the start of the model. The model however develops horizontal gradients and baroclinicity rather quickly during the simulation.

The limitations of the 3D model chosen are twofold. First, a nonhydrostatic model (for example, Walker and Watts, 1995) more accurately simulates the deep convection processes. Second, the vertical component of the Coriolis acceleration normally ignored in ocean and lake models may be important in lakes such as Lake Baikal because of the weak stratification of its deep waters. Nevertheless, the inaccuracies resulting from these limitations are overwhelmed by those due to even small errors in surface forcing in deep dimictic lakes, so that in the final analysis, it is the accuracy with which one can specify the surface forcing that constrains the fidelity of the model simulations.

The resolution of the model is 0.5 km in the horizontal (21 × 18 grid). There are 38 sigma levels in the vertical, distributed to resolve the upper layers to a

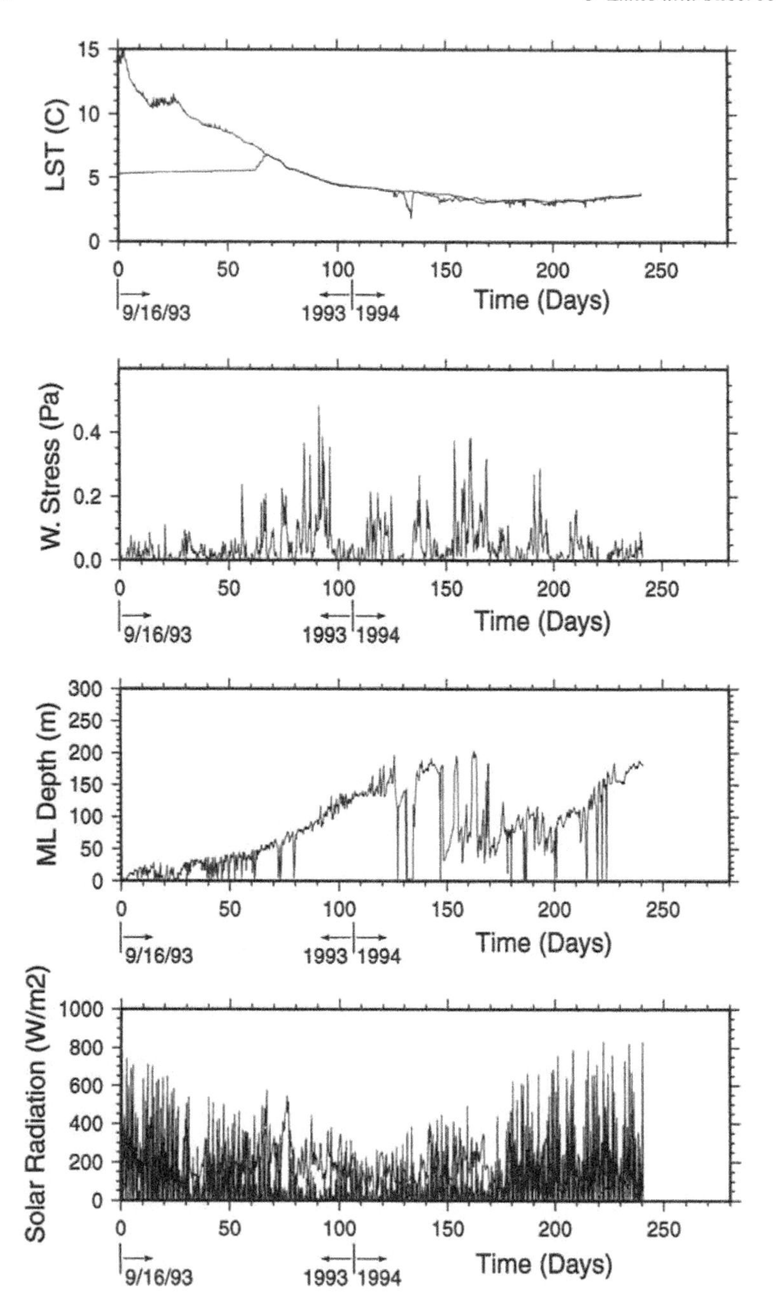

Figure 8.5.5. Time series plots of the (a) lake temperatures at 0 and 50 m, (b) wind stress magnitude measured, (c) the mixed layer depth, and (d) the solar insolation during the winter of 1993–1994 in the North Basin from a 1-D mixing model (from Kantha and Crawford, 1999a).

depth of 350 to 20 m. Good resolution in the vertical, especially at the base of the mixed layer, is essential to simulating the TBI accurately.

Surface forcing for the 3D model is identical to that used in the 1D model. This is simply because meteorological observations have been made at only one point on the lake and once again horizontal homogeneity needs to be invoked for simplicity. This may introduce some errors, since it is known that the wind whips around the caldera often, and it is likely that there are horizontal gradients in meteorological forcing. Nevertheless, as a first approximation, homogeneity might be adequate.

The bottom boundary conditions are slightly different from those used in the 1D model. We prescribe the same Dirichlet boundary conditions on temperature and salinity at grid points where the depth exceeds 400 m and therefore there is likelihood of a layer of sediment through which heat and dissolved solids can percolate into the water column. At grid points, where the bottom depth is less, we use Neumann conditions; we assume that the heat and salt fluxes at the bottom are zero.

Figure 8.5.6 shows the temperature at the deepest point in the North Basin during the winter of 1993/1994. Comparison with observed temperature profiles (Figure 8.5.2) shows that the 3D simulations are far more realistic and tend to produce deep convection that was notably absent in the 1D simulations. The presence of lake-wide gyre-like circulation may be responsible for this difference.

Numerical models such as these along with continued monitoring of the lake are a powerful combination in studies of chemical and nutrient exchanges between the upper and the lower layers of the lake.

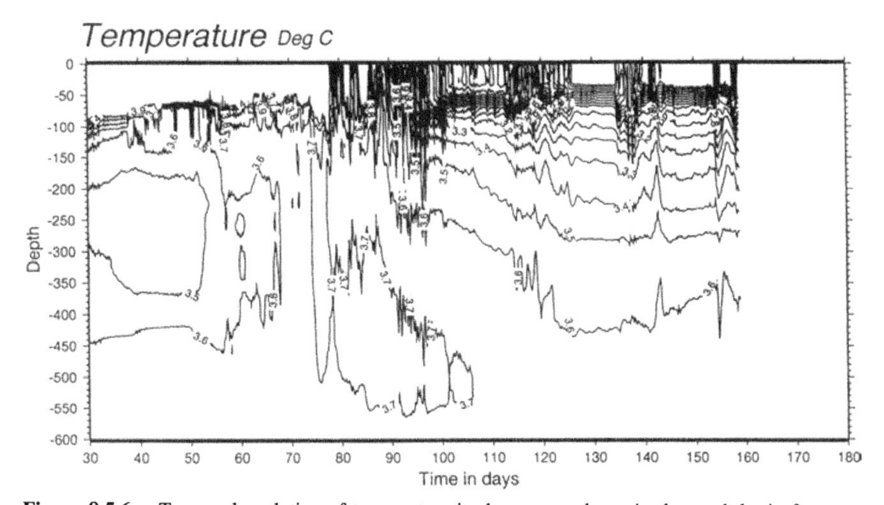

Figure 8.5.6. Temporal evolution of temperature in the water column in the north basin from a 3-D circulation model during the winter of 1993–1994. Mixing reaches deep into the water column. Only temperature contours less than or equal to 4 °C are plotted; contour interval is 0.1 °C.

8.6 LAKE BAIKAL

Lake Baikal in Siberia is a deep freshwater lake—the deepest and largest (by volume) in the world. Its importance lies in the fact that it holds about 20% of the fresh water on the globe [approximately 266,600 km^3, which is only 2.5% of all water on the globe; nearly two-thirds of this fresh water is locked up in glaciers and icecaps–see Postel et al. (1996)]. The sediments underneath the lake store the climatic record for the past 16 million years. Unfortunately, the lake is being increasingly polluted by industries situated around it.

Lake Baikal is a crescent-shaped lake that sits on top of a rift which is 9000 m deep and has collected some 7000 m of sediments over the past 16 of its 25 million year history (Figure 8.6.1). The lake consists of three deep basins, the central one being the deepest, 1637 m deep at its deepest point. The northern basin is comparatively shallow, 920 m deep, and the southern one is 1433 m deep. The sills between the southern and central, and the central and northern, basins are about 400 m deep, and therefore the deep waters in the various basins are essentially isolated from one another. The lake has an average width of 80 km and a length of roughly 635 km, and its volume is roughly 23,000 km^3. Of the several rivers flowing into Lake Baikal, the Selenga River that empties into the lake between the south and central basins is the most important since it supplies half the water flowing into the lake. The Upper Angara River that empties into the northernmost part of the lake is the next in importance. There is an average outflow of around 1800 $m^3 s^{-1}$ from the lake through the Angara River that flows out of the western shores of the southern basin. Most of the pollution in the lake is through Selenga River inflow and from industries around the town of Irkutsk, situated at the mouth of the Angara River, and hence pretty much confined to the southern portions of the lake. The circulation in the lake, especially the exchange between the southern and central basins, is therefore of considerable importance, from the point of view of pollutant transport.

For dynamical purposes, the concentration of dissolved matter in the lake is negligibly small and only the temperature in the water column is important. However, the decrease of the temperature of the density maximum with depth plays a very important role in mixing in the lake. The deep water in the lake is ventilated enough to have dissolved oxygen concentrations as high as 9 mg l^{-1} (Shimarev et al., 1993), and the mean age of waters below the upper 250 m is about 8 years (Weiss et al., 1991). Alternative estimates put the renewal timescale for deep water at around 11 years. However, the dynamical mechanisms responsible for this deep ventilation are unclear and still being sorted out.

Weiss et al. (1991) report that in July 1988, the near-bottom waters of the lake had higher chlorofluorocarbon concentrations than the intermediate waters, suggesting that there was direct ventilation of the deep. The observations of

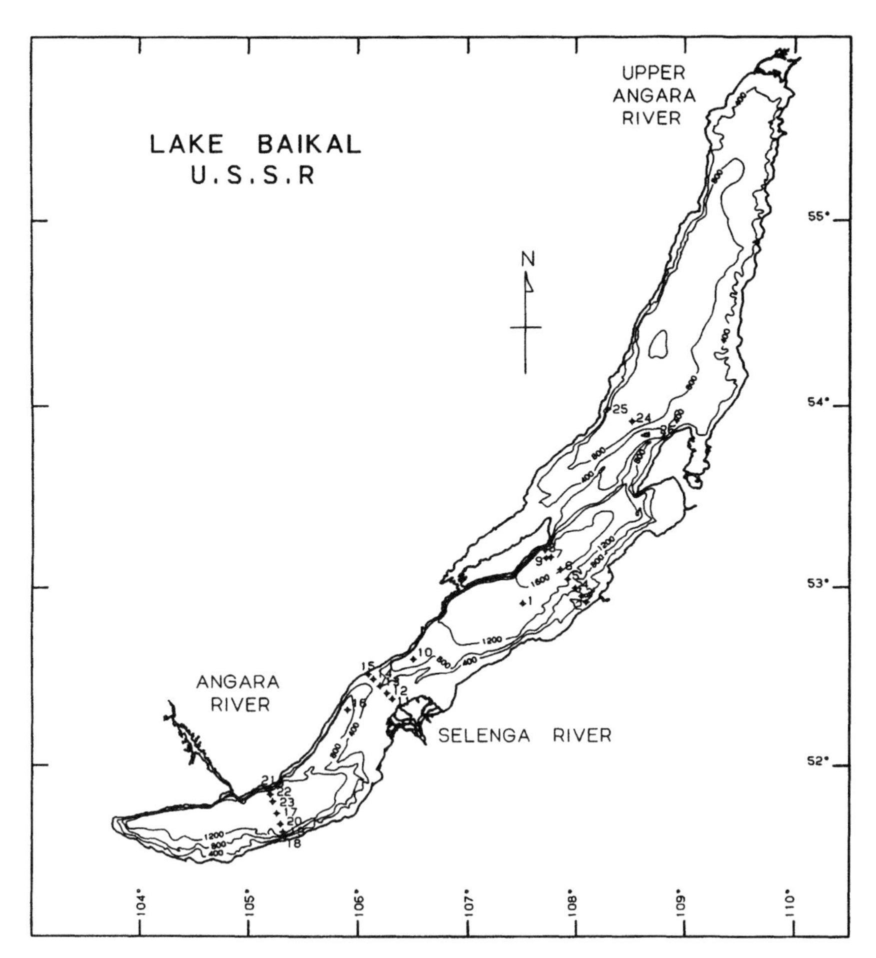

Figure 8.6.1. A sketch of Lake Baikal and its topography (reprinted from Carmack and Weiss, Deep Convection and Deep Water Formation in the Oceans, Copyright 1991, Page No. 223, with permission from Elsevier Science).

Shimarev et al. (1993) suggest that cabbeling instability near thermal bars might be responsible for initiating deep convection. Thermal bars are especially prevalent near river inflows and have timescales of a few weeks, enough to initiate deep mixing events through CI. Shimarev et al. (1993) observed such an instability in June 1991 in a thermal bar that exists near the outflow of the Selenga River in May to June. The near-shore temperatures ranged between 5 and 14°C, whereas the deep lake temperatures were between 3 and 3.8°C, conditions conducive to CI and formation of a cold dense water plume on the open side of the lake (see Figure 8.3.1). Once this mixed water begins to sink, it can reach depths where its temperature exceeds the T_{md} and TBI ensues, causing the water to sink deeper, down to the bottom, where it can move along the slope to depths where its density equals that of the surrounding waters (Shimarev et al., 1993). The convective plume may not always reach the lake bottom, however; in June 1991 it did not, while in 1988 it apparently did. Shimarev et al. suggest that since large spring thermal bars are a regular feature in Lake Baikal, the above mechanism must cause significant deep mixing in Lake Baikal.

Because of the cold air temperatures during winter (average of –20°C), ice forms in the lake and grows to a thickness of about a meter. The lake extends over 5° in latitude and this latitudinal variation makes the ice form first in the North Basin around November, and the thaw begins a month later there than in the South. The ice begins to thaw in April in the South. Strong westerlies, which tend to pile up ice along the eastern shores, help disperse and melt the ice cover. By mid-May, the ice is gone, although it lingers till early June in the North Basin. The presence of ice cover during spring heating has important consequences to mixing. The ice cover tends to shield the water from wind forcing and therefore even though the reverse stratification present holds the potential for TBI, the penetrative convection due to solar heating through the ice cover of the upper layers may not be able to push the base of the mixed layer below the compensation depth to initiate TBI , whereas during autumn, once reverse stratification develops, strong mixing events can lower the mixed layer base to the compensation depth and initiate TBI (Carmack and Weiss, 1991). Therefore, deep mixing events due to TBI are more likely during autumn cooling (see Figure 8.3.2), if we exclude those due to spring thermal bars.

The temperatures in the North Basin are generally higher, and those in the Central Basin lower, than the rest, presumably due to stronger winds and more intense deep convection in the latter, and low winds in the former.

Figure 8.6.2 shows an idealized 1D model simulation (see Sections 8.4 and 8.5 for model details) of the evolution of temperature in a deep freshwater lake during winter. The model simulations start from a stratified condition with the mixed layer depth close to the compensation depth. A steady wind stress and a steady heat loss are imposed at the surface. Two cases are studied, one with a net cooling rate of 25 W m^{-2} and the other with 50 W m^{-2}. All other conditions are kept the same and the evolution of the temperature structure is modeled. Figure 8.6.2 shows the temperature profiles at 5-day intervals for both cases; deep mixing occurs in the first case but not in the second. Instead reverse stratification builds up quickly for the second case and prevents deep convection from happening. Thus if the rate of cooling is not too strong when the lake passes through the critical overturning phase, TBI ensues and mixes the entire water column. Thus the rate of cooling (and the intensity of wind mixing) during the critical overturning phase of the lake evolution can influence the occurrence of deep convection and mixing events in a deep freshwater lake. These simulations highlight the importance of TBI in a deep freshwaterlake such as Lake Baikal.

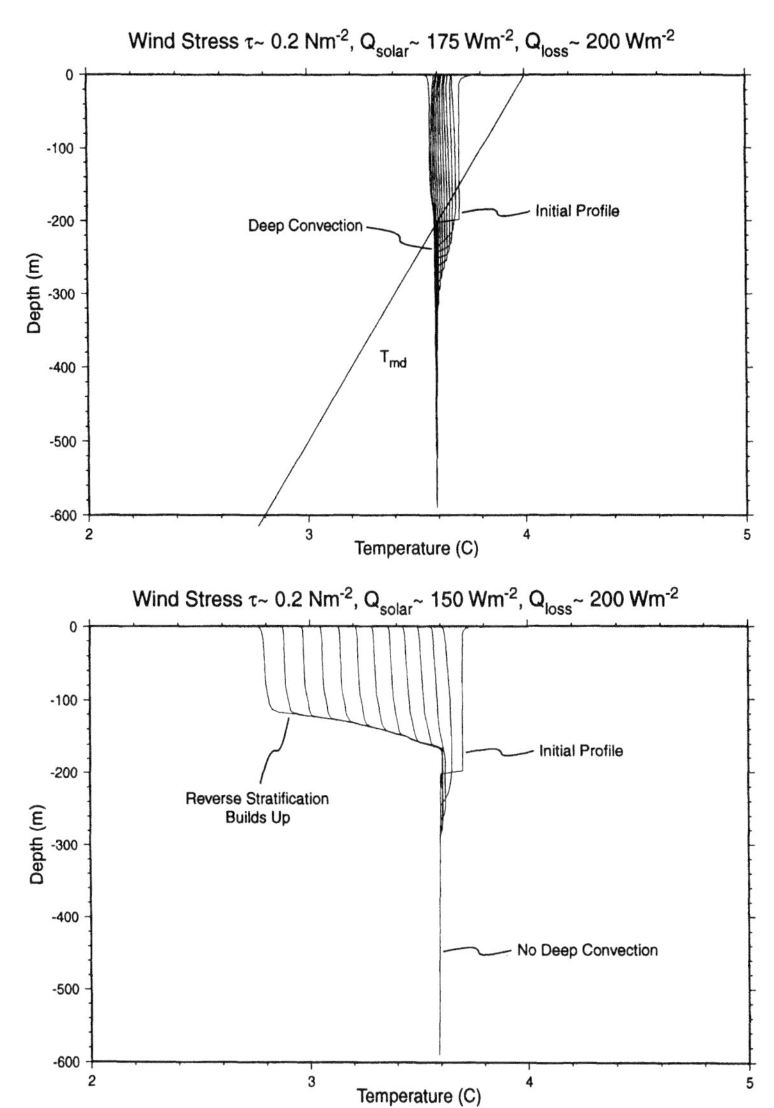

Figure 8.6.2. Temperature profiles from a 1-D mixing model (Kantha and Crawford, 1999b) for (top) a weak cooling rate and (bottom) a strong cooling rate. Note the deep mixing event in the former and a rapid buildup of reverse stratification in the latter.

LIST OF SYMBOLS

θ	Latitude
ρ	Density of water
a	Internal Rossby radius of deformation (c/f)
b	Lake width
c	Internal gravity wave speed
g	Acceleration due to gravity
h	Surface layer thickness
k_T, k_S	Kinematic molecular diffusivities of heat and salt
t	Time
H	Lake depth
L	Lake length
T	Temperature
T_{md}	Temperature of maximum density of fresh water
U	Horizontal velocity
W_e	Wedderburn number

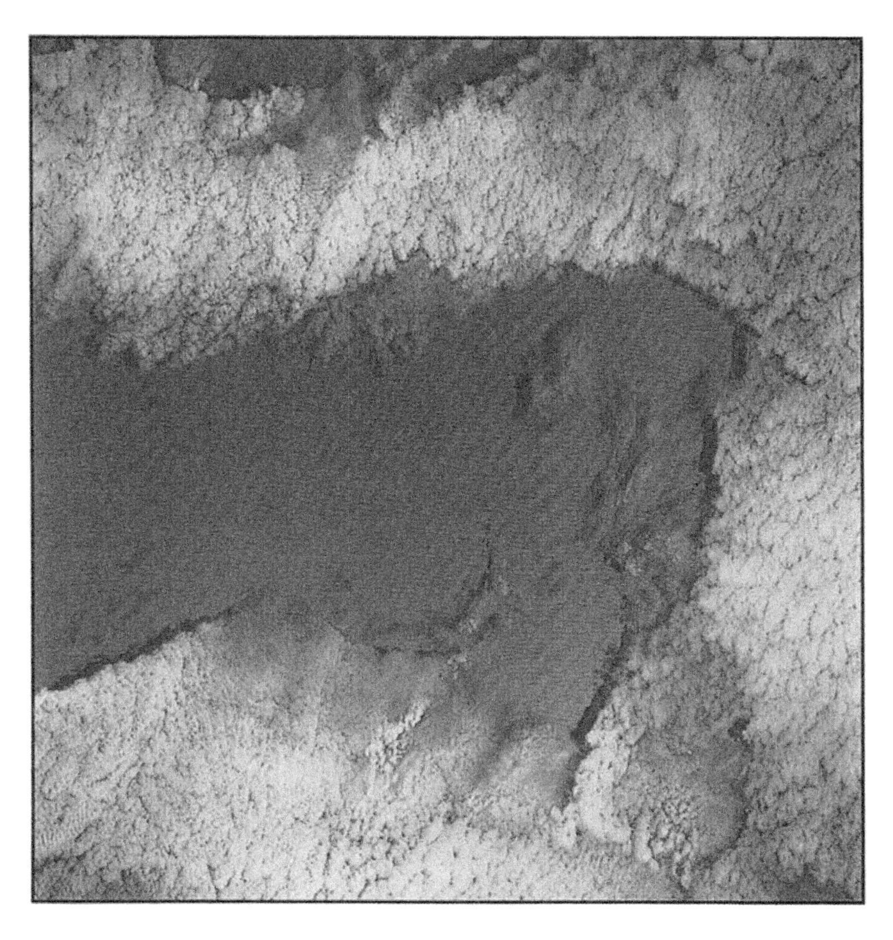

Plate I. A swell in the northeast Pacific, as seen from the space shuttle Challenger in 1984.

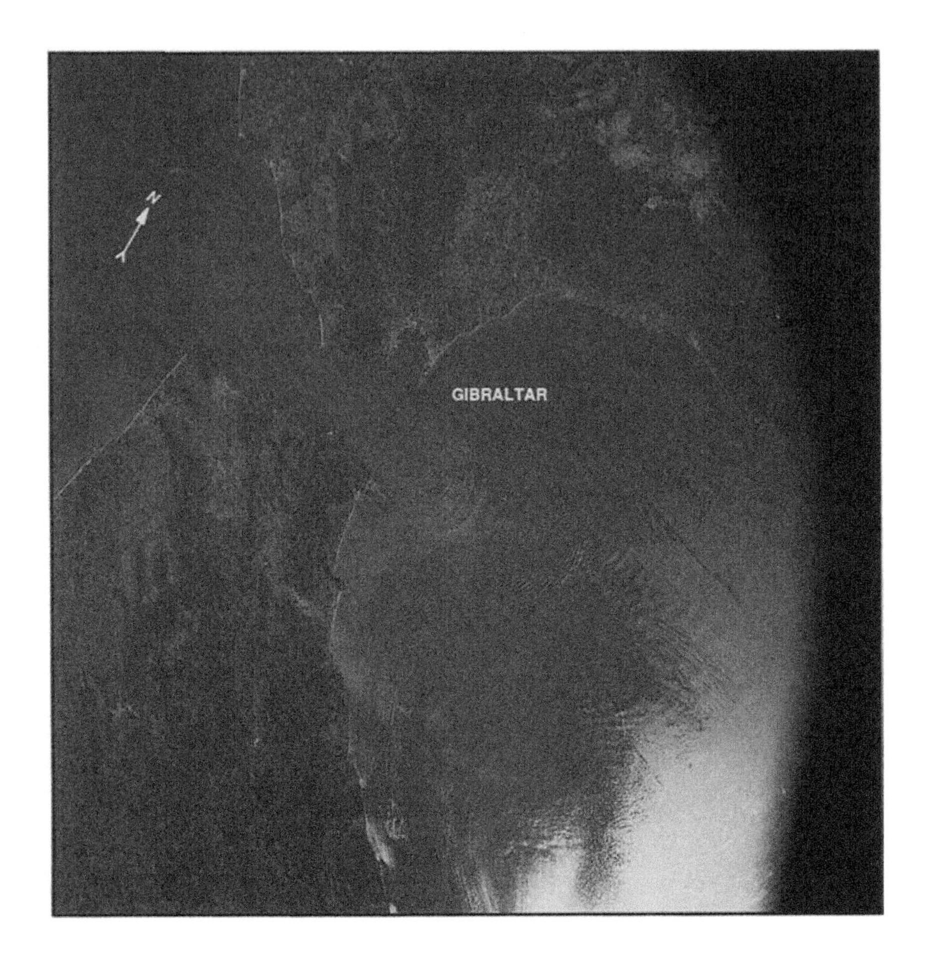

Plate II. Solitans, over the eastern part of the Alboran Sea, as seen from the space shuttle Challenger in 1984.

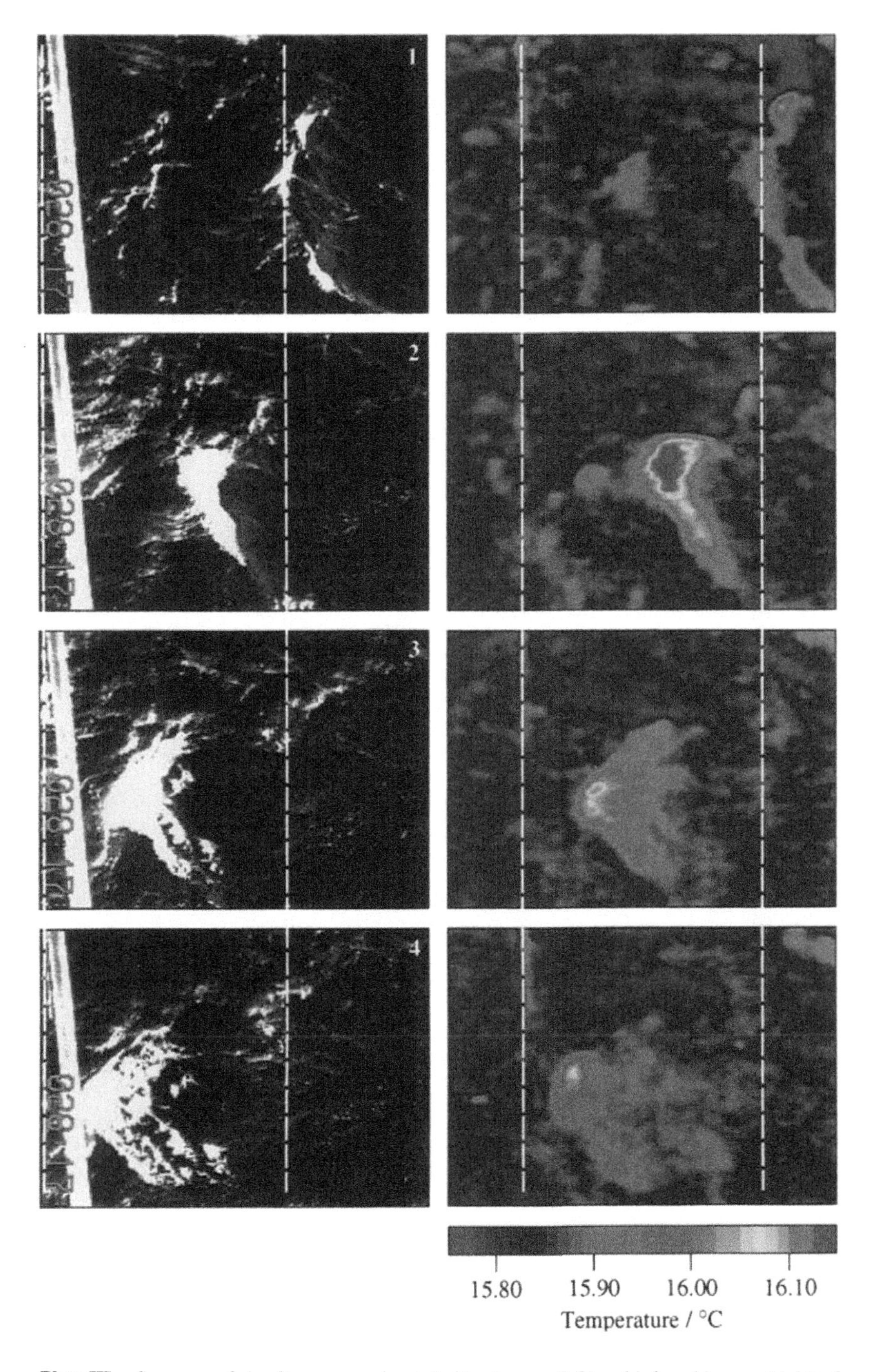

Plate III. Sequence of simultaneous, co-located video images (left) and infrared images (right) of a breaking wave in the open ocean. Image size is approximately 5m x 10m. The breaking wave is propagating from right to left; time increases down the page and spans 1.6 s. The whitecap in the video images appears as the warmest region in the infrared images. The infrared signature of the turbulent wake results from the disruption of the cool skin layer by the breaking wave.

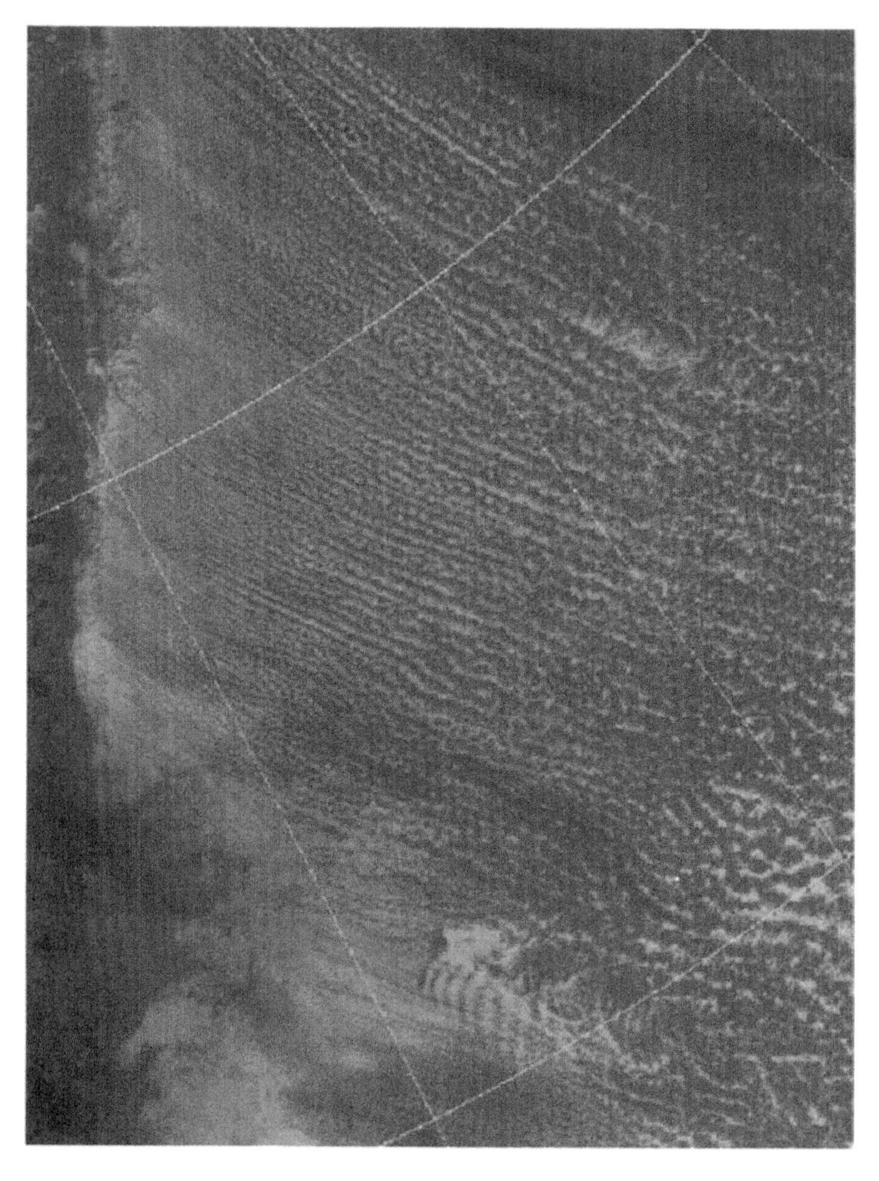

Plate IV. This NOAA AVHRR IR image depicts cloud structure associated with horizontal roll vortices in the atmospheric boundary layer. The mean roll spacing is about 5 km. (Figure courtesy of J. A. Johannessen.)

Appendix A

Units

A.1 THE BUCKINGHAM PI THEOREM

As in most areas of scientific endeavor, we seek quantitative functional relationships between various parameters governing a small scale process. We humans prefer to think of such relationships in terms of dimensional parameters. For example, what is the phase speed in meters per second of a deep water surface wave with a frequency of 10 cycles per second? There exists, of course, an answer in dimensional units. However, Nature does not recognize dimensional quantities in functional relationships governing physical or other processes. After all, dimensional standards are defined by humans and agreed to universally (a variable standard is not very useful), but without Nature's consent and subject always to potential revisions. That is why only functional relationships involving dimensionless quantities are meaningful and useful. Also constants in a functional relationship should be dimensionless to be of much utility. And the first thing to check in an equation is whether the terms on both sides of the equation have the right units.

In studying small scale processes we often look at scaling laws. These laws should involve functional relationships between dimensionless quantities. An excellent way of deriving such relationships is by the use of the Buckingham PI theorem, which states that if a physical process involves N dimensional quantities involving M units, it is possible to define (M − N) dimensionless quantities that should of course be related to each other functionally. It is a powerful tool for analyzing physical processes and organizing observational data. The PI theorem cannot provide the constants or the form of functional relationships between the dimensionless quantities. However, if the functional relationship involves power law relationships, it often provides answers with only the constants that need to be determined empirically or theoretically. The following examples demonstrate the utility of the theorem.

Take the case of a surface water wave. We need to find the frequency n of the wave as a function of its other characteristics. Since we suspect that gravitational restoring forces are involved, gravitational acceleration g is a parameter in the problem as well as the density difference across the air–sea interface, which involves ρ_a and ρ_w. For sufficiently small waves, surface tension effects cannot be neglected and therefore surface tension γ (its kinematic value) is a parameter in the problem. The only other parameters are its wavelength λ (or equivalently wavenumber k), its amplitude a , and the water depth d. So there are three units involved, length [L] , mass [M], and time [T], and eight quantities, n (dimension $[T^{-1}]$), k (dimension $[L^{-1}]$), g (dimension $[LT^{-2}]$), d (dimension [L]) and a (dimension [L]), γ (dimension $[L^3T^{-2}]$), ρ_a (dimension $[ML^{-3}]$), and ρ_w (dimension $[ML^{-3}]$). It is possible to find five dimensionless quantities that should be related to each other. It is easily verified that they are $n^2/gk, \gamma k^2/g, \rho_a/\rho_w$, kd, and ka , and therefore

$$n^2/gk = f(\gamma k^2/g, \rho_a/\rho_w, kd, ka) \tag{A1}$$

is the so-called dispersion relationship that relates the wave frequency to its wavenumber and other parameters. Of course the PI theorem (or dimensional analysis) cannot provide the functional relationship or the constants involved. For that one has to appeal to a mathematical theory. But observational data on waves could be organized in the above fashion to determine both the constant and the functional relationship. However, we can simplify and seek limiting cases. If we recognize that the density of air is three orders of magnitude smaller than that of water and it is the air–sea density difference that matters, then ρ_a/ρ_w is irrelevant since $(\rho_w - \rho_a)/\rho_w \sim 1$ and hence

$$(n^2/gk) = f(\gamma k^2/g, kd, ka) \tag{A2}$$

If we now restrict our attention to infinitesimal waves, the amplitude drops out of the relationship and we have

$$(n^2/gk) = f(\gamma k^2/g, kd) \tag{A3}$$

Theory (see Chapter 5) gives

$$(n^2/gk) = [1 + (\gamma k^2/g)]\tanh(kd) \tag{A4}$$

If we now restrict our attention to waves long enough so that surface tension is not a parameter in the problem, then the nondimensional constant involving γ

drops out of the functional relationship and we have

$$(n^2 / gk) = f(kd) \tag{A5}$$

Now let us look at deep water waves. We suspect then that the depth d should not be a parameter in the functional relationship. Then

$$(n^2 / gk) = c_1 \tag{A6}$$

This is indeed the dispersion relationship for deep water waves and the constant $c_1 = 1$ from theory. If, on the other hand, we look at shallow water waves, we suspect then that only d should be relevant in the relationship for phase speed $c = n/k$ and not k. If we rearrange the functional relationship replacing n by c,

$$(c^2 / gd) = f(kd) \tag{A7}$$

and since k should drop out we get

$$(c^2 / gd) = c_2 \tag{A8}$$

where $c_2 = 1$ by theory. Shallow water waves are nondispersive and their phase speed is dependent only on the water depth and is independent of their wavenumber. For surface-tension-dominated (capillary) waves, only γ should be relevant and not g, since restoring forces are due only to surface tension and not gravitational forces. Also the water depth is irrelevant. Then in the modified functional relationship (A3)

$$(n^2 / gk) = f(\gamma k^2 / g) \tag{A9}$$

g should drop out. This happens if

$$(n^2 / gk) = c_3(\gamma k^2 / g) \tag{A10}$$

This of course gives

$$n^2 = c_3(\gamma k^3) \tag{A11}$$

as the dispersion relation for capillary waves, where $c_3 = 1$ by theory.

Dimensional analysis is therefore a very powerful tool in investigating natural processes. It provides a good idea of relevant nondimensional parameters and often the relationship to the extent of an unknown constant! It is always a good idea to explore what nondimensional parameters govern a particular process by employing the PI theorem. If Galileo had knowledge of this powerful tool, he could have easily deduced that the period of a simple pendulum T would not depend on its mass because of the disparate units and should depend on only its length l and of course the gravitational constant g!

$$T^2 \sim (l / g) \tag{A12}$$

Salient results in as complex a phenomenon as turbulence, such as the celebrated Kolmogoroff –5/3 law for the turbulence spectrum in the inertial subrange, were originally derived not by using sophisticated mathematical theories or the most powerful supercomputers but simply by physical intuition and the PI theorem. For a delightful look at how nondimensional numbers such as the Froude number can be used to explore the constraints physical laws impose on biological organisms, see Vogel (1988, 1998).

A.2 USEFUL QUANTITIES

A.2.1 SI (INTERNATIONAL SYSTEM) UNITS AND CONVENTIONS

The basic SI units for our applications are

Length	–	meter (m)
Mass	–	kilogram (kg)
Time	–	second (s)
Temperature	–	Kelvin (K)

From these, units for all other quantities can be derived. The derived units are

$1\ \text{rad} = 1\ \text{m m}^{-1}$	– Angle
$1\ \text{Hz} = 1\ \text{s}^{-1}$	– Frequency
$1\ \text{N} = 1\ \text{kg m s}^{-2}$	– Force
$1\ \text{Pa} = 1\ \text{N m}^{-2}$	– Pressure (also stress)
$1\ \text{J} = 1\ \text{N m} = 1\ \text{kg m}^2\ \text{s}^{-2}$	– Energy (also work, heat)
$1\ \text{W} = 1\ \text{J s}^{-1} = 1\ \text{kg m}^2\ \text{s}^{-3}$	– Power (also heat flux)

Prefixes are as follows:

10^{-1}	deci (d)	10^{1}	deka (da)
10^{-2}	centi (c)	10^{2}	hecto (h)
10^{-3}	milli (m)	10^{3}	kilo (k)
10^{-6}	micro (u)	10^{6}	mega (M)
10^{-9}	nano (n)	10^{9}	giga (G)

10^{-12}	pico (p)	10^{12}	tera (T)
10^{-15}	femto (f)	10^{15}	peta (P)
10^{-18}	atto (a)	10^{18}	exa (C)
10^{-21}	zepto (z)	10^{21}	zetta (Z)
10^{-24}	yocto (y)	10^{24}	yotta (Y)

Symbols for physical quantities are italicized, but units are not (for example, temperature T in degrees Celsius, °C). Unit symbols are lower cased, unless derived from proper names (for example, 1 watt is 1 W and 1 kilogram is 1 kg), not pluralized (for example, 3 seconds is 3 s), and not followed by a period. When prefixes are used, no space is allowed in between (for example, 3 nanoseconds is 3 ns, not 3 n s), compound prefixes are to be avoided (for example, 1 terawatt is 1 TW, not 1 MMW), and when written out fully, should begin with a lower case (for example, millijoules). Multiplication of units may be indicated by the use of a space in between, and division by a negative exponent (for example, 1 newton meter is 1 N m and 1 newton per meter squared per second is 1 N m^{-2} s^{-1}). When written out in full, no mathematical operations should be commingled (for example, 1 watt per meters squared, not 1 watt/meter^2), and a product denoted by a space in between (for example, 1 newton meter). Large numbers are to be divided by spaces between each three digit groups (for example, 237 865 540) and a decimal marker is to be always preceded if not preceded by a zero (for example, 0.8 kg, not .8 kg). Units and the numbers are to be divided by a space (for example, 10 kg, not 10kg). Prefixes are to be used whenever possible to bring the numerical value between 0.1 and 1000 (for example 180 kW, not 180 000 W). The word "degree" or its symbol is not used in conjunction with the unit kelvin (for example, 215 kelvin or 215 K, not 215 degrees kelvin or 215°K), but only with Celsius (degrees Celsius or °C).

Following the above conventions (Nelson, 1982) avoids needless proof-reading corrections by the copy editors. Also in the era of electronic self-publishing on the Internet, it is important to adhere strictly to standards that have been so painstakingly arrived at after decades of experimentation.

A.2.2 Useful Conversion Factors

1 pound (mass) = 0.453 593 kg
1 inch = 2.54 cm
1 foot = 0.3048 m
1 fathom = 6 ft = 1.8288 m
1 mile = 1.609 344 km
1 nautical mile = 1.85318 km
1 acre = 4047 m^2
1 gal = 3.786 liter
1 bbl = 31.5 gal = 119.26 liter

1 mile hr^{-1} = 0.447041 m s^{-1} = 1.609 344 km hr^{-1}
1 ft s^{-1} = 0.3048 m s^{-1}
1 knot = 0.51477 m s^{-1} = 1.853 172 km hr^{-1}
1 pound (force) = 4.4482 N
1 bar = 10^5 Pa = 10^5 N m^{-2} = 10^6 dyn cm^{-2}
1 psi = 6894.724 Pa
1 horsepower = 746 W
°F = 32.0 + 1.8 × °C

1 solar day = 86,400 s
1 sidereal day = 86,164 s
1 degree = 0.01745 (= 2π/360) rad
1 Hz = 2π rad s^{-1}
1 cm s^{-1} = 0.864 km day^{-1}
1 N = 10^5 dynes
1 Pa = 1 N m^{-2} = 10 dyn cm^{-2} = 10^{-2} mb
1 mb = 10^2 Pa; 1 m of water ~ 10.1 kPa
1 standard atmosphere = 101.325 kPa (14.7 psi, 76 cm Hg)
1 J = 1 N m = 1 W s = 10^7 ergs = 6.24×10^{18} ev = 0.2389 cal
1 W = 1 J s^{-1}
1 petawatt = 10^{15} W
1 ly day^{-1} = 0.484 W m^{-2}
1 metric ton (t) = 1000 kg
1 hectare (ha) = 10^4 m^2
0°C = 273.15 K

A.2.3 Useful Universal Constants

π =	3.141 592 653 589 793 238 462 643
e =	2.718 281 828 459 045 235 360 287
γ =	0.577 215 664 901 532 860 606 512 (Euler constant)

Universal gas constant R = 8.314 51 J mol^{-1} K^{-1}
Gas constant for dry air = 287.05 J kg^{-1} K^{-1}
Gas constant for water vapor = 461.53 J kg^{-1} K^{-1}
Molecular weight of dry air = 28.97
Molecular weight of water vapor = 18.016
Avagadro constant = 6.02252×10^{23} mol^{-1}
Speed of light = 2.9979246×10^8 m s^{-1}
Newton's gravitational constant G = $6.672\ 590 \times 10^{-11}$ N m^2 kg^{-2}
Solar constant = 1376 W m^{-2} (in 200- to 4000-nm range; 40% of it in visible 400- to 670-nm range, 1.78×10^{17} W)

Stefan–Boltzmann constant $\sigma = 5.670\,51 \times 10^{-8}$ W m^{-2} K^{-4}
Boltzmann constant k = 1.38×10^{-23} J K^{-4}
Planck's constant = 6.6262×10^{-34} J s

A.2.4 Useful Geophysical Constants

Radius a = 6371 km (6378.139 km, equatorial; 6356.754 km, polar; Hydrostatic flattening, 1/299.638)
Radius of the core (equatorial) = 3486 km (flattening = 1/392.7)
Mass of Earth = 5.977×10^{24} kg (mantle = 4.05×10^{24} kg, core = 1.90×10^{24} kg, crust =2.5×10^{22} kg)
Average density ρ_E = 5517 kg m^{-3}
Average density of the upper mantle = 3330 kg m^{-3}
Average density of the crust = 2850 kg m^{-3}
Polar moment of inertia of the entire Earth C ~ 8.0376×10^{37} kg m^2
Equatorial moment of inertia of the entire Earth A ~ 8.0115×10^{37} kg m^2 (C – A = 2.610×10^{35} kg m^2)
Polar moment of inertia of the core C_c ~ 0.9140×10^{37} kg m^2
Equatorial moment of inertia of the core A_c ~ 0.9117×10^{37} kg m^2 ($C_c - A_c$ = 2.328×10^{34} kg m^2)
Polar moment of inertia of the mantle C_m ~ 7.1236×10^{37} kg m^2
Equatorial moment of inertia of the mantle A_m ~ 7.1000×10^{37} kg m^2 ($C_m - A_m$ = 2.377×10^{34} kg m^2)
Orbital dates: spring (vernal) equinox = March 20, autumnal equinox = September 22, summer solstice = June 20, winter solstice = December 21, perihelion = January 3, aphelion = July 4
Obliquity of Ecliptic = 23° 26′
Rotation rate = 7.292115×10^{-5} rad s^{-1}
Mean orbital velocity = 29.77 km s^{-1}
Chandler wobble period = 433.3 days
Average surface temperature = 288 K
Average albedo = 0.33
Sidereal day = 23 hr 56 m 4.09 s
Sidereal year = 365.25 days
Tropical year = 365.2422 days
Current angular momentum of the Earth–Moon system = 3.5×10^{34} Kg m^2 s^{-1}
Current energy in the Earth–Moon system = 2.5412×10^{29} J
Current energy dissipation rate in the Earth–Moon system = 3.17×10^{12} W

A.2.5 Useful Physical Constants

Mass $M_E = 5.977 \times 10^{24}$ kg
Mean mass of the atmosphere = 5.1352×10^{18} kg
(dry air = 5.122×10^{18} kg, water vapor = 1.32×10^{16} kg)
Mass of the ocean =1.35×10^{21} kg
Mass of ice = 2.2×10^{19} kg
Mass of water in lakes and rivers = 5.0×10^{17} kg (in sediments = 2×10^{20} kg)

Surface area of Earth = 5.10×10^{14} m^2
Ocean surface area = 3.611×10^{14} m^2 (71%, includes sea ice)
Deep ocean (>1000 m) surface area = 3.05×10^{14} m^2
Land surface area = 1.489×10^{14} m^2
Surface area of sea ice = 0.261×10^{14} m^2 (7% of ocean area) average, 0.11×10^{14} km^2 in the northern hemisphere (8–15 $\times 10^6$ km^2, Arctic, sub-Arctic Seas); and 0.14×10^{14} km^2 in the southern hemisphere (4–21 $\times 10^6$ km^2)

Mean ocean depth = 3795 m (~4.2 km excluding the shelf)
Mean sea ice thickness = 3 m (Arctic), 0.7 m (Antarctic)
Standard sea level pressure = 101.325 kPa
Mean sea level pressure = 98.44 kPa (dry air = 98.19, water vapor = 0.25 kPa)
Standard sea level temperature = 288.15 K
Typical density of dry air at 20°C and 100 kPa (1 bar) = 1.210 kg m^{-3}
Typical density of seawater = 1026 kg m^{-3}
Typical density of fresh water = 1000 kg m^{-3}
Typical density of ice = 917 kg m^{-3}
Typical density of snow = 330 kg m^{-3}
Average density of seawater = 1035 kg m^{-3}
Latent heat of vaporization of water at 0°C = 2.501×10^6 J kg^{-1}
Latent heat of vaporization of water at 100°C = 2.250×10^6 J kg^{-1}
Latent heat of sublimation of ice at 0°C = 2.835×10^6 J kg^{-1}
Latent heat of fusion for water at 0°C = 3.347×10^5 J kg^{-1}
Specific heat at constant pressure for dry air = 1005.6 J kg^{-1} K^{-1}
Specific heat of water vapor at 0°C = 1859 J kg^{-1} K^{-1}
Specific heat at constant pressure for water at 0°C = 4217.4 J kg^{-1} K^{-1}
Specific heat of ice and snow at 0°C = 2099 J kg^{-1} K^{-1}
Ratio of specific heats at constant pressure and volume for dry air = 1.4
Kinematic viscosity of seawater at 10°C, 35 psu = 1.8×10^{-6} m^2 s^{-1}
Molecular heat diffusivity for seawater = 1.39×10^{-7} m^2 s^{-1}
Molecular salt diffusivity for seawater = 9.0×10^{-10} m^2 s^{-1}
Kinematic viscosity of air = 1.53×10^{-5} m^2 s^{-1} [$1.35 \times 10^{-5} + 10^{-7} (T_a - 273.15)$]

Heat diffusivity for air = 2.16×10^{-5} m^2 s^{-1} [$1.9 \times 10^{-5} + 1.26 \times 10^{-7}$ (T_a–273.15)]
Water vapor diffusivity for air = 2.57×10^{-5} m^2 s^{-1} [$2.26 \times 10^{-5} + 1.51 \times 10^{-7}$ (T_a–273.15)]
Molecular salt diffusivity for seawater = 9.0×10^{-10} m^2 s^{-1}
Prandtl number for air = 0.71
Schmidt number for water vapor in air = 0.595
Prandtl number for water = 8 to 13
Schmidt number for salt diffusion in water ~ 2000
Typical emissivity of water, ice, and snow surfaces = 0.97
Typical albedo of water surface = 0.1
Typical albedo of ice surface = 0.7
Typical albedo of fresh snow = 0.85
Typical thermal conductivity of air = 0.024 W m^{-1} K^{-1}
[equal to $0.0238+7.12 \times 10^{-5}$ (T_a–273.15) at temperature T_a in K]
Typical thermal conductivity of water = 0.6 W m^{-1} K^{-1}
Typical thermal conductivity of ice = 2.034 W m^{-1} K^{-1}
Typical thermal conductivity of snow = 0.3097 W m^{-1} K^{-1}
Typical sound speed at sea level in air = 320 m s^{-1}
Typical sound speed at sea level in water = 1500 m s^{-1}
Typical sound speed in ice = 4000 m s^{-1}
Typical volumetric expansion coefficient of seawater for heat ~2×10^{-4} K^{-1}
Typical volumetric expansion coefficient of seawater for salinity ~8×10^{-4} psu^{-1}
Typical surface tension for water = 7.4×10^{-5} N m^{-1}
Typical geothermal heat flux at sea bottom = 0.05 W m^{-2} (at midocean spreading centers, this value can be an order of magnitude or more higher)

A.2.6 USEFUL DYNAMICAL QUANTITIES

Von Karman constant κ = 0.40–0.41
Charnock constant = 0.011
Kolmogoroff constant 1D (3D) = 0.50 (1.62)
Batchelor constant 1D (3D) = 0.42 (1.36)
Phillips constant = 0.015
Toba constant = 0.11

Acceleration due to gravity g = 9.806 65 (midlatitude), 9.780 32 (equator), and 9.832 04 m s^{-2} (poles) at sea level, with 9.797 6 m s^{-2} as the surface average, but at arbitrary latitude θ and altitude z,

$$g = (9.78032 + 5.172 \times 10^{-2} \sin^2\theta - 6.0 \times 10^{-5} \sin^2\theta)(1 + z/a)^{-2}$$

Earth's rotation rate Ω = 7.292 116 × 10 m s^{-25} s m s^{-21}

1 inertial period (IP) = $T/(2\sin\theta)$, where θ is the latitude and T is the rotation period (1/2 sidereal day at the poles, and 1 sidereal day at 30° latitude)

Value of $f = 2\Omega\sin\theta$: equatorial, 0; midlatitude (30°), 7.29 × 10^{-5} s^{-1}; poles (90°), 1.458 × 10^{-4} s^{-1}

Beta (β) = (2Ω/a) cos θ, where a is Earth's radius, and θ is the latitude, 2.289 159 × 10^{-11} cos θ m^{-1} s^{-1} (2.2891 × 10^{-11} m^{-1} s^{-1} at 0°, 2.0 × 10^{-11} m^{-1} s^{-1} at 30, and 1.618 68 × 10^{-11} at 45°, 0 at 90° latitude

Rossby radius of deformation a = C/f for the first baroclinic mode: 330 km at 3°, 100 km at 10°, 40 km at 30°, 20 km at 45°, and 5 km at the poles

Equatorial Rossby radius of deformation a = $(C/\beta)^{1/2}$ for the first baroclinic mode: 330 km. Second baroclinic mode: 256 km

Equatorial Rossby radius = 1200 km (atmosphere)

Extratropical long Rossby wave speed $C = \beta a^2$: 22 cm s^{-1} at 10°, and 1.5 cm s^{-1} at 30°

Equatorial Kelvin wave speed = 2.5 m s^{-1} (first baroclinic mode), 1.5 m s^{-1} (second), and 30 m s^{-1} (atmosphere)

Equatorial Rossby wave speed: mode 1 of the first baroclinic mode ~ 0.8 m s^{-1}, and mode 2 ~ 0.5 m s^{-1}

Baroclinic midlatitude gravity/Kelvin wave speed: first mode ~ 2.5 m s^{-1}, and second mode ~ 1.5 m s^{-1}

The values quoted above for some physical properties as typical values are suitable as rough estimates. For more precise values, which are often functions of many other parameters such as temperature, see appropriate tables.

Appendix B

Equations of State

B.1 EQUATION OF STATE FOR THE OCEAN

The equation of state that is now commonly used is presented by Millero and Poisson (1981) and given in UNESCO Technical Paper in Marine Science Number 36 (UNESCO, 1981). The density of seawater (in kg m^{-3}) as a function of temperature T (°C), salinity S (psu), and pressure p (bars) over the typical oceanic range of –2 to 40°C temperature, 0 to 42 psu salinity, and 0 to 1000 bars pressure is given to a precision of 3.5×10^{-3} kg m^{-3} by

$$\rho(T,S,p) = \rho(T,S,0)[1 - p/K(T,S,p)]^{-1} \qquad \text{(B1)}$$

where K(T,S,p) is the secant bulk modulus, and the density of seawater at one standard atmosphere pressure (p = 0) is given by (Pond and Pickard, 1989; Gill, 1982)

$$\begin{aligned}
\rho(T,S,0) = \; & + 999.842\,594 && + 6.793\,952 \times 10^{-2}\, T \\
& - 9.095\,290 \times 10^{-3}\, T^2 && + 1.001\,685 \times 10^{-4}\, T^3 \\
& - 1.120\,083 \times 10^{-6}\, T^4 && + 6.536\,332 \times 10^{-9}\, T^5 \\
& + 8.244\,93 \times 10^{-1}\, S && - 4.089\,9 \times 10^{-3}\, TS \\
& + 7.643\,80 \times 10^{-5}\, T^2S && - 8.246\,7 \times 10^{-7}\, T^3S \\
& + 5.387\,50 \times 10^{-9}\, T^4S && - 5.724\,66 \times 10^{-3}\, S^{1.5} \\
& + 1.022\,70 \times 10^{-4}\, TS^{1.5} && - 1.654\,6 \times 10^{-6}\, T^2S^{1.5} \\
& + 4.831\,40 \times 10^{-4}\, S^2 && \qquad \text{(B2)}
\end{aligned}$$

$$
\begin{aligned}
K(T,S,p) =\ & \\
& +19\,652.21 \\
& +148.420\,6\ T \quad -2.327\,105\ T^2 \\
& +1.360\,477 \times 10^{-2}\ T^3 \quad -5.155\,288 \times 10^{-5}\ T^4 \\
& +3.239\,908\ p \quad +1.437\,13 \times 10^{-3}\ Tp \\
& +1.160\,92 \times 10^{-4}\ T^2 p \quad -5.779\,05 \times 10^{-7}\ T^3 p \\
& +8.509\,35 \times 10^{-5}\ p^2 \quad -6.122\,93 \times 10^{-6}\ Tp^2 \\
& +5.278\,7 \times 10^{-8}\ T^2 p^2 \\
& +54.674\,6\ S \quad -0.603\,459\ TS \\
& +1.099\,87 \times 10^{-2}\ T^2 S \quad -6.167\,0 \times 10^{-5}\ T^3 S \\
& +7.944 \times 10^{-2}\ S^{1.5} \quad +1.648\,3 \times 10^{-2}\ TS^{1.5} \\
& -5.300\,9 \times 10^{-4}\ T^2 S^{1.5} \quad +2.283\,8 \times 10^{-3}\ pS \\
& -1.098\,1 \times 10^{-5}\ TpS \quad -1.607\,8 \times 10^{-6}\ T^2 pS \\
& +1.910\,75 \times 10^{-4}\ pS^{1.5} \quad -9.934\,8 \times 10^{-7}\ p^2 S \\
& +2.081\,6 \times 10^{-3}\ Tp^2 S \quad +9.169\,7 \times 10^{-10}\ T^2 p^2 S
\end{aligned}
\tag{B3}
$$

Test values for various quantities above are as follows (Pond and Pickard, 1989):

At $T = 5°C$, $S = 0$, $p = 0$ bars:	$\rho = 999.966\,75$ kg m^{-3}
At $T = 5°C$, $S = 0$, $p = 1000$ bars:	$\rho = 1044.128\,02$ kg m^{-3}
At $T = 25°C$, $S = 0$, $p = 0$ bars:	$\rho = 997.047\,96$ kg m^{-3}
At $T = 25°C$, $S = 0$, $p = 1000$ bars:	$\rho = 1037.902\,04$ kg m^{-3}
At $T = 5°C$, $S = 35$, $p = 0$ bars:	$\rho = 1027.675\,47$ kg m^{-3}
At $T = 5°C$, $S = 35$, $p = 1000$ bars:	$\rho = 1069.489\,14$ kg m^{-3}
At $T = 25°C$, $S = 35$, $p = 0$ bars:	$\rho = 1023.343\,06$ kg m^{-3}
At $T = 25°C$, $S = 35$, $p = 1000$ bars:	$\rho = 1062.538\,17$ kg m^{-3}

The specific volume is given by

$$\alpha(T,S,p) = \alpha(T,S,0)[1 - p/K(T,S,p)]^{-1} \tag{B4}$$

The adiabatic lapse rate can be calculated using

$$\Gamma = g\alpha T/c_p \tag{B5}$$

The static stability can be calculated by

$$N^2 = g\alpha\left(\Gamma + \frac{dT}{dz}\right) - g\beta\frac{dS}{dz} = g^2\alpha^2 T/c_p + g\alpha\frac{dT}{dz} - g\beta\frac{dS}{dz} \tag{B6}$$

where N is the Brunt–Vaisala (buoyancy) frequency and g the gravitational constant, and

$$\alpha = \frac{1}{\rho}\left[\frac{\partial \rho}{\partial T}\right]_{p,S} ; \beta = \frac{1}{\rho}\left[\frac{\partial \rho}{\partial S}\right]_{p,T} . \qquad \text{(B7)}$$

Potential temperature $\theta(T,S,p)$ to a 0.001°C precision in the range 2 to 30°C, 30 to 40 psu, 0 to 1000 bars, is given by

$$\begin{aligned}
\theta(T,S,p) = T \\
& - 3.650\,4 \times 10^{-4}\ p & & - 8.319\,8 \times 10^{-5}\ Tp \\
& + 5.406\,5 \times 10^{-7}\ T^2p & & - 4.027\,4 \times 10^{-9}\ T^3p \\
& - 8.930\,9 \times 10^{-7}\ p^2 & & + 3.162\,8 \times 10^{-8}\ Tp^2 \\
& - 2.198\,7 \times 10^{-10} T^2p^2 \\
& + 1.605\,6 \times 10^{-10} p^3 & & - 5.048\,4 \times 10^{-12}\ Tp^3 \\
& - 1.743\,9 \times 10^{-5} p(S-35) \\
& + 2.977\,8 \times 10^{-7}\ Tp(S-35) \\
& + 4.105\,7 \times 10^{-9}\ p^2(S-35) & & \qquad \text{(B8)}
\end{aligned}$$

A test value for T = 10°C, S = 25 psu, p = 1000 bars gives a θ of 8.467 851 6.

The dependence of the freezing point (°C) on salinity (0 to 40 psu) and pressure (0 to 50 bars) to a 0.004°C precision is

$$T_f(S,p) = -0.057\,5\ S + 1.710\,523 \times 10^{-3}\ S^{1.5} - 2.154\,996 \times 10^{-4}\ S^2$$

$$- 7.53 \times 10^{-3}\ p \qquad \text{(B9)}$$

This equation will be of particular interest in studying the formation of supercooled waters and ice formation on the bottom of ice shelves.

B.2 EQUATION OF STATE FOR FRESHWATER LAKES

Equations are given below (from Chen and Millero, 1986) to calculate the following properties of water in a freshwater lake (nominal salt content of 0.15 g kg^{-1}) over the typical limnological range of 0–30°C temperature, 0–0.6 salinity, and 0–180 bars pressure: density, coefficient of thermal expansion, temperature of maximum density, maximum density, specific heat at constant pressure, adiabatic temperature gradient, sound speed, freezing point, and static stability.

The equation of state to a precision of better than 2×10^{-3} kg m^{-3} is

$$
\begin{aligned}
\rho(T,S_l,0) = \\
&+ 999.8395 && + 6.7914 \times 10^{-2}\,T \\
&- 9.0894 \times 10^{-3}\,T^2 && + 1.0171 \times 10^{-4}\,T^3 \\
&- 1.2846 \times 10^{-6}\,T^4 && + 1.1592 \times 10^{-8}\,T^3 \\
&- 5.0125 \times 10^{-11}\,T^6 \\
&+ 8.181 \times 10^{-1}\,S_l && - 3.850 \times 10^{-3}\,TS_l \\
&+ 4.960 \times 10^{-5}\,T^2S_l
\end{aligned}
\qquad (B10)
$$

$$
\begin{aligned}
K(T,S_l,p) = \\
&+ 19\,652.17 \\
&+ 148.113\,T && - 2.293\,T^2 \\
&+ 1.256 \times 10^{-2}\,T^3 && - 4.180 \times 10^{-5}\,T^4 \\
&+ 3.2726\,p && - 2.147 \times 10^{-4}\,Tp \\
&+ 1.128 \times 10^{-4}\,T^2p \\
&+ 53.238\,S_l && - 0.313\,TS_l \\
&+ 5.728 \times 10^{-3}\,PS_l
\end{aligned}
\qquad (B11)
$$

where T is the temperature in °C, and S is the salinity defined as the total grams of dissolved salt in 1 kg of lake water. The lake water salinity is related to the salinity commonly used in oceanography by

$$S_l = 1.004\,885\,S \qquad (B12)$$

The coefficient of thermal expansion $\alpha(T,S_l,p) = -\frac{1}{\rho}\left[\frac{\partial \rho}{\partial T}\right]_{p,S_l}$ to a 0.3×10^{-6} $^{\circ}C^{-1}$ precision is given by

$$
\begin{aligned}
\alpha(T,S_l,p) = \\
&+ 68.00 \times 10^{-6} \\
&+ 1.82091 \times 10^{-5}\,T && - 3.0866 \times 10^{-7}\,T^2 \\
&+ 5.3445 \times 10^{-9}\,T^3 && - 6.0721 \times 10^{-11}\,T^4 \\
&+ 3.1441 \times 10^{-13}\,T^5 \\
&+ 3.682 \times 10^{-6}\,p && - 1.520 \times 10^{-8}\,Tp \\
&+ 1.910 \times 10^{-10}\,T^2p \\
&+ 4.599 \times 10^{-6}\,S_l && - 1.999 \times 10^{-7}\,TS_l \\
&+ 2.790 \times 10^{-9}\,T^2S_l \\
&- 4.613 \times 10^{-9}\,pS_l
\end{aligned}
\qquad (B13)
$$

As pure water at 0°C is warmed, it increases in density until a maximum of 999.972 kg m^{-3} is reached at 3.9839°C. At 1 atm, a nominal amount of salt (0.15 g kg^{-1}) in a freshwater lake contributes 0.033°C to the depression of the temperature of maximum density. With increasing salinity or pressure the temperature of the maximum density decreases.

$$
\begin{aligned}
T_{MD}(S_l,p) = {} & + 3.9839 \\
& - 1.9911 \times 10^{-2}\, p - 5.822 \times 10^{-6}\, p^2 \\
& - 0.2219\, S_l - 1.106 \times 10^{-4}\, pS_l
\end{aligned} \qquad \text{(B14)}
$$

$$
\begin{aligned}
\rho_{max}(S_l,p) = {} & + 999.972 \\
& - 4.94686 \times 10^{-2}\, p - 2.0918 \times 10^{-6}\, p^2 \\
& + 8.0357 \times 10^{-1}\, S_l + 1.0 \times 10^{-4}\, pS_l
\end{aligned} \qquad \text{(B15)}
$$

The specific heat at constant pressure c_p (in J kg^{-1} $°C^{-1}$) to a precision of 0.4 J kg^{-1} $°C^{-1}$ is given by

$$
\begin{aligned}
c_p(T,S_l,p) = {} & \\
& + 4217.4 \\
& - 3.6608\, T + 1.3129 \times 10^{-1}\, T^2 \\
& - 2.210 \times 10^{-3}\, T^3 + 1.508 \times 10^{-5}\, T^4 \\
& - 4.917 \times 10^{-1}\, p + 1.335 \times 10^{-2}\, Tp \\
& - 2.177 \times 10^{-4}\, T^2 p \\
& + 1.50 \times 10^{-4}\, p^2 \\
& - 6.616 \times 10^{-2}\, S_l + 9.28 \times 10^{-3}\, TS_l \\
& - 2.39 \times 10^{-5}\, T^2 S_l
\end{aligned} \qquad \text{(B16)}
$$

As a result of the compressibility of water, pressure changes are accompanied by adiabatic temperature changes. Although these effects are relatively small, they are nevertheless significant in the static stability of the water column. If a water sample is raised adiabatically from the deeper layers to the surface, the pressure decreases and the water sample expands in volume, thus performing work against the external pressure. The result is a drop in temperature. The resulting temperature after adiabatic movement of the water parcel is the potential temperature, and the rate of change of temperature with pressure, called the adiabatic temperature gradient (in °C bar^{-1}), is given to a precision of

$5 \times 10^{-6}\,°\mathrm{C\ bar}^{-1}$ by

$$
\left[\frac{\partial T}{\partial p}\right]_a =
\begin{aligned}
&-4.407 \times 10^{-4} \\
&+1.159 \times 10^{-4}\, T - 1.447 \times 10^{-6}\, T^2 \\
&+2.202 \times 10^{-8}\, T^3 - 1.588 \times 10^{-10}\, T^4 \\
&+2.422 \times 10^{-6}\, p - 8.169 \times 10^{-8}\, Tp \\
&+8.651 \times 10^{-10}\, T^2 p \\
&-4.55 \times 10^{-10}\, p^2 \\
&+2.725 \times 10^{-5}\, S_l - 1.025 \times 10^{-6}\, TS_l \\
&+2.351 \times 10^{-8}\, T^2 S_l
\end{aligned}
\tag{B17}
$$

The adiabatic lapse rate can be calculated using either

$$
\Gamma = \left[\frac{\partial T}{\partial p}\right]_a \frac{dp}{dz} = -\rho g \left[\frac{\partial T}{\partial p}\right]_a \tag{B18}
$$

or

$$
\Gamma = g\alpha T / c_p \tag{B19}
$$

The static stability in the water column can be calculated by

$$
N^2 = g\alpha\left(\Gamma + \frac{dT}{dz}\right) - g\beta\frac{dS_l}{dz} = g^2\alpha^2 T / c_p + g\alpha\frac{dT}{dz} - g\beta\frac{dS_l}{dz} \tag{B20}
$$

where N is the Brunt–Vaisala (buoyancy) frequency and g the gravitational constant, and

$$
\alpha = \frac{1}{\rho}\left[\frac{\partial \rho}{\partial T}\right]_{p,S_l} ; \quad \beta = \frac{1}{\rho}\left[\frac{\partial \rho}{\partial S_l}\right]_{p,T} \tag{B21}
$$

The equation of sound speed in lake waters to $0.04\ \mathrm{m\ s^{-1}}$ precision is as follows:

$$
C_S(T, S_l, p) =
\begin{aligned}
&+1402.388 \\
&+5.037\,1 T - 5.808\,5 \times 10^{-2}\, T^2 \\
&+3.342 \times 10^{-4}\, T^3 - 1.478 \times 10^{-6}\, T^4 \\
&+3.146 \times 10^{-9}\, T^5 \\
&+1.556\,4 \times 10^{-1}\, p + 4.046 \times 10^{-4}\, Tp \\
&-8.15 \times 10^{-7}\, T^2 p \\
&+1.593 \times 10^{-5}\, p^2 \\
&+1.322\, S_l - 7.01 \times 10^{-3}\, TS_l \\
&+4.9 \times 10^{-5}\, T^2 S_l \\
&-5.58 \times 10^{-5}\, p S_l
\end{aligned}
\tag{B22}
$$

At 0°C temperature, 0.15 S_l, and 1 atm, the sound speed is 0.2 m s^{-1} higher in a freshwater lake than in pure water. This effect is important for echo soundings in deep lakes. The dependence of the freezing point on salinity (0 to 0.65) and pressure (0 to 50 bars) is

$$T_f = -0.0137 - 0.0525\, S_l - 7.48 \times 10^{-3}\, p \tag{B23}$$

This equation will be of particular interest in studying the formation of supercooled waters and ice formation on the bottom of ice shelves. Test values for various quantities above (Chen and Millero, 1986) are as follows:

At T = 10°C, $S_l = 0$, p = 100 bars,

ρ = 1004.4277 kg m^{-3}, α = 111.52 × 10^{-6} °C^{-1}, T_{MD} = 1.9346°C, ρ_{max} = 1004.8979 kg m^{-3}, c_p = 4155.36 J kg^{-1} °C^{-1}, C_S = 1463.39 m s^{-1}, $[\partial T/\partial p]_a$ = 7.586 4°C bar^{-1}, T_f = −0.7617°C.

At T = 10°C, $S_l = 0.5$, p = 100 bars,

ρ = 1005.0920 kg m^{-3}, α = 112.63 × 10^{-6} °C^{-1}, T_{MD} = 1.8181°C, ρ_{max} = 1005.3047 kg m^{-3}, c_p = 4155.55 J kg^{-1} °C^{-1}, C_S = 1464.016 m s^{-1}, $[\partial T/\partial p]_a$ = 7.6643°C bar^{-1}, T_f = −0.7877°C.

B.2.1 EQUATION OF STATE WITH DISSOLVED CO_2

If C is the concentration of dissolved CO_2 (in grams per kilogram), then the density (from Giggenbach, 1990; see also Weiss, 1974) is

$$\rho_2 = (1000 + C)/\,[0.75 \times 10^{-3}\, C + (1000/\rho_1)] \tag{B24}$$

where ρ_1 is the water density without the dissolved gas that can be obtained from relations above. One can also use

$$\rho_2 = \rho_1\,(1. + 0.266 \times 10^{-6} \rho_1\, C) \tag{B25}$$

If the water contains dissolved solids as well and if S is the concentration of dissolved matter (in grams per kilogram), the density is

$$\rho_3 = \rho_2 + 0.8\, S \tag{B26}$$

The saturation value of the CO_2 concentration at depth D (in meters) is given by (Giggenbach, 1990)

$$C_s = 244\ D/\ H \tag{B27}$$

where H is Henry's law constant:

$$H = 1640 + 95\ (25 - T). \tag{B28}$$

Therefore the depth D_s (in meters) at which a parcel brought up from a depth D (in meters), temperature T (in degrees Celsius), and concentration C (in grams per kilogram) would be saturated is given by (ignoring the small adiabatic cooling of the water parcel due to decrease in depth)

$$D_s = C\ H/244.$$

B.3 EQUATION OF STATE FOR THE ATMOSPHERE

Air is a mixture of gases, the primary constituents being in nearly constant proportions: N_2 (78.1% by volume), O_2 (21%), and Ar (0.9%). Dry air can be treated as an ideal gas,

$$p_{da} = \rho_{da} R_a T \tag{B29}$$

where R_a = 287.04 J kg^{-1} K^{-1}. T is the air temperature, ρ_{da} is the density, and p_{da} is the pressure of dry air. Water vapor can also be treated as an ideal gas,

$$p_V = \rho_V R_V T \tag{B30}$$

where R_v = 461.50 J kg^{-1} K^{-1}. T is the temperature, ρ_v is the density of water vapor (also called absolute humidity, the mass of vapor per unit volume of moist air), and p_v is the partial pressure of the water vapor. The total pressure is the sum of the two partial pressures, assuming air is a mixture of ideal gases:

$$p = p_{da} + p_v \tag{B31}$$

If q is the specific humidity, mass of water vapor per unit mass of moist air,

$$\rho_V = q\rho_a;\ \rho_a = \rho_{da} + \rho_V \tag{B32}$$

$$\frac{p_v}{p} = \frac{q}{\varepsilon + (1-\varepsilon)q} = \frac{r}{r+\varepsilon} \tag{B33}$$

where $\varepsilon = \frac{R_a}{R_v} = 0.62197$, $r = \frac{q}{1-q}$ is the mixing ratio. The saturation value q_s, the maximum possible value for q, depends strongly on temperature: 0.0038 (r = 0.0038) at 0°C and 0.04 (r = 0.042) at 37°C.

The equation of state for the moist air can therefore be written as

$$\rho_a = \frac{p}{RT_v} \tag{B34}$$

where T_v is the virtual temperature given by

$$T_V = T(1 - q - q/\varepsilon) = T(1 + 0.6078q) \tag{B35}$$

Since air is mostly diatomic ($c_{pda} = 3.5\ R_{da}$) and water vapor is triatomic ($c_{pv} = 4\ R_v$),

$$c_p = c_{pda}(1 - q + 8q/7\varepsilon) = 1004.6(1 + 0.8375q) \tag{B36}$$

$$c_V = c_{vda}(1 - q + 6q/5\varepsilon) = 714.3(1 + 0.9294q) \tag{B37}$$

The Clausius–Clapeyron equation provides the rate of change of saturation vapor pressure p_{vs} over a water surface with temperature,

$$\frac{dp_{vs}}{dT} = \frac{L_v}{T\left(\frac{1}{\rho_v} - \frac{1}{\rho_w}\right)} \sim \frac{L_v \rho_v}{T} = \frac{L_v p_{vs}}{R_v T^2} \tag{B38}$$

where ρ_V is the density of water vapor and ρ_W is the density of liquid water. Expressions for L_v, the latent heat of vaporization (in J kg^{-1}), and saturation vapor pressure p_{vs} (mb) of pure water vapor over a water surface (to 0.2% precision between –40°C and +40°C) can be written as

$$L_v(T) = 2.5008 \times 10^6 - 2.3 \times 10^3\ T \tag{B39}$$

$$p_{vs}(T) = 10^{(0.7859 + 0.03477T)/(1 + 0.00412T)} \tag{B40}$$

In air, the saturation vapor pressure is slightly higher by a factor of 1 to 1.006,

$$p_{vs} = p'_{vs}\left[1+10^{-6}p(4.5+0.0006T^2)\right], \tag{B41}$$

where p is the total pressure in millibars (mb). Over seawater, this value must be multiplied by a factor of 0.98. Knowing T and p, one can therefore compute p_{vs} and hence the saturation specific humidity q_s, the saturation mixing ratio r_s, and the relative humidity RH:

$$RH = \frac{r}{r_s} = \frac{q(1-q_s)}{q_s(1-q)} \tag{B42}$$

For evaporation over ice the expressions are different, and to a 0.3% precision in the –40°C and 0°C range the saturation vapor pressure is given by

$$p'_{vsi}(T) = p'_{vs}(T)10^{0.00422T} \tag{B43}$$

and

$$L_{vi}(T) = 2.839 \times 10^6 - 3.6\,(T+35)^2 \tag{B44}$$

The standard psychrometer equation that relates wet bulb temperature T_{wb} to air temperature is

$$T_{wb} = T_a - \frac{L_E k_v}{c_{pa} k_H}(q_s(T_{wb}) - q_a) \tag{B45}$$

where k_v and k_H are water vapor and heat diffusivities and q_s is the saturation specific humidity.

To within 0.3%, the dry lapse rate for the atmosphere is $\Gamma = g/c_p$, about 10 K km^{-1}. The moist adiabatic lapse rate (in K km^{-1}) is given by

$$\Gamma_m = 6.4 - 0.12T + 2.5\text{x}10^{-5}T^3 + [-2.4 + 10^{-3}(T-5)^2](1-10^{-3}p) \tag{B46}$$

Appendix C

Important Scales and Nondimensional Quantities

C.1 LENGTH SCALES

Buoyancy scale, $\ell_b = (\overline{w^2})^{1/2} N^{-1}$, is the maximum turbulence length scale associated with ambient stratification. A looser definition would be $\ell_b = (q/N)$. It has been found to be proportional to the Thorpe scale ℓ_T and Ozmidov scale ℓ_O. It is indicative of the vertical displacement at which all the kinetic energy of the eddy is converted to potential energy by working against buoyancy forces. It is also related to the density distribution via $\ell_b = (\overline{\rho'^2})^{1/2} \left(\frac{\partial \bar{\rho}}{\partial z} \right)^{1}$.

Deep convection length scale, $l_{dc} = (B_o / f^3)^{1/2}$, where B_0 is the buoyancy flux, characterizes the size of the convective plumes in rotation-dominated deep convection process. It also denotes the depth beyond which rotational effects become important in the convective layer and the plumes assume sinewy rope-like shapes.

Displacement scale, d, characterizes the displacement of the zero velocity point due to the influence of canopy or other surface roughness elements.

Ekman scale, $\ell_E \sim (u_* / f)$, where f is the Coriolis Parameter, characterizes the size of the friction-influenced layer adjacent to a boundary due to an applied shear stress. It sets the upper bound on the depth of the OML and ABL under neutral and stably stratified conditions.

Energy-containing scale, ℓ, indicates the size of the turbulent eddies that contain a majority of the TKE. It corresponds to the peak of the turbulence energy spectrum. It is proportional to the integral scale or macroscale of turbulence.

Kolmogoroff microscale, $\eta\left(\frac{\nu^3}{\varepsilon}\right)^{1/4}$, also called the dissipative scale, is indicative of the passive viscous scales at which energy handed down the spectrum from large eddies is dissipated by viscosity. It is also the smallest scale possible in a turbulent flow. The turbulence is isotropic and the spectrum universal around this scale.

Monin–Obukhoff (Obukhoff) scale, $L=\left(\frac{u_*^3}{\kappa Q_b}\right)$, is the length scale that characterizes gravitational stratification and is indicative of the height at which shear production of turbulence to the buoyancy production (destruction) is comparable. Below that height, shear production dominates; above it buoyancy production (destruction) dominates.

Ozmidov scale, $\ell_O=\left(\frac{\varepsilon}{N^3}\right)^{1/2}$, determines the maximum scale of the turbulent eddies in a stably stratified flow. It has been found to be proportional to the Thorpe scale ℓ_T and buoyancy scale ℓ_b.

Rotation length scale, $\ell_r=\left(\frac{q}{f,f_y}\right)$, is the characteristic rotation scale. It is proportional to the Ekman scale in neutrally stratified shear flow.

Roughness scale, z_0, characterizes the roughness elements and determines the resulting friction exerted on the flow. It is usually 1/30 of the average physical height of the roughness elements. It is a function only of the characteristics such as size, density, and distribution of roughness elements for immobile surfaces, but for mobile surfaces, such as the sea surface and sand, z_0 is determined dynamically by the flow.

Taylor microscale, $\lambda=\left(\frac{5\nu q^2}{\varepsilon}\right)^{1/2}$, is the intermediate scale between the large eddy scale ℓ and the Kolmogoroff microscale. Turbulence strain rate is proportional to q/λ. Vorticity fluctuations in a turbulent flow are concentrated at this scale.

Thorpe scale, ℓ_T, is indicative of the scale of the overturns in a turbulent stratified fluid, and is unaffected by the presence of internal waves. The Thorpe scale is the local distance over which the net displacement of parcels integrate to zero. It is determined by sorting potential density (more conveniently, potential temperature) profile measurements, which may contain inversions, into a profile with density increasing monotonically with depth. The RMS of vertical displacements of parcels needed to accomplish this is the Thorpe scale, usually O(1 m).

C.2 TIMESCALES

Buoyancy timescale, $T_b \sim \frac{1}{N}$, is the timescale of oscillation of a vertically displaced parcel in a stably stratified fluid.

(Turbulent) diffusion timescale, $t_b \sim \frac{\theta^2}{\varepsilon_0}$, is the timescale for the temperature (density) fluctuations to decay due to turbulent diffusion.

Eddy turnover timescale, $t_c = \frac{\ell}{q}$, is the timescale over which large energy-containing eddies lose (transfer) a significant portion of their energy to dissipation scales through the nonlinear cascade process. This is also the timescale over which turbulence decays when it is cut off from its source. It is inversely proportional to the turbulence frequency $f_t = \frac{q}{\ell}$.

Eddy viscous timescale, $t_v = \frac{\nu}{q^2}$, is the timescale over which direct action of viscosity damps eddy motions.

Inertial period, $t_v = 2\pi/f$, where f is the Coriolis parameter, is characteristic of the influence of planetary rotation.

Kolmogoroff timescale, $t_\eta \sim \left(\frac{\nu}{\varepsilon}\right)^{1/2}$, is the timescale associated with dissipation scales.

C.3 VELOCITY SCALES

(Turbulent) buoyancy velocity scale, $v_b = (\varepsilon/N)^{1/2}$, is the turbulence velocity scale influenced by buoyancy forces in a stably stratified turbulent flow, and is associated with the Ozmidov scale.

Deardorff convective velocity scale, $w_* = (q_b D)^{1/3}$, is the characteristic velocity scale in the bulk of the convective mixed layer.

Deep convection velocity scale, $u_{dc} = (B_o/f)^{1/2}$, where B_0 is the buoyancy flux, characterizes the vertical velocity associated with convective plumes in rotation-dominated deep convection processes.

Friction velocity, $u_* = (\tau/\rho)^{1/2}$, is the characteristic velocity scale for turbulent shear flows. It determines the velocity scale of turbulence and mixing in the boundary layer.

Kolmogoroff velocity scale, $v_\eta = (\nu\varepsilon)^{1/4}$, is the viscous velocity scale characteristic of the high wavenumber end of the turbulence spectrum, the velocity scale associated with the Kolmogoroff range of the spectrum.

Wave phase speed, $c_p = (g_e H)^{1/2}$, or NH, where g_e is the effective gravitational acceleration and H is the depth of the fluid, determines the characteristic wave propagation velocity scale.

C.4 NONDIMENSIONAL QUANTITIES

(Surface wave field) Age $\left(A_w = \frac{c_p}{u_*}\right)$ is indicative of the maturity of a surface wave field. A fully developed wave field corresponds to high values of wave age (~30). Its importance lies in the fact that it determines the roughness of the sea and hence the air–sea exchanges. Young seas (low wave age) are rougher than older ones.

Buoyancy flux ratio, $R = \left(\frac{q}{q_0}\right)$, is indicative of penetrative convection; it is the ratio of entrainment flux at the inversion to applied buoyancy flux at the surface in the ABL. Its usual value is ~0.2, although for very strong inversions, it can be much smaller.

Burger number, Bu = NH/fL, is the square of the ratio of the Rossby radius of deformation to the horizontal scale of the flow, indicative of the importance of baroclinicity in the flow.

Cox number, $C_T = \left(\dfrac{\left\langle \dfrac{\partial T}{\partial x_i} \dfrac{\partial T}{\partial x_i} \right\rangle}{\left(\dfrac{\partial T}{\partial z} \right)^2} \right)$, an important parameter in microscale turbulence, is of importance to microstructure measurements.

Dalton number, $Da = c_E = \left(\dfrac{E}{\rho L_E U_a (q_s - q_a)} \right)$, is the coefficient of water vapor transfer across the air–sea interface.

Deep convection Rayleigh number, $Ra = B_0 D^4/(\nu k^2)$, where B_0 is the buoyancy flux and D the convective layer depth, characterizes the Rayleigh number associated with the deep convection process.

Deep convection Rossby number, $R_{dc} = l_{cd}/D = [B_0/(f^3 D^2)]^{1/2}$, where B_0 is the buoyancy flux and D the convective layer depth, denotes the importance of rotation. For small values of R_{dc}, the rotational effects are important.

Density ratio, $R_\rho = \left(\dfrac{\beta_T \dfrac{\partial T}{\partial z}}{\beta_S \dfrac{\partial S}{\partial z}} \right) = \left(\dfrac{\beta_T \Delta T}{\beta_S \Delta S} \right)$, indicates the relative contribution of thermal or solute stratification to buoyancy stratification of the water column. An important parameter in double diffusion, could be called Turner-Stern number.

Double-diffusion number, $D_D = \left(\dfrac{F_{S,T}}{k_{S,T} \dfrac{\partial (S,T)}{\partial z}} \right) = \dfrac{K^e_{S,T}}{k_{S,T}}$, is the ratio of effective diffusivity to molecular diffusivity of salt (heat) in salt finger (diffusive convection) processes. Denotes the effectiveness or efficiency of double-diffusive processes.

Flux ratio, $R_F = \left(\dfrac{\beta_T F_T}{\beta_S F_S} \right)$, indicates the relative contribution of thermal and solute fluxes to the buoyancy flux in double-diffusive processes. It is a function of the density ratio.

(External) Froude number, $F_e = U/(gD)^{1/2}$, analogous to the Mach number in compressible flows, is the ratio of the fluid velocity to the shallow water wave speed, and is indicative of the importance of inertial forces relative to gravitational forces. The flow is supercritical for $F_e > 1$, and subcritical for $F_e < 1$.

(Internal) Froude number, $F = \left(\frac{U}{NH}\right)$ or $\left(\frac{U}{\sqrt{D\Delta b}}\right)$, denotes stratification effects in, for example, flow over topography (H is the mountain height, U is the flow velocity, D is the mixed layer height, and Δb is the buoyancy change at its top). Often an inverse Froude number is used. For high inverse Froude numbers, the flow tends to go around the mountain, rather than over it.

(Rotational turbulence) Froude number, $F_t^r = \left(\frac{q}{(f, f_y)\ell}\right) = \frac{t_r}{t_e} \sim \frac{f}{f, f_y}$, is the ratio of the time-scale associated with rotation to the eddy turnover time-scale. It is more appropriately a Rossby number.

(Turbulence) Froude number, $F_t = \left(\frac{a}{N\ell}\right) = \frac{t_b}{t_e} \sim \frac{f_t}{N}$, is the ratio of the time-scale associated with stable stratification to the eddy turnover time-scale.

Hoenikker number, $H_O = \left(\frac{Q_b}{k u_*^2 V_{SO}}\right)$, where Q_b is the buoyancy flux, k is the wave-number, and V_{SO} is the Stokes velocity, is important in Langmuir circulation.

Kuelegan number, $Ke = \left(\frac{u_*^3}{\nu g}\right)$, denotes the importance of wave breaking for the molecular skin of the ocean.

Langmuir cell number, $N_S = \left(\frac{V_{SO}}{u_*}\right)$, indicates the importance of Langmuir circulation relative to wind mixing. It is the ratio of the Stokes velocity of surface waves to the frictional velocity.

Langmuir production ratio, $R_S = \left(\frac{\overline{u_1 u_3}\frac{\partial V_S}{\partial x_3}}{\overline{u_1 u_3}\frac{\partial U_1}{\partial x_3} + \overline{u_2 u_3}\frac{\partial U_2}{\partial x_3}}\right)$, is the ratio of TKE production due to Langmuir cell circulation to that by conventional shear production.

Lewis number, $Le = \left(\frac{k_T}{k_S}\right)$, is the ratio of thermal diffusivity to solute diffusivity; it is a counterpart of the Prandtl number.

Mixing efficiency, $\gamma_m = \frac{Ri_f}{1 + Ri_f}$, indicates the efficiency of conversion of TKE into potential energy of the system.

Nusselt number, $Nu = \left(\frac{q_T}{k_T \frac{\Delta T}{d}}\right) = \left(\frac{q_b}{k_T \frac{\Delta b}{d}}\right)$, is the ratio of the convective heat (buoyancy) flux (kinematic) q_T (q_b) to a purely conductive value, for a given temperature (buoyancy) difference maintained between two levels distance d apart.

Peclet number, $Pe = \left(\frac{UL}{k_t}\right)$, is the ratio of advection to thermal diffusion; it is a counterpart of the Reynolds number.

(Turbulent) Prandtl number, $Pr_t = \left(\frac{K_M}{K_H}\right)$, is the ratio of the momentum diffusivity (kinematic viscosity) to the thermal diffusivity. Its value is around 0.85. A molecular Prandtl number can also be defined. Its value is 0.73 for air and 7–12 for seawater, depending on the temperature.

Rayleigh number, $Ra = \left(\frac{g\beta_T \Delta d^3}{\nu k_T}\right) = \left(\frac{\Delta b d^3}{\nu k_T}\right)$, governs free convection. It is a counterpart of the Reynolds number. The flow heated from the bottom or cooled from the top becomes gravitationally unstable above a certain value of Ra and free convection ensues. For fully turbulent convection, $Nu \sim Ra^{1/3}$ holds approximately at large Ra numbers typical of geophysical flows.

(Flow) Reynolds number, $R_N = \left(\frac{UL}{\nu}\right)$, is the ratio of inertial forces to viscous forces in a flow. Low R_N flows are laminar. Transition occurs at a certain value of R_N in a particular sheared flow.

(Turbulence) Reynolds number, $R_t = \left(\frac{q\ell}{\nu}\right) \sim \frac{t_\nu}{t_e}$, is the ratio of the eddy viscosity to the molecular viscosity (equivalently the ratio of the eddy viscous time-scale to the eddy turnover time-scale). The larger the value of R_t, the lesser the molecular effects. Asymptotic invariance with respect to molecular effects demands that this number be large.

(Bulk) Richardson number, $Ri_b = \left(\frac{D\Delta b}{u_*^2}\right)$, denotes the stability of a buoyancy interface and hence its resistance to turbulent entrainment across it.

(Flux) Richardson number, $Ri_f = \left(\frac{Q_b}{-\overline{u_i u_j}\frac{\partial U_i}{\partial x_j}}\right)$, is the ratio of the buoyancy production (destruction) to shear production of TKE. When positive, it is an indicator of the mixing efficiency of turbulence.

(Gradient) Richardson number, $Ri_g = \left(\frac{g}{\bar{\rho}}\frac{\partial\bar{\rho}}{\partial z}\right)\left(\frac{\partial U_i}{\partial z}\frac{\partial U_i}{\partial z}\right)^{-1} = \left(\frac{N}{f_s}\right) = \left(\frac{t_e}{t_b}\right)^2$, is indicative of static stability of a stratified sheared flow. The classical Miles–Howard theorem states that the flow is stable if $Ri_g > 0.25$. It can be thought of as the square of the ratio of the time-scale associated with mean shear and the buoyancy period.

(Rotation gradient) Richardson number, $Ri_R = (f,f_y)\left(\frac{\partial U_i}{\partial z}\frac{\partial U_i}{\partial z}\right)^{-1/2} = \left(\frac{f,f_y}{f_s}\right)$, is indicative of static stability of a rotating sheared flow.

(Curvature gradient) Richardson number, $Ri_c = \left(\frac{\frac{U}{R}}{\left(1+\frac{z}{R}\right)}\right)\left(\frac{\partial U}{\partial z}\right)^{-1}$, where z is the distance from the curved surface, is indicative of static stability of a sheared flow with streamline curvature.

(Skin) Richardson number, $Ri_{cs} = \left(\frac{g\beta Q_H \nu}{u_*^2} \right)$, denotes the relative importance of convection to shear in the molecular skin layer of the ocean.

(Stratification) Richardson number, $Ri_N = \left(\frac{ND}{u_*, w_*} \right)$, denotes the stability of a linearly stratified fluid to vertical turbulent perturbations.

Rossby number, $Ro = \left(\frac{U}{fL} \right)$, characterizes the relative importance of rotation on flow dynamics. For small Rossby numbers typical of large scale geophysical flows, rotational effects dominate.

(Friction) Rossby number, $Ro_* = \left(\frac{u_*}{fz_0} \right)$, characterizes the relative size of the roughness layer.

Schmidt number, $Sc_t = \left(\frac{K_M}{k_S} \right)$, is the ratio of momentum diffusivity (kinematic viscosity) to scalar diffusivity. Its value is around 0.85. A molecular Schmidt number can also be defined. Its value is 0.65 for water vapor diffusion in air and 2000–2500 for salt diffusion in seawater.

(Turbulence) shear number, $Ri_t = \left(\frac{\ell}{q} \frac{\partial U}{\partial z} \right) = \frac{t_e}{t_s}$, is the ratio of the turbulence timescale to the scale of mean shear.

Stanton number, $St = c_H = \left(\frac{H_s}{\rho C_p U_a (T_s - T_a)} \right)$, is the coefficient of heat transfer across the air–sea interface.

Stern number, $S_N = \left(\frac{F_b}{\nu N^2} \right)$, where F_b is the buoyancy flux, is the ratio of effective diffusivity to kinematic viscosity; equivalent to an effective Reynolds number in salt fingers.

Strouhal number, $S_l = nD/U$, is the ratio of the advection time-scale to the timescale of oscillations in a fluid. A particular example is the oscillation of cables and columns from shedding of Karman Street eddies in their wake.

Turbulence activity, $A_t = \left(\frac{\varepsilon}{\nu N^2}\right) = \left(\frac{t_b}{t_v}\right)^2$, is often used to characterize "fossil" turbulence. It is the square of the ratio of the stratification time-scale to the Kolmogoroff time-scale.

Turner angle, $Tu = \tan^{-1}\left(\frac{R_\rho + 1}{R_\rho - 1}\right)$, delineates stability boundaries for salt fingers and double diffusion.

Wedderburn Number, $We = \left(\frac{g'h^2}{u_*^2 L}\right)$, where h is the surface layer thickness, and L is the length of the lake, is an important parameter in lake dynamics.

Appendix D

Wave Motions

A linear wave in multidimensional space can be characterized by its amplitude a, the wavenumber vector k_i, and its intrinsic frequency n. Unlike turbulence, in wave motions, n cannot assume arbitrary values. It is often externally (for example, in tidal processes) or dynamically (in internal-inertial waves) constrained to selected portions of the frequency space. It is related to k_i by the dispersion relation that is derived from the governing dynamical equations. If there is a mean current U_i, the intrinsic frequency n is related to the actual frequency ω by $n = \omega - k_i U_i$; the frequency ω is Doppler shifted from n by the amount $k_i U_i$. When the scale of the variations in the background medium, in which the wave is propagating, is large compared to the period and the wavelength of the wave, WKB methods can be used (Whitham, 1974).

A group of waves, a wave train, travels along a ray path with slow changes in ω and k_i at a velocity given by the sum of the background velocity in the medium U_i, and the group velocity c_{gi} of the waves given by

$$c_{gi} = dn / dk_i \tag{D1}$$

This is also the velocity at which the energy in the wave packet travels. The phase velocity of the waves composing the wave train is

$$c_i = \frac{n}{(k_i k_i)^{1/2}} k_i \tag{D2}$$

This is the velocity at which the crests and troughs, the phase lines, travel. Only for nondispersive waves are the group and phase velocities equal. If the wave can be represented by its property ζ (where ζ is, say, displacement), for narrowband wave fields, which require that the background medium vary slowly along the propagation path of the wave,

$$\zeta \sim \exp(i\chi) \tag{D3}$$

where $\chi = \chi(x_i, t)$ is the phase function, slowly varying in time and space. The wavenumber k_i and frequency ω can be defined in terms of χ as

$$k_i = \frac{\partial \chi}{\partial x_i}, \quad \omega = -\frac{\partial \chi}{\partial t} \tag{D4}$$

These satisfy the kinematic conservation equation for wave crests

$$\frac{\partial k_i}{\partial t} + \frac{\partial \omega}{\partial x_i} = 0 \tag{D5}$$

In vectorial notation, Eqs. (D5) and (D4) reduce to

$$\frac{\partial \mathbf{k}}{\partial t} + \nabla(n + \mathbf{k} \cdot \mathbf{U}) = 0; \ \nabla \times \mathbf{k} = 0 \tag{D6}$$

These equations define the ray trajectories along which the waves travel. In addition, a dynamical equation for conservation of wave action can be written as

$$\frac{\partial A}{\partial t} + \frac{\partial}{\partial x_i}\left[\left(U_i + c_{g_i}\right)A\right] = S \tag{D7}$$

where A is the action density given by

$$A = \frac{E}{n} = \frac{E}{\omega - k_i U_i} \tag{D8}$$

where E is the energy density of the waves averaged over the wave period. E includes both the kinetic and the potential energy. S represents the effect of sources and sinks due to generation and dissipation. Equation (D8) with S = 0 holds for nondissipative waves, which conserve their energy. It is quite useful in dealing with small amplitude wave motions in a background flow, where the energy density is not conserved, but the action density is.

References

Agarwal, Y. C., E. A. Terray, M. A. Donelan, P. A. Hwang, A. J. Williams III, W. M. Drennan, K. K. Kahma, and S. A. Kitaigorodskii. 1992. Enhanced dissipation of kinetic energy beneath surface waves. *Nature* **359,** 219–220.

Albrecht, B. A., R. S. Penc, and W. H. Schubert. 1985. An observational study of cloud-topped mixed layers. *J. Atmos. Sci.* **42,** 800–822.

Albrecht, B. A., D. A. Randall, and S. Nicholls. 1988. Observations of marine stratocumulus clouds during FIRE. *Bull. Am. Meteorol. Soc.* **69,** 618–626.

Albrecht, B. A., C. S. Bretherton, D. Johnson, W. H. Schubert, and A. S. Frisch. 1995. The Atlantic Stratocumulus Transition Experiment—ASTEX. *Bull. Am. Meteorol. Soc.* **76,** 889–904.

Allen, J. S. 1980. Models of wind-driven currents on the continental shelf. *Annu. Rev. Fluid Mech.* **12,** 389–433.

Al-Zanaidi, M. A., and W. H. Hui. 1984. Turbulent air flow over water waves. *J. Fluid Mech.* **148,** 225–246.

Andre, J. C., G. De Moor, P. Lacarrere, G. Therry, and R. du Vachat. 1978. Modeling the 24-hour evolution of the mean and turbulent structures of the planetary boundary layer. *J. Atmos. Sci.* **35,** 1861–1883.

Andre, J. C., and L. Mahrt. 1982. The nocturnal surface inversion and influence of clear-air radiative cooling. *J. Atmos. Sci.* **39,** 864–878.

Andre, J. C., and P. Lacarrere. 1985. Mean and turbulent structures of the oceanic surface layer as determined from one-dimensional third order simulations *J. Phys. Oceanogr.* **15,** 121–132.

Andreas, E. L. 1987a. On the Kolmogoroff constants for the temperature-humidity cospectrum and the refractive index spectrum. *J. Atmos. Sci.* **44,** 2399–2406.

Andreas, E. L. 1987b. A theory for the scalar roughness and the scalar transfer coefficients over snow and sea ice. *Boundary Layer Meteorol.* **38,** 159–184.

Andreas, E. L. 1992. Sea spray and the turbulent air-sea heat fluxes. *J. Geophys. Res.* **97,** 11429–11441.

Andreas, E. L. 1996. *The Atmospheric Boundary Layer over Polar Marine Surfaces.* U.S. Army Corps of Engineers Cold Regions Research and Engineering Laboratory Monograph, Vol. 96-2. 38 pp.

Andreas, E. L. 1998. A new sea spray generation function for wind speeds up to 32 m s^{-1}. *J. Phys. Oceanogr.* **28,** 2175–2184,

Andreas, E. L., and S. F. Ackley. 1982. On the difference in ablation seasons of Arctic and Antarctic sea ice. *J. Atmos. Sci.* **39,** 440–447.

Andren, A., and C.-H. Moeng. 1993. Single point closures in a neutrally stratified boundary layer. *J. Atmos. Sci.* **50,** 3366–3379.

Andren, A. 1995. The structure of stably stratified atmospheric boundary layers: A large-eddy simulation study. *Q. J. R. Meteorol. Soc.* **121,** 961–985.

Andren, A., A. R. Brown, J. Graf, P. J. Mason, C.-H. Moeng, F. T. M. Nieuwstadt, and U. Schumann. 1994. Large-eddy simulation of a neutrally stratified boundary layer: A comparison of four computer codes. *Q. J. R. Meteorol. Soc.* **120,** 1457–1484.

Andrews, D. G., J. R. Holton, and C. B. Leovy. 1987. *Middle Atmosphere Dynamics.* Academic Press, San Diego.

Angevine, W. M., A. W. Grimsdell, L. M. Hartten, and A. C. Delany. 1998. The Flatland Boundary Layer Experiments. *Bull. Am. Meteorol. Soc.* **79,** 419–431.

Anis, A., and J. N. Moum. 1992. The superadiabatic surface layer of the ocean during convection. *J. Phys. Oceanogr.* **22,** 1221–1227.

Anis, A., and J. N. Moum. 1995. Surface wave-turbulence interactions: Scaling $\varepsilon(z)$ near the sea surface. *J. Phys. Oceanogr.* **25,** 2025–2045.

Apel, J. R. 1987. *Principles of Ocean Physics.* Academic Press, San Diego. 631 pp.

Apel, J. R., J. R. Holbrook, A. K. Liu, and J. J. Tsai. 1985. The Sulu Sea Internal Soliton Experiment. *J. Phys. Oceanogr.* **15,** 1625–1651.

Armi, L., and R. C. Millard. 1976. The bottom boundary layer of the deep ocean. *J. Geophys. Res.* **81,** 4983–4990.

Armi, L., D. Hebert, N. Oakey, J. F. Price, P. L. Richardson, H. T. Rossby, and B. Ruddick. 1989. Two years in the life of a Mediterranean salt lens. *J. Phys. Oceanogr.* **19,** 354–370.

Arnone, R. A., and R. W. Gould. 1998. Coastal monitoring using ocean color. *Sea Technol.,* Sept. 18–27.

Arya, S. P. S. 1988. *Introduction to Micrometeorology.* Academic Press, New York. 303 pp.

Atakturk, S. S., and K. B. Katsaros. 1999. Wind stress and surface waves observed on Lake Washington. *J. Phys. Oceanogr.* **29,** 633–650.

Atkinson, B. W., and J. Wu. Zhang. 1996. Mesoscale shallow convection in the atmosphere. *Rev. Geophys.* **34,** 403–431.

Atlas, D. (Ed.). 1990. *Radar in Meteorology.* American Meteorological Society, Boston, MA.

Avissar, R., and M. M. Verstraete. 1990. The representation of continental surface processes in atmospheric models. *Rev. Geophys.* **28,** 35–52.

Bacmeister, J. T., and R. T. Pierrehumbert. 1988. On high-drag states of nonlinear stratified flow over an obstacle.

Bacon, C. R., and M. A. Lanphere. 1990. The geologic setting of Crater Lake. In *Crater Lake: An Ecosystem Study,* edited by E. T. Drake, G. L. Larson, J. Dymond, and R. Collier, pp. 19–28. AAAS, Pacific Division, San Francisco, CA.

Baines, P. G. 1987. Upstream blocking and airflow over mountains. *Annu. Rev. Fluid Mech.* **19,** 75–97.

Baines, P. G., and K. P. Hoinka. 1985. Stratified flow over two-dimensional topography in fluid of infinite depth: A laboratory simulation. *J. Atmos. Sci.* **42,** 1614–1630.

Balk, A. M. 1996. A Lagrangian for water waves. *Phys. Fluids* **8,** 416–420.

Ball, F. K. 1960. Control of inversion height by surface heating. *Q. J. R. Meteorol. Soc.* **86,** 483–494.

Bang, I.-K., J.-K. Choi, L. Kantha, C. Horton, M. Clifford, M.-S. Suk, K.-I. Chang, S.-Y. Nam, and H.-J. Lie. 1996. A hindcast experiment in the East Sea (sea of Japan). *La Mer,* **34,** 108–130.

Banner, M. 1990a. Equilibrium spectra of wind waves. *J. Phys. Oceanogr.* **20,** 966–984.

Banner, M. 1990b. The influence of wave breaking on the surface pressure distribution in wind-wave interactions. *J. Fluid Mech.* **211,** 463–495.

Banner, M. L., and R. H. J. Grimshaw (Eds.). 1992. *Breaking Waves*. Springer-Verlag, Berlin/New York.

Banner, M. L., and D. H. Peregrine. 1993. Wave breaking in deep water. *Annu. Rev. Fluid Mech.* **25,** 373–397.

Bardina, J., J. H. Ferziger, and R. S. Rogallo. 1985. Effect of rotation on isotropic turbulence: Computation and modelling. *J. Fluid Mech.* **154,** 321–336.

Barenblatt, G. I. 1978. Strong interaction of gravity waves and turbulence. *Izv. Acad. Sci. USSR, Atmos. Oceanic Phys.* **13,** 581–583.

Barenblatt, G. I., and N. Goldenfeld. 1995. Does fully developed turbulence exist? Reynolds number independence versus asymptotic covariance. *Phys. Fluids* **7,** 3078–3082.

Barry, R. G., M. C. Serreze, J. A. Maslanik, and R. H. Preller. 1993. The Arctic sea ice-climate system: Observations and modeling. *Rev. Geophys.* **31,** 397–422.

Barstow, S. 1996. Validating TOPEX/Poseidon significant wave heights. *AVISO Newsletter* **4,** 23.

Batchelor, G. K. 1953. *The Theory of Homogeneous Turbulence*. Cambridge Univ. Press, London. 197 pp.

Batchelor, G. K. 1959. Small scale variation of convected quantities like temperature in a turbulent fluid, part 1. *J. Fluid Mech.* **5,** 113.

Batchelor, G. K. 1967. *An Introduction to Fluid Mechanics*. Cambridge Univ. Press, London. 615 pp.

Batchelor, G. K. 1996. *The Life and Legacy of G. I. Taylor*. Cambridge Univ. Press, Cambridge.

Behrenfeld, M. J., and Z. S. Kolber. 1999. Widespread iron limitation of phytoplankton in the south Pacific Ocean. *Science* **283,** 840–843.

Beljaars, A. C. M., and A. A. M. Holtslag. 1991. Flux parameterization over land surfaces for atmospheric models. *J. Appl. Meteorol.* **30,** 327–341.

Bell, T. H., Jr. 1975. Topographically generated internal waves in the deep ocean. *J. Geophys. Res.* **80,** 320–327.

Bell, T. H., Jr. 1978. Radiation damping of inertial oscillations in the upper ocean. *J. Fluid Mech.* **88,** 289–308,

Bennett, T. J. 1982. A coupled atmosphere-sea ice model study of the role of sea ice in climate predictability. *J. Atmos. Sci.* **39,** 1456–1465.

Benoit, R. 1977. On the integral of the surface layer proile-gradient functions. *J. Appl. Meteorol.* **16,** 859–860.

Benoit, R. M., M. Desgagne, P. Pellerin, S. Pellerin, Y. Chartier, and S. Desjardins. 1997. The Canadian MC2: A semi-Lagrangian, semi-implicit wide-band atmospheric model suited for finescale process studies and simulation. *Mon. Weather Rev.* **125,** 2382–2415.

Belcher, S. E., and J. C. R. Hunt. 1998. Turbulent flow over hills and waves. *Annu. Rev. Fluid Mech.* **30,** 507–538.

Bergstrom, H., and A. Smedman. 1994. Stably stratified flow in a marine surface layer. *Boundary Layer Meteorol.* **72,** 239–265.

Berliand, M. E., and T. G. Berliand. 1952. Determining the net long-wave radiation of the Earth with consideration of the effects of cloudiness. *Izv. Nauk. SSSR Ser. Geofiz.* **1.**

Berliand, T. G. 1960. Methods of climatological computation of total incoming solar radiation (in Russian). *Meteorol. Gidrol.* **6,** 9–12.

Betts, A. K. 1973. Non-precipitating cumulus convection and its parameterization. *Q. J. R. Meteorol. Soc.* **99,** 178–196.

Bignami, F., S. Marullo, R. Santoleri, and M. E. Schiano. 1995. Longwave radiation budget in the Mediterranean Sea. *J. Geophys. Res.* **100,** 2501–2514.

Birch, K. G., and J. A. Ewing. 1986. *Observations of Wind Waves on a Reservoir*. IOS Report no. 234, Wormley, UK. 37 pp.

Bishop, J. K. B., and W. B. Rossow. 1991. Spatial and temporal variability of global surface solar irradiance. *J. Geophys. Res.* **96,** 16839–16858.

Blackader, A. K. 1962. The vertical distribution of wind and turbulence exchange in a neutral atmosphere. *J. Geophys. Res.* **67,** 3095–3102.

Blackadar, A. K. 1979. High resolution models of the planetary boundary layer. *Adv. Environ. Sci. Eng.* **1,** 50–85.

Blanc, T. V. 1985. Variation of bulk-derived surface flux, stability and roughness results due to the use of different transfer coefficient schemes. *J. Phys. Oceanogr.* **15,** 650–659.

Blumberg, A. F., and G. L. Mellor. 1987. A description of a three-dimensional coastal ocean circulation model. In *Three-Dimensional Coastal Ocean Models,* Vol. 4, pp. 208. American Geophysical Union, Washington, DC.

Bohle-Carbonell, M. 1986. Currents in Lake Geneva. *Limnol. Oceanogr.* **31,** 1255–1266.

Bond, N. A., and M. J. McPhaden. 1995. An indirect estimate of the diurnal cycle in the upper ocean turbulent fluxes at the equator, 140 °W. *J. Geophys. Res.* **100,** 18369–18378.

Bonmarin, P. 1989. The geometric properties of deep-water breaking waves. *J. Fluid Mech.* **209,** 405–433.

Booker, J. R., and F. P. Bretherton. 1967. The critical layer for internal gravity waves in a shear flow. *J. Fluid Mech.* **27,** 513–539.

Bougeault, P., A. Jansa Clar, B. Benech, B. Carissimo, J. Pelon, and E. Richard. 1990. Momentum budget over the Pyrenees: The PYREX experiment. *Bull. Am. Meteorol. Soc.* **71,** 806–818.

Bourassa, M. A., D. G. Vincent, and W. L. Wood. 1999. A flux parameterization including the effects of capillary waves and sea state. *J. Atmospheric Sci.* **56,** 1123–1139.

Bouws, E. 1986. *Provisional Results of a Wind Wave Experiment in a Shallow Lake(Lake Marken, The Netherlands).* KNMI Afdeling Oceanografisch Onderzoek Memo OO-86-21, De Bilt. 15 pp.

Bowden, K. F. 1978. Physical problems of the benthic boundary layer. *Geophys. Survey* **3,** 255–296.

Boyce, F. M. 1975. Internal waves in the Straits of Gibraltar. *Deep-Sea Res.* **22,** 597–610.

Bradley, E. F. 1968. A micrometeorological study of velocity profiles and surface drag in the region modified by a change in surface roughness. *Q. J. R. Meteorol. Soc.* **94,** 361–379.

Bradley, E. F., P. A. Coppin, and J. S. Godfrey. 1991. Measurements of sensible and latent heat flux in the western equatorial Pacific Ocean. *J. Geophys. Res.* **96,** 3375–3389.

Bradley, E. F., J. S. Godfrey, P. A. Coppin, and J. A. Butt. 1993. Observations of net heat flux into the surface mixed layer of the western equatorial Pacific Ocean. *J. Geophys. Res.* **98,** 22521–22532.

Bradshaw, P. 1969. The analogy between streamline curvature and buoyancy in turbulent shear flow. *J. Fluid Mech.* **36,** 177–191.

Brainard, R. E., M. J. McPhaden, and R. W. Garwood, Jr. 1994. Observations of the diurnal cycle of high frequency internal waves and their relationship to mixing in the central equatorial Pacific. *Eos Trans. AGU.* **75,** 121.

Brandt, A., and H. J. S. Fernando (Eds.). 1996. *Double Diffusive Convection.* American Geophysical Union, Washington, DC.

Bretherton, F. P. 1966. The propagation of groups of internal gravity waves in a shear flow. *Q. J. R. Meteorol. Soc.* **92,** 466–480.

Brewer, P. G. 1983. Carbon dioxide and the oceans. In *Changing Climate.* National Academy Press, Washington, DC.

Broecker, W. S., and T. Takahashi. 1978. The relationship between lysocline depth and in situ carbonate ion concentration. *Deep-Sea Res.* **25,** 65–95.

Broecker, W. S., and T.-H. Peng. 1982. *Tracers in the Sea.* Eldigio, New York. 690 pp.

Broecker, W. S., *et al.* 1986. Isotopic versus micro meteorological ocean CO_2 fluxes: A serious conflict. *J. Geophys. Res.* **91,** 10517–10527.

Brost, R. A., and J. C. Wyngaard. 1978. A model study of the stably stratified planetary boundary layer. *J. Atmos. Sci.* **35,** 1427–1440.

Brown, A. R., S. H. Derbyshire, and P. J. Mason. 1994. Large-eddy simulation of stable atmospheric boundary layer with a revised stochastic subgrid model. *Q. J. R. Meteorol. Soc.* **120,** 1485–1512.

Brown, R. A. 1990. Meteorology. In *Polar Oceanography*, edited by W. O. Smith, Jr., pp. 1–46. Academic Press, San Diego.

Brunt, D. 1932. Notes on radiation in the atmosphere. *Q. J. R. Meteorol. Soc.* **58,** 389–420.

Brutsaert, W. 1975. The roughness length for water vapor, sensible heat and other scalars. *J. Atmos. Sci.* **32,** 2028–2031.

Brutsaert, W. 1975. A theory for local evaporation (or heat transfer) from rough and smooth surfaces at ground level. *Water Resour. Res.* **11,** 543–550.

Brutsaert, W. 1982. *Evaporation into the Atmosphere: Theory, History and Applications.* Reidel, MA. 229 pp.

Brutsaert, W. 1991. The formulation of evaporation from land surfaces. In *Recent Advances in the Modeling of Hydrologic Systems,* edited by D. S. Bowles and P. E. OÕConnell, pp. 67–84. Kluwer Academic, Norwell, MA.

Brutsaert, W., and G. H. Jerka (Eds.). 1984. *Gas Transfer at Water Surfaces.* Reidel, Dordrecht.

Budyko, M. I. 1974. *Climate and Life.* Academic Press, San Diego. 508 pp.

Bunker, A. F. 1976. Computations of surface energy flux and annual air-sea interaction cycles of the North Atlantic Ocean. *Mon. Weather Rev.* **104,** 1122–1140.

Burchard, H., O. Petersen, and T. P. Rippeth. 1998. Comparing the performance of the Mellor-Yamada and the k-ε two-equation turbulence models. *J. Geophys. Res.* **103,** 10543–10554.

Businger, J. A. 1988. A note on the Businger-Dyer profiles. *Boundary Layer Meteorol.* **42,** 145–151.

Businger, J. A., J. C. Wyngaard, Y. Izumi, and E. F. Bradley. 1971. Flux-profile relationships in the atmospheric surface layer. *J. Atmos. Sci.* **28,** 181–189.

Byrne, J. V. 1965. Morphometry of Crater Lake, Oregon. *Limnol. Oceanogr.* **10,** 462–465.

Cahalan, R. F., D. Silberstein, and J. B. Snider. 1995. Liquid water path and plane-parallel albedo bias during ASTEX. *J. Atmos. Sci.* **52,** 3002–3012.

Cairns, J. L. 1975. Internal wave measurements from a mid-water float. *J. Geophys. Res.* **80,** 299–306.

Cairns, J. L., and G. O. Williams. 1976. Internal wave observations from a midwater float, 2. *J. Geophys. Res.* **81,** 1943–1950.

Caldwell, D. R. 1978. Variability in the bottom mixed layer on the Oregon shelf. *Deep-Sea Res.* **25,** 1235–1243.

Caldwell, D. R., and W. P. Elliott. 1971. Surface stresses produced by rainfall. *J. Phys. Oceanogr.* **2,** 145–148.

Canuto, V. M., and M. S. Dubovikov. 1996a. A dynamical model for turbulence. I. General formulism. *Phys. Fluids.* **8,** 571–586.

Canuto, V. M., and M. S. Dubovikov. 1996b. A dynamical model for turbulence. II. Shear-driven flows. *Phys. Fluids* **8,** 587–598.

Canuto, V. M., and M. S. Dubovikov. 1996c. A dynamical model for turbulence. III. Numerical results. *Phys. Fluids* **8,** 599–613.

Carlson, H., K. Richter, and H. Walden. 1967. Messungen der statistischen verteilung der auslenkung der meeresoberflache im seegang. *Dt. Hydrogr. Z.* **20,** 59–64.

Carmack, E. C. 1986. Circulation and mixing in ice-covered waters. In *The Geophysics of Sea Ice,* ed. N. Untersteiner, pp. 641–712. Plenum, New York.

Carmack, E. C. 1990. Large-scale physical oceanography of polar oceans. In *Polar Oceanography,* ed. W. O. Smith, Jr., pp. 171–222. Academic Press, San Diego.

Carmack, E. C., and T. D. Foster. 1975. On the flow of water out of the Weddell Sea. *Deep-Sea Res.* **22,** 711–724.

Carmack, E. C., and D. M. Farmer. 1982. Cooling processes in deep temperate lakes: A review with examples from two lakes in British columbia. *J. Mar. Res.* **40,** 85–111.

Carmack, E. C., R. C. Wiegand, R. J. Daley, C. B. J. Gray, S. Jasper, and C. H. Pharo. 1986. Mechanisms influencing the circulation and distribution of water mass in a medium residence-time lake. *Limnol. Oceanogr.* **31,** 249–265.

Carmack, E. C., and R. F. Weiss. 1991. Convection in Lake Baikal: An example of thermobaric instability. In *Deep Convection and Deep Water Formation in the Oceans,* ed. P. C. Chu and J. C. Gascard. Elsevier, Amsterdam/New York.

Carmack, E. C., K. Aagaard, J. H. Swift, R. G. Perkin, F. A. McLaughlin, R. W. Macdonald, and E. P. Jones. 1995. Thermohaline transitions. In *Physical Processes in Lakes and Oceans,* ed. J. Imberger. International Union of Theoretical and Applied Mechanics.

Carson, D. J. 1973. The development of a dry inversion-capped convectively unstable boundary layer. *Q. J. R. Meteorol. Soc.* **99,** 450–467.

Cartwright, D. E. 1963. The use of directional spectra in studying the output of a wave recorder on a moving ship. In *Ocean Wave Spectra.* Prentice-Hall, Englewood Cliffs, NJ, 203–218.

Caughey, S. J. 1982. Observed characteristics of the atmospheric boundary layer. In *Atmospheric Turbulence and Air Pollution Modelling,* ed. F. T. M. Nieuwstadt and H. van Dop, pp. 107–158. Riedel, Dordrecht.

Caughey, S. J., and S. G. Palmer. 1979. Some aspects of turbulence structure through the depth of the convective boundary layer. *Q. J. R. Meteorol. Soc.* **105,** 811–827.

Caughey, S. J., and J. C. Wyngaard. 1979. The turbulent kinetic energy budget in convective conditions. *Q. J. R. Meteorol. Soc.* **105,** 231–239.

Caughey, S. J., J. C. Wyngaard, and J. C. Kaimal. 1979. Turbulence in the evolving stable boundary layer. *J. Atmos. Sci.* **36,** 1041–1052.

Champagne, F. 1978. The fine scale structure of the turbulent velocity field. *J. Fluid Mech.* **86,** 67–108.

Charleson, R. J., J. E. Lovelock, M. O. Andreae, and S. G. Warren. 1987. Ocean phytoplankton, atmospheric sulfur, cloud albedo and climate. *Nature* **326,** 655–661.

Charnock, H. 1955. Wind stress on a water surface. *Q. J. R. Meteorol. Soc.* **81,** 639–640.

Chasnov, J. R. 1991. Simulation of the Kolmogorov inertial subrange using an improved subgrid model. *Phys. Fluids A* **3,** 188–200.

Chen, C. F., and D. H. Johnson. 1984. Double-diffusive convection: A report on an engineering Foundation Conference. *J. Fluid Mech.* **138,** 405–416.

Chen, C. T., and F. J. Millero. 1986. Precise thermodynamic properties for natural waters covering only the limnological range. *Limnol. Oceanogr.* **31,** 657–662.

Choi, J. K., L. H. Kantha, and R. R. Leben. 1995. A nowcast/forecast experiment using TOPEX/POSEIDON and ERS-1 altimetric data assimilation into a three-dimensional circulation model of the Gulf of Mexico. IAPSO Abstract, XXI General Assembly.

Chollet, J.-P., and M. Lesieur. 1981. Parameterization of small scale of three-dimensional isotropic turbulence utilizing spectral closures. *J. Atmos. Sci.* **38,** 2747–2757.

Chou, S.-H., and E.-N. Yeh. 1987. Airborne measurements of surface layer turbulence over the ocean during cold air outbreaks. *J. Atmos. Sci.* **44,** 3721–3733.

Clark, N. E., L. Eber, R. M. Laurs, J. A. Renner, and J. F. T. Saur. 1974. *Heat Exchange between Ocean and Atmosphere in the Eastern North Pacific for 1961–1971.* NOAA Technical Rep. NMFS SSRF-682, U.S. Dept. of Commerce, Washington, DC.

Clark, T. L., and W. R. Peltier. 1984. Critical level reflection and the resonant growth of nonlinear mountain waves. *J. Atmos. Sci.* **41,** 3122–3134.

Clark, T. L., W. D. Hall, and R. M. Banta. 1994. Two- and three-dimensional simulations of the 9 January 1989 severe Boulder windstorm: Comparison with observations. *J. Atmos. Sci.* **51,** 2317–2343.

Clarke, R. H., A. J. Dyer, R. R. Brook, D. G. Reid, and A. J. Troup. 1971. *The Wangara Experiment: Boundary Layer Data.* Technical Paper 19, Div. Meteor. Phys., CSIRO Australia. 363 pp. [NTIS N71-37838]

Clayson, C. A. 1995. Modeling the mixed layer in the tropical Pacific and air-sea interactions: Application to a westerly wind burst. Ph.D. dissertation, Dept. Aerospace Engineering Sciences, University of Colorado, Boulder.

Clayson, C. A., and J. A. Curry. 1996. Determination of surface turbulent fluxes for TOGA COARE: Comparison of satellite retrievals and in situ measurements. *J. Geophys. Res.* **101,** 28,515–28,528.

Clayson, C. A., J. A. Curry, and C. W. Fairall. 1996. Evaluation of turbulent fluxes at the ocean surface using surface renewal theory. *J. Geophys. Res.* **101,** 28,503–28,513.

Clayson, C. A., A. Chen, L. H. Kantha, and P. J. Webster. 1997. Numerical simulations of the equatorial Pacific during the TOGA/COARE IOP. AMS abstract, 22nd Conference on Hurricanes and Tropical Meteorology, May 19–23, 1997, Fort Collins, CO, pp. 600–601.

Clayson, C. A., and L. H. Kantha. 1999. Turbulent kinetic energy and dissipation rate in the equatorial mixed layer. *J. Phys. Oceanogr.* **29,** 2146–2166.

Clayson, C. A., and L. H. Kantha. 2000. On the effect of surface gravity waves on mixing in an oceanic mixed layer model. (in review)

Coates, M. J., G. N. Ivey, and J. R. Taylor. 1995. Unsteady turbulent convection into rotating, linearly stratified fluid: Modeling deep ocean convection. *J. Phys. Oceanogr.* **25,** 3032–3050.

Codiga, D. L. 1997a. Physics and observational signatures of free, forced, and frictional stratified seamount-trapped waves. *J. Geophys. Res.* **102,** 23009–23024.

Codiga, D. L. 1997b. Trapped wave modification and critical surface formation by mean flow at a seamount with applications at Fieberling Guyot. *J. Geophys. Res.* **102,** 23025–23040.

Codiga, D. L., and C. C. Eriksen. 1997. Observations of low-frequency circulation and amplified subinertial tidal currents at Cobb Seamount. *J. Geophys. Res.* **102,** 22993–23008.

Coleman, G. N., J. H. Ferziger, and P. R. Spalart. 1990. A numerical study of the turbulent Ekman layer. *J. Fluid Mech.* **213,** 313–348.

Coleman, G. N., J. H. Ferziger, and P. R. Spalart. 1992. Direct simulation of the stably stratified turbulent Ekman layer. *J. Fluid Mech.* **244,** 677–712.

Coles, D. 1956. The law of the wake in the turbulent boundary layer. *J. Fluid Mech.* **1,** 191.

Comte Bellot, G., and S. Corrsin. 1971. Simple Eulerian time correlation of full and narrow velocity signals in grid-generated isotropic turbulence. *J. Fluid Mech.* **48,** 273.

Coppin, P. A., E. F. Bradley, I. J. Barton, and J. S. Godfrey. 1991. Simultaneous observations of sea surface temperature in the Western Equatorial Pacific Ocean by bulk, radiative and satellite methods. *J. Geophys. Res.* **96,** 3401–3409.

Corrsin, S. 1951. On the spectrum of isotropic temperature fluctuations in isotropic turbulence. *J. Appl. Phys.* **22,** 469–473.

Corrsin, S., and A. L. Kistler. 1954. *The Free Stream Boundaries of Turbulent Flows.* National Advisory Committee on Aeronautics Technical Note NACA TN 3133.

Corrsin, S. 1974. Limitations of gradient transport models in random walk and in turbulence. *Adv. Geophys. A* **18,** 25–71.

Cotton, P. D., and D. J. T. Carter. 1994. Cross calibration of TOPEX, ERS-1, and Geosat wave heights. *J. Geophys. Res.* **99,** 25025–25033.

Cox, C. S., and W. H. Munk. 1954. Statistics of the sea surface derived from Sun glitter. *J. Mar. Res.* **13,** 198–227.

Craig, P. D., and M. L. Banner. 1994. Modeling wave-enhanced turbulence in the ocean surface layer. *J. Phys. Oceanogr.* **24,** 2546–2559.

Craik, A. D. D. 1970. A wave-interaction model for the generation of windrows. *J. Fluid Mech.* **41,** 802–822.

Craik, A. D. D. 1977. The generation of Langmuir circulations by an instability mechanism. *J. Fluid Mech.* **81,** 209–223.

Craik, A. D. D. 1982. The drift velocity of water waves. *J. Fluid Mech.* **116,** 187–205.

Craik, A. D. D., and S. Leibovich. 1976. A rational model for Langmuir circulation. *J. Fluid Mech.* **73,** 401–426.

Crawford, G. B., and R. W. Collier. 1997. Observation of a deep mixing event in Crater Lake, Oregon. *Limnol. Oceanogr.* **42,** 299–306.

Crawford, G. B., R. W. Collier, and K. T. Redmond. In preparation. On the heat and water budgets of Crater Lake, Oregon.

Crawford, G. C. 1993. Upper ocean response to storms—A resonant system. Ph.D. dissertation, Univ. of British Columbia, Vancouver.

Crawford, W. R., and T. R. Osborn. 1981. Control of equatorial ocean currents by turbulent dissipation. *Science* **212,** 539–540.

Crow, S. C. 1968. Viscoelastic properties of fine-grained incompressible turbulence. *J. Fluid Mech.* **33,** 1–20.

Csanady, G. T. 1974. Equilibrium theory of the planetary boundary layer with an inversion lid. *Boundary Layer Meteorol.* **6,** 63–79.

Csanady, G. T. 1985. Hydrodynamics of large lakes. *Annu. Rev. Fluid Mech.* **17,** 357–386.

Csanady, G. T. 1990. The role of breaking wavelets in air-sea gas transfer. *J. Geophys. Res.* **95,** 749–759.

Cummins, P. F. 1995. Numerical simulations of upstream bores and solitons in a two-layer flow past an obstacle. *J. Phys. Oceanogr.* **25,** 1504–1515.

Cummins, P. F., and M. Li. 1998. Comment on "Energetics of borelike internal waves" by Frank S. Henyey and Antje Hoering. *J. Geophys. Res.* **103,** 3339–3341.

Curry, J. A., C. A. Clayson, W. B. Rossow, Y. Zhang, and P. J. Webster. 1993. Determination of the tropical sea surface energy balance from satellite. In *Proceedings, Twentieth Conference on Hurricanes and Tropical Meteorology, 10–14 May 1993,* pp. 591–594. American Meteorological Society, Providence, RI.

Dabberdt, W. F., D. H. Lenschow, T. W. Horst, P. R. Zimmerman, S. P. Oncley, and A. C. Delany. 1993. Atmosphere-surface exchange measurements. *Science* **260,** 1472–1480.

Dahl, P. H., and A. T. Jessup. 1995. On bubble clouds produced by breaking waves: An event analysis of ocean acoustic measurements. *J. Geophys. Res.* **100,** 5007–5020.

Dalu, G. A., and R. Purini. 1981. The diurnal thermocline due to buoyant convection. *Q. J. R. Meteorol. Soc.* **108,** 929–935.

D'Asaro, E. A. 1985. The energy flux from the wind to near-intertial motions in the surface mixed layer. *J. Phys. Oceanogr.* **15,** 1043–1059.

D'Asaro, E. A., D. M. Farmer, J. Osse, and G. T. Dairiki. 1996. A Lagrangian float. *J. Atmos. Ocean. Tech.* **13,** 1230–1246.

Davis, R. E., R. DeSzoeke, and P. P. Niiler. 1981a. Variability in the upper ocean during MILE. I. The heat and momentum balances. *Deep-Sea Res.* **28,** 1427–1451.

Davis, R. E., R. DeSzoeke, and P. P. Niiler. 1981b. Variability in the upper ocean during MILE. II. Modeling the mixed layer response. *Deep-Sea Res.* **28,** 1453–1475.

Dawson, T. H. 1997. Group structure in random seas. *J. Atmos. Oceanic Phys.* **14,** 741–747.

Deardorff, J. W. 1966. The counter-gradient heat flux in the lower atmosphere and in the laboratory. *J. Atmos. Sci.* **23,** 503–506.

Deardorff, J. W. 1972a. Theoretical expression for the countergradient vertical heat flux. *J. Geophys. Res.* **77,** 5900–5904.

Deardorff, J. W. 1972b. Parameterization of the planetary boundary layer for use in general circulation models. *Mon. Weather Rev.* **100,** 93–106.

Deardorff, J. W. 1973. The use of subgrid transport equations in a three-dimensional model of atmospheric turbulence. *J. Fluids Engineering, Trans. ASME* **95,** 429–438.

Deardorff, J. W. 1980a. Progress in understanding entrainment at the top of a mixed layer. In *A Workshop on the Planetary Boundary Layer,* ed. J. C. Wyngaard, pp. 36–66. American Meteorological Society, Providence, RI.

Deardorff, J. W. 1980b. Cloud-top entrainment instability. *J. Atmos. Sci.* **37,** 131–147.

Deardorff, J. W. 1985. Sub-grid-scale turbulence modeling. *Adv. Geophys. B* **28,** 337–343.

Deardorff, J. W., G. E. Willis, and D. K. Lilly. 1969. Laboratory investigation of non-steady penetrative convection. *J. Fluid Mech.* **35,** 7–31.

Deardorff, J. W., and G. E. Willis. 1985. Further results from a laboratory model of the convective planetary boundary layer. *Boundary Layer Meteorol.* **32,** 205–236.

DeCosmo, J., K. B. Katsaros, S. D. Smith, R. J. Anderson, W. A. Oost, K. Bumke, and H. Chadwick. 1996. Air-sea exchange of water vapor and sensible heat: The humidity exchange over the sea (HEXOS) results. *J. Geophys. Res.* **101,** 12001–12016.

Delage, Y. 1988. A parameterization of the stable atmospheric boundary layer. *Boundary Layer Meteorol.* **43,** 365–381.

Denbo, D. W., and E. D. Skyllingstad. 1996. An ocean large-eddy simulation model with application to deep convection in the Greenland Sea. *J. Geophys. Res.* **101,** 1095–1110.

Denman, K. L. 1973. A time-dependent model of the upper ocean. *J. Phys. Oceanogr.* **3,** 173–184.

Denman, K. L., and A. E. Gargett. 1995. Biological-physical interactions in the upper ocean: The role of vertical and small scale transport processes. *Annu. Rev. Fluid Mech.* **27,** 225–255.

Denmead, O. T., and E . F. Bradley. 1987. On scalar transport in plant canopies. *Irrig. Sci.* **8,** 131–149.

Derbyshire, S. H. 1990. Nieuwstadt's stable boundary layer revisited. *Q. J. R. Meteorol. Soc.* **116,** 127–158.

Derbyshire, S. H. 1994. A "balanced" approach to stable boundary layer dynamics. *J. Atmos. Sci.* **51,** 3486–3504.

Dietrich, G., K. Kalle, W. Krauss, and G. Siedler. 1980. *General Oceanography*, pp. 626. Wiley, New York.

Dillon, T. M., J. N. Moum, T. K. Chereskin, and D. R. Caldwell. 1989. Zonal momentum balance at the equator *J. Phys. Oceanogr.* **19,** 561–570.

Ding, L., and D. M. Farmer. 1994. Observations of breaking surface wave statistics. *J. Phys. Oceanogr.* **24,** 1368–1387.

Dobson, F. W., and S. D. Smith. 1988. Bulk models of solar radiation at sea. *Q. J. R. Meteorol. Soc.* **114,** 165–182.

Domaradzki, J. A. 1988. Analysis of energy transfer in direct numerical simulations of isotropic turbulence. *Phys. Fluids* **31,** 2747–2758.

Domaradzki, J. A., and R. S. Rogallo. 1990. Local energy transfer and nonlocal interactions in homogeneous isotropic turbulence. *Phys. Fluids* **2,** 413–425.

Domaradzki, J. A., W. Liu, and M. E. Brachet. 1993. An analysis of subgrid scale interactions in numerically simulated isotropic turbulence. *Phys. Fluids A* **5,** 1747–1759.

Domaradzki, J. A., W. Liu, C. Hartel, and L. Kleiser. 1994. Energy transfer in numerically simulated wall-bounded flows. *Phys. Fluids* **6,** 1583–1593.

Donelan, M. A. 1979. On the fraction of wind momentum retained by waves. In *Marine Forecasting: Prediction and Modelling in Ocean Hydrodynamics,* ed. J. Nihoul, pp. 141–159. Elsevier, Amsterdam/New York.

Donelan, M. A. 1990. Air-sea interaction. In *The Sea: Ocean Engineering Science*, eds. B. LeMehaute and D. Hanes, Vol. 9, pp. 239–292. Wiley, New York.

Donelan, M. A., K. N. Birch, and D. C. Beesley. 1974. Generalized Profiles of Wind Speed, Temperature, and Humidity. *Proceedings of the 17th Conference on Great Lakes Research,* International Association for Great Lakes Research, 369–388.

Donelan, M. A., M. S. Longuet-Higgins, and J. S. Turner. 1972. Periodicity in whitecaps. *Nature* **239,** 449–451.

Donelan, M. A., J. Hamilton, and W. H. Hui. 1985. Directional spectra of wind generated waves. *Philos. Trans. R. Soc. London, Ser. A* **315,** 509–562.

Donelan, M. A., and W. J. Pierson. 1987. Radar scattering and equilibrium ranges in wind-generated waves with applications to scatterometry. *J. Geophys. Res.* **92,** 4971–5029.

Donelan, M. A., and W. H. Hui. 1990. Mechanics of ocean surface waves. In G. L. Geernaert and W. J. Plant (Eds.), *Surface Waves and Fluxes,* Vol. I, pp. 209–246. Kluwer Academic, Dordrecht/Norwell, MA.

Donelan, M. A., F. W. Dobson, S. D. Smith, and R. J. Anderson. 1993. On the dependence of sea surface roughness on wave development. *J. Phys. Oceanogr.* **23,** 2143–2149.

Donelan, M. A., F. W. Dobson, S. D. Smith, and R. J. Anderson. 1995. Reply. *J. Phys. Oceanogr.* **25,** 1908–1909.

Donelan, M. A., W. M. Drennan, and A. K. Magnusson. 1996. Nonstationary analysis of the directional properties of propagating waves. *J. Phys. Oceanogr.* **26,** 1901–1914.

Donelan, M. A., W. M. Drennan, and K. B. Katsaros. 1997. The air-sea momentum flux in conditions of wind sea and swell. *J. Phys. Oceanogr.* **27,** 2087–2099.

Doney, S. C., R. G. Najjar, and S. Stewart. 1995. Photochemistry, mixing and diurnal cycles in the upper ocean. *J. Mar. Res.* **53,** 341–369.

Donlon, C. J., and I. S. Robinson. 1997. Observations of the oceanic thermal skin in the Atlantic Ocean. *J. Geophys. Res.* **102,** 18585–18606.

Doyle, J. D., and T. W. Warner. 1990. Mesoscale coastal processes during GALE IOP2. *Mon. Weather Rev.* **118,** 283–308.

Drennan, W. M., W. H. Hui, and G. Tenti. 1988. Accurate calculation of Stokes wave near breaking. In C. Graham and S. K. Malik (Eds.), *Continuum Mechanics and Its Applications.* Hemisphere, Washington, DC/New York.

Drennan, W. M., K. K. Kahma, E. A. Terray, M. A. Donelan, and S. A. Kitaigorodskii. 1992. Observations of the enhancement of the kinetic energy dissipation beneath breaking wind waves. In M. L. Banner and R. H. J. Grimshaw (Eds.), *Breaking Waves,* pp. 95–101. Springer-Verlag, Berlin/New York.

Drennan, W. M., M. A. Donelan, E. A. Terray, and K. B. Katsaros. 1996. Oceanic turbulence dissipation measurements in SWADE. *J. Phys. Oceanogr.* **26,** 808–815.

Driedonks, A. G. M., and P. G. Duynkerke. 1989. Current problems in the stratocumulus-topped atmospheric boundary layer. *Boundary Layer Meteorol.* **46,** 275–304.

Duncan, J. H. 1981. An investigation of breaking waves produced by a towed airfoil. *Proc. R. Soc. London, A* **377,** 331–348.

Duncan, J. H. 1994. The formation of spilling breaking water waves. *Phys. Fluids* **6,** 2558–2560.

Dupuis, H., P. K. Taylor, A. Weill, and K. Katsaros. 1997. Inertial dissipation method applied to derive turbulent fluxes over the ocean during the Surface of the Ocean, Fluxes and Interactions with the Atmosphere/Atlantic Stratocumulus Transition Experiment (SOFIA/ASTEX) and Structure des Echanges Mer-Atmosphere. Proprietes des Heterogeneites Oceaniques: Recherche

Experimentale (SEMAPHORE) experiments with low to moderate wind speeds. *J. Geophys. Res.* **102,** 21115–21130.

Duynkerke, P. G., H. Zhang, and P. J. Jonker. 1995. Microphysical and turbulent structure of nocturnal stratocumulus as observed during ASTEX. *J. Atmos. Sci.* **52,** 2763–2777.

Dyer, A. J. 1974. A review of flux-profile relationships. *Boundary Layer Meteorol.* **7,** 363–372.

Dyer, A. J., and T. V. Crawford. 1965. Observations of climate at a leading edge. *Q. J. R. Meteorol. Soc.* **91,** 345–348.

Dymond, J., R. W. Collier, J. McManus, and G. Larson. 1996. Unbalanced particle flux budgets in Crater Lake, Oregon: Implications for edge effects and sediment focusing in lakes. *Limnol. Oceanogr.* **41,** 732–743.

Eddington, L. W., J. J. O'Brien, and D. W. Stuart. 1992. Forced mesoscale variability in a well-mixed marine layer. *Mon. Weather Rev.* **12,** 2881–2896.

Efimova, N. A. 1961. On methods of calculating monthly values of net longwave radiation. *Meteorol. Gidrol.* **10,** 28–33.

Ekman, V. W. 1905. On the influence of the earth's rotation on ocean currents. *Ark. Mat. Astron. Syst.* **2,** 1–52.

Elfouhaily, T., B. Chapron, K. Katsaros, and D. Vandemark. 1997. A unified directional spectrum for long and short wind-driven waves. *J. Geophys. Res.* **102,** 15781–15796.

Eliassen, A., and S. Thorsteinsson. 1984. Numerical studies of stratified air flow over a mountain ridge on the rotating earth. *Tellus A* **36,** 172–186.

Enriquez, A. G., and C. A. Friehe. 1993. Measurement and parameterization of fluxes of momemtum, sensible heat and moisture over a coastal region. In *Proceedings, Tenth Symposium on Turbulence and Diffusion, 1993,* pp. 15–18.

Enriquez , A. G., and C. A. Friehe. 1997. Bulk parameterization of momentum, heat and moisture fluxes over a coastal upwelling area. *J. Geophys. Res.* **102,** 5781–5798.

Eriksen, C. C. 1985a. The TROPIC HEAT program: An overview. *EOS, Trans. Am. Geophys. Union* **66,** 50–52.

Eriksen, C. C. 1985b. Implications of ocean bottom reflection for internal wave spectra and mixing. *J. Phys. Oceanogr.* **15,** 1145–1156.

Etling, D. 1993. Turbulence collapse in stably stratified flows: Application to the atmosphere. In S. D. Mobbs and J. C. King (Eds.), *Waves and Turbulence in Stably Stratified Flows*, pp. 1–21. Clarendon, Oxford.

Evans, W. C., L. D. White, M. L. Tuttle, G. W. Kling, G. Tanyileke, and R. L. Michel. 1994. Six years of change at Lake Nyos, Cameroon, yield clues to the past and cautions for the future. *Geochem. J.* **28,** 139–162.

Ewing, G., and E. D. McAlister. 1960. On the thermal boundary layer of the ocean. *Science* **131,** 1374–1376.

Eynk, G. L. 1994. The RNG method in statistical hydrodynamics. *Phys. Fluids* **6,** 3063–3075.

Fairall, C. W., and S. E. Larsen. 1986. Inertial dissipation methods and turbulent fluxes at the air-ocean interface. *Boundary Layer Meteorol.* **34,** 287–301.

Fairall, C. W., J. B. Ebson, S. E. Larsen, and P. G. Mestayer. 1990. Inertial-dissipation air-sea flux measurements: A prototype system using real time spectral computations. *J. Atmos. Oceanic Technol.* **7,** 425–453.

Fairall, C. W., E. F. Bradley, D. P. Rogers, J. B. Edson, and G. S. Young. 1996a. Bulk parameterization of air-sea fluxes for Tropical Ocean-Global Atmosphere Coupled-Ocean Atmosphere Response Experiment. *J. Geophys. Res.* **101,** 3747–3764.

Fairall, C. W., E. F. Bradley, J. S. Godfrey, G. A. Wick, J. B. Edson, and G. S. Young. 1996b. Cool-skin and warm-layer effects on sea surface temperature. *J. Geophys. Res.* **101,** 1295–1308.

Fairall, C. W., J. D. Kepert, and G. J. Holland. 1998. The effect of sea spray on surface energy transports over the ocean. *Global Atmos. Ocean System* **2,** 121–142.

Falkowski, P. G., R. T. Barber, and V. Smetacek. 1998. Biogeochemical controls and feedbacks on ocean primary production. *Science* **281,** 200–206,

Farge, M. 1992a. Wavelet transforms and their applications to turbulence. *Annu. Rev. Fluid Mech.* **24,** 395–457.

Farge, M. 1992b. The continuous wavelet transform of two-dimensional turbulent flows. In M. B. Ruskai *et al.* (Eds.), *Wavelets and Their Applications,* pp. 275–302. Jones and Barlett, Boston, MA.

Farge, M., N. Kevlahan, V. Perrier, and E. Goirand. 1996. Wavelets and turbulence. *Proc. IEEE* **84,** 639–669.

Farmer, D. M. 1975. Penetrative convection in the absence of mean shear. *Q. J. R. Meteorol. Soc.* **101,** 868–891.

Farmer, D. M. 1998. Observing the ocean side of the air-sea interface. *Oceanography* **10,** 106–110.

Farmer, D. M., and E. C. Carmack. 1981. Wind mixing and restratification in a lake near the temperature of maximum density. *J. Phys. Oceanogr.* **11,** 1516–1533.

Farmer, D. M., C. McNeil, and B. Johnson. 1993. Evidence for the importance of bubbles to the enhancement of air-sea gas flux. *Nature* **361,** 620–623.

Farmer, D. M., and M. Li. 1995. Patterns of bubble clouds organized by Langmuir circulation. *J. Phys. Oceanogr.* **25,** 1426–1440.

Fasham, M., H. W. Ducklow, and S. M. McKelvie. 1990. A nitrogen-based model of phytoplankton dynamics in the oceanic mixed layer. *J. Mar. Res.* **48,** 591–639.

Federov, K. N. 1978. *The Thermohaline Finestructure of the Ocean.* Pergamon, Elmsford, NY.

Federovich, E., R. Kaiser, M. Rau, and E. Plate. 1996. Wind tunnel study of turbulent flow structure in the convective boundary layer capped by a temperature inversion. *J. Atmos. Sci.* **53,** 1273–1289.

Fernando, H. J. S. 1987. The formation of layered structure when a stable salinity gradient is heated from below. *J. Fluid Mech.* **182,** 425–442.

Fernando, H. J. S. 1991. Turbulent mixing in stratified fluids. *Annu. Rev. Fluid Mech.* **23,** 455–493.

Fernando, H. J. S., and C. Ching. 1991. An experimental study on thermohaline staircases. In R. W. Schmitt (Ed.), *Double Diffusion in Oceanography*, pp. 91–20. WHOI report.

Fernando, H. J. S., R. Chen, and D. L. Boyer. 1991. Effects of rotation on convective turbulence. *J. Fluid Mech.* **228,** 513–547.

Fernando, H. J. S., and C. Y. Ching. 1994. Effects of background rotation on turbulent plumes. *J. Phys. Oceanogr.* **23,** 2115–2129.

Field, C. B., M. J. Behrenfeld, J. T. Randerson, and P. Falkowski. 1998. Primary production of the biosphere: Integrating terrestrial and oceanic components. *Science* **281,** 237–240.

Finnigan, J. J. 1979a. Turbulence in waving wheat. I. Mean statistics and honami. *Boundary Layer Meteorol.* **16,** 181–211.

Finnigan, J. J. 1979b. Turbulence in waving wheat. II. Structure of momentum transfer. *Boundary Layer Meteorol.* **16,** 213–236.

Finnigan, J. J. 1985. Turbulence transport in flexible plant canopies. In B. A. Hutchinson and B. B. Hicks (Eds.), *The Forest-Atmosphere Interaction,* pp. 443–480. Reidel, Dordrecht.

Flato, G. M., and W. D. Hibler III. 1996. Ridging and strength in modeling the thickness distribution of Arctic sea ice. *J. Geophys. Res.* **101,** 18611–18626.

Fleury, M., and R. G. Lueck. 1994. Direct heat flux estimates using a towed vehicle. *J. Phys. Oceanogr.* **24,** 801–818.

Flugge-Lotz, I., and W. Flugge. 1973. Ludwig Prandtl in the nineteen-thirties: Reminiscences. *Annu. Rev. Fluid Mech.* **5,** 1–8.

Ford, D.E. and Stefan, H., 1980: Stratification variability in three morphologically different lakes under identical meteorological forcing, *Water Resour. Bull,* **16**, 243–247.

Forrer, J., and M. W. Rotach. 1997. On the turbulence structure in the stable boundary layer over the Greenland ice sheet. *Boundary Layer Meteorol.* **85,** 111–136.

Forristall, G. Z. 1981. Measurements of a saturated range in ocean wave spectra. *J. Geophys. Res.* **86,** 8075–8084.

Foster, D., D. Nelson, and M. Stephen. 1977. Large-distance and long-time properties of a randomly stirred fluid. *Phys. Rev. A* **16,** 732–756.

Foster, T. D. 1972. An analysis of the cabbeling instability in sea water. *J. Phys. Oceanogr.* **2,** 294–301.

Fox, M. J. H. 1976. On thenonlinear transfer of energy in the peak of a gravity-wave spectrum. II. *Proc. R. Soc. A* **348,** 467–483.

Freeth, S. J. 1992. The Lake Nyos disaster. In S. J. Freeth, C. O. Ofoegbu, and K. M. Onuoha (Eds.), *Natural Hazards in West and Central Africa*, pp. 63–82. Vieweg, Wiesbaden.

Freeth, S. J., and R. L. F. Kay. 1987. The Lake Nyos gas disaster. *Nature* **325,** 104–105.

Freeth, S. J., G. W. Kling, M. Kusakabe, J. Maley, F. M. Tchoua and K. Tietze. 1990. Conclusions from Lake Nyos disaster. *Nature* **348,** 201.

Frenzen, P. and C. A. Vogel, 1992. A new study of the TKE budget in the surface layer. Part 1: The von Karman constant and rates of TKE production, Abstract, AMS Tenth Symposium on Turbulence and Diffusion, Portland, OR, 1992, 157–159

Friehe, C. A. 1986. Fine scale measurements of velocity, temperature and humidity in the atmospheric boundary layer. In D. H. Lenschow (Ed.), *Probing the Atmospheric Boundary Layer*, pp. 29–38. Am. Meteorol. Soc., Boston.

Friehe, C. A., and K. F. Schmitt. 1976. Parameterization of air-sea interface fluxes of sensible heat and moisture by the bulk aerodynamic formulas. *J. Phys. Oceanogr.* **6,** 801–809.

Frisch, U. 1995. *Turbulence, the Legacy of A. N. Kolmogorov*. Cambridge Univ. Press, Cambridge.

Frisch, A. S., D. H. Lenschow, C. W. Fairall, W. H. Schubert, and J. S. Gibson. 1995. Doppler radar measurements of turbulence in marine stratiform cloud during ASTEX. *J. Atmos. Sci.* **52,** 2800–2808.

Frouin, R., C. Gautier, K. B. Katsaros, and R. J. Lind. 1988. A comparison of satellite and empirical formula techniques for estimating insolation over the oceans. *J. Appl. Meteorol.* **12,** 1016–1023.

Fu, L.-L. 1981. Observations and models of inertial waves in the deep ocean. *Rev. Geophys.* **19,** 141–170.

Gal-Chen, T., M. Xu, and W. L. Eberhard. 1992. Estimations of atmospheric boundary layer fluxes and other turbulence parameters from Doppler lidar data. *J. Geophys. Res.* **97,** 18409–18423.

Galperin, B., L. H. Kantha, S. Hassid, and A. Rosati. 1988. A quasi-equilibrium turbulent energy model for geophysical flows. *J. Atmos. Sci.* **45,** 55–62.

Galperin, B., and L. H. Kantha. 1989. A turbulence model for rotating flows. *AIAA J.* **27,** 750–757.

Galperin, B., A. Rosati, L. H. Kantha, and G. L. Mellor. 1989. Modeling rotating stratified turbulent flows with application to oceanic mixed layers. *J. Phys. Oceanogr.* **19,** 901–916.

Galperin, B., and S. A. Orzag (Eds.). 1993. *Large Eddy Simulation of Complex Engineering and Geophysical Flows,* pp. 617. Cambridge University Press, New York.

Gardner, W. D. 1998. Quantifying vertical fluxes from turbulence in the ocean. *Oceanography* **10,** 116–121.

Gargett, A. E. 1989. Ocean turbulence. *Annu. Rev. Fluid Mech.* **21,** 419–451.

Gargett, A. E., P. J. Hendricks, T. B. Sanford, T. R. Osborn, and A. J. Williams, III, 1981. A composite spectrum of vertical shear in the upper ocean, *J. Phys. Oceanogr.*, **11**, 1258–1271.

Gargett, A. E., and R. W. Schmitt. 1982. Observations of salt fingers in the central waters of the eastern North Pacific. *J. Geophys. Res.* **87,** 8017–8029.

Gargett, A. E., and G. Holloway. 1984. Dissipation and diffusion by internal wave breaking. *J. Mar. Res.* **42,** 15–27.

Gargett, A. E., T. R. Osborn, and P. Nasmyth. 1984. Local isotropy and decay of turbulence in a stratified turbulence. *J. Fluid Mech.* **144,** 231–280.

Gargett, A. E., and G. Holloway. 1992. Sensitivity of the GFDL ocean model to different diffusivities for heat and salt. *J. Phys. Oceanogr.* **22,** 1158–1177.

Gargett, A. E., and B. Ferron. 1996. The effects of differential vertical diffusion of T and S in a box model of thermohaline circulation. *J. Mar. Res.* **54,** 827–866.

Garratt, J. R. 1977. Review of drag coefficients over oceans and continents. *Mon. Weather Rev.* **105,** 915–929.

Garratt, J. R. 1990. The internal boundary layer—A review. *Boundary Layer Meteorol.* **50,** 171–203.

Garratt, J. R. 1992., *The Atmospheric Boundary Layer.*, Cambridge Univ. Press, Cambridge. 316 pp.

Garratt, J. R., and B. F. Ryan. 1989. The structure of stably stratified internal boundary layer in offshore flow over the sea. *Boundary Layer Meteorol.* **47,** 17–40.

Garrett, C., R. Outerbridge, and K. Thompson. 1993. Interannual variability in Mediterranean heat and buoyancy fluxes. *J. Clim.* **6,** 900–910.

Garrett, C. J., and W. Munk. 1972. Space-time scales of internal waves. *Geophys. Astrophys. Fluid Dyn.* **2,** 255–264.

Garrett, C. J., and W. Munk. 1975. Space-time scales of internal waves: A progress report. *J. Geophys. Res.* **80,** 291–297.

Garrett, C. J. R., and W. H. Munk. 1979. Internal waves in the ocean. *Annu. Rev. Fluid Mech.* **11,** 339–369.

Garwood, R. W., Jr. 1977. An oceanic mixed layer model capable of simulating cyclic states. *J. Phys. Oceanogr.* **7,** 455–468.

Garwood, R. W., Jr., P. C. Gallacher, and P. Muller. 1985a. Wind direction and equilibrium mixed layer depth: General theory. *J. Phys. Oceanogr.* **15,** 1325–1331.

Garwood, R. W., Jr., P. Muller, and P. C. Gallacher. 1985b. Wind direction and equilibrium mixed layer depth in the tropical Pacific Ocean. *J. Phys. Oceanogr.* **15,** 1332–1338.

Garwood, R. W., P. C. Chu, P. Muller, and N. Schneider. 1989. Equatorial entrainment zone: The diurnal cycle. In J. Picaut *et al.* (Eds.), *Western Pacific International Meeting and Workshop on TOGA/COARE, Proceedings,* pp. 435–443. Centre ORSTOM de Noumea, New Caledonia.

Garwood, R. W., S. M. Isakari, and P. C. Gallacher. 1994. Thermobaric convection. In *The Polar Oceans,* pp. 199–209. Am. Geophys. Union.

Gaspar, P. 1988. Modeling the seasonal cycle of the upper ocean. *J. Phys. Oceanogr.* **18,** 161–180.

Gaspar, P., Y. Gregoris, and J.-M. Lefevre. 1990. A simple eddy kinetic energy model for simulations of the oceanic vertical mixing: Tests at station Papa and Long-Term Upper Ocean Study Site. *J. Geophys. Res.* **95,** 16179–16193.

Gautier, C., and K. B. Katsaros. 1984. Insolation during STREX. 1. Comparison between surface measurements and satellite estimates. *J. Geophys. Res.* **39,** 11778–11788.

Gaynor, J. E., D. E. Wolfe, and J.-P. Ye, Turbulence structure of the atmospheric surface layer over Arctic ice and near a lead, Abstract, AMS Tenth Symposium on Turbulence and Diffusion, Portland, OR, 1992, (J3) 37–39.

Geernaert, G. L. 1990. Bulk parameterizations for the wind stress and heat fluxes. In G. L. Geernaert and W. J. Plant (Eds.), *Surface Waves and Fluxes,* Vol. 1, pp. 91–172. Kluwer Academic, Dordrecht/Norwell, MA.

Geernaert, G. L., S. E. Larsen, and F. Hansen. 1987. Measurements of the wind stress, heat flux, and turbulent intensity during storm conditions over the North Sea. *J. Geophys. Res.* **92,** 13127–13139.

Gemmrich, J. R., T. D. Mudge, and V. D. Polocichko. 1994. On the energy input from wind to surface waves. *J. Phys. Oceanogr.* **24,** 2413–2417.

Gemmrich, J. R., and D. M. Farmer. 1999. Near-surface turbulence and thermal structure in a wind-driven sea. *J. Phys. Oceanogr.* **29,** 480–499.

Germano, M. 1992. Turbulence: The filtering approach. *J. Fluid Mech.* **238,** 325.
Germano, M., U. Piomelli, P. Moin, and W. H. Cabot. 1991. A dynamic subgrid-scale eddy viscosity model. *Phys. Fluids A* **3,** 1760.
Gerz, T., U. Schumann, and S. E. Elghobashi. 1989. Direct numerical simulation of stratified homogeneous turbulent shear flows. *J. Fluid Mech.* **200,** 563–594.
Gerz, T., and J. M. L. M. Palma. 1994. Sheared and stably stratified homogeneous turbulence: Comparison of DNS and LES. In P. R. Voke, L. Kleiser, and J.-P. Chollet (Eds.), *Direct and Large Eddy Simulation I,* pp. 145–156. Kluwer Academic, Dordrecht/Norwell, MA.
Gibson, M. M., and B. A. Younis. 1986. Calculation of swirling jets with Reynolds stress closure. *Phys. Fluids* **29,** 38–48.
Giggenbach, W. F. 1990. Water and gas chemistry of Lake Nyos and its bearing on the eruptive process. *J. Volcanol. Geothermal Res.* **42,** 337–362.
Gill, A. E. 1982. *Atmosphere-Ocean Dynamics.* Academic Press, New York.
Gilman, C., and C. Garrett. 1994. Heat flux parameterizations for the Mediterranean Sea: The role of aerosols and constraints from the water budget. *J. Geophys. Res.* **99,** 5119–5134.
Gnanadesikan, A. 1994. Dynamics of Langmuir circulations in oceanic surface layers. Ph.D. dissertation, MIT/Woods Hole Joint Program in Physical Oceanography, Woods Hole, MA. 350 pp.
Gnanadesikan, A. 1996. Modelling the diurnal cycle of carbon monoxide: Sensitivity to physics, chemistry, biology and optics. *J. Geophys. Res.* **101,** 12177–12191.
Godfrey, J. S., and A. C. M. Beljaars. 1991. On the turbulent fluxes of buoyancy, heat and moisture at the air-sea interface at low wind speeds. *J. Geophys. Res.* **96,** 22043–22048.
Godfrey, J. S., M. Nunez, E. F. Bradley, P. A. Coppin, and E. J. Lindstrom. 1991. On the net surface heat flux into the western equatorial Pacific. *J. Geophys. Res.* **96,** 3391–3400.
Gosnell, R., C. W. Fairall, and P. J. Webster. 1995. The sensible heat of rainfall in the tropical ocean. *J. Geophys. Res.* **100,** 18437–18442.
Gossard, E. E., and W. H. Hooke. 1975. *Waves in the Atmosphere.* Elsevier, Amsterdam/New York.
Grant, H. L., R. W. Stewart,and A. Molliet. 1962. Turbulence spectra from a tidal channel. *J. Fluid Mech.* **12,** 241–268.
Grant, W. D., and O. S. Madsen. 1986. The continental-shelf bottom boundary layer. *Annu. Rev. Fluid Mech.* **18,** 265–305.
Grassl, H. 1976. The dependence of the measured cool skin of the ocean on wind stress and total heat flux. *Boundary Layer Meteorol.* **10,** 465–474.
Gregg, M. C. 1976. Temperature and salinity microstructure in the Pacific equatorial undercurrent. *J. Geophys. Res.* **81,** 1180–1196.
Gregg, M. C. 1987. Diapycnal mixing in a thermocline: A review. *J. Geophys. Res.* **92,** 5249–5286.
Gregg, M. C. 1989. Scaling of turbulent dissipation in the thermocline. *J. Geophys. Res.* **94,** 9686–9697.
Gregg, M. C., H. Peters, J. C. Wesson, N. S. Oakey, and T. S. Shay. 1985. Intensive measurements of turbulence and shear in the equatorial undercurrent. *Nature* **318,** 140–144.
Gregg, M. C., and T. B. Sanford. 1987. Shear and turbulence in thermohaline staircases. *Deep-Sea Res.* **34,** 1689–1696.
Griffiths, R. W., and B. R. Ruddick. 1980. Accurate fluxes across a salt-sugar finger interface deduced from direct density measurements. *J. Fluid Mech.* **99,** 85–95.
Gryning, S.-E., and E. Batchvarova. 1996. A model for the height of the internal boundary layer over an area with irregular coastline. *Boundary Layer Meteorol.* **78,** 405–413.
Hadlock, R., and C. W. Kreitzberg. 1988. The experiment on rapidly intensifying cyclones over the Atlantic (ERICA) field study: Objectives and plans. *Bull. Am. Meteorol. Soc.* **69,** 1309–1320.

Hakkinen, S. 1990. Models and their applications to polar oceanography. In W. O. Smith, Jr. (Ed.), *Polar Oceanography,* pp. 335–384. Academic Press, San Diego.

Hakkinen, S., and G. L. Mellor. 1990. One hundred years of Arctic ice cover variations as simulated by a one-dimensional, ice-ocean model. *J. Geophys. Res.* **95,** 15959–15969.

Hakkinen, S., and G. L. Mellor. 1994. A review of coupled ice-ocean models. In *The Polar Oceans and Their Role in Shaping the Global Environment,* pp. 21–31. American Geophysical Union, Washington, DC.

Haney, R. L. 1991. On the pressure gradient force over step topography in sigma-coordinate models. *J. Phys. Oceanogr.* **21,** 610–619.

Hanjalic, K., and B. E. Launder. 1972. A Reynolds stress model of turbulence and its application to thin shear flows. *J. Fluid Mech.* **52,** 609–638.

Hanson, H. P., 1987. Radiative/turbulent transfer interactions in layer clouds. *J. Atmos. Sci.*, **44,** 1287–1295.

Hasselmann, K. 1962. On the nonlinear energy transfer in gravity wave spectrum, part I. *J. Fluid Mech.* **12,** 481–500.

Hasselmann, K., T. P. Barnett, E. Bouws, H. Carlson, D. E. Cartwright, K. Enke, J. A. Ewing, H. Gienapp, D. E. Hasselmann, P. Kruseman, A. Meerburg, P. Muller, D. J. Olbers, K. Richter, W. Sell, and H. Walden. 1973. Measurements of wind-wave growth and swell decay during the Joint North Sea Wave Project (JONSWAP). *Dtsch. Hydrogr. Z.* **12**(suppl.), 95.

Hasselmann, S., and K. Hasselmann. 1985. Computations and parameterizations of the nonlinear energy transfer in a gravity wave spectrum, part I: A new method for efficient computations of the exact nonlinear transfer integral. *J. Phys. Oceanogr.* **15,** 1369–1377.

Hassid, S., and B. Galperin. 1983. A turbulent energy model for geophysical flows. *Boundary Layer Meteorol.* **26,** 397–412.

Hassid, S., and B. Galperin. 1994. Modeling rotating flows with neutral and unstable stratification. *J. Geophys. Res.* **99,** 12533–12548.

Haugen, D. A. (Ed.). 1973. *Workshop on Micrometeorology.* Am. Meteorol. Soc., Boston. 392 pp.

Hebert, D., N. Oakey, and B. R. Ruddick. 1990. Evolution of a Mediterranean salt lens. *J. Phys. Oceanogr.* **20,** 1468–1483.

Hebert, D., J. N. Moum, C. A. Paulson, and D. R. Caldwell. 1992. Turbulence and internal waves at the equator, part II: Details of a single event. *J. Phys. Oceanogr.* **22,** 1346–1356.

Heidt, F. D. 1977. The growth of a mixed layer in a stratified fluid. *Boundary Layer Meteorol.* **12,** 439–461.

Heimbach, P., S. Hasselmann, and K. Hasselmann. 1998. Statistical analysis and intercomparison of WAM model data with global ERS-1 SAR wave mode spectral retrievals over 3 years. *J. Geophys. Res.* **103,** 7931–7977.

Henderson-Sellers, B. 1984. *Engineering Limnology.* Pitman Adv. Publ., Boston.

Henderson-Sellers, B., and A. M. Davies. 1989. Thermal stratification modeling for oceans and lakes. *Annu. Rev. Numerical Fluid Mech.,* 86–156.

Henyey, F. S. 1991. Scaling of internal wave model predictions for ε. In P. Muller and D. Henderson (Eds.), *Dynamics of Ocaenic Internal Gravity Waves. Aha Huliko'a Hawaiian Winter Workshop,* pp. 233–236. University of Hawaii at Manoa.

Henyey, F. S., J. Wright, and S. M. Flatte. 1986. Energy and action flow through the internal wave field: An eikonal approach. *J. Geophys. Res.* **91,** 8487–8495.

Henyey, F. S., and A. Hoering. 1997. Energetics of borelike internal waves. *J. Geophys. Res.* **102,** 3323–3330.

Herring, J. R. 1979. Subgrid scale modeling—An introduction and an overview. *Turb. Shear Flows* **1,** 347–352.

Hibler, W. D. III. 1979. A dynamic-thermodynamic sea ice model. *J. Phys. Oceanogr.* **9,** 815–846.

Hibler, W. D., III, and K. Bryan. 1987. A diagnostic ice-ocean model. *J. Phys. Oceanogr.* **17,** 987–1015.

Hicks, B. B., P. Hyson, and C. J. Moore, 1975. A study of eddy fluxes over a forest. *J. Appl. Meteor.*, **14**, 58–66.

Hinze, J. O. 1976. *Turbulence*, 2nd ed. McGraw Hill, New York.

Hogstrom, U. 1988. Non-dimensional wind and temperature profiles in the atmospheric surface layer: A reevaluation. *Boundary Layer Meteorol.* **42,** 55–78.

Hogstrom, U. 1996. Review of some basic characteristics of the atmospheric surface layer. *Boundary Layer Meteorol.* **78,** 215–246.

Holland, D. M., L. A. Mysak, D. K. Manak, and J. M. Oberhuber. 1993. Sensitivity study of a dynamic-thermodynamic sea-ice model. *J. Geophys. Res.* **98,** 2561–2586.

Holland, D. M., L. A. Mysak, and J. M. Oberhuber. 1996. Simulation of the mixed-layer circulation in the Arctic Ocean. *J. Geophys. Res.* **101,** 1111–1128.

Holland, J. Z., 1981. Atmospheric Boundary Layer. In IFYGL Ð The International Field Year for the Great Lakes. Edited by E. J. Aubert and T. L. Richards. Published by NOAA, Ann Arbor, MI.

Holloway, G. 1980. Oceanic internal waves are not weak waves. *J. Phys. Oceanogr.*, **10**, 906–914.

Holt, S. E., J. R. Koseff, and J. H. Ferziger. 1992. A numerical study of the evolution and structure of homogeneous stably startified sheared turbulence. *J. Fluid Mech.* **237,** 499–539.

Holt, T., and S. Raman. 1988. A review and comparative evaluation of multi-level boundary layer parameterizations for first-order and turbulent kinetic energy closure schemes. *Rev. Geophys.* **26,** 761–780.

Holtslag, A. A. M., and C.-H. Moeng. 1991. Eddy diffusivity and countergradient transport in the convective atmospheric boundary layer. *J. Atmos. Sci.* **48,** 1690–1698.

Hooke, W. H. and R. M. Jones, 1986. Dissipative waves excited by gravity-wave encounters with the stably stratified planetary boundary layer. *J. Atmos. Sci.*, **43**, 2048–2060.

Hopfinger, E. J. 1987. Turbulence in stratified flows: A review. *J. Geophys. Res.* **92,** 5287–5303.

Hopfinger, E. J., and J.-A. Toly. 1976. Spatially decaying turbulence and its relation to mixing across density interfaces. *J. Fluid Mech.* **78,** 155–175.

Horn, W., C. H. Mortimer, and D. J. Schwab. 1986. Wind-induced internal seiches in Lake Zurich observed and modeled. *Limnol. Oceanogr.* **31,** 1232–1254.

Horton, C., M. Clifford, J. Schmitz, and L. Kantha. 1997. A real-time oceanographic nowcast/forecast system for the Mediterranean Sea. *J. Geophys. Res.* **102,** 25123–25156.

Houze, R. A., Jr. 1993. *Cloud Dynamics*. Academic Press, San Diego.

Howard, L. N. 1961. Note on a paper of John W. Miles. *J. Fluid Mech.* **10,** 509.

Howard, L. N., and G. Veronis. 1987. The salt finger zone. *J. Fluid Mech.* **183,** 1–24.

Howell, J. F., and L. Mahrt. 1994. An adaptive decomposition application to turbulence. In E. Foufoula-Georgiou and P. Kumar (Eds.), *Wavelets in Geophysics*, pp. 107–128. Academic Press, San Diego.

Huang, R. X. 1999. Mixing and energetics of the oceanic thermohaline circulation. *J. Phys. Oceanogr.* **29,** 727–746.

Hubbard, B. B. 1996. *The World According to Wavelets*. Peters, Wellseley, MA.

Hudgins, L. H., C. A. Friehe, and M. E. Mayer. 1993. Wavelet transform and atmospheric turbulence. *Phys. Rev. Lett.* **71,** 3279–3282.

Hunkins, K. 1975. The oceanic boundary layer and stress beneath a drifting ice floe. *J. Geophys. Res.* **80,** 3425–3433.

Hunt, J. C. R., D. D. Stretch, and R. E. Britter. 1988. Length scales in stably stratified turbulent flows and their use in turbulence models. In J. S. Puttock (Ed.), *Stably Stratified Flow and Dense Gas Dispersion*. Clarendon, Oxford.

Hunt, J. C. R., G. J. Shutts, and S. H. Derbyshire. 1996. Stably stratified flows in meteorology. *Dyn. Atm. Oceans* **23,** 63–79.

Huppert, H. E. 1986. The intrusion of fluid mechanics into geology. *J. Fluid Mech.* **173,** 557–594.
Huppert, H., and J. W. Miles. 1969. Lee waves in a stratified flow, part 3: Semi-elliptical obstacles. *J. Fluid Mech.* **35,** 481–496.
Huppert, H. E., and P. C. Manins. 1973. Limiting conditions for salt-fingering at an interface. *Deep-Sea Res.* **20,** 315–323.
Huppert, H. E., and P. F. Linden. 1979. On heating a stable salinity gradient from below. *J. Fluid Mech.* **95,** 431–464.
Huppert, H. E., and J. S. Turner. 1981. Double-diffusive convection. *J. Fluid Mech.* **106,** 299–329.
Idso, S. B. 1981. A set of equations for full spectrum and 8–14 microns and 10.5–12.5 microns thermal radiation from cloudless skies. *Water Resour. Res.* **17,** 295–304.
Idso, S. B., and R. D. Jackson. 1969. Thermal radiation from the atmosphere. *J. Geophys. Res.* **74,** 5397–5403.
Imberger, J. 1985. The diurnal mixed layer. *Limnol. Oceanogr.* **30,** 737–770.
Imberger, J. (Ed.). 1995. *Physical Processes in Lakes and Oceans.* International Union of Theoretical and Applied Mechanics.
Imberger, J., and J. C. Patterson. 1981. A dynamic reservoir simulation model—DYRESM:5. In *Transport Models for Inland and Coastal Waters,* pp. 310–360. Academic Press, New York.
Imberger, J., and P. F. Hamblin. 1982. Dynamics of lakes, reservoirs, and cooling ponds. *Annu. Rev. Fluid Mech.* **14,** 153–187.
Imberger, J., and J. C. Patterson. 1990. Physical limnology. *Adv. Appl. Mech.* **27,** 303–475.
Ingham, M. C. 1966. The salinity extrema of the world ocean. Ph.D. dissertation, Oregon State University, Corvallis. 63 pp.
Isemer, H. J., and L. Hasse. 1987. *The Bunker Climate Atlas of the North Atlantic Ocean: 2. Air-Sea Interactions.* Springer-Verlag, Berlin/New York. 256 pp.
Ivanoff, A. 1977. Oceanic absorption of solar energy. In E. B. Kraus (Eds.), *Modelling and Prediction of the Upper Layers of the Ocean.* Pergamon, New York. 326 pp.
Ivlev, V. S. 1955. *Experimental Ecology of the Feeding of Fishes.* Pischepromazat, Moscow. 302 pp.
Jacobs, J. D. 1978. Radiation climate of Broughton Island. In R. G. Barry and J. D. Jacobs (Eds.), *Energy Budget Studies in Relation to Fast-Ice Breakup Processes in Davis Strait,* pp. 105–120. Occasional Paper 26, Inst. Arctic Alpine Res., Univ. of Colorado, Boulder.
Jacobs, S. 1984. Mass transport in a turbulent boundary layer under a progressive water wave. *J. Fluid Mech.* **146,** 303–312.
Jahne, B., and H. Haubecker. 1998. Air-water gas exchange. *Annu. Rev. Fluid Mech.* **30,** 443–468.
Jahne, B., and K. S. Riemer. 1990. Two-dimensional wave number spectra of small-scale water surface waves. *J. Geophys. Res.* **95,** 11531–11546.
Janssen, P. A. E. M. 1999. On the effect of ocean waves on the kinetic energy balance and consequences for the inertial dissipation technique. *J. Phys. Oceanogr.* **29,** 530–534.
Jeffreys, H. 1924. On the formation of waves by wind. *Proc. R. Soc. London A* **107,** 189–206.
Jenkins, W. J. 1980. Tritium and ^{3}He in the Sargasso Sea. *J. Mar. Res.* **38,** 533–569.
Jerlov, N. G. 1976. *Marine Optics.* Elsevier, New York. 231 pp.
Jessup, A. T. 1995. The infrared signature of breaking waves. In M. A. Donelan, W. J. Plant, and W. H. Hui (Eds.), *Proceedings, Symposium on Air-Sea Interaction, Marseilles, France, 1993.* Univ. Toronto Press, Toronto.
Jessup, A. T., C. J. Zappa, and H. Yeh. 1997a. Defining and quantifying microscale wave breaking with infrared imagery. *J. Geophys. Res.* **102,** 23145–23154.
Jessup, A. T., C. J. Zappa, and V. Hesany. 1997b. Infrared remote sensing of breaking waves. *Nature* **385,** 52–55.
Jobson, H. E., and W. W. Sayre. 1970. Vertical mass transfer in open channel flow. *J. Hydraul. Div. Am. Soc. Civ. Eng.* **96**(HY3), 703–724.

Johnson, E. S., and D. S. Luther. 1994. Mean zonal momentum balance in the upper and central equatorial Pacific Ocean. *J. Geophys. Res.* **99,** 7689–7706.

Johnson, G. C., T. B. Sanford, and M. O. Baringer. 1994. Stress on the Mediterranean outflow plume: Part I. Velocity and water property measurements. *J. Phys. Oceanogr.* **24,** 2072–2083.

Johnson, H. K., J. Hojstrup, H. J. Vested, and S. E. Larsen. 1998. On the dependence of sea surface roughness on wind waves. *J. Phys. Oceanogr.* **28,** 1702–1716.

Jones, H., and J. Marshall. 1993. Convection with rotation in a neutral ocean: A study of open-ocean deep convection. *J. Phys. Oceanogr.* **23,** 1009–1039.

Jones, I. S. F. 1985. Turbulence below wind waves. In Y. Toba and H. Mitsuyasu (Eds.), *The Ocean Surface*, pp. 437–442. Reidel, Dordrecht.

Jones, I. S. F., and Y. Toba. 1995. Comments on "On the dependence of sea surface roughness on wave development." *J. Phys. Oceanogr.* **25,** 1905–1907.

Jones, W. L. 1968. Reflexion and stability of of waves in stably stratified fluids with shear flow: A numerical study. *J. Fluid Mech.* **34,** 609–624.

Josey, S. A., D. Oakley, and R. W. Pascal. 1997. On estimating the atmospheric longwave flux at the ocean surface from ship meteorological reports. *J. Geophys. Res.* **102,** 27961–27972.

Julien, K., S. Legg, J. C. McWilliams, and J. Werne. 1996. Penetrative convection in rapidly rotating flows: Preliminary results from numerical simulation. *Dyn. Atmos. Oceans* **24,** 237–249.

Kader, B. A. 1981. Temperature and concentration profiles in fully turbulent boundary layers. *Int. J. Heat Mass Transfer* **24,** 1541–1544.

Kader, B. A., and A. M. Yaglom. 1972. Heat and mass transfer laws for fully turbulent wall flows. *Int. J. Heat Mass Transfer* **15,** 2329–2351.

Kagan, B. A. 1995. *Ocean-Atmosphere Interaction and Climate Modeling*. Cambridge Univ. Press, Cambridge. 377 pp.

Kahma, K. K. 1981a. A study of the growth of the wave spectrum with fetch. *J. Phys. Oceanogr.* **11,** 1503–1515.

Kahma, K. K. 1981b. On the wind speed dependence of the saturation range of the wave spectrum. In M. Lepparanta (Ed.), *X Geofysiikan paiavat Helsingissa.* Seura, Helsinki.

Kahma, K. K., and M. A. Donelan. 1988. A laboratory study of the minimum wind speed for wind wave generation. *J. Fluid Mech.* **192,** 339–364.

Kahma, K. K., and C. J. Calkoen. 1992. Reconciling discrepancies in the observed growth of wind-generated waves. *J. Phys. Oceanogr.* **22,** 1389–1405.

Kaimal, J. C., J. C. Wyngaard, Y. Izumi, and R. Cote. 1972. Spectral characteristics of surface-layer turbulence. *Q. J. R. Meteorol. Soc.* **98,** 563–589.

Kaimal, J. C., J. C. Wyngaard, D. A. Haugen, O. R. Cote, Y. Izumi, S. J. Caughey, and C. J. Readings. 1976. Turbulence structure in the convective boundary layer. *J. Atmos. Sci.* **33,** 2152–2169.

Kaimal, J. C., and J. J. Finnigan. 1994. *Atmospheric Boundary Layer Flows. Their Structure and Measurement.* Oxford University Press, London. 289 pp.

Kallstrand, B., and A.-S. Smedman. 1997. A case study of the near-neutral coastal internal boundary-layer growth: Aircraft measurements compared with different model estimates. *Boundary Layer Meteorol.* **85,** 1–33.

Kaltenbach, H.-J., T. Gerz, and U. Schumann. 1991. Transport of passive scalars in neutrally and stably stratified homogeneous turbulent shear flows. In *Advances in Turbulence,* Vol. 3, pp. 327–334. Springer, Berlin.

Kaltenbach, H.-J., T. Gerz, and U. Schumann. 1994. Large eddy simulation of homogeneous turbulence and diffusion in stably stratified shear flow. *J. Fluid Mech.* **280,** 1–40.

Kampf, J., and J. O. Backhaus. 1998. Shallow, brine-driven free convection in polar oceans: Nonhydrostatic numerical process studies. *J. Geophys. Res.* **103,** 5577–5594.

Kanari, S.-I. 1989. An inference on the process of gas outburst from Lake Nyos, Cameroon. *J. Volcanol. Geothermal Res.* **39,** 135–149.

Kantha, L. H. 1977. Note on the role of internal waves in thermocline erosion. In E. B. Kraus (Eds.), *Modelling and Prediction of the Upper Layers of the Ocean*, pp. 73–178. Pergamon, New York.

Kantha, L. H. 1979a. On generation of internal waves by turbulence in the mixed layer. *Dyn. Atmos. Oceans* **3,** 39–46.

Kantha, L. H. 1979b. On leaky modes on a buoyancy interface. *Dyn. Atmos. Oceans* **3,** 47–54.

Kantha, L. H. 1980a. Experimental simulation of the "retreat" of the seasonal thermocline by surface heating. In H. J. Freeland, D. M. Farmer, and C. D. Levings (Eds.), *Fjord Oceanography,* pp. 197–204. Plenum Press, New York.

Kantha, L. H. 1980b. Turbulent entrainment at a buoyancy interface due to convective turbulence. In H. J. Freeland, D. M. Farmer, and C. D. Levings (Eds.), *Fjord Oceanography,* pp. 205–214. Plenum Press, New York.

Kantha, L. H. 1980c. Laboratory experiments on attenuation of internal waves by turbulence in the mixed layer. In *Proceedings, Second International Symposium on Stratified Flows, Trondheim, Norway,* pp. 731–741. IABH.

Kantha, L. H. 1981. "Basalt fingers"—Origin of columnar joints? *Geol. Mag.* **118,** 251–264.

Kantha, L. H. 1985. On some aspects of second moment closure. Geophysical Fluid Dynamics Laboratory, Princeton, NJ. Unpublished report, 151 pp.

Kantha, L. H. 1995a. A numerical model of Arctic leads. *J. Geophys. Res.* **100,** 4653–4672.

Kantha, L. H. 1995b. Barotropic tides in the global ocean from a nonlinear tidal model assimilating altimetric tides. 1. Model description and results. *J. Geophys. Res.* **100,** 25283–25308.

Kantha, L. H. 1998. Tides—A modern perspective. *Mar. Geodesy* **21,** 275–297.

Kantha, L. H., O. M. Phillips, and R. S. Azad. 1977. On turbulent entrainment at a stable density interface. *J. Fluid Mech.* **79,** 753–768.

Kantha, L. H., and G. L. Mellor. 1989a. A numerical model of the atmospheric boundary layer over a marginal ice zone. *J. Geophys. Res.* **94,** 4959–4970.

Kantha, L. H., and G. L. Mellor. 1989b. A two-dimensional coupled ice-ocean model of the Bering Sea marginal ice zone. *J. Geophys. Res.* **94,** 10924–10936.

Kantha, L. H., A. Rosati, and B. Galperin. 1989. Effect of rotation on vertical mixing and associated turbulence in stratified fluids. *J. Geophys. Res.* **94,** 4843–4854.

Kantha, L. H., and A. Rosati. 1990. The effect of curvature on turbulence in stratified fluids. *J. Geophys. Res.* **95,** 20313–20330.

Kantha, L., and S. Piacsek. 1993. Ocean models. In *Computing Science Education Project.* Department of Energy, Washington, DC. 90 pp.

Kantha, L. H., and C. A. Clayson. 1994. An improved mixed layer model for geophysical applications. *J. Geophys. Res.* **99,** 25235–25266.

Kantha, L. H., C. Tierney, J. W. Lopez, S. D. Desai, M. E. Parke, and L. Drexler. 1995. Barotropic tides in the global oceans from a nonlinear tidal model assimilating altimetric tides. 2. Altimetric and geophysical implications. *J. Geophys. Res.* **100,** 25309–25317.

Kantha, L. H., G. C. Crawford, R. W. Collier, and S. J. Freeth. 1996. A numerical model of the Crater Lake. Presented at the Chapman Conference on Crater Lake, Oregon, Sept. 1996.

Kantha, L. H., and S. J. Freeth. 1996. A numerical model of the evolution of temperature and CO_2 in lake Nyos since the outgassing event. *J. Geophys. Res.* **101,** 8127–8203.

Kantha, L. H., and S. Piacsek. 1997. Computational ocean modeling. In A. B. Tucker, Jr. (Ed.), *The Computer Science and Engineering Handbook*, pp. 934–958. CRC Press, Boca Raton, FL.

Kantha, L. H., and C. C. Tierney. 1997. Global baroclinic tides. *Prog. Oceanogr.,* **40,** 163–178.

Kantha, L. H., and C. A. Clayson. 2000. *Numerical Models of Oceans and Oceanic Processes.* Academic Press, San Diego.

Karman, T. von. 1930. Mechanische Ahnlichkeit und turbulenz. *Nachr. Ges. Wiss. Gottingen, Math. Phys.* **K1,** 58–76.

Karman, T. von, and L. Howarth. 1938. On the statistical theory of isotropic turbulence. *Proc. R. Soc. London A,* **164,** 192.

Katsaros, K. B. 1980. The aqeous thermal boundary layer. *Boundary Layer Meteorol.* **18,** 107–127.

Katsaros, K. B. 1990. Parameterization schemes and models for estimating the surface radiation budget. In G. L. Geernaert and W. J. Plant (Eds.), *Surface Waves and Fluxes,* Vol. 3, pp. 339–368. Kluwer Academic, Dordrecht/Norwell, MA.

Katul, G., C-I. Hsieh, R. Oren, D. Ellsworth, and N. Phillips. 1996. Latent and sensible heat flux predictions from a uniform pine forest using surface renewal and flux variance methods. *Boundary Layer Meteorol.* **80,** 249–282.

Katz, E. J. 1975. Tow spectra from MODE. *J. Geophys. Res.* **80,** 1163–1167.

Kawai, S., K. Okeda, and Y. Toba. 1977. Field data support of three-seconds power law and $gu_*\sigma^{-4}$ spectral form for growing wind waves. *J. Oceanogr. Soc. Jpn.* **33,** 137–150.

Kazanski, A. B., and A. S. Monin. 1960. A turbulent regime above the ground atmospheric layer. *Izv. Geophys. Ser.* **1,** 110–112.

Keen, R. A. 1988. Equatorial westerlies and the southern oscillation. In R. Lukas and P. Webster (Eds.), *Proceedings of the U.S. TOGA Western Pacific Air-Sea Interaction Workshop,* pp. 121–140. Technical Report USTOGA-8, UCAR, Boulder, CO.

Kelley, D. E. 1984. Effective diffusivities within oceanic thermohaline staircases. *J. Geophys. Res.* **89,** 10484–10488.

Kelley, D. 1987. Interface migration in thermohaline staircases. *J. Phys. Oceanogr.* **17,** 1633–1639.

Kelley, D. E. 1990. Fluxes through diffusive interfaces: A new formulation. *J. Geophys. Res.* **95,** 3365–3371.

Kerr, R. M., J. A. Domaradzki, and G. Barbier. 1996. Small-scale properties of nonlinear interactions and subgrid-scale energy transfer in isotropic turbulence. *Phys. Fluids* **8,** 197–208.

Key, J. R., R. A. Silcox, and R. S. Stone. 1996. Evaluation of surface radiative flux parameterizations for use in sea ice models. *J. Geophys. Res.* **101,** 3839–3849.

Kibby, H. V., J. R. Donaldson, and C. E. Bond. 1968. Temperature and current observations in Crater Lake, Oregon. *Limnol. Oceanogr.* **13,** 363–366.

Kiefer, D. A., and D. G. Mitchell. 1983. A simple steady state description of phytoplankton growth based on absorption cross section and quantum efficiency. *Limnol. Oceanogr.* **28,** 770–776.

Kiehl, J. T., and K. E. Trenberth. 1997. Earth's annual global mean energy budget. *Bull. Am. Meteorol. Soc.* **78,** 197–208.

Killworth, P. D. 1983. Deep convection in the world ocean. *Rev. Geophys. Space Phys.* **21,** 1–26.

Kim, Y.-J., and A. Arakawa. 1995. Improvement of orographic gravity wave parameterization using mesoscale gravity wave model. *J. Atmos. Sci.* **52,** 1875–1902.

Kimball, H. H. 1928. Amount of solar radiation that reaches the surface of the earth on the land and on the sea and methods by which it is measured. *Mon. Weather Rev.* **56,** 393–399.

Kinsman, B. 1984. *Wind Waves.* Dover, New York.

Kitaigorodskii, S. A. 1962. Application of the theory of similarity to the analysis of wind-generated water waves as a stochastic process. *Bull. Acad. Sci. USSR Geophys. Ser.* **1,** 105–117.

Kitaigorodskii, S. A. 1983. On the theory of the equilibrium range in the spectrum of wind-generated gravity waves. *J. Phys. Oceanogr.* **13,** 960–972.

Kitaigorodskii, S. A. 1984. On the fluid dynamical theory of turbulent gas transfer across an air-sea interface in the presence of breaking wind-waves. *J. Phys. Oceanogr.* **14,** 960–972.

Kitaigorodskii, S. A. 1986. The equilibrium ranges in wind-wave spectra. In O. M. Phillips and K. Hasselmann (Eds.), *Wave Dynamics and Radio Probing of the Ocean Surface,* pp. 9–40. Plenum, New York.

Kitaigorodskii, S. A., V. P. Krasitskii, and M. M. Zaslavskii. 1975. On Phillips' theory of equilibrium range in the spectra of wind-generated gravity waves. *J. Phys. Oceanogr.* **5,** 410–420.

Kitaigorodskii, S. A., M. A. Donelan, J. L. Lumley, and E. A. Terray. 1983. Wave-turbulence interactions in the upper ocean. Part II. Statistical characteristics of wave and turbulent components of the random velocity field in the marine surface layer. *J. Phys. Oceanogr.* **13,** 1988–1999.

Kitaigorodskii, S. A., and M. A. Donelan. 1984. Wind-wave effects on gas transfer. In W. Brutsaert and G. H. Jirka (Eds.), *Gas Transfer at Water Surfaces*, pp. 147–170.

Klemp, J. B., and D. K. Lilly. 1975. The dynamics of wave-induced downslope winds. *J. Atmos. Sci.* **32,** 320–339.

Kling, G. W. 1987. Seasonal mixing and catastrophic degassing in tropical lakes, Cameroon, West Africa. *Science* **237,** 1022–1024.

Kling, G. W., M. A. Clark, H. R. Compton, J. D. Levine, W. C. Evans, A. M. Humphrey, E. J. Koenigsberg, J. P. Lockwood, M. L. Tuttle, and G. N. Wagner. 1987. The 1986 Lake Nyos gas disaster in Cameroon, West Africa. *Science* **236,** 169–175.

Kling, G. W., M. L. Tuttle, and W. C. Evans. 1989. The evolution of thermal structure and water chemistry in Lake Nyos. *J. Volcanol. Geothermal Res.* **39,** 151–165.

Kogan, Y. L., M. P. Khairoutdinov, D. K. Lilly, Z. N. Kogan, and Q. Liu. 1995. Modeling of stratocumulus cloud layers in a large eddy simulation model with explicit microphysics. *J. Atmos. Sci.* **52,** 2923–2940.

Kolmogoroff, A. N. 1941a. Local structure of turbulence in an incompressible fluid at very high Reynolds numbers (in Russian). *Dokl. Akad. Nauk. SSSR.* **30,** 299–303.

Kolmogoroff, A. N. 1941b. Energy dissipation in locally isotropic turbulence (in Russian). *Dokl. Akad. Nauk. SSSR.* **32,** 19–21.

Kolmogoroff, A. N. 1942. Equations of turbulent motion of an incompressible fluid (in Russian). *Izv. Akad. Nauk. USSR. Ser. Phys.* **6,** 56–58.

Kolmogoroff, A. N. 1962. A refinement of previous hypotheses concerning the local structure of turbulence in viscous incompressible fluid at high Reynolds number. *J. Fluid Mech.* **13,** 82–85.

Komen, G. J., S. Hasselmann, and K. Hasselmann. 1984. On the existence of a fully developed wind-sea spectrum. *J. Phys. Oceanogr.* **14,** 1271–1285.

Komen, G. J., L. Cavaleri, M. Donelan, K. Hasselmann, S. Hasselmann, and P. A. E. M. Janssen. 1994. *Dynamics and Modelling of Ocean Waves*. Cambridge University Press, Cambridge. 532 pp.

Kondo, J. 1975. Air-sea bulk transfer coefficients in diabetic conditions. *Boundary Layer Meteorol.* **9,** 91–112.

Kondo, J. and H. Yamazawa, 1986. Aerodynamic roughness over an inhomogeneous ground surface. *Boundary-Layer Meteorol.*, **35**, 331–348.

Koracin, D., and R. Berkowicz. 1988. Nocturnal boundary-layer height: Observations by acoustic sounders and predictions in terms of surface-layer parameters. *Boundary Layer Meteorol.* **43,** 65–83.

Kosovic, B. 1996. Subgrid-scale modeling for the large eddy simulation of stably stratified boundary layers. Ph.D. dissertation, Department of Aerospace Engineering Sciences, University of Colorado, Boulder.

Kraichnan, R. H. 1964. Direct interaction approximation for shear and thermally driven turbulence. *Phys. Fluids* **7,** 1048–1062.

Kraichnan, R. H. 1971. Inertial range transfer in two- and three-dimensional turbulence. *J. Fluid Mech.* **47,** 525–535.

Kraichnan, R. H. 1976. Eddy viscosity in two and three dimensions. *J. Atmos. Sci.* **33,** 1521–1536.

Kraus, E. B., and J. A. Businger. 1994. *Atmosphere-Ocean Interaction*. Oxford Univ. Press, New York. 352 pp.

Kudryavtsev, V. N., V. K. Makin, and B. Chapron. 1999. Coupled sea surface-atmosphere model. 2. Spectrum of short wind waves. *J. Geophys. Res.* **104,** 7625–7640.

Kumar, P., and E. Foufoula-Georgiou. 1994. Wavelet analysis in geophysics: An introduction. In E. Foufoula-Georgiou and P. Kumar (Eds.), *Wavelets in Geophysics*, pp. 1–43. Academic Press, San Diego.

Kumar, P., and E. Foufoula-Georgiou. 1997. Wavelet analysis for geophysical applications. *Rev. Geophys.* **35,** 385–412.

Kundu, K. P. 1980. A numerical investigation of mixed-layer dynamics. *J. Phys. Oceanogr.* **10,** 220–236.

Kunze, E. 1985. Near-inertial wave propagation in geostrophic shear. *J. Phys. Oceanogr.* **14,** 544–565.

Kunze, E. 1987. Limits on growing, finite length salt fingers: A Richardson number constraint. *J. Mar. Res.* **45,** 533–556.

Kunze, E. 1991. Behavior of salt fingers in shear. In R. W. Schmitt (Ed.), *Double Diffusion in Oceanography,* pp. 61–74. WHOI report 91-20, Woods Hole Oceanographic Institution, MA.

Kunze, E. 1994. A proposed constraint for salt fingers in shear. *J. Mar. Res.* **52,** 999–1016.

Kunze, E. 1996. Quantifying salt-fingering fluxes in the ocean. In A. Brandt and H. J. S. Fernando (Eds.), *Double Diffusive Convection*. American Geophysical Union, Washington, DC.

Kunze, E., and T. B. Sanford. 1996. Abyssal mixing: Where it is not. *J. Phys. Oceanogr.* **26,** 2286–2296.

Kusakabe, M., T. Ohsumi, and S. Aramaki. 1989. The Lake Nyos disaster: Chemical and isotopic evidence in waters and dissolved gases from three Cameroonian crater lakes, Nyos, Monoun and Wum. *J. Volcanol. Geotherm. Res.* **39,** 167–185.

Kustas, W. P., D. C. Goodrich, M. S. Moran, S. A. Amer, L. B. Bach, J. H. Blanford, A. Chehbouni, H. Claassen, W. E. Clements, P. C. Doraiswamy, P. Dubois, T. R. Clarke, C. S.S.T. Daughtry, D. I. Gellman, T. A. Grant, L. E. Hipps, A. R. Huete, K. S. Humes, T. J. Jackson, T. O. Keefer, W. D. Nichols, R. Parry, E. M. Perry, R. T. Pinker, P. J. Pinter, Jr., J. Qi, A. C. Riggs, T. J. Schmugge, A. M. Shutko, D. I. Stannard, E. Swiatek, J. D. van Leeuwen, J. van Zyl, A. Vidal, J. Washburne, and M. A. Weltz, 1991. An Interdiciplinary Field Study of the Energy and Water Fluxes in the Atmosphere-Biosphere system over semiarid rangelands: description and some preliminary results. *Bull. Am. Meteorol. Soc.,* **72,** 1683–1705.

Lab Sea Group. 1998. The Labrador Sea Deep Convection Experiment. *Bull. Am. Meteorol. Soc.* **79,** 2033–2058.

Laevastu, T. 1960. Factors affecting the temperature of the surface layer of the sea. *Comment. Phys. Math.* **25,** 1–136.

Lam, S. H. 1992. On the RNG theory of turbulence. *Phys. Fluids A* **4,** 1007–1017.

Lamb, H. 1945. *Hydrodynamics,* 6th ed. Dover, New York. 738 pp.

Landau, L. D., and E. M. Lifshitz. 1959. *Fluid Mechanics.* Pergamon, 536 pp.

Langmuir, I. 1938. Surface motion of water induced by wind. *Science* **87,** 119–123.

Large, W. G., and S. Pond. 1982. Sensible and latent heat fluxes over the ocean. *J. Phys. Oceanogr.* **12,** 464–482.

Large, W. G., J. C. McWilliams, and P. P. Niiler. 1986. Upper ocean thermal response to strong autumnal forcing of the northeast Pacific. *J. Phys. Oceanogr.* **16,** 1524–1550.

Large, W. G., J. C. McWilliams, and S. C. Doney. 1994. Oceanic vertical mixing: A review and a model with nonlocal boundary layer parameterization. *Rev. Geophys.* **32,** 363–403.

Large, W. G., and G. B. Crawford. 1995. Observations and simulations of upper-ocean response to wind events during the Ocean Storms. *J. Phys. Oceanogr.* **25,** 2959–2971.

Larson, D. W. 1972. Temperature, transparency, and phytoplankton productivity in Crater Lake, Oregon. *Limnol. Oceanogr.* **17,** 410–417.
Launder, B. E. 1975. On the effects of gravitational field on the turbulent transport of heat and momentum. *J. Fluid Mech.* **67,** 569–581.
Launder, B. E., D. P. Tselepidakis, and B. A. Younis. 1987. A second moment closure study of rotating channel flow. *J. Fluid Mech.* **183,** 63–75.
Lavoie, R. L. 1972. A mesoscale model of lake-effect storms. *J. Atmos. Sci.* **29,** 1025–1040.
Lawrence, G. A. 1993. The hydraulics of steady two-layer flow over a fixed obstacle. *J. Fluid Mech.* **254,** 605–633.
LeBlond, P. H., and L. A. Mysak. 1978. *Waves in the Ocean.* Elsevier, Amsterdam/New York.
Ledwell, J. R., A. J. Watson, and C. S. Law. 1993. Evidence of slow mixing across the pycnocline from an open-ocean tracer-release experiment. *Nature* **364,** 701–703.
Lee, D.-K., and P. P. Niiler. 1998. The inertial chimney: The near-inertial energy drainage from the ocean surface to the deep layer. *J. Geophys. Res.* **103,** 7579–7592.
Lee, K., F. J. Millero, and R. Wanninkhof. 1997. The carbon dioxide system in the Atlantic Ocean. *J. Geophys. Res.* **102,** 15693–15707.
Leetmaa, A., and M. Ji. 1989. Operational hindcasting of the tropical Pacific. *Dyn. Atmos. Oceans* **13,** 465–490.
Leibovich, S. 1977. Convective instability of stably stratified water in the ocean. *J. Fluid Mech.* **82,** 561–585.
Leibovich, S. 1980. On wave-current interaction theories of Langmuir circulations. *J. Fluid Mech.* **99,** 715–724.
Leibovich, S. 1983. The form and dynamics of Langmuir circulations. *Annu. Rev. Fluid Mech.* **15,** 391–427.
Leibovich, S., and A. Tandon. 1993. Three-dimensional Langmuir circulation instability in a stratified layer. *J. Geophys. Res.* **98,** 16501–16507.
Leith, C. E. 1971. Atmospheric predictability and two-dimensional turbulence. *J. Atmos. Sci.* **28,** 145–156.
Lemmin, U., and C. H. Mortimer. 1986. Tests of an extension to internal seiches of Defant's procedure for determination of surface seiche characteristics in real lakes. *Limnol. Oceanogr.* **31,** 1207–1231.
Lenschow, D. H. (Ed.). 1986. *Probing the Atmospheric Boundary Layer.* American Meteorological Society, Boston, MA.
Lenschow, D. H. 1994. In P. Matson and R. Harriss (Eds.), *Methods in Ecology: Trace Gases.* Blackwell, MA.
Lenschow, D. H., J. C. Wyngaard, and W. T. Pennel. 1980. Mean field and second-moment budgets in the baroclinic, convective boundary layer. *J. Atmos. Sci.* **37,** 1313–1326.
Lenschow, D. H., X. S. Li, C. J. Zhu, and B. B. Stankov. 1988. The stably stratified boundary layer over the great plains. *Boundary Layer Meteorol.* **42,** 95–121.
Leonard, A. 1974. Energy cascade in large-eddy simulations of turbulent fluid flows. *Adv. Geophys. A* **18,** 237–243.
Leonard, C. L., C. R. McClain, R. Murtugudde, E. E. Hofmann, and L. W. Harding, Jr. 1999. An iron-based ecosystem model of the central equatorial Pacific. *J. Geophys. Res.* **104,** 1325–1341.
Leonov, A. I., and Yu. Z. Miropolskiy. 1973. Resonant excitation of internal gravity waves in the ocean by atmospheric pressure fluctuations. *Izv. Acad. Sci. Atmos. USSR Oceanic Phys.* **9,** 480–485.
Lesieur, M. 1990. *Turbulence in Fluids,* 2nd ed. Kluwer Academic, Dordrecht. 412 pp.
Lesieur, M., and O. Metais. 1996. New trends in large-eddy simulations of turbulence. *Annu. Rev. Fluid Mech.* **28,** 45–82.

Leslie, D. C. 1973. *Developments in the Theory of Turbulence.* Clarendon Press, Oxford.

Leslie, D. C., and G. L. Quarini. 1979. The application of turbulence theory to the formulation of subgrid modelling procedures. *J. Fluid Mech.* **91,** 65–91.

Lettau, H. H. 1979. Wind and temperature profile prediction for diabatic surface layers including strong inversion cases. *Boundary Layer Meteorol.* **17,** 443–464.

Levine, M. D. 1986. Internal waves in the ocean: A review. *Rev. Geophys.* **21,** 1206–1216.

Levine, M. D. 1990. Internal waves under the Arctic pack ice during the Arctic Internal Wave Experiment: The coherence structure. *J. Geophys. Res.* **95,** 7347–7357.

Levine, M. D., R. A. deSzoeke, and P. P. Niiler. 1983. Internal waves in the upper ocean during MILE. *J. Phys. Oceanogr.* **13,** 240–257.

Levine, M. D., C. A. Paulson, and J. H. Morison. 1985. Internal waves in the Arctic Ocean: Comparison with lower-latitude observations. *J. Phys. Oceanogr.* **15,** 800–809.

Levine, M. D., J. D. Irish, T. E. Ewart, and S. A. Reynolds. 1986. Simultaneous spatial and temporal measurements of the internal wave field during MATE. *J. Geophys. Res.* **91,** 9709–9719.

Levine, M. D., C. A. Paulson, and J. H. Morison. 1987. Observations of internal gravity waves under the Arctic pack ice. *J. Geophys. Res.* **92,** 779–782.

Levine, M. D., L. Padman, R. D. Muench, and J. H. Morison. 1997. Internal waves and tides in the western Weddell Sea: Observations from Ice Station Weddell. *J. Geophys. Res.* **102,** 1073–1090.

Li, M., and C. Garrett. 1995. Is Langmuir circulation driven by surface waves or surface cooling? *J. Phys. Oceanogr.* **25,** 64–76.

Li, X., C.-H. Sui, D. Adamec, and K.-M. Lau. 1998. Impacts of precipitation in the upper ocean in the western Pacific warm pool during TOGA-COARE. *J. Geophys. Res.* **103,** 5347–5360.

Lien, R.-C., D. R. Caldwell, M. C. Gregg, and J. N. Moum. 1995. Turbulence variability at the equator in the central Pacific at the beginning of the 1991–1993 El Nino. *J. Geophys. Res.* **100,** 6881–6898.

Lien, R.-C., M. J. McPhaden, and M. C. Gregg. 1996. High-frequency internal waves at 0°, 140 °W and their possible relationship to deep-cycle turbulence. *J. Phys. Oceanogr.* **26,** 581–600.

Lighthill, J. 1978. *Waves in Fluids.* Cambridge Univ. Press, New York. 504 pp.

Lilly, D. K. 1967. The representation of small scale turbulence in numerical simulation experiments. Proc. IBM Sci. Comput. Symp. Environ. Sci., IBM Form 320-1951, 195–202.

Lilly, D. K. 1968. Models of cloud-topped mixed layers under a strong inversion. *Q. J. R. Meteorol. Soc.* **94,** 292–309.

Lilly, D. K. 1978. A severe downslope windstorm and aircraft turbulence event induced by a mountain wave. *J. Atmos. Sci.* **35,** 59–77.

Lilly, D. K. 1992. A proposed modification of the Germano subgrid scale closure method. *Phys. Fluids A* **4,** 633–635.

Lilly, D. K., and E. J. Zipser. 1972. The front range windstorm of 11 January 1972—A meteorological narrative. *Weatherwise* **25,** 56–63.

Lin, R. Q., and W. Perrie. 1997. A new coastal wave model. Part V. Five-wave interactions. *J. Phys. Oceanogr.* **27,** 2169–2186.

Lind, R. J., and K. B. Katsaros. 1982. A model of long wave irradiance for use with surface observations. *J. Appl. Meteorol.* **12,** 1015–1023.

Lind, R. J., K. B. Katsaros, and M. Gube. 1984. Radiation budget components and their parameterization in JASIN. *Q. J. R. Meteorol. Soc.* **110,** 1061–1071.

Linden, P. F. 1973. On the structure of salt fingers. *Deep-Sea Res.* **20,** 325–340.

Linden, P. F. 1976. The formation and destruction of fine-structure by double-diffusive processes. *Deep-Sea Res.* **23,** 895–908.

Linden, P. F. 1978. The formation of banded salt finger structure. *J. Geophys. Res.* **83,** 2902–2912.

Lindstrom, E., R. Lukas, R. Fine, E. Firing, S. Godfrey, G. Meyers, and M. Tsuchiya. 1987. The Western Equatorial Pacific Ocean Circulation Study. *Nature* **330,** 533–537.

Lindzen, R. S. 1990. *Dynamics in Atmospheric Physics.* Cambridge Univ. Press, Cambridge.
Ling, S. C. 1993. Effect of breaking waves on the transport of heat and vapor fluxes from the ocean. *J. Phys. Oceanogr.* **23,** 2360–2372.
List, R. J. 1958. *Smithsonian Meteorological Tables,* 6th ed. Smithsonian Institute, Washington, DC.
Liu, A. K., J. R. Holbrook, and J. R. Apel. 1985. Nonlinear internal wave evolution in the Sulu Sea. *J. Phys. Oceanogr.* **15,** 1613–1624.
Liu, A. K., Y. S. Chang, M.-K. Hsu, and N. K. Liang. 1998. Evolution of nonlinear internal waves in the East and South China Seas. *J. Geophys. Res.* **103,** 7995–8008.
Liu, W. T. 1986. Statistical relation between monthly mean precipitable water and surface-level humidity over global oceans. *Mon. Weather Rev.* **114,** 1591–1602.
Liu, W. T. 1990. Remote sensing of surface turbulence heat flux. In G. L. Geernaert and W. J. Plant (Eds.), *Surface Waves and Fluxes,* Vol. 2, pp. 293–309. Kluwer Academic, Dordrecht/Norwell, MA.
Liu, W. T., K. B. Katsaros, and J. A. Businger. 1979. Bulk parameterization of air-sea exchanges of heat and water vapor including the molecular constraints at the interface. *J. Atmos. Sci.* **36,** 1722–1735.
Liu, W. T., and P. P. Niiler. 1984. Determination of monthly mean humidity in the atmospheric surface layer over oceans from satellite data. *J. Phys. Oceanogr.* **14,** 1451–1457.
Lombardo, C. P., and M. C. Gregg. 1989. Similarity scaling of viscous and thermal dissipation in a convecting surface boundary layer. *J. Geophys. Res.* **94,** 6273–6284.
Long, R. R. 1953. Some aspects of stratified fluids. I. A theoretical investigation *Tellus,* **5,** 42–58.
Long, R. R. 1954. Some aspects of stratified fluids. II. Experiments with a two-fluid system. *Tellus* **6,** 97–115.
Long, R. R. 1955. Some aspects of stratified fluids. III. Continuous density gradients. *Tellus* **7,** 341–357.
Long, R. R. 1970. Blocking effects in flow over obstacles. *Tellus* **22,** 471–479.
Long, R. R. 1972. Finite amplitude disturbances in the flow of inviscidrotating and stratified fluids over obstacles. *Annu. Rev. Fluid Mech.* **4,** 69–92.
Longuet-Higgins, M. S. 1969. On wave breaking and the equilibrium spectrum of wind-generated waves. *Proc. R. Soc. London A* **310,** 151–159.
Longuet-Higgins, M. S. 1972. On the form of the highest progressive and standing waves in deep water. *Proc. R. Soc. London A* **331,** 445–456.
Longuet-Higgins, M. S. 1976. On the nonlinear transfer of energy in the peak of a gravity-wave spectrum: A simplified model. *Proc. R. Soc. London A* **347,** 311–328.
Longuet-Higgins, M. S. 1979. Why is a water wave like a grandfather clock? *Phys. Fluids* **22,** 1828–1829.
Longuet-Higgins, M. S. 1984. Statistical properties of wave groups in a random sea state. *Phil. Trans. R. Soc. London A* **312,** 219–250.
Longuet-Higgins, M. S. 1985a. Accelerations in steep gravity waves. *J. Phys. Oceanogr.* **15,** 1570–1579.
Longuet-Higgins, M. S. 1985b. A new way to calculate steep gravity waves. In Y. Toba and H. Mitsuyasu (Eds.), *The Ocean Surface,* pp. 1–15. Reidel, Dordrecht.
Longuet-Higgins, M. S. 1986. Eulerian and Lagrangian aspects of surface waves. *J. Fluid Mech.* **173,** 683–707.
Longuet-Higgins, M. S. 1992. Capillary rollers and bores. *J. Fluid Mech.* **240,** 659–679.
Longuet-Higgins, M. S. 1995. Parasitic capillary waves: A direct calculation. *J. Fluid Mech.* **301,** 79–107.
Longuet-Higgins, M. S., and R. W. Stewart. 1964. Radiation stresses in water waves: A physical discussion with applications. *Deep-Sea Res.* **11,** 529–562.

Longuet-Higgins, M. S., and E. D. Cokelet. 1976. The deformation of steep surface waves on water. I. A numerical method of computation. *Proc. R. Soc. London A* **350,** 1–26.

Longuet-Higgins, M. S., and D. G. Dommermuth. 1997. Crest instabilities of gravity waves. Part 3. Nonlinear development and breaking. *J. Fluid Mech.* **336,** 33–50.

Lott, F., and M. Miller. 1997. A new sub-grid scale orographic drag parameterization: Its formulation and testing. *Q. J. R. Meteorol. Soc.* **123,** 101–127.

Lozier, M. S., W. B. Owens, and R. G. Curry. 1995. The climatology of the North Atlantic. *Prog. Oceanogr.* **36,** 1–44.

Lueck, R. 1987. Microstructure measurements in a thermocline staircase. *Deep-Sea Res.* **34,** 1677–1688.

Lukas, R., and E. Lindstrom. 1991. The mixed layer of the western equatorial Pacific Ocean. *J. Geophys. Res.* **96,** 3343–3358.

Lumb, F. E. 1964. The influence of cloud on hourly amount of total solar radiation at the sea surface. *Q. J. R. Meteorol. Soc.* **90,** 43–56.

Lumley, J. L. 1978. Computational modelling of turbulent flows. *Adv. Appl. Mech.* **18,** 124–176.

Lumley, J. L. 1983. Turbulence modeling. *J. Appl. Mech.* **50,** 1097–1103.

Lumley, J. L., and H. A. Panofsky. 1964. *The Structure of Atmospheric Turbulence.* Wiley, New York. 239 pp.

Lumley, J. L., and E. A. Terray. 1983. Kinematics of turbulence convected by a random wave field. *J. Phys. Oceanogr.* **13,** 2000–2077.

Luyten, J. E., J. Pedlosky, and H. Stommel. 1983. The ventilated thermocline. *J. Phys. Oceanogr.* **13,** 292–309.

Lyra, G. 1940. Uber den Einfluss von Bodener hebungen auf die Stromung einer stabil geschichteten Atmosphare. *Beitr. Phys. Freien Atmos.* **26,** 197–206.

Maat, N., C. Kraan, and W. A. Oost. 1991. The roughness of wind waves. *Boundary Layer Meteorol.* **54,** 89–103.

McAlister, E. D., and W. McLeish. 1969. Heat transfer in the top millimeter of the ocean. *J. Geophys. Res.* **74,** 3408–3414.

McClain, C. R., K. Arrigo, K.-S. Tai, and D. Turk. 1996. Observations and simulations of physical and biological processes at ocean weather station P, 1951–1980. *J. Geophys. Res.* **101,** 3697–3714.

McComas, C. H., and F. P. Bretherton. 1977. Resonant interactions of oceanic internal waves. *J. Geophys. Res.* **82,** 1397–1412.

McComas, C. H., and P. Muller. 1981. The dynamic balance of internal waves. *J. Phys. Oceanogr.* **11,** 970–986.

McComb, W. D. 1990. *The Physics of Fluid Turbulence.* Oxford, New York.

McDougall, T. J. 1983. Double-diffusive convection caused by coupled molecular diffusion. *J. Fluid Mech.* **126,** 379–397.

McDougall, T. J. 1987. Thermobaricity, cabelling and water mass conversion. *J. Geophys. Res.* **92,** 5448–5464.

McDougall, T. J. 1991. Interfacial advection in the thermohaline staircase east of Barbados. *Deep-Sea Res.* **38,** 367–370.

McDougall, T. J., and J. R. Taylor. 1984. Flux measurements across a finger interface at low values of the stability ratio. *J. Mar. Res.* **42,** 1–14.

McFarlane, N. A. 1987. The effect of orographically excited gravity wave drag on the general circulation of the lower stratosphere and troposphere. *J. Atmos. Sci.* **44,** 1775–1800.

McGillicuddy, D. J., Jr., J. J. McCarthy, and A. R. Robinson. 1995a. Coupled physical and biological modeling of the spring bloom in the North Atlantic. I. Model formulation and one-dimensional bloom processes. *Deep-Sea Res.* **42,** 1313–1357.

McGillicuddy, D. J., Jr., A. R. Robinson, and J. J. McCarthy. 1995b. Coupled physical and biological modeling of the spring bloom in the North Atlantic II. Three-dimensional bloom and post-bloom effects. *Deep-Sea Res.* **42,** 1358–1397.

McGoldrick, L. F. 1965. Resonant interactions among capillary-gravity waves. *J. Fluid Mech.* **21,** 305–332.

McGoldrick, L. F., O. M. Phillips, N. Huang, and T. Hodgson. 1966. Measurement of resonant wave interactions. *J. Fluid Mech.* **25,** 437–456.

Mack, S. A. 1985. Two-dimensional measurements of ocean microstructure: The role of double diffusion. *J. Phys. Oceanogr.* **15,** 1581–1604.

Mack, S. A. 1989. Towed chain measurement of ocean microstructure. *J. Phys. Oceanogr.* **15,** 1581–1604.

Mack, S. A., and H. C. Schoeberlein. 1993. Discriminating salt fingering from turbulence-induced microstructure: Analysis of towed temperature-conductivity chain data. *J. Phys. Oceanogr.* **23,** 2073–2106.

McLean, J. W. 1982. Instabilities of finite-amplitude gravity waves on water of finite depth. *J. Fluid Mech.* **114,** 331–341.

McManus, J., R. W. Collier, and J. Dymond. 1993. Mixing processes in Crater Lake, Oregon. *J. Geophys. Res.* **98,** 18295–18307.

McManus, J., R. W. Collier, J. Dymond, C. G. Wheat, and G. Larson. 1996. Spatial and temporal distribution of dissolved oxygen in Crater Lake, Oregon. *Limnol. Oceanogr.* 722–731.

McPhaden, M. J., H. P. Freitag, S. P. Hayes, B. A. Taft, Z. Chen, and K. Wyrtki. 1988. The response of the equatorial Pacific to a westerly wind burst in May 1986. *J. Geophys. Res.* **93,** 10589–10603.

McPhaden, M. J., and H. Peters. 1992. On the diurnal cycle of internal wave variability in the eastern equatorial Pacific: Results from moored observations. *J. Phys. Oceanogr.* **22,** 1317–1329.

McPhaden, M. J., F. Bahr, Y. Du Penhoat, S. P. Hayes, P. P. Niiler, P. L. Richardson, and J. M. Toole. 1992. The response of the western equatorial Pacific Ocean to westerly wind bursts during November 1989 to January 1990. *J. Geophys. Res.* **97,** 14289–14303.

McPhee, M. G. 1979. The effect of the oceanic boundary layer on the mean drift of pack ice: Application of a simple model. *J. Phys. Oceanogr.* **9,** 388–400.

McPhee, M. G. 1980. A study of oceanic boundary-layer characteristics including inertial oscillation at three drifting stations in the Arctic Ocean. *J. Phys. Oceanogr.* **10,** 870–884.

McPhee, M. G. 1983. Turbulent heat and momentum transfer in the oceanic boundary layer under melting pack ice. *J. Geophys. Res.* **88,** 2827–2835.

McPhee, M. G. 1986. The upper ocean. In N. Untersteiner (Ed.), *The Geophysics of Sea Ice,* Plenum Publ. Corp., New York, pp. 339–394.

McPhee, M. G. 1990. Small scale processes. In W. O. Smith, Jr. (Ed.), *Polar Oceanography,* pp. 287–334. Academic Press, San Diego.

McPhee, M. G. 1991. A quasi-analytical model for the under-ice boundary layer. *Ann. Geol.* **15,** 148–154.

McPhee, M. G. 1992. Turbulent heat flux in the upper ocean under sea ice. *J. Geophys. Res.* **97,** 5365–5379.

McPhee, M. G. 1994. On the turbulent mixing length in the oceanic boundary layer. *J. Phys. Oceanogr.* **24,** 2014–2031.

McPhee, M. G., and J. D. Smith. 1976. Measurements of the turbulent boundary layer under pack ice. *J. Phys. Oceanogr.* **6,** 696–711.

McPhee, M. G., and L. H. Kantha. 1989. Generation of internal waves by sea ice. *J. Geophys. Res.* **94,** 3287–3302.

McPhee, M. G., and D. G. Martinson. 1994. Turbulent mixing under drifting pack ice in the Weddell Sea. *Science* **263,** 218–221.

McWilliams, J. C. 1993. Modeling the oceanic planetary boundary layer. In B. galperin and S. A. Orszag (Eds.), *Large Eddy Simulation of Complex Engineering and Geophysical Flows*, pp. 489–509. Cambridge University Press, New York.

Mahrt, L. 1982. Momentum balance of gravity flows. *J. Atmos. Sci.* **39,** 2701–2711.

Mahrt, L., D. Vickers, J. Howell, J. Hojstrup, J. M. Wilczak, J. Edson, and J. Hare. 1996. Sea surface drag coefficients in the Riso Air Sea Experiment. *J. Geophys. Res.* **101,** 14327–14335.

Mailhot, J., J. W. Strapp, J. I. MacPherson, R. Benoit, S. Belair, N. R. Donaldson, F. Froude, M. Benjamin, I. Zawadzki, and R. R. Rogers. 1998. The Montreal-96 Experiment on Regional Mixing and Ozone (MERMOZ): An overview and some preliminary results. *Bull. Am. Meteorol. Soc.* **79,** 433–442.

Makin, V. K., and V. N. Kudryavtsev. 1999. Coupled sea surface-atmosphere model. 1. Wind over waves coupling, *J. Geophys. Res.* **104,** 7613–7624.

Markatos, N. C. 1986. The mathematical modelling of turbulent flows. *Appl. Math. Modelling* **10,** 190–220.

Marmorino, G. O., and D. R. Caldwell. 1976. Heat and salt transport through a diffusive thermohaline interface. *Deep-Sea Res.* **23,** 59–67.

Marmorino, G. O., W. K. Brown, and W. D. Morris. 1987. Two-dimensional temperature structure in the C-SALT thermohaline staircase. *Deep-Sea Res.* **34,** 1667–1675.

Marra, J. 1995. Bioluminescence and optical variability in the ocean: An overview of the Marine Light-Mixed Layers Program. *J. Geophys. Res.* **100,** 6521–6525.

Marra, J., C. Langdon, and C. A. Knudson. 1995. Primary production, water column changes, and the demise of a Phaeocytis bloom at the Marine Light-Mixed Layers site (59 °N, 21 °W) in the northeast Atlantic Ocean. *J. Geophys. Res.* **100,** 6633–6643.

Marshall, J., and F. Schott. 1998. *Open-Ocean Convection: Observations, Theory and Models.* Center for Global Change Science Report 52, MIT, Cambridge, MA. 158 pp.

Marshunova, M. S. 1966. Principal characteristics of the radiation balance of the underlying surface. In J. O. Fletcher *et al.* (Eds.), *Soviet Data on the Arctic Heat Budget and Its Climate Influence.* Report R.M. 5003-PR, Rand Corp., Santa Monica, CA.

Martin, P. J. 1985. Simulation of the mixed layer at OWS November and Papa with several models. *J. Geophys. Res.* **90,** 903–916.

Martin, P. J. 1986. *Testing and Comparison of Several Mixed-Layer Models.* Naval Oceanographic Research and Development Agency (NORDA) Report 143, Naval Research Laboratory, Stennis Space Center, MS. 27 pp.

Martin, P. J., and R. A. Allard. 1993. *The Calculation of Solar Extinction from Satellite Color for Ocean Modeling.* Report NRL/MR/7322-93-7021, Naval Research Laboratory, Washington, DC. 70 pp.

Martin, G. M., D. W. Johnson, and D. A. Hegg. 1995. Observations of the interaction between cumulus clouds and warm stratocumulus clouds in the marine boundary layer during ASTEX. *J. Atmospheric Sci.* **52,** 2902–2922.

Maslowe, S. A. 1986. Critical layers in shear flows. *Annu. Rev. Fluid Mech.* **18,** 405–432.

Mason, P. J. 1994. Large eddy simulation: A critical review of the technique. *Q. J. R. Meteorol. Soc.* **120,** 1–35.

Mason, P. J., and S. H. Derbyshire. 1990. Large eddy simulation of the stably-stratified atmospheric boundary layer. *Boundary Layer Meteorol.* **53,** 117–162.

Mass, C. F., and D. P. Dempsey. 1985. A one-level, mesoscale model for diagnosing surface winds in mountainous and coastal regions. *Mon. Weather Rev.* **113,** 565–578.

Maxworthy, T., and S. Narimousa. 1994. Unsteady, turbulent convection into a homogeneous, rotating fluid, with oceanographic application. *J. Phys.Oceanogr.* **24,** 865–887.

May, B. D., and D. E. Kelley. 1997. Effect of baroclinicity on double-diffusive interleaving. *J. Phys. Oceanogr.* **27,** 1997–2008.

Maykut, G. A., and D. K. Perovich. 1987. The role of shortwave radiation in the summer decay of a sea ice cover. *J. Geophys. Res.* **92,** 7032–7044.

Maykut, G. A., and M. G. McPhee. 1995. Solar heating of the Arctic mixed layer. *J. Geophys. Res.* **100,** 24691–24703.

Melas, D. 1989. The temperature structure in a stably stratified internal boundary layer over a cold sea. *Boundary Layer Meteorol.* **48,** 361–375.

Mellor, G. L. 1973. Analytic prediction of the properties of stratified planetary surface layers. *J. Atmos. Sci.* **30,** 1061–1069.

Mellor, G. L. 1989. Retrospect on oceanic boundary layer modeling and second moment closure. In P. Muller and D. Henderson (Eds.), *Parameterization of Small-Scale Processes,* Proceedings of the Aha Hulikoa Hawaiian Winter Workshop.

Mellor, G. L. 1991. *User's Guide for a Three-Dimensional, Primitive Equation, Numerical Ocean Model.* Program in Atmospheric and Oceanic Sciences Report, Princeton University, Princeton, NJ. 35 pp.

Mellor, G. L., and T. Yamada. 1974. A hierarchy of turbulence closure models for planetary boundary layers. *J. Atmos. Sci.* **31,** 1791–1806.

Mellor, G. L., and T. Yamada. 1982. Development of a turbulence closure model for geophysical fluid problems. *Rev. Geophys. Space Phys.* **20,** 851–875.

Mellor, G. L., and L. H. Kantha. 1989. An ice-ocean coupled model. *J. Geophys. Res.* **94,** 10937–10954.

Melville, W. K. 1994. Energy dissipation by breaking waves. *J. Phys. Oceanogr.* **24,** 2041–2049.

Melville, W. K. 1996. The role of surface-wave breaking in air-sea interaction. *Annu. Rev. Fluid Mech.* **28,** 279–321.

Melville, W. K., and R. J. Rapp. 1985. Momentum flux in breaking waves. *Nature* **317,** 514–516.

Miles, J. W. 1957. On the generation of surface waves by shear flows. Part I. *J. Fluid Mech.* **3,** 185–204.

Miles, J. W. 1961. On the stability of heterogeneous shear flows. *J. Fluid Mech.* **10,** 496–515.

Miles, J. W. 1963. On the stability of heterogeneous shear flows. Part 2. *J. Fluid Mech.* **16,** 209–227.

Miles, J. W. 1993. Surface-wave generation revisited. *J. Fluid Mech.* **256,** 427–441.

Miller, D. K., and K. B. Katsaros. 1992. Satellite-derived surface latent heat fluxes in a rapidly intensifying marine cyclone. *Mon. Weather Rev.* **120,** 1093–1107.

Miller, M. A., and B. A. Albrecht. 1995. Surface-based observations of mesoscale cumulus-stratocumulus interaction during ASTEX. *J. Atmos. Sci.* **52,** 2809–2826.

Miller, M. J., A. C. M. Beljaars, and T. N. Palmer. 1992. The sensitivity of the ECMWF model to the parameterization of evaporation from tropical oceans. *J. Clim.* **5,** 418–434.

Millero, F. J., and A. Poisson. 1981. International one-atmosphere equation of state of seawater. *Deep-Sea Res.* **27A,** 255–264.

Millikan, C. B. 1939. A critical discussion of turbulent flow in channels and circular tubes. In *Proc. Fifth Intern. Cong. Appl. Mech.,* pp. 386–392. John Wiley, New York.

Mitsuyasu, H. 1985. A note on momentum transfer from wind to waves. *J. Geophys. Res.* **90,** 3343–3345.

Moeng, C.-H. 1984. A large-eddy simulation model for the study of planetary boundary layer turbulence. *J. Atmos. Sci.* **41,** 2052–2062.

Moeng, C.-H., and J. C. Wyngaard. 1986. An analysis of closures for pressure-scalar covariances in the convective boundary layer. *J. Atmos. Sci.* **43,** 2499–2513.

Moeng, C.-H., and J. C. Wyngaard. 1989. Evaluation of turbulent transport and dissipation closures in second-order modeling. *J. Atmos. Sci.* **46,** 2311–2330.

Moeng, C.-H., and P. P. Sullivan. 1994. A comparison of shear- and buoyancy-driven planetary boundary layer flows. *J. Atmos. Sci.* **51,** 999–1022.

Moeng, C.-H., W. R. Cotton, C. Bretherton, A. Chlond, M. Khairoutdinov, S. Krueger, W. S. Lewellen, M. K. MacVean, J. R. M. Pasquier, H. A. Rand, A. P. Siebesma, B. Stevens, and R. I. Sykes. 1996. Simulation of a stratocumulous-topped planetary boundary layer: Intercomparison among different numerical codes. *Bull. Am. Meteorol. Soc.* **77,** 261–278.

Moin, P., and J. Kim. 1982. Numerical investigation of turbulent channel flow. *J. Fluid Mech.* **118,** 341–377.

Moin, P., and K. Mahesh.1998. Direct numerical simulation: A tool in turbulence research. *Annu. Rev. Fluid Mech.* **30,** 539–578.

Moisan, J. R., and E. E. Hofmann. 1996. Modeling nutrient and plankton processes in the California coastal transition zone. 1. A time- and depth-dependent model. *J. Geophys. Res.* **101,** 22647–22676.

Monin, A. S., and A. M. Obukhoff. 1954. Basic laws of turbulent mixing in the ground layer of the atmosphere. *Trans. Akad. Nauk SSSR. Geofiz. Inst.* **151,** 163–187.

Monin, A. S., and A. M. Yaglom. 1971. *Statistical Fluid Mechanics: Mechanics of Turbulence*, Vols. 1 and 2, pp. 769 and 874. MIT Press, Cambridge, MA.

Monteith, J. L. 1975. *Vegetation and the Atmosphere*, Vol. 1. *Principles.* Academic Press, New York. 278 pp.

Monteith, J. L. 1976. *Vegetation and the Atmosphere*, Vol. 2. *Case Studies.* Academic Press, New York. 439 pp.

Morel, A. 1988. Optical modeling of the upper ocean in relation to its biogenous content (case I waters). *J. Geophys. Res.* **93,** 10749–10768.

Morel, A., and D. Antoine. 1994. Heating rate within the upper ocean in relation to its bio-optical state. *J. Phys. Oceanogr.* **24,** 1652–1665.

Morison, J. H., M. G. McPhee, and G. A. Maykut. 1987. Boundary layer, upper ocean, and ice observations in the Greenland Sea marginal ice zone. *J. Geophys. Res.* **92,** 6987–7011.

Morison, J. H., M. G. McPhee, T. B. Curtin, and C. A. Paulson. 1992. The oceanography of winter leads. *J. Geophys. Res.* **97,** 11199–11218.

Mortimer, C.H., 1974, Lake hydrodynamics, *Mitt. Internat. Verein. Limnol.*, **20**, 124–197.

Moum, J. N. 1990. The quest for K_ρ—Preliminary results from direct measurements of turbulent fluxes in the ocean. *J. Phys. Oceanogr.* **20,** 1980–1984.

Moum, J. N. 1996a. Efficiency of mixing in the main thermocline. *J. Geophys. Res.* **101,** 16500–16509.

Moum, J. N. 1996b. Energy-containing scales of turbulence in the ocean thermocline. *J. Geophys. Res.* **101,** 14905–14109.

Moum, J. N. 1998. The flux of particles to the deep sea: Methods, measurements and mechanisms. *Oceanography* **10,** 111–115.

Moum, J. N., and D. R. Caldwell. 1985. Local influences on shear flow turbulence in the equatorial ocean. *Science* **230,** 315–316.

Moum, J. N., and T. R. Osborn. 1986. Mixing in the main thermocline. *J. Phys. Oceanogr.* **16,** 1250–1259.

Moum, J. N., and D. R. Caldwell. 1989. Mixing in the equatorial surface layer. *J. Geophys. Res.* **94,** 2005–2021.

Moum, J. N., D. R. Caldwell, and C. A. Paulson. 1989. Mixing in the equtorial surface layer and thermocline. *J. Geophys. Res.* **94,** 2005–2021.

Moum, J. N., D. Hebert, C. A. Paulson, and D. R. Caldwell. 1992a. Turbulence and internal waves at the equator. Part I. Statistics from towed thermistors and a microstructure profiler. *J. Phys. Oceanogr.* **22,** 1330–1345.

Moum, J. N., M. J. McPhaden, D. Hebert, H. Peters, C. A. Paulson, and D. R. Caldwell. 1992b. Internal waves, dynamical instabilities and turbulence in the equatorial thrmocline: An introduction to the three papers in this issue. *J. Phys. Oceanogr.* **22,** 1357–1359.

Moum, J. N., and D. R. Caldwell. 1994. Experiment explores the dynamics of ocean mixing. *EOS Trans.* **75,** 489–495.

Moum, J. N., M. C. Gregg, R.-C. Lien, and M.-E. Carr. 1995. Comparison of turbulent kinetic energy dissipation rate estimates from two ocean microstructure profilers. *J. Atmos. Oceanic Technol.* **12,** 346–366.

Mowbray, D., and B. S. H. Rarity. 1967. A theoretical and experimental investigation of the phase configuration of internal waves of small amplitude in a density stratified fluid. *J. Fluid Mech.* **28,** 1–16.

Muench, R. D., H. J. S. Fernando, and G. R. Stegan. 1990. Temperature and salinity staircases in the northwestern Weddell Sea. *J. Phys. Oceanogr.* **20,** 295–306.

Mulhearn, P. J. 1977. Relations between surface fluxes and mean profiles of velocity, temperature, and concentration downwind of a change in surface roughness. *Q. J. R. Meteorol. Soc.* **103,** 785–802.

Mulhearn, P. J. 1981. On the formation of a stably stratified internal boundary layer by advection of warm air over a cooler sea. *Boundary Layer Meteorol.* **21,** 247–254.

Muller, G. 1998. Starch columns: Analog model for bsalt columns. *J. Geophys. Res.* **103,** 15239–15253.

Muller, P. 1976. *Parameterization of One-Dimensional Wind Wave Spectra and Their Dependence on the State of Development.* Hamburger Geophysikalishe Einzelschriften, Heft 31. 177 pp.

Muller, P., and D. Henderson (Eds.). *Dynamics of the Oceanic Surface Mixed Layer. Proceedings of the Aha Huliko'a Hawaaian Winter Workshop.* Hawaii Institute of Geophysics Special Publication.

Muller, P., D. J. Olbers, and J. Willebrand. 1978. IWEX spectrum. *J. Geophys. Res.* **83,** 479–500.

Muller, P., G. Holloway, F. Henyey, and N. Pomphrey. 1986. Nonlinear interactions among internal gravity waves. *Rev. Geophys.* **24,** 493–536.

Muller, P., and D. Henderson (Eds.). 1991. *Dynamics of Oceanic Internal Gravity Waves. Aha Huliko'a Hawaiian Winter Workshop.* University of Hawaii at Manoa.

Mullineaux, C. W. 1999. The plankton and the planet. *Science* **283,** 801–802.

Munk, W. 1966. Abyssal recipes. *Deep-Sea Res.* **13,** 707–730.

Munk, W. 1981. Internal wave and small scale processes. In B. A. Warren and C. Wunsch (Eds.), *Evolution of Physical Oceanography*, pp. 264–291. MIT Press, Cambridge, MA.

Munk, W. H. 1968. Once again—Tidal friction. *Q. J. R. Astron. Soc.* **9,** 352–375.

Munk, W. H. 1997. Once again—Once again—Tidal friction. *Prog. Oceanogr.,* **40,** 7–36.

Munk, W. H., G. R. Miller, F. E. Snodgrass, and N. F. Barber. 1963. Directional recording of swell from distant storms. *Philos. Trans. R. Soc. London Ser. A* **255,** 505–584.

Munk, W. H., and C. Wunsch. 1998. The moon and mixing: Abyssal recipes II: Energetics of tidal and wind mixing. *Deep-Sea Res.* **I45,** 1977–2010.

Murthy, B. S., T. Dharmaraj, and K. G. Vernekar. 1996. Sodar observations of the nocturnal boundary layer at Kharagpur, India. *Boundary Layer Meteorol.* **81,** 201–209.

Najjar, R. G. 1992. Marine biogeochemistry. In K. Trenberth (Ed.), *Climate System Modeling,* pp. 214–280. Cambridge University Press, Cambridge.

Najjar, R. G., D. J. Erickson III, and S. Madronich. 1994. Modeling the air-sea fluxes of gases from the decomposition of dissolved organic matter: Carbonyl sulfide and carbon monoxide. In R. Zepp and C. Sonntag (Eds.), *The Role of Non-living Organic Matter in the Earth's Carbon Cycle.* Wiley, New York.

Nakagawa, H., I. Nezu, and H. Ueda. 1975. Turbulence in open channel flow over smooth and rough beds. *Proc. Jpn. Soc. Civ. Eng.* **241,** 155–168.
Nappo, C. J., Jr., 1977. Mesoscale flow over complex terrain during the Eastern Tennessee Trajectory Experiment (ETTEX). *J. Appl. Meteor.*, **15**, 1186–1196.
Neal, V. T., S. Neshyba, and W. Denner. 1969. Thermal stratification in the Arctic Ocean. *Science* **166,** 373–374.
Neal, V. T., S. J. Neshyba, and W. W. Denner. 1972. Vertical temperature structure in Crater Lake, Oregon. *Limnol. Oceanogr.* **17,** 451–454.
Nelkin, M. 1994. Universality and scaling in fully developed turbulence. *Adv. Phys.* **43,** 143–181.
Nelson, R. A. 1982. *SI: The International System of Units,* 2nd ed. American Association of Physics Teachers, College Park, MD.
Neshyba, S., V. T. Neal, and W. Denner. 1971. Temperature and conductivity measurements under ice island T-3. *J. Geophys. Res.* **76,** 8107–8120.
Nevison, C. D., R. F. Weiss, and D. J. Erickson III. 1995. Global oceanic emissions of nitrous oxide. *J. Geophys. Res.* **100,** 15809, 15820.
Newman, F. C. 1976. Temperature steps in Lake Kivu, a bottom heated saline lake. *J. Phys. Oceanogr.* **6,** 157–163.
Nicholls, J. M. 1973. *The Airflow over Mountains: Research 1958–1972.* WMO Tech. Note no. 127.
Nicholls, S. 1984. The dynamics of stratocumulus: Aircraft observations and comparisons with a mixed layer model. *Q. J. R. Meteorol. Soc.* **110,** 783–820.
Nicholls, S., 1989. The structure of radiatively driven convection in stratocumulus, *Q. J. R. Meteorol. Soc.*, **115**, 487–511.
Nicholls, S. and J. Leighton, 1986. An observational study of the structure of stratiform cloud sheets: Part I. Structure, *Q. J. R. Meteorol. Soc.*, **112**, 431–460.
Niebauer, H. J. 1991. Bio-physical oceanographic interactions at the edge of the Arctic ice pack. *J. Mar. Systems* **2,** 209–232.
Nield, D. A. 1967. The thermohaline Rayleigh-Jeffries problem. *J. Fluid Mech.* **29,** 545–558.
Nieuwstadt, F. T. M. 1984. The turbulent structure of the stable nocturnal boundary layer. *J. Atmos. Sci.* **41,** 2202–2216.
Nieuwstadt, F. T. M., and H. Tennekes. 1981. A rate equation for the nocturnal boundary layer height. *J. Atmos. Sci.* **38,** 1418–1428.
Nieuwstadt, F. T. M., and H. van Dop (Eds.). 1982. *Atmospheric Turbulence and Air Pollution Modelling*. Reidel, Dordrecht. 358 pp.
Nieuwstadt, F. T. M., P. J. Mason, C.-H. Moeng, and U. Schumann. 1991. Large eddy simulation of the convective boundary layer: A comparison of four computer codes. In *Proceedings of the Eighth Symposium on Turbulent Shear Flows.* Springer-Verlag, Berlin/New York.
Niiler, P. P. 1975. Deepening of the wind-mixed layer. *J. Mar. Res.* **33,** 405–422.
Niiler, P. P. 1992. The ocean circulation. In K. E. Trenberth (Ed.), *Climate System Modeling,* pp. 117–148. Cambridge Univ. Press, New York.
Niiler, P. P., and E. B. Kraus. 1977. One-dimensional models of the upper ocean. In E. B. Kraus (Ed.), *Modelling and Prediction of the Upper Layers of the Ocean,* pp. 143–172. Pergamon, New York.
Nokes, R. I. 1988. On the entrainment rate across a density interface. *J. Fluid Mech.* **188,** 185–204.
Nowell,A. R. M. 1983. The benthic boundary layer and sediment transport. *Rev. Geophys. Space Phys.* **21,** 1181–1192.
Oberhuber, J. M. 1993a. Simulation of the Atlantic circulation with a coupled sea ice-mixed layer-isopycnal general circulation model. I. Model description. *J. Phys. Oceanogr.* **23,** 808–829.
Oberhuber, J. M. 1993b. Simulation of the Atlantic circulation with a coupled sea ice-mixed layer-isopycnal general circulation model. II. Model experiment. *J. Phys. Oceanogr.* **23,** 830–845.

Obukhoff, A. M. Turbulence in the atmosphere with nonuniform temperature. *Trans. Akad. Nauk SSSR Inst. Teoret. Geofiz.* **1.** Also *Boundary Layer Meteorol.* **2,** 7–29)

Obukhoff, A. M. 1949. The structure of the temperature field in a turbulent flow. *Izv. Akad. Nauk. SSSR. Ser. Geogr. Geophys.* **13,** 58.

Oguz, T., H. Ducklow, P. Malanotte-Rizzoli, S. Tugrul, N. P. Nezlin, and U. Unluata. 1996. Simulation of annual plankton productivity cycle in the Black Sea by a one-dimensional physical-biological model. *J. Geophys. Res.* **101,** 16585–16599.

Ohlmann, J. C., D. A. Siegel, and L. Washburn. 1998. Radiant heating of the western equatorial Pacific during TOGA-COARE. *J. Geophys. Res.* **103,** 5379–5398.

Oke, T. R. 1978. *Boundary Layer Climates*. Halsted, New York.

Olbers, D. J. 1983. Models of the oceanic internal wave field. *Rev. Geophys. Space Phys.* **21,** 1567–1606.

Oncley, S. P., 1992. TKE dissipation measurements during the FLAT experiment, Abstract, AMS Tenth Symposium on Turbulence and Diffusion, Portland, OR, 1992, 165–166.

Oncley, S. P., J. A. Businger, C. A. Friehe, J. C. La Rue, E. C. Itsweire, and S. S. Chang. 1990. Surface layer profiles and turbulence measurements over uniform land under near-neutral conditions. In *Proceedings, Ninth Symposium on Turbulence and Diffusion, April 30–May 3, 1990,* pp. 237–240.

Orszag, S. A. 1970. Analytical theories of turbulence. *J. Fluids Mech.* **41,** 363–386.

Osborn, T. R. 1980. Estimates of local rate of vertical diffusion from dissipation measurements. *J. Phys. Oceanogr.* **10,** 83–89.

Osborn, T. R., and C. S. Cox. 1972. Oceanic fine structure. *Geophys. Fluid Dyn.* **3,** 321–345.

Osborn, T., D. M. Farmer, S. Vagle, S. Thorpe, and M. Cure. 1992. Measurements of bubble plumes and turbulence from a submarine. *Atmos.-Ocean* **30,** 419–440.

Osborne, A. R., and T. L. Birch. 1980. Internal solitons in the Andaman Sea. *Science* **208,** 451–460.

Ostrovsky, L. A., and Y. A. Stepanyants. 1989. Do internal solitons exist in the ocean? *Rev. Geophys.* **27,** 293–310.

Ostrovsky, L. A., V. I. Kazakov, P. A. Matusov, and D. Zaborskikh. 1996. Experimental study of the internal wave damping on small-scale turbulence. *J. Phys. Oceanogr.* **26,** 398–405.

Ostrovsky, L. A., and D. Zaborskikh. 1996. Damping of internal gravity waves by small-scale turbulence. *J. Phys. Oceanogr.* **26,** 388–397.

Ostwatitsch, K., and K. Wieghardt. 1987. Ludwig Prandtl and his Kaiser-Wilhelm-Institut. *Annu. Rev. Fluid Mech.* **19,** 1–25.

Overland, J. E. 1985. Atmospheric boundary layer structure and drag coefficients over sea ice. *J. Geophys. Res.* **90,** 9029–9049.

Overland, J. E., M. H. Hitchman, and Y. J. Han. 1979. *A regional surface wind model for mountainous coastal areas.* NOAA Technical Report ERL 407-PMEL 32, U.S. Dept. of Commerce, Washington, DC. 34 pp.

Overland, J. E., R. M. Reynolds, and C. H. Pease. 1983. A model of the atmospheric boundary layer over the marginal ice zone. *J. Geophys. Res.* **88,** 2836–2840.

Overland, J. E., and K. L. Davidson. 1992. Geostrophic drag coefficients over sea ice. *Tellus* **44,** 54–66.

Pacanowski, R. C., and S. G. H. Philander. 1981. Parameterization of vertical mixing in numerical models of the tropical oceans. *J. Phys. Oceanogr.* **11,** 1443–1451.

Padman, L. 1995. Small scale processes in the Arctic Ocean. In W. O. Smith, Jr., and J. M. Grebmeier (Eds.), *Arctic Oceanography: Marginal Ice Zones and Continental Shelves,* pp. 97–129. AGU, Washington, DC.

Padman, L., and T. M. Dillon. 1987. Vertical fluxes through the Beaufort Sea thermohaline staircase. *J. Geophys. Res.* **92,** 799–806.

Padman, L., and T. M. Dillon. 1988. On the horizontal extent of the Canada Basin thermohaline steps. *J. Phys. Oceanogr.* **18,** 1458–1462.

Padman, L., and T. M. Dillon. 1989. Thermal microstructure and internal waves in the Canada Basin diffusive staircase. *Deep-Sea Res.* **36,** 531–542.

Padman, L., and T. M. Dillon. 1991. Turbulent mixing near the Yermak Plateau during the Coordinated Eastern Arctic Experiment. *J. Geophys. Res.* **96,** 4769–4782.

Palmer, T. N., G. J. Shutts, and R. Swinbank. 1986. Alleviation of a systematic westerly bias in general circulation and numerical weather prediction models through an orgraphic gravity wave drag parameterization. *Q. J. R. Meteorol. Soc.* **112,** 1001–1039.

Panofsky, H. A., and J. A. Dutton. 1984. *Atmospheric Turbulence, Models and Methods for Engineering Applications.* Wiley, New York. 397 pp.

Parkinson, C. L., and W. M. Washington. 1979. A large scale numerical model of sea ice. *J. Geophys. Res.* **84,** 311–337.

Parlange, M. B., W. E. Eichinger, and J. D. Albertson. 1995. Regional scale evaporation and the atmospheric boundary layer. *Rev. Geophys.* **33,** 99–124.

Paulson, C. A. 1970. The mathematical representation of wind speed and temperature profiles in the unstable atmospheric surface layer. *J. Appl. Meteorol.* **9,** 857–860.

Paulson, C. A., and T. W. Parker. 1972. Cooling of a water surface by evaporation, radiation and heat transfer. *J. Geophys. Res.* **77,** 491–495.

Paulson, C. A., and J. J. Simpson. 1977. Irradiance measurements in the upper ocean. *J. Phys. Oceanogr.* **7,** 952–956.

Paulson, C. A., and J. J. Simpson. 1981. The temperature difference across the cool skin of the ocean. *J. Geophys. Res.* **86,** 11044–11054.

Payne, R. E. 1972. Albedo of the sea surface. *J. Atmos. Sci.* **29,** 959–970.

Peeters, F., G. Piepke, R. Kipfer, R. Hohmann, and D. M. Imboden. 1996. Description of stability and neutrally buoyant transport in freshwater lakes. *Limnol. Oceanogr.* **41,** 1711–1724.

Peixoto, J. P., and A. H. Oort. 1992. *The Physics of Climate.* American Institute of Physics, New York.

Peng, T. H., W. S. Broecker, G. C. Matheieu, Y. H. Li, and A. E. Bainbrige. 1979. Radon evasion rates in the Atlantic and Pacific oceans as determined during GEOSECS program. *J. Geophys. Res.* **84,** 2471–2486.

Penman, H. L., and I. F. Long. 1960. Weather in wheat: An essay in micrometeorology. *Q. J. R. Meteorol. Soc.* **86,** 16–50.

Peters, H., M. C. Gregg, and J. M. Toole. 1988. On the parameterization of equatorial turbulence. *J. Geophys. Res.* **93,** 1199–1218.

Peters, H., M. C. Gregg, and J. M. Toole. 1989. Meridional variability of turbulence through the equatorial undercurrent. *J. Geophys. Res.* **94,** 18003–18009.

Peters, H., M. C. Gregg, and T. B. Sanford. 1991. Equatorial and off-equatorial fine-scale and large-scale shear variability at 140 W. *J. Geophys. Res.* **96,** 16913–16928.

Peters, H., M. C. Gregg, and T. B. Sanford. 1994. The diurnal cycle of the upper equatorial ocean: Turbulence, fine-scale shear, and mean shear. *J. Geophys. Res.* **99,** 7707–7724.

Peters, H., M. C. Gregg, and T. B. Sanford. 1995a. On the parameterization of equatorial turbulence: Effect of fine scale varaiations below the range of the diurnal cycle. *J. Geophys. Res.* **100,** 18333–18348.

Peters, H., M. C. Gregg, and T. B. Sanford. 1995b. Detail and scaling of turbulent overturns in the Pacific equatorial undercurrent. *J. Geophys. Res.* **100,** 18349–18368.

Phillips, O. M. 1955. The irrotational motion outside a free turbulent boundary. *Proc. Cambridge Philos. Soc.* **51,** 220–229.

Phillips, O. M. 1957. On the generation of waves by turbulent wind. *J. Fluid Mech.* **2,** 417–445.

Phillips, O. M. 1958. The equilibrium range in the spectrum of wind-generated ocean waves. *J. Fluid Mech.* **4,** 426–434.

Phillips, O. M. 1960. On the dynamics of unsteady gravity waves of finite amplitude. Part I. *J. Fluid Mech.* **9,** 193–217.

Phillips, O. M. 1967. *Wave Motions in Fluids: Stony Brook Lectures.* Johns Hopkins University Report, Baltimore, MD.

Phillips, O. M. 1971. On spectra measured in undulating layered medium. *J. Phys. Oceanogr.* **1,** 1–16.

Phillips, O. M. 1974. Nonlinear dispersive waves. *Annu. Rev. Fluid Mech.,* **6,** 93–109.

Phillips, O. M. 1977. *The Dynamics of the Upper Ocean,* 2nd ed. Cambridge Univ. Press, Cambridge. 309 pp.

Phillips, O. M. 1985. Spectral and statistical properties of the equilibrium range in wind-generated gravity waves. *J. Fluid Mech.* **156,** 505–531.

Phillips, O. M., and M. L. Banner. 1974. Wave breaking in the presence of wind drift and swell. *J. Fluid Mech.* **66,** 625–640.

Pierrehumbert, R. T., and B. Wyman. 1985. Upstream effects of mesoscale mountains. *J. Atmos. Sci.* **42,** 977–1003.

Pierson, W. J., and L. Moskowitz. 1964. A proposed spectral form for fully-developed wind seas based on the similarity theory of S. A. Kitaigorodskii. *J. Geophys. Res.* **69,** 5191–5203.

Pinkel, R., M. Merrifield, M. McPhaden, J. Picaut, S. Rutledge, D. Siegel, and L. Washburn. 1997. Solitary waves in the western equatorial Pacific Ocean. *Geophys. Res. Lett.* **24,** 1603–1606.

Piomelli, U., W. H. Cabot, P. Moin, and S. Lee. 1991. Subgrid-scale back scatter in turbulent and transitional flows. *Phys. Fluids A* **3,** 1766–1776.

Piomelli, U., Y. Yu, and R. J. Adrian. 1996. Subgrid-scale energy transfer and near-wall turbulence structure. *Phys. Fluids* **8,** 215–224.

Plant, W. J. 1982. A relation between wind stress and wave slope. *J. Geophys. Res.* **87,** 1961–1967.

Plate, E. J. 1971. *Aerodynamic Characteristics of Atmospheric Boundary Layers.* AEC Critical Review Series, Atomic Energy Commission. 190 pp.

Platnick, S., and F. P. J. Valero. 1995. A validation of a satellite cloud retrieval during ASTEX. *J. Atmos. Sci.* **52,** 2985–3001.

Plueddemann, A. J. 1992. Internal wave observations from the Arctic Environmental Drifting Buoy. *J. Geophys. Res.* **97,** 12619–12638.

Plueddemann, A. J., R. A. Weller, M. Stramska, T. D. Dickey, and J. Marra. 1995. Vertical structure of the upper ocean during Marine Light-Mixed Layers experiment. *J. Geophys. Res.* **100,** 6605–6619.

Plueddemann, A. J., J. A. Smith, D. M. Farmer, R. A. Weller, W. R. Crawford, R. Pinkel, S. Vagle, and A. Gnanadesikan. 1996. Structure and variability of Langmuir circulation during the Surface Wave Processes Program. *J. Geophys. Res.* **101,** 3525–3543.

Pollard, R. T. 1977. Observations and theories of Langmuir circulations and their role in near surface mixing. In M. Angel (Ed.), *A Voyage of Discovery: G. Deacon 70th Anniversary Volume,* pp. 235–251. Pergamon, Elmsford, NY.

Pollard, R. T., R. B. Rhines, and R. O. R. Y. Thompson. 1973. The deepening of the wind-mixed layer. *Geophys. Fluid Dyn.* **3,** 381–404.

Polzin, K., J. M. Toole, and R. W. Schmitt. 1995. Fine scale parameterizations of turbulent dissipation. *J. Phys. Oceanogr.* **25,** 306–328.

Polzin, K., K. Speer, J. Toole, and R. Schmitt. 1996. Intense mixing of Antarctic bottom water in the equatorial Atlantic. *Nature* **380,** 54–57.

Pond, S., and G. L. Pickard. 1989. *Introductory Dynamical Oceanography,* 2nd ed. Pergamon, Elmsford, NY. 329 pp.

Pope, S. B. 1975. A more general effective-viscosity hypothesis. *J. Fluid Mech.* **72,** 331–340.

Postel, S. L., G. C. Daily, and P. R. Ehrlich. 1996. Human appropriation of renewable fresh water. *Science* **271,** 785–788.

Potts, M. A. 1998. A study of convective processes in polar and subpolar seas using non-hydrostatic models. Doctoral dissertation, Department of Aerospace Engineering Sciences, University of Colorado, Boulder.

Prandtl, L. 1905. Ueber Flussigkeitsbewegung bei Sehr Kleiner Reibung, *Verh. III, Proc. Third Intern. Cong. Mathematicians* (Heidelberg 1904), pp. 484–491.

Prandtl, L. 1925. Bericht uber untersuchungen zur ausgebildeten turbulenz. *Z. Angew. Math. Mech.* **5,** 136–139.

Prandtl, L. 1932. Zur turbulenten stromung in rohren und langs platten. *Ergebn. Aerodyn. Versuchsanst Gottingen* **4,** 18–29.

Prandtl, L. 1945. Uber ein neues formelsystem fur die ausgebildete turbulenz. *Nachr. Akad. Wiss. Gottingen Math. Phys.* **K1,** 6–19.

Prandtl, L. 1952. *Essentials of Fluid Mechanics*. Hafner, New York. 452 pp.

Praskovsky, A., and S. Oncley. 1994. Measurements of the Kolmogorov constant and intermittency exponent at very high Reynolds numbers. *Phys. Fluids* **6,** 2886.

Preston, B. N., T. C. Laurent, W. D. Comper, and G. J. Checkley. 1980. Rapid polymer transport in concentrated solutions through the formation of ordered structures. *Nature* **287,** 499–503.

Price, J. F. 1979. Observations of a rain-formed mixed layer. *J. Phys. Oceanogr.* **9,** 643–649.

Price, J. F., C. N. K. Mooers, and J. C. Van Leer. 1978. Observation and simulation of storm-induced mixed layer deepening. *J. Phys. Oceanogr.* **8,** 582–599.

Price, J. F., R. A. Weller, and R. R. Schudlich, 1987. Wind-driven currents and Ekman transport. *Science* **238**, 1534–1538.

Price, J. F., R. A. Weller, and R. Pinkel. 1986. Diurnal cycling: Observations and models of the upper ocean response to diurnal heating, cooling and wind mixing. *J. Geophys. Res.* **91,** 8411–8427.

Price, J. F., R. A. Weller, C. K. Bowers, and M. G. Briscoe. 1987. Diurnal response of sea surface temperature observed at the Long-Term Upper Ocean Study (34°N, 70°W) in the Sargasso Sea. *J. Geophys. Res.* **92,** 14480–14490.

Priscu, J. C., R. H. Spigel, M. M. Gibbs, and M. T. Downes. 1986. A numerical analysis of hypolimnetic nitrogen and phosphorous transformations in Lake Rotoiti, New Zealand: A geothermally influenced lake. *Limnol. Oceanogr.* **31,** 812–831.

Purtell, L. P., P. S. Klebanoff, and F. T. Buckley. 1981. Turbulent boundary layer at low Reynolds number. *Phys. Fluids* **24,** 802–811.

Queney, P. 1948. The problem of air flow over mountains: A summary of theoretical studies. *Bull. Am. Meteorol. Soc.* **29,** 16–26.

Queney, P., G. Corby, N. Gerbier, H. Koschmieder, and J. Zierep. 1960. *The Airflow over Mountains.* WMO Technical Note no. 34.

Raasch, S., and D. Etling. 1998. Modeling deep ocean convection: Large eddy simulation in comparison with laboratory experiments. *J. Phys. Oceanogr.* **28,** 1786–1802.

Ramanathan, V., R. D. Cess, E. F. Harrison, P. Minnis, B. R. Barkstrom, E. Ahmad, and D. Hartmann. 1989. Cloud-radiative forcing and climate: Results from the Earth Radiation Budget Experiment. *Science* **243,** 57–63.

Ramp, S. R., R. W. Garwood, C. O. Davis, and R. L. Snow. 1991. Surface heating and patchiness in the coastal ocean off central California during a wind relaxation event. *J. Geophys. Res.* **96,** 14947–14957.

Randall, D. A. 1980. Conditional instability of the first kind upside down. *J. Atmos. Sci.* **37,** 125–130.

Randall, D. A. 1995. Atlantic Stratocumulus Transition Experiment (Editorial). *J. Atmos. Sci.* **52,** 2705.

Randall, D. A., J. A. Coakley, Jr., C. W. Fairall, R. A. Kropfli, and D. H. Lenschow. 1984. Outlook for research on subtropical marine stratiform clouds. *Bull. Am. Meteorol. Soc.* **65,** 1290–1301.

Randall, D., J. Curry, D. Battisti, G. Flato, R. Gumbine, S. Hakkinen, D. Martinson, R. Preller, J. Walsh, and J. Weatherly. 1998. Status of and outlook for large-scale modeling of atmosphere-ice-ocean interactions in the Arctic. *Bull. Am. Meteorol. Soc.* **79,** 197–219.

Rapp, R. J., and W. K. Melville. 1990. Laboratory measurements of deep-water breaking waves. *Philos. Trans. R. Soc. London Ser. A* **331,** 735–800.

Raupach, M. R. 1981. Conditional statistics of Reynolds stress in rough wall and smooth wall boundary layers. *J. Fluid Mech.* **108,** 309–322.

Raupach, M. R. 1988. Canopy transport processes. In W. L. Steffen and O. T. Denmead (Eds.), *Flow and Transport in the Natural Environment: Advances and Applications,* pp. 95–127. Springer-Verlag, Berlin.

Raupach, M. R. 1991. Vegetation-atmosphere interaction in homogeneous and heterogeneous terrain: Some implications of mixed layer dynamics. *Vegetation* **91,** 105–120.

Raupach, M. R., and A. S. Thom. 1981. Turbulence in and above plant canopies. *Annu. Rev. Fluid Mech.* **13,** 97–129.

Raupach, M. R., R. A. Antonia, and S. Rajagopalan. 1991. Rough wall turbulent boundary layers. *Appl. Mech. Rev.* **44,** 1–25.

Rayleigh, L. 1883. Investigation of the character of the equilibrium of an incompressible heavy fluid of variable density. *Proc. London Math. Soc.* **14,** 170–178.

Redmond, K. T. 1990. Crater Lake climate and lake level variability. In E. T. Drake, G. L. Larson, J. Dymond, and R. Collier (Eds.), *Crater Lake: An Ecosystem Study,* pp. 127–142. AAAS, Pacific Division, San Francisco, CA.

Reed, R. K. 1977. On estimating insolation over the ocean. *J. Phys. Oceanogr.* **7,** 482–485.

Reed, R. K. 1982. Comparison of measured and estimated insolation over the eastern Pacific Ocean. *J. Appl. Meteorol.* **21,** 339–341.

Reed, R. K., and R. E. Brainard. 1983. A comparison of computed and observed insolation under clear skies over the Pacific Ocean. *J. Clim. Appl. Meteorol.* **22,** 1125–1128.

Reynolds, O. 1895. On the dynamical theory of incompressible viscous fluids and the determination of the criterion. *Philos. Trans. R. Soc. London Ser. A* **186,** 123–164.

Reynolds, W. C. 1990. The potential and limitations of direct and large eddy simulations. *Lect. Notes Phys.* **357,** 313–343.

Richardson, P. L., D. Walsh, L. Armi, M. Schroder, and J. F. Price. 1989. Tracking three Meddies with SOFAR floats. *J. Phys. Oceanogr.* **19,** 371–383.

Robinson, I. S., N. C. Wells, and H. Charnock. 1984. The sea surface thermal boundary layer and its relevance to the measurement of sea surface temperature by airborne and spaceborne radiometers. *Int. J. Remote Sensing* **5,** 19–45.

Rodi, W. 1987. Examples of calculation methods for flow and mixing in stratified fluids. *J. Geophys. Res.* **92,** 5305–5328.

Rogallo, R. S., and P. Moin. 1984. Numerical simulation of turbulent flows. *Annu. Rev. Fluid Mech.* **16,** 99–137.

Rogers, D. P., 1989. The Marine Boundary Layer in the Vicinity of an Ocean Front. *J. Atmos. Sci.*, **46**, 2044–2062.

Rogers, D. P., X. Yang, P. M. Norris, D. W. Johnson, G. M. Martin, C. A. Friehe, and B. W. Berger. 1995. Diurnal evolution of the cloud-topped marine boundary layer. Part I: Nocturnal stratocumulus development. *J. Atmos. Sci.* **52,** 2953–2966.

Rogers, K. R., G. A. Lawrence, and P. F. Hamblin. 1995. Observations and numerical simulation of a shallow ice-covered mid-latitude lake. *Limnol. Oceanogr.* **40,** 374–385.

Rossow, W. B., and R. A. Schiffer. 1991. ISCCP cloud data products. *Bull. Am. Meteorol. Soc.* **72,** 2–20.

Rotta, J. C. 1951. Statistische Theorie nichthomogener Turbulenz, 1. *Z. Phys.* **129,** 547–572.

Rowland, J. R. and A. Arnold, 1975. Vertical velocity structure and geometry of clear air convective elements, in Preprints of 16th Radar Meteorology Conference, Houston, TX, pp. 296–303. American Meteorological Society, Boston, MA.

Rubinstein, R., and J. M. Barton. 1990. Nonlinear Reynolds stress models and renormalization group. *Phys. Fluids A* **2,** 1472–1476.

Rubinstein, R., and J. M. Barton. 1992. Renormalization group analysis of the Reynolds stress transport equation. *Phys. Fluids A* **4,** 1759–1766.

Ruddick, B. 1983. A practical indicator of the stability of the water column to double-diffusive activity. *Deep-Sea Res.* **30,** 1105–1107.

Ruddick, B. 1992. Intrusive mixing in a Mediterranean salt lens—Intrusion slopes and dynamical mechanisms. *J. Phys. Oceanogr.* **22,** 1274–1285.

Ruddick, B. 1998. Differential fluxes of heat and salt: Implications for circulation and ecosystem modeling. *Oceanography* **10,** 122–127.

Ruddick, B. R., and J. S. Turner. 1979. The vertical length scale of double-diffusive intrusions. *Deep-Sea Res.* **26A,** 903–913.

Ruddick, B., and D. Hebert. 1988. The mixing of Meddy "Sharon". In J. C. J. Nihoul and B. M. Jamart (Eds.), *Small Scale Mixing in the Ocean.* Elsevier, New York. 541 pp.

Ruddick, B., and D. Walsh. 1996. Observations of the density perturbations which drive thermohaline intrusions. In A. Brandt and H. J. S. Fernando (Eds.), *Double-Diffusive Convection.* Am. Geophys. Union, Washington, DC. 334 pp.

Rudnick, D. L., and R. Ferrari. 1999. Compensation of horizontal temperature and salinity gradients in the ocean mixed layer. *Science* **283,** 526–529,

Saddoughi, S. G., and S. V. Veeravalli. 1994. Local isotropy in turbulent boundary layer at high Reynolds number. *J. Fluid Mech.* **268,** 333.

Said, F., and A. Druilhet. 1991. Study of atmospheric marine boundary layer. *Boundary-Layer Meteorol.* **54,** 225–247.

Sandstrom, H., and N. S. Oakey. 1995. Dissipation in internal tides and solitary waves. *J. Phys. Oceanogr.* **25,** 604–614.

Sarmiento, J. L. 1992. Biogeochemical ocean models. In K. Trenberth (Ed.), *Climate System Modeling*, pp. 519–551. Cambridge Univ. Press, Cambridge.

Sarmiento, J. L., H. W. Feely, W. S. Moore, A. E. Bainbridge, and W. S. Broeker. 1976. The relationship between vertical eddy diffusion and buoyancy gradient in the deep sea. *Earth Planet. Sci. Lett.* **32,** 357–370.

Sathyendranath, S., and T. Platt. 1988. The spectral irradiance field at the surface and in the interior of the ocean: A model for applications in oceanography and remote sensing. *J. Geophys. Res.* **93,** 9270–9280.

Saunders, P. M. 1967. The temperature at the ocean-air interface. *J. Atmos. Sci.* **24,** 269–273.

Schiano, M. E. 1996. Insolation over the western Mediterranean Sea: A comparison of direct measurements and Reed's formula. *J. Geophys. Res.* **101,** 3831–3838.

Schlichting, H. 1977. *Boundary Layer Theory.* McGraw Hill, New York. 747 pp.

Schluessel, P., H.-Y. Shin, W. J. Emery, and H. Grassl. 1987. Comparison of satellite-derived sea surface temperatures with in situ skin measurements. *J. Geophys. Res.* **92,** 2859–2874.

Schluessel, P., W. J. Emery, H. Grassl, and T. Mammen. 1990. On the bulk-skin temperature difference andits impact on satellite remote sensing of sea surface temperature. *J. Geophys. Res.* **95,** 13341–13356.

Schluessel, P., A. V. Soloviev, and W. J. Emery, 1997. Cool and freshwater skin of the ocean during rainfall, *Boundary-Layer Meteorol.*, **82**, 437–472.

Schmetz, P., J. Schmetz, and E. Raschke. 1986. Estimation of daytime downward longwave radiation at the surface from satellite and grid point data. *Theor. Appl. Climatol.* **37,** 136–149.

Schmidt, H., and U. Schumann. 1989. Coherent structures of the convective boundary layer derived from large-eddy simulations. *J. Fluid Mech.* **200,** 511–562.

Schmitt, R. W. 1979a. Flux measurements on salt fingers at an interface. *J. Mar. Res.* **37,** 419–436.

Schmitt, R. W. 1979b. The growth rate of supercritical salt fingers. *Deep-Sea. Res.* **26A,** 23–44.

Schmitt, R. W. 1983. The characteristics of salt fingers in a variety of fluid systems, including stellar interiors, liquid metals, oceans and magmas. *Phys. Fluids* **26,** 2373–2377.

Schmitt, R. W. 1990. On the density ratio balance in Central Water. *J. Phys. Oceanogr.* **20,** 900–906.

Schmitt, R. W. 1991. *Double Diffusion in Oceanography: Proceedings of a Meeting, Sep. 26–29, 1989.* WHOI Technical Report WHOI 91-20.

Schmitt, R. W. 1994. Double diffusion in oceanography. *Annu. Rev. Fluid Mech.* **26,** 255–285.

Schmitt, R. W. 1995. The salt finger experiments of Jevons (1857) and Rayleigh (1880). *J. Phys. Oceanogr.* **25,** 8–17.

Schmitt, R. W. 1999. Spice and the demon. *Science* **283,** 498–499.

Schmitt, R. W., and D. T. Georgi. 1982. Fine structure and microstructure in the North Atlantic current *J. Mar. Res.* **40,** 659–705.

Schott, F. A., M. Visbeck, and J. Fischer. 1993. Observations of vertical currents and convection in the Greenland Sea during the winter of 1988/1989. *J. Geophys. Res.* **98,** 14401–14422.

Schott, F. A., M. Visbeck, U. Send, J. Fischer, L. Stramma, and Y. Desaubies. 1996. Observations of deep convection in the Gulf of Lions, northern Mediterranen, during winter of 1991/1992. *J. Phys. Oceanogr.* **26,** 505–524.

Schudlich, R. R., and J. F. Price. 1992. Diurnal cycles of current, temperature, and turbulent dissipation in a model of the equatorial upper ocean. *J. Geophys. Res.* **97,** 5409–5422.

Schudlich, R. R., and J. F. Price. 1998. Observations of seasonal variation in the Ekman layer. *J. Phys. Oceanogr.* **28,** 1187–1204.

Schumann, U. 1975. Subgrid scale model for finite difference simulation of turbulent flows in plane channels and annuli. *J. Comput. Phys.* **18,** 376–404.

Schumann, U. 1996. Direct and large eddy simulations of stratified homogeneous shear flows. *Dyn. Atm. Oceans* **23,** 81–97.

Schumann, U., and Gerz, T. 1995. Turbulent mixing in stably stratified shear flows. *J. Appl. Meteorol.* **34,** 33–48.

Scorer, R. S. 1949. The theory of waves in the lee of mountains. *Q. J. R. Meteorol. Soc.* **75,** 41–56.

Seckel, G. R., and F. H. Beaudry. 1973. The radiation from the sun and sky over the North Pacific Ocean (abstract). *EOS Trans.* **54,** 1114.

Serra, Y. L., D. P. Rogers, D. E. Hagan, C. A. Friehe, R. L. Grossman, R. A. Weller, and S. Anderson. 1997. Atmospheric boundary layer over the central and western equatorial Pacific Ocean observed during COARE and CEPEX. *J. Geophys. Res.* **102,** 23217–23238.

Shay, T. J., and M. C. Gregg. 1986. Convectively driven turbulent mixing in the upper ocean. *J. Phys. Oceanogr.* **16,** 1777–1798.

Shen, C. Y. 1991. Numerical modeling of salt fingers at a density interface. In R. W. Schmitt (Ed.), *Double Diffusion in Oceanography,* pp. 75–86. WHOI Report 91-20.

Shen, C. Y. 1993. Heat-salt finger fluxes across a density interface. *Phys. Fluids A* **5,** 2633–2643.

Shen, C. Y., and G. Veronis. 1997. Numerical simulation of two-dimensional salt fingers. *J. Geophys. Res.* **102,** 23131–23144.

Shimaraev, M. N., N. G. Granin, and A. A. Zhadanov. 1993. Deep ventilation of Lake Baikal waters due to spring thermal bars. *Limnol. Oceanogr.* **38,** 1068–1072.

Shine, K. P. 1984. Parameterization of shortwave flux over high albedo surfaces as a function of cloud thickness and surface albedo. *Q. J. R. Meteorol. Soc.* **110,** 747–764.

Shirasawa, K., and R. G. Ingram. 1991. Characteristics of the turbulent oceanic boundary layer under sea ice. Part 1: A review of the ice-ocean boundary layer. *J. Mar. Systems* **2,** 153–160.

Shirtcliffe, T. G. L. 1967. Thermosolutal convection: Observation of an overstable mode. *Nature* **213,** 489–490.

Shuttleworth, W. J., 1989. Micrometeorology of temperate and tropical forest. *Phil. Trans. Roy. Soc. London B*, **324**, 299–334.

Shutts, G. J. 1995. Gravity wave drag parameterization over complex terrain: The effect of critical level absorption in directional wind shear. *Q. J. R. Meterol. Soc.* **121,** 1005–1022.

Siegel, D. A., and T. D. Dickey. 1987. On the parameterization of irradiance for open ocean photoprocesses. *J. Geophys. Res.* **92,** 14648–14662.

Siggia, E. D. 1994. High Rayleigh number convection. *Annu. Rev. Fluid Mech.* **26,** 137–168.

Sigvaldason, G. E. 1989. International conference on Lake Nyos disaster, Yaounde, Cameroon 16–20 March 1987: Conclusions and recommendations. *J. Volcanol. Geotherm. Res.* **39,** 97–107.

Simmons, W. F. 1969. A variational method for weak resonant wave interactions. *Proc. R. Soc. A* **309,** 551–575.

Simpson, H. J. 1970. Tritium in Crater Lake, Oregon. *J. Geophys. Res.* **75,** 5195–5207.

Simpson, J. H., W. R. Crawford, T. P. Rippeth, A. R. Campbell, and J. V. S. Cheok. 1996. The vertical structure of turbulent dissipation in shelf seas. *J. Phys. Oceanogr.* **26,** 1579–1590.

Simpson, J. J., and C. A. Paulson. 1979. Mid-ocean observations of atmospheric radiation. *Q. J. R. Meteorol. Soc.* **105,** 487–502.

Simpson, J. J., and T. D. Dickey. 1981a. The relationship between downward irradiance and upper ocean structure. *J. Phys. Oceanogr.* **11,** 309–323.

Simpson, J. J., and T. D. Dickey. 1981b. Alternative parameterizations of downward irradiance and their dynamical significance. *J. Phys. Oceanogr.* **11,** 876–882.

Skyllingstad, E. D., and D. W. Denbo. 1994. The role of internal gravity waves in the equatorial current system. *J. Phys. Oceanogr.* **24,** 2093–2110.

Skyllingstad, E. D., and D. W. Denbo. 1995. An ocean large-eddy simulation of Langmuir circulations and convection in the surface layer. *J. Geophys. Res.* **100,** 8501–8522.

Slingo, A., Brown, R. and C. L. Wrench, 1982a. A field study of nocturnal stratocumulus: III. High resolution radiative and microphysical observations, *Quart. J. R. Met. Soc.*, **108**, 145–165.

Slingo, A., S. Nicholls, and J. Schmetz, 1982b. Aircraft observations of marine stratocumulus during JASIN, *Quart. J. R. Met. Soc.*, **108**, 833–856.

Smagorinsky, J. S. 1963. General circulation experiments with primitive equations. I. The basic experiment. *Mon. Weather Rev.* **91,** 99–164.

Smedman-Hogstrom, A.-S., and U. Hogstrom, 1978. A practical method for determining wind frequency distributions for the lowest 200 m from routine meteorological data, *J. Appl. Meteor.* **17**, 942–954.

Smedman, A.-S., H. Bergstrom, and B. Grisogono. 1997. Evolution of stable internal boundary layers over a cold sea. *J. Geophys. Res.* **102,** 1091–1100.

Smith, D., and J. Morison. 1993. Numerical study of haline convection beneath leads in sea ice. *J. Geophys. Res.* **98,** 10069–10083.

Smith, D., and J. Morison. 1998. Numerical study of haline convection beneath leads in sea ice. *J. Geophys. Res.* **103,** 10069–10083.

Smith, J. A. 1990. Modulation of short wind waves by long waves. In G. L. Geernaert and W. J. Plant (Eds.), *Surface Waves and Fluxes*, Vol. 1, pp. 247–284. Kluwer Academic, Dordrecht/Norwell, MA.

Smith, J. A. 1992. Observed growth of Langmuir circulations. *J. Geophys. Res.* **97,** 5651–5664.

Smith, J. A. 1998. Evolution of Langmuir circulation during a storm. *J. Geophys. Res.* **103,** 12649–12668.

Smith, J. A., R. A. Weller, and R. Pinkel. 1987. Velocity structure in the mixed layer during MILDEX. *J. Phys. Oceanogr.* **17,** 425–439.

Smith, J. D. 1977. Modeling sediment transport on continental shelves. In E. D. Goldberg, I. N. McCave, J. J. O'Brien, and J. H. Steele (Eds.), *The Sea,* Vol. 6, pp. 539–577. Wiley–Interscience, New York.

Smith, L. M., and S. L. Woodruff.1998. Renormalization-Group Analysis of turbulence. *Annu. Rev. Fluid Mech.* **30,** 275–310.

Smith, L. M., and W. C. Reynolds. 1992. On the Yakhot-Orszag renormalization group method for deriving turbulence statistics and models. *Phys. Fluids A* **4,** 364–390.

Smith, M. H., P. M. Park, and I. E. Consterdine. 1993. Marine aerosol concentrations and estimated fluxes over the sea. *Q. J. R. Meteorol. Soc.* **119,** 809–824.

Smith, N. R. 1993. Ocean modeling in a global ocean observing system. *Rev. Geophys.* **31,** 281–317.

Smith, N. R., and G. D. Hess. 1993. A comparison of vertical eddy mixing parameterizations for equatorial ocean models. *J. Phys. Oceanogr.* **23,** 1823–1830.

Smith, R. B. 1976. The generation of lee waves by the Blue Ridge. *J. Atmos. Sci.* **33,** 507–519.

Smith, R. B. 1978. A measurement of mountain drag. *J. Atmos. Sci.* **35,** 1644–1654.

Smith, R. B. 1979. The influence of mountains on the atmosphere. *Adv. Geophys.* **21,** 87–230.

Smith, R. B. 1985. On severe downslope winds. *J. Atmos. Sci.* **42,** 2597–2603.

Smith, R. B. 1987. Aerial observations of the Yugoslavian bora. *J. Atmos. Sci.* **44,** 269–297.

Smith, R. B. 1989. Hydrostatic airflow over mountains. *Adv. Geophys.* **31,** 1–41.

Smith, R. C., J. E. Tyler, and C. R. Goldman. 1973. Optical properties and color of Lake Tahoe and Crater Lake. *Limnol. Oceanogr.* **18,** 189–199.

Smith, S. D. 1988. Coefficients for sea surface wind stress, heat flux and wind profiles as a function of wind speed and temperature. *J. Geophys. Res.* **93,** 15467–15472.

Smith, S. D. 1989. Water vapor flux at the sea surface (review paper). *Boundary Layer Meteorol.* **47,** 277–293.

Smith, S. D., R. D. Muench, and C. H. Pease. 1990. Polynyas and leads: An overview of physical processes and environment. *J. Geophys. Res.* **95,** 9461–9479.

Smith, S. D., R. J. Anderson, W. A. Oost, C. Kraan, N. Maat, J. DeCosmo, K. B. Katsaros, K. L. Davidson, K. Bumke, L. Hasse, and H. M. Cadwick. 1992. Sea surface wind stress and drag coefficients: The HEXOS results. *Boundary Layer Meteorol.* **60,** 109–142.

Smith, S. D., C. W. Fairall, G. L. Geernaert, and L. Hasse. 1996a. Air-sea fluxes: 25 years of progress. *Boundary Layer Meteorol.* **78,** 247–290.

Smith, S. D., K. B. Katsaros, W. B. Oost, and P. G. Metsayer. 1996b. The impact of the HEXOS programme. *Boundary Layer Meteorol.* **78,** 121–141.

Smyth, W. D., D. Hebert, and J. N. Moum. 1996a. Local ocean response to a multiphase westerly wind burst. 1. Dynamic response. *J. Geophys. Res.* **101,** 22495–22512.

Smyth, W. D., D. Hebert, and J. N. Moum. 1996b. Local ocean response to a multiphase westerly wind burst. 2. Thermal and freshwater response. *J. Geophys. Res.* **101,** 22513–22533.

Snodgrass, F. E., G. W. Groves, K. F. Hasselmann, G. R. Miller, W. H. Munk, and W. H. Powers. 1966. Propagation of ocean swell across the Pacific. *Philos. Trans. R. Soc. London* **259,** 431–497.

Snyder, R. L., F. W. Dobson, J. A. Elliott, and R. B. Long. 1981. Array measurements of atmospheric pressure fluctuations above surface gravity waves. *J. Fluid Mech.* **102,** 1–59.

Sobey, R. J. 1986. Wind-wave prediction. *Annu. Rev. Fluid Mech.* **18,** 149–172.

Soloviev, A. V., and P. Schluessel. 1994. Parameterization of the cool skin of the ocean and of the air-ocean gas transfer on the basis of modelling surface renewal. *J. Phys. Oceanogr.* **24,** 1339–1346.

Soloviev, A. V., and P. Schluessel. 1996. Evolution of cool-skin and direct air-sea gas transfer coefficient during daytime. *Boundary Layer Meteorol.* **77,** 45–68.

Sommer, T. P., and R. M. C. So. 1995. On the modeling of homogeneous turbulence in a stably stratified flow. *Phys. Fluids* **7,** 2766–2777.

Song, X., and C. A. Friehe. 1997. Surface air-sea fluxes and upper ocean heat budget at 156 °E, 4 °S during the Tropical Ocean-Global Atmosphere Coupled Ocean Atmosphere Response Experiment. *J. Geophys. Res.* **102,** 23109–23130.

Sorbjan, Z. 1988a. Structure of the stably-stratified boundary layer during the SESAME–1979 experiment. *Boundary Layer Meteorol.* **44,** 255–266.

Sorbjan, Z. 1988b. Local similarity in the atmospheric boundary layer. In *Proceedings, Eighth Symposium on Turbulence and Diffusion,* pp. 353–356. American Meteorological Society.

Sorbjan, Z. 1989. *Structure of the Atmospheric Boundary Layer.* Prentice Hall, New York. 317 pp.

Sorbjan, Z. 1991. Evaluation of local similarity functions in the convective boundary layer. *J. Appl. Meteorol.* **30,** 1565–1583.

Spaziale, C. G. 1991. Analytical methods for the development of Reynolds-stress closures in turbulence. *Annu. Rev. Fluid Mech.* **23,** 107–158.

Spigel, R. H., J. Imberger, and K. N. Rayner. 1986. Modeling the diurnal mixed layer. *Limnol. Oceanogr.* **31,** 533–556.

Sreenivasan, K. R. 1995. On the universality of the Kolmogorov constant. *Phys. Fluids* **7,** 2778–2784.

Sreenivasan, K. R. 1996. The passive scalar spectrum and the Obukhov-Corrsin constant. *Phys. Fluids* **8,** 189–196.

Sreenivasan, K. R., and R. A. Antonia. 1997. The phenomenology of small-scale turbulence. *Annu. Rev. Fluid Mech.* **29,** 435–472.

Srokosz, M. A. 1990. Wave statistics. In G. L. Geernaert and W. J. Plant (Eds.), *Surface Waves and Fluxes, Vol. 1—Current Theory,* pp. 285–332. Kluwer Academic, Dordrecht/Norwell, MA.

Stacey, M. W., and S. Pond. 1997. On the Mellor-Yamada turbulence closure scheme: The surface boundary condition for q^2. *J. Phys. Oceanogr.* **27,** 2081–2086.

Stanisic, M. M. 1985. *Mathematical Theory of Turbulence.* Springer-Verlag, New York. 429 p.

Starr, V. P. 1945. Water transport of surface waves. *J. Meteorol.* **2,** 129–131.

Stefan, H. G., X. Fang, D. Wright, J. G. Eaton, and J. H. McCormick. 1995. Simulation of dissolved oxygen profiles in a transparent, dimictic lake *Limnol. Oceanogr.* **40,** 105–118.

Stefan, H. G., and X. Fang. 1994. Model simulations of dissolved oxygen characteristics of Minnesota lakes: Past and future. *Environmental Management.* **18,** 73–92.

Steinhorn, I. 1985. The disappearance of the long-term meromictic stratification of the Dead Sea. *Limnol. Oceanogr.* **30,** 451–472.

Stern, M. E. 1960. The salt fountain and thermohaline convection. *Tellus* **12,** 172–175.

Stern, M. E. 1967. Lateral mixing of water masses. *Deep-Sea Res.* **14**, 747–753.

Stern, M. E. 1969. Collective instability of salt fingers. *J. Fluid Mech.* **35,** 209–218.

Stern, M. E. 1975. *Ocean Circulation Physics.* Academic Press, New York.

Stern, M. E. 1976. Maximum buoyancy flux across a salt finger interface. *J. Marine Res.* **34,** 95–110.

Stern, M. E., and J. S. Turner. 1969. Salt fingers and convecting layers. *Deep-Sea Res.* **34,** 95–110.

Steele, J. 1974. Spatial heterogeneity and population stability. *Nature* **248,** 83.

Stoens, A., C. Menkes, M.-H. Radenac, Y. Dandonneau, N. Grima, G. Eldin, L. Memery, C. Navarette, J.-M. Andre, T. Moutin, and P. Raimbault. 1999. The coupled physical-new production system in the equatorial Pacific during the 1992–1995 El Nino. *J. Geophys. Res.* **104,** 3323–3339.

Stokes, G. G. 1847. On the theory of oscillating waves. *Trans. Cambridge Philos. Soc.* **8,** 441–455.

Stommel, H. M., A. B. Arons, and D. Blanchard. 1956. An oceanographic curiosity: The perpetual salt fountain. *Deep-Sea Res.* **3,** 152–153.

Stramma, L., P. Cornillon, R. A. Weller, J. F. Price, and M. G. Briscoe. 1986. Large diurnal sea surface temperature variability: Satellite and in situ measurements. *J. Phys. Oceanogr.* **16,** 827–837.

Stramska, M., T. D. Dickey, A. J. Plueddemann, R. A. Weller, C. Langdon, and J. Marra. 1995. Bio-optical variability associated with phytoplankton dynamics in the North Atlantic Ocean during spring and summer of 1991. *J. Geophys. Res.* **100,** 6621–6632.

Straskraba, M. 1980. The effects of physical variables on freshwater production: Analyses based on models. In E. D. Le Cren and R. H. McConnell (Eds.), *The Functioning of Freshwater Ecosystems,* pp. 13–84. Cambridge Univ. Press, Cambridge.

Stull, R. B. 1973. An inversion rise model based on penetrative convection. *J. Atmos. Sci.* **30,** 1092–1099.

Stull, R. B. 1976. Internal gravity waves generated by penetrative convection. *J. Atmos. Sci.* **33,** 1279–1286.

Stull, R. B. 1988. *An Introduction to Boundary Layer Meteorology.* Kluwer Academic, Hingham, MA. 666 pp.

Sullivan, P. P., and C.-H. Moeng. 1993. An evaluation of the dynamic subgrid-scale model in buoyancy driven flows. Presented at AMS 10th Symp. on Boundary layers and Turbulence, Portland, OR.

Sullivan, P. P., J. C. McWilliams, and C.-H. Moeng. 1994. A subgrid-scale model for large eddy simulation of planetary boundary layer flows. *Boundary Layer Meteorol.* **71,** 247–276.

Sutton, O. G. 1953. *Micrometeorology.* McGraw-Hill, New York. 333 pp.

Sverdrup, H. U., and W. H. Munk. 1947. *Wind, Sea and Swell: Theory of Relations for Forecasting.* U.S. Navy Hydrogr. Off. Publ. 601.

SWAMP Group. 1985. *Ocean Wave Modeling.* Plenum, New York. 256 pp.

Swift, C. T. 1990. Passive microwave remote sensing of ocean surface wind speed. In G. L. Geernaert and W. J. Plant (Eds.), *Surface Waves and Fluxes*, Vol. 2, pp. 265–292. Kluwer Academic, Dordrecht/Norwell, MA.

Swinbank, W. C. 1963. Longwave radiation from clear skies. *Q. J. R. Meteorol. Soc.* **89,** 339–348.

Tabata, S. 1964. Insolation in relation to cloud amount and sunÕs altitude. In K. Yoshida (Ed.), *Studies in Oceanography,* pp. 202–210. Univ. Tokyo Press, Tokyo.

Tabata, S. 1965. Variability of oceanographic conditions at Ocean Station P in the northeast Pacific Ocean. *Trans. R. Soc. Canada.* **3**(Ser. 4), 367–418.

Takahashi, T., W. S. Broecker, and A. E. Bainbridge, 1979. The alkalinity and total carbon dioxide concentration in the world's oceans. In *Carbon Cycle Modeling*, ed. by B. Bolin, John Wiley, 159–199.

Takao, S., and U. Narusawa. 1980. An experimental study of heat and mass transfer across a diffusive interface. *Int. J. Heat Mass Transfer* **23,** 1283–1285.

Taylor, G. I. 1915. Eddy motion in the atmosphere. *Philos. Trans. R. Soc. London Ser. A* **215,** 1.

Taylor, G. I. 1935. Statistical theory of turbulence I-IV. *Proc. R. Soc. London Ser. A* **151,** 421.

Taylor, G. I. 1938. Spectrum of turbulence. *Proc. R. Soc. London Ser. A* **164,** 476.

Taylor, J., and P. Bucens. 1989. Laboratory experiments on the structure of salt fingers. *Deep-Sea Res.* **36,** 1675–1704.

Taylor, J. R., and G. Veronis. 1996. Experiments on double-diffusive sugar-salt fingers at high stability ratio. *J. Fluid Mech.* **321,** 315–333.

Taylor, P. A., P. J. Mason, and E. F. Bradley. 1987. Boundary-layer flow over low hills. *Boundary-Layer Meteorol.* **39,** 107–132.

Taylor, P. A., and H. W. Teunissen. 1987. The Askervein Hill Project: Overview and background data. *Boundary-Layer Meteorol.* **39,** 15–39.

Tazieff, H. 1989. Mechanisms of the Nyos carbon dioxide disaster and of so-called phreatic steam eruptions. *J. Volcanol. Geothermal Res.* **39,** 109–116.

Telbany, M. M. M. E., and A. J. Reynolds. 1982. The structure of turbulent plane Couette flow. *J. Fluids Eng.* **104,** 367–372.

Tennekes, H. 1968. Outline of a second order theory for turbulent pipe flow. *AIAA J.* **6,** 1735.

Tennekes, H. 1973. The logarithmic wind profile. *J. Atmos. Sci.* **30,** 234–238.

Tennekes, H., and J. L. Lumley. 1982. *A First Course in Turbulence,* 2nd ed. MIT Press, Cambridge, MA. 300 pp.

Terray, E. A., M. A. Donelan, Y. C. Agarwal, W. M. Drennan, K. K. Kahma, A. J. Williams III, P. A. Hwang, and S. A. Kitaigorodskii. 1996. Estimates of kinetic energy dissipation under breaking waves. *J. Phys. Oceanogr.* **26,** 792–807.

Teunissen, H. W., M. E. Shokr, A. J. Bowen, C. J. Wood, and D. W. R. Green. 1987. The Askervein Hill project: Wind-tunnel simulations at three length scales. *Boundary-Layer Meteorol.* **40,** 1–29.

Therry, G., and P. Lacarrere. 1983. Improving the eddy-kinetic-energy model for planetary boundary layer description. *Boundary Layer Meteorol.* **25,** 63–88.

Thom, A. S. 1971. Momentum absorption by vegetation. *Q. J. R. Meteorol. Soc.* **97,** 414–428.

Thom, A. S. 1975. Momentum, mass and heat exchange of plant communities. In J. L. Monteith (Ed.), *Vegetation and the Atmosphere,* pp. 57–110. Academic Press, London.

Thompson, R. S., 1978. Note on the aerodynamic roughness length for complex terrain, *J. Appl. Meteor.*, **17**, 1402–1403.

Thorpe, S. A. 1975. The excitation, dissipation, and interaction of internal waves in the deep ocean. *J. Geophys. Res.* **80,** 328–338.

Thorpe, S. A. 1977. Turbulence and mixing in a Sottish loch. *Philos. Trans. R. Soc. London Ser. A* **286,** 125–181.

Thorpe, S. A. 1979. On the shape and breaking of finite amplitude internal gravity waves in a shear flow. *J. Fluid Mech.* **85,** 7–31.

Thorpe, S. A. 1984. The effect of Langmuir circulation on the distribution of submerged bubbles caused by breaking wind waves. *J. Fluid Mech.* **142,** 151–170.

Thorpe, S. A. 1992a. The breakup of Langmuir circulation and the instability of an array of vortices. *J. Phys. Oceanogr.* **16,** 1462–1478.

Thorpe, S. A. 1992b. Bubble clouds and the dynamics of the upper ocean. *Q. J. R. Meteorol. Soc.* **118,** 1–22.

Thorpe, S. A. 1993. Energy loss by breaking waves. *J. Phys. Oceanogr.* **23,** 2498–2502.

Thorpe, S. A. 1995. Dynamical processes of transfer at the sea interface. *Prog. Oceanogr.* **35,** 315–352.

Thorsteinsson, S. 1988. Finite amplitude stratified air flow past isolated mountains on an f-plane. *Tellus* **40A,** 220–236.

Tietze, K. 1992. Cyclic gas bursts: Are they a 'usual' feature of Lake Nyos and other gas-bearing lakes? In S. J. Freeth, K. M. Onuoha, and C. S. Ofoegbu (Eds.), *Natural Hazards in West and Central Africa*, pp. 97–108. Earth Evolution Series, International Monograph Series on Interdisciplinary Earth Sciences Research and Applications, Vieweg, Braunschweig.

Tjernstroem, M. and A-S. Smedman, 1993. The vertical turbulence structure of the coastal marine atmospheric boundary layer. *J. Geophys. Res.*, **98**, 4809–4826.

Toba, Y. 1973. Local balance in the air-sea boundary processes. III. On the spectrum of wind waves. *J. Oceanogr. Soc. Jpn.* **29,** 209–220.

Toba, Y., and I. S. F. Jones. 1992. The influence of sea state on atmospheric drag. *EOS Trans. Am. Geophys. Union* **73,** 306.

Tolman, H. L., and D. Chalikov, 1996. Source terms in a third generation wind wave model, *J. Phys. Oceanogr.*, **26,** 2497-2518.

Toole, J. M., and D. T. Georgi. 1981. On the dynamics and effects of double-difffusively driven intrusions. *Prog. Oceanogr.* **10,** 123–145.

Toole, J. M., K. L. Polzin, and R. W. Schmitt. 1994. Estimates of diapycnal mixing in the abyssal ocean. *Science* **264,** 1120–1123.

Toole, J. M., R. W. Schmitt, and K. L. Polzin. 1997. Near-boundary mixing above the flanks of a mid-latitude seamount. *J. Geophys. Res.* **102,** 947–959.

Townsend, A. A. 1966. The flow in a turbulent boundary layer after a change in surface roughness. *J. Fluid Mech.* **26,** 255–266.

Townsend, A. A. 1976. *The Structure of Turbulent Shear Flow*, 2nd ed. Cambridge Univ. Press, Cambridge. 429 pp.

Trasvina, A., E. D. Barton, J. Brown, H. S. Velez, P. M. Kosro, and R. L. Smith. 1995. Offshore wind forcing in the Gulf of Tehuantepec, Mexico: The asymmetric circulation. *J. Geophys. Res.* **100,** 20649–20663.

Trenberth, K. E. (Ed.). 1992. *Climate System Modeling.* Cambridge Univ. Press, New York. 788 pp.

Troen, I. B., and L. Mahrt. 1986. A simple model of the atmospheric boundary layer: Sensitivity to surface evaporation. *Boundary Layer Meteorol.* **37,** 129–148.

Turner, J. S. 1965. The coupled turbulent transports of salt and heat across a sharp density interface. *Int. J. Heat Mass Transfer* **38,** 375–400.

Turner, J. S. 1967. Salt fingers across a density interface. *Deep-Sea Res.* **14,** 599–611.

Turner, J. S. 1968. The behavior of a stable density gradient heated from below. *J. Fluid Mech.* **33,** 183–200.

Turner, J. S. 1973. *Buoyancy Effects in Fluids.* Cambridge Univ. Press, Cambridge. 367 pp.

Turner, J. S. 1974. Double-diffusive phenomena. *Annu. Rev. Fluid Mech.* **6,** 37–56.

Turner, J. S. 1978. The behavior of a stable density gradient heated from below. *J. Geophys. Res.* **83,** 2887–2901.

Turner, J. S. 1981. Small scale mixing processes. In B. A. Warren and C. Wunsch (Eds.), *Evolution of Physical Oceanography*, pp. 236–262. MIT Press, Cambridge, MA.

Turner, J. S. 1985. Multicomponent convection. *Annu. Rev. Fluid Mech.* **17,** 11–44.

Turner, J. S. 1996. Laboratory models of double-diffusive processes. In A. Brandt and H. J. S. Fernando (Eds.), *Double Diffusive Convection.* American Geophysical Union, Washington, DC.

Turner, J. S. 1997. G. I. Taylor in his later years. *Annu. Rev. Fluid Mech.* **29,** 1–25.

Turner, J. S., and H. Stommel. 1964. A new case of convection in the presence of combined vertical salinity and temperature gradients. *Proc. Natl. Acad. Sci.* **52,** 49–53.

Tuttle, M. L., M. A. Clark, H. R. Compton, J. D. Devine, W. C. Evans, A. M. Humphrey, G. W. Kling, E. J. Koenigsberg, J. P. Lockwood, and G. N. Wagner. 1987. *The 21 August 1986 Lake Nyos gas disaster, Cameroon.* U.S. Geological Survey Open-File Report no. 87–97, USGS, Washington, DC. 58 pp.

Ueda, H., R. Moller, S. Komori, and T. Mizushina. 1977. Eddy diffusivity near the free surface of an open channel flow. *Int. J. Heat Mass Transfer* **20,** 1127–1136.

UNESCO. 1981. *Tenth Report of the Joint Panel on Oceanographic Tables and Standards.* UNESCO Technical Papers in Marine Science no. 36, UNESCO. 24 pp.

Vickers, D., and L. Mahrt. 1997. Fetch limited drag coefficients. *Boundary Layer Meteorol.* **85,** 53–79.

Visbeck, M., J. Marshall, and H. Jones. 1996. Dynamics of isolated convective regions in the ocean. *J. Phys. Oceanogr.* **26,** 1721–1734.

Vogel C. A. and P. Frenzen, 1992. A new study of the TKE budget in the surface layer. Part 2: The dissipation function and divergent transport terms, Abstract, AMS Tenth Symposium on Turbulence and Diffusion, Portland, OR, 1992, 161–164.

Vogel, S. 1988. *Life's Devices*. Princeton University Press, Princeton, NJ.

Vogel, S. 1998. Exposing life's limits with dimensionless numbers. *Phys. Today*, Nov., 22–27.

Wallace, D. W. R., and C. D. Wirick. 1992. Large air-sea fluxes associated with breaking waves. *Nature* **356,** 694–696.

Walker, S. J., and R. G. Watts. 1995. A three-dimensional numerical model of deep ventilation in temperate lakes. *J. Geophys. Res.* **100,** 22711–22728.

Walsh, J. J., D. A. Dieterle, F. E. Muller-Karger, R. Bohrer, W. P. Bissett, R. J. Varela, R. Aparicio, R. Diaz, R. Thunell, G. T. Taylor, M. I. Scranton, K. A. Fanning, and E. T. Peltzer. 1999. Simulation of carbon-nitrogen cycling during spring upwelling in the Cariaco Basin. *J. Geophys. Res.* **104,** 7807–7825.

WAMDI Group. 1988. The WAM model—A third generation ocean wave prediction model. *J. Phys. Oceanogr.* **18,** 1775–1810.

Wang, D., W. G. Large, and J. C. McWilliams. 1996. Large eddy simulation of the equatorial ocean boundary layer: Diurnal cycling, eddy viscosity, and horizontal rotation. *J. Geophys. Res.* **101,** 3649–3662.

Wanninkhof, R. 1992. Relationship between wind speed and gas exchange over the ocean. *J. Geophys. Res.* **97,** 7373–7382.

Wayland, R. J., and S. Raman. 1994. Structure of the marine atmospheric boundary layer during two cold air outbreaks of varying intensities: GALE 86. *Boundary Layer Meteorol.* **71,** 43–66.

Weatherly, G. L., and P. J. Martin. 1978. On the structure and dynamics of the oceanic bottom boundary layer. *J. Phys. Oceanogr.* **8,** 557–570.

Webster, P. J., and R. Lukas. 1992. TOGA COARE: The Coupled Ocean-Atmosphere Response Experiment. *Bull. Am. Meteorol. Soc.* **73,** 1377–1416.

Webster, P. J., C. A. Clayson, and J. A. Curry. 1996. Clouds, radiation and the diurnal cycle of sea surface temperature in the tropical western Pacific. *J. Clim.* **9,** 1712–1730.

Wehausen, J., and E. Laitone. 1960. Surface waves. In *Handbuch der Physik IX,* pp. 446–778. Springer-Verlag, Berlin.

Weiss, R. F. 1974. Carbon dioxide in water and sea water. The solubility of a non-ideal gas. *Mar. Chem.* **2,** 203–215.

Weiss, R. F., E. C. Carmack, and V. M. Koropalov. 1991. Deep water renewal and biological production in Lake Baikal. *Nature* **349,** 665–669.

Weller, R. A., J. P. Dean, J. Marra, J. F. Price, E. A. Francis, and D. C. Boardman, 1985. Three-dimensional flow in the upper ocean. *Science*, **227**, 1552–1556.

Weller, R. A., and J. F. Price. 1988. Langmuir circulation within the oceanic mixed layer. *Deep-Sea Res.* **35,** 711–747.

Weller, R. A., M. A. Donelan, M. G. Briscoe, and N. E. Huang. 1991. Riding the crest: A tale of two wave experiments. *Bull. Am. Meteorol. Soc.* **72,** 163–183.

Wentz, F. J., L. A. Mattox, and S. Peteherych. 1986. New algorithms for microwave measurements of ocean winds: Applications to Seasat and the Special Sensor Microwave Imagery. *J. Geophys. Res.* **91,** 2289–2307.

Wesson, J. C., and M. C. Gregg. 1994. Mixing in Camarinal Sill in the Strait of Gibraltar. *J. Geophys. Res.* **99,** 9847–9878.

Westphal, D. L., *et al.* 1996. Initialization and validation of a simulation of cirrus using FIRE-II data. *J. Atmos. Sci.* **53,** 3397–3429.

Wetzel, R.G., 1975, *Limnology*, W.B. Saunders, Philadelphia, Pennsylvania, 743 pp.

Whitham, G. B., 1970. Two-timing, variational principles and waves, *J. Fluid Mech.*, **44**, 373–395.

Whitham, G. 1974. Dispersive waves and variational principles. In S. Leibovich and A. R. Seebass (Eds.), *Nonlinear Waves,* pp. 139–169. Cornell Univ. Press, Ithaca, NY.

Wick, G. 1995. Evaluation of the variability and predictability of the bulk-skin sea surface temperature difference with applications to satellite-measured sea surface temperature. Ph.D. dissertation, University of Colorado, Boulder. 140 pp.

Wick, G. A., W. J. Emery, and P. Schluessel. 1992. A comprehensive comparison between satellite-measured skin and multichannel sea surface temperature. *J. Geophys. Res.* **97,** 5569–5595.

Wick, G., W. J. Emery, L. H. Kantha, and P. Schluessel. 1996. The behavior of the bulk-skin sea surface temperature difference under varying wind speed and heat flux. *J. Phys. Oceanogr.* **26,** 1969–1988.

Wick, G. A., and A. T. Jessup. 1998. Simulation of ocean skin temperature modulation by swell waves. *J. Geophys. Res.* **103,** 3149–3161.

Wijesekera, H., L. Padman, T. Dillon, M. Levine, C. Paulson, and R. Pinkel. 1993. The application of internal-wave dissipation models to a region of strong mixing. *J. Phys. Oceanogr.* **23,** 269–286.

Wijesekera, H. W., and M. C. Gregg. 1996. Surface layer response to weak winds, westerly wind bursts, and rain squalls in the western Pacific warm pool. *J. Geophys. Res.* **101,** 977–997.

Wilczak, J. M., and J. W. Glendening. 1988. Observations and mixed layer modeling of a terrain-induced mesoscale gyre: The Denver cyclone. *Mon. Weather Rev.* **116,** 2688–2711.

Wilczak, J. M., E. E. Gossard, W. D. Neff, and W. L. Eberhard. 1996. Ground-based remote sensing of the atmospheric boundary layer: 25 years of progress. *Boundary Layer Meteorol.* **78,** 321–349.

Williams, A. J. 1975. Images of ocean microstructure. *Deep-Sea Res.* **22,** 811–829.

Williams, D. L., and R. P. Von Herzen. 1983. On the terrestrial heat flow and physical limnology of Crater Lake, Oregon. *J. Geophys. Res.* **88,** 1094–1104.

Willis, G. E., and J. W. Deardorff. 1974. A laboratory model of the unstable boundary layer. *J. Atmos. Sci.* **31,** 1297–1307.

Wilson, K. G. 1971. Renormalization group and critical phenomena. *Phys. Rev. B* **4,** 3174–3187.

Wimbush, M., and W. H. Munk. 1970. The benthic boundary layer. In A. Maxwell (Ed.), *The Sea,* Vol. 4, pp. 731–758. Wiley–Interscience, New York.

Winant, C. D. 1980. Coastal circulation and wind-induced currents. *Annu. Rev. Fluid Mech.* **12,** 271–301.

Wittich, K.-P. 1991. The nocturnal boundary layer over northern Germany: An observational study. *Boundary Layer Meteorol.* **55,** 47–66.

Wood, E. F. 1991. Global scale hydrology: Advances in land surface modeling. *Rev. Geophys.* **29,** 193–201.

Woods, J. D. 1968. An investigation of some physical processes associated with the vertical flow of heat through the upper ocean. *Met. Mag.* **97,** 65–72.

Woods, J. D. 1980. Diurnal and seasonal variation of convection in the wind mixed layer of the ocean. *Q. J. R. Meteorol. Soc.* **106,** 379–394.

Woods, J. D., and R. L. Wiley. 1972. Billow turbulence and ocean microstructure. *Deep-Sea Res.* **19,** 87–121.

Wu, J. 1974. Evaporation due to spray. *J. Geophys. Res.* **79,** 4107–4109.

Wu, J. 1975. Wind-induced drift currents. *J. Fluid Mech.* **68,** 49–70.

Wu, J. 1985. On the cool skin of the ocean. *Boundary Layer Meteorol.* **31,** 203–207.

Wu, J. 1988. Variations of whitecap coverage with wind stress and water temperature. *J. Phys. Oceanogr.* **18,** 1448–1453.

Wu, J. 1990. On parameterization of sea spray. *J. Geophys. Res.* **95,** 18269–18279.

Wu, J. 1992. On moisture flux across the sea surface. *Boundary Layer Meteorol.* **60,** 361–374.

Wu, J. 1994. The sea surface is aerodynamically rough even under light winds. *Boundary Layer Meteorol.* **69,** 149–158.

Wu, J. 1995. Small-scale wave breaking: A widespread sea surface phenomenon and its consequence for air-sea exchanges. *J. Phys. Oceanogr.* **25,** 407–412.

Wu, J. 1998. Insignificant evaporation from escaping sea spray droplets. *J. Geophys. Res.* **103,** 3163–3165.

Wunsch, C., 1972. Temperature microstructure on the Bermuda slope, with application to the mean flow. *Tellus* **24**, 350–367.

Wunsch, C. 1975. Internal tides in the ocean. *Rev. Geophys. Space Phys.* **13,** 167–182.

Wurtele, M. G., R. D. Sharman, and A. Datta. 1996. Atmospheric lee waves. *Annu. Rev. Fluid Mech.* **28,** 429–476.

Wyngaard, J. C. 1980. *Workshop on the Planetary Boundary Layer.* Am. Meteorol. Soc., Boston. 322 pp.

Wyngaard, J. C. 1973. On surface-layer turbulence. In D. Haugen (Ed.), *Workshop on Micrometeorology,* pp. 101–149. Am. Meteorol. Soc., Boston.

Wyngaard, J. C. 1982. Boundary-layer modeling. In F. T. M. Nieuwstadt and D. Van Dop (Eds.), *Atmospheric Turbulence and Air Pollution Modeling*, pp. 69–106. Reidel, Boston.

Wyngaard, J. C. 1992. Atmospheric turbulence. *Annu. Rev. Fluid Mech.* **24,** 205–233.

Wyngaard, J. C., I. Izumi, and S. A. Collins. 1971. Behavior of the refractive index structure parmeter near the ground. *J. Opt. Soc. Am.* **61,** 1646–1650.

Wyngaard, J. C., and C.-H. Moeng. 1993. Large eddy simulation in geophysical turbulence parameterization. In *Large Eddy Simulation of Complex Engineering and Geophysical Flows.* Cambridge Univ. Press, New York.

Yaglom, A. M. 1979. Similarity laws for constant pressure and pressure-gradient turbulent wall flows. *Annu. Rev. Fluid Mech.* **11,** 505–540.

Yaglom, A. M. 1981. Laws of small scale turbulence in atmosphere and ocean (in commemoration of the 40th anniversary of theory of locally isotropic turbulence). *Izv. Atmos. Oceanic Phys.* **17,** 919–935.

Yaglom, A. M. 1994. A. N. Kolmogoroff as a fluid mechanician and founder of a school in turbulence research. *Annu. Rev. Fluid Mech.* **26,** 1–22.

Yaglom, A. M., and B. A. Kader. 1974. Heat and mass transfer between a rough wall and turbulent fluid flow at high Reynolds and Peclet numbers. *J. Fluid Mech.* **62,** 601–623.

Yakhot, V., and S. A. Orszag. 1986. Renormalization group analysis of turbulence. I. Basic theory. *J. Sci. Computing* **1,** 3–51.

Yakhot, V., and S. A. Orszag. 1987. Relation between the Kolmogorov and Batchelor constants. *Phys. Fluids* **30,** 3.

Yakhot, A., S. A. Orszag, V. Yakhot, and M. Israeli. 1989. Renormalization group formulation of large-eddy simulations. *J. Sci. Computing* **4,** 139–158.

Yakhot, V., and S. A. Orszag. 1992. Development of turbulence models for shear flows by a double expansion technique. *Phys. Fluids A* **4,** 1510–1520.

Yakhot, V., and L. M. Smith. 1992. The renormalization group, the e-expansion and derivation of turbulence models. *J. Sci. Comput.* **7,** 35–65.

Yamada, T. 1976. On the similarity functions A, B and C of the planetary boundary layer. *J. Atmos. Sci.* **33,** 781–793.

Yamada, T. 1983. Simulations of nocturnal drainage flows by a q^2l turbulence closure model. *J. Atmos. Sci.* **40,** 91–106.

Yardanov, D. 1976. On the universal functions in the resistance law of a baroclinic planetary boundary layer. *Izv. Atmos. Oceanic Phys.* **12,** 769–772.

Yelland, M. J., B. I. Moat, P. K. Taylor, R. W. Pascal, J. Hutchings, and V. C. Cornell. 1998. Wind stress measurements from the open ocean corrected for airflow distortion by the ship. *J. Phys. Oceanogr.* **28,** 1511–1526.

Yokoyama, O., M. Gamo, and S. Yamamoto. 1979. The vertical profiles of the turbulence quantities in the atmospheric boundary layer. *J. Meteorol. Soc. Jpn.* **57,** 264–272.

You, Y. 1998. Rain-formed barrier layer of the western equatorial Pacific warm pool: A case study. *J. Geophys. Res.* **103,** 5361–5378.

Young, G. S. 1987. Mixed layer spectra from aircraft measurements. *J. Atmos. Sci.* **44,** 1251–1256.

Young, G. S., D. V. Ledvina, and C. W. Fairall. 1992. Influence of precipitating convection on the surface budget observed during a Tropical Ocean Global Atmosphere pilot cruise in the tropical western Pacific Ocean. *J. Geophys. Res.* **97,** 9595–9603.

Young, G. S., and Holland. 1996. *Atlas of the Oceans: Wind and Wave Climate* (with CD ROM). Pergamon, Elmsford, NY.

Young, W. R. 1994. Subinertial mixed layer approximation. *J. Phys. Oceanogr.* **24,** 1812–1826.

Zakharov, V. E. 1968. Stability of periodic waves of finite amplitude on the surface of a deep fluid. *Zh. Pril. Mekh. Tekh. Fiz.* **3,** 80–94.

Zakharov, V. E. 1992. Inverse and direct cascade in the wind-driven surface wave turbulence and wave-breaking. In M. L. Banner and R. H. Grimshaw (Eds.), *Breaking Waves,* pp. 69–91. Springer-Verlag, Berlin/New York.

Zedel, L., and D. Farmer. 1991. Organized structures in subsurface bubble clouds: Langmuir circulation in the open ocean. *J. Geophys. Res.* **91,** 8889–8900.

Zeller, K. F., 1992. Vertical ozone gradient behavior and similarity relationship above a flat shortgrass prairie site, Abstract, AMS Tenth Symposium on Turbulence and Diffusion, Portland, OR, 226–229.

Zeman, O., and N. O. Jensen. 1987. Modification of turbulence characteristics in flow over hills. *Q. J. R. Meteorol. Soc.* **113,** 55–80.

Zhang, J., R. W. Schmitt, and R. X. Huang. 1998. Sensitivity of the GFDL Modular Ocean Model to parameterization of double-diffusive processes. *J. Phys. Oceanogr.* **28,** 589–605.

Zhang, S. F., S. P. Oncley, and J. A. Businger. 1988. A critical evaluation of the von Karman constant from a new atmospheric surface layer experiment. Paper presented at the Eighth Symposium on Turbulence and Diffusion, Am. Meteorol. Soc., San Diego, April 26–29.

Zhang, X. 1995. Capillary-gravity and capillary waves generated in a wind wave tank: Observations and theories. *J. Fluid Mech.* **289,** 51–82.

Zilitinkevich, S. S. 1972. On the determination of the height of the Ekman boundary layer. *Boundary Layer Meteorol.* **3,** 141–145.

Zilitinkevich, S. S. 1991. *Turbulent Penetrative Convection.* Avebury Technical, Aldershot, UK. 179 pp.

Zillman, J. W. 1972. *A Study of Some Aspects of the Radiation and Heat Budgets of the Southern Hemisphere Oceans.* Meteorol. Stud. Report 26, Bur. Meteorol., Dept. Interior, Canberra.

Zipser, E. J., and A. J. Bedard. 1982. Front range eindstorms revisited, small scale differences amid large scale similarities. *Weatherwise* **35,** 82–85.

Zubov, N. N. 1945. *Arctic Ice.* Glavsevmorput, Moscow. 360 pp.

Biographies

Ludwig Prandtl (1875–1953) of Germany, along with G. I. Taylor (or simply GI to his admirers) of Great Britain, Andrei Nikolayevich Kolmogoroff of Russia, and von Karman of the United States, is a towering personality in turbulence theory. When the history of fluid mechanics in the 20th century gets written, three names will stand out: Ludwig Prandtl, Geoffrey Ingram Taylor of Great Britain, and Theodore von Karman of the United States. All three made important contributions to turbulence. Prandtl of course made the most seminal contribution to fluid mechanics: the boundary layer theory. If Nobel prizes awarded included the area of fluid mechanics, Prandtl would probably have received one. His contributions include mixing length theory of turbulence (an admittedly "engineering" approach, that for the first time enabled turbulence data to be normalized properly), the lifting line theory, affine transformations, and Prandtl–Meyer expansions. He even made contributions to solid mechanics. In addition, many of his students and colleagues themselves made seminal contributions to turbulence and fluid mechanics. The famous examples are von Karman, Blassius, Betz, and Busemann.

Prandtl, while an extremely competent scientist, was quite at loss in many other matters. One story that illustrates this is the manner in which he got married, as told by von Karman in his memoirs. Prandtl became quite famous due to his boundary layer theory (which was incidentally published as a seven-page article in a rather obscure applied mathematics journal!) in his early twenties and he became a professor at Gottingen at a very young age. He, however, did not get married till his thirties. One day out of the blue, he decided to get married and settle into the social life around Gottingen. It was fashionable in those days to marry the daughter of one's professor. So one day he approached his professor's wife and asked her daughter's hand in marriage. After the usual small talk, he thanked his professor and his wife and left. Only then did the professor's wife realize that Professor Prandtl forgot to mention which one—you see they had two daughters. An urgent family conference was convened and it was decided that the eldest should marry the professor after all. That is how Professor Prandtl came to be married. Fascinating accounts of Prandtl's life and research can be found in Ostwatitsch and Wieghardt (1987; see also Flugge-Lotz and Flugge, 1973).

Prandtl had an insatiable curiosity. Once, walking along with some visitors on the top floor of a building, his visitors were amazed to discover that the dignified Professor Prandtl had suddenly stopped and had started bouncing vigorously. He later explained to his visiting dignitaries that he was mentally calculating the natural frequency of the structure and verifying it experimentally!

Prandtl never quite received the honors he so richly deserved for his contributions to fluid mechanics. His reputation was tainted by his indirect association with Nazis during the Second World War. While he himself was not a Nazi, his laboratory did extremely valuable fundamental research on aerodynamics, funded by the German government, and this indirectly helped its war effort. Prandtl never became famous in this country, unlike von Karman, who migrated to this country from Germany before the War and who did. In fact, there was intense competition between the master and the pupil, and in at least one instance, the pupil won—the law of the wall is named after von Karman. After the war, von Karman, who was an honorary colonel in the U.S. Army, was sent to Gottingen at the head of a scientific team to assess the scientific advances made in Germany during the war. The irony is that when he visited Gottingen, he sat in Prandtl's chair and interviewed Prandtl. (Photograph used with permission, from the Annual Review of Fluid Mechanics, Volume 5, 1973, by Annual Reviews http://www.AnnualReviews.org.)

G. I. Taylor pioneered the more respectable "scientific" approach to turbulence, using stochastic theory. G. K. Batchelor, himself a prominent fluid dynamicist, was his student and a life-long colleague. GI is held in awe by his admirers. With a keen physical insight, he was able to make many contributions to fluid mechanics. He was also an experimentalist, and took delight in devising simple experiments to verify his theoretical deductions and elucidate fluid flow processes. Owen Phillips, one of his students at Cambridge, used to remark that GI's laboratory was just as disorderly looking as LHK's laboratory at Johns Hopkins! Fascinating accounts of GI's life and contributions can be found in Batchelor (1996) and Turner (1997). (Photograph used with permission, from the Annual Review of Fluid Mechanics, Volume 6, 1974, by Annual Reviews http://www.AnnualReviews.org.)

Andrei Nikolaevich Kolmogoroff (1903–1987) was a great mathematician. He is regarded as one among the select group of brilliant mathematicians of the 20th century that includes Poincare, Hilbert, and von Neumann. His seminal contribution to statistical fluid mechanics is one of the cornerstones of turbulence (Yaglom, 1994). His students included such masters as A. M. Obukhoff, Andrei S. Monin, M. D. Millionshchikoff, and A. M. Yaglom, who themselves made seminal contributions to fluid mechanics and turbulence. (Photograph used with permission, from the Annual Review of Fluid Mechanics, Volume 26, 1994, by Annual Reviews http://www.AnnualReviews.org. John Lumley, photographer.)

Von Karman, born in Hungary, came to work with Prandtl at Gottingen. While there, he discovered the Karman vortex street. He is well-known for his discovery of the universal law of the wall. He emigrated to the United States, joined CalTech, and continued to make contributions to a variety of topics, including the statistical theory of turbulence. His memoirs detail his life and contributions to fluid mechanics.

Deardorff received his PhD in Meteorology at the University of Washington in 1959, with interests at that time in air-sea interaction. During 1962 to 1978 he was a senior scientist at NCAR, where he and Glen Willis researched turbulent thermal convection in the laboratory. In 1963 his interests expanded to include numerical modeling of the same, which soon extended to modeling of the planetary boundary layer, and later to numerical and laboratory modeling of pollutant diffusion. In 1978 he took a research professor position in the Department of Atmospheric Sciences (now part of the College of Oceanic and Atmospheric Sciences) at Oregon State University. He is well-known for his pioneering studies of large eddy simulations. By 1986 his interests had shifted towards UFOs and he took early retirement to research that field. He has written a book on the UFO topic, and two on the related topic of the origins of Christianity and its Gospels.

Owen M. Phillips belongs to the select group of brilliant theoreticians of this century. A student of G. K. Batchelor and none other than G. I. Taylor himself, he is one of the Cambridge scholars. With a keen analytical intellect and a penetrative physical insight, Phillips manages to bring out the essence of the process in any topic he investigates. Elected to the Royal Society at a very young age, Phillips has been on the forefront of turbulence and surface wave theory with numerous contributions. He has spent most of his professional life at the Johns Hopkins University. His monograph, *The Dynamics of the Upper Ocean*, known simply as DUO to many of his students and colleagues, has been a standard reference source for decades. He has lately been interested in flow through porous media with applications to geological fluid mechanics. LHK considers it to be a privilege to have been able to work with one of the "masters" in this field.

Michael Longuet-Higgins, also from Cambridge University, a close friend and colleague of Phillips, is one of the true giants in this field. There isn't a single area of surface wave dynamics that he has not touched and influenced profoundly one way or another over the past few decades, as can be seen from the list of references. His keen physical insight and considerable mathematical skills have greatly advanced the state-of-art in surface gravity wave theory and computation.

Sergei Kitaigorodskii has also been a major contributor to the field of surface waves. He headed a laboratory in the former Soviet Union until the early seventies, when he found himself increasingly at odds with his administration over issues of classified research. His marriage to a Finnish woman led to increasing conflicts with his superiors and he left the Soviet Union to work at the Johns Hopkins University. He returned to Russia in the early nineties and is now at the Shirshov Institute of Oceanology in Moscow. Armed with a keen physical insight, he has made many contributions to the theory of the wind-wave spectrum and gas transfer across the air-sea interface.

Klaus Hasselmann is well known for his contributions to wave research, especially in the area of wave-wave interactions. He was instrumental in the development of the popular, widely used third-generation wave model WAM. It is to people like Longuet-Higgins, Phillips, Hasselmann and Kitaigorodskii that we owe much in this difficult, yet fascinating field of surface gravity waves.

Walter Munk of the Scripps Institute of Oceanography in San Diego, is one of the "founding fathers" of modern oceanography (Henry Stommel of the Woods Hole Oceanographic Institution is another). Working out of the Scripps Institute for more than five decades, Munk has made seminal contributions to oceanography as well as geophysics, such as internal waves, tides, Earth's rotation, acoustic tomography, and many other diverse topics. His brilliant insight into oceanic processes and his infectious enthusiasm for the subject are legendary. LHK savors the rare privilege of his interactions with this true giant of oceanography.

The name of **J. Stewart Turner** stands out in the field of double-diffusive processes. Although he did not discover these processes himself, he has made numerous contributions, principally through strikingly beautiful laboratory experiments. Like Owen Phillips, he is a fellow of the Royal Society and a Cambridge scholar and was a student of G. K. Batchelor and G. I. Taylor.

Solitons or solitary waves have a fascinating history. The first documented observation of a solitary wave was made in 1834 by John Scott Russell, an engineer hired by a barge company to investigate the possibility of replacing horses by steam power. In those days, canals were heavily used for transporting cargo, and while Russell was observing a heavily loaded barge being pulled by a pair of horses along a narrow canal, the rope broke and the barge was suddenly stopped. As the barge settled into the water, it spawned a large solitary hump in the water roughly 10 m long and about 1/2 m high that propagated away from the barge and traveled at about 4.5 m s^{-1} along the canal for 2 to 3 km without much change of shape or speed. Russell diligently followed it on horseback until he lost sight of it around a bend. Subsequently, he made many more observations of this strange phenomenon and reported on them in 1844. However, his discoveries were not received well, and the possibility of a solitary wave was dismissed as physically impossible by none other than the royal astronomer George Airy and Stokes! It was only after 1895, when Danish scientists Korteweg and de Vries derived an equation for finite-amplitude shallow water waves from Navier–Stokes equations, that the possibility was widely conceded. Solitary waves remained, however, a curiosity till 1965, when Zabusky and Kruskal described numerical solutions which showed that two colliding solitary waves retain their shape and speed and emerge unchanged after the collision. This particle-like behavior made them coin the word "soliton" to describe these waves. Until then, it had been assumed that solitary waves would interact in a strongly nonlinear fashion, resulting in their eventual destruction, and therefore they were not of great importance in nonlinear wave physics. In 1967, Gardner, Green, Kruskal, and Miura obtained an analytical solution to the initial value problem governed by the K-dV equation and showed analytically that an initial waveform evolves into a train of one or more solitons with a dispersive tail. Since then, interest in solitons has grown by leaps and bounds, and they are now found in many branches of science and are invoked to explain diverse processes including transmission of signals through nerves.

Internal wave solitons in the oceans often make their presence known by their spectacular surface manifestations. For a long time, fishermen and sailors in the Orient knew of a phenomenon they often witnessed that spawned fear and superstition in them. When the sea was otherwise reasonably calm and peaceful, they would observe a long crest of very choppy water, many, many kilometers long, pass under their boats with a deafening roar, to be followed several minutes later by another and another. Since there was no apparent cause such as a sudden burst of wind, this phenomenon was quite scary when first encountered by an unsuspecting sailor or fisherman. However, the fishermen learned quickly that this choppy water was a good fishing locale and, once the fear wore off, began making use of the strange phenomenon. Like bioluminescence in these waters, the phenomenon was routinely observed but remained a mystery for a long time.

The advent of satellite photography showed the ubiquitous nature of tidally generated internal wave solitons in the global oceans and pointed out their possible important role in tidal dissipation. Internal wave solitons became important to naval operations when it was suspected, without solid proof, of course, that the loss of the U.S. nuclear submarine Thresher with all hands might have been due to its unexpected encounter with a large IW soliton in the Gulf of Maine, which might have taken it down suddenly and unexpectedly below its designed operational depth!

Like many scientific discoveries, the history of double diffusion took a serendipitous route. Discovery of double-diffusive phenomena is rightly attributed to Melvin Stern, then at Woods Hole Oceanographic Institution. In 1960, he showed that the differing diffusivities of heat and salt would lead to instability and ensuing fluid motions. Interestingly, Henry Stommel, also at Woods Hole, narrowly missed discovering double diffusion. Although he suggested that if a pipe that would prevent salt transfer but allow heat to be conducted were inserted into the ocean, a perpetual salt fountain would result, he missed recognizing that the enormous disparity between heat and salt diffusion (two orders of magnitude) would accomplish essentially the same objective, with the ocean forming its own "pipes." Stern went on to make seminal contributions, including calculation of the structure of salt fingers and the possibility of diffusive convection. The existence of salt fingers was demonstrated by Turner and Stommel in 1964. They were first observed in the oceans by Williams in 1975.

However, it was Jevons (1857) who, in experimenting on cloud forms in the laboratory, discovered salt fingers when he poured cold fresh water underneath warm salty water. He correctly attributed the process to the difference in diffusivities of heat and salt, but mistakenly tried to apply it to clouds. Lord Rayleigh apparently repeated Jevons' experiments but failed to follow up, and discovered instead the Brunt–Vaisala frequency, the frequency of natural oscillations in stratified fluids (Rayleigh, 1883), by neglecting molecular processes entirely! It took nearly 75 years for the phenomenon to be rediscovered by Stommel and Stern! A fascinating account of this tortuous path that led to the discovery of double diffusion can be found in Schmitt (1995).

Index

S

International Geophysics Series

EDITED BY

RENATA DMOWSKA
Division of Applied Science
Harvard University
Cambridge, Massachusetts

JAMES R. HOLTON
Department of Atmospheric Sciences
University of Washington
Seattle, Washington

H. THOMAS ROSSBY
Graduate School of Oceanography
University of Rhode Island
Narragansett, Rhode Island

Volume 1 B. GUTENBERG. Physics of the Earth's Interior. 1959*

Volume 2 J.W. CHAMBERLAIN. Physics of the Aurora and Airglow. 1961*

Volume 3 S.K. RUNCORN (ed.). Continental Drift. 1962*

Volume 4 C.E. JUNGE. Air Chemistry and Radioactivity. 1963*

Volume 5 R.G. FLEAGLE AND J.A. BUSINGER. An Introduction to Atmospheric Physics. 1963*

Volume 6 L. DEFOUR AND R. DEFAY. Thermodynamics of Clouds. 1963*

Volume 7 H. U. ROLL. Physics of the Marine Atmosphere. 1965*

Volume 8 R.A. CRAIG. The Upper Atmosphere: Meteorology and Physics. 1965*

* Out of print.

Volume 9 W.L. Webb. Structure of the Stratosphere and Mesosphere. 1966*

Volume 10 M. Caputo. The Gravity Field of the Earth from Classical and Modern Methods. 1967*

Volume 11 S. Matsushita and W.H. Campbell (eds.). Physics of Geomagnetic Phenomena (In two volumes). 1967*

Volume 12 K.Y. Kondratyev. Radiation in the Atmosphere. 1969*

Volume 13 E.P. Men and C.W. Newton. Atmospheric Circulation Systems: Their Structure and Physical Interpretation. 1969*

Volume 14 H. Rishbeth and O.K. Garriott. Introduction to Ionospheric Physics. 1969*

Volume 15 C.S. Ramage. Monsoon Meteorology. 1971*

Volume 16 J.R. Holton. An Introduction to Dynamic Meteorology. 1972*

Volume 17 K.C. Yeh and C.H. Liu. Theory of Ionospheric Waves. 1972*

Volume 18 M.I. Budyko. Climate and Life. 1974*

Volume 19 M.E. Stern. Ocean Circulation Physics. 1975

Volume 20 J.A. Jacobs. The Earth's Core. 1975*

Volume 21 D.H. Miller. Water at the Surface of the Earth: An Introduction to Ecosystem Hydrodynamics. 1977

Volume 22 J.W. Chamberlain. Theory of Planetary Atmospheres: An Introduction to Their Physics and Chemistry. 1978*

Volume 23 J.R. Holton. An Introduction to Dynamic Meteorology, Second Edition. 1979*

Volume 24 A.S. Dennis. Weather Modification by Cloud Seeding. 1980

Volume 25 R.G. Fleagle and J.A. Businger. An Introduction to Atmospheric Physics, Second Edition. 1980

Volume 26 K. Liou. An Introduction to Atmospheric Radiation. 1980

Volume 27 D.H. Miller. Energy at the Surface of the Earth: An Introduction to the Energetics of Ecosystems. 1981

Volume 28 H.G. Landsberg. The Urban Climate. 1991

Volume 29 M.I. Budkyo. The Earth's Climate: Past and Future. 1982*

Volume 30 A.E. Gill. Atmosphere-Ocean Dynamics. 1982

Volume 31 P. Lanzano. Deformations of an Elastic Earth. 1982*

Volume 32 R. Merrill McElhinny. The Earth's Magnetic Field: Its History, Origin, and Planetary Perspective. 1983

Volume 33 J.S. Lewis and R.G. Prinn. Planets and Their Atmospheres: Origin and Evolution. 1983

Volume 34 R. Meissner. The Continental Crust: A Geophysical Approach. 1986

Volume 35 SAGITOV, BODKI, NAZARENKO, AND TADZHIDINOV. Lunar Gravimetry. 1986

Volume 36 CHAMBERLAIN AND HUNTEN. Theory of Planetary Atmospheres, 2nd Edition. 1987

Volume 37 J.A. JACOBS. The Earth's Core, 2nd Edition. 1987

Volume 38 J.R. APEL. Principles of Ocean Physics. 1987

Volume 39 M.A. UMAN. The Lightning Discharge. 1987

Volume 40 ANDREWS, Holton, and Leovy. Middle Atmosphere Dynamics. 1987

Volume 41 P. WARNECK. Chemistry of the Natural Atmosphere. 1988

Volume 42 S.P. ARYA. Introduction to Micrometeorology. 1988

Volume 43 M.C. KELLEY. The Earth's Ionosphere. 1989

Volume 44 W.R. COTTON AND R.A. ANTHES. Storm and Cloud Dynamics. 1989

Volume 45 W. MENKE. GEOPHYSICAL DATA ANALYSIS: Discrete Inverse Theory, Revised Edition. 1989

Volume 46 S.G. PHILANDER. El Niño, La Niña, and the Southern Oscillation. 1990

Volume 47 R.A. BROWN. Fluid Mechanics of the Atmosphere. 1991

Volume 48 J.R. HOLTON. An Introduction to Dynamic Meteorology, Third Edition. 1992

Volume 49 A. KAUFMAN. Geophysical Field Theory and Method.

Part A: Gravitational, Electric, and Magnetic Fields. 1992

Part B: Electromagnetic Fields I. 1994

Part C: Electromagnetic Fields II. 1994

Volume 50 BUTCHER, ORIANS, CHARLSON, AND WOLFE. Global Biogeochemical Cycles. 1992

Volume 51 B. EVANS AND T. WONG. Fault Mechanics and Transport Properties of Rocks. 1992

Volume 52 R.E. HUFFMAN. Atmospheric Ultraviolet Remote Sensing. 1992

Volume 53 R.A. HOUZE, JR. Cloud Dynamics. 1993

Volume 54 P.V. HOBBS. Aerosol-Cloud-Climate Interactions. 1993

Volume 55 S. J. GIBOWICZ AND A. KIJKO. An Introduction to Mining Seismology. 1993

Volume 56 D.L. HARTMANN. Global Physical Climatology. 1994

Volume 57 M.P. RYAN. Magmatic Systems. 1994

Volume 58 T. LAY AND T.C. WALLACE. Modern Global Seismology. 1995

Volume 59 D.S. WILKS. Statistical Methods in the Atmospheric Sciences. 1995

Volume 60 F. NEBEKER. Calculating the Weather. 1995

Volume 61 M.L. SALBY. Fundamentals of Atmospheric Physics. 1996

Volume 62 J.P. McCalpin. Paleoseismology. 1996

Volume 63 R. Merrill, M. McElhinny, and P. McFadden. The Magnetic Field of the Earth: Paleomagnetism, the Core, and the Deep Mantle. 1996

Volume 64 N.D. Opdyke and J. Channell. Magnetic Stratigraphy. 1996

Volume 65 J.A. Curry and P.J. Webster. Thermodynamics of Atmospheres and Oceans. 1998

Volume 66 L.H. Kantha and C.A. Clayson. Numerical Models of Oceans and Oceanic Processes. 2000

Volume 67 L.H. Kantha and C.A. Clayson. Small Scale Processes in Geophysical Fluid Flows. 2000

Printed and bound by CPI Group (UK) Ltd, Croydon, CR0 4YY

08/05/2025

01864897-0002